Glossary of Notation

Symbol	Meaning	Page
$C(X)$	Set of all functions continuous on X	2
$C^n(X)$	Set of all functions having n continuous derivatives on X	4
$C^\infty(X)$	Set of all functions having derivatives of all orders on X	4
$0.\overline{3}$	A decimal in which the numeral 3 repeats indefinitely	11
$\mathbb{R}$	Set of real numbers	11
$fl(y)$	Floating-point form of the real number y	19
$O(\cdot)$	Order of convergence	36
Δ	Forward difference	88
$\bar{z}$	Complex conjugate of the complex number z	96
$\binom{n}{k}$	The kth binomial coefficient of order n	123
$f[\cdot]$	Divided difference of the function f	124
∇	Backward difference	128
$\mathbb{R}^n$	Set of ordered n-tuples of real numbers	254
τ_i	Local truncation error at the ith step	269
$\rightarrow$	Equation replacement	351
$\leftrightarrow$	Equation interchange	351
(a_{ij})	Matrix with a_{ij} as the entry in the ith row and jth column	353
$\mathbf{x}$	Column vector or element of $\mathbb{R}^n$	353
$[A, \mathbf{b}]$	Augmented matrix	354
O_n	$n \times n$ matrix with all zero entries	376
δ_{ij}	Kronecker delta, 1 if $i = j$, 0 if $i \neq j$	378
I_n	$n \times n$ identity matrix	378
A^{-1}	Inverse matrix of the matrix A	380
A^t	Transpose matrix of the matrix A	383
M_{ij}	Minor of a matrix	389
$\det A$	Determinant of the matrix A	389
$\mathbf{0}$	Vector with all zero entries	392
$\|\mathbf{x}\|$	Arbitrary norm of the vector $\mathbf{x}$	424
$\|\mathbf{x}\|_2$	The l_2 norm of the vector $\mathbf{x}$	425
$\|\mathbf{x}\|_\infty$	The l_∞ norm of the vector $\mathbf{x}$	425
$\|A\|$	Arbitrary norm of the matrix A	431
$\|A\|_2$	The l_2 norm of the matrix A	432
$\|A\|_\infty$	The l_∞ norm of the matrix A	432
$\rho(A)$	The spectral radius of the matrix A	439
$K(A)$	The condition number of the matrix A	463
Π_n	Set of all polynomials of degree n or less	490
$\tilde{\Pi}_n$	Set of all monic polynomials of degree n	500
$\text{sign}(x)$	Sign of the number x, 1 if $x > 0$, -1 if $x < 0$	518
$\mathcal{T}_n$	Set of all trigonometric polynomials of degree n or less	519
C	Set of complex numbers	545
$\mathbf{F}$	Function mapping $\mathbb{R}^n$ into $\mathbb{R}^n$	590
$A(\mathbf{x})$	Matrix whose entries are functions form $\mathbb{R}^n$ into $\mathbb{R}$	599
$J(\mathbf{x})$	Jacobian matrix	601
∇g	Gradient of the function g	615

Numerical Analysis

SIXTH EDITION

Numerical Analysis

SIXTH EDITION

Richard L. Burden
Youngstown State University

J. Douglas Faires
Youngstown State University

Brooks/Cole Publishing Company
ITP® An International Thomson Publishing Company

Pacific Grove • Albany • Belmont • Bonn • Boston • Cincinnati • Detroit
Johannesburg • London • Madrid • Melbourne • Mexico City • New York
Paris • Singapore • Tokyo • Toronto • Washington

Sponsoring Editor: *Steve Quigley, Gary Ostedt*
Marketing Team: *Maureen Riopelle, Deborah Petit*
Editorial Associate: *Carol Benedict*
Production Coordinator: *Kirk Bomont*
Manuscript Editor: *Linda Thompson*
Interior Design: *Julia Gecha*
Cover Design: *Cheryl Carrington*
Interior Illustration: *Scientific Illustrators*
Cover Photo: *David Bishop/Phototake*
Project Management and Typesetting:
Integre Technical Publishing Co., Inc.
Cover Printing: *Phoenix Color Corporation, Inc.*
Printing and Binding: *Quebecor Printing Fairfield*

For more information, contact:

BROOKS/COLE PUBLISHING COMPANY
511 Forest Lodge Road
Pacific Grove, CA 93950
USA

International Thomson Publishing Europe
Berkshire House 168–173
High Holborn
London WC1V 7AA
England

Thomas Nelson Australia
102 Dodds Street
South Melbourne, 3205
Victoria, Australia

Nelson Canada
1120 Birchmount Road
Scarborough, Ontario
Canada M1K 5G4

International Thomson Editores
Seneca 53
Col. Polanco
11560 México, D.F., México

International Thomson Publishing GmbH
Königswinterer Strasse 418
53227 Bonn
Germany

International Thomson Publishing Asia
221 Henderson Road
#05–10 Henderson Building
Singapore 0315

International Thomson Publishing Japan
Hirakawacho Kyowa Building, 3F
2-2-1 Hirakawacho
Chiyoda-ku, Tokyo 102
Japan

Printed in the United States of America

10 9 8 7 6 5 4 3 2

Library of Congress Cataloging in Publication Data
Burden, Richard L.
Numerical analysis / Richard Burden, J. Douglas Faires. — 6th ed.
p. cm.
Includes bibliographical references (p. –) and index.
ISBN 0-534-95532-0
1. Numerical analysis. I. Faires, J. Douglas. II. Title.
QA297.B84 1997
519.4—dc20 96-27594
CIP

Contents

Contents

Preface

About the Text

We have developed the material in this text for a sequence of courses in the theory and application of numerical approximation techniques. The text is designed primarily for junior-level mathematics, science, and engineering majors who have completed at least the first year of the standard college calculus sequence and have some knowledge of a high-level programming language. Familiarity with the fundamentals of matrix algebra and differential equations is also useful, but adequate introductory material on these topics is presented in the text so that courses in these subjects should not be needed as prerequisites.

Previous editions of the book have been used in a wide variety of situations. In some cases, the mathematical analysis underlying the development of approximation techniques is emphasized rather than the methods themselves; in others, the emphasis is reversed. The book has also been used as the core reference for courses at the beginning graduate level in engineering and computer science programs, as the basis for the actuarial examination in numerical methods, where self-study is common, and in first-year courses in introductory analysis offered at international universities. We have tried to adapt the book to fit these diverse requirements without compromising our original purpose:

> *To give an introduction to modern approximation techniques; to explain how, why, and when they can be expected to work; and to provide a firm basis for future study in numerical analysis and scientific computing.*

The book contains sufficient material for a full year of study, but we expect many readers to use the text for only a single-term course. In such a course, students learn to identify the type of problems that require numerical techniques for their solution and see examples of error propagation that can occur when numerical methods are applied. They accurately approximate the solutions of problems that cannot be solved exactly and learn techniques for estimating bounds for the error in the approximations. The remainder of the text serves as a reference for methods that are not considered in the course. Either the full-year or single-course treatment is consistent with the aims of the text.

Virtually every concept in the text is illustrated by example, and this edition contains more than 2000 class-tested exercises. These exercises range from elementary applications of methods and algorithms to generalizations and extensions of theory. In addition, the exercise sets include a large number of applied problems from diverse areas of engineering, as well as from the physical, computer, biological, and social sciences. The applications chosen concisely demonstrate how numerical methods can be, and often must be, applied in real-life situations.

New for This Edition

In the past decade a number of software packages have been developed to produce symbolic mathematical computations. Predominant among them in the academic environment are Derive, Maple, and Mathematica. There are versions of the software packages for most common computer systems, and student versions of some are available at very reasonable prices. Although there are significant differences among the packages, both in performance and price, they all can perform standard algebra and calculus operations.

Having a symbolic computation package available can be very useful in the study of approximation techniques. The results in most of our examples and exercises have been generated using problems for which exact values can be determined, since this permits the performance of the approximation method to be monitored. Exact solutions can often be obtained quite easily using symbolic computation. In addition, many numerical techniques have error bounds that require bounding a higher ordinary or partial derivative of a function. This can be a tedious task and one that is not particularly instructive once the techniques of calculus have been mastered. These derivatives can be quickly obtained symbolically, and a little insight will often permit a symbolic computation to aid in the bounding process as well.

We have chosen Maple as our standard package because of its wide distribution, but Derive or Mathematica could be substituted in its place with only minor modifications. In addition, we have added examples and exercises whenever we felt that a computer algebra system would be of significant benefit and discussed the approximation methods that Maple employs when it is unable to solve a problem exactly.

Another addition for this edition is the inclusion of a number of quickly worked exercises of the type included on the actuarial examination in numerical methods. These problems are designed to be solved in 5 to 10 minutes, and to get to the heart of the technique being considered. To accommodate this and still keep the book to a reasonable length, we have had to omit some of the exercises that were included in earlier editions. We hope none of your favorites have been included in this culling, and if they have, please let us know so that perhaps they can be resurrected in the future.

There are relatively few places in this edition that contain substantial text changes from the fifth edition. The section on Romberg integration in Chapter 4 has been rewritten and moved earlier in the chapter. This places it closer to the discussion of Richardson extrapolation, on which the technique is based. The material in Section 5.5 on variable step-size one-step methods leading to the Runge-Kutta-Fehlberg method has been rewritten for better understanding, as has the introduction to Gragg extrapolation in Section 5.8.

More discussion on positive definite matrices has been included in Section 6.6, and conditions are presented that permit easier verification of positive definiteness in matrices of small size. The theory, in Section 7.2, underlying the convergence of iterative methods for solving linear systems has been rewritten to more closely parallel the discussion of fixed-point methods in Chapter 2. In Chapter 8, we have rewritten the discussion of Fourier series and moved this material from Section 8.2 to Section 8.5 so that the discussion of continuous and discrete trigonometric approximation can be more easily paralleled. The discussion of Chebyshev polynomials has also been redone and simplified in Section 8.3.

More of the theory of linear algebra is presented in the first section of Chapter 9, and Aitken's Δ^2 technique is used in all the methods in Section 9.2 to improve convergence. The presentation in Section 11.4 on finite-difference techniques for nonlinear problems has been redone to simplify the notation, which should lead to a better understanding of this topic.

Although the major changes in this edition are quite small, those familiar with our past editions will find that virtually every page has been modified in some way. All the references have been updated and revised. The exercises have all been reworked and often reordered or simplified. A number of the theoretical and application exercises that we felt were not being extensively used have been replaced by new problems that incorporate the use of technology.

We hope you will find that these changes have been beneficial to the teaching and study of numerical analysis. Almost all of them have been motivated by changes in the presentation of the material to our students.

Algorithms

As in the previous editions, we give a detailed, structured algorithm without program listing for each method in the text. The algorithms are in a form that students with even limited programming experience can code.

With this edition we have included a disk containing programs for solutions to representative exercises using the algorithms. The programs for each algorithm are written in the programming languages Fortran, Pascal, and C. In addition, we have also coded the programs using the computer algebra systems Maple and Mathematica, which should ensure that a set of programs is available for most common computing systems.

A *Student Study Guide* is available with this edition that illustrates the calls required for these programs, but most of this information is quite intuitive for those with programming experience. The study guide also contains worked out solutions to many of the problems, including all of those that we feel would provide good practice for students preparing for the actuarial examination in numerical methods.

The publisher can provide instructors with an *Instructor's Manual* that provides answers and solutions to all the exercises in the book. All the computation results in the *Instructor's Manual* were regenerated for this edition using the programs on the disk to ensure compatibility between the various programming systems.

The algorithms in the text lead to programs that give correct results for the examples and exercises in the text, but no attempt was made to write general-purpose professional software. In particular, the algorithms are not always written in the form that leads to the most efficient program in terms of either time or storage requirements. When a conflict occurred between writing an extremely efficient algorithm and writing a slightly different one illustrating the important features of the method, the latter path was invariably taken.

Suggested Course Outlines

The text has been designed to allow instructors flexibility in the choice of topics as well as in the level of theoretical rigor and in the emphasis on applications. In line with these aims, we provide detailed references for the results that are not demonstrated in the text as well as for the applications that are used to indicate the practical importance of the methods. The text references cited are those most likely to be available in college libraries and have been updated to reflect the most recent edition at the time this book was placed into production. For the most recent methods we also include quotations from original research papers when we feel this material is accessible to our audience.

The following flow chart indicates chapter prerequisites. The only deviation from this chart is described in the footnote at the bottom of the first page of Section 3.4. Most of the possible sequences that can be generated from this chart have been taught by the authors at Youngstown State University.

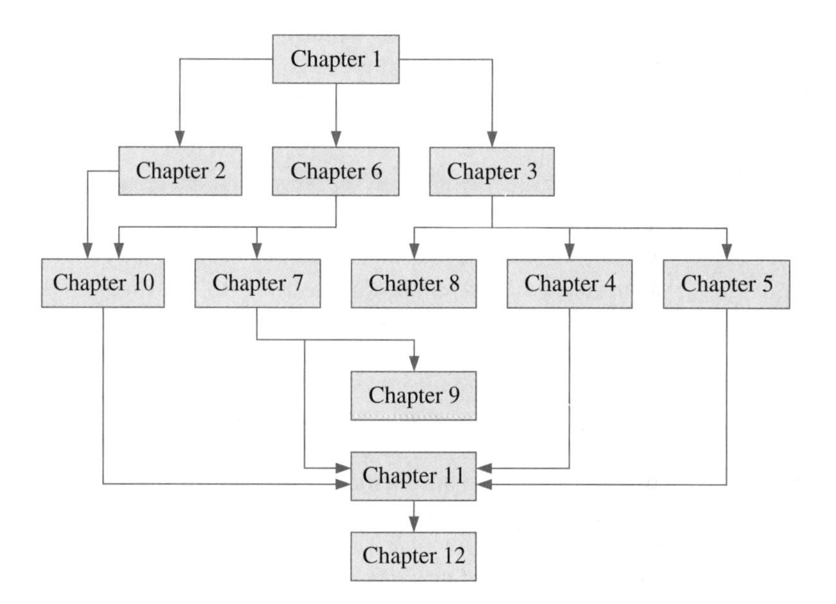

Acknowledgments

We feel most fortunate to have had so many of our students and colleagues communicate with us regarding their impressions of the earlier editions of this book. All of these comments are taken very seriously; we try to include all the suggestions that are in line with the philosophy of the book, and are extremely grateful to all those that have taken the time to contact us and inform us of improvements we can make in subsequent versions.

In previous editions of *Numerical Analysis*, errors seemed to creep in in the dead of night during the production of the book. No matter how often we proofread, there were still digits transposed or subscripts wrong. It is our hope that you will find this edition to be virtually error free. For the first time we have been able to prepare a TEX version that was used directly in the production of the book. This gave us much better control over the final version of the material in the book. No longer can we shift the blame to an unknown typesetter, because in this edition we did most of the typesetting.

In addition to the many unsolicited comments on the previous edition, we have had the benefit of receiving reviews from the following, whose efforts we greatly appreciate.

J. R. Dorroh, Louisiana State University
Mahmoud Fath El-Den, Ft. Hays State University
David R. Hill, Temple University
Israel Koltracht, University of Connecticut-Storrs
Michael Ratliff, Northern Arizona University
Larry Snyder, Ohio University
Charles Waters, Mankato State University

As has been our practice in past editions of the book, we have used student help at Youngstown State University in preparing the sixth edition. The extensive effort of converting the previous edition into a TEX document was very capably performed by Christy Conn, a junior mathematics and computer science major, to whom we owe a great deal of thanks. The Pascal programs for the algorithms were translated into C language by Douglas Knickerbocker, who has recently received his MS degree in mathematics. The Maple and Mathematica procedures were written primarily by two students in our University Scholars program, Robert Komara, a junior mathematics and physics major, and John Slanina, a sophomore with majors in mathematics and mechanical engineering. We are grateful to all of you, not only for your good work, but also for your good humor during some trying times.

We would also like to express gratitude to our colleagues on the faculty and administration of Youngstown State University for providing us the opportunity and facilities to complete this work, and particularly to John Buoni, the Chair of Mathematics and Statistics, for his support.

Finally, we would like to thank all those who have used and adopted the various editions of *Numerical Analysis* over the years. It has been most gratifying to hear from so many students, and new faculty, who used our book for their first exposure to the study of numerical methods. We hope this edition continues the trend and adds to the enjoyment of students studying numerical analysis. If you have any suggestions for improvements that can be incorporated into future editions of the book, we would be most grateful for your comments. We can be contacted most easily by electronic mail at the addresses listed below.

Richard L. Burden
burden@math.ysu.edu

J. Douglas Faires
faires@math.ysu.edu

CHAPTER

1

Mathematical Preliminaries

■ ■ ■

IN beginning chemistry courses, students are confronted with the *ideal gas law*,

$$PV = NRT,$$

which relates the pressure P, volume V, temperature T, and number of moles N of an "ideal" gas. In this equation R is a constant that depends on the measurement system.

Suppose two experiments are conducted to test this law, using the same gas in each case. In the first experiment,

$$P = 1.00 \text{ atm}, \qquad V = 0.100 \text{ m}^3,$$
$$N = 0.00420 \text{ mol}, \qquad R = 0.08206.$$

Using the ideal gas law, we predict the temperature of the gas to be

$$T = \frac{PV}{NR} = \frac{(1.00)(0.100)}{(0.00420)(0.08206)} = 290.15 \text{ K} = 17^\circ \text{C}.$$

When we measure the temperature of the gas, we find that the true temperature is 15° C.

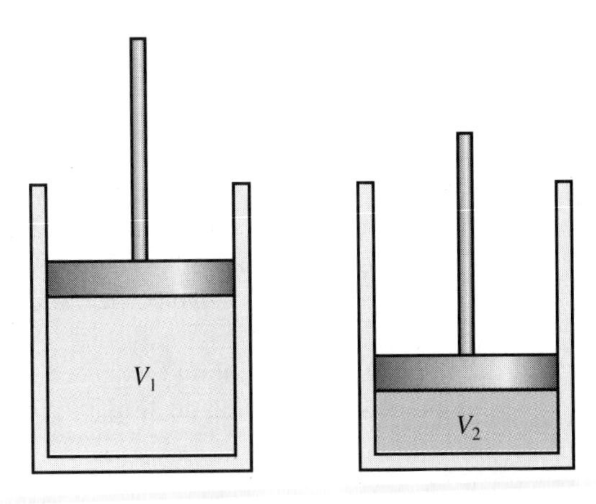

We then repeat the experiment using the same values of R and N, but we increase the pressure by a factor of two and reduce the volume by the same factor. Since the product PV remains the same, the predicted temperature is still 17° C, but now we find that the actual temperature of the gas is 19° C.

Clearly, the ideal gas law is suspect when an error of this magnitude is obtained. Before concluding that the law is invalid in this situation, however, we should examine the data to determine whether the error can be attributed to the experimental results. If so, we could determine how much more accurate our experimental results would need to be to ensure that an error of this magnitude could not occur.

Analysis of the error involved in calculations is an important topic in numerical analysis and is introduced in Section 1.2. This particular application is considered in Exercise 28 of that section.

This chapter contains a short review of those topics from elementary single-variable calculus that will be needed in later chapters, together with an introduction to convergence, error analysis, and the machine representation of numbers.

1.1 Review of Calculus

The concepts of *limit* and *continuity* of a function are fundamental to the study of calculus.

Definition 1.1 Let f be a function defined on a set X of real numbers. Then f has the **limit** L at x_0, written $\lim_{x\to x_0} f(x) = L$, if, given any real number $\epsilon > 0$, there exists a real number $\delta > 0$ such that $|f(x) - L| < \epsilon$ whenever $x \in X$ and $0 < |x - x_0| < \delta$. (See Figure 1.1.) ■

Definition 1.2 Let f be a function defined on a set X of real numbers and $x_0 \in X$. Then f is **continuous** at x_0 if $\lim_{x\to x_0} f(x) = f(x_0)$. The function f is continuous on the set X if it is continuous at each number in X. ■

$C(X)$ denotes the set of all functions that are continuous on X. When X is an interval of the real line, the parentheses in this notation are omitted. For example, the set of all functions continuous on the closed interval $[a, b]$ is denoted $C[a, b]$.

The *limit of a sequence* of real or complex numbers is defined in a similar manner.

Definition 1.3 Let $\{x_n\}_{n=1}^{\infty}$ be an infinite sequence of real or complex numbers. The sequence **converges** to a number x (called the **limit**) if, for any $\epsilon > 0$, there exists a positive integer $N(\epsilon)$ such

Figure 1.1

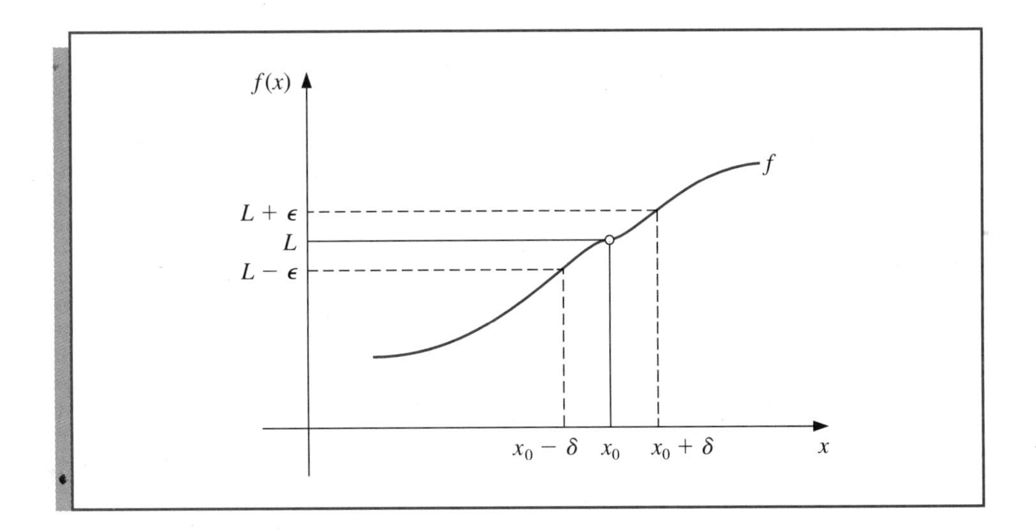

that $n > N(\epsilon)$ implies $|x_n - x| < \epsilon$. The notation $\lim_{n\to\infty} x_n = x$, or $x_n \to x$ as $n \to \infty$, means that the sequence $\{x_n\}_{n=1}^{\infty}$ converges to x. ■

The following theorem relates the concepts of convergence and continuity.

Theorem 1.4 If f is a function defined on a set X of real numbers and $x_0 \in X$, then the following are equivalent:

a. f is continuous at x_0;

b. If $\{x_n\}_{n=1}^{\infty}$ is any sequence in X converging to x_0, then $\lim_{n\to\infty} f(x_n) = f(x_0)$. ■

The functions we consider when discussing numerical methods are assumed to be continuous since this is a minimal requirement for predictable behavior. Functions that are not continuous can skip over points of interest, which causes difficulty when we attempt to approximate a solution to a problem. More sophisticated assumptions about a function generally lead to better approximation results. For example, a function with a smooth graph will normally behave more predictably than one with numerous jagged features. The analytic definition of smoothness relies on the concept of the derivative.

Definition 1.5 If f is a function defined in an open interval containing x_0, then f is **differentiable** at x_0 if

$$f'(x_0) = \lim_{x\to x_0} \frac{f(x) - f(x_0)}{x - x_0}$$

exists. The number $f'(x_0)$ is called the **derivative** of f at x_0. A function that has a derivative at each number in a set X is **differentiable** on X. The derivative of f at x_0 is the slope of the tangent line to the graph of f at $(x_0, f(x_0))$, as shown in Figure 1.2. ■

Theorem 1.6 If the function f is differentiable at x_0, then f is continuous at x_0. ■

Figure 1.2

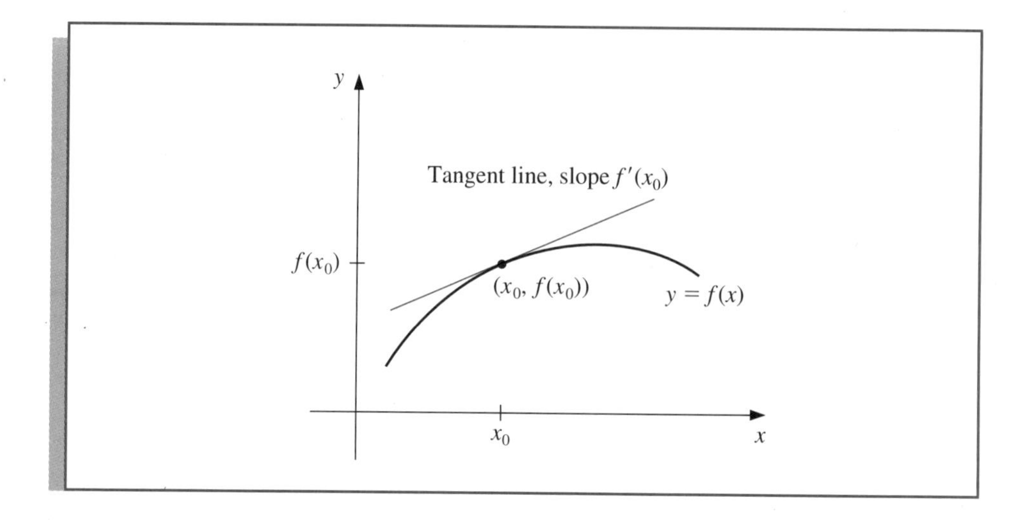

The set of all functions that have n continuous derivatives on X is denoted $C^n(X)$, and the set of functions that have derivatives of all orders on X is denoted $C^\infty(X)$. Polynomial, rational, trigonometric, exponential, and logarithmic functions are in $C^\infty(X)$, where X consists of all numbers at which the functions are defined. When X is an interval of the real line, we will again omit the parentheses in this notation.

The next theorems are of fundamental importance in deriving methods for error estimation. The proofs of these theorems and the other unreferenced results in this section can be found in any standard calculus text.

Theorem 1.7 **(Rolle's Theorem)**

Suppose $f \in C[a, b]$ and f is differentiable on (a, b). If $f(a) = f(b) = 0$, then a number c in (a, b) exists with $f'(c) = 0$. (See Figure 1.3.) ■

Figure 1.3

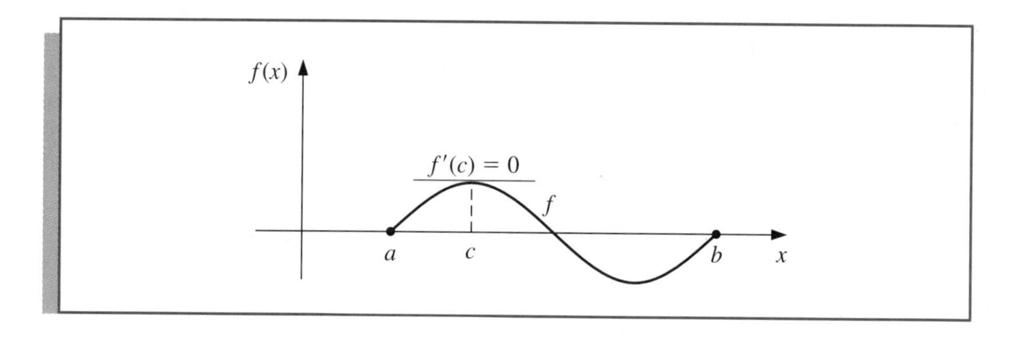

Theorem 1.8 **(Mean Value Theorem)**

If $f \in C[a, b]$ and f is differentiable on (a, b), then a number c in (a, b) exists with

$$f'(c) = \frac{f(b) - f(a)}{b - a}. \quad \text{(See Figure 1.4.)}$$ ■

Figure 1.4

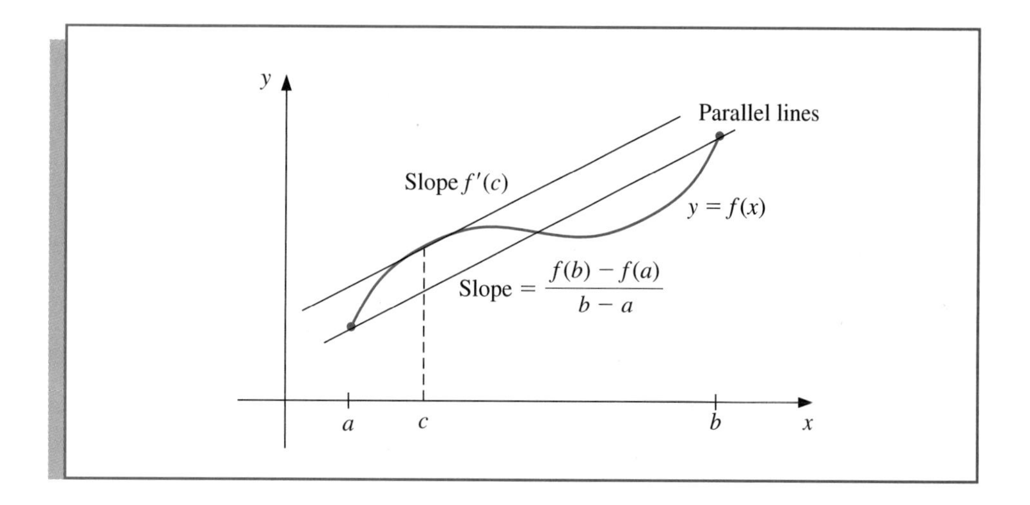

Theorem 1.9 **(Extreme Value Theorem)**

If $f \in C[a, b]$, then $c_1, c_2 \in [a, b]$ exist with $f(c_1) \leq f(x) \leq f(c_2)$ for each $x \in [a, b]$. If, in addition, f is differentiable on (a, b), then the numbers c_1 and c_2 occur either at the endpoints of $[a, b]$ or where f' is zero. (See Figure 1.5.) ■

Figure 1.5

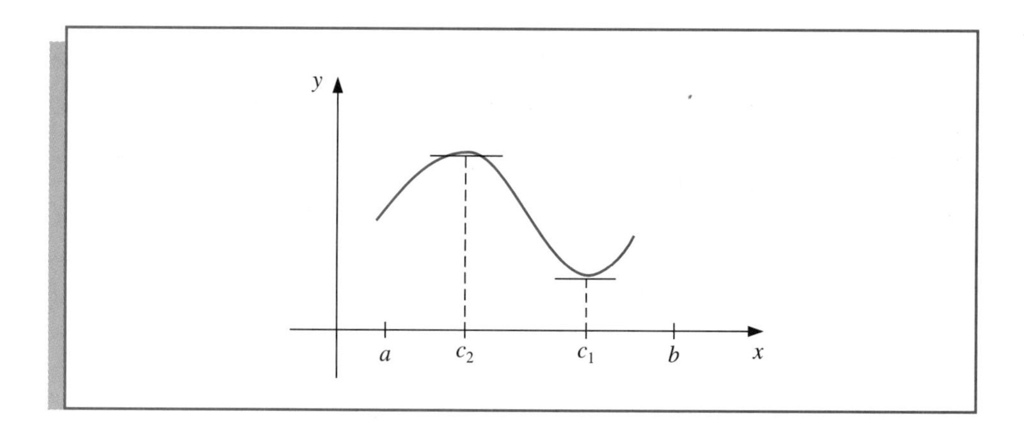

As mentioned in the preface, we will use the computer algebra system Maple whenever appropriate. This package is particularly useful for symbolic differentiation and plotting graphs. Both techniques are illustrated in the next example.

EXAMPLE 1 Find $\max_{a \leq x \leq b} |f(x)|$ for

$$f(x) = 160 \cos 2x - 64x \sin 2x$$

on the intervals $[1, 2]$ and $[0.5, 1]$.

We first illustrate the graphing capabilities of Maple. To access the graphing package enter the command

```
>with(plots);
```

The commands within the package are then displayed. We define f by entering

```
>f:= 160*cos(2*x)-64*x*sin(2*x);
```

The response from Maple is

$$f := 160\cos(2x) - 64x\sin(2x)$$

To graph f on the interval [0.5, 1] requires the command

```
>plot(f,x=0.5..1);
```

The graph appears as shown in Figure 1.6, and we can determine the coordinates of any point of the graph by moving the mouse pointer to the desired point and clicking the left mouse button. This technique can by used to estimate extrema of functions and where a graph crosses the axes.

Figure 1.6

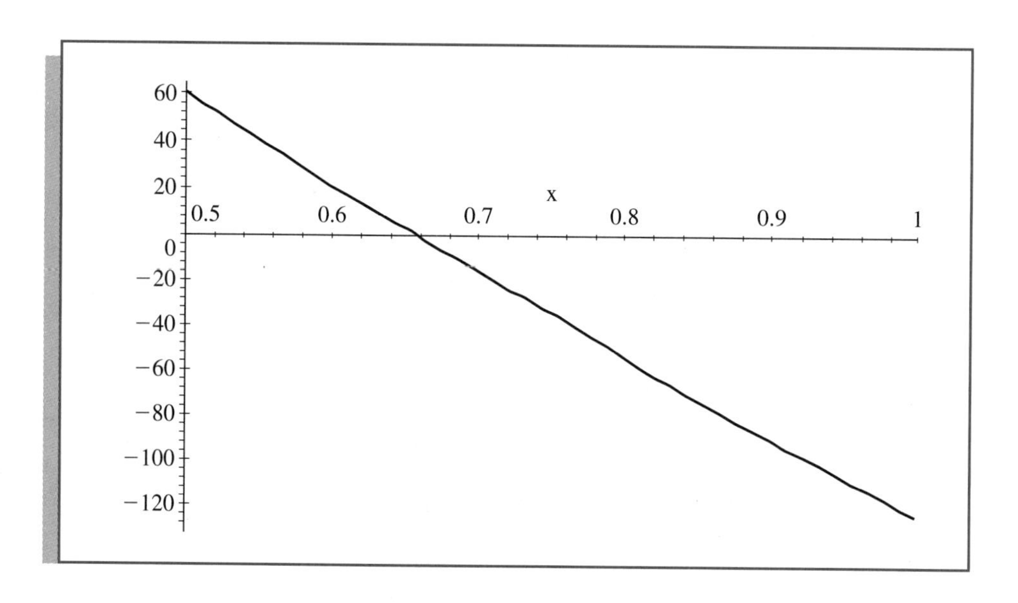

We complete the example using the Extreme Value Theorem. First consider the interval [1, 2]. To obtain the first derivative $g = f'$, enter

```
>g:=diff(f,x);
```

Maple gives

$$g := -384\sin(2x) - 128x\cos(2x)$$

We can solve $g(x) = 0$ for $1 \leq x \leq 2$ with the command

```
>fsolve(g,x,1..2);
```

to obtain 1.358229874.

We can compute $f(1.358229874)$ using

```
>evalf(subs(x=1.358229874,f));
```

Maple responds with -181.6096428.

Since

$$f(1) = -124.7785292 \quad \text{and} \quad f(2) = -7.71225990,$$

we have

$$\max_{1\leq x\leq 2} |160\cos 2x - 64x\sin 2x| = 181.6096428.$$

For the interval $[a, b] = [0.5, 1]$ we enter

```
>fsolve(g,x,0.5..1);
```

Maple responds with

$$\text{fsolve}(-384\sin(2x) - 128x\cos(2x), x, .5..1)$$

to indicate that it could not find a solution to

$$-384\sin 2x - 128x\cos 2x = 0$$

in $[0.5, 1]$. If you graph g you will see that there is no solution in this interval. Using the fact that $f(0.5) = 59.52129743$ and $f(1.0) = -124.7785292$ gives

$$\max_{0.5\leq x\leq 1} |160\cos 2x - 64x\sin 2x| = 124.7785292.$$

■

The other basic concept of calculus that will be used extensively is the Riemann integral.

Definition 1.10 The **Riemann integral** of the function f on the interval $[a, b]$ is the following limit, provided it exists:

$$\int_a^b f(x)\,dx = \lim_{\max \Delta x_i \to 0} \sum_{i=1}^{n} f(z_i)\,\Delta x_i,$$

where the numbers $x_0, x_1, \ldots, x_n$ satisfy $a = x_0 \leq x_1 \leq \cdots \leq x_n = b$, and where $\Delta x_i = x_i - x_{i-1}$ and z_i is arbitrarily chosen in the interval $[x_{i-1}, x_i]$ for each $i = 1, 2, \ldots, n$.

■

A function f that is continuous on an interval $[a, b]$ is Riemann integrable on the interval. This permits us to choose, for computational convenience, the points x_i to be equally spaced in $[a, b]$, and for each $i = 1, 2, \ldots, n$, to choose $z_i = x_i$. In this case,

$$\int_a^b f(x)\,dx = \lim_{n\to\infty} \frac{b-a}{n} \sum_{i=1}^{n} f(x_i),$$

where the numbers shown in Figure 1.7 as x_i are $x_i = a + i(b-a)/n$.

Figure 1.7

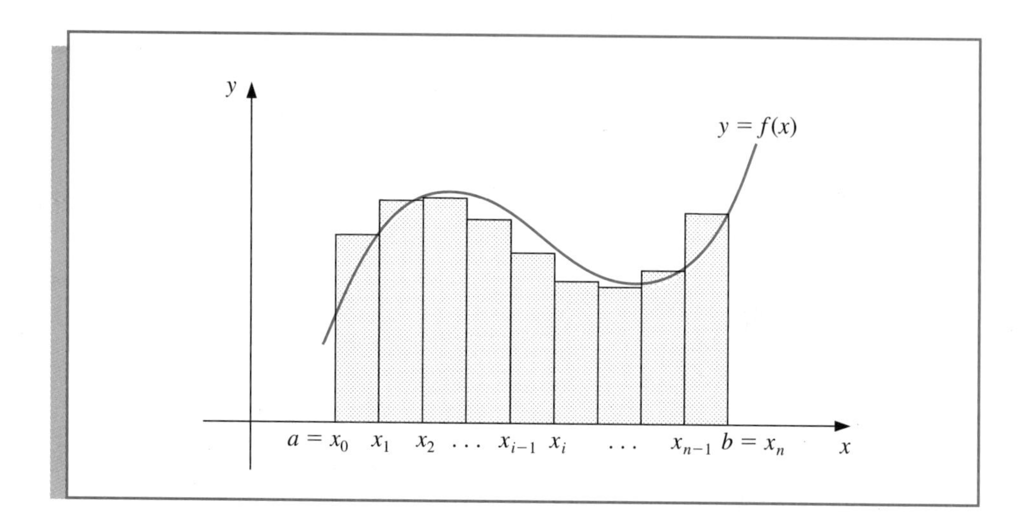

Two other results will be needed in our study of numerical methods. The first is a generalization of the usual Mean Value Theorem for Integrals.

Theorem 1.11 **(Weighted Mean Value Theorem for Integrals)**

If $f \in C[a, b]$, the Riemann integral of g exists on $[a, b]$, and $g(x)$ does not change sign on $[a, b]$, then there exists a number c in (a, b) with

$$\int_a^b f(x)g(x)\,dx = f(c)\int_a^b g(x)\,dx.$$ ■

When $g(x) \equiv 1$, Theorem 1.11 gives the **average value** of the function f over the interval $[a, b]$. (See Figure 1.8.) This average value is

$$f(c) = \frac{1}{b-a}\int_a^b f(x)\,dx.$$

The proof of Theorem 1.11 is not generally given in a basic calculus course but can be found in most analysis texts (see, for example, [Fu, p. 162]).

The other theorem we will need that is not generally presented in a basic calculus course is derived by applying Rolle's Theorem successively to $f, f', \ldots$, and, finally, to $f^{(n-1)}$.

Figure 1.8

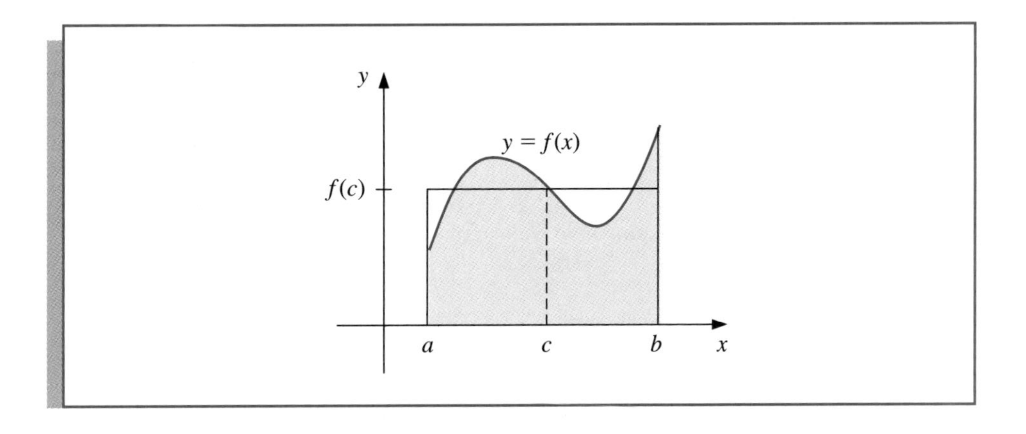

Theorem 1.12 **(Generalized Rolle's Theorem)**

Suppose $f \in C[a, b]$ is n times differentiable on (a, b). If $f(x)$ is zero at the $n + 1$ distinct numbers $x_0, \ldots, x_n$ in $[a, b]$, then a number c in (a, b) exists with $f^{(n)}(c) = 0$. ■

The next theorem is the Intermediate Value Theorem. Although its statement is intuitively clear, the proof is beyond the scope of the usual calculus course. The proof can be found in most analysis texts (see, for example, [Fu, p. 67]).

Theorem 1.13 **(Intermediate Value Theorem)**

If $f \in C[a, b]$ and K is any number between $f(a)$ and $f(b)$, then there exists a number c in (a, b) for which $f(c) = K$. ■

Figure 1.9 shows one choice for the number that is guaranteed by the Intermediate Value Theorem. In this example there are two other possibilities.

Figure 1.9

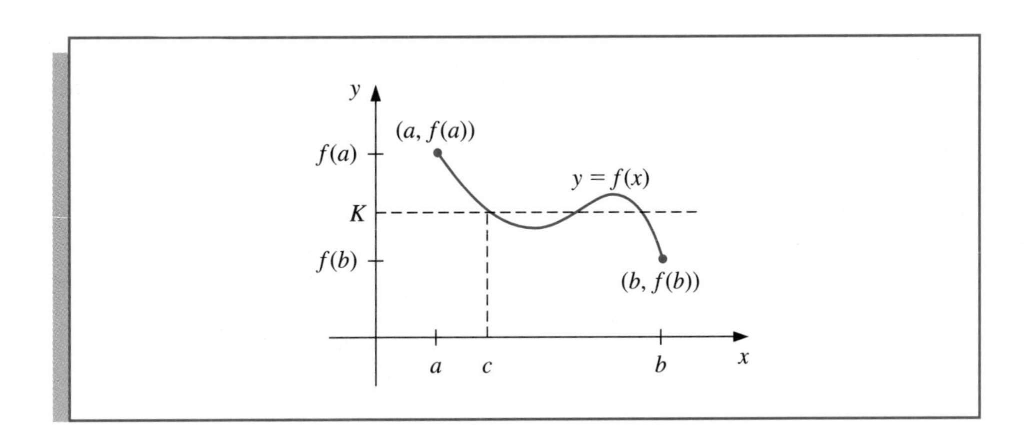

EXAMPLE 2 Show that $x^5 - 2x^3 + 3x^2 - 1 = 0$ has a solution in the interval $[0, 1]$.

Consider $f(x) = x^5 - 2x^3 + 3x^2 - 1$. Since f is a polynomial function, it is continuous on $[0, 1]$. Also

$$f(0) = -1 < 0 < 1 = f(1),$$

so the Intermediate Value Theorem implies that there is a number x in $(0, 1)$ with $x^5 - 2x^3 + 3x^2 - 1 = 0$. ■

As seen in Example 2, the Intermediate Value Theorem is important as an aid to determine when solutions to certain problems exist. It does not, however, give a means for finding these solutions. This topic is considered in Chapter 2.

The final theorem in this review from calculus describes the development of the Taylor polynomials. The importance of the Taylor polynomials to the study of numerical analysis cannot be overemphasized, and the following result will be used repeatedly.

Theorem 1.14 **(Taylor's Theorem)**

Suppose $f \in C^n[a, b]$, that $f^{(n+1)}$ exists on $[a, b]$, and $x_0 \in [a, b]$. For every $x \in [a, b]$ there exists a number $\xi(x)$ between x_0 and x with

$$f(x) = P_n(x) + R_n(x),$$

where

$$P_n(x) = f(x_0) + f'(x_0)(x - x_0) + \frac{f''(x_0)}{2!}(x - x_0)^2 + \cdots + \frac{f^{(n)}(x_0)}{n!}(x - x_0)^n$$
$$= \sum_{k=0}^{n} \frac{f^{(k)}(x_0)}{k!}(x - x_0)^k$$

and

$$R_n(x) = \frac{f^{(n+1)}(\xi(x))}{(n + 1)!}(x - x_0)^{n+1}.$$ ■

Here $P_n(x)$ is called the ***n*th Taylor polynomial** for f about x_0 and $R_n(x)$ is called the **remainder term** (or **truncation error**) associated with $P_n(x)$. The infinite series obtained by taking the limit of $P_n(x)$ as $n \to \infty$ is called the **Taylor series** for f about x_0. In the case $x_0 = 0$, the Taylor polynomial is often called a **Maclaurin polynomial**, and the Taylor series is called a **Maclaurin series**.

The term **truncation error** refers to the error involved in using a truncated, or finite, summation to approximate the sum of an infinite series. This terminology will be reintroduced in subsequent chapters.

EXAMPLE 3 Determine (a) the second and (b) the third Taylor polynomials for $f(x) = \cos x$ about $x_0 = 0$, and use these polynomials to approximate $\cos(0.01)$. (c) Use the third Taylor polynomial and its remainder term to approximate $\int_0^{0.1} \cos x \, dx$.

Since $f \in C^\infty(\mathbb{R})$, Taylor's Theorem can be applied for any $n > 0$. Also, $f'(x) = -\sin x$, $f''(x) = -\cos x$, $f'''(x) = \sin x$, and $f^{(4)}(x) = \cos x$, so $f(0) = 1$, $f'(0) = 0$, $f''(0) = -1$ and $f'''(0) = 0$.

a. For $n = 2$ and $x_0 = 0$, we have

$$\cos x = 1 - \frac{1}{2}x^2 + \frac{1}{6}x^3 \sin \xi(x),$$

where $\xi(x)$ is a number between 0 and x. (See Figure 1.10.)

Figure 1.10

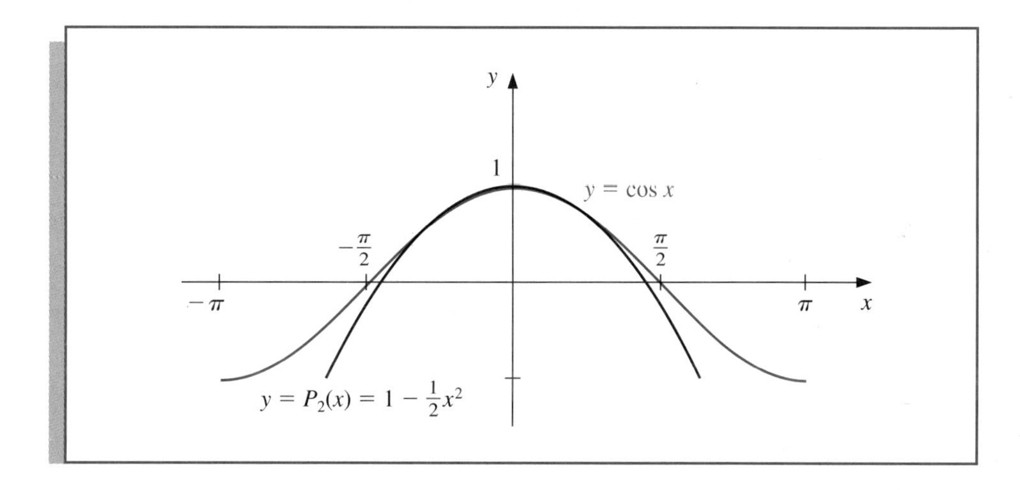

With $x = 0.01$, the Taylor polynomial and remainder term are

$$\begin{aligned}\cos 0.01 &= 1 - \frac{1}{2}(0.01)^2 + \frac{1}{6}(0.01)^3 \sin \xi(x) \\ &= 0.99995 + (0.1\overline{6}) \times 10^{-6} \sin \xi(x),\end{aligned}$$

where $0 < \xi(x) < 0.01$. (The bar over the 6 in 0.16 is used to indicate that this digit repeats indefinitely.) Since $|\sin \xi(x)| < 1$ for all x, we have

$$|\cos 0.01 - 0.99995| \leq 0.1\overline{6} \times 10^{-6},$$

so the approximation 0.99995 matches at least the first five digits of $\cos 0.01$. Using standard tables we find that $\cos 0.01 = 0.99995000042$, so the approximation actually gives agreement through the first nine digits.

The error bound is much larger than the actual error. This is due in part to the poor bound we used for $|\sin \xi(x)|$. It can be shown that for all values of x, we have $|\sin x| \leq |x|$. Since $0 < \xi < 0.01$, we could have used the fact that $|\sin \xi(x)| \leq 0.01$ in the error formula, producing the bound $0.1\overline{6} \times 10^{-8}$.

b. Since $f'''(0) = 0$, the third Taylor polynomial and remainder term about $x_0 = 0$ are

$$\cos x = 1 - \frac{1}{2}x^2 + \frac{1}{24}x^4 \cos \tilde{\xi}(x),$$

where $0 < \tilde{\xi}(x) < 0.01$. The approximating polynomial remains the same, and the approximation is still 0.99995, but we now have a much better accuracy assurance. Since $|\cos \tilde{\xi}(x)| \leq 1$ for all x, we have

$$\left| \frac{1}{24}x^4 \cos \tilde{\xi}(x) \right| \leq \frac{1}{24}(0.01)^4(1) \approx 4.2 \times 10^{-10}.$$

The first two parts of the example illustrate the two objectives of numerical analysis. The first is to find an approximation, which the Taylor polynomials in both parts provide. The second is to determine the accuracy of the approximation. In this case the third Taylor polynomial was much more informative than the second, even though both polynomials gave the same approximation.

c. Since

$$\cos x = 1 - \frac{1}{2}x^2 + \frac{1}{24}x^4 \cos \tilde{\xi}(x),$$

we have

$$\begin{aligned}
\int_0^{0.1} \cos x \, dx &= \int_0^{0.1} \left(1 - \frac{1}{2}x^2\right) dx + \frac{1}{24} \int_0^{0.1} x^4 \cos \tilde{\xi}(x) \, dx \\
&= \left[x - \frac{1}{6}x^3 \right]_0^{0.1} + \frac{1}{24} \int_0^{0.1} x^4 \cos \tilde{\xi}(x) \, dx \\
&= 0.1 - \frac{1}{6}(0.1)^3 + \frac{1}{24} \int_0^{0.1} x^4 \cos \tilde{\xi}(x) \, dx.
\end{aligned}$$

So

$$\int_0^{0.1} \cos x \, dx \approx 0.1 - \frac{1}{6}(0.1)^3 = 0.0998\overline{3}.$$

A bound for the error in this approximation is determined from the integral of the Taylor remainder term and the fact that $|\cos \tilde{\xi}(x)| \leq 1$ for all x:

$$\begin{aligned}
\frac{1}{24} \left| \int_0^{0.1} x^4 \cos \tilde{\xi}(x) \, dx \right| &\leq \frac{1}{24} \int_0^{0.1} x^4 |\cos \tilde{\xi}(x)| \, dx \\
&\leq \frac{1}{24} \int_0^{0.1} x^4 \, dx = 8.\overline{3} \times 10^{-8}.
\end{aligned}$$

Since the true value of this integral is

$$\int_0^{0.1} \cos x\,dx = \sin x\Big]_0^{0.1} = \sin 0.1 \approx 0.099833417,$$

the error for this approximation is within the bound. ■

We could also use Maple in Example 2. Define f by

```
>f:=cos(x);
```

We obtain the third Taylor polynomial with

```
>s3:=taylor(f,x=0,4): p3:=convert(s3, polynom);
```

The first part computes the Taylor series with four terms (degree 3) and remainder expanded about $x_0 = 0$. The second part converts the series `s3` to the polynomial `p3` by dropping the remainder. To obtain 11 decimal digits of display we enter

```
>Digits:=11;
```

To evaluate $f(0.5)$, $P_3(0.5)$, and $|f(0.5) - P_3(0.5)|$, we enter the following commands.

```
>y1:=evalf(subs(x=0.01,f));
>y2:=evalf(subs(x=0.01,p3));
>error:=abs(y1-y2);
```

These commands produce the values

$$y1 := .99995000042,\ y2 := .99995000000, \quad \text{and error } := .42 \times 10^{-9}.$$

To obtain a graph similar to Figure 1.10, enter

```
>plot({ f,p3 },x=-Pi..Pi);
```

The commands and responses for the integrals are

```
>q1:=int(f,x=0..0.1);
>q2:=int(p3,x=0..0.1);
>error:=abs(q1-q2);
```

These give the values

$$q1 := .099833416647,\ q2 := .099833333333, \quad \text{and error } := .83314 \times 10^{-7}.$$

EXERCISE SET 1.1

1. Show that the following equations have at least one solution in the given intervals.
   a. $x\cos x - 2x^2 + 3x - 1 = 0$, $[0.2, 0.3]$ and $[1.2, 1.3]$
   b. $(x-2)^2 - \ln x = 0$, $[1, 2]$ and $[e, 4]$

c. $2x\cos(2x) - (x-2)^2 = 0$, $[2, 3]$ and $[3, 4]$

d. $x - (\ln x)^x = 0$, $[4, 5]$

2. Find intervals containing solutions to the following equations.

a. $x - 3^{-x} = 0$

b. $4x^2 - e^x = 0$

c. $x^3 - 2x^2 - 4x + 3 = 0$

d. $x^3 + 4.001x^2 + 4.002x + 1.101 = 0$

3. Show that the first derivatives of the following functions are zero at least once in the given intervals.

a. $f(x) = 1 - e^x + (e-1)\sin((\pi/2)x)$, $[0, 1]$

b. $f(x) = (x-1)\tan x + x\sin \pi x$, $[0, 1]$

c. $f(x) = x\sin \pi x - (x-2)\ln x$, $[1, 2]$

d. $f(x) = (x-2)\sin x \ln(x+2)$, $[-1, 3]$

4. Find $\max_{a \le x \le b} |f(x)|$ for the following functions and intervals.

a. $f(x) = (2 - e^x + 2x)/3$, $[0, 1]$

b. $f(x) = (4x-3)/(x^2 - 2x)$, $[0.5, 1]$

c. $f(x) = 2x\cos(2x) - (x-2)^2$, $[2, 4]$

d. $f(x) = 1 + e^{-\cos(x-1)}$, $[1, 2]$

5. Use the Intermediate Value Theorem and Rolle's Theorem to show that the graph of $f(x) = x^3 + 2x + k$ crosses the x-axis exactly once, regardless of the value of the constant k.

6. Suppose $f \in C[a, b]$ and $f'(x)$ exists on (a, b). Show that if $f'(x) \neq 0$ for all x in (a, b), then there can exist at most one number p in $[a, b]$ with $f(p) = 0$.

7. Let $f(x) = x^3$.

a. Find the second Taylor polynomial $P_2(x)$ about $x_0 = 0$.

b. Find $R_2(0.5)$ and the actual error in using $P_2(0.5)$ to approximate $f(0.5)$.

c. Repeat part (a) using $x_0 = 1$.

d. Repeat part (b) using the polynomial from part (c).

8. Find the third Taylor polynomial $P_3(x)$ for the function $f(x) = \sqrt{x+1}$ about $x_0 = 0$. Approximate $\sqrt{0.5}$, $\sqrt{0.75}$, $\sqrt{1.25}$, and $\sqrt{1.5}$ using $P_3(x)$, and find the actual errors.

9. Find the second Taylor polynomial $P_2(x)$ for the function $f(x) = e^x \cos x$ about $x_0 = 0$.

a. Use $P_2(0.5)$ to approximate $f(0.5)$. Find an upper bound for the error $|f(0.5) - P_2(0.5)|$ using the error formula, and compare it to the actual error.

b. Find a bound for the error $|f(x) - P_2(x)|$ in using $P_2(x)$ to approximate $f(x)$ on the interval $[0, 1]$.

c. Approximate $\int_0^1 f(x)\,dx$ using $\int_0^1 P_2(x)\,dx$.

d. Find an upper bound for the error in (c) using $\int_0^1 |R_2(x)|\,dx$, and compare the bound to the actual error.

10. Repeat Exercise 9 using $x_0 = \pi/6$.

11. Find the third Taylor polynomial $P_3(x)$ for the function $f(x) = (x-1)\ln x$ about $x_0 = 1$.

a. Use $P_3(0.5)$ to approximate $f(0.5)$. Find an upper bound for the error $|f(0.5) - P_3(0.5)|$ using the error formula, and compare it to the actual error.

b. Find a bound for the error $|f(x) - P_3(x)|$ in using $P_3(x)$ to approximate $f(x)$ on the interval $[0.5, 1.5]$.

c. Approximate $\int_{0.5}^{1.5} f(x)\,dx$ using $\int_{0.5}^{1.5} P_3(x)\,dx$.

d. Find an upper bound for the error in (c) using $\int_{0.5}^{1.5} |R_3(x)|\,dx$, and compare the bound to the actual error.

12. Let $f(x) = 2x\cos(2x) - (x-2)^2$ and $x_0 = 0$.

a. Find the third Taylor polynomial $P_3(x)$ and use it to approximate $f(0.4)$.

b. Use the error formula in Taylor's Theorem to find an upper bound for the error $|f(0.4) - P_3(0.4)|$. Compute the actual error.

c. Find the fourth Taylor polynomial $P_4(x)$ and use it to approximate $f(0.4)$.

d. Use the error formula in Taylor's Theorem to find an upper bound for the error $|f(0.4) - P_4(0.4)|$. Compute the actual error.

13. Find the fourth Taylor polynomial $P_4(x)$ for the function $f(x) = xe^{x^2}$ about $x_0 = 0$.

a. Find an upper bound for $|f(x) - P_4(x)|$ for $0 \le x \le 0.4$.

b. Approximate $\int_0^{0.4} f(x)\,dx$ using $\int_0^{0.4} P_4(x)\,dx$.

c. Find an upper bound for the error in (b) using $\int_0^{0.4} P_4(x)\,dx$.

d. Approximate $f'(0.2)$ using $P_4'(0.2)$, and find the error.

14. Use the error term of a Taylor polynomial to estimate the error involved in using $\sin x \approx x$ to approximate $\sin 1^\circ$.

15. Use a Taylor polynomial about $\pi/4$ to approximate $\cos 42^\circ$ to an accuracy of 10^{-6}.

16. Let $f(x) = e^{x/2}\sin\frac{x}{3}$. Use Maple to determine the following.

a. The third Maclaurin polynomial $P_3(x)$.

b. $f^{(4)}(x)$ and a bound for the error $|f(x) - P_3(x)|$ on $[0, 1]$.

17. Let $f(x) = \ln(x^2 + 2)$. Use Maple to determine the following.

a. The Taylor polynomial $P_3(x)$ for f expanded about $x_0 = 1$.

b. The maximum error $|f(x) - P_3(x)|$ for $0 \le x \le 1$.

c. The Maclaurin polynomial $\tilde{P}_3(x)$ for f.

d. The maximum error $|f(x) - \tilde{P}_3(x)|$ for $0 \le x \le 1$.

e. Does $P_3(0)$ approximate $f(0)$ better than $\tilde{P}_3(1)$ approximates $f(1)$?

18. Let $f(x) = (1-x)^{-1}$ and $x_0 = 0$. Find the nth Taylor polynomial $P_n(x)$ for $f(x)$ about x_0. Find a value of n necessary for $P_n(x)$ to approximate $f(x)$ to within 10^{-6} on $[0, 0.5]$.

19. Let $f(x) = e^x$ and $x_0 = 0$. Find the nth Taylor polynomial $P_n(x)$ for $f(x)$ about x_0. Find a value of n necessary for $P_n(x)$ to approximate $f(x)$ to within 10^{-6} on $[0, 0.5]$.

20. Find the nth Maclaurin polynomial $P_n(x)$ for $f(x) = \arctan x$.

21. The polynomial $P_2(x) = 1 - \frac{1}{2}x^2$ is to be used to approximate $f(x) = \cos x$ in $[-\frac{1}{2}, \frac{1}{2}]$. Find a bound for the maximum error.

22. The nth Taylor polynomial for a function f at x_0 is sometimes referred to as the polynomial of degree at most n that "best" approximates f near x_0.

a. Explain why this description is accurate.

b. Find the quadratic polynomial that best approximates a function f near $x_0 = 1$ if the tangent line at $x_0 = 1$ has equation $y = 4x - 1$, and if $f''(1) = 6$.

23. A Maclaurin polynomial for e^x is used to give the approximation 2.5 to e. The error bound in this approximation is established to be $E = \frac{1}{6}$. Find a bound for the error in E.

24. The error function defined by

$$\operatorname{erf}(x) = \frac{2}{\sqrt{\pi}} \int_0^x e^{-t^2}\,dt$$

gives the probability that any one of a series of trials will lie within x units of the mean, assuming that the trials have a normal distribution with mean 0 and standard deviation $\sqrt{2}/2$. This integral cannot be evaluated in terms of elementary functions, so an approximating technique must be used.

a. Integrate the Maclaurin series for e^{-t^2} to show that

$$\operatorname{erf}(x) = \frac{2}{\sqrt{\pi}} \sum_{k=0}^{\infty} \frac{(-1)^k x^{2k+1}}{(2k+1)k!}.$$

b. The error function can also be expressed in the form

$$\operatorname{erf}(x) = \frac{2}{\sqrt{\pi}} e^{-x^2} \sum_{k=0}^{\infty} \frac{2^k x^{2k+1}}{1 \cdot 3 \cdot 5 \cdots (2k+1)}.$$

Verify that the two series agree for $k = 1, 2, 3$, and 4. [*Hint:* Use the Maclaurin series for e^{-x^2}.]

c. Use the series in part (a) to approximate erf(1) to within 10^{-7}.

d. Use the same number of terms used in part (c) to approximate erf(1) with the series in part (b).

e. Explain why difficulties occur using the series in part (b) to approximate erf(x).

25. A function $f : [a, b] \to \mathbb{R}$ is said to satisfy a *Lipschitz condition* with Lipschitz constant L on $[a, b]$ if, for every $x, y \in [a, b]$, we have $|f(x) - f(y)| \leq L|x - y|$.

a. Show that if f satisfies a Lipschitz condition with Lipschitz constant L on an interval $[a, b]$, then $f \in C[a, b]$.

b. Show that if f has a derivative that is bounded on $[a, b]$ by L, then f satisfies a Lipschitz condition with Lipschitz constant L on $[a, b]$.

c. Give an example of a function that is continuous on a closed interval but does not satisfy a Lipschitz condition on the interval.

26. Suppose $f \in C[a, b]$, that x_1 and x_2 are in $[a, b]$, and that c_1 and c_2 are positive constants. Show that a number ξ exists between x_1 and x_2 with

$$f(\xi) = \frac{c_1 f(x_1) + c_2 f(x_2)}{c_1 + c_2}.$$

1.2 Roundoff Errors and Computer Arithmetic

The arithmetic performed by a calculator or computer is different from the arithmetic that we use in our algebra and calculus courses. From our past experiences we expect that we will always have as true statements such things as $2 + 2 = 4$, $4^2 = 16$, and $(\sqrt{3})^2 = 3$. In standard computational arithmetic we will have the first two but not the third. To understand why this is true we must explore the world of finite-digit arithmetic.

In our traditional mathematical world we permit numbers with an infinite number of nonperiodic digits. The arithmetic we use in this world *defines* $\sqrt{3}$ as the unique positive number that when multiplied by itself produces the integer 3. In the computational world, however, each representable number has only a fixed, finite number of digits. Since $\sqrt{3}$

does not have a finite-digit representation, it is given an approximate representation within the machine, one whose square will not be precisely 3, although it will likely be sufficiently close to 3 to be acceptable in most situations. In most cases this machine representation and arithmetic is satisfactory and passes without notice or concern. But we must be aware that this is not always true and be alert to the problems that it can produce.

Roundoff error occurs when a calculator or computer is used to perform real-number calculations. This error arises because the arithmetic performed in a machine involves numbers with only a finite number of digits, with the result that calculations are performed with approximate representations of the actual numbers. In a typical computer, only a relatively small subset of the real number system is used for the representation of all the real numbers. This subset contains only rational numbers, both positive and negative, and stores a fractional part, called the **mantissa**, together with an exponential part, called the **characteristic**. For example, a single-precision **floating-point** number used in the IBM 3000 series consists of a 1-binary-digit (**bit**) sign indicator, a 7-bit exponent with a base of 16, and a 24-bit mantissa.

Since 24 binary digits correspond to between 6 and 7 decimal digits, we can assume that this number has at least 6 decimal digits of precision for the floating-point number system. The exponent of 7 binary digits gives a range of 0000000 = 0 to 1111111 = 127. However, using only positive integers for the characteristic does not permit an adequate representation of numbers with small magnitude. To ensure that numbers with small magnitude are equally representable, 64 is subtracted from the characteristic, so the range of the exponential part is actually from -64 to $+63$.

Consider, for example, the machine number

0	1000010	101100110000010000000000

.

The leftmost bit is a zero, which indicates that the number is positive. The next seven bits, 1000010, are equivalent to the decimal number

$$1 \cdot 2^6 + 0 \cdot 2^5 + 0 \cdot 2^4 + 0 \cdot 2^3 + 0 \cdot 2^2 + 1 \cdot 2^1 + 0 \cdot 2^0 = 66$$

and are used to describe the characteristic, 16^{66-64}. The final 24 bits indicate that the mantissa is

$$1 \cdot \left(\frac{1}{2}\right)^1 + 1 \cdot \left(\frac{1}{2}\right)^3 + 1 \cdot \left(\frac{1}{2}\right)^4 + 1 \cdot \left(\frac{1}{2}\right)^7 + 1 \cdot \left(\frac{1}{2}\right)^8 + 1 \cdot \left(\frac{1}{2}\right)^{14}.$$

As a consequence, this machine number precisely represents the decimal number

$$+\left[\left(\frac{1}{2}\right)^1 + \left(\frac{1}{2}\right)^3 + \left(\frac{1}{2}\right)^4 + \left(\frac{1}{2}\right)^7 + \left(\frac{1}{2}\right)^8 + \left(\frac{1}{2}\right)^{14}\right] 16^{66-64} = 179.015625.$$

However, the next smallest machine number is

0	1000010	101100110000001111111111

= 179.0156097412109375,

and the next largest machine number is

0	1000010	101100110000010000000001	= 179.0156402587890625.

So, our original machine number represents not only 179.015625, but many real numbers that are between this number and its nearest machine number neighbors. To be precise, the original machine number is used to represent any real number in the interval

$$[179.01561737060546875, 179.01563262939453125).$$

To ensure uniqueness of representation and obtain all the available precision of the system, a normalization is imposed by requiring that at least one of the four leftmost bits of the mantissa of a machine number is a 1. Consequently, 15×2^{28} numbers of the form

$$\pm 0.d_1 d_2 \ldots d_{24} \times 16^{e_1 e_2 \ldots e_7}$$

are used by this system to represent all real numbers. With this representation, the number of binary machine numbers used to represent $[16^n, 16^{n+1}]$ is independent of n within the limit of the machine—that is, for $-64 \leq n \leq 63$. This requirement also implies that the smallest normalized, positive machine number that can be represented is

0	0000000	000100000000000000000000	$= 16^{-65} \approx 10^{-78}$,

whereas the largest is

0	1111111	111111111111111111111111	$\approx 16^{63} \approx 10^{76}$.

Numbers occurring in calculations that have a magnitude of less than 16^{-65} result in what is called **underflow** and are often set to zero, and numbers greater than 16^{63} result in **overflow** and cause the computations to halt.

The arithmetic used on microcomputers differs somewhat from that used on mainframe computers. In 1985, the IEEE (Institute for Electrical and Electronic Engineers) published a report called *Binary Floating Point Arithmetic Standard 754–1985*. In this report, formats were specified for single, double, and extended precisions, and these standards are generally followed by microcomputer manufacturers using floating-point hardware. For example, the numerical coprocessor for IBM-compatible microcomputers implements a 64-bit representation for a real number, called a *long real*. The first bit is a sign indicator, denoted s. This is followed by an 11-bit exponent c and a 52-bit mantissa f. The base for the exponent is 2 and, to obtain numbers with both large and small magnitude, the actual exponent is $c - 1023$. In addition, a normalization is imposed that requires that the units digit be 1, and this digit is not stored as part of the 52-bit mantissa. Using this system gives a floating-point number of the form

$$(-1)^s * 2^{c-1023} * (1 + f),$$

which provides between 15 and 16 decimal digits of precision and a range of approximately 10^{-308} to 10^{308}. This is the form that most compilers use when a mathematical coprocessor is available.

The use of binary digits tends to conceal the computational difficulties that occur when a finite collection of machine numbers is used to represent all the real numbers. To explain the problems that can arise, we will now assume, for simplicity, that machine numbers are represented in the normalized decimal floating-point form

$$\pm 0.d_1d_2\ldots d_k \times 10^n, \quad 1 \le d_1 \le 9, \quad 0 \le d_i \le 9, \quad \text{for each } i = 2, \ldots, k.$$

We call numbers of this form *decimal machine numbers*.

Any positive real number y can be normalized to

$$y = 0.d_1d_2\ldots d_kd_{k+1}d_{k+2}\ldots \times 10^n.$$

If y is within the numerical range of the machine, the floating-point form of y, denoted $fl(y)$, is obtained by terminating the mantissa of y at k decimal digits. There are two ways of performing this termination. One method, called **chopping**, is to simply chop off the digits $d_{k+1}d_{k+2}\ldots$ to obtain

$$fl(y) = 0.d_1d_2\ldots d_k \times 10^n.$$

The other method, called **rounding**, adds $5 \times 10^{n-(k+1)}$ to y and then chops to obtain a number of the form

$$fl(y) = 0.\delta_1\delta_2\ldots\delta_k \times 10^n.$$

In this method, if $d_{k+1} \ge 5$, we add 1 to d_k to obtain $fl(y)$; that is, we round up. If $d_{k+1} < 5$, we merely chop off all but the first k digits; so we round down.

EXAMPLE 1 The number π has an infinite decimal expansion of the form $\pi = 3.14159265\ldots$. Written in normalized decimal form, we have

$$\pi = 0.314159265\ldots \times 10^1.$$

The five-digit floating-point form of π using chopping is

$$fl(\pi) = 0.31415 \times 10^1 = 3.1415.$$

Since the sixth digit of the decimal expansion of π is a 9, the floating-point form of π using five-digit rounding is

$$fl(\pi) = (0.31415 + 0.00001) \times 10^1 = 3.1416.$$

■

The error that results from replacing a number with its floating-point form is called **roundoff error** (regardless of whether the rounding or chopping method is used). The following definition describes two methods for measuring approximation errors.

Definition 1.15 If p^* is an approximation to p, the **absolute error** is $|p - p^*|$, and the **relative error** is $\frac{|p - p^*|}{|p|}$, provided that $p \neq 0$. ■

Consider the absolute and relative errors in representing p by p^* in the following example.

EXAMPLE 2

a. If $p = 0.3000 \times 10^1$ and $p^* = 0.3100 \times 10^1$, the absolute error is 0.1 and the relative error is $0.333\overline{3} \times 10^{-1}$.

b. If $p = 0.3000 \times 10^{-3}$ and $p^* = 0.3100 \times 10^{-3}$, the absolute error is 0.1×10^{-4} and the relative error is $0.333\overline{3} \times 10^{-1}$.

c. If $p = 0.3000 \times 10^4$ and $p^* = 0.3100 \times 10^4$, the absolute error is 0.1×10^3 and the relative error is $0.333\overline{3} \times 10^{-1}$.

This example shows that the same relative error, $0.333\overline{3} \times 10^{-1}$, occurs for widely varying absolute errors. As a measure of accuracy, the absolute error may be misleading and the relative error more meaningful. ■

The following definition uses relative error to give a measure of significant digits of accuracy for an approximation.

Definition 1.16 The number p^* is said to approximate p to t **significant digits** (or figures) if t is the largest nonnegative integer for which

$$\frac{|p - p^*|}{|p|} < 5 \times 10^{-t}.$$ ■

Table 1.1 illustrates the continuous nature of significant digits by listing, for the various values of p, the least upper bound of $|p - p^*|$, denoted max $|p - p^*|$, when p^* agrees with p to four significant digits.

Table 1.1

p	0.1	0.5	100	1000	5000	9990	10000
$\max \lvert p - p^* \rvert$	0.00005	0.00025	0.05	0.5	2.5	4.995	5.

Returning to the machine representation of numbers, we see that the floating-point representation $fl(y)$ for the number y has the relative error

$$\left| \frac{y - fl(y)}{y} \right|.$$

If k decimal digits and chopping are used for the machine representation of

$$y = 0.d_1 d_2 \ldots d_k d_{k+1} \ldots \times 10^n,$$

then

$$\left|\frac{y - fl(y)}{y}\right| = \left|\frac{0.d_1d_2 \ldots d_kd_{k+1} \ldots \times 10^n - 0.d_1d_2 \ldots d_k \times 10^n}{0.d_1d_2 \ldots \times 10^n}\right|$$

$$= \left|\frac{0.d_{k+1}d_{k+2} \ldots \times 10^{n-k}}{0.d_1d_2 \ldots \times 10^n}\right| = \left|\frac{0.d_{k+1}d_{k+2} \ldots}{0.d_1d_2 \ldots}\right| \times 10^{-k}.$$

Since $d_1 \neq 0$, the minimal value of the denominator is 0.1. The numerator is bounded above by 1. As a consequence,

$$\left|\frac{y - fl(y)}{y}\right| \leq \frac{1}{0.1} \times 10^{-k} = 10^{-k+1}.$$

In a similar manner, a bound for the relative error when using k-digit rounding arithmetic is $0.5 \times 10^{-k+1}$. (See Exercise 24.)

Note that the bounds for the relative error using k-digit arithmetic are independent of the number being represented. This is due to the manner in which the machine numbers are distributed along the real line. Because of the exponential form of the characteristic, the same number of decimal machine numbers is used to represent each of the intervals $[0.1, 1]$, $[1, 10]$, and $[10, 100]$. In fact, within the limits of the machine, the number of decimal machine numbers in $[10^n, 10^{n+1}]$ is constant for all integers n.

In addition to inaccurate representation of numbers, the arithmetic performed in a computer is not exact. The arithmetic involves manipulating binary digits by various shifting, or logical, operations. Since the actual mechanics of these operations are not pertinent to this presentation, we shall devise our own approximation to computer arithmetic. Although our arithmetic will not give the exact picture, it suffices to explain the problems that occur. (For an explanation of the manipulations actually involved, the reader is urged to consult more technically oriented computer science texts, such as [Ma], *Computer system architecture*.)

Assume that the floating-point representations $fl(x)$ and $fl(y)$ are given for the real numbers x and y and that the symbols $\oplus, \ominus, \otimes, \oslash$ represent machine addition, subtraction, multiplication, and division operations, respectively. We will assume a finite-digit arithmetic given by

$$x \oplus y = fl(fl(x) + fl(y)), \quad x \otimes y = fl(fl(x) \times fl(y)),$$

$$x \ominus y = fl(fl(x) - fl(y)), \quad x \oslash y = fl(fl(x) \div fl(y)).$$

This arithmetic corresponds to performing exact arithmetic on the floating-point representations of x and y and then converting the exact result to its finite-digit floating-point representation.

Rounding arithmetic is easily implemented in Maple. The command

```
>Digits:=t;
```

causes all arithmetic to be rounded to t digits. For example, $fl(fl(x) + fl(y))$ is performed using t-digit rounding arithmetic by

```
>evalf(evalf(x)+evalf(y));
```

Implementing t-digit chopping arithmetic in Maple is more difficult and requires a sequence of steps or a procedure. Exercise 27 explores this problem.

EXAMPLE 3 Suppose that $x = \frac{1}{3}$, $y = \frac{5}{7}$, and that five-digit chopping is used for arithmetic calculations involving x and y. Table 1.2 lists the values of these computer-type operations on $fl(x) = 0.33333 \times 10^0$ and $fl(y) = 0.71428 \times 10^0$. ■

Table 1.2

Operation	Result	Actual value	Absolute error	Relative error
$x \oplus y$	0.10476×10^1	$22/21$	0.190×10^{-4}	0.182×10^{-4}
$x \ominus y$	0.38095×10^0	$8/21$	0.238×10^{-5}	0.625×10^{-5}
$x \otimes y$	0.23809×10^0	$5/21$	0.524×10^{-5}	0.220×10^{-4}
$x \oplus\!\!\!\div y$	0.21428×10^1	$15/7$	0.571×10^{-4}	0.267×10^{-4}

Since the maximum relative error for the operations in Example 3 is 0.267×10^{-4}, the arithmetic produces satisfactory five-digit results. Suppose, however, that we also have $u = 0.714251$, $v = 98765.9$, and $w = 0.111111 \times 10^{-4}$, so that $fl(u) = 0.71425 \times 10^0$, $fl(v) = 0.98765 \times 10^5$, and $fl(w) = 0.11111 \times 10^{-4}$. (These numbers were chosen to illustrate some problems that can arise with finite-digit arithmetic.)

In Table 1.3, $y \ominus u$ results in a small absolute error but a large relative error. The subsequent division by the small number w or multiplication by the large number v magnifies the absolute error without modifying the relative error. The addition of the large and small numbers u and v produces large absolute error but not large relative error.

Table 1.3

Operation	Result	Actual value	Absolute error	Relative error
$y \ominus u$	0.30000×10^{-4}	0.34714×10^{-4}	0.471×10^{-5}	0.136
$(y \ominus u) \oplus\!\!\!\div w$	0.27000×10^1	0.31243×10^1	0.424	0.136
$(y \ominus u) \otimes v$	0.29629×10^1	0.34285×10^1	0.465	0.136
$u \oplus v$	0.98765×10^5	0.98766×10^5	0.161×10^1	0.163×10^{-4}

One of the most common error-producing calculations involves the cancellation of significant digits due to the subtraction of nearly equal numbers. Suppose two nearly equal numbers x and y, with $x > y$, have the k-digit representations

$$fl(x) = 0.d_1 d_2 \ldots d_p \alpha_{p+1} \alpha_{p+2} \ldots \alpha_k \times 10^n,$$

and

$$fl(y) = 0.d_1 d_2 \ldots d_p \beta_{p+1} \beta_{p+2} \ldots \beta_k \times 10^n.$$

The floating-point form of $x - y$ is

$$fl(fl(x) - fl(y)) = 0.\sigma_{p+1} \sigma_{p+2} \ldots \sigma_k \times 10^{n-p}$$

where

$$0.\sigma_{p+1}\sigma_{p+2}\ldots\sigma_k = 0.\alpha_{p+1}\alpha_{p+2}\ldots\alpha_k - 0.\beta_{p+1}\beta_{p+2}\ldots\beta_k.$$

The floating-point number used to represent $x - y$ has at most $k - p$ digits of significance. However, in most calculation devices, $x - y$ will be assigned k digits, with the last p being either zero or randomly assigned. Any further calculations involving $x - y$ retain the problem of having only $k - p$ digits of significance, since a chain of calculations cannot be expected to be more accurate than its weakest portion.

If a finite-digit representation or calculation introduces an error, further enlargement of the error occurs when dividing by a number with small magnitude (or, equivalently, when multiplying by a number with large magnitude). Suppose, for example, that the number z has the finite-digit approximation $z + \delta$, where the error δ is introduced by representation or by previous calculation. Dividing by $\varepsilon \neq 0$ results in the approximation

$$\frac{z}{\varepsilon} \approx fl\left(\frac{z+\delta}{fl(\varepsilon)}\right).$$

Suppose $\varepsilon = 10^{-n}$, where $n > 0$. Then

$$\frac{z}{\varepsilon} = z \times 10^n$$

and

$$fl\left(\frac{z+\delta}{fl(\varepsilon)}\right) = (z+\delta) \times 10^n.$$

Thus, the absolute error in this approximation, $|\delta| \times 10^n$, is the original absolute error, $|\delta|$, multiplied by the factor 10^n.

EXAMPLE 4 Let $p = 0.54617$ and $q = 0.54601$. The exact value of $r = p - q$ is $r = 0.00016$. Suppose the subtraction is performed using four-digit arithmetic. Rounding p and q to four digits gives $p^* = 0.5462$ and $q^* = 0.5460$, respectively, and $r^* = p^* - q^* = 0.0002$ is the four-digit approximation to r. Since

$$\frac{|r - r^*|}{|r|} = \frac{|0.00016 - 0.0002|}{|0.00016|} = 0.25,$$

the result has only one significant digit, whereas p^* and q^* were accurate to four and five significant digits, respectively.

If a simple chopping procedure is used to obtain the four digits, the four-digit approximations to p, q, and r are $p^* = 0.5461$, $q^* = 0.5460$, and $r^* = p^* - q^* = 0.0001$. This gives

$$\frac{|r - r^*|}{|r|} = \frac{|0.00016 - 0.0001|}{|0.00016|} = 0.375,$$

which also results in only one significant digit of accuracy. ■

The loss of accuracy due to roundoff error can often be avoided by a careful sequencing of operations or reformulation of the problem, as illustrated in the next two examples.

EXAMPLE 5 The quadratic formula states that the roots of $ax^2 + bx + c = 0$, when $a \neq 0$, are

$$x_1 = \frac{-b + \sqrt{b^2 - 4ac}}{2a} \quad \text{and} \quad x_2 = \frac{-b - \sqrt{b^2 - 4ac}}{2a}. \tag{1.1}$$

Consider

$$x^2 + 62.10x + 1 = 0,$$

a quadratic equation with approximate roots

$$x_1 = -0.01610723 \quad \text{and} \quad x_2 = -62.08390.$$

In this equation, b^2 is much larger than $4ac$, so the numerator in the calculation for x_1 involves the *subtraction* of nearly equal numbers. Suppose we perform the calculations for x_1 using four-digit rounding arithmetic. First, we have

$$\sqrt{b^2 - 4ac} = \sqrt{(62.10)^2 - 4.000} = \sqrt{3856. - 4.000} = \sqrt{3852.} = 62.06.$$

So

$$fl(x_1) = \frac{-b + \sqrt{b^2 - 4ac}}{2a} = \frac{-62.10 + 62.06}{2.000} = \frac{-0.04000}{2.000} = -0.02000$$

is an approximation to $x_1 = -0.01611$, whose relative error is large:

$$\frac{|-0.01611 + 0.02000|}{|-0.01611|} \approx 2.4 \times 10^{-1}.$$

On the other hand, the calculation for x_2 involves the *addition* of the nearly equal numbers $-b$ and $-\sqrt{b^2 - 4ac}$ and presents no problem:

$$fl(x_2) = \frac{-b - \sqrt{b^2 - 4ac}}{2a} = \frac{-62.10 - 62.06}{2.000} = \frac{-124.2}{2.000} = -62.10.$$

This approximation to $x_2 = -62.08$ has relative error

$$\frac{|-62.08 + 62.10|}{|-62.08|} \approx 3.2 \times 10^{-4}.$$

To obtain a more accurate four-digit rounding approximation for x_1, we change the form of the quadratic formula by *rationalizing the numerator*:

$$x_1 = \frac{-b + \sqrt{b^2 - 4ac}}{2a} \left(\frac{-b - \sqrt{b^2 - 4ac}}{-b - \sqrt{b^2 - 4ac}} \right) = \frac{b^2 - (b^2 - 4ac)}{2a(-b - \sqrt{b^2 - 4ac})},$$

which simplifies to

$$x_1 = \frac{-2c}{b + \sqrt{b^2 - 4ac}}. \tag{1.2}$$

Using (1.2) gives

$$fl(x_1) = \frac{-2.000}{62.10 + 62.06} = \frac{-2.000}{124.2} = -0.01610,$$

with the small relative error 6.2×10^{-4}. ■

The rationalization technique can also be applied to give the following alternative form for x_2:

$$x_2 = \frac{-2c}{b - \sqrt{b^2 - 4ac}}. \tag{1.3}$$

This is the form to use if b is a negative number. In our problem, however, the use of this formula results in not only the subtraction of nearly equal numbers, but also the division by the small result of this subtraction. The inaccuracy that this combination produces,

$$fl(x_2) = \frac{-2c}{b - \sqrt{b^2 - 4ac}} = \frac{-2.000}{62.10 - 62.06} = \frac{-2.000}{0.04000} = -50.00,$$

has the large relative error 1.9×10^{-1}.

EXAMPLE 6 Evaluate $f(x) = x^3 - 6.1x^2 + 3.2x + 1.5$ at $x = 4.71$ using three-digit arithmetic.

Table 1.4 gives the intermediate results in the calculations. Carefully verify these results to be sure that your notion of finite-digit arithmetic is correct. Note that the three-digit chopping values simply retain the leading three digits, with no rounding involved, and differ significantly from the three-digit rounding values.

Table 1.4

	x	x^2	x^3	$6.1x^2$	$3.2x$
Exact	4.71	22.1841	104.487111	135.32301	15.072
Three-digit (chopping)	4.71	22.1	104.	134.	15.0
Three-digit (rounding)	4.71	22.2	105.	135.	15.1

$$\text{Exact:} \quad f(4.71) = 104.487111 - 135.32301 + 15.072 + 1.5$$
$$= -14.263899;$$
$$\text{Three-digit (chopping):} \quad f(4.71) = ((104. - 134.) + 15.0) + 1.5 = -13.5;$$
$$\text{Three-digit (rounding):} \quad f(4.71) = ((105. - 135.) + 15.1) + 1.5 = -13.4.$$

The relative errors for the three-digit methods are

$$\left|\frac{-14.263899 + 15.0}{-14.263899}\right| \approx 0.05 \quad \text{for chopping}$$

and

$$\left|\frac{-14.263899 + 15.1}{-14.263899}\right| \approx 0.06 \quad \text{for rounding.}$$

As an alternative approach, $f(x)$ can be written in a **nested** manner as

$$f(x) = x^3 - 6.1x^2 + 3.2x + 1.5 = ((x - 6.1)x + 3.2)x + 1.5.$$

This gives

$$\text{Three-digit (chopping)}: \quad f(4.71) = ((4.71 - 6.1)4.71 + 3.2)4.71 + 1.5 = -14.2$$

and a three-digit rounding answer of -14.3. The new relative errors are

$$\text{Three-digit (chopping):} \quad \left|\frac{-14.263899 + 14.2}{-14.263899}\right| \approx 0.0045;$$

$$\text{Three-digit (rounding):} \quad \left|\frac{-14.263899 + 14.3}{-14.263899}\right| \approx 0.0025.$$

Nesting has reduced the relative error for the chopping approximation to less than one-tenth that obtained initially. For the rounding approximation the improvement has been even more dramatic; the error in this case has been reduced by more than 95%. ■

Polynomials should *always* be expressed in nested form before performing an evaluation, because this form minimizes the number of required arithmetic calculations. The decreased error in Example 6 is due to the reduction in computations from four multiplications and three additions to two multiplications and three additions. One way to reduce roundoff error is to reduce the number of error-producing computations.

EXERCISE SET 1.2

1. Compute the absolute error and relative error in approximations of p by p^*.

a. $p = \pi,\ p^* = 22/7$

b. $p = \pi,\ p^* = 3.1416$

c. $p = e,\ p^* = 2.718$

d. $p = \sqrt{2},\ p^* = 1.414$

e. $p = e^{10},\ p^* = 22000$

f. $p = 10^{\pi},\ p^* = 1400$

g. $p = 8!,\ p^* = 39900$

h. $p = 9!,\ p^* = \sqrt{18\pi}(9/e)^9$

2. Find the largest interval in which p^* must lie to approximate p with relative error at most 10^{-4} for each value of p.

a. π **b.** e **c.** $\sqrt{2}$ **d.** $\sqrt[3]{7}$

3. Suppose p^* must approximate p with relative error at most 10^{-3}. Find the largest interval in which p^* must lie for each value of p.

a. 150 **b.** 900 **c.** 1500 **d.** 90

4. Perform the following computations (i) exactly, (ii) using three-digit chopping arithmetic, and (iii) using three-digit rounding arithmetic. (iv) Compute the relative errors in parts (ii) and (iii).

a. $\dfrac{4}{5} + \dfrac{1}{3}$ **b.** $\dfrac{4}{5} \cdot \dfrac{1}{3}$

c. $\left(\dfrac{1}{3} - \dfrac{3}{11}\right) + \dfrac{3}{20}$ **d.** $\left(\dfrac{1}{3} + \dfrac{3}{11}\right) - \dfrac{3}{20}$

5. Use three-digit rounding arithmetic to perform the following calculations. Compute the absolute error and relative error with the exact value determined to at least five digits.

a. $133 + 0.921$ **b.** $133 - 0.499$

c. $(121 - 0.327) - 119$ **d.** $(121 - 119) - 0.327$

e. $\dfrac{\frac{13}{14} - \frac{6}{7}}{2e - 5.4}$ **f.** $-10\pi + 6e - \dfrac{3}{62}$

g. $\left(\dfrac{2}{9}\right) \cdot \left(\dfrac{9}{7}\right)$ **h.** $\dfrac{\pi - \frac{22}{7}}{\frac{1}{17}}$

6. Repeat Exercise 5 using four-digit rounding arithmetic.

7. Repeat Exercise 5 using three-digit chopping arithmetic.

8. Repeat Exercise 5 using four-digit chopping arithmetic.

9. The first three nonzero terms of the Maclaurin series for the arctangent function are given by $P(x) = x - \frac{1}{3}x^3 + \frac{1}{5}x^5$. Compute the absolute error and relative error in the following approximations of π using the polynomial $P(x)$ in place of the arctangent:

a. $4\left[\arctan\left(\dfrac{1}{2}\right) + \arctan\left(\dfrac{1}{3}\right)\right]$ **b.** $16 \arctan\left(\dfrac{1}{5}\right) - 4 \arctan\left(\dfrac{1}{239}\right)$

10. The number e can be defined by $e = \sum_{n=0}^{\infty} \frac{1}{n!}$, where $n! = n(n-1)\cdots 2 \cdot 1$ for $n \neq 0$ and $0! = 1$. Compute the absolute error and relative error in the following approximations of e:

a. $\displaystyle\sum_{n=0}^{5} \frac{1}{n!}$ **b.** $\displaystyle\sum_{n=0}^{10} \frac{1}{n!}$

11. Let

$$f(x) = \frac{x \cos x - \sin x}{x - \sin x}.$$

a. Find $\lim_{x \to 0} f(x)$.

b. Use four-digit rounding arithmetic to evaluate $f(0.1)$.

c. Replace each trigonometric function with its third Maclaurin polynomial, and repeat part (b).

d. The actual value is $f(0.1) = -1.99899998$. Find the relative error for the values obtained in parts (b) and (c).

12. Let

$$f(x) = \frac{e^x - e^{-x}}{x}.$$

a. Find $\lim_{x \to 0} f(x)$.

b. Use three-digit rounding arithmetic to evaluate $f(0.1)$.

c. Replace each exponential function with its third Maclaurin polynomial, and repeat part (b).

d. The actual value is $f(0.1) = 2.003335000$. Find the relative error for the values obtained in parts (a) and (b).

13. Use four-digit rounding arithmetic and the formulas of Example 5 to find the most accurate approximations to the roots of the following quadratic equations. Compute the absolute errors and relative errors.

a. $\frac{1}{3}x^2 - \frac{123}{4}x + \frac{1}{6} = 0$

b. $\frac{1}{3}x^2 + \frac{123}{4}x - \frac{1}{6} = 0$

c. $1.002x^2 - 11.01x + 0.01265 = 0$

d. $1.002x^2 + 11.01x + 0.01265 = 0$

14. Repeat Exercise 13 using four-digit chopping arithmetic.

15. Use the IBM mainframe format to find the decimal equivalent of the following floating-point machine numbers.

a. 0 1000011 101010010011000000000000

b. 1 1000011 101010010011000000000000

c. 0 0111111 010001111000000000000000

d. 0 0111111 010001111000000000000001

16. Find the next largest and smallest machine numbers in decimal form for the numbers given in Exercise 15.

17. Suppose two points (x_0, y_0) and (x_1, y_1) are on a straight line with $y_1 \neq y_0$. Two formulas are available to find the x-intercept of the line:

$$x = \frac{x_0 y_1 - x_1 y_0}{y_1 - y_0} \quad \text{and} \quad x = x_0 - \frac{(x_1 - x_0)y_0}{y_1 - y_0}.$$

a. Show that both formulas are algebraically correct.

b. Use the data $(x_0, y_0) = (1.31, 3.24)$ and $(x_1, y_1) = (1.93, 4.76)$ and three-digit rounding arithmetic to compute the x-intercept both ways. Which method is better and why?

18. The Taylor polynomial of degree n for $f(x) = e^x$ is $\sum_{i=0}^{n} \frac{x^i}{i!}$. Use the Taylor polynomial of degree nine and three-digit chopping arithmetic to find an approximation to e^{-5} by each of the following methods.

a. $e^{-5} \approx \sum_{i=0}^{9} \frac{(-5)^i}{i!} = \sum_{i=0}^{9} \frac{(-1)^i 5^i}{i!}$

b. $e^{-5} = \frac{1}{e^5} \approx \frac{1}{\sum_{i=0}^{9} \frac{5^i}{i!}}$.

An approximate value of e^{-5} correct to three digits is 6.74×10^{-3}. Which formula, (a) or (b), gives the most accuracy, and why?

19. The two-by-two linear system

$$ax + by = e,$$

$$cx + dy = f,$$

where a, b, c, d, e, f are given, can be solved for x and y as follows:

$$\text{set } m = \frac{c}{a}, \quad \text{provided } a \neq 0;$$

$$d_1 = d - mb;$$

$$f_1 = f - me;$$

$$y = \frac{f_1}{d_1};$$

$$x = \frac{e - by}{a}.$$

Use this procedure and four-digit rounding arithmetic to solve the following linear systems.

a. $1.130x - 6.990y = 14.20$
$8.110x + 12.20y = -0.1370$

b. $1.013x - 6.099y = 14.22$
$-18.11x + 112.2y = -0.1376$

20. Repeat Exercise 19 using four-digit chopping arithmetic.

21. **a.** Show that the polynomial nesting technique described in Example 6 can also be applied to the evaluation of

$$f(x) = 1.01e^{4x} - 4.62e^{3x} - 3.11e^{2x} + 12.2e^{x} - 1.99.$$

b. Use three-digit rounding arithmetic, the assumption that $e^{1.53} = 4.62$, and the fact that $e^{nx} = (e^x)^n$ to evaluate $f(1.53)$ as given in part (a).

c. Redo the calculation in part (b) by first nesting the calculations.

d. Compare the approximations in parts (b) and (c) to the true three-digit result $f(1.53) = -7.61$.

22. A rectangular parallelepiped has sides 3 cm, 4 cm, and 5 cm, measured to the nearest centimeter. What are the best upper and lower bounds for the volume of this parallelepiped? What are the best upper and lower bounds for the surface area?

23. Let $P_n(x)$ be the Maclaurin polynomial of degree n for the arctangent function. Use Maple carrying 75 decimal digits to find the value of n required to approximate π to within 10^{-25} using the following formulas.

a. $4\left[P_n\left(\frac{1}{2}\right) + P_n\left(\frac{1}{3}\right)\right]$

b. $16P_n\left(\frac{1}{5}\right) - 4P_n\left(\frac{1}{239}\right)$

24. Suppose that $fl(y)$ is a k-digit rounding approximation to y. Show that

$$\left|\frac{y - fl(y)}{y}\right| \leq 0.5 \times 10^{-k+1}.$$

[*Hint:* If $d_{k+1} < 5$, then $fl(y) = 0.d_1d_2\ldots d_k \times 10^n$. If $d_{k+1} \geq 5$, then $fl(y) = 0.d_1d_2\ldots d_k \times 10^n + 10^{n-k}$.]

25. The binomial coefficient

$$\binom{m}{k} = \frac{m!}{k!\,(m-k)!}$$

describes the number of ways of choosing a subset of k objects from a set of m elements.

a. Suppose decimal machine numbers are of the form

$$\pm 0.d_1d_2d_3d_4 \times 10^n,$$

with $1 \leq d_1 \leq 9$, $0 \leq d_i \leq 9$, if $i = 2, 3, 4$ and $|n| \leq 15$. What is the largest value of m for which the binomial coefficient $\binom{m}{k}$ can be computed for all k by the definition without causing overflow?

b. Show that $\binom{m}{k}$ can also be computed by

$$\binom{m}{k} = \binom{m}{k}\binom{m-1}{k-1}\cdots\binom{m-k+1}{1}.$$

c. What is the largest value of m for which the binomial coefficient $\binom{m}{3}$ can be computed by the formula in part (b) without causing overflow?

d. Use four-digit chopping arithmetic to compute the number of possible 5-card hands in a 52-card deck. Compute the actual and relative errors.

26. Let $f \in C[a, b]$ be a function whose derivative exists on (a, b). Suppose f is to be evaluated at x_0 in (a, b), but instead of computing the actual value $f(x_0)$, the approximate value, $\tilde{f}(x_0)$, is the actual value of f at $x_0 + \epsilon$, that is, $\tilde{f}(x_0) = f(x_0 + \epsilon)$.

a. Use the Mean Value Theorem to estimate the absolute error $|f(x_0) - \tilde{f}(x_0)|$ and the relative error $|f(x_0) - \tilde{f}(x_0)|/|f(x_0)|$, assuming $f(x_0) \neq 0$.

b. If $\epsilon = 5 \times 10^{-6}$ and $x_0 = 1$, find bounds for the absolute and relative errors for

(i) $f(x) = e^x$ **(ii)** $f(x) = \sin x$

c. Repeat part (b) with $\epsilon = (5 \times 10^{-6})x_0$ and $x_0 = 10$.

27. The following Maple procedure chops a floating-point number x to t digits.

```
chop:=proc(x,t);
    if x=0 then 0
    else
          e:=trunc(evalf(log10(abs(x))));
          if abs(x)>1 then e:=e+1 fi;
          x2:=evalf(trunc(x*10^(t-e))*10^(e-t));
    fi
end;
```

Verify the procedure works for the following values.

a. $x = 124.031,\ t = 5$

b. $x = 124.036,\ t = 5$

c. $x = -124.031,\ t = 5$

d. $x = -124.036,\ t = 5$

e. $x = 0.00653,\ t = 2$

f. $x = 0.00656,\ t = 2$

g. $x = -0.00653,\ t = 2$

h. $x = -0.00656,\ t = 2$

28. The opening example to this chapter described a physical experiment involving the temperature of a gas under pressure. In this application, we were given $P = 1.00$ atm, $V = 0.100$ m^3, $N = 0.00420$ mol, and $R = 0.08206$. Solving for T in the ideal gas law gives

$$T = \frac{PV}{NR} = \frac{(1.00)(0.100)}{(0.00420)(0.08206)} = 290.15 \text{ K} = 17^\circ\text{C}.$$

In the laboratory, it was found that T was 15°C under these conditions, and when the pressure was doubled and the volume halved, T was 19°C. Assume that the data are rounded values accurate to the places given, and show that both laboratory figures are within the bounds of accuracy for the ideal gas law.

1.3 Algorithms and Convergence

The examples in Section 1.2 demonstrate ways that machine calculations involving approximations can result in the growth of roundoff errors. Throughout the text we will be examining approximation procedures, called *algorithms*, involving sequences of calculations. An **algorithm** is a procedure that describes, in an unambiguous manner, a finite sequence of steps to be performed in a specified order. The object of the algorithm is to implement a procedure to solve a problem or approximate a solution to the problem.

A **pseudocode** is used for describing the algorithms. This pseudocode specifies the form of the input to be supplied and the form of the desired output. Not all numerical procedures give satisfactory output for arbitrarily chosen input. As a consequence, a stopping technique independent of the numerical technique is incorporated into each algorithm so that infinite loops are unlikely to occur.

Two punctuation symbols are used in the algorithms: the period (.) indicates the termination of a step, and the semicolon (;) separates tasks within a step. Indentation is used to indicate that groups of statements are to be treated as a single entity.

Looping techniques in the algorithms are either counter-controlled, such as,

For $\quad i = 1, 2, \ldots, n$

Set $\quad x_i = a_i + i \cdot h$

or condition-controlled, such as

While $i < N$ do Steps 3–6.

To allow for conditional execution, we use the standard

If... then

or

If... then

else

constructions.

The steps in the algorithms generally follow the rules of structured program construction. They have been arranged so that there should be minimal difficulty translating pseudocode into any programming language suitable for scientific applications. However, many of the more advanced topics in the latter half of the book require large matrices or complex arithmetic which may be difficult to program in certain languages.

The algorithms are liberally laced with comments. These are written in italics and contained within parentheses to distinguish them from the algorithmic statements.

EXAMPLE 1 An algorithm to compute

$$\sum_{i=1}^{N} x_i = x_1 + x_2 + \cdots + x_N,$$

where N and the numbers $x_1, x_2, \ldots x_N$ are given, is described by the following:

INPUT $\quad N, x_1, x_2, \ldots, x_n$.

OUTPUT $\quad SUM = \sum_{i=1}^{N} x_i$.

Step 1 Set $SUM = 0$. (*Initialize accumulator.*)

Step 2 For $i = 1, 2, \ldots, N$ do
set $SUM = SUM + x_i$. (*Add next term.*)

Step 3 OUTPUT (SUM);
STOP. ■

EXAMPLE 2 The Nth Taylor polynomial for $f(x) = \ln x$ expanded about $x_0 = 1$ is

$$P_N(x) = \sum_{i=1}^{N} \frac{(-1)^{i+1}}{i}(x-1)^i,$$

and the value of ln 1.5 to eight decimal places is 0.40546511. Suppose we want to compute the minimal value of N required for

$$|\ln 1.5 - P_N(1.5)| < 10^{-5}$$

without using the Taylor polynomial truncation error formula. From calculus we know that if $\sum_{n=1}^{\infty} a_n$ is an alternating series with limit A whose terms decrease in magnitude, then A and the Nth partial sum $A_N = \sum_{n=1}^{N} a_n$ differ by less than the magnitude of the $(N+1)$st term; that is,

$$|A - A_N| \le |a_{N+1}|.$$

An algorithm that uses this result to solve the problem is as follows.

INPUT value x, tolerance TOL, maximum number of iterations M.

OUTPUT degree N of the polynomial or a message of failure.

Step 1 Set $N = 1$;
$y = x - 1$;
$SUM = 0$;
$POWER = y$;
$TERM = y$;
$SIGN = -1$. (*Used to implement alternation of signs.*)

Step 2 While $N \le M$ do Steps 3–5.

Step 3 Set $SIGN = -SIGN$; (*Alternate signs.*)
$SUM = SUM + SIGN \cdot TERM$; (*Accumulate terms.*)
$POWER = POWER \cdot y$;
$TERM = POWER/(N+1)$. (*Calculate next term.*)

Step 4 If $|TERM| < TOL$ then (*Test for accuracy.*)
OUTPUT (N);
STOP. (*Procedure completed successfully.*)

Step 5 Set $N = N + 1$. (*Prepare for next iteration.*)

Step 6 OUTPUT ('Method Failed'); (*Procedure completed unsuccessfully.*)
STOP.

The input for our problem is $x = 1.5$, $TOL = 10^{-5}$, and perhaps $M = 15$. This choice of M provides an upper bound for the number of calculations we are willing to have performed, recognizing that the algorithm is likely to fail if this bound is exceeded. Whether the output is a value for N or the failure message depends on the precision of the computational device being used. ∎

We are interested in choosing methods that will produce dependably accurate results for a wide range of problems. One criterion we will impose on an algorithm whenever possible is that small changes in the initial data produce correspondingly small changes in the final results. An algorithm that satisfies this property is called **stable**; it is **unstable** when this criterion is not fulfilled. Some algorithms are stable for certain, but not all, choices of initial data. We will characterize the stability properties of algorithms whenever possible.

To consider further the subject of roundoff error growth and its connection to algorithm stability, suppose an error with magnitude E_0 is introduced at some stage in the calculations and that the magnitude of the error after n subsequent operations is denoted by E_n. The two cases that arise most often in practice are defined as follows.

Definition 1.17 Suppose that E_n represents the magnitude of an error after n subsequent operations. If $E_n \approx CnE_0$, where C is a constant independent of n, the growth of error is said to be **linear**. If $E_n \approx C^n E_0$, for some $C > 1$, the growth of error is called **exponential**. ∎

Linear growth of error is usually unavoidable, and when C and E_0 are small the results are generally acceptable. Exponential growth of error should be avoided, since the term C^n becomes large for even relatively small values of n. This leads to unacceptable inaccuracies, regardless of the size of E_0. As a consequence, an algorithm that exhibits linear growth of error is stable, whereas an algorithm exhibiting exponential error growth is unstable. (See Figure 1.11.)

Figure 1.11

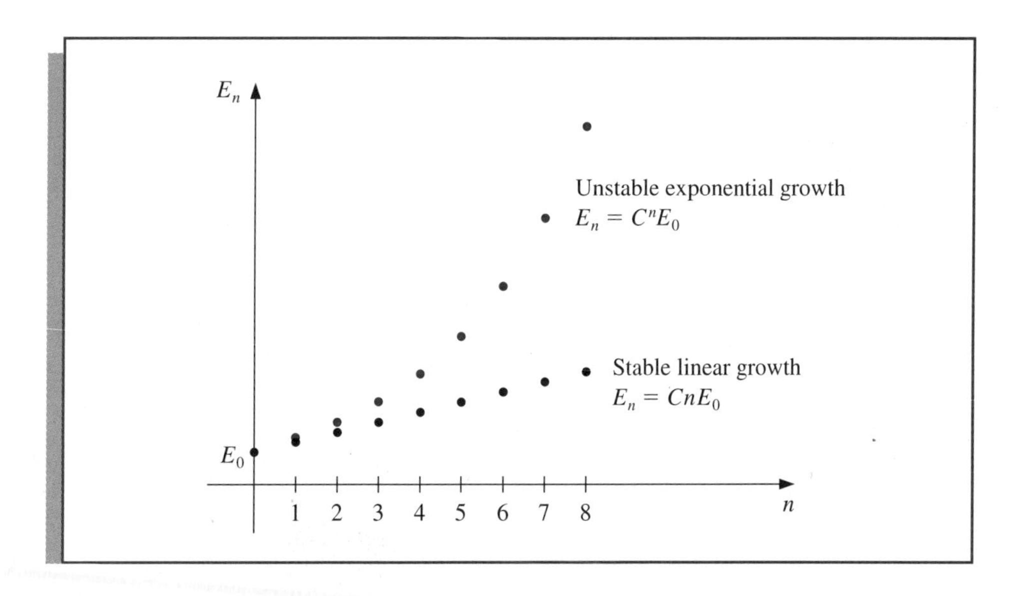

EXAMPLE 3 The recursive equation

$$p_n = \frac{10}{3}p_{n-1} - p_{n-2}, \quad \text{for } n = 2, 3, \ldots$$

has the solution

$$p_n = c_1 \left(\frac{1}{3}\right)^n + c_2 3^n$$

for any constants c_1 and c_2 since

$$\begin{aligned}
\frac{10}{3}p_{n-1} - p_{n-2} &= \frac{10}{3}\left[c_1\left(\frac{1}{3}\right)^{n-1} + c_2 3^{n-1}\right] - \left[c_1\left(\frac{1}{3}\right)^{n-2} + c_2 3^{n-2}\right] \\
&= c_1\left(\frac{1}{3}\right)^{n-2}\left[\frac{10}{3}\cdot\frac{1}{3} - 1\right] + c_2 3^{n-2}\left[\frac{10}{3}\cdot 3 - 1\right] \\
&= c_1\left(\frac{1}{3}\right)^{n-2}\left(\frac{1}{9}\right) + c_2 3^{n-2}(9) = c_1\left(\frac{1}{3}\right)^n + c_2 3^n = p_n.
\end{aligned}$$

If $p_0 = 1$ and $p_1 = \frac{1}{3}$, we have $c_1 = 1$ and $c_2 = 0$, so $p_n = \left(\frac{1}{3}\right)^n$ for all n. Suppose that five-digit rounding arithmetic is used to compute the terms of the sequence given by this equation. Then $p_0 = 1.0000$ and $p_1 = 0.33333$, which requires modifying the constants to $c_1 = 1.0000$ and $c_2 = -0.12500 \times 10^{-5}$. The sequence $\{\hat{p}_n\}_{n=0}^{\infty}$ generated is then given by

$$\hat{p}_n = 1.0000\left(\frac{1}{3}\right)^n - 0.12500 \times 10^{-5}(3)^n,$$

and the roundoff error

$$p_n - \hat{p}_n = 0.12500 \times 10^{-5}(3^n)$$

grows exponentially with n. This is reflected in the extreme inaccuracies found in Table 1.5.

Table 1.5

n	Computed $\hat{p}_n$	Correct p_n	Relative Error
0	0.10000×10^1	0.10000×10^1	
1	0.33333×10^0	0.33333×10^0	
2	0.11110×10^0	0.11111×10^0	9×10^{-5}
3	0.37000×10^{-1}	0.37037×10^{-1}	1×10^{-3}
4	0.12230×10^{-1}	0.12346×10^{-1}	9×10^{-3}
5	0.37660×10^{-2}	0.41152×10^{-2}	8×10^{-2}
6	0.32300×10^{-3}	0.13717×10^{-2}	8×10^{-1}
7	-0.26893×10^{-2}	0.45725×10^{-3}	7×10^0
8	-0.92872×10^{-2}	0.15242×10^{-3}	6×10^1

The equation

$$p_n = 2p_{n-1} - p_{n-2}, \quad \text{for } n = 2, 3, \ldots$$

has the solution $p_n = c_1 + c_2 n$ for any constants c_1 and c_2 because

$$\begin{aligned} 2p_{n-1} - p_{n-2} &= 2(c_1 + c_2(n-1)) - (c_1 + c_2(n-2)) \\ &= c_1(2-1) + c_2(2n - 2 - n + 2) = c_1 + c_2 n = p_n. \end{aligned}$$

If $p_0 = 1$ and $p_1 = \frac{1}{3}$, the constants in this equation become $c_1 = 1$ and $c_2 = -\frac{2}{3}$, so $p_n = 1 - \frac{2n}{3}$. Five-digit rounding arithmetic in this case results in $\hat{p}_0 = 1.0000$, and $\hat{p}_1 = 0.33333$; consequently, $c_1 = 1.0000$ and $c_2 = -0.66667$. Thus,

$$\hat{p}_n = 1.0000 - 0.66667n.$$

The roundoff error is

$$p_n - \hat{p}_n = \left(0.66667 - \frac{2}{3}\right) n,$$

which grows linearly with n. This is reflected in the stability found in Table 1.6. ■

Table 1.6

n	Computed $\hat{p}_n$	Correct p_n	Relative Error
0	0.10000×10^1	0.10000×10^1	
1	0.33333×10^0	0.33333×10^0	
2	-0.33330×10^0	-0.33333×10^0	9×10^{-5}
3	-0.10000×10^1	-0.10000×10^1	0
4	-0.16667×10^1	-0.16667×10^1	0
5	-0.23334×10^1	-0.23333×10^1	4×10^{-5}
6	-0.30000×10^1	-0.30000×10^1	0
7	-0.36667×10^1	-0.36667×10^1	0
8	-0.43334×10^1	-0.43333×10^1	2×10^{-5}

The effects of roundoff error can be reduced by using high-order-digit arithmetic such as the double- or multiple-precision option available on most digital computers. Disadvantages to using multiple-precision arithmetic are that it takes more computation time and the growth of roundoff error is not eliminated but is only postponed until subsequent computations are performed.

One approach to estimating roundoff error is to use interval arithmetic (that is, to retain the largest and smallest possible values at each step) so that, in the end, we obtain an interval that contains the true value. Unfortunately, we may have to find a very small interval for reasonable implementation.

Since iterative techniques involving sequences are often used, the section concludes with a brief discussion of some terminology used to describe the rate at which convergence occurs when employing a numerical technique. In general, we would like the technique to converge as rapidly as possible. The following definition is used to compare the convergence rates of various methods.

Definition 1.18 Suppose $\{\beta_n\}_{n=1}^{\infty}$ is a sequence known to converge to zero and $\{\alpha_n\}_{n=1}^{\infty}$ converges to a number α. If a positive constant K exists with

$$|\alpha_n - \alpha| \le K|\beta_n| \quad \text{for large } n,$$

then we say that $\{\alpha_n\}_{n=1}^{\infty}$ converges to α with **rate of convergence** $O(\beta_n)$. (This expression is read "big oh of β_n.") This situation is indicated by writing $\alpha_n = \alpha + O(\beta_n)$ or $\alpha_n \to \alpha$ with rate of convergence $O(\beta_n)$. ■

Although Definition 1.18 permits $\{\alpha_n\}_{n=1}^{\infty}$ to be compared with an arbitrary sequence $\{\beta_n\}_{n=1}^{\infty}$, in nearly every situation we have

$$\beta_n = \frac{1}{n^p}$$

for some number $p > 0$. We consider the largest value of p so that $\alpha_n = \alpha + O(1/n^p)$ to describe the rate at which α_n converges to α.

EXAMPLE 4 Suppose that the sequences $\{\alpha_n\}$ and $\{\hat{\alpha}_n\}$ are described by $\alpha_n = (n+1)/n^2$ and $\hat{\alpha}_n = (n+3)/n^3$ for each integer $n \ge 1$. Although $\lim_{n\to\infty} \alpha_n = 0$ and $\lim_{n\to\infty} \hat{\alpha}_n = 0$, the sequence $\{\hat{\alpha}_n\}$ converges to this limit much faster than the sequence $\{\alpha_n\}$.

In fact, using five-digit rounding arithmetic gives the entries in Table 1.7.

Table 1.7

n	1	2	3	4	5	6	7
α_n	2.00000	0.75000	0.44444	0.31250	0.24000	0.19444	0.16327
$\hat{\alpha}_n$	4.00000	0.62500	0.22222	0.10938	0.064000	0.041667	0.029155

If we let $\beta_n = 1/n$ and $\hat{\beta}_n = 1/n^2$ for each n, we see that

$$|\alpha_n - 0| = \frac{n+1}{n^2} \le \frac{n+n}{n^2} = 2 \cdot \frac{1}{n} = 2\beta_n$$

and

$$|\hat{\alpha}_n - 0| = \frac{n+3}{n^3} \le \frac{n+3n}{n^3} = 4 \cdot \frac{1}{n^2} = 4\hat{\beta}_n,$$

so

$$\alpha_n = 0 + O\left(\frac{1}{n}\right) \quad \text{and} \quad \hat{\alpha}_n = 0 + O\left(\frac{1}{n^2}\right).$$

The rate of convergence of $\{\alpha_n\}$ to zero is similar to the convergence of $\{1/n\}$ to zero, whereas $\{\hat{\alpha}_n\}$ converges to zero at a rate similar to the more rapidly convergent sequence $\{1/n^2\}$. ■

We also use the "big oh" notation to describe the rate at which functions converge.

Definition 1.19 Suppose that $\lim_{h\to 0} G(h) = 0$ and $\lim_{h\to 0} F(h) = L$. If a positive constant K exists with

$$|F(h) - L| \leq K|G(h)| \quad \text{for sufficiently small } h,$$

then we write $F(h) = L + O(G(h))$. ■

Generally, the comparison function G has the form $G(h) = h^p$ for some number $p > 0$. We consider the largest value of p for which $F(h) = L + O(h^p)$ to describe the rate of convergence of $F(h)$ to L.

EXAMPLE 5 From Example 3(b) of Section 1.1 we know that by using a third Taylor polynomial,

$$\cos h = 1 - \frac{1}{2}h^2 + \frac{1}{24}h^4 \cos \tilde{\xi}(h)$$

for some number $\tilde{\xi}(h)$ between zero and h.

Consequently,

$$\cos h + \frac{1}{2}h^2 = 1 + \frac{1}{24}h^4 \cos \tilde{\xi}(h).$$

This result implies that

$$\cos h + \frac{1}{2}h^2 = 1 + O(h^4),$$

since $|(\cos h + \frac{1}{2}h^2) - 1| = |\frac{1}{24} \cos \tilde{\xi}(h)|h^4 \leq \frac{1}{24}h^4$. The implication is that $\cos h + \frac{1}{2}h^2$ converges to its limit, 1, about as fast as h^4 converges to zero. ■

EXERCISE SET 1.3

1. **a.** Use three-digit chopping arithmetic to compute the sum $\sum_{i=1}^{10} \frac{1}{i^2}$ first by $\frac{1}{1} + \frac{1}{4} + \cdots + \frac{1}{100}$ and then by $\frac{1}{100} + \frac{1}{81} + \cdots + \frac{1}{1}$. Which method is more accurate, and why?

   b. Write an algorithm to sum the finite series $\sum_{i=1}^{N} x_i$ in reverse order.

2. The number e is defined by $e = \sum_{n=0}^{\infty} \frac{1}{n!}$ where $n! = n(n-1)\cdots 2 \cdot 1$ for $n \neq 0$ and $0! = 1$. Use four-digit chopping arithmetic to compute the following approximations to e. Also compute absolute and relative errors.

 a. $e \approx \sum_{n=0}^{5} \frac{1}{n!}$
 b. $e \approx \sum_{j=0}^{5} \frac{1}{(5-j)!}$
 c. $e \approx \sum_{n=0}^{10} \frac{1}{n!}$
 d. $e \approx \sum_{j=0}^{10} \frac{1}{(10-j)!}$

3. The Maclaurin series for the arctangent function converges for $-1 < x \leq 1$ and is given by

$$\arctan x = \lim_{n\to\infty} P_n(x) = \lim_{n\to\infty} \sum_{i=1}^{n} (-1)^{i+1} \frac{x^{2i-1}}{(2i-1)}.$$

 a. Use the fact that $\tan \pi/4 = 1$ to determine the number n of terms of the series that need to be summed to ensure that $|4P_n(1) - \pi| < 10^{-3}$.
 b. The C programming language requires the value of π to be within 10^{-10}. How many terms of the series would we need to sum to obtain this degree of accuracy?

4. Exercise 3 details a rather inefficient means of obtaining an approximation to π. The method can be improved substantially by observing that $\pi/4 = \arctan \frac{1}{2} + \arctan \frac{1}{3}$ and evaluating the series for the arctangent at $\frac{1}{2}$ and at $\frac{1}{3}$. Determine the number of terms that must be summed to ensure an approximation to π to within 10^{-3}.

5. Another formula for computing π can be deduced from the identity $\pi/4 = 4\arctan \frac{1}{5} - \arctan \frac{1}{239}$. Determine the number of terms that must be summed to ensure an approximation to π to within 10^{-3}.

6. Find the rates of convergence of the following sequences as $n \to \infty$.

 a. $\lim_{n\to\infty} \sin\left(\frac{1}{n}\right) = 0$
 b. $\lim_{n\to\infty} \sin\left(\frac{1}{n^2}\right) = 0$
 c. $\lim_{n\to\infty} \left[\sin\left(\frac{1}{n}\right)\right]^2 = 0$
 d. $\lim_{n\to\infty} [\ln(n+1) - \ln(n)] = 0$

7. Find the rates of convergence of the following functions as $h \to 0$.

 a. $\lim_{h\to 0} \frac{\sin h - h\cos h}{h} = 0$
 b. $\lim_{h\to 0} \frac{1 - e^h}{h} = -1$
 c. $\lim_{h\to 0} \frac{\sin h}{h} = 1$
 d. $\lim_{h\to 0} \frac{1 - \cos h}{h} = 0$

8. **a.** How many multiplications and additions are required to determine a sum of the form

$$\sum_{i=1}^{n} \sum_{j=1}^{i} a_i b_j?$$

 b. Modify the sum in part (a) to an equivalent form that reduces the number of computations.

9. Let $P(x) = a_n x^n + a_{n-1}x^{n-1} + \cdots + a_1 x + a_0$ be a polynomial and let x_0 be given. Construct an algorithm to evaluate $P(x_0)$ using nested multiplication.

10. Example 5 of Section 1.2 gives alternative formulas for the roots x_1 and x_2 of $ax^2 + bx + c = 0$. Construct an algorithm with input a, b, c and output x_1, x_2 that computes the roots x_1 and x_2 (which may be equal or be complex conjugates) in order to employ the best formula for each.

11. Construct an algorithm that has as input an integer $n \geq 1$, numbers $x_0, x_1, \ldots, x_n$, and a number x and that produces as output the product $(x - x_0)(x - x_1)\cdots(x - x_n)$.

12. Assume that

$$\frac{1-2x}{1-x+x^2} + \frac{2x-4x^3}{1-x^2+x^4} + \frac{4x^3-8x^7}{1-x^4+x^8} + \cdots = \frac{1+2x}{1+x+x^2}$$

for $x < 1$. Let $x = 0.25$. Write and execute an algorithm that determines the number of terms needed on the left side of the equation so that the left side differs from the right side by less than 10^{-6}.

13. **a.** Suppose that $0 < q < p$ and that $\alpha_n = \alpha + O(n^{-p})$. Show that $\alpha_n = \alpha + O(n^{-q})$.

b. Make a table listing $1/n$, $1/n^2$, $1/n^3$, and $1/n^4$ for $n = 5, 10, 100$, and 1000, and discuss the varying rates of convergence of these sequences as n becomes large.

14. **a.** Suppose that $0 < q < p$ and that $F(h) = L + O(h^p)$. Show that $F(h) = L + O(h^q)$.

b. Make a table listing h, h^2, h^3, and h^4 for $h = 0.5, 0.1, 0.01$, and 0.001, and discuss the varying rates of convergence of these powers of h as h approaches zero.

15. Suppose that as x approaches zero,

$$F_1(x) = L_1 + O(x^\alpha) \quad \text{and} \quad F_2(x) = L_2 + O(x^\beta).$$

Let c_1 and c_2 be nonzero constants and define

$$F(x) = c_1F_1(x) + c_2F_2(x) \quad \text{and} \quad G(x) = F_1(c_1x) + F_2(c_2x).$$

Show that if $\gamma = \text{minimum}\{\alpha, \beta\}$, then as x approaches zero,

a. $F(x) = c_1L_1 + c_2L_2 + O(x^\gamma)$ **b.** $G(x) = L_1 + L_2 + O(x^\gamma)$.

16. The sequence $\{F_n\}$ described by $F_0 = 1$, $F_1 = 1$, and $F_{n+2} = F_n + F_{n+1}$, if $n \geq 0$, is called a *Fibonacci sequence*. Its terms occur naturally in many botanical species, particularly those with petals or scales arranged in the form of a logarithmic spiral. Consider the sequence $\{x_n\}$, where $x_n = F_{n+1}/F_n$. Assuming that $\lim_{n\to\infty} x_n = x$ exists, show that $x = (1 + \sqrt{5})/2$. This number is called the *golden ratio*.

17. The Fibonacci sequence also satisfies the equation

$$F_n \equiv \tilde{F}_n = \frac{1}{\sqrt{5}}\left[\left(\frac{1+\sqrt{5}}{2}\right)^n - \left(\frac{1-\sqrt{5}}{2}\right)^n\right].$$

a. Write a Maple procedure to calculate F_{100}.

b. Use Maple with the default value of `Digits` followed by `evalf` to calculate $\tilde{F}_{100}$.

c. Why is the result from part (a) more accurate than the result from part (b)?

d. Why is the result from part (b) obtained more rapidly than the result from part (a)?

e. What results when you use the command `simplify` instead of `evalf` to compute $\tilde{F}_{100}$?

18. The harmonic series $1 + \frac{1}{2} + \frac{1}{3} + \frac{1}{4} + \cdots$ diverges, but the sequence $\gamma_n = 1 + \frac{1}{2} + \cdots + \frac{1}{n} - \ln n$ converges, since $\{\gamma_n\}$ is a bounded, nonincreasing sequence. The limit $\gamma = 0.5772156649\ldots$ of the sequence $\{\gamma_n\}$ is called Euler's constant.

a. Use the default value of `Digits` in Maple to determine the value of n for $\gamma_n - \gamma$ to be less than 10^{-2}.

b. Use the default value of `Digits` in Maple to determine the value of n for $\gamma_n - \gamma$ to be less than 10^{-3}.

c. What happens if you use the default value of `Digits` in Maple to determine the value of n for γ to be within 10^{-4}?

1.4 Numerical Software

Computer software for approximating the numerical solutions to problems is available in many forms. With this book we have provided programs written in C, FORTRAN, Maple, Mathematica, and Pascal that can be used to solve the problems given in the examples and exercises. These programs will give satisfactory results for most problems that you will likely need to solve, but they are what we call *special-purpose* programs. We use this term to distinguish these programs from those available in the standard mathematical subroutine libraries. The programs in these packages will be called *general purpose*.

The programs in general-purpose software packages differ in their intent from the algorithms and programs provided with this book. General-purpose software packages consider ways to reduce errors due to machine rounding, underflow, and overflow. They also describe the range of input that will lead to results of a certain specified accuracy. Since these are machine-dependent characteristics, general-purpose software packages need to include parameters that describe the floating-point characteristics of the machine being used for computations.

To illustrate some differences between programs included in a general-purpose package and a program that we would provide for use in this book, let us consider an algorithm that computes the Euclidean norm of an n-dimensional vector $\mathbf{x} = (x_1, x_2, \ldots, x_n)^t$. This norm is often required within larger programs and is defined by

$$\|\mathbf{x}\| = \left[\sum_{i=1}^{n} x_i^2\right]^{1/2}.$$

The norm gives a measure for the distance from the vector $\mathbf{x}$ to the zero vector. For example, the vector $\mathbf{x} = (2, 1, 3, -2, -1)^t$ has

$$\|\mathbf{x}\| = [2^2 + 1^2 + 3^2 + (-2)^2 + (-1)^2]^{1/2} = \sqrt{19},$$

so its distance from $\mathbf{0} = (0, 0, 0, 0, 0)^t$ is $\sqrt{19} \approx 4.36$.

An algorithm of the type we would present for this problem is given here. It includes no machine-dependent parameters and provides no accuracy assurances, but it will give accurate results "most of the time."

INPUT $n, x_1, x_2, \ldots, x_n$.

OUTPUT $NORM$.

Step 1 Set $SUM = 0$.

Step 2 For $i = 1, 2, \ldots, n$ set $SUM = SUM + x_i^2$.

Step 3 Set $NORM = SUM^{1/2}$.

Step 4 OUTPUT ($NORM$);
STOP.

A program based on this algorithm is easy to write and understand. However, the program could fail to give sufficient accuracy for a number of reasons. For example, the magnitude of some of the numbers might be too large or too small to be accurately represented in the computer's floating-point system. Also, the normal order for performing the calculations might not produce the most accurate results, or the standard software square-root routine might not be the best available for the problem. Matters of this type are considered by algorithm designers when writing programs for general-purpose software. These programs are often used as subprograms for solving larger problems, so they must incorporate controls that we will not need.

Let us now consider an algorithm for a general-purpose software program for computing the Euclidean norm. First, it is possible that although a component x_i of the vector is within the range of the machine, the square of the component is not. This can occur when, for some i, $|x_i|$ is so small that x_i^2 causes underflow or when $|x_i|$ is so large that x_i^2 causes overflow. It is also possible for all these terms to be within the range of the machine with overflow occurring from the addition of a square of one of the terms to the previously computed sum.

Since accuracy criteria depend on the machine on which the calculations are being performed, machine-dependent parameters must be incorporated into the algorithm. Suppose we are working on a hypothetical computer with base 10, having $t \geq 4$ digits of precision, a minimum exponent *emin*, and a maximum exponent *emax*. Then the set of floating-point numbers in this machine consists of 0 and the numbers of the form

$$x = f \cdot 10^e, \quad \text{where} \quad f = \pm(f_1 10^{-1} + f_2 10^{-2} + \cdots + f_t 10^{-t}),$$

$1 \leq f_1 \leq 9$ and $0 \leq f_i \leq 9$, for each $i = 2, \ldots, t$, and where $emin \leq e \leq emax$. These constraints imply that the smallest positive number represented in the machine is $\sigma = 10^{emin-1}$, so any computed number x with $|x| < \sigma$ causes underflow and results in x being set to 0. The largest positive number is $\lambda = (1 - 10^{-t})10^{emax}$, and any computed number x with $|x| > \lambda$ causes overflow. When underflow occurs, the program will continue, often without a significant loss of accuracy. If overflow occurs, the program will fail.

The algorithm assumes that the floating-point characteristics of the machine are described using parameters N, s, S, y, and Y. The maximum number of entries that can be summed with at least $t/2$ digits of accuracy is given by N. This implies the algorithm will proceed to find the norm of a vector $\mathbf{x} = (x_1, x_2, \ldots, x_n)^t$ only if $n \leq N$. To resolve the underflow-overflow problem, the nonzero floating-point numbers are partitioned into three groups: small-magnitude numbers x, those satisfying $0 < |x| < y$; medium-magnitude numbers x, where $y \leq |x| < Y$; and large-magnitude numbers x, where $Y \leq |x|$. The parameters y and Y are chosen so that there will be no underflow-overflow problem in squaring and summing the medium-magnitude numbers. Squaring small-magnitude numbers can cause underflow, so a scale factor S much greater than 1 is used with the result that $(Sx)^2$ avoids the underflow even when x^2 does not. Summing and squaring numbers having a large magnitude can cause overflow, so in this case a positive scale factor s much smaller than 1 is used to ensure that $(sx)^2$ does not cause overflow when calculated or incorporated into a sum, even though x^2 would.

To avoid unnecessary scaling, y and Y are chosen so that the range of medium-magnitude numbers is as large as possible. The algorithm is a modification of one described in [Brow, W, p. 471]. It incorporates a procedure for scaling the components of the vector that are small in magnitude until a component with medium magnitude is encountered. It then unscales the previous sum and continues by squaring and summing small and medium numbers until a component with a large magnitude is encountered. Once a component with large magnitude appears, the algorithm scales the previous sum and proceeds to scale, square, and sum the remaining numbers. The algorithm assumes that, in transition from small to medium numbers, unscaled small numbers are negligible when compared to medium numbers. Similarly, in transition from medium to large numbers, unscaled medium numbers are negligible when compared to large numbers. Thus, the choices of the scaling parameters must be made so that numbers are equated to 0 only when they are truly negligible. Typical relationships between the machine characteristics as described by t, σ, λ, *emin*, *emax*, and the algorithm parameters N, s, S, y, and Y are given after the algorithm.

The algorithm uses three flags to indicate the various stages in the summation process. These flags are given initial values in Step 3 of the algorithm. FLAG 1 is 1 until a medium or large component is encountered; then it is changed to 0. FLAG 2 is 0 while small numbers are being summed, changes to 1 when a medium number is first encountered, and changes back to 0 when a large number is found. FLAG 3 is initially 0 and changes to 1 when a large number is first encountered. Step 3 also introduces the flag DONE, which is 0 until the calculations are complete and then changes to 1.

INPUT $N, s, S, y, Y, \lambda, n, x_1, x_2, \ldots, x_n$.

OUTPUT *NORM* or an appropriate error message.

Step 1 If $n \leq 0$ then OUTPUT ('The integer n must be positive.');
STOP.

Step 2 If $n \geq N$ then OUTPUT ('The integer n is too large.');
STOP.

Step 3 Set $SUM = 0$;
$FLAG1 = 1$; (*Small numbers are being summed.*)
$FLAG2 = 0$;
$FLAG3 = 0$;
$DONE = 0$;
$i = 1$.

Step 4 While ($i \leq n$ and $FLAG1 = 1$) do Step 5.

Step 5 If $|x_i| < y$ then set $SUM = SUM + (Sx_i)^2$;
$i = i + 1$
else set $FLAG1 = 0$. (*A non-small number encountered.*)

Step 6 If $i > n$ then set $NORM = (SUM)^{1/2}/S$;
$DONE = 1$
else set $SUM = (SUM/S)/S$; (*Scale for larger numbers.*)
$FLAG2 = 1$.

Step 7 While ($i \leq n$ and $FLAG2 = 1$) do Step 8. (*Sum medium-sized numbers.*)

Step 8 If $|x_i| < Y$ then set $SUM = SUM + x_i^2$;
$i = i + 1$
else set $FLAG2 = 0$. (*A large number has been encountered.*)

Step 9 If $DONE = 0$ then
if $i > n$ then set $NORM = (SUM)^{1/2}$;
$DONE = 1$
else set $SUM = ((SUM)s)s$; *(Scale for large numbers.)*
$FLAG3 = 1$.

Step 10 While ($i \leq n$ and $FLAG3 = 1$) do Step 11.

Step 11 Set $SUM = SUM + (sx_i)^2$; *(Sum large numbers.)*
$i = i + 1$.

Step 12 If $DONE = 0$ then
if $SUM^{1/2} < \lambda s$ then set $NORM = (SUM)^{1/2}/s$;
$DONE = 1$
else set $SUM = \lambda$. *(Norm is too large.)*

Step 13 If $DONE = 1$ then OUTPUT ('Norm is', $NORM$)
else OUTPUT ('Norm $\geq$', $NORM$, 'overflow').

Step 14 STOP.

The relationships between the machine characteristics t, σ, λ, $emin$, $emax$, and the algorithm parameters N, s, S, y, and Y were chosen by Brown in [Brow, W, p. 471] as follows:

$N = 10^{e_N}$, where $e_N = \lfloor (t-2)/2 \rfloor$, the greatest integer less than or equal to $(t-2)/2$;

$s = 10^{e_s}$, where $e_s = \lfloor -(emax + e_N)/2 \rfloor$;

$S = 10^{e_S}$, where $e_S = \lceil (1 - emin)/2 \rceil$, the smallest integer greater than or equal to $(1 - emin)/2$;

$y = 10^{e_y}$, where $e_y = \lceil (emin + t - 2)/2 \rceil$;

$Y = 10^{e_Y}$, where $e_Y = \lfloor (emax - e_N)/2 \rfloor$.

The reliability built into this general-purpose algorithm has greatly increased the complexity compared to the special-purpose algorithm given earlier in the section.

There are many forms of general-purpose numerical software available commercially and in the public domain. Most of the early software was written for mainframe computers. A good reference for this software is *Sources and development of mathematic software*, edited by Wayne Cowell [Co]. Now that the desktop computer has become sufficiently powerful, much of the standard numerical software is available for personal computers and workstations. Most of this numerical software is written in FORTRAN, although some packages are written in C and Pascal.

ALGOL procedures were presented for matrix computations in 1971 in [WR]. A package of FORTRAN subroutines based mainly on the ALGOL procedures was then developed into the EISPACK routines. These routines are documented in the manuals published by Springer-Verlag as part of their Lecture Notes in Computer Science series [Sm,B] and

[Gar]. The FORTRAN subroutines are used to compute eigenvalues and eigenvectors for a variety of types of matrices. The EISPACK project was the first large-scale numerical software package to be made available in the public domain and led the way for many packages to follow.

LINPACK is a package of FORTRAN subroutines for analyzing and solving systems of linear equations and solving linear least squares problems. The documentation for this package is contained in [DBMS]. A step-by-step introduction to LINPACK, EISPACK, and BLAS (Basic Linear Algebra Subroutines) is given in [CV].

The LAPACK package, first available in 1992, is a library of FORTRAN subroutines that supercedes LINPACK and EISPACK by integrating these two sets of algorithms into a unified and updated package. The software has been restructured to achieve greater efficiency on vector processors and other high-performance or shared-memory multiprocessors. LAPACK is expanded in depth and breadth in version 2.0, which is available in FORTRAN and C. The package BLAS is not a part of LAPACK, but the FORTRAN 77 code for BLAS is distributed with LAPACK. The *LAPACK User's Guide, 2nd ed.* [An], is available from SIAM or over the Internet at http://www.netlib.org. The complete LAPACK or individual routines from LAPACK can also be obtained through netlib at netlib@ornl.gov or netlib@research.att.com.

Other packages for solving specific types of problems are available in the public domain. Information about these programs can be obtained through electronic mail by sending the line "help" to one of the following Internet addresses: netlib@research.att.com, netlib@ornl.gov, netlib@nac.no, or netlib@draci.cs.uow.edu.au or to the uucp address uunet!research!netlib. As an alternative to netlib, you can use Xnetlib to search the database and retrieve software. More information can be found in the article *Software distribution using Xnetlib* by Dongarra, Roman, and Wade [DRW].

These software packages are highly efficient, accurate, and reliable. They are thoroughly tested, and documentation is readily available. Although the packages are portable, it is a good idea to investigate the machine dependence and read the documentation thoroughly. The programs test for almost all special contingencies that might result in error and failures. At the end of each chapter we will discuss some of the appropriate general-purpose packages.

Commercially available packages also represent the state of the art in numerical methods. Their contents are often based on the public-domain packages but include methods for almost every type of problem.

IMSL (International Mathematical Software Library) consists of the libraries MATH, STAT, and SFUN for numerical mathematics, statistics, and special functions, respectively. These libraries contain more than 700 subroutines written in FORTRAN 77, which solve most common numerical analysis problems. In 1970 IMSL became the first large-scale scientific library for mainframes. Since that time the libraries have been made available for computer systems ranging from supercomputers to personal computers. The libraries are available commercially from IMSL, 2500 Park West Tower One, 2500 City West Boulevard, Houston, TX 77042-3020. The packages are delivered in compiled form with extensive documentation. There is an example program for each routine as well as background reference information. IMSL contains methods for linear systems, eigensystem analysis, interpolation and approximation, integration and differentiation, differential equations, transforms, nonlinear equations, optimization, and basic matrix/vector operations. The library also contains extensive statistical routines.

The Numerical Algorithms Group (NAG) has been in existence in the United Kingdom since 1971. NAG offers more than 1100 subroutines in a FORTRAN 77 library for more than 90 different computers. Subsets of their library are available for IBM personal computers (the PC50 Library consists of 50 of the most frequently used routines) and workstations (the Workstation Library contains 172 routines). The NAG C Library offers many of the same routines as the FORTRAN Library. The NAG user's manual includes instructions and examples, along with sample output for each of the routines. A useful introduction to the NAG routines is [Ph]. The NAG library contains routines to perform most standard numerical analysis tasks in a manner similar to those in the IMSL. It also includes some statistical routines and a set of graphic routines. The library is commercially available from Numerical Algorithms Group, Inc., 1400 Opus Place, Suite 200, Downers Grove, IL 60515–5702.

The IMSL and NAG packages are designed for the mathematician, scientist, or engineer who wishes to call high-quality FORTRAN subroutines from within a program. The documentation available with the commercial packages illustrates the typical driver program required to use the library routines. The next two software packages are stand-alone environments. When activated, the user enters commands to cause the package to solve a problem. However, each package allows programming within the command language. Although the incorporated language resembles Pascal, it is not possible to call external routines that consist of compiled FORTRAN or C subprograms.

MATLAB is a matrix laboratory that was originally a Fortran program published by Cleve Moler [Mo]. The laboratory is based mainly on the EISPACK and LINPACK subroutines, although functions such as nonlinear systems, numerical integration, cubic splines, curve fitting, optimization, ordinary differential equations, and graphical tools have been incorporated. MATLAB is currently written in C and assembler, and the PC version of this package requires a numeric coprocessor. The basic structure is to perform matrix operations, such as finding the eigenvalues of a matrix entered from the command line or from an external file via function calls. This is a powerful self-contained system that is especially useful for instruction in an applied linear algebra course. MATLAB has been available since 1985 and can be purchased from The MathWorks Inc., Cochituate Place, 24 Prime Park Way, Natick, MA 01760. The e-mail address of The Mathworks is info@mathworks.com, and the web address is http://www.mathworks.com. The MATLAB software is designed to run on many computers, including IBM PC compatibles, APPLE Macintosh, and SUN workstations. A student version of MATLAB has been recently released that does not require a coprocessor but will use one if it is available.

The second package is GAUSS, a mathematical and statistical system for IBM personal computers produced by Lee E. Ediefson and Samuel D. Jones in 1985. It is coded mainly in assembler and based primarily on EISPACK and LINPACK. As in the case of MATLAB, integration/differentiation, nonlinear systems, fast Fourier transforms, and graphics are available. GAUSS is oriented less toward instruction in linear algebra and more toward statistical analysis of data. This package also uses a numeric coprocessor if one is available. It can be purchased from Aptech Systems, Inc., 23804 S.E. Kent-Kangley Road, Maple Valley, WA 98038 (info@aptech.com).

There are numerous packages available that can be classified as supercalculator packages for the PC. These should not be confused, however, with the general-purpose software listed here. If you have an interest in one of these packages, you should read *Supercalculators on the PC* by B. Simon and R. M. Wilson [SW].

Additional information about software and software libraries can be found in the books by Cody and Waite [CW] and by Köckler [Ko], and in the 1995 article by Dongarra and Walker [DW]. More information about floating-point computation can be found in the book by Chaitin-Chatelin and Fraysse [CF] and the article by Goldberg [Go].

Books which address the application of numerical techniques on parallel computers include those by Schendel [Sche],Ortega [Or1], Phillips and Freeman [PF], and Golub and Ortega [GO].

CHAPTER

2

Solutions of Equations in One Variable

■ ■ ■

The growth of a large population can be modeled over short periods of time by assuming that the population grows continuously with time at a rate proportional to the number of individuals present at that time. If we let $N(t)$ denote the number of individuals at time t and λ denote the constant birth rate of the population, the population satisfies the differential equation

$$\frac{dN(t)}{dt} = \lambda N(t).$$

The solution to this equation is $N(t) = N_0 e^{\lambda t}$, where N_0 denotes the initial population.

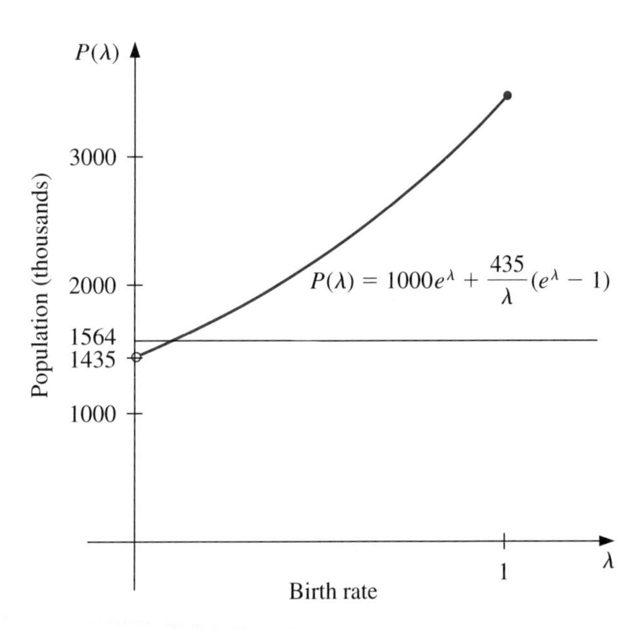

This model is valid only when the population is isolated, with no immigration from outside the community. If immigration is permitted at a constant rate v, the differential equation governing the situation becomes

$$\frac{dN(t)}{dt} = \lambda N(t) + v,$$

whose solution is

$$N(t) = N_0 e^{\lambda t} + \frac{v}{\lambda}(e^{\lambda t} - 1).$$

Suppose a certain population contains one million individuals initially, that 435,000 individuals immigrate into the community in the first year, and that 1,564,000 individuals are present at the end of 1 year. To determine the birth rate of this population, we must solve for λ in the equation

$$1{,}564{,}000 = 1{,}000{,}000 e^{\lambda} + \frac{435{,}000}{\lambda}(e^{\lambda} - 1).$$

The numerical methods discussed in this chapter are used to find approximations to solutions of equations of this type, when the exact solutions cannot be obtained by algebraic methods. The solution to this particular problem is considered in Exercise 20 of Section 2.3.

2.1 The Bisection Method

In this chapter we consider one of the basic problems of numerical approximation, the root-finding problem. It involves finding a *root* x of an equation of the form $f(x) = 0$, for a given function f. (The number x is also called a *zero* of f.) This is one of the oldest known approximation problems, yet research continues in this area at the present time. The problem of finding an approximation to the root of an equation can be traced back at least as far as 1700 B.C. A cuneiform table in the Yale Babylonian Collection dating from that period gives approximations to $\sqrt{2}$, approximations that can essentially be found by applying a technique we will see in Section 2.3.

The first technique, based on the Intermediate Value Theorem, is called the **Bisection**, or **Binary-search, method**. Suppose f is a continuous function defined on the interval $[a, b]$, with $f(a)$ and $f(b)$ of opposite sign. By the Intermediate Value Theorem, there exists a number p in (a, b) with $f(p) = 0$. Although the procedure will work for the case when $f(a)$ and $f(b)$ have opposite signs and there is more than one root in the interval (a, b), we assume for simplicity that the root in this interval is unique. The method calls for a repeated halving of subintervals of $[a, b]$ and, at each step, locating the half containing p.

To begin, set $a_1 = a$ and $b_1 = b$, and let p_1 be the midpoint of $[a, b]$; that is,

$$p_1 = \frac{1}{2}(a_1 + b_1).$$

If $f(p_1) = 0$, then $p = p_1$; if not, then $f(p_1)$ has the same sign as either $f(a_1)$ or $f(b_1)$. If $f(p_1)$ and $f(a_1)$ have the same sign, then $p \in (p_1, b_1)$, and we set $a_2 = p_1$ and $b_2 = b_1$. If $f(p_1)$ and $f(a_1)$ have opposite signs, then $p \in (a_1, p_1)$, and we set $a_2 = a_1$, and $b_2 = p_1$. We then reapply the process to the interval $[a_2, b_2]$. This produces the method described in Algorithm 2.1. (See Figure 2.1.)

Figure 2.1

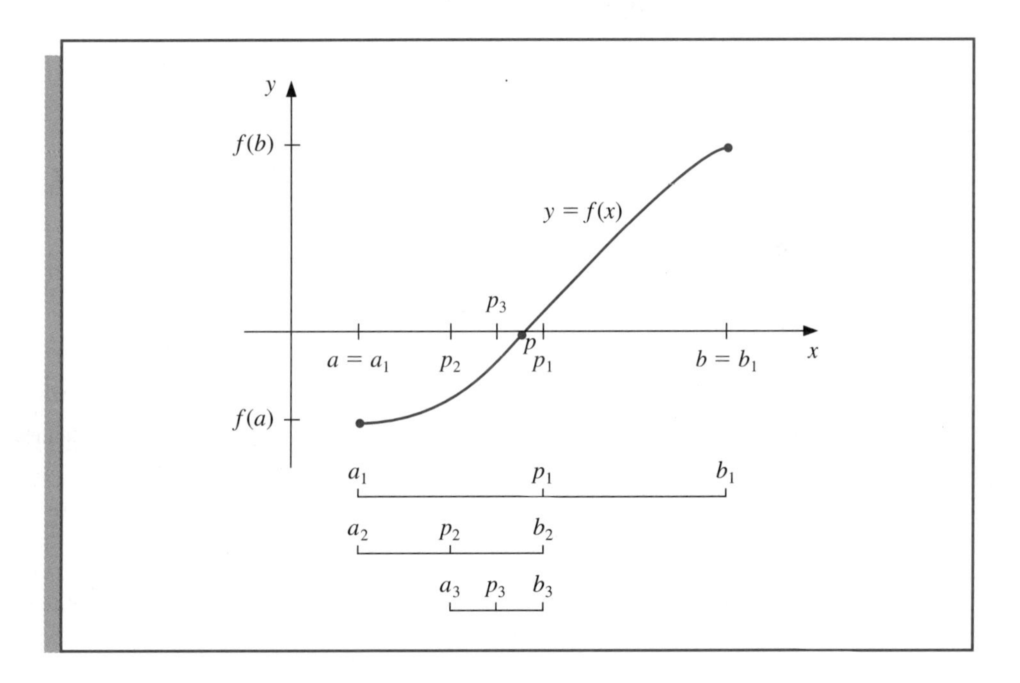

ALGORITHM 2.1

Bisection

To find a solution to $f(x) = 0$ given the continuous function f on the interval $[a, b]$, where $f(a)$ and $f(b)$ have opposite signs:

INPUT endpoints a, b; tolerance TOL; maximum number of iterations N_0.

OUTPUT approximate solution p or message of failure.

Step 1 Set $i = 1$;
$FA = f(a)$.

Step 2 While $i \le N_0$ do Steps 3–6.

Step 3 Set $p = a + (b - a)/2$; (*Compute* p_i.)
$FP = f(p)$.

Step 4 If $FP = 0$ or $(b - a)/2 < TOL$ then
OUTPUT (p); *(Procedure completed successfully.)*
STOP.

Step 5 Set $i = i + 1$.

Step 6 If $FA \cdot FP > 0$ then set $a = p$; *(Compute a_i, b_i.)*
$FA = FP$
else set $b = p$.

Step 7 OUTPUT ('Method failed after N_0 iterations, N_0 =', N_0);
(Procedure completed unsuccessfully.)
STOP.

Listed next are some other stopping procedures that can be applied in Step 4 of Algorithm 2.1 or in any of the iterative techniques in this chapter. Select a tolerance $\epsilon > 0$ and generate $p_1, \ldots, p_N$ until one of the following conditions is met:

(2.1) $$|p_N - p_{N-1}| < \epsilon,$$

(2.2) $$\frac{|p_N - p_{N-1}|}{|p_N|} < \epsilon, \quad p_N \neq 0, \quad \text{or}$$

(2.3) $$|f(p_N)| < \epsilon.$$

Unfortunately, difficulties can arise using any of these stopping criteria. For example, there exist sequences $\{p_n\}$ with the property that the differences $p_n - p_{n-1}$ converge to zero while the sequence itself diverges. (See Exercise 15.) It is also possible for $f(p_n)$ to be close to zero while p_n differs significantly from p. (See Exercise 14.) Without additional knowledge about f or p, Inequality (2.2) is the best stopping criterion to apply because it tests relative error.

When using a computer to generate approximations, it is good practice to set an upper bound on the number of iterations performed. This will eliminate entering an infinite loop, a possibility that can arise when the sequence diverges (and also when the program is incorrectly coded). This was done in Algorithm 2.1 where the bound N_0 was set and the procedure terminated if $i > N_0$.

Note that to start the Bisection Algorithm, an interval $[a, b]$ must be found with $f(a) \cdot f(b) < 0$. At each step the length of the interval known to contain a zero of f is reduced by a factor of 2; hence it is advantageous to choose the initial interval $[a, b]$ as small as possible. For example, if $f(x) = 2x^3 - x^2 + x - 1$, then

$$f(-4) \cdot f(4) < 0 \quad \text{and} \quad f(0) \cdot f(1) < 0,$$

so the Bisection Algorithm could be used on either of the intervals $[-4, 4]$ or $[0, 1]$. Starting the Bisection Algorithm on $[0, 1]$ instead of $[-4, 4]$ will reduce by 3 the number of iterations required to achieve a specified accuracy.

The following example illustrates the Bisection Algorithm. The iteration in this example is terminated when the relative error is less than 0.0001—that is, when

$$\frac{|p - p_n|}{|p|} < 10^{-4}.$$

EXAMPLE 1 The equation $f(x) = x^3 + 4x^2 - 10 = 0$ has a root in $[1, 2]$ since $f(1) = -5$ and $f(2) = 14$. The Bisection Algorithm gives the values in Table 2.1.

Table 2.1

n	a_n	b_n	p_n	$f(p_n)$
1	1.0	2.0	1.5	2.375
2	1.0	1.5	1.25	−1.79687
3	1.25	1.5	1.375	0.16211
4	1.25	1.375	1.3125	−0.84839
5	1.3125	1.375	1.34375	−0.35098
6	1.34375	1.375	1.359375	−0.09641
7	1.359375	1.375	1.3671875	0.03236
8	1.359375	1.3671875	1.36328125	−0.03215
9	1.36328125	1.3671875	1.365234375	0.000072
10	1.36328125	1.365234375	1.364257813	−0.01605
11	1.364257813	1.365234375	1.364746094	−0.00799
12	1.364746094	1.365234375	1.364990235	−0.00396
13	1.364990235	1.365234375	1.365112305	−0.00194

After 13 iterations, $p_{13} = 1.365112305$ approximates the root p with an error

$$|p - p_{13}| < |b_{14} - a_{14}| = |1.365234375 - 1.365112305| = 0.000122070.$$

Since $|a_{14}| < |p|$,

$$\frac{|p - p_{13}|}{|p|} < \frac{|b_{14} - a_{14}|}{|a_{14}|} \leq 9.0 \times 10^{-5},$$

so the approximation is correct to at least four significant digits. The correct value of p, to nine decimal places, is $p = 1.365230013$. Note that p_9 is closer to p than is the final approximation p_{13}. You might suspect this is true since $|f(p_9)| < |f(p_{13})|$, but we cannot verify this unless the true answer is known. ■

The Bisection method, though conceptually clear, has significant drawbacks. It is slow to converge (that is, N may become quite large before $|p - p_N|$ is sufficiently small), and a good intermediate approximation can be inadvertently discarded. However, the method has the important property that it always converges to a solution and for that reason is often used as a starter for the more efficient methods presented later in this chapter.

Theorem 2.1 Suppose that $f \in C[a, b]$ and $f(a) \cdot f(b) < 0$. The Bisection method as given in Algorithm 2.1 generates a sequence $\{p_n\}$ approximating a zero p of f with

$$|p_n - p| \leq \frac{b - a}{2^n}, \quad n \geq 1.$$

Proof For each $n \geq 1$, we have

$$b_n - a_n = \frac{1}{2^{n-1}}(b-a) \quad \text{and} \quad p \in (a_n, b_n).$$

Since $p_n = \frac{1}{2}(a_n + b_n)$ for all $n \geq 1$, it follows that

$$|p_n - p| \leq \frac{1}{2}(b_n - a_n) = \frac{b-a}{2^n}. \quad \blacksquare \blacksquare \blacksquare$$

According to Definition 1.18, this inequality implies that $\{p_n\}_{\infty}^{n=1}$ converges to p with rate of convergence $O(1/2^n)$ since

$$|p_n - p| \leq (b-a)\frac{1}{2^n}$$

implies that

$$p_n = p + O\left(\frac{1}{2^n}\right).$$

It is important to realize that Theorem 2.1 gives only a bound for the approximation error and that this bound may be quite conservative. For example, this bound applied to the problem in Example 1 ensures only that

$$|p - p_9| \leq \frac{2-1}{2^9} \approx 2 \times 10^{-3},$$

but the actual error is much smaller:

$$|p - p_9| = |1.365230013 - 1.365234375| \approx 4.4 \times 10^{-6}.$$

EXAMPLE 2 To determine the number of iterations necessary to solve $f(x) = x^3 + 4x^2 - 10 = 0$ with accuracy 10^{-3} using $a_1 = 1$ and $b_1 = 2$ requires finding an integer N that satisfies

$$|p_N - p| \leq 2^{-N}(b-a) = 2^{-N} < 10^{-3}.$$

To determine N we will use logarithms. Although logarithms to any base would suffice, we will use base-10 logarithms since the tolerance is given as a power of 10. Since $2^{-N} < 10^{-3}$ implies that $\log_{10} 2^{-N} < \log_{10} 10^{-3} = -3$, we have

$$-N \log_{10} 2 < -3 \quad \text{or} \quad N > \frac{3}{\log_{10} 2} \approx 9.96.$$

Hence, about ten iterations are required to ensure an approximation accurate to within 10^{-3}. Table 2.1 shows that the value of $p_9 = 1.365234375$ is accurate to within 10^{-4}. It is important to keep in mind that the error analysis gives only a bound for the number of iterations necessary, and in many cases this bound is much larger than the actual number required. ■

EXERCISE SET 2.1

1. Use the Bisection method to find p_3 for $f(x) = \sqrt{x} - \cos x$ on $[0, 1]$.
2. Let $f(x) = 3(x + 1)(x - \frac{1}{2})(x - 1)$. Use the Bisection method on the following intervals to find p_3.
   a. $[-2, 1.5]$ **b.** $[-1.25, 2.5]$
3. Use the Bisection method to find solutions accurate to within 10^{-2} for $x^3 - 7x^2 + 14x - 6 = 0$ on each interval.
   a. $[0, 1]$ **b.** $[1, 3.2]$ **c.** $[3.2, 4]$
4. Use the Bisection method to find solutions accurate to within 10^{-2} for $x^4 - 2x^3 - 4x^2 + 4x + 4 = 0$ on each interval.
   a. $[-2, -1]$ **b.** $[0, 2]$ **c.** $[2, 3]$ **d.** $[-1, 0]$
5. Use the Bisection method to find a solution accurate to within 10^{-3} for $x = \tan x$ on $[4, 4.5]$.
6. Use the Bisection method to find a solution accurate to within 10^{-3} for $2 + \cos(e^x - 2) - e^x = 0$ on $[0.5, 1.5]$.
7. Use the Bisection method to find solutions accurate to within 10^{-5} for the following problems.
   a. $x - 2^{-x} = 0$ for $0 \le x \le 1$
   b. $e^x - x^2 + 3x - 2 = 0$ for $0 \le x \le 1$
   c. $2x\cos(2x) - (x + 1)^2 = 0$ for $-3 \le x \le -2$ and for $-1 \le x \le 0$
   d. $x\cos x - 2x^2 + 3x - 1 = 0$ for $0.2 \le x \le 0.3$ and for $1.2 \le x \le 1.3$
8. Let $f(x) = (x + 2)(x + 1)^2x(x - 1)^3(x - 2)$. To which zero of f does the Bisection method converge for the following intervals?
   a. $[-1.5, 2.5]$ **b.** $[-0.5, 2.4]$ **c.** $[-0.5, 3]$ **d.** $[-3, -0.5]$
9. Let $f(x) = (x + 2)(x + 1)x(x - 1)^3(x - 2)$. To which zero of f does the Bisection method converge for the following intervals?
   a. $[-3, 2.5]$ **b.** $[-2.5, 3]$ **c.** $[-1.75, 1.5]$ **d.** $[-1.5, 1.75]$
10. Find an approximation to $\sqrt{3}$ correct to within 10^{-4} using the Bisection Algorithm. [*Hint:* Consider $f(x) = x^2 - 3$.]
11. Find an approximation to $\sqrt[3]{25}$ correct to within 10^{-4} using the Bisection Algorithm.
12. Use Theorem 2.1 to find a bound for the number of iterations needed to achieve an approximation with accuracy 10^{-3} to the solution of $x^3 + x - 4 = 0$ lying in the interval $[1, 4]$. Find an approximation to the root with this degree of accuracy.
13. Use Theorem 2.1 to find a bound for the number of iterations needed to achieve an approximation with accuracy 10^{-4} to the solution of $x^3 - x - 1 = 0$ lying in the interval $[1, 2]$. Find an approximation to the root with this degree of accuracy.
14. Let $f(x) = (x - 1)^{10}$, $p = 1$, and $p_n = 1 + 1/n$. Show that $|f(p_n)| < 10^{-3}$ whenever $n > 1$ but that $|p - p_n| < 10^{-3}$ requires that $n > 1000$.
15. Let $\{p_n\}$ be the sequence defined by $p_n = \sum_{k=1}^{n} 1/k$. Show that $\{p_n\}$ diverges even though $\lim_{n\to\infty}(p_n - p_{n-1}) = 0$.

16. The function defined by $f(x) = \sin \pi x$ has zeros at every integer. Determine an interval $[a, b]$ with $a < 0$ and $b > 2$ for which the Bisection method converges to

a. 0 **b.** 2 **c.** 1

17. A trough of length L has a cross section in the shape of a semicircle with radius r (See the accompanying figure.) When filled with water to within a distance h of the top, the volume V of water is

$$V = L\left[0.5\pi r^2 - r^2 \arcsin(h/r) - h(r^2 - h^2)^{1/2}\right]$$

Suppose $L = 10$ ft, $r = 1$ ft, and $V = 12.4$ ft^3. Find the depth of water in the trough to within 0.01 ft.

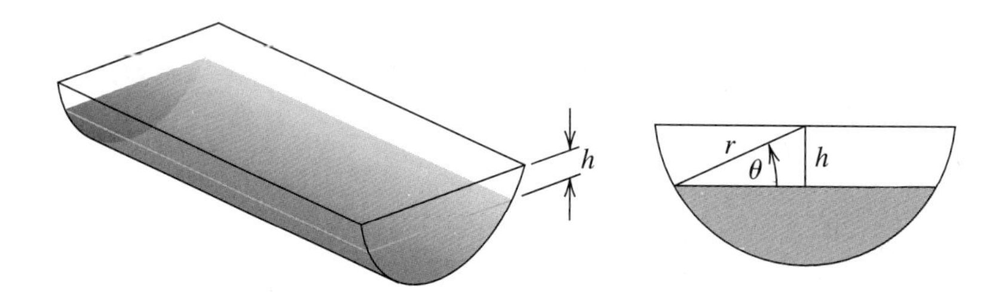

18. A particle starts at rest on a smooth inclined plane whose angle θ is changing at a constant rate

$$\frac{d\theta}{dt} = \omega < 0.$$

At the end of t seconds, the position of the object is given by

$$x(t) - \frac{g}{2\omega^2}\left(\frac{e^{wt} - e^{-wt}}{2} - \sin \omega t\right).$$

Suppose the particle has moved 1.7 ft in 1 s. Find, to within 10^{-5}, the rate ω at which θ changes. Assume that $g = -32.17$ ft/s^2.

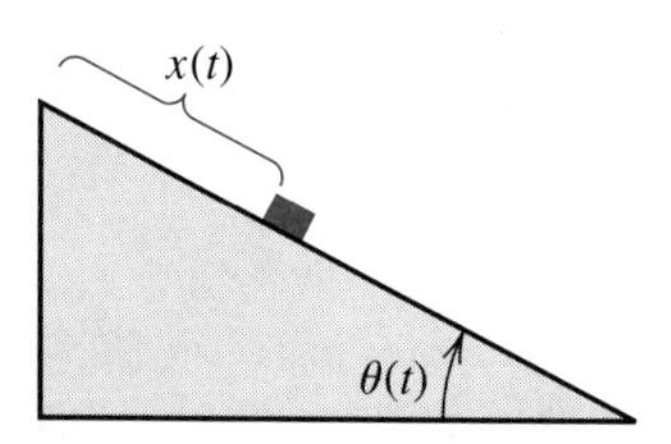

2.2 Fixed-Point Iteration

A **fixed point** for a given function g is a number p for which $g(p) = p$. In this section we consider the problem of finding solutions to fixed-point problems and the connection between these problems and the root-finding problems we wish to solve.

Root-finding problems and fixed-point problems are equivalent classes in the following sense:

> Given a root-finding problem $f(p) = 0$, we can define a function g with a fixed point at p in a number of ways, for example, as $g(x) = x - f(x)$ or as $g(x) = x + 3f(x)$. Conversely, if the function g has a fixed point at p, then the function defined by $f(x) = x - g(x)$ has a zero at p.

Although the problems we wish to solve are in the root-finding form, the fixed-point form is easier to analyze, and certain fixed-point choices lead to very powerful root-finding techniques.

Our first task is to become comfortable with this new type of problem and to decide when a function has a fixed point and how the fixed points can be approximated to within a specified accuracy.

EXAMPLE 1 The function $g(x) = x^2 - 2$, for $-2 \le x \le 3$, has fixed points at $x = -1$ and $x = 2$, since

$$g(-1) = (-1)^2 - 2 = -1 \quad \text{and} \quad g(2) = 2^2 - 2 = 2.$$

This can be seen in Figure 2.2. ∎

Figure 2.2

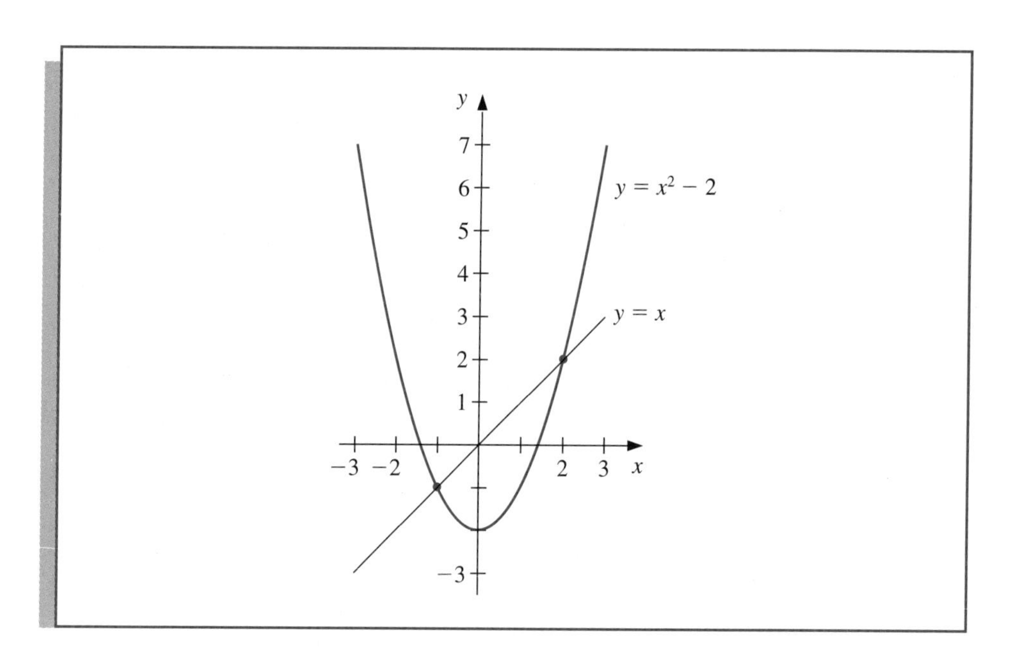

The following theorem gives sufficient conditions for the existence and uniqueness of a fixed point.

Theorem 2.2

a. If $g \in C[a, b]$ and $g(x) \in [a, b]$ for all $x \in [a, b]$, then g has a fixed point in $[a, b]$.

b. If, in addition, $g'(x)$ exists on (a, b) and a positive constant $k < 1$ exists with

$$|g'(x)| \leq k, \quad \text{for all } x \in (a, b),$$

then the fixed point in $[a, b]$ is unique. (See Figure 2.3.)

Figure 2.3

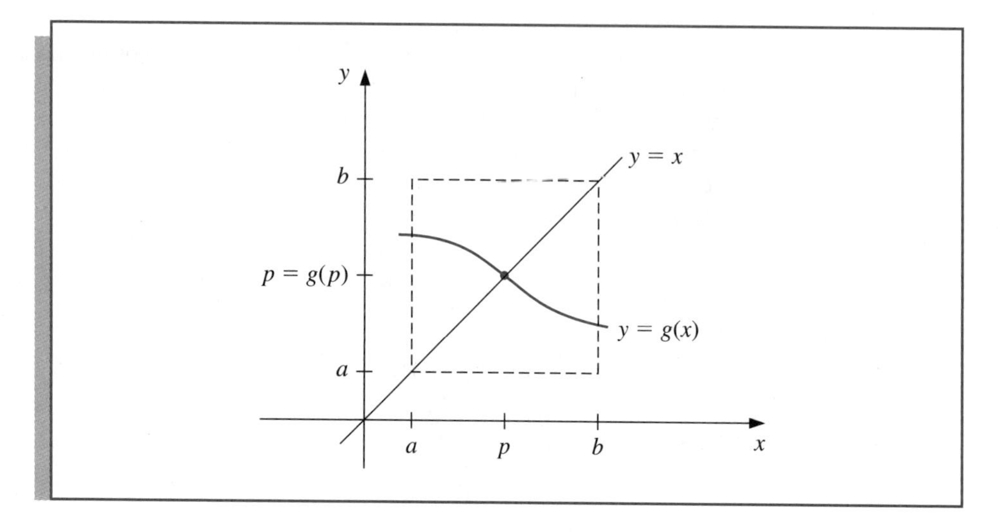

Proof

a. If $g(a) = a$ or $g(b) = b$, then g has a fixed point at an endpoint. Suppose not; then it must be true that $g(a) > a$ and $g(b) < b$. The function $h(x) = g(x) - x$ is continuous on $[a, b]$, and we have

$$h(a) = g(a) - a > 0 \quad \text{and} \quad h(b) = g(b) - b < 0.$$

The Intermediate Value Theorem implies that there exists $p \in (a, b)$ for which $h(p) = 0$. Thus, $g(p) - p = 0$, and p is a fixed point of g.

b. Suppose, in addition, that $|g'(x)| \leq k < 1$ and that p and q are both fixed points in $[a, b]$ with $p \neq q$. By the Mean Value Theorem, a number ξ exists between p and q, and hence in $[a, b]$, with

$$\frac{g(p) - g(q)}{p - q} = g'(\xi).$$

Then

$$|p - q| = |g(p) - g(q)| = |g'(\xi)||p - q| \leq k|p - q| < |p - q|,$$

which is a contraction. This contradiction must come from the only supposition, $p \neq q$. Hence, $p = q$ and the fixed point in $[a, b]$ is unique. ▪ ▪ ▪

EXAMPLE 2

a. Let $g(x) = (x^2 - 1)/3$ on $[-1, 1]$. The Extreme Value Theorem implies that the absolute minimum of g occurs at $x = 0$ and $g(0) = -\frac{1}{3}$. Similarly, the absolute maximum of g occurs at $x = \pm 1$ and has the value $g(\pm 1) = 0$. Moreover, g is continuous and

$$|g'(x)| = \left|\frac{2x}{3}\right| \le \frac{2}{3}, \quad \text{for all } x \in [1, 1],$$

so g satisfies all the hypotheses of Theorem 2.2 and has a unique fixed point in $[-1, 1]$.

In this example, the unique fixed point p in the interval $[-1, 1]$ can be determined algebraically. If

$$p = g(p) = \frac{p^2 - 1}{3}, \quad \text{then} \quad p^2 - 3p - 1 = 0,$$

which, by the quadratic formula, implies that

$$p = \frac{1}{2}\left(3 - \sqrt{13}\right).$$

Note that g also has a unique fixed point $p = \frac{1}{2}\left(3 + \sqrt{13}\right)$ for the interval $[3, 4]$. However, $g(4) = 5$ and $g'(4) = \frac{8}{3} > 1$, so g does not satisfy the hypotheses of Theorem 2.2 on $[3, 4]$. This shows that the hypotheses of Theorem 2.2 are sufficient to guarantee a unique fixed point but are not necessary. (See Figure 2.4.)

Figure 2.4

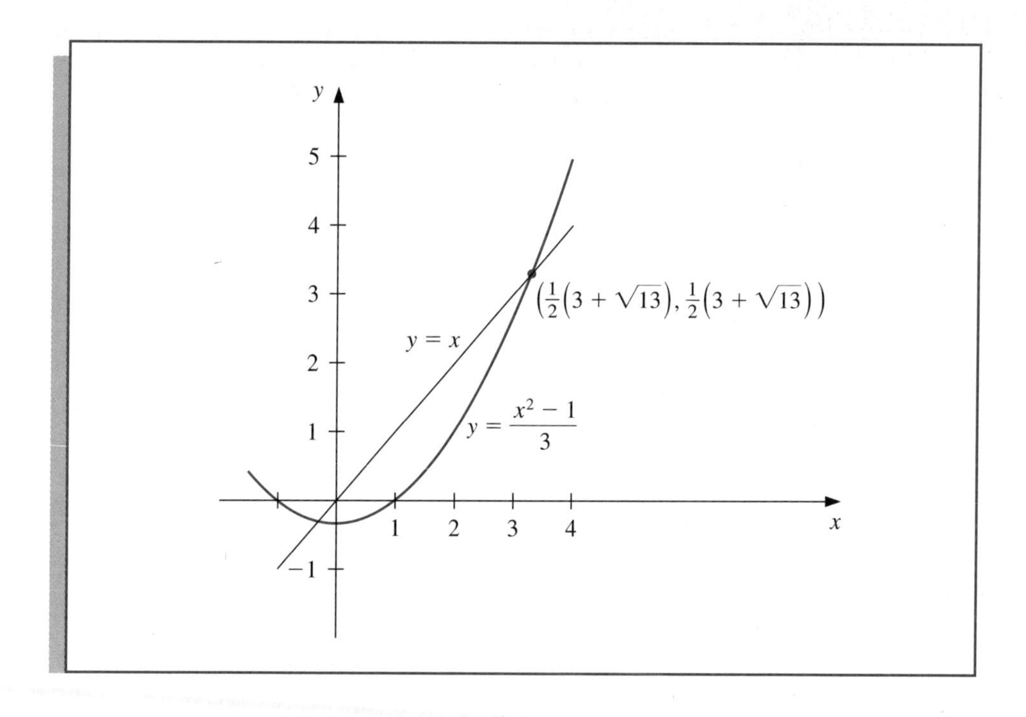

b. Let $g(x) = 3^{-x}$. Since $g'(x) = -3^{-x} \ln 3 < 0$ on $[0, 1]$, the function g is decreasing on $[0, 1]$. So, $g(1) = \frac{1}{3} \le g(x) \le 1 = g(0)$ for $0 \le x \le 1$. Thus, for $x \in [0, 1]$, we have $g(x) \in [0, 1]$, and g has a fixed point in $[0, 1]$. Since

$$g'(0) = -\ln 3 = -1.098612289,$$

$|g'(x)| \not\le 1$ on $[0, 1]$, and Theorem 2.2 cannot be used to determine uniqueness. However, g is decreasing, and it is clear from Figure 2.5 that the fixed point must be unique. ■

Figure 2.5

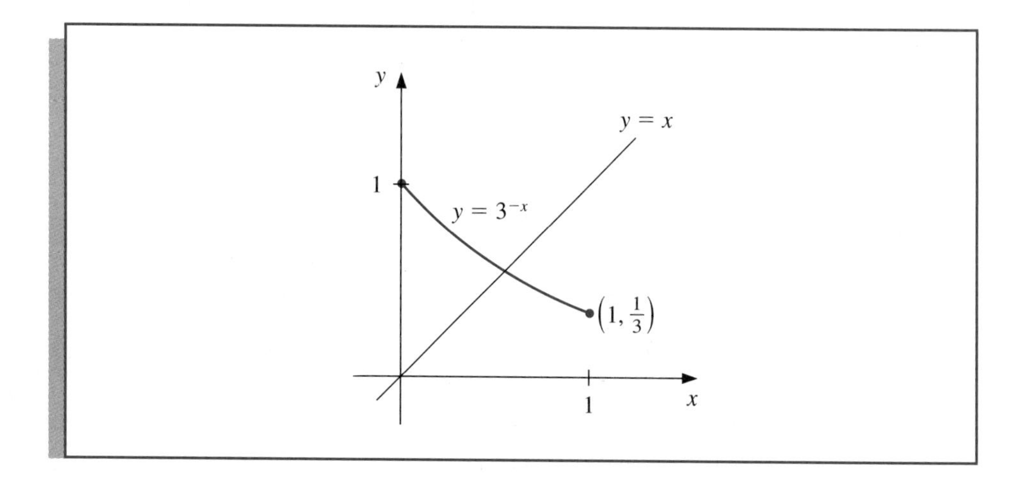

To approximate the fixed point of a function g, we choose an initial approximation p_0 and generate the sequence $\{p_n\}_{n=0}^{\infty}$ by letting $p_n = g(p_{n-1})$ for each $n \ge 1$. If the sequence converges to p and g is continuous, then, by Theorem 1.4,

$$p = \lim_{n\to\infty} p_n = \lim_{n\to\infty} g(p_{n-1}) = g\left(\lim_{n\to\infty} p_{n-1}\right) = g(p),$$

and a solution to $x = g(x)$ is obtained. This technique is called **fixed-point iteration**, or **functional iteration**. The procedure is detailed in Algorithm 2.2 and illustrated in Figure 2.6.

ALGORITHM **2.2**

Fixed-Point Iteration

To find a solution to $p = g(p)$ given an initial approximation p_0:

INPUT initial approximation p_0; tolerance TOL; maximum number of iterations N_0.

OUTPUT approximate solution p or message of failure.

Step 1 Set $i = 1$.

Step 2 While $i \le N_0$ do Steps 3–6.

Step 3 Set $p = g(p_0)$. (*Compute p_i.*)

Step 4 If $|p - p_0| < TOL$ then
OUTPUT (p); (*Procedure completed successfully.*)
STOP.

Step 5 Set $i = i + 1$.

Step 6 Set $p_0 = p$. (*Update p_0.*)

Step 7 OUTPUT ('Method failed after N_0 iterations, N_0 =', N_0);
(*Procedure completed unsuccessfully.*)
STOP.

Figure 2.6

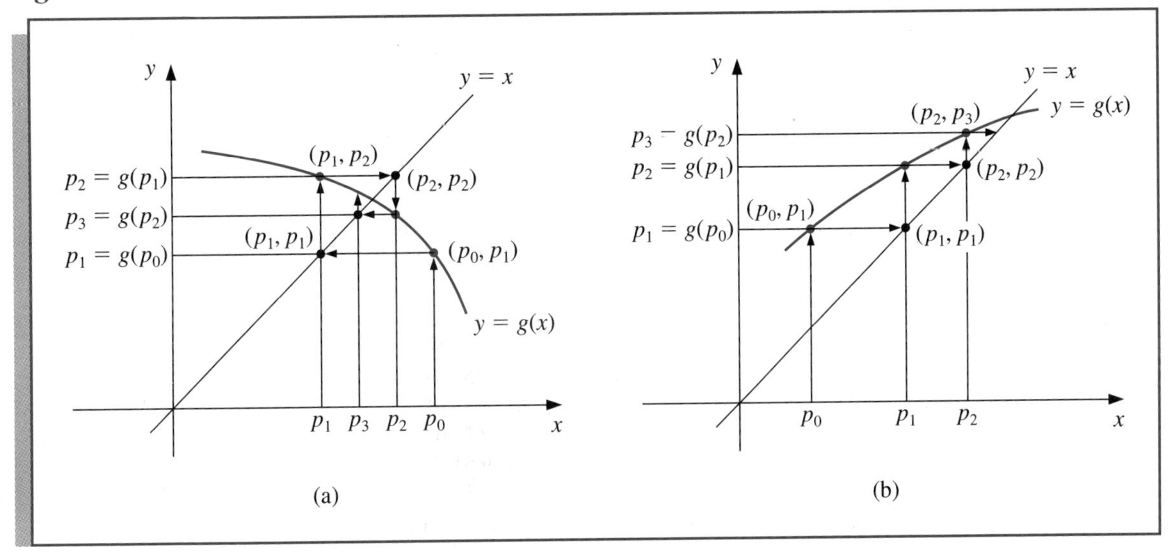

The following example illustrates the technique of functional iteration.

EXAMPLE 3 The equation $x^3 + 4x^2 - 10 = 0$ has a unique root in [1, 2]. There are many ways to change the equation to the form $x = g(x)$ by simple algebraic manipulation. For example, to obtain the function g described in (c), we can manipulate the equation $x^3 + 4x^2 - 10 = 0$ as follows:

$$4x^2 = 10 - x^3, \quad \text{so} \quad x^2 = \frac{1}{4}(10 - x^3)$$

and

$$x = \pm\frac{1}{2}(10 - x^3)^{1/2}.$$

To obtain a positive solution, $g_3(x)$ is chosen as shown. It is not important to derive the functions shown here, but you should verify that the fixed point of each is actually a solution to the original equation, $x^3 + 4x^2 - 10 = 0$.

a. $x = g_1(x) = x - x^3 - 4x^2 + 10$

b. $x = g_2(x) = \left(\frac{10}{x} - 4x\right)^{1/2}$

c. $x = g_3(x) = \frac{1}{2}\left(10 - x^3\right)^{1/2}$

d. $x = g_4(x) = \left(\frac{10}{4 + x}\right)^{1/2}$

e. $x = g_5(x) = x - \frac{x^3 + 4x^2 - 10}{3x^2 + 8x}$

With $p_0 = 1.5$, Table 2.2 lists the results of the fixed-point iteration method for all five choices of g.

Table 2.2

n	(a)	(b)	(c)	(d)	(e)
0	1.5	1.5	1.5	1.5	1.5
1	-0.875	0.8165	1.286953768	1.348399725	1.373333333
2	6.732	2.9969	1.402540804	1.367376372	1.365262015
3	-469.7	$(-8.65)^{1/2}$	1.345458374	1.364957015	1.365230014
4	1.03×10^8		1.375170253	1.365264748	1.365230013
5			1.360094193	1.365225594	
6			1.367846968	1.365230576	
7			1.363887004	1.365229942	
8			1.365916734	1.365230022	
9			1.364878217	1.365230012	
10			1.365410062	1.365230014	
15			1.365223680	1.365230013	
20			1.365230236		
25			1.365230006		
30			1.365230013		

The actual root is 1.365230013, as was noted in Example 1 of Section 2.1. Comparing the results to the Bisection Algorithm given in that example, it can be seen that excellent results have been obtained for choices (c), (d), and (e), since the Bisection method requires 27 iterations for such accuracy. It is interesting to note that choice (a) led to divergence and that (b) became undefined because it involved the square root of a negative number. ■

Even though the various functions in this example are fixed-point problems for the same root-finding problem, they differ vastly as techniques for approximating the solution to the root-finding problem. Their purpose is to illustrate the true question that needs to be answered:

How can we find a fixed-point problem that produces a sequence that *rapidly* converges to a solution to a given root-finding problem?

The following result gives us some clues concerning the paths we should pursue and, perhaps more importantly, some we should reject.

Theorem 2.3 **(Fixed-Point Theorem)**

Let $g \in C[a, b]$ be such that $g(x) \in [a, b]$ for all x in $[a, b]$. Suppose, in addition, that g' exists on (a, b) and a positive constant $k < 1$ exists with

$$|g'(x)| \leq k, \quad \text{for all } x \in (a, b).$$

Then for any number p_0 in $[a, b]$, the sequence defined by

$$p_n = g(p_{n-1}), \quad n \geq 1,$$

converges to the unique fixed point p in $[a, b]$.

Proof Theorem 2.2 implies that a unique fixed point exists in $[a, b]$. Since g maps $[a, b]$ into itself, the sequence $\{p_n\}_{n=0}^{\infty}$ is defined for all $n \geq 0$ and $p_n \in [a, b]$ for all n. Using the fact that $|g'(x)| \leq k$ and the Mean Value Theorem, we have

$$|p_n - p| = |g(p_{n-1}) - g(p)| = |g'(\xi)||p_{n-1} - p| \leq k|p_{n-1} - p|,$$

where $\xi \in (a, b)$. Applying this inequality inductively gives

$$\textbf{(2.4)} \qquad |p_n - p| \leq k|p_{n-1} - p| \leq k^2|p_{n-2} - p| \leq \cdots \leq k^n|p_0 - p|.$$

Since $k < 1$,

$$\lim_{n\to\infty} |p_n - p| \leq \lim_{n\to\infty} k^n|p_0 - p| = 0,$$

and $\{p_n\}_{n=0}^{\infty}$ converges to p. ■ ■ ■

Corollary 2.4 If g satisfies the hypotheses of Theorem 2.3, bounds for the error involved in using p_n to approximate p are given by

$$|p_n - p| \leq k^n \max\{p_0 - a, b - p_0\}$$

and

$$|p_n - p| \leq \frac{k^n}{1-k}|p_1 - p_0|, \quad \text{for all } n \geq 1.$$

Proof The first bound follows from Inequality (2.4):

$$|p_n - p| \leq k^n|p_0 - p| \leq k^n \max\{p_0 - a, b - p_0\}$$

since $p \in [a, b]$.

For $n \geq 1$, the procedure used in the proof of Theorem 2.3 implies that

$$|p_{n+1} - p_n| = |g(p_n) - g(p_{n-1})| \leq k|p_n - p_{n-1}| \leq \cdots \leq k^n|p_1 - p_0|.$$

Thus, for $m > n \geq 1$,

$$\begin{aligned}|p_m - p_n| &= |p_m - p_{m-1} + p_{m-1} - p_{m-2} + \cdots + p_{n+1} - p_n| \\ &\leq |p_m - p_{m-1}| + |p_{m-1} - p_{m-2}| + \cdots + |p_{n+1} - p_n| \\ &\leq k^{m-1}|p_1 - p_0| + k^{m-2}|p_1 - p_0| + \cdots k^n|p_1 - p_0| \\ &= k^n(1 + k + k^2 + \cdots + k^{m-n-1})|p_1 - p_0|.\end{aligned}$$

By Theorem 2.3, $\lim_{m\to\infty} p_m = p$, so

$$|p - p_n| = \lim_{m\to\infty} |p_m - p_n| \leq k^n|p_1 - p_0| \sum_{i=0}^{\infty} k^i.$$

But $\sum_{i=0}^{\infty} k^i$ is a geometric series with ratio k, and $0 < k < 1$. This sequence converges to $1/(1 - k)$, which gives the second bound:

$$|p - p_n| \leq \frac{k^n}{1 - k}|p_1 - p_0|.$$

▪ ▪ ▪

Both inequalities in the corollary relate the rate at which $\{p_n\}$ converges to the bound k on the first derivative. The rate of convergence depends on the factor k^n. The smaller the value of k, the faster the convergence, which may be very slow if k is close to 1. In the following example, the fixed-point methods in Example 3 are reconsidered in light of the results described in Theorem 2.3.

EXAMPLE 4

a. When $g_1(x) = x - x^3 - 4x^2 + 10$, $g_1'(x) = 1 - 3x^2 - 8x$. There is no interval $[a, b]$ containing p for which $|g_1'(x)| < 1$. Although Theorem 2.3 does not guarantee that the method must fail for this choice of g, there is no reason to expect convergence.

b. With $g_2(x) = [(10/x) - 4x]^{1/2}$, we can see that g_2 does not map $[1, 2]$ into $[1, 2]$, and the sequence $\{p_n\}_{n=0}^{\infty}$ is not defined with $p_0 = 1.5$. Moreover, there is no interval containing p such that

$$|g_2'(x)| < 1, \quad \text{since} \quad |g_2'(p)| \approx 3.4.$$

c. For the function $g_3(x) = \frac{1}{2}(10 - x^3)^{1/2}$,

$$g_3'(x) = -\frac{3}{4}x^2(10 - x^3)^{-1/2} < 0 \quad \text{on } [1, 2],$$

so g_3 is strictly decreasing on $[1, 2]$. However, $|g_3'(2)| \approx 2.12$, so the condition $|g_3'(x)| \leq k < 1$ fails on $[1, 2]$. A closer examination of the sequence $\{p_n\}_{n=0}^{\infty}$ starting with $p_0 = 1.5$ shows that it suffices to consider the interval $[1, 1.5]$ instead of $[1, 2]$. On this interval it is still true that $g_3'(x) < 0$ and g_3 is strictly decreasing, but, additionally,

$$1 < 1.28 \approx g_3(1.5) \le g_3(x) \le g_3(1) = 1.5$$

for all $x \in [1, 1.5]$. This shows that g_3 maps the interval $[1, 1.5]$ into itself. Since it is also true that $|g_3'(x)| \le |g_3'(1.5)| \approx 0.66$ on this interval, Theorem 2.3 confirms the convergence of which we were already aware.

d. For $g_4(x) = (10/(4+x))^{1/2}$ we have

$$|g_4'(x)| = \left|\frac{-5}{\sqrt{10}(4+x)^{3/2}}\right| \le \frac{5}{\sqrt{10}(5)^{3/2}} < 0.15, \quad \text{for all} \quad x \in [1, 2].$$

The bound on the magnitude of $g_4'(x)$ is much smaller than the bound on the magnitude of $g_3'(x)$, which explains the more rapid convergence using g_4. The other part of Example 3 can be handled in a similar manner. ■

EXERCISE SET 2.2

1. Use algebraic manipulation to show that each of the following functions has a fixed point at p precisely when $f(p) = 0$, where $f(x) = x^4 + 2x^2 - x - 3$.

a. $g_1(x) = \left(3 + x - 2x^2\right)^{1/4}$

b. $g_2(x) = \left(\dfrac{x + 3 - x^4}{2}\right)^{1/2}$

c. $g_3(x) = \left(\dfrac{x+3}{x^2+2}\right)^{1/2}$

d. $g_4(x) = \dfrac{3x^4 + 2x^2 + 3}{4x^3 + 4x - 1}$

2. a. Perform four iterations, if possible, on each of the functions g defined in Exercise 1. Let $p_0 = 1$ and $p_{n+1} = g(p_n)$ for $n = 0, 1, 2, 3$.

b. Which function do you think gives the best approximation to the solution?

3. The following three methods are proposed to compute $21^{1/3}$. Rank them in order, based on their apparent speed of convergence, assuming $p_0 = 1$.

a. $p_n = \dfrac{20p_{n-1} + 21/p_{n-1}^2}{21}$

b. $p_n = p_{n-1} - \dfrac{p_{n-1}^3 - 21}{3p_{n-1}^2}$

c. $p_n = p_{n-1} - \dfrac{p_{n-1}^4 - 21p_{n-1}}{p_{n-1}^2 - 21}$

d. $p_n = \left(\dfrac{21}{p_{n-1}}\right)^{1/2}$

4. The following four methods are proposed to compute $7^{1/5}$. Rank them in order, based on their apparent speed of convergence, assuming $p_0 = 1$.

a. $p_n = \left(1 + \dfrac{7 - p_{n-1}^3}{p_{n-1}^2}\right)^{1/2}$

b. $p_n = p_{n-1} - \dfrac{p_{n-1}^5 - 7}{p_{n-1}^2}$

c. $p_n = p_{n-1} - \dfrac{p_{n-1}^5 - 7}{5p_{n-1}^4}$

d. $p_n = p_{n-1} - \dfrac{p_{n-1}^5 - 7}{12}$

5. Use a fixed-point iteration method to determine a solution accurate to within 10^{-2} for $x^4 - 3x^2 - 3 = 0$ on $[1, 2]$. Use $p_0 = 1$.

6. Use a fixed-point iteration method to determine a solution accurate to within 10^{-2} for $x^3 - x - 1 = 0$ on $[1, 2]$. Use $p_0 = 1$.

7. Use Theorem 2.2 to show that $g(x) = \pi + 0.5\sin(x/2)$ has a unique fixed point on $[0, 2\pi]$. Use fixed-point iteration to find an approximation to the fixed point that is accurate to within 10^{-2}. Use Corollary 2.4 to estimate the number of iterations required to achieve 10^{-2} accuracy, and compare this theoretical estimate to the number actually needed.

8. Use Theorem 2.2 to show that $g(x) = 2^{-x}$ has a unique fixed point on $[\frac{1}{3}, 1]$. Use fixed-point iteration to find an approximation to the fixed point accurate to within 10^{-4}. Use Corollary 2.4 to estimate the number of iterations required to achieve 10^{-4} accuracy, and compare this theoretical estimate to the number actually needed.

9. Use a fixed-point iteration method to find an approximation to $\sqrt{3}$ that is accurate to within 10^{-4}. Compare your result and the number of iterations required with the answer obtained in Exercise 10 of Section 2.1.

10. Use a fixed-point iteration method to find an approximation to $\sqrt[3]{25}$ that is accurate to within 10^{-4}. Compare your result and the number of iterations required with the answer obtained in Exercise 11 of Section 2.1.

11. For each of the following equations, determine an interval $[a, b]$ on which fixed-point iteration will converge. Estimate the number of iterations necessary to obtain approximations accurate to within 10^{-5}, and perform the calculations.

    a. $x = \dfrac{2 - e^x + x^2}{3}$ **b.** $x = \dfrac{5}{x^2} + 2$

    c. $x = (e^x/3)^{1/2}$ **d.** $x = 5^{-x}$

    e. $x = 6^{-x}$ **f.** $x = 0.5(\sin x + \cos x)$

12. For each of the following equations, determine a function g and an interval $[a, b]$ on which fixed-point iteration will converge to a positive solution of the equation.

    a. $3x^2 - e^x = 0$ **b.** $x - \cos x = 0$

    Find the solutions to within 10^{-5}.

13. Find all the zeros of $f(x) = x^2 + 10\cos x$ by using the fixed-point iteration method for an appropriate iteration function g. Find the zeros accurate to within 10^{-4}.

14. Use a fixed-point iteration method to determine a solution accurate to within 10^{-4} for $x = \tan x$, for x in $[4, 5]$.

15. Use a fixed-point iteration method to determine a solution accurate to within 10^{-2} for $2\sin \pi x + x = 0$ on $[1, 2]$. Use $p_0 = 1$.

16. Let A be a given positive constant and $g(x) = 2x - Ax^2$.

    a. Show that if fixed point iteration converges to a nonzero limit, then the limit is $p = 1/A$, so the inverse of a number can be found using only multiplications and subtractions.

    b. Find an interval about $1/A$ for which fixed-point iteration converges, provided p_0 is in that interval.

17. Find a function g defined on $[0, 1]$ that satisfies none of the hypotheses of Theorem 2.2 but still has a unique fixed point on $[0, 1]$.

18. **a.** Show that Theorem 2.2 is true if the inequality $|g'(x)| \le k$ is replaced by $g'(x) \le k$ for all $x \in (a, b)$. [*Hint:* Only uniqueness is in question.]

    b. Show that Theorem 2.3 may not hold if inequality $|g'(x)| \le k$ is replaced by $g'(x) \le k$. [*Hint:* Show that $g(x) = 1 - x^2$, for x in $[0, 1]$, provides a counterexample.]

19. **a.** Use Theorem 2.3 to show that the sequence defined by

$$x_n = \frac{1}{2}x_{n-1} + \frac{1}{x_{n-1}}, \quad \text{for } n \ge 1,$$

    converges to $\sqrt{2}$ whenever $x_0 > \sqrt{2}$.

b. Use the fact that $0 < \left(x_0 - \sqrt{2}\right)^2$ whenever $x_0 \neq \sqrt{2}$ to show that if $0 < x_0 < \sqrt{2}$, then $x_1 > \sqrt{2}$.

c. Use the results of parts (a) and (b) to show that the sequence in (a) converges to $\sqrt{2}$ whenever $x_0 > 0$.

20. **a.** Show that if A is any positive number, then the sequence defined by

$$x_n = \frac{1}{2}x_{n-1} + \frac{A}{2x_{n-1}}, \quad \text{for } n \geq 1,$$

converges to $\sqrt{A}$ whenever $x_0 > 0$.

b. What happens if $x_0 < 0$?

21. Replace the assumption in Theorem 2.3 that "a positive number $k < 1$ exists with $|g'(x)| \leq k$" with "g satisfies a Lipschitz condition on the interval $[a, b]$ with Lipschitz constant $L < 1$." (See Exercise 25, Section 1.1.) Show that the conclusions of this theorem are still valid.

22. Suppose that g is continuously differentiable on some interval (c, d) that contains the fixed point p of g. Show that if $|g'(p)| < 1$, then there exists a $\delta > 0$ such that the fixed-point iteration converges for any initial approximation p_0, whenever $|p_0 - p| \leq \delta$.

23. An object falling vertically through the air is subjected to viscous resistance as well as to the force of gravity. Assume that an object with mass m is dropped from a height y_0, and that the height of the object after t seconds is

$$y(t) = y_0 + \frac{mg}{k}t - \frac{m^2 g}{k^2}(1 - e^{-kt/m}),$$

where $g = -32.17$ ft/s^2 and k represents the coefficient of air resistance in lb-s/ft. Suppose $y_0 = 300$ ft, $m = 0.25$ lb, and $k = 0.1$ lb-s/ft. Find, to within 0.01 s, the time it takes this quarter-pounder to hit the ground.

24. Let $g \in C^1[a, b]$ and p be in (a, b) with $g(p) = p$ and $|g'(p)| > 1$. Show that there exists a $\delta > 0$ such that if $0 < |p_0 - p| < \delta$, then $|p_0 - p| < |p_1 - p|$. Thus, no matter how close the initial approximation p_0 is to p, the next iterate p is farther away, so the fixed-point iteration does not converge if $p_0 \neq p$.

2.3 The Newton-Raphson Method

The **Newton-Raphson** (or simply **Newton's**) **method** is one of the most powerful and well-known numerical methods for solving a root-finding problem $f(x) = 0$. There are many ways of introducing Newton's method. The most common is to consider the technique graphically. Another possibility is to derive Newton's method as a technique to obtain faster convergence than offered by other types of functional iteration. This is done in Section 2.4. A third means of introducing Newton's method, which is discussed next, is based on Taylor polynomials.

Suppose that $f \in C^2[a, b]$. Let $\overline{x} \in [a, b]$ be an approximation to p such that $f'(\overline{x}) \neq 0$ and $|\overline{x} - p|$ is "small." Consider the first Taylor polynomial for $f(x)$ expanded about $\overline{x}$,

$$f(x) = f(\overline{x}) + (x - \overline{x})f'(\overline{x}) + \frac{(x - \overline{x})^2}{2}f''(\xi(x)),$$

where $\xi(x)$ lies between x and $\overline{x}$. Since $f(p) = 0$, this equation, with $x = p$, gives

$$0 = f(\overline{x}) + (p - \overline{x})f'(\overline{x}) + \frac{(p - \overline{x})^2}{2} f''(\xi(p)).$$

Newton's method is derived by assuming that since $|p - \overline{x}|$ is small the term involving $(p - \overline{x})^2$ is much smaller and that

$$0 \approx f(\overline{x}) + (p - \overline{x})f'(\overline{x}).$$

Solving for p in this equation gives

$$p \approx x - \frac{f(\overline{x})}{f'(\overline{x})}.$$

This sets the stage for the Newton-Raphson method, which starts with an initial approximation p_0 and generates the sequence $\{p_n\}$ defined by

(2.5)
$$p_n = p_{n-1} - \frac{f(p_{n-1})}{f'(p_{n-1})}, \quad \text{for } n \geq 1.$$

Figure 2.7 illustrates how the approximations are obtained using successive tangents. (Also see Exercise 11.) Starting with the initial approximation p_0, the approximation p_1 is the x-intercept of the tangent line to the graph of f at $(p_0, f(p_0))$. The approximation p_2 is the x-intercept of the tangent line to the graph of f at $(p_1, f(p_1))$ and so on. Algorithm 2.3 follows this procedure.

Figure 2.7

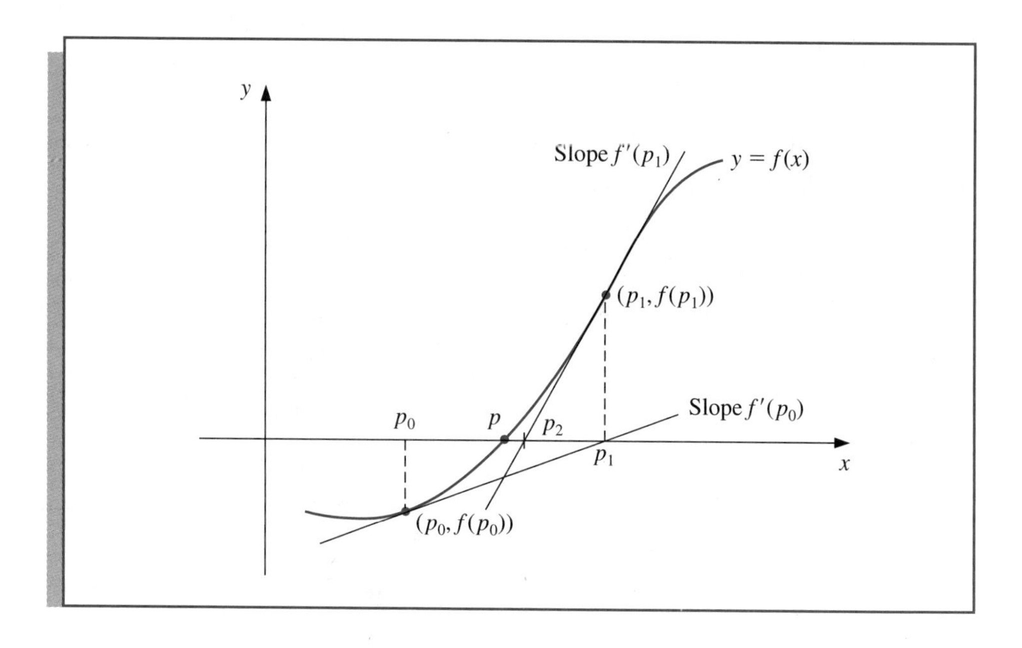

ALGORITHM 2.3

Newton-Raphson

To find a solution to $f(x) = 0$ given the differentiable function f and an initial approximation p_0:

INPUT initial approximation p_0; tolerance TOL; maximum number of iterations N_0.

OUTPUT approximate solution p or message of failure.

Step 1 Set $i = 1$.

Step 2 While $i \leq N_0$ do Steps 3–6.

> *Step 3* Set $p = p_0 - f(p_0)/f'(p_0)$. *(Compute p_i.)*
>
> *Step 4* If $|p - p_0| < TOL$ then
> OUTPUT (p); *(Procedure completed successfully.)*
> STOP.
>
> *Step 5* Set $i = i + 1$.
>
> *Step 6* Set $p_0 = p$. *(Update p_0.)*

Step 7 OUTPUT ('Method failed after N_0 iterations, N_0 =', N_0);
(Procedure completed unsuccessfully.)
STOP.

The stopping-technique inequalities given with the Bisection method are applicable to Newton's method. That is, select a tolerance $\varepsilon > 0$ and construct $p_1, \ldots p_N$ until

(2.6)
$$|p_N - p_{N-1}| < \varepsilon,$$

(2.7)
$$\frac{|p_N - p_{N-1}|}{|p_N|} < \varepsilon, \quad p_N \neq 0,$$

or

(2.8)
$$|f(p_N)| < \varepsilon.$$

A form of Inequality (2.6) is used in Step 4 of Algorithm 2.3. Note that inequality (2.8) may not give much information about the actual error $|p_N - p|$. (See Exercise 14 in Section 2.1.)

Newton's method is a functional iteration technique of the form $p_n = g(p_{n-1})$, for which

$$g(p_{n-1}) = p_{n-1} - \frac{f(p_{n-1})}{f'(p_{n-1})}, \quad \text{for } n \geq 1.$$

It is clear from this equation that Newton's method cannot be continued if $f'(p_{n-1}) = 0$ for some n. We will see that the method is most effective when f' is bounded away from zero near p.

EXAMPLE 1

a. To approximate a solution to the equation $x = \cos x$, let $f(x) = \cos x - x$. Then

$$f\left(\frac{\pi}{2}\right) = -\frac{\pi}{2} < 0 < 1 = f(0),$$

and, by the Intermediate Value Theorem, there exists a zero of f in $[0, \pi/2]$. The graphs of the equations $y = x$ and $y = \cos x$ appear in Figure 2.8; their intersection is the fixed point of $g(x) = \cos x$.

Figure 2.8

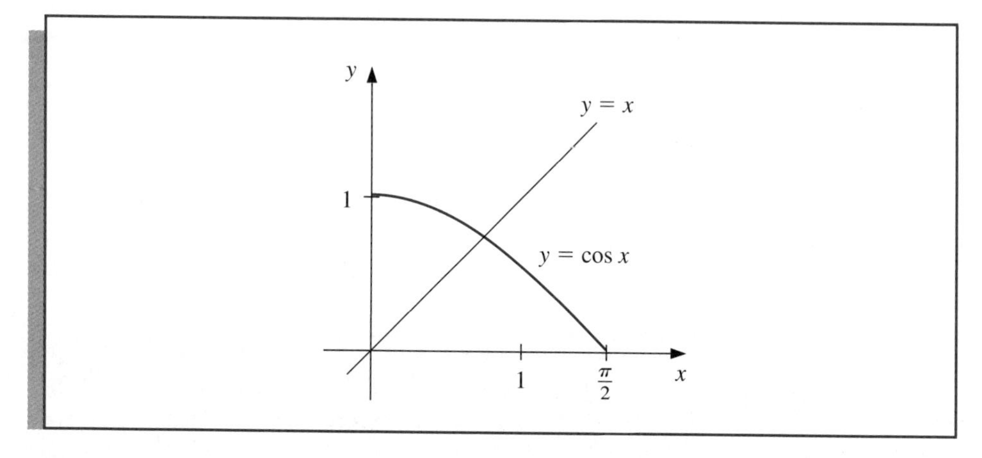

The graph shows that $f(x) = 0$ has a unique solution in $[0, \pi/2]$. Since $f'(x) = -\sin x - 1$, Newton's method has the form

$$p_n = p_{n-1} - \frac{\cos p_{n-1} - p_{n-1}}{-\sin p_{n-1} - 1}, \quad \text{for } n \geq 1,$$

where p_0 is yet to be selected. For some problems it suffices to let p_0 be arbitrary, whereas for others it is important to select a good initial approximation. For the problem under consideration, the graph in Figure 2.8 suggests $p_0 = \pi/4$ as an initial approximation. With $p_0 = \pi/4$, the approximations in Table 2.3 are generated. An excellent approximation is obtained with $n = 3$.

Table 2.3

n	p_n
0	0.7853981635
1	0.7395361337
2	0.7390851781
3	0.7390851332
4	0.7390851332

b. To obtain the unique solution to $x^3 + 4x^2 - 10 = 0$ on the interval $[1, 2]$ by Newton's method, generate the sequence $\{p_n\}_{n=1}^{\infty}$ given by

$$p_n = p_{n-1} - \frac{p_{n-1}^3 + 4p_{n-1}^2 - 10}{3p_{n-1}^2 + 8p_{n-1}}, \quad \text{for } n \geq 1.$$

Selecting $p_0 = 1.5$ (see Figure 2.9) produces the results of Example 3(e) of Section 2.2, in which $p_3 = 1.36523001$ is correct to the eighth decimal place. ■

Figure 2.9

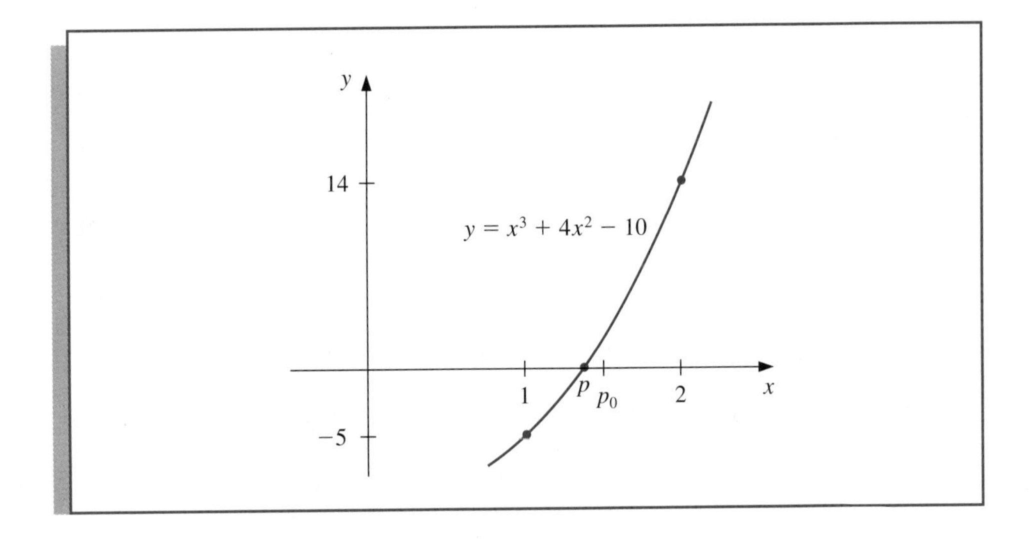

The Taylor series derivation of Newton's method at the beginning of the section points out the importance of an accurate initial approximation. The crucial assumption is that the term involving $(p - \bar{x})^2$ is, by comparison, so small that it can be deleted. This will clearly be false unless $\bar{x}$ is a good approximation to p. In particular, if p_0 is not sufficiently close to the actual root, Newton's method may not converge to the root. This, however, is not always the case. (Exercises 12 and 16 illustrate some of the possibilities that can occur.)

The following convergence theorem for Newton's method illustrates the theoretical importance of the choice of p_0.

Theorem 2.5 Let $f \in C^2[a, b]$. If $p \in [a, b]$ is such that $f(p) = 0$ and $f'(p) \neq 0$, then there exists a $\delta > 0$ such that Newton's method generates a sequence $\{p_n\}_{n=1}^{\infty}$ converging to p for any initial approximation $p_0 \in [p - \delta, p + \delta]$.

Proof The proof is based on analyzing Newton's method as the functional iteration scheme $p_n = g(p_{n-1})$, for $n \geq 1$, with

$$g(x) = x - \frac{f(x)}{f'(x)}.$$

Let k be any number in $(0, 1)$. We first find an interval $[p - \delta, p + \delta]$ that g maps into itself, and $|g'(x)| \leq k$ for all $x \in (p - \delta, p + \delta)$.

Since $f'(p) \neq 0$ and f' is continuous, there exists $\delta_1 > 0$ such that $f'(x) \neq 0$ for $x \in [p - \delta_1, p + \delta_1] \subset [a, b]$. Thus, g is defined and continuous on $[p - \delta_1, p + \delta_1]$. Also,

$$g'(x) = 1 - \frac{f'(x)f'(x) - f(x)f''(x)}{[f'(x)]^2} = \frac{f(x)f''(x)}{[f'(x)]^2}$$

for $x \in [p - \delta_1, p + \delta_1]$, and since $f \in C^2[a, b]$, we have $g \in C^1[p - \delta_1, p + \delta_1]$.

By assumption, $f(p) = 0$, so

$$g'(p) = \frac{f(p)f''(p)}{[f'(p)]^2} = 0.$$

Since g' is continuous and $0 < k < 1$, there exists a δ, with $0 < \delta < \delta_1$, and

$$|g'(x)| \leq k \quad \text{for all } x \in [p - \delta, p + \delta].$$

It remains to show that $g : [p - \delta, p + \delta] \to [p - \delta, p + \delta]$. If $x \in [p - \delta, p + \delta]$, the Mean Value Theorem implies that, for some number ξ between x and p, $|g(x) - g(p)| = |g'(\xi)||x - p|$. So

$$|g(x) - p| = |g(x) - g(p)| = |g'(\xi)||x - p| \leq k|x - p| < |x - p|.$$

Since $x \in [p - \delta, p + \delta]$, it follows that $|x - p| < \delta$ and that $|g(x) - p| < \delta$. This result implies that $g : [p - \delta, p + \delta] \to [p - \delta, p + \delta]$.

All the hypotheses of the Fixed-Point Theorem are now satisfied for $g(x) = x - f(x)/f'(x)$, so the sequence $\{p_n\}_{n=1}^{\infty}$ defined by

$$p_n = g(p_{n-1}), \quad \text{for } n \geq 1,$$

converges to p for any $p_0 \in [p - \delta, p + \delta]$. ▪ ▪ ▪

Theorem 2.5 states that, under reasonable assumptions, Newton's method converges provided a sufficiently accurate initial approximation is chosen. It also implies that the constant k that bounds the derivative of g, and, consequently, indicates the speed of convergence of the method, decreases to zero as the procedure continues.

Newton's method is an extremely powerful technique, but it has a major difficulty: the need to know the value of the derivative of f at each approximation. Frequently $f'(x)$ is more difficult to determine and needs more arithmetic operations to calculate than $f(x)$.

To circumvent the problem of the derivative evaluation in Newton's method, we derive a slight variation. By definition,

$$f'(p_{n-1}) = \lim_{x \to p_{n-1}} \frac{f(x) - f(p_{n-1})}{x - p_{n-1}}.$$

Letting $x = p_{n-2}$, we have

$$f'(p_{n-1}) \approx \frac{f(p_{n-2}) - f(p_{n-1})}{p_{n-2} - p_{n-1}} = \frac{f(p_{n-1}) - f(p_{n-2})}{p_{n-1} - p_{n-2}}.$$

Using this approximation for $f'(p_{n-1})$ in Newton's formula gives

(2.9)
$$p_n = p_{n-1} - \frac{f(p_{n-1})(p_{n-1} - p_{n-2})}{f(p_{n-1}) - f(p_{n-2})}.$$

The technique using this formula is called the **Secant method** and is presented in Algorithm 2.4. (See Figure 2.10.) Starting with the two initial approximations p_0 and p_1, the approximation p_2 is the x-intercept of the line joining $(p_0, f(p_0))$ and $(p_1, f(p_1))$. The approximation p_3 is the x-intercept of the line joining $(p_1, f(p_1))$ and $(p_2, f(p_2))$, and so on.

Figure 2.10

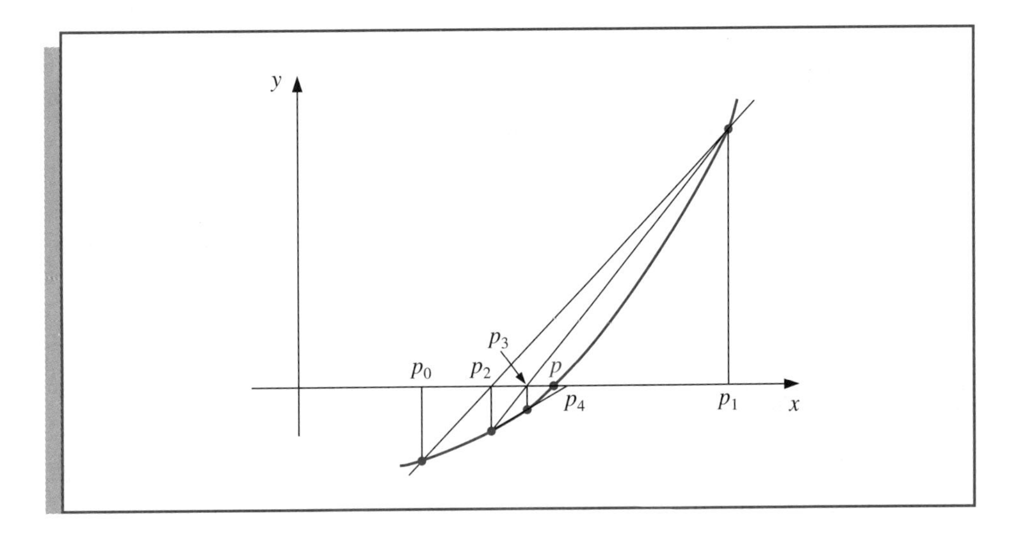

ALGORITHM 2.4

Secant

To find a solution to $f(x) = 0$ given the continuous function f and initial approximations p_0 and p_1:

INPUT initial approximations p_0, p_1; tolerance TOL; maximum number of iterations N_0.

OUTPUT approximate solution p or message of failure.

Step 1 Set $i = 2$;
$q_0 = f(p_0)$;
$q_1 = f(p_1)$.

Step 2 While $i \le N_0$ do Steps 3–6.

Step 3 Set $p = p_1 - q_1(p_1 - p_0)/(q_1 - q_0)$. *(Compute p_i.)*

Step 4 If $|p - p_1| < TOL$ then
OUTPUT (p); *(Procedure completed successfully.)*
STOP.

Step 5 Set $i = i + 1$.

Step 6 Set $p_0 = p_1$; *(Update p_0, q_0, p_1, q_1.)*
$q_0 = q_1$;

$p_1 = p;$
$q_1 = f(p).$

Step 7 OUTPUT ('Method failed after N_0 iterations, N_0 =', N_0);
(*Procedure completed unsuccessfully.*)
STOP.

The next example involves a problem considered in Example 1(a), where we used Newton's method with $p_0 = \pi/4$.

EXAMPLE 2 Use the Secant method to find a zero of $f(x) = \cos x - x$. In Example 1 we used the initial approximation $p_0 = \pi/4$. Here we need two initial approximations. Table 2.4 lists the calculations with $p_0 = 0.5$, $p_1 = \pi/4$, and the formula

$$p_n = p_{n-1} - \frac{(p_{n-1} - p_{n-2})(\cos p_{n-1} - p_{n-1})}{(\cos p_{n-1} - p_{n-1}) - (\cos p_{n-2} - p_{n-2})}, \quad \text{for } n \geq 2,$$

from Algorithm 2.4. ■

Table 2.4

n	p_n
0	0.5
1	0.7853981635
2	0.7363841388
3	0.7390581392
4	0.7390851493
5	0.7390851332

By comparing the results here with those from Example 1 given in Table 2.3, we see that p_5 is accurate to the tenth decimal place. Note that the convergence of the Secant method is slightly slower in this example than that of Newton's method, which obtained this degree of accuracy with p_3. This result is generally true. (See Theorem 2.8 and Exercise 12 of Section 2.4.)

Newton's method (or the Secant method) is often used to refine an answer obtained by another technique, such as the Bisection method. Since Newton's method requires a good first approximation but generally gives rapid convergence, it serves this purpose well.

Each successive pair of approximations in the Bisection method brackets a root p of the equation; that is, for each positive integer n, a root lies between p_n and p_{n-1}. This implies that for each n the Bisection method iterations satisfy

$$|p_n - p| < |p_n - p_{n-1}|,$$

which provides an easily calculated error bound for the approximations. Root bracketing is not guaranteed for either Newton's method or the Secant method. Table 2.3 contains results from Newton's method applied to $f(x) = \cos x - x$, where an approximate root was found to be 0.7390851332. Notice that this root is not bracketed by either p_0, p_1 or p_1, p_2. The Secant method approximations for this problem are given in Table 2.4. The initial approximations p_0 and p_1 were chosen to bracket the root, but the pair of approximations p_3 and p_4 fail to do so.

The **method of False Position** (also called the *Regula Falsi method*) generates approximations in the same manner as the Secant method, but it provides a test to ensure that the root is bracketed between successive iterations. Although it is not a method we generally recommend, it illustrates the way that bracketing can be incorporated.

We first choose initial approximations p_0 and p_1 with $f(p_0) \cdot f(p_1) < 0$. The approximation p_2 is chosen in the same manner as in the Secant method, as the x-intercept of the line joining $(p_0, f(p_0))$ and $(p_1, f(p_1))$. To decide which secant line to use to compute p_3, we check $f(p_2) \cdot f(p_1)$. If this value is negative, then p_1, p_2 bracket a root, and we choose p_3 as the x-intercept of the line joining $(p_1, f(p_1))$ and $(p_2, f(p_2))$. If not, we choose p_3 as the x-intercept of the line joining $(p_0, f(p_0))$ and $(p_2, f(p_2))$ and then interchange the indices on p_0 and p_1. In a similar manner, once p_3 is found, the sign of $f(p_3) \cdot f(p_2)$ determines whether we use p_2 and p_3 or p_3 and p_1 to compute p_4. In the latter case a relabeling of p_2 and p_1 is performed. The relabeling ensures that the root is bracketed between successive iterations. The process is described in Algorithm 2.5, and Figure 2.11 shows how the iterations can differ from those of the Secant method. In this illustration, the first three approximations are the same, but the fourth approximations differ.

Figure 2.11

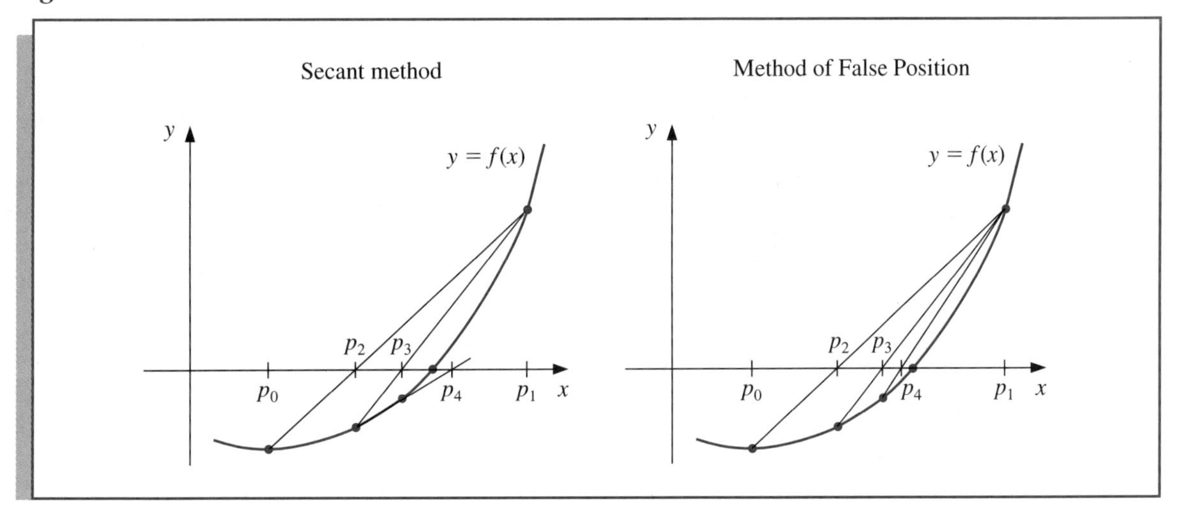

ALGORITHM **2.5**

Method of False Position

To find a solution to $f(x) = 0$ given the continuous function f on the interval $[p_0, p_1]$ where $f(p_0)$ and $f(p_1)$ have opposite signs:

INPUT initial approximations p_0, p_1; tolerance TOL; maximum number of iterations N_0.

OUTPUT approximate solution p or message of failure.

Step 1 Set $i = 2$;
$q_0 = f(p_0)$;
$q_1 = f(p_1)$.

Step 2 While $i \leq N_0$ do Steps 3–7.

Step 3 Set $p = p_1 - q_1(p_1 - p_0)/(q_1 - q_0)$. (*Compute* p_i.)

Step 4 If $|p - p_1| < TOL$ then
OUTPUT (p); (*Procedure completed successfully.*)
STOP.

Step 5 Set $i = i + 1$;
$q = f(p)$.

Step 6 If $q \cdot q_1 < 0$ then set $p_0 = p_1$;
$q_0 = q_1$.

Step 7 Set $p_1 = p$;
$q_1 = q$.

Step 8 OUTPUT ('Method failed after N_0 iterations, N_0 =', N_0);
(*Procedure completed unsuccessfully.*)
STOP.

EXAMPLE 3 Table 2.5 shows the results of the method of False Position applied to $f(x) = \cos x - x$ with the same initial approximations we used for the Secant method in Example 2. Notice that the approximations agree through p_3 and that the method of False Position requires an additional iteration to obtain the same accuracy as the Secant method. ■

Table 2.5

n	p_n
0	0.5
1	0.7353981635
2	0.7363841388
3	0.7390581392
4	0.7390848638
5	0.7390851305
6	0.7390851332

The added insurance of the method of False Position commonly requires more calculation than the Secant method, just as the simplification that the Secant method provides over Newton's method usually comes at the expense of additional iterations. Further examples of the positive and negative features of these methods can be seen by working Exercises 13 and 14.

EXERCISE SET 2.3

1. Let $f(x) = x^2 - 6$ and $p_0 = 1$. Use Newton's method to find p_2.
2. Let $f(x) = -x^3 - \cos x$ and $p_0 = -1$. Use Newton's method to find p_2. Could $p_0 = 0$ be used?
3. Let $f(x) = x^2 - 6$. With $p_0 = 3$ and $p_1 = 2$ find p_3.
   a. Use the Secant method.
   b. Use the method of False Position.
   c. Which of (a) or (b) is closer to $\sqrt{6}$?
4. Let $f(x) = -x^3 - \cos x$. With $p_0 = -1$ and $p_1 = 0$ find p_3.
   a. Use the Secant method.
   b. Use the method of False Position.
   c. Which of (a) or (b) is closer to $\sqrt{6}$?
5. Use Newton's method to find solutions accurate to within 10^{-4} for the following problems.
   a. $x^3 - 2x^2 - 5 = 0, \quad [1, 4]$
   b. $x^3 + 3x^2 - 1 = 0, \quad [-3, -2]$
   c. $x - \cos x = 0, \quad [0, \pi/2]$
   d. $x - 0.8 - 0.2\sin x = 0, \quad [0, \pi/2]$
6. Use Newton's method to find solutions accurate to within 10^{-5} for the following problems.
   a. $e^x + 2^{-x} + 2\cos x - 6 = 0 \quad \text{for } 1 \le x \le 2$
   b. $\ln(x - 1) + \cos(x - 1) = 0 \quad \text{for } 1.3 \le x \le 2$
   c. $2x\cos 2x - (x - 2)^2 = 0 \quad \text{for } 2 \le x \le 3 \text{ and for } 3 \le x \le 4$
   d. $(x - 2)^2 - \ln x = 0 \quad \text{for } 1 \le x \le 2 \text{ and for } e \le x \le 4$
   e. $e^x - 3x^2 = 0 \quad \text{for } 0 \le x \le 1 \text{ and for } 3 \le x \le 5$
   f. $\sin x - e^{-x} = 0 \quad \text{for } 0 \le x \le 1, \text{ for } 3 \le x \le 4, \text{ and for } 6 \le x \le 7$
7. Repeat Exercise 5 using (i) the Secant method and (ii) the method of False Position.
8. Repeat Exercise 6 using (i) the Secant method and (ii) the method of False Position.
9. Use Newton's method to approximate, to within 10^{-4}, the value of x that produces the point on the graph of $y = x^2$ that is closest to $(1, 0)$. [*Hint:* Minimize $[d(x)]^2$, where $d(x)$ represents the distance from (x, x^2) to $(1, 0)$.]
10. Use Newton's method to approximate, to within 10^{-4}, the value of x that produces the point on the graph of $y = 1/x$ that is closest to $(2, 1)$.
11. The following describes Newton's method graphically: Suppose that $f'(x)$ exists and is nonzero on $[a, b]$. Further, suppose there exists one $p \in [a, b]$ such that $f(p) = 0$, and let $p_0 \in [a, b]$ be arbitrary. Let p_1 be the point at which the tangent line to f at $(p_0, f(p_0))$ crosses the x-axis. For each $n \ge 1$ let p_n be the x-intercept of the line tangent to f at $(p_{n-1}, f(p_{n-1}))$. Derive the formula describing this method.
12. Use Newton's method to solve the equation

$$0 = \frac{1}{2} + \frac{1}{4}x^2 - x\sin x - \frac{1}{2}\cos 2x, \quad \text{with } p_0 = \frac{\pi}{2}.$$

    Iterate using Newton's method until an accuracy of 10^{-5} is obtained. Explain why the result seems unusual for Newton's method. Also, solve the equation with $p_0 = 5\pi$ and $p_0 = 10\pi$.
13. The fourth-degree polynomial

$$f(x) = 230x^4 + 18x^3 + 9x^2 - 221x - 9$$

    has two real zeros, one in $[-1, 0]$ and the other in $[0, 1]$. Attempt to approximate these zeros to within 10^{-6} using the

a. Method of False Position

b. Secant method

c. Newton's method

Use the endpoints of each interval as the initial approximations in (a) and (b) and the midpoints as the initial approximation in (c).

14. The function $f(x) = \tan \pi x - 6$ has a zero at $(1/\pi) \arctan 6 \approx 0.447431543$. Let $p_0 = 0$ and $p_1 = 0.48$ and use ten iterations of each of the following methods to approximate this root. Which method is most successful and why?

a. Bisection method

b. Method of False Position

c. Secant method

15. The iteration equation for the Secant method can be written in the simpler form

$$p_n = \frac{f(p_{n-1})p_{n-2} - f(p_{n-2})p_{n-1}}{f(p_{n-1}) - f(p_{n-2})}.$$

Explain why, in general, this iteration equation is likely to be less accurate than the one given in Algorithm 2.4.

16. The equation $x^2 - 10 \cos x = 0$ has two solutions, ± 1.3793646. Use Newton's method to approximate the solutions to within 10^{-5} with the following values of p_0.

a. $p_0 = -100$ **c.** $p_0 = -25$ **e.** $p_0 = 50$

b. $p_0 = -50$ **d.** $p_0 = 25$ **f.** $p_0 = 100$

17. Use Maple to determine how many iterations of Newton's method with $p_0 = \pi/4$ are needed to find a zero of $f(x) = \cos x - x$ to within 10^{-100}.

18. Repeat Exercise 17 with $p_0 = 1/2$, $p_1 = \pi/4$, and the Secant method.

19. The function described by $f(x) = \ln(x^2 + 1) - e^{0.4x} \cos \pi x$ has an infinite number of zeros.

a. Determine, within 10^{-6}, the only negative zero.

b. Determine, within 10^{-6}, the four smallest positive zeros.

c. Determine a reasonable initial approximation to find the nth smallest positive zero of f. [*Hint:* Sketch an approximate graph of f.]

d. Use part (c) to determine, within 10^{-6}, the 25th smallest positive zero of f.

20. Find an approximation for λ, accurate to within 10^{-4}, for the population equation

$$1{,}564{,}000 = 1{,}000{,}000 e^{\lambda} + \frac{435{,}000}{\lambda}(e^{\lambda} - 1),$$

discussed in the introduction to this chapter. Use this value to predict the population at the end of the second year, assuming that the immigration rate during this year remains at 435,000 individuals per year.

21. The sum of two numbers is 20. If each number is added to its square root, the product of the two sums is 155.55. Determine the two numbers to within 10^{-4}.

22. The accumulated value of a savings account based on regular periodic payments can be determined from the *annuity due equation,*

$$A = \frac{P}{i}[(1 + i)^n - 1].$$

In this equation A is the amount in the account, P is the amount regularly deposited, and i is the rate of interest per period for the n deposit periods. An engineer would like to have a savings account valued at \$750,000 upon retirement in 20 years and can afford to put \$1500 per month toward this goal. What is the minimal interest rate at which this amount can be invested, assuming that the interest is compounded monthly?

23. Problems involving the amount of money required to pay off a mortgage over a fixed period of time involve the formula

$$A = \frac{P}{i}[1 - (1 + i)^{-n}],$$

known as an *ordinary annuity equation.* In this equation A is the amount of the mortgage, P is the amount of each payment, and i is the interest rate per period for the n payment periods. Suppose that a 30-year home mortgage in the amount of \$135,000 is needed and that the borrower can afford house payments of at most \$1000 per month. What is the maximal interest rate the borrower can afford to pay?

24. A drug administered to a patient produces a concentration in the blood stream given by $c(t) = Ate^{-t/3}$ milligrams per milliliters, t hours after A units have been injected. The maximum safe concentration is 1 mg/ml.

a. What amount should be injected to reach this maximum safe concentration and when does this maximum occur?

b. An additional amount of this drug is to be administered to the patient after the concentration falls to 0.25 mg/ml. Determine, to the nearest minute, when this second injection should be given.

c. Assuming that the concentration from consecutive injections is additive and that 75% of the amount originally injected is administered in the second injection, when is it time for the third injection?

25. Let $f(x) = 3^{3x+1} - 7 \cdot 5^{2x}$.

a. Use the Maple commands `solve` and `fsolve` to try to find all the zeros of f.

b. Plot $f(x)$ to find initial approximations to the zeros of f.

c. Use Newton's method to find the zeros of f to within 10^{-16}.

d. Find the exact solutions of $f(x) = 0$ algebraically.

26. Repeat Exercise 25 using $f(x) = 2^{x^2} - 3 \cdot 7^{x+1}$.

27. The logistic population growth model is described by an equation of the form

$$P(t) = \frac{P_L}{1 - ce^{-kt}},$$

where P_L, c, and $k > 0$ are constants and $P(t)$ is the population at time t. P_L represents the limiting value of the population since $\lim_{t\to\infty} P(t) = P_L$. Use the census data for the years 1950, 1960, and 1970 listed in the table on page 104 to determine the constants P_L, c, and k for a logistic growth model. Use the logistic model to predict the population of the United States in 1980 and in 2000, assuming $t = 0$ at 1950. Compare the 1980 prediction to the actual value.

28. The Gompertz population growth model is described by

$$P(t) = P_L e^{-ce^{-kt}},$$

where P_L, c, and $k > 0$ are constants and $P(t)$ is the population at time t. Repeat Exercise 27 using the Gompertz growth model in place of the logistic model.

29. Player A will shut out (win by a score of 21–0) a player B in a game of racquetball with probability

$$P = \frac{1+p}{2}\left(\frac{p}{1-p+p^2}\right)^{21},$$

where p denotes the probability A will win any specific rally (independent of the server). (See [K, J, p. 267].) Determine, to within 10^{-3}, the minimal value of p that will ensure that A will shut out B in at least half the matches they play.

30. In the design of all-terrain vehicles, it is necessary to consider the failure of the vehicle when attempting to negotiate two types of obstacles. One type of failure is called *hang-up failure* and occurs when the vehicle attempts to cross an obstacle that causes the bottom of the vehicle to touch the ground. The other type of failure is called *nose-in failure* and occurs when the vehicle descends into a ditch and its nose touches the ground.

The accompanying figure, adapted from [Bek], shows the components associated with the nose-in failure of a vehicle. In that reference it is shown that the maximum angle α that can be negotiated by a vehicle when β is the maximum angle at which hang-up failure does *not* occur satisfies the equation

$$A \sin\alpha \cos\alpha + B \sin^2\alpha - C \cos\alpha - E \sin\alpha = 0,$$

where

$$A = l \sin\beta_1, \quad B = l \cos\beta_1, \quad C = (h + 0.5D)\sin\beta_1 - 0.5D \tan\beta_1,$$

$$\text{and} \quad E = (h + 0.5D)\cos\beta_1 - 0.5D.$$

a. It is stated that when $l = 89$ in., $h = 49$ in., $D = 55$ in., and $\beta_1 = 11.5°$, angle α is approximately $33°$. Verify this result.

b. Find α for the situation when l, h, and β_1 are the same as in part (a) but $D = 30$ in.

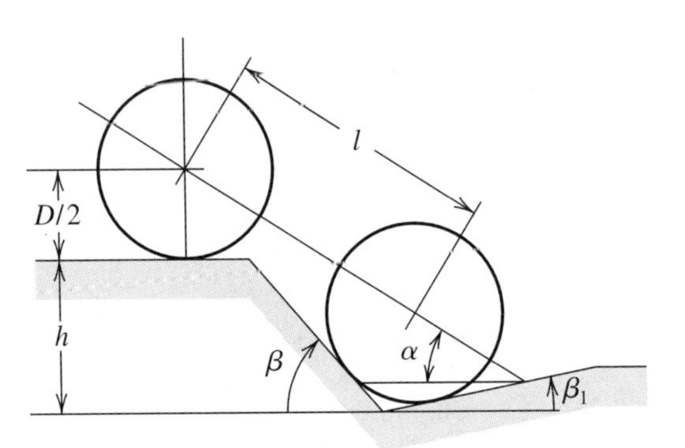

2.4 Error Analysis for Iterative Methods

In this section we investigate the order of convergence of functional iteration schemes and, as a means of obtaining rapid convergence, rediscover Newton's method. We also consider ways of accelerating the convergence of Newton's method in special circumstances. First we need a new procedure for measuring how rapidly a sequence converges.

Definition 2.6 Suppose $\{p_n\}_{n=0}^{\infty}$ is a sequence that converges to p, with $p_n \neq p$ for all n. If positive constants λ and α exist with

$$\lim_{n\to\infty} \frac{|p_{n+1} - p|}{|p_n - p|^{\alpha}} = \lambda,$$

then $\{p_n\}_{n=0}^{\infty}$ **converges to p of order α, with asymptotic error constant λ.** ■

An iterative technique of the form $p_n = g(p_{n-1})$ is said to be of **order α** if the sequence $\{p_n\}$ converges to the solution $p = g(p)$ of order α.

In general, a sequence with a high order of convergence converges more rapidly than a sequence with a lower order. The asymptotic constant affects the speed of convergence but is not as important as the order. Two cases of order are given special attention.

1. If $\alpha = 1$, the sequence is **linearly convergent**.
2. If $\alpha = 2$, the sequence is **quadratically convergent**.

The next example compares a linearly convergent sequence to one that is quadratically convergent and demonstrates why we will be trying to find methods that produce higher-order convergent sequences.

EXAMPLE 1 Suppose $\{p_n\}$ and $\{\tilde{p}_n\}$ converge to zero, that $\{p_n\}$ is linear with

$$\lim_{n\to\infty} \frac{|p_{n+1}|}{|p_n|} = 0.5,$$

and that $\{\tilde{p}_n\}$ is quadratic with the same asymptotic error constant,

$$\lim_{n\to\infty} \frac{|\tilde{p}_{n+1}|}{|\tilde{p}_n|^2} = 0.5.$$

Suppose also, for simplicity, that

$$\frac{|p_{n+1}|}{|p_n|} \approx 0.5 \quad \text{and} \quad \frac{|\tilde{p}_{n+1}|}{|\tilde{p}_n|^2} \approx 0.5.$$

For the linearly convergent scheme, this assumption means that

$$|p_n - 0| = |p_n| \approx 0.5|p_{n-1}| \approx (0.5)^2|p_{n-2}| \approx \cdots \approx (0.5)^n|p_0|,$$

whereas the quadratically convergent procedure has

$$\begin{aligned} |\tilde{p}_n - 0| = |\tilde{p}_n| &\approx 0.5|p_{n-1}|^2 \approx (0.5)\left[0.5|\tilde{p}_{n-2}|^2\right]^2 = (0.5)^3|\tilde{p}_{n-2}|^4 \\ &\approx (0.5)^3\left[(0.5)|\tilde{p}_{n-3}|^2\right]^4 = (0.5)^7|\tilde{p}_{n-3}|^8 \approx \cdots \approx (0.5)^{2^n-1}|\tilde{p}_0|^{2^n}. \end{aligned}$$

Table 2.6 illustrates the relative speed of convergence of the sequences to zero when $|p_0| = |\tilde{p}_0| = 1$.

Table 2.6

n	Linear Convergence Sequence $\{p_n\}$ $(0.5)^n$	Quadratic Convergence Sequence $\{\tilde{p}_n\}$ $(0.5)^{2^n-1}$
1	5.0000×10^{-1}	5.0000×10^{-1}
2	2.5000×10^{-1}	1.2500×10^{-1}
3	1.2500×10^{-1}	7.8125×10^{-3}
4	6.2500×10^{-2}	3.0518×10^{-5}
5	3.1250×10^{-2}	4.6566×10^{-10}
6	1.5625×10^{-2}	1.0842×10^{-19}
7	7.8125×10^{-3}	5.8775×10^{-39}

The quadratically convergent sequence is within 10^{-38} of zero by the seventh term. At least 126 terms are needed to ensure this accuracy for the linearly convergent sequence. ■

Quadratically convergent sequences generally converge much more quickly than those that converge only linearly, but many techniques that generate convergent sequences do so only linearly.

Theorem 2.7 Let $g \in C[a, b]$ be such that $g(x) \in [a, b]$ for all $x \in [a, b]$. Suppose, in addition, that g' is continuous on (a, b) and a positive constant $k < 1$ exists with

$$|g'(x)| \leq k, \quad \text{for all } x \in (a, b).$$

If $g'(p) \neq 0$, then for any number p_0 in $[a, b]$ the sequence

$$p_n = g(p_{n-1}), \quad \text{for } n \geq 1,$$

converges only linearly to the unique fixed point p in $[a, b]$.

Proof We know from the Fixed-Point Theorem 2.3 in Section 2.2 that the sequence converges to p. Since g' exists on $[a, b]$, we can apply the Mean Value Theorem to g to show that for any n,

$$p_{n+1} - p = g(p_n) - g(p) = g'(\xi_n)(p_n - p),$$

where ξ_n is between p_n and p. Since $\{p_n\}_{n=0}^{\infty}$ converges to p, $\{\xi_n\}_{n=0}^{\infty}$ also converges to p. Since g' is continuous on $[a, b]$, we have

$$\lim_{n\to\infty} g'(\xi_n) = g'(p).$$

Thus,

$$\lim_{n\to\infty} \frac{p_{n+1} - p}{p_n - p} = \lim_{n\to\infty} g'(\xi_n) = g'(p) \quad \text{and} \quad \lim_{n\to\infty} \frac{|p_{n+1} - p|}{|p_n - p|} = |g'(p)|.$$

Hence, fixed-point iteration exhibits linear convergence with asymptotic error constant $|g'(p)|$ whenever $g'(p) \neq 0$. ■ ■ ■

Theorem 2.7 implies that higher-order convergence for fixed-point methods can occur only when $g'(p) = 0$. The next result describes additional conditions that ensure the quadratic convergence we seek.

Theorem 2.8 Let p be a solution of the equation $x = g(x)$. Suppose that $g'(p) = 0$ and g'' is continuous and strictly bounded by M on an open interval I containing p. Then there exists a $\delta > 0$ such that, for $p_0 \in [p - \delta, p + \delta]$, the sequence defined by $p_n = g(p_{n-1})$, when $n \geq 1$, converges at least quadratically to p. Moreover, for sufficiently large values of n,

$$|p_{n+1} - p| < \frac{M}{2}|p_n - p|^2.$$

Proof Choose k in $(0, 1)$ and $\delta > 0$ such that on the interval $[p - \delta, p + \delta]$, contained in I, we have $|g'(x)| \leq k$ and g'' continuous. Since $|g'(x)| \leq k < 1$, the argument used in the proof of Theorem 2.5 in Section 2.3 shows that the terms of the sequence $\{p_n\}_{n=0}^{\infty}$ are contained in $[p - \delta, p + \delta]$. Expanding $g(x)$ in a linear Taylor polynomial for $x \in [p - \delta, p + \delta]$ gives

$$g(x) = g(p) + g'(p)(x - p) + \frac{g''(\xi)}{2}(x - p)^2,$$

where ξ lies between x and p. The hypotheses $g(p) = p$ and $g'(p) = 0$ imply that

$$g(x) = p + \frac{g''(\xi)}{2}(x - p)^2.$$

In particular, when $x = p_n$,

$$p_{n+1} = g(p_n) = p + \frac{g''(\xi_n)}{2}(p_n - p)^2,$$

with ξ_n between p_n and p. Thus,

$$p_{n+1} - p = \frac{g''(\xi_n)}{2}(p_n - p)^2.$$

Since $|g'(x)| \leq k < 1$ on $[p - \delta, p + \delta]$ and g maps $[p - \delta, p + \delta]$ into itself, it follows from the Fixed-Point Theorem that $\{p_n\}_{n=0}^{\infty}$ converges to p. But ξ_n is between p and p_n for each n, so $\{\xi_n\}_{n=0}^{\infty}$ also converges to p, and, since g'' is continuous,

$$\lim_{n\to\infty} \frac{|p_{n+1} - p|}{|p_n - p|^2} = \frac{|g''(p)|}{2}.$$

This result implies that the sequence $\{p_n\}_{n=0}^{\infty}$ is quadratically convergent if $g''(p) \neq 0$ and of higher-order convergence if $g''(p) = 0$.

Since g'' is strictly bounded by M on the interval $[p - \delta, p + \delta]$, this also implies that for sufficiently large values of n,

$$|p_{n+1} - p| < \frac{M}{2}|p_n - p|^2.$$

■ ■ ■

Theorems 2.7 and 2.8 tell us that our search for quadratically convergent fixed-point methods should point in the direction of functions whose derivatives are zero at the fixed point.

The easiest way to construct a fixed-point problem associated with a root-finding problem $f(x) = 0$ is to subtract a multiple of $f(x)$ (which vanishes at the root) from x. So let us consider a scheme of the form

$$p_n = g(p_{n-1}), \quad \text{for } n \geq 1,$$

for g in the form

$$g(x) = x - \phi(x)f(x),$$

where ϕ is a differentiable function that will be chosen later.

For the iterative procedure derived from g to be quadratically convergent, we need to have $g'(p) = 0$. Since

$$g'(x) = 1 - \phi'(x)f(x) - f'(x)\phi(x),$$

we have $g'(p) = 1 - f'(p)\phi(p)$, and $g'(p) = 0$ if and only if $\phi(p) = 1/f'(p)$.

A reasonable approach is to let $\phi(x) = 1/f'(x)$, which will ensure that $\phi(p) = 1/f'(p)$. The natural procedure for producing quadratic convergence then becomes

$$p_n = g(p_{n-1}) = p_{n-1} - \frac{f(p_{n-1})}{f'(p_{n-1})},$$

which is Newton's method.

In the preceding discussion the restriction was made that $f'(p) \neq 0$, where p is the solution to $f(x) = 0$. From the definition of Newton's method, it is clear that difficulties might occur if $f'(p_n)$ goes to zero simultaneously with $f(p_n)$. In particular, Newton's method and the Secant method will generally give problems if $f'(p) = 0$ when $f(p) = 0$. To examine these difficulties in more detail, we make the following definition.

Definition 2.9 A solution p of $f(x) = 0$ is a **zero of multiplicity m** of f if for $x \neq p$, we can write $f(x) = (x - p)^m q(x)$, where $\lim_{x \to p} q(x) \neq 0$. ■

In essence, $q(x)$ represents that portion of $f(x)$ that does not contribute to the zero of f. The following result gives a means to easily identify **simple** zeros of a function, those that have multiplicity one.

Theorem 2.10 $f \in C^1[a, b]$ has a simple zero at p in (a, b) if and only if $f(p) = 0$, but $f'(p) \neq 0$.

Proof If f has a simple zero at p, then $f(p) = 0$ and $f(x) = (x - p)q(x)$, where $\lim_{x \to p} q(x) \neq 0$. Since $f \in C^1[a, b]$,

$$f'(p) = \lim_{x \to p} f'(x) = \lim_{x \to p}[q(x) + (x-p)q'(x)] = \lim_{x \to p} q(x) \neq 0.$$

Conversely, if $f(p) = 0$, but $f'(p) \neq 0$, expand f in a zeroth Taylor polynomial about p. Then

$$f(x) = f(p) + f'(\xi(x))(x-p) = (x-p)f'(\xi(x)),$$

where $\xi(x)$ is between x and p. Since $f \in C^1[a,b]$,

$$\lim_{x \to p} f'(\xi(x)) = f'\left(\lim_{x \to p} \xi(x)\right) = f'(p) \neq 0.$$

Letting $q = f' \circ \xi$ gives $f(x) = (x-p)q(x)$, where $\lim_{x \to p} q(x) \neq 0$. Thus f has a simple zero at p. ■ ■ ■

The following generalization of Theorem 2.10 is considered in Exercise 10.

Theorem 2.11 The function $f \in C^m[a,b]$ has a zero of multiplicity m at p if and only if

$$0 = f(p) = f'(p) = f''(p) = \cdots = f^{(m-1)}(p), \quad \text{but} \quad f^{(m)}(p) \neq 0. \quad ■$$

The result in Theorem 2.10 implies that an interval about p exists such that Newton's method converges quadratically to p for any initial approximation $p_0 = p$, provided that p is a simple zero. The following example shows that quadratic convergence may not occur if the zero is not simple.

EXAMPLE 2 The function described by $f(x) = e^x - x - 1$ has a zero of multiplicity two at $p = 0$, since $f(0) = e^0 - 0 - 1 = 0$ and $f'(0) = e^0 - 1 = 0$, but $f''(0) = e^0 = 1$. In fact, $f(x)$ can be expressed in the form

$$f(x) = (x-0)^2 \frac{e^x - x - 1}{x^2},$$

where, by L'Hôpital's rule,

$$\lim_{x \to 0} \frac{e^x - x - 1}{x^2} = \lim_{x \to 0} \frac{e^x - 1}{2x} = \lim_{x \to 0} \frac{e^x}{2} = \frac{1}{2} \neq 0.$$

The terms generated by Newton's method applied to f with $p_0 = 1$ are shown in Table 2.7. The sequence is clearly converging, but not quadratically, to zero. The graph of f is shown in Figure 2.12. ■

Table 2.7

n	p_n	n	p_n
0	1.0	9	2.7750×10^{-3}
1	0.58198	10	1.3881×10^{-3}
2	0.31906	11	6.9411×10^{-4}
3	0.16800	12	3.4703×10^{-4}
4	0.08635	13	1.7416×10^{-4}
5	0.04380	14	8.8041×10^{-5}
6	0.02206	15	4.2610×10^{-5}
7	0.01107	16	1.9142×10^{-6}
8	0.005545		

Figure 2.12

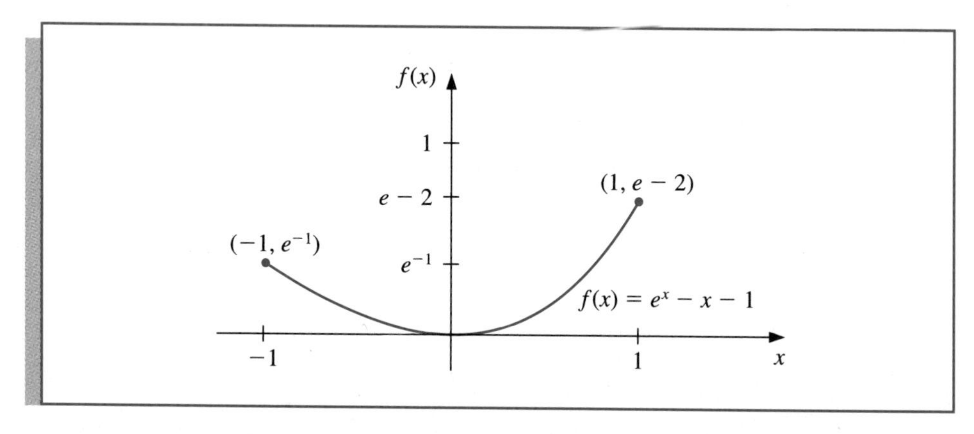

One method of handling the problem of multiple roots is to define a function μ by

$$\mu(x) = \frac{f(x)}{f'(x)}.$$

If p is a zero of f of multiplicity m and $f(x) = (x-p)^m q(x)$, then

$$\begin{aligned}\mu(x) &= \frac{(x-p)^m q(x)}{m(x-p)^{m-1}q(x) + (x-p)^m q'(x)} \\ &= (x-p)\frac{q(x)}{mq(x) + (x-p)q'(x)}\end{aligned}$$

also has a zero at p. However, since $q(p) \neq 0$,

$$\frac{q(p)}{mq(p) + (p-p)q'(p)} = \frac{1}{m} \neq 0,$$

so p is a zero of multiplicity 1 of $\mu(x)$. Newton's method can then be applied to the function μ to give

$$g(x) = x - \frac{\mu(x)}{\mu'(x)} = x - \frac{f(x)/f'(x)}{\{[f'(x)]^2 - [f(x)][f''(x)]\}/[f'(x)]^2}$$

or

(2.10)
$$g(x) = x - \frac{f(x)f'(x)}{[f'(x)]^2 - f(x)f''(x)}.$$

If g has the required continuity conditions, functional iteration applied to g will be quadratically convergent regardless of the multiplicity of the zero of f. Theoretically, the only drawback to this method is the additional calculation of $f''(x)$ and the more laborious procedure of calculating the iterates. In practice, however, the presence of a multiple zero can cause serious roundoff problems because the denominator of (2.10) consists of the difference of two numbers that are close to zero.

EXAMPLE 3 Table 2.8 lists the approximations to the double zero at $x = 0$ of $f(x) = e^x - x - 1$ using (2.10) and a calculator with ten digits of precision. The initial approximation of $p_0 = 1$ was chosen so that the entries can be compared with those in Table 2.7. What Table 2.8 does not show is that no improvement to the zero approximation $-2.8085217 \times 10^{-7}$ occurs in subsequent computations using (2.10) and this calculator since both the numerator and the denominator approach zero. ■

Table 2.8

n	p_n
1	$-2.3421061 \times 10^{-1}$
2	$-8.4582788 \times 10^{-3}$
3	$-1.1889524 \times 10^{-5}$
4	$-6.8638230 \times 10^{-6}$
5	$-2.8085217 \times 10^{-7}$

EXAMPLE 4 In Example 3 of Section 2.2 we solved $f(x) = x^3 + 4x^2 - 10 = 0$ for the zero $p = 1.36523001$. To compare convergence for a zero of multiplicity one by Newton's method and the modified Newton's method listed in Eq. (2.10), let

(i)
$$p_n = p_{n-1} - \frac{p_{n-1}^3 + 4p_{n-1}^2 - 10}{3p_{n-1}^2 + 8p_{n-1}}, \quad \text{from Newton's method}$$

and, from Eq. (2.10),

(ii)
$$p_n = p_{n-1} - \frac{(p_{n-1}^3 + 4p_{n-1}^2 - 10)(3p_{n-1}^2 + 8p_{n-1})}{(3p_{n-1}^2 + 8p_{n-1})^2 - (p_{n-1}^3 + 4p_{n-1}^2 - 10)(6p_{n-1} + 8)}.$$

With $p_0 = 1.5$, the first three iterates for (i) and (ii) are shown in Table 2.9. The results illustrate the rapid convergence of both methods in the case of a simple zero. ■

Table 2.9

	(i)	(ii)
p_1	1.37333333	1.35689898
p_2	1.36526201	1.36519585
p_3	1.36523001	1.36523001

EXERCISE SET 2.4

1. Use Newton's method to find solutions accurate to within 10^{-5} to the following problems.
   a. $x^2 - 2xe^{-x} + e^{-2x} = 0, \quad \text{for } 0 \le x \le 1$
   b. $\cos\left(x + \sqrt{2}\right) + x\left(x/2 + \sqrt{2}\right) = 0, \quad \text{for } -2 \le x \le -1$
   c. $x^3 - 3x^2(2^{-x}) + 3x(4^{-x}) - 8^{-x} = 0, \quad \text{for } 0 \le x \le 1$
   d. $e^{6x} + 3(\ln 2)^2 e^{2x} - e^{4x}\ln 8 - (\ln 2)^3 = 0, \quad \text{for } -1 \le x \le 0$
2. Repeat Exercise 1 using the modified Newton-Raphson method described in Eq. (2.10). Is there an improvement in speed or accuracy over Exercise 1?
3. Use Newton's method and the modified Newton-Raphson method described in Eq. (2.10) to find a solution accurate to within 10^{-5} to the problem
$$e^{6x} + 1.441e^{2x} - 2.079e^{4x} - 0.3330 = 0 \quad \text{for } -1 \le x \le 0.$$
   This is the same problem as 1(d) with the coefficients replaced by their four-digit approximations. Compare the solutions to the results in 1(d) and 2(d).
4. Show that the following sequences $\{p_n\}$ converge linearly to $p = 0$. How large must n be before $|p_n - p| \le 5 \times 10^{-2}$?
   a. $p_n = \dfrac{1}{n}, \quad n \ge 1$
   b. $p_n = \dfrac{1}{n^2}, \quad n \ge 1$
5. Show that for any positive integer k, the sequence defined by $p_n = 1/n^k$ converges linearly to $p = 0$. For each pair of integers k and m, determine a number N for which $1/N^k < 10^{-m}$.
6. **a.** Show that the sequence $p_n = 10^{-2^n}$ converges quadratically to zero.
   b. Show that the sequence $p_n = 10^{-n^k}$ does not converge to zero quadratically, regardless of the size of the exponent $k > 1$.
7. **a.** Construct a sequence that converges to zero of order 3.
   b. Suppose $\alpha > 1$. Construct a sequence that converges to zero of order α.
8. Suppose p is a zero of multiplicity m of f, where f''' is continuous on an open interval containing p. Show that the following fixed-point method has $g'(p) = 0$:
$$g(x) = x - \frac{mf(x)}{f'(x)}.$$
9. Show that the Bisection Algorithm 2.1 gives a sequence with an error bound that converges linearly to zero.
10. Suppose that f has m continuous derivatives. Modify the proof of Theorem 2.10 to show that f has a zero of multiplicity m at p if and only if
$$0 = f(p) = f'(p) = \cdots = f^{(m-1)}(p) \quad \text{but} \quad f^{(m)}(p) \ne 0.$$
11. The iterative method to solve $f(x) = 0$, given by the fixed-point method $g(x) = x$, where
$$p_n = g(p_{n-1}) = p_{n-1} - \frac{f(p_{n-1})}{f'(p_{n-1})} - \frac{f''(p_{n-1})}{2f'(p_{n-1})}\left[\frac{f(p_{n-1})}{f'(p_{n-1})}\right]^2, \quad \text{for } n = 1, 2, 3, \ldots,$$
   has $g'(p) = g''(p) = 0$. This will generally yield cubic ($\alpha = 3$) convergence. Use the analysis of Example 1 to compare quadratic and cubic convergence.
12. It can be shown (see, for example, [DaB, pp. 228–229]) that if $\{p_n\}_{n=0}^{\infty}$ are convergent Secant method approximations to p, the solution to $f(x) = 0$, then a constant C exists with $|p_{n+1} - p| \approx C\,|p_n - p|\,|p_{n-1} - p|$ for sufficiently large values of n. Assume $\{p_n\}$ converges to p of order α and show that $\alpha = \left(1 + \sqrt{5}\right)/2$. (*Note:* This implies that the order of convergence of the Secant method is approximately 1.62).

2.5 Accelerating Convergence

Because it is rare to have the luxury of quadratic convergence, we now consider a technique called **Aitken's Δ^2 method** that can be used to accelerate the convergence of a sequence that is linearly convergent, regardless of its origin or application.

Suppose $\{p_n\}_{n=0}^{\infty}$ is a linearly convergent sequence with limit p. To motivate the construction of a sequence $\{\hat{p}_n\}$ that converges more rapidly to p than does $\{p_n\}$, let us first assume that the signs of $p_n - p$, $p_{n+1} - p$, and $p_{n+2} - p$ agree and that n is sufficiently large that

$$\frac{p_{n+1} - p}{p_n - p} \approx \frac{p_{n+2} - p}{p_{n+1} - p}.$$

Then

$$(p_{n+1} - p)^2 \approx (p_{n+2} - p)(p_n - p),$$

so

$$p_{n+1}^2 - 2p_{n+1}p + p^2 \approx p_{n+2}p_n - (p_n + p_{n+2})p + p^2$$

and

$$(p_{n+2} + p_n - 2p_{n+1})p \approx p_{n+2}p_n - p_{n+1}^2.$$

Solving for p gives

$$p \approx \frac{p_{n+2}p_n - p_{n+1}^2}{p_{n+2} - 2p_{n+1} + p_n}.$$

Adding and subtracting the terms p_n^2 and $2p_np_{n+1}$ in the numerator permits us to rewrite this expression as

$$\begin{aligned} p &\approx \frac{p_n^2 + p_np_{n+2} + 2p_np_{n+1} - 2p_np_{n+1} - p_n^2 - p_{n+1}^2}{p_{n+2} - 2p_{n+1} + p_n} \\ &= \frac{(p_n^2 + p_np_{n+2} + 2p_np_{n+1}) - (p_n^2 - 2p_np_{n+1} - p_{n+1}^2)}{p_{n+2} - 2p_{n+1} + p_n} \\ &= p_n - \frac{(p_{n+1} - p_n)^2}{p_{n+2} - 2p_{n+1} + p_n}. \end{aligned}$$

Aitken's Δ^2 method is based on the assumption that the sequence $\{\hat{p}_n\}_{n=0}^{\infty}$ defined by

$$\hat{p}_n = p_n - \frac{(p_{n+1} - p_n)^2}{p_{n+2} - 2p_{n+1} + p_n} \tag{2.11}$$

converges more rapidly to p than does the original sequence $\{p_n\}_{n=0}^{\infty}$.

EXAMPLE 1 The sequence $\{p_n\}_{n=1}^{\infty}$, where $p_n = \cos(1/n)$, converges linearly to $p = 1$. The first few terms of the sequences $\{p_n\}_{n=1}^{\infty}$ and $\{\hat{p}_n\}_{n=1}^{\infty}$ are given in Table 2.10. It certainly appears that $\{\hat{p}_n\}_{n=1}^{\infty}$ converges more rapidly to $p = 1$ than does $\{p_n\}_{n=1}^{\infty}$. ■

Table 2.10

n	p_n	$\hat{p}_n$
1	0.54030	0.96178
2	0.87758	0.98213
3	0.94496	0.98979
4	0.96891	0.99342
5	0.98007	0.99541
6	0.98614	
7	0.98981	

The Δ notation associated with this technique has its origin in the following definition.

Definition 2.12 Given the sequence $\{p_n\}_{n=0}^{\infty}$, the **forward difference** Δp_n is defined by

$$\Delta p_n = p_{n+1} - p_n, \quad \text{for } n \geq 0.$$

Higher powers $\Delta^k p_n$ are defined recursively by

$$\Delta^k p_n = \Delta(\Delta^{k-1} p_n), \quad \text{for } k \geq 2. \quad ■$$

The definition implies that

$$\Delta^2 p_n = \Delta(p_{n+1} - p_n) = \Delta p_{n+1} - \Delta p_n = (p_{n+2} - p_{n+1}) - (p_{n+1} - p_n).$$

So

$$\Delta^2 p_n = p_{n+2} - 2p_{n+1} + p_n,$$

and the formula for $\hat{p}_n$ given in Eq. (2.11) can be written as

$$\hat{p} = p_n - \frac{(\Delta p_n)^2}{\Delta^2 p_n}, \quad \text{for } n \geq 0. \tag{2.12}$$

To this point in our discussion of Aitken's Δ^2 method, we have stated that the sequence $\{\hat{p}_n\}_{n=0}^{\infty}$, converges to p more rapidly than does the original sequence $\{p_n\}_{n=0}^{\infty}$, but we have not said what is meant by the term "more rapid" convergence. Theorem 2.13 explains and justifies this terminology. The proof of this theorem is considered in Exercise 14.

Theorem 2.13 Suppose that $\{p_n\}$ is a sequence that converges linearly to the limit p and that for all sufficiently large values of n we have $(p_n - p)(p_{n+1} - p) > 0$. Then the sequence $\{\hat{p}_n\}_{n=0}^{\infty}$

converges to p faster than $\{p_n\}_{n=0}^{\infty}$ in the sense that

$$\lim_{n\to\infty} \frac{\hat{p}_n - p}{p_n - p} = 0. \qquad \blacksquare$$

By applying a modified Aitken's Δ^2 method to a linearly convergent sequence obtained from fixed-point iteration, we can accelerate the convergence to quadratic. This procedure is known as Steffensen's method and differs slightly from applying Aitken's Δ^2 method directly to the linearly convergent fixed-point iteration sequence. Aitken's Δ^2 method would construct the terms in order:

$$p_0, \quad p_1 = g(p_0), \quad p_2 = g(p_1), \quad \hat{p}_0 = \{\Delta^2\}(p_0), \quad p_3 = g(p_2), \quad \hat{p}_1 = \{\Delta^2\}(p_1), \ldots,$$

where $\{\Delta^2\}$ indicates that Eq. (2.12) is used. Steffensen's method constructs the same first four terms, p_0, p_1, p_2, and $\hat{p}_0$. However, at this step it assumes that $\hat{p}_0$ is a better approximation to p than is p_2 and applies fixed-point iteration to $\hat{p}_0$ instead of p_2. Using this notation the sequence generated is

$$p_0^{(0)}, \quad p_1^{(0)} = g(p_0^{(0)}), \quad p_2^{(0)} = g(p_1^{(0)}), \quad p_0^{(1)} = \{\Delta^2\}(p_0^{(0)}), \quad p_1^{(1)} = g(p_0^{(1)}), \ldots.$$

Every third term is generated by Eq. (2.12); the others use fixed-point iteration on the previous term. The process is described in Algorithm 2.6.

ALGORITHM 2.6

Steffensen's

To find a solution to $p = g(p)$ given an initial approximation p_0:

INPUT initial approximation p_0; tolerance TOL; maximum number of iterations N_0.

OUTPUT approximate solution p or message of failure.

Step 1 Set $i = 1$.

Step 2 While $i \le N_0$ do Steps 3–6.

Step 3 Set $p_1 = g(p_0)$; (*Compute* $p_1^{(i-1)}$.)
$p_2 = g(p_1)$; (*Compute* $p_2^{(i-1)}$.)
$p = p_0 - (p_1 - p_0)^2/(p_2 - 2p_1 + p_0)$. (*Compute* $p_0^{(i)}$.)

Step 4 If $|p - p_0| < TOL$ then
OUTPUT (p); (*Procedure completed successfully.*)
STOP.

Step 5 Set $i = i + 1$.
Step 6 Set $p_0 = p$. (*Update* p_0.)

Step 7 OUTPUT ('Method failed after N_0 iterations, N_0 =', N_0);
(*Procedure completed unsuccessfully.*)
STOP.

Note that $\Delta^2 p_n$ may be zero, which would introduce a zero in the denominator of the next iterate. If this occurs, we terminate the sequence and select $p_2^{(n-1)}$ as the approximate answer.

EXAMPLE 2 To solve $x^3 + 4x^2 - 10 = 0$ using Steffensen's method, let $x^3 + 4x^2 = 10$ and solve for x by dividing by $x + 4$. This procedure produces the fixed-point method

$$g(x) = \left(\frac{10}{x+4}\right)^{1/2},$$

and $x = g(x)$ implies that $x^3 + 4x^2 - 10 = 0$.

Steffensen's procedure with $p_0 = 1.5$ gives the values in Table 2.11. The iterate $p_0^{(2)} = 1.365230013$ is accurate to the ninth decimal place. In this example, Steffensen's method gave about the same rate of convergence as Newton's method (see Example 4 in Section 2.4). ■

Table 2.11

k	$p_0^{(k)}$	$p_1^{(k)}$	$p_2^{(k)}$
0	1.5	1.348399725	1.367376372
1	1.365265224	1.365225534	1.365230583
2	1.365230013		

From Example 2, it appears that Steffensen's method gives quadratic convergence without evaluating a derivative, and Theorem 2.14 verifies that this is the case. The proof of this theorem can be found in [He2, pp. 90–92] or [IK, pp. 103–107].

Theorem 2.14 Suppose that $x = g(x)$ has the solution p with $g'(p) \neq 1$. If there exists a $\delta > 0$ such that $g \in C^3[p - \delta, p + \delta]$, then Steffensen's method gives quadratic convergence for any $p_0 \in [p - \delta, p + \delta]$. ■

EXERCISE SET 2.5

1. The following sequences are linearly convergent. Generate the first five terms of the sequence $\{\hat{p}_n\}$ using Aitken's Δ^2 method.
   a. $p_0 = 0.5, \quad p_n = (2 - e^{p_{n-1}} + p_{n-1}^2)/3, \quad n \geq 1$
   b. $p_0 = 0.75, \quad p_n = (e^{p_{n-1}}/3)^{1/2}, \quad n \geq 1$
   c. $p_0 = 0.5, \quad p_n = 3^{-p_{n-1}}, \quad n \geq 1$
   d. $p_0 = 0.5, \quad p_n = \cos p_{n-1}, \quad n \geq 1$
2. Consider the function $f(x) = e^{6x} + 3(\ln 2)^2 e^{2x} - \ln 8 e^{4x} - (\ln 2)^3$. Use Newton's method with $p_0 = 0$ to approximate a zero of f. Generate terms until $|p_{n+1} - p_n| < 0.0002$. Construct the sequence $\{\hat{p}_n\}$. Is the convergence improved?
3. Let $g(x) = \cos(x - 1)$ and $p_0^{(0)} = 2$. Use Steffensen's method to find $p_0^{(1)}$.
4. Let $g(x) = 1 + (\sin x)^2$ and $p_0^{(0)} = 1$. Use Steffensen's method to find $p_0^{(1)}$ and $p_0^{(2)}$.

5. Steffensen's method is applied to a function g using $p_0^{(0)} = 1$ and $p_2^{(0)} = 3$ to obtain $p_0^{(1)} = 0.75$. What could $p_1^{(0)}$ be?

6. Steffensen's method is applied to a function g using $p_0^{(0)} = 1$ and $p_1^{(0)} = \sqrt{2}$ to obtain $p_0^{(1)} = 2.7802$. What is $p_2^{(0)}$?

7. Solve $x^3 - x - 1 = 0$ for the root in $[1, 2]$ to an accuracy of 10^{-4} using Steffensen's method and compare to the results of Exercise 6 of Section 2.2.

8. Solve $x - 2^{-x} = 0$ for the root in $[0, 1]$ to an accuracy of 10^{-4} using Steffensen's method, and compare to the results of Exercise 8 of Section 2.2.

9. Use Steffensen's method with $p_0 = 2$ to compute an approximation to $\sqrt{3}$ accurate to within 10^{-4}. Compare this result with those obtained in Exercise 9 of Section 2.2 and Exercise 10 of Section 2.1.

10. Use Steffensen's method to approximate the solutions of the following equations to within 10^{-5}.

a. $x = (2 - e^x + x^2)/3$, where g is the function in Exercise 11(a) of Section 2.2.

b. $x = 0.5(\sin x + \cos x)$, where g is the function in Exercise 11(f) of Section 2.2.

c. $3x^2 - e^x = 0$, where g is the function in Exercise 12(a) of Section 2.2.

d. $x - \cos x = 0$, where g is the function in Exercise 12(b) of Section 2.2.

11. The following sequences converge to 0. Use Aitken's Δ^2 method to generate $\{\hat{p}_n\}$ until $|\hat{p}_n| \leq 5 \times 10^{-2}$:

a. $p_n = \dfrac{1}{n}, \quad n \geq 1$ **b.** $p_n = \dfrac{1}{n^2}, \quad n \geq 1$

12. A sequence $\{p_n\}$ is said to be **superlinearly convergent** to p if

$$\lim_{n\to\infty} \frac{|p_{n+1} - p|}{|p_n - p|} = 0.$$

a. Show that if $p_n \to p$ of order α for $\alpha > 1$, then $\{p_n\}$ is superlinearly convergent to p.

b. Show that $p_n = 1/n^n$ is superlinearly convergent to zero but does not converge to zero of order α for any $\alpha > 1$.

13. Suppose that $\{p_n\}$ is superlinearly convergent to p. Show that

$$\lim_{n\to\infty} \frac{|p_{n+1} - p_n|}{|p_n - p|} = 1.$$

14. Prove Theorem 2.13. [*Hint:* Let $\delta_n = (p_{n+1} - p)/(p_n - p) - \lambda$ and show that $\lim_{n\to\infty} \delta_n = 0$. Then express $(\hat{p}_n - p)/(p_n - p)$ in terms of δ_n, δ_{n+1}, and λ.]

15. Let $P_n(x)$ be the nth Taylor polynomial for $f(x) = e^x$ expanded about $x_0 = 0$.

a. For fixed x, show that $p_n = P_n(x)$ satisfies the hypotheses of Theorem 2.13.

b. Let $x = 1$. Use Aitken's Δ^2 method to generate the sequence $\hat{p}_0, \ldots, \hat{p}_8$.

c. Does Aitken's method accelerate convergence in this situation?

2.6 Zeros of Polynomials and Müller's Method

A *polynomial of degree n* has the form

$$P(x) = a_n x^n + a_{n-1} x^{n-1} + \cdots + a_1 x + a_0,$$

where the a_i's, called the *coefficients* of $P(x)$, are constants and $a_n \neq 0$. The zero function, $P(x) = 0$ for all values of x, is considered a polynomial but is assigned no degree.

Theorem 2.15 **(Fundamental Theorem of Algebra)**

If $P(x)$ is a polynomial of degree $n \geq 1$, then $P(x) = 0$ has at least one (possibly complex) root. ■

Although Theorem 2.15 is basic to any study of elementary functions, the usual proof requires techniques from the study of complex-function theory. The reader is referred to [SaS, p. 155], for the culmination of a systematic development of the topics needed to prove Theorem 2.15.

An important consequence of Theorem 2.15 is the following corollary.

Corollary 2.16 If $P(x)$ is a polynomial of degree $n \geq 1$, then there exist unique constants $x_1, x_2, \ldots, x_k$, possibly complex, and positive integers $m_1, m_2, \ldots, m_k$ such that $\sum_{i=1}^{k} m_i = n$ and

$$P(x) = a_n(x - x_1)^{m_1}(x - x_2)^{m_2} \cdots (x - x_k)^{m_k}.$$ ■

Corollary 2.16 states that the collection of zeros of a polynomial is unique and that, if each zero x_i is counted as many times as its multiplicity m_i, a polynomial of degree n has exactly n zeros.

The following corollary of the Fundamental Theorem of Algebra will be used often in this section and in later chapters.

Corollary 2.17 Let $P(x)$ and $Q(x)$ be polynomials of degree at most n. If $x_1, x_2, \ldots, x_k$, with $k > n$, are distinct numbers with $P(x_i) = Q(x_i)$ for $i = 1, 2, \ldots, k$, then $P(x) = Q(x)$ for all values of x. ■

To use the Newton-Raphson procedure to locate approximate zeros of a polynomial $P(x)$, we need to evaluate $P(x)$ and its derivative at specified values. Since both $P(x)$ and its derivative are polynomials, computational efficiency requires that the evaluation of these functions be done in the nested manner discussed in Section 1.2. Horner's method incorporates this nesting technique, and, as a consequence, requires only n multiplications and n additions to evaluate an arbitrary nth-degree polynomial.

Theorem 2.18 **(Horner's Method)**

Let

$$P(x) = a_n x^n + a_{n-1}x^{n-1} + \cdots + a_1 x + a_0.$$

If $b_n = a_n$ and

$$b_k = a_k + b_{k+1}x_0, \quad \text{for } k = n-1, n-2, \cdots, 1, 0,$$

then $b_0 = P(x_0)$. Moreover, if

$$Q(x) = b_n x^{n-1} + b_{n-1}x^{n-2} + \cdots + b_2 x + b_1,$$

then

$$P(x) = (x - x_0)Q(x) + b_0.$$

Proof By the definition of $Q(x)$,

$$\begin{aligned}(x - x_0)Q(x) + b_0 &= (x - x_0)(b_n x^{n-1} + \cdots + b_2 x + b_1) + b_0\\ &= (b_n x^n + b_{n-1}x^{n-1} + \cdots + b_2 x^2 + b_1 x)\\ &\quad - (b_n x_0 x^{n-1} + \cdots + b_2 x_0 x + b_1 x_0) + b_0\\ &= b_n x^n + (b_{n-1} - b_n x_0)x^{n-1} + \cdots + (b_1 - b_2 x_0)x + (b_0 - b_1 x_0).\end{aligned}$$

By the hypothesis, $b_n = a_n$ and $b_k - b_{k+1}x_0 = a_k$, so

$$(x - x_0)Q(x) + b_0 = P(x) \quad \text{and} \quad b_0 = P(x_0). \qquad \blacksquare\ \blacksquare\ \blacksquare$$

EXAMPLE 1 Use Horner's method to evaluate $P(x) = 2x^4 - 3x^2 + 3x - 4$ at $x_0 = -2$.

When we use hand calculation in Horner's method, we first construct a table, which suggests the *synthetic division* name often applied to the technique. For this problem the table appears as follows:

	Coefficient of x^4	Coefficient of x^3	Coefficient of x^2	Coefficient of x	Constant term
$x_0 = -2$	$a_4 = 2$	$a_3 = 0$	$a_2 = -3$	$a_1 = 3$	$a_0 = -4$
		$b_4x_0 = -4$	$b_3x_0 = 8$	$b_2x_0 = -10$	$b_1x_0 = 14$
	$b_4 = 2$	$b_3 = -4$	$b_2 = 5$	$b_1 = -7$	$b_0 = 10$

So

$$P(x) = (x + 2)(2x^3 - 4x^2 + 5x - 7) + 10 \quad \text{and} \quad P(-2) = 10. \qquad \blacksquare$$

An additional advantage of using the Horner (or synthetic-division) procedure is that, since

$$P(x) = (x - x_0)Q(x) + b_0,$$

where

$$Q(x) = b_n x^{n-1} + b_{n-1}x^{n-2} + \cdots + b_2 x + b_1,$$

differentiating with respect to x gives

(2.13) $$P'(x) = Q(x) + (x - x_0)Q'(x) \quad \text{and} \quad P'(x_0) = Q(x_0).$$

When the Newton-Raphson method is being used to find an approximate zero of a polynomial, $P(x)$ and $P'(x)$ can be evaluated in the same manner. Algorithm 2.7 computes $P(x_0)$ and $P'(x_0)$ using Horner's method.

ALGORITHM 2.7

Horner's

To evaluate the polynomial

$$P(x) = a_n x^n + a_{n-1}x^{n-1} + \cdots + a_1 x + a_0$$

and its derivative at x_0:

INPUT degree n; coefficients $a_0, a_1, \ldots, a_n$; x_0.

OUTPUT $y = P(x_0)$; $z = P'(x_0)$.

Step 1 Set $y = a_n$; (*Compute b_n for P.*)
$z = a_n$. (*Compute b_{n-1} for Q.*)

Step 2 For $j = n-1, n-2, \ldots, 1$
set $y = x_0 y + a_j$; (*Compute b_j for P.*)
$z = x_0 z + y$. (*Compute b_{j-1} for Q.*)

Step 3 Set $y = x_0 y + a_0$. (*Compute b_0 for P.*)

Step 4 OUTPUT (y, z);
STOP.

EXAMPLE 2 Find an approximation to one of the zeros of

$$P(x) = 2x^4 - 3x^2 + 3x - 4$$

using the Newton-Raphson procedure and synthetic division to evaluate $P(x_n)$ and $P'(x_n)$ for each iterate x_n.

With $x_0 = -2$ as an initial approximation, we obtained $P(-2)$ in Example 1 by

$$\begin{array}{r|rrrrrl} x_0 = -2 & 2 & 0 & -3 & 3 & -4 & \\ & & -4 & 8 & -10 & 14 & \\ \hline & 2 & -4 & 5 & -7 & 10 & = P(-2). \end{array}$$

Using Theorem 2.18 and Eq. (2.13),

$$Q(x) = 2x^3 - 4x^2 + 5x - 7 \quad \text{and} \quad P'(-2) = Q(-2);$$

so $P'(-2)$ can be found by evaluating $Q(-2)$ in a similar manner:

$$\begin{array}{r|rrrr} x_0 = -2 & 2 & -4 & 5 & -7 \\ & & -4 & 16 & -42 \\ \hline & 2 & -8 & 21 & -49 \end{array} \quad = Q(-2) = P'(-2)$$

and

$$x_1 = x_0 - \frac{P(x_0)}{P'(x_0)} = -2 - \frac{10}{-49} \approx -1.796.$$

Repeating the procedure to find x_2,

$$\begin{array}{r|rrrrr} -1.796 & 2 & 0 & -3 & 3 & -4 \\ & & -3.592 & 6.451 & -6.197 & 5.742 \\ \hline & 2 & -3.592 & 3.451 & -3.197 & 1.742 \quad = P(x_1) \\ & & -3.592 & 12.902 & -29.368 & \\ \hline & 2 & -7.184 & 16.353 & -32.565 & = Q(x_1) \quad = P'(x_1). \end{array}$$

So $P(-1.796) = 1.742$, $P'(-1.796) = -32.565$, and

$$x_2 = -1.796 - \frac{1.742}{-32.565} \approx -1.7425.$$

In a similar manner, $x_3 = -1.73897$. An actual zero to five decimal places is -1.73896. ■

Note that the polynomial $Q(x)$ depends on the approximation being used and changes from iterate to iterate.

If the Nth iterate, x_N, in the Newton-Raphson procedure is an approximate zero for P, then

$$P(x) = (x - x_N)Q(x) + b_0 = (x - x_N)Q(x) + P(x_N) \approx (x - x_N)Q(x);$$

so $x - x_N$ is an approximate factor of $P(x)$. Letting $\hat{x}_1 = x_N$ be the approximate zero of P and $Q_1(x) \equiv Q(x)$ be the approximate factor gives

$$P(x) \approx (x - \hat{x}_1)Q_1(x).$$

We can find a second approximate zero of P by applying the Newton-Raphson procedure to $Q_1(x)$. If $P(x)$ is an nth-degree polynomial with n real zeros, this procedure applied repeatedly will eventually result in $(n - 2)$ approximate zeros of P and an approximate quadratic factor $Q_{n-2}(x)$. At this stage, $Q_{n-2}(x) = 0$ can be solved by the quadratic formula to find the last two approximate zeros of P. Although this method can be used to find all the approximate zeros, it depends on repeated use of approximations and can lead to very inaccurate results.

The procedure just described is called **deflation**. The accuracy difficulty with deflation is due to the fact that, when we obtain the approximate zeros of $P(x)$, the Newton-Raphson procedure is used on the reduced polynomial $Q_k(x)$, that is, the polynomial having the property that

$$P(x) \approx (x - \hat{x}_1)(x - \hat{x}_2) \cdots (x - \hat{x}_k)Q_k(x).$$

An approximate zero $\hat{x}_{k+1}$ of Q_k will generally not approximate a root of $P(x) = 0$ as well as it does a root of the reduced equation $Q_k(x) = 0$, and inaccuracy increases as k increases. One way to eliminate this difficulty is to use the reduced equations to find approximations $\hat{x}_2, \hat{x}_3, \ldots, \hat{x}_k$ to the zeros of P and then improve these approximations by applying Newton's method to the original polynomial $P(x)$.

It has been previously noted that the success of Newton's method often depends on obtaining a good initial approximation. One way to find approximate zeros of P is as follows:

> Evaluate $P(x)$ at x_i for $i = 1, 2, \ldots, k$. If $P(x_i)P(x_j) < 0$, then P has a zero between x_i and x_j.

The problem is to choose the x_i's so that the chance of missing a change of sign is minimized, while keeping the number of x_i's reasonably small. To illustrate the possible difficulty of this problem, consider the polynomial

$$P(x) = 16x^4 - 40x^3 + 5x^2 + 20x + 6.$$

For any integer x_i we have $P(x_i) > 0$. So, evaluating $P(x)$ at even this infinite number of x_i's would not locate an interval (x_i, x_j) containing a zero of P. However, P does have real zeros, which we will find in Example 3. It happened that this particular choice of the x_i's is inappropriate for this polynomial because of the location of the roots.

Another problem with applying Newton's method to polynomials concerns the possibility of the polynomial having complex roots when all the coefficients are real numbers. If the initial approximation using Newton's method is a real number, all subsequent approximations will also be real numbers. One way to overcome this difficulty is to begin with a complex initial approximation and do all the computations using complex arithmetic. An alternative approach has its basis in the following theorem.

Theorem 2.19 If $z = a + bi$ is a complex zero of multiplicity m of the polynomial $P(x)$, then $\bar{z} = a - bi$ is also a zero of multiplicity m of the polynomial $P(x)$ and $(x^2 - 2ax + a^2 + b^2)^m$ is a factor of $P(x)$. ∎

A synthetic division involving quadratic polynomials can be devised to approximately factor the polynomial so that one term will be a quadratic polynomial whose complex roots are approximations to the roots of the original polynomial. This technique was described in some detail in our second edition [BFR].

Instead of proceeding along these lines, we now consider a method first presented by D. E. Müller [Mu] in 1956. This technique can be used for any root-finding problem, but is particularly useful for approximating the roots of polynomials.

Müller's method is an extension of the Secant method. The Secant method begins with two initial approximations x_0 and x_1 and determines the next approximation x_2 as the intersection of the x-axis with the line through $(x_0, f(x_0))$ and $(x_1, f(x_1))$. (See Figure 2.13(a).) Müller's method uses three initial approximations, x_0, x_1, and x_2, and determines the next approximation x_3 by considering the intersection of the x-axis with the parabola through $(x_0, f(x_0))$, $(x_1, f(x_1))$, and $(x_2, f(x_2))$. (See Figure 2.13(b).)

Figure 2.13

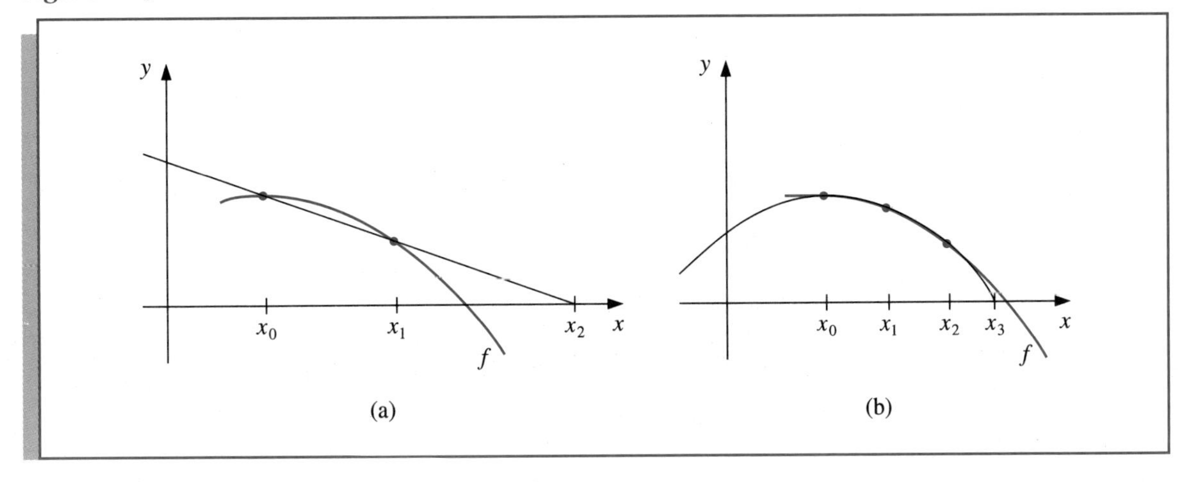

The derivation of Müller's method begins by considering the quadratic polynomial

$$P(x) = a(x - x_2)^2 + b(x - x_2) + c$$

that passes through $(x_0, f(x_0))$, $(x_1, f(x_1))$, and $(x_2, f(x_2))$. The constants a, b, and c can be determined from the conditions

$$f(x_0) = a(x_0 - x_2)^2 + b(x_0 - x_2) + c,$$
$$f(x_1) = a(x_1 - x_2)^2 + b(x_1 - x_2) + c,$$

and

$$f(x_2) = a \cdot 0^2 + b \cdot 0 + c$$

to be

(2.14)
$$c = f(x_2),$$
$$b = \frac{(x_0 - x_2)^2[f(x_1) - f(x_2)] - (x_1 - x_2)^2[f(x_0) - f(x_2)]}{(x_0 - x_2)(x_1 - x_2)(x_0 - x_1)},$$

and

$$a = \frac{(x_1 - x_2)[f(x_0) - f(x_2)] - (x_0 - x_2)[f(x_1) - f(x_2)]}{(x_0 - x_2)(x_1 - x_2)(x_0 - x_1)}.$$

To determine x_3, a zero of P, we apply the quadratic formula to $P(x) = 0$. Because of roundoff error problems caused by the subtraction of nearly equal numbers, however, we apply the formula in the manner prescribed in Example 5 of Section 1.2:

$$x_3 - x_2 = \frac{-2c}{b \pm \sqrt{b^2 - 4ac}}.$$

This formula gives two possibilities for x_3, depending on the sign preceding the radical term. In Müller's method, the sign is chosen to agree with the sign of b. Chosen in this manner, the denominator will be the largest in magnitude and will result in x_3 being selected as the zero of P that is closest to x_2. Thus,

$$x_3 = x_2 - \frac{2c}{b + \text{sign}(b)\sqrt{b^2 - 4ac}},$$

where a, b, and c are given in Eq. (2.14).

Once x_3 is determined, the procedure is reinitialized using x_1, x_2, and x_3 in place of x_0, x_1, and x_2 to determine the next approximation, x_4. The method continues until a satisfactory conclusion is obtained. Since, at each step, the method involves the radical $\sqrt{b^2 - 4ac}$, the method can approximate complex roots when $b^2 - 4ac < 0$. Algorithm 2.8 implements this procedure.

ALGORITHM **2.8**

Müller's

To find a solution to $f(x) = 0$ given three approximations, x_0, x_1, and x_2:

INPUT x_0, x_1, x_2; tolerance TOL; maximum number of iterations N_0.

OUTPUT approximate solution p or message of failure.

Step 1 Set $h_1 = x_1 - x_0$;
$h_2 = x_2 - x_1$;
$\delta_1 = (f(x_1) - f(x_0))/h_1$;
$\delta_2 = (f(x_2) - f(x_1))/h_2$;
$d = (\delta_2 - \delta_1)/(h_2 + h_1)$;
$i = 3$.

Step 2 While $i \le N_0$ do Steps 3–7.

Step 3 $b = \delta_2 + h_2 d$;
$D = (b^2 - 4f(x_2)d)^{1/2}$. (*Note: May require complex arithmetic.*)

Step 4 If $|b - D| < |b + D|$ then set $E = b + D$
else set $E = b - D$.

Step 5 Set $h = -2f(x_2)/E$;
$p = x_2 + h$.

Step 6 If $|h| < TOL$ then
OUTPUT (p); (*Procedure completed successfully.*)
STOP.

Step 7 Set $x_0 = x_1$; *(Prepare for next iteration.)*
$x_1 = x_2$;
$x_2 = p$;
$h_1 = x_1 - x_0$;
$h_2 = x_2 - x_1$;
$\delta_1 = (f(x_1) - f(x_0))/h_1$;
$\delta_2 = (f(x_2) - f(x_1))/h_2$;
$d = (\delta_2 - \delta_1)/(h_2 + h_1)$;
$i = i + 1$.

Step 8 OUTPUT ('Method failed after N_0 iterations, N_0 =', N_0);
(Procedure completed unsuccessfully.)
STOP.

EXAMPLE 3 Consider the polynomial $P(x) = 16x^4 - 40x^3 + 5x^2 + 20x + 6$. Using Algorithm 2.8 with $TOL = 10^{-5}$ and different values of x_0, x_1, and x_2 produces the results in Table 2.12.

Table 2.12

a.

	$x_0 = 0.5,\quad x_1 = -0.5,\quad x_2 = 0$	
i	x_i	$P(x_i)$
3	$-0.555556 + 0.598352i$	$-29.4007 - 3.89872i$
4	$-0.435450 + 0.102101i$	$1.33223 - 1.19309i$
5	$-0.390631 + 0.141852i$	$0.375057 - 0.670164i$
6	$-0.357699 + 0.169926i$	$-0.146746 - 0.00744629i$
7	$-0.356051 + 0.162856i$	$-0.183868 \times 10^{-2} + 0.539780 \times 10^{-3}i$
8	$-0.356062 + 0.162758i$	$0.286102 \times 10^{-5} + 0.953674 \times 10^{-6}i$

b.

	$x_0 = 0.5,\quad x_1 = 1.0,\quad x_2 = 1.5$	
i	x_i	$P(x_i)$
3	1.28785	-1.37624
4	1.23746	0.126941
5	1.24160	0.219440×10^{-2}
6	1.24168	0.257492×10^{-4}
7	1.24168	0.257492×10^{-4}

c.

	$x_0 = 2.5,\quad x_1 = 2.0,\quad x_2 = 2.25$	
i	x_i	$P(x_i)$
3	1.96059	-0.611255
4	1.97056	0.748825×10^{-2}
5	1.97044	-0.295639×10^{-4}
6	1.97044	-0.259639×10^{-4}

The actual values for the roots of the equation are 1.241677, 1.970446, and $-0.356062 \pm 0.162758i$, which demonstrates the accuracy of the approximations from Müller's method. ■

Example 3 illustrates that Müller's method can approximate the roots of polynomials with a variety of starting values. In fact, Müller's method generally converges to the root of a polynomial for any initial approximation choice, although problems can be constructed for which convergence will not occur for certain choices of initial approximations. This can occur, for example, if for some i we have $P(x_i) = P(x_{i+1}) = P(x_{i+2}) \neq 0$. The quadratic equation then reduces to a nonzero constant function and never intersects the x-axis. This is not usually the case, however, and general-purpose software packages using Müller's method request only one initial approximation per root and will even supply this approximation as an option.

EXERCISE SET 2.6

1. Find the approximations to within 10^{-4} to all the real zeros of the following polynomials using Newton's method.
   - **a.** $P(x) = x^3 - 2x^2 - 5$
   - **b.** $P(x) = x^3 + 3x^2 - 1$
   - **c.** $P(x) = x^3 - x - 1$
   - **d.** $P(x) = x^4 + 2x^2 - x - 3$
   - **e.** $P(x) = x^3 + 4.001x^2 + 4.002x + 1.101$
   - **f.** $P(x) = x^5 - x^4 + 2x^3 - 3x^2 + x - 4$
2. Find approximations to within 10^{-5} to all the zeros of each of the following polynomials by first finding the real zeros using Newton's method and then reducing to polynomials of lower degree to determine any complex zeros.
   - **a.** $P(x) = x^4 + 5x^3 - 9x^2 - 85x - 136$
   - **b.** $P(x) = x^4 - 2x^3 - 12x^2 + 16x - 40$
   - **c.** $P(x) = x^4 + x^3 + 3x^2 + 2x + 2$
   - **d.** $P(x) = x^5 + 11x^4 - 21x^3 - 10x^2 - 21x - 5$
   - **e.** $P(x) = 16x^4 + 88x^3 + 159x^2 + 76x - 240$
   - **f.** $P(x) = x^4 - 4x^2 - 3x + 5$
   - **g.** $P(x) = x^4 - 2x^3 - 4x^2 + 4x + 4$
   - **h.** $P(x) = x^3 - 7x^2 + 14x - 6$
3. Repeat Exercise 1 using Müller's method.
4. Repeat Exercise 2 using Müller's method.
5. Use Newton's method to find, within 10^{-3}, the zeros and critical points of the following functions. Use this information to sketch the graph of P.
   - **a.** $P(x) = x^3 - 9x^2 + 12$
   - **b.** $P(x) = x^4 - 2x^3 - 5x^2 + 12x - 5$
6. $P(x) = 10x^3 - 8.3x^2 + 2.295x - 0.21141 = 0$ has a root at $x = 0.29$. Use Newton's method with an initial approximation $x_0 = 0.28$ to attempt to find this root. Explain what happens.
7. Use Maple to find the exact roots of the polynomial $P(x) = x^3 + 4x - 4$.

8. Use Maple to find the exact roots of the polynomial $P(x) = x^3 - 2x - 5$.

9. Use each of the following methods to find a solution accurate to within 10^{-4} for the problem

$$600x^4 - 550x^3 + 200x^2 - 20x - 1 = 0, \quad \text{for } 0.1 \le x \le 1.$$

a. Bisection method

b. Newton's method

c. Secant method

d. method of False Position

e. Müller's method

10. Two ladders crisscross an alley of width W. Each ladder reaches from the base of one wall to some point on the opposite wall. The ladders cross at a height H above the pavement. Find W given that the lengths of the ladders are $x_1 = 20$ ft and $x_2 = 30$ ft and that $H = 8$ ft.

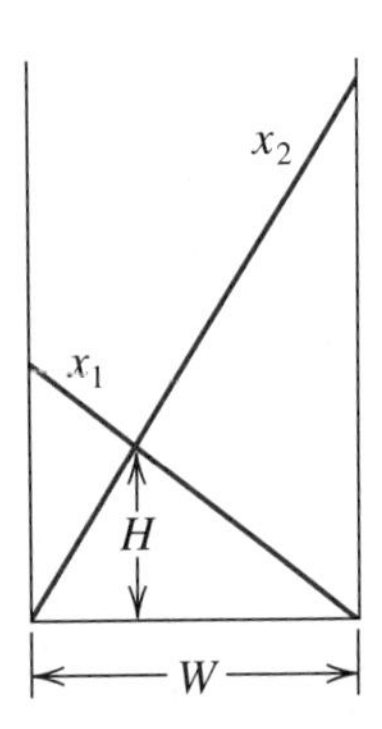

11. A can in the shape of a right circular cylinder is to be constructed to contain 1000 cm^3. The circular top and bottom of the can must have a radius of 0.25 cm more than the radius of the can so that the excess can be used to form a seal with the side. The sheet of material being formed into the side of the can must also be 0.25 cm longer than the circumference of the can so that a seal can be formed. Find, to within 10^{-4}, the minimal amount of material needed to construct the can.

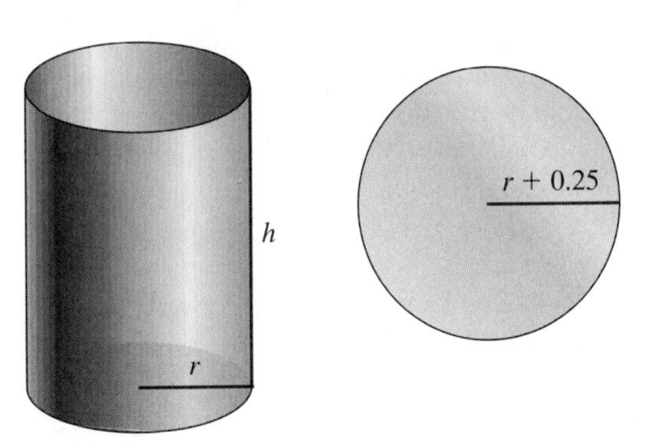

12. In 1224 Leonardo of Pisa, better known as Fibonacci, answered a mathematical challenge of John of Palermo in the presence of Emperor Frederick II. His challenge was to find a root of the equation $x^3 + 2x^2 + 10x = 20$. He first showed that the equation had no rational roots and no Euclidean irrational root—that is, no root in one of the forms $a \pm \sqrt{b}$, $\sqrt{a} \pm \sqrt{b}$, $\sqrt{a \pm \sqrt{b}}$, or $\sqrt{\sqrt{a} \pm \sqrt{b}}$, where a and b are rational numbers. He then approximated the only real root, probably using an algebraic technique of Omar Khayyam involving the intersection of a circle and a parabola. His answer was given in the base-60 number system as

$$1 + 22\left(\frac{1}{60}\right) + 7\left(\frac{1}{60}\right)^2 + 42\left(\frac{1}{60}\right)^3 + 33\left(\frac{1}{60}\right)^4 + 4\left(\frac{1}{60}\right)^5 + 40\left(\frac{1}{60}\right)^6.$$

How accurate was his approximation?

2.7 Survey of Methods and Software

In this chapter we have considered the problem of solving the equation $f(x) = 0$, where f is a given continuous function. All the methods begin with an initial approximation and generate a sequence that converges to a root of the equation if the method is successful. If $[a, b]$ is an interval on which $f(a)$ and $f(b)$ are of opposite sign, then the Bisection method and the method of False Position will converge. However, the convergence of these methods may be slow. Faster convergence is generally obtained using the Secant method or the Newton-Raphson method. Good initial approximations are required for these methods, two for the Secant method and one for the Newton-Raphson method, so the Bisection or the False Position method can be used as starter methods for the Secant or Newton-Raphson method.

Müller's method will give rapid convergence without a particularly good initial approximation. Müller's method is not quite as efficient as Newton's method; its order of convergence near a root is approximately $\alpha = 1.84$ (see [YG, p. 156]), compared to the quadratic, $\alpha = 2$, of Newton's method. However, it is better than the Secant method, whose order is approximately $\alpha = 1.62$.

Deflation is generally used with Müller's method once an approximate root of a polynomial has been determined. After an approximation to the root of the deflated equation has been determined, use either Müller's method or Newton's method in the original polynomial with this root as the initial approximation. This procedure will ensure that the root being approximated is a solution to the true equation, not to the deflated equation. We recommended Müller's method for finding all the zeros of polynomials, real or complex. Müller's method can also be used for an arbitrary continuous function.

Other high-order methods are available for determining the roots of polynomials. If this topic is of particular interest, we recommend that consideration be given to Laguerre's method, which gives cubic convergence and also approximates complex roots (see [Ho, pp. 176–179] for a complete discussion), the Jenkins-Traub method (see [JT]), and Brent's method (see [Bre]), which is based on the Bisection and False-position methods.

Another method of interest, Cauchy's method, is similar to Müller's method but avoids the failure problem of Müller's method when $f(x_i) = f(x_{i+1}) = f(x_{i+2})$, for some i. For an interesting discussion of this method, as well as more detail on Müller's method, we recommend [YG, Sections 4.10, 4.11, and 5.4].

Given a specified function f and a tolerance, an efficient program should produce an approximation to one or more solutions of $f(x) = 0$, each having an absolute or relative error within the tolerance, and the results should be generated in a reasonable amount of time. If the program cannot accomplish this task, it should at least give meaningful explanations of why success was not obtained and an indication of how to remedy the cause of failure.

The IMSL FORTRAN subroutine ZANLY uses Müller's method with deflation to approximate a number of roots of $f(x) = 0$. The routine ZBREN, due to R. P. Brent, uses a combination of linear interpolation, an inverse quadratic interpolation similar to Müller's method, and the Bisection method. It requires specifying an interval $[a, b]$ that contains a root. The IMSL routine ZREAL is based on a variation of Müller's method and approximates zeros of a real function f when only poor initial approximations are available. Routines for finding zeros of polynomials are ZPORC, which uses the Jenkins-Traub method for finding zeros of a real polynomial; ZPLRC, which uses Laguerre's method to find zeros of a real polynomial; and ZPOCC, which uses the Jenkins-Traub method to find zeros of a complex polynomial.

The NAG FORTRAN subroutine C05ADF uses the Bisection method in conjunction with a method based on inverse linear interpolation, similar to the False Position and Secant methods, to approximate a real zero of $f(x) = 0$ in the interval $[a, b]$. The subroutine C05AZF is similar to C05ADF but requires a single starting value instead of an interval. The subroutine C05AGF will find an interval containing a root on its own. NAG also supplies subroutines C02AEF and C02ADF to approximate all zeros of a real polynomial or complex polynomial, respectively. The subroutine C02AGF uses a modified Laguerre method to find the roots of a real polynomial.

Within MATLAB, the function ROOTS is used to compute all the roots, both real and complex, of a polynomial. For an arbitrary function, FZERO computes a root near a specified initial approximation to within a specified tolerance.

Maple has the procedure `fsolve` to find roots of equations. For example,

```
>fsolve(x^2 - x - 1);
```

returns the numbers $-.6180339887$ and 1.618033989. You can also specify a particular variable and interval to search. For example,

```
>fsolve(x^2 - x - 1,x,1..2);
```

returns only the number 1.618033989. `fsolve` uses a variety of specialized techniques that depend on the particular form of the equation or system of equations.

Notice that in spite of the diversity of methods, the professionally written packages are based primarily on the methods and principles discussed in this chapter. You should be able to use these packages by reading the manuals accompanying the packages to better understand the parameters and the specifications of the results that are obtained.

There are three books that we consider to be classics on the solution of nonlinear equations, those by Traub [Tr], by Ostrowski [Os], and by Householder [Ho]. In addition, the book by Brent [Bre] served as the basis for many of the currently used root-finding methods.

CHAPTER

3

Interpolation and Polynomial Approximation

■ ■ ■

A census of the population of the United States is taken every 10 years. The following table lists the population, in thousands of people, from 1940 to 1990.

Year	1940	1950	1960	1970	1980	1990
Population (in thousands)	132,165	151,326	179,323	203,302	226,542	249,633

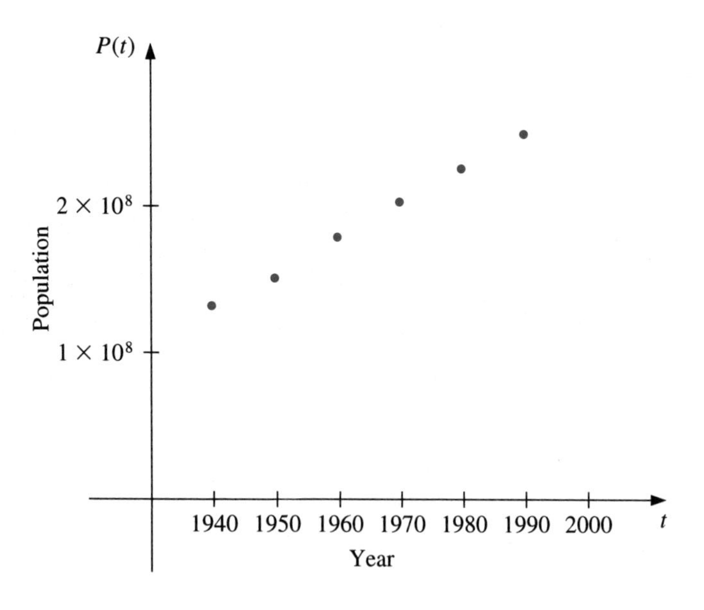

In reviewing these data, we might ask whether they could be used to provide a reasonable estimate of the population, say, in 1965 or even in the year 2000. Predictions of this type can be obtained by using a function that fits the given data. This process is called *interpolation* and is the subject of this chapter. This population problem is considered throughout the chapter and in Exercises 24 of Section 3.1, 14 of Section 3.2, and 24 of Section 3.4.

One of the most useful and well-known classes of functions mapping the set of real numbers into itself is the class of *algebraic polynomials*, the set of functions of the form

$$P_n(x) = a_n x^n + a_{n-1}x^{n-1} + \cdots + a_1 x + a_0,$$

where n is a nonnegative integer and $a_0, \ldots, a_n$ are real constants. One reason for their importance is that they uniformly approximate continuous functions. Given any function, defined and continuous on a closed and bounded interval, there exists a polynomial that is as "close" to the given function as desired. This result is expressed precisely in the following theorem. (See Figure 3.1.)

Figure 3.1

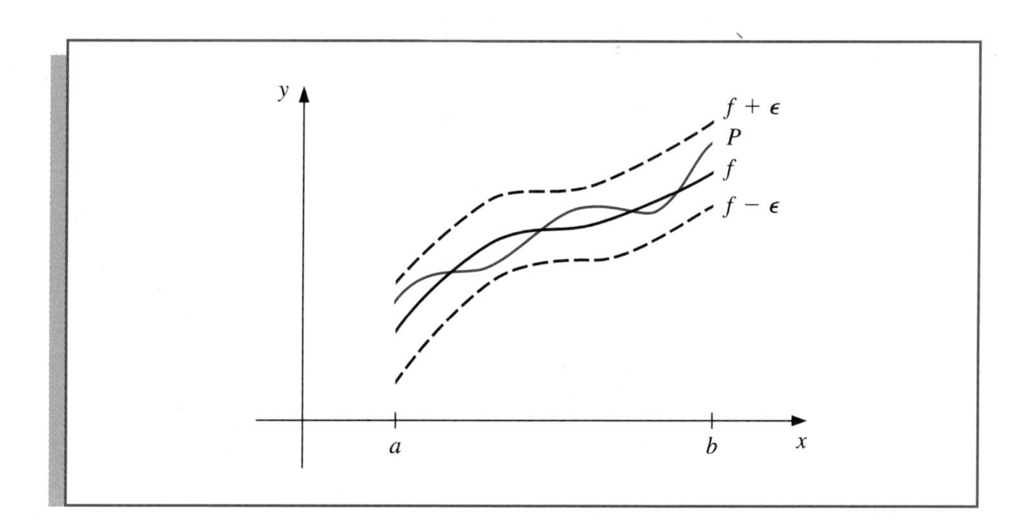

Theorem 3.1 **(Weierstrass Approximation Theorem)**

Suppose f is defined and continuous on $[a, b]$. For each $\varepsilon > 0$ there exists a polynomial $P(x)$, defined on $[a, b]$, with the property that

$$|f(x) - P(x)| < \varepsilon, \quad \text{for all } x \text{ in } [a, b]. \qquad \blacksquare$$

The proof of this theorem can be found in any elementary text on real analysis (see, for example, [Bart, pp. 165–172]).

Another important reason for considering the class of polynomials in the approximation of functions is that the derivative and indefinite integral of a polynomial are easy to determine and are also polynomials. For these reasons, polynomials are often used for approximating continuous functions.

The Taylor polynomials were introduced in the first section of the book, where they were described as one of the fundamental building blocks of numerical analysis. Given this prominence, you might assume that polynomial interpolation would make heavy use of these functions. However, this is not the case. The Taylor polynomials agree as closely as possible with a given function at a specific point, but they concentrate their accuracy near that point. A good interpolation polynomial needs to provide a relatively accurate approximation over an entire interval, and Taylor polynomials do not generally do this. For example, suppose we calculate the first six Taylor polynomials about $x_0 = 0$ for $f(x) = e^x$. Since the derivatives of $f(x)$ are all e^x, which evaluated at $x_0 = 0$ gives 1, the Taylor polynomials are

$$P_0(x) = 1, \quad P_1(x) = 1 + x, \quad P_2(x) = 1 + x + \frac{x^2}{2}, \quad P_3(x) = 1 + x + \frac{x^2}{2} + \frac{x^3}{6},$$

$$P_4(x) = 1 + x + \frac{x^2}{2} + \frac{x^3}{6} + \frac{x^4}{24}, \quad \text{and} \quad P_5(x) = 1 + x + \frac{x^2}{2} + \frac{x^3}{6} + \frac{x^4}{24} + \frac{x^5}{120}.$$

Table 3.1

x	$P_0(x)$	$P_1(x)$	$P_2(x)$	$P_3(x)$	$P_4(x)$	$P_5(x)$	e^x
−2.0	1.00000	−1.00000	1.00000	−0.33333	0.33333	0.06667	0.13534
−1.5	1.00000	−0.50000	0.62500	0.06250	0.27344	0.21016	0.22313
−1.0	1.00000	0.00000	0.50000	0.33333	0.37500	0.36667	0.36788
−0.5	1.00000	0.50000	0.62500	0.60417	0.60677	0.60651	0.60653
0.0	1.00000	1.00000	1.00000	1.00000	1.00000	1.00000	1.00000
0.5	1.00000	1.50000	1.62500	1.64583	1.64844	1.64870	1.64872
1.0	1.00000	2.00000	2.50000	2.66667	2.70833	2.71667	2.71828
1.5	1.00000	2.50000	3.62500	4.18750	4.39844	4.46172	4.48169
2.0	1.00000	3.00000	5.00000	6.33333	7.00000	7.26667	7.38906

Table 3.1 lists the values of the Taylor polynomial for various values of x. Notice that even for the higher-degree polynomials, the error becomes progressively worse as we move away from zero. This result can also be seen from the graphs of the polynomials shown in Figure 3.2.

Although better approximations are obtained for this problem if higher-degree Taylor polynomials are used, this is not always the case. Consider, as an extreme example, using Taylor polynomials of various degrees for $f(x) = 1/x$ expanded about $x_0 = 1$ to approximate $f(3) = \frac{1}{3}$. Since $f(x) = x^{-1}$, $f'(x) = -x^{-2}$, $f''(x) = (-1)^2 2 \cdot x^{-3}$, and, in general, $f^{(k)}(x) = (-1)^k k!\, x^{-k-1}$, the nth Taylor polynomial is

$$P_n(x) = \sum_{k=0}^{n} \frac{f^{(k)}(1)}{k!}(x-1)^k = \sum_{k=0}^{n} (-1)^k (x-1)^k.$$

Figure 3.2

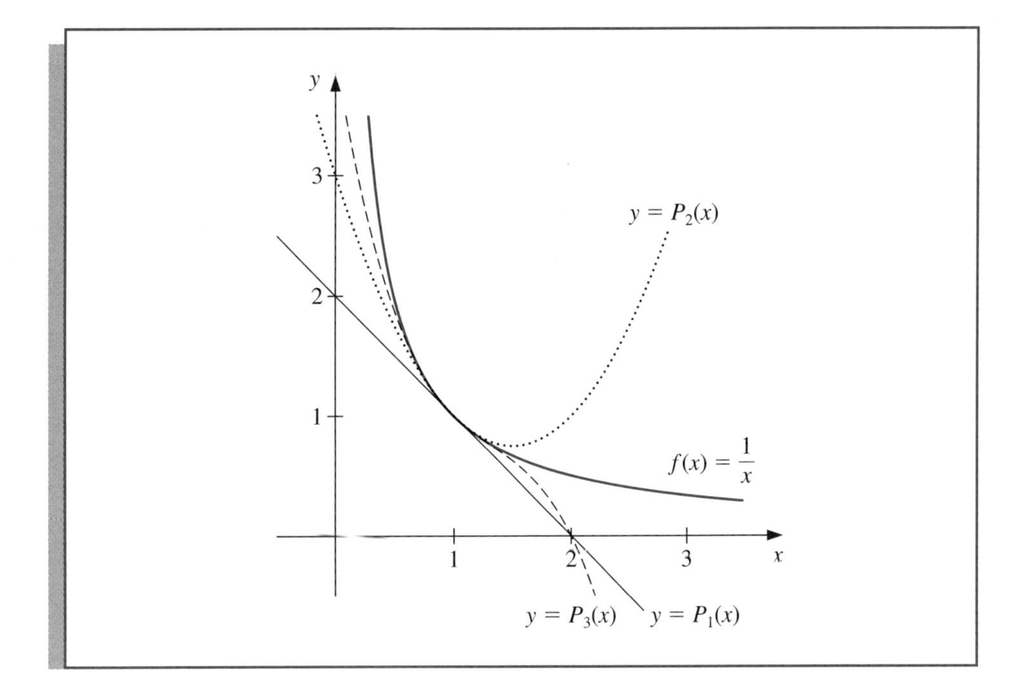

To approximate $f(3) = \frac{1}{3}$ by $P_n(3)$ for increasing values of n, we obtain the values in Table 3.2—rather a dramatic failure!

Table 3.2

n	0	1	2	3	4	5	6	7
$P_n(3)$	1	-1	3	-5	11	-21	43	-85

Since the Taylor polynomials have the property that all the information used in the approximation is concentrated at the single point x_0, the type of difficulty that occurs here is quite common. This problem generally limits Taylor polynomial approximation to the situation in which approximations are needed only at points close to x_0. For ordinary computational purposes it is more efficient to use methods that include information at various points, which we consider in the remainder of this chapter. The primary use of Taylor polynomials in numerical analysis is not for approximation purposes but for the derivation of numerical techniques.

3.1 Interpolation and the Lagrange Polynomial

Since the Taylor polynomials are not appropriate for interpolation, alternative methods are needed. In this section we find approximating polynomials that are determined simply by specifying certain points on the plane through which they must pass.

The problem of determining a polynomial of degree one that passes through the distinct points (x_0, y_0) and (x_1, y_1) with $x_0 \neq x_1$ is the same as approximating a function f for which $f(x_0) = y_0$ and $f(x_1) = y_1$ by means of a first-degree polynomial interpolating, or agreeing with, the values of f at the given points.

Consider the linear polynomial

$$P(x) = \frac{x - x_1}{x_0 - x_1} y_0 + \frac{x - x_0}{x_1 - x_0} y_1.$$

When $x = x_0$,

$$P(x_0) = 1 \cdot y_0 + 0 \cdot y_1 = y_0 = f(x_0),$$

and when $x = x_1$,

$$P(x_1) = 0 \cdot y_0 + 1 \cdot y_1 = y_1 = f(x_1),$$

so $P(x)$ has the required properties. (See Figure 3.3.)

Figure 3.3

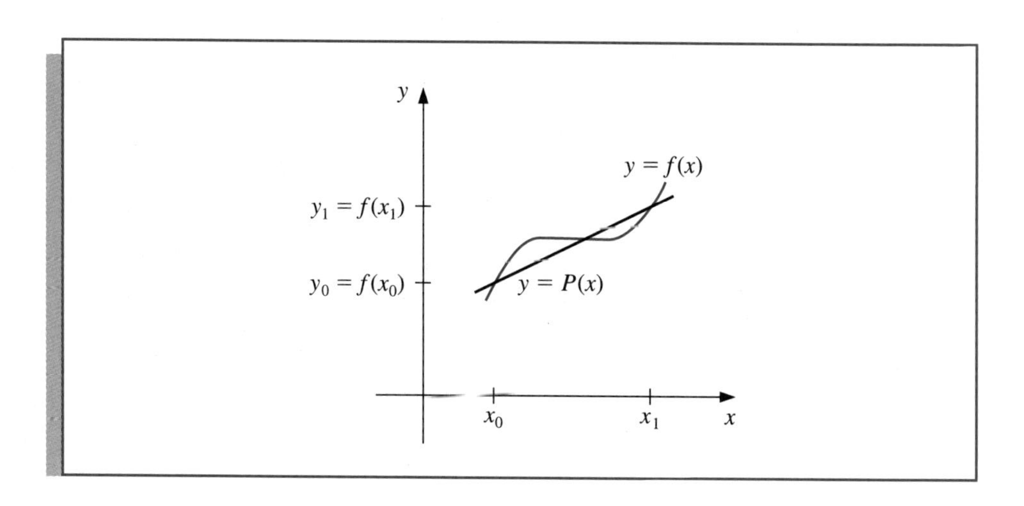

To generalize the concept of linear interpolation, consider the construction of a polynomial of degree at most n that passes through the $n + 1$ points $(x_0, f(x_0)), (x_1, f(x_1)), \ldots, (x_n, f(x_n))$. (See Figure 3.4.)

The linear polynomial passing through $(x_0, f(x_0))$ and $(x_1, f(x_1))$ was constructed using the quotients

$$L_0(x) = \frac{x - x_1}{x_0 - x_1} \quad \text{and} \quad L_1(x) = \frac{x - x_0}{x_1 - x_0}.$$

When $x = x_0$, $L_0(x_0) = 1$ and $L_1(x_0) = 0$. When $x = x_1$, $L_0(x_1) = 0$ and $L_1(x_1) = 1$.

Figure 3.4

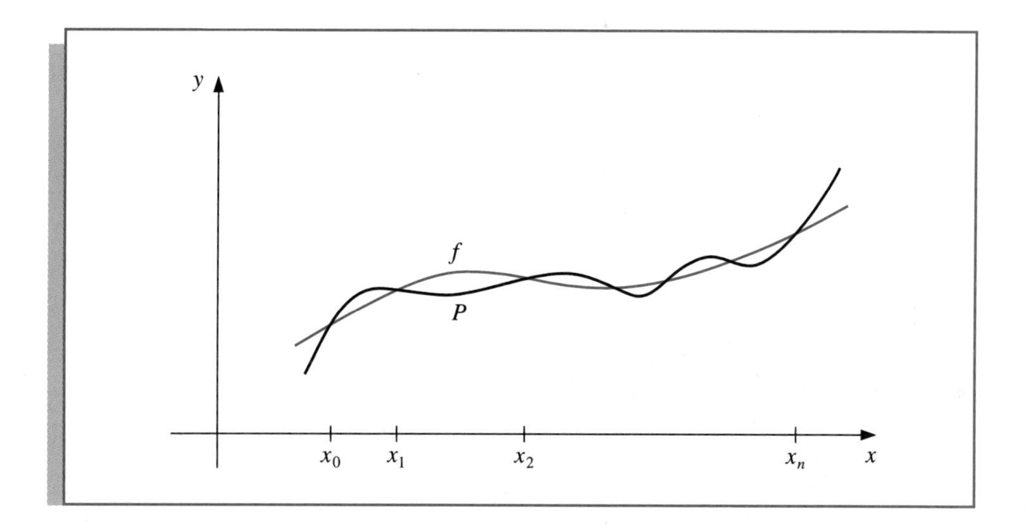

For the general case we construct, for each $k = 0, 1, \ldots, n$, a quotient $L_{n,k}(x)$ with the property that $L_{n,k}(x_i) = 0$ when $i \neq k$ and $L_{n,k}(x_k) = 1$. To satisfy $L_{n,k}(x_i) = 0$ for each $i \neq k$ requires that the numerator of $L_{n,k}(x)$ contain the term

(3.1) $$(x - x_0)(x - x_1)\cdots(x - x_{k-1})(x - x_{k+1})\cdots(x - x_n).$$

To satisfy $L_{n,k}(x_k) = 1$, the denominator of $L_k(x)$ must agree with (3.1) when it is evaluated at $x = x_k$. That is,

$$L_{n,k}(x) = \frac{(x - x_0)\cdots(x - x_{k-1})(x - x_{k+1})\cdots(x - x_n)}{(x_k - x_0)\cdots(x_k - x_{k-1})(x_k - x_{k+1})\cdots(x_k - x_n)} = \prod_{\substack{i=0 \\ i \neq k}}^{n} \frac{(x - x_i)}{(x_k - x_i)}.$$

A sketch of the graph of a typical $L_{n,k}$ (in the case when n is even and k is odd) is shown in Figure 3.5.

Figure 3.5

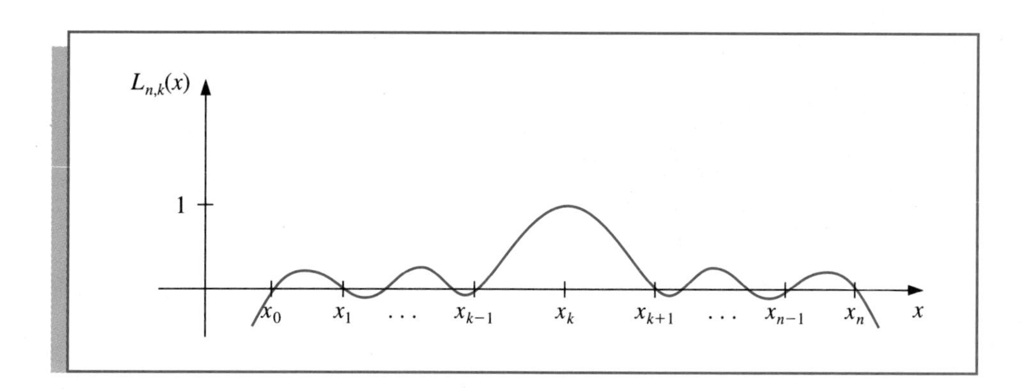

The interpolating polynomial is easily described now that the form of $L_{n,k}(x)$ is known. This polynomial, called the **nth Lagrange interpolating polynomial**, is defined in the following theorem.

Theorem 3.2 If $x_0, x_1, \ldots, x_n$ are $n+1$ distinct numbers and f is a function whose values are given at these numbers, then there exists a unique polynomial $P(x)$ of degree at most n with the property that

$$f(x_k) = P(x_k) \quad \text{for each } k = 0, 1, \ldots, n.$$

This polynomial is given by

$$P(x) = f(x_0)L_{n,0}(x) + \cdots + f(x_n)L_{n,n}(x) = \sum_{k=0}^{n} f(x_k)L_{n,k}(x), \tag{3.2}$$

where

$$L_{n,k}(x) = \frac{(x-x_0)(x-x_1)\cdots(x-x_{k-1})(x-x_{k+1})\cdots(x-x_n)}{(x_k-x_0)(x_k-x_1)\cdots(x_k-x_{k-1})(x_k-x_{k+1})\cdots(x_k-x_n)} = \prod_{\substack{i=0\\ i\neq k}}^{n} \frac{(x-x_i)}{(x_k-x_i)}, \tag{3.3}$$

for each $k = 0, 1, \ldots, n$. ■

We will write $L_{n,k}(x)$ simply as $L_k(x)$ when there is no confusion as to its degree.

EXAMPLE 1 Using the numbers (or nodes) $x_0 = 2$, $x_1 = 2.5$, and $x_2 = 4$ to find the second interpolating polynomial for $f(x) = 1/x$ requires that we determine the coefficient polynomials $L_0(x)$, $L_1(x)$, and $L_2(x)$:

$$L_0(x) = \frac{(x-2.5)(x-4)}{(2-2.5)(2-4)} = (x-6.5)x + 10,$$

$$L_1(x) = \frac{(x-2)(x-4)}{(2.5-2)(2.5-4)} = \frac{(-4x+24)x-32}{3},$$

and

$$L_2(x) = \frac{(x-2)(x-2.5)}{(4-2)(4-2.5)} = \frac{(x-4.5)x+5}{3}.$$

Since $f(x_0) = f(2) = 0.5$, $f(x_1) = f(2.5) = 0.4$, and $f(x_2) = f(4) = 0.25$, we have

$$
\begin{aligned}
P(x) &= \sum_{k=0}^{2} f(x_k)L_k(x) \\
&= 0.5((x - 6.5)x + 10) + 0.4\frac{(-4x + 24)x - 32}{3} + 0.25\frac{(x - 4.5)x + 5}{3} \\
&= (0.05x - 0.425)x + 1.15.
\end{aligned}
$$

An approximation to $f(3) = \frac{1}{3}$ (see Figure 3.6) is

$$f(3) \approx P(3) = 0.325.$$

Figure 3.6

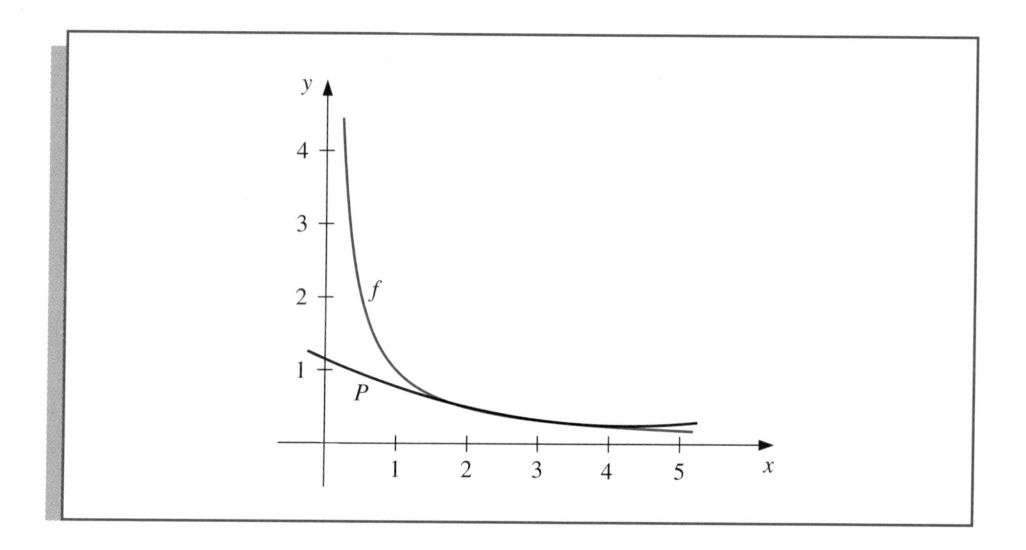

Compare this to Table 3.2, where no Taylor polynomial (expanded about $x_0 = 1$) could be used to reasonably approximate $f(3) = \frac{1}{3}$. ■

The next step is to calculate a remainder term or bound for the error involved in approximating a function by an interpolating polynomial. This is done in the following theorem.

Theorem 3.3 Suppose $x_0, x_1, \ldots, x_n$ are distinct numbers in the interval $[a, b]$ and $f \in C^{n+1}[a, b]$. Then, for each x in $[a, b]$, a number $\xi(x)$ in (a, b) exists with

(3.4)
$$f(x) = P(x) + \frac{f^{(n+1)}(\xi(x))}{(n+1)!}(x - x_0)(x - x_1)\cdots(x - x_n),$$

where $P(x)$ is the interpolating polynomial given in Eq. (3.2).

Proof Note first that if $x = x_k$ for $k = 0, 1, \ldots, n$, then $f(x_k) = P(x_k)$, and choosing $\xi(x_k)$ arbitrarily in (a, b) yields Eq. (3.4). If $x \neq x_k$ for any $k = 0, 1, \ldots, n$, define the function g for t in $[a, b]$ by

$$\begin{aligned} g(t) &= f(t) - P(t) - [f(x) - P(x)]\frac{(t - x_0)(t - x_1)\cdots(t - x_n)}{(x - x_0)(x - x_1)\cdots(x - x_n)} \\ &= f(t) - P(t) - [f(x) - P(x)]\prod_{i=0}^{n}\frac{(t - x_i)}{(x - x_i)}. \end{aligned}$$

Since $f \in C^{n+1}[a, b]$, $P \in C^{\infty}[a, b]$, and $x \neq x_k$ for any k, it follows that $g \in C^{n+1}[a, b]$. For $t = x_k$ we have

$$\begin{aligned} g(x_k) &= f(x_k) - P(x_k) - [f(x) - P(x)]\prod_{i=0}^{n}\frac{(x_k - x_i)}{(x - x_i)} \\ &= 0 - [f(x) - P(x)] \cdot 0 = 0. \end{aligned}$$

Moreover,

$$g(x) = f(x) - P(x) - [f(x) - P(x)]\prod_{i=0}^{n}\frac{(x - x_i)}{(x - x_i)} = f(x) - P(x) - [f(x) - P(x)] = 0.$$

Thus, $g \in C^{n+1}[a, b]$ and g vanishes at the $n + 2$ distinct numbers $x, x_0, x_1, \ldots, x_n$. By the Generalized Rolle's Theorem, there exists ξ in (a, b) for which $g^{(n+1)}(\xi) = 0$. So,

$$\textbf{(3.5)} \quad 0 = g^{(n+1)}(\xi) = f^{(n+1)}(\xi) - P^{(n+1)}(\xi) - [f(x) - P(x)]\frac{d^{n+1}}{dt^{n+1}}\left[\prod_{i=0}^{n}\frac{(t - x_i)}{(x - x_i)}\right]_{t=\xi}.$$

Since $P(x)$ is a polynomial of degree at most n, its $(n + 1)$st derivative, $P^{(n+1)}(x)$, is identically zero. Also, $\prod_{i=0}^{n}[(t - x_i)/(x - x_i)]$ is a polynomial of degree $(n + 1)$, so

$$\prod_{i=0}^{n}\frac{(t - x_i)}{(x - x_i)} = \left[\frac{1}{\prod_{i=0}^{n}(x - x_i)}\right]t^{n+1} + \text{(lower-degree terms in } t\text{)},$$

and

$$\frac{d^{n+1}}{dt^{n+1}}\prod_{i=0}^{n}\frac{(t - x_i)}{(x - x_i)} = \frac{(n + 1)!}{\prod_{i=0}^{n}(x - x_i)}.$$

Equation (3.5) now becomes

$$0 = f^{(n+1)}(\xi) - 0 - [f(x) - P(x)]\frac{(n + 1)!}{\prod_{i=0}^{n}(x - x_i)},$$

and, upon solving for $f(x)$, we have

$$f(x) = P(x) + \frac{f^{(n+1)}(\xi)}{(n+1)!}\prod_{i=0}^{n}(x - x_i).$$ ■ ■ ■

The error formula in Theorem 3.3 is an important theoretical result because Lagrange polynomials are used extensively for deriving numerical differentiation and integration methods. Error bounds for these techniques are obtained from the Lagrange error formula.

Note that the error form for the Lagrange polynomial is quite similar to that for the Taylor polynomial. The nth Taylor polynomial about x_0 concentrates all the known information at x_0 and has an error term of the form

$$\frac{f^{(n+1)}(\xi(x))}{(n+1)!}(x - x_0)^{n+1}.$$

The Lagrange polynomial of degree n uses information at the distinct numbers $x_0, x_1, \ldots, x_n$ and, in place of $(x - x_0)^n$, its error formula uses a product of the $n + 1$ terms $(x - x_0)$, $(x - x_1), \ldots, (x - x_n)$:

$$\frac{f^{(n+1)}(\xi(x))}{(n+1)!}(x - x_0)(x - x_1)\cdots(x - x_n).$$

The specific use of this error formula is restricted to those functions whose derivatives have known bounds.

EXAMPLE 2 Suppose a table is to be prepared for the function $f(x) = e^x$, for x in $[0, 1]$. Assume the number of decimal places to be given per entry is $d \geq 8$ and that the difference between adjacent x-values, the step size, is h. What should h be for linear interpolation (that is, the Lagrange polynomial of degree 1) to give an absolute error of at most 10^{-6}?

Let $x \in [0, 1]$, $x_0, x_1, \ldots$ be the numbers at which f is evaluated and suppose j satisfies $x_j \leq x \leq x_{j+1}$. Eq. (3.4) implies that the error in linear interpolation is

$$|f(x) - p(x)| = \left|\frac{f^{(2)}(\xi)}{2!}(x - x_j)(x - x_{j+1})\right| = \frac{|f^{(2)}(\xi)|}{2}|(x - x_j)(x - x_{j+1})|.$$

Since the step size is h, it follows that $x_j = jh$, $x_{j+1} = (j+1)h$, and

$$|f(x) - P(x)| \leq \frac{|f^{(2)}(\xi)|}{2!}|(x - jh)(x - (j+1)h|.$$

Hence,

$$\begin{aligned}|f(x) - P(x)| &\leq \frac{1}{2}\max_{\xi\in[0,1]} e^{\xi}\max_{x_j\leq x\leq x_{j+1}}|(x - jh)(x - (j+1)h)|\\ &\leq \frac{1}{2}e\max_{x_j\leq x\leq x_{j+1}}|(x - jh)(x - (j+1)h)|.\end{aligned}$$

By considering $g(x) = (x - jh)(x - (j+1)h)$ for $jh \leq x \leq (j+1)h$ and using techniques of calculus (see Exercise 28) we find that

$$\max_{x_j \leq x \leq x_{j+1}} |(x - jh)(x - (j+1)h| = \max_{x_j \leq x \leq x_{j+1}} |g(x)| = \left| g\left(\left(j + \frac{1}{2}\right) h\right) \right| = \frac{h^2}{4}.$$

Consequently, the error in linear interpolation is bounded by

$$|f(x) - P(x)| \leq \frac{eh^2}{8},$$

and it is sufficient for h to be chosen so that

$$\frac{eh^2}{8} \leq 10^{-6}, \quad \text{which implies that} \quad h < 1.72 \times 10^{-3}.$$

Since $n = (1 - 0)/h$ must be an integer, one logical choice for the step size is $h = 0.001$. ■

The next example illustrates interpolation for a situation when the error portion of Eq. (3.4) cannot be used.

EXAMPLE 3 Table 3.3 lists values of a function at various points. The approximations to $f(1.5)$ obtained by various Lagrange polynomials will be compared.

Table 3.3

x	$f(x)$
1.0	0.7651977
1.3	0.6200860
1.6	0.4554022
1.9	0.2818186
2.2	0.1103623

Since 1.5 is between 1.3 and 1.6, the linear polynomial will use $x_0 = 1.3$ and $x_1 = 1.6$. The value of the interpolating polynomial at 1.5 is

$$P_1(1.5) = \frac{(1.5 - 1.6)}{(1.3 - 1.6)}(0.6200860) + \frac{(1.5 - 1.3)}{(1.6 - 1.3)}(0.4554022) = 0.5102968.$$

Two polynomials of degree two can reasonably be used, one by letting $x_0 = 1.3$, $x_1 = 1.6$, and $x_2 = 1.9$, which gives

$$\begin{aligned} P_2(1.5) = {} & \frac{(1.5 - 1.6)(1.5 - 1.9)}{(1.3 - 1.6)(1.3 - 1.9)}(0.6200860) + \frac{(1.5 - 1.3)(1.5 - 1.9)}{(1.6 - 1.3)(1.6 - 1.9)}(0.4554022) \\ & + \frac{(1.5 - 1.3)(1.5 - 1.6)}{(1.9 - 1.3)(1.9 - 1.6)}(0.2818186) \\ = {} & 0.5112857, \end{aligned}$$

and the other by letting $x_0 = 1.0$, $x_1 = 1.3$, and $x_2 = 1.6$, in which case

$$\hat{P}_2(1.5) = 0.5124715.$$

In the third-degree case there are also two choices for the polynomial. One is with $x_0 = 1.3$, $x_1 = 1.6$, $x_2 = 1.9$, and $x_3 = 2.2$, which gives

$$P_3(1.5) = 0.5118302.$$

The other is obtained by letting $x_0 = 1.0$, $x_1 = 1.3$, $x_2 = 1.6$, and $x_3 = 1.9$, giving

$$\hat{P}_3(1.5) = 0.5118127.$$

The fourth-degree Lagrange polynomial uses all the entries in the table. With $x_0 = 1.0$, $x_1 = 1.3$, $x_2 = 1.6$, $x_3 = 1.9$, and $x_4 = 2.2$, it can be shown that

$$P_4(1.5) = 0.5118200.$$

Since $P_3(1.5)$, $\hat{P}_3(1.5)$, and $P_4(1.5)$ all agree to within 2×10^{-5} units, we expect this degree of accuracy for these approximations. We also expect $P_4(1.5)$ to be the most accurate approximation, since it uses more of the given data.

The function we are approximating is the Bessel function of the first kind of order zero, whose value at 1.5 is known to be 0.5118277, so the true accuracies of the approximations are as follows:

$$\begin{aligned}
|P_1(1.5) - f(1.5)| &\approx 1.5 \times 10^{-3}, \\
|P_2(1.5) - f(1.5)| &\approx 5.4 \times 10^{-4}, \\
|\hat{P}_2(1.5) - f(1.5)| &\approx 6.4 \times 10^{-4}, \\
|P_3(1.5) - f(1.5)| &\approx 2.5 \times 10^{-6}, \\
|\hat{P}_3(1.5) - f(1.5)| &\approx 1.5 \times 10^{-5}, \\
|P_4(1.5) - f(1.5)| &\approx 7.7 \times 10^{-6}.
\end{aligned}$$

Notice that $P_3(1.5)$ is the most accurate approximation. However, with no knowledge of the actual value of $f(1.5)$, we would accept $P_4(1.5)$ as the best approximation since it uses more of the given data. The error, or remainder, term derived in Theorem 3.3 cannot be applied here because we do not know the fourth derivative of f. Unfortunately, this is generally the case. ■

There are two immediate difficulties with using Lagrange polynomials, as illustrated in Example 3. First, the error term is difficult to apply, so the degree of the polynomial needed for the desired accuracy is generally not known until computations are determined. The usual practice is to compute the results given from various polynomials until appropriate agreement is obtained. In addition, the work done in calculating the approximation by the second polynomial does not lessen the work needed to calculate the third approximation; nor is the fourth approximation easier to obtain once the third approximation is known.

We will now derive these approximating polynomials in a manner that uses the previous calculations to greater advantage.

Definition 3.4 Let f be a function defined at $x_0, x_1, x_2, \ldots, x_n$, and suppose that $m_1, m_2, \ldots, m_k$ are k distinct integers with $0 \leq m_i \leq n$ for each i. The Lagrange polynomial that agrees with f at the k points $x_{m_1}, x_{m_2}, \ldots, x_{m_k}$ is denoted $P_{m_1,m_2,\ldots,m_k}(x)$. ■

EXAMPLE 4 If $x_0 = 1$, $x_1 = 2$, $x_2 = 3$, $x_3 = 4$, $x_4 = 6$, and $f(x) = x^3$, then $P_{1,2,4}(x)$ is the polynomial that agrees with f at $x_1 = 2$, $x_2 = 3$, and $x_4 = 6$; that is,

$$P_{1,2,4}(x) = \frac{(x-3)(x-6)}{(2-3)(2-6)}(8) + \frac{(x-2)(x-6)}{(3-2)(3-6)}(27) + \frac{(x-2)(x-3)}{(6-2)(6-3)}(216).$$ ■

The next result describes a method for recursively generating Lagrange polynomial approximations.

Theorem 3.5 Let f be defined at $x_0, x_1, \ldots, x_k$, and x_j and x_i be two distinct numbers in this set. Then

$$P(x) = \frac{(x - x_j)P_{0,1,\ldots,j-1,j+1,\ldots,k}(x) - (x - x_i)P_{0,1,\ldots,i-1,i+1,\ldots,k}(x)}{(x_i - x_j)}$$

describes the kth Lagrange polynomial that interpolates f at the $k+1$ points $x_0, x_1, \ldots, x_k$.

Proof For ease of notation, let $Q \equiv P_{0,1,\ldots,i-1,i+1,\ldots,k}$ and $\hat{Q} \equiv P_{0,1,\ldots,j-1,j+1,\ldots,k}$. Since $Q(x)$ and $\hat{Q}(x)$ are polynomials of degree $k-1$ or less, $P(x)$ is of degree at most k. If $0 \leq r \leq k$ and $r \neq i, j$, then $Q(x_r) = \hat{Q}(x_r) = f(x_r)$, so

$$P(x_r) = \frac{(x_r - x_j)\hat{Q}(x_r) - (x_r - x_i)Q(x_r)}{x_i - x_j} = \frac{(x_i - x_j)}{(x_i - x_j)}f(x_r) = f(x_r).$$

Moreover,

$$P(x_i) = \frac{(x_i - x_j)\hat{Q}(x_i) - (x_i - x_i)Q(x_i)}{x_i - x_j} = \frac{(x_i - x_j)}{(x_i - x_j)}f(x_i) = f(x_i)$$

and, similarly, $P(x_j) = f(x_j)$. But, by definition, $P_{0,1,\ldots,k}(x)$ is the unique polynomial of degree at most k which agrees with f at $x_0, x_1, \ldots, x_k$. Thus, $P \equiv P_{0,1,\ldots,k}$. ■ ■ ■

Theorem 3.5 implies that the interpolating polynomials can be generated recursively. For example, they can be generated in the manner shown in Table 3.4, where each row is completed before the succeeding rows are begun.

This procedure is called **Neville's method**. The P notation used in Table 3.4 is cumbersome because of the number of subscripts used to represent the entries. Note, however, that as an array is being constructed, only two subscripts are needed. Proceeding down the table corresponds to using consecutive points x_i with larger i, and proceeding to the right corresponds to increasing the degree of the interpolating polynomial. Since the points appear consecutively in each entry, we need to describe only a starting point and the number of additional points used in constructing the approximation.

Let $Q_{i,j}(x)$, for $0 \leq i \leq j$, denote the interpolating polynomial of degree j on the $j+1$ numbers $x_{i-j}, x_{i-j+1}, \ldots, x_{i-1}, x_i$; that is,

$$Q_{i,j} = P_{i-j,i-j+1,\ldots,i-1,i}.$$

Using this notation for Neville's method provides the Q notation array in Table 3.4.

Table 3.4

x_0	$P_0 = Q_{0,0}$				
x_1	$P_1 = Q_{1,0}$	$P_{0,1} = Q_{1,1}$			
x_2	$P_2 = Q_{2,0}$	$P_{1,2} = Q_{2,1}$	$P_{0,1,2} = Q_{2,2}$		
x_3	$P_3 = Q_{3,0}$	$P_{2,3} = Q_{3,1}$	$P_{1,2,3} = Q_{3,2}$	$P_{0,1,2,3} = Q_{3,3}$	
x_4	$P_4 = Q_{4,0}$	$P_{3,4} = Q_{4,1}$	$P_{2,3,4} = Q_{4,2}$	$P_{1,2,3,4} = Q_{4,3}$	$P_{0,1,2,3,4} = Q_{4,4}$

EXAMPLE 5 Values of various interpolating polynomials at $x = 1.5$ were obtained in Example 3 using the data shown in the first two columns of Table 3.5. In this example, we approximate $f(1.5)$ using the result in Theorem 3.5. If $x_0 = 1.0$, $x_1 = 1.3$, $x_2 = 1.6$, $x_3 = 1.9$, and $x_4 = 2.2$, then $Q_{0,0} = f(1.0)$, $Q_{1,0} = f(1.3)$, $Q_{2,0} = f(1.6)$, $Q_{3,0} = f(1.9)$, and $Q_{4,0} = f(2.2)$. These are the five polynomials of degree zero (constants) that approximate $f(1.5)$.

Calculating the first-degree approximation $Q_{1,1}(1.5)$ gives

$$\begin{aligned} Q_{1,1}(1.5) &= \frac{(x - x_0)Q_{1,0} - (x - x_1)Q_{0,0}}{x_1 - x_0} \\ &= \frac{(1.5 - 1.0)Q_{1,0} - (1.5 - 1.3)Q_{0,0}}{1.3 - 1.0} \\ &= \frac{0.5(0.6200860) - 0.2(0.7651977)}{0.3} = 0.5233449. \end{aligned}$$

Similarly,

$$Q_{2,1}(1.5) = \frac{(1.5 - 1.3)(0.4554022) - (1.5 - 1.6)(0.600860)}{1.6 - 1.3} = 0.5102968,$$

$$Q_{3,1}(1.5) = 0.5132634, \quad \text{and} \quad Q_{4,1}(1.5) = 0.5104270.$$

The best linear approximation is expected to be $Q_{2,1}$ since 1.5 is between $x_1 = 1.3$ and $x_2 = 1.6$.

In a similar manner, approximations using higher-degree polynomials are given by

$$Q_{2,2}(1.5) = \frac{(1.5 - 1.0)(0.5102968) - (1.5 - 1.6)(0.5233449)}{1.6 - 1.0} = 0.5124715,$$

$$Q_{3,2}(1.5) = 0.5112857, \quad \text{and} \quad Q_{4,2}(1.5) = 0.5137361.$$

The higher-degree approximations are generated in a similar manner and are shown in Table 3.5. ■

Table 3.5

1.0	0.7651977				
1.3	0.6200860	0.5233449			
1.6	0.4554022	0.5102968	0.5124715		
1.9	0.2818186	0.5132634	0.5112857	0.5118127	
2.2	0.1103623	0.5104270	0.5137361	0.5118302	0.5118200

If the latest approximation, $Q_{4,4}$, is not as accurate as desired, another node, x_5, can be selected, and another row added to the table:

$$x_5 \quad Q_{5,0} \quad Q_{5,1} \quad Q_{5,2} \quad Q_{5,3} \quad Q_{5,4} \quad Q_{5,5}.$$

Then $Q_{4,4}$, $Q_{5,4}$, and $Q_{5,5}$ can be compared to determine further accuracy.

In our example, the function is the Bessel function of the first kind of order zero, whose value at 2.5 is -0.0483838. Using this we can construct a new row of approximations to $f(1.5)$:

$$2.5 \quad -0.0483838 \quad 0.4807699 \quad 0.5301984 \quad 0.5119070 \quad 0.5118430 \quad 0.5118277.$$

The final new entry, 0.5118277, is actually correct to seven decimal places.

EXAMPLE 6 Table 3.6 lists the values of $f(x) = \ln x$ accurate to the places given

Table 3.6

i	x_i	$\ln x_i$
0	2.0	0.6931
1	2.2	0.7885
2	2.3	0.8329

We will use Neville's method to approximate $f(2.1) = \ln 2.1$. Completing the table gives the entries in Table 3.7.

Table 3.7

i	x_i	$x - x_i$	Q_{i0}	Q_{i1}	Q_{i2}
0	2.0	0.1	0.6931		
1	2.2	-0.1	0.7885	0.7410	
2	2.3	-0.2	0.8329	0.7441	0.7420

Thus, $P_2(2.1) = Q_{22} = 0.7420$. Since $f(2.1) = \ln 2.1 = 0.7419$ to four decimal places, the absolute error is

$$|f(2.1) - P_2(2.1)| = |0.7419 - 0.7420| = 10^{-4}.$$

However, $f'(x) = 1/x$, $f''(x) = -1/x^2$, and $f'''(x) = 2/x^3$, so the error formula (3.4) gives an error bound

$$|f(2.1) - P_2(2.1)| = \left|\frac{f'''(\xi)}{3!}(x - x_0)(x - x_1)(x - x_2)\right|$$

$$= \left|\frac{1}{3\xi^3}(0.1)(-0.1)(-0.2)\right| \leq 8.\overline{3} \times 10^{-5}.$$

Notice that the actual error, 10^{-4}, exceeds the error bound, $8.\overline{3} \times 10^{-5}$. This apparent contradiction is a consequence of finite-digit computations. We used four-digit approximations, and the error formula (3.4) assumes infinite-digit arithmetic. This is what caused our actual errors to exceed the theoretical error estimate. ■

Algorithm 3.1 constructs the entries in Neville's method by rows.

ALGORITHM 3.1

Neville's Iterated Interpolation

To evaluate the interpolating polynomial P on the $n + 1$ distinct numbers $x_0, \ldots, x_n$ at the number x for the function f:

INPUT numbers $x, x_0, x_1, \ldots, x_n$; values $f(x_0), f(x_1), \ldots, f(x_n)$ as the first column $Q_{0,0}, Q_{1,0}, \ldots, Q_{n,0}$ of Q.

OUTPUT the table Q with $P(x) = Q_{n,n}$.

Step 1 For $i = 1, 2, \ldots, n$
for $j = 1, 2, \ldots, i$

$$\text{set } Q_{i,j} = \frac{(x - x_{i-j})Q_{i,j-1} - (x - x_i)Q_{i-1,j-1}}{x_i - x_{i-j}}.$$

Step 2 OUTPUT (Q);
STOP.

The algorithm can be modified to allow for the addition of new interpolating nodes. For example, the inequality

$$|Q_{i,i} - Q_{i-1,i-1}| < \varepsilon$$

can be used as a stopping criterion, where ε is a prescribed error tolerance. If the inequality is true, $Q_{i,i}$ is a reasonable approximation to $f(x)$. If the inequality is false, a new interpolation point, x_{i+1}, is added.

EXERCISE SET 3.1

1. For the given functions $f(x)$, let $x_0 = 0$, $x_1 = 0.6$, and $x_2 = 0.9$. Construct interpolation polynomials of degree at most one and at most two to approximate $f(0.45)$, and find the actual error.

a. $f(x) = \cos x$ **b.** $f(x) = \sqrt{1+x}$

c. $f(x) = \ln(x+1)$ **d.** $f(x) = \tan x$

2. Use Theorem 3.3 to find an error bound for the approximations in Exercise 1.

3. Use appropriate Lagrange interpolating polynomials of degrees one, two, and three to approximate each of the following:

a. $f(8.4)$ if $f(8.1) = 16.94410$, $f(8.3) = 17.56492$, $f(8.6) = 18.50515$, $f(8.7) = 18.82091$

b. $f\left(-\frac{1}{3}\right)$ if $f(-0.75) = -0.07181250$, $f(-0.5) = -0.02475000$, $f(-0.25) = 0.33493750$, $f(0) = 1.10100000$

c. $f(0.25)$ if $f(0.1) = -0.62049958$, $f(0.2) = -0.28398668$, $f(0.3) = 0.00660095$, $f(0.4) = 0.24842440$

d. $f(0.9)$ if $f(0.6) = -0.17694460$, $f(0.7) = 0.01375227$, $f(0.8) = 0.22363362$, $f(1.0) = 0.65809197$

4. Use Neville's method to obtain the approximations for Exercise 3.

5. Use Neville's method to approximate $\sqrt{3}$ with the function $f(x) = 3^x$ and the values $x_0 = -2$, $x_1 = -1$, $x_2 = 0$, $x_3 = 1$, and $x_4 = 2$.

6. Use Neville's method to approximate $\sqrt{3}$ with the function $f(x) = \sqrt{x}$ and the values $x_0 = 0$, $x_1 = 1$, $x_2 = 2$, $x_3 = 4$, and $x_4 = 5$. Compare the accuracy with that of Exercise 5.

7. The data for Exercise 3 were generated using the following functions. Use the error formula to find a bound for the error and compare the bound to the actual error for the cases $n = 1$ and $n = 2$.

a. $f(x) = x \ln x$

b. $f(x) = x^3 + 4.001x^2 + 4.002x + 1.101$

c. $f(x) = x \cos x - 2x^2 + 3x - 1$

d. $f(x) = \sin(e^x - 2)$

8. Let $f(x) = \sqrt{x - x^2}$ and $P_2(x)$ be the interpolation polynomial on $x_0 = 0$, x_1 and $x_2 = 1$. Find the largest value of x_1 in $(0, 1)$ for which $f(0.5) - P_2(0.5) = -0.25$.

9. Let $P_3(x)$ be the interpolating polynomial for the data $(0, 0)$, $(0.5, y)$, $(1, 3)$, and $(2, 2)$. Find y if the coefficient of x^3 in $P_3(x)$ is 6.

10. Use the Lagrange interpolating polynomial of degree three or less and four-digit chopping arithmetic to approximate $\cos 0.750$ using the following values. Find an error bound for the approximation.

$$\cos 0.698 = 0.7661 \quad \cos 0.733 = 0.7432 \quad \cos 0.768 = 0.7193 \quad \cos 0.803 = 0.6946$$

The actual value of $\cos 0.750$ is 0.7317 (to four decimal places). Explain the discrepancy between the actual error and the error bound.

11. Use the following values and four-digit rounding arithmetic to construct a third Lagrange polynomial approximation to $f(1.09)$. The function being approximated is $f(x) = \log_{10}(\tan x)$. Use this knowledge to find a bound for the error in the approximation.

$$f(1.00) = 0.1924 \quad f(1.05) = 0.2414 \quad f(1.10) = 0.2933 \quad f(1.15) = 0.3492$$

12. Repeat Exercise 11 using Maple and ten-digit rounding arithmetic.

13. Neville's method is used to approximate $f(0.5)$, giving the following table.

$x_0 = 0$	$P_0 = 0$		
$x_1 = 0.4$	$P_1 = 2.8$	$P_{01} = 3.5$	
$x_2 = 0.7$	P_2	P_{12}	$P_{012} = \frac{27}{7}$

Determine $P_2 = f(0.7)$.

14. Neville's method is used to approximate $f(0.4)$, giving the following table.

$x_0 = 0$	$P_0 = 1$			
$x_1 = 0.25$	$P_1 = 2$	$P_{01} = 2.6$		
$x_2 = 0.5$	P_2	P_{12}	P_{012}	
$x_3 = 0.75$	$P_3 = 8$	$P_{23} = 2.4$	$P_{123} = 2.96$	$P_{0123} = 3.016$

Determine $P_2 = f(0.5)$.

15. Construct the Lagrange interpolating polynomials for the following functions, and find a bound for the absolute error on the interval $[x_0, x_n]$.

a. $f(x) = e^{2x} \cos 3x, \quad x_0 = 0, x_1 = 0.3, x_2 = 0.6, n = 2$

b. $f(x) = \sin(\ln x), \quad x_0 = 2.0, x_1, = 2.4, x_2 = 2.6, n = 2$

c. $f(x) = \ln x, \quad x_0 = 1, x_1 = 1.1, x_2 = 1.3, x_3 = 1.4, n = 3$

d. $f(x) = \cos x + \sin x, \quad x_0 = 0, x_1 = 0.25, x_2 = 0.5, x_3 = 1.0, n = 3$

16. Let $f(x) = e^x$, for $0 \leq x \leq 2$.

a. Approximate $f(0.25)$ using linear interpolation with $x_0 = 0$ and $x_1 = 0.5$.

b. Approximate $f(0.75)$ using linear interpolation with $x_0 = 0.5$ and $x_1 = 1$.

c. Approximate $f(0.25)$ and $f(0.75)$ using the second interpolating polynomial with $x_0 = 0$, $x_1 = 1$, and $x_2 = 2$.

d. Which approximations are better and why?

17. Suppose you need to construct eight-decimal-place tables for the common, or base-10, logarithm function from $x = 1$ to $x = 10$ in such a way that linear interpolation is accurate to within 10^{-6}. Determine a bound for the step size for this table. What choice of step size would you make to ensure that $x = 10$ is included in the table?

18. Suppose $x_j = j$ for $j = 0, 1, 2, 3$ and it is known that

$$P_{0,1}(x) = x + 1, \quad P_{1,2}(x) = 3x - 1, \quad \text{and} \quad P_{1,2,3}(1.5) = 4.$$

Find $P_{0,1,2,3}(1.5)$.

19. Suppose $x_j = j$ for $j = 0, 1, 2, 3$ and it is known that

$$P_{0,1}(x) = 2x + 1, \quad P_{0,2}(x) = x + 1, \quad \text{and} \quad P_{1,2,3}(2.5) = 3.$$

Find $P_{0,1,2,3}(2.5)$.

20. Neville's Algorithm is used to approximate $f(0)$ using $f(-2)$, $f(-1)$, $f(1)$, and $f(2)$. Suppose $f(-1)$ was overstated by 2 and $f(1)$ was understated by 3. Determine the error in the original calculation of the value of the interpolating polynomial to approximate $f(0)$.

21. Construct a sequence of interpolating values y_n to $f(1 + \sqrt{10})$, where $f(x) = (1 + x^2)^{-1}$ for $-5 \leq x \leq 5$, as follows: For each $n = 1, 2, \ldots, 10$, let $h = 10/n$ and $y_n = P_n(1+\sqrt{10})$, where $P_n(x)$ is the interpolating polynomial for $f(x)$ at the nodes $x_0^{(n)}, x_1^{(n)}, \ldots, x_n^{(n)}$ and $x_j^{(n)} = -5 + jh$ for each $j = 0, 1, 2, \ldots, n$. Does the sequence $\{y_n\}$ appear to converge to $f(1 + \sqrt{10})$?

Inverse Interpolation Suppose $f \in C^1[a, b]$, $f'(x) \neq 0$ on $[a, b]$ and f has one zero p in $[a, b]$. Let $x_0, \ldots, x_n$, be $n + 1$ distinct numbers in $[a, b]$ with $f(x_k) = y_k$ for each $k = 0, 1, \ldots, n$. To approximate p construct the interpolating polynomial of degree n on the nodes $y_0, \ldots, y_n$ for f^{-1}. Since $y_k = f(x_k)$ and $0 = f(p)$, it follows that $f^{-1}(y_k) = x_k$ and $p = f^{-1}(0)$. Using iterated interpolation to approximate $f^{-1}(0)$ is called *iterated inverse interpolation*.

22. Use iterated inverse interpolation to find an approximation to the solution of $x - e^{-x} = 0$, using the data

x	0.3	0.4	0.5	0.6
e^{-x}	0.740818	0.670320	0.606531	0.548812

23. Construct an algorithm that can be used for inverse interpolation.

24. **a.** The introduction to this chapter included a table listing the population of the United States from 1940 to 1990. Use Lagrange interpolation to approximate the population in the years 1930, 1965, and 2000.

b. The population in 1930 was approximately 123,203,000. How accurate do you think your 1965 and 2000 figures are?

25. It is suspected that the high amounts of tannin in mature oak leaves inhibit the growth of the winter moth (*Operophtera bromata L., Geometridae*) larvae that extensively damage these trees in certain years. The following table lists the average weight of two samples of larvae at times in the first 28 days after birth. The first sample was reared on young oak leaves, whereas the second sample was reared on mature leaves from the same tree.

a. Use Lagrange interpolation to approximate the average weight curve for each sample.

b. Find an approximate maximum average weight for each sample by determining the maximum of the interpolating polynomial.

Day	0	6	10	13	17	20	28
Sample 1 average weight (mg)	6.67	17.33	42.67	37.33	30.10	29.31	28.74
Sample 2 average weight (mg)	6.67	16.11	18.89	15.00	10.56	9.44	8.89

26. In Exercise 24 of Section 1.1 a Maclaurin series was integrated to approximate erf(1), where erf(x) is the normal distribution error function defined by

$$\operatorname{erf}(x) = \frac{2}{\sqrt{\pi}} \int_0^x e^{-t^2}\, dt.$$

a. Use the Maclaurin series to construct a table for erf(x) that is accurate to within 10^{-4} for erf(x_i), where $x_i = 0.2i$, for $i = 0, 1, \ldots, 5$.

b. Use both linear interpolation and quadratic interpolation to obtain an approximation to erf($\frac{1}{3}$). Which approach seems most feasible?

27. Prove Theorem 1.14 by following the procedure in the proof of Theorem 3.3. [*Hint*: Let

$$g(t) = f(t) - P(t) - [f(x) - P(x)] \cdot \frac{(t - x_0)^{n+1}}{(x - x_0)^{n+1}},$$

where P is the nth Taylor polynomial, and use Theorem 1.12.]

28. Show that $\max_{x_j \le x \le x_{j+1}} |g(x)| = h^2/4$, where $g(x) = (x - jh)(x - (j+1)h)$.

29. The Bernstein polynomial of degree n for $f \in C[0, 1]$ is given by

$$B_n(x) = \sum_{k=0}^{n} \binom{n}{k} f\left(\frac{k}{n}\right) x^k (1 - x)^{n-k},$$

where $\binom{n}{k}$ denotes $\dfrac{n!}{k!\,(n-k)!}$. These polynomials can be used in a constructive proof of the Weierstrass Approximation Theorem 3.1 (see [Bart]) since $\lim_{n\to\infty} \max_{x\in[0,1]} |B_n(x) - f(x)| = 0$.

a. Find $B_3(x)$ for the functions

(i) $f(x) = x$ **(ii)** $f(x) = 1$

b. Show that for each $k \leq n$,

$$\binom{n-1}{k-1} = \left(\frac{k}{n}\right)\binom{n}{k}.$$

c. Use part (b) and the fact, from (ii) in part (a), that

$$1 = \sum_{k=0}^{n} \binom{n}{k} x^k (1-x)^{n-k}, \quad \text{for each } n,$$

to show that, for $f(x) = x^2$,

$$B_n(x) = \left(\frac{n-1}{n}\right) x^2 + \frac{1}{n}x.$$

d. Use part (c) to estimate the value of n necessary for $\left|B_n(x) - x^2\right| \leq 10^{-6}$ to hold for all x in $[0, 1]$.

3.2 Divided Differences

Iterated interpolation was used in the previous section to generate successively higher-degree polynomial approximations at a specific point. Divided-difference methods introduced in this section are used to successively generate the polynomials themselves. Our treatment of divided-difference methods will be brief, since the results in this section will not be used extensively in subsequent material. Most older texts on numerical analysis have extensive treatments of divided-difference methods. If a more comprehensive treatment is needed, the book by Hildebrand [Hild] is a particularly good reference.

Suppose that $P_n(x)$ is the nth Lagrange polynomial that agrees with the function f at the distinct numbers $x_0, x_1, \ldots, x_n$. The divided differences of f with respect to $x_0, x_1, \ldots, x_n$ are derived to express $P_n(x)$ in the form

(3.6)
$$P_n(x) = a_0 + a_1(x - x_0) + a_2(x - x_0)(x - x_1) + \cdots + a_n(x - x_0)(x - x_1)\cdots(x - x_{n-1})$$

for appropriate constants $a_0, a_1, \ldots, a_n$.

To determine the first of these constants, a_0, note that if $P_n(x)$ is written in the form of Eq. (3.6), then evaluating $P_n(x)$ at x_0 leaves only the constant term a_0; that is,

$$a_0 = P_n(x_0) = f(x_0).$$

Similarly, when $P(x)$ is evaluated at x_1, the only nonzero terms in the evaluation of $P_n(x_1)$ are the constant and linear terms,

$$f(x_0) + a_1(x_1 - x_0) = P_n(x_1) = f(x_1);$$

so

$$a_1 = \frac{f(x_1) - f(x_0)}{x_1 - x_0}. \tag{3.7}$$

We now need to introduce the divided-difference notation, which is reminiscent of the Aitken's Δ^2 notation used in Section 2.5. The **zeroth divided difference** of the function f with respect to x_i, denoted $f[x_i]$, is simply the value of f at x_i:

$$f[x_i] = f(x_i). \tag{3.8}$$

The remaining divided differences are defined inductively; the first divided difference of f with respect to x_i and x_{i+1} is denoted $f[x_i, x_{i+1}]$ and defined as

$$f[x_i, x_{i+1}] = \frac{f[x_{i+1}] - f[x_i]}{x_{i+1} - x_i}. \tag{3.9}$$

The second divided difference $f[x_i, x_{i+1}, x_{i+2}]$ is defined as

$$f[x_i, x_{i+1}, x_{i+2}] = \frac{f[x_{i+1}, x_{i+2}] - f[x_i, x_{i+1}]}{x_{i+2} - x_i}.$$

Similarly, after the $(k-1)$st divided differences,

$$f[x_i, x_{i+1}, x_{i+2}, \ldots, x_{i+k-1}] \quad \text{and} \quad f[x_{i+1}, x_{i+2}, \ldots, x_{i+k-1}, x_{i+k}],$$

have been determined, the **kth divided difference** relative to $x_i, x_{i+1}, x_{i+2}, \ldots, x_{i+k}$ is given by

$$f[x_i, x_{i+1}, \ldots, x_{i+k-1}, x_{i+k}] = \frac{f[x_{i+1}, x_{i+2}, \ldots, x_{i+k}] - f[x_i, x_{i+1}, \ldots, x_{i+k-1}]}{x_{i+k} - x_i}. \tag{3.10}$$

With this notation, Eq. (3.7) can be reexpressed as $a_1 = f[x_0, x_1]$, and the interpolating polynomial in Eq. (3.6) is

$$\begin{aligned} P_n(x) = f[x_0] + f[x_0, x_1](x - x_0) + a_2(x - x_0)(x - x_1) \\ + \cdots + a_n(x - x_0)(x - x_1) \cdots (x - x_{n-1}). \end{aligned}$$

As might be expected from the evaluation of a_0 and a_1, the required constants are

$$a_k = f[x_0, x_1, x_2, \ldots, x_k],$$

for each $k = 0, 1, \ldots, n$. So $P_n(x)$ can be rewritten as (see [Hild, pp. 43–47])

(3.11)
$$P_n(x) = f[x_0] + \sum_{k=1}^{n} f[x_0, x_1, \ldots, x_k](x - x_0)\cdots(x - x_{k-1}).$$

The value of $f[x_0, x_1, \ldots, x_k]$ is independent of the order of the numbers $x_0, x_1, \ldots, x_k$ as is shown in Exercise 17. This equation is known as **Newton's interpolatory divided-difference formula**. The determination of the divided differences from tabulated data points is outlined in Table 3.8. Two fourth and one fifth difference could also be determined from these data.

Table 3.8

x	$f(x)$	First divided differences	Second divided differences	Third divided differences
x_0	$f[x_0]$			
		$f[x_0, x_1] = \dfrac{f[x_1] - f[x_0]}{x_1 - x_0}$		
x_1	$f[x_1]$		$f[x_0, x_1, x_2] = \dfrac{f[x_1, x_2] - f[x_0, x_1]}{x_2 - x_0}$	
		$f[x_1, x_2] = \dfrac{f[x_2] - f[x_1]}{x_2 - x_1}$		$f[x_0, x_1, x_2, x_3] = \dfrac{f[x_1, x_2, x_3] - f[x_0, x_1, x_2]}{x_3 - x_0}$
x_2	$f[x_2]$		$f[x_1, x_2, x_3] = \dfrac{f[x_2, x_3] - f[x_1, x_2]}{x_3 - x_1}$	
		$f[x_2, x_3] = \dfrac{f[x_3] - f[x_2]}{x_3 - x_2}$		$f[x_1, x_2, x_3, x_4] = \dfrac{f[x_2, x_3, x_4] - f[x_1, x_2, x_3]}{x_4 - x_1}$
x_3	$f[x_3]$		$f[x_2, x_3, x_4] = \dfrac{f[x_3, x_4] - f[x_2, x_3]}{x_4 - x_2}$	
		$f[x_3, x_4] = \dfrac{f[x_4] - f[x_3]}{x_4 - x_3}$		$f[x_2, x_3, x_4, x_5] = \dfrac{f[x_3, x_4, x_5] - f[x_2, x_3, x_4]}{x_5 - x_2}$
x_4	$f[x_4]$		$f[x_3, x_4, x_5] = \dfrac{f[x_4, x_5] - f[x_3, x_4]}{x_5 - x_3}$	
		$f[x_4, x_5] = \dfrac{f[x_5] - f[x_4]}{x_5 - x_4}$		
x_5	$f[x_5]$			

Newton's interpolatory divided-difference formula can be implemented using Algorithm 3.2. The form of the output can be modified to produce all the divided differences, as done in Example 1.

ALGORITHM **3.2**

Newton's Interpolatory Divided-Difference Formula

To obtain the divided-difference coefficients of the interpolatory polynomial P on the $(n+1)$ distinct numbers $x_0, x_1, \ldots, x_n$ for the function f:

INPUT numbers $x_0, x_1, \ldots, x_n$; values $f(x_0), f(x_1), \ldots, f(x_n)$ as $F_{0,0}, F_{1,0}, \ldots, F_{n,0}$.

OUTPUT the numbers $F_{0,0}, F_{1,1}, \ldots, F_{n,n}$ where

$$P(x) = \sum_{i=0}^{n} F_{i,i} \prod_{j=0}^{i-1}(x - x_j).$$

Step 1 For $i = 1, 2, \ldots, n$
For $j = 1, 2, \ldots, i$
set $F_{i,j} = \dfrac{F_{i,j-1} - F_{i-1,j-1}}{x_i - x_{i-j}}$.

Step 2 OUTPUT $(F_{0,0}, F_{1,1}, \ldots, F_{n,n})$; ($F_{i,i}$ *is* $f[x_0, x_1, \ldots, x_i]$.)
STOP.

EXAMPLE 1 In Example 3 of Section 3.1, various interpolating polynomials were used to approximate $f(1.5)$, using the data in the first three columns of Table 3.9. The remaining entries of Table 3.9 contain divided differences computed using Algorithm 3.2.

The coefficients of the Newton forward divided-difference form of the interpolatory polynomial are along the diagonal in the table. The polynomial is

$$\begin{aligned} P_4(x) = {} & 0.7651977 - 0.4837057(x - 1.0) - 0.1087339(x - 1.0)(x - 1.3) \\ & + 0.0658784(x - 1.0)(x - 1.3)(x - 1.6) \\ & + 0.0018251(x - 1.0)(x - 1.3)(x - 1.6)(x - 1.9). \end{aligned}$$

Notice that the value $P_4(1.5) = 0.5118200$ agrees with the result in Section 3.1, Example 3, as it must because the polynomials are the same. ■

Table 3.9

i	x_i	$f[x_i]$	$f[x_{i-1}, x_i]$	$f[x_{i-2}, x_{i-1}, x_i]$	$f[x_{i-3}, \ldots, x_i]$	$f[x_{i-4}, \ldots, x_i]$
0	1.0	0.7651977				
			−0.4837057			
1	1.3	0.6200860		−0.1087339		
			−0.5489460		0.0658784	
2	1.6	0.4554022		−0.0494433		0.0018251
			−0.5786120		0.0680685	
3	1.9	0.2818186		0.0118183		
			−0.5715210			
4	2.2	0.1103623				

The Mean Value Theorem applied to Eq. (3.9) when $i = 0$,

$$f[x_0, x_1] = \frac{f(x_1) - f(x_0)}{x_1 - x_0},$$

implies that when f' exists, $f[x_0, x_1] = f'(\xi)$ for some number ξ between x_0 and x_1. The following theorem generalizes this result.

Theorem 3.6 Suppose that $f \in C^n[a, b]$ and $x_0, x_1, \ldots, x_n$ are distinct numbers in $[a, b]$. Then a number ξ exists in (a, b) with

$$f[x_0, x_1, \ldots, x_n] = \frac{f^{(n)}(\xi)}{n!}.$$

Proof Let

$$g(x) = f(x) - P_n(x).$$

Since $f(x_i) = P_n(x_i)$ for each $i = 0, 1, \ldots, n$, the function g has $n+1$ distinct zeros in $[a, b]$. The Generalized Rolle's Theorem implies that a number ξ in (a, b) exists with $g^{(n)}(\xi) = 0$, so

$$0 = f^{(n)}(\xi) - P_n^{(n)}(\xi).$$

Since $P_n(x)$ is a polynomial of degree n whose leading coefficient is $f[x_0, x_1, \ldots, x_n]$,

$$P_n^{(n)}(x) = f[x_0, x_1, \ldots, x_n] \cdot n!$$

As a consequence,

$$f[x_0, x_1, \ldots, x_n] = \frac{f^{(n)}(\xi)}{n!}.$$

■ ■ ■

Newton's interpolatory divided-difference formula can be expressed in a simplified form when $x_0, x_1, \ldots, x_n$ are arranged consecutively with equal spacing. Introducing the notation $h = x_{i+1} - x_i$ for each $i = 0, 1, \ldots, n-1$ and $x = x_0 + sh$, the difference $x - x_i$ can be written as $x - x_i = (s - i)h$. So Eq. (3.11) becomes

$$\begin{aligned} P_n(x) = P_n(x_0 + sh) &= f[x_0] + shf[x_0, x_1] + s(s-1)h^2 f[x_0, x_1, x_2] \\ &\quad + \cdots + s(s-1)(s-n+1)h^n f[x_0, x_1, \ldots, x_n] \\ &= \sum_{k=0}^{n} s(s-1)\cdots(s-k+1)h^k f[x_0, x_1, \ldots, x_k]. \end{aligned}$$

Using binomial-coefficient notation,

$$\binom{s}{k} = \frac{s(s-1)\cdots(s-k+1)}{k!},$$

we can express $P_n(x)$ compactly as

(3.12) $$P_n(x) = P_n(x_0 + sh) = \sum_{k=0}^{n} \binom{s}{k} k!\, h^k f[x_0, x_1, \ldots, x_k].$$

This formula is called the **Newton forward divided-difference formula**. Another form, called the **Newton forward-difference formula**, is constructed by making use of the forward difference notation Δ introduced in Aitken's Δ^2 method. With this notation,

$$f[x_0, x_1] = \frac{f(x_1) - f(x_0)}{x_1 - x_0} = \frac{1}{h}\Delta f(x_0)$$

$$f[x_0, x_1, x_2] = \frac{1}{2h}\left[\frac{\Delta f(x_1) - \Delta f(x_0)}{h}\right] = \frac{1}{2h^2}\Delta^2 f(x_0),$$

and, in general,

$$f[x_0, x_1, \ldots, x_k] = \frac{1}{k!\,h^k}\Delta^k f(x_0).$$

Then, Eq. (3.12) has the following formula.

Newton Forward-Difference Formula

(3.13)
$$P_n(x) = \sum_{k=0}^{n} \binom{s}{k} \Delta^k f(x_0)$$

If the interpolating nodes are reordered as $x_n, x_{n-1}, \ldots, x_0$, a formula similar to Eq. (3.11) results:

$$\begin{aligned} P_n(x) = {} & f[x_n] + f[x_n, x_{n-1}](x - x_n) + f[x_n, x_{n-1}, x_{n-2}](x - x_n)(x - x_{n-1}) \\ & + \cdots + f[x_n, \ldots, x_0](x - x_n)(x - x_{n-1}) \cdots (x - x_1). \end{aligned}$$

If the nodes are equally spaced with $x = x_n + sh$ and $x_i = x_n - (n - i)h$ for each $i = 0, 1, \ldots, n$, then

$$\begin{aligned} P_n(x) &= P_n(x_n + sh) \\ &= f[x_n] + shf[x_n, x_{n-1}] + s(s+1)h^2 f[x_n, x_{n-1}, x_{n-2}] + \cdots \\ &\quad + s(s+1)\cdots(s+n-1)h^n f[x_n, \ldots, x_0]. \end{aligned}$$

This form is called the **Newton backward divided-difference formula**. It is used to derive a more commonly applied formula known as the **Newton backward-difference formula**. To discuss this formula, we need the following definition.

Definition 3.7 Given the sequence $\{p_n\}_{n=0}^{\infty}$, define the backward difference ∇p_n (read *nabla* p_n) by

$$\nabla p_n = p_n - p_{n-1}, \quad \text{for } n \geq 1.$$

Higher powers are defined recursively by

$$\nabla^k p_n = \nabla(\nabla^{k-1} p_n), \quad \text{for } k \geq 2.$$

■

Definition 3.7 implies that

$$f[x_n, x_{n-1}] = \frac{1}{h}\nabla f(x_n), \quad f[x_n, x_{n-1}, x_{n-2}] = \frac{1}{2h^2}\nabla^2 f(x_n),$$

and, in general,

$$f[x_n, x_{n-1}, \ldots, x_{n-k}] = \frac{1}{k!\,h^k}\nabla^k f(x_n).$$

Consequently,

$$P_n(x) = f[x_n] + s\nabla f(x_n) + \frac{s(s+1)}{2}\nabla^2 f(x_n) + \cdots + \frac{s(s+1)\cdots(s+n-1)}{n!}\nabla^n f(x_n).$$

We extend the binomial coefficient notation to include all real values of s by letting

$$\binom{-s}{k} = \frac{-s(-s-1)\cdots(-s-k+1)}{k!} = (-1)^k\frac{s(s+1)\cdots(s+k-1)}{k!}.$$

Then

$$P_n(x) = f(x_n)+(-1)^1\binom{-s}{1}\nabla f(x_n)+(-1)^2\binom{-s}{2}\nabla^2 f(x_n)+\cdots+(-1)^n\binom{-s}{n}\nabla^n f(x_n).$$

This gives the following result.

Newton Backward–Difference Formula

(3.14)
$$P_n(x) = \sum_{k=0}^{n}(-1)^k\binom{-s}{k}\nabla^k f(x_n)$$

EXAMPLE 2 The divided-difference Table 3.10 corresponds to the data in Example 1.

Table 3.10

		First divided differences	Second divided differences	Third divided differences	Fourth divided differences
1.0	0.7651977				
		−0.4837057			
1.3	0.6200860		−0.1087339		
		−0.5489460		0.0658784	
1.6	0.4554022		−0.0494433		0.0018251
		−0.5786120		0.0680685	
1.9	0.2818186		0.0118183		
		−0.5715210			
2.2	0.1103623				

If an approximation to $f(1.1)$ is required, the reasonable choice for the nodes would be $x_0 = 1.0$, $x_1 = 1.3$, $x_2 = 1.6$, $x_3 = 1.9$, and $x_4 = 2.2$, since this choice makes the earliest possible use of the data points closest to $x = 1.1$, and also makes use of the fourth divided difference. This implies that $h = 0.3$ and $s = \frac{1}{3}$, so the Newton forward divided-difference formula is used with the divided differences that have a *solid* underscore in Table 3.10:

$$
\begin{aligned}
P_4(1.1) = P_4\left(1.0 + \frac{1}{3}(0.3)\right)
&= 0.7651997 + \frac{1}{3}(0.3)(-0.4837057) + \frac{1}{3}\left(-\frac{2}{3}\right)(0.3)^2(-0.1087339) \\
&\quad + \frac{1}{3}\left(-\frac{2}{3}\right)\left(-\frac{5}{3}\right)(0.3)^3(0.0658784) \\
&\quad + \frac{1}{3}\left(-\frac{2}{3}\right)\left(-\frac{5}{3}\right)\left(-\frac{8}{3}\right)(0.3)^4(0.0018251) \\
&= 0.7196480.
\end{aligned}
$$

To approximate a value when x is close to the end of the tabulated values, say, $x = 2.0$, we would again like to make the earliest use of the data points closest to x. This requires using the Newton backward divided-difference formula with $s = -\frac{2}{3}$ and the divided differences in Table 3.10 that have a *dashed* underscore:

$$
\begin{aligned}
P_4(2.0) = P_4\left(2.2 - \frac{2}{3}(0.3)\right)
&= 0.1103623 - \frac{2}{3}(0.3)(-0.5715210) - \frac{2}{3}\left(\frac{1}{3}\right)(0.3)^2(0.0118183) \\
&\quad - \frac{2}{3}\left(\frac{1}{3}\right)\left(\frac{4}{3}\right)(0.3)^3(0.0680685) - \frac{2}{3}\left(\frac{1}{3}\right)\left(\frac{4}{3}\right)\left(\frac{7}{3}\right)(0.3)^4(0.0018251) \\
&= 0.2238754.
\end{aligned}
$$

■

The Newton formulas are not appropriate for approximating a value x that lies near the center of the table since employing either the backward or forward method in such a way that the highest-order difference is involved will not allow x_0 to be close to x. A number of divided-difference formulas are available in this instance, each of which has situations when it can be used to maximum advantage. These methods are known as **centered-difference formulas**. There are a number of such methods, but we will present only one, Stirling's method, and again refer the interested reader to [Hild] for a complete presentation.

For the centered-difference formulas we choose x_0 near the point being approximated and label the nodes directly below x_0 as $x_1, x_2, \ldots$ and those directly above as $x_{-1}, x_{-2}, \ldots$. With this convention, **Stirling's formula** is given by

(3.15)
$$
\begin{aligned}
P_n(x) = P_{2m+1}(x) &= f[x_0] + \frac{sh}{2}(f[x_{-1}, x_0] + f[x_0, x_1]) + s^2h^2 f[x_{-1}, x_0, x_1] \\
&\quad + \frac{s(s^2-1)h^3}{2}(f[x_{-2}, x_{-1}, x_0, x_1] + f[x_{-1}, x_0, x_1, x_2]) \\
&\quad + \cdots + s^2(s^2-1)(s^2-4)\cdots(s^2-(m-1)^2)h^{2m} f[x_{-m}, \ldots, x_m] \\
&\quad + \frac{s(s^2-1)\cdots(s^2-m^2)h^{2m+1}}{2}(f[x_{-m-1}, \ldots, x_m] + f[x_{-m}, \ldots, x_{m+1}])
\end{aligned}
$$

if $n = 2m + 1$ is odd. If $n = 2m$ is even, we use the same formula but delete the last line. The entries used for this formula are underscored in Table 3.11.

Table 3.11

x	$f(x)$	First divided differences	Second divided differences	Third divided differences	Fourth divided differences
x_{-2}	$f[x_{-2}]$				
		$f[x_{-2}, x_{-1}]$			
x_{-1}	$f[x_{-1}]$		$f[x_{-2}, x_{-1}, x_0]$		
		$\underline{f[x_{-1}, x_0]}$		$\underline{f[x_{-2}, x_{-1}, x_0, x_1]}$	
x_0	$\underline{f[x_0]}$		$\underline{f[x_{-1}, x_0, x_1]}$		$\underline{f[x_{-2}, x_{-1}, x_0, x_1, x_2]}$
		$\underline{f[x_0, x_1]}$		$\underline{f[x_{-1}, x_0, x_1, x_2]}$	
x_1	$f[x_1]$		$f[x_0, x_1, x_2]$		
		$f[x_1, x_2]$			
x_2	$f[x_2]$				

EXAMPLE 3 Consider the table of data that was given in the previous examples. To use Stirling's formula to approximate $f(1.5)$ with $x_0 = 1.6$, we use the underscored entries in the difference Table 3.12.

Table 3.12

x	$f(x)$	First divided differences	Second divided differences	Third divided differences	Fourth divided differences
1.0	0.7651977				
		−0.4837057			
1.3	0.6200860		−0.1087339		
		$\underline{-0.5489460}$		$\underline{0.0658784}$	
1.6	$\underline{0.4554022}$		$\underline{-0.0494433}$		$\underline{0.0018251}$
		$\underline{-0.5786120}$		$\underline{0.0680685}$	
1.9	0.2818186		0.0118183		
		−0.5715210			
2.2	0.1103623				

The formula, with $h = 0.3$, $x_0 = 1.6$, and $s = -\frac{1}{3}$, becomes

$$f(1.5) \approx P_4\left(1.6 + \left(-\frac{1}{3}\right)(0.3)\right)$$

$$= 0.4554022 + \left(-\frac{1}{3}\right)\left(\frac{0.3}{2}\right)((-0.5489460) - (0.5786120))$$

$$+ \left(-\frac{1}{3}\right)^2 (0.3)^2(-0.0494433)$$

$$+\frac{1}{2}\left(-\frac{1}{3}\right)\left(\left(-\frac{1}{3}\right)^2-1\right)(0.3)^3(0.0658784+0.0680685)$$

$$+\left(-\frac{1}{3}\right)^2\left(\left(-\frac{1}{3}\right)^2-1\right)(0.3)^4(0.0018251)$$

$$=0.5118200.$$

■

EXERCISE SET 3.2

1. Use Newton's interpolatory divided-difference formula or Algorithm 3.2 to construct interpolating polynomials of degree one, two, and three for the following data. Approximate the specified value using each of the polynomials.

a. $f(8.4)$ if $f(8.1) = 16.94410$, $f(8.3) = 17.56492$, $f(8.6) = 18.50515$, $f(8.7) = 18.82091$

b. $f(0.9)$ if $f(0.6) = -0.17694460$, $f(0.7) = 0.01375227$, $f(0.8) = 0.22363362$, $f(1.0) = 0.65809197$

2. Use Newton's forward-difference formula to construct interpolating polynomials of degree one, two, and three for the following data. Approximate the specified value using each of the polynomials.

a. $f\left(-\frac{1}{3}\right)$ if $f(-0.75) = -0.07181250$, $f(-0.5) = -0.02475000$, $f(-0.25) = 0.33493750$, $f(0) = 1.10100000$

b. $f(0.25)$ if $f(0.1) = -0.62049958$, $f(0.2) = -0.28398668$, $f(0.3) = 0.00660095$, $f(0.4) = 0.24842440$

3. Use Newton's backward-difference formula to construct interpolating polynomials of degree one, two, and three for the following data. Approximate the specified value using each of the polynomials.

a. $f\left(-\frac{1}{3}\right)$ if $f(-0.75) = -0.07181250$, $f(-0.5) = -0.02475000$, $f(-0.25) = 0.33493750$, $f(0) = 1.10100000$

b. $f(0.25)$ if $f(0.1) = -0.62049958$, $f(0.2) = -0.28398668$, $f(0.3) = 0.00660095$, $f(0.4) = 0.24842440$

4. a. Use Algorithm 3.2 to construct the interpolating polynomial of degree four for the unequally spaced points given in the following table:

x	$f(x)$
0.0	−6.00000
0.1	−5.89483
0.3	−5.65014
0.6	−5.17788
1.0	−4.28172

b. Add $f(1.1) = -3.99583$ to the table and construct the interpolating polynomial of degree five.

5. **a.** Approximate $f(0.05)$ using the following data and the Newton forward divided-difference formula:

x	0.0	0.2	0.4	0.6	0.8
$f(x)$	1.00000	1.22140	1.49182	1.82212	2.22554

b. Use the Newton backward divided-difference formula to approximate $f(0.65)$.

c. Use Stirling's formula to approximate $f(0.43)$.

6. Show that the polynomial interpolating the following data has degree 3.

x	-2	-1	0	1	2	3
$f(x)$	1	4	11	16	13	-4

7. **a.** Show that the Newton forward divided-difference polynomials

$$P(x) = 3 - 2(x+1) + 0(x+1)(x) + (x+1)(x)(x-1)$$

and

$$Q(x) = -1 + 4(x+2) - 3(x+2)(x+1) + (x+2)(x+1)(x)$$

both interpolate the data

x	-2	-1	0	1	2
$f(x)$	-1	3	1	-1	3

b. Why does part (a) not violate the uniqueness property of interpolating polynomials?

8. A fourth-degree polynomial $P(x)$ satisfies $\Delta^4 P(0) = 24$, $\Delta^3 P(0) = 6$, and $\Delta^2 P(0) = 0$, where $\Delta P(x) = P(x+1) - P(x)$. Compute $\Delta^2 P(10)$.

9. The following data are given for a polynomial $P(x)$ of unknown degree.

x	0	1	2
$P(x)$	2	-1	4

Determine the coefficient of x^2 in $P(x)$ if all third-order forward differences are 1.

10. The following data are given for a polynomial $P(x)$ of unknown degree.

x	0	1	2	3
$P(x)$	4	9	15	18

Determine the coefficient of x^3 in $P(x)$ if all fourth-order forward differences are 1.

11. The Newton forward divided-difference formula is used to approximate $f(0.3)$ given the following data.

x	0.0	0.2	0.4	0.6
$f(x)$	15.0	21.0	30.0	51.0

Suppose it is discovered that $f(0.4)$ was understated by 10 and $f(0.6)$ was overstated by 5. By what amount should the approximation to $f(0.3)$ be changed?

12. For a function f the Newton's interpolatory divided-difference formula gives the interpolating polynomial

$$P_3(x) = 1 + 4x + 4x(x - 0.25) + \frac{16}{3}x(x - 0.25)(x - 0.5)$$

on the nodes $x_0 = 0$, $x_1 = 0.25$, $x_2 = 0.5$ and $x_3 = 0.75$. Find $f(0.75)$.

13. For a function f the forward divided-differences are given by

$x_0 = 0.0$	$f[x_0]$		
		$f[x_0, x_1]$	
$x_1 = 0.4$	$f[x_1]$		$f[x_0, x_1, x_2] = \frac{50}{7}$
		$f[x_1, x_2] = 10$	
$x_2 = 0.7$	$f[x_2] = 6$		

Determine the missing entries in the table.

14. **a.** The introduction to this chapter included a table listing the population of the United States from 1940 to 1990. Use appropriate divided differences to approximate the population in the years 1930, 1965, and 2000.

    b. The population in 1930 was approximately 123,203,000. How accurate do you think your 1965 and 2000 figures are?

15. Given

$$\begin{aligned} P_n(x) = f[x_0] &+ f[x_0, x_1](x - x_0) + a_2(x - x_0)(x - x_1) \\ &+ a_3(x - x_0)(x - x_1)(x - x_2) + \cdots \\ &+ a_n(x - x_0)(x - x_1)\cdots(x - x_{n-1}). \end{aligned}$$

Use $P_n(x_2)$ to show that $a_2 = f[x_0, x_1, x_2]$.

16. Show that

$$f[x_0, x_1, \ldots, x_n, x] = \frac{f^{(n+1)}(\xi(x))}{(n+1)!}$$

for some $\xi(x)$. [*Hint:* From Eq. (3.4)

$$f(x) = P_n(x) + \frac{f^{(n+1)}(\xi(x))}{(n+1)!}(x - x_0)\cdots(x - x_n).$$

Considering the interpolation polynomial of degree $n + 1$ on $x_0, x_1, \ldots, x_n, x$, we have

$$f(x) = P_{n+1}(x) = P_n(x) + f[x_0, x_1, \ldots, x_n, x](x - x_0)\cdots(x - x_n).]$$

17. Let $i_0, i_1, \ldots, i_n$ be a rearrangement of the integers $0, 1, \ldots, n$. Show that $f[x_{i_0}, x_{i_1}, \ldots, x_{i_n}] = f[x_0, x_1, \ldots, x_n]$. [*Hint:* Consider the leading coefficient of the nth Lagrange polynomial on the data $\{x_0, x_1, \ldots, x_n\} = \{x_{i_0}, x_{i_1}, \ldots, x_{i_n}\}$.]

3.3 Hermite Interpolation

Osculating polynomials generalize both the Taylor polynomials and the Lagrange polynomials. Given $n + 1$ distinct numbers $x_0, x_1, \ldots, x_n$ and nonnegative integers $m_0, m_1, \ldots, m_n$, the osculating polynomial approximating a function $f \in C^m[a, b]$, where $m = \max\{m_0, m_1, \ldots, m_n\}$ and $x_i \in [a, b]$ for each $i = 0, \ldots, n$, is the polynomial of least degree with the property that it agrees with the function f and all its derivatives of order less than or equal to m_i at x_i for each $i = 0, 1, \ldots, n$. The degree of this osculating polynomial is at most

$$M = \sum_{i=0}^{n} m_i + n$$

since the number of conditions to be satisfied is $\sum_{i=0}^{n}(m_i + 1) = \sum_{i=0}^{n} m_i + (n + 1)$, and a polynomial of degree M has $M + 1$ coefficients that can be used to satisfy these conditions.

Definition 3.8 Let $x_0, x_1, \ldots, x_n$ be $n + 1$ distinct numbers in $[a, b]$ and m_i be a nonnegative integer associated with x_i for $i = 0, 1, \ldots, n$. Suppose that $f \in C^m[a, b]$ and $m = \max_{0 \le i \le n} m_i$. The **osculating polynomial** approximating f is the polynomial $P(x)$ of least degree such that

$$\frac{d^k P(x_i)}{dx^k} = \frac{d^k f(x_i)}{dx^k} \quad \text{for each } i = 0, 1, \ldots, n \text{ and } k = 0, 1, \ldots, m_i. \qquad \blacksquare$$

Note that when $n = 0$, the osculating polynomial approximating f is simply the m_0th Taylor polynomial for f at x_0. When $m_i = 0$ for each i, the osculating polynomial is the nth Lagrange polynomial interpolating f on $x_0, x_1, \ldots, x_n$.

The case when $m_i = 1$ for each $i = 0, 1, \ldots, n$ gives a class called the **Hermite polynomials**. For a given function f, these polynomials agree with f at $x_0, x_1, \ldots, x_n$. In addition, since their first derivatives agree with those of f, they have the same "shape" as the function at $(x_i, f(x_i))$ in the sense that the *tangent lines* to the polynomial and to the function agree. We will restrict our study of osculating polynomials to this situation and consider first a theorem that describes precisely the form of the Hermite polynomials.

Theorem 3.9 If $f \in C^1[a, b]$ and $x_0, \ldots, x_n \in [a, b]$ are distinct, the unique polynomial of least degree agreeing with f and f' at $x_0, \ldots, x_n$ is the Hermite polynomial of degree at most $2n + 1$ given by

$$H_{2n+1}(x) = \sum_{j=0}^{n} f(x_j) H_{n,j}(x) + \sum_{j=0}^{n} f'(x_j) \hat{H}_{n,j}(x),$$

where

$$H_{n,j}(x) = [1 - 2(x - x_j) L'_{n,j}(x_j)] L^2_{n,j}(x)$$

and

$$\hat{H}_{n,j}(x) = (x - x_j) L^2_{n,j}(x).$$

In this context, $L_{n,j}(x)$ denotes the jth Lagrange coefficient polynomial of degree n defined in Eq. (3.3).

Moreover, if $f \in C^{2n+2}[a, b]$, then

$$f(x) = H_{2n+1}(x) + \frac{(x - x_0)^2 \cdots (x - x_n)^2}{(2n + 2)!} f^{(2n+2)}(\xi),$$

for some ξ with $a < \xi < b$.

Proof First recall that

$$L_{n,j}(x_i) = \begin{cases} 0, & \text{if } i \neq j, \\ 1, & \text{if } i = j. \end{cases}$$

Hence when $i \neq j$,

$$H_{n,j}(x_i) = 0 \quad \text{and} \quad \hat{H}_{n,j}(x_i) = 0,$$

whereas

$$H_{n,i}(x_i) = [1 - 2(x_i - x_i)L'_{n,i}(x_i)] \cdot 1 = 1$$

and

$$\hat{H}_{n,i}(x_i) = (x_i - x_i) \cdot 1^2 = 0.$$

As a consequence,

$$H_{2n+1}(x_i) = \sum_{\substack{j=0 \\ j \neq i}}^{n} f(x_j) \cdot 0 + f(x_i) \cdot 1 + \sum_{j=0}^{n} f'(x_j) \cdot 0 = f(x_i),$$

so H_{2n+1} agrees with f at $x_0, x_1, \ldots, x_n$.

To show the agreement of H'_{2n+1} with f' at the nodes, first note that $L_{n,j}(x)$ is a factor of $H'_{n,j}(x)$, so $H'_{n,j}(x_i) = 0$ when $i \neq j$. In addition, when $i = j$,

$$\begin{aligned} H'_{n,i}(x_i) &= -2L'_{n,i}(x_i) \cdot L^2_{n,i}(x_i) + [1 - 2(x_i - x_i)L'_{n,i}(x_i)]2L_{n,i}(x_i)L'_{n,i}(x_i) \\ &= -2L'_{n,i}(x_i) + 2L'_{n,i}(x_i) = 0. \end{aligned}$$

Hence $H'_{n,j}(x_i) = 0$ for all i and j.

Finally,

$$\begin{aligned} \hat{H}'_{n,j}(x_i) &= L^2_{n,j}(x_i) + (x_i - x_j)2L_{n,j}(x_i)L'_{n,j}(x_i) \\ &= L_{n,j}(x_i)[L_{n,j}(x_i) + 2(x_i - x_j)L'_{n,j}(x_i)], \end{aligned}$$

so $\hat{H}_{n,j}(x_i) = 0$ if $i \neq j$ and $\hat{H}_{n,i}(x_i) = 1$. Combining these facts we have

$$H'_{2n+1}(x_i) = \sum_{j=0}^{n} f(x_j) \cdot 0 + \sum_{\substack{j=0 \\ j \neq i}}^{n} f'(x_j) \cdot 0 + f'(x_i) \cdot 1 = f'(x_i).$$

Therefore, H_{2n+1} agrees with f and H'_{2n+1} with f' at $x_0, x_1, \ldots, x_n$.

The uniqueness of this polynomial and the error formula are considered in Exercise 8.

■ ■ ■

EXAMPLE 1 Use the Hermite polynomial that agrees with the data listed in Table 3.13 to find an approximation of $f(1.5)$.

Table 3.13

k	x_k	$f(x_k)$	$f'(x_k)$
0	1.3	0.6200860	−0.5220232
1	1.6	0.4554022	−0.5698959
2	1.9	0.2818186	−0.5811571

First compute the Lagrange polynomials and their derivatives:

$$L_{2,0}(x) = \frac{(x - x_1)(x - x_2)}{(x_0 - x_1)(x_0 - x_2)} = \frac{50}{9}x^2 - \frac{175}{9}x + \frac{152}{9}, \qquad L'_{2,0}(x) = \frac{100}{9}x - \frac{175}{9};$$

$$L_{2,1}(x) = \frac{(x - x_0)(x - x_2)}{(x_1 - x_0)(x_1 - x_2)} = \frac{-100}{9}x^2 + \frac{320}{9}x - \frac{247}{9}, \quad L'_{2,1}(x) = \frac{-200}{9}x + \frac{320}{9};$$

and

$$L_{2,2} = \frac{(x - x_0)(x - x_1)}{(x_2 - x_0)(x_2 - x_1)} = \frac{50}{9}x^2 - \frac{145}{9}x + \frac{104}{9}, \qquad L'_{2,2}(x) = \frac{100}{9}x - \frac{145}{9}.$$

The polynomials $H_{2,j}(x)$ and $\hat{H}_{2,j}(x)$ are then

$$\begin{aligned} H_{2,0}(x) &= [1 - 2(x - 1.3)(-5)] \left(\frac{50}{9}x^2 - \frac{175}{9}x + \frac{152}{9} \right)^2 \\ &= (10x - 12) \left(\frac{50}{9}x^2 - \frac{175}{9}x + \frac{152}{9} \right)^2, \end{aligned}$$

$$H_{2,1}(x) = 1 \cdot \left(\frac{-100}{9}x^2 + \frac{320}{9}x - \frac{247}{9} \right)^2,$$

$$H_{2,2}(x) = 10(2 - x) \left(\frac{50}{9}x^2 - \frac{145}{9}x + \frac{104}{9} \right)^2,$$

$$\hat{H}_{2,0}(x) = (x - 1.3)\left(\frac{50}{9}x^2 - \frac{175}{9}x + \frac{152}{9}\right)^2,$$

$$\hat{H}_{2,1}(x) = (x - 1.6)\left(\frac{-100}{9}x^2 + \frac{320}{9}x - \frac{247}{9}\right)^2,$$

and

$$\hat{H}_{2,2}(x) = (x - 1.9)\left(\frac{50}{9}x^2 - \frac{145}{9}x + \frac{104}{9}\right)^2.$$

Finally,

$$\begin{aligned} H_5(x) = {} & 0.6200860H_{2,0}(x) + 0.4554022H_{2,1}(x) + 0.2818186H_{2,2}(x) \\ & - 0.5220232\hat{H}_{2,0}(x) - 0.5698959\hat{H}_{2,1}(x) - 0.5811571\hat{H}_{2,2}(x) \end{aligned}$$

and

$$\begin{aligned} H_5(1.5) &= 0.6200860\left(\frac{4}{27}\right) + 0.4554022\left(\frac{64}{81}\right) + 0.2818186\left(\frac{5}{81}\right) \\ &\quad - 0.5220232\left(\frac{4}{405}\right) - 0.5698959\left(\frac{-32}{405}\right) - 0.5811571\left(\frac{-2}{405}\right) \\ &= 0.5118277, \end{aligned}$$

a result that is accurate to the places listed. ∎

Although Theorem 3.9 provides a complete description of the Hermite polynomials, it is clear from Example 1 that the need to determine and evaluate the Lagrange polynomials and their derivatives makes the procedure tedious even for small values of n. An alternative method for generating Hermite approximations has as its basis the Newton interpolatory divided-difference formula (3.11) for the Lagrange polynomial at $x_0, x_1, \ldots, x_n$,

$$P_n(x) = f[x_0] + \sum_{k=1}^{n} f[x_0, x_1, \ldots, x_k](x - x_0)\cdots(x - x_{k-1}),$$

and the connection between the nth divided difference and the nth derivative of f, as outlined in Theorem 3.6 in Section 3.2.

Suppose that the n distinct numbers $x_0, x_1, \ldots, x_n$ are given together with the values of f and f' at these numbers. Define a new sequence $z_0, z_1, \ldots, z_{2n+1}$ by

$$z_{2i} = z_{2i+1} = x_i, \quad \text{for each } i = 0, 1, \ldots, n,$$

and construct the divided difference table in the form of Table 3.8 that uses $z_0, z_1, \ldots, z_{2n+1}$.

Since $z_{2i} = z_{2i+1} = x_i$ for each i, we cannot define $f[z_{2i}, z_{2i+1}]$ by the basic relation (3.9). If we assume, based on Theorem 3.6, that the reasonable substitution in this situation is $f[z_{2i}, z_{2i+1}] = f'(z_{2i}) = f'(x_i)$, we can use the entries

$$f'(x_0), f'(x_1), \ldots, f'(x_n)$$

in place of the undefined first divided differences

$$f[z_0, z_1], f[z_2, z_3], \ldots, f[z_{2n}, z_{2n+1}].$$

The remaining divided differences are produced as usual, and the appropriate divided differences are employed in Newton's interpolatory divided-difference formula. Table 3.14 shows the entries that are used for the first three divided-difference columns when determining the Hermite polynomial $H_5(x)$ for x_0, x_1, and x_2. The remaining entries are generated in the same manner as in Table 3.8. The Hermite polynomial is given by

$$H_{2n+1}(x) = f[z_0] + \sum_{k=1}^{2n+1} f[z_0, \ldots, z_k](x - z_0)(x - z_1) \cdots (x - z_{k-1}).$$

A proof of this fact can be found in [Po, p. 56].

Table 3.14

z	$f(z)$	First divided differences	Second divided differences
$z_0 = x_0$	$f[z_0] = f(x_0)$		
		$f[z_0, z_1] = f'(x_0)$	
$z_1 = x_0$	$f[z_1] = f(x_0)$		$f[z_0, z_1, z_2] = \dfrac{f[z_1, z_2] - f[z_0, z_1]}{z_2 - z_0}$
		$f[z_1, z_2] = \dfrac{f[z_2] - f[z_1]}{z_2 - z_1}$	
$z_2 = x_1$	$f[z_2] = f(x_1)$		$f[z_1, z_2, z_3] = \dfrac{f[z_2, z_3] - f[z_1, z_2]}{z_3 - z_1}$
		$f[z_2, z_3] = f'(x_1)$	
$z_3 = x_1$	$f[z_3] = f(x_1)$		$f[z_2, z_3, z_4] = \dfrac{f[z_3, z_4] - f[z_2, z_3]}{z_4 - z_2}$
		$f[z_3, z_4] = \dfrac{f[z_4] - f[z_3]}{z_4 - z_3}$	
$z_4 = x_2$	$f[z_4] = f(x_2)$		$f[z_3, z_4, z_5] = \dfrac{f[z_4, z_5] - f[z_3, z_4]}{z_5 - z_3}$
		$f[z_4, z_5] = f'(x_2)$	
$z_5 = x_2$	$f[z_5] = f(x_2)$		

EXAMPLE 2 The entries in Table 3.15 use the data given in Example 1. The underlined entries are the given data; the remainder are generated by the standard divided-difference formula (3.10).

$$\begin{aligned} H_5(1.5) &= 0.6200860 + (1.5 - 1.3)(-0.5220232) + (1.5 - 1.3)^2(-0.0897427) \\ &\quad + (1.5 - 1.3)^2(1.5 - 1.6)(0.0663657) + (1.5 - 1.3)^2(1.5 - 1.6)^2(0.0026663) \\ &\quad + (1.5 - 1.3)^2(1.5 - 1.6)^2(1.5 - 1.9)(-0.0027738) \\ &= 0.5118277. \end{aligned}$$

■

Table 3.15

1.3	0.6200860					
		−0.5220232				
1.3	0.6200860		−0.0897427			
		−0.5489460		0.0663657		
1.6	0.4554022		−0.0698330		0.0026663	
		−0.5698959		0.0679655		−0.0027738
1.6	0.4554022		−0.0290537		0.0010020	
		−0.5786120		0.0685667		
1.9	0.2818186		−0.0084837			
		−0.5811571				
1.9	0.2818186					

The technique used in Algorithm 3.3 can be extended for use in determining other osculating polynomials. A concise discussion of the procedures can be found in [Po, pp. 53–57].

ALGORITHM 3.3

Hermite Interpolation

To obtain the coefficients of the Hermite interpolating polynomial $H(x)$ on the $(n + 1)$ distinct numbers $x_0, \ldots, x_n$ for the function f:

INPUT numbers $x_0, x_1, \ldots, x_n$; values $f(x_0), \ldots, f(x_n)$ and $f'(x_0), \ldots, f'(x_n)$.

OUTPUT the numbers $Q_{0,0}, Q_{1,1}, \ldots, Q_{2n+1,2n+1}$ where

$$\begin{aligned} H(x) = {} & Q_{0,0} + Q_{1,1}(x - x_0) + Q_{2,2}(x - x_0)^2 + Q_{3,3}(x - x_0)^2(x - x_1) \\ & + Q_{4,4}(x - x_0)^2(x - x_1)^2 + \cdots \\ & + Q_{2n+1,2n+1}(x - x_0)^2(x - x_1)^2 \cdots (x - x_{n-1})^2(x - x_n). \end{aligned}$$

Step 1 For $i = 0, 1, \ldots, n$ do Steps 2 and 3.

Step 2 Set $z_{2i} = x_i$;
$z_{2i+1} = x_i$;
$Q_{2i,0} = f(x_i)$;
$Q_{2i+1,0} = f(x_i)$;
$Q_{2i+1,1} = f'(x_i)$.

Step 3 If $i \neq 0$ then set

$$Q_{2i,1} = \frac{Q_{2i,0} - Q_{2i-1,0}}{z_{2i} - z_{2i-1}}.$$

Step 4 For $i = 2, 3, \ldots, 2n + 1$

for $j = 2, 3, \ldots, i$ set $Q_{i,j} = \dfrac{Q_{i,j-1} - Q_{i-1,j-1}}{z_i - z_{i-j}}$.

Step 5 OUTPUT $(Q_{0,0}, Q_{1,1}, \ldots, Q_{2n+1,2n+1})$;
STOP

EXERCISE SET 3.3

1. Use Theorem 3.9 or Algorithm 3.3 to construct an approximating polynomial for the following data.

a.

x	$f(x)$	$f'(x)$
8.3	17.56492	3.116256
8.6	18.50515	3.151762

b.

x	$f(x)$	$f'(x)$
0.8	0.22363362	2.1691753
1.0	0.65809197	2.0466965

c.

x	$f(x)$	$f'(x)$
-0.5	-0.0247500	0.7510000
-0.25	0.3349375	2.1890000
0	1.1010000	4.0020000

d.

x	$f(x)$	$f'(x)$
0.1	-0.62049958	3.58502082
0.2	-0.28398668	3.14033271
0.3	0.00660095	2.66668043
0.4	0.24842440	2.16529366

2. The data in Exercise 1 were generated using the following functions. Use the polynomials constructed in Exercise 1 for the given value of x to approximate $f(x)$, and calculate the actual error.

a. $f(x) = x \ln x$; approximate $f(8.4)$.

b. $f(x) = \sin(e^x - 2)$; approximate $f(0.9)$.

c. $f(x) = x^3 + 4.001x^2 + 4.002x + 1.101$; approximate $f(-\frac{1}{3})$.

d. $f(x) = x \cos x - 2x^2 + 3x - 1$; approximate $f(0.25)$.

3. **a.** Use the following values and five-digit rounding arithmetic to construct a Hermite interpolating polynomial to approximate $\sin 0.34$.

x	$\sin x$	$D_x \sin x = \cos x$
0.30	0.29552	0.95534
0.32	0.31457	0.94924
0.35	0.34290	0.93937

b. Determine an error bound for the approximation in part (a) and compare to the actual error.

c. Add $\sin 0.33 = 0.32404$ and $\cos 0.33 = 0.94604$ to the data and redo the calculations.

4. Let $f(x) = 3xe^x - e^{2x}$.

a. Approximate $f(1.03)$ by the Hermite interpolating polynomial of degree at most three using $x_0 = 1$ and $x_1 = 1.05$. Compare the actual error to the error bound.

b. Repeat (a) with the Hermite interpolating polynomial of degree at most five, using $x_0 = 1$, $x_1 = 1.05$, and $x_2 = 1.07$.

5. Use the error formula and Maple to find a bound for the errors in the approximations of $f(x)$ in parts (a) and (c) of Exercise 2.

6. The following table lists data for the function described by $f(x) = e^{0.1x^2}$. Approximate $f(1.25)$ by using $H_5(1.25)$ and $H_3(1.25)$, where H_5 uses the nodes $x_0 = 1$, $x_1 = 2$, and $x_2 = 3$ and H_3 uses the nodes $\bar{x}_0 = 1$ and $\bar{x}_1 = 1.5$. Find error bounds for these approximations.

x	$f(x) = e^{0.1x^2}$	$f'(x) = 0.2xe^{0.1x^2}$
$x_0 = \bar{x}_0 = 1$	1.105170918	0.2210341836
$\bar{x}_1 = 1.5$	1.252322716	0.3756968148
$x_1 = 2$	1.491824698	0.5967298792
$x_2 = 3$	2.459603111	1.475761867

7. A car traveling along a straight road is clocked at a number of points. The data from the observations are given in the following table, where the time is in seconds, the distance is in feet, and the speed is in feet per second.

Time	0	3	5	8	13
Distance	0	225	383	623	993
Speed	75	77	80	74	72

a. Use a Hermite polynomial to predict the position of the car and its speed when $t = 10$ s.

b. Use the derivative of the Hermite polynomial to determine whether the car ever exceeds a 55-mi/h speed limit on the road. If so, what is the first time the car exceeds this speed?

c. What is the predicted maximum speed for the car?

8. **a.** Show that $H_{2n+1}(x)$ is the unique polynomial of least degree agreeing with f and f' at $x_0, \ldots, x_n$. [*Hint*: Assume that $P(x)$ is another such polynomial and consider $D = H_{2n+1} - P$ and D' at $x_0, x_1, \ldots, x_n$.]

b. Derive the error term in Theorem 3.9. [*Hint*: Use the same method as in the Lagrange error derivation, Theorem 3.3, defining

$$g(t) = f(t) - H_{2n+1}(t) - \frac{(t - x_0)^2 \cdots (t - x_n)^2}{(x - x_0)^2 \cdots (x - x_n)^2}[f(x) - H_{2n+1}(x)]$$

and showing that $g'(t)$ has $(2n + 2)$ distinct zeros in $[a, b]$.]

9. Let $z_0 = x_0$, $z_1 = x_0$, $z_2 = x_1$, and $z_3 = x_1$. Form the following divided-difference table.

$z_0 = x_0$	$f[z_0] = f(x_0)$			
		$f[z_0, z_1] = f'(x_0)$		
$z_1 = x_0$	$f[z_1] = f(x_0)$		$f[z_0, z_1, z_2]$	
		$f[z_1, z_2]$		$f[z_0, z_1, z_2, z_3]$
$z_2 = x_1$	$f[z_2] = f(x_1)$		$f[z_1, z_2, z_3]$	
		$f[z_2, z_3] = f'(x_1)$		
$z_3 = x_1$	$f[z_3] = f(x_1)$			

Show that the cubic Hermite polynomial $H_3(x)$ can also be written as $f[z_0] + f[z_0, z_1](x - x_0) + f[z_0, z_1, z_2](x - x_0)^2 + f[z_0, z_1, z_2, z_3](x - x_0)^2(x - x_1)$.

3.4 Cubic Spline Interpolation*

The previous sections of this chapter concerned the approximation of an arbitrary function on a closed interval by a polynomial. However, the oscillatory nature of high-degree polynomials and the property that a fluctuation over a small portion of the interval can induce large fluctuations over the entire range restricts their use. We will see a good example of this at the end of this section. (See Figure 3.12.)

An alternative approach is to divide the interval into a collection of subintervals and construct a (generally) different approximating polynomial on each subinterval. Approximation by functions of this type is called **piecewise polynomial approximation**.

The simplest piecewise polynomial approximation is piecewise linear interpolation, which consists of joining a set of data points

$$\{(x_0, f(x_0)), (x_1, f(x_1)), \ldots, (x_n, f(x_n))\}$$

by a series of straight line segments, such as those shown in Figure 3.7.

Figure 3.7

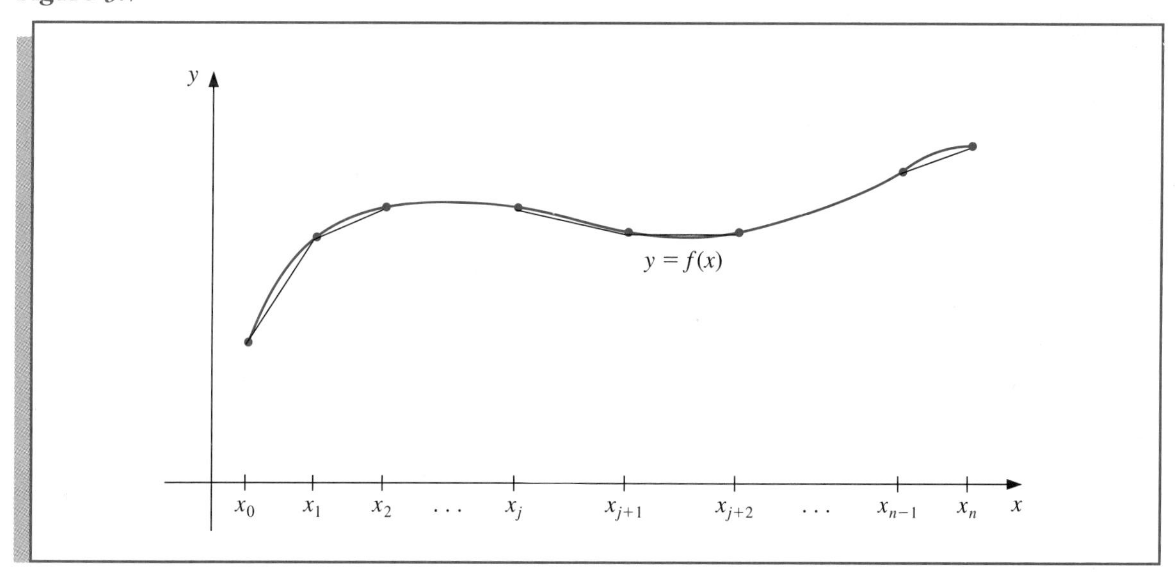

A disadvantage of linear function approximation is that there is no assurance of differentiability at each of the endpoints of the subintervals, which, in a geometrical context, means that the interpolating function is not "smooth" at these points. Often it is clear from physical conditions that such a smoothness condition is required and that the approximating function must be continuously differentiable.

*The proofs of the theorems in this section rely on results in Chapter 6.

An alternative procedure is to use a piecewise polynomial of Hermite type. For example, if the values of the function f and of f' are known at each of the points $x_0 < x_1 < \cdots < x_n$, a Hermite polynomial of degree three can be used on each of the subintervals $[x_0, x_1], [x_1, x_2], \ldots, [x_{n-1}, x_n]$ to obtain a function that is continuously differentiable on the interval $[x_0, x_n]$. To determine the appropriate Hermite cubic polynomial on a given interval is simply a matter of computing $H_3(x)$ for that interval. Since the Lagrange interpolating polynomials needed to determine H_3 are of first degree, this can be accomplished without great difficulty. However, to use Hermite piecewise polynomials for general interpolation, we need to know the derivative of the function being approximated, which is frequently not available.

The remainder of this section considers approximation using piecewise polynomials that require no derivative information, except perhaps at the endpoints of the interval on which the function is being approximated.

The simplest type of differentiable piecewise polynomial function on an entire interval $[x_0, x_n]$ is the function obtained by fitting a quadratic polynomial between each successive pair of nodes. This is done by constructing a quadratic on $[x_0, x_1]$ agreeing with the function at x_0 and x_1, another quadratic on $[x_1, x_2]$ agreeing with the function at x_1 and x_2, and so on. Since a general quadratic polynomial has three arbitrary constants—the constant term, the coefficient of x, and the coefficient of x^2—and only two conditions are required to fit the data at the endpoints of each subinterval, flexibility exists that allows the quadratic to be chosen so that the interpolant has a continuous derivative on $[x_0, x_n]$. The difficulty with this procedure arises when there is a need to specify conditions about the derivative of the interpolant at the endpoints x_0 and x_n. There is not a sufficient number of constants to ensure that the conditions will be satisfied. (See Exercise 22.)

The most common piecewise polynomial approximation uses cubic polynomials between each successive pair of nodes and is called **cubic spline interpolation**. A general cubic polynomial involves four constants, so there is sufficient flexibility in the cubic spline procedure to ensure that the interpolant is not only continuously differentiable on the interval, but also has a continuous second derivative on the interval. The construction of the cubic spline does not, however, assume that the derivatives of the interpolant agree with those of the function, even at the nodes. (See Figure 3.8.)

Definition 3.10 Given a function f defined on $[a, b]$ and a set of nodes $a = x_0 < x_1 < \cdots < x_n = b$, a **cubic spline interpolant** S for f is a function that satisfies the following conditions:

a. $S(x)$ is a cubic polynomial, denoted $S_j(x)$, on the subinterval $[x_j, x_{j+1}]$ for each $j = 0, 1, \ldots, n-1$;

b. $S(x_j) = f(x_j)$ for each $j = 0, 1, \ldots, n$;

c. $S_{j+1}(x_{j+1}) = S_j(x_{j+1})$ for each $j = 0, 1, \ldots, n-2$;

d. $S'_{j+1}(x_{j+1}) = S'_j(x_{j+1})$ for each $j = 0, 1, \ldots, n-2$;

e. $S''_{j+1}(x_{j+1}) = S''_j(x_{j+1})$ for each $j = 0, 1, \ldots, n-2$;

f. One of the following set of boundary conditions is satisfied:

(i) $S''(x_0) = S''(x_n) = 0$ (**free or natural boundary**);

(ii) $S'(x_0) = f'(x_0)$ and $S'(x_n) = f'(x_n)$ (**clamped boundary**). ■

Although cubic splines are defined with other boundary conditions, the conditions given are sufficient for our purposes. When the free boundary conditions occur, the spline

Figure 3.8

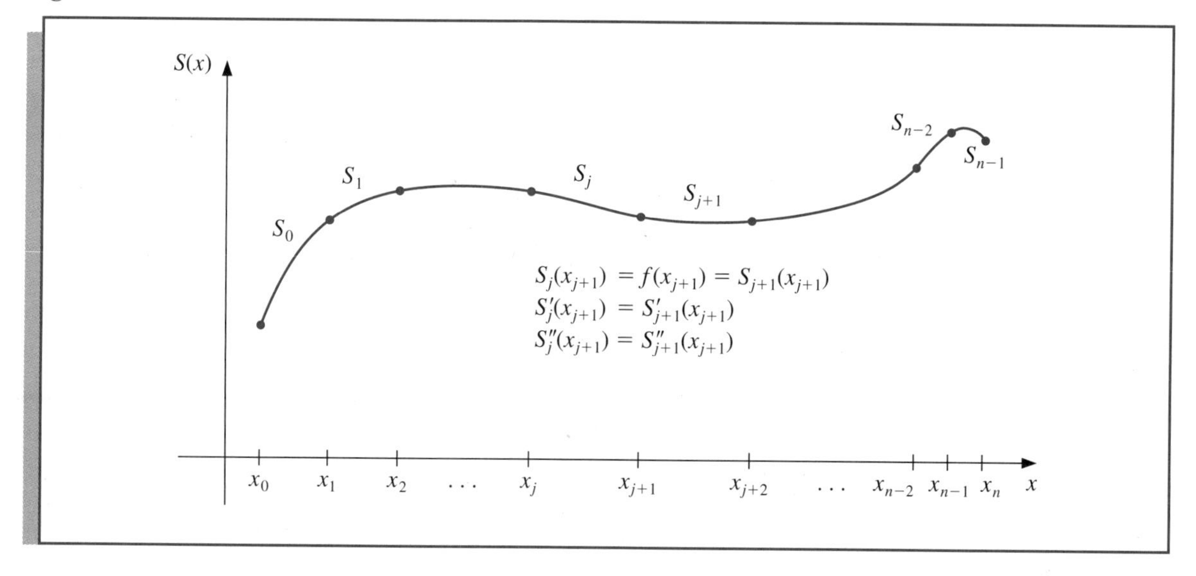

is called a **natural spline**, and its graph approximates the shape that a long flexible rod would assume if forced to go through each of the data points $\{(x_0, f(x_0)), (x_1, f(x_1)), \ldots, (x_n, f(x_n))\}$.

In general, clamped boundary conditions lead to more accurate approximations since they include more information about the function. However, for this type of boundary condition to hold, it is necessary to have either the values of the derivative at the endpoints or an accurate approximation to those values.

To construct the cubic spline interpolant for a given function f, the conditions in the definition are applied to the cubic polynomials

$$S_j(x) = a_j + b_j(x - x_j) + c_j(x - x_j)^2 + d_j(x - x_j)^3$$

for each $j = 0, 1, \ldots, n - 1$.

Clearly,

$$S_j(x_j) = a_j = f(x_j),$$

and if condition (c) is applied,

$$a_{j+1} = S_{j+1}(x_{j+1}) = S_j(x_{j+1}) = a_j + b_j(x_{j+1} - x_j) + c_j(x_{j+1} - x_j)^2 + d_j(x_{j+1} - x_j)^3$$

for each $j = 0, 1, \ldots, n - 2$.

Since terms $(x_{j+1} - x_j)$ will be used repeatedly in this development, it is convenient to introduce the simpler notation

$$h_j = x_{j+1} - x_j$$

for each $j = 0, 1, \ldots, n-1$. If we also define $a_n = f(x_n)$, then the equation

$$a_{j+1} = a_j + b_j h_j + c_j h_j^2 + d_j h_j^3 \tag{3.16}$$

holds for each $j = 0, 1, \ldots, n-1$.

In a similar manner, define $b_n = S'(x_n)$ and observe that

$$S_j'(x) = b_j + 2c_j(x - x_j) + 3d_j(x - x_j)^2$$

implies $S_j'(x_j) = b_j$ for each $j = 0, 1, \ldots, n-1$. Applying condition (d) gives

$$b_{j+1} = b_j + 2c_j h_j + 3d_j h_j^2 \tag{3.17}$$

for each $j = 0, 1, \ldots, n-1$.

Another relation between the coefficients of the S_j is obtained by defining $c_n = S''(x_n)/2$ and applying condition (e). In this case,

$$c_{j+1} = c_j + 3d_j h_j, \tag{3.18}$$

for each $j = 0, 1, \ldots, n-1$.

Solving for d_j in Eq. (3.18) and substituting this value into Eqs. (3.16) and (3.17) gives the new equations

$$a_{j+1} = a_j + b_j h_j + \frac{h_j^2}{3}(2c_j + c_{j+1}) \tag{3.19}$$

and

$$b_{j+1} = b_j + h_j(c_j + c_{j+1}) \tag{3.20}$$

for each $j = 0, 1, \ldots, n-1$.

The final relationship involving the coefficients is obtained by solving the appropriate equation in the form of equation (3.19), first for b_j,

$$b_j = \frac{1}{h_j}(a_{j+1} - a_j) - \frac{h_j}{3}(2c_j + c_{j+1}), \tag{3.21}$$

and then, with a reduction of the index, for b_{j-1}:

$$b_{j-1} = \frac{1}{h_{j-1}}(a_j - a_{j-1}) - \frac{h_{j-1}}{3}(2c_{j-1} + c_j).$$

Substituting these values into the equation derived from Eq. (3.20), with the index reduced by 1, gives the linear system of equations

$$h_{j-1}c_{j-1} + 2(h_{j-1} + h_j)c_j + h_j c_{j+1} = \frac{3}{h_j}(a_{j+1} - a_j) - \frac{3}{h_{j-1}}(a_j - a_{j-1}), \tag{3.22}$$

for each $j = 1, 2, \ldots, n-1$. This system involves only $\{c_j\}_{j=0}^{n}$ as unknowns since the values of $\{h_j\}_{j=0}^{n-1}$ and $\{a_j\}_{j=0}^{n}$ are given by the spacing of the nodes $\{x_j\}_{j=0}^{n}$ and the values of f at the nodes.

Note that once the values of $\{c_j\}_{j=0}^{n}$ are known, it is a simple matter to find the remainder of the constants $\{b_j\}_{j=0}^{n-1}$ from Eq. (3.21) and $\{d_j\}_{j=0}^{n-1}$ from Eq. (3.18) and to construct the cubic polynomials $\{S_j(x)\}_{j=0}^{n-1}$.

The major question that arises in connection with this construction is whether the values of $\{c_j\}_{j=0}^{n}$ can be found using the system of equations given in (3.22) and, if so, whether these values are unique. The following theorems indicate that this is the case when either of the boundary conditions given in part (f) of the definition are imposed. The proofs of these theorems require material from linear algebra, which is discussed in Chapter 6.

Theorem 3.11 If f is defined at $a = x_0 < x_1 < \cdots < x_n = b$, then f has a unique natural spline interpolant on the nodes $x_0, x_1, \ldots, x_n$, that is, a spline interpolant that satisfies the boundary conditions $S''(a) = 0$ and $S''(b) = 0$.

Proof The boundary conditions in this case imply that $c_n = S''(x_n)/2 = 0$ and that

$$0 = S''(x_0) = 2c_0 + 6d_0(x_0 - x_0);$$

so $c_0 = 0$.

The two equations $c_0 = 0$ and $c_n = 0$ together with the equations in (3.22) produce a linear system described by the vector equation $A\mathbf{x} = \mathbf{b}$, where A is the $(n+1)$-by-$(n+1)$ matrix

$$A = \begin{bmatrix} 1 & 0 & 0 & \cdots & \cdots & \cdots & 0 \\ h_0 & 2(h_0+h_1) & h_1 & \ddots & & & \vdots \\ 0 & h_1 & 2(h_1+h_2) & h_2 & \ddots & & \vdots \\ \vdots & \ddots & \ddots & \ddots & \ddots & \ddots & 0 \\ \vdots & & & \ddots & h_{n-2} & 2(h_{n-2}+h_{n-1}) & h_{n-1} \\ 0 & \cdots & \cdots & \cdots & 0 & 0 & 1 \end{bmatrix},$$

and $\mathbf{b}$ and $\mathbf{x}$ are the vectors

$$\mathbf{b} = \begin{bmatrix} 0 \\ \frac{3}{h_1}(a_2 - a_1) - \frac{3}{h_0}(a_1 - a_0) \\ \vdots \\ \frac{3}{h_{n-1}}(a_n - a_{n-1}) - \frac{3}{h_{n-2}}(a_{n-1} - a_{n-2}) \\ 0 \end{bmatrix} \quad \text{and} \quad \mathbf{x} = \begin{bmatrix} c_0 \\ c_1 \\ \vdots \\ c_n \end{bmatrix}.$$

The matrix A is strictly diagonally dominant, so it satisfies the hypotheses of Theorem 6.20 in Section 6.6. Therefore, the linear system has a unique solution for $c_0, c_1, \ldots, c_n$.

■ ■ ■

The solution to the cubic spline problem with the boundary conditions $S''(x_0) = S''(x_n) = 0$ can be obtained by applying Algorithm 3.4.

ALGORITHM 3.4

Natural Cubic Spline

To construct the cubic spline interpolant S for the function f, defined at the numbers $x_0 < x_1 < \cdots < x_n$, satisfying $S''(x_0) = S''(x_n) = 0$:

INPUT $n; x_0, x_1, \ldots, x_n; a_0 = f(x_0), a_1 = f(x_1), \ldots, a_n = f(x_n)$.

OUTPUT a_j, b_j, c_j, d_j for $j = 0, 1, \ldots, n-1$. (*Note:* $S(x) = S_j(x) = a_j + b_j(x - x_j) + c_j(x - x_j)^2 + d_j(x - x_j)^3$ *for* $x_j \le x \le x_{j+1}$.)

Step 1 For $i = 0, 1, \ldots, n-1$ set $h_i = x_{i+1} - x_i$.

Step 2 For $i = 1, 2, \ldots, n-1$ set

$$\alpha_i = \frac{3}{h_i}(a_{i+1} - a_i) - \frac{3}{h_{i-1}}(a_i - a_{i-1}).$$

Step 3 Set $l_0 = 1$; (*Steps 3, 4, 5, and part of Step 6 solve a tridiagonal linear system using a method described in Algorithm 6.7.*)
$\mu_0 = 0$;
$z_0 = 0$.

Step 4 For $i = 1, 2, \ldots, n-1$
set $l_i = 2(x_{i+1} - x_{i-1}) - h_{i-1}\mu_{i-1}$;
$\mu_i = h_i/l_i$;
$z_i = (\alpha_i - h_{i-1}z_{i-1})/l_i$.

Step 5 Set $l_n = 1$;
$z_n = 0$;
$c_n = 0$.

Step 6 For $j = n-1, n-2, \ldots, 0$
set $c_j = z_j - \mu_j c_{j+1}$;
$b_j = (a_{j+1} - a_j)/h_j - h_j(c_{j+1} + 2c_j)/3$;
$d_j = (c_{j+1} - c_j)/(3h_j)$.

Step 7 OUTPUT (a_j, b_j, c_j, d_j for $j = 0, 1, \ldots, n-1$);
STOP.

A result similar to the Theorem 3.11 holds in the case of clamped boundary conditions.

Theorem 3.12 If f is defined at $a = x_0 < x_1 < \cdots < x_n = b$ and differentiable at a and b, then f has a unique clamped spline interpolant on the nodes $x_0, x_1, \ldots, x_n$, that is, a spline interpolant that satisfies the boundary conditions $S'(a) = f'(a)$ and $S'(b) = f'(b)$.

Proof It can be seen, using the fact that $S'(a) = S'(x_0) = b_0$, that Eq. (3.21) with $j = 0$ implies

$$f'(a) = \frac{a_1 - a_0}{h_0} - \frac{h_0}{3}(2c_0 + c_1).$$

Consequently,

$$2h_0c_0 + h_0c_1 = \frac{3}{h_0}(a_1 - a_0) - 3f'(a).$$

Similarly,

$$f'(b) = b_n = b_{n-1} + h_{n-1}(c_{n-1} + c_n),$$

so Eq. (3.21) with $j = n - 1$ implies that

$$\begin{aligned} f'(b) &= \frac{a_n - a_{n-1}}{h_{n-1}} - \frac{h_{n-1}}{3}(2c_{n-1} + c_n) + h_{n-1}(c_{n-1} + c_n) \\ &= \frac{a_n - a_{n-1}}{h_{n-1}} + \frac{h_{n-1}}{3}(c_{n-1} + 2c_n), \end{aligned}$$

and

$$h_{n-1}c_{n-1} + 2h_{n-1}c_n = 3f'(b) - \frac{3}{h_{n-1}}(a_n - a_{n-1}).$$

Equations (3.22), together with the equations

$$2h_0c_0 + h_0c_1 = \frac{3}{h_0}(a_1 - a_0) - 3f'(a)$$

and

$$h_{n-1}c_{n-1} + 2h_{n-1}c_n = 3f'(b) - \frac{3}{h_{n-1}}(a_n - a_{n-1}),$$

determine the linear system $A\mathbf{x} = \mathbf{b}$, where

$$A = \begin{bmatrix} 2h_0 & h_0 & 0 & \cdots & \cdots & \cdots & 0 \\ h_0 & 2(h_0 + h_1) & h_1 & \ddots & & & \vdots \\ 0 & h_1 & 2(h_1 + h_2) & h_2 & \ddots & & \vdots \\ \vdots & \ddots & \ddots & \ddots & \ddots & \ddots & 0 \\ \vdots & & & \ddots & h_{n-2} & 2(h_{n-2} + h_{n-1}) & h_{n-1} \\ 0 & \cdots & \cdots & \cdots & 0 & h_{n-1} & 2h_{n-1} \end{bmatrix},$$

$$\mathbf{b} = \begin{bmatrix} \frac{3}{h_0}(a_1 - a_0) - 3f'(a) \\ \frac{3}{h_1}(a_2 - a_1) - \frac{3}{h_0}(a_1 - a_0) \\ \vdots \\ \frac{3}{h_{n-1}}(a_n - a_{n-1}) - \frac{3}{h_{n-2}}(a_{n-1} - a_{n-2}) \\ 3f'(b) - \frac{3}{h_{n-1}}(a_n - a_{n-1}) \end{bmatrix} \quad \text{and} \quad \mathbf{x} = \begin{bmatrix} c_0 \\ c_1 \\ \vdots \\ c_n \end{bmatrix}.$$

The matrix A is strictly diagonally dominant, so it satisfies the conditions of Theorem 6.20. Therefore, the linear system has a unique solution for $c_0, c_1, \ldots, c_n$. ▪ ▪ ▪

The solution to the cubic spline problem with the boundary conditions $S'(x_0) = f'(x_0)$ and $S'(x_n) = f'(x_n)$ can be obtained by applying Algorithm 3.5.

ALGORITHM 3.5

Clamped Cubic Spline

To construct the cubic spline interpolant S for the function f defined at the numbers $x_0 < x_1 < \cdots < x_n$, satisfying $S'(x_0) = f'(x_0)$ and $S'(x_n) = f'(x_n)$:

INPUT n; $x_0, x_1, \ldots, x_n$; $a_0 = f(x_0)$, $a_1 = f(x_1), \ldots, a_n = f(x_n)$; $FPO = f'(x_0)$; $FPN = f'(x_n)$.

OUTPUT a_j, b_j, c_j, d_j for $j = 0, 1, \ldots, n-1$.
(*Note:* $S(x) = S_j(x) = a_j + b_j(x - x_j) + c_j(x - x_j)^2 + d_j(x - x_j)^3$ *for* $x_j \le x \le x_{j+1}$.)

Step 1 For $i = 0, 1, \ldots, n-1$ set $h_i = x_{i+1} - x_i$.

Step 2 Set $\alpha_0 = 3(a_1 - a_0)/h_0 - 3FPO$;
$\alpha_n = 3FPN - 3(a_n - a_{n-1})/h_{n-1}$.

Step 3 For $i = 1, 2, \ldots, n-1$

$$\text{set } \alpha_i = \frac{3}{h_i}(a_{i+1} - a_i) - \frac{3}{h_{i-1}}(a_i - a_{i-1}).$$

Step 4 Set $l_0 = 2h_0$; (*Steps 4,5,6, and part of Step 7 solve a tridiagonal linear system using a method described in Algorithm 6.7.*)
$\mu_0 = 0.5$;
$z_0 = \alpha_0/l_0$.

Step 5 For $i = 1, 2, \ldots, n-1$
set $l_i = 2(x_{i+1} - x_{i-1}) - h_{i-1}\mu_{i-1}$;
$\mu_i = h_i/l_i$;
$z_i = (\alpha_i - h_{i-1}z_{i-1})/l_i$.

Step 6 Set $l_n = h_{n-1}(2 - \mu_{n-1})$;
$z_n = (\alpha_n - h_{n-1}z_{n-1})/l_n$;
$c_n = z_n$.

Step 7 For $j = n-1, n-2, \ldots, 0$
set $c_j = z_j - \mu_j c_{j+1}$;
$b_j = (a_{j+1} - a_j)/h_j - h_j(c_{j+1} + 2c_j)/3$;
$d_j = (c_{j+1} - c_j)/(3h_j)$.

Step 8 OUTPUT (a_j, b_j, c_j, d_j for $j = 0, 1, \ldots, n-1$);
STOP.

EXAMPLE 1 Figure 3.9 shows a ruddy duck in flight. To approximate the top profile of the duck, we have chosen points along the curve through which we want the approximating curve to pass.

Figure 3.9

Table 3.16 lists the coordinates of 21 data points relative to the superimposed coordinate system shown in Figure 3.10.

Notice that more points are used when the curve is changing rapidly than when it is changing more slowly.

Table 3.16

x	0.9	1.3	1.9	2.1	2.6	3.0	3.9	4.4	4.7	5.0	6.0	7.0	8.0	9.2	10.5	11.3	11.6	12.0	12.6	13.0	13.3
$f(x)$	1.3	1.5	1.85	2.1	2.6	2.7	2.4	2.15	2.05	2.1	2.25	2.3	2.25	1.95	1.4	0.9	0.7	0.6	0.5	0.4	0.25

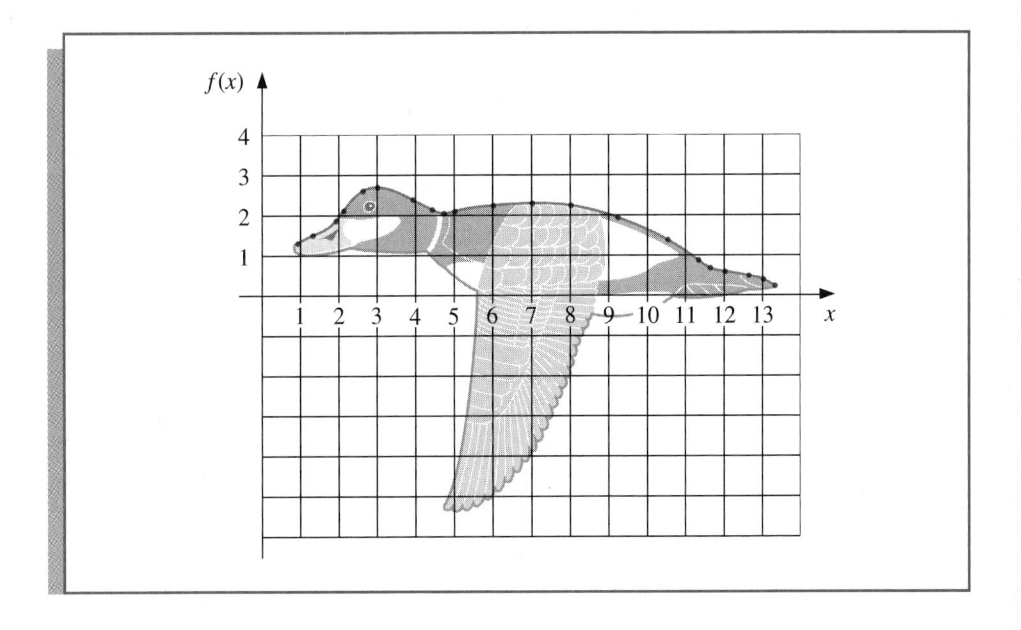

Figure 3.10

Using Algorithm 3.4 to generate the free cubic spline for this data produces the coefficients shown in Table 3.17. This spline curve is nearly identical to the profile, as shown in Figure 3.11.

Table 3.17

j	x_j	a_j	b_j	c_j	d_j
0	0.9	1.3	5.40	0.00	−0.25
1	1.3	1.5	0.42	−0.30	0.95
2	1.9	1.85	1.09	1.41	−2.96
3	2.1	2.1	1.29	−0.37	−0.45
4	2.6	2.6	0.59	−1.04	0.45
5	3.0	2.7	−0.02	−0.50	0.17
6	3.9	2.4	−0.50	−0.03	0.08
7	4.4	2.15	−0.48	0.08	1.31
8	4.7	2.05	−0.07	1.27	−1.58
9	5.0	2.1	0.26	−0.16	0.04
10	6.0	2.25	0.08	−0.03	0.00
11	7.0	2.3	0.01	−0.04	−0.02
12	8.0	2.25	−0.14	−0.11	0.02
13	9.2	1.95	−0.34	−0.05	−0.01
14	10.5	1.4	−0.53	−0.10	−0.02
15	11.3	0.9	−0.73	−0.15	1.21
16	11.6	0.7	−0.49	0.94	−0.84
17	12.0	0.6	−0.14	−0.06	0.04
18	12.6	0.5	−0.18	0.00	−0.45
19	13.0	0.4	−0.39	−0.54	0.60
20	13.3	0.25			

Figure 3.11

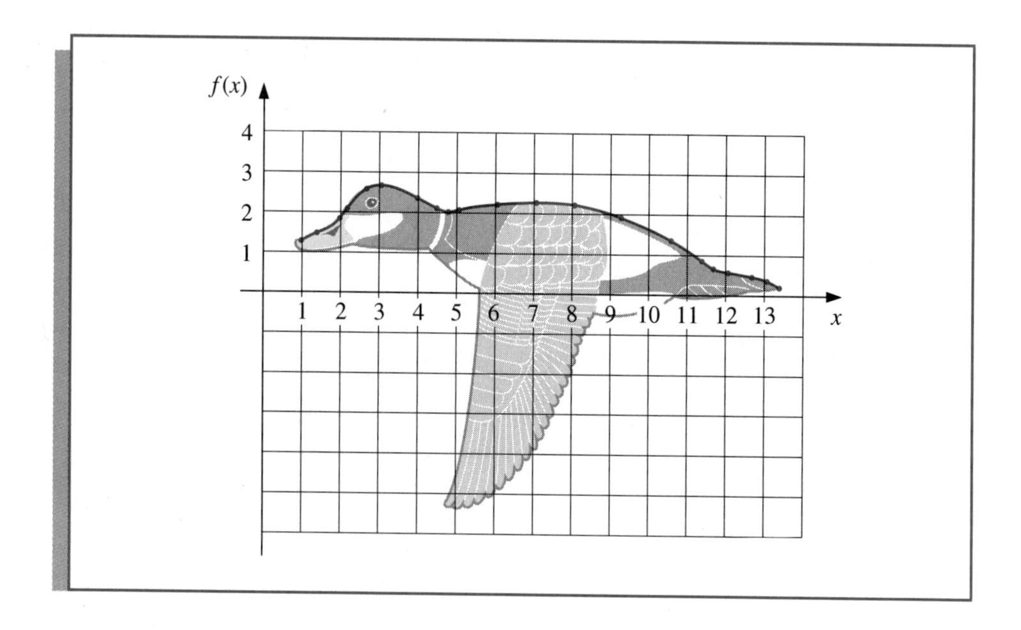

For comparison purposes, Figure 3.12 gives an illustration of the curve that is generated using a Lagrange interpolating polynomial to fit the data given in Table 3.16. This produces a very strange illustration of the back of a duck, in flight or otherwise. The interpolating polynomial in this case is of degree 20 and oscillates wildly, except when contained between data points that are in close proximity.

Figure 3.12

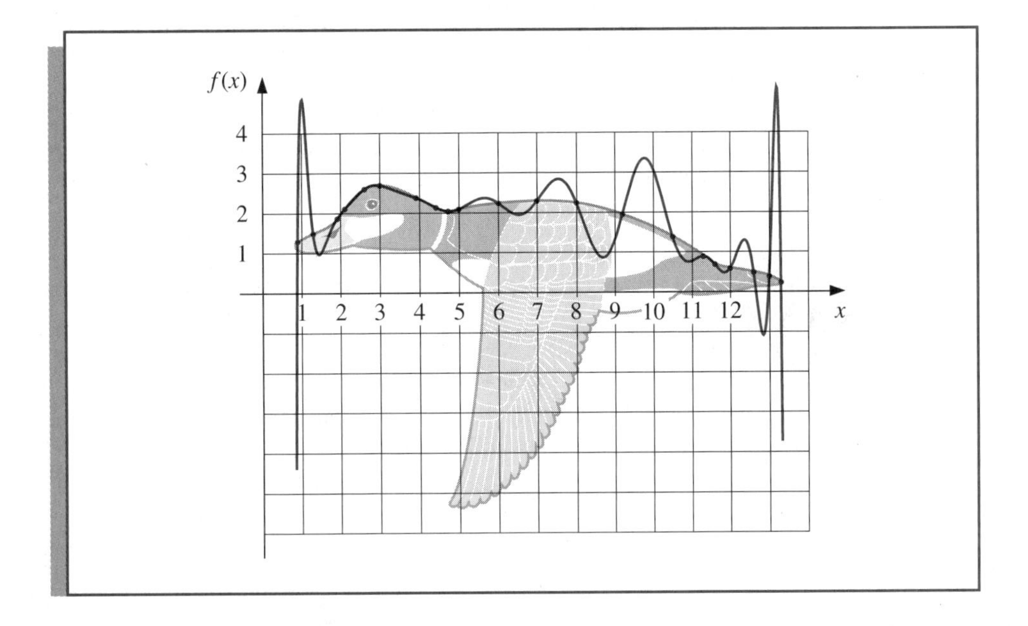

To use a clamped spline to approximate this curve we would need derivative approximations for the endpoints. Even if these approximations were available, we could expect little improvement because of the close agreement of the free cubic spline to the curve of the top profile. ■

Constructing a cubic spline to approximate the lower profile of the ruddy duck would be much more difficult since the curve for this portion cannot be expressed as a function of x, and at certain points the curve does not appear to be smooth. These problems can be resolved by using separate splines to represent various portions of the curve, but a more effective approach to curves of this type is considered in the next section.

The clamped boundary conditions are generally preferred when approximating functions by cubic splines, so the derivative of the function must be estimated at the endpoints of the interval. In the case where the nodes are equally spaced near both endpoints, approximations can be obtained by using Eq. (4.7) or any of the other appropriate formulas given in Sections 4.1 and 4.2. In the case of unequally spaced nodes, the problem is considerably more difficult.

To conclude this section, we list an error-bound formula for the cubic spline with clamped boundary conditions. The proof of this result can be found in [Schul, pp. 57–58].

Theorem 3.13 Let $f \in C^4[a, b]$ with $\max_{a\le x\le b} |f^{(4)}(x)| = M$. If S is the unique clamped cubic spline interpolant to f with respect to the nodes $a = x_0 < x_1 < \cdots < x_n = b$, then

$$\max_{a\le x\le b} |f(x) - S(x)| \le \frac{5M}{384} \max_{0\le j\le n-1} (x_{j+1} - x_j)^4. \qquad \blacksquare$$

A fourth-order error-bound result also holds in the case of free boundary conditions, but it is more difficult to express. (See [BD, pp. 827–835].)

The free boundary conditions will generally give less accurate results than the clamped conditions near the ends of the interval $[x_0, x_n]$ unless the function f happens to nearly satisfy $f''(x_0) = f''(x_n) = 0$. An alternative to the free boundary condition that does not require knowledge of the derivative of f is the *not-a-knot* condition, (see [Deb, pp. 55-56]). This condition requires that $S'''(x)$ be continuous at x_1 and at x_{n-1}.

EXERCISE SET 3.4

1. Determine the free cubic spline S that interpolates the data $f(0) = 0$, $f(1) = 1$, and $f(2) = 2$.
2. Determine the clamped cubic spline s that interpolates the data $f(0) = 0$, $f(1) = 1$, $f(2) = 2$ and satisfies $s'(0) = s'(2) = 1$.
3. Construct the free cubic spline for the following data.

a.

x	$f(x)$
8.3	17.56492
8.6	18.50515

b.

x	$f(x)$
0.8	0.22363362
1.0	0.65809197

c.

x	$f(x)$
−0.5	−0.0247500
−0.25	0.3349375
0	1.1010000

d.

x	$f(x)$
0.1	−0.62049958
0.2	−0.28398668
0.3	0.00660095
0.4	0.24842440

4. The data in Exercise 3 were generated using the following functions. Use the cubic splines constructed in Exercise 3 for the given value of x to approximate $f(x)$ and $f'(x)$, and calculate the actual error.
   - **a.** $f(x) = x \ln x$; approximate $f(8.4)$ and $f'(8.4)$.
   - **b.** $f(x) = \sin(e^x - 2)$; approximate $f(0.9)$ and $f'(0.9)$.
   - **c.** $f(x) = x^3 + 4.001x^2 + 4.002x + 1.101$; approximate $f(-\frac{1}{3})$ and $f'(-\frac{1}{3})$.
   - **d.** $f(x) = x \cos x - 2x^2 + 3x - 1$; approximate $f(0.25)$ and $f'(0.25)$.
5. Construct the clamped cubic spline using the data of Exercise 3 and the fact that
   - **a.** $f'(8.3) = 3.116256$ and $f'(8.6) = 3.151762$
   - **b.** $f'(0.8) = 2.1691753$ and $f'(1.0) = 2.0466965$
   - **c.** $f'(-0.5) = 0.7510000$ and $f'(0) = 4.0020000$
   - **d.** $f'(0.1) = 3.58502082$ and $f'(0.4) = 2.16529366$
6. Repeat Exercise 4 using the cubic splines constructed in Exercise 5.

7. A natural cubic spline S on $[0, 2]$ is defined by

$$S(x) = \begin{cases} S_0(x) = 1 + 2x - x^3, & \text{if } 0 \le x < 1, \\ S_1(x) = a + b(x-1) + c(x-1)^2 + d(x-1)^3, & \text{if } 1 \le x \le 2. \end{cases}$$

Find a, b, c, and d.

8. A clamped cubic spline s for a function f is defined on $[1, 3]$ by

$$s(x) = \begin{cases} s_0(x) = 3(x-1) + 2(x-1)^2 - (x-1)^3, & \text{if } 1 \le x < 2, \\ s_1(x) = a + b(x-2) + c(x-2)^2 + d(x-2)^3, & \text{if } 2 \le x \le 3. \end{cases}$$

Given $f'(1) = f'(3)$, find a, b, c, and d.

9. A natural cubic spline S is defined by

$$S(x) = \begin{cases} S_0(x) = 1 + B(x-1) - D(x-1)^3, & \text{if } 1 \le x < 2, \\ S_1(x) = 1 + b(x-2) - \frac{3}{4}(x-2)^2 + d(x-2)^3, & \text{if } 2 \le x \le 3. \end{cases}$$

If S interpolates the data $(1, 1)$, $(2, 1)$, and $(3, 0)$, find B, D, b, and d.

10. A clamped cubic spline s for a function f is defined by

$$s(x) = \begin{cases} s_0(x) = 1 + Bx + 2x^2 - 2x^3, & \text{if } 0 \le x < 1, \\ s_1(x) = 1 + b(x-1) - 4(x-1)^2 + 7(x-1)^3, & \text{if } 1 \le x \le 2. \end{cases}$$

Find $f'(0)$ and $f'(2)$.

11. Construct a free cubic spline to approximate $f(x) = \cos \pi x$ by using the values given by $f(x)$ at $x = 0, 0.25, 0.5, 0.75$, and 1.0. Integrate the spline over $[0, 1]$, and compare the result to $\int_0^1 \cos \pi x \, dx = 0$. Use the derivatives of the spline to approximate $f'(0.5)$ and $f''(0.5)$. Compare these approximations to the actual values.

12. Construct a free cubic spline to approximate $f(x) = e^{-x}$ by using the values given by $f(x)$ at $x = 0, 0.25, 0.75$, and 1.0. Integrate the spline over $[0, 1]$, and compare the result to $\int_0^1 e^{-x} \, dx = 1 - 1/e$. Use the derivatives of the spline to approximate $f'(0.5)$ and $f''(0.5)$. Compare the approximations to the actual values.

13. Repeat Exercise 11, constructing instead the clamped cubic spline with $f'(0) = f'(1) = 0$.

14. Repeat Exercise 12, constructing instead the clamped cubic spline with $f'(0) = -1$, $f'(1) = -e^{-1}$.

15. Suppose that $f(x)$ is a polynomial of degree 3 on the interval $[a, b]$. Show that $f(x)$ is its own clamped cubic spline, but that it cannot be its own free cubic spline.

16. Suppose the data $\{(x_i, f(x_i))\}_{i=1}^n$ lie on a straight line. What can be said about the free and clamped cubic splines for the function f? [*Hint*: Take a cue from the results of Exercises 1 and 2.]

17. Given the partition $x_0 = 0$, $x_1 = 0.05$, and $x_2 = 0.1$ of $[0, 0.1]$, find the piecewise linear interpolating function F for $f(x) = e^{2x}$. Approximate $\int_0^{0.1} e^{2x} \, dx$ with $\int_0^{0.1} F(x) \, dx$ and compare the results to the actual value.

18. Let $f \in C^2[a, b]$, and let the nodes $a = x_0 < x_1 < \cdots < x_n = b$ be given. Derive an error estimate similar to that in Theorem 3.13 for the piecewise linear interpolating function F. Use this estimate to derive error bounds for Exercise 17.

19. Extend Algorithms 3.4 and 3.5 to include as output the first and second derivatives of the spline at the nodes.

20. Extend Algorithms 3.4 and 3.5 to include as output the integral of the spline over the interval $[x_0, x_n]$.

21. Given the partition $x_0 = 0$, $x_1 = 0.05$, $x_2 = 0.1$ of $[0, 0.1]$ and $f(x) = e^{2x}$:

a. Find the cubic spline s with clamped boundary conditions that interpolates f.

b. Find an approximation for $\int_0^{0.1} e^{2x} \, dx$ by evaluating $\int_0^{0.1} s(x) \, dx$.

c. Use Theorem 3.13 to estimate $\max_{0\le x\le 0.1}|f(x)-s(x)|$ and

$$\left|\int_0^{0.1} f(x)\,dx - \int_0^{0.1} s(x)\,dx\right|.$$

d. Determine the cubic spline S with free boundary conditions and compare $S(0.02)$, $s(0.02)$, and $e^{0.04} = 1.04081077$.

22. Let f be defined on $[a, b]$, and let the nodes $a = x_0 < x_1 < x_2 = b$ be given. A quadratic spline interpolating function S consists of the quadratic polynomial

$$S_0(x) = a_0 + b_0(x - x_0) + c_0(x - x_0)^2 \quad \text{on } [x_0, x_1]$$

and the quadratic polynomial

$$S_1(x) = a_1 + b_1(x - x_1) + c_1(x - x_1)^2 \quad \text{on } [x_1, x_2]$$

such that

(i) $S(x_0) = f(x_0)$, $S(x_1) = f(x_1)$, and $S(x_2) = f(x_2)$,

(ii) $S \in C^1[x_0, x_2]$.

Show that conditions (i) and (ii) lead to five equations in the six unknowns a_0, b_0, c_0, a_1, b_1, and c_1. The problem is to decide what additional condition to impose to make the solution unique. Does the condition

$$S \in C^2[x_0, x_2]$$

lead to a meaningful solution?

23. Determine a quadratic spline s that interpolates the data $f(0) = 0$, $f(1) = 1$, $f(2) = 2$ and satisfies $s'(0) = 2$.

24. **a.** The introduction to this chapter included a table listing the population of the United States from 1940 to 1990. Use free cubic spline interpolation to approximate the population in the years 1930, 1965, and 2000.

b. The population in 1930 was approximately 123,203,000. How accurate do you think your 1965 and 2000 figures are?

25. A car traveling along a straight road is clocked at a number of points. The data from the observations are given in the following table, where the time is in seconds, the distance is in feet, and the speed is in feet per second.

Time	0	3	5	8	13
Distance	0	225	383	623	993
Speed	75	77	80	74	72

a. Use a clamped cubic spline to predict the position of the car and its speed when $t = 10$ s.

b. Use the derivative of the spline to determine whether the car ever exceeds a 55-mi/h speed limit on the road; if so, what is the first time the car exceeds this speed?

c. What is the predicted maximum speed for the car?

26. The 1995 Kentucky Derby was won by a horse named Thunder Gulch in a time of $2{:}01\frac{1}{5}$ (2 min and $1\frac{1}{5}$ s) for the $1\frac{1}{4}$-mi race. Times at the quarter-mile, half-mile, and mile poles were $22\frac{2}{5}$, $45\frac{4}{5}$, and $1{:}35\frac{3}{5}$.

a. Use these values together with the starting time to construct a free cubic spline for Thunder Gulch's race.

b. Use the spline to predict the time at the three-quarter-mile pole, and compare this to the actual time of $1{:}10\frac{1}{5}$.

c. Use the spline to approximate Thunder Gulch's starting speed and speed at the finish line.

27. It is suspected that the high amounts of tannin in mature oak leaves inhibit the growth of the winter moth (*Operophtera bromata L., Geometridae*) larvae that extensively damage these trees in certain years. The following table lists the average weight of two samples of larvae at times in the first 28 days after birth. The first sample was reared on young oak leaves, whereas the second sample was reared on mature leaves from the same tree.

a. Use a free cubic spline to approximate the average weight curve for each sample.

b. Find an approximate maximum average weight for each sample by determining the maximum of the spline.

Day	0	6	10	13	17	20	28
Sample 1 average weight (mg)	6.67	17.33	42.67	37.33	30.10	29.31	28.74
Sample 2 average weight (mg)	6.67	16.11	18.89	15.00	10.56	9.44	8.89

28. The upper portion of the noble beast shown here is to be approximated using clamped cubic spline interpolants. The curve is drawn on a grid from which the table is constructed. Use Algorithm 3.5 to construct the three clamped cubic splines.

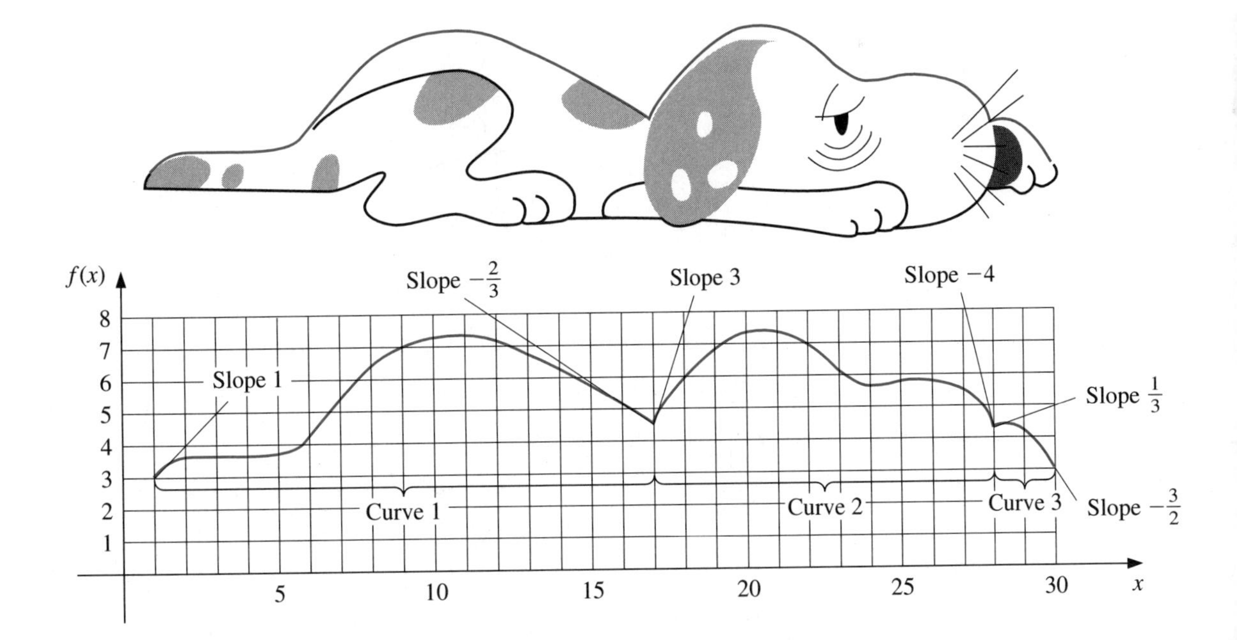

Curve 1				Curve 2				Curve 3			
i	x_i	$f(x_i)$	$f'(x_i)$	i	x_i	$f(x_i)$	$f'(x_i)$	i	x_i	$f(x_i)$	$f'(x_i)$
0	1	3.0	1.0	0	17	4.5	3.0	0	27.7	4.1	0.33
1	2	3.7		1	20	7.0		1	28	4.3	
2	5	3.9		2	23	6.1		2	29	4.1	
3	6	4.2		3	24	5.6		3	30	3.0	−1.5
4	7	5.7		4	25	5.8					
5	8	6.6		5	27	5.2					
6	10	7.1		6	27.7	4.1	−4.0				
7	13	6.7									
8	17	4.5	−0.67								

29. Repeat Exercise 28, constructing three natural splines using Algorithm 3.4.

3.5 Parametric Curves

None of the techniques we have developed can be used to generate curves of the form shown in Figure 3.13 since this curve cannot be expressed as a function of one coordinate variable in terms of the other. In this section we will see how to represent general curves by using a parameter to express both the x- and y-coordinate variables. This technique can be extended to represent general curves and surfaces in space.

Figure 3.13

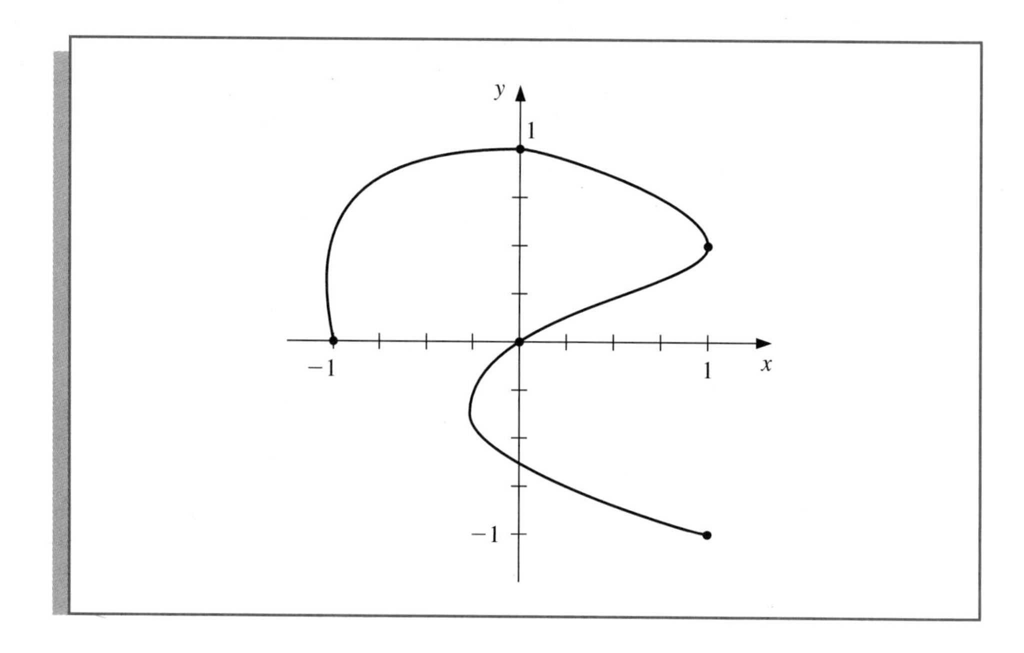

A straightforward parametric technique for determining a polynomial or piecewise polynomial to connect the points $(x_0, y_0), (x_1, y_1), \ldots, (x_n, y_n)$ in the order given is to use a parameter t on an interval $[t_0, t_n]$, with $t_0 < t_1 < \cdots < t_n$, and construct approximation functions with

$$x_i = x(t_i) \quad \text{and} \quad y_i = y(t_i) \quad \text{for each } i = 0, 1, \ldots, n.$$

The following example demonstrates the technique in the case where both approximating functions are Lagrange interpolating polynomials:

EXAMPLE 1 Construct a pair of Lagrange polynomials to approximate the curve shown in Figure 3.13, using the data points shown on the curve.

There is flexibility in choosing the parameter, and we will choose the points $\{t_i\}$ equally spaced in [0, 1]. In this case, we have the data in Table 3.18.

Table 3.18

i	0	1	2	3	4
t_i	0	0.25	0.5	0.75	1
x_i	-1	0	1	0	1
y_i	0	1	0.5	0	-1

This produces the interpolating polynomials

$$x(t) = \left(\left(\left(64t - \frac{352}{3}\right)t + 60\right)t - \frac{14}{3}\right)t - 1$$

and

$$y(t) = \left(\left(\left(-\frac{64}{3}t + 48\right)t - \frac{116}{3}\right)t + 11\right)t$$

Plotting this parametric system produces the graph in Figure 3.14. Although it passes through the required points and has the same basic shape, it is quite a crude approximation to the original curve. A more accurate approximation would require additional nodes, with the accompanying increase in computation. ■

Hermite and spline curves can be generated in a similar manner, but again we have extensive computational effort.

Applications in computer graphics require the rapid generation of smooth curves that can be easily and quickly modified. In addition, changing one portion of these curves should have little or no effect on other portions of the curves. This eliminates the use of interpolating polynomials and splines since changing one portion of these curves affects the whole curve, which is undesirable both from an aesthetic and computational standpoint.

The choice of curve for use in computer graphics is generally a form of the piecewise cubic Hermite polynomial. Each portion of a cubic Hermite polynomial is completely

Figure 3.14

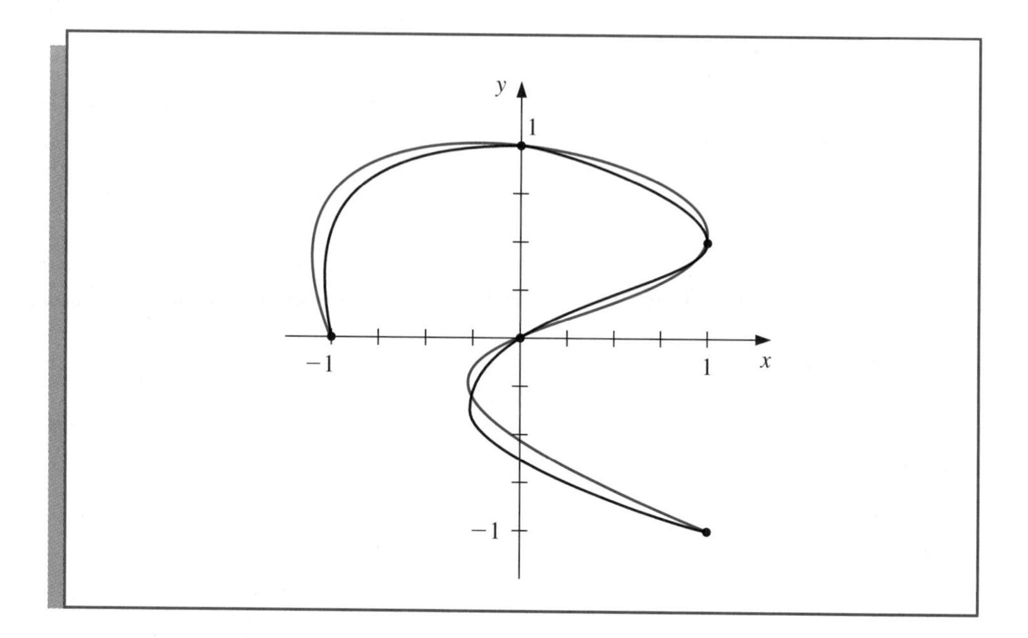

determined by specifying its endpoints and the derivatives at these endpoints. As a consequence, one portion of the curve can be changed while leaving most of the curve the same. Only the adjacent portions must be modified to ensure smoothness at the endpoints. The computations can be performed quickly and the curve can be modified a section at a time.

The problem with Hermite interpolation is the need to specify the derivatives at the endpoints of each section of the curve. Suppose the curve has $n + 1$ data points $(x_0, y_0), \ldots, (x_n, y_n)$ and we wish to parameterize the cubic to allow complex features. Then we must specify $x'(t_i)$ and $y'(t_i)$ for each $i = 0, 1, \ldots, n$, where $(x_i, y_i) = (x(t_i), y(t_i))$. This is not as difficult as it would first appear, however, since each portion can be generated independently, provided that we ensure that the derivatives at the endpoints of each portion match those in the adjacent portion.

Essentially, then, we can simplify the process to one of determining, on each section of the curve, a pair of cubic Hermite polynomials in the parameter t, where $t_0 = 0$ and $t_1 = 1$, given the endpoint data $(x(0), y(0))$, $(x(1), y(1))$ and the derivatives dy/dx (at $t = 0$) and dy/dx (at $t = 1$). Notice, however, that we are specifying only six conditions, and each cubic polynomial has four parameters, for a total of eight. This situation provides considerable flexibility in choosing the pair of cubic Hermite polynomials to satisfy these conditions. The reason this flexibility occurs is that the natural form for determining $x(t)$ and $y(t)$ requires that we specify $x'(0)$, $x'(1)$, $y'(0)$, and $y'(1)$. The explicit Hermite curve in x and y requires specifying only the quotients

$$\frac{dy}{dx}(t = 0) = \frac{y'(0)}{x'(0)} \quad \text{and} \quad \frac{dy}{dx}(t = 1) = \frac{y'(1)}{x'(1)}.$$

By multiplying $x'(0)$ and $y'(0)$ by a common scaling factor, the tangent line to the curve at $(x(0), y(0))$ remains the same, but the shape of the curve varies. The larger the scaling factor, the closer the curve comes to approximating the tangent line near $(x(0), y(0))$. A similar situation exists at the other endpoint $(x(1), y(1))$.

To further simplify the process, the derivative at an endpoint is specified graphically by describing a second point, called a *guidepoint*, on the desired tangent line. The farther the guidepoint is from the node, the larger the scaling factor and the more closely the curve approximates the tangent line near the node.

In Figure 3.15, the nodes occur at (x_0, y_0) and (x_1, y_1), the guidepoint for (x_0, y_0) is $(x_0 + \alpha_0, y_0 + \beta_0)$, and the guidepoint for (x_1, y_1) is $(x_1 - \alpha_1, y_1 - \beta_1)$. The cubic Hermite polynomial $x(t)$ on $[0, 1]$ must satisfy

$$x(0) = x_0, \quad x(1) = x_1, \quad x'(0) = \alpha_0, \quad \text{and} \quad x'(1) = \alpha_1.$$

Figure 3.15

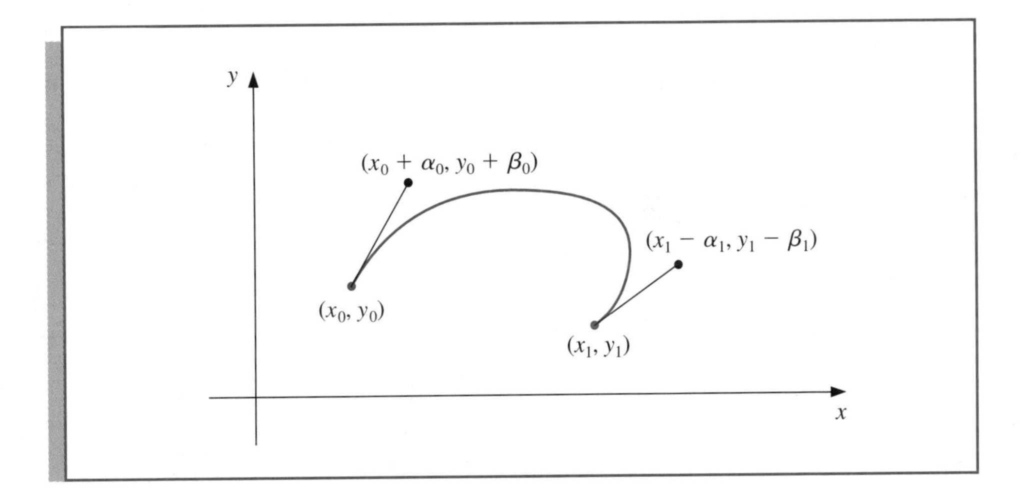

It is easily verified that the unique cubic polynomial satisfying these conditions is

$$x(t) = [2(x_0 - x_1) + (\alpha_0 + \alpha_1)]t^3 + [3(x_1 - x_0) - (\alpha_1 + 2\alpha_0)]t^2 + \alpha_0 t + x_0. \tag{3.23}$$

In a similar manner, the unique cubic polynomial satisfying

$$y(0) = y_0, \quad y(1) = y_1, \quad y'(0) = \beta_0, \quad \text{and} \quad y'(1) = \beta_1$$

is

$$y(t) = [2(y_0 - y_1) + (\beta_0 + \beta_1)]t^3 + [3(y_1 - y_0) - (\beta_1 + 2\beta_0)]t^2 + \beta_0 t + y_0. \tag{3.24}$$

EXAMPLE 2 The graphs in Figure 3.16 show some possibilities of the curves produced by Eqs. (3.23) and (3.24) when the nodes are $(0, 0)$ and $(1, 0)$ and the slopes at these nodes are 1 and -1, respectively. The specification of the slope at the endpoints requires only that $\alpha_0 = \beta_0$ and $\alpha_1 = -\beta_1$, since the ratios $\alpha_0/\beta_0 = 1$ and $\alpha_1/\beta_1 = -1$ give the slopes at the left and right endpoints, respectively. ■

Figure 3.16

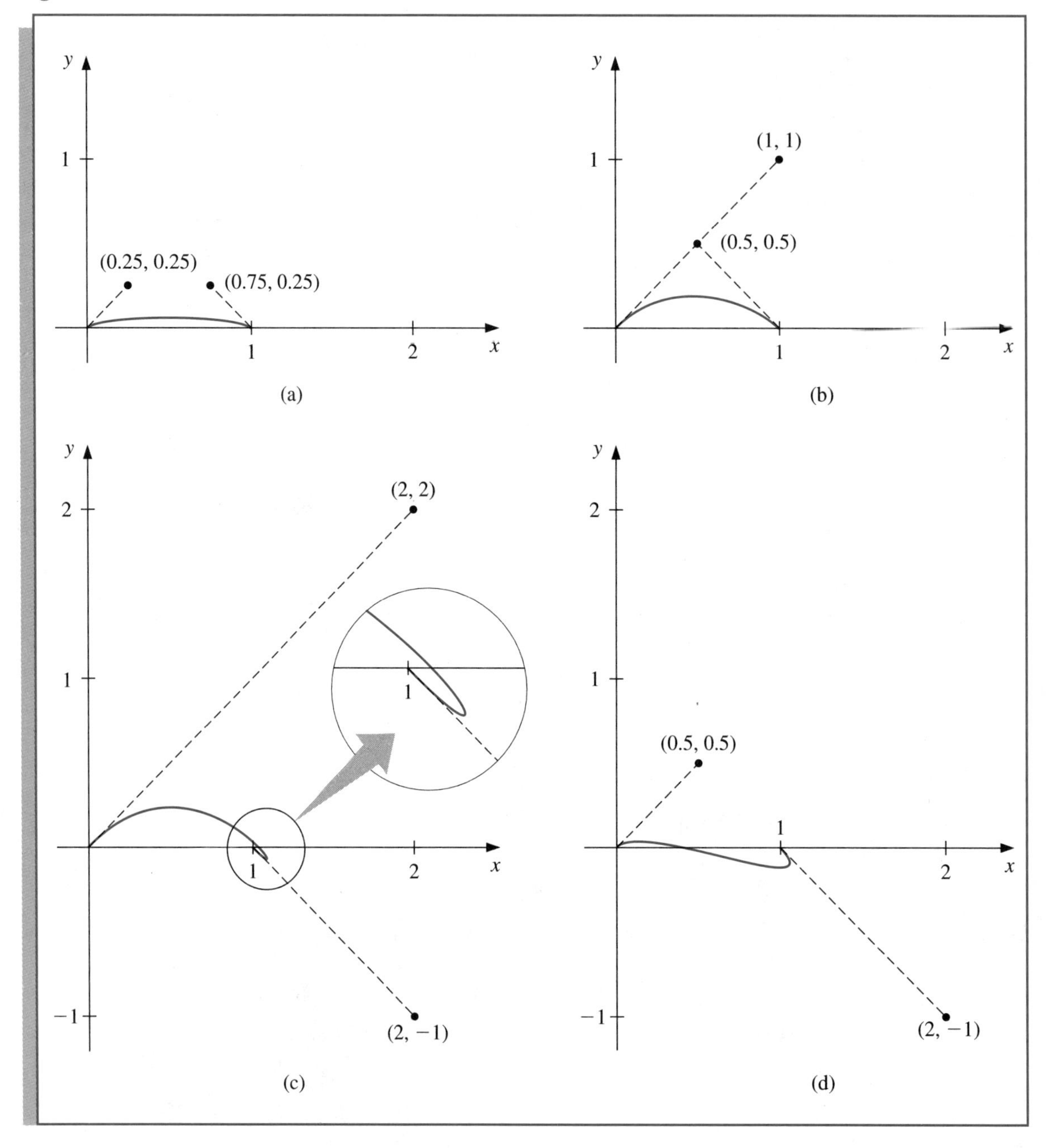

The standard procedure for determining curves in an interactive graphics mode is to first use an input device, such as a mouse or trackball, to set the nodes appropriately. The guidepoints are then placed to generate a first approximation to the desired curve. These can be set manually, but most graphics systems permit you to use your input device to draw the curve on the screen freehand and will select appropriate nodes and guidepoints for your freehand curve.

The nodes and guidepoints can then be manipulated into a position that produces an aesthetically satisfying curve. Since the computation is minimal, the curve can generally be determined so quickly that the resulting change is seen immediately. Moreover, all the data needed to compute the curves are imbedded in the coordinates of the nodes and guidepoints, so no analytical knowledge is required of the user of the system.

Popular graphics programs use this type of system for their freehand graphic representations but in a slightly modified form. The Hermite cubics are described as Bézier polynomials, which incorporate a scaling factor of 3 when computing the derivatives at the endpoints. This modifies the parametric equations to

$$\textbf{(3.25)}\quad x(t) = [2(x_0 - x_1) + 3(\alpha_0 + \alpha_1)]t^3 + [3(x_1 - x_0) - 3(\alpha_1 + 2\alpha_0)]t^2 + 3\alpha_0 t + x_0,$$

and

$$\textbf{(3.26)}\quad y(t) = [2(y_0 - y_1) + 3(\beta_0 + \beta_1)]t^3 + [3(y_1 - y_0) - 3(\beta_1 + 2\beta_0)]t^2 + 3\beta_0 t + y_0,$$

for $0 \le t \le 1$, but this change is transparent to the user of the system.

Algorithm 3.6 constructs a set of Bézier curves based on the parametric equations in (3.25) and (3.26).

ALGORITHM **3.6**

Bézier Curve

To construct the cubic Bézier curves $C_0, \ldots, C_{n-1}$ in parametric form, where C_i is represented by

$$(x_i(t), y_i(t)) = (a_0^{(i)} + a_1^{(i)}t + a_2^{(i)}t^2 + a_3^{(i)}t^3, b_0^{(i)} + b_1^{(i)}t + b_2^{(i)}t^2 + b_3^{(i)}t^3),$$

for $0 \le t \le 1$, as determined by the left endpoint (x_i, y_i), left guidepoint (x_i^+, y_i^+), right endpoint (x_{i+1}, y_{i+1}), and right guidepoint (x_{i+1}^-, y_{i+1}^-) for each $i = 0, 1, \ldots, n-1$;

INPUT n; $(x_0, y_0), \ldots, (x_n, y_n)$; $(x_0^+, y_0^+), \ldots, (x_{n-1}^+, y_{n-1}^+)$; $(x_1^-, y_1^-), \ldots, (x_n^-, y_n^-)$.

OUTPUT coefficients $\{a_0^{(i)}, a_1^{(i)}, a_2^{(i)}, a_3^{(i)}, b_0^{(i)}, b_1^{(i)}, b_2^{(i)}, b_3^{(i)}$, for $0 \le i \le n-1\}$.

Step 1 For each $i = 0, 1, \ldots, n-1$ do Steps 2 and 3.

Step 2 Set $a_0^{(i)} = x_i$;
$b_0^{(i)} = y_i$;
$a_1^{(i)} = 3(x_i^+ - x_i)$;
$b_1^{(i)} = 3(y_i^+ - y_i)$;
$a_2^{(i)} = 3(x_i + x_{i+1}^- - 2x_i^+)$;
$b_2^{(i)} = 3(y_i + y_{i+1}^- - 2y_i^+)$;
$a_3^{(i)} = x_{i+1} - x_i + 3x_i^+ - 3x_{i+1}^-$;
$b_3^{(i)} = y_{i+1} - y_i + 3y_i^+ - 3y_{i+1}^-$;

Step 3 OUTPUT $(a_0^{(i)}, a_1^{(i)}, a_2^{(i)}, a_3^{(i)}, b_0^{(i)}, b_1^{(i)}, b_2^{(i)}, b_3^{(i)})$.

Step 4 STOP.

Three-dimensional curves are generated in a similar manner by additionally specifying third components $z_0, z_1, \ldots, z_n$ for the nodes and $z_0^+, z_1^+, \ldots, z_{n-1}^+$ and $z_1^-, z_2^-, \ldots, z_n^-$ for the guidepoints. The more difficult problem involving the representation of three-dimensional curves concerns the loss of the third dimension when the curve is projected onto a two-dimensional medium such as a computer screen or printer paper. Various projection techniques are used, but this topic lies within the realm of computer graphics. For an introduction to this topic and ways that the technique can be modified for surface representations, see one of the many books on computer graphics methods, such as [Hill,F].

EXERCISE SET 3.5

1. Let $(x_0, y_0) = (0, 0)$ and $(x_1, y_1) = (5, 2)$ be the endpoints of a curve. Use the given guidepoints to construct parametric cubic Hermite approximations $(x(t), y(t))$ to the curve and graph the approximations.

   a. (1, 1) and (6, 1) **b.** (0.5, 0.5) and (5.5, 1.5)

   c. (1, 1) and (6, 3) **d.** (2, 2) and (7, 0)

2. Repeat Exercise 1 using cubic Bézier polynomials.

3. Construct and graph the cubic Bézier polynomials given the following points and guidepoints.

   a. Point (1, 1) with guidepoint (1.5, 1.25) to point (6, 2) with guidepoint (7, 3)

   b. Point (1, 1) with guidepoint (1.25, 1.5) to point (6, 2) with guidepoint (5, 3)

   c. Point (0, 0) with guidepoint (0.5, 0.5) to point (4, 6) with entering guidepoint (3.5, 7) and exiting guidepoint (4.5, 5) to point (6, 1) with guidepoint (7, 2)

   d. Point (0, 0) with guidepoint (0.5, 0.25) to point (2, 1) with entering guidepoint (3, 1) and exiting guidepoint (3, 1) to point (4, 0) with entering guidepoint (5, 1) and exiting guidepoint (3, −1) to point (6, −1) with guidepoint (6.5, −0.25)

4. Use the data in the following table and Algorithm 3.6 to approximate the letter $\mathcal{N}$.

i	x_i	y_i	α_i	β_i	α_i'	β_i'
0	3	6	3.3	6.5		
1	2	2	2.8	3.0	2.5	2.5
2	6	6	5.8	5.0	5.0	5.8
3	5	2	5.5	2.2	4.5	2.5
4	6.5	3			6.4	2.8

5. Suppose a cubic Bézier polynomial is placed through (u_0, v_0) and (u_3, v_3) with guidepoints (u_1, v_1) and (u_2, v_2), respectively.

   a. Derive the parametric equations for $u(t)$ and $v(t)$ assuming that

$$u(0) = u_0, \quad u(1) = u_3, \quad u'(0) = u_1 - u_0, \quad u'(1) = u_3 - u_2.$$

and

$$v(0) = v_0, \quad v(1) = v_3, \quad v'(0) = v_1 - v_0, \quad v'(1) = v_3 - v_2.$$

b. Let $f(i/3) = u_i$ for $i = 0, 1, 2, 3$ and $g(i/3) = v_i$ for $i = 0, 1, 2, 3$. Show that the Bernstein polynomial of degree three in t for f is $u(t)$, and the Bernstein polynomial of degree three in t for g is $v(t)$. (See Exercise 29 of Section 3.1.)

3.6 Survey of Methods and Software

In this chapter we have considered approximating a function using polynomials and piecewise polynomials. The function can be specified by a given defining equation or by providing points in the plane through which the graph of the function passes. A set of nodes $x_0, x_1, \ldots, x_n$ is given in each case, and more information, such as the value of various derivatives, may also be required. The problem is to find an approximating function that satisfies the conditions specified by these data.

The interpolating polynomial $P(x)$ is the polynomial of least degree that satisfies, for a function f,

$$P(x_i) = f(x_i) \quad \text{for each } i = 0, 1, \ldots, n.$$

Although there is a unique interpolating polynomial, it can take many different forms. The Lagrange form is most often used for interpolating tables when n is small and for deriving formulas for approximating derivatives and integrals. Neville's method is used for evaluating several interpolating polynomials at the same value of x. Newton's forms of the polynomial are more appropriate for computation and are also used extensively for deriving formulas for solving differential equations. However, polynomial interpolation has the inherent weakness of oscillation, particularly if the number of nodes is large. In this case there are other methods that can be better applied.

The Hermite polynomials interpolate a function and its derivative at the nodes. They can be very accurate but require more information about the function being approximated. The Hermite polynomials also exhibit oscillation weakness when there are a large number of nodes.

The most commonly used form of interpolation is piecewise polynomial interpolation. If function and derivative values are available, piecewise cubic Hermite interpolation is recommended. This is the preferred method for interpolating values of a function that is the solution to a differential equation. When only the function values are available, free cubic spline interpolation can be used. This forces the second derivative of the spline to be zero at the endpoints. Other cubic splines require additional data. For example, the clamped cubic spline needs values of the derivative of the function at the endpoints of the interval.

There are other methods of interpolation that are commonly used. Trigonometric interpolation is used with large amounts of data when the function has a periodic nature. In particular, the Fast Fourier Transform discussed in Chapter 8 is employed. Interpolation by rational functions is also used. If the data are suspected to be inaccurate, smoothing techniques can be applied, and some form of least squares fit of data is recommended. Polynomials, trigonometric functions, rational functions, and splines can be used in least squares fitting of data. We consider these topics in Chapter 8.

Interpolation routines included in the IMSL Library are based on the book *A Practical Guide to Splines* by Carl de Boor [Deb] and use interpolation by cubic splines. The subroutine CSDEC is for interpolation by cubic splines with user-supplied end conditions, CSPER is for interpolation by cubic splines with periodic end conditions, and CSHER is for interpolation by quasi-Hermite piecewise polynomials. The subroutine CSDEC incorporates Algorithms 3.4 and 3.5. The subroutine CSINT uses the not-a-knot condition mentioned at the end of Section 3.4. There are also cubic splines to minimize oscillations or to preserve concavity. Methods for two-dimensional interpolation by bicubic splines are also included.

The NAG library contains the subroutines E01AEF for polynomial and Hermite interpolation, E01BAF for cubic spline interpolation, and E01BEF for piecewise cubic Hermite interpolation. The subroutine E01ABF is used to interpolate data at equally spaced points. The routine E01AAF is applied if the data are given at unequally spaced points. NAG also contains subroutines for interpolating functions of two variables.

The MATLAB function POLYFIT can be used to find an interpolating function of degree at most n that passes through $n + 1$ specified points. Cubic splines can be produced with the function SPLINE.

Maple constructs an interpolating polynomial with the command

```
>interp(X,Y,x);
```

where `X` is the list `[x[0],x[1],...,x[n]]`, `Y` is the list `[f(x[0]),f(x[1]),...,f(x[n])]`, and `x` is the variable to be used. For example, the interpolating polynomial in Example 1 of Section 3.1 can be constructed with the command

```
>p:=interp([2,2.5,4], [0.5,0.4,0.25],x);
```

The result is

$$p := .0500000000x^2 - .4250000000x + 1.150000000.$$

To approximate $f(3) = \frac{1}{3}$ enter

```
>subs(x=3,p);
```

to obtain .3250000000.

The natural cubic spline can also be constructed with Maple. First enter

```
>readlib(spline);
```

to make the package available. With `X` and `Y` as in the preceding paragraph the command

```
>spline(X,Y,x,3);
```

constructs the natural cubic spline interpolating `X` = [x[0],..., x[n]] and `Y` = [y[0],..., y[n]], where `x` is the variable and `3` refers to the degree of the cubic spline. Linear and quadratic splines can also be created.

General references to the methods in this chapter are the books by Powell [Po] and by Davis [Da2]. The seminal paper on splines is due to Schoenberg [Scho]. Important books on splines are by Schultz [Schul], De Boor [Deb], and Schumaker [Schum]. A recent book by Dierckx [Di] is also recommended for those needing more information about splines.

CHAPTER

4

Numerical Differentiation and Integration

■ ■ ■

A sheet of corrugated roofing is constructed using a machine that presses a flat sheet of aluminum into one whose cross section has the form of a sine wave.

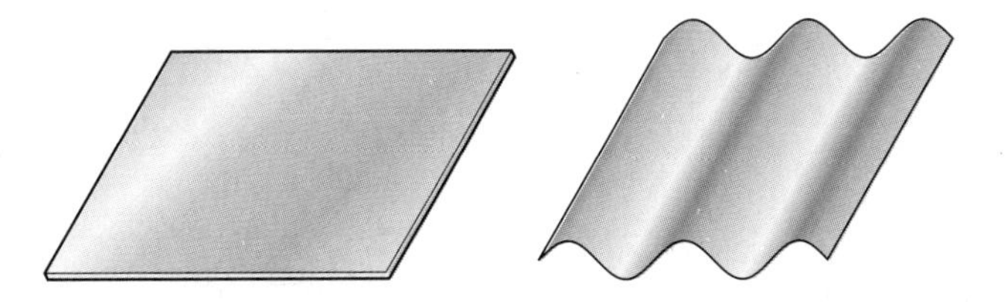

A corrugated sheet 4 ft long is needed, the height of each wave is 1 in. from the center line, and each wave has a period of approximately 2π in. The problem of finding the length of the initial flat sheet is one of determining the length of the curve given by $f(x) = \sin x$ from $x = 0$ in. to $x = 48$ in. From calculus we know that this length is

$$L = \int_0^{48} \sqrt{1 + \left(\frac{df(x)}{dx}\right)^2}\, dx = \int_0^{48} \sqrt{1 + (\cos x)^2}\, dx,$$

so the problem reduces to evaluating this integral. Although the sine function is one of the most common mathematical functions, the calculation of its length gives rise to an elliptic integral of the second kind, which cannot be evaluated by ordinary methods. Approximation methods are

developed in this chapter to reduce problems of this type to elementary exercises. This particular problem is considered in Exercise 21 of Section 4.4 and Exercise 10 of Section 4.5.

We mentioned in the introduction to Chapter 3 that one reason for using algebraic polynomials to approximate an arbitrary set of data is that given any continuous function defined on a closed interval, there exists a polynomial that is arbitrarily close to the function at every point in the interval. Also, the derivatives and integrals of polynomials are easily obtained and evaluated. It should not be surprising, then, that most procedures for approximating integrals and derivatives use the polynomials that approximate the function.

4.1 Numerical Differentiation

The derivative of the function f at x_0 is

$$f'(x_0) = \lim_{h\to 0} \frac{f(x_0+h) - f(x_0)}{h}.$$

To approximate this number, suppose first that $x_0 \in (a, b)$, where $f \in C^2[a, b]$, and that $x_1 = x_0 + h$ for some $h \neq 0$ that is sufficiently small to ensure that $x_1 \in [a, b]$. We construct the first Lagrange polynomial $P_{0,1}(x)$ for f determined by x_0 and x_1, with its error term:

$$\begin{aligned} f(x) &= P_{0,1}(x) + \frac{(x-x_0)(x-x_1)}{2!} f''(\xi(x)). \\ &= \frac{f(x_0)(x-x_0-h)}{-h} + \frac{f(x_0+h)(x-x_0)}{h} \\ &\quad + \frac{(x-x_0)(x-x_0-h)}{2} f''(\xi(x)) \end{aligned}$$

for some $\xi(x)$ in $[a, b]$. Differentiating gives

$$\begin{aligned} f'(x) &= \frac{f(x_0+h) - f(x_0)}{h} + D_x\left[\frac{(x-x_0)(x-x_0-h)}{2} f''(\xi(x))\right] \\ &= \frac{f(x_0+h) - f(x_0)}{h} + \frac{2(x-x_0)-h}{2} f''(\xi(x)) \\ &\quad + \frac{(x-x_0)(x-x_0-h)}{2} D_x(f''(\xi(x))), \end{aligned}$$

so

$$f'(x) \approx \frac{f(x_0 + h) - f(x_0)}{h}.$$

One difficulty with this formula for approximating $f'(x)$ for arbitrary values of x is that we have no information about $D_x f''(\xi(x)) = f'''(\xi(x)) \cdot \xi'(x)$, so the truncation error cannot be estimated. When x is x_0, however, the coefficient of $D_x f''(\xi(x))$ is zero, and the formula simplifies to

$$\textbf{(4.1)} \qquad f'(x_0) = \frac{f(x_0 + h) - f(x_0)}{h} - \frac{h}{2} f''(\xi).$$

For small values of h, the difference quotient $[f(x_0 + h) - f(x_0)]/h$ can be used to approximate $f'(x_0)$ with an error bounded by $Mh/2$, where M is a bound on $f''(x)$ for $x \in [a, b]$. This formula is known as the **forward-difference formula** if $h > 0$ (see Figure 4.1) and the **backward-difference formula** if $h < 0$.

Figure 4.1

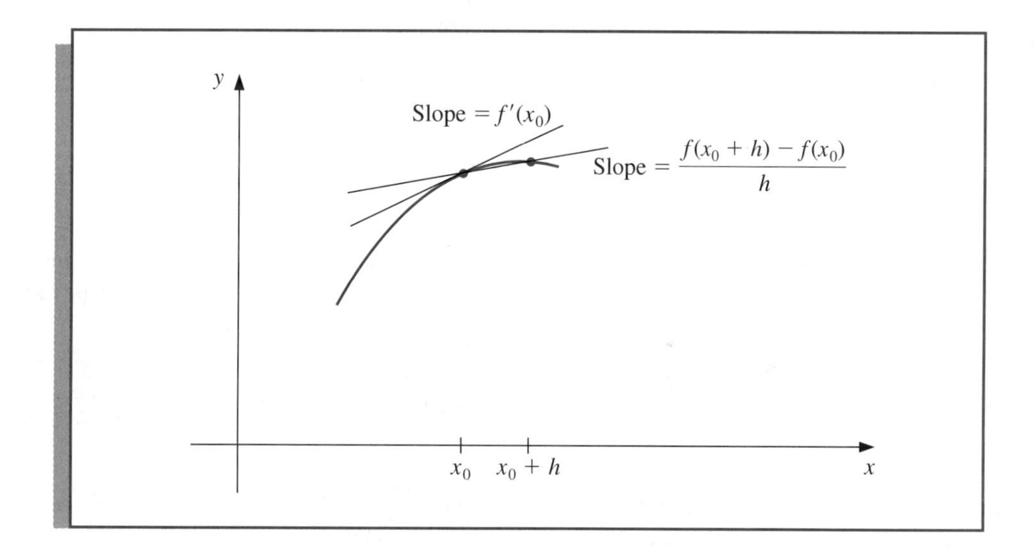

EXAMPLE 1 Let $f(x) = \ln x$ and $x_0 = 1.8$. The quotient

$$\frac{f(1.8 + h) - f(1.8)}{h}, \quad \text{for } h > 0,$$

is used to approximate $f'(1.8)$ with error

$$\frac{|hf''(\xi)|}{2} = \frac{|h|}{2\xi^2} \leq \frac{|h|}{2(1.8)^2}, \quad \text{where} \quad 1.8 < \xi < 1.8 + h.$$

The results in Table 4.1 are produced when $h = 0.1, 0.01$, and 0.001.

Table 4.1

h	$f(1.8+h)$	$\dfrac{f(1.8+h)-f(1.8)}{h}$	$\dfrac{\lvert h\rvert}{2(1.8)^2}$
0.1	0.64185389	0.5406722	0.0154321
0.01	0.59332685	0.5540180	0.0015432
0.001	0.58834207	0.5554013	0.0001543

Since $f'(x) = 1/x$, the exact value of $f'(1.8)$ is $0.55\overline{5}$, and the error bounds are appropriate. ■

To obtain more general derivative approximation formulas, suppose that $\{x_0, x_1, \ldots, x_n\}$ are $(n+1)$ distinct numbers in some interval I and that $f \in C^{n+1}(I)$. From Theorem 3.3,

$$f(x) = \sum_{k=0}^{n} f(x_k)L_k(x) + \frac{(x-x_0)\cdots(x-x_n)}{(n+1)!} f^{(n+1)}(\xi(x))$$

for some $\xi(x)$ in I, where $L_k(x)$ denotes the kth Lagrange coefficient polynomial for f at $x_0, x_1, \ldots, x_n$. Differentiating this expression gives

$$f'(x) = \sum_{k=0}^{n} f(x_k)L_k'(x) + D_x\left[\frac{(x-x_0)\cdots(x-x_n)}{(n+1!)}\right] f^{(n+1)}(\xi(x))$$
$$+ \frac{(x-x_0)\cdots(x-x_n)}{(n+1)!} D_x[f^{(n+1)}(\xi(x))].$$

Again, we have a problem estimating the truncation error unless x is one of the numbers x_j. In this case, the term involving $D_x[f^{(n+1)}(\xi(x))]$ is zero, and the formula becomes

$$f'(x_j) = \sum_{k=0}^{n} f(x_k)L_k'(x_j) + \frac{f^{(n+1)}(\xi(x_j))}{(n+1)!} \prod_{\substack{k=0\\k\neq j}}^{n} (x_j - x_k). \tag{4.2}$$

Equation (4.2) is called an **$(n+1)$-point formula** to approximate $f'(x_j)$, since a linear combination of the $n+1$ values $f(x_k)$ is used for $k = 0, 1, \ldots, n$.

In general, using more evaluation points in Eq. (4.2) produces greater accuracy, although the number of functional evaluations and growth of roundoff error discourages this somewhat. The most common formulas are those involving three and five evaluation points.

We first derive some useful three-point formulas and consider aspects of their errors. Since

$$L_0(x) = \frac{(x-x_1)(x-x_2)}{(x_0-x_1)(x_0-x_2)}, \quad \text{we have} \quad L_0'(x) = \frac{2x-x_1-x_2}{(x_0-x_1)(x_0-x_2)}.$$

Similarly,

$$L_1'(x) = \frac{2x-x_0-x_2}{(x_1-x_0)(x_1-x_2)} \quad \text{and} \quad L_2'(x) = \frac{2x-x_0-x_1}{(x_2-x_0)(x_2-x_1)}.$$

Hence, from Eq. (4.2),

(4.3)
$$f'(x_j) = f(x_0)\left[\frac{2x_j - x_1 - x_2}{(x_0 - x_1)(x_0 - x_2)}\right] + f(x_1)\left[\frac{2x_j - x_0 - x_2}{(x_1 - x_0)(x_1 - x_2)}\right]$$
$$+ f(x_2)\left[\frac{2x_j - x_0 - x_1}{(x_2 - x_0)(x_2 - x_1)}\right] + \frac{1}{6}f^{(3)}(\xi_j)\prod_{\substack{k=0\\k\neq j}}^{2}(x_j - x_k),$$

for each $j = 0, 1, 2$, where the notation ξ_j indicates that this point depends on x_j.

The three formulas from Eq. (4.3) become especially useful if the nodes are equally spaced, that is, when

$$x_1 = x_0 + h \quad \text{and} \quad x_2 = x_0 + 2h, \qquad \text{for some } h \neq 0.$$

We will assume equally spaced nodes throughout the remainder of this section.

Using Eq. (4.3) with $x_j = x_0$, $x_1 = x_0 + h$, and $x_2 = x_0 + 2h$ gives

$$f'(x_0) = \frac{1}{h}\left[-\frac{3}{2}f(x_0) + 2f(x_1) - \frac{1}{2}f(x_2)\right] + \frac{h^2}{3}f^{(3)}(\xi_0).$$

Doing the same for $x_j = x_1$ gives

$$f'(x_1) = \frac{1}{h}\left[-\frac{1}{2}f(x_0) + \frac{1}{2}f(x_2)\right] - \frac{h^2}{6}f^{(3)}(\xi_1),$$

and for $x_j = x_2$,

$$f'(x_2) = \frac{1}{h}\left[\frac{1}{2}f(x_0) - 2f(x_1) + \frac{3}{2}f(x_2)\right] + \frac{h^2}{3}f^{(3)}(\xi_2).$$

Since $x_1 = x_0 + h$ and $x_2 = x_0 + 2h$, these formulas can also be expressed as

$$f'(x_0) = \frac{1}{h}\left[-\frac{3}{2}f(x_0) + 2f(x_0 + h) - \frac{1}{2}f(x_0 + 2h)\right] + \frac{h^2}{3}f^{(3)}(\xi_0),$$

$$f'(x_0 + h) = \frac{1}{h}\left[-\frac{1}{2}f(x_0) + \frac{1}{2}f(x_0 + 2h)\right] - \frac{h^2}{6}f^{(3)}(\xi_1), \quad \text{and}$$

$$f'(x_0 + 2h) = \frac{1}{h}\left[\frac{1}{2}f(x_0) - 2f(x_0 + h) + \frac{3}{2}f(x_0 + 2h)\right] + \frac{h^2}{3}f^{(3)}(\xi_2).$$

As a matter of convenience, the variable substitution x_0 for $x_0 + h$ is used in the middle equation to change this formula to an approximation for $f'(x_0)$. A similar change, x_0 for

$x_0 + 2h$, is used in the last equation. This gives three formulas for approximating $f'(x_0)$:

$$f'(x_0) = \frac{1}{2h}[-3f(x_0) + 4f(x_0+h) - f(x_0+2h)] + \frac{h^2}{3}f^{(3)}(\xi_0),$$

$$f'(x_0) = \frac{1}{2h}[-f(x_0-h) + f(x_0+h)] - \frac{h^2}{6}f^{(3)}(\xi_1), \quad \text{and}$$

$$f'(x_0) = \frac{1}{2h}[f(x_0-2h) - 4f(x_0-h) + 3f(x_0)] + \frac{h^2}{3}f^{(3)}(\xi_2)$$

Finally, note that since the last of these equations can be obtained from the first by simply replacing h with $-h$, there are actually only two formulas:

(4.4)
$$f'(x_0) = \frac{1}{2h}[-3f(x_0) + 4f(x_0+h) - f(x_0+2h)] + \frac{h^2}{3}f^{(3)}(\xi_0),$$

where ξ_0 lies between x_0 and $x_0 + 2h$, and

(4.5)
$$f'(x_0) = \frac{1}{2h}[f(x_0+h) - f(x_0-h)] - \frac{h^2}{6}f^{(3)}(\xi_1),$$

where ξ_1 lies between $(x_0 - h)$ and $(x_0 + h)$.

The error in Eq. (4.5) is approximately half the error in Eq. (4.4). This is because Eq. (4.5) uses data on both sides of x_0, and Eq. (4.4) uses data on only one side. Note also that f needs to be evaluated at only two points in Eq. (4.5), whereas in Eq. (4.4) three evaluations are needed. Figure 4.2 gives an illustration of the approximation produced from Eq. (4.5).

Figure 4.2

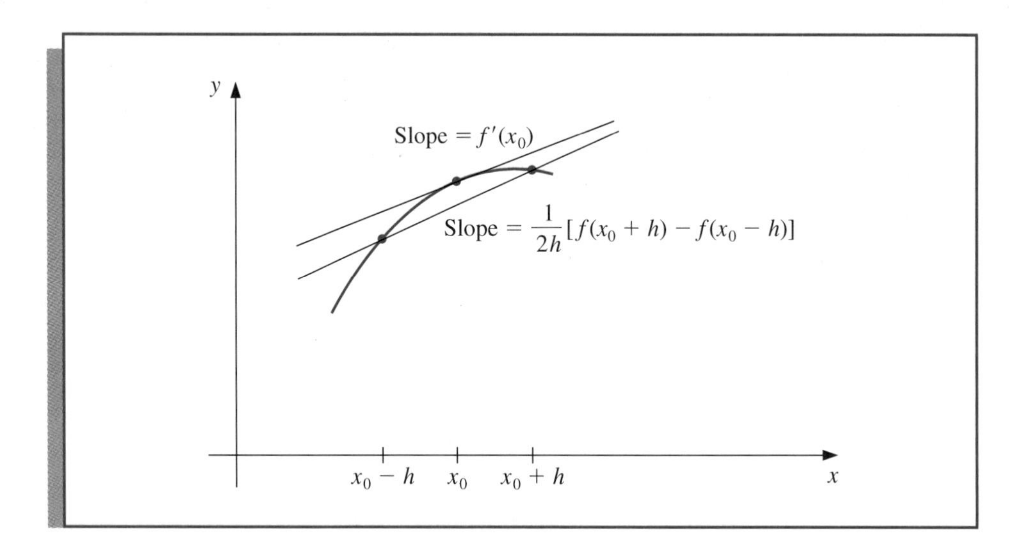

The approximation in Eq. (4.4) is useful near the ends of the interval I, since information about f outside the interval may not be available.

The methods presented in Eqs. (4.4) and (4.5) are called **three-point formulas** (even though the third point $f(x_0)$ does not appear in Eq. (4.5)). Similarly, there are methods known as **five-point formulas** that involve evaluating the function at two more points but whose error term is of the form $O(h^4)$. A derivation of Eq. (4.6) is considered in Section 4.2.

$$\textbf{(4.6)}\quad f'(x_0) = \frac{1}{12h}[f(x_0 - 2h) - 8f(x_0 - h) + 8f(x_0 + h) - f(x_0 + 2h)] + \frac{h^4}{30}f^{(5)}(\xi).$$

Another five-point formula that is useful, particularly with regard to the clamped cubic spline interpolation of Section 3.4, is

$$\textbf{(4.7)}\quad \begin{aligned} f'(x_0) = \frac{1}{12h}[&-25f(x_0) + 48f(x_0 + h) - 36f(x_0 + 2h) \\ &+ 16f(x_0 + 3h) - 3f(x_0 + 4h)] + \frac{h^4}{5}f^{(5)}(\xi), \end{aligned}$$

where ξ lies between x_0 and $x_0 + 4h$. Left-endpoint approximations can be found using this formula with $h > 0$ and right-endpoint approximations, with $h < 0$.

EXAMPLE 2 Values for $f(x) = xe^x$ are given in Table 4.2.

Table 4.2

x	$f(x)$
1.8	10.889365
1.9	12.703199
2.0	14.778112
2.1	17.148957
2.2	19.855030

Since $f'(x) = (x + 1)e^x$, $f'(2.0) = 22.167168$. Approximating $f'(2.0)$ using the various three- and five-point formulas produces the following results.

Three-Point Formulas

Using (4.4) with $h = 0.1$: $\frac{1}{0.2}[-3f(2.0) + 4f(2.1) - f(2.2)] = 22.032310$,

Using (4.4) with $h = -0.1$: $\frac{1}{-0.2}[-3f(2.0) + 4f(1.9) - f(1.8)] = 22.054525$,

Using (4.5) with $h = 0.1$: $\frac{1}{0.2}[f(2.1) - f(1.9)] = 22.228790$,

Using (4.5) with $h = 0.2$: $\frac{1}{0.4}[f(2.2) - f(1.8)] = 22.414163$.

Five-Point Formula

Using (4.6) with $h = 0.1$ (the only five-point formula applicable):

$$\frac{1}{1.2}[f(1.8) - 8f(1.9) + 8f(2.1) - f(2.2)] = 22.166999.$$

The errors in the formulas are approximately

$$1.35 \times 10^{-1}, \quad 1.13 \times 10^{-1}, \quad -6.16 \times 10^{-2}, \quad -2.47 \times 10^{-1}, \quad \text{and} \quad 1.69 \times 10^{-4},$$

respectively. Clearly, the five-point formula gives the superior result. Note also that the error from Eq. (4.5) with $h = 0.1$ is approximately half of the magnitude of the error produced using Eq. (4.4) with either $h = 0.1$ or $h = -0.1$. ■

Methods can also be derived to find approximations to higher derivatives of a function using only tabulated values of the function at various points. The derivation is algebraically tedious, however, so only a representative procedure will be presented.

Expand a function f in a third Taylor polynomial about a point x_0 and evaluate at $x_0 + h$ and $x_0 - h$. Then

$$f(x_0 + h) = f(x_0) + f'(x_0)h + \frac{1}{2}f''(x_0)h^2 + \frac{1}{6}f'''(x_0)h^3 + \frac{1}{24}f^{(4)}(\xi_1)h^4$$

and

$$f(x_0 - h) = f(x_0) - f'(x_0)h + \frac{1}{2}f''(x_0)h^2 - \frac{1}{6}f'''(x_0)h^3 + \frac{1}{24}f^{(4)}(\xi_{-1})h^4,$$

where $x_0 - h < \xi_{-1} < x_0 < \xi_1 < x_0 + h$.

If we add these equations, we obtain

$$f(x_0 + h) + f(x_0 - h) = 2f(x_0) + f''(x_0)h^2 + \frac{1}{24}[f^{(4)}(\xi_1) + f^{(4)}(\xi_{-1})]h^4.$$

Solving this equation for $f''(x_0)$ gives

$$\textbf{(4.8)} \quad f''(x_0) = \frac{1}{h^2}[f(x_0 - h) - 2f(x_0) + f(x_0 + h)] - \frac{h^2}{24}[f^{(4)}(\xi_1) + f^{(4)}(\xi_{-1})].$$

Suppose $f^{(4)}$ is continuous on $[x_0 - h, x_0 + h]$. Since $\frac{1}{2}[f^{(4)}(\xi_1) + f^{(4)}(\xi_{-1})]$ is between $f^{(4)}(\xi_1)$ and $f^{(4)}(\xi_{-1})$, the Intermediate Value Theorem implies that a number ξ exists between ξ_1 and ξ_{-1}, and hence in $(x_0 - h, x_0 + h)$, with

$$f^{(4)}(\xi) = \frac{1}{2}\left[f^{(4)}(\xi_1) + f^{(4)}(\xi_{-1})\right].$$

This permits us to rewrite Eq. (4.8) as

(4.9)
$$f''(x_0) = \frac{1}{h^2}[f(x_0 - h) - 2f(x_0) + f(x_0 + h)] - \frac{h^2}{12}f^{(4)}(\xi)$$

for some ξ, where $x_0 - h < \xi < x_0 + h$.

EXAMPLE 3 For the data given in Example 2 for $f(x) = xe^x$ we can use Eq. (4.9) to approximate $f''(2.0)$. Since $f''(x) = (x+2)e^x$ the exact value is $f''(2.0) = 29.556224$. Using (4.9) with $h = 0.1$ gives

$$f''(2.0) \approx \frac{1}{0.01}[f(1.9) - 2f(2.0) + f(2.1)] = 29.593200.$$

Using (4.9) with $h = 0.2$ gives

$$f''(2.0) \approx \frac{1}{0.04}[f(1.8) - 2f(2.0) + f(2.2)] = 29.704275.$$

The errors are approximately -3.70×10^{-2} and -1.48×10^{-1}, respectively. ■

A particularly important subject in the study of numerical differentiation is the effect roundoff error plays in the approximation. Let us examine Eq. (4.5):

$$f'(x_0) = \frac{1}{2h}[f(x_0 + h) - f(x_0 - h)] - \frac{h^2}{6}f^{(3)}(\xi_1)$$

more closely. Suppose that in evaluating $f(x_0 + h)$ and $f(x_0 - h)$ we encounter roundoff errors $e(x_0 + h)$ and $e(x_0 - h)$. Then our computed values $\tilde{f}(x_0 + h)$ and $\tilde{f}(x_0 - h)$ are related to the true values $f(x_0 + h)$ and $f(x_0 - h)$ by the formulas

$$f(x_0 + h) = \tilde{f}(x_0 + h) + e(x_0 + h)$$

and

$$f(x_0 - h) = \tilde{f}(x_0 - h) + e(x_0 - h).$$

The total error in the approximation,

$$f'(x_0) - \frac{\tilde{f}(x_0 + h) - \tilde{f}(x_0 - h)}{2h} = \frac{e(x_0 + h) - e(x_0 - h)}{2h} - \frac{h^2}{6}f^{(3)}(\xi_1),$$

will have a part due to roundoff error and a part due to truncation error. If we assume that the roundoff errors $e(x_0 \pm h)$ are bounded by some number $\varepsilon > 0$ and that the third derivative of f is bounded by a number $M > 0$, then

$$\left| f'(x_0) - \frac{\tilde{f}(x_0 + h) - \tilde{f}(x_0 - h)}{2h} \right| \leq \frac{\varepsilon}{h} + \frac{h^2}{6}M.$$

To reduce the truncation error, $h^2M/6$, we must reduce h. But as h is reduced, the roundoff error ε/h grows. In practice, then, it is seldom advantageous to let h be too small since the roundoff error will dominate the calculations.

EXAMPLE 4 Consider using the values in Table 4.3 to approximate $f'(0.900)$, where $f(x) = \sin x$. The true value is $\cos 0.900 = 0.62161$.

Table 4.3

x	$\sin x$	x	$\sin x$
0.800	0.71736	0.901	0.78395
0.850	0.75128	0.902	0.78457
0.880	0.77074	0.905	0.78643
0.890	0.77707	0.910	0.78950
0.895	0.78021	0.920	0.79560
0.898	0.78208	0.950	0.81342
0.899	0.78270	1.000	0.84147

Using the formula

$$f'(0.900) \approx \frac{f(0.900 + h) - f(0.900 - h)}{2h}$$

with different values of h gives the approximations in Table 4.4.

Table 4.4

h	Approximation to $f'(0.900)$	Error
0.001	0.62500	0.00339
0.002	0.62250	0.00089
0.005	0.62200	0.00039
0.010	0.62150	−0.00011
0.020	0.62150	−0.00011
0.050	0.62140	−0.00021
0.100	0.62055	−0.00106

The optimal choice for h appears to lie between 0.005 and 0.05. If we perform some analysis on the error term,

$$e(h) = \frac{\varepsilon}{h} + \frac{h^2}{6}M,$$

we can use calculus to verify (see Exercise 23) that a minimum for e occurs at $h = \sqrt[3]{3\varepsilon/M}$, where

$$M = \max_{x\in[0.800,1.00]} |f'''(x)| = \max_{x\in[0.800,1.00]} |\cos x| = \cos 0.8 \approx 0.69671.$$

Since values of f are given to five decimal places, it is reasonable to assume that the roundoff error is bounded by $\varepsilon = 0.000005$. Therefore, the optimal choice of h is approximately

$$h = \sqrt[3]{\frac{3(0.000005)}{0.69671}} \approx 0.028,$$

which is consistent with the results in Table 4.4.

In practice, we cannot compute an optimal h to use in approximating the derivative since we have no knowledge of the third derivative of the function. But we must remain aware that reducing the step size will not always improve the approximation. ■

Although we have considered only the roundoff-error problems that are presented by the three-point formula Eq. (4.5), similar difficulties occur with all the differentiation formulas. The reason can be traced to the need to divide by a power of h. As we found in Section 1.2 (see, in particular, Example 3), division by small numbers tends to exaggerate roundoff error and should be avoided if possible. In the case of numerical differentiation, it is impossible to avoid the problem entirely, although the higher-order methods reduce the difficulty.

Keep in mind that as an approximation method, numerical differentiation is *unstable* since the small values of h needed to reduce truncation error also cause the roundoff error to grow. This is the first class of unstable methods we have encountered, and these techniques would be avoided if it were possible. However, in addition to being used for computational purposes, the formulas we have derived are needed for approximating the solutions of ordinary and partial-differential equations.

EXERCISE SET 4.1

1. Use the forward-difference formulas and backward-difference formulas to determine approximations that will complete the following tables.

a.

x	$f(x)$	$f'(x)$
0.5	0.4794	
0.6	0.5646	
0.7	0.6442	

b.

x	$f(x)$	$f'(x)$
0.0	0.00000	
0.2	0.74140	
0.4	1.3718	

2. The data in Exercise 1 were taken from the following functions. Compute the actual errors in Exercise 1 and find error bounds using the error formulas.

a. $f(x) = \sin x$ **b.** $f(x) = e^x - 2x^2 + 3x - 1$

3. Use the most appropriate three-point formula to determine approximations that will complete the following tables.

a.

x	$f(x)$	$f'(x)$
1.1	9.025013	
1.2	11.02318	
1.3	13.46374	
1.4	16.44465	

b.

x	$f(x)$	$f'(x)$
8.1	16.94410	
8.3	17.56492	
8.5	18.19056	
8.7	18.82091	

c.

x	$f(x)$	$f'(x)$
2.9	−4.827866	
3.0	−4.240058	
3.1	−3.496909	
3.2	−2.596792	

d.

x	$f(x)$	$f'(x)$
2.0	3.6887983	
2.1	3.6905701	
2.2	3.6688192	
2.3	3.6245909	

4. The data in Exercise 3 were taken from the following functions. Compute the actual errors in Exercise 3 and find error bounds using the error formulas.

a. $f(x) = e^{2x}$

b. $f(x) = x \ln x$

c. $f(x) = x \cos x - x^2 \sin x$

d. $f(x) = 2(\ln x)^2 + 3 \sin x$

5. Use the most accurate formula possible to determine approximations that will complete the following tables.

a.

x	$f(x)$	$f'(x)$
2.1	−1.709847	
2.2	−1.373823	
2.3	−1.119214	
2.4	−0.9160143	
2.5	−0.7470223	
2.6	−0.6015966	

b.

x	$f(x)$	$f'(x)$
−3.0	9.367879	
−2.8	8.233241	
−2.6	7.180350	
−2.4	6.209329	
−2.2	5.320305	
−2.0	4.513417	

6. The data in Exercise 5 were taken from the given functions. Compute the actual errors in Exercise 5 and find error bounds using the error formulas and Maple.

a. $f(x) = \tan x$

b. $f(x) = e^{x/3} + x^2$

7. Use the following data and the knowledge that the first five derivatives of f were bounded on $[1, 5]$ by 2, 3, 6, 12 and 23, respectively, to approximate $f'(3)$ as accurately as possible. Find a bound for the error.

x	1	2	3	4	5
$f(x)$	2.4142	2.6734	2.8974	3.0976	3.2804

8. Repeat Exercise 7, assuming instead that the third derivative of f is bounded on $[1, 5]$ by 4.

9. Repeat Exercise 1 using four-digit rounding arithmetic, and compare the errors to those in Exercise 2.

10. Repeat Exercise 3 using four-digit chopping arithmetic, and compare the errors to those in Exercise 4.

11. Repeat Exercise 5 using four-digit rounding arithmetic, and compare the errors to those in Exercise 6.

12. Consider the following table of data.

x	0.2	0.4	0.6	0.8	1.0
$f(x)$	0.9798652	0.9177710	0.8080348	0.6386093	0.3843735

a. Use all appropriate formulas to approximate $f'(0.4)$ and $f''(0.4)$.

b. Use all appropriate formulas to approximate $f'(0.6)$ and $f''(0.6)$.

13. Let $f(x) = \cos \pi x$. Use Eq. (4.9) and the values of $f(x)$ at $x = 0.25$, 0.5, and 0.75 to approximate $f''(0.5)$. Compare this result to the exact value and to the approximation found in Exercise 11 of Section 3.4. Explain why this method is particularly accurate for this problem. Find a bound for the error.

14. Let $f(x) = 3xe^x - \cos x$. Use the following data and Eq. (4.9) to approximate $f''(1.3)$ with $h = 0.1$ and $h = 0.01$.

x	1.20	1.29	1.30	1.31	1.40
$f(x)$	11.59006	13.78176	14.04276	14.30741	16.86187

Compare your results to $f''(1.3)$.

15. Consider the following table of data:

x	0.2	0.4	0.6	0.8	1.0
$f(x)$	0.9798652	0.9177710	0.8080348	0.6386093	0.3843735

a. Use Eq. (4.7) to approximate $f'(0.2)$.

b. Use Eq. (4.7) to approximate $f'(1.0)$.

c. Use Eq. (4.6) to approximate $f'(0.6)$.

16. Derive an $O(h^4)$ five-point formula to approximate $f'(x_0)$ that uses $f(x_0 - h)$, $f(x_0)$, $f(x_0 + h)$, $f(x_0 + 2h)$, and $f(x_0 + 3h)$. [*Hint*: Consider the expression $Af(x_0 - h) + Bf(x_0 + h) + Cf(x_0 + 2h) + Df(x_0 + 3h)$. Expand in fifth Taylor polynomials and choose A, B, C, and D appropriately.]

17. Use the formula derived in Exercise 16 and the data of Exercise 15 to approximate $f'(0.4)$ and $f'(0.8)$.

18. Analyze the roundoff errors, as in Example 4, for the formula

$$f'(x_0) = \frac{f(x_0 + h) - f(x_0)}{h} - \frac{h}{2}f''(\xi_0).$$

Find an optimal $h > 0$ in terms of M, a bound for f'' on $(x_0, x_0 + h)$.

19. In Exercise 7 of Section 3.3 data were given describing a car traveling on a straight road. That problem asked to predict the position and speed of the car when $t = 10$s. Use the following times and positions to predict the speed at each time listed.

Time	0	3	5	8	10	13
Distance	0	225	383	623	742	993

20. In a circuit with impressed voltage $\mathcal{E}(t)$ and inductance L, Kirchhoff's first law gives the relationship

$$\mathcal{E} = L\frac{di}{dt} + Ri,$$

where R is the resistance in the circuit and i is the current. Suppose we measure the current for several values of t and obtain:

t	1.00	1.01	1.02	1.03	1.0
i	3.10	3.12	3.14	3.18	3.24

where t is measured in seconds, i is in amperes, the inductance L is a constant 0.98 henries, and the resistance is 0.142 ohms. Approximate the voltage $\mathcal{E}$ at the values $t = 1.00, 1.01, 1.02, 1.03$, and 1.04.

21. All calculus students know that the derivative of a function f at x can be defined as

$$f'(x) = \lim_{h\to 0}\frac{f(x + h) - f(x)}{h}.$$

Choose your favorite function f, nonzero number x, and computer or calculator. Generate approximations $f_n'(x)$ to $f'(x)$ by

$$f_n'(x) = \frac{f(x + 10^{-n}) - f(x)}{10^{-n}}$$

for $n = 1, 2, \ldots, 20$ and describe what happens.

22. Derive a method for approximating $f'''(x_0)$, whose error term is of order h^2, by expanding the function f in a fifth Taylor polynomial about x_0 and evaluating at $x_0 \pm h$ and $x_0 \pm 2h$.

23. Consider the function

$$e(h) = \frac{\varepsilon}{h} + \frac{h^2}{6}M,$$

where M is a bound for the third derivative of a function. Show that $e(h)$ has a minimum at $\sqrt[3]{3\varepsilon/M}$.

4.2 Richardson's Extrapolation

Richardson's Extrapolation is used to generate high-accuracy results while using low-order formulas. Although the name attached to the method refers to a paper written by L. F. Richardson and J. A. Gaunt [RG] in 1927, the idea behind the technique is much older. An interesting article regarding the history and application of extrapolation can be found in [Joy].

Extrapolation can be applied whenever it is known that the approximation technique has an error term with a predictable form, one that depends on a parameter, usually the step size h. Suppose that for each number $h \neq 0$ we have a formula $N(h)$ that approximates an unknown value M and that the truncation error involved with the approximation has the form

$$M - N(h) = K_1 h + K_2 h^2 + K_3 h^3 + \cdots$$

for some collection of unknown, but nonzero, constants $K_1, K_2, K_3, \ldots$.

Since the truncation error is $O(h)$, we would expect, for example, that

$$M - N(0.1) \approx 0.1K_1, \quad M - N(0.01) \approx 0.01K_1,$$

and, in general, $M - N(h) \approx K_1 h$, unless there was a large variation in magnitude among the constants $K_1, K_2, K_3, \ldots$.

The object of extrapolation is to find an easy way to combine the rather inaccurate $O(h)$ approximations in an appropriate way to produce formulas with a higher-order truncation error. Suppose, for example, we could combine the $N(h)$ formulas so as to produce an $O(h^2)$ approximation formula, $\hat{N}(h)$, for M with

$$M - \hat{N}(h) = \hat{K}_2 h^2 + \hat{K}_3 h^3 + \cdots,$$

for some, again unknown, collection of constants $\hat{K}_1, \hat{K}_2, \ldots$. Then we would have

$$M - \hat{N}(0.1) \approx 0.01\hat{K}_2, \quad M - \hat{N}(0.01) \approx 0.0001\hat{K}_2,$$

and so on. If the constants K_1 and $\hat{K}_2$ are roughly of the same magnitude, and there is no reason to expect that they are not, then the $\hat{N}(h)$ approximations would be much better that the corresponding $N(h)$ approximations. The extrapolation continues by combining the $\hat{N}(h)$ approximations in a manner that produces formulas with $O(h^3)$ truncation error, and so on.

To see specifically how we can generate these higher-order formulas, let us consider the formula for approximating M of the form

(4.10) $$M = N(h) + K_1 h + K_2 h^2 + K_3 h^3 + \cdots.$$

Since the formula is assumed to hold for all positive h, consider the result when we replace the parameter h by half its value. Then we have the formula

$$M = N\left(\frac{h}{2}\right) + K_1\frac{h}{2} + K_2\frac{h^2}{4} + K_3\frac{h^3}{8} + \cdots.$$

Subtracting (4.10) from twice this equation eliminates the term involving K_1 and gives

$$M = \left[N\left(\frac{h}{2}\right) + \left(N\left(\frac{h}{2}\right) - N(h)\right)\right] + K_2\left(\frac{h^2}{2} - h^2\right) + K_3\left(\frac{h^3}{4} - h^3\right) + \cdots.$$

To facilitate the discussion, we define $N_1(h) \equiv N(h)$ and

$$N_2(h) = N_1\left(\frac{h}{2}\right) + \left[N_1\left(\frac{h}{2}\right) - N_1(h)\right].$$

Then we have the $O(h^2)$ approximation formula for M:

(4.11) $$M = N_2(h) - \frac{K_2}{2}h^2 - \frac{3K_3}{4}h^3 - \cdots.$$

If we now replace h by $h/2$ in this formula, we have

(4.12) $$M = N_2\left(\frac{h}{2}\right) - \frac{K_2}{8}h^2 - \frac{3K_3}{32}h^3 - \cdots.$$

This can be combined with Eq. (4.11) to eliminate the h^2 term. Specifically, subtracting (4.11) from 4 times Eq. (4.12) gives

$$3M = 4N_2\left(\frac{h}{2}\right) - N_2(h) + \frac{3K_3}{8}h^3 + \cdots,$$

which simplifies to the $O(h^3)$ formula for approximating M:

$$M = \left[N_2\left(\frac{h}{2}\right) + \frac{N_2(h/2) - N_2(h)}{3}\right] + \frac{K_3}{8}h^3 + \cdots.$$

By defining

$$N_3(h) \equiv N_2\left(\frac{h}{2}\right) + \frac{N_2(h/2) - N_2(h)}{3},$$

we have the $O(h^3)$ formula:

$$M = N_3(h) + \frac{K_3}{8}h^3 + \cdots.$$

The process is continued by constructing the $O(h^4)$ approximation

$$N_4(h) = N_3\left(\frac{h}{2}\right) + \frac{N_3(h/2) - N_3(h)}{7},$$

the $O(h^5)$ approximation

$$N_5(h) = N_4\left(\frac{h}{2}\right) + \frac{N_4(h/2) - N_4(h)}{15},$$

and so on. In general, if M can be written in the form

(4.13)
$$M = N(h) + \sum_{j=1}^{m-1} K_j h^j + O(h^m),$$

then for each $j = 2, 3, \ldots, m$, we have an $O(h^j)$ approximation of the form

(4.14)
$$N_j(h) = N_{j-1}\left(\frac{h}{2}\right) + \frac{N_{j-1}(h/2) - N_{j-1}(h)}{2^{j-1} - 1}.$$

These approximations are generated by rows in the order indicated by the numbered entries in Table 4.5. This is done to take best advantage of the highest-order formulas.

Table 4.5

$O(h)$	$O(h^2)$	$O(h^3)$	$O(h^4)$
1 $N_1(h) \equiv N(h)$			
2 $N_1\left(\frac{h}{2}\right) \equiv N\left(\frac{h}{2}\right)$	**3** $N_2(h)$		
4 $N_1\left(\frac{h}{4}\right) \equiv N\left(\frac{h}{4}\right)$	**5** $N_2\left(\frac{h}{2}\right)$	**6** $N_3(h)$	
7 $N_1\left(\frac{h}{8}\right) \equiv N\left(\frac{h}{8}\right)$	**8** $N_2\left(\frac{h}{4}\right)$	**9** $N_3\left(\frac{h}{2}\right)$	**10** $N_4(h)$

Extrapolation can be applied whenever the truncation error for a formula has the form

$$\sum_{j=1}^{m-1} K_j h^{\alpha_j} + O(h^{\alpha_m}),$$

for a collection of constants K_j and when $\alpha_1 < \alpha_2 < \alpha_3 < \cdots < \alpha_m$. In the next example we have $\alpha_j = 2j$.

EXAMPLE 1 The centered difference formula in Eq. (4.5) to approximate $f'(x_0)$ can be expressed with an error formula:

$$f'(x_0) = \frac{1}{2h}[f(x_0+h) - f(x_0-h)] - \frac{h^2}{6}f'''(x_0) - \frac{h^4}{120}f^{(5)}(x_0) - \cdots.$$

Since this error formula contains only even powers of h, extrapolation is more effective than as outlined in the opening discussion. In this case we have the $O(h^2)$ approximation

$$f'(x_0) = N_1(h) - \frac{h^2}{6}f'''(x_0) - \frac{h^4}{120}f^{(5)}(x_0) - \cdots, \tag{4.15}$$

where

$$N_1(h) \equiv N(h) = \frac{1}{2h}[f(x_0+h) - f(x_0-h)].$$

Replacing h by $h/2$ in this formula gives the approximation

$$f'(x_0) = N_1\left(\frac{h}{2}\right) - \frac{h^2}{24}f'''(x_0) - \frac{h^4}{1920}f^{(5)}(x_0) - \cdots.$$

Subtracting (4.15) from 4 times this equation eliminates the $O(h^2)$ term that involves $f'''(x_0)$ and gives

$$3f'(x_0) = 4N_1\left(\frac{h}{2}\right) - N_1(h) + \frac{h^4}{160}f^{(5)}(x_0) + \cdots.$$

Dividing by 3 provides the $O(h^4)$ formula

$$f'(x_0) = N_2(h) + \frac{h^4}{480}f^{(5)}(x_0) + \cdots,$$

where

$$N_2(h) = N_1\left(\frac{h}{2}\right) + \frac{N_1(h/2) - N_1(h)}{3}.$$

Continuing this procedure gives, for each $j = 2, 3, \ldots,$ the $O(h^{2j})$ approximation

$$N_j(h) = N_{j-1}\left(\frac{h}{2}\right) + \frac{N_{j-1}(h/2) - N_{j-1}(h)}{4^{j-1} - 1}.$$

Notice that the denominator of the quotient is now $4^{j-1} - 1$ instead of $2^{j-1} - 1$ because we are now eliminating powers of h^2 instead of powers of h. Since $(h/2)^2 = h^2/4$, the multipliers used to eliminate the powers of h^2 are powers of 4 instead of 2.

Suppose that $x_0 = 2.0$, $h = 0.2$, and $f(x) = xe^x$. Then

$$N_1(0.2) = N(0.2) = \frac{1}{0.4}[f(2.2) - f(1.8)] = 22.414160,$$

$$N_1(0.1) = N(0.1) = 22.228786,$$

and

$$N_1(0.05) = N(0.05) = 22.182564.$$

The extrapolation table for these data is shown in Table 4.6. The exact value of $f'(x) = xe^x + e^x$ at $x_0 = 2.0$ to six decimal places is 22.167168. ■

Table 4.6

$N_1(0.2) = 22.414160$		
$N_1(0.1) = 22.228786$	$N_2(0.2) = N_1(0.1) + \frac{N_1(0.1) - N_1(0.2)}{3}$	
	$= 22.166995$	
$N_1(0.05) = 22.182564$	$N_2(0.1) = N_1(0.05) + \frac{N_1(0.05) - N_1(0.1)}{3}$	$N_3(0.2) = N_2(0.1) + \frac{N_2(0.1) - N_2(0.2)}{15}$
	$= 22.167157$	$= 22.167168$

Since each column beyond the first in the extrapolation table is obtained by a simple averaging process, the technique can produce high-order approximations with minimal computational cost and roundoff error. However, as k increases, the roundoff error in $N_1(h/2^k)$ will generally increase because of the instability of numerical differentiation.

In Section 4.1, we discussed both three- and five-point methods for approximating $f'(x_0)$ given various functional values of f. The three-point methods were derived by differentiating a Lagrange interpolating polynomial for f. The five-point methods can be obtained in a similar manner, but the derivation is tedious. Extrapolation can be used to derive these formulas more easily.

Suppose we expand the function f in a fourth Taylor polynomial about x_0. Then

$$f(x) = f(x_0) + f'(x_0)(x - x_0) + \frac{1}{2}f''(x_0)(x - x_0)^2 + \frac{1}{6}f'''(x_0)(x - x_0)^3$$
$$+ \frac{1}{24}f^{(4)}(x_0)(x - x_0)^4 + \frac{1}{120}f^{(5)}(\xi)(x - x_0)^5$$

for some number ξ between x and x_0. Evaluating f at $x_0 + h$ and $x_0 - h$ gives

(4.16)
$$f(x_0 + h) = f(x_0) + f'(x_0)h + \frac{1}{2}f''(x_0)h^2 + \frac{1}{6}f'''(x_0)h^3 + \frac{1}{24}f^{(4)}(x_0)h^4 + \frac{1}{120}f^{(5)}(\xi_1)h^5$$

and

(4.17)
$$f(x_0 - h) = f(x_0) - f'(x_0)h + \frac{1}{2}f''(x_0)h^2 - \frac{1}{6}f'''(x_0)h^3 + \frac{1}{24}f^{(4)}(x_0)h^4 - \frac{1}{120}f^{(5)}(\xi_2)h^5,$$

where $x_0 - h < \xi_2 < x_0 < \xi_1 < x_0 + h$. Subtracting Eq. (4.17) from Eq. (4.16) produces

(4.18)
$$f(x_0 + h) - f(x_0 - h) = 2hf'(x_0) + \frac{h^3}{3}f'''(x_0) + \frac{h^5}{120}[f^{(5)}(\xi_1) + f^{(5)}(\xi_2)].$$

If $f^{(5)}$ is continuous on $[x_0 - h, x_0 + h]$, the Intermediate Value Theorem implies that a number $\tilde{\xi}$ in $(x_0 - h, x_0 + h)$ exists with

$$f^{(5)}(\tilde{\xi}) = \frac{1}{2}[f^{(5)}(\xi_1) + f^{(5)}(\xi_2)].$$

As a consequence, Eq. (4.18) can be solved for $f'(x_0)$ to give the $O(h^2)$ approximation

(4.19)
$$f'(x_0) = \frac{1}{2h}[f(x_0 + h) - f(x_0 - h)] - \frac{h^2}{6}f'''(x_0) - \frac{h^4}{120}f^{(5)}(\tilde{\xi}).$$

Although the approximation in Eq. (4.19) is the same as that given in the three-point formula in Eq. (4.5), the unknown evaluation point occurs now in $f^{(5)}$, rather than in f'''. Extrapolation takes advantage of this by first replacing h in Eq. (4.19) with $2h$ to give the new formula

(4.20)
$$f'(x_0) = \frac{1}{4h}[f(x_0 + 2h) - f(x_0 - 2h)] - \frac{4h^2}{6}f'''(x_0) - \frac{16h^4}{120}f^{(5)}(\hat{\xi}),$$

where $\hat{\xi}$ is between $x_0 - 2h$ and $x_0 + 2h$.

Multiplying Eq. (4.19) by 4 and subtracting Eq. (4.20) produces

$$3f'(x_0) = \frac{2}{h}[f(x_0+h)-f(x_0-h)] - \frac{1}{4h}[f(x_0+2h)-f(x_0-2h)] - \frac{h^4}{30}f^{(5)}(\tilde{\xi}) + \frac{2h^4}{15}f^{(5)}(\hat{\xi}).$$

If $f^{(5)}$ is continuous on $[x_0 - 2h, x_0 + 2h]$, an alternative method can be used to show that $f^{(5)}(\tilde{\xi})$ and $f^{(5)}(\hat{\xi})$ can be replaced by a common value $f^{(5)}(\xi)$. Using this result and dividing by 3 produces the five-point formula

$$f'(x_0) = \frac{1}{12h}[f(x_0 - 2h) - 8f(x_0 - h) + 8f(x_0 + h) - f(x_0 + 2h)] + \frac{h^4}{30}f^{(5)}(\xi),$$

which is the five-point formula given as Eq. (4.6). Other formulas for first and higher derivatives can be derived in a similar manner. Some of these formulas are considered in the exercises.

The technique of extrapolation is used throughout the text. The most prominent applications occur in approximating integrals in Section 4.5 and for determining approximate solutions to differential equations in Section 5.8.

EXERCISE SET 4.2

1. Apply the extrapolation process described in Example 1 to determine $N_3(h)$, an approximation to $f'(x_0)$, for the following functions and stepsizes.

   a. $f(x) = \ln x$, $x_0 = 1.0$, $h = 0.4$
   b. $f(x) = x + e^x$, $x_0 = 0.0$, $h = 0.4$
   c. $f(x) = 2^x \sin x$, $x_0 = 1.05$, $h = 0.4$
   d. $f(x) = x^3 \cos x$, $x_0 = 2.3$, $h = 0.4$

2. Add another line to the extrapolation table in Exercise 1 to obtain the approximation $N_4(h)$.
3. Repeat Exercise 1 using four-digit rounding arithmetic.
4. Repeat Exercise 2 using four-digit rounding arithmetic.
5. The following data give approximations to the integral

$$M = \int_0^{\pi} \sin x \, dx.$$

$$N_1(h) = 1.570796, \quad N_1\left(\frac{h}{2}\right) = 1.896119, \quad N_1\left(\frac{h}{4}\right) = 1.974232, \quad N_1\left(\frac{h}{8}\right) = 1.993570.$$

   Assuming $M = N_1(h) + K_1h^2 + K_2h^4 + K_3h^6 + K_4h^8 + O(h^{10})$, construct an extrapolation table to determine $N_4(h)$.

6. The following data can be used to approximate the integral

$$M = \int_0^{3\pi/2} \cos x \, dx.$$

$$N_1(h) = 2.356194, \qquad N_1\left(\frac{h}{2}\right) = -0.4879837,$$

$$N_1\left(\frac{h}{4}\right) = -0.8815732, \quad N_1\left(\frac{h}{8}\right) = -0.9709157.$$

   Assume a formula exists of the type given in Exercise 5 and determine $N_4(h)$.

7. Show that the five-point formula in Eq. (4.6) applied to $f(x) = xe^x$ at $x_0 = 2.0$ gives $N_2(0.2)$ in Table 4.6 when $h = 0.1$ and $N_2(0.1)$ when $h = 0.05$.
8. The forward-difference formula can be expressed as

$$f'(x_0) = \frac{1}{h}[f(x_0 + h) - f(x_0)] - \frac{h}{2}f''(x_0) - \frac{h^2}{6}f'''(x_0) + O(h^3).$$

   Use extrapolation to derive an $O(h^3)$ formula for $f'(x_0)$.

9. Suppose that $N(h)$ is an approximation to M for every $h > 0$ and that

$$M = N(h) + K_1h + K_2h^2 + K_3h^3 + \cdots$$

for some constants $K_1, K_2, K_3, \ldots$. Use the values $N(h)$, $N\left(\frac{h}{3}\right)$, and $N\left(\frac{h}{9}\right)$ to produce an $O(h^3)$ approximation to M.

10. Suppose that $N(h)$ is an approximation to M for every $h > 0$ and that

$$M = N(h) + K_1h^2 + K_2h^4 + K_3h^6 + \cdots$$

for some constants $K_1, K_2, K_3, \ldots$. Use the values $N(h)$, $N\left(\frac{h}{3}\right)$, and $N\left(\frac{h}{9}\right)$ to produce an $O(h^6)$ approximation to M.

11. We learn in calculus that $e = \lim_{h\to 0}(1 + h)^{1/h}$.

a. Determine approximations to e corresponding to $h = 0.04$, 0.02, and 0.01.

b. Use extrapolation on the approximations, assuming that constants $K_1, K_2, \ldots,$ exist with

$$e = (1 + h)^{1/h} + K_1h + K_2h^2 + K_3h^3 + \cdots$$

to produce an $O(h^3)$ approximation to e, where $h = 0.04$.

c. Do you think that the assumption in part (b) is correct?

12. a. Show that

$$\lim_{h\to 0}\left(\frac{2+h}{2-h}\right)^{1/h} = e.$$

b. Compute approximations to e using the formula $N(h) = \left(\frac{2+h}{2-h}\right)^{1/h}$ for $h = 0.04$, 0.02, and 0.01.

c. Assume that $e = N(h) + K_1h + K_2h^2 + K_3h^3 + \cdots$. Use extrapolation, with at least 16 digits of precision, to compute an $O(h^3)$ approximation to e with $h = 0.04$. Do you think the assumption is correct?

d. Show that $N(-h) = N(h)$.

e. Use part (d) to show that $K_1 = K_3 = K_5 = \cdots = 0$ in the formula

$$e = N(h) + K_1h + K_2h^2 + K_3h^3 + K_4h^4 + K_5h^5 + \cdots,$$

so that the formula reduces to

$$e = N(h) + K_2h^2 + K_4h^4 + K_6h^6 + \cdots.$$

f. Use the results of part (e) and extrapolation to compute an $O(h^6)$ approximation to e with $h = 0.04$.

13. Suppose the following extrapolation table has been constructed to approximate the number M with $M = N_1(h) + K_1h^2 + K_2h^4 + K_3h^6$:

$N_1(h)$		
$N_1\left(\frac{h}{2}\right)$	$N_2(h)$	
$N_1\left(\frac{h}{4}\right)$	$N_2\left(\frac{h}{2}\right)$	$N_3(h)$

a. Show that the linear interpolating polynomial $P_{0,1}(h)$ through $(h^2, N_1(h))$ and $(h^2/4, N_1(h/2))$ satisfies $P_{0,1}(0) = N_2(h)$. Similarly, show that $P_{1,2}(0) = N_2(h/2)$.

b. Show that the linear interpolating polynomial $P_{0,2}(h)$ through $(h^4, N_2(h))$ and $(h^4/16, N_2(h/2))$ satisfies $P_{0,2}(0) = N_3(h)$.

14. Suppose that $N_1(h)$ is a formula that produces $O(h)$ approximations to a number M and that

$$M = N_1(h) + K_1 h + K_2 h^2 + \cdots$$

for a collection of positive constants $K_1, K_2, \ldots$. Then $N_1(h), N_1(h/2), N_1(h/4), \ldots$ are all lower bounds for M. What can be said about the extrapolated approximations $N_2(h), N_3(h), \ldots$?

15. The semiperimeters of regular polygons with k sides that inscribe and circumscribe the unit circle were used by Archimedes as early as 200 B.C. to approximate π, the circumference of a semicircle. Geometry can be used to show that the sequence of inscribed and circumscribed semiperimeters $\{p_k\}$ and $\{p_k\}$, respectively, satisfy

$$p_k = k \sin\left(\frac{\pi}{k}\right) \quad \text{and} \quad P_k = k \tan\left(\frac{\pi}{k}\right),$$

with $p_k < \pi < P_k$ whenever $k \geq 4$.

a. Show that $p_4 = 2\sqrt{2}$ and $P_4 = 4$.

b. Show that for $k \geq 4$, the sequences satisfy the recurrence relations

$$P_{2k} = \frac{2p_k P_k}{p_k + P_k} \quad \text{and} \quad p_{2k} = \sqrt{p_k P_{2k}}.$$

c. Approximate π to within 10^{-4} by computing p_k and P_k until $P_k - p_k < 10^{-4}$.

d. Use Taylor Series to show that

$$\pi = p_k + \frac{\pi^3}{3!}\left(\frac{1}{k}\right)^2 - \frac{\pi^5}{5!}\left(\frac{1}{k}\right)^4 + \cdots$$

and

$$\pi = P_k - \frac{\pi^3}{3}\left(\frac{1}{k}\right)^2 - \frac{2\pi^5}{15}\left(\frac{1}{k}\right)^4 - \cdots.$$

e. Use extrapolation with $h = 1/k$ to better approximate π.

4.3 Elements of Numerical Integration

The need often arises for evaluating the definite integral of a function that has no explicit antiderivative or whose antiderivative is not easy to obtain. The basic method involved in approximating $\int_a^b f(x)\,dx$ is called **numerical quadrature** and uses a sum of the type

$$\sum_{i=0}^{n} a_i f(x_i)$$

to approximate $\int_a^b f(x)\,dx$.

The methods of quadrature in this section are based on the interpolation polynomials given in Chapter 3. We first select a set of distinct nodes $\{x_0, \ldots, x_n\}$ from the interval $[a, b]$. Then we integrate the Lagrange interpolating polynomial

$$P_n(x) = \sum_{i=0}^{n} f(x_i) L_i(x)$$

and its truncation error term over $[a, b]$ to obtain

$$\int_a^b f(x)\,dx = \int_a^b \sum_{i=0}^{n} f(x_i)L_i(x)\,dx + \int_a^b \prod_{i=0}^{n}(x - x_i)\frac{f^{(n+1)}(\xi(x))}{(n+1)!}\,dx$$

$$= \sum_{i=0}^{n} a_i f(x_i) + \frac{1}{(n+1)!}\int_a^b \prod_{i=0}^{n}(x - x_i)f^{(n+1)}(\xi(x))\,dx,$$

where $\xi(x)$ is in $[a, b]$ for each x and

$$a_i = \int_a^b L_i(x)\,dx \quad \text{for each } i = 0, 1, \ldots, n.$$

The quadrature formula is, therefore,

$$\int_a^b f(x)\,dx \approx \sum_{i=0}^{n} a_i f(x_i),$$

with error given by

$$E(f) = \frac{1}{(n+1)!}\int_a^b \prod_{i=0}^{n}(x - x_i)f^{(n+1)}(\xi(x))\,dx.$$

Before discussing the general situation of quadrature formulas, let us consider formulas produced by using first and second Lagrange polynomials with equally spaced nodes. This gives the **Trapezoidal rule** and **Simpson's rule**, which are commonly introduced in calculus courses.

To derive the Trapezoidal rule for approximating $\int_a^b f(x)\,dx$, let $x_0 = a$, $x_1 = b$, $h = b - a$ and use the linear Lagrange polynomial:

$$P_1(x) = \frac{(x - x_1)}{(x_0 - x_1)}f(x_0) + \frac{(x - x_0)}{(x_1 - x_0)}f(x_1).$$

Then,

(4.21)
$$\int_a^b f(x)\,dx = \int_{x_0}^{x_1}\left[\frac{(x - x_1)}{(x_0 - x_1)}f(x_0) + \frac{(x - x_0)}{(x_1 - x_0)}f(x_1)\right]dx$$

$$+ \frac{1}{2}\int_{x_0}^{x_1} f''(\xi(x))(x - x_0)(x - x_1)\,dx.$$

Since $(x - x_0)(x - x_1)$ does not change sign on $[x_0, x_1]$, the Weighted Mean Value Theorem for Integrals can be applied to the error term to give

$$\int_{x_0}^{x_1} f''(\xi(x))(x - x_0)(x - x_1)\,dx = f''(\xi)\int_{x_0}^{x_1}(x - x_0)(x - x_1)\,dx$$

$$= f''(\xi)\left[\frac{x^3}{3} - \frac{(x_1 + x_0)}{2}x^2 + x_0x_1x\right]_{x_0}^{x_1}$$

$$= -\frac{h^3}{6}f''(\xi).$$

Consequently, Eq. (4.21) implies that

$$\int_a^b f(x)\,dx = \left[\frac{(x - x_1)^2}{2(x_0 - x_1)}f(x_0) + \frac{(x - x_0)^2}{2(x_1 - x_0)}f(x_1)\right]_{x_0}^{x_1} - \frac{h^3}{12}f''(\xi)$$

$$= \frac{(x_1 - x_0)}{2}[f(x_0) + f(x_1)] - \frac{h^3}{12}f''(\xi).$$

Since $h = x_1 - x_0$, we have the following rule:

Trapezoidal Rule:

$$\int_a^b f(x)\,dx = \frac{h}{2}[f(x_0) + f(x_1)] - \frac{h^3}{12}f''(\xi).$$

This formula is called the Trapezoidal rule because when f is a function with positive values, $\int_a^b f(x)\,dx$ is approximated by the area in a trapezoid, as shown in Figure 4.3.

Figure 4.3

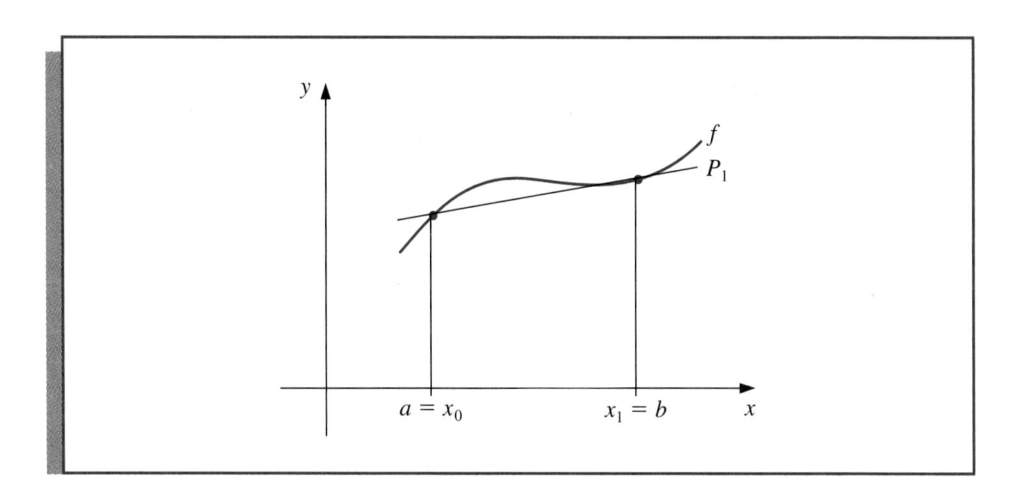

Since the error term for the Trapezoidal rule involves f'', the rule gives the exact result when applied to any function whose second derivative is identically zero, that is, any polynomial of degree 1 or less.

Simpson's rule results from integrating over $[a, b]$ the second Lagrange polynomial with nodes $x_0 = a$, $x_2 = b$, and $x_1 = a + h$, where $h = (b - a)/2$. (See Figure 4.4.)

Figure 4.4

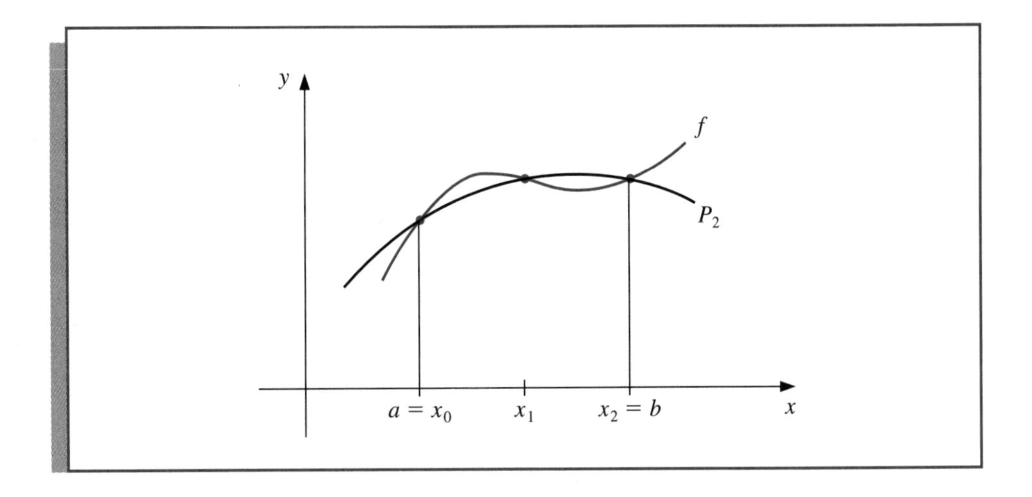

Therefore,

$$\int_a^b f(x)\,dx = \int_{x_0}^{x_2} \left[\frac{(x-x_1)(x-x_2)}{(x_0-x_1)(x_0-x_2)} f(x_0) + \frac{(x-x_0)(x-x_2)}{(x_1-x_0)(x_1-x_2)} f(x_1) + \frac{(x-x_0)(x-x_1)}{(x_2-x_0)(x_2-x_1)} f(x_2) \right] dx + \int_{x_0}^{x_2} \frac{(x-x_0)(x-x_1)(x-x_2)}{6} f^{(3)}(\xi(x))\,dx.$$

Deriving Simpson's rule in this manner, however, provides only an $O(h^4)$ error term involving $f^{(3)}$. By approaching the problem in another way, a higher-order term involving $f^{(4)}$ can be derived.

To illustrate this alternative formula, suppose that f is expanded in the third Taylor polynomial about x_1. Then for each x in $[x_0, x_2]$, a number $\xi(x)$ in (x_0, x_2) exists with

$$f(x) = f(x_1) + f'(x_1)(x-x_1) + \frac{f''(x_1)}{2}(x-x_1)^2 + \frac{f'''(x_1)}{6}(x-x_1)^3 + \frac{f^{(4)}(\xi(x))}{24}(x-x_1)^4$$

and

$$\textbf{(4.22)} \quad \int_{x_0}^{x_2} f(x)\,dx = \left[f(x_1)(x-x_1) + \frac{f'(x_1)}{2}(x-x_1)^2 + \frac{f''(x_1)}{6}(x-x_1)^3 + \frac{f'''(x_1)}{24}(x-x_1)^4 \right]_{x_0}^{x_2} + \frac{1}{24}\int_{x_0}^{x_2} f^{(4)}(\xi(x))(x-x_1)^4\,dx.$$

Since $(x-x_1)^4$ is never negative on $[x_0, x_2]$, the Weighted Mean Value Theorem for Integrals implies that

$$\frac{1}{24}\int_{x_0}^{x_2} f^{(4)}(\xi(x))(x-x_1)^4\,dx = \frac{f^{(4)}(\xi_1)}{24}\int_{x_0}^{x_2}(x-x_1)^4\,dx = \frac{f^{(4)}(\xi_1)}{120}(x-x_1)^5\Big]_{x_0}^{x_2}$$

for some number ξ_1 in (x_0, x_2).

However, $h = x_2 - x_1 = x_1 - x_0$, so

$$(x_2-x_1)^2 - (x_0-x_1)^2 = (x_2-x_1)^4 - (x_0-x_1)^4 = 0,$$

whereas

$$(x_2-x_1)^3 - (x_0-x_1)^3 = 2h^3 \quad \text{and} \quad (x_2-x_1)^5 - (x_0-x_1)^5 = 2h^5.$$

Consequently, Eq. (4.22) can be rewritten as

$$\int_{x_0}^{x_2} f(x)\,dx = 2hf(x_1) + \frac{h^3}{3}f''(x_1) + \frac{f^{(4)}(\xi_1)}{60}h^5.$$

If we now replace $f''(x_1)$ by the approximation given in Eq. (4.9) of Section 4.1, we have

$$\begin{aligned}\int_{x_0}^{x_2} f(x)\,dx &= 2hf(x_1) + \frac{h^3}{3}\left\{\frac{1}{h^2}[f(x_0) - 2f(x_1) + f(x_2)] - \frac{h^2}{12}f^{(4)}(\xi_2)\right\} + \frac{f^{(4)}(\xi_1)}{60}h^5 \\ &= \frac{h}{3}[f(x_0) + 4f(x_1) + f(x_2)] - \frac{h^5}{12}\left[\frac{1}{3}f^{(4)}(\xi_2) - \frac{1}{5}f^{(4)}(\xi_1)\right].\end{aligned}$$

It can be shown by alternative methods (see Exercise 18) that the values ξ_1 and ξ_2 in this expression can be replaced by a common value ξ in (x_0, x_2). This gives Simpson's rule.

Simpson's Rule:

$$\int_{x_0}^{x_2} f(x)\,dx = \frac{h}{3}[f(x_0) + 4f(x_1) + f(x_2)] - \frac{h^5}{90}f^{(4)}(\xi).$$

Since the error term involves the fourth derivative of f, Simpson's rule gives exact results when applied to any polynomial of degree three or less.

EXAMPLE 1 The Trapezoidal rule for a function f on the interval $[0, 2]$ is

$$\int_0^2 f(x)\,dx \approx f(0) + f(2),$$

and Simpson's rule for f on [0,2] is

$$\int_0^2 f(x)\,dx \approx \frac{1}{3}[f(0) + 4f(1) + f(2)].$$

The results to three places for some elementary functions are summarized in Table 4.7. Notice that in each instance Simpson's Rule is significantly better. ■

Table 4.7

$f(x)$	x^2	x^4	$1/(x+1)$	$\sqrt{1+x^2}$	$\sin x$	e^x
Exact value	2.667	6.400	1.099	2.958	1.416	6.389
Trapezoidal	4.000	16.000	1.333	3.326	0.909	8.389
Simpson's	2.667	6.667	1.111	2.964	1.425	6.421

The standard derivation of quadrature error formulas is based on determining the class of polynomials for which these formulas produce exact results. The next definition is used to facilitate the discussion of this derivation.

Definition 4.1 The **degree of accuracy**, or **precision**, of a quadrature formula is the largest positive integer n such that the formula is exact for x^k, when $k = 0, 1, \ldots, n$. ■

Definition 4.1 implies that the Trapezoidal and Simpson's rules have degree of precision one and three, respectively.

Integration and summation are linear operations; that is,

$$\int_a^b (\alpha f(x) + \beta g(x))\,dx = \alpha \int_a^b f(x)\,dx + \beta \int_a^b g(x)\,dx$$

and

$$\sum_{i=0}^{n} (\alpha f(x_i) + \beta g(x_i)) = \alpha \sum_{i=0}^{n} f(x_i) + \beta \sum_{i=0}^{n} g(x_i)$$

for each pair of integrable functions f and g and each pair of real constants α and β. This implies (see Exercise 19) that the degree of precision of a quadrature formula is n if and only if the error $E(P(x)) = 0$ for all polynomials $P(x)$ of degree $k = 0, 1, \ldots, n$, but $E(P(x)) \neq 0$ for some polynomial $P(x)$ of degree $n + 1$.

The Trapezoidal and Simpson's rules are examples of a class of methods known as Newton-Cotes formulas. There are two types of Newton-Cotes formulas, open and closed.

The $(n+1)$*-point closed Newton-Cotes formula* uses nodes $x_i = x_0 + ih$, for $i = 0, 1, \ldots, n$, where $x_0 = a$, $x_n = b$ and $h = (b - a)/n$. (See Figure 4.5.) It is called closed because the endpoints of the closed interval $[a, b]$ are included as nodes. The formula assumes the form

$$\int_a^b f(x)\,dx \approx \sum_{i=0}^{n} a_i f(x_i),$$

Figure 4.5

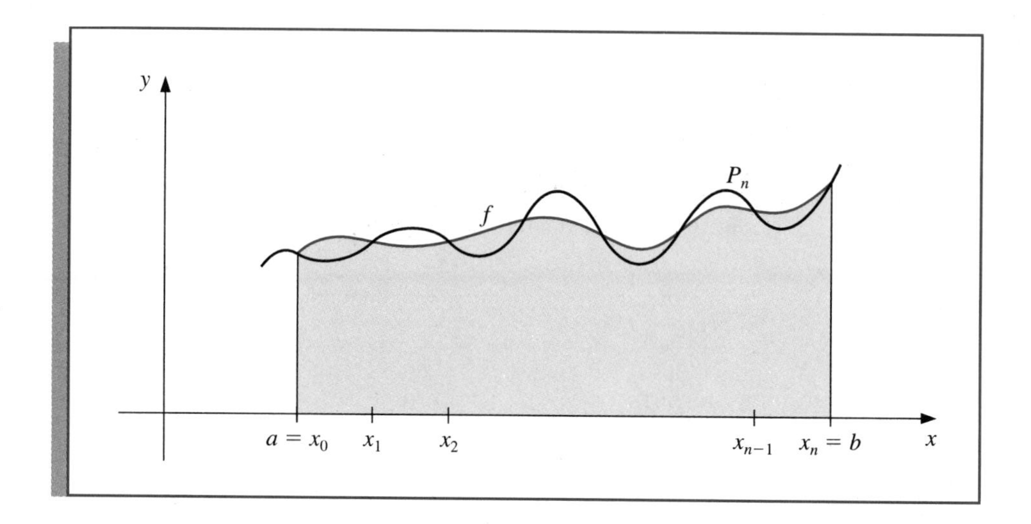

where

$$a_i = \int_{x_0}^{x_n} L_i(x)\,dx = \int_{x_0}^{x_n} \prod_{\substack{j=0 \\ j\neq i}}^{n} \frac{(x - x_j)}{(x_i - x_j)}\,dx.$$

The following theorem details the error analysis associated with the closed Newton-Cotes formulas. For a proof of this theorem, see [IK, p. 313].

Theorem 4.2 Suppose that $\sum_{i=0}^{n} a_i f(x_i)$ denotes the $(n+1)$-point closed Newton-Cotes formula with $x_0 = a$, $x_n = b$, and $h = (b-a)/n$. There exists $\xi \in (a, b)$ for which

$$\int_a^b f(x)\,dx = \sum_{i=0}^{n} a_i f(x_i) + \frac{h^{n+3} f^{(n+2)}(\xi)}{(n+2)!} \int_0^n t^2(t-1)\cdots(t-n)\,dt$$

if n is even and $f \in C^{n+2}[a, b]$, and

$$\int_a^b f(x)\,dx = \sum_{i=0}^{n} a_i f(x_i) + \frac{h^{n+2} f^{(n+1)}(\xi)}{(n+1)!} \int_0^n t(t-1)\cdots(t-n)\,dt$$

if n is odd and $f \in C^{n+1}[a, b]$. ■

Note that when n is an even integer, the degree of precision is $n+1$, although the interpolation polynomial is of degree at most n. In case n is odd, the second part of the theorem shows that the degree of precision is only n.

Some of the common **closed Newton-Cotes formulas** with their error terms are as follows:

$n = 1$: Trapezoidal Rule

$$\text{(4.23)} \quad \int_{x_0}^{x_1} f(x)\,dx = \frac{h}{2}[f(x_0) + f(x_1)] - \frac{h^3}{12}f''(\xi), \quad \text{where} \quad x_0 < \xi < x_1.$$

$n = 2$: Simpson's Rule

$$\text{(4.24)} \quad \int_{x_0}^{x_2} f(x)\,dx = \frac{h}{3}[f(x_0) + 4f(x_1) + f(x_2)] - \frac{h^5}{90}f^{(4)}(\xi), \quad \text{where} \quad x_0 < \xi < x_2.$$

$n = 3$: Simpson's Three-Eighths Rule

$$\text{(4.25)} \quad \int_{x_0}^{x_3} f(x)\,dx = \frac{3h}{8}[f(x_0) + 3f(x_1) + 3f(x_2) + f(x_3)] - \frac{3h^5}{80}f^{(4)}(\xi),$$

$$\text{where} \quad x_0 < \xi < x_3.$$

$n = 4$:

$$\text{(4.26)} \quad \int_{x_0}^{x_4} f(x)\,dx = \frac{2h}{45}[7f(x_0) + 32f(x_1) + 12f(x_2) + 32f(x_3) + 7f(x_4)] - \frac{8h^7}{945}f^{(6)}(\xi),$$

$$\text{where} \quad x_0 < \xi < x_4.$$

In the *open Newton-Cotes formulas*, the nodes $x_i = x_0 + ih$ are used for each $i = 0, 1, \ldots, n$, where $h = (b - a)/(n + 2)$ and $x_0 = a + h$. This implies that $x_n = b - h$, so we label the endpoints by setting $x_{-1} = a$ and $x_{n+1} = b$, as shown in Figure 4.6. Open

Figure 4.6

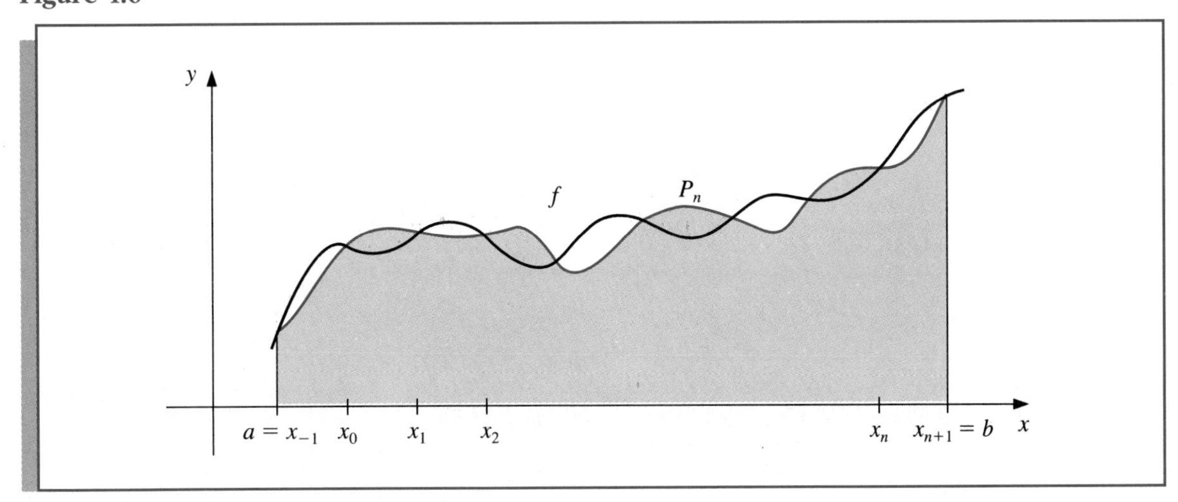

formulas contain all the nodes used for approximation within the open interval (a, b). The formulas become

$$\int_a^b f(x)\,dx = \int_{x_{-1}}^{x_{n+1}} f(x)\,dx \approx \sum_{i=0}^{n} a_i f(x_i),$$

where again

$$a_i = \int_a^b L_i(x)\,dx.$$

The following theorem is analogous to Theorem 4.2; its proof is contained in [IK, p. 314].

Theorem 4.3 Suppose that $\sum_{i=0}^{n} a_i f(x_i)$ denotes the $(n+1)$-point open Newton-Cotes formula with $x_{-1} = a$, $x_{n+1} = b$, and $h = (b-a)/(n+2)$. There exists $\xi \in (a, b)$ for which

$$\int_a^b f(x)\,dx = \sum_{i=0}^{n} a_i f(x_i) + \frac{h^{n+3} f^{(n+2)}(\xi)}{(n+2)!} \int_{-1}^{n+1} t^2(t-1)\cdots(t-n)\,dt$$

if n is even and $f \in C^{n+2}[a, b]$, and

$$\int_a^b f(x)\,dx = \sum_{i=0}^{n} a_i f(x_i) + \frac{h^{n+2} f^{(n+1)}(\xi)}{(n+1)!} \int_{-1}^{n+1} t(t-1)\cdots(t-n)\,dt$$

if n is odd and $f \in C^{n+1}[a, b]$. ■

Some of the common **open Newton-Cotes** formulas with their error terms are as follows:

$n = 0$: Midpoint Rule

(4.27)
$$\int_{x_{-1}}^{x_1} f(x)\,dx = 2hf(x_0) + \frac{h^3}{3} f''(\xi), \quad \text{where} \quad x_{-1} < \xi < x_1.$$

$n = 1$:

(4.28)
$$\int_{x_{-1}}^{x_2} f(x)\,dx = \frac{3h}{2}[f(x_0) + f(x_1)] + \frac{3h^3}{4} f''(\xi), \quad \text{where} \quad x_{-1} < \xi < x_2.$$

$n = 2$:

(4.29)
$$\int_{x_{-1}}^{x_3} f(x)\,dx = \frac{4h}{3}[2f(x_0) - f(x_1) + 2f(x_2)] + \frac{14h^5}{45} f^{(4)}(\xi),$$

$$\text{where} \quad x_{-1} < \xi < x_3.$$

$n = 3$:

$$\int_{x_{-1}}^{x_4} f(x)\,dx = \frac{5h}{24}[11f(x_0) + f(x_1) + f(x_2) + 11f(x_3)] + \frac{95}{144}h^5 f^{(4)}(\xi), \tag{4.30}$$

where $x_{-1} < \xi < x_4$.

EXAMPLE 2 Using the closed and open Newton-Cotes formulas listed as (4.23)–(4.26) and (4.27)–(4.30) to approximate $\int_0^{\pi/4} \sin x\,dx = 1 - \sqrt{2}/2$ gives the results in Table 4.8. ■

Table 4.8

n	0	1	2	3	4
Closed formulas		0.27768018	0.29293264	0.29291070	0.29289318
Error		0.01521303	0.00003942	0.00001748	0.00000004
Open formulas	0.30055887	0.29798754	0.29285866	0.29286923	
Error	0.00766565	0.00509432	0.00003456	0.00002399	

EXERCISE SET 4.3

1. Approximate the following integrals using the Trapezoidal rule.

   a. $\int_{0.5}^{1} x^4\,dx$

   b. $\int_{0}^{0.5} \frac{2}{x-4}\,dx$

   c. $\int_{1}^{1.5} x^2 \ln x\,dx$

   d. $\int_{0}^{1} x^2 e^{-x}\,dx$

   e. $\int_{1}^{1.6} \frac{2x}{x^2-4}\,dx$

   f. $\int_{1}^{1.6} \frac{2}{x^2-4}\,dx$

   g. $\int_{0}^{\pi/4} x \sin x\,dx$

   h. $\int_{0}^{\pi/4} e^{3x} \sin 2x\,dx$

2. Find a bound for the error in Exercise 1 using the error formula and compare to the actual error.
3. Repeat Exercise 1 using Simpson's rule.
4. Repeat Exercise 2 using Simpson's rule and the results of Exercise 3.
5. Repeat Exercise 1 using the Midpoint rule.
6. Repeat Exercise 2 using the Midpoint rule and the results of Exercise 5.
7. The Trapezoidal rule applied to $\int_0^2 f(x)\,dx$ gives the value 4, and Simpson's rule gives the value 2. What is $f(1)$?
8. The Trapezoidal rule applied to $\int_0^2 f(x)\,dx$ gives the value 5, and the Midpoint rule gives the value 4. What value does Simpson's rule give?
9. Find the degree of precision of the quadrature formula

$$\int_{-1}^{1} f(x)\,dx = f\left(-\frac{\sqrt{3}}{3}\right) + f\left(\frac{\sqrt{3}}{3}\right).$$

10. Let $h = (b - a)/3$, $x_0 = a$, $x_1 = a + h$, and $x_2 = b$. Find the degree of precision of the quadrature formula

$$\int_a^b f(x)\,dx = \frac{9}{4}hf(x_1) + \frac{3}{4}hf(x_2).$$

11. The quadrature formula $\int_{-1}^{1} f(x)\,dx = c_0 f(-1) + c_1 f(0) + c_2 f(1)$ is exact for all polynomials of degree less than or equal to 2. Determine c_0, c_1, and c_2.

12. The quadrature formula $\int_{-1}^{1} f(x)\,dx = c_0 f(0) + c_1 f(1) + c_2 f(2)$ is exact for all polynomials of degree less than or equal to 2. Determine c_0, c_1, and c_2.

13. Find the constants c_0, c_1, and x_1 so that the quadrature formula

$$\int_0^1 f(x)\,dx = c_0 f(0) + c_1 f(x_1)$$

has the highest possible degree of precision.

14. Find the constants x_0, x_1, and c_1 so that the quadrature formula

$$\int_0^1 f(x)\,dx = \frac{1}{2}f(x_0) + c_1 f(x_1)$$

has the highest possible degree of precision.

15. Approximate the following integrals using formulas (4.23) through (4.30). Are the accuracies of the approximations consistent with the error formulas? Which of parts (d) and (e) give the better approximation?

a. $\int_0^{0.1} \sqrt{1 + x}\,dx$

b. $\int_0^{\pi/2} (\sin x)^2\,dx$

c. $\int_{1.1}^{1.5} e^x\,dx$

d. $\int_1^{10} \frac{1}{x}\,dx$

e. $\int_1^{5.5} \frac{1}{x}\,dx + \int_{5.5}^{10} \frac{1}{x}\,dx$

f. $\int_0^1 x^{1/3}\,dx$

16. Given the function f at the following values:

x	1.8	2.0	2.2	2.4	2.6
$f(x)$	3.12014	4.42569	6.04241	8.03014	10.46675

Approximate $\int_{1.8}^{2.6} f(x)\,dx$ using all the quadrature formulas of this section that can be applied.

17. Suppose the data of Exercise 16 have roundoff errors given by the following table:

x	1.8	2.0	2.2	2.4	2.6
Error in $f(x)$	2×10^{-6}	-2×10^{-6}	-0.9×10^{-6}	-0.9×10^{-6}	2×10^{-6}

Calculate the errors due to roundoff in Exercise 16.

18. Derive Simpson's rule with error term by using

$$\int_{x_0}^{x_2} f(x)\,dx = a_0 f(x_0) + a_1 f(x_1) + a_2 f(x_2) + k f^{(4)}(\xi).$$

Find a_0, a_1, and a_2 from the fact that Simpson's rule is exact for $f(x) = x^n$ when $n = 1$, 2, and 3. Then find k by applying the integration formula with $f(x) = x^4$.

19. Prove the statement following Definition 4.1; that is, show that a quadrature formula has degree of precision n if and only if the error $E(P(x)) = 0$ for all polynomials $P(x)$ of degree $k = 0, 1, \ldots, n$, but $E(P(x)) \neq 0$ for some polynomial $P(x)$ of degree $n + 1$.

20. Derive Simpson's three-eighths rule, Eq. (4.25), with error term by the use of Theorem 4.2.

21. Derive Eq. (4.28) with error term by the use of Theorem 4.3.

4.4 Composite Numerical Integration

The Newton-Cotes formulas are generally unsuitable for use over large integration intervals. High-degree formulas would be required, and the values of the coefficients in these formulas are difficult to obtain. Also, the Newton-Cotes formulas are based on interpolatory polynomials that use equally spaced nodes, a procedure that is inaccurate over large intervals because of the oscillatory nature of high-degree polynomials. In this section, we discuss a *piecewise* approach to numerical integration that uses the low-order Newton-Cotes formulas. These are the techniques most often applied.

Consider finding an approximation to $\int_0^4 e^x\,dx$. Simpson's rule with $h = 2$ gives

$$\int_0^4 e^x\,dx \approx \frac{2}{3}(e^0 + 4e^2 + e^4) = 56.76958.$$

Since the exact answer in this case is $e^4 - e^0 = 53.59815$, the error -3.17143 is far larger than we would normally accept.

To apply a piecewise technique to this problem, divide $[0, 4]$ into $[0, 2]$ and $[2, 4]$ and use Simpson's rule twice with $h = 1$:

$$\begin{aligned}\int_0^4 e^x\,dx &= \int_0^2 e^x\,dx + \int_2^4 e^x\,dx\\ &\approx \frac{1}{3}[e^0 + 4e + e^2] + \frac{1}{3}[e^2 + 4e^3 + e^4]\\ &= \frac{1}{3}[e^0 + 4e + 2e^2 + 4e^3 + e^4]\\ &= 53.86385.\end{aligned}$$

The error has been reduced to -0.26570. Encouraged by our results, we subdivide the intervals $[0, 2]$ and $[2, 4]$ and use Simpson's rule with $h = \frac{1}{2}$, giving

$$\begin{aligned}\int_0^4 e^x\,dx &= \int_0^1 e^x\,dx + \int_1^2 e^x\,dx + \int_2^3 e^x\,dx + \int_3^4 e^x\,dx\\ &\approx \frac{1}{6}[e_0 + 4e^{1/2} + e] + \frac{1}{6}[e + 4e^{3/2} + e^2]\\ &\quad + \frac{1}{6}[e^2 + 4e^{5/2} + e^3] + \frac{1}{6}[e^3 + 4e^{7/2} + e^4]\end{aligned}$$

$$= \frac{1}{6}[e^0 + 4e^{1/2} + 2e + 4e^{3/2} + 2e^2 + 4e^{5/2} + 2e^3 + 4e^{7/2} + e^4]$$
$$= 53.61622.$$

The error for this approximation is -0.01807.

To generalize this procedure, choose an even integer n. Subdivide the interval $[a, b]$ into n subintervals, and apply Simpson's rule on each consecutive pair of subintervals. (See Figure 4.7.)

Figure 4.7

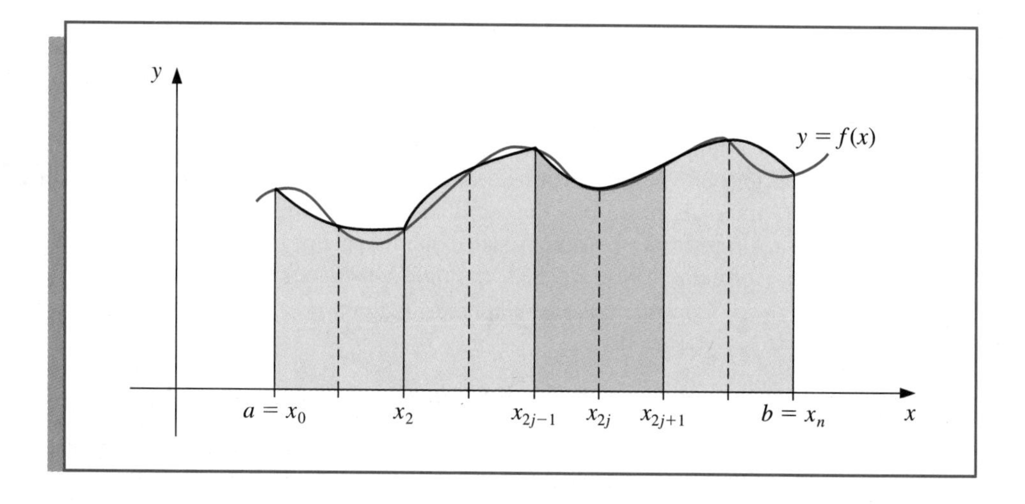

With $h = (b-a)/n$ and $x_j = a + jh$ for each $j = 0, 1, \ldots, n$, we have

$$\int_a^b f(x)\,dx = \sum_{j=1}^{n/2} \int_{x_{2j-2}}^{x_{2j}} f(x)\,dx$$
$$= \sum_{j=1}^{n/2} \left\{ \frac{h}{3}[f(x_{2j-2}) + 4f(x_{2j-1}) + f(x_{2hj})] - \frac{h^5}{90} f^{(4)}(\xi_j) \right\}$$

for some ξ_j with $x_{2j-2} < \xi_j < x_{2j}$, provided that $f \in C^4[a, b]$. Using the fact that for each $j = 1, 2, \ldots, (n/2) - 1$, we have $f(x_{2j})$ appearing in the term corresponding to the interval $[x_{2j-2}, x_{2j}]$ and also in the term corresponding to the interval $[x_{2j}, x_{2j+2}]$, we can reduce this sum to

$$\int_a^b f(x)\,dx = \frac{h}{3}\left[f(x_0) + 2\sum_{j=1}^{(n/2)-1} f(x_{2j}) + 4\sum_{j=1}^{n/2} f(x_{2j-1}) + f(x_n) \right] - \frac{h^5}{90}\sum_{j=1}^{n/2} f^{(4)}(\xi_j).$$

The error associated with this approximation is

$$E(f) = -\frac{h^5}{90}\sum_{j=1}^{n/2} f^{(4)}(\xi_j),$$

where $x_{2j-2} < \xi_j < x_{2j}$ for each $j = 1, 2, \ldots, n/2$. If $f \in C^4[a, b]$, the Extreme Value Theorem implies that $f^{(4)}$ assumes its maximum and minimum in $[a, b]$. Since

$$\min_{x\in[a,b]} f^{(4)}(x) \le f^{(4)}(\xi_j) \le \max_{x\in[a,b]} f^{(4)}(x),$$

we have

$$\frac{n}{2}\min_{x\in[a,b]} f^{(4)}(x) \le \sum_{j=1}^{n/2} f^{(4)}(\xi_j) \le \frac{n}{2}\max_{x\in[a,b]} f^{(4)}(x),$$

and

$$\min_{x\in[a,b]} f^{(4)}(x) \le \frac{2}{n}\sum_{j=1}^{n/2} f^{(4)}(\xi_j) \le \max_{x\in[a,b]} f^{(4)}(x).$$

By the Intermediate Value Theorem, there is a $\mu \in (a, b)$ such that

$$f^{(4)}(\mu) = \frac{2}{n}\sum_{j=1}^{n/2} f^{(4)}(\xi_j).$$

Thus,

$$E(f) = -\frac{h^5}{180} n f^{(4)}(\mu),$$

or, since $n = (b - a)/h$,

$$E(f) = -\frac{b-a}{180} h^4 f^{(4)}(\mu).$$

These observations produce the following result.

Theorem 4.4 Let $f \in C^4[a, b]$, n be even, $h = (b - a)/n$, and $x_j = a + jh$ for each $j = 0, 1, \ldots, n$. There exists a $\mu \in (a, b)$ for which the **Composite Simpson's rule** for n subintervals can be written with its error term as

$$\int_a^b f(x)\,dx = \frac{h}{3}\left[f(a) + 2\sum_{j=1}^{(n/2)-1} f(x_{2j}) + 4\sum_{j=1}^{n/2} f(x_{2j-1}) + f(b)\right] - \frac{b-a}{180} h^4 f^{(4)}(\mu). \quad \blacksquare$$

Algorithm 4.1 uses the Composite Simpson's rule on n subintervals. This is the most frequently used general-purpose quadrature algorithm.

ALGORITHM 4.1

Composite Simpson's Rule

To approximate the integral $I = \int_a^b f(x)\,dx$:

INPUT endpoints a, b; even positive integer n.

OUTPUT approximation XI to I.

Step 1 Set $h = (b - a)/n$.

Step 2 Set $XI0 = f(a) + f(b)$;
$XI1 = 0$; (*Summation of* $f(x_{2i-1})$.)
$XI2 = 0$. (*Summation of* $f(x_{2i})$.)

Step 3 For $i = 1, \ldots, n - 1$ do Steps 4 and 5.

Step 4 Set $X = a + ih$.

Step 5 If i is even then set $XI2 = XI2 + f(X)$
else set $XI1 = XI1 + f(X)$.

Step 6 Set $XI = h(XI0 + 2 \cdot XI2 + 4 \cdot XI1)/3$.

Step 7 OUTPUT (XI);
STOP.

The subdivision approach can be applied to any of the lower-order formulas. The extensions of the Trapezoidal (see Figure 4.8) and Midpoint rules are given without proof. Since the Trapezoidal rule requires only one interval for each application, the integer n can be either odd or even.

Figure 4.8

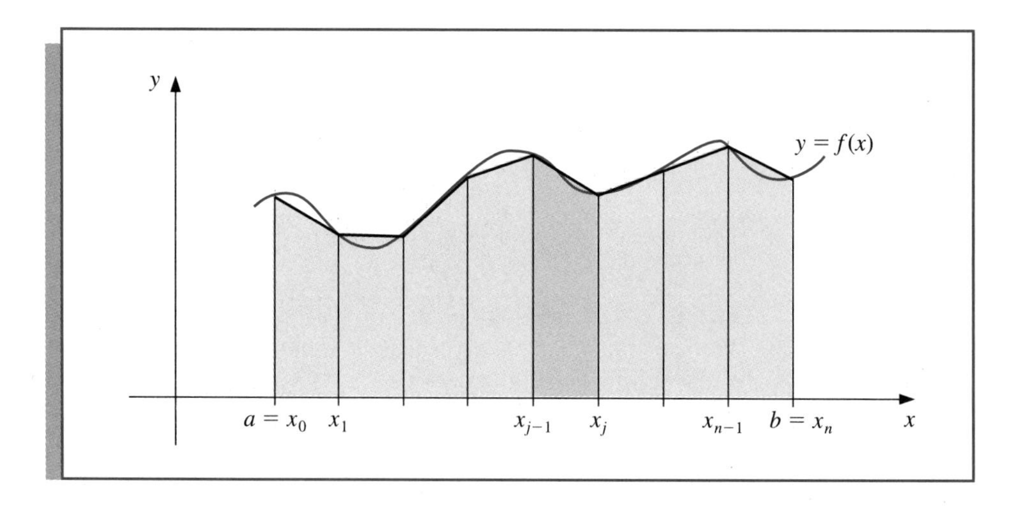

Theorem 4.5 Let $f \in C^2[a, b]$, $h = (b - a)/n$, and $x_j = a + jh$ for each $j = 0, 1, \ldots, n$. There exists a $\mu \in (a, b)$ for which the **Composite Trapezoidal rule** for n subintervals can be written

with its error term as

$$\int_a^b f(x)\,dx = \frac{h}{2}\left[f(a) + 2\sum_{j=1}^{n-1} f(x_j) + f(b)\right] - \frac{b-a}{12}h^2 f''(\mu). \qquad \blacksquare$$

For the Composite Midpoint rule, n must again be even. (See Figure 4.9.)

Theorem 4.6 Let $f \in C^2[a,b]$, n be even, $h = (b-a)/(n+2)$, and $x_j = a + (j+1)h$ for each $j = -1, 0, \ldots, n+1$. There exists a $\mu \in (a,b)$ for which the **Composite Midpoint rule** for $n+2$ subintervals can be written with its error term as

$$\int_a^b f(x)\,dx = 2h\sum_{j=0}^{n/2} f(x_{2j}) + \frac{b-a}{6}h^2 f''(\mu). \qquad \blacksquare$$

Figure 4.9

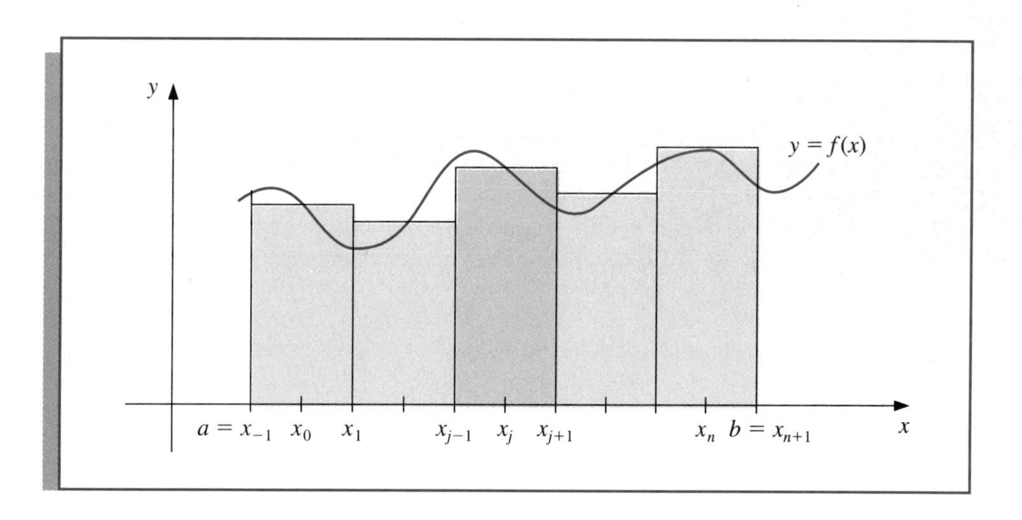

EXAMPLE 1 Consider approximating $\int_0^\pi \sin x\,dx$ with an absolute error less than 0.00002, using the Composite Simpson's rule. The Composite Simpson's rule gives

$$\int_0^\pi \sin x\,dx = \frac{h}{3}\left[2\sum_{j=1}^{(n/2)-1} \sin x_{2j} + 4\sum_{j=1}^{n/2} \sin x_{2j-1}\right] - \frac{\pi h^4}{180}\sin\mu.$$

Since the absolute error is to be less than 0.00002, the inequality

$$\left|\frac{\pi h^4}{180}\sin\mu\right| \le \frac{\pi h^4}{180} = \frac{\pi^5}{180 n^4} < 0.00002$$

is used to determine n and h. Completing these calculations gives $n \geq 18$. If $n = 20$, then $h = \pi/20$, and the formula gives

$$\int_0^{\pi} \sin x\,dx \approx \frac{\pi}{60}\left[2\sum_{j=1}^{9}\sin\left(\frac{j\pi}{10}\right) + 4\sum_{j=1}^{10}\sin\left(\frac{(2j-1)\pi}{20}\right)\right] = 2.000006.$$

To be assured of this degree of accuracy using the Composite Trapezoidal rule requires that

$$\left|\frac{\pi h^2}{12}\sin\mu\right| \leq \frac{\pi h^2}{12} = \frac{\pi^3}{12n^2} < 0.00002,$$

or that $n \geq 360$. Since this is many more calculations than are needed for the Composite Simpson's rule, we would not want to use the Composite Trapezoidal rule on this problem.

For comparison purposes, the Composite Trapezoidal rule with $n = 20$ and $h = \pi/20$ gives

$$\int_0^{\pi} \sin x\,dx \approx \frac{\pi}{40}\left[2\sum_{j=1}^{19}\sin\left(\frac{j\pi}{20}\right) + \sin 0 + \sin\pi\right] = \frac{\pi}{40}\left[2\sum_{j=1}^{19}\sin\left(\frac{j\pi}{20}\right)\right]$$
$$= 1.9958860.$$

The exact answer is 2, so Simpson's rule with $n = 20$ gave an answer well within the required error bound, whereas the Trapezoidal rule with $n = 20$ clearly did not. ■

Maple incorporates both the Composite Simpson's rule and the Composite Trapezoidal Rule. To obtain access to the library where they are defined, enter

```
>with(student);
```

The calls for the methods are `trapezoid(f,x=a..b,n)` and `simpson(f,x=a..b,n)`. For our example the following commands can be used:

```
>f:=sin(x);
```

$$f := \sin(x)$$

```
>trapezoid(f,x=0..Pi,20);
```

$$\frac{1}{20}\pi\left(\sum_{i=1}^{19}\sin\left(\frac{1}{20}i\pi\right)\right)$$

```
>evalf(");
```

$$1.995885974$$

```
>evalf(simpson(f,x=0..Pi,20));
```

$$2.000006785$$

An important property shared by all the Composite Newton-Cotes integration techniques is a stability with respect to roundoff error. To demonstrate this feature, suppose we apply the Composite Simpson's rule with n subintervals to a function f on $[a, b]$ and determine the maximum bound for the roundoff error. Assume that $f(x_i)$ is approximated by $\tilde{f}(x_i)$ and that

$$f(x_i) = \tilde{f}(x_i) + e_i, \quad \text{for each } i = 0, 1, \ldots, n,$$

where e_i denotes the roundoff error associated with using $\tilde{f}(x_i)$ to approximate $f(x_i)$. Then the accumulated error, $e(h)$, in the Composite Simpson's rule is

$$\begin{aligned} e(h) &= \left| \frac{h}{3} \left[e_0 + 2 \sum_{j=1}^{(n/2)-1} e_{2j} + 4 \sum_{j=1}^{n/2} e_{2j-1} + e_n \right] \right| \\ &\leq \frac{h}{3} \left[|e_0| + 2 \sum_{j=1}^{(n/2)-1} |e_{2j}| + 4 \sum_{j=1}^{n/2} |e_{2j-1}| + |e_n| \right]. \end{aligned}$$

If the roundoff errors are uniformly bounded by ε, then

$$e(h) \leq \frac{h}{3} \left[\varepsilon + 2 \left(\frac{n}{2} - 1 \right) \varepsilon + 4 \left(\frac{n}{2} \right) \varepsilon + \varepsilon \right] = \frac{h}{3} 3n\varepsilon = nh\varepsilon.$$

But $nh = b - a$, so

$$e(h) \leq (b - a)\varepsilon,$$

a bound independent of h. This result implies that the procedure is stable as h approaches zero. Recall that this was not true of the numerical differentiation procedures considered at the beginning of this chapter.

EXERCISE SET 4.4

1. Use the Composite Trapezoidal rule with the indicated values of n to approximate the following integrals.

   a. $\displaystyle\int_1^2 x \ln x \, dx, \quad n = 4$

   b. $\displaystyle\int_{-2}^2 x^3 e^x \, dx, \quad n = 4$

   c. $\displaystyle\int_0^2 \frac{2}{x^2 + 4} \, dx, \quad n = 6$

   d. $\displaystyle\int_0^\pi x^2 \cos x \, dx, \quad n = 6$

   e. $\displaystyle\int_0^2 e^{2x} \sin 3x \, dx, \quad n = 8$

   f. $\displaystyle\int_1^3 \frac{x}{x^2 + 4} \, dx, \quad n = 8$

   g. $\displaystyle\int_3^5 \frac{1}{\sqrt{x^2 - 4}} \, dx, \quad n = 8$

   h. $\displaystyle\int_0^{3\pi/8} \tan x \, dx, \quad n = 8$

2. Use the Composite Simpson's rule to approximate the integrals in Exercise 1.
3. Use the Composite Midpoint rule with $n + 2$ subintervals to approximate the integrals in Exercise 1.

4. Approximate $\int_0^2 x^2 e^{-x^2}\,dx$ using $h = 0.25$.

a. Use the Composite Trapezoidal rule.

b. Use the Composite Simpson's rule.

c. Use the Composite Midpoint rule.

5. Suppose that $f(0.25) = f(0.75) = \alpha$. Find α if the Composite Trapezoidal rule with $n = 2$ gives the value 2 for $\int_0^1 f(x)\,dx$ and with $n = 4$ gives the value 1.75.

6. The Midpoint rule for approximating $\int_{-1}^1 f(x)\,dx$ gives the value 12, the Composite Midpoint rule with $n = 2$ gives 5, and the Composite Simpson's rule gives 6. Use the fact that $f(-1) = f(1)$ and $f(-0.5) = f(0.5) - 1$ to determine $f(-1)$, $f(-0.5)$, $f(0)$, $f(0.5)$, and $f(1)$.

7. Determine the values of n and h required to approximate

$$\int_0^2 e^{2x} \sin 3x\,dx$$

to within 10^{-4}.

a. Use the Composite Trapezoidal rule.

b. Use the Composite Simpson's rule.

c. Use the Composite Midpoint rule.

8. Repeat Exercise 7 for the integral $\int_0^\pi x^2 \cos x\,dx$.

9. Determine the values of n and h required to approximate

$$\int_0^2 \frac{1}{x+4}\,dx$$

to within 10^{-5} and compute the approximation.

a. Use the Composite Trapezoidal rule.

b. Use the Composite Simpson's rule.

c. Use the Composite Midpoint rule.

10. Repeat Exercise 9 for the integral $\int_1^2 x \ln x\,dx$.

11. Let f be defined by

$$f(x) = \begin{cases} x^3 + 1, & 0 \le x \le 0.1, \\ 1.001 + 0.03(x-0.1) + 0.3(x-0.1)^2 + 2(x-0.1)^3, & 0.1 \le x \le 0.2, \\ 1.009 + 0.15(x-0.2) + 0.9(x-0.2)^2 + 2(x-0.2)^3, & 0.2 \le x \le 0.3. \end{cases}$$

a. Investigate the continuity of the derivatives of f.

b. Use the Composite Trapezoidal rule with $n = 6$ to approximate $\int_0^{0.3} f(x)\,dx$, and estimate the error using the error bound.

c. Use the Composite Simpson's rule with $n = 6$ to approximate $\int_0^{0.3} f(x)\,dx$. Are the results more accurate than in part (b)?

12. Show that the error $E(f)$ for Composite Simpson's rule can be approximated by

$$-\frac{h^4}{180}[f'''(b) - f'''(a)].$$

[*Hint:* $\sum_{j=1}^{n/2} f^{(4)}(\xi_j)(2h)$ is a Riemann Sum for $\int_a^b f^{(4)}(x)\,dx$.]

13. **a.** Derive an estimate for $E(f)$ in the Composite Trapezoidal rule using the method in Exercise 12.

b. Repeat part (a) for the Composite Midpoint rule.

14. Use the error estimates of Exercises 12 and 13 to estimate the errors in Exercise 8.

15. Use the error estimates of Exercises 12 and 13 to estimate the errors in Exercise 10.

16. In multivariable calculus and in statistics courses it is shown that

$$\int_{-\infty}^{\infty} \frac{1}{\sigma\sqrt{2\pi}} e^{-(1/2)(x/\sigma)^2}\, dx = 1$$

for any positive σ. The function

$$f(x) = \frac{1}{\sigma\sqrt{2\pi}} e^{-(1/2)(x/\sigma)^2}$$

is the *normal density function* with *mean* $\mu = 0$ and *standard deviation* σ. The probability that a randomly chosen value described by this distribution lies in $[a, b]$ is given by $\int_a^b f(x)\, dx$. Approximate to within 10^{-5} the probability that a randomly chosen value described by this distribution will lie in

a. $[-2\sigma, 2\sigma]$ **b.** $[-3\sigma, 3\sigma]$

17. Determine to within 10^{-6} the length of the graph of the ellipse with equation $4x^2 + 9y^2 = 36$.

18. A car laps a race track in 84 s. The speed of the car at each 6-second interval is determined using a radar gun and is given from the beginning of the lap, in feet/second, by the entries in the following table:

Time	0	6	12	18	24	30	36	42	48	54	60	66	72	78	84
Speed	124	134	148	156	147	133	121	109	99	85	78	89	104	116	123

How long is the track?

19. A particle of mass m moving through a fluid is subjected to a viscous resistance R, which is a function of the velocity v. The relationship between the resistance R, velocity v, and time t is given by the equation

$$t = \int_{v(t_0)}^{v(t)} \frac{m}{R(u)}\, du.$$

Suppose that $R(v) = -v\sqrt{v}$ for a particular fluid, where R is in newtons and v is in meters/second. If $m = 10$ kg and $v(0) = 10$ m/s, approximate the time required for the particle to slow to $v = 5$ m/s.

20. To simulate the thermal characteristics of disk brakes (see the following figure), D. A. Secrist and R. W. Hornbeck [SH] needed to approximate numerically the "area averaged lining temperature," T, of the brake pad from the equation

$$T = \frac{\int_{r_e}^{r_0} T(r) r\theta_p\, dr}{\int_{r_e}^{r_0} r\theta_p\, dr},$$

where r_e represents the radius at which the pad-disk contact begins, r_0 represents the outside radius of the pad-disk contact, θ_p represents the angle subtended by the sector brake pads, and $T(r)$ is the temperature at each point of the pad, obtained numerically from analyzing the heat equation (see Section 12.2). If $r_e = 0.308$ ft, $r_0 = 0.478$ ft, $\theta_p = 0.7051$ radians, and the temperatures given in the following table have been calculated at the various points on the disk, find an approximation for T.

r (ft)	$T(r)$ (°F)	r (ft)	$T(r)$ (°F)	r (ft)	$T(r)$ (°F)
0.308	640	0.376	1034	0.444	1204
0.325	794	0.393	1064	0.461	1222
0.342	885	0.410	1114	0.478	1239
0.359	943	0.427	1152		

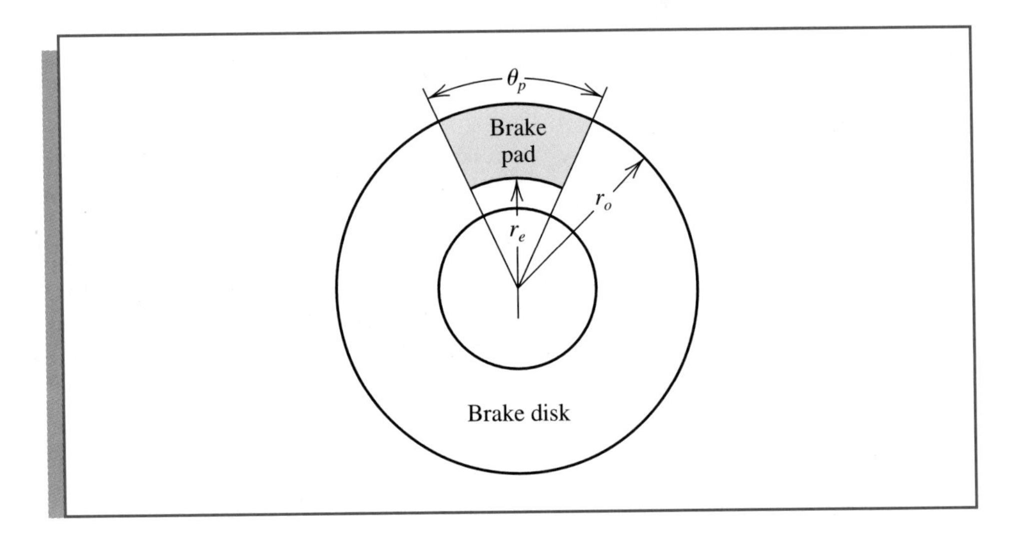

21. Find an approximation to within 10^{-4} of the value of the integral considered in the application opening this chapter:

$$\int_0^{48} \sqrt{1 + (\cos x)^2}\, dx$$

22. The equation

$$\int_0^x \frac{1}{\sqrt{2\pi}} e^{-t^2/2}\, dt = 0.45$$

can be solved for x by using Newton's method with

$$f(x) = \int_0^x \frac{1}{\sqrt{2\pi}} e^{-t^2/2}\, dt - 0.45$$

and

$$f'(x) = \frac{1}{\sqrt{2\pi}} e^{-x^2/2}.$$

To evaluate f at the approximation p_k, we need a quadrature formula to approximate

$$\int_0^{p_k} \frac{1}{\sqrt{2\pi}} e^{-t^2/2}\, dt.$$

a. Find a solution to $f(x) = 0$ accurate to within 10^{-5} using Newton's method with $p_0 = 0.5$ and the Composite Simpson's rule.

b. Repeat (a) using the Composite Trapezoidal rule in place of the Composite Simpson's rule.

4.5 Romberg Integration

Romberg integration uses the Composite Trapezoidal rule to give preliminary approximations and then applies the Richardson extrapolation process to obtain improvements of the approximations. Recall from Section 4.2 that Richardson extrapolation can be performed on any approximation procedure of the form

$$M - N(h) = K_1 h + K_2 h^2 + \cdots + K_n h^n,$$

where the $K_1, K_2, \ldots, K_n$ are constants and $N(h)$ is an approximation to the unknown value M. The truncation error in this formula is dominated by $K_1 h$ when h is small, so this formula gives $O(h)$ approximations. Richardson's extrapolation uses an averaging technique to produce formulas with higher-order truncation error. In Section 4.2 we saw how this could be used to produce derivative approximations. In this section we will use extrapolation to approximate definite integrals.

To begin the presentation of the Romberg integration scheme, recall that the Composite Trapezoidal rule for approximating the integral of a function f on an interval $[a, b]$ using m subintervals is

$$\int_a^b f(x)\,dx = \frac{h}{2}\left[f(a) + f(b) + 2\sum_{j=1}^{m-1} f(x_j)\right] - \frac{(b-a)}{12}h^2 f''(\mu),$$

where $a < \mu < b$, $h = (b-a)/m$ and $x_j = a + jh$ for each $j = 0, 1, \ldots, m$.

We first obtain Composite Trapezoidal rule approximations with $m_1 = 1$, $m_2 = 2$, $m_3 = 4, \ldots,$ and $m_n = 2^{n-1}$, where n is a positive integer. The values of the step size h_k corresponding to m_k are $h_k = (b-a)/m_k = (b-a)/2^{k-1}$. With this notation the Trapezoidal rule becomes

$$\int_a^b f(x)\,dx = \frac{h_k}{2}\left[f(a) + f(b) + 2\left(\sum_{i=1}^{2^{k-1}-1} f(a + ih_k)\right)\right] - \frac{(b-a)}{12}h_k^2 f''(\mu_k), \tag{4.31}$$

where μ_k is a number in (a, b).

If the notation $R_{k,1}$ is introduced to denote the portion of Eq. (4.31) used for the trapezoidal approximation, then:

$$R_{1,1} = \frac{h_1}{2}[f(a) + f(b)] = \frac{(b-a)}{2}[f(a) + f(b)];$$

$$R_{2,1} = \frac{h_2}{2}[f(a) + f(b) + 2f(a + h_2)]$$

$$= \frac{(b-a)}{4}\left[f(a) + f(b) + 2f\left(a + \frac{(b-a)}{2}\right)\right]$$

$$= \frac{1}{2}[R_{1,1} + h_1 f(a + h_2)];$$

$$R_{3,1} = \frac{1}{2}\{R_{2,1} + h_2[f(a + h_3) + f(a + 3h_3)]\};$$

and, in general (see Figure 4.10),

(4.32)
$$R_{k,1} = \frac{1}{2}\left[R_{k-1,1} + h_{k-1}\sum_{i=1}^{2^{k-2}} f(a + (2i-1)h_k)\right],$$

for each $k = 2, 3, \ldots, n$. (See Exercises 12 and 13.)

Figure 4.10

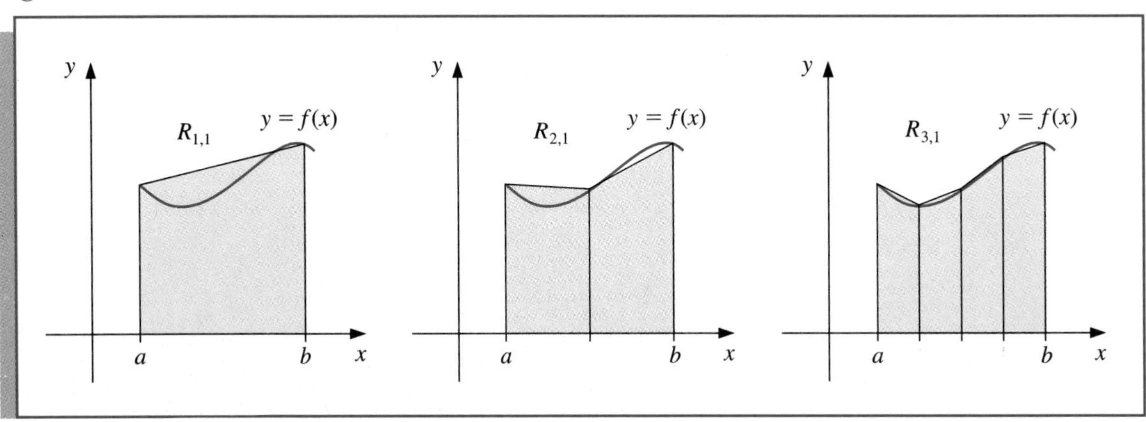

EXAMPLE 1 Using Eq. (4.32) to perform the first step of the Romberg integration scheme for approximating $\int_0^\pi \sin x\,dx$ with $n = 6$ leads to:

$$R_{1,1} = \frac{\pi}{2}[\sin 0 + \sin \pi] = 0;$$

$$R_{2,1} = \frac{1}{2}\left[R_{1,1} + \pi \sin \frac{\pi}{2}\right] = 1.57079633;$$

$$R_{3,1} = \frac{1}{2}\left[R_{2,1} + \frac{\pi}{2}\left(\sin \frac{\pi}{4} + \sin \frac{3\pi}{4}\right)\right] = 1.89611890;$$

$$R_{4,1} = \frac{1}{2}\left[R_{3,1} + \frac{\pi}{4}\left(\sin \frac{\pi}{8} + \sin \frac{3\pi}{8} + \sin \frac{5\pi}{8} + \sin \frac{7\pi}{8}\right)\right] = 1.97423160;$$

$$R_{5,1} = 1.99357034, \quad \text{and} \quad R_{6,1} = 1.99839336.$$

■

Since the correct value for the integral in Example 1 is 2, it appears that, although the calculations involved are not difficult, the convergence is slow. Richardson extrapolation will be used to speed the convergence.

It can be shown, although not easily (see [RR, pp. 136–138]), that if $f \in C^\infty[a, b]$, the Composite Trapezoidal rule can be written with an alternative error term in the form

(4.33)
$$\int_a^b f(x)\,dx - R_{k,1} = \sum_{i=1}^{\infty} K_i h_k^{2i} = K_1 h_k^2 + \sum_{i=2}^{\infty} K_i h_k^{2i},$$

where K_i for each i is independent of h_k and depends only on $f^{(2i-1)}(a)$ and $f^{(2i-1)}(b)$.

With the Composite Trapezoidal rule in this form, we can eliminate the term involving h_k^2 by combining this equation with its counterpart with h_k replaced by $h_{k+1} = h_k/2$:

(4.34)
$$\int_a^b f(x)\,dx - R_{k+1,1} = \sum_{i=1}^{\infty} K_i h_{k+1}^{2i} = \sum_{i=1}^{\infty} \frac{K_i h_k^{2i}}{2^{2i}} = \frac{K_1 h_k^2}{4} + \sum_{i=2}^{\infty} \frac{K_i h_k^{2i}}{4^i}.$$

Subtracting Eq. (4.33) from 4 times (4.34) and simplifying gives the $O(h_k^4)$ formula

$$\int_a^b f(x)\,dx - \left[R_{k+1,1} + \frac{R_{k+1,1} - R_{k,1}}{3}\right] = \sum_{i=2}^{\infty} \frac{K_i}{3}\left(\frac{h_k^{2i}}{4^{i-1}} - h_k^{2i}\right)$$
$$= \sum_{i=2}^{\infty} \frac{K_i}{3}\left(\frac{1 - 4^{i-1}}{4^{i-1}}\right) h_k^{2i}.$$

Extrapolation can now be applied to this formula to obtain an $O(h_k^6)$ result, and so on. To simplify the notation we define

$$R_{k,2} = R_{k,1} + \frac{R_{k,1} - R_{k-1,1}}{3},$$

for each $k = 2, 3, \ldots, n$, and apply the Richardson extrapolation procedure to these values. Continuing this notation, we have, for each $k = 2, 3, 4, \ldots, n$ and $j = 2, \ldots, k$, an $O(h_k^{2j})$ approximation formula defined by

(4.35)
$$R_{k,j} = R_{k,j-1} + \frac{R_{k,j-1} - R_{k-1,j-1}}{4^{j-1} - 1}.$$

The results that are generated from these formulas are shown in Table 4.9.

Table 4.9

$R_{1,1}$					
$R_{2,1}$	$R_{2,2}$				
$R_{3,1}$	$R_{3,2}$	$R_{3,3}$			
$R_{4,1}$	$R_{4,2}$	$R_{4,3}$	$R_{4,4}$		
$\vdots$	$\vdots$	$\vdots$	$\vdots$	$\ddots$	
$R_{n,1}$	$R_{n,2}$	$R_{n,3}$	$R_{n,4}$	$\cdots$	$R_{n,n}$

The Romberg technique has the additional desirable feature that it allows an entire new row in the table to be calculated by simply doing one additional application of the Composite Trapezoidal rule and then using the previously calculated values to obtain the succeeding entries in the row. The method used to construct a table of this type calculates the entries row by row, that is, in the order $R_{1,1}$, $R_{2,1}$, $R_{2,2}$, $R_{3,1}$, $R_{3,2}$, $R_{3,3}$, etc. Algorithm 4.2 describes this technique in detail.

ALGORITHM 4.2

Romberg

To approximate the integral $I = \int_a^b f(x)\,dx$, select an integer $n > 0$.

INPUT endpoints a, b; integer n.

OUTPUT an array R. *(Compute R by rows; only the last 2 rows are saved in storage.)*

Step 1 Set $h = b - a$;

$$R_{1,1} = \frac{h}{2}(f(a) + f(b)).$$

Step 2 OUTPUT $(R_{1,1})$.

Step 3 For $i = 2, \ldots, n$ do Steps 4–8.

Step 4 Set $$R_{2,1} = \frac{1}{2}\left[R_{1,1} + h\sum_{k=1}^{2^{i-2}} f(a + (k - 0.5)h)\right].$$

(Approximation from Trapezoidal method.)

Step 5 For $j = 2, \ldots, i$

set $$R_{2,j} = R_{2,j-1} + \frac{R_{2,j-1} - R_{1,j-1}}{4^{j-1} - 1}.$$ *(Extrapolation.)*

Step 6 OUTPUT ($R_{2,j}$ for $j = 1, 2, \ldots, i$).

Step 7 Set $h = h/2$.

Step 8 For $j = 1, 2, \ldots, i$ set $R_{1,j} = R_{2,j}$. *(Update row 1 of R.)*

Step 9 STOP.

EXAMPLE 2 In Example 1, the values for $R_{1,1}$ through $R_{6,1}$ were obtained for approximating $\int_0^\pi \sin x\,dx$. With Algorithm 4.2, the Romberg table is shown in Table 4.10. ■

Table 4.10

0					
1.57079633	2.09439511				
1.89611890	2.00455976	1.99857073			
1.97423160	2.00026917	1.99998313	2.00000555		
1.99357034	2.00001659	1.99999975	2.00000001	1.99999999	
1.99839336	2.00000103	2.00000000	2.00000000	2.00000000	2.00000000

Algorithm 4.2 requires a preset integer n to determine the number of rows to be generated. It is often more useful to prescribe an error tolerance for the approximation and generate n, within some upper bound, until consecutive diagonal entries $R_{n-1,n-1}$ and $R_{n,n}$ agree to within the tolerance. To guard against the possibility that two consecutive row elements agree with each other but not with the value of the integral being approximated, we generate approximations until not only $|R_{n-1,n-1} - R_{n,n}|$ is within the tolerance, but also $|R_{n-2,n-2} - R_{n-1,n-1}|$. Although not a universal safeguard, this will ensure that two differently generated sets of approximations agree within the specified tolerance before $R_{n,n}$, is accepted as sufficiently accurate.

Romberg integration applied to f on $[a, b]$ relies on the assumption that the Composite Trapezoidal rule has an error term that can be expressed in the form of Eq. (4.33); that is, we must have $f \in C^{2k+2}[a, b]$ for the kth row to be generated. General-purpose algorithms using Romberg integration include a check at each stage to ensure that this assumption is fulfilled. These methods are known as *cautious Romberg algorithms* and are described in [Joh]. This reference also describes methods for using the Romberg technique as an adaptive procedure, similar to the adaptive Simpson's rule that will be discussed in Section 4.6.

EXERCISE SET 4.5

1. Use Romberg integration to compute $R_{3,3}$ for the following integrals.

a. $\displaystyle\int_1^{1.5} x^2 \ln x\, dx$

b. $\displaystyle\int_0^{1} x^2 e^{-x}\, dx$

c. $\displaystyle\int_0^{0.35} \frac{2}{x^2-4}\, dx$

d. $\displaystyle\int_0^{\pi/4} x^2 \sin x\, dx$

e. $\displaystyle\int_0^{\pi/4} e^{3x} \sin 2x\, dx$

f. $\displaystyle\int_1^{1.6} \frac{2x}{x^2-4}\, dx$

g. $\displaystyle\int_3^{3.5} \frac{x}{\sqrt{x^2-4}}\, dx$

h. $\displaystyle\int_0^{\pi/4} (\cos x)^2\, dx$

2. Calculate $R_{4,4}$ for the integrals in Exercise 1.

3. Use Romberg integration to approximate the integrals in Exercise 1 to within 10^{-6}. Compute the Romberg table until $|R_{n-1,n-1} - R_{n,n}| < 10^{-6}$, or until $n = 10$. Compare your results to the exact values of the integrals.

4. Apply Romberg integration to the following integrals until $R_{n-1,n-1}$ and $R_{n,n}$ agree to within 10^{-4}.

a. $\displaystyle\int_0^{1} x^{1/3}\, dx$

b. $\displaystyle\int_0^{0.3} f(x)\, dx,$ where

$$f(x) = \begin{cases} x^3 + 1, & 0 \le x \le 0.1 \\ 1.001 + 0.03(x-0.1) + 0.3(x-0.1)^2 + 2(x-0.1)^3, & 0.1 < x \le 0.2, \\ 1.009 + 0.15(x-0.2) + 0.9(x-0.2)^2 + 2(x-0.2)^3, & 0.2 < x \le 0.3. \end{cases}$$

5. Use the following data to approximate $\int_1^5 f(x)\,dx$ as accurately as possible.

x	1	2	3	4	5
$f(x)$	2.4142	2.6734	2.8974	3.0976	3.2804

6. Romberg integration is used to approximate

$$\int_0^1 \frac{x^2}{1+x^3}\,dx.$$

If $R_{11} = 0.250$ and $R_{22} = 0.2315$, what is R_{21}?

7. Romberg integration is used to approximate

$$\int_2^3 f(x)\,dx.$$

If $f(2) = 0.51342$, $f(3) = 0.36788$, $R_{31} = 0.43687$, and $R_{33} = 0.43662$, find $f(2.5)$.

8. Romberg integration for approximating $\int_0^1 f(x)\,dx$ gives $R_{11} = 4$ and $R_{22} = 5$. Find $f\left(\frac{1}{2}\right)$.

9. Romberg integration for approximating $\int_a^b f(x)\,dx$ gives $R_{11} = 8$, $R_{22} = \frac{16}{3}$, and $R_{33} = \frac{208}{45}$. Find R_{31}.

10. Use Romberg integration to compute the following approximations to

$$\int_0^{48} \sqrt{1 + (\cos x)^2}\,dx.$$

[*Note*: The results in this exercise are most interesting if you are using a device with between seven- and nine-digit arithmetic.]

a. Determine $R_{1,1}$, $R_{2,1}$, $R_{3,1}$, $R_{4,1}$, and $R_{5,1}$ and use these approximations to predict the value of the integral.

b. Determine $R_{2,2}$, $R_{3,3}$, $R_{4,4}$, and $R_{5,5}$ and modify your prediction.

c. Determine $R_{6,1}$, $R_{6,2}$, $R_{6,3}$, $R_{6,4}$, $R_{6,5}$, and $R_{6,6}$ and modify your prediction.

d. Determine $R_{7,7}$, $R_{8,8}$, $R_{9,9}$, and $R_{10,10}$ and make a final prediction.

e. Explain why this integral causes difficulty with Romberg integration and how it can be reformulated to more easily determine an accurate approximation.

11. Show that the approximation obtained from $R_{k,2}$ is the same as that given by the Composite Simpson's rule described in Theorem 4.4 with $h = h_k$.

12. Show that, for any k,

$$\sum_{i=1}^{2^{k-1}-1} f\left(a + \frac{i}{2}h_{k-1}\right) = \sum_{i=1}^{2^{k-2}} f\left(a + \left(i - \frac{1}{2}\right)h_{k-1}\right) + \sum_{i=1}^{2^{k-2}-1} f(a + ih_{k-1}).$$

13. Use the result of Exercise 12 to verify Eq. (4.32); that is, show that for all k,

$$R_{k,1} = \frac{1}{2}\left[R_{k-1,1} + h_{k-1}\sum_{i=1}^{2^{k-2}} f\left(a + \left(i - \frac{1}{2}\right)h_{k-1}\right)\right].$$

14. In Exercise 24 of Section 1.1 a Maclaurin series was integrated to approximate erf(1), where erf(x) is the normal distribution error function defined by

$$\text{erf}(x) = \frac{2}{\sqrt{\pi}}\int_0^x e^{-t^2}\,dt.$$

Approximate erf(1) to within 10^{-7}.

4.6 Adaptive Quadrature Methods

The composite formulas require the use of equally spaced nodes. This is inappropriate when integrating a function on an interval that contains both regions with large functional variation and regions with small functional variation. If the approximation error is to be evenly distributed, a smaller step size is needed for the large-variation regions than for those with less variation. An efficient technique for this type of problem should predict the amount of functional variation and adapt the step size to the varying requirements. These methods are called **Adaptive quadrature methods**. The method we discuss is based on the Composite Simpson's rule, but the technique is easily modified to use other composite procedures.

Suppose that we want to approximate $\int_a^b f(x)\,dx$ to within a specified tolerance $\varepsilon > 0$. The first step in the procedure is to apply Simpson's rule with step size $h = (b-a)/2$. This procedure results in the following (see Figure 4.11):

(4.36)
$$\int_a^b f(x)\,dx = S(a,b) - \frac{h^5}{90}f^{(4)}(\mu), \quad \text{for some } \mu \text{ in } (a,b),$$

where

$$S(a,b) = \frac{h}{3}[f(a) + 4f(a+h) + f(b)].$$

Figure 4.11

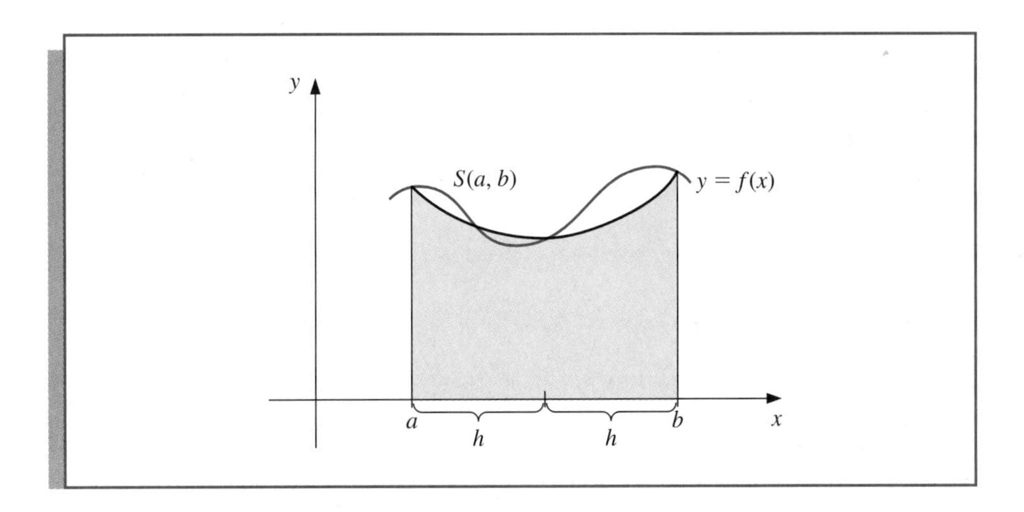

The next step is to determine a way to estimate the accuracy of our approximation, in particular, one that does not require determining $f^{(4)}(\mu)$. To do this, we first apply the Composite Simpson's rule to the problem with $n = 4$ and step size $(b-a)/4 = h/2$, giving

$$\textbf{(4.37)} \quad \int_a^b f(x)\,dx = \frac{h}{6}\left[f(a) + 4f\left(a + \frac{h}{2}\right) + 2f(a+h) + 4f\left(a + \frac{3h}{2}\right) + f(b)\right] - \left(\frac{h}{2}\right)^4 \frac{(b-a)}{180} f^{(4)}(\tilde{\mu}),$$

for some $\tilde{\mu}$ in (a, b). To simplify notation, let

$$S\left(a, \frac{a+b}{2}\right) = \frac{h}{6}\left[f(a) + 4f\left(a + \frac{h}{2}\right) + f(a+h)\right]$$

and

$$S\left(\frac{a+b}{2}, b\right) = \frac{h}{6}\left[f(a+h) + 4f\left(a + \frac{3h}{2}\right) + f(b)\right].$$

Then Eq. (4.37) can be rewritten (see Figure 4.12) as

$$\textbf{(4.38)} \quad \int_a^b f(x)\,dx = S\left(a, \frac{a+b}{2}\right) + S\left(\frac{a+b}{2}, b\right) - \frac{1}{16}\left(\frac{h^5}{90}\right) f^{(4)}(\tilde{\mu}).$$

The error estimation is derived by assuming that $\mu \approx \tilde{\mu}$ or, more precisely, that $f^{(4)}(\mu) \approx f^{(4)}(\tilde{\mu})$. The success of the technique depends on the accuracy of this assumption. If it is accurate, then equating the integrals in Eqs. (4.36) and (4.38) implies that

$$S\left(a, \frac{a+b}{2}\right) + S\left(\frac{a+b}{2}, b\right) - \frac{1}{16}\left(\frac{h^5}{90}\right) f^{(4)}(\tilde{\mu}) \approx S(a, b) - \frac{h^5}{90} f^{(4)}(\mu),$$

Figure 4.12

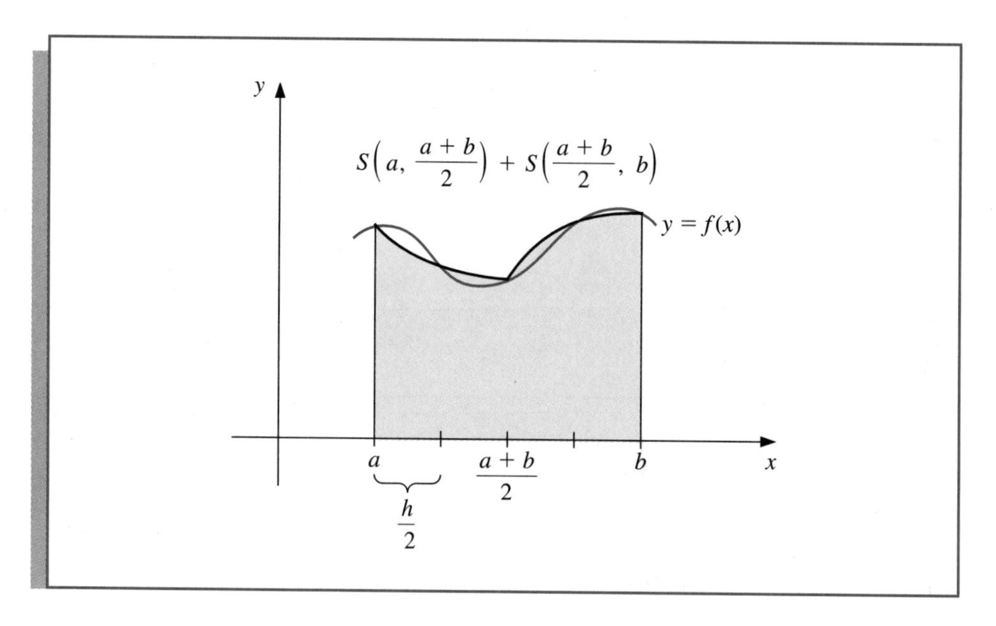

so

$$\frac{h^5}{90} f^{(4)}(\mu) \approx \frac{16}{15}\left[S(a,b) - S\left(a, \frac{a+b}{2}\right) - S\left(\frac{a+b}{2}, b\right)\right].$$

Using this estimate in Eq. (4.38) produces the error estimation

$$\left|\int_a^b f(x)\,dx - S\left(a, \frac{a+b}{2}\right) - S\left(\frac{a+b}{2}, b\right)\right| \approx \frac{1}{15}\left|S(a,b) - S\left(a, \frac{a+b}{2}\right) - S\left(\frac{a+b}{2}, b\right)\right|.$$

This result means that $S(a, (a+b)/2) + S((a+b)/2, b)$ approximates $\int_a^b f(x)\,dx$ about 15 times better than it agrees with the known value $S(a, b)$. Thus, if

$$\left|S(a,b) - S\left(a, \frac{a+b}{2}\right) - S\left(\frac{a+b}{2}, b\right)\right| < 15\varepsilon, \tag{4.39}$$

we expect to have

$$\left|\int_a^b f(x)\,dx - S\left(a, \frac{a+b}{2}\right) - S\left(\frac{a+b}{2}, b\right)\right| < \varepsilon. \tag{4.40}$$

So if ε is an acceptable error tolerance,

$$S\left(a, \frac{a+b}{2}\right) + S\left(\frac{a+b}{2}, b\right)$$

is assumed to be a sufficiently accurate approximation to $\int_a^b f(x)\,dx$.

EXAMPLE 1 To check the accuracy of the error estimate given in (4.39) and (4.40), consider its application to the integral

$$\int_0^{\pi/2} \sin x\,dx = 1.$$

In this case,

$$S\left(0, \frac{\pi}{2}\right) = \frac{\pi/4}{3}\left(\sin 0 + 4\sin\frac{\pi}{4} + \sin\frac{\pi}{2}\right) = \frac{\pi}{12}(2\sqrt{2}+1) = 1.002279878$$

and

$$S\left(0, \frac{\pi}{4}\right) + S\left(\frac{\pi}{4}, \frac{\pi}{2}\right) = \frac{\pi/8}{3}\left(\sin 0 + 4\sin\frac{\pi}{8} + 2\sin\frac{\pi}{4} + 4\sin\frac{3\pi}{8} + \sin\frac{\pi}{2}\right)$$
$$= 1.000134585.$$

So

$$\frac{1}{15}\left|S\left(0,\frac{\pi}{2}\right)-S\left(0,\frac{\pi}{4}\right)-S\left(\frac{\pi}{4},\frac{\pi}{2}\right)\right| = 0.000143020.$$

This closely approximates the actual error,

$$\left|\int_0^{\pi/2} \sin dx - 1.000134585\right| = 0.000134585,$$

even though $D_x^4 \sin x = \sin x$ varies significantly in the interval $(0, \pi/2)$. ■

When the error estimate given in (4.39) does not hold, we apply the Simpson's rule technique individually to the subintervals $[a, (a+b)/2]$ and $[(a+b)/2, b]$. Then we use the error estimation procedure to determine if the approximation to the integral on each subinterval is within a tolerance of $\varepsilon/2$. If so, we sum the approximations to produce an approximation to $\int_a^b f(x)\,dx$ within the tolerance ε.

If the approximation on one of the subintervals fails to be within the tolerance $\varepsilon/2$, that subinterval is itself subdivided and the procedure is reapplied to the two subintervals to determine if the approximation on each subinterval is accurate to within $\varepsilon/4$. This halving procedure is continued until each portion is within the required tolerance. Although problems can be constructed for which this tolerance will never be met, the technique is usually successful, because each subdivision typically increases the accuracy of the approximation by a factor of 16 while requiring an increased accuracy factor of only 2.

Algorithm 4.3 details this Adaptive quadrature procedure for Simpson's rule, although some technical difficulties arise that require the implementation of the method to differ slightly from the preceding discussion. For example, in Step 1 the tolerance has been set at 10ε rather than the 15ε figure in Inequality (4.39). This bound is chosen conservatively to compensate for error in the assumption $f^{(4)}(\mu) \approx f^{(4)}(\tilde{\mu})$. In problems where $f^{(4)}$ is known to be widely varying, you should lower this bound even further.

The procedure listed in the algorithm first approximates the integral on the leftmost subinterval in a subdivision. This requires introducing a procedure for efficiently storing and recalling previously computed functional evaluations for the nodes in the right half subintervals. Steps 3, 4, and 5 contain a stacking procedure with an indicator to keep track of the data that will be required for calculating the approximation on the subinterval immediately adjacent and to the right of the subinterval on which the approximation is being generated. The method is easier to implement on a computer if a programming language is used that permits recursion, such as C or Pascal.

ALGORITHM **4.3**

Adaptive Quadrature

To approximate the integral $I = \int_a^b f(x)\,dx$ to within a given tolerance:

INPUT endpoints a, b; tolerance TOL; limit N to number of levels.

OUTPUT approximation APP or message that N is exceeded.

Step 1 Set $APP = 0$;
$i = 1$;
$TOL_i = 10\ TOL$;
$a_i = a$;
$h_i = (b - a)/2$;
$FA_i = f(a)$;
$FC_i = f(a + h_i)$;
$FB_i = f(b)$;
$S_i = h_i(FA_i + 4FC_i + FB_i)/3$; *(Approximation from Simpson's method for entire interval.)*
$L_i = 1$.

Step 2 While $i > 0$ do Steps 3–5.

Step 3 Set $FD = f(a_i + h_i/2)$;
$FE = f(a_i + 3h_i/2)$;
$S1 = h_i(FA_i + 4FD + FC_i)/6$; *(Approximations from Simpson's method for halves of subintervals.)*
$S2 = h_i(FC_i + 4FE + FB_i)/6$;
$v_1 = a_i$; *(Save data at this level.)*
$v_2 = FA_i$;
$v_3 = FC_i$;
$v_4 = FB_i$;
$v_5 = h_i$;
$v_6 = TOL_i$;
$v_7 = S_i$;
$v_8 = L_i$.

Step 4 Set $i = i - 1$. *(Delete the level.)*

Step 5 If $|S1 + S2 - v_7| < v_6$
then set $APP = APP + (S1 + S2)$
else
if $(v_8 \geq N)$
then
OUTPUT ('LEVEL EXCEEDED'); *(Procedure fails.)*
STOP.
else *(Add one level.)*
set $i = i + 1$; *(Data for right half subinterval.)*
$a_i = v_1 + v_5$;
$FA_i = v_3$;
$FC_i = FE$;
$FB_i = v_4$;
$h_i = v_5/2$;
$TOL_i = v_6/2$;
$S_i = S2$;
$L_i = v_8 + 1$;
set $i = i + 1$; *(Data for left half subinterval.)*
$a_i = v_1$;
$FA_i = v_2$;

$FC_i = FD;$
$FB_i = v_3;$
$h_i = h_{i-1};$
$TOL_i = TOL_{i-1};$
$S_i = S1;$
$L_i = L_{i-1}.$

Step 6 OUTPUT (APP); (APP *approximates* I *to within* TOL.)
STOP.

EXAMPLE 2 The graph of the function $f(x) = (100/x^2)\sin(10/x)$ for x in $[1, 3]$ is shown in Figure 4.13. Using the Adaptive Quadrature Algorithm 4.3 with tolerance 10^{-4} to approximate $\int_1^3 f(x)\,dx$ produces -1.426014, a result that is accurate to within 1.1×10^{-5}. The approximation required that Simpson's rule with $n = 4$ be performed on the 23 subintervals whose endpoints are shown on the horizontal axis in Figure 4.13. The total number of functional evaluations required for this approximation is 93.

Figure 4.13

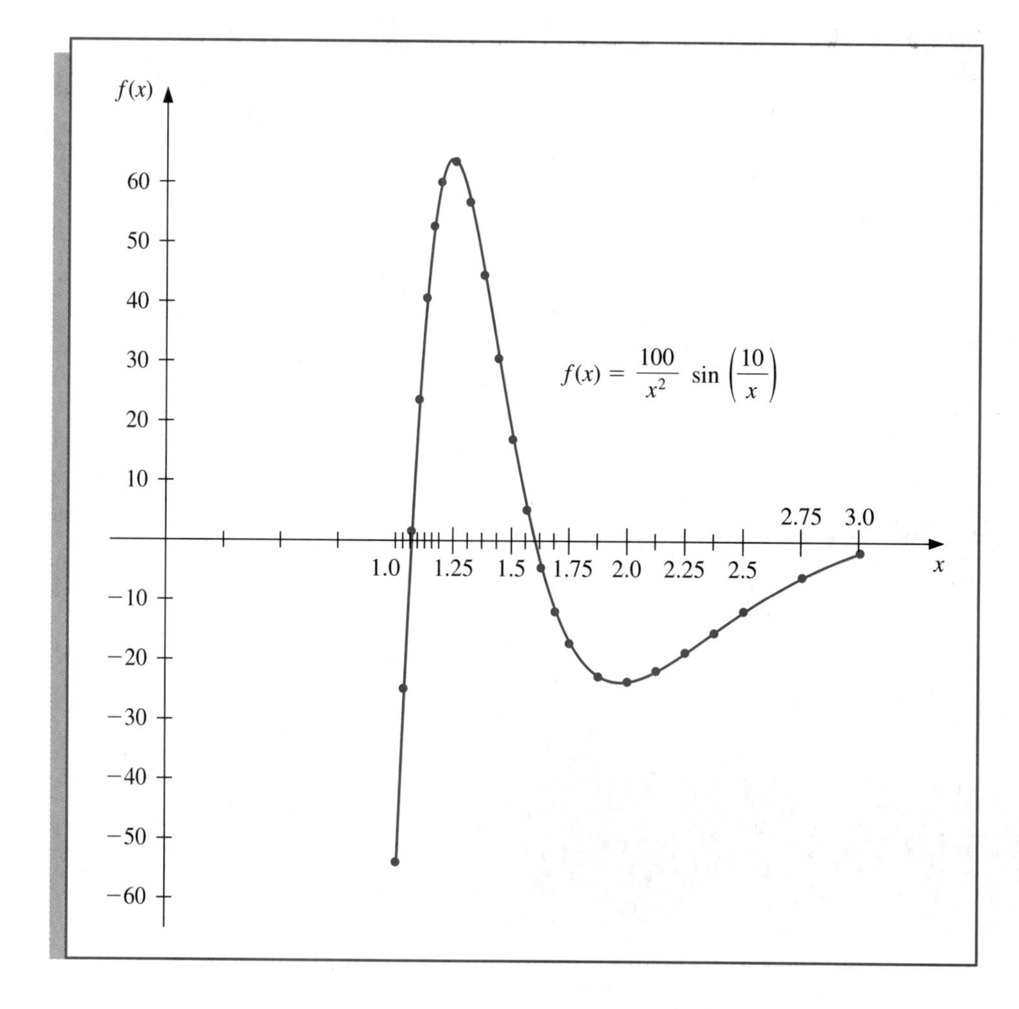

The largest value of h for which the standard Composite Simpson's rule gives 10^{-4} accuracy is $h = \frac{1}{88}$. This application requires 177 function evaluations, nearly twice as many as the adaptive technique. ■

EXERCISE SET 4.6

1. Compute the Simpson's rule approximations $S(a, b)$, $S(a, (a+b)/2)$, and $S((a+b)/2, b)$ for the following integrals, and verify the estimate given in the approximation formula.

   a. $\int_1^{1.5} x^2 \ln x \, dx$

   b. $\int_0^1 x^2 e^{-x} \, dx$

   c. $\int_0^{0.35} \frac{2}{x^2 - 4} \, dx$

   d. $\int_0^{\pi/4} x^2 \sin x \, dx$

   e. $\int_0^{\pi/4} e^{3x} \sin 2x \, dx$

   f. $\int_1^{1.6} \frac{2x}{x^2 - 4} \, dx$

   g. $\int_3^{3.5} \frac{x}{\sqrt{x^2 - 4}} \, dx$

   h. $\int_0^{\pi/4} \cos^2 x \, dx$

2. Use Adaptive quadrature to find approximations to within 10^{-3} for the integrals in Exercise 1. Do not use a computer program to generate these results.

3. Use Adaptive quadrature to approximate the following integrals to within 10^{-5}.

   a. $\int_1^3 e^{2x} \sin 3x \, dx$

   b. $\int_1^3 e^{3x} \sin 2x \, dx$

   c. $\int_0^5 \left[2x \cos(2x) - (x-2)^2\right] dx$

   d. $\int_0^5 \left[4x \cos(2x) - (x-2)^2\right] dx$

4. Use Simpson's Composite rule with $n = 4, 6, 8, \ldots$, until successive approximations to the following integrals agree to within 10^{-6}. Determine the number of nodes required. Use the Adaptive Quadrature Algorithm to approximate the integral to within 10^{-6} and count the number of nodes. Did Adaptive quadrature produce any improvement?

   a. $\int_0^{\pi} x \cos x^2 \, dx$

   b. $\int_0^{\pi} x \sin x^2 \, dx$

   c. $\int_0^{\pi} x^2 \cos x \, dx$

   d. $\int_0^{\pi} x^2 \sin x \, dx$

5. Sketch the graphs of $\sin(1/x)$ and $\cos(1/x)$ on $[0.1, 2]$. Use Adaptive quadrature to approximate the integrals

$$\int_{0.1}^{2} \sin \frac{1}{x} \, dx \quad \text{and} \quad \int_{0.1}^{2} \cos \frac{1}{x} \, dx$$

   to within 10^{-3}.

6. Let $T(a, b)$ and $T(a, \frac{a+b}{2}) + T(\frac{a+b}{2}, b)$ be the single and double applications of the Trapezoidal rule to $\int_a^b f(x)\,dx$. Derive the relationship between

$$\left| T(a, b) - T\left(a, \frac{a+b}{2}\right) - T\left(\frac{a+b}{2}, b\right) \right|$$

   and

$$\left| \int_a^b f(x)\,dx - T\left(a, \frac{a+b}{2}\right) - T\left(\frac{a+b}{2}, b\right) \right|.$$

7. The differential equation

$$mu''(t) + ku(t) = F_0 \cos \omega t$$

describes a spring-mass system with mass m, spring constant k, and no applied damping. The term $F_0 \cos \omega t$ describes a periodic external force applied to the system. The solution to the equation when the system is initially at rest ($u'(0) = u(0) = 0$) is

$$u(t) = \frac{2F_0}{m(\omega_0^2 - \omega^2)} \sin \frac{(\omega_0 - \omega)}{2} t \sin \frac{(\omega_0 + \omega)}{2} t, \quad \text{where} \quad \omega_0 = \sqrt{\frac{k}{m}} \neq \omega.$$

Sketch the graph of u when $m = 1$, $k = 9$, $F_0 = 1$, $\omega = 2$, and $t \in [0, 2\pi]$. Approximate $\int_0^{2\pi} u(t)\,dt$ to within 10^{-4}.

8. If the term $cu'(t)$ is added to the left side of the motion equation in Exercise 7, the resulting differential equation describes a spring-mass system that is damped with damping constant c. The solution to this equation when the solution is initially at rest is

$$u(t) = c_1 e^{r_1 t} + c_2 e^{r_2 t} + \frac{F_0}{\sqrt{m^2(\omega_0^2 - \omega^2)^2 + c^2\omega^2}} \cos(\omega t - \delta),$$

where

$$r_1 = \frac{-c + \sqrt{c^2 - 4\omega_0^2 m^2}}{2m}, \quad r_2 = \frac{-c - \sqrt{c^2 - 4\omega_0^2 m^2}}{2m},$$

and

$$\delta = \arctan\left(\frac{c\omega}{m(\omega_0^2 - \omega^2)}\right).$$

Sketch the graph of u when $m = 1, k = 9, F_0 = 1, c = 10, \omega = 2$ and $t \in [0, 2\pi]$. Approximate $\int_0^{2\pi} u(t)\,dt$ to within 10^{-4}.

9. The study of light diffraction at a rectangular aperture involves the Fresnel integrals

$$c(t) = \int_0^t \cos \frac{\pi}{2} w^2\,dw \quad \text{and} \quad s(t) = \int_0^t \sin \frac{\pi}{2} w^2\,dw.$$

Construct a table of values for $c(t)$ and $s(t)$ that is accurate to within 10^{-4} for values of $t = 0.1, 0.2, \ldots, 1.0$.

4.7 Gaussian Quadrature

The Newton-Cotes formulas were derived by integrating interpolating polynomials. Since the error term in the interpolating polynomial of degree n involves the $(n + 1)$st derivative of the function being approximated, a formula of this type is exact when approximating any polynomial of degree less than or equal to n.

All the Newton-Cotes formulas use values of the function at equally spaced points. This practice is convenient when the formulas are combined to form the composite rules we considered earlier, but this restriction can significantly decrease the accuracy of the approximation. Consider, for example, the Trapezoidal rule applied to determine the integrals of the functions shown in Figure 4.14.

Figure 4.14

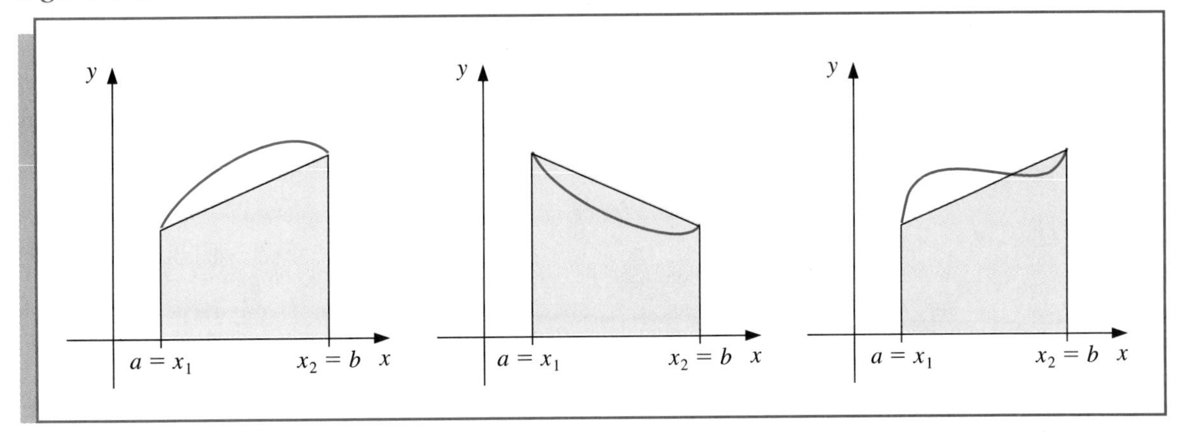

The Trapezoidal rule approximates the integral of the function by integrating the linear function that joins the endpoints of the graph of the function. But this is not likely the best line for approximating the integral. Lines such as those shown in Figure 4.15 would likely give much better approximations in most cases.

Figure 4.15

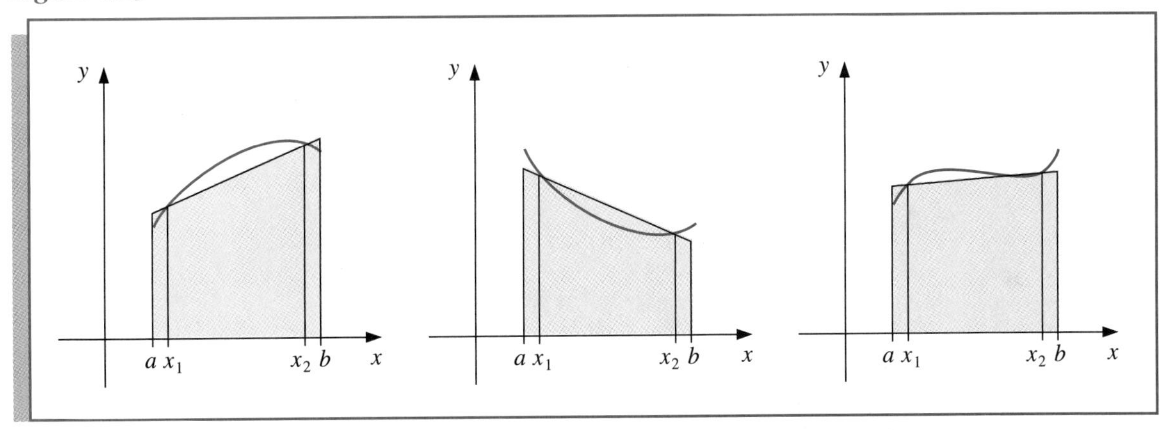

Gaussian quadrature chooses the points for evaluation in an optimal, rather than equally spaced, way. The nodes $x_1, x_2, \ldots, x_n$ in the interval $[a, b]$ and coefficients $c_1, c_2, \ldots, c_n$, are chosen to minimize the expected error obtained in performing the approximation

$$\int_a^b f(x)\,dx \approx \sum_{i=1}^{n} c_i f(x_i)$$

for an arbitrary function f. To measure this accuracy, we assume that the best choice of these values is that producing the exact result for the largest class of polynomials, that is, the choice which gives the greatest degree of precision.

The coefficients $c_1, c_2, \ldots, c_n$ in the approximation formula are arbitrary, and the nodes $x_1, x_2, \ldots, x_n$ are restricted only by the specification that they lie in $[a, b]$, the interval of integration. This gives us $2n$ parameters to choose. If the coefficients of a polynomial are considered as parameters, the class of polynomials of degree at most $(2n-1)$ also contains $2n$ parameters. This, then, is the largest class of polynomials for which it is possible to expect the formula to be exact. For the proper choice of the values and constants exactness on this set can be obtained.

To illustrate the procedure for choosing the appropriate parameters, we will show how to select the coefficients and nodes when $n = 2$ and the interval of integration is $[-1, 1]$. We will then discuss the more general situation for an arbitrary choice of nodes and coefficients and show how the technique is modified when integrating over an arbitrary interval $[a, b]$.

Suppose we want to determine c_1, c_2, x_1, and x_2 so that the integration formula

$$\int_{-1}^{1} f(x)\,dx \approx c_1 f(x_1) + c_2 f(x_2)$$

gives the exact result whenever $f(x)$ is a polynomial of degree $2(2) - 1 = 3$ or less, that is, when

$$f(x) = a_0 + a_1 x + a_2 x^2 + a_3 x^3,$$

for some collection of constants, a_0, a_1, a_2, and a_3. Because

$$\int (a_0 + a_1 x + a_2 x^2 + a_3 x^3)\,dx = a_0 \int 1\,dx + a_1 \int x\,dx + a_2 \int x^2\,dx + a_3 \int x^3\,dx,$$

this is equivalent to showing that the formula gives exact results when $f(x)$ is 1, x, x^2, and x^3. Hence, we need c_1, c_2, x_1, and x_2, so that

$$c_1 \cdot 1 + c_2 \cdot 1 = \int_{-1}^{1} 1\,dx = 2, \qquad c_1 \cdot x_1 + c_2 \cdot x_2 = \int_{-1}^{1} x\,dx = 0,$$

$$c_1 \cdot x_1^2 + c_2 \cdot x_2^2 = \int_{-1}^{1} x^2\,dx = \frac{2}{3}, \quad \text{and} \quad c_1 \cdot x_1^3 + c_2 \cdot x_2^3 = \int_{-1}^{1} x^3\,dx = 0.$$

A little algebra shows that this system of equations has the unique solution

$$c_1 = 1, \quad c_2 = 1, \quad x_1 = -\frac{\sqrt{3}}{3}, \quad \text{and} \quad x_2 = \frac{\sqrt{3}}{3},$$

which gives the approximation formula

$$\int_{-1}^{1} f(x)\,dx \approx f\left(\frac{-\sqrt{3}}{3}\right) + f\left(\frac{\sqrt{3}}{3}\right). \tag{4.41}$$

This formula has degree of precision three, that is, it produces the exact result for every polynomial of degree three or less.

This technique could be used to determine the nodes and coefficients for formulas that give exact results for higher-degree polynomials, but an alternative method can be used to obtain them more easily. In Sections 8.2 and 8.3 we will consider various collections of orthogonal polynomials, functions that have the property that a particular definite integral of the product of any two of them is zero. The set that is relevant to our problem is the set of Legendre polynomials, a collection $\{p_0(x), P_1(x), \ldots, P_n(x), \ldots,\}$ with properties:

1. For each n, $P_n(x)$ is a polynomial of degree n.
2. $\int_{-1}^{1} P(x)P_n(x)\,dx = 0$ whenever $P(x)$ is a polynomial of degree less than n.

The first few Legendre polynomials are

$$P_0(x) = 1, \quad P_1(x) = x, \quad P_2(x) = x^2 - \frac{1}{3},$$

$$P_3(x) = x^3 - \frac{3}{5}x, \quad \text{and} \quad P_4(x) = x^4 - \frac{6}{7}x^2 + \frac{3}{35}.$$

The roots of these polynomials are distinct, lie in the interval $(-1, 1)$, have a symmetry with respect to the origin, and, most importantly, are the correct choice for determining the parameters that solve our problem. The nodes $x_1, x_2, \ldots, x_n$ needed to produce an integral approximation formula that gives exact results for any polynomial of degree less than $2n$ are the roots of the nth-degree Legendre polynomial. This is established by the following result.

Theorem 4.7 Suppose that $x_1, x_2, \ldots, x_n$ are the roots of the nth Legendre polynomial $P_n(x)$ and that for each $i = 1, 2, \ldots, n$, the numbers c_i are defined by

$$c_i = \int_{-1}^{1} \prod_{\substack{j=1 \\ j \neq i}}^{n} \frac{x - x_j}{x_i - x_j}\,dx.$$

If $P(x)$ is any polynomial of degree less than $2n$, then

$$\int_{-1}^{1} P(x)\,dx = \sum_{i=1}^{n} c_i P(x_i).$$

Proof Let us first consider the situation for a polynomial $R(x)$ of degree less than n. Rewrite $R(x)$ as an $(n-1)$st Lagrange polynomial with nodes at the roots of the nth Legendre polynomial $P_n(x)$. This representation of $R(x)$ is exact, since the error term involves the nth derivative of R and the nth derivative of R is zero. Hence,

$$\int_{-1}^{1} R(x)\,dx = \int_{-1}^{1} \left[\sum_{i=1}^{n} \prod_{\substack{j=1 \\ j \neq i}}^{n} \frac{x - x_j}{x_i - x_j} R(x_i) \right] dx$$

$$= \sum_{i=1}^{n} \left[\int_{-1}^{1} \prod_{\substack{j=1 \\ j \neq i}}^{n} \frac{x - x_j}{x_i - x_j}\, dx \right] R(x_i) = \sum_{i=1}^{n} c_i R(x_i).$$

This verifies the result for polynomials of degree less than n.

If the polynomial $P(x)$ of degree less than $2n$ is divided by the nth Legendre polynomial $P_n(x)$, then two polynomials $Q(x)$ and $R(x)$ of degree less than n are produced with

$$P(x) = Q(x)P_n(x) + R(x).$$

We now invoke the unique power of the Legendre polynomials. First, the degree of the polynomial $Q(x)$ is less than n, so (by property 2),

$$\int_{-1}^{1} Q(x)P_n(x)\, dx = 0.$$

Next, since x_i is a root of $P_n(x)$ for each $i = 1, 2, \ldots, n$, we have

$$P(x_i) = Q(x_i)P_n(x_i) + R(x_i) = R(x_i).$$

Finally, since $R(x)$ is a polynomial of degree less than n, the opening argument implies that

$$\int_{-1}^{1} R(x)\, dx = \sum_{i=1}^{n} c_i R(x_i).$$

Putting these facts together verifies that the formula is exact for the polynomial $P(x)$:

$$\begin{aligned}\int_{-1}^{1} P(x)\, dx &= \int_{-1}^{1} [Q(x)P_n(x) + R(x)]\, dx \\ &= \int_{-1}^{1} R(x)\, dx = \sum_{i=1}^{n} c_i R(x_i) = \sum_{i=1}^{n} c_i P(x_i).\end{aligned}$$

▪ ▪ ▪

The constants c_i needed for the quadrature rule can be generated from the equation in Theorem 4.7, but both these constants and the roots of the Legendre polynomials are extensively tabulated. Table 4.11 lists these values for $n = 2$, 3, 4, and 5. Others can be found in [StS].

An integral $\int_a^b f(x)\, dx$ over an arbitrary $[a, b]$ can be transformed into an integral over $[-1, 1]$ by using the change of variables (See Figure 4.16):

$$t = \frac{2x - a - b}{b - a} \iff x = \frac{1}{2}[(b - a)t + a + b].$$

Table 4.11

n	Roots $r_{n,i}$	Coefficients $c_{n,i}$
2	0.5773502692	1.0000000000
	−0.5773502692	1.0000000000
3	0.7745966692	0.5555555556
	0.0000000000	0.8888888889
	−0.7745966692	0.5555555556
4	0.8611363116	0.3478548451
	0.3399810436	0.6521451549
	−0.3399810436	0.6521451549
	−0.8611363116	0.3478548451
5	0.9061798459	0.2369268850
	0.5384693101	0.4786286705
	0.0000000000	0.5688888889
	−0.5384693101	0.4786286705
	−0.9061798459	0.2369268850

Figure 4.16

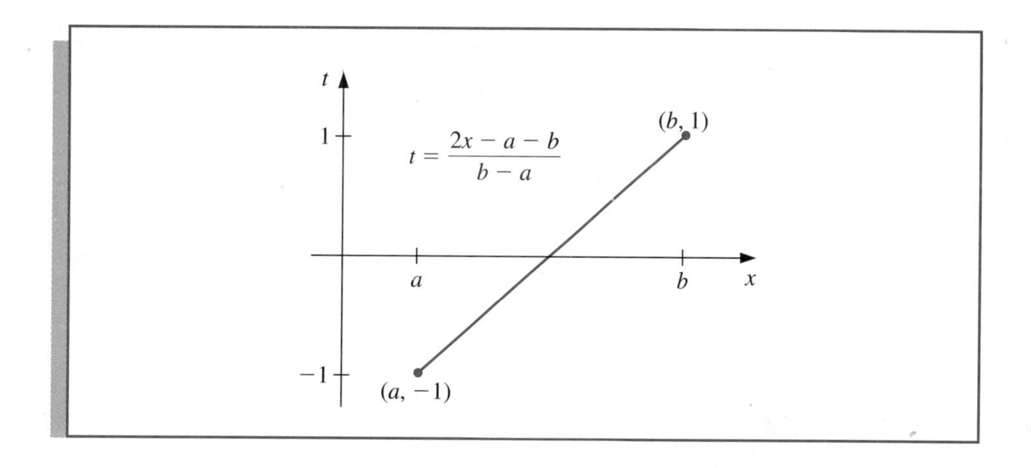

This permits Gaussian quadrature to be applied to

(4.42)
$$\int_a^b f(x)\,dx = \int_{-1}^{1} f\left(\frac{(b-a)t + (b+a)}{2}\right)\frac{(b-a)}{2}\,dt.$$

EXAMPLE 1 Consider the problem of finding approximations to $\int_1^{1.5} e^{-x^2}\,dx$. Table 4.12 lists the values for the Newton-Cotes formulas given in Section 4.3. The exact value of the integral to seven decimal places is 0.1093643.

Table 4.12

n	0	1	2	3	4
Closed formulas		0.1183197	0.1093104	0.1093404	0.1093643
Open formulas	0.1048057	0.1063473	0.1094116	0.1093971	

The Gaussian quadrature procedure applied to this problem requires that the integral first be transformed into a problem whose interval of integration is $[-1, 1]$. Using Eq. (4.42), we have

$$\int_1^{1.5} e^{-x^2}\,dx = \frac{1}{4}\int_{-1}^{1} e^{(-(t+5)^2/16)}\,dt.$$

Using the values in Table 4.11 gives the improved Gaussian quadrature approximations for this problem

$n = 2$:

$$\int_1^{1.5} e^{-x^2}\,dx \approx \frac{1}{4}[e^{-(5+0.5773502692)^2/16} + e^{-(5-0.5773502692)^2/16}] = 0.1094003$$

$n = 3$:

$$\int_1^{1.5} e^{-x^2}\,dx \approx \frac{1}{4}[(0.5555555556)e^{-(5+0.7745966692)^2/16} + (0.8888888889)e^{-(5)^2/16}$$
$$+ (0.5555555556)e^{-(5-0.7745966692)^2/16}]$$
$$= 0.1093642.$$

For comparison, the values obtained using the Romberg procedure with $n = 4$ are listed in Table 4.13. ■

Table 4.13

0.1183197			
0.1115627	0.1093104		
0.1099114	0.1093610	0.1093643	
0.1095009	0.1093641	0.1093643	0.1093643

EXERCISE SET 4.7

1. Approximate the following integrals using Gaussian quadrature with $n = 2$ and compare your results to the exact values of the integrals.

   a. $\int_1^{1.5} x^2 \ln x\,dx$

   b. $\int_0^1 x^2 e^{-x}\,dx$

   c. $\int_0^{0.35} \frac{2}{x^2-4}\,dx$

   d. $\int_0^{\pi/4} x^2 \sin x\,dx$

   e. $\int_0^{\pi/4} e^{3x} \sin 2x\,dx$

   f. $\int_1^{1.6} \frac{2x}{x^2-4}\,dx$

   g. $\int_3^{3.5} \frac{x}{\sqrt{x^2-4}}\,dx$

   h. $\int_0^{\pi/4} (\cos x)^2\,dx$

2. Repeat Exercise 1 with $n = 3$.

3. Repeat Exercise 1 with $n = 4$.

4. Repeat Exercise 1 with $n = 5$.

5. Determine constants a, b, c, and d that will produce a quadrature formula

$$\int_{-1}^{1} f(x)\,dx = af(-1) + bf(1) + cf'(-1) + df'(1)$$

that has degree of precision 3.

6. Determine constants a, b, c, and d that will produce a quadrature formula

$$\int_{-1}^{1} f(x)\,dx = af(-1) + bf(0) + cf(1) + df'(-1) + ef'(1)$$

that has degree of precision 4.

7. Verify the entries for the values of $n = 2$ and 3 in Table 4.11 by finding the roots of the respective Legendre polynomials and using the equations preceding this table to find the coefficients associated with the values.

8. Show that the formula $Q(P) = \sum_{i=1}^{n} c_i P(x_i)$ cannot have degree of precision greater than $2n - 1$, regardless of the choice of $c_1, \ldots, c_n$ and $x_1, \ldots, x_n$. [*Hint*: Construct a polynomial that has a double root at each of the x_i's.]

4.8 Multiple Integrals

The techniques discussed in the previous sections can be modified in a straightforward manner for use in the approximation of multiple integrals. Consider the double integral

$$\iint_R f(x, y)\,dA,$$

where R is a rectangular region in the plane;

$$R = \{(x, y) \mid a \le x \le b, c \le y \le d\},$$

for some constants a, b, c, and d. (See Figure 4.17.) To illustrate the approximation technique, we employ the Composite Simpson's rule, although any other composite formula could be used in its place.

To apply the Composite Simpson's rule, we divide the region R by partitioning both $[a, b]$ and $[c, d]$ into an even number of subintervals. To simplify the notation we choose integers n and m and partition $[a, b]$ and $[c, d]$ with the evenly spaced mesh points $x_0, x_1, \ldots, x_{2n}$ and $y_0, y_1, \ldots, y_{2m}$, respectively. These subdivisions determine step sizes $h =$

Figure 4.17

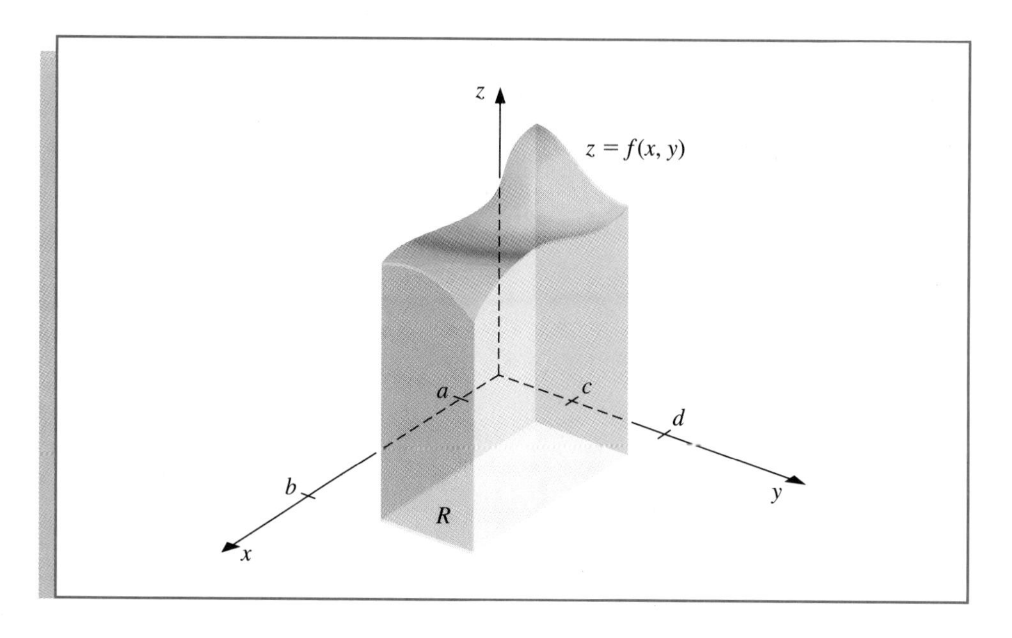

$(b-a)/2n$ and $k=(d-c)/2m$. Writing the double integral as the iterated integral

$$\iint_R f(x,y)\,dA = \int_a^b \left(\int_c^d f(x,y)\,dy \right) dx,$$

we first use the Composite Simpson's rule to evaluate

$$\int_c^d f(x,y)\,dy,$$

treating x as a constant. Let $y_j = c + jk$ for each $j = 0, 1, \ldots, 2m$. Then

$$\int_c^d f(x,y)\,dy = \frac{k}{3}\left[f(x,y_0) + 2\sum_{j=1}^{m-1} f(x,y_{2j}) + 4\sum_{j=1}^{m} f(x,y_{2j-1}) + f(x,y_{2m}) \right] - \frac{(d-c)k^4}{180}\frac{\partial^4 f(x,\mu)}{\partial y^4}$$

for some μ in (c,d). Thus,

$$\int_a^b \int_c^d f(x,y)\,dy\,dx = \frac{k}{3}\left[\int_a^b f(x,y_0)\,dx + 2\sum_{j=1}^{m-1}\int_a^b f(x,y_{2j})\,dx + 4\sum_{j=1}^{m}\int_a^b f(x,y_{2j-1})\,dx + \int_a^b f(x,y_{2m})\,dx \right] - \frac{(d-c)k^4}{180}\int_a^b \frac{\partial^4 f(x,\mu)}{\partial y^4}\,dx.$$

The Composite Simpson's rule is now employed on the integrals in this equation. Let $x_i = a + ih$ for each $i = 0, 1, \ldots, 2n$. Then for each $j = 0, 1, \ldots, 2m$ we have

$$\int_a^b f(x, y_j)\,dx = \frac{h}{3}\left[f(x_0, y_j) + 2\sum_{i=1}^{n-1} f(x_{2i}, y_j) + 4\sum_{i=1}^{n} f(x_{2i-1}, y_j) + f(x_{2n}, y_j)\right] - \frac{(b-a)h^4}{180}\frac{\partial^4 f}{\partial x^4}(\xi_j, y_j)$$

for some ξ_j in (a, b). The resulting approximation has the form

$$\begin{aligned}
\int_a^b \int_c^d f(x, y)\,dy\,dx \approx \frac{hk}{9}\Bigg\{&\left[f(x_0, y_0) + 2\sum_{i=1}^{n-1} f(x_{2i}, y_0) + 4\sum_{i=1}^{n} f(x_{2i-1}, y_0) + f(x_{2n}, y_0)\right] \\
&+ 2\left[\sum_{j=1}^{m-1} f(x_0, y_{2j}) + 2\sum_{j=1}^{m-1}\sum_{i=1}^{n-1} f(x_{2i}, y_{2j}) \right. \\
&\left. + 4\sum_{j=1}^{m-1}\sum_{i=1}^{n} f(x_{2i-1}, y_{2j}) + \sum_{j=1}^{m-1} f(x_{2n}, y_{2j})\right] \\
&+ 4\left[\sum_{j=1}^{m} f(x_0, y_{2j-1}) + 2\sum_{j=1}^{m}\sum_{i=1}^{n-1} f(x_{2i}, y_{2j-1}) \right. \\
&\left. + 4\sum_{j=1}^{m}\sum_{i=1}^{n} f(x_{2i-1}, y_{2j-1}) + \sum_{j=1}^{m} f(x_{2n}, y_{2j-1})\right] \\
&+ \left[f(x_0, y_{2m}) + 2\sum_{i=1}^{n-1} f(x_{2i}, y_{2m}) + 4\sum_{i=1}^{n} f(x_{2i-1}, y_{2m}) \right. \\
&\left. + f(x_{2n}, y_{2m})\right]\Bigg\}.
\end{aligned}$$

The error term E is given by

$$\begin{aligned}
E = \frac{-k(b-a)h^4}{540}&\left[\frac{\partial^4 f(\xi_0, y_0)}{\partial x^4} + 2\sum_{j=1}^{m-1}\frac{\partial^4 f(\xi_{2j}, y_{2j})}{\partial x^4} + 4\sum_{j=1}^{m}\frac{\partial^4 f(\xi_{2j-1}, y_{2j-1})}{\partial x^4}\right. \\
&\left. + \frac{\partial^4 f(\xi_{2m}, y_{2m})}{\partial x^4}\right] - \frac{(d-c)k^4}{180}\int_a^b \frac{\partial^4 f(x, \mu)}{\partial y^4}\,dx.
\end{aligned}$$

If $\partial^4 f/\partial x^4$ is continuous, the Intermediate Value Theorem can be repeatedly applied to show that the evaluation of the partial derivatives with respect to x can be replaced by a common value and that

$$E = \frac{-k(b-a)h^4}{540}\left[6m\frac{\partial^4 f}{\partial x^4}(\overline{\eta}, \overline{\mu})\right] - \frac{(d-c)k^4}{180}\int_a^b \frac{\partial^4 f(x, \mu)}{\partial y^4}\,dx$$

for some $(\overline{\eta}, \overline{\mu})$ in R. If $\partial^4 f/\partial y^4$ is also continuous, the Weighted Mean Value Theorem for Integrals implies that

$$\int_a^b \frac{\partial^4 f(x,\mu)}{\partial y^4}\,dx = (b-a)\frac{\partial^4 f}{\partial y^4}(\hat{\eta}, \hat{\mu})$$

for some $(\hat{\eta}, \hat{\mu})$ in R. Since $2m = (d-c)/k$, the error term has the form

$$\begin{aligned} E &= \frac{-k(b-a)h^4}{540}\left[6m\frac{\partial^4 f}{\partial x^4}(\overline{\eta}, \overline{\mu})\right] - \frac{(d-c)(b-a)}{180}k^4\frac{\partial^4 f}{\partial y^4}(\hat{\eta}, \hat{\mu}) \\ &= -\frac{(d-c)(b-a)}{180}\left[h^4\frac{\partial^4 f}{\partial x^4}(\overline{\eta}, \overline{\mu}) + k^4\frac{\partial^4 f}{\partial y^4}(\hat{\eta}, \hat{\mu})\right] \end{aligned}$$

for some $(\overline{\eta}, \overline{\mu})$ and $(\hat{\eta}, \hat{\mu})$ in R.

EXAMPLE 1 The Composite Simpson's rule applied to approximate

$$\int_{1.4}^{2.0}\int_{1.0}^{1.5} \ln(x+2y)\,dy\,dx$$

with $n = 2$ and $m = 1$ uses the step sizes $h = 0.15$ and $k = 0.25$. The region of integration R is shown in Figure 4.18, together with the nodes (x_i, y_j), where $i = 0, 1, 2, 3, 4$ and $j = 0, 1, 2$ and $w_{i,j}$, which are the coefficients of $f(x_i, y_j) = \ln(x_i + 2y_j)$ in the sum.

Figure 4.18

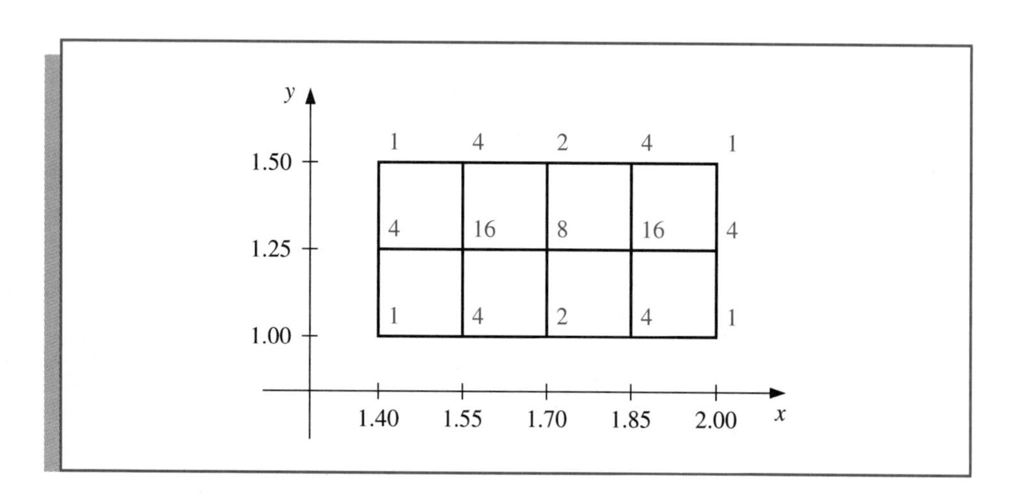

The approximation is

$$\begin{aligned} \int_{1.4}^{2.0}\int_{1.0}^{1.5} \ln(x+2y)\,dy\,dx &\approx \frac{(0.15)(0.25)}{9}\sum_{i=0}^{4}\sum_{j=0}^{2} w_{i,j}\ln(x_i+2y_j) \\ &= 0.4295524387. \end{aligned}$$

Since

$$\frac{\partial^4 f}{\partial x^4}(x, y) = \frac{-6}{(x+2y)^4} \quad \text{and} \quad \frac{\partial^4 f}{\partial y^4}(x, y) = \frac{-96}{(x+2y)^4},$$

the error is bounded by

$$\begin{aligned} |E| &\le \frac{(0.5)(0.6)}{180}\left[(0.15)^4 \max_{(x,y)\text{ in }R} \frac{6}{(x+2y)^4} + (0.25)^4 \max_{(x,y)\text{ in }R} \frac{96}{(x+2y)^4}\right] \\ &\le 4.72 \times 10^{-6}. \end{aligned}$$

The actual value of the integral to ten decimal places is

$$\int_{1.4}^{2.0} \int_{1.0}^{1.5} \ln(x+2y)\, dy\, dx = 0.4295545265;$$

so the approximation is accurate to within 2.1×10^{-6}. ∎

The same techniques can be applied for the approximation of triple integrals as well as higher integrals for functions of more than three variables. The number of functional evaluations required for the approximation is the product of the number of functional evaluations required when the method is applied to each variable. To reduce the number of functional evaluations, more efficient methods such as Gaussian quadrature, Romberg integration, or Adaptive quadrature can be incorporated in place of the Newton-Cotes formulas. The following example illustrates the use of Gaussian quadrature for the integral considered in Example 1.

EXAMPLE 2 Consider the double integral given in Example 1. Before employing a Gaussian quadrature technique to approximate this integral, we transform the region of integration

$$R = \{(x, y) | 1.4 \le x \le 2.0,\ 1.0 \le y \le 1.5\}$$

into

$$\hat{R} = \{(u, v) | -1 \le u \le 1,\ -1 \le v \le 1\}.$$

The linear transformations that accomplish this are

$$u = \frac{1}{2.0 - 1.4}(2x - 1.4 - 2.0) \quad \text{and} \quad v = \frac{1}{1.5 - 1.0}(2y - 1.0 - 1.5),$$

or, equivalently, $x = 0.3u + 1.7$ and $y = 0.25v + 1.25$. Employing this change of variables gives an integral on which Gaussian quadrature can be applied:

$$\int_{1.4}^{2.0} \int_{1.0}^{1.5} \ln(x+2y)\, dy\, dx = 0.075 \int_{-1}^{1} \int_{-1}^{1} \ln(0.3u + 0.5v + 4.2)\, dv\, du.$$

The Gaussian quadrature formula for $n = 3$ in both u and v requires that we use the nodes

$$u_1 = v_1 = r_{3,2} = 0, \quad u_0 = v_0 = r_{3,1} = -0.7745966692,$$

and

$$u_2 = v_2 = r_{3,3} = 0.7745966692.$$

The associated weights are $c_{3,2} = 0.8888888889$ and $c_{3,1} = c_{3,3} = 0.5555555556$. (See Table 4.11 on page 227.) Thus,

$$\int_{1.4}^{2.0} \int_{1.0}^{1.5} \ln(x + 2y)\, dy\, dx \approx 0.075 \sum_{i=1}^{3} \sum_{j=1}^{3} c_{3,i} c_{3,j} \ln(0.3 r_{3,i} + 0.5 r_{3,j} + 4.2)$$

$$= 0.4295545313.$$

Although this result requires only 9 functional evaluations compared to 15 for the Composite Simpson's rule considered in Example 1, this result is accurate to within 4.8×10^{-9}, compared to 2.1×10^{-6} accuracy in Example 1. ■

The use of approximation methods for double integrals is not limited to integrals with rectangular regions of integration. The techniques previously discussed can be modified to approximate double integrals of the form

(4.43)
$$\int_a^b \int_{c(x)}^{d(x)} f(x, y)\, dy\, dx$$

or

(4.44)
$$\int_c^d \int_{a(y)}^{b(y)} f(x, y)\, dx\, dy$$

In fact, integrals on regions not of this type can also be approximated by performing appropriate partitions of the region. (See Exercise 10.)

To describe the technique involved with approximating an integral in the form

$$\int_a^b \int_{c(x)}^{d(x)} f(x, y)\, dy\, dx,$$

we will use the basic Simpson's rule to integrate with respect to both variables. The step size for the variable x is $h = (b - a)/2$, but the step size for y varies with x (see Figure 4.19) and is written

$$k(x) = \frac{d(x) - c(x)}{2}.$$

Figure 4.19

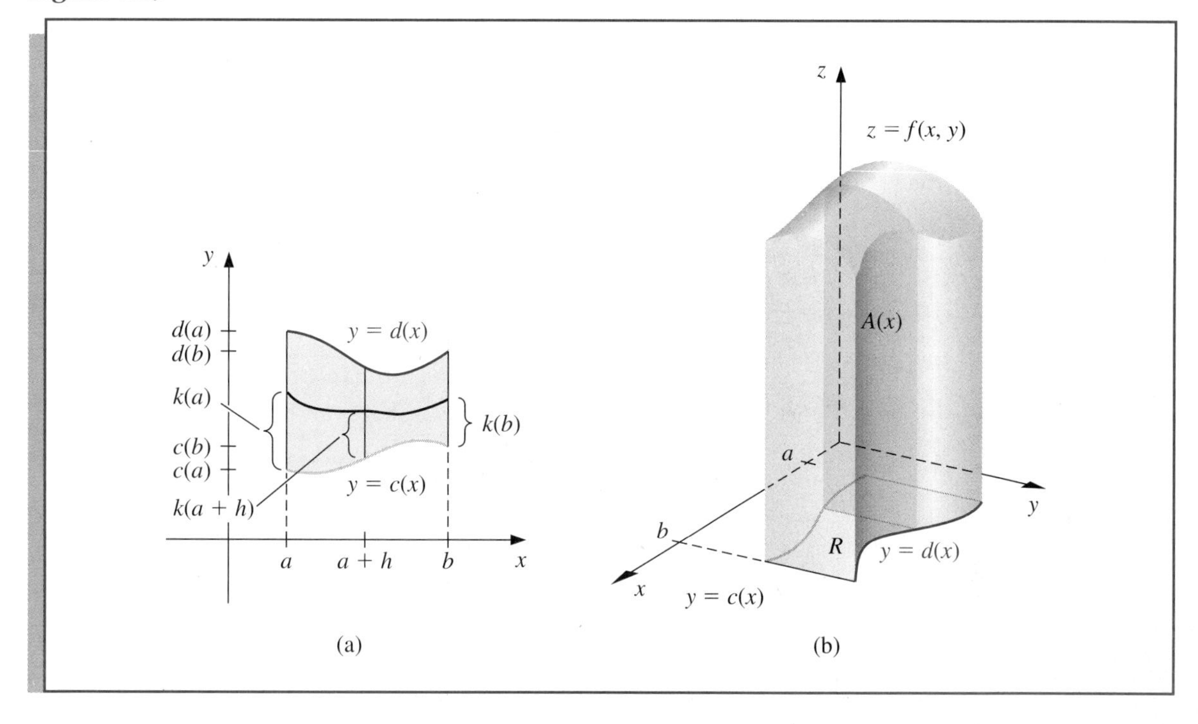

Consequently,

$$\int_a^b \int_{c(x)}^{d(x)} f(x, y)\, dy\, dx \approx \int_a^b \frac{k(x)}{3}[f(x, c(x)) + 4f(x, c(x) + k(x)) + f(x, d(x))]\, dx$$

$$\approx \frac{h}{3}\left\{ \frac{k(a)}{3}[f(a, c(a)) + 4f(a, c(a) + k(a)) + f(a, d(a))] \right.$$

$$+ \frac{4k(a+h)}{3}[f(a+h, c(a+h)) + 4f(a+h, c(a+h)$$

$$+ k(a+h)) + f(a+h, d(a+h))]$$

$$\left. + \frac{k(b)}{3}[f(b, c(b)) + 4f(b, c(b) + k(b)) + f(b, d(b))] \right\}.$$

Algorithm 4.4 applies the Composite Simpson's rule to an integral in the form (4.43). Integrals in the form (4.44) can, of course, be handled similarly.

ALGORITHM 4.4

Simpson's Double Integral

To approximate the integral $I = \int_a^b \int_{c(x)}^{d(x)} f(x, y)\,dy\,dx$:

INPUT endpoints a, b: positive integers m, n.

OUTPUT approximation J to I.

Step 1 Set $h = (b - a)/(2n)$;
$J_1 = 0$; *(End terms.)*
$J_2 = 0$; *(Even terms.)*
$J_3 = 0$. *(Odd terms.)*

Step 2 For $i = 0, 1, \ldots, 2n$ do Steps 3–8.

Step 3 Set $x = a + ih$; *(Composite Simpson's method for x.)*
$HX = (d(x) - c(x))/(2m)$;
$K_1 = f(x, c(x)) + f(x, d(x))$; *(End terms.)*
$K_2 = 0$; *(Even terms.)*
$K_3 = 0$. *(Odd terms.)*

Step 4 For $j = 1, 2, \ldots, 2m - 1$ do Step 5 and 6.

Step 5 Set $y = c(x) + jHX$;
$Q = f(x, y)$.

Step 6 If j is even then set $K_2 = K_2 + Q$
else set $K_3 = K_3 + Q$.

Step 7 Set $L = (K_1 + 2K_2 + 4K_3)HX/3$.

$$\left(L \approx \int_{c(x_i)}^{d(x_i)} f(x_i, y)\,dy \quad \textit{by the Composite Simpson's method.}\right)$$

Step 8 If $i = 0$ or $i = 2n$ then set $J_1 = J_1 + L$
else if i is even then set $J_2 = J_2 + L$
else set $J_3 = J_3 + L$.

Step 9 Set $J = h(J_1 + 2J_2 + 4J_3)/3$.

Step 10 OUTPUT (J);
STOP.

To apply Gaussian quadrature to

$$\int_a^b \int_{c(x)}^{d(x)} f(x, y)\,dy\,dx$$

first requires transforming, for each x in $[a, b]$, the interval $[c(x), d(x)]$ to $[-1, 1]$ and then applying Gaussian quadrature. This results in the formula

$$\int_a^b \int_{c(x)}^{d(x)} f(x, y)\, dy\, dx$$
$$\approx \int_a^b \frac{d(x) - c(x)}{2} \sum_{j=1}^{n} c_{n,j} f\left(x, \frac{(d(x) - c(x))r_{n,j} + d(x) + c(x)}{2}\right) dx,$$

where, as before, the roots $r_{n,j}$ and coefficients $c_{n,j}$ come from Table 4.11. Now the interval $[a, b]$ is tranformed to $[-1, 1]$ and Gaussian quadrature is applied to approximate the integral on the right side of this equation. The details are given in Algorithm 4.5.

ALGORITHM 4.5

Gaussian Double Integral

To approximate the integral $\int_a^b \int_{c(x)}^{d(x)} f(x, y)\, dy\, dx$:

INPUT endpoints a, b; positive integers m, n.
(The roots $r_{i,j}$ and coefficients $c_{i,j}$ need to be available for $i = \max\{m, n\}$ and for $1 \le j \le i$.)

OUTPUT approximation J to I.

Step 1 Set $h_1 = (b - a)/2$;
$h_2 = (b + a)/2$;
$J = 0$.

Step 2 For $i = 1, 2, \ldots, m$ do Steps 3–5.

Step 3 Set $JX = 0$;
$x = h_1 r_{m,i} + h_2$;
$d_1 = d(x)$;
$c_1 = c(x)$;
$k_1 = (d_1 - c_1)/2$;
$k_2 = (d_1 + c_1)/2$.

Step 4 For $j = 1, 2, \ldots, n$ do
set $y = k_1 r_{n,j} + k_2$;
$Q = f(x, y)$;
$JX = JX + c_{n,j} Q$.

Step 5 Set $J = J + c_{m,i} k_1 JX$.

Step 6 Set $J = h_1 J$.

Step 7 OUTPUT (J);
STOP.

EXAMPLE 3 The volume of the solid in Figure 4.20 is approximated by applying Simpson's Double Integral Algorithm with $n = m = 5$ to

$$\int_{0.1}^{0.5} \int_{x^3}^{x^2} e^{y/x}\, dy\, dx.$$

This requires 121 evaluations of the function $f(x, y) = e^{y/x}$ and produces the result 0.0333054, which is accurate to nearly seven decimal places. Applying the Gaussian Quadrature Algorithm with $n = m = 5$ requires only 25 function evaluations and gives the approximation 0.03330556611, which is accurate to 11 decimal places. ■

Figure 4.20

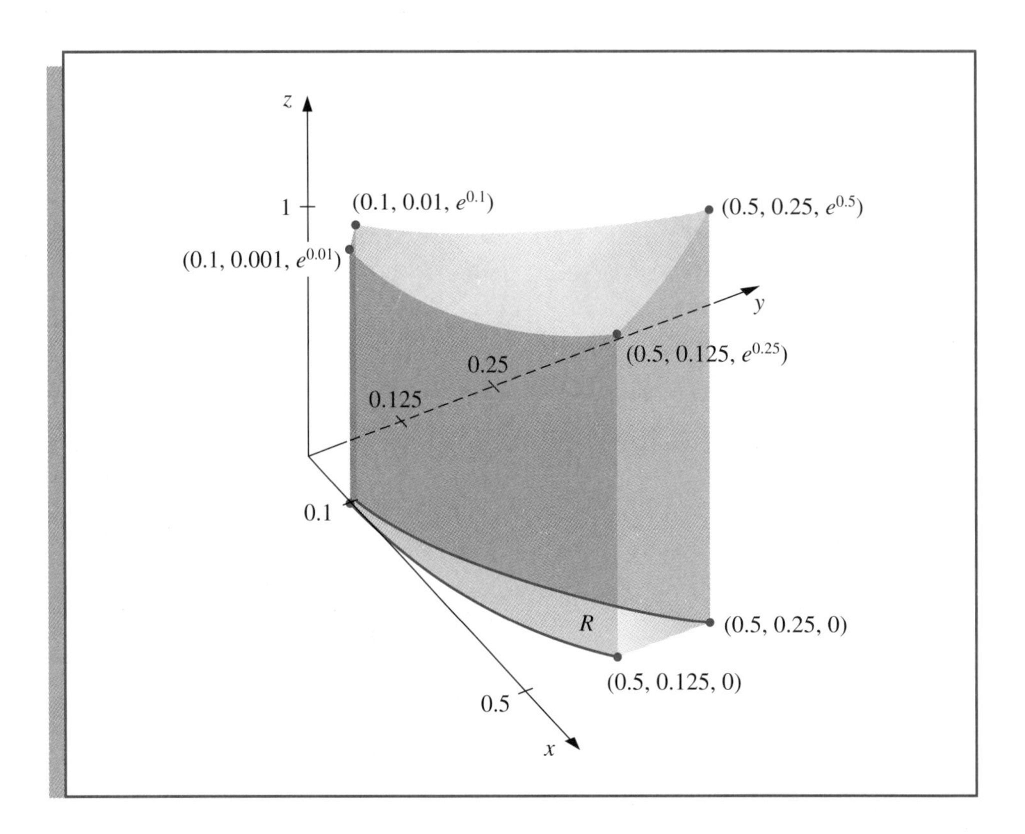

Triple integrals of the form

$$\int_a^b \int_{c(x)}^{d(x)} \int_{\alpha(x,y)}^{\beta(x,y)} f(x, y, z)\, dz\, dy\, dx$$

(see Figure 4.21) are approximated in a similar manner. Because of the number of calculations involved, Gaussian quadrature is the method of choice. Algorithm 4.6 implements this procedure.

Figure 4.21

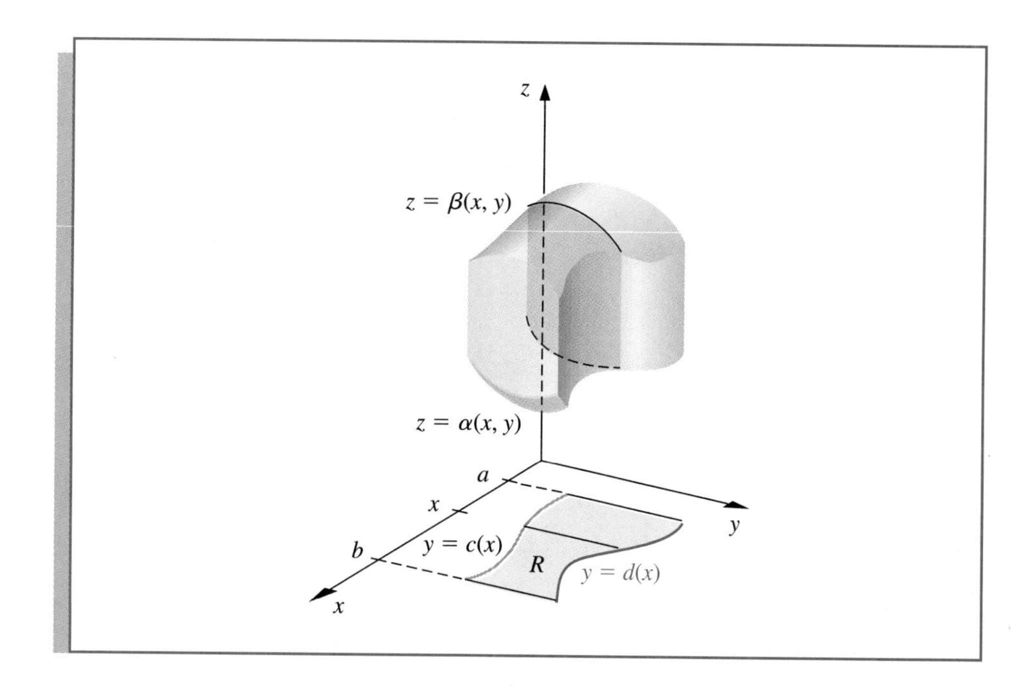

ALGORITHM 4.6

Gaussian Triple Integral

To approximate the integral $\int_a^b \int_{c(x)}^{d(x)} \int_{\alpha(x,y)}^{\beta(x,y)} f(x, y, z)\, dz\, dy\, dx$:

INPUT endpoints a, b; positive integers m, n, p.
(The roots $r_{i,j}$ and coefficients $c_{i,j}$ need to be available for $i = \max\{n, m, p\}$ and for $1 \le j \le i$.)

OUTPUT approximation J to I.

Step 1 Set $h_1 = (b - a)/2$;
$h_2 = (b + a)/2$;
$J = 0$.

Step 2 For $i = 1, 2, \ldots, m$ do Steps 3–8.

Step 3 Set $JX = 0$;
$x = h_1 r_{m,i} + h_2$;
$d_1 = d(x)$;
$c_1 = c(x)$;
$k_1 = (d_1 - c_1)/2$;
$k_2 = (d_1 + c_1)/2$.

Step 4 For $j = 1, 2, \ldots, n$ do Steps 5–7.

Step 5 Set $JY = 0$;
$y = k_1 r_{n,j} + k_2$;
$\beta_1 = \beta(x, y)$;
$\alpha_1 = \alpha(x, y)$;
$l_1 = (\beta_1 - \alpha_1)/2$;
$l_2 = (\beta_1 + \alpha_1)/2$.

Step 6 For $k = 1, 2, \ldots, p$ do
set $z = l_1 r_{p,k} + l_2$;
$Q = f(x, y, z)$;
$JY = JY + c_{p,k} Q$.

Step 7 Set $JX = JX + c_{n,j} l_1 JY$.

Step 8 Set $J = J + c_{m,i} k_1 JX$.

Step 9 Set $J = h_1 J$.

Step 10 OUTPUT (J);
STOP.

The following example requires the evaluation of four triple integrals.

EXAMPLE 4 The center of a mass of a solid region D with density function σ occurs at

$$(\overline{x}, \overline{y}, \overline{z}) = \left(\frac{M_{yz}}{M}, \frac{M_{xz}}{M}, \frac{M_{xy}}{M} \right),$$

where

$$M_{yz} = \iiint_D x\sigma(x, y, z)\, dV, \quad M_{xz} = \iiint_D y\sigma(x, y, z)\, dV$$

and

$$M_{xy} = \iiint_D z\sigma(x, y, z)\, dV$$

are the moments about the coordinate planes and

$$M = \iiint_D \sigma(x, y, z) dV$$

is the mass. The solid shown in Figure 4.22 is bounded by the upper nappe of the cone $z^2 = x^2 + y^2$ and the plane $z = 2$ and has density function given by

$$\sigma(x, y, z) = \sqrt{x^2 + y^2}.$$

Figure 4.22

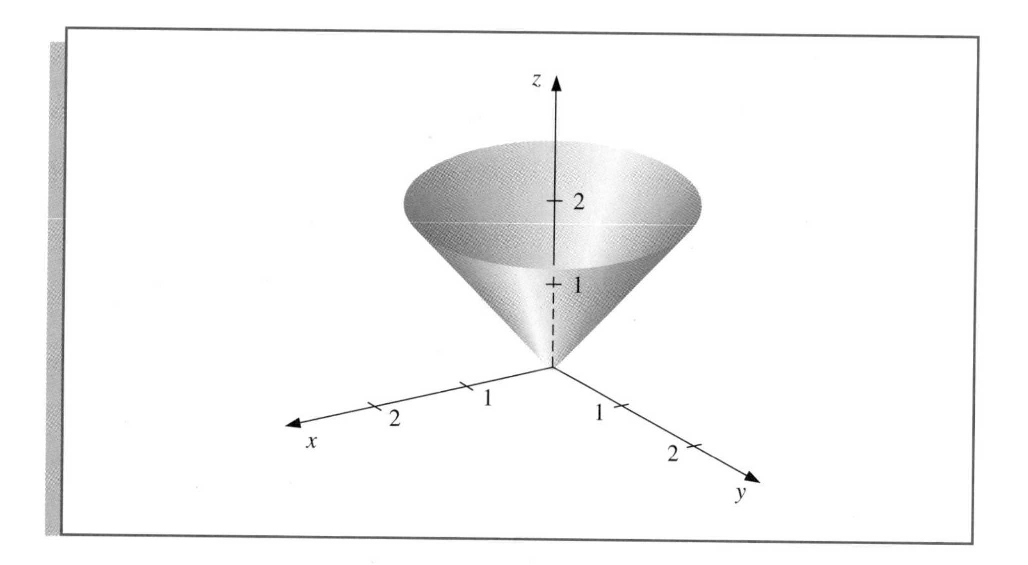

Applying the Gaussian Triple Integral Algorithm 4.6 with $n = m = p = 5$ requires 125 function evaluations per integral and gives the following approximations:

$$M = \int_{-2}^{2} \int_{-\sqrt{4-x^2}}^{\sqrt{4-x^2}} \int_{\sqrt{x^2+y^2}}^{2} \sqrt{x^2 + y^2}\, dz\, dy\, dx$$

$$= 4 \int_{0}^{2} \int_{0}^{\sqrt{4-x^2}} \int_{\sqrt{x^2+y^2}}^{2} \sqrt{x^2 + y^2}\, dz\, dy\, dx \approx 8.37504476,$$

$$M_{yz} = \int_{-2}^{2} \int_{-\sqrt{4-x^2}}^{\sqrt{4-x^2}} \int_{\sqrt{x^2+y^2}}^{2} x\sqrt{x^2 + y^2}\, dz\, dy\, dx \approx -5.55111512 \times 10^{-17},$$

$$M_{xz} = \int_{-2}^{2} \int_{-\sqrt{4-x^2}}^{\sqrt{4-x^2}} \int_{\sqrt{x^2+y^2}}^{2} y\sqrt{x^2 + y^2}\, dz\, dy\, dx \approx -8.01513675 \times 10^{-17},$$

and

$$M_{xy} = \int_{-2}^{2} \int_{-\sqrt{4-x^2}}^{\sqrt{4-x^2}} \int_{\sqrt{x^2+y^2}}^{2} z\sqrt{x^2 + y^2}\, dz\, dy\, dx \approx 13.40038156.$$

This implies that the approximate location of the center of mass is

$$(\overline{x}, \overline{y}, \overline{z}) = (0, 0, 1.60003701).$$

By direct evaluation of the integrals, the center of mass can be shown to occur at (0, 0, 1.6).

■

EXERCISE SET 4.8

1. Use Algorithm 4.4 with $n = m = 2$ to approximate the following double integrals, and compare the results to the exact answers.

   a. $\int_{2.1}^{2.5} \int_{1.2}^{1.4} xy^2\, dy\, dx$

   b. $\int_{0}^{0.5} \int_{0}^{0.5} e^{y-x}\, dy\, dx$

   c. $\int_{2}^{2.2} \int_{x}^{2x} (x^2 + y^3)\, dy\, dx$

   d. $\int_{1}^{1.5} \int_{0}^{x} (x^2 + \sqrt{y})\, dy\, dx$

2. Find the smallest values for $n = m$ so that Algorithm 4.4 can be used to approximate the integrals in Exercise 1 to within 10^{-6} of the actual value.

3. Use Algorithm 4.4 with (i) $n = 2, m = 4$, (ii) $n = 4, m = 2$, and (iii) $n = m = 3$ to approximate the following double integrals, and compare the results to the exact answers.

   a. $\int_{0}^{\pi/4} \int_{\sin x}^{\cos x} \left(2y \sin x + (\cos x)^2\right)\, dy\, dx$

   b. $\int_{1}^{e} \int_{1}^{x} \ln xy\, dy\, dx$

   c. $\int_{0}^{1} \int_{x}^{2x} (x^2 + y^3)\, dy\, dx$

   d. $\int_{0}^{1} \int_{x}^{2x} (y^2 + x^3)\, dy\, dx$

   e. $\int_{0}^{\pi} \int_{0}^{x} \cos x\, dy\, dx$

   f. $\int_{0}^{\pi} \int_{0}^{x} \cos y\, dy\, dx$

   g. $\int_{0}^{\pi/4} \int_{0}^{\sin x} \frac{1}{\sqrt{1 - y^2}}\, dy\, dx$

   h. $\int_{-\pi}^{3\pi/2} \int_{0}^{2\pi} (y \sin x + x \cos y)\, dy\, dx$

4. Find the smallest values for $n = m$ so that Algorithm 4.4 can be used to approximate the integrals in Exercise 3 to within 10^{-6} of the actual value.

5. Use Algorithm 4.5 with $n = m = 2$ to approximate the integrals in Exercise 1, and compare the results to those obtained in Exercise 1.

6. Find the smallest values of $n = m$ so that Algorithm 4.5 can be used to approximate the integrals in Exercise 1 to within 10^{-6}. Do not continue beyond $n = m = 5$. Compare the number of functional evaluations required to the number required in Exercise 2.

7. Use Algorithm 4.5 with (i) $n = m = 3$, (ii) $n = 3, m = 4$, (iii) $n = 4, m = 3$, and (iv) $n = m = 4$ to approximate the integrals in Exercise 3.

8. Use Algorithm 4.5 with $n = m = 5$ to approximate the integrals in Exercise 3. Compare the number of functional evaluations required to the number required in Exercise 4.

9. Use Algorithm 4.4 with $n = m = 7$ and Algorithm 4.5 with $n = m = 4$ to approximate

$$\iint_R e^{-(x+y)}\, dA$$

   for the region R in the plane bounded by the curves $y = x^2$ and $y = \sqrt{x}$.

10. Use Algorithm 4.4 to approximate

$$\iint_R \sqrt{xy + y^2}\, dA,$$

   where R is the region in the plane bounded by the lines $x + y = 6$, $3y - x = 2$, and $3x - y = 2$. First partition R into two regions R_1 and R_2 on which Algorithm 4.4 can be applied. Use $n = m = 3$ on both R_1 and R_2.

11. A plane lamina is defined to be a thin sheet of continuously distributed mass. If σ is a function describing the density of a lamina having the shape of a region R in the xy-plane, then the center of the mass of the lamina $(\bar{x}, \bar{y})$ is defined by

$$\bar{x} = \frac{\iint_R x\sigma(x, y)\, dA}{\iint_R \sigma(x, y)\, dA}, \quad \bar{y} = \frac{\iint_R y\sigma(x, y)\, dA}{\iint_R \sigma(x, y)\, dA}.$$

Use Algorithm 4.4 with $n = m = 7$ to find the center of mass of the lamina described by $R = \{(x, y) \mid 0 \leq x \leq 1,\ 0 \leq y \leq \sqrt{1 - x^2}\}$ with the density function $\sigma(x, y) = e^{-(x^2+y^2)}$. Compare the approximation to the exact result.

12. Repeat Exercise 11 using Algorithm 4.5 with $n = m = 5$.

13. The area of the surface described by $z = f(x, y)$ for (x, y) in R is given by

$$\iint_R \sqrt{[f_x(x, y)]^2 + [f_y(x, y)]^2 + 1}\, dA.$$

Use Algorithm 4.4 with $n = m = 4$ to find an approximation to the area of the surface on the hemisphere $x^2 + y^2 + z^2 = 9$, $z \geq 0$ that lies above the region in the plane described by $R = \{(x, y) \mid 0 \leq x \leq 1,\ 0 \leq y \leq 1\}$.

14. Repeat Exercise 13 using Algorithm 4.5 with $n = m = 4$.

15. Use Algorithm 4.6 with $n = m = p = 2$ to approximate the following triple integrals, and compare the results to the exact answers.

a. $\displaystyle\int_0^1 \int_1^2 \int_0^{0.5} e^{x+y+z}\, dz\, dy\, dx$

b. $\displaystyle\int_0^1 \int_x^1 \int_0^y y^2 z\, dz\, dy\, dx$

c. $\displaystyle\int_0^1 \int_{x^2}^x \int_{x-y}^{x+y} y\, dz\, dy\, dx$

d. $\displaystyle\int_0^1 \int_{x^2}^x \int_{x-y}^{x+y} z\, dz\, dy\, dx$

e. $\displaystyle\int_0^{\pi} \int_0^x \int_0^{xy} \frac{1}{y} \sin\frac{z}{y}\, dz\, dy\, dx$

f. $\displaystyle\int_0^1 \int_0^1 \int_{-xy}^{xy} e^{x^2+y^2}\, dz\, dy\, dx$

16. Repeat Exercise 15 using $n = m = p = 3$.

17. Repeat Exercise 15 using $n = m = p = 4$ and $n = m = p = 5$.

18. Use Algorithm 4.6 with $n = m = p = 4$ to approximate

$$\iiint_S xy \sin(yz)\, dV,$$

where S is the solid bounded by the coordinate planes and the planes $x = \pi$, $y = \pi/2$, $z = \pi/3$. Compare this approximation to the exact result.

19. Use Algorithm 4.6 with $n = m = p = 5$ to approximate

$$\iiint_S \sqrt{xyz}\, dV,$$

where S is the region in the first octant bounded by the cylinder $x^2 + y^2 = 4$, the sphere $x^2 + y^2 + z^2 = 4$, and the plane $x + y + z = 8$. How many functional evaluations are required for the approximation?

4.9 Improper Integrals

Improper integrals result when the notion of integration is extended either to an interval of integration on which the function is unbounded or to an interval with one or more infinite endpoints. In either circumstance, the normal rules of integral approximation must be modified.

We will first consider the situation when the integrand is unbounded at the left endpoint of the interval of integration, as shown in Figure 4.23. We will then show that by a suitable manipulation, the other improper integrals can be reduced to problems of this form.

Figure 4.23

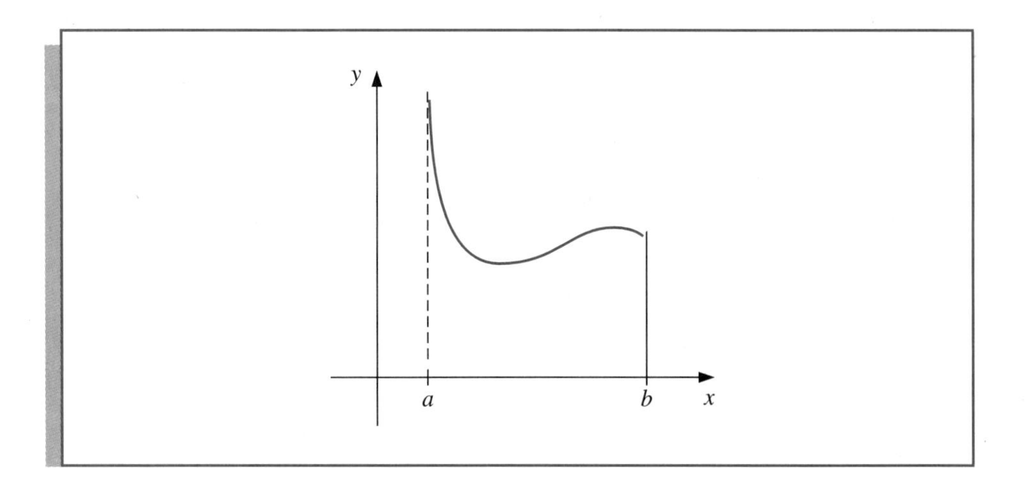

It is shown in calculus that the improper integral with a singularity at the left endpoint,

$$\int_a^b \frac{dx}{(x-a)^p},$$

converges if and only if $0 < p < 1$, and in this case, we define

$$\int_a^b \frac{dx}{(x-a)^p} = \frac{(b-a)^{1-p}}{1-p}.$$

If f is a function that can be written in the form

$$f(x) = \frac{g(x)}{(x-a)^p},$$

where $0 < p < 1$ and g is continuous on $[a, b]$, then the improper integral

$$\int_a^b f(x)\,dx$$

also exists. We will approximate this integral using the Composite Simpson's rule. If $g \in C^5[a, b]$ we can construct the fourth Taylor polynomial, $P_4(x)$, for g about a,

$$P_4(x) = g(a) + g'(a)(x-a) + \frac{g''(a)}{2!}(x-a)^2 + \frac{g'''(a)}{3!}(x-a)^3 + \frac{g^{(4)}(a)}{4!}(x-a)^4,$$

and write

$$\int_a^b f(x)\,dx = \int_a^b \frac{g(x) - P_4(x)}{(x-a)^p}\,dx + \int_a^b \frac{P_4(x)}{(x-a)^p}\,dx. \tag{4.45}$$

We can exactly determine the value of

$$\begin{aligned} \int_a^b \frac{P_4(x)}{(x-a)^p}\,dx &= \sum_{k=0}^{4} \int_a^b \frac{g^{(k)}(a)}{k!}(x-a)^{k-p}\,dx \\ &= \sum_{k=0}^{4} \frac{g^{(k)}(a)}{k!(k+1-p)}(b-a)^{k+1-p}. \end{aligned} \tag{4.46}$$

This is generally the dominant portion of the approximation, especially when the Taylor polynomial $P_4(x)$ agrees closely with $g(x)$ throughout the interval $[a, b]$. To approximate the integral of f then, we need to add this value to the approximation of

$$\int_a^b \frac{g(x) - P_4(x)}{(x-a)^p}\,dx.$$

First define

$$G(x) = \begin{cases} \dfrac{g(x) - P_4(x)}{(x-a)^p}, & \text{if } a < x \le b, \\ 0, & \text{if } x = a. \end{cases}$$

Since $0 < p < 1$ and $P_4^{(k)}(a)$ agrees with $g^{(k)}(a)$ for each $k = 0, 1, 2, 3, 4$, we have $G \in C^4[a, b]$. This implies that the Composite Simpson's rule can be applied to approximate the integral of G on $[a, b]$ and the error term for this rule will be valid. Adding this approximation to the value in Eq. (4.46) gives an approximation to the improper integral of f on $[a, b]$, within the accuracy of the Composite Simpson's rule approximation.

EXAMPLE 1 To approximate the values of the improper integral

$$\int_0^1 \frac{e^x}{\sqrt{x}}\,dx,$$

we will use the Composite Simpson's rule with $h = 0.25$. Since the fourth Taylor polynomial for e^x about $x = 0$ is

$$P_4(x) = 1 + x + \frac{x^2}{2} + \frac{x^3}{6} + \frac{x^4}{24},$$

we have

$$\int_0^1 \frac{P_4(x)}{\sqrt{x}}\,dx = \int_0^1 \left(x^{-1/2} + x^{1/2} + \frac{1}{2}x^{3/2} + \frac{1}{6}x^{5/2} + \frac{1}{24}x^{7/2}\right) dx$$
$$= \lim_{M\to 0^+}\left[2x^{1/2} + \frac{2}{3}x^{3/2} + \frac{1}{5}x^{5/2} + \frac{1}{21}x^{7/2} + \frac{1}{108}x^{9/2}\right]_M^1$$
$$= 2 + \frac{2}{3} + \frac{1}{5} + \frac{1}{21} + \frac{1}{108} \approx 2.9235450.$$

Table 4.14 lists the approximate values of

$$G(x) = \begin{cases} \dfrac{e^x - P_4(x)}{\sqrt{x}}, & \text{if } 0 < x \le 1, \\ 0, & \text{if } x = 0. \end{cases}$$

Table 4.14

x	$G(x)$
0.00	0
0.25	0.0000170
0.50	0.0004013
0.75	0.0026026
1.00	0.0099485

Applying the Composite Simpson's rule to G using these data gives

$$\int_0^1 G(x)\,dx \approx \frac{0.25}{3}[0 + 4(0.0000170) + 2(0.0004013)$$
$$+ 4(0.0026026) + 0.0099485] = 0.0017691.$$

Hence,

$$\int_0^1 \frac{e^x}{\sqrt{x}}\,dx \approx 2.9235450 + 0.0017691 = 2.9253141.$$

This result is accurate within the accuracy of the Composite Simpson's rule approximation for the function G. Since $|G^{(4)}(x)| < 1$ on $[0, 1]$, the error is bounded by

$$\frac{1-0}{180}(0.25)^4 = 0.0000217.$$

Applying Maple to this improper integral gives the 9-decimal-place result 2.925303492. The actual error of our approximation is 1.06×10^{-5}, well within the error bound. ■

To approximate the improper integral with a singularity at the right endpoint, we simply apply the technique we used above but expand in terms of the right endpoint b instead of the left endpoint a. Alternatively, we could make the substitution

$$z = -x, \qquad dz = -\,dx$$

to change the improper integral into one of the form

$$\int_a^b f(x)\,dx = \int_{-b}^{-a} f(-z)\,dz, \tag{4.47}$$

which has its singularity at the left endpoint. (See Figure 4.24.)

Figure 4.24

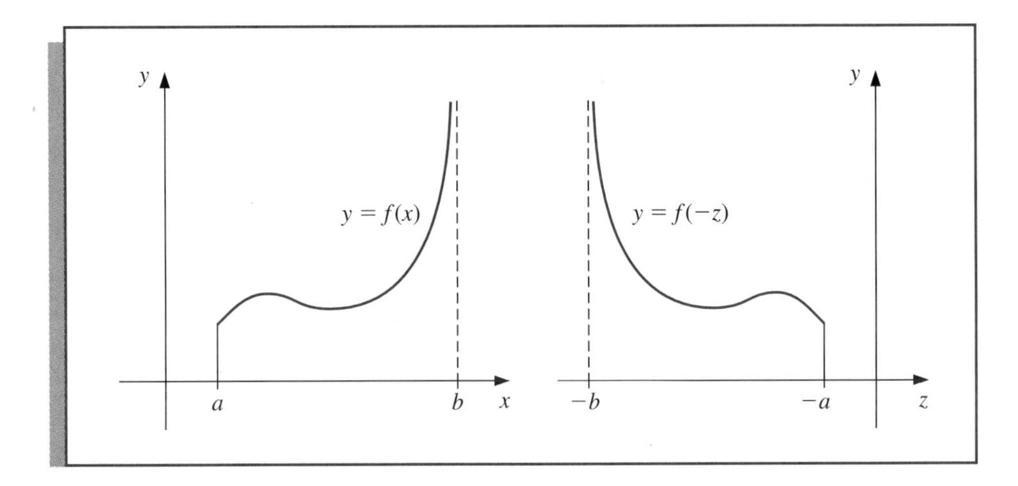

Improper integrals with interior singularities (for example, at c, where $a < c < b$) are treated as the sum of improper integrals with endpoint singularities, since

$$\int_a^b f(x)\,dx = \int_a^c f(x)\,dx + \int_c^b f(x)\,dx.$$

The other type of improper integrals involves infinite limits of integration. The basic integral of this type has the form

$$\int_a^\infty \frac{1}{x^p}\,dx,$$

which is converted to an integral with left endpoint singularity by making the integration substitution

$$t = x^{-1}, \quad dt = -x^{-2}\,dx, \quad \text{so} \quad dx = -x^2\,dt = -t^{-2}\,dt.$$

Then

$$\int_a^\infty \frac{1}{x^p}\,dx = \int_{1/a}^{0} -\frac{t^p}{t^2}\,dt = \int_0^{1/a} \frac{1}{t^{2-p}}\,dt.$$

In a similar manner, the variable change $t = x^{-1}$ converts the improper integral $\int_a^\infty f(x)\,dx$ into one that has a left endpoint singularity at zero:

(4.48)
$$\int_a^\infty f(x)\,dx = \int_0^{1/a} t^{-2} f\left(\frac{1}{t}\right)\,dt.$$

It can now be approximated using a quadrature formula of the type described earlier.

EXAMPLE 2 To approximate the value of the improper integral

$$I = \int_1^\infty x^{-3/2} \sin\frac{1}{x}\,dx,$$

we make the change of variable $t = x^{-1}$ to obtain

$$I = \int_0^1 t^{-1/2} \sin t\,dt,$$

The fourth Taylor polynomial, $P_4(t)$, for $\sin t$ about 0 is

$$P_4(t) = t - \frac{1}{6}t^3,$$

so we have

$$\begin{aligned} I &= \int_0^1 \frac{\sin t - t + \frac{1}{6}t^3}{t^{1/2}}\,dt + \int_0^1 \left(t^{1/2} - \frac{1}{6}t^{5/2}\right)\,dt \\ &= \int_0^1 \frac{\sin t - t + \frac{1}{6}t^3}{t^{1/2}}\,dt + \left[\frac{2}{3}t^{3/2} - \frac{1}{21}t^{7/2}\right]_0^1 \\ &= \int_0^1 \frac{\sin t - t + \frac{1}{6}t^3}{t^{1/2}}\,dt + 0.61904761. \end{aligned}$$

Applying the Composite Simpson's rule with $n = 16$ to the remaining integral gives

$$I = 0.0014890097 + 0.61904761 = 0.62053661,$$

which is accurate to within 4.0×10^{-8}. ■

EXERCISE SET 4.9

1. Use Simpson's Composite rule and the given values of n to approximate the following improper integrals:

   a. $\int_0^1 x^{-1/4} \sin x\,dx : \quad n = 4$

   b. $\int_0^1 \frac{e^{2x}}{\sqrt[5]{x^2}}\,dx : \quad n = 6$

c. $\displaystyle\int_1^2 \frac{\ln x}{(x-1)^{1/5}}\,dx: \quad n = 8$ d. $\displaystyle\int_0^1 \frac{\cos 2x}{x^{1/3}}\,dx: \quad n = 6$

2. Use the Composite Simpson's rule and the given values of n to approximate the following improper integrals:

a. $\displaystyle\int_0^1 \frac{e^{-x}}{\sqrt{1-x}}\,dx: \quad n = 6$ b. $\displaystyle\int_0^2 \frac{xe^x}{\sqrt[3]{(x-1)^2}}\,dx: \quad n = 8$

3. Use the transformation $t = x^{-1}$ and then the Composite Simpson's rule and the given values of n to approximate the following improper integrals:

a. $\displaystyle\int_1^\infty \frac{1}{x^2+9}\,dx: \quad n = 4$ c. $\displaystyle\int_1^\infty \frac{\cos x}{x^3}\,dx: \quad n = 6$

b. $\displaystyle\int_1^\infty \frac{1}{1+x^4}\,dx: \quad n = 4$ d. $\displaystyle\int_1^\infty x^{-4}\sin x\,dx: \quad n = 6$

4. The improper integral $\int_0^\infty f(x)\,dx$ cannot be converted into an integral with finite limits using the substitution $t = 1/x$ because the limit at zero becomes infinite. The problem is resolved by first writing $\int_0^\infty f(x)\,dx = \int_0^1 f(x)\,dx + \int_1^\infty f(x)\,dx$. Apply this technique to approximate the following improper integrals to within 10^{-6}.

a. $\displaystyle\int_0^\infty \frac{1}{1+x^4}\,dx$ b. $\displaystyle\int_0^\infty \frac{1}{(1+x^2)^3}\,dx$

5. The Laguerre polynomials, given by $L_0(x) = 1$, $L_1(x) = 1 - x$, $L_2(x) = x^2 - 4x + 2$, $L_3(x) = -x^3 + 9x^2 - 18x + 6$ are shown in Exercise 15 of Section 8.2 to satisfy $\int_0^\infty e^{-x}L_i(x)L_j(x)\,dx = 0$ for each $i = 0, 1, 2, 3$ and $j = 0, 1, 2, 3$, when $i \neq j$. These polynomials can be used to give approximations to $\int_0^\infty e^{-x}f(x)\,dx$, provided this improper integral exists. The derivation parallels that given for the Legendre polynomials presented in the proof of Theorem 4.7. Show that this set of polynomials gives a formula with degree of precision at least five to approximate

$$\int_0^\infty e^{-x}f(x)\,dx.$$

6. Find the roots of the Laguerre polynomials discussed in Exercise 5, and use the corresponding coefficients obtained from the equations in Theorem 4.7 to find approximations to

$$\int_0^\infty e^{-x}\sin x\,dx \quad \text{when } n = 2 \text{ and } n = 3.$$

7. Use the Laguerre polynomials to approximate $\int_0^\infty \sqrt{x}e^{-x}\,dx$.

8. Use the Laguerre polynomials to approximate $\int_{-\infty}^\infty (1+x^2)^{-1}\,dx$.

9. Suppose a body of mass m is traveling vertically upward starting at the surface of the earth. If all resistance except gravity is neglected, the escape velocity v is given by

$$v^2 = 2gR\int_R^\infty z^{-2}\,dz, \quad \text{where } z = \frac{x}{R},$$

$R = 3960$ mi is the radius of the earth, and $g = 0.00609$ mi/s^2 is the force of gravity at the earth's surface. Approximate the escape velocity v.

4.10 Survey of Methods and Software

In this chapter we considered approximating integrals of functions of one, two, or three variables and approximating the derivatives of a function of a single real variable.

The Midpoint rule, Trapezoidal rule, and Simpson's rule were studied to introduce the techniques and error analysis of quadrature methods. Composite Simpson's rule is easy to use and produces accurate approximations unless the function oscillates in a subinterval of the interval of integration. Adaptive quadrature can be used if the function is suspected of oscillatory behavior. To minimize the number of nodes and increase the degree of precision, we used Gaussian quadrature. Romberg integration was introduced to take advantage of the easily applied Composite Trapezoidal rule and extrapolation.

Most software for integrating a function of a single real variable is based on the adaptive approach or extremely accurate Gaussian formulas. Cautious Romberg integration is an adaptive technique that includes a check to make sure that the integrand is smoothly behaved over subintervals of the interval of integration. This method has been successfully used in software libraries. Multiple integrals are generally approximated by extending good adaptive methods to higher dimensions. Gaussian-type quadrature is also recommended to decrease the number of function evaluations.

The main routines in both the IMSL and NAG Libraries are based on *QUADPACK: A subroutine package for automatic integration* by R. Piessens, E. de Doncker-Kapenga, C. W. Überhuber, and D. K. Kahaner published by Springer-Verlag in 1983 [PDUK]. The routines are also available as public domain software.

The IMSL Library contains the function QDAGS, which is an adaptive integration scheme based on the 21-point Gaussian-Kronrod rule using a 10-point Gaussian rule for error estimation. The Gaussian rule uses the ten points $x_1, \ldots, x_{10}$ and weights $w_1, \ldots, w_{10}$ to give the quadrature formula $\sum_{i=1}^{10} w_i f(x_i)$ to approximate $\int_a^b f(x)\,dx$. The additional points $x_{11}, \ldots, x_{21}$, and the new weights $v_1, \ldots, v_{21}$, are then used in the Kronrod formula $\sum_{i=1}^{21} v_i f(x_i)$. The results of the two formulas are compared to eliminate error. The advantage in using $x_1, \ldots, x_{10}$ in each formula is that f needs to be evaluated only at 21 points. If independent 10- and 21-point Gaussian rules were used, 31 function evaluations would be needed. This procedure permits endpoint singularities in the integrand.

Other IMSL subroutines are QDAGP, which allows user-specified singularities; QDAGI, which allows infinite intervals of integration; and QDNG, which is a nonadaptive procedure for smooth functions. The subroutine TWODQ uses the Gauss-Kronrod rules to integrate a function of two variables. There is also a subroutine QAND to use Gaussian quadrature to integrate a function of n variables over n intervals of the form $[a_i, b_i]$.

The NAG Library includes the subroutine D01AJF to compute the integral of f over the interval $[a, b]$ using an adaptive method based on Gaussian Quadrature using Gauss 10-point and Kronrod 21-point rules. The subroutine D01AHF is used to approximate $\int_a^b f(x)\,dx$ with a family of Gaussian-type formulas based on 1, 3, 5, 7, 15, 31, 63, 127, and 255 nodes. These interlacing high-precision rules are due to Patterson [Pat] and are used in an adaptive manner. The subroutine D01GBF is for multiple integrals and D01GAF approximates an integral given only data points instead of the function f. NAG includes many other subroutines for approximating integrals.

The Maple function call

```
>int(f,x=a..b);
```

computes the definite integral $\int_a^b f(x)\,dx$. The numerical method employed by Maple applies singularity handling routines and then uses Clenshaw-Curtis quadrature, which is described in [CC]. If this fails, an adaptive Newton-Cotes formula is applied. The method attempts to achieve a relative error tolerance $0.5 \times 10^{(1-\texttt{Digits})}$, where `Digits` is the variable in Maple that specifies the number of digits of rounding Maple uses for numerical calculation. The default value for `Digits` is 10, but it can be changed to any positive integer `n` by the command

```
>Digits:=n;
```

Although numerical differentiation is unstable, derivative approximation formulas are needed for solving differential equations. The NAG Library includes the subroutine D04AAF for the numerical differentiation of a function of one real variable with differentiation to the fourteenth derivative being possible. The IMSL function DERIV uses an adaptive change in step size for finite differences to approximate a derivative of f at x to within a given tolerance. IMSL also includes the subroutine QDDER to compute the derivatives of a function defined on a set of points using quadratic interpolation. Both packages allow the differentiation and integration of interpolatory cubic splines constructed by the subroutines mentioned in Section 3.4.

For further reading on numerical integration we recommend the books by Engels [E] and by Davis and Rabinowitz [DR]. For more information on Gaussian quadrature see Stroud and Secrest [StS]. Books on multiple integrals include those by Stroud [Stro] and the recent book by Sloan and Joe [SJ].

CHAPTER

5

Initial-Value Problems for Ordinary Differential Equations

■ ■ ■

The motion of a swinging pendulum under certain simplifying assumptions is described by the second-order differential equation

$$\frac{d^2\theta}{dt^2} - \frac{g}{L}\sin\theta = 0$$

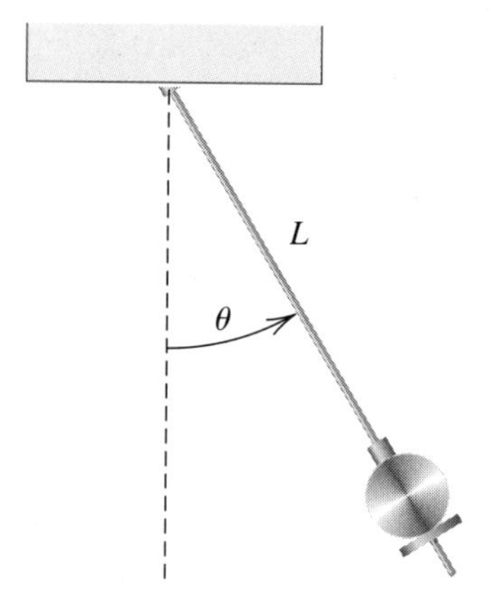

where L is the length of the pendulum, g is the gravitational constant of the earth, and θ is the angle the pendulum makes with the vertical or equilibrium position. If, in addition, we specify the position of the pendulum when the motion begins, $\theta(t_0) = \theta_0$, and its velocity at that time, $\theta'(t_0) = \theta_0'$, we have what is called an initial-value problem.

For small values of θ, the approximation $\theta \approx \sin\theta$ can be used to simplify this problem to the linear initial-value problem

$$\frac{d^2\theta}{dt^2} - \frac{g}{L}\theta = 0, \quad \theta(t_0) = \theta_0,\ \theta'(t_0) = \theta_0'.$$

This problem can be solved by a standard differential-equation technique. For larger values of θ, approximation methods must be used. A problem of this type is considered in Exercise 6 of Section 5.9.

Any textbook on ordinary differential equations details a number of methods for explicitly finding solutions to first-order initial-value problems. In practice, however, few of the problems originating from the study of physical phenomena can be solved exactly.

The first part of this chapter is concerned with approximating the solution $y(t)$ to a problem of the form

$$\frac{dy}{dt} = f(t, y), \quad \text{for } a \le t \le b,$$

subject to an initial condition

$$y(a) = \alpha.$$

Later in the chapter we deal with the extension of these methods to a system of first-order differential equations in the form

$$\begin{aligned}
\frac{dy_1}{dt} &= f_1(t, y_1, y_2, \ldots, y_n),\\
\frac{dy_2}{dt} &= f_2(t, y_1, y_2, \ldots, y_n),\\
&\vdots\\
\frac{dy_n}{dt} &= f_n(t, y_1, y_2, \ldots, y_n),
\end{aligned}$$

for $a \le t \le b$, subject to the initial conditions

$$y_1(a) = \alpha_1, \quad y_2(a) = \alpha_2, \quad \ldots, \quad y_n(a) = \alpha_n.$$

and the relationship of a system of this type to the general nth-order initial-value problem of the form

$$y^{(n)} = f(t, y, y', y'', \ldots, y^{(n-1)})$$

for $a \le t \le b$, subject to the initial conditions

$$y(a) = \alpha_1, \quad y'(a) = \alpha_2, \quad \ldots, \quad y^{n-1}(a) = \alpha_n.$$

5.1 The Elementary Theory of Initial-Value Problems

Differential equations are used to model problems in science and engineering that involve the change of some variable with respect to another. Most of these problems require the solution to an initial-value problem, that is, the solution to a differential equation that satisfies a given initial condition.

In most real-life situations the differential equation that models the problem is too complicated to solve exactly, and one of two approaches is taken to approximate the solution. The first approach is to simplify the differential equation to one that can be solved exactly and then use the solution of the simplified equation to approximate the solution to the original equation. The other approach, which we will examine in this chapter, uses methods for approximating the solution of the original problem. This is the approach that is most commonly taken, since the approximation methods give more accurate results and realistic error information.

The methods we consider in this chapter do not produce a continuous approximation to the solution of the initial-value problem. Rather, approximations are found at certain specified, and often equally spaced, points. Some method of interpolation, commonly Hermite interpolation, is used if intermediate values are needed.

We need some definitions and results from the theory of ordinary differential equations before considering methods for approximating the solutions to initial-value problems. Initial-value problems obtained by observing physical phenomena generally only approximate the true situation, so we need to know whether small changes in the statement of the problem introduce correspondingly small changes in the solution. This is also important because of the introduction of roundoff error when numerical methods are used.

Definition 5.1 A function $f(t, y)$ is said to satisfy a **Lipschitz condition** in the variable y on a set $D \subset \mathbb{R}^2$ if a constant $L > 0$ exists with the property that

$$|f(t, y_1) - f(t, y_2)| \leq L|y_1 - y_2|$$

whenever $(t, y_1), (t, y_2) \in D$. The constant L is called a **Lipschitz constant** for f. ■

EXAMPLE 1 If $D = \{(t, y) \mid 1 \leq t \leq 2, -3 \leq y \leq 4\}$ and $f(t, y) = t\,|y|$, then for each pair of points (t, y_1) and (t, y_2) in D we have

$$|f(t, y_1) - f(t, y_2)| = \big|t\,|y_1| - t\,|y_2|\big| = |t|\,\big||y_1| - |y_2|\big| \leq 2|y_1 - y_2|.$$

Thus, f satisfies a Lipschitz condition on D in the variable y with Lipschitz constant 2. The smallest value possible for the Lipschitz constant for this problem is $L = 2$, since, for example,

$$|f(2, 1) - f(2, 0)| = |2 - 0| = 2|1 - 0|.$$ ■

Definition 5.2 A set $D \subset \mathbb{R}^2$ is said to be **convex** if whenever (t_1, y_1) and (t_2, y_2) belong to D, the point $((1 - \lambda)t_1 + \lambda t_2, (1 - \lambda)y_1 + \lambda y_2)$ also belongs to D for each λ in $[0,1]$. ■

In geometric terms, Definition 5.2 states that a set is convex provided that whenever two points belong to the set, the entire straight-line segment between the points also belongs to the set. (See Figure 5.1.) The sets we consider in this chapter are generally of the form $D = \{(t, y) \mid a \leq t \leq b, -\infty < y < \infty\}$ for some constants a and b. It is easy to verify (see Exercise 5) that such sets are convex.

Figure 5.1

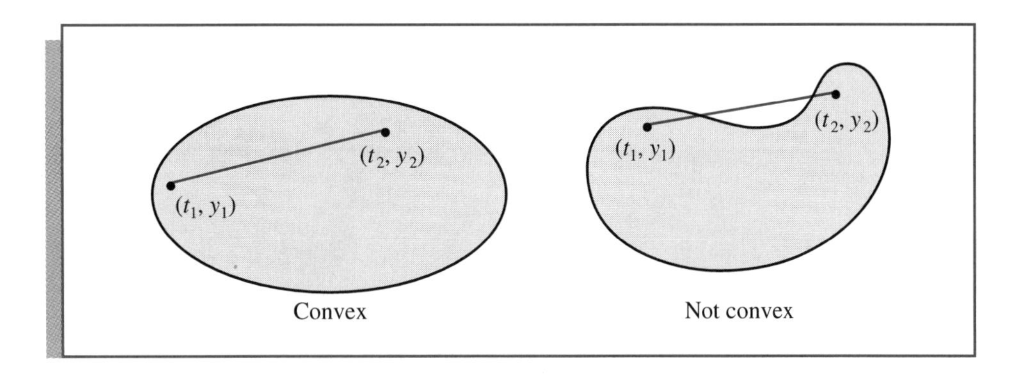

Theorem 5.3 Suppose $f(t, y)$ is defined on a convex set $D \subset \mathbb{R}^2$. If a constant $L > 0$ exists with

(5.1)
$$\left|\frac{\partial f}{\partial y}(t, y)\right| \leq L, \quad \text{for all } (t, y) \in D,$$

then f satisfies a Lipschitz condition on D in the variable y with Lipschitz constant L. ■

The proof of Theorem 5.3 is discussed in Exercise 4; it is similar to the proof of the corresponding result for functions of one variable discussed in Exercise 25 of Section 1.1.

As the next theorem will show, it is often of significant interest to determine whether the function involved in an initial-value problem satisfies a Lipschitz condition in its second variable, and condition (5.1) is generally easier to apply than the definition. We should note, however, that Theorem 5.3 gives only sufficient conditions for a Lipschitz condition to hold; a reexamination of Example 1 will demonstrate that these conditions are definitely *not necessary*.

The following theorem is a version of the fundamental existence and uniqueness theorem for first-order ordinary differential equations. Although the theorem can be proved with the hypothesis reduced somewhat, this form of the theorem is sufficient for our purposes. (The proof of the theorem, in approximately this form, can be found in [BiR, pp. 142–155].)

Theorem 5.4 Suppose that $D = \{(t, y) \mid a \leq t \leq b, \ -\infty < y < \infty\}$ and that $f(t, y)$ is continuous on D. If f satisfies a Lipschitz condition on D in the variable y, then the initial-value problem

$$y'(t) = f(t, y), \quad a \leq t \leq b, \quad y(a) = \alpha,$$

has a unique solution $y(t)$ for $a \leq t \leq b$. ■

EXAMPLE 2 Consider the initial-value problem

$$y' = 1 + t\sin(ty), \quad 0 \le t \le 2, \quad y(0) = 0.$$

Holding t constant and applying the Mean Value Theorem to the function

$$f(t, y) = 1 + t\sin(ty),$$

we find that whenever $y_1 < y_2$, a number ξ in (y_1, y_2) exists with

$$\frac{f(t, y_2) - f(t, y_1)}{y_2 - y_1} = \frac{\partial}{\partial y} f(t, \xi) = t^2 \cos(t\xi).$$

Thus,

$$|f(t, y_2) - f(t, y_1)| = |y_2 - y_1||t^2 \cos(t\xi)| \le 4|y_2 - y_1|,$$

and f satisfies a Lipschitz condition in the variable y with Lipschitz constant $L = 4$. Since, additionally, $f(t, y)$ is continuous when $0 \le t \le 2$ and $-\infty < y < \infty$, Theorem 5.4 implies that a unique solution exists to this initial-value problem.

If you have completed a course in differential equations you might try to find the exact solution to this problem. ■

Now that we have, to some extent, taken care of the question of when initial-value problems have unique solutions, we can move to the other question posed earlier in the section:

> How do we determine whether a particular problem has the property that small changes, or perturbations, in the statement of the problem introduce correspondingly small changes in the solution?

As usual, we first need to give a workable definition to express this concept.

Definition 5.5 The initial-value problem

(5.2)
$$\frac{dy}{dt} = f(t, y), \quad a \le t \le b, \quad y(a) = \alpha,$$

is said to be a **well-posed problem** if:

1. A unique solution, $y(t)$, to the problem exists;
2. For any $\varepsilon > 0$, there exists a positive constant $k(\varepsilon)$ with the property that, whenever $|\varepsilon_0| < \varepsilon$ and $\delta(t)$ is continuous with $|\delta(t)| < \varepsilon$ on $[a, b]$, a unique solution, $z(t)$, to the problem

   (5.3)
   $$\frac{dz}{dt} = f(t, z) + \delta(t), \quad a \le t \le b, \quad z(a) = \alpha + \varepsilon_0,$$

   exists with

   $$|z(t) - y(t)| < k(\varepsilon)\varepsilon, \quad \text{for all } a \le t \le b.$$ ■

The problem specified by Eq. (5.3) is called a **perturbed problem** associated with the original problem (5.2). It assumes the possibility of an error $\delta(t)$ being introduced in the statement of the differential equation, as well as an error ε_0 being present in the initial condition.

Numerical methods will always be concerned with solving a perturbed problem, since any roundoff error introduced in the representation perturbs the original problem. Unless the original problem is well-posed, there is little reason to expect that the numerical solution to a perturbed problem will accurately approximate the solution to the original problem.

The following theorem specifies conditions that ensure that an initial-value problem is well-posed. The proof of this theorem can be found in [BiR, pp. 142–147].

Theorem 5.6 Suppose $D = \{(t, y) \mid a \leq t \leq b \text{ and } -\infty < y < \infty\}$. If f is continuous and satisfies a Lipschitz condition in the variable y on the set D, then the initial-value problem

$$\frac{dy}{dt} = f(t, y), \quad a \leq t \leq b, \quad y(a) = \alpha$$

is well-posed. ■

EXAMPLE 3 Let $D = \{(t, y) \mid 0 \leq t \leq 1, \ -\infty < y < \infty\}$ and consider the initial-value problem

$$\frac{dy}{dt} = y - t^2 + 1, \quad 0 \leq t \leq 2, \quad y(0) = 0.5. \tag{5.4}$$

Since

$$\left|\frac{\partial(y - t^2 + 1)}{\partial y}\right| = |1| = 1,$$

Theorem 5.3 implies that $f(t, y) = y - t^2 + 1$ satisfies a Lipschitz condition on D with Lipschitz constant 1. Since f is continuous on D, Theorem 5.6 implies that the problem is well-posed.

Consider the perturbed problem

$$\frac{dz}{dt} = z - t^2 + 1 + \delta, \quad 0 \leq t \leq 2, \quad z(0) = 0.5 + \varepsilon_0, \tag{5.5}$$

where δ and ε_0 are constants. The solutions to Eqs. (5.4) and (5.5) are

$$y(t) = (t + 1)^2 - 0.5e^t \quad \text{and} \quad z(t) = (t + 1)^2 + (\delta + \varepsilon_0 - 0.5)e^t - \delta,$$

respectively. It is easy to verify that if $|\delta| < \varepsilon$ and $|\varepsilon_0| < \varepsilon$, then

$$|y(t) - z(t)| = |(\delta + \varepsilon_0)e^t - \delta| \leq |\delta + \varepsilon_0|e^2 + |\delta| \leq (2e^2 + 1)\varepsilon$$

for all t. This agrees with the result obtained by the use of Theorem 5.6. ■

Maple can be used to solve many initial-value problems. Consider the problem

$$\frac{dy}{dt} = y - t^2 + 1, \quad 0 \le t \le 2, \quad y(0) = 0.5.$$

To define the differential equation, enter

```
>deq:=D(y)(t)=y(t)-t*t+1;
```

and the initial condition

```
>init:=y(0)=0.5;
```

The names deq and init are chosen by the user. The command to solve the initial-value problems is

```
>deqsol:=dsolve({deq,init},y(t));
```

The response is

$$\text{deqsol} := y(t) = t^2 + 2t + 1 - .5000000000e^t$$

To use the solution to obtain $y(1.5)$, we enter

```
>q:=rhs(deqsol);
>evalf(subs(t=1.5,q));
```

with the result 4.009155465.

The function rhs is used to assign the solution of the initial-value problem to the function q, which we then evaluate at $t = 1.5$. The function dsolve can fail if an explicit solution to the initial-value problem cannot be found. For example, the command

```
>deqsol2:=dsolve({D(y)(t)=1+t*sin(t*y(t)),y(0)=0},y(t));
```

does not succeed because an explicit solution cannot be found. In this case a numerical method must be used.

EXERCISE SET 5.1

1. Use Theorem 5.4 to show that each of the following initial-value problems has a unique solution and find the solution.

   a. $y' = y\cos t, \quad 0 \le t \le 1, \quad y(0) = 1.$

   b. $y' = \dfrac{2}{t}y + t^2e^2, \quad 1 \le t \le 2, \quad y(1) = 0.$

   c. $y' = -\dfrac{2}{t}y + t^2e^t, \quad 1 \le t \le 2, \quad y(1) = \sqrt{2}e.$

   d. $y' = \dfrac{4t^3y}{1+t^4}, \quad 0 \le t \le 1, \quad y(0) = 1.$

2. For each choice of $f(t, y)$ given in parts (a)–(d):
   (i) Does f satisfy a Lipschitz condition on $D = \{(t, y) \mid 0 \le t \le 1, -\infty < y < \infty\}$?
   (ii) Can Theorem 5.6 be used to show that the initial-value problem

$$y' = f(t, y), \quad 0 \le t \le 1, \quad y(0) = 1,$$

   is well-posed?

   a. $f(t, y) = t^2 y + 1$ **b.** $f(t, y) = ty$
   c. $f(t, y) = 1 - y$ **d.** $f(t, y) = -ty + \dfrac{4t}{y}$

3. For the following initial-value problems, show that the given equation implicitly defines a solution. Approximate $y(2)$, using Newton's method.
   a. $y' = -\dfrac{y^3 + y}{(3y^2 + 1)t}$, $1 \le t \le 2$, $y(1) = 1$; $y^3 t + yt = 2$
   b. $y' = -\dfrac{y \cos t + 2te^y}{\sin t + t^2 e^y + 2}$, $1 \le t \le 2$, $y(1) = 0$; $y \sin t + t^2 e^y + 2y = 1$

4. Prove Theorem 5.3 by applying the Mean Value Theorem to $f(t, y)$, holding t fixed.

5. Show that, for any constants a and b, the set $D = \{(t, y) \mid a \le t \le b, -\infty < y < \infty\}$ is convex.

6. Suppose the perturbation $\delta(t)$ is proportional to t, that is, $\delta(t) = \delta t$ for some constant δ. Show directly that the following initial-value problems are well-posed.
   a. $y' = 1 - y$, $0 \le t \le 2$, $y(0) = 0$
   b. $y' = t + y$, $0 \le t \le 2$, $y(0) = -1$
   c. $y' = \dfrac{2}{t} y + t^2 e^t$, $1 \le t \le 2$, $y(1) = 0$
   d. $y' = -\dfrac{2}{t} y + t^2 e^t$, $1 \le t \le 2$, $y(1) = \sqrt{2}e$

7. Picard's method for solving the initial-value problem

$$y' = f(t, y), \quad a \le t \le b, \quad y(a) = \alpha$$

   is described as follows: Let $y_0(t) = \alpha$ for each t in $[a, b]$. Define a sequence $\{y_k(t)\}$ of functions by

$$y_k(t) = \alpha + \int_a^t f(\tau, y_{k-1}(\tau))\, d\tau, \quad k = 1, 2, \ldots.$$

   a. Integrate $y' = f(t, y(t))$ and use the initial condition to derive Picard's method.
   b. Generate $y_0(t)$, $y_1(t)$, $y_2(t)$, and $y_3(t)$ for the initial-value problem

$$y' = -y + t + 1, \quad 0 \le t \le 1, \quad y(0) = 1.$$

   c. Compare the result in part (b) to the Maclaurin series of the actual solution $y(t) = t + e^{-t}$.

5.2 Euler's Method

In this section we will consider Euler's method. Although Euler's method is seldom used in practice, the simplicity of its derivation can be used to illustrate the techniques involved in the construction of some of the more advanced techniques, without the cumbersome algebra that accompanies these constructions.

The object of the method is to obtain an approximation to the well-posed initial-value problem

$$\frac{dy}{dt} = f(t, y), \quad a \le t \le b, \quad y(a) = \alpha. \tag{5.6}$$

In actuality, a continuous approximation to the solution $y(t)$ will not be obtained; instead, approximations to $y(t)$ will be generated at various values, called **mesh points**, in the interval $[a, b]$. Once the approximate solution is obtained at the points, the approximate solution at other points in the interval can be obtained by interpolation.

We first make the stipulation that the mesh points are equally distributed throughout the interval $[a, b]$. This condition is ensured by choosing a positive integer N and selecting the mesh points $\{t_0, t_1, t_2, \ldots, t_N\}$, where

$$t_i = a + ih, \quad \text{for each } i = 0, 1, 2, \ldots, N.$$

The common distance between the points $h = (b - a)/N$ is called the **step size**.

We will use Taylor's Theorem to derive Euler's method.

Suppose that $y(t)$, the unique solution to (5.6), has two continuous derivatives on $[a, b]$, so that for each $i = 0, 1, 2, \ldots, N - 1$,

$$y(t_{i+1}) = y(t_i) + (t_{i+1} - t_i)y'(t_i) + \frac{(t_{i+1} - t_i)^2}{2}y''(\xi_i)$$

for some number ξ_i in (t_i, t_{i+1}). If $h = t_{i+1} - t_i$, then

$$y(t_{i+1}) = y(t_i) + hy'(t_i) + \frac{h^2}{2}y''(\xi_i),$$

and, since $y(t)$ satisfies the differential equation (5.6),

$$y(t_{i+1}) = y(t_i) + hf(t_i, y(t_i)) + \frac{h^2}{2}y''(\xi_i). \tag{5.7}$$

Euler's method constructs $w_i \approx y(t_i)$ for each $i = 1, 2, \ldots, N$, by deleting the remainder term. Thus,

$$\begin{aligned} w_0 &= \alpha, \\ w_{i+1} &= w_i + hf(t_i, w_i), \quad \text{for each } i = 0, 1, \ldots, N - 1. \end{aligned} \tag{5.8}$$

Equation (5.8) is called the **difference equation** associated with Euler's method. As we will see later in this chapter, the theory and solution of difference equations parallel, in many ways, the theory and solution of differential equations. Algorithm 5.1 implements Euler's method.

ALGORITHM 5.1

Euler's

To approximate the solution of the initial-value problem

$$y' = f(t, y), \quad a \le t \le b, \quad y(a) = \alpha,$$

at $(N + 1)$ equally spaced numbers in the interval $[a, b]$:

INPUT endpoints a, b; integer N; initial condition α.

OUTPUT approximation w to y at the $(N + 1)$ values of t.

Step 1 Set $h = (b - a)/N$;
$t = a$;
$w = \alpha$;
OUTPUT (t, w).

Step 2 For $i = 1, 2, \ldots, N$ do Steps 3, 4.

Step 3 Set $w = w + hf(t, w)$; (*Compute* w_i.)
$t = a + ih$. (*Compute* t_i.)

Step 4 OUTPUT (t, w).

Step 5 STOP.

To interpret Euler's method geometrically, note that when w_i is a close approximation to $y(t_i)$, the assumption that the problem is well-posed implies that

$$f(t_i, w_i) \approx y'(t_i) = f(t_i, y(t_i)).$$

The graph of the function highlighting $y(t_i)$ is shown in Figure 5.2(a). One step in Euler's method appears in Figure 5.2(b), and a series of steps appears in Figure 5.3.

Figure 5.2

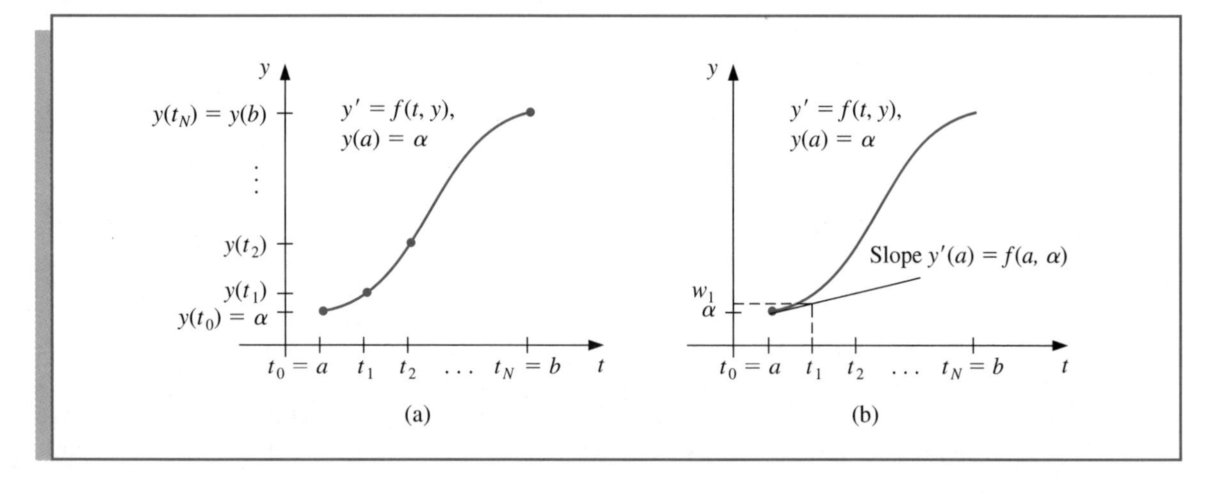

Figure 5.3

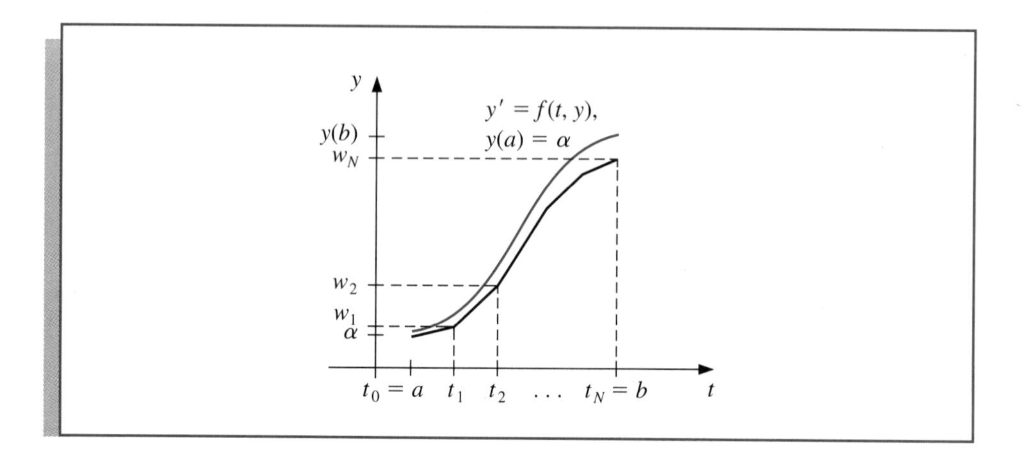

EXAMPLE 1 Suppose Euler's method is used to approximate the solution to the initial-value problem

$$y' = y - t^2 + 1, \quad 0 \le t \le 2, \quad y(0) = 0.5,$$

with $N = 10$. Then $h = 0.2$, $t_i = 0.2i$, $w_0 = 0.5$, and

$$w_{i+1} = w_i + h(w_i - t_i^2 + 1) = w_i + 0.2[w_i - 0.04i^2 + 1] = 1.2w_i - 0.008i^2 + 0.2,$$

for $i = 0, 1, \ldots, 9$. The exact solution is $y(t) = (t + 1)^2 - 0.5e^t$. Table 5.1 shows the comparison between the approximate values at t_i and the actual values. ■

Table 5.1

t_i	w_i	$y_i = y(t_i)$	$\lvert y_i - w_i \rvert$
0.0	0.5000000	0.5000000	0.0000000
0.2	0.8000000	0.8292986	0.0292986
0.4	1.1520000	1.2140877	0.0620877
0.6	1.5504000	1.6489406	0.0985406
0.8	1.9884800	2.1272295	0.1387495
1.0	2.4581760	2.6408591	0.1826831
1.2	2.9498112	3.1799415	0.2301303
1.4	3.4517734	3.7324000	0.2806266
1.6	3.9501281	4.2834838	0.3333557
1.8	4.4281538	4.8151763	0.3870225
2.0	4.8657845	5.3054720	0.4396874

Note that the error grows slightly as the value of t increases. This controlled error growth is a consequence of the stability of Euler's method, which implies that the error is expected to grow in no worse than a linear manner.

Although Euler's method is not accurate enough to warrant its use in practice, it is sufficiently elementary to analyze the error that is produced from its application. The error analysis for the more accurate methods that we consider in subsequent sections follows the same pattern but is more complicated.

To derive an error bound for Euler's method, we first consider two computational lemmas.

Lemma 5.7 For all $x \geq -1$ and any positive m, we have $0 \leq (1+x)^m \leq e^{mx}$.

Proof Applying Taylor's Theorem with $f(x) = e^x$, $x_0 = 0$, and $n = 1$ gives

$$e^x = 1 + x + \frac{1}{2}x^2 e^{\xi},$$

where ξ is between x and zero. Thus,

$$0 \leq 1 + x \leq 1 + x + \frac{1}{2}x^2 e^{\xi} = e^x$$

and, since $1 + x \geq 0$,

$$0 \leq (1+x)^m \leq (e^x)^m = e^{mx}. \qquad \blacksquare\ \blacksquare\ \blacksquare$$

Lemma 5.8 If s and t are positive real numbers, $\{a_i\}_{i=0}^{k}$ is a sequence satisfying $a_0 \geq -t/s$, and

(5.9) $$a_{i+1} \leq (1+s)a_i + t, \quad \text{for each } i = 0, 1, 2, \ldots, k,$$

then

$$a_{i+1} \leq e^{(i+1)s}\left(a_0 + \frac{t}{s}\right) - \frac{t}{s}.$$

Proof For a fixed integer i, Inequality (5.9) implies that

$$\begin{aligned}
a_{i+1} &\leq (1+s)a_i + t \\
&\leq (1+s)[(1+s)a_{i-1} + t] + t \\
&\leq (1+s)\{(1+s)[(1+s)a_{i-2} + t] + t\} + t \\
&\ \ \vdots \\
&\leq (1+s)^{i+1}a_0 + \left[1 + (1+s) + (1+s)^2 + \cdots + (1+s)^i\right] t.
\end{aligned}$$

But

$$1 + (1+s) + (1+s)^2 + \cdots + (1+s)^i = \sum_{j=0}^{i}(1+s)^j$$

is a geometric series with ratio $(1+s)$ and, as such, sums to

$$\frac{1 - (1+s)^{i+1}}{1 - (1+s)} = \frac{1}{s}\left[(1+s)^{i+1} - 1\right].$$

Thus,

$$a_{i+1} \leq (1+s)^{i+1} a_0 + \frac{(1+s)^{i+1} - 1}{s} t = (1+s)^{i+1} \left(a_0 + \frac{t}{s} \right) - \frac{t}{s},$$

and, by Lemma 5.7,

$$a_{i+1} \leq e^{(i+1)s} \left(a_0 + \frac{t}{s} \right) - \frac{t}{s}. \qquad \blacksquare\ \blacksquare\ \blacksquare$$

Theorem 5.9 Suppose f is continuous and satisfies a Lipschitz condition with constant L on

$$D = \{(t, y) \mid a \leq t \leq b, -\infty < y < \infty\}$$

and that a constant M exists with the property that

$$|y''(t)| \leq M, \quad \text{for all } t \in [a, b].$$

Let $y(t)$ denote the unique solution to the initial-value problem

$$y' = f(t, y), \quad a \leq t \leq b, \quad y(a) = \alpha,$$

and $w_0, w_1, \ldots, w_N$ be the approximations generated by Euler's method for some positive integer N. Then, for each $i = 0, 1, 2, \ldots, N$,

$$|y(t_i) - w_i| \leq \frac{hM}{2L} \left[e^{L(t_i - a)} - 1 \right]. \tag{5.10}$$

Proof When $i = 0$ the result is clearly true, since $y(t_0) = w_0 = \alpha$.

From Eq. (5.7), we have for $i = 0, 1, \ldots, N-1$,

$$y(t_{i+1}) = y(t_i) + hf(t_i, y(t_i)) + \frac{h^2}{2} y''(\xi_i);$$

and from the equations in (5.8),

$$w_{i+1} = w_i + hf(t_i, w_i).$$

Consequently, using the notation $y_i = y(t_i)$ and $y_{i+1} = y(t_{i+1})$, we have

$$y_{i+1} - w_{i+1} = y_i - w_i + h[f(t_i, y_i) - f(t_i, w_i)] + \frac{h^2}{2} y''(\xi_i)$$

and

$$|y_{i+1} - w_{i+1}| \leq |y_i - w_i| + h|f(t_i, y_i) - f(t_i, w_i)| + \frac{h^2}{2} |y''(\xi_i)|.$$

Since f satisfies a Lipschitz condition in the second variable with constant L and $|y''(t)| \le M$, we have

$$|y_{i+1} - w_{i+1}| \le (1 + hL)|y_i - w_i| + \frac{h^2M}{2}.$$

Referring to Lemma 5.8 and letting $a_j = |y_j - w_j|$ for each $j = 0, 1, \ldots, N$, while $s = hL$ and $t = h^2M/2$, we see that

$$|y_{i+1} - w_{i+1}| \le e^{(i+1)hL}\left(|y_0 - w_0| + \frac{h^2M}{2hL}\right) - \frac{h^2M}{2hL}.$$

Since $|y_0 - w_0| = 0$ and $(i + 1)h = t_{i+1} - t_0 = t_{i+1} - a$, we have

$$|y_{i+1} - w_{i+1}| \le \frac{hM}{2L}\left[e^{(t_{i+1}-a)L} - 1\right],$$

for each $i = 0, 1, \ldots, N - 1$. ■ ■ ■

The weakness of Theorem 5.9 lies in the requirement that a bound be known for the second derivative of the solution. Although this condition often prohibits us from obtaining a realistic error bound, it should be noted that if $\partial f/\partial t$ and $\partial f/\partial y$ both exist, the chain rule for partial differentiation implies that

$$y''(t) = \frac{dy'}{dt}(t) = \frac{df}{dt}(t, y(t)) = \frac{\partial f}{\partial t}(t, y(t)) + \frac{\partial f}{\partial y}(t, y(t)) \cdot f(t, y(t)).$$

So it is at times possible to obtain an error bound for $y''(t)$ without explicitly knowing $y(t)$.

EXAMPLE 2 Returning to the initial-value problem

$$y' = y - t^2 + 1, \quad 0 \le t \le 2, \quad y(0) = 0.5,$$

considered in Example 1, we see that since $f(t, y) = y - t^2 + 1$, we have $\partial f(t, y)/\partial y = 1$ for all y, so $L = 1$. For this problem, the exact solution is $y(t) = (t + 1)^2 - \frac{1}{2}e^t$, so $y''(t) = 2 - 0.5e^t$ and

$$|y''(t)| \le 0.5e^2 - 2, \quad \text{for all } t \in [0, 2].$$

Using the inequality in the error bound for Euler's method with $h = 0.2$, $L = 1$, and $M = 0.5e^2 - 2$ gives the error bound

$$|y_i - w_i| \le 0.1(0.5e^2 - 2)(e^{t_i} - 1).$$

Table 5.2 lists the actual error found in Example 1, together with this error bound. Note that even though the true bound for the second derivative of the solution was used, the error bound is considerably larger than the actual error. ■

Table 5.2

t_i	0.2	0.4	0.6	0.8	1.0	1.2	1.4	1.6	1.8	2.0
Actual Error	0.02930	0.06209	0.09854	0.13875	0.18268	0.23013	0.28063	0.33336	0.38702	0.43969
Error Bound	0.03752	0.08334	0.13931	0.20767	0.29117	0.39315	0.51771	0.66985	0.85568	1.08264

The principal importance of the error-bound formula given in Theorem 5.9 is that the bound depends linearly on the step size h. Consequently, diminishing the step size should give correspondingly greater accuracy to the approximations.

Neglected in the result of Theorem 5.9 is the effect that roundoff error plays in the choice of step size. As h becomes smaller, more calculations are necessary and more roundoff error is expected. In actuality then, the difference-equation form

$$w_0 = \alpha,$$
$$w_{i+1} = w_i + hf(t_i, w_i), \quad \text{for each } i = 0, 1, \ldots, N-1,$$

is not used to calculate the approximation to the solution y_i at a mesh point t_i. We use instead an equation of the form

(5.11)
$$u_0 = \alpha + \delta_0,$$
$$u_{i+1} = u_i + hf(t_i, u_i) + \delta_{i+1}, \quad \text{for each } i = 0, 1, \ldots, N-1.$$

where δ_i denotes the roundoff error associated with u_i. Using methods similar to those in the proof of Theorem 5.9, we can produce an error bound for the finite-digit approximations to $y_i(t)$ given by Euler's method.

Theorem 5.10 Let $y(t)$ denote the unique solution to the initial-value problem

(5.12)
$$y' = f(t, y), \quad a \le t \le b, \quad y(a) = \alpha$$

and $u_0, u_1, \ldots, u_N$ be the approximations obtained using (5.11). If $|\delta_i| < \delta$ for each $i = 0, 1, \ldots, N$ and the hypotheses of Theorem 5.9 hold for (5.12), then

(5.13)
$$|y(t_i) - u_i| \le \frac{1}{L}\left(\frac{hM}{2} + \frac{\delta}{h}\right)\left[e^{L(t_i-a)} - 1\right] + |\delta_0| e^{L(t_i-a)},$$

for each $i = 0, 1, \ldots, N$. ■

The error bound (5.13) is no longer linear in h. In fact, since

$$\lim_{h\to 0}\left(\frac{hM}{2} + \frac{\delta}{h}\right) = \infty,$$

the error would be expected to become large for sufficiently small values of h. Calculus can be used to determine a lower bound for the step size h. Letting $E(h) = (hM/2) + (\delta/h)$ implies that $E'(h) = (M/2) - (\delta/h^2)$.

$$\text{If} \quad h < \sqrt{2\delta/M}, \text{ then} \quad E'(h) < 0 \text{ and } E(h) \text{ is decreasing.}$$

$$\text{If} \quad h > \sqrt{2\delta/M}, \text{ then} \quad E'(h) > 0 \text{ and } E(h) \text{ is increasing.}$$

The minimal value of $E(h)$ occurs when

$$h = \sqrt{\frac{2\delta}{M}}. \tag{5.14}$$

Decreasing h beyond this value tends to increase the total error in the approximation. Normally, however, the value of δ is sufficiently small that this lower bound for h does not affect the operation of Euler's method.

EXERCISE SET 5.2

1. Use Euler's method to approximate the solutions for each of the following initial-value problems.
   - **a.** $y' = te^{3t} - 2y, \quad 0 \le t \le 1, \quad y(0) = 0$, with $h = 0.5$
   - **b.** $y' = 1 + (t - y)^2, \quad 2 \le t \le 3, \quad y(2) = 1$, with $h = 0.5$
   - **c.** $y' = 1 + y/t, \quad 1 \le t \le 2, \quad y(1) = 2$, with $h = 0.25$
   - **d.** $y' = \cos 2t + \sin 3t, \quad 0 \le t \le 1 \quad y(0) = 1$, with $h = 0.25$
2. The actual solutions to the initial-value problems in Exercise 1 are given here. Compare the actual error at each step to the error bound.
   - **a.** $y(t) = \dfrac{1}{5}te^{3t} - \dfrac{1}{25}e^{3t} + \dfrac{1}{25}e^{-2t}$
   - **b.** $y(t) = t + \dfrac{1}{1-t}$
   - **c.** $y(t) = t \ln t + 2t$
   - **d.** $y(t) = \dfrac{1}{2}\sin 2t - \dfrac{1}{3}\cos 3t + \dfrac{4}{3}$
3. Use Euler's method to approximate the solutions for each of the following initial-value problems.
   - **a.** $y' = \dfrac{y}{t} - \left(\dfrac{y}{t}\right)^2, \quad 1 \le t \le 2, \quad y(1) = 1$, with $h = 0.1$
   - **b.** $y' = 1 + \dfrac{y}{t} + \left(\dfrac{y}{t}\right)^2, \quad 1 \le t \le 3, \quad y(1) = 0$, with $h = 0.2$
   - **c.** $y' = -(y + 1)(y + 3), \quad 0 \le t \le 2, \quad y(0) = -2$, with $h = 0.2$
   - **d.** $y' = -5y + 5t^2 + 2t, \quad 0 \le t \le 1, \quad y(0) = \frac{1}{3}$, with $h = 0.1$
4. The actual solutions to the initial-value problems in Exercise 3 are given here. Compute the actual error in the approximations of Exercise 3.
   - **a.** $y(t) = \dfrac{t}{1 + \ln t}$
   - **b.** $y(t) = t\tan(\ln t)$
   - **c.** $y(t) = -3 + \dfrac{2}{1 + e^{-2t}}$
   - **d.** $y(t) = t^2 + \dfrac{1}{3}e^{-5t}$
5. Given the initial-value problem

$$y' = \frac{2}{t}y + t^2e^t, \quad 1 \le t \le 2, \quad y(1) = 0$$

   with exact solution $y(t) = t^2(e^t - e)$:
   - **a.** Use Euler's method with $h = 0.1$ to approximate the solution and compare it with the actual values of y.

b. Use the answers generated in part (a) and linear interpolation to approximate the following values of y and compare them to the actual values.

(i) $y(1.04)$ **(ii)** $y(1.55)$ **(iii)** $y(1.97)$

c. Compute the value of h necessary for $|y(t_i) - w_i| \leq 0.1$, using Eq. (5.10).

6. Given the initial-value problem

$$y' = \frac{1}{t^2} - \frac{y}{t} - y^2, \quad 1 \leq t \leq 2, \quad y(1) = -1$$

with exact solution $y(t) = -1/t$:

a. Use Euler's method with $h = 0.05$ to approximate the solution and compare it with the actual values of y.

b. Use the answers generated in part (a) and linear interpolation to approximate the following values of y and compare them to the actual values.

(i) $y(1.052)$ **(ii)** $y(1.555)$ **(iii)** $y(1.978)$

c. Compute the value of h necessary for $|y(t_i) - w_i| \leq 0.05$ using eq. (5.10).

7. Given the initial-value problem

$$y' = -y + t + 1, \quad 0 \leq t \leq 5, \quad y(0) = 1$$

with exact solution $y(t) = e^{-t} + t$:

a. Approximate $y(5)$ using Euler's method with $h = 0.2$, $h = 0.1$, and $h = 0.05$.

b. Determine the optimal value of h to use in computing $y(5)$, assuming $\delta = 10^{-6}$ and that Eq. (5.14) is valid.

8. Use the results of Exercise 3 and linear interpolation to approximate the following values of $y(t)$. Compare the approximations obtained to the actual values obtained using the functions given in Exercise 4.

a. $y(1.25)$ and $y(1.93)$ **b.** $y(2.1)$ and $y(2.75)$

c. $y(1.4)$ and $y(1.93)$ **d.** $y(0.54)$ and $y(0.94)$

9. Let $E(h) = \dfrac{hM}{2} + \dfrac{\delta}{h}$.

a. For the initial-value problem

$$y' = -y + 1, \quad 0 \leq t \leq 1, \quad y(0) = 0,$$

compute the value of h to minimize $E(h)$. Assume $\delta = 5 \times 10^{-(n+1)}$ if you are using n-digit arithmetic in part (c).

b. For the optimal h computed in part (a), use Eq. (5.13) to compute the minimal error obtainable.

c. Compare the actual error obtained using $h = 0.1$ and $h = 0.01$ to the minimal error in part (b). Can you explain the results?

10. Consider the initial-value problem

$$y' = -10y, \quad 0 \leq t \leq 2, \quad y(0) = 1,$$

which has solution $y(t) = e^{-10t}$. What happens when Euler's method is applied to this problem with $h = 0.1$? Does this behavior violate Theorem 5.9?

11. In a book entitled *Looking at History Through Mathematics*. Rashevsky [Ra, pp. 103–110] considers a model for a problem involving the production of nonconformists in society. Suppose that a society has a population of $x(t)$ individuals at time t, in years, and that all nonconformists who mate with other nonconformists have offspring who are also nonconformists, while a fixed proportion r of all other offspring are also nonconformist. If the birth and death rates for all individuals are assumed to be the constants b and d, respectively, and if conformists and nonconformists mate at random, the problem can be expressed by the differential equations

$$\frac{dx(t)}{dt} = (b - d)x(t) \quad \text{and} \quad \frac{dx_n(t)}{dt} = (b - d)x_n(t) + rb(x(t) - x_n(t)),$$

where $x_n(t)$ denotes the number of nonconformists in the population at time t.

a. If the variable $p(t) = x_n(t)/x(t)$ is introduced to represent the proportion of nonconformists in the society at time t, show that these equations can be combined and simplified to the single differential equation

$$\frac{dp(t)}{dt} = rb(1 - p(t)).$$

b. Assuming that $p(0) = 0.01$, $b = 0.02$, $d = 0.015$, and $r = 0.1$, approximate the solution $p(t)$ from $t = 0$ to $t = 50$ when the step size is $h = 1$ year.

c. Solve the differential equation for $p(t)$ exactly, and compare your result in part (b) when $t = 50$ with the exact value at that time.

12. In a circuit with impressed voltage $\mathcal{E}$ having resistance R, inductance L, and capacitance C in parallel, the current i satisfies the differential equation

$$\frac{di}{dt} = C\frac{d^2\mathcal{E}}{dt^2} + \frac{1}{R}\frac{d\mathcal{E}}{dt} + \frac{1}{L}\mathcal{E}.$$

Suppose $C = 0.3$ farads, $R = 1.4$ ohms, $L = 1.7$ henries, and the voltage is given by

$$\mathcal{E}(t) = e^{-0.06\pi t}\sin(2t - \pi).$$

If $i(0) = 0$, find the current i for the values $t = 0.1j$, where $j = 0, 1, \ldots, 100$.

5.3 Higher-Order Taylor Methods

Since the object of numerical techniques is to determine sufficiently accurate approximations with minimal effort, we need a means for comparing the efficiency of various approximation methods. The first device we consider is called the *local truncation error* of the method. The local truncation error at a specified step measures the amount by which the exact solution to the differential equation fails to satisfy the difference equation being used for the approximation.

Definition 5.11 The difference method

$$w_0 = \alpha$$

$$w_{i+1} = w_i + h\phi(t_i, w_i), \quad \text{for each } i = 0, 1, \ldots, N - 1,$$

has **local truncation error** given by

$$\tau_{i+1}(h) = \frac{y_{i+1} - (y_i + h\phi(t_i, y_i))}{h} = \frac{y_{i+1} - y_i}{h} - \phi(t_i, y_i),$$

for each $i = 0, 1, \ldots, N - 1$, where, as usual, $y_i = y(t_i)$ denotes the exact value of the solution at t_i. ■

For Euler's method, the local truncation error at the ith step for the problem

$$y' = f(t, y), \quad a \le t \le b, \quad y(a) = \alpha,$$

is

$$\tau_{i+1}(h) = \frac{y_{i+1} - y_i}{h} - f(t_i, y_i) \quad \text{for each } i = 0, 1, \ldots, N-1.$$

This error is a *local error* because it measures the accuracy of the method at a specific step, assuming that the method was exact at the previous step. As such, it depends on the differential equation, the step size, and the particular step in the approximation.

By considering Eq. (5.7) in the previous section, we see that Euler's method has

$$\tau_{i+1}(h) = \frac{h}{2} y''(\xi_i), \quad \text{for some } \xi_i \text{ in } (t_i, t_{i+1}).$$

When $y''(t)$ is known to be bounded by a constant M on $[a, b]$, this implies

$$|\tau_{i+1}(h)| \le \frac{h}{2} M,$$

so the local truncation error in Euler's method is $O(h)$. One way to select difference-equation methods for solving ordinary differential equations is in such a manner that their local truncation errors are $O(h^p)$ for as large a value of p as possible, while keeping the number and complexity of calculations of the methods within a reasonable bound.

Since Euler's method was derived by using Taylor's Theorem with $n = 1$ to approximate the solution of the differential equation, our first attempt to find methods for improving the convergence properties of difference methods is to extend this technique of derivation to larger values of n.

Suppose the solution $y(t)$ to the initial-value problem

$$y' = f(t, y), \quad a \le t \le b, \quad y(a) = \alpha$$

has $(n+1)$ continuous derivatives. If we expand the solution, $y(t)$, in terms of its nth Taylor polynomial about t_i, we obtain

(5.15) $$y(t_{i+1}) = y(t_i) + hy'(t_i) + \frac{h^2}{2} y''(t_i) + \cdots + \frac{h^n}{n!} y^{(n)}(t_i) + \frac{h^{n+1}}{(n+1)!} y^{(n+1)}(\xi_i),$$

for some ξ_i in (t_i, t_{i+1}).

Successive differentiation of the solution, $y(t)$, gives

$$y'(t) = f(t, y(t)),$$
$$y''(t) = f'(t, y(t)),$$

and, in general,

$$y^{(k)}(t) = f^{(k-1)}(t, y(t)).$$

Substituting these results into Eq. (5.15) gives

(5.16)
$$y(t_{i+1}) = y(t_i) + hf(t_i, y(t_i)) + \frac{h^2}{2}f'(t_i, y(t_i)) + \cdots + \frac{h^n}{n!}f^{(n-1)}(t_i, y(t_i)) + \frac{h^{n+1}}{(n+1)!}f^{(n)}(\xi_i, y(\xi_i)).$$

The difference-equation method corresponding to Eq. (5.16) is called the *Taylor Method of order n* and is obtained by deleting the remainder term involving ξ_i.

Taylor Method of Order n:

(5.17)
$$w_0 = \alpha,$$
$$w_{i+1} = w_i + hT^{(n)}(t_i, w_i) \quad \text{for each } i = 0, 1, \ldots, N-1,$$

where

$$T^{(n)}(t_i, w_i) = f(t_i, w_i) + \frac{h}{2}f'(t_i, w_i) + \cdots + \frac{h^{n-1}}{n!}f^{(n-1)}(t_i, w_i).$$

Note that Euler's method is Taylor's method of order one.

EXAMPLE 1 To apply Taylor's method of orders two and four to the initial-value problem

$$y' = y - t^2 + 1, \quad 0 \le t \le 2, \quad y(0) = 0.5,$$

which was studied in the previous sections, we must find the first three derivatives of $f(t, y(t)) = y(t) - t^2 + 1$ with respect to the variable t:

$$f'(t, y(t)) = \frac{d}{dt}(y - t^2 + 1) = y' - 2t = y - t^2 + 1 - 2t,$$
$$f''(t, y(t)) = \frac{d}{dt}(y - t^2 + 1 - 2t) = y' - 2t - 2$$
$$= (y - t^2 + 1) - 2t - 2 = y - t^2 - 2t - 1,$$

and

$$f'''(t, y(t)) = \frac{d}{dt}(y - t^2 - 2t - 1) = y' - 2t - 2 = y - t^2 - 2t - 1.$$

So

$$T^{(2)}(t_i, w_i) = f(t_i, w_i) + \frac{h}{2}f'(t_i, w_i) = w_i - t_i^2 + 1 + \frac{h}{2}(w_i - t_i^2 - 2t_i + 1)$$
$$= \left(1 + \frac{h}{2}\right)(w_i - t_i^2 + 1) - ht_i$$

and

$$\begin{aligned}
T^{(4)}(t_i, w_i) &= f(t_i, w_i) + \frac{h}{2} f'(t_i, w_i) + \frac{h^2}{6} f''(t_i, w_i) + \frac{h^3}{24} f'''(t_i, w_i) \\
&= w_i - t_i^2 + 1 + \frac{h}{2}(w_i - t_i^2 - 2t_i + 1) + \frac{h^2}{6}(w_i - t_i^2 - 2t_i - 1) \\
&\quad + \frac{h^3}{24}(w_i - t_i^2 - 2t_i - 1) \\
&= \left(1 + \frac{h}{2} + \frac{h^2}{6} + \frac{h^3}{24}\right)(w_i - t_i^2) - \left(1 + \frac{h}{3} + \frac{h^2}{12}\right) ht_i \\
&\quad + 1 + \frac{h}{2} - \frac{h^2}{6} - \frac{h^3}{24}.
\end{aligned}$$

The Taylor methods of orders two and four are, consequently,

$$\begin{aligned}
w_0 &= 0.5, \\
w_{i+1} &= w_i + h\left[\left(1 + \frac{h}{2}\right)(w_i - t_i^2 + 1) - ht_i\right]
\end{aligned}$$

and

$$\begin{aligned}
w_0 &= 0.5, \\
w_{i+1} &= w_i + h\left[\left(1 + \frac{h}{2} + \frac{h^2}{6} + \frac{h^3}{24}\right)(w_i - t_i^2) - \left(1 + \frac{h}{3} + \frac{h^2}{12}\right) ht_i \right. \\
&\quad \left. + 1 + \frac{h}{2} - \frac{h^2}{6} - \frac{h^3}{24}\right],
\end{aligned}$$

for $i = 0, 1, \ldots, N - 1$.

If $h = 0.2$, then $N = 10$ and $t_i = 0.2i$ for each $i = 1, 2, \ldots, 10$. Thus, the second-order method becomes

$$\begin{aligned}
w_0 &= 0.5, \\
w_{i+1} &= w_i + 0.2\left[\left(1 + \frac{0.2}{2}\right)(w_i - 0.04i^2 + 1) - 0.04i\right] \\
&= 1.22w_i - 0.0088i^2 - 0.008i + 0.22,
\end{aligned}$$

and the fourth-order method becomes

$$\begin{aligned}
w_{i+1} &= w_i + 0.2\left[\left(1 + \frac{0.2}{2} + \frac{0.04}{6} + \frac{0.008}{24}\right)(w_i - 0.04i^2) \right. \\
&\quad \left. - \left(1 + \frac{0.2}{3} + \frac{0.04}{12}\right)(0.04i) + 1 + \frac{0.2}{2} - \frac{0.04}{6} - \frac{0.008}{24}\right] \\
&= 1.2214w_i - 0.008856i^2 - 0.00856i + 0.2186,
\end{aligned}$$

for each $i = 0, 1, \ldots, 9$.

Table 5.3 lists the actual values of the solution $y(t) = (t+1)^2 - 0.5e^t$, the results from the Taylor methods of orders two and four, and the actual errors involved with these methods.

Table 5.3

t_i	Exact $y(t_i)$	Taylor Order 2 w_i	Error $\lvert y(t_i) - w_i\rvert$	Taylor Order 4 w_i	Error $\lvert y(t_i) - w_i\rvert$
0.0	0.5000000	0.5000000	0	0.5000000	0
0.2	0.8292986	0.8300000	0.0007014	0.8293000	0.0000014
0.4	1.2140877	1.2158000	0.0017123	1.2140910	0.0000034
0.6	1.6489406	1.6520760	0.0031354	1.6489468	0.0000062
0.8	2.1272295	2.1323327	0.0051032	2.1272396	0.0000101
1.0	2.6408591	2.6486459	0.0077868	2.6408744	0.0000153
1.2	3.1799415	3.1913480	0.0114065	3.1799640	0.0000225
1.4	3.7324000	3.7486446	0.0162446	3.7324321	0.0000321
1.6	4.2834838	4.3061464	0.0226626	4.2835285	0.0000447
1.8	4.8151763	4.8462986	0.0311223	4.8152377	0.0000615
2.0	5.3054720	5.3476843	0.0422123	5.3055554	0.0000834

Suppose we need to determine an approximation to an intermediate point in the table, for example at $t = 1.25$. If we use linear interpolation on the Taylor method of order four approximations at $t = 1.2$ and $t = 1.4$ we have

$$y(1.25) \approx \left(\frac{1.25 - 1.4}{1.2 - 1.4}\right) 3.1799640 + \left(\frac{1.25 - 1.2}{1.4 - 1.2}\right) 3.7324321 = 3.3180810.$$

Since $y(1.25) = 3.3173285$, this approximation has an error of 0.0007525, which is nearly 30 times the average of the approximation errors at 1.2 and 1.4.

To improve the approximation to $y(1.25)$ we can use cubic Hermite interpolation. This requires approximations to $y'(1.2)$ and $y'(1.4)$ as well as approximations to $y(1.2)$ and $y(1.4)$. But the derivative approximations are available from the differential equation since $y'(t) = f(t, y(t))$. In our example that means that $y'(t) = y(t) - t^2 + 1$, so

$$y'(1.2) = y(1.2) - (1.2)^2 + 1 \approx 3.1799640 - 1.44 + 1 = 2.7399640$$

and

$$y'(1.4) = y(1.4) - (1.4)^2 + 1 \approx 3.7324327 - 1.96 + 1 = 2.7724321.$$

Following the divided-difference procedure in Section 3.3, we have the information in Table 5.4. The underlined entries come from the data, and the other entries use the divided-difference formulas.

Table 5.4

1.2	3.1799640			
		2.7399640		
1.2	3.1799640		0.1118825	
		2.7623405		−0.3071225
1.4	3.7324321		0.0504580	
		2.7724321		
1.4	3.7324321			

The cubic Hermite polynomial is

$$y(t) \approx 3.1799640 + (t - 1.2)2.7399640 + (t - 1.2)^2 0.1118825 + (t - 1.2)^2(t - 1.4)(-0.3071225),$$

so

$$y(1.25) \approx 3.1799640 + 0.1369982 + 0.0002797 + 0.0001152 = 3.3173571,$$

a result that is accurate to within 0.0000286. This is about the average of the error at 1.2 and 1.4, which is less than 4% of the error obtained using linear interpolation. ■

As might be expected from our study of Euler's method, Taylor's method of order n has local truncation error $O(h^n)$, provided that the solution of the differential equation is sufficiently well-behaved. This is seen by noting that Eq. (5.16) can be rewritten

$$y_{i+1} - y_i - hf(t_i, y_i) - \frac{h^2}{2}f'(t_i, y_i) - \cdots - \frac{h^n}{n!}f^{(n-1)}(t_i, y_i) = \frac{h^{n+1}}{(n+1)!}f^{(n)}(\xi_i, y(\xi_i)),$$

for some ξ_i in (t_i, t_{i+1}), so the local truncation error is

$$\tau_{i+1}(h) = \frac{y_{i+1} - y_i}{h} - T^{(n)}(t_i, y_i) = \frac{h^n}{(n+1)!}f^{(n)}(\xi_i, y(\xi_i)),$$

for each $i = 0, 1, \ldots, N-1$. If $y \in C^{n+1}[a, b]$, this implies that $y^{(n+1)}(t) = f^{(n)}(t, y(t))$ is bounded on $[a, b]$ and that $\tau_i = O(h^n)$ for each $i = 1, 2, \ldots, N$.

EXERCISE SET 5.3

1. Use Taylor's method of order two to approximate the solutions for each of the following initial-value problems.
   a. $y' = te^{3t} - 2y, \quad 0 \le t \le 1, \quad y(0) = 0$, with $h = 0.5$
   b. $y' = 1 + (t - y)^2, \quad 2 \le t \le 3, \quad y(2) = 1$, with $h = 0.5$
   c. $y' = 1 + \dfrac{y}{t}, \quad 1 \le t \le 2, \quad y(1) = 2$, with $h = 0.25$
   d. $y' = \cos 2t + \sin 3t, \quad 0 \le t \le 1, \quad y(0) = 1$, with $h = 0.25$
2. Repeat Exercise 1 using Taylor's method of order four.

3. Use Taylor's method of order two and four to approximate the solution for each of the following initial-value problems.

a. $y' = \dfrac{y}{t} - \left(\dfrac{y}{t}\right)^2, \quad 1 \le t \le 1.2, \quad y(1) = 1$, with $h = 0.1$

b. $y' = \sin t + e^{-t}, \quad 0 \le t \le 1, \quad y(0) = 0$, with $h = 0.5$

c. $y' = \dfrac{1}{t}(y^2 + y), \quad 1 \le t \le 3, \quad y(1) = -2$, with $h = 0.5$

d. $y' = -ty + \dfrac{4t}{y}, \quad 0 \le t \le 1, \quad y(0) = 1$, with $h = 0.25$

4. Use the Taylor method of order two with $h = 0.1$ to approximate the solution to

$$y' = 1 + t\sin(ty), \quad 0 \le t \le 2, \quad y(0) = 0.$$

5. Given the initial-value problem

$$y' = \frac{2}{t}y + t^2 e^t, \quad 1 \le t \le 2, \quad y(1) = 0$$

with exact solution $y(t) = t^2(e^t - e)$:

a. Use Taylor's method of order two with $h = 0.1$ to approximate the solution and compare it with the actual values of y.

b. Use the answers generated in part (a) and linear interpolation to approximate y at the following values and compare them to the actual values of y.

(i) $y(1.04)$ **(ii)** $y(1.55)$ **(iii)** $y(1.97)$

c. Use Taylor's method of order four with $h = 0.1$ to approximate the solution and compare it with the actual values of y.

d. Use the answers generated in part (c) and piecewise cubic Hermite interpolation to approximate y at the following values and compare them to the actual values of y.

(i) $y(1.04)$ **(ii)** $y(1.55)$ **(iii)** $y(1.97)$

6. Given the initial-value problem

$$y' = \frac{1}{t^2} - \frac{y}{t} - y^2, \quad 1 \le t \le 2, \quad y(1) = -1$$

with exact solution $y(t) = -1/t$:

a. Use Taylor's method of order two with $h = 0.05$ to approximate the solution and compare it with the actual values of y.

b. Use the answers generated in part (a) and linear interpolation to approximate the following values of y and compare them to the actual values.

(i) $y(1.052)$ **(ii)** $y(1.555)$ **(iii)** $y(1.978)$

c. Use Taylor's method of order four with $h = 0.05$ to approximate the solution and compare it with the actual values of y.

d. Use the answers generated in part (c) and piecewise cubic Hermite interpolation to approximate the following values of y and compare them to the actual values.

(i) $y(1.052)$ **(ii)** $y(1.555)$ **(iii)** $y(1.978)$

7. A projectile of mass $m = 0.11$ kg shot vertically upward with initial velocity $v(0) = 8$ m/s is slowed due to the force of gravity $F_g = mg$ and due to air resistance $F_r = -kv|v|$, where $g = -9.8$ m/s^2 and $k = 0.002$ kg/m. The differential equation for the velocity v is given by

$$mv' = mg - kv|v|.$$

a. Find the velocity after $0.1, 0.2, \ldots, 1.0$ s.

b. To the nearest tenth of a second, determine when the projectile reaches its maximum height and begins falling.

5.4 Runge-Kutta Methods

The Taylor methods outlined in the previous section have the desirable property of high-order local truncation error, but the disadvantage of requiring the computation and evaluation of the derivatives of $f(t, y)$. This is a complicated and time-consuming procedure for most problems, so the Taylor methods are seldom used in practice.

Runge-Kutta methods have the high-order local truncation error of the Taylor methods while eliminating the need to compute and evaluate the derivatives of $f(t, y)$. Before presenting the ideas behind their derivation, we need to state Taylor's Theorem in two variables. The proof of this result can be found in any standard book on advanced calculus (see, for example, [Fu, p. 331]).

Theorem 5.12 Suppose that $f(t, y)$ and all its partial derivatives of order less than or equal to $n + 1$ are continuous on $D = \{(t, y) \mid a \le t \le b, c \le y \le d\}$, and let $(t_0, y_0) \in D$. For every $(t, y) \in D$, there exists ξ between t and t_0 and μ between y and y_0 with

$$f(t, y) = P_n(t, y) + R_n(t, y),$$

where

$$\begin{aligned} P_n(t, y) = f(t_0, y_0) &+ \left[(t - t_0)\frac{\partial f}{\partial t}(t_0, y_0) + (y - y_0)\frac{\partial f}{\partial y}(t_0, y_0)\right] \\ &+ \left[\frac{(t - t_0)^2}{2}\frac{\partial^2 f}{\partial t^2}(t_0, y_0) + (t - t_0)(y - y_0)\frac{\partial^2 f}{\partial t \partial y}(t_0, y_0)\right. \\ &+ \left.\frac{(y - y_0)^2}{2}\frac{\partial^2 f}{\partial y^2}(t_0, y_0)\right] + \cdots \\ &+ \left[\frac{1}{n!}\sum_{j=0}^{n}\binom{n}{j}(t - t_0)^{n-j}(y - y_0)^j\frac{\partial^n f}{\partial t^{n-j}\partial y^j}(t_0, y_0)\right] \end{aligned}$$

and

$$R_n(t, y) = \frac{1}{(n + 1)!}\sum_{j=0}^{n+1}\binom{n + 1}{j}(t - t_0)^{n+1-j}(y - y_0)^j\frac{\partial^{n+1} f}{\partial t^{n+1-j}\partial y^j}(\xi, \mu).$$ ■

The function $P_n(t, y)$ is called the ***n*th Taylor polynomial in two variables** for the function f about (t_0, y_0), and $R_n(t, y)$ is the remainder term associated with $P_n(t, y)$.

EXAMPLE 1 Figure 5.4 shows the graph of the function

$$f(t, y) = \exp\left[-\frac{(t - 2)^2}{4} - \frac{(y - 3)^2}{4}\right]\cos(2t + y - 7)$$

together with the second Taylor polynomial of f about $(2, 3)$, the polynomial in two variables

$$P_2(t, y) = 1 - \frac{9}{4}(t-2)^2 - 2(t-2)(y-3) - \frac{3}{4}(y-3)^2.$$

The differentiation required to determine this polynomial would be tedious to do by hand. Fortunately, there is a Maple procedure to do the work for us. First we need to initiate the multiple variable Taylor polynomial procedure by entering the command

```
>readlib(mtaylor);
```

which produces the response

proc() . . . end

The Taylor polynomial we need in this example is found by issuing the command

```
>mtaylor(exp(-(t-2)^2/4-(y-3)^2/4)*cos(2*t+y-7),[t=2,y=3],3);
```

The final parameter in this command indicates that we want the second multivariate Taylor polynomial, that is, the quadratic polynomial. If this parameter is 2, we get the linear polynomial, and if it is 1, we get the constant polynomial. When this parameter is omitted, it defaults to 6 and gives the fifth Taylor polynomial.

Figure 5.4

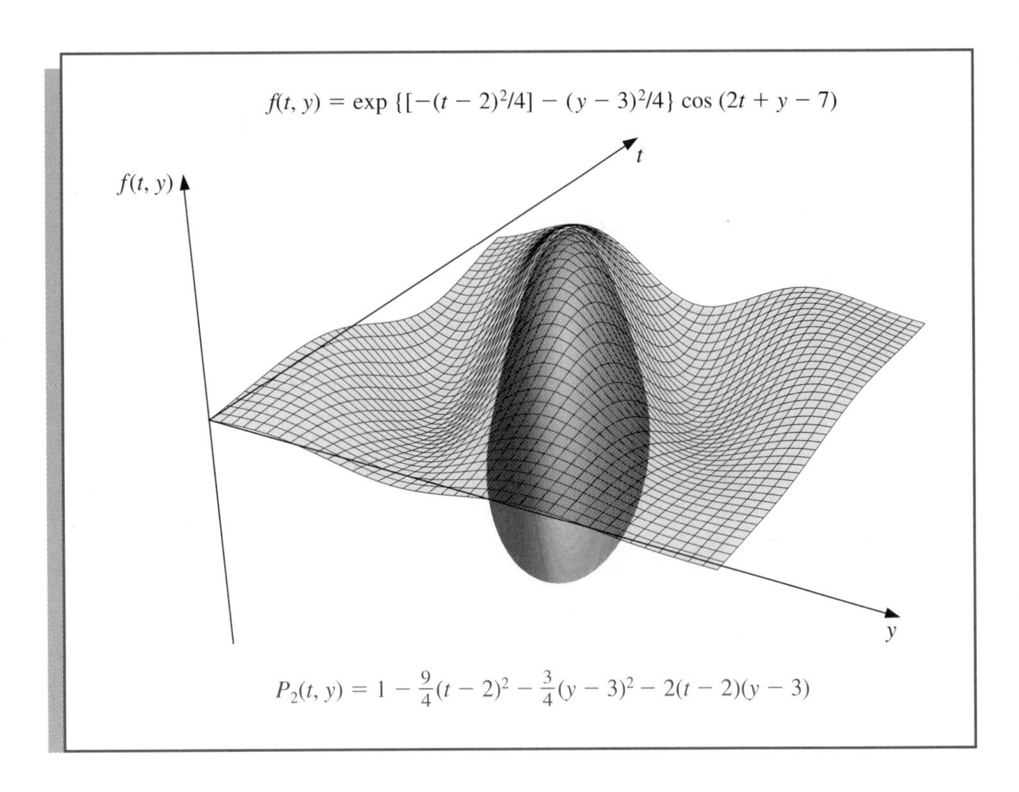

The response from this Maple command is the polynomial

$$1 - \frac{9}{4}(t-2)^2 - 2(t-2)(y-3) - \frac{3}{4}(y-3)^2. \qquad \blacksquare$$

The first step in deriving a Runge-Kutta method is to determine values for a_1, α_1, and β_1 with the property that $a_1 f(t+\alpha_1, y+\beta_1)$ approximates

$$T^{(2)}(t,y) = f(t,y) + \frac{h}{2}f'(t,y)$$

with error no greater than $O(h^2)$, the local truncation error for the Taylor method of order two. Since

$$f'(t,y) = \frac{df}{dt}(t,y) = \frac{\partial f}{\partial t}(t,y) + \frac{\partial f}{\partial y}(t,y)\cdot y'(t) \quad \text{and} \quad y'(t) = f(t,y),$$

this implies

$$\textbf{(5.18)} \qquad T^{(2)}(t,y) = f(t,y) + \frac{h}{2}\frac{\partial f}{\partial t}(t,y) + \frac{h}{2}\frac{\partial f}{\partial y}(t,y)\cdot f(t,y).$$

Expanding $f(t+\alpha_1, y+\beta_1)$ in its Taylor polynomial of degree one about (t, y) gives

$$\textbf{(5.19)} \qquad a_1 f(t+\alpha_1, y+\beta_1) = a_1 f(t,y) + a_1\alpha_1 \frac{\partial f}{\partial t}(t,y) + a_1\beta_1\frac{\partial f}{\partial y}(t,y) + a_1\cdot R_1(t+\alpha_1, y+\beta_1),$$

where

$$\textbf{(5.20)} \qquad R_1(t+\alpha_1, y+\beta_1) = \frac{\alpha_1^2}{2}\frac{\partial^2 f}{\partial t^2}(\xi,\mu) + \alpha_1\beta_1\frac{\partial^2 f}{\partial t\partial y}(\xi,\mu) + \frac{\beta_1^2}{2}\frac{\partial^2 f}{\partial y^2}(\xi,\mu),$$

for some ξ between t and $t+\alpha_1$ and μ between y and $y+\beta_1$.

Matching the coefficients of f and its derivatives in Eqs. (5.18) and (5.19) gives the three equations

$$f(t,y): \quad a_1 = 1; \qquad \frac{\partial f}{\partial t}(t,y): \quad a_1\alpha_1 = \frac{h}{2};$$

and

$$\frac{\partial f}{\partial y}(t,y): \quad a_1\beta_1 = \frac{h}{2}f(t,y).$$

The parameters a_1, α_1, and β_1 are uniquely determined to be

$$a_1 = 1, \quad \alpha_1 = \frac{h}{2}, \quad \text{and} \quad \beta_1 = \frac{h}{2} f(t, y);$$

so

$$T^{(2)}(t, y) = f\left(t + \frac{h}{2}, y + \frac{h}{2} f(t, y)\right) - R_1\left(t + \frac{h}{2}, y + \frac{h}{2} f(t, y)\right),$$

and from Eq. (5.20),

$$R_1\left(t + \frac{h}{2}, y + \frac{h}{2} f(t, y)\right) = \frac{h^2}{8}\frac{\partial^2 f}{\partial t^2}(\xi, \mu) + \frac{h^2}{4} f(t, y)\frac{\partial^2 f}{\partial t \partial y}(\xi, \mu) + \frac{h^2}{8}(f(t, y))^2 \frac{\partial^2 f}{\partial y^2}(\xi, \mu).$$

If all the second-order partial derivatives of f are bounded, then

$$R_1\left(t + \frac{h}{2}, y + \frac{h}{2} f(t, y)\right)$$

is $O(h^2)$, the order of the local truncation error of Taylor method of order two. As a consequence, using the new procedure in place of the Taylor method of order two might add some error, but it does not increase the order of the error.

The difference-equation method resulting from replacing $T^{(2)}(t, y)$ in Taylor's method of order two by $f(t + (h/2), y + (h/2) f(t, y))$ is a specific Runge-Kutta method known as the *Midpoint method.*

Midpoint Method:

$$w_0 = \alpha,$$

$$w_{i+1} = w_i + hf\left(t_i + \frac{h}{2}, w_i + \frac{h}{2} f(t_i, w_i)\right), \quad \text{for each } i = 0, 1, \ldots, N-1.$$

Since only three parameters are present in $a_1 f(t + \alpha_1, y + \beta_1)$ and all are needed in the match of $T^{(2)}$, we need a more complicated form to satisfy the conditions required for any of the higher-order Taylor methods.

The most appropriate four-parameter form for approximating

$$T^{(3)}(t, y) = f(t, y) + \frac{h}{2} f'(t, y) + \frac{h^2}{6} f''(t, y)$$

is

$$a_1 f(t, y) + a_2 f(t + \alpha_2, y + \delta_2 f(t, y)), \tag{5.21}$$

and even with this there is insufficient flexibility to match the term

$$\frac{h^2}{6}\left[\frac{\partial f}{\partial y}(t, y)\right]^2 f(t, y)$$

resulting from the expansion of $(h^2/6)f''(t, y)$. Consequently, the best that can be obtained from using (5.21) are methods with $O(h^2)$ local truncation error. The fact that (5.21) has four parameters, however, gives a flexibility in their choice, so a number of $O(h^2)$ methods can be derived. One of the most important is the *Modified Euler method*, which corresponds to choosing $a_1 = a_2 = \frac{1}{2}$ and $\alpha_2 = \delta_2 = h$ and has the following difference-equation form.

Modified Euler Method:

$$w_0 = \alpha,$$

$$w_{i+1} = w_i + \frac{h}{2}[f(t_i, w_i) + f(t_{i+1}, w_i + hf(t_i, w_i))], \text{ for each } \quad i = 0, 1, 2, \ldots, N-1.$$

The other important $O(h^2)$ method is *Heun's method*, which corresponds to $a_1 = \frac{1}{4}$, $a_2 = \frac{3}{4}$, and $\alpha_2 = \delta_2 = \frac{2}{3}h$, and has the following difference-equation form.

Heun's Method:

$$w_0 = \alpha,$$

$$w_{i+1} = w_i + \frac{h}{4}\left[f(t_i, w_i) + 3f\left(t_i + \frac{2}{3}h, w_i + \frac{2}{3}hf(t_i, w_i)\right)\right],$$

$$\text{for each} \quad i = 0, 1, 2, \ldots, N-1.$$

Both are classified as Runge-Kutta methods of order two, the order of their local truncation error.

EXAMPLE 2 Suppose we apply the Runge-Kutta methods of order two to our usual example,

$$y' = y - t^2 + 1, \quad 0 \le t \le 2, \quad y(0) = 0.5,$$

with $N = 10$, $h = 0.2$, $t_i = 0.2i$, and $w_0 = 0.5$ in each case. The difference equations are

$$\text{Midpoint method:} \quad w_{i+1} = 1.22w_i - 0.0088i^2 - 0.008i + 0.218;$$

$$\text{Modified Euler method:} \quad w_{i+1} = 1.22w_i - 0.0088i^2 - 0.008i + 0.216;$$

$$\text{Heun's method:} \quad w_{i+1} = 1.22w_i - 0.0088i^2 - 0.008i + 0.217\overline{3},$$

for each $i = 0, 1, \ldots, 9$. Table 5.5 lists the results of these calculations. ∎

Although $T^{(3)}(t, y)$ can be approximated with error $O(h^3)$ by an expression of the form

$$f(t + \alpha_1, y + \delta_1 f(t + \alpha_2, y + \delta_2 f(t, y))),$$

Table 5.5

t_i	$y(t_i)$	Midpoint Method	Error	Modified Euler Method	Error	Heun's Method	Error
0.0	0.5000000	0.5000000	0	0.5000000	0	0.5000000	0
0.2	0.8292986	0.8280000	0.0012986	0.8260000	0.0032986	0.8273333	0.0019653
0.4	1.2140877	1.2113600	0.0027277	1.2069200	0.0071677	1.2098800	0.0042077
0.6	1.6489406	1.6446592	0.0042814	1.6372424	0.0116982	1.6421869	0.0067537
0.8	2.1272295	2.1212842	0.0059453	2.1102357	0.0169938	2.1176014	0.0096281
1.0	2.6408591	2.6331668	0.0076923	2.6176876	0.0231715	2.6280070	0.0128521
1.2	3.1799415	3.1704634	0.0094781	3.1495789	0.0303627	3.1635019	0.0164396
1.4	3.7324000	3.7211654	0.0112346	3.6936862	0.0387138	3.7120057	0.0203944
1.6	4.2834838	4.2706218	0.0128620	4.2350972	0.0483866	4.2587802	0.0247035
1.8	4.8151763	4.8009586	0.0142177	4.7556185	0.0595577	4.7858452	0.0293310
2.0	5.3054720	5.2903695	0.0151025	5.2330546	0.0724173	5.2712645	0.0342074

involving four parameters, the algebra involved in the determination of α_1, δ_1, α_2, and δ_2 is quite involved and will not be presented. In fact, the Runge-Kutta method of order three resulting from this expression is not generally used. The most common Runge-Kutta method in use is of order four and, in difference-equation form, is given by the following.

Runge-Kutta Order Four:

$$
\begin{aligned}
w_0 &= \alpha, \\
k_1 &= hf(t_i, w_i), \\
k_2 &= hf\left(t_i + \frac{h}{2}, w_i + \frac{1}{2}k_1\right), \\
k_3 &= hf\left(t_i + \frac{h}{2}, w_i + \frac{1}{2}k_2\right), \\
k_4 &= hf\left(t_{i+1}, w_i + k_3\right), \\
w_{i+1} &= w_i + \frac{1}{6}(k_1 + 2k_2 + 2k_3 + k_4),
\end{aligned}
$$

for each $i = 0, 1, \ldots, N-1$. This method has local truncation error $O(h^4)$, provided the solution $y(t)$ has five continuous derivatives. The reason for introducing the notation k_1, k_2, k_3, k_4 into the method is to eliminate the need for successive nesting in the second variable of $f(t, y)$ (see Exercise 17). Algorithm 5.2 implements the Runge-Kutta method of order four.

ALGORITHM **5.2**

Runge-Kutta (Order Four)

To approximate the solution of the initial-value problem

$$y' = f(t, y), \quad a \le t \le b, \quad y(a) = \alpha,$$

at $(N + 1)$ equally spaced numbers in the interval $[a, b]$:

INPUT endpoints a, b; integer N; initial condition α.

OUTPUT approximation w to y at the $(N + 1)$ values of t.

Step 1 Set $h = (b - a)/N$;
$t = a$;
$w = \alpha$;
OUTPUT (t, w).

Step 2 For $i = 1, 2, \ldots, N$ do Steps 3–5.

Step 3 Set $K_1 = hf(t, w)$;
$K_2 = hf(t + h/2, w + K_1/2)$;
$K_3 = hf(t + h/2, w + K_2/2)$;
$K_4 = hf(t + h, w + K_3)$.

Step 4 Set $w = w + (K_1 + 2K_2 + 2K_3 + K_4)/6$; (*Compute* w_i.)
$t = a + ih$. (*Compute* t_i.)

Step 5 OUTPUT (t, w).

Step 6 STOP.

EXAMPLE 3 Using the Runge-Kutta method of order four to obtain approximations to the solution of the initial-value problem

$$y' = y - t^2 + 1, \quad 0 \leq t \leq 2, \quad y(0) = 0.5,$$

with $h = 0.2$, $N = 10$, and $t_i = 0.2i$ gives the results and errors listed in Table 5.6. ∎

Table 5.6

t_i	Exact $y_i = y(t_i)$	Runge-Kutta Order Four w_i	Error $\lvert y_i - w_i \rvert$
0.0	0.5000000	0.5000000	0
0.2	0.8292986	0.8292933	0.0000053
0.4	1.2140877	1.2140762	0.0000114
0.6	1.6489406	1.6489220	0.0000186
0.8	2.1272295	2.1272027	0.0000269
1.0	2.6408591	2.6408227	0.0000364
1.2	3.1799415	3.1798942	0.0000474
1.4	3.7324000	3.7323401	0.0000599
1.6	4.2834838	4.2834095	0.0000743
1.8	4.8151763	4.8150857	0.0000906
2.0	5.3054720	5.3053630	0.0001089

The main computational effort in applying the Runge-Kutta methods is the evaluation of f. In the second-order methods, the local truncation error is $O(h^2)$, and the cost is two functional evaluations per step. The Runge-Kutta method of order four requires four evaluations per step and the local truncation error is $O(h^4)$. Butcher (see [But] for a summary) has established the relationship between the number of evaluations per step and the order of the local truncation error shown in Table 5.7. This table indicates why the methods of order less than five with smaller step size are used in preference to the higher-order methods using a larger step size.

Table 5.7

Evaluations per step	2	3	4	$5 \leq n \leq 7$	$8 \leq n \leq 9$	$10 \leq n$
Best possible local error	$O(h^2)$	$O(h^3)$	$O(h^4)$	$O(h^{n-1})$	$O(h^{n-2})$	$O(h^{n-3})$

One measure of comparing the lower-order Runge-Kutta methods is described as follows:

> Since the Runge-Kutta method of order four requires four evaluations per step, it should give more accurate answers than Euler's method with one-quarter the step size if it is to be superior. Similarly, if the Runge-Kutta method of order four is to be superior to the second-order Runge-Kutta methods, it should give more accuracy with step size h than a second-order method with step size $\frac{1}{2}h$, because the fourth-order method requires twice as many evaluations per step.

An illustration of the superiority of the Runge-Kutta fourth-order method by this measure is shown in the following example.

EXAMPLE 4 For the problem

$$y' = y - t^2 + 1, \quad 0 \leq t \leq 2, \quad y(0) = 0.5,$$

Euler's method with $h = 0.025$, the Midpoint method with $h = 0.05$, and the Runge-Kutta fourth-order method with $h = 0.1$ are compared at the mesh points 0.1, 0.2, 0.3, 0.4, and 0.5. Each of these techniques requires 20 functional evaluations to determine the values listed in Table 5.8 to approximate $y(0.5)$. In this example, the fourth-order method is clearly superior. ■

Table 5.8

t_i	Exact	Euler $h = 0.025$	Modified Euler $h = 0.05$	Runge-Kutta Order Four $h = 0.1$
0.0	0.5000000	0.5000000	0.5000000	0.5000000
0.1	0.6574145	0.6554982	0.6573085	0.6574144
0.2	0.8292986	0.8253385	0.8290778	0.8292983
0.3	1.0150706	1.0089334	1.0147254	1.0150701
0.4	1.2140877	1.2056345	1.2136079	1.2140869
0.5	1.4256394	1.4147264	1.4250141	1.4256384

Euler's method, the Modified Euler method, and the Runge-Kutta Order Four method can be used in Maple. To implement one of these methods we enter the command

```
>with(share);
>readshare(ODE,plots);
```

To see how the Runge-Kutta Order Four column in Table 5.8 is created using Maple, define the function $f(t, y)$ by

```
>eq:=(t,y)->y-t*t+1;
```

Now invoke the Runge-Kutta method with

```
>eqrk:=rungekutta(eq,[0,0.5],0.1,5);
```

$$
\begin{aligned}
\text{eqrk} := \text{array}\,(0..5, [\\
(0) &= [0, .5]\\
(1) &= [.1, .6574143750]\\
(2) &= [.2, .8292982760]\\
(3) &= [.3, 1.015070059]\\
(4) &= [.4, 1.214086906]\\
(5) &= [.5, 1.425638396]\\
&])
\end{aligned}
$$

In the function call [0, 0.5] refers to the initial condition $y(0) = 0.5$, 0.1 refers to h, and 5 is the number of steps n. The response is an array where the jth item in the list is the pair (t_j, w_j). Accessing an entry in the list is illustrated by the computation of $|y(t_3) - w_3|$ with

```
>evalf(subs(t=0.3,(t+1)^2-0.5*exp(t))-eqrk[3][2]);
```

The first part computes $y(0.3)$ and the last part accesses the second coordinate of the item labeled (3).

Euler's method can be used in a similar manner with the command `firsteuler`, and the Modified Euler method is called with the command `impeuler`.

EXERCISE SET 5.4

1. Use the Modified Euler method to approximate the solutions to each of the following initial-value problems, and compare the results to the actual values.
   a. $y' = te^{3t} - 2y$, $0 \le t \le 1$, $y(0) = 0$, with $h = 0.5$; actual solution $y(t) = \frac{1}{5}te^{3t} - \frac{1}{25}e^{3t} + \frac{1}{25}e^{-2t}$.
   b. $y' = 1 + (t - y)^2$, $2 \le t \le 3$, $y(2) = 1$, with $h = 0.5$; actual solution $y(t) = t + 1/(1 - t)$.
   c. $y' = 1 + \frac{y}{t}$, $1 \le t \le 2$, $y(1) = 2$, with $h = 0.25$; actual solution $y(t) = t \ln t + 2t$.
   d. $y' = \cos 2t + \sin 3t$, $0 \le t \le 1$, $y(0) = 1$, with $h = 0.25$; actual solution $y(t) = \frac{1}{2}\sin 2t - \frac{1}{3}\cos 3t + \frac{4}{3}$.

2. Repeat Exercise 1 using Heun's method.

3. Repeat Exercise 1 using the Midpoint method.

4. Use the Modified Euler method to approximate the solutions to each of the following initial-value problems, and compare the results to the actual values.

a. $y' = \frac{y}{t} - \left(\frac{y}{t}\right)^2$, $1 \le t \le 2$, $y(1) = 1$, with $h = 0.1$; actual solution $y(t) = t/(1 + \ln t)$.

b. $y' = 1 + \frac{y}{t} + \left(\frac{y}{t}\right)^2$, $1 \le t \le 3$, $y(1) = 0$, with $h = 0.2$; actual solution $y(t) = t\tan(\ln t)$.

c. $y' = -(y + 1)(y + 3)$, $0 \le t \le 2$, $y(0) = -2$, with $h = 0.2$; actual solution $y(t) = -3 + 2(1 + e^{-2t})^{-1}$.

d. $y' = -5y + 5t^2 + 2t$, $0 \le t \le 1$, $y(0) = \frac{1}{3}$, with $h = 0.1$; actual solution $y(t) = t^2 + \frac{1}{3}e^{-5t}$.

5. Use the results of Exercise 4 and linear interpolation to approximate values of $y(t)$, and compare the results to the actual values.

a. $y(1.25)$ and $y(1.93)$

b. $y(2.1)$ and $y(2.75)$

c. $y(1.3)$ and $y(1.93)$

d. $y(0.54)$ and $y(0.94)$

6. Repeat Exercise 4 using Heun's method.

7. Repeat Exercise 5 using the results of Exercise 6.

8. Repeat Exercise 4 using the Midpoint method.

9. Repeat Exercise 5 using the results of Exercise 8.

10. Repeat Exercise 1 using the Runge-Kutta method of order four.

11. Repeat Exercise 4 using the Runge-Kutta method of order four.

12. Use the results of Exercise 11 and Cubic Hermite interpolation to approximate values of $y(t)$ and compare the approximations to the actual values.

a. $y(1.25)$ and $y(1.93)$

b. $y(2.1)$ and $y(2.75)$

c. $y(1.3)$ and $y(1.93)$

d. $y(0.54)$ and $y(0.94)$

13. Show that the Midpoint method, the Modified Euler method, and Heun's method give the same approximations to the initial-value problem

$$y' = -y + t + 1, \quad 0 \le t \le 1, \quad y(0) = 1,$$

for any choice of h. Why is this true?

14. Water flows from an inverted conical tank with circular orifice at the rate

$$\frac{dx}{dt} = -0.6\pi r^2 \sqrt{-2g}\frac{\sqrt{x}}{A(x)},$$

where r is the radius of the orifice, x is the height of the liquid level from the vertex of the cone, and $A(x)$ is the area of the cross section of the tank x units above the orifice. Suppose $r = 0.1$ ft, $g = -32.17$ ft/s^2, and the tank has an initial water level of 8 ft and initial volume of $512(\pi/3)$ ft^3.

a. Compute the water level after 10 min with $h = 20$ s.

b. Determine, to within 1 min, when the tank will be empty.

15. The irreversible chemical reaction in which two molecules of solid potassium dichromate ($K_2Cr_2O_7$), two molecules of water (H_2O), and three atoms of solid sulfur (S) combine to yield three molecules of the gas sulfur dioxide (SO_2), four molecules of solid potassium hydroxide (KOH), and two molecules of solid chromic oxide (Cr_2O_3) can be represented symbolically by the stoichiometric equation:

$$2K_2Cr_2O_7 + 2H_2O + 3S \longrightarrow 4KOH + 2Cr_2O_3 + 3SO_2.$$

If n_1 molecules of $K_2Cr_2O_7$, n_2 molecules of H_2O, and n_3 molecules of S are originally available, the following differential equation describes the amount $x(t)$ of KOH after time t:

$$\frac{dx}{dt} = k\left(n_1 - \frac{x}{2}\right)^2\left(n_2 - \frac{x}{2}\right)^2\left(n_3 - \frac{3x}{4}\right)^3,$$

where k is the velocity constant of the reaction. If $k = 6.22 \times 10^{-19}$, $n_1 = n_2 = 2 \times 10^3$, and $n_3 = 3 \times 10^3$, how many units of potassium hydroxide will have been formed after 0.2 s?

16. Show that the difference method

$$w_0 = \alpha,$$

$$w_{i+1} = w_i + a_1 f(t_i, w_i) + a_2 f(t_i + \alpha_2, w_i + \delta_2 f(t_i, w_i)),$$

for each $i = 0, 1, \ldots, N-1$, cannot have local truncation error $O(h^3)$ for any choice of constants a_1, a_2, α_2, and δ_2.

17. The Runge-Kutta method of order four can be written in the form

$$w_0 = \alpha,$$

$$\begin{aligned} w_{i+1} = w_i &+ \frac{h}{6} f(t_i, w_i) + \frac{h}{3} f(t_i + \alpha_1 h, w_i + \delta_1 f(t_i, w_i)) \\ &+ \frac{h}{3} f(t_i + \alpha_2 h, w_i + \delta_2 h f(t_i + \gamma_2 h, w_i + \gamma_3 h f(t_i, w_i))) \\ &+ \frac{h}{6} f(t_i + \alpha_3 h, w_i + \delta_3 h f(t_i + \gamma_4 h, w_i + \gamma_5 h f(t_i + \gamma_6 h, w_i + \gamma_7 h f(t_i, w_i)))). \end{aligned}$$

Find the values of the constants

$$\alpha_1,\ \alpha_2,\ \alpha_3,\ \delta_1,\ \delta_2,\ \delta_3,\ \gamma_2,\ \gamma_3,\ \gamma_4,\ \gamma_5,\ \gamma_6,\ \text{and } \gamma_7.$$

5.5 Error Control and the Runge-Kutta-Fehlberg Method

The appropriate use of varying step size was seen in Section 4.6 to produce integral approximating methods that are efficient in the amount of computation required. In itself, this might not be sufficient to favor these methods due to the increased complication of applying them. However, they have another feature that makes them worthwhile. They incorporate in the step-size procedure an estimate of the truncation error that does not require the approximation of the higher derivatives of the function. These methods are called *adaptive* because they adapt the number and position of the nodes used in the approximation to ensure that the truncation error is kept within a specified bound.

There is a close connection between the problem of approximating the value of a definite integral and that of approximating the solution to an initial-value problem. It is not surprising, then, that there are adaptive methods for approximating the solutions to initial-value problems and that these methods are not only efficient, but also incorporate the control of error.

An ideal difference-equation method

$$w_{i+1} = w_i + h_i\phi(t_i, w_i, h_i), \quad i = 0, 1, \ldots, N-1,$$

for approximating the solution, $y(t)$, to the initial-value problem

$$y' = f(t, y), \quad a \le t \le b, \quad y(a) = \alpha$$

would have the property that, given a tolerance $\varepsilon > 0$, the minimal number of mesh points would be used to ensure that the global error, $|y(t_i) - w_i|$, would not exceed ε for any $i = 0, 1, \ldots, N$. Having a minimal number of mesh points and also controlling the global error of a difference method is, not surprisingly, inconsistent with the points being equally spaced in the interval. In this section we examine techniques used to control the error of a difference-equation method in an efficient manner by the appropriate choice of mesh points.

Although we cannot generally determine the global error of a method, we will see in Section 5.10 that there is a close connection between the local truncation error and the global error. By using methods of differing order we can predict the local truncation error and, using this prediction, choose a step size that will keep the global error in check.

To illustrate the technique, suppose that we have two approximation techniques. The first is an nth-order method obtained from an nth-order Taylor method of the form

$$y(t_{i+1}) = y(t_i) + h\phi(t_i, y(t_i), h) + O(h^{n+1}),$$

producing approximations

$$\begin{aligned} w_0 &= \alpha \\ w_{i+1} &= w_i + h\phi(t_i, w_i, h), \quad \text{for } i > 0, \end{aligned}$$

with local truncation error $\tau_{i+1}(h) = O(h^n)$. In general, the method is generated by applying a Runge-Kutta modification to the Taylor method, but the specific derivation is unimportant here.

The second method is similar but of higher order. For example, let us suppose it comes from an $(n+1)$st-order Taylor method of the form

$$y(t_{i+1}) = y(t_i) + h\tilde{\phi}(t_i, y(t_i), h) + O(h^{n+2}),$$

producing approximations

$$\begin{aligned} \tilde{w}_0 &= \alpha \\ \tilde{w}_{i+1} &= \tilde{w}_i + h\tilde{\phi}(t_i, \tilde{w}_i, h), \quad \text{for } i > 0, \end{aligned}$$

with local truncation error, $\tilde{\tau}_{i+1}(h) = O(h^{n+1})$.

We first make the assumption that $w_i \approx y(t_i) \approx \tilde{w}_i$ and choose a fixed step size h to generate the approximations w_{i+1} and $\tilde{w}_{i+1}$ to $y(t_{i+1})$. Then

$$\tau_{i+1}(h) = \frac{y(t_{i+1}) - y(t_i)}{h} - \phi(t_i, y(t_i), h)$$

$$= \frac{y(t_{i+1}) - w_i}{h} - \phi(t_i, w_i, h)$$

$$= \frac{y(t_{i+1}) - [w_i + h\phi(t_i, w_i, h)]}{h}$$

$$= \frac{1}{h}(y(t_{i+1}) - w_{i+1}).$$

In a similar manner

$$\tilde{\tau}_{i+1}(h) = \frac{1}{h}(y(t_{i+1}) - \tilde{w}_{i+1}).$$

As a consequence,

$$\tau_{i+1}(h) = \frac{1}{h}(y(t_{i+1}) - w_{i+1})$$

$$= \frac{1}{h}\left[(y(t_{i+1}) - \tilde{w}_{i+1}) + (\tilde{w}_{i+1} - w_{i+1})\right]$$

$$= \tilde{\tau}_{i+1}(h) + \frac{1}{h}(\tilde{w}_{i+1} - w_{i+1}).$$

But $\tau_{i+1}(h)$ is $O(h^n)$ and $\tilde{\tau}_{i+1}(h)$ is $O(h^{n+1})$, so the significant portion of $\tau_{i+1}(h)$ must come from $(1/h)[\tilde{w}_{i+1} - w_{i+1}]$. This gives us an easily computed approximation for the local truncation error of the $O(h^n)$ method:

$$\tau_{i+1}(h) \approx \frac{1}{h}[\tilde{w}_{i+1} - w_{i+1}].$$

The object, however, is not simply to estimate the local truncation error but to adjust the step size to keep it within a specified bound. To do this we now assume that since $\tau_{i+1}(h)$ is $O(h^n)$, a number K, independent of h, exists with

$$\tau_{i+1}(h) \approx Kh^n.$$

Then the local truncation error produced by applying the nth-order method with a new step size qh can be estimated using the original approximations w_{i+1} and $\tilde{w}_{i+1}$:

$$\tau_{i+1}(qh) \approx K(qh)^n = q^n(Kh^n) \approx q^n\tau_{i+1}(h) \approx \frac{q^n}{h}(\tilde{w}_{i+1} - w_{i+1}).$$

To bound $\tau_{i+1}(qh)$ by ε, we choose q so that

$$\frac{q^n}{h}|\tilde{w}_{i+1} - w_{i+1}| \approx |\tau_{i+1}(qh)| \leq \varepsilon,$$

that is, so that

$$q \leq \left(\frac{\varepsilon h}{|\tilde{w}_{i+1} - w_{i+1}|}\right)^{1/n}.$$

One popular technique that uses this inequality for error control is called the **Runge-Kutta-Fehlberg method**. (See [Fe].) This technique consists of using a Runge-Kutta method with local truncation error of order five,

$$\tilde{w}_{i+1} = w_i + \frac{16}{135}k_1 + \frac{6656}{12825}k_3 + \frac{28561}{56430}k_4 - \frac{9}{50}k_5 + \frac{2}{55}k_6,$$

to estimate the local error in a Runge-Kutta method of order four given by

$$w_{i+1} = w_i + \frac{25}{216}k_1 + \frac{1408}{2565}k_3 + \frac{2197}{4104}k_4 - \frac{1}{5}k_5,$$

where

$$k_1 = hf(t_i, w_i),$$

$$k_2 = hf\left(t_i + \frac{h}{4}, w_i + \frac{1}{4}k_1\right),$$

$$k_3 = hf\left(t_i + \frac{3h}{8}, w_i + \frac{3}{32}k_1 + \frac{9}{32}k_2\right),$$

$$k_4 = hf\left(t_i + \frac{12h}{13}, w_i + \frac{1932}{2197}k_1 - \frac{7200}{2197}k_2 + \frac{7296}{2197}k_3\right),$$

$$k_5 = hf\left(t_i + h, w_i + \frac{439}{216}k_1 - 8k_2 + \frac{3680}{513}k_3 - \frac{845}{4104}k_4\right),$$

$$k_6 = hf\left(t_i + \frac{h}{2}, w_i - \frac{8}{27}k_1 + 2k_2 - \frac{3544}{2565}k_3 + \frac{1859}{4104}k_4 - \frac{11}{40}k_5\right).$$

An advantage to this method is that only six evaluations of f are required per step. Arbitrary Runge-Kutta methods of order four and five used together require (see Table 5.7 in Section 5.4) at least four evaluations of f for the fourth-order method and an additional six for the fifth-order method, for a total of at least ten functional evaluations.

In the error-control theory, an initial value of h at the ith step was used to find the first values of w_{i+1} and $\tilde{w}_{i+1}$, which led to the determination of q for that step, and then the calculations were repeated. This procedure requires twice the number of functional evaluations per step as without the error control. In practice, the value of q to be used is chosen somewhat differently in order to make the increased functional-evaluation cost worthwhile. The value of q determined at the ith step is used for two purposes:

1. To reject, if necessary, the initial choice of h at the ith step and repeat the calculations using qh, and
2. To predict an appropriate initial choice of h for the $(i + 1)$st step.

Because of the penalty that must be paid in terms of functional evaluations if the steps are repeated, q tends to be chosen conservatively; in fact, for the Runge-Kutta-Fehlberg method with $n = 4$, the usual choice is

$$q = \left(\frac{\varepsilon h}{2|\tilde{w}_{i+1} - w_{i+1}|}\right)^{1/4} = 0.84\left(\frac{\varepsilon h}{|\tilde{w}_{i+1} - w_{i+1}|}\right)^{1/4}.$$

In Algorithm 5.3 for the Runge-Kutta-Fehlberg method, Step 9 is added to eliminate large modifications in step size. This is done to avoid spending too much time with small step sizes in regions with irregularities in the derivatives of y and to avoid large step sizes, which can result in skipping sensitive regions between the steps. In some instances the step-size increase procedure is omitted completely from the algorithm, and the step-size decrease procedure is modified to be incorporated only when needed to bring the error under control.

ALGORITHM 5.3

Runge-Kutta-Fehlberg

To approximate the solution of the initial-value problem

$$y' = f(t, y), \quad a \le t \le b, \quad y(a) = \alpha,$$

with local truncation error within a given tolerance:

INPUT endpoints a, b; initial condition α; tolerance TOL; maximum step size $hmax$; minimum step size $hmin$.

OUTPUT t, w, h where w approximates $y(t)$ and the step size h was used, or a message that the minimum step size was exceeded.

Step 1 Set $t = a$;
$w = \alpha$;
$h = hmax$;
$FLAG = 1$;
OUTPUT (t, w).

Step 2 While $(FLAG = 1)$ do Steps 3–11.

Step 3 Set $K_1 = hf(t, w)$;

$$K_2 = hf\left(t + \tfrac{1}{4}h, w + \tfrac{1}{4}K_1\right);$$

$$K_3 = hf\left(t + \tfrac{3}{8}h, w + \tfrac{3}{32}K_1 + \tfrac{9}{32}K_2\right);$$

$$K_4 = hf\left(t + \tfrac{12}{13}h, w_i + \tfrac{1932}{2197}K_1 - \tfrac{7200}{2197}K_2 + \tfrac{7296}{2197}K_3\right);$$

$$K_5 = hf\left(t + h, w_i + \tfrac{439}{216}K_1 - 8K_2 + \tfrac{3680}{513}K_3 - \tfrac{845}{4104}K_4\right);$$

$$K_6 = hf\left(t + \tfrac{1}{2}h, w - \tfrac{8}{27}K_1 + 2K_2 - \tfrac{3544}{2565}K_3 + \tfrac{1859}{4104}K_4 - \tfrac{11}{40}K_5\right).$$

Step 4 Set $R = \frac{1}{h}\left|\frac{1}{360}K_1 - \frac{128}{4275}K_3 - \frac{2197}{75240}K_4 + \frac{1}{50}K_5 + \frac{2}{55}K_6\right|$.

(*Note:* $R = \frac{1}{h}|\tilde{w}_{i+1} - w_{i+1}|$.)

Step 5 If $R \le TOL$ then do Steps 6 and 7.

Step 6 Set $t = t + h$; (*Approximation accepted.*)

$$w = w + \tfrac{25}{216}K_1 + \tfrac{1408}{2565}K_3 + \tfrac{2197}{4104}K_4 - \tfrac{1}{5}K_5.$$

Step 7 OUTPUT (t, w, h).

Step 8 Set $\delta = 0.84(TOL/R)^{1/4}$.

Step 9 If $\delta \leq 0.1$ then set $h = 0.1h$
else if $\delta \geq 4$ then set $h = 4h$
else set $h = \delta h$. *(Calculate new h.)*

Step 10 If $h > hmax$ then set $h = hmax$.

Step 11 If $t \geq b$ then set $FLAG = 0$
else if $t + h > b$ then set $h = b - t$
else if $h < hmin$ then
set $FLAG = 0$;
OUTPUT ('*minimum h exceeded*').
(*Procedure completed unsuccessfully.*)

Step 12 (*The procedure is complete.*)
STOP.

EXAMPLE 1 Algorithm 5.3 will be used to approximate the solution to the initial-value problem

$$y' = y - t^2 + 1, \quad 0 \leq t \leq 2, \quad y(0) = 0.5,$$

which has solution $y(t) = (t + 1)^2 - 0.5e^t$. The input consists of tolerance $TOL = 10^{-5}$, a maximum step size $hmax = 0.25$, and a minimum step size $hmin = 0.01$. The results are shown in Table 5.9. The last two columns in Table 5.9 show the results of the fifth-order method. For small values of t the error is less than the error in the fourth-order method, but the error exceeds that of the fourth-order method when t increases. ■

Table 5.9

		RKF-4				RKF-5	
t_i	$y_i = y(t_i)$	w_i	h_i	R_i	$\lvert y_i - w_i \rvert$	$\hat{w}_i$	$\lvert y_i - \hat{w}_i \rvert$
0	0.5	0.5				0.5	
0.2500000	0.9204873	0.9204886	0.2500000	6.2×10^{-6}	1.3×10^{-6}	0.9204870	2.424×10^{-7}
0.4865522	1.3964884	1.3964910	0.2365522	4.5×10^{-6}	2.6×10^{-6}	1.3964900	1.510×10^{-6}
0.7293332	1.9537446	1.9537488	0.2427810	4.3×10^{-6}	4.2×10^{-6}	1.9537477	3.136×10^{-6}
0.9793332	2.5864198	2.5864260	0.2500000	3.8×10^{-6}	6.2×10^{-6}	2.5864251	5.242×10^{-6}
1.2293332	3.2604520	3.2604605	0.2500000	2.4×10^{-6}	8.5×10^{-6}	3.2604599	7.895×10^{-6}
1.4793332	3.9520844	3.9520955	0.2500000	7×10^{-7}	1.11×10^{-5}	3.9520954	1.096×10^{-5}
1.7293332	4.6308127	4.6308268	0.2500000	1.5×10^{-6}	1.41×10^{-5}	4.6308272	1.446×10^{-5}
1.9793332	5.2574687	5.2574861	0.2500000	4.3×10^{-6}	1.73×10^{-5}	5.2574871	1.839×10^{-5}
2.0000000	5.3054720	5.3054896	0.0206668		1.77×10^{-5}	5.3054896	1.768×10^{-5}

An implementation of the Runge-Kutta-Fehlberg method is available in Maple using the `dsolve` command with the numeric option. Consider the initial-value problem of Example 1. The command

```
>g:=dsolve({D(y)(t)=y(t)-t*t+1,y(0)=0.5},y(t),numeric);
```

returns the procedure

$$g := \text{proc}(rkf45\_x) \ldots \text{end}$$

We can evaluate y as shown by example, using

```
>g(2.0);
```

which gives

$$[t = 2., y(t) = 5.305471958400194]$$

EXERCISE SET 5.5

1. Use the Runge-Kutta-Fehlberg method with tolerance $TOL = 10^{-4}$, $hmax = 0.25$, and $hmin = 0.05$ to approximate the solutions to the following initial-value problems. Compare the results to the actual values.
   a. $y' = te^{3t} - 2y, \quad 0 \le t \le 1, \quad y(0) = 0$; actual solution $y(t) = \frac{1}{5}te^{3t} - \frac{1}{25}e^{3t} + \frac{1}{25}e^{-2t}$.
   b. $y' = 1 + (t - y)^2, \quad 2 \le t \le 3, \quad y(2) = 1$; actual solution $y(t) = t + 1/(1 - t)$.
   c. $y' = 1 + \frac{y}{t}, \quad 1 \le t \le 2, \quad y(1) = 2$; actual solution $y(t) = t \ln t + 2t$.
   d. $y' = \cos 2t + \sin 3t, \quad 0 \le t \le 1, \quad y(0) = 1$; actual solution $y(t) = \frac{1}{2}\sin 2t - \frac{1}{3}\cos 3t + \frac{4}{3}$.
2. Use the Runge-Kutta Fehlberg Algorithm with tolerance $TOL = 10^{-4}$ to approximate the solution to the following initial-value problems.
   a. $y' = \left(\frac{y}{t}\right)^2 + \frac{y}{t}, \quad 1 \le t \le 1.2, \quad y(1) = 1$, with $hmax = 0.05$ and $hmin = 0.02$.
   b. $y' = \sin t + e^{-t}, \quad 0 \le t \le 1, \quad y(0) = 0$, with $hmax = 0.25$ and $hmin = 0.02$.
   c. $y' = \frac{1}{t}(y^2 + y), \quad 1 \le t \le 3, \quad y(1) = -2$, with $hmax = 0.5$ and $hmin = 0.02$.
   d. $y' = t^2, \quad 0 \le t \le 2, \quad y(0) = 0$, with $hmax = 0.5$ and $hmin = 0.02$.
3. Use the Runge-Kutta-Fehlberg method with tolerance $TOL = 10^{-6}$, $hmax = 0.5$, and $hmin = 0.05$ to approximate the solutions to the following initial-value problems. Compare the results to the actual values.
   a. $y' = \frac{y}{t} - \frac{y^2}{t^2}, \quad 1 \le t \le 4, \quad y(1) = 1$; actual solution $y(t) = \frac{t}{1 + \ln t}$.
   b. $y' = 1 + \frac{y}{t} + \left(\frac{y}{t}\right)^2, \quad 1 \le t \le 3, \quad y(1) = 0$; actual solution $y(t) = t\tan(\ln t)$.
   c. $y' = -(y + 1)(y + 3), \quad 0 \le t \le 3, \quad y(0) = -2$; actual solution $y(t) = -3 + 2(1 + e^{-2t})^{-1}$.
   d. $y' = (t + 2t^3)y^3 - ty, \quad 0 \le t \le 2, \quad y(0) = \frac{1}{3}$; actual solution $y(t) = (3 + 2t^2 + 6e^{t^2})^{-1/2}$.
4. The Runge-Kutta-Verner method is based on the formulas

$$w_{i+1} = w_i + \frac{13}{160}k_1 + \frac{2375}{5984}k_3 + \frac{5}{16}k_4 + \frac{12}{85}k_5 + \frac{3}{44}k_6 \quad \text{and}$$

$$\tilde{w}_{i+1} = w_i + \frac{3}{40}k_1 + \frac{875}{2244}k_3 + \frac{23}{72}k_4 + \frac{264}{1955}k_5 + \frac{125}{11592}k_7 + \frac{43}{616}k_8,$$

where

$$k_1 = hf(t_i, w_i),$$

$$k_2 = hf\left(t_i + \frac{h}{6}, w_i + \frac{1}{6}k_1\right),$$

$$k_3 = hf\left(t_i + \frac{4h}{15}, w_i + \frac{4}{75}k_1 + \frac{16}{75}k_2\right),$$

$$k_4 = hf\left(t_i + \frac{2h}{3}, w_i + \frac{5}{6}k_1 - \frac{8}{3}k_2 + \frac{5}{2}k_3\right),$$

$$k_5 = hf\left(t_i + \frac{5h}{6}, w_i - \frac{165}{64}k_1 + \frac{55}{6}k_2 - \frac{425}{64}k_3 + \frac{85}{96}k_4\right),$$

$$k_6 = hf\left(t_i + h, w_i + \frac{12}{5}k_1 - 8k_2 + \frac{4015}{612}k_3 - \frac{11}{36}k_4 + \frac{88}{255}k_5\right),$$

$$k_7 = hf\left(t_i + \frac{h}{15}, w_i - \frac{8263}{15000}k_1 + \frac{124}{75}k_2 - \frac{643}{680}k_3 - \frac{81}{250}k_4 + \frac{2484}{10625}k_5\right),$$

$$k_8 = hf\left(t_i + h, w_i + \frac{3501}{1720}k_1 - \frac{300}{43}k_2 + \frac{297275}{52632}k_3 - \frac{319}{2322}k_4 + \frac{24068}{84065}k_5 + \frac{3850}{26703}k_7\right).$$

The sixth-order method $\tilde{w}_{i+1}$ is used to estimate the error in the fifth-order method w_{i+1}. Construct an algorithm similar to the Runge-Kutta-Fehlberg Algorithm and repeat Exercise 3 using this new method.

5. In the theory of the spread of contagious disease (see [Ba1] or [Ba2]), a relatively elementary differential equation can be used to predict the number of infective individuals in the population at any time, provided appropriate simplification assumptions are made. In particular, let us assume that all individuals in a fixed population have an equally-likely chance of being infected and once infected remain in that state. If we let $x(t)$ denote the number of susceptible individuals at time t and $y(t)$ denote the number of infectives, it is reasonable to assume that the rate at which the number of infectives changes is proportional to the product of $x(t)$ and $y(t)$ since the rate depends on both the number of infectives and the number of susceptibles present at that time. If the population is large enough to assume that $x(t)$ and $y(t)$ are continuous variables, the problem can be expressed

$$\frac{dy}{dt}(t) = kx(t)y(t),$$

where k is a constant and $x(t) + y(t) = m$, the total population. This equation can be rewritten involving only $y(t)$ as

$$\frac{dy}{dt}(t) = k(m - y(t))y(t).$$

a. Assuming that $m = 100,000$, $y(0) = 1000$, $k = 2 \times 10^{-6}$, and that time is measured in days, find an approximation to the number of infective individuals at the end of 30 days.

b. The differential equation in part (a) is called a *Bernoulli equation* and can be transformed into a linear differential equation in $u(t)$ by letting $u(t) = (y(t))^{-1}$. Use this technique to find the exact solution to the equation, under the same assumptions as in part (a), and compare the true value of $y(t)$ to the approximation given there. What is $\lim_{t\to\infty} y(t)$? Does this agree with your intuition?

6. In the previous exercise, all infected individuals remained in the population to spread the disease. A more realistic proposal is to introduce a third variable $z(t)$ to represent the number of individuals who are removed from the affected population at a given time t by isolation, recovery and consequent immunity, or death. This quite naturally complicates the problem, but it can be shown (see [Ba2]) that an approximate solution can be given in the form

$$x(t) = x(0)e^{-(k_1/k_2)z(t)} \quad \text{and} \quad y(t) = m - x(t) - z(t),$$

where k_1 is the infective rate, k_2 is the removal rate, and $z(t)$ is determined from the differential equation

$$\frac{dz}{dt}(t) = k_2\left(m - z(t) - x(0)e^{-(k_1/k_2)z(t)}\right).$$

The authors are not aware of any technique for solving this problem directly, so a numerical procedure must be applied. Find an approximation to $z(30)$, $y(30)$, and $x(30)$, assuming that $m = 100,000$, $x(0) = 99,000$, $k_1 = 2 \times 10^{-6}$, and $k_2 = 10^{-4}$.

5.6 Multistep Methods

The methods discussed to this point in the chapter are called **one-step methods** because the approximation for the mesh point t_{i+1} involves information from only one of the previous mesh points, t_i. Although these methods can use functional evaluation information at points between t_i and t_{i+1}, they do not retain that information for direct use in future approximations. All the information used by these methods is obtained within the subinterval over which the solution is being approximated.

Since the approximate solution is available at each of the mesh points $t_0, t_1, \ldots, t_i$ before the approximation at t_{i+1} is obtained and because the error $|w_j - y(t_j)|$ tends to increase with j, it seems reasonable to develop methods that use these more accurate previous data when approximating the solution at t_{i+1}.

Methods using the approximation at more than one previous mesh point to determine the approximation at the next point are called *multistep* methods. The precise definition of these methods follows, together with the definition of the two types of multistep methods.

Definition 5.13 An ***m*-step multistep method** for solving the initial-value problem

$$y' = f(t, y), \quad a \le t \le b, \quad y(a) = \alpha \tag{5.22}$$

is one whose difference equation for finding the approximation w_{i+1} at the mesh point t_{i+1} can be represented by the following equation, where m is an integer greater than 1:

$$\begin{aligned} w_{i+1} &= a_{m-1}w_i + a_{m-2}w_{i-1} + \cdots + a_0 w_{i+1-m} \\ &\quad + h[b_m f(t_{i+1}, w_{i+1}) + b_{m-1} f(t_i, w_i) \\ &\quad + \cdots + b_0 f(t_{i+1-m}, w_{i+1-m})], \end{aligned} \tag{5.23}$$

for $i = m-1, m, \ldots, N-1$, where $h = (b-a)/N$, the $a_0, a_1, \ldots, a_{m-1}$ and $b_0, b_1, \ldots, b_m$ are constants, and the starting values

$$w_0 = \alpha, \quad w_1 = \alpha_1, \quad w_2 = \alpha_2, \quad \ldots, \quad w_{m-1} = \alpha_{m-1}$$

are specified. ■

When $b_m = 0$, the method is called **explicit**, or **open**, since Eq. (5.23) then gives w_{i+1} explicitly in terms of previously determined values. When $b_m \neq 0$, the method is called **implicit**, or **closed**, since w_{i+1} occurs on both sides of Eq. (5.23) and is specified only implicitly.

EXAMPLE 1 The equations

(5.24) $$w_0 = \alpha, \quad w_1 = \alpha_1, \quad w_2 = \alpha_2, \quad w_3 = \alpha_3,$$

$$w_{i+1} = w_i + \frac{h}{24}\left[55f(t_i, w_i) - 59f(t_{i-1}, w_{i-1}) + 37f(t_{i-2}, w_{i-2}) - 9f(t_{i-3}, w_{i-3})\right]$$

for each $i = 3, 4, \ldots, N-1$, define an explicit four-step method known as the **fourth-order Adams-Bashforth technique**. The equations

(5.25) $$w_0 = \alpha, \quad w_1 = \alpha_1, \quad w_2 = \alpha_2,$$

$$w_{i+1} = w_i + \frac{h}{24}\left[9f(t_{i+1}, w_{i+1}) + 19f(t_i, w_i) - 5f(t_{i-1}, w_{i-1}) + f(t_{i-2}, w_{i-2})\right],$$

for each $i = 2, 3, \ldots, N-1$, define an implicit three-step method known as the **fourth-order Adams-Moulton technique**. ■

The starting values in either (5.24) or (5.25) must be specified, generally by assuming $w_0 = \alpha$ and generating the remaining values by either a Runge-Kutta method or some other one-step technique.

To apply an implicit method such as (5.25) directly, we must solve the implicit equation for w_{i+1}. It is not clear that this can be done in general or that a unique solution for w_{i+1} will always be obtained.

To begin the derivation of the multistep methods, note that the solution to the initial-value problem (5.22), if integrated over the interval $[t_i, t_{i+1}]$, has the property that

$$y(t_{i+1}) - y(t_i) = \int_{t_i}^{t_{i+1}} y'(t)\,dt = \int_{t_i}^{t_{i+1}} f(t, y(t))\,dt.$$

Consequently,

(5.26) $$y(t_{i+1}) = y(t_i) + \int_{t_i}^{t_{i+1}} f(t, y(t))\,dt.$$

Since we cannot integrate $f(t, y(t))$ without knowing $y(t)$, the solution to the problem, we instead integrate an interpolating polynomial $P(t)$ to $f(t, y(t))$ that is determined by some of the previously obtained data points $(t_0, w_0), (t_1, w_1), \ldots, (t_i, w_i)$. When we assume, in addition, that $y(t_i) \approx w_i$, Eq. (5.26) becomes

(5.27) $$y(t_{i+1}) \approx w_i + \int_{t_i}^{t_{i+1}} P(t)\,dt.$$

Although any form of the interpolating polynomial can be used for the derivation, it is most convenient to use the Newton backward-difference formula.

To derive an Adams-Bashforth explicit m-step technique, we form the backward-difference polynomial $P_{m-1}(t)$ through $(t_i, f(t_i, y(t_i)))$, $(t_{i-1}, f(t_{i-1}, y(t_{i-1})))$, $\ldots$, $(t_{i+1-m}, f(t_{i+1-m}, y(t_{i+1-m})))$. Since $P_{m-1}(t)$ is an interpolatory polynomial of degree $m-1$, some number ξ_i in (t_{i+1-m}, t_i) exists with

$$f(t, y(t)) = P_{m-1}(t) + \frac{f^{(m)}(\xi_i, y(\xi_i))}{m!}(t - t_i)(t - t_{i-1}) \cdots (t - t_{i+1-m}).$$

Introducing the variable substitution $t = t_i + sh$, with $dt = h\,ds$ into $P_{m-1}(t)$ and the error term implies that

$$\begin{aligned}
\int_{t_i}^{t_{i+1}} f(t, y(t))\,dt &= \int_{t_i}^{t_{i+1}} \sum_{k=0}^{m-1} (-1)^k \binom{-s}{k} \nabla^k f(t_i, y(t_i))\,dt \\
&\quad + \int_{t_i}^{t_{i+1}} \frac{f^{(m)}(\xi_i, y(\xi_i))}{m!}(t - t_i)(t - t_{i-1}) \cdots (t - t_{i+1-m})\,dt \\
&= \sum_{k=0}^{m-1} \nabla^k f(t_i, y(t_i)) h (-1)^k \int_0^1 \binom{-s}{k} ds \\
&\quad + \frac{h^{m+1}}{m!} \int_0^1 s(s+1) \cdots (s+m-1) f^{(m)}(\xi_i, y(\xi_i))\,ds.
\end{aligned}$$

The integrals $(-1)^k \int_0^1 \binom{-s}{k} ds$ for various values of k are easily evaluated and are listed in Table 5.10. For example, when $k = 3$,

$$\begin{aligned}
(-1)^3 \int_0^1 \binom{-s}{3} ds &= -\int_0^1 \frac{(-s)(-s-1)(-s-2)}{1 \cdot 2 \cdot 3}\,ds \\
&= \frac{1}{6} \int_0^1 (s^3 + 3s^2 + 2s)\,ds \\
&= \frac{1}{6}\left[\frac{s^4}{4} + s^3 + s^2\right]_0^1 = \frac{1}{6}\left(\frac{9}{4}\right) = \frac{3}{8}.
\end{aligned}$$

Table 5.10

k	0	1	2	3	4	5
$(-1)^k \int_0^1 \binom{-s}{k} ds$	1	$\frac{1}{2}$	$\frac{5}{12}$	$\frac{3}{8}$	$\frac{251}{720}$	$\frac{95}{288}$

As a consequence,

$$\text{(5.28)}\quad \int_{t_i}^{t_{i+1}} f(t, y(t))\,dt = h\left[f(t_i, y(t_i)) + \frac{1}{2}\nabla f(t_i, y(t_i)) + \frac{5}{12}\nabla^2 f(t_i, y(t_i)) + \cdots\right]$$
$$+ \frac{h^{m+1}}{m!}\int_0^1 s(s+1)\cdots(s+m-1)f^{(m)}(\xi_i, y(\xi_i))\,ds.$$

Since $s(s+1)\cdots(s+m-1)$ does not change sign on $[0, 1]$, the Weighted Mean Value Theorem for Integrals can be used to deduce that for some number μ_i, where $t_{i+1-m} < \mu_i < t_{i+1}$, the error term in Eq. (5.28) becomes

$$\frac{h^{m+1}}{m!}\int_0^1 s(s+1)\cdots(s+m-1)f^{(m)}(\xi_i, y(\xi_i))\,ds$$
$$= \frac{h^{m+1}f^{(m)}(\mu_i, y(\mu_i))}{m!}\int_0^1 s(s+1)\cdots(s+m-1)\,ds$$

or

$$\text{(5.29)}\qquad h^{m+1}f^{(m)}(\mu_i, y(\mu_i))(-1)^m\int_0^1 \binom{-s}{m}\,ds.$$

Since $y(t_{i+1}) - y(t_i) = \int_{t_i}^{t_{i+1}} f(t, y(t))\,dt$, Eq. (5.28) can be written as

$$\text{(5.30)}\quad y(t_{i+1}) = y(t_i) + h\left[f(t_i, y(t_i)) + \frac{1}{2}\nabla f(t_i, y(t_i)) + \frac{5}{12}\nabla^2 f(t_i, y(t_i)) + \cdots\right]$$
$$+ h^{m+1}f^{(m)}(\mu_i, y(\mu_i))(-1)^m\int_0^1 \binom{-s}{m}\,ds.$$

EXAMPLE 2 To derive the three-step Adams-Bashforth technique, consider Eq. (5.30) with $m = 3$:

$$y(t_{i+1}) \approx y(t_i) + h\left[f(t_i, y(t_i)) + \frac{1}{2}\nabla f(t_i, y(t_i)) + \frac{5}{12}\nabla^2 f(t_i, y(t_i))\right]$$
$$= y(t_i) + h\left\{f(t_i, y(t_i)) + \frac{1}{2}[f(t_i, y(t_i)) - f(t_{i-1}, y(t_{i-1}))]\right.$$
$$\left. + \frac{5}{12}[f(t_i, y(t_i)) - 2f(t_{i-1}, y(t_{i-1})) + f(t_{i-2}, y(t_{i-2}))]\right\}$$
$$= y(t_i) + \frac{h}{12}[23f(t_i, y(t_i)) - 16f(t_{i-1}, y(t_{i-1})) + 5f(t_{i-2}, y(t_{i-2}))].$$

The three-step Adams-Bashforth method is, consequently,

$$w_0 = \alpha, \quad w_1 = \alpha_1, \quad w_2 = \alpha_2,$$
$$w_{i+1} = w_i + \frac{h}{12}[23f(t_i, w_i) - 16f(t_{i-1}, w_{i-1}) + 5f(t_{i-2}, w_{i-2})]$$

for $i = 2, 3, \ldots, N-1$. ■

Multistep methods can also be derived by using Taylor series. An example of the procedure involved is considered in Exercise 10. A derivation using a Lagrange interpolating polynomial is discussed in Exercise 9.

The local truncation error for multistep methods is defined analogously to that of one-step methods. As in the case of one-step methods, the local truncation error provides a measure of how the solution to the differential equation fails to solve the difference equation.

Definition 5.14 If $y(t)$ is the solution to the initial-value problem

$$y' = f(t, y), \quad a \leq t \leq b, \quad y(a) = \alpha,$$

and

$$\begin{aligned} w_{i+1} &= a_{m-1}w_i + a_{m-2}w_{i-1} + \cdots + a_0 w_{i+1-m} \\ &\quad + h[b_m f(t_{i+1}, w_{i+1}) + b_{m-1}f(t_i, w_i) + \cdots + b_0 f(t_{i+1-m}, w_{i+1-m})] \end{aligned}$$

is the $(i+1)$st step in a multistep method, the **local truncation error** at this step is

$$\begin{aligned} \tau_{i+1}(h) &= \frac{y(t_{i+1}) - a_{m-1}y(t_i) - \cdots - a_0 y(t_{i+1-m})}{h} \\ &\quad - [b_m f(t_{i+1}, y(t_{i+1})) + \cdots + b_0 f(t_{i+1-m}, y(t_{i+1-m}))], \end{aligned} \qquad \textbf{(5.31)}$$

for each $i = m-1, m, \ldots, N-1$. ■

EXAMPLE 3 To determine the local truncation error for the three-step Adams-Bashforth method derived in Example 2, consider the form of the error given in Eq. (5.29):

$$h^4 f^{(3)}(\mu_i, y(\mu_i))(-1)^3 \int_0^1 \binom{-s}{3} ds = \frac{3h^4}{8} f^{(3)}(\mu_i, y(\mu_i)).$$

Using the fact that $f^{(3)}(\mu_i, y(\mu_i)) = y^{(4)}(\mu_i)$ and the difference equation derived in Example 2, we have

$$\begin{aligned} \tau_{i+1}(h) &= \frac{y(t_{i+1}) - y(t_i)}{h} - \frac{1}{12}[23f(t_i, y(t_i)) - 16f(t_{i-1}, y(t_{i-1})) + 5f(t_{i-2}, y(t_{i-2}))] \\ &= \frac{1}{h}\left[\frac{3h^4}{8} f^{(3)}(\mu_i, y(\mu_i))\right] = \frac{3h^3}{8} y^{(4)}(\mu_i), \quad \text{for some } \mu_i \in (t_{i-2}, t_{i+1}). \end{aligned}$$ ■

Some of the explicit multistep methods together with their required starting values and local truncation errors are as follows. The derivation of these techniques is similar to the procedure in Examples 2 and 3.

Adams-Bashforth Two-Step Method:

$$\begin{aligned} w_0 &= \alpha, \quad w_1 = \alpha_1, \\ w_{i+1} &= w_i + \frac{h}{2}[3f(t_i, w_i) - f(t_{i-1}, w_{i-1})], \end{aligned} \qquad \textbf{(5.32)}$$

where $i = 1, 2, \ldots, N - 1$. The local truncation error is $\tau_{i+1}(h) = \frac{5}{12}y'''(\mu_i)h^2$, for some $\mu_i \in (t_{i-1}, t_{i+1})$.

Adams-Bashforth Three-Step Method:

(5.33)
$$w_0 = \alpha, \quad w_1 = \alpha_1, \quad w_2 = \alpha_2,$$
$$w_{i+1} = w_i + \frac{h}{12}[23f(t_i, w_i) - 16f(t_{i-1}, w_{i-1}) + 5f(t_{i-2}, w_{i-2})],$$

where $i = 2, 3, \ldots, N - 1$. The local truncation error is $\tau_{i+1}(h) = \frac{3}{8}y^{(4)}(\mu_i)h^3$, for some $\mu_i \in (t_{i-2}, t_{i+1})$.

Adams-Bashforth Four-Step Method:

(5.34)
$$w_0 = \alpha, \quad w_1 = \alpha_1, \quad w_2 = \alpha_2, \quad w_3 = \alpha_3,$$
$$w_{i+1} = w_i + \frac{h}{24}[55f(t_i, w_i) - 59f(t_{i-1}, w_{i-1}) + 37f(t_{i-2}, w_{i-2}) - 9f(t_{i-3}, w_{i-3})],$$

where $i = 3, 4, \ldots, N - 1$. The local truncation error is $\tau_{i+1}(h) = \frac{251}{720}y^{(5)}(\mu_i)h^4$, for some $\mu_i \in (t_{i-3}, t_{i+1})$.

Adams-Bashforth Five-Step Method:

(5.35)
$$w_0 = \alpha, \quad w_1 = \alpha_1, \quad w_2 = \alpha_2, \quad w_3 = \alpha_3, \quad w_4 = \alpha_4,$$
$$w_{i+1} = w_i + \frac{h}{720}[1901f(t_i, w_i) - 2774f(t_{i-1}, w_{i-1})$$
$$+ 2616f(t_{i-2}, w_{i-2}) - 1274f(t_{i-3}, w_{i-3}) + 251f(t_{i-4}, w_{i-4})],$$

where $i = 4, 5, \ldots, N - 1$. The local truncation error is $\tau_{i+1}(h) = \frac{95}{288}y^{(6)}(\mu_i)h^5$, for some $\mu_i \in (t_{i-4}, t_{i+1})$.

Implicit methods are derived by using $(t_{i+1}, f(t_{i+1}, y(t_{i+1})))$ as an additional interpolation node in the approximation of the integral

$$\int_{t_i}^{t_{i+1}} f(t, y(t))\, dt.$$

Some of the more common implicit methods are as follows.

Adams-Moulton Two-Step Method:

(5.36)
$$w_0 = \alpha, \quad w_1 = \alpha_1,$$
$$w_{i+1} = w_i + \frac{h}{12}[5f(t_{i+1}, w_{i+1}) + 8f(t_i, w_i) - f(t_{i-1}, w_{i-1})],$$

where $i = 1, 2, \ldots, N - 1$. The local truncation error is $\tau_{i+1}(h) = -\frac{1}{24}y^{(4)}(\mu_i)h^3$, for some $\mu_i \in (t_{i-1}, t_{i+1})$.

Adams-Moulton Three-Step Method:

(5.37)
$$w_0 = \alpha, \quad w_1 = \alpha_1, \quad w_2 = \alpha_2,$$
$$w_{i+1} = w_i + \frac{h}{24}[9f(t_{i+1}, w_{i+1}) + 19f(t_i, w_i) - 5f(t_{i-1}, w_{i-1}) + f(t_{i-2}, w_{i-2})],$$

where $i = 2, 3, \ldots, N-1$. The local truncation error is $\tau_{i+1}(h) = -\frac{19}{720}y^{(5)}(\mu_i)h^4$, for some $\mu_i \in (t_{i-2}, t_{i+1})$.

Adams-Moulton Four-Step Method:

(5.38)
$$w_0 = \alpha, \quad w_1 = \alpha_1, \quad w_2 = \alpha_2, \quad w_3 = \alpha_3,$$
$$w_{i+1} = w_i + \frac{h}{720}[251f(t_{i+1}, w_{i+1}) + 646f(t_i, w_i) - 264f(t_{i-1}, w_{i-1}) + 106f(t_{i-2}, w_{i-2}) - 19f(t_{i-3}, w_{i-3})],$$

where $i = 3, 4, \ldots, N-1$. The local truncation error is $\tau_{i+1}(h) = -\frac{3}{160}y^{(6)}(\mu_i)h^5$, for some $\mu_i \in (t_{i-3}, t_{i+1})$.

It is interesting to compare an m-step Adams-Bashforth explicit method to an $(m-1)$-step Adams-Moulton implicit method. Both involve m evaluations of f per step, and both have the terms $y^{(m+1)}(\mu_i)h^m$ in their local truncation errors. In general, the coefficients of the terms involving f and in the local truncation error are smaller for the implicit methods than for the explicit methods. This leads to greater stability and smaller roundoff errors for the implicit methods.

EXAMPLE 4 Consider the initial-value problem

$$y' = y - t^2 + 1, \quad 0 \le t \le 2, \quad y(0) = 0.5,$$

and the approximations given by the Adams-Bashforth four-step method and the Adams-Moulton three-step method, both using $h = 0.2$.

The Adams-Bashforth method has the difference equation

$$w_{i+1} = w_i + \frac{h}{24}[55f(t_i, w_i) - 59f(t_{i-1}, w_{i-1}) + 37f(t_{i-2}, w_{i-2}) - 9f(t_{i-3}, w_{i-3})]$$

for $i = 3, 4, \ldots, 9$, which, when simplified using $f(t, y) = y - t^2 + 1$, $h = 0.2$, and $t_i = 0.2i$, becomes

$$w_{i+1} = \frac{1}{24}[35w_i - 11.8w_{i-1} + 7.4w_{i-2} - 1.8w_{i-3} - 0.192i^2 - 0.192i + 4.736].$$

The Adams-Moulton method has the difference equation

$$w_{i+1} = w_i + \frac{h}{24}[9f(t_{i+1}, w_{i+1}) + 19f(t_i, w_i) - 5f(t_{i-1}, w_{i-1}) + f(t_{i-2}, w_{i-2})],$$

for $i = 2, 3, \ldots, 9$, which reduces to

$$w_{i+1} = \frac{1}{24}[1.8w_{i+1} + 27.8w_i - w_{i-1} + 0.2w_{i-2} - 0.192i^2 - 0.192i + 4.736].$$

To use this method explicitly, we solve for w_{i+1}, which gives

$$w_{i+1} = \frac{1}{22.2}[27.8w_i - w_{i-1} + 0.2w_{i-2} - 0.192i^2 - 0.192i + 4.736].$$

for $i = 2, 3, \ldots, 9$.

The results in Table 5.11 were obtained using the exact values from $y(t) = (t+1)^2 - 0.5e^t$ for α, α_1, α_2, and α_3 in the Adams-Bashforth case and for α, α_1, and α_2 in the Adams-Moulton case. ■

Table 5.11

t_i	Exact	Adams-Bashforth w_i	Error	Adams-Moulton w_i	Error
0.0	0.5000000				
0.2	0.8292986				
0.4	1.2140877				
0.6	1.6489406			1.6489341	0.0000065
0.8	2.1272295	2.1273124	0.0000828	2.1272136	0.0000160
1.0	2.6408591	2.6410810	0.0002219	2.6408298	0.0000293
1.2	3.1799415	3.1803480	0.0004065	3.1798937	0.0000478
1.4	3.7324000	3.7330601	0.0006601	3.7323270	0.0000731
1.6	4.2834838	4.2844931	0.0010093	4.2833767	0.0001071
1.8	4.8151763	4.8166575	0.0014812	4.8150236	0.0001527
2.0	5.3054720	5.3075838	0.0021119	5.3052587	0.0002132

In Example 4 the implicit Adams-Moulton method gave better results than the explicit Adams-Bashforth method of the same order. Although this is generally the case, the implicit methods have the inherent weakness of first having to convert the method algebraically to an explicit representation for w_{i+1}. This procedure is not always possible, as can be seen by considering the elementary initial-value problem

$$y' = e^y, \quad 0 \le t \le 0.25, \quad y(0) = 1.$$

Since $f(t, y) = e^y$, the three-step Adams-Moulton method has

$$w_{i+1} = w_i + \frac{h}{24}[9e^{w_{i+1}} + 19e^{w_i} - 5e^{w_{i-1}} + e^{w_{i-2}}]$$

as its difference equation, and this equation cannot be solved explicitly for w_{i+1}.

In practice, implicit multistep methods are not used as described here. Rather, they are used to improve approximations obtained by explicit methods. The combination of an explicit and implicit technique is called a **predictor-corrector method**. The explicit method predicts an approximation, and the implicit method corrects this prediction.

Consider the following fourth-order method for solving an initial-value problem. The first step is to calculate the starting values w_0, w_1, w_2, and w_3 for the four-step Adams-Bashforth method. To do this, we use a fourth-order one-step method, the Runge-Kutta method of order four. The next step is to calculate an approximation, $w_4^{(0)}$, to $y(t_4)$ using the Adams-Bashforth method as predictor:

$$w_4^{(0)} = w_3 + \frac{h}{24}[55f(t_3, w_3) - 59f(t_2, w_2) + 37f(t_1, w_1) - 9f(t_0, w_0)].$$

This approximation is improved by inserting $w_4^{(0)}$ in the right side of the three-step Adams-Moulton method and using that method as a corrector:

$$w_4^{(1)} = w_3 + \frac{h}{24}[9f(t_4, w_4^{(0)}) + 19f(t_3, w_3) - 5f(t_2, w_2) + f(t_1, w_1)].$$

The only new function evaluation required in this procedure is $f(t_4, w_4^{(0)})$ in the corrector equation. All the other values of f have been calculated for earlier approximations.

The value $w_4^{(1)}$ is then used as the approximation to $y(t_4)$, and the technique of using the Adams-Bashforth method as a predictor and the Adams-Moulton method as a corrector is repeated to find $w_5^{(0)}$ and $w_5^{(1)}$, the initial and final approximations to $y(t_5)$, etc.

Improved approximations to $y(t_{i+1})$ can be obtained by iterating the Adams-Moulton formula

$$w_{i+1}^{(k+1)} = w_i + \frac{h}{24}[9f(t_{i+1}, w_{i+1}^{(k)}) + 19f(t_i, w_i) - 5f(t_{i-1}, w_{i-1}) + f(t_{i-2}, w_{i-2})].$$

However, $\{w_{i+1}^{(k+1)}\}$ converges to the approximation given by the implicit formula rather than to the solution $y(t_{i+1})$, and it is usually more efficient to use a reduction in the step size if improved accuracy is needed.

Algorithm 5.4 is based on the fourth-order Adams-Bashforth method as predictor and one iteration of the Adams-Moulton method as corrector, with the starting values obtained from the fourth-order Runge-Kutta method.

ALGORITHM 5.4

Adams Fourth-Order Predictor-Corrector

To approximate the solution of the initial-value problem

$$y' = f(t, y), \quad a \le t \le b, \quad y(a) = \alpha,$$

at $(N + 1)$ equally spaced numbers in the interval $[a, b]$:

INPUT endpoints a, b; integer N; initial condition α.

OUTPUT approximation w to y at the $(N + 1)$ values of t.

Step 1 Set $h = (b - a)/N$;
$t_0 = a$;
$w_0 = \alpha$;
OUTPUT (t_0, w_0).

Step 2 For $i = 1, 2, 3$, do Steps 3–5. *(Compute starting values using Runge-Kutta method.)*

Step 3 Set $K_1 = hf(t_{i-1}, w_{i-1})$;
$K_2 = hf(t_{i-1} + h/2, w_{i-1} + K_1/2)$;
$K_3 = hf(t_{i-1} + h/2, w_{i-1} + K_2/2)$;
$K_4 = hf(t_{i-1} + h, w_{i-1} + K_3)$.

Step 4 Set $w_i = w_{i-1} + (K_1 + 2K_2 + 2K_3 + K_4)/6$;
$t_i = a + ih$.

Step 5 OUTPUT (t_i, w_i).

Step 6 For $i = 4, \ldots, N$ do Steps 7–10.

Step 7 Set $t = a + ih$;
$w = w_3 + h[55f(t_3, w_3) - 59f(t_2, w_2) + 37f(t_1, w_1) - 9f(t_0, w_0)]/24$; *(Predict* w_i.*)*
$w = w_3 + h[9f(t, w) + 19f(t_3, w_3) - 5f(t_2, w_2) + f(t_1, w_1)]/24$. *(Correct* w_i.*)*

Step 8 OUTPUT (t, w).

Step 9 For $j = 0, 1, 2$
set $t_j = t_{j+1}$; *(Prepare for next iteration.)*
$w_j = w_{j+1}$.

Step 10 Set $t_3 = t$;
$w_3 = w$.

Step 11 STOP.

EXAMPLE 5 Table 5.12 lists the results obtained by using Algorithm 5.4 for the initial-value problem

$$y' = y - t^2 + 1, \quad 0 \le t \le 2, \quad y(0) = 0.5$$

with $N = 10$. ∎

Other multistep methods can be derived using integration of interpolating polynomials over intervals of the form $[t_j, t_{i+1}]$, for $j \le i - 1$, to obtain an approximation to $y(t_{i+1})$. When an interpolating polynomial is integrated over $[t_{i-3}, t_{i+1}]$, the result is an explicit technique known as **Milne's method**,

$$w_{i+1} = w_{i-3} + \frac{4h}{3}[2f(t_i, w_i) - f(t_{i-1}, w_{i-1}) + 2f(t_{i-2}, w_{i-2})],$$

which has local truncation error $\frac{14}{45}h^4 y^{(5)}(\xi_i)$, for some $\xi_i \in (t_{i-3}, t_{i+1})$.

Table 5.12

t_i	$y_i = y(t_i)$	w_i	Error $\lvert y_i - w_i \rvert$
0.0	0.5000000	0.5000000	0
0.2	0.8292986	0.8292933	0.0000053
0.4	1.2140877	1.2140762	0.0000114
0.6	1.6489406	1.6489220	0.0000186
0.8	2.1272295	2.1272056	0.0000239
1.0	2.6408591	2.6408286	0.0000305
1.2	3.1799415	3.1799026	0.0000389
1.4	3.7324000	3.7323505	0.0000495
1.6	4.2834838	4.2834208	0.0000630
1.8	4.8151763	4.8150964	0.0000799
2.0	5.3054720	5.3053707	0.0001013

This method is occasionally used as a predictor for the implicit **Simpson's method**,

$$w_{i+1} = w_{i-1} + \frac{h}{3}[f(t_{i+1}, w_{i+1}) + 4f(t_i, w_i) + f(t_{i-1}, w_{i-1})],$$

which has local truncation error $-(h^4/90)y^{(5)}(\xi_i)$, for some $\xi_i \in (t_{i-1}, t_{i+1})$, and is obtained by integrating an interpolating polynomial over $[t_{i-1}, t_{i+1}]$.

The local truncation error involved with a predictor-corrector method of the Milne-Simpson type is generally smaller than that of the Adams-Bashforth-Moulton method. But the technique has limited use because of problems of stability, which do not occur with the Adams procedure. Elaboration on this difficulty is given in Section 5.10.

EXERCISE SET 5.6

1. Use all the Adams-Bashforth methods to approximate the solutions to the following initial-value problems. In each case use exact starting values and compare the results to the actual values.
   a. $y' = te^{3t} - 2y$, $0 \le t \le 1$, $y(0) = 0$, with $h = 0.2$; actual solution $y(t) = \frac{1}{5}te^{3t} - \frac{1}{25}e^{3t} + \frac{1}{25}e^{-2t}$.
   b. $y' = 1 + (t - y)^2$, $2 \le t \le 3$, $y(2) = 1$, with $h = 0.2$; actual solution $y(t) = t + 1/(1 - t)$.
   c. $y' = 1 + \frac{y}{t}$, $1 \le t \le 2$, $y(1) = 2$, with $h = 0.2$; actual solution $y(t) = t \ln t + 2t$.
   d. $y' = \cos 2t + \sin 3t$, $0 \le t \le 1$, $y(0) = 1$, with $h = 0.2$; actual solution $y(t) = \frac{1}{2}\sin 2t - \frac{1}{3}\cos 3t + \frac{4}{3}$.
2. Use all the Adams-Moulton methods to approximate the solutions to the Exercises 1(a), 1(c), and 1(d). In each case use exact starting values and explicitly solve for w_{i+1}. Compare the results to the actual values.

3. Use each of the Adams-Bashforth methods to approximate the solutions to the following initial-value problems. In each case use starting values obtained from the Runge-Kutta method of order four. Compare the results to the actual values.

a. $y' = \frac{y}{t} - \left(\frac{y}{t}\right)^2$, $1 \le t \le 2$, $y(1) = 1$, with $h = 0.1$; actual solution $y(t) = \frac{t}{1 + \ln t}$.

b. $y' = 1 + \frac{y}{t} + \left(\frac{y}{t}\right)^2$, $1 \le t \le 3$, $y(1) = 0$, with $h = 0.2$; actual solution $y(t) = t \tan(\ln t)$.

c. $y' = -(y + 1)(y + 3)$, $0 \le t \le 2$, $y(0) = -2$, with $h = 0.1$; actual solution $y(t) = -3 + \frac{2}{1 + e^{-2t}}$.

d. $y' = -5y + 5t^2 + 2t$, $0 \le t \le 1$, $y(0) = 1/3$, with $h = 0.1$; actual solution $y(t) = t^2 + \frac{1}{3}e^{-5t}$.

4. Use Algorithm 5.4 to approximate the solutions to the initial-value problems in Exercise 1.

5. Use Algorithm 5.4 to approximate the solutions to the initial-value problems in Exercise 3.

6. Change Algorithm 5.4 so that the corrector can be iterated for a given number p iterations. Repeat Exercise 5 with $p = 2, 3$, and 4 iterations. Which choice of p gives the best answer for each initial-value problem?

7. The initial-value problem

$$y' = e^y, \quad 0 \le t \le 0.20, \quad y(0) = 1$$

has solution

$$y(t) = 1 - \ln(1 - et).$$

Applying the three-step Adams-Moulton method to this problem is equivalent to finding the fixed point w_{i+1} of

$$g(w) = w_i + \frac{h}{24}[9e^w + 19e^{w_i} - 5e^{w_{i-1}} + e^{w_{i-2}}].$$

a. With $h = 0.01$, obtain w_{i+1} by functional iteration for $i = 2, \ldots, 19$ using exact starting values w_0, w_1, and w_2. At each step use w_i to initially approximate w_{i+1}.

b. Will Newton's method speed the convergence over functional iteration?

8. Use the Milne-Simpson Predictor-Corrector method to approximate the solutions to the initial-value problems in Exercise 3.

9. **a.** Derive Eq. (5.32) by using the Lagrange form of the interpolating polynomial.

b. Derive Eq. (5.34) by using Newton's backward-difference form of the interpolating polynomial.

10. Derive Eq. (5.33) by the following method. Set

$$y(t_{i+1}) = y(t_i) + ahf(t_i, y(t_i)) + bhf(t_{i-1}, y(t_{i-1})) + chf(t_{i-2}, y(t_{i-2}))$$

Expand $y(t_{i+1})$, $f(t_{i-2}, y(t_{i-2}))$, and $f(t_{i-1}, y(t_{i-1}))$ in Taylor series about $(t_i, y(t_i))$, and equate the coefficients of h, h^2 and h^3 to obtain a, b, and c.

11. Derive Eq. (5.36) and its local truncation error by using an appropriate form of an interpolating polynomial.

12. Derive Simpson's method by applying Simpson's rule to the integral

$$y(t_{i+1}) - y(t_{i-1}) = \int_{t_{i-1}}^{t_{i+1}} f(t, y(t))\, dt.$$

13. Derive Milne's method by applying the open Newton-Cotes formula (4.29) to the integral

$$y(t_{i+1}) - y(t_{i-3}) = \int_{t_{i-3}}^{t_{i+1}} f(t, y(t))\, dt.$$

14. Verify the entries in Table 5.10.

5.7 Variable Step-Size Multistep Methods

The Runge-Kutta-Fehlberg method is used for error control because at each step it provides, at little additional cost, *two* approximations that can be compared and related to the local error. Predictor-corrector techniques always generate two approximations at each step, so they are natural candidates for error-control adaptation.

To demonstrate the error-control procedure, we will construct a variable step-size predictor-corrector method using the four-step Adams-Bashforth method as predictor and the three-step Adams-Moulton method as corrector.

The Adams-Bashforth four-step method comes from the relation

$$y(t_{i+1}) = y(t_i) + \frac{h}{24}[55f(t_i, y(t_i)) - 59f(t_{i-1}, y(t_{i-1})) + 37f(t_{i-2}, y(t_{i-2})) - 9f(t_{i-3}, y(t_{i-3}))] + \frac{251}{720}y^{(5)}(\hat{\mu}_i)h^5,$$

for some $\hat{\mu}_i \in (t_{i-3}, t_{i+1})$. The assumption that the approximations $w_0, w_1, \ldots, w_i$ are all exact implies that the Adams-Bashforth truncation error is

$$\frac{y(t_{i+1}) - w_{i+1}^{(0)}}{h} = \frac{251}{720}y^{(5)}(\hat{\mu}_i)h^4. \tag{5.39}$$

A similar analysis of the Adams-Moulton three-step method, which comes from

$$y(t_{i+1}) = y(t_i) + \frac{h}{24}[9f(t_{i+1}, y(t_{i+1})) + 19f(t_i, y(t_i)) - 5f(t_{i-1}, y(t_{i-1})) + f(t_{i-2}, y(t_{i-2}))] - \frac{19}{720}y^{(5)}(\tilde{\mu}_i)h^4,$$

for some $\tilde{\mu}_i \in (t_{i-2}, t_{i+1})$ leads to the local truncation error

$$\frac{y(t_{i+1}) - w_{i+1}}{h} = -\frac{19}{720}y^{(5)}(\tilde{\mu}_i)h^4. \tag{5.40}$$

To proceed further, we must make the assumption that for small values of h,

$$y^{(5)}(\hat{\mu}_i) \approx y^{(5)}(\tilde{\mu}_i).$$

The effectiveness of the error-control technique depends directly on this assumption. If we subtract Eq. (5.40) from Eq. (5.39), we have

$$\frac{w_{i+1} - w_{i+1}^{(0)}}{h} = \frac{h^4}{720}\left[251y^{(5)}(\hat{\mu}_i) + 19y^{(5)}(\tilde{\mu}_i)\right] \approx \frac{3}{8}h^4 y^{(5)}(\tilde{\mu}_i),$$

so

(5.41)
$$y^{(5)}(\tilde{\mu}_i) \approx \frac{8}{3h^5}(w_{i+1} - w_{i+1}^{(0)}).$$

Using this result to eliminate the term involving $h^4 y^{(5)}(\tilde{\mu}_i)$ from (5.40) gives the approximation to the error

$$|\tau_{i+1}(h)| = \frac{|y(t_{i+1}) - w_{i+1}|}{h} \approx \frac{19h^4}{720} \cdot \frac{8}{3h^5}|w_{i+1} - w_{i+1}^{(0)}| = \frac{19|w_{i+1} - w_{i+1}^{(0)}|}{270h}.$$

Suppose we now reconsider (5.40) with a new step size qh generating new approximations $\hat{w}_{i+1}^{(0)}$ and $\hat{w}_{i+1}$. The object is to choose q so that the local truncation error given in (5.40) is bounded by a prescribed tolerance ε. If we assume that the value $y^{(5)}(\mu)$ in (5.40) associated with qh is also approximated using (5.41), then

$$\frac{|y(t_i + qh) - \hat{w}_{i+1}|}{qh} = \frac{19}{720}|y^{(5)}(\mu)|q^4h^4 \approx \frac{19}{720}\left[\frac{8}{3h^5}|w_{i+1} - w_{i+1}^{(0)}|\right] q^4h^4,$$

and we need to choose q so that

$$\frac{|y(t_i + qh) - \hat{w}_{i+1}|}{qh} \approx \frac{19}{270}\frac{|w_{i+1} - w_{i+1}^{(0)}|}{h}q^4 < \varepsilon.$$

That is, we choose q so that

$$q < \left(\frac{270}{19}\frac{h\varepsilon}{|w_{i+1} - w_{i+1}^{(0)}|}\right)^{1/4} \approx 2\left(\frac{h\varepsilon}{|w_{i+1} - w_{i+1}^{(0)}|}\right)^{1/4}.$$

A number of approximation assumptions have been made in this development, so in practice q is chosen conservatively, usually as

$$q = 1.5\left(\frac{h\varepsilon}{|w_{i+1} - w_{i+1}^{(0)}|}\right)^{1/4}.$$

A change in step size for a multistep method is more costly in terms of function evaluations than for a one-step method since new, equally-spaced starting values must be computed. As a consequence, it is common practice to ignore the step-size change whenever the local truncation error is between $\varepsilon/10$ and ε, that is, when

$$\frac{\varepsilon}{10} < |\tau_{i+1}(h)| = \frac{|y(t_{i+1}) - w_{i+1}|}{h} \approx \frac{19|w_{i+1} - w_{i+1}^{(0)}|}{270h} < \varepsilon.$$

In addition, q is generally given an upper bound to ensure that a single unusually accurate approximation does not result in too large a step size. Algorithm 5.5 incorporates this safeguard with an upper bound of 4.

Remember that since the multistep methods require equal step sizes for the starting values, any change in step size necessitates recalculating new starting values at that point. In Algorithm 5.5 this is done by calling a Runge-Kutta subalgorithm (Algorithm 5.2).

ALGORITHM **5.5**

Adams Variable Step-Size Predictor-Corrector

To approximate the solution of the initial-value problem

$$y' = f(t, y), \quad a \leq t \leq b, \quad y(a) = \alpha$$

with local truncation error within a given tolerance:

INPUT endpoints a, b; initial condition α; tolerance TOL; maximum step size $hmax$; minimum step size $hmin$.

OUTPUT i, t_i, w_i, h where at the ith step w_i approximates $y(t_i)$ and the step size h was used, or a message that the minimum step size was exceeded.

Step 1 Set up a subalgorithm for the Runge-Kutta fourth-order method to be called $RK4(h, v_0, x_0, v_1, x_1, v_2, x_2, v_3, x_3)$ that accepts as input a step size h and starting values $v_0 \approx y(x_0)$ and returns $\{(x_j, v_j) | j = 1, 2, 3\}$ defined by the following:

for $j = 1, 2, 3$

set $K_1 = hf(x_{j-1}, v_{j-1})$;

$K_2 = hf(x_{j-1} + h/2, v_{j-1} + K_1/2)$

$K_3 = hf(x_{j-1} + h/2, v_{j-1} + K_2/2)$

$K_4 = hf(x_{j-1} + h, v_{j-1} + K_3)$

$v_j = v_{j-1} + (K_1 + 2K_2 + 2K_3 + K_4)/6$;

$x_j = x_0 + jh$.

Step 2 Set $t_0 = a$;

$w_0 = \alpha$;

$h = hmax$;

$FLAG = 1$; (*FLAG will be used to exit the loop in Step 4.*)

$LAST = 0$; (*LAST will indicate when the last value is calculated.*)

OUTPUT (t_0, w_0).

Step 3 Call $RK4(h, w_0, t_0, w_1, t_1, w_2, t_2, w_3, t_3)$;

Set $NFLAG = 1$; (*Indicates computation from RK4.*)

$i = 4$;

$t = t_3 + h$.

Step 4 While ($FLAG = 1$) do Steps 5–20.

Step 5 Set $WP = w_{i-1} + \frac{h}{24}[55f(t_{i-1}, w_{i-1}) - 59f(t_{i-2}, w_{i-2})$

$+ 37f(t_{i-3}, w_{i-3}) - 9f(t_{i-4}, w_{i-4})]$; (*Predict w_i.*)

$$WC = w_{i-1} + \frac{h}{24}[9f(t, WP) + 19f(t_{i-1}, w_{i-1})$$
$$- 5f(t_{i-2}, w_{i-2}) + f(t_{i-3}, w_{i-3})]; \quad (\textit{Correct } w_i.)$$
$$\sigma = 19|WC - WP|/(270h).$$

Step 6 If $\sigma \leq TOL$ then do Steps 7–16 (*Result accepted.*)
else do Steps 17–19. (*Result rejected.*)

Step 7 Set $w_i = WC$; (*Result accepted.*)
$t_i = t$.

Step 8 If $NFLAG = 1$ then for $j = i - 3, i - 2, i - 1, i$
OUTPUT (j, t_j, w_j, h);
(*Previous results also accepted.*)
else OUTPUT (i, t_i, w_i, h).
(*Previous results already accepted.*)

Step 9 If $LAST = 1$ then set $FLAG = 0$ (*Next step is* 20.)
else do Steps 10–16.

Step 10 Set $i = i + 1$;
$NFLAG = 0$.

Step 11 If $\sigma \leq 0.1\ TOL$ or $t_{i-1} + h > b$ then do Steps 12–16.
(*Increase h if it is more accurate than required or decrease h to include b as a mesh point.*)

Step 12 Set $q = (TOL/(2\sigma))^{1/4}$.

Step 13 If $q > 4$ then set $h = 4h$
else set $h = qh$.

Step 14 If $h > hmax$ then set $h = hmax$.

Step 15 If $t_{i-1} + 4h > b$ then
set $h = (b - t_{i-1})/4$;
$LAST = 1$.

Step 16 Call $RK4(h, w_{i-1}, t_{i-1}, w_i, t_i, w_{i+1}, t_{i+1}, w_{i+2}, t_{i+2})$;
Set $NFLAG = 1$;
$i = i + 3$. (*True branch completed. Next step is 20.*)

Step 17 Set $q = (TOL/(2\sigma))^{1/4}$. (*False branch from Step 6: Result rejected.*)

Step 18 If $q < 0.1$ then set $h = 0.1h$
else set $h = qh$.

Step 19 If $h < hmin$ then set $FLAG = 0$;
OUTPUT ('*hmin* exceeded')
else
if $NFLAG = 1$ then set $i = i - 3$;
(*Previous results also rejected.*)
Call $RK4(h, w_{i-1}, t_{i-1}, w_i, t_i, w_{i+1}, t_{i+1}, w_{i+2}, t_{i+2})$;

set $i = i + 3$;
$NFLAG = 1$.

Step 20 Set $t = t_{i-1} + h$.

Step 21 STOP.

EXAMPLE 1 Table 5.13 lists the results obtained using Algorithm 5.5 to find approximations to the solution of the initial-value problem

$$y' = y - t^2 + 1, \quad 0 \le t \le 2, \quad y(0) = 0.5,$$

which has solution $y(t) = (t+1)^2 - 0.5e^t$. Included in the input is the tolerance $TOL = 10^{-5}$, maximum step size $hmax = 0.25$, and minimum step size $hmin = 0.01$. ■

Table 5.13

t_i	$y(t_i)$	w_i	h_i	σ_i	$\lvert y(t_i) - w_i \rvert$
0	0.5	0.5			
0.1257017	0.7002323	0.7002318	0.1257017	4.051×10^{-6}	0.0000005
0.2514033	0.9230960	0.9230949	0.1257017	4.051×10^{-6}	0.0000011
0.3771050	1.1673894	1.1673877	0.1257017	4.051×10^{-6}	0.0000017
0.5028066	1.4317502	1.4317480	0.1257017	4.051×10^{-6}	0.0000022
0.6285083	1.7146334	1.7146306	0.1257017	4.610×10^{-6}	0.0000028
0.7542100	2.0142869	2.0142834	0.1257017	5.210×10^{-6}	0.0000035
0.8799116	2.3287244	2.3287200	0.1257017	5.913×10^{-6}	0.0000043
1.0056133	2.6556930	2.6556877	0.1257017	6.706×10^{-6}	0.0000054
1.1313149	2.9926385	2.9926319	0.1257017	7.604×10^{-6}	0.0000066
1.2570166	3.3366642	3.3366562	0.1257017	8.622×10^{-6}	0.0000080
1.3827183	3.6844857	3.6844761	0.1257017	9.777×10^{-6}	0.0000097
1.4857283	3.9697541	3.9697433	0.1030100	7.029×10^{-6}	0.0000108
1.5887383	4.2527830	4.2527711	0.1030100	7.029×10^{-6}	0.0000120
1.6917483	4.5310269	4.5310137	0.1030100	7.029×10^{-6}	0.0000133
1.7947583	4.8016639	4.8016488	0.1030100	7.029×10^{-6}	0.0000151
1.8977683	5.0615660	5.0615488	0.1030100	7.760×10^{-6}	0.0000172
1.9233262	5.1239941	5.1239764	0.0255579	3.918×10^{-8}	0.0000177
1.9488841	5.1854932	5.1854751	0.0255579	3.918×10^{-8}	0.0000181
1.9744421	5.2460056	5.2459870	0.0255579	3.918×10^{-8}	0.0000186
2.0000000	5.3054720	5.3054529	0.0255579	3.918×10^{-8}	0.0000191

EXERCISE SET 5.7

1. Use the Adams Variable Step-Size Predictor-Corrector Algorithm with tolerance $TOL = 10^{-4}$, $hmax = 0.25$, and $hmin = 0.025$ to approximate the solutions to the given initial-value problems. Compare the results to the actual values.
   a. $y' = te^{3t} - 2y, \quad 0 \le t \le 1, \quad y(0) = 0;$ actual solution $y(t) = \frac{1}{5}te^{3t} - \frac{1}{25}e^{3t} + \frac{1}{25}e^{-2t}$.
   b. $y' = 1 + (t - y)^2, \quad 2 \le t \le 3, \quad y(2) = 1;$ actual solution $y(t) = t + 1/(1 - t)$.

c. $y' = 1 + y/t, \quad 1 \leq t \leq 2, \quad y(1) = 2;$ actual solution $y(t) = t \ln t + 2t$.

d. $y' = \cos 2t + \sin 3t, \quad 0 \leq t \leq 1, \quad y(0) = 1;$ actual solution $y(t) = \frac{1}{2}\sin 2t - \frac{1}{3}\cos 3t + \frac{4}{3}$.

2. Use the Adams Variable Step-Size Predictor-Corrector Algorithm with $TOL = 10^{-4}$ to approximate the solutions to the following initial-value problems:

a. $y' = \left(\frac{y}{t}\right)^2 + \left(\frac{y}{t}\right), \quad 1 \leq t \leq 1.2, \quad y(1) = 1$, with $hmax = 0.05$ and $hmin = 0.01$.

b. $y' = \sin t + e^{-t}, \quad 0 \leq t \leq 1, \quad y(0) = 0$, with $hmax = 0.2$ and $hmin = 0.01$.

c. $y' = \frac{1}{t}(y^2 + y), \quad 1 \leq t \leq 3, \quad y(1) = -2$, with $hmax = 0.4$ and $hmin = 0.01$.

d. $y' = -ty + \frac{4t}{y}, \quad 0 \leq t \leq 1, \quad y(0) = 1$, with $hmax = 0.2$ and $hmin = 0.01$.

3. Use the Adams Variable Step-Size Predictor-Corrector Algorithm with tolerance $TOL = 10^{-6}$, $hmax = 0.5$, and $hmin = 0.02$ to approximate the solutions to the given initial-value problems. Compare the results to the actual values.

a. $y' = \frac{y}{t} - \frac{y^2}{t^2}, \quad 1 \leq t \leq 4, \quad y(1) = 1;$ actual solution $y(t) = t/(1 + \ln t)$.

b. $y' = 1 + \frac{y}{t} + \left(\frac{y}{t}\right)^2, \quad 1 \leq t \leq 3, \quad y(1) = 0;$ actual solution $y(t) = t \tan(\ln t)$.

c. $y' = -(y + 1)(y + 3), \quad 0 \leq t \leq 3, \quad y(0) = -2;$ actual solution $y(t) = -3 + 2(1 + e^{-2t})^{-1}$.

d. $y' = (t + 2t^3)y^3 - ty, \quad 0 \leq t \leq 2, \quad y(0) = \frac{1}{3};$ actual solution $y(t) = (3 + 2t^2 + 6e^{t^2})^{-1/2}$.

4. Construct an Adams Variable Step-Size Predictor-Corrector Algorithm based on the Adams-Bashforth five-step method and the Adams-Moulton four-step method. Repeat Exercise 3 using this new method.

5. An electrical circuit consists of a capacitor of constant capacitance $C = 1.1$ farads in series with a resistor of constant resistance $R_0 = 2.1$ ohms. A voltage $\mathcal{E}(t) = 110 \sin t$ is applied at time $t = 0$. When the resistor heats up, the resistance becomes a function of the current i,

$$R(t) = R_0 + ki, \quad \text{where } k = 0.9,$$

and the differential equation for i becomes

$$\left(1 + \frac{2k}{R_0}i\right)\frac{di}{dt} + \frac{1}{R_0 C}i = \frac{1}{R_0}\frac{d\mathcal{E}}{dt}.$$

Find the current i after 2 s, assuming $i(0) = 0$.

5.8 Extrapolation Methods

Extrapolation was used in Section 4.5 for the approximation of definite integrals, where we found that by correctly averaging relatively inaccurate trapezoidal approximations we could produce new approximations that are exceedingly accurate. In this section we will apply extrapolation to increase the accuracy of approximations to the solution of initial-value problems. As we have previously seen, the original approximations must have an error expansion of a specific form for the procedure to be successful.

To apply extrapolation to solve initial-value problems, we use a technique based on the Midpoint method:

(5.42)
$$w_{i+1} = w_{i-1} + 2hf(t_i, w_i), \quad \text{for } i \geq 1.$$

This technique requires two starting values, since both w_0 and w_1 are needed before the first midpoint approximation, w_2, can be determined. As usual, we use the initial condition for $w_0 = y(a) = \alpha$. To determine the second starting value, w_1, we apply Euler's method. Subsequent approximations are obtained from (5.42). After a series of approximations of this type are generated ending at a value t, an endpoint correction is performed that involves the final two midpoint approximations. This produces an approximation $w(t, h)$ to $y(t)$ that has the form

(5.43)
$$y(t) = w(t, h) + \sum_{k=1}^{\infty} \delta_k h^{2k},$$

where the δ_k are constants related to the derivatives of the solution $y(t)$. The important point is that the δ_k do not depend on the step size h. The details of this procedure can be found in the paper by Gragg [Gr].

To illustrate the extrapolation technique for solving

$$y'(t) = f(t, y), \quad a \leq t \leq b, \quad y(a) = \alpha,$$

let us assume that we have a fixed step size h and that we wish to approximate $y(t_1) = y(a + h)$.

For the first extrapolation step we let $h_0 = h/2$ and use Euler's method with $w_0 = \alpha$ to approximate $y(a + h_0) = y(a + h/2)$ as

$$w_1 = w_0 + h_0 f(a, w_0).$$

We then apply the Midpoint method with $t_{i-1} = a$ and $t_i = a + h_0 = a + h/2$ to produce a first approximation to $y(a + h) = y(a + 2h_0)$,

$$w_2 = w_0 + 2h_0 f(a + h_0, w_1).$$

The endpoint correction is applied to obtain the final approximation to $y(a + h)$ for the step size h_0. This results in the $O(h_0^2)$ approximation to $y(t_1)$

$$y_{1,1} = \frac{1}{2}[w_2 + w_1 + h_0 f(a + 2h_0, w_2)].$$

We save the approximation $y_{1,1}$ and discard the intermediate results w_1 and w_2.

To obtain the next approximation, $y_{2,1}$, to $y(t_1)$, we let $h_1 = h/4$ and use Euler's method with $w_0 = \alpha$ to obtain an approximation to $y(a + h_1) = y(a + h/4)$ that we will call w_1:

$$w_1 = w_0 + h_1 f(a, w_0).$$

Next we produce approximations w_2 to $y(a+2h_1) = y(a+h/2)$ and w_3 to $y(a+3h_1) = y(a+3h/4)$ given by

$$w_2 = w_0 + 2h_1 f(a+h_1, w_1) \quad \text{and} \quad w_3 = w_1 + 2h_1 f(a+2h_1, w_2).$$

Then we produce the approximation w_4 to $y(a+4h_1) = y(t_1)$ given by

$$w_4 = w_2 + 2h_1 f(a+3h_1, w_3).$$

The endpoint correction is now applied to w_3 and w_4 to produce the improved $O(h_1^2)$ approximation to $y(t_1)$,

$$y_{2,1} = \frac{1}{2}[w_4 + w_3 + h_1 f(a+4h_1, w_4)].$$

Because of the form of the error given in (5.43), the two approximations to $y(a+h)$ have the property that

$$y(a+h) = y_{1,1} + \delta_1 \left(\frac{h}{2}\right)^2 + \delta_2 \left(\frac{h}{2}\right)^4 + \cdots = y_{1,1} + \delta_1 \frac{h^2}{4} + \delta_2 \frac{h^4}{16} + \cdots,$$

and

$$y(a+h) = y_{2,1} + \delta_1 \left(\frac{h}{4}\right)^2 + \delta_2 \left(\frac{h}{4}\right)^4 + \cdots = y_{2,1} + \delta_1 \frac{h^2}{16} + \delta_2 \frac{h^4}{256} + \cdots.$$

We can eliminate the $O(h^2)$ portion of this truncation error by averaging these two formulas appropriately. Specifically, if we subtract the first from 4 times the second and divide the result by 3, we have

$$y(a+h) = y_{2,1} + \frac{1}{3}(y_{2,1} - y_{1,1}) - \delta_2 \frac{h^4}{64} + \cdots.$$

So the approximation

$$y_{2,2} = y_{2,1} + \frac{1}{3}(y_{2,1} - y_{1,1})$$

has error of order $O(h^4)$.

Continuing in this manner, we next let $h_2 = h/6$ and apply Euler's method once followed by the Midpoint method five times. Then we use the endpoint correction to determine the h^2 approximation, $y_{3,1}$, to $y(a+h)$. This approximation can be averaged with $y_{2,1}$ to produce a second $O(h^4)$ approximation that we denote $y_{3,2}$. Then $y_{3,2}$ and $y_{2,2}$ are averaged to eliminate the $O(h^4)$ error terms and produce an approximation with error of order $O(h^6)$. Higher-order formulas are generated by continuing the process.

The only significant difference between the extrapolation performed here and that used for Romberg integration in Section 4.5 results from the way the subdivisions are chosen. In Romberg integration there is a convenient formula for representing the Composite

Trapezoidal rule approximations that uses consecutive divisions of the step size by the integers 1, 2, 4, 8, 16, 32, 64, ... This procedure permits the averaging process to proceed in an easily followed manner.

We do not have a means for easily producing refined approximations for initial-value problems, so the divisions for the extrapolation technique are chosen to minimize the number of required function evaluations. The averaging procedure arising from this choice of subdivision, shown in Table 5.14, is not as elementary, but, other than that, the process is the same as that used for Romberg integration.

Table 5.14

$y_{1,1} = w(t, h_0)$		
$y_{2,1} = w(t, h_1)$	$y_{2,2} = y_{2,1} + \dfrac{h_1^2}{h_0^2 - h_1^2}(y_{2,1} - y_{1,1})$	
$y_{3,1} = w(t, h_2)$	$y_{3,2} = y_{3,1} + \dfrac{h_2^2}{h_1^2 - h_2^2}(y_{3,1} - y_{2,1})$	$y_{3,3} = y_{3,2} + \dfrac{h_2^2}{h_0^2 - h_2^2}(y_{3,2} - y_{2,2})$

Algorithm 5.6 uses the extrapolation technique with the sequence of integers $q_0 = 2$, $q_1 = 4$, $q_2 = 6$, $q_3 = 8$, $q_4 = 12$, $q_5 = 16$, $q_6 = 24$, and $q_7 = 32$. A basic step size h is selected, and the method progresses by using $h_i = h/q_i$, for each $i = 0, \ldots, 7$, to approximate $y(t + h)$. The error is controlled by requiring that the approximations $y_{1,1}, y_{2,2}, \ldots$ be computed until $|y_{i,i} - y_{i-1,i-1}|$ is less than a given tolerance. If the tolerance is not achieved by $i = 8$, then h is reduced, and the process is reapplied. If $y_{i,i}$ is found to be acceptable, then w_1 is set to $y_{i,i}$ and computations begin again to determine w_2, which will approximate $y(t_2) = y(a + 2h)$. The process is repeated until the approximation w_N to $y(b)$ is determined.

ALGORITHM 5.6

Extrapolation

To approximate the solution of the initial-value problem

$$y' = f(t, y), \quad a \le t \le b, \quad y(a) = \alpha,$$

with local truncation error within a given tolerance:

INPUT endpoints a, b; initial condition α; tolerance TOL; maximum step size $hmax$; minimum step size $hmin$.

OUTPUT T, W, h where W approximates $y(t)$ and step size h was used, or a message that minimum step size was exceeded.

Step 1 Initialize the array $NK = (2, 4, 6, 8, 12, 16, 24, 32)$.

Step 2 Set $TO = a$;
$WO = \alpha$;
$h = hmax$;
$FLAG = 1$. (*FLAG is used to exit the loop in Step* 4.)

Step 3 For $i = 1, 2, \ldots, 7$
for $j = 1, \ldots, i$
set $Q_{i,j} = (NK_{i+1}/NK_j)^2$. (*Note* : $Q_{i,j} = h_j^2/h_{i+1}^2$.)

Step 4 While ($FLAG = 1$) do Steps 5–20.

Step 5 Set $k = 1$;
$NFLAG = 0$. (*When desired accuracy is achieved, NFLAG is set to* 1.)

Step 6 While ($k \leq 8$ and $NFLAG = 0$) do Steps 7–14.

Step 7 Set $HK = h/NK_k$;
$T = TO$;
$W2 = WO$;
$W3 = W2 + HK \cdot f(T, W2)$; (*Euler's first step.*)
$T = TO + HK$.

Step 8 For $j = 1, \ldots, NK_k - 1$
set $W1 = W2$;
$W2 = W3$;
$W3 = W1 + 2HK \cdot f(T, W2)$; (*Midpoint method.*)
$T = TO + (j + 1) \cdot HK$.

Step 9 Set $y_k = [W3 + W2 + HK \cdot f(T, W3)]/2$.
(*End-point correction to compute* $y_{k,1}$.)

Step 10 If $k \geq 2$ then do Steps 11–13.
(*Note:* $y_{k-1} \equiv y_{k-1,1}, y_{k-2} \equiv y_{k-2,2}, \ldots, y_1 \equiv y_{k-1,k-1}$ *since only the previous row of the table is saved.*)

Step 11 Set $j = k$;
$v = y_1$. (*Save* $y_{k-1,k-1}$.)

Step 12 While ($j \geq 2$) do

$$\text{set } y_{j-1} = y_j + \frac{y_j - y_{j-1}}{Q_{k-1,j-1} - 1};$$

(*Extrapolation to compute* $y_{j-1} \equiv y_{k,k-j+2}$.)

$$\left(\textit{Note}: \quad y_{j-1} = \frac{h_{j-1}^2 y_j - h_k^2 y_{j-1}}{h_{j-1}^2 - h_k^2}. \right)$$

$j = j - 1$.

Step 13 If $|y_1 - v| \leq TOL$ then set $NFLAG = 1$.
(y_1 *is accepted as the new* w.)

Step 14 Set $k = k + 1$.

Step 15 Set $k = k - 1$.

Step 16 If $NFLAG = 0$ then do Steps 17 and 18 (*Result rejected.*)
else do Steps 19 and 20. (*Result accepted.*)

Step 17 Set $h = h/2$. *(New value for w rejected, decrease h.)*

Step 18 If $h < hmin$ then
OUTPUT ('*hmin* exceeded');
Set $FLAG = 0$.
(True branch completed, next step is back to Step 4.)

Step 19 Set $WO = y_1$; *(New value for w accepted.)*
$TO = TO + h$;
OUTPUT (TO, WO, h).

Step 20 If $TO \geq b$ then set $FLAG = 0$
(Procedure completed successfully.)
else if $TO + h > b$ then set $h = b - TO$
(Terminate at $t = b$.)
else if ($k \leq 3$ and $h < 0.5(hmax)$) then set $h = 2h$.
(Increase step size if possible.)

Step 21 STOP.

EXAMPLE 1 Consider the initial-value problem

$$y' = y - t^2 + 1, \quad 0 \leq t \leq 2, \quad y(0) = 0.5,$$

which has solution $y(t) = (t + 1)^2 - 0.5e^t$. The Extrapolation Algorithm will be applied to this problem with $h = 0.25$, $TOL = 10^{-10}$, $hmax = 0.25$, and $hmin = 0.01$. Table 5.15 is obtained in the computation of w_1. ■

Table 5.15

$y_{1,1} = 0.9187011719$				
$y_{2,1} = 0.9200379848$	$y_{2,2} = 0.9204835892$			
$y_{3,1} = 0.9202873689$	$y_{3,2} = 0.9204868761$	$y_{3,3} = 0.9204872870$		
$y_{4,1} = 0.9203747896$	$y_{4,2} = 0.9204871876$	$y_{4,3} = 0.9204872914$	$y_{4,4} = 0.9204872917$	
$y_{5,1} = 0.9204372763$	$y_{5,2} = 0.9204872656$	$y_{5,3} = 0.9204872916$	$y_{5,4} = 0.9204872917$	$y_{5,5} = 0.9204872917$

The computations stopped with $w_1 = y_{5,5}$ because $|y_{5,5} - y_{4,4}| \leq 10^{-10}$ and $y_{5,5}$ is accepted as the approximation to $y(t_1) = y(0.25)$. The complete set of approximations accurate to the places listed is given in Table 5.16.

The proof that the method presented in Algorithm 5.6 converges involves results from summability theory; it can be found in the original paper of Gragg [Gr]. A number of other extrapolation procedures are available, some of which use the variable step-size techniques. For additional procedures based on the extrapolation process, see the Bulirsch and Stoer papers [BS1], [BS2], [BS3] or the text by Stetter [Stet]. The methods used by Bulirsch and Stoer involve interpolation with rational functions instead of the polynomial interpolation used in the Gragg procedure.

Table 5.16

t_i	$y_i = y(t_i)$	w_i	h_i	k
0.25	0.9204872917	0.9204872917	0.25	5
0.50	1.4256393646	1.4256393646	0.25	5
0.75	2.0039999917	2.0039999917	0.25	5
1.00	2.6408590858	2.6408590858	0.25	5
1.25	3.3173285213	3.3173285212	0.25	4
1.50	4.0091554648	4.0091554648	0.25	3
1.75	4.6851986620	4.6851986619	0.25	3
2.00	5.3054719505	5.3054719505	0.25	3

EXERCISE SET 5.8

1. Use the Extrapolation Algorithm with tolerance $TOL = 10^{-4}$, $hmax = 0.25$, and $hmin = 0.05$ to approximate the solutions to the following initial-value problems. Compare the results to the actual values.

   a. $y' = te^{3t} - 2y$, $0 \le t \le 1$, $y(0) = 0$; actual solution $y(t) = \frac{1}{5}te^{3t} - \frac{1}{25}e^{3t} + \frac{1}{25}e^{-2t}$.

   b. $y' = 1 + (t - y)^2$, $2 \le t \le 3$, $y(2) = 1$; actual solution $y(t) = t + 1/(1 - t)$.

   c. $y' = 1 + y/t$, $1 \le t \le 2$, $y(1) = 2$; actual solution $y(t) = t \ln t + 2t$.

   d. $y' = \cos 2t + \sin 3t$, $0 \le t \le 1$, $y(0) = 1$; actual solution $y(t) = \frac{1}{2}\sin 2t - \frac{1}{3}\cos 3t + \frac{4}{3}$.

2. Use the Extrapolation Algorithm with $TOL = 10^{-4}$ to approximate the solutions to the following initial-value problems:

   a. $y' = \left(\frac{y}{t}\right)^2 + \left(\frac{y}{t}\right)$, $1 \le t \le 1.2$, $y(1) = 1$, with $hmax = 0.05$ and $hmin = 0.02$.

   b. $y' = \sin t + e^{-t}$, $0 \le t \le 1$, $y(0) = 0$, with $hmax = 0.25$ and $hmin = 0.02$.

   c. $y' = \frac{1}{t}(y^2 + y)$, $1 \le t \le 3$, $y(1) = -2$, with $hmax = 0.5$ and $hmin = 0.02$.

   d. $y' = -ty + \frac{4t}{y}$, $0 \le t \le 1$, $y(0) = 1$, with $hmax = 0.25$ and $hmin = 0.02$.

3. Use the Extrapolation Algorithm with tolerance $TOL = 10^{-6}$, $hmax = 0.5$, and $hmin = 0.05$ to approximate the solutions to the following initial-value problems. Compare the results to the actual values.

   a. $y' = \frac{y}{t} - \frac{y^2}{t^2}$, $1 \le t \le 4$, $y(1) = 1$; actual solution $y(t) = t/(1 + \ln t)$.

   b. $y' = 1 + \frac{y}{t} + \left(\frac{y}{t}\right)^2$, $1 \le t \le 3$, $y(1) = 0$; actual solution $y(t) = t\tan(\ln t)$.

   c. $y' = -(y + 1)(y + 3)$, $0 \le t \le 3$, $y(0) = -2$; actual solution $y(t) = -3 + 2(1 + e^{-2t})^{-1}$.

   d. $y' = (t + 2t^3)y^3 - ty$, $0 \le t \le 2$, $y(0) = \frac{1}{3}$; actual solution $y(t) = (3 + 2t^2 + 6e^{t^2})^{-1/2}$.

4. Let $P(t)$ be the number of individuals in a population at time t, measured in years. If the average birth rate b is constant and the average death rate d is proportional to the size of the population

(due to overcrowding), then the growth rate of the population is given by the **logistic equation**

$$\frac{dP(t)}{dt} = bP(t) - k[P(t)]^2$$

where $d = kP(t)$. Suppose $P(0) = 50,976$, $b = 2.9 \times 10^{-2}$, and $k = 1.4 \times 10^{-7}$. Find the population after 5 years.

5.9 Higher-Order Equations and Systems of Differential Equations

This section contains an introduction to the numerical solution of higher-order differential equations subject to initial conditions. The techniques discussed are limited to those that transform a higher-order equation into a system of first-order differential equations. Before discussing the transformation procedure, some remarks are needed concerning systems that involve first-order differential equations.

An ***m*th-order system** of first-order initial-value problems can be expressed in the form

(5.44)

$$\begin{aligned}\frac{du_1}{dt} &= f_1(t, u_1, u_2, \ldots, u_m),\\ \frac{du_2}{dt} &= f_2(t, u_1, u_2, \ldots, u_m),\\ &\vdots\\ \frac{du_m}{dt} &= f_m(t, u_1, u_2, \ldots, u_m),\end{aligned}$$

for $a \le t \le b$, with the initial conditions

(5.45)

$$u_1(a) = \alpha_1, \quad u_2(a) = \alpha_2, \quad \ldots, \quad u_m(a) = \alpha_m.$$

The object is to find m functions $u_1, u_2, \ldots, u_m$ that satisfy the system of differential equations together with all the initial conditions.

To discuss existence and uniqueness of solutions to systems of equations, we need to extend the definition of the Lipschitz condition to functions of several variables.

Definition 5.15 The function $f(t, y_1, \ldots, y_m)$, defined on the set

$$D = \{(t, u_1, \ldots, u_m) \mid a \le t \le b, -\infty < u_i < \infty, \text{ for each } i = 1, 2, \ldots, m\}$$

is said to satisfy a **Lipschitz condition** on D in the variables $u_1, u_2, \ldots, u_m$ if a constant $L > 0$ exists with the property that

(5.46)

$$|f(t, u_1, \ldots, u_m) - f(t, z_1, \ldots, z_m)| \le L \sum_{j=1}^{m} |u_j - z_j|,$$

for all $(t, u_1, \ldots, u_m)$ and $(t, z_1, \ldots, z_m)$ in D. ■

By using the Mean Value Theorem, it can be shown that if f and its first partial derivatives are continuous on D and if

$$\left| \frac{\partial f(t, u_1, \ldots, u_m)}{\partial u_i} \right| \le L,$$

for each $i = 1, 2, \ldots, m$ and all $(t, u_1, \ldots, u_m)$ in D, then f satisfies a Lipschitz condition on D with Lipschitz constant L (see [BiR, p. 141]). A basic existence and uniqueness theorem follows. Its proof can be found in [BiR, pp. 152–154].

Theorem 5.16 Suppose

$$D = \{(t, u_1, u_2, \ldots, u_m) \mid a \le t \le b, -\infty < u_i < \infty, \quad \text{for each } i = 1, 2, \ldots, m\},$$

and let $f_i(t, u_1, \ldots, u_m)$, for each $i = 1, 2, \ldots, m$, be continuous on D and satisfy a Lipschitz condition there. The system of first-order differential equations (5.44), subject to the initial conditions (5.45), has a unique solution $u_1(t), \ldots, u_m(t)$ for $a \le t \le b$. ■

Methods to solve systems of first-order differential equations are generalizations of the methods for a single first-order equation presented earlier in this chapter. For example, the classical Runge-Kutta method of order four given by

$$\begin{aligned}
w_0 &= \alpha, \\
k_1 &= hf(t_i, w_i), \\
k_2 &= hf\left(t_i + \frac{h}{2}, w_i + \frac{1}{2}k_1\right), \\
k_3 &= hf\left(t_i + \frac{h}{2}, w_i + \frac{1}{2}k_2\right), \\
k_4 &= hf(t_{i+1}, w_i + k_3),
\end{aligned}$$

and

$$w_{i+1} = w_i + \frac{1}{6}[k_1 + 2k_2 + 2k_3 + k_4], \quad \text{for each } i = 0, 1, \ldots, N-1,$$

used to solve the first-order initial-value problem

$$y' = f(t, y), \quad a \le t \le b, \quad y(a) = \alpha,$$

is generalized as follows.

Let an integer $N > 0$ be chosen and set $h = (b - a)/N$. Partition the interval $[a, b]$ into N subintervals with the mesh points

$$t_j = a + jh \quad \text{for each } j = 0, 1, \ldots, N.$$

Use the notation w_{ij} to denote an approximation to $u_i(t_j)$ for each $j = 0, 1, \ldots, N$ and $i = 1, 2, \ldots, m$. That is, w_{ij} approximates the ith solution $u_i(t)$ of (5.44) at the jth mesh point t_j. For the initial conditions, set

(5.47) $$w_{1,0} = \alpha_1, \quad w_{2,0} = \alpha_2, \quad \ldots, \quad w_{m,0} = \alpha_m.$$

Suppose that the values $w_{1,j}, w_{2,j}, \ldots, w_{m,j}$ have been computed. We obtain $w_{1,j+1}, w_{2,j+1}, \ldots, w_{m,j+1}$ by first calculating

(5.48) $$k_{1,i} = hf_i(t_j, w_{1,j}, w_{2,j}, \ldots, w_{m,j}), \quad \text{for each } i = 1, 2, \ldots, m;$$

(5.49) $$k_{2,i} = hf_i\left(t_j + \frac{h}{2}, w_{1,j} + \frac{1}{2}k_{1,1}, w_{2,j} + \frac{1}{2}k_{1,2}, \ldots, w_{m,j} + \frac{1}{2}k_{1,m}\right),$$

for each $i = 1, 2, \ldots, m$;

(5.50) $$k_{3,i} = hf_i\left(t_j + \frac{h}{2}, w_{1,j} + \frac{1}{2}k_{2,1}, w_{2,j} + \frac{1}{2}k_{2,2}, \ldots, w_{m,j} + \frac{1}{2}k_{2,m}\right),$$

for each $i = 1, 2, \ldots, m$;

(5.51) $$k_{4,i} = hf_i(t_j + h, w_{1,j} + k_{3,1}, w_{2,j} + k_{3,2}, \ldots, w_{m,j} + k_{3,m}),$$

for each $i = 1, 2, \ldots, m$; and then

(5.52) $$w_{i,j+1} = w_{i,j} + \frac{1}{6}[k_{1,i} + 2k_{2,i} + 2k_{3,i} + k_{4,i}],$$

for each $i = 1, 2, \ldots m$. Note that all the values $k_{1,1}, k_{1,2}, \ldots, k_{1,m}$ must be computed before any of the terms of the form $k_{2,i}$ can be determined. In general, each $k_{l,1}, k_{l,2}, \ldots, k_{l,m}$ must be computed before any of the expressions $k_{l+1,i}$.

Algorithm 5.7 implements the Runge-Kutta fourth-order method for systems of initial-value problems.

ALGORITHM **5.7**

Runge-Kutta Method for Systems of Differential Equations

To approximate the solution of the mth-order system of first-order initial-value problems

$$u_j' = f_j(t, u_1, u_2, \ldots, u_m), \quad a \le t \le b, \quad \text{with} \quad u_j(a) = \alpha_j,$$

for $j = 1, 2, \ldots, m$ at $(N + 1)$ equally spaced numbers in the interval $[a, b]$:

INPUT endpoints a, b; number of equations m; integer N; initial conditions $\alpha_1, \ldots, \alpha_m$.

OUTPUT approximations w_j to $u_j(t)$ at the $(N + 1)$ values of t.

Step 1 Set $h = (b - a)/N$;
$t = a$.

Step 2 For $j = 1, 2, \ldots, m$ set $w_j = \alpha_j$.

Step 3 OUTPUT $(t, w_1, w_2, \ldots, w_m)$.

Step 4 For $i = 1, 2, \ldots, N$ do steps 5–11.

Step 5 For $j = 1, 2, \ldots, m$ set
$k_{1,j} = hf_j(t, w_1, w_2, \ldots, w_m)$.

Step 6 For $j = 1, 2, \ldots, m$ set

$$k_{2,j} = hf_j\left(t + \tfrac{h}{2}, w_1 + \tfrac{1}{2}k_{1,1}, w_2 + \tfrac{1}{2}k_{1,2}, \ldots, w_m + \tfrac{1}{2}k_{1,m}\right).$$

Step 7 For $j = 1, 2, \ldots, m$ set

$$k_{3,j} = hf_j\left(t + \tfrac{h}{2}, w_1 + \tfrac{1}{2}k_{2,1}, w_2 + \tfrac{1}{2}k_{2,2}, \ldots, w_m + \tfrac{1}{2}k_{2,m}\right).$$

Step 8 For $j = 1, 2, \ldots, m$ set
$k_{4,j} = hf_j(t + h, w_1 + k_{3,1}, w_2 + k_{3,2}, \ldots, w_m + k_{3,m})$.

Step 9 For $j = 1, 2, \ldots, m$ set
$w_j = w_j + (k_{1,j} + 2k_{2,j} + 2k_{3,j} + k_{4,j})/6$.

Step 10 Set $t = a + ih$.

Step 11 OUTPUT $(t, w_1, w_2, \ldots, w_m)$.

Step 12 STOP.

EXAMPLE 1 Kirchhoff's Law states that the sum of all instantaneous voltage changes around a closed circuit is zero. This law implies that the current $I(t)$ in a closed circuit containing a resistance of R ohms, a capacitance of C farads, an inductance of L henries, and a voltage source of $\mathcal{E}(t)$ volts satisfies the equation

$$LI'(t) + RI(t) + \frac{1}{C}\int I(t)\,dt = \mathcal{E}(t).$$

The currents $I_1(t)$ and $I_2(t)$ in the left and right loops, respectively, of the circuit shown in Figure 5.5 are the solutions to the system of equations

$$2I_1(t) + 6[I_1(t) - I_2(t)] + 2I_1'(t) = 12,$$

$$\frac{1}{0.5}\int I_2(t)\,dt + 4I_2(t) + 6[I_2(t) - I_1(t)] = 0.$$

Figure 5.5

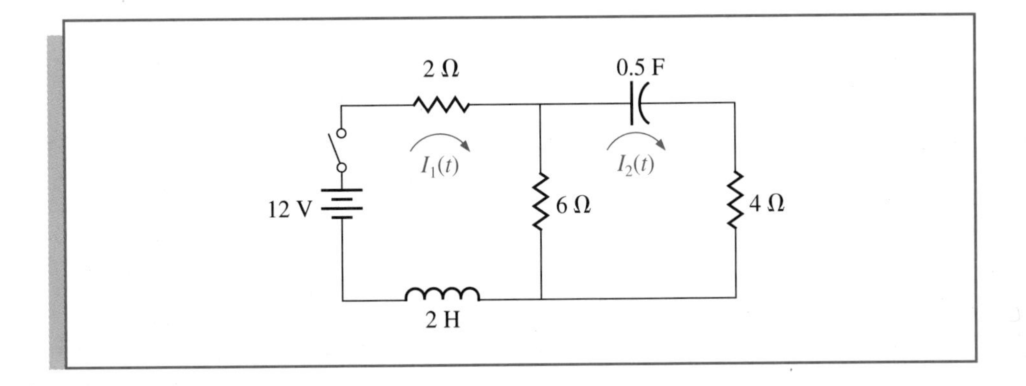

Suppose that the switch in the circuit is closed at time $t = 0$. Then $I_1(0) = 0$ and $I_2(0) = 0$. Differentiating the second equation and substituting the first equation into the result gives the system

$$I_1' = f_1(t, I_1, I_2) = -4I_1 + 3I_2 + 6, \; I_1(0) = 0,$$
$$I_2' = f_2(t, I_1, I_2) = 0.6I_1' - 0.2I_2 = -2.4I_1 + 1.6I_2 + 3.6, \; I_2(0) = 0.$$

The exact solution to this system is

$$I_1(t) = -3.375e^{-2t} + 1.875e^{-0.4t} + 1.5,$$
$$I_2(t) = -2.25e^{-2t} + 2.25e^{-0.4t}.$$

We will apply the Runge-Kutta method of order four to this system with $h = 0.1$. Since $w_{1,0} = I_1(0) = 0$ and $w_{2,0} = I_2(0) = 0$,

$$k_{1,1} = hf_1(t_0, w_{1,0}, w_{2,0}) = 0.1\, f_1(0, 0, 0) = 0.1[-4(0) + 3(0) + 6] = 0.6,$$
$$k_{1,2} = hf_2(t_0, w_{1,0}, w_{2,0}) = 0.1\, f_2(0, 0, 0) = 0.1[-2.4(0) + 1.6(0) + 3.6] = 0.36,$$
$$k_{2,1} = hf_1\left(t_0 + \frac{1}{2}h, w_{1,0} + \frac{1}{2}k_{1,1}, w_{2,0} + \frac{1}{2}k_{1,2}\right) = 0.1\, f_1(0.05, 0.3, 0.18)$$
$$= 0.1[-4(0.3) + 3(0.18) + 6] = 0.534,$$
$$k_{2,2} = hf_2\left(t_0 + \frac{1}{2}h, w_{1,0} + \frac{1}{2}k_{1,1}, w_{2,0} + \frac{1}{2}k_{1,2}\right) = 0.1\, f_2(0.05, 0.3, 0.18)$$
$$= 0.1[-2.4(0.3) + 1.6(0.18) + 3.6] = 0.3168.$$

Generating the remaining entries in a similar manner produces

$$k_{3,1} = (0.1)f_1(0.05, 0.267, 0.1584) = 0.54072,$$
$$k_{3,2} = (0.1)f_2(0.05, 0.267, 0.1584) = 0.321264,$$
$$k_{4,1} = (0.1)f_1(0.1, 0.54072, 0.321264) = 0.4800912,$$

and

$$k_{4,2} = (0.1)f_2(0.1, 0.54072, 0.321264) = 0.28162944.$$

As a consequence,

$$I_1(0.1) \approx w_{1,1} = w_{1,0} + \frac{1}{6}[k_{1,1} + 2k_{2,1} + 2k_{3,1} + k_{4,1}]$$
$$= 0 + \frac{1}{6}[0.6 + 2(0.534) + 2(0.54072) + 0.4800912] = 0.5382552$$

and

$$I_2(0.1) \approx w_{2,1} = w_{2,0} + \frac{1}{6}[k_{1,2} + 2k_{2,2} + 2k_{3,2} + k_{4,2}] = 0.3196263.$$

The remaining entries in Table 5.17 are generated in a similar manner. ■

Table 5.17

t_j	$w_{1,j}$	$w_{2,j}$	$\lvert I_1(t_j) - w_{1,j}\rvert$	$\lvert I_2(t_j) - w_{2,j}\rvert$
0.0	0	0	0	0
0.1	0.5382550	0.3196263	0.8285×10^{-5}	0.5803×10^{-5}
0.2	0.9684983	0.5687817	0.1514×10^{-4}	0.9596×10^{-5}
0.3	1.310717	0.7607328	0.1907×10^{-4}	0.1216×10^{-4}
0.4	1.581263	0.9063208	0.2098×10^{-4}	0.1311×10^{-4}
0.5	1.793505	1.014402	0.2193×10^{-4}	0.1240×10^{-4}

Maple's command `dsolve` can be used to solve systems of first-order differential equations. The system in Example 1 is defined with

```
>sys2:=D(u1)(t)=-4*u1(t)+3*u2(t)+6,
        D(u2)(t)=-2.4*u1(t)+1.6*u2(t)+3.6;
```

and the initial conditions with

```
>init2:=u1(0)=0,u2(0)=0;
```

The system is solved with the command

```
>sol2:=dsolve({sys2,init2},{u1(t),u2(t)});
```

to obtain

$$\text{sol2} := \{u2(t) = 2.250000000e^{(-.4000000000t)} - 2.250000000e^{(-2.t)},$$
$$u1(t) = 1.500000000 - 3.375000000e^{(-2.t)} + 1.875000000e^{(-.4000000000t)}\}$$

To isolate the solution in function form use

```
>r1:=rhs(sol2[2]);
```

$$r1 := 1.500000000 - 3.375000000e^{(-2.t)} + 1.875000000e^{(-.4000000000t)}\}$$

and

```
>r2:=rhs(sol2[1]);
```

which gives a similar response.

To evaluate $u_1(0.5)$ and $u_2(0.5)$, use

```
>evalf(subs(t=0.5,r1));evalf(subs(t=0.5,r2);
```

to get 1.793527048 and 1.014415451.

The command `dsolve` will fail if an explicit solution cannot be found. In that case we can use the numeric option in `dsolve`, which applies the Runge-Kutta-Fehlberg technique. For example,

```
>g:=dsolve({sys2,init2},{u1(t),u2(t)},numeric);
```

returns the procedure

$$g := \text{proc}(rkf45\_x)\ldots\text{end}$$

To approximate the solutions at $t = 0.5$, enter

```
>g(0.5);
```

to obtain

$$[t = .5,\ u2(t) = 1.014415454702917,\ u1(t) = 1.793527052437666]$$

The `rungekutta` command in Maple can also be used for systems, without invoking `dsolve`. We illustrate by applying the method to the system in Example 1. First, define the functions f_1 and f_2 by

```
>f1:=(t,u1,u2) -> -4*u1+3*u2+6;
>f2:=(t,u1,u2) -> -2.4*u1+1.6*u2+3.6;
```

We need to access the differential equation solving procedure, so we enter

```
>with(share);
>readshare(ODE,plots);
```

The procedure is called with

```
>rhp:=rungekutta([f1,f2],[0,0,0],0.1,5);
```

The second parameter refers to the initial conditions $[a, \alpha_1, \alpha_2]$, and the third and fourth refer to $h = 0.1$ and $n = 5$. Maple responds with the array

$$\begin{aligned}
&\text{rhp} := \text{array}\ (0..5, [\\
&\quad (0) = [0, 0, 0]\\
&\quad (1) = [.1, .5382552001, .3196262401]\\
&\quad (2) = [.2, .9684987377, .5687821731]\\
&\quad (3) = [.3, 1.310719039, .7607331320]\\
&\quad (4) = [.4, 1.581265239, .9063206181]\\
&\quad (5) = [.5, 1.793507490, 1.014402417]\\
&\quad])
\end{aligned}$$

To access the entry with subscript 3 enter

```
>rhp[3][2]; rhp[3][3];
```

which gives the results 1.310719039 and .767331320.

A general mth-order initial-value problem

$$y^{(m)}(t) = f(t, y, y', \ldots, y^{(m-1)}), \quad a \le t \le b,$$

with initial conditions $y(a) = \alpha_1, y'(a) = \alpha_2, \ldots, y^{(m-1)}(a) = \alpha_m$ can be converted into a system of equations in the form (5.44) and (5.45).

Let $u_1(t) = y(t), u_2(t) = y'(t), \ldots$, and $u_m(t) = y^{(m-1)}(t)$. This produces the first-order system

$$\begin{aligned}
\frac{du_1}{dt} &= \frac{dy}{dt} = u_2,\\
\frac{du_2}{dt} &= \frac{dy'}{dt} = u_3,\\
&\vdots\\
\frac{du_{m-1}}{dt} &= \frac{dy^{(m-2)}}{dt} = u_m,
\end{aligned}$$

and

$$\frac{du_m}{dt} = \frac{dy^{(m-1)}}{dt} = y^{(m)} = f(t, y, y', \ldots, y^{(m-1)}) = f(t, u_1, u_2, \ldots, u_m),$$

with initial conditions

$$u_1(a) = y(a) = \alpha_1, \quad u_2(a) = y'(a) = \alpha_2, \quad \ldots, \quad u_m(a) = y^{(m-1)}(a) = \alpha_m.$$

EXAMPLE 2 Consider the second-order initial-value problem

$$y'' - 2y' + 2y = e^{2t}\sin t, \qquad \text{for } 0 \le t \le 1, \quad \text{with } y(0) = -0.4,\ y'(0) = -0.6.$$

With $u_1(t) = y(t)$ and $u_2(t) = y'(t)$, this equation is transformed into the system

$$\begin{aligned}
u_1'(t) &= u_2(t),\\
u_2'(t) &= e^{2t}\sin t - 2u_1(t) + 2u_2(t),
\end{aligned}$$

with initial conditions

$$u_1(0) = -0.4, \quad u_2(0) = -0.6.$$

The Runge-Kutta fourth-order method will be used to approximate the solution to this problem using $h = 0.1$. The initial conditions give $w_{1,0} = -0.4$ and $w_{2,0} = -0.6$.

Eqs. (5.48) through (5.51) with $j = 0$ give

$$k_{1,1} = hf_1(t_0, w_{1,0}, w_{2,0}) = hw_{2,0} = -0.06,$$

$$k_{1,2} = hf_2(t_0, w_{1,0}, w_{2,0}) = h\left[e^{2t_0}\sin t_0 - 2w_{1,0} + 2w_{2,0}\right] = -0.04,$$

$$k_{2,1} = hf_1\left(t_0 + \frac{h}{2}, w_{1,0} + \frac{1}{2}k_{1,1}, w_{2,0} + \frac{1}{2}k_{1,2}\right) = h\left[w_{2,0} + \frac{1}{2}k_{1,2}\right] = -0.062,$$

$$k_{2,2} = hf_2\left(t_0 + \frac{h}{2}, w_{1,0} + \frac{1}{2}k_{1,1}, w_{2,0} + \frac{1}{2}k_{1,2}\right)$$

$$= h\left[e^{2(t_0+0.05)}\sin(t_0 + 0.05) - 2\left(w_{1,0} + \frac{1}{2}k_{1,1}\right) + 2\left(w_{2,0} + \frac{1}{2}k_{1,2}\right)\right]$$

$$= -0.03247644757,$$

$$k_{3,1} = h\left[w_{2,0} + \frac{1}{2}k_{2,2}\right] = -0.06162382238,$$

$$k_{3,2} = h\left[e^{2(t_0+0.05)}\sin(t_0 + 0.05) - 2\left(w_{1,0} + \frac{1}{2}k_{2,1}\right) + 2\left(w_{2,0} + \frac{1}{2}k_{2,2}\right)\right]$$

$$= -0.03152409237,$$

$$k_{4,1} = h[w_{2,0} + k_{3,2}] = -0.06315240924,$$

and

$$k_{4,2} = h[e^{2(t_0+0.1)}\sin(t_0 + 0.1) - 2(w_{1,0} + k_{3,1}) + 2(w_{2,0} + k_{3,2})] = -0.02178637298.$$

So

$$w_{1,1} = w_{1,0} + \frac{1}{6}[k_{1,1} + 2k_{2,1} + 2k_{3,1} + k_{4,1}] = -0.4617333423 \quad \text{and}$$

$$w_{2,1} = w_{2,0} + \frac{1}{6}[k_{1,2} + 2k_{2,2} + 2k_{3,2} + k_{4,2}] = -0.6316312421.$$

The value $w_{1,1}$ approximates $u_1(0.1) = y(0.1) = 0.2e^{2(0.1)}[\sin 0.1 - 2\cos 0.1]$, and $w_{2,1}$ approximates $u_2(0.1) = y'(0.1) = 0.2e^{2(0.1)}[4\sin 0.1 - 3\cos 0.1]$.

The set of values $w_{1,j}$ and $w_{2,j}$ for $j = 0, 1, \ldots, 10$, are presented in Table 5.18 and compared to the actual values of $u_1(t) = 0.2e^{2t}(\sin t - 2\cos t)$ and $u_2(t) = u_1'(t) = 0.2e^{2t}(4\sin t - 3\cos t)$. ■

We can also use `dsolve` from Maple on higher-order equations. Note that the nth derivative $y^{(n)}(t)$ is specified by `(D@@n)(y)(t)`. To define the differential equation of Example 2, use

```
>def2:=(D@@2)(y)(t)-2*D(y)(t)+2*y(t)=exp(2*t)*sin(t);
```

and to specify the initial conditions use

```
>init2:=y(0)=-0.4, D(y)(0)=-0.6;
```

Table 5.18

t_j	$y(t_j) = u_1(t_j)$	$w_{1,j}$	$y'(t_j) = u_2(t_j)$	$w_{2,j}$	$\lvert y(t_j) - w_{1,j}\rvert$	$\lvert y'(t_j) - w_{2,j}\rvert$
0.0	−0.40000000	−0.40000000	−6.0000000	−0.60000000	0	0
0.1	−0.46173297	−0.46173334	−0.6316304	−0.63163124	3.7×10^{-7}	7.75×10^{-7}
0.2	−0.52555905	−0.52555988	−0.6401478	−0.64014895	8.3×10^{-7}	1.01×10^{-6}
0.3	−0.58860005	−0.58860144	−0.6136630	−0.61366381	1.39×10^{-6}	8.34×10^{-7}
0.4	−0.64661028	−0.64661231	−0.5365821	−0.53658203	2.03×10^{-6}	1.79×10^{-7}
0.5	−0.69356395	−0.69356666	−0.3887395	−0.38873810	2.71×10^{-6}	5.96×10^{-7}
0.6	−0.72114849	−0.72115190	−0.1443834	−0.14438087	3.41×10^{-6}	7.75×10^{-7}
0.7	−0.71814890	−0.71815295	0.2289917	0.22899702	4.05×10^{-6}	2.03×10^{-6}
0.8	−0.66970677	−0.66971133	0.7719815	0.77199180	4.56×10^{-6}	5.30×10^{-6}
0.9	−0.55643814	−0.55644290	1.534764	0.15347815	4.76×10^{-6}	9.54×10^{-6}
1.0	−0.35339436	−0.35339886	2.578741	0.25787663	4.50×10^{-6}	1.34×10^{-5}

The solution is obtained by the command

```
>sol2:=dsolve({def2,init2},y(t));
```

to obtain

$$\text{sol2} := y(t) = \frac{1}{10}e^{(2t)}\sin(t)\sin(2t) - \frac{1}{5}e^{(2t)}\sin(t)\cos(2t) - \frac{3}{5}e^{(2t)}\cos(t) + \frac{1}{5}e^{(2t)}\cos(t)^3 + \frac{2}{5}e^{(2t)}\cos(t)^2\sin(t).$$

We isolate the solution in function form using

```
>y:=rhs(sol2);
```

To obtain $y(1.0)$ enter

```
>evalf(subs(t=1.0,y));
```

which gives the result $-.3533943558$.

Runge-Kutta-Fehlberg is also available for higher-order equations via the `dsolve` command with the numeric option. We enter the command

```
>g:=dsolve({def2,init2},y(t),numeric);
```

with the Maple response

$$g := \text{proc}(rkf45\_x)\ldots\text{end}$$

We can approximate $y(1.0)$ using the command

```
>g(1.0);
```

to give

$$\left[t = 1., \ y(t) = -.3533943468075346, \ \frac{\partial}{\partial t} y(t) = 2.578746659404821 \right].$$

The other one-step methods can be extended to systems in a similar way. If methods such as the Runge-Kutta-Fehlberg method are extended with error control, then each component of the numerical solution $(w_{1j}, w_{2j}, \ldots, w_{mj})$ must be examined for accuracy. If any of the components fail to be sufficiently accurate, the entire numerical solution $(w_{1j}, w_{2j}, \ldots, w_{mj})$ must be recomputed.

The multistep methods and predictor-corrector techniques can also be extended to systems. Again, if error control is used each component must be accurate. The extension of the extrapolation technique to systems can also be done, but the notation becomes quite involved. If this topic is of interest see [HNW1].

Convergence theorems and error estimates for systems are similar to those considered in Section 5.10 for the single equations, except that the bounds are given in terms of vector norms, a topic considered in Chapter 7. (A good reference for these theorems is [Ge1, pp. 45–72].)

EXERCISE SET 5.9

1. Use the Runge-Kutta method for systems to approximate the solutions of the following systems of first-order differential equations and compare the results to the actual solutions.

a. $u_1' = 3u_1 + 2u_2 - (2t^2 + 1)e^{2t}, \quad 0 \le t \le 1, \quad u_1(0) = 1;$
$u_2' = 4u_1 + u_2 + (t^2 + 2t - 4)e^{2t}, \quad 0 \le t \le 1, \quad u_2(0) = 1;$
$h = 0.2;$ actual solutions $u_1(t) = \frac{1}{3}e^{5t} - \frac{1}{3}e^{-t} + e^{2t}$ and $u_2(t) = \frac{1}{3}e^{5t} + \frac{2}{3}e^{-t} + t^2e^{2t}$.

b. $u_1' = -4u_1 - 2u_2 + \cos t + 4\sin t, \quad 0 \le t \le 2, \quad u_1(0) = 0;$
$u_2' = 3u_1 + u_2 - 3\sin t, \quad 0 \le t \le 2, \quad u_2(0) = -1;$
$h = 0.1;$ actual solutions $u_1(t) = 2e^{-t} - 2e^{-2t} + \sin t$ and $u_2(t) = -3e^{-t} + 2e^{-2t}$.

c. $u_1' = u_2, \quad 0 \le t \le 2, \quad u_1(0) = 1;$
$u_2' = -u_1 - 2e^t + 1, \quad 0 \le t \le 2, \quad u_2(0) = 0;$
$u_3' = -u_1 - e^t + 1, \quad 0 \le t \le 2, \quad u_3(0) = 1;$
$h = 0.5;$ actual solutions $u_1(t) = \cos t + \sin t - e^t + 1, \quad u_2(t) = -\sin t + \cos t - e^t$, and $u_3(t) = -\sin t + \cos t$.

d. $u_1' = u_2 - u_3 + t, \quad 0 \le t \le 1, \quad u_1(0) = 1;$
$u_2' = 3t^2, \quad 0 \le t \le 1, \quad u_2(0) = 1;$
$u_3' = u_2 + e^{-t}, \quad 0 \le t \le 1, \quad u_3(0) = -1;$
$h = 0.1;$ actual solutions $u_1(t) = -0.05t^5 + 0.25t^4 + t + 2 - e^{-t}, \quad u_2(t) = t^3 + 1$, and $u_3(t) = 0.25t^4 + t - e^{-t}$.

2. Use the Runge-Kutta for Systems Algorithm to approximate the solutions of the following higher-order differential equations and compare the results to the actual solutions.

a. $y'' - 2y' + y = te^t - t, \quad 0 \le t \le 1, \quad y(0) = y'(0) = 0$, with $h = 0.1$; actual solution $y(t) = \frac{1}{6}t^3e^t - te^t + 2e^t - t - 2$.

b. $t^2y'' - 2ty' + 2y = t^3 \ln t, \quad 1 \le t \le 2, \quad y(1) = 1, \quad y'(1) = 0$, with $h = 0.1$; actual solution $y(t) = \frac{7}{4}t + \frac{1}{2}t^3 \ln t - \frac{3}{4}t^3$.

c. $y''' + 2y'' - y' - 2y = e^t, \quad 0 \le t \le 3, \quad y(0) = 1, \quad y'(0) = 2, \quad y''(0) = 0$, with $h = 0.2$; actual solution $y(t) = \frac{43}{36}e^t + \frac{1}{4}e^{-t} - \frac{4}{9}e^{-2t} + \frac{1}{6}te^t$.

d. $t^3y''' - t^2y'' + 3ty' - 4y = 5t^3 \ln t + 9t^3$, $1 \le t \le 2$, $y(1) = 0$, $y'(1) = 1$, $y''(1) = 3$, with $h = 0.1$; actual solution $y(t) = -t^2 + t\cos(\ln t) + t\sin(\ln t) + t^3 \ln t$.

3. Change the Adams Fourth-Order Predictor-Corrector Algorithm to obtain approximate solutions to systems of first-order equations.

4. Repeat Exercise 1 using the algorithm developed in Exercise 3.

5. Repeat Exercise 2 using the algorithm developed in Exercise 3.

6. Suppose the swinging pendulum described in the lead example of this chapter is 2 ft long and that $g = -32.17$ ft/s^2. With $h = 0.1$ s, compare the angle θ obtained for the following two initial-value problems at $t = 0$, 1, and 2 s:

a. $\dfrac{d^2\theta}{dt^2} - \dfrac{g}{L}\sin\theta = 0, \quad \theta(0) = \dfrac{\pi}{6}, \quad \theta'(0) = 0,$

b. $\dfrac{d^2\theta}{dt^2} - \dfrac{g}{L}\theta = 0, \quad \theta(0) = \dfrac{\pi}{6}, \quad \theta'(0) = 0.$

7. The study of mathematical models for predicting the population dynamics of competing species has its origin in independent works published in the early part of this century by A. J. Lotka and V. Volterra. Consider the problem of predicting the population of two species, one of which is a predator, whose population at time t is $x_2(t)$, feeding on the other, which is the prey, whose population is $x_1(t)$. We will assume that the prey always has an adequate food supply and that its birth rate at any time is proportional to the number of prey alive at that time; that is, birth rate (prey) is $k_1x_1(t)$. The death rate of the prey depends on both the number of prey and predators alive at that time. For simplicity, we assume death rate (prey) $= k_2x_1(t)x_2(t)$. The birth rate of the predator, on the other hand, depends on its food supply, $x_1(t)$, as well as on the number of predators available for reproduction purposes. For this reason, we assume that the birth rate (predator) is $k_3x_1(t)x_2(t)$. The death rate of the predator will be taken as simply proportional to the number of predators alive at the time; that is, death rate (predator) $= k_4x_2(t)$.

Since $x_1'(t)$ and $x_2'(t)$ represent the change in the prey and predator populations, respectively, with respect to time, the problem is expressed by the system of nonlinear differential equations

$$x_1'(t) = k_1x_1(t) - k_2x_1(t)x_2(t) \quad \text{and} \quad x_2'(t) = k_3x_1(t)x_2(t) - k_4x_2(t).$$

Solve this system for $0 \le t \le 4$, assuming that the initial population of the prey is 1000 and of the predators is 500 and that the constants are $k_1 = 3$, $k_2 = 0.002$, $k_3 = 0.0006$, and $k_4 = 0.5$. Sketch a graph of the solutions to this problem, plotting both populations with time, and describe the physical phenomena represented. Is there a stable solution to this population model? If so, for what values x_1 and x_2 is the solution stable?

8. In Exercise 7 we considered the problem of predicting the population in a predator-prey model. Another problem of this type is concerned with two species competing for the same food supply. If the numbers of species alive at time t are denoted by $x_1(t)$ and $x_2(t)$, it is often assumed that, although the birth rate of each of the species is simply proportional to the number of species alive at that time, the death rate of each species depends on the population of both species. We will assume that the population of a particular pair of species is described by the equations

$$\frac{dx_1(t)}{dt} = x_1(t)[4 - 0.0003x_1(t) - 0.0004x_2(t)]$$

and

$$\frac{dx_2(t)}{dt} = x_2(t)[2 - 0.0002x_1(t) - 0.0001x_2(t)].$$

If it is known that the initial population of each species is 10,000, find the solution to this system for $0 \le t \le 4$. Is there a stable solution to this population model? If so, for what values of x_1 and x_2 is the solution stable?

5.10 Stability

A number of methods have been presented in this chapter for approximating the solution to an initial-value problem. Although numerous other techniques are available, we have chosen the methods described here because they generally satisfied three criteria:

1. Their development is clear enough that the first-year student in numerical analysis can understand how and why they work.
2. One or more of the methods will give satisfactory results for most of the problems that are encountered by students in science and engineering.
3. Most of the more advanced and complex techniques are based on one or a combination of the procedures described here.

In this section, we discuss why these methods give satisfactory results when some similar methods do not.

Before we begin this discussion, we need to present two definitions concerned with the convergence of one-step difference-equation methods to the solution of the differential equation as the step size decreases.

Definition 5.17 A one-step difference-equation method with local truncation error $\tau_i(h)$ at the ith step is said to be **consistent** with the differential equation it approximates if

$$\lim_{h\to 0}\max_{1\le i\le N}|\tau_i(h)| = 0. \qquad \blacksquare$$

Note that this definition is a *local* definition, since, for each of the values $\tau_i(h)$, we are comparing the exact value $f(t_i, y_i)$ to the difference-equation approximation of y' at t_i. A more realistic means of analyzing the effects of making h small is to determine the *global* effect of the method. This is the maximum error of the method over the entire range of the approximation, assuming only that the method gives the exact result at the initial value.

Definition 5.18 A one-step difference-equation method is said to be **convergent** with respect to the differential equation it approximates if

$$\lim_{h\to 0}\max_{1\le i\le N}|w_i - y(t_i)| = 0,$$

where $y_i = y(t_i)$ denotes the exact value of the solution of the differential equation and w_i is the approximation obtained from the difference method at the ith step. $\blacksquare$

Examining Inequality (5.10) of Section 5.2 in the error-bound formula for Euler's method, it can be said that under the hypotheses of Theorem 5.9,

$$\max_{1\le i\le N}|w_i - y(t_i)| \le \frac{Mh}{2L}|e^{L(b-a)} - 1|,$$

so Euler's method is convergent with respect to a differential equation satisfying the conditions of this theorem, and the rate of convergence is $O(h)$.

A one-step method is consistent precisely when the difference equation for the method approaches the differential equation when the step size goes to zero; that is, the local truncation error approaches zero as the step size approaches zero. The definition of convergence has a similar connotation. A method is convergent precisely when the solution to the difference equation approaches the solution to the differential equation as the step size goes to zero.

The other error-bound type of problem that exists when using difference methods to approximate solutions to differential equations is a consequence of not using exact results. In practice, neither the initial conditions nor the arithmetic that is subsequently performed is represented exactly because of the roundoff error associated with finite-digit arithmetic. In Section 5.2 we saw that this consideration can lead to difficulties even for the convergent Euler's method. To analyze this situation, at least partially, we will try to determine which methods are stable, in the sense that small changes or perturbations in the initial conditions produce correspondingly small changes in the subsequent approximations; that is, a **stable** method is one whose results depend *continuously* on the initial data.

Since the concept of stability of a one-step difference equation is somewhat analogous to the condition of a differential equation being well-posed, it is not surprising that the Lipschitz condition appears here, as it did in the corresponding theorem for differential equations, Theorem 5.6.

Part (i) of the following theorem concerns the stability of a one-step method. The proof of this result is not difficult and is considered in Exercise 1. Part (ii) of Theorem 5.19 concerns sufficient conditions for a consistent method to be convergent. Part (iii) justifies the remark made in Section 5.5 about controlling the global error of a method by controlling its local truncation error and implies that when the local truncation error has the rate of convergence $O(h^n)$, the global error will have the same rate of convergence. The proofs of parts (ii) and (iii) are more difficult than that of part (i), and can be found within the material presented in [Ge1, pp. 57–58].

Theorem 5.19 Suppose the initial-value problem

$$y' = f(t, y), \quad a \leq t \leq b, \quad y(a) = \alpha,$$

is approximated by a one-step difference method in the form

$$\begin{aligned} w_0 &= \alpha, \\ w_{i+1} &= w_i + h\phi(t_i, w_i, h). \end{aligned}$$

Suppose also that a number $h_0 > 0$ exists and that $\phi(t, w, h)$ is continuous and satisfies a Lipschitz condition in the variable w with Lipschitz constant L on the set

$$D = \{(t, w, h) \mid a \leq t \leq b, -\infty < w < \infty, 0 \leq h \leq h_0\}.$$

Then

(i) The method is stable;

(ii) The difference method is convergent if and only if it is consistent—that is, if and only if

$$\phi(t, y, 0) = f(t, y), \quad \text{for all } a \le t \le b;$$

(iii) If a function τ exists, and for each $i = 1, 2, \ldots, N$, the local truncation error $\tau_i(h)$ satisfies $|\tau_i(h)| \le \tau(h)$ whenever $0 \le h \le h_0$, then

$$|y(t_i) - w_i| \le \frac{\tau(h)}{L} e^{L(t_i - a)}. \qquad \blacksquare$$

EXAMPLE 1 Consider the Modified Euler method given by

$$w_0 = \alpha,$$

$$w_{i+1} = w_i + \frac{h}{2}[f(t_i, w_i) + f(t_{i+1}, w_i + hf(t_i, w_i))], \quad \text{for } i = 0, 1, \ldots, N - 1.$$

We will verify that this method satisfies the hypothesis of Theorem 5.19. For this method,

$$\phi(t, w, h) = \frac{1}{2}f(t, w) + \frac{1}{2}f(t + h, w + hf(t, w)).$$

If f satisfies a Lipschitz condition on $\{(t, w) | a \le t \le b, -\infty < w < \infty\}$ in the variable w with constant L, then, since

$$\begin{aligned}\phi(t, w, h) - \phi(t, \overline{w}, h) &= \frac{1}{2}f(t, w) + \frac{1}{2}f(t + h, w + hf(t, w)) \\ &\quad - \frac{1}{2}f(t, \overline{w}) - \frac{1}{2}f(t + h, \overline{w} + hf(t, \overline{w})),\end{aligned}$$

the Lipschitz condition on f leads to

$$\begin{aligned}|\phi(t, w, h) - \phi(t, \overline{w}, h)| &\le \frac{1}{2}L|w - \overline{w}| + \frac{1}{2}L|w + hf(t, w) - \overline{w} - hf(t, \overline{w})| \\ &\le L|w - \overline{w}| + \frac{1}{2}L|hf(t, w) - hf(t, \overline{w})| \\ &\le L|w - \overline{w}| + \frac{1}{2}hL^2|w - \overline{w}| \\ &= \left(L + \frac{1}{2}hL^2\right)|w - \overline{w}|.\end{aligned}$$

Therefore, ϕ satisfies a Lipschitz condition in w on the set

$$\{(t, w, h) \mid a \le t \le b, -\infty < w < \infty, 0 \le h \le h_0\}.$$

for any $h_0 > 0$ with constant

$$L' = L + \frac{1}{2}h_0L^2.$$

Finally, if f is continuous on $\{(t, w) \mid a \le t \le b, -\infty < w < \infty\}$, then ϕ is continuous on

$$\{(t, w, h) \mid a \le t \le b, -\infty < w < \infty, 0 \le h \le h_0\};$$

so Theorem 5.19 implies that the Modified Euler method is stable. Letting $h = 0$, we have

$$\phi(t, w, 0) = \frac{1}{2}f(t, w) + \frac{1}{2}f(t + 0, w + 0 \cdot f(t, w)) = f(t, w),$$

so the consistency condition expressed in Theorem 5.19, part (ii), holds. Thus, the method is convergent. Moreover, we have seen that for this method the local truncation error is $O(h^2)$, so the convergence of the Modified Euler method also has rate $O(h^2)$. ■

For multistep methods, the problems involved with consistency, convergence, and stability are compounded because of the number of approximations involved at each step. In the one-step methods, the approximation w_{i+1} depends directly only on the previous approximation w_i, whereas the multistep methods use at least two of the previous approximations, and the usual methods that are employed involve more.

The general multistep method for approximating the solution to the initial-value problem

(5.53)
$$y' = f(t, y), \quad a \le t \le b, \quad y(a) = \alpha,$$

can be written in the form

$$w_0 = \alpha, \quad w_1 = \alpha_1, \quad \ldots, \quad w_{m-1} = \alpha_{m-1},$$

(5.54)
$$w_{i+1} = a_{m-1}w_i + a_{m-2}w_{i-1} + \cdots + a_0w_{i+1-m} + hF(t_i, h, w_{i+1}, w_i, \ldots, w_{i+1-m}),$$

for each $i = m - 1, m, \ldots, N - 1$, where $a_0, a_1, \ldots, a_{m+1}$ are constants and, as usual, $h = (b - a)/N$ and $t_i = a + ih$.

The local truncation error for a multistep method expressed in this form is

$$\tau_{i+1}(h) = \frac{y(t_{i+1}) - a_{m-1}y(t_i) - \cdots - a_0y(t_{i+1-m})}{h} - F(t_i, h, y(t_{i+1}), y(t_i), \ldots, y(t_{i+1-m})),$$

for each $i = m - 1, m, \ldots, N - 1$. As in the one-step methods, the local truncation error measures how the solution $y(t)$ to the differential equation fails to satisfy the difference equation.

For the four-step Adams-Bashforth method, we have seen that

$$\tau_{i+1}(h) = \frac{251}{720}y^{(5)}(\mu_i)h^4, \quad \text{for some } \mu_i \in (t_{i-3}, t_{i+1}),$$

whereas the three-step Adams-Moulton method has

$$\tau_{i+1}(h) = -\frac{19}{720}y^{(5)}(\mu_i)h^4, \quad \text{for some } \mu_i \in (t_{i-2}, t_{i+1}),$$

provided, of course, that $y \in C^5[a, b]$.

Throughout the analysis, two assumptions will be made concerning the function F:

1. If $f \equiv 0$ (that is, if the differential equation is homogeneous), then $F \equiv 0$ also.
2. F satisfies a Lipschitz condition with respect to $\{w_j\}$, in the sense that a constant L exists and, for every pair of sequences $\{v_j\}_{j=0}^N$ and $\{\tilde{v}_j\}_{j=0}^N$ and for $i = m-1, m, \ldots, N-1$, we have

$$|F(t_i, h, v_{i+1}, \ldots, v_{i+1-m}) - F(t_i, h, \tilde{v}_{i+1}, \ldots, \tilde{v}_{i+1-m})| \le L\sum_{j=0}^{m} |v_{i+1-j} - \tilde{v}_{i+1-j}|.$$

The Adams-Bashforth and Adams-Moulton methods satisfy both of these conditions, provided f satisfies a Lipschitz condition. (See Exercise 2.)

The concept of convergence for multistep methods is the same as that for one-step methods; a multistep method is **convergent** if the solution to the difference equation approaches the solution to the differential equation as the step size approaches zero. This means that

$$\lim_{h\to 0} \max_{0\le i\le N} |w_i - y(t_i)| = 0.$$

For consistency, however, a slightly different situation occurs. Again, we want a multistep method to be consistent provided that the difference equation approaches the differential equation as the step size approaches zero; that is, the local truncation error must approach zero at each step as the step size approaches zero. The additional condition occurs because of the number of starting values required for multistep methods. Since usually only the first starting value, $w_0 = \alpha$, is exact, we need to require that the errors in all the starting values $\{\alpha_i\}$ approach zero as the step size approaches zero. So, both

(5.55) $$\lim_{h\to 0} |\tau_i(h)| = 0, \quad \text{for all } i = m, m+1, \ldots, N \quad \text{and}$$

(5.56) $$\lim_{h\to 0} |\alpha_i - y(t_i)| = 0, \quad \text{for all } i = 1, 2, \ldots, m-1,$$

must be true for a multistep method in the form (5.54) to be **consistent**. Note that (5.56) implies that a multistep method will not be consistent unless the one-step method generating the starting values is also consistent.

The following theorem for multistep methods is similar to Theorem 5.19, part (iii), and gives a relationship between the local truncation error and global error of a multistep method. It provides the theoretical justification for attempting to control global error by controlling local truncation error. The proof of a slightly more general form of this theorem can be found in [IK, pp. 387–388].

Theorem 5.20 Suppose the initial-value problem

$$y' = f(t, y), \quad a \le t \le b, \quad y(a) = \alpha,$$

is approximated by an Adams predictor-corrector method with an m-step Adams-Bashforth predictor equation

$$w_{i+1} = w_i + h[b_{m-1} f(t_i, w_i) + \cdots + b_0 f(t_{i+1-m}, w_{i+1-m})]$$

with local truncation error $\tau_{i+1}(h)$, and an $(m-1)$-step Adams-Moulton corrector equation

$$w_{i+1} = w_i + h[\tilde{b}_{m-1} f(t_i, w_{i+1}) + \tilde{b}_{m-2} f(t_i, w_i) + \cdots + \tilde{b}_0 f(t_{i+2-m}, w_{i+2-m})]$$

with local truncation error $\tilde{\tau}_{i+1}(h)$. In addition, suppose that $f(t, y)$ and $f_y(t, y)$ are continuous on $D = \{(t, y) \mid a \le t \le b \text{ and } -\infty < y < \infty\}$ and that f_y is bounded. Then the local truncation error $\sigma_{i+1}(h)$ of the predictor-corrector method is

$$\sigma_{i+1}(h) = \tilde{\tau}_{i+1}(h) + h\tau_{i+1}(h)\tilde{b}_{m-1} \frac{\partial f}{\partial y}(t_{i+1}, \theta_{i+1}),$$

where θ_{i+1} is a number between zero and $h\tau_{i+1}(h)$.

Moreover, there exist constants k_1 and k_2 such that

$$|w_i - y(t_i)| \le \left[\max_{0 \le j \le m-1} |w_j - y(t_j)| + k_1 \sigma(h)\right] e^{k_2(t_i - a)},$$

where $\sigma(h) = \max_{m \le j \le N} |\sigma_j(h)|$. ■

Before discussing connections between consistency, convergence, and stability for multistep methods, we need to consider in more detail the difference equation for a multistep method. In doing so, we will discover the reason for choosing the Adams methods as our standard multistep methods.

Associated with the difference equation (5.54) given at the beginning of this discussion,

$$w_0 = \alpha, \quad w_1 = \alpha_1, \quad \ldots, \quad w_{m-1} = \alpha_{m-1},$$
$$w_{i+1} = a_{m-1} w_i + a_{m-2} w_{i-1} + \cdots + a_0 w_{i+1-m} + hF(t_i, h, w_{i+1}, w_i, \ldots, w_{i+1-m}),$$

is the **characteristic equation** of the method, given by

$$\lambda^m - a_{m-1}\lambda^{m-1} - a_{m-2}\lambda^{m-2} - \cdots - a_1\lambda - a_0 = 0. \tag{5.57}$$

The magnitudes of the roots of the characteristic equation of a multistep method are associated with the stability of the method with respect to roundoff error. To see this, consider applying the standard multistep method (5.54) to the trivial initial-value problem

$$y' \equiv 0, \quad y(a) = \alpha, \quad \text{where } \alpha \ne 0. \tag{5.58}$$

This problem has exact solution $y(t) \equiv \alpha$. By examining Eqs. (5.26) and (5.27) in Section 5.6, we can see that any multistep method will, in theory, produce the exact solution $w_n = \alpha$ for all n. The only deviation from the exact solution is due to the inherent roundoff error associated with the calculations involved in the method.

The right side of the differential equation in (5.58) has $f(t, y) \equiv 0$, so by assumption (1), we have $F(t_i, h, w_{i+1}, w_{i+2}, \ldots, w_{i+1-m}) = 0$ in the difference equation (5.54). As a consequence, the standard form of the difference equation becomes

(5.59)
$$w_{i+1} = a_{m-1}w_i + a_{m-2}w_{i-1} + \cdots + a_0 w_{i+1-m}.$$

Suppose λ is one of the roots of the characteristic equation associated with (5.54). Then $w_n = \lambda^n$ for each n is a solution to (5.59) since

$$\lambda^{i+1} - a_{m-1}\lambda^i - a_{m-2}\lambda^{i-1} - \cdots - a_0\lambda^{i+1-m} = \lambda^{i+1-m}[\lambda^m - a_{m-1}\lambda^{m-1} - \cdots - a_0] = 0.$$

In fact, if $\lambda_1, \lambda_2, \ldots, \lambda_m$ are distinct roots of the characteristic equation for (5.54), it can be shown that *every* solution to (5.59) can be expressed in the form

(5.60)
$$w_n = \sum_{i=1}^{m} c_i \lambda_i^n,$$

for some unique collection of constants $c_1, c_2, \ldots, c_m$.

Since the exact solution to (5.58) is $y(t) = \alpha$, the choice $w_n = \alpha$, for all n, is a solution to (5.59). Using this fact in (5.58) gives

$$0 = \alpha - \alpha a_{m-1} - \alpha a_{m-2} - \cdots - \alpha a_0 = \alpha[1 - a_{m-1} - a_{m-2} - \cdots - a_0].$$

This implies that $\lambda = 1$ is one of the solutions of the characteristic equation (5.57). We will assume that in the representation (5.60) this solution is described by $\lambda_1 = 1$ and $c_1 = \alpha$, so all solutions to (5.59) are expressed as

(5.61)
$$w_n = \alpha + \sum_{i=2}^{m} c_i \lambda_i^n.$$

If all the calculations were exact, the constants $c_2, c_3, \ldots, c_m$ would all be zero. In practice, the constants $c_2, c_3, \ldots, c_m$ are not zero due to roundoff error. In fact, the roundoff error grows exponentially unless $|\lambda_i| \leq 1$ for each of the roots $\lambda_2, \lambda_3, \ldots, \lambda_m$. The smaller the magnitude of these roots, the more stable the method will be with respect to the growth of roundoff error.

We made the simplifying assumption in deriving (5.61) that the roots of the characteristic equation are distinct. The situation is similar when multiple roots occur. For example, if $\lambda_k = \lambda_{k+1} = \cdots = \lambda_{k+p}$ for some k and p, it simply requires replacing the sum

$$c_k\lambda_k^n + c_{k+1}\lambda_{k+1}^n + \cdots + c_{k+p}\lambda_{k+p}^n$$

od
is

in (5.61) by

(5.62)
$$c_k\lambda_k^n + c_{k+1}n\lambda_k^{n-1} + c_{k+2}n(n-1)\lambda_k^{n-2} + \cdots + c_{k+p}[n(n-1)\cdots(n-p+1)]\lambda_k^{n-p}.$$

(See [He2, pp. 119–145].) Although the form of the solution is modified, the roundoff effect if $|\lambda_k| > 1$ remains the same.

Although we have considered only the special case of approximating initial-value problems of the form (5.58), the stability characteristics for this equation determine the stability for the situation when $f(t, y)$ is not identically zero. This is due to the fact that the solution to the homogeneous equation (5.58) is embedded in the solution to any equation. The following definitions are motivated by this discussion.

Definition 5.21 Let $\lambda_1, \lambda_2, \ldots, \lambda_m$ denote the (not necessarily distinct) roots of the characteristic equation

$$\lambda^m - a_{m-1}\lambda^{m-1} - \cdots - a_1\lambda - a_0 = 0$$

associated with the multistep difference method

$$w_0 = \alpha, \quad w_1 = \alpha_1, \quad \ldots, \quad w_{m-1} = \alpha_{m-1}$$

and

$$w_{i+1} = a_{m-1}w_i + a_{m-2}w_{i-1} + \cdots + a_0w_{i+1-m} + hF(t_i, h, w_{i+1}, w_i, \ldots, w_{i+1-m}).$$

If $|\lambda_i| \leq 1$ for each $i = 1, 2, \ldots, m$ and all roots with absolute value 1 are simple roots, then the difference method is said to satisfy the **root condition**. ■

Definition 5.22

(i) Methods that satisfy the root condition and have $\lambda = 1$ as the only root of the characteristic equation of magnitude one are called **strongly stable**.

(ii) Methods that satisfy the root condition and have more than one distinct root with magnitude one are called **weakly stable**.

(iii) Methods that do not satisfy the root condition are called **unstable**. ■

Consistency and convergence of a multistep method are closely related to the roundoff stability of the method. The next theorem details these connections. For the proof of this result and the theory on which it is based, see [IK, pp. 410–417].

Theorem 5.23 A multistep method of the form

$$w_0 = \alpha, \quad w_1 = \alpha_1, \quad \ldots, \quad w_{m-1} = \alpha_{m-1},$$

where

$$w_{i+1} = a_{m-1}w_i + a_{m-2}w_{i-1} + \cdots + a_0w_{i+1-m} + hF(t_i, h, w_{i+1}, w_i, \ldots, w_{i+1-m})$$

is stable if and only if it satisfies the root condition. Moreover, if the difference meth is consistent with the differential equation, then the method is stable if and only if it convergent. ■

EXAMPLE 2 We have seen that the fourth-order Adams-Bashforth method can be expressed as

$$w_{i+1} = w_i + hF(t_i, h, w_{i+1}, w_i, \ldots, w_{i-3}),$$

where

$$F(t_i, h, w_{i+1}, w_i, \ldots, w_{i-3}) = \frac{1}{24}[55f(t_i, w_i) - 59f(t_{i-1}, w_{i-1}) + 37f(t_{i-2}, w_{i-2}) - 9f(t_{i-3}, w_{i-3})];$$

so $m = 4$, $a_0 = 0$, $a_1 = 0$, $a_2 = 0$, and $a_3 = 1$.

The characteristic equation for this Adams-Bashforth method is, consequently,

$$\lambda^4 - \lambda^3 = \lambda^3(\lambda - 1) = 0,$$

which has roots $\lambda_1 = 1$, $\lambda_2 = 0$, $\lambda_3 = 0$, and $\lambda_4 = 0$. It satisfies the root condition and is strongly stable.

The Adams-Moulton method has a similar characteristic equation, $\lambda^3 - \lambda^2 = 0$, and is also strongly stable. ■

EXAMPLE 3 The explicit multistep method given by

$$w_{i+1} = w_{i-3} + \frac{4h}{3}[2f(t_i, w_i) - f(t_{i-1}, w_{i-1}) + 2f(t_{i-2}, w_{i-2})]$$

was introduced in Section 5.6 as the fourth-order Milne's method. Since the characteristic equation for this method, $\lambda^4 - 1 = 0$, has four roots with magnitude one: $\lambda_1 = 1$, $\lambda_2 = -1$, $\lambda_3 = i$, and $\lambda_4 = -i$, the method satisfies the root condition, but it is only weakly stable.

Consider the initial-value problem

$$y' = -6y + 6, \quad 0 \leq t \leq 1, \quad y(0) = 2,$$

which has the exact solution $y(t) = 1 + e^{-6t}$. For comparison purposes, the strongly stable explicit fourth-order Adams-Bashforth method and Milne's method are used to approximate the solution to this problem with $h = 0.1$, with exact values for the starting values. The results in Table 5.19 show the effects of a weakly stable method versus a strongly stable method for this problem.

The reason for choosing the Adams-Bashforth-Moulton as our standard fourth-order predictor-corrector technique in Section 5.6 over the Milne-Simpson method of the same order is that both the Adams-Bashforth and Adams-Moulton methods are strongly stable. They are more likely to give accurate approximations to a wider class of problems than is the predictor-corrector based on the Milne and Simpson techniques, both of which are weakly stable. ■

Table 5.19

t_i	Exact $y(t_i)$	Adams-Bashforth Method w_i	Error $\lvert y_i - w_i \rvert$	Milne's Method w_i	Error $\lvert y_i - w_i \rvert$
0.10000000		1.5488116		1.5488116	
0.20000000		1.3011942		1.3011942	
0.30000000		1.1652989		1.1652989	
0.40000000	1.0907180	1.0996236	8.906×10^{-3}	1.0983785	7.661×10^{-3}
0.50000000	1.0497871	1.0513350	1.548×10^{-3}	1.0417344	8.053×10^{-3}
0.60000000	1.0273237	1.0425614	1.524×10^{-2}	1.0486438	2.132×10^{-2}
0.70000000	1.0149956	1.0047990	1.020×10^{-2}	0.9634506	5.154×10^{-2}
0.80000000	1.0082297	1.0359090	2.768×10^{-2}	1.1289977	1.208×10^{-1}
0.90000000	1.0045166	0.9657936	3.872×10^{-2}	0.7282684	2.762×10^{-1}
1.00000000	1.0024788	1.0709304	6.845×10^{-2}	1.6450917	6.426×10^{-1}

EXERCISE SET 5.10

1. To prove Theorem 5.19, part (i), show that the hypotheses imply that there exists a constant $K > 0$ such that

$$|u_i - v_i| \leq K|u_0 - v_0|, \quad \text{for each } 1 \leq i \leq N,$$

whenever $\{u_i\}_{i=1}^N$ and $\{v_i\}_{i=1}^N$ satisfy the difference equation $w_{i+1} = w_i + h\phi(t_i, w_i, h)$.

2. For the Adams-Bashforth and Adams-Moulton methods of order four,

   a. Show that if $f = 0$, then

$$F(t_i, h, w_{i+1}, \ldots, w_{i+1-m}) = 0.$$

   b. Show that if f satisfies a Lipschitz condition with constant L, then a constant C exists with

$$|F(t_i, h, w_{i+1}, \ldots, w_{i+1-m}) - F(t_i, h, v_{i+1}, \ldots, v_{i+1-m})| \leq C\sum_{j=0}^{m} |w_{i+1-j} - v_{i+1-j}|.$$

3. Use the results of Exercise 17 in Section 5.4 to show that the Runge-Kutta fourth-order method is consistent.

4. Consider the differential equation

$$y' = f(t, y), \quad a \leq t \leq b, \quad y(a) = \alpha.$$

   a. Show that

$$y'(t_i) = \frac{-3y(t_i) + 4y(t_{i+1}) - y(t_{i+2})}{2h} + \frac{h^2}{3}y'''(\xi_i),$$

   for some ξ_i, where $t_i < \xi_i < t_{i+2}$.

   b. Part (a) suggests the difference method

$$w_{i+2} = 4w_{i+1} - 3w_i - 2hf(t_i, w_i), \quad \text{for } i = 0, 1, \ldots, N-2.$$

Use this method to solve

$$y' = 1 - y, \quad 0 \le t \le 1, \quad y(0) = 0,$$

with $h = 0.1$. Use the starting values $w_0 = 0$ and $w_1 = y(t_1) = 1 - e^{-0.1}$.

c. Repeat part (b) with $h = 0.01$ and $w_1 = 1 - e^{-0.01}$.

d. Analyze this method for consistency, stability, and convergence.

5. Given the multistep method

$$w_{i+1} = -\frac{3}{2}w_i + 3w_{i-1} - \frac{1}{2}w_{i-2} + 3hf(t_i, w_i), \quad \text{for } i = 2, \ldots, N-1,$$

with starting values w_0, w_1, w_2:

a. Find the local truncation error.

b. Comment on consistency, stability, and convergence.

6. Obtain an approximate solution to the differential equation

$$y' = -y, \quad 0 \le t \le 10, \quad y(0) = 1$$

using Milne's method with $h = 0.1$ and then $h = 0.01$, with starting values $w_0 = 1$ and $w_1 = e^{-h}$ in both cases. How does decreasing h from $h = 0.1$ to $h = 0.01$ affect the number of correct digits in the approximate solutions at $t = 1$ and $t = 10$?

7. Investigate stability for the difference method

$$w_{i+1} = -4w_i + 5w_{i-1} + 2h[f(t_i, w_i) + 2hf(t_{i-1}, w_{i-1})],$$

for $i = 1, 2, \ldots, N-1$, with starting values w_0, w_1.

8. Consider the problem $y' = 0$, for $0 \le t \le 10$, with $y(0) = 0$, which has the solution $y \equiv 0$. If the difference method of Exercise 4 is applied to the problem, then

$$w_{i+1} = 4w_i - 3w_{i-1}, \quad \text{for } i = 1, 2, \ldots, N-1,$$

$$w_0 = 0, \quad \text{and} \quad w_1 = \alpha_1.$$

Suppose $w_1 = \alpha_1 = \varepsilon$, where ε is a small roundoff error. Compute w_i exactly for $i = 2, 3, \ldots, 6$ to find how the error ε is propagated.

5.11 Stiff Differential Equations

Significant difficulties can occur when standard numerical techniques are applied to approximate the solution of a differential equation when the exact solution contains terms of the form $e^{\lambda t}$, where λ is a complex number with negative real part. This term decays to zero with increasing t, but its approximation generally will not have this property unless restrictions are placed on the step size of the method. The problem is particularly acute when the exact solution consists of a steady-state term that does not grow significantly with t, together with a transient term that decays rapidly to zero. In such a problem, the numerical method should approximate the steady-state portion of the solution, but unless care is taken, the error associated with the decaying transient portion will dominate the calculations and produce meaningless results.

Problems involving rapidly decaying transient solutions occur naturally in a wide variety of applications, including the study of spring and damping systems, the analysis of control systems, and problems in chemical kinetics. These are all examples of a class of problems called **stiff systems** of differential equations, due to their application in analyzing the motion of spring and mass systems with large spring constants.

EXAMPLE 1 The system of initial-value problems

$$u_1' = 9u_1 + 24u_2 + 5\cos t - \frac{1}{3}\sin t, \quad u_1(0) = \frac{4}{3}$$

$$u_2' = -24u_1 - 51u_2 - 9\cos t + \frac{1}{3}\sin t, \quad u_2(0) = \frac{2}{3}$$

has the unique solution

$$u_1(t) = 2e^{-3t} - e^{-39t} + \frac{1}{3}\cos t,$$

$$u_2(t) = -e^{-3t} + 2e^{-39t} - \frac{1}{3}\cos t.$$

The transient term e^{-39t} in the solution causes this system to be stiff. Applying Algorithm 5.7, the Runge-Kutta Fourth-Order Method for Systems, gives results listed in Table 5.20. When $h = 0.05$ stability results and the approximations are accurate. Increasing the step size to $h = 0.1$, however, leads to the disastrous results shown in the table. ■

Table 5.20

t	$u_1(t)$	$w_1(t)$ $h = 0.05$	$w_1(t)$ $h = 0.1$	$u_2(t)$	$w_2(t)$ $h = 0.05$	$w_2(t)$ $h = 0.1$
0.1	1.793061	1.712219	−2.645169	−1.032001	−0.8703152	7.844527
0.2	1.423901	1.414070	−18.45158	−0.8746809	−0.8550148	38.87631
0.3	1.131575	1.130523	−87.47221	−0.7249984	−0.7228910	176.4828
0.4	0.9094086	0.9092763	−934.0722	−0.6082141	−0.6079475	789.3540
0.5	0.7387877	9.7387506	−1760.016	−0.5156575	−0.5155810	3520.999
0.6	0.6057094	0.6056833	−7848.550	−0.4404108	−0.4403558	15697.84
0.7	0.4998603	0.4998361	−34989.63	−0.3774038	−0.3773540	69979.87
0.8	0.4136714	0.4136490	−155979.4	−0.3229535	−0.3229078	311959.5
0.9	0.3416143	0.3415939	−695332.0	−0.2744088	−0.2743673	1390664.
1.0	0.2796748	0.2796568	−3099671.	−0.2298877	−0.2298511	6199352.

Although stiffness is usually associated with systems of differential equations, the approximation characteristics of a particular numerical method applied to a stiff system can be predicted by examining the error produced when the method is applied to a simple *test equation*,

(5.63) $$y' = \lambda y, \quad y(0) = \alpha, \quad \text{where } \lambda < 0.$$

The solution to this equation is $y(t) = \alpha e^{\lambda t}$, which contains the transient term $e^{\lambda t}$ and the steady-state term is zero, so the approximation characteristics of a method are easy to determine. (A more complete discussion of the roundoff error associated with stiff systems requires examining the test equation when λ is a complex number with negative imaginary part; see [Ge1, p. 222].)

First consider Euler's method applied to the test equation. Letting $h = (b - a)/N$ and $t_j = jh$, for $j = 0, 1, 2, \ldots, N$, Eq. (5.8) implies that

$$w_0 = \alpha,$$

and

$$w_{j+1} = w_j + h(\lambda w_j) = (1 + h\lambda)w_j,$$

so

(5.64) $$w_{j+1} = (1 + h\lambda)^{j+1}w_0 = (1 + h\lambda)^{j+1}\alpha, \quad \text{for } j = 0, 1, \ldots, N - 1.$$

Since the exact solution is $y(t) = \alpha e^{\lambda t}$, the absolute error is

$$|y(t_j) - w_j| = |e^{jh\lambda} - (1 + h\lambda)^j||\alpha| = |(e^{h\lambda})^j - (1 + h\lambda)^j|\,|\alpha|,$$

and the accuracy is determined by how well the term $1 + h\lambda$ approximates $e^{h\lambda}$. When $\lambda < 0$, the exact solution $(e^{h\lambda})^j$ decays to zero as j increases, but by (5.64), the approximation will have this property only if $|1 + h\lambda| < 1$. This effectively restricts the step size h for Euler's method to satisfy $h < 2/|\lambda|$.

Suppose now that a roundoff error δ_0 is introduced in the initial condition for Euler's method,

$$w_0 = \alpha + \delta_0.$$

At the jth step the roundoff error is

$$\delta_j = (1 + h\lambda)^j\delta_0.$$

Since $\lambda < 0$, the condition for the control of the growth of roundoff error is the same as the condition for controlling the absolute error, $h < 2/|\lambda|$.

The situation is similar for other one-step methods. In general, a function Q exists with the property that the difference method, when applied to the test equation, gives

(5.65) $$w_{i+1} = Q(h\lambda)w_i.$$

The accuracy of the method depends upon how well $Q(h\lambda)$ approximates $e^{h\lambda}$, and the error will grow without bound if $|Q(h\lambda)| > 1$. An nth-order Taylor method, for example, will have stability with regard to both the growth of roundoff error and absolute error, provided h is chosen to satisfy

$$\left|1 + h\lambda + \left(\frac{1}{2}\right)h^2\lambda^2 + \cdots + \left(\frac{1}{n!}\right)h^n\lambda^n\right| < 1.$$

Exercise 6 examines the specific case when the method is the classical fourth-order Runge-Kutta method, a Taylor method of order four.

When a multistep method of the form (5.23) is applied to the test equation, the result is

$$w_{j+1} = a_{m-1}w_j + \cdots + a_0 w_{j+1-m} + h\lambda(b_m w_{j+1} + b_{m-1}w_j + \cdots + b_0 w_{j+1-m}),$$

for $j = m-1, \ldots, N-1$, or

$$(1 - h\lambda b_m)w_{j+1} - (a_{m-1} + h\lambda b_{m-1})w_j - \cdots - (a_0 + h\lambda b_0)w_{j+1-m} = 0.$$

Associated with this homogeneous difference equation is a **characteristic polynomial**

$$Q(z, h\lambda) = (1 - h\lambda b_m)z^m - (a_{m-1} + h\lambda b_{m-1})z^{m-1} - \cdots - (a_0 + h\lambda b_0).$$

This polynomial is similar to the polynomial in the characteristic equation (5.57) but it also incorporates the test equation. The theory here parallels the stability discussion in Section 5.10.

Suppose $w_0, \ldots, w_{m-1}$ are given, and, for fixed $h\lambda$, let $\beta_1, \ldots, \beta_m$ be the roots of the equation $Q(z, h\lambda) = 0$. If $\beta_1, \ldots, \beta_m$ are distinct, then $c_1, \ldots, c_m$ exist with

(5.66) $$w_j = \sum_{k=1}^{m} c_k(\beta_k)^j, \quad \text{for } j = 0, \ldots, N.$$

If $Q(z, h\lambda) = 0$ has multiple roots, w_j is similarly defined. (See Eq. (5.62) in Section 5.10.) If w_j is to accurately approximate $y(t_j) = e^{jh\lambda} = (e^{h\lambda})^j$, then all roots β_k must satisfy $|\beta_k| < 1$; otherwise, certain choices of α will result in $c_k \neq 0$, and the term $c_k(\beta_k)^j$ will not decay to zero.

EXAMPLE 2 The test differential equation

$$y' = -30y, \quad 0 \le t \le 1.5, \quad y(0) = \frac{1}{3}$$

has exact solution $y = \frac{1}{3}e^{-30t}$. Using $h = 0.1$ for Euler's Algorithm 5.1, Runge-Kutta Fourth-Order Algorithm 5.2, and the Adams Predictor-Corrector Algorithm 5.4 gives the results at $t = 1.5$ in Table 5.21. ■

Table 5.21

Exact solution	9.54173×10^{-21}
Euler's method	-1.09225×10^{4}
Runge-Kutta method	3.95730×10^{1}
Predictor-corrector method	8.03840×10^{5}

The inaccuracies in Example 2 are due to the fact that $|Q(h\lambda)| > 1$ for Euler's method and the Runge-Kutta method and that $Q(z, h\lambda) = 0$ has roots with modulus exceeding 1 for the predictor-corrector method. To apply these methods to this problem, the step size must be reduced. The following definition is used to describe the amount of step-size reduction that is required.

Definition 5.24 The **region *R* of absolute stability** for a one-step method is $R = \{h\lambda \in C \mid |Q(h\lambda)| < 1\}$, and for a multistep method, it is $R = \{h\lambda \in C \mid |\beta_k| < 1 \text{ for all roots } \beta_k \text{ of } Q(z, h\lambda) = 0\}$. ■

Equations (5.65) and (5.66) imply that a method can be applied effectively to a stiff equation only if $h\lambda$ is in the region of absolute stability of the method, which for a given problem places a restriction on the size of h. Even though the exponential term in the exact solution decays quickly to zero, λh must remain within the region of absolute stability throughout the interval of t values for the approximation to decay to zero and the growth of error to be under control. This means that although h could normally be increased because of truncation error considerations, the absolute stability criterion forces h to remain small. Variable step-size methods are especially vulnerable to this problem, since an examination of the local truncation error might indicate that the step size could increase, which would inadvertently result in λh being outside the region of absolute stability.

Since the region of absolute stability of a method is generally the critical factor in producing accurate approximations for stiff systems, numerical methods are sought with as large a region of absolute stability as possible. A numerical method is said to be ***A*-stable** if its region R of absolute stability contains the entire left half-plane.

The **Implicit Trapezoidal method**, given by

(5.67)
$$\begin{aligned} w_0 &= \alpha, \\ w_{j+1} &= w_j + \frac{h}{2}[f(t_{j+1}, w_{j+1}) + f(t_j, w_j)], \quad 0 \le j \le N-1, \end{aligned}$$

is an A-stable method (see Exercise 9) and is the only A-stable multistep method. Although the Trapezoidal method does not give accurate approximations for large step sizes, the approximations do not grow exponentially like those of the Runge-Kutta method.

The techniques commonly used for stiff systems are implicit multistep methods. Generally, w_{i+1} is obtained by solving a nonlinear equation or nonlinear system iteratively, often by Newton's method. Consider, for example, the Implicit Trapezoidal method

$$w_{j+1} = w_j + \frac{h}{2}[f(t_{j+1}, w_{j+1}) + f(t_j, w_j)].$$

Having computed t_j, t_{j+1}, and w_j, we need to determine $w = w_{j+1}$, the solution to

(5.68)
$$F(w) = w - w_j - \frac{h}{2}[f(t_{j+1}, w) + f(t_j, w_j)] = 0.$$

To approximate this solution, select $w_{j+1}^{(0)}$, usually as w_j, and generate $w_{j+1}^{(k)}$ by applying Newton's method to (5.68),

$$w_{j+1}^{(k)} = w_{j+1}^{(k-1)} - \frac{F(w_{j+1}^{(k-1)})}{F'(w_{j+1}^{(k-1)})}$$

$$= w_{j+1}^{(k-1)} - \frac{w_{j+1}^{(k-1)} - w_j - \frac{h}{2}[f(t_j, w_j) + f(t_{j+1}, w_{j+1}^{(k-1)})]}{1 - \frac{h}{2} f_y(t_{j+1}, w_{j+1}^{(k-1)})}$$

until $|w_{j+1}^{(k)} - w_{j+1}^{(k-1)}|$ is sufficiently small. This is the procedure that is used in Algorithm 5.8. Normally only three or four iterations per step are required.

The Secant method can be used as an alternative to Newton's method in Eq. (5.68), but then two distinct initial approximations to w_{j+1} are required. To employ the Secant method, the usual practice is to let $w_{j+1}^{(0)} = w_j$ and obtain $w_{j+1}^{(1)}$ from some explicit multistep method. When a system of stiff equations is involved, a generalization is required for either Newton's or the Secant method. These topics are considered in Chapter 10.

ALGORITHM **5.8**

Trapezoidal with Newton Iteration

To approximate the solution of the initial-value problem

$$y' = f(t, y), \quad a \le t \le b, \quad y(a) = \alpha$$

at $(N + 1)$ equally spaced numbers in the interval $[a, b]$:

INPUT endpoints a, b; integer N; initial condition α; tolerance TOL; maximum number of iterations M at any one step.

OUTPUT approximation w to y at the $(N + 1)$ values of t, or a message of failure.

Step 1 Set $h = (b - a)/N$;
$t = a$;
$w = \alpha$;
OUTPUT (t, w).

Step 2 For $i = 1, 2, \ldots, N$ do Steps 3–7.

Step 3 Set $k_1 = w + \frac{h}{2} f(t, w)$;
$w_0 = k_1$;
$j = 1$;
$FLAG = 0$.

Step 4 While $FLAG = 0$ do Steps 5–6.

Step 5 Set $w = w_0 - \dfrac{w_0 - \frac{h}{2} f(t + h, w_0) - k_1}{1 - \frac{h}{2} f_y(t + h, w_0)}$.

Step 6 If $|w - w_0| < TOL$ then set $FLAG = 1$
else set $j = j + 1$;
$w_0 = w$;
if $j > M$ then
OUTPUT ('MAXIMUM NUMBER OF ITERATIONS EXCEEDED');
STOP.

Step 7 Set $t = a + ih$;
OUTPUT (t, w).

Step 8 STOP.

EXAMPLE 3 The stiff initial-value problem

$$y' = 5e^{5t}(y - t)^2 + 1, \quad 0 \le t \le 1, \quad y(0) = -1$$

has solution $y(t) = t - e^{-5t}$. To show the effects of stiffness, the Implicit Trapezoidal method and the Runge-Kutta fourth-order method are applied both with $N = 4$ and $h = 0.25$, and with $N = 5$ and $h = 0.20$. The Trapezoidal method performs well in both cases using $M = 10$ and $TOL = 10^{-6}$, as does Runge-Kutta with $h = 0.2$. However, $h = 0.25$ is outside the region of absolute stability of the Runge-Kutta method, which is evident from the results in Table 5.22. ■

Table 5.22

	Runge–Kutta Method		Trapezoidal Method	
	$h = 0.2$		$h = 0.2$	
t_i	w_i	$\|y(t_i) - w_i\|$	w_i	$\|y(t_i) - w_i\|$
0.0	−1.0000000	0	−1.0000000	0
0.2	−0.1488521	1.9027×10^{-2}	−0.1414969	2.6383×10^{-2}
0.4	0.2684884	3.8237×10^{-3}	0.2748614	1.0197×10^{-2}
0.6	0.5519927	1.7798×10^{-3}	0.5539828	3.7700×10^{-3}
0.8	0.7822857	6.0131×10^{-4}	0.7830720	1.3876×10^{-3}
1.0	0.9934905	2.2845×10^{-4}	0.9937726	5.1050×10^{-4}
	$h = 0.25$		$h = 0.25$	
t_i	w_i	$\|y(t_i) - w_i\|$	w_i	$\|y(t_i) - w_i\|$
0.0	−1.0000000	0	−1.0000000	0
0.25	0.4014315	4.37936×10^{-1}	0.0054557	4.1961×10^{-2}
0.5	3.4374753	3.01956×10^{0}	0.4267572	8.8422×10^{-3}
0.75	1.44639×10^{23}	1.44639×10^{23}	0.7291528	2.6706×10^{-3}
1.0	Overflow		0.9940199	7.5790×10^{-4}

We have presented here only a small amount of what the reader frequently encountering stiff differential equations should know. We recommend for further details that [Ge2], [Lam], or [SGe] be consulted.

EXERCISE SET 5.11

1. Solve the following stiff initial-value problems using Euler's method and compare the results with the actual solution.
   a. $y' = -9y, \quad 0 \le t \le 1, \quad y(0) = e$, with $h = 0.1$; actual solution $y(t) = e^{1-9t}$.
   b. $y' = -20(y - t^2) + 2t, \quad 0 \le t \le 1, \quad y(0) = \frac{1}{3}$, with $h = 0.1$; actual solution $y(t) = t^2 + \frac{1}{3}e^{-20t}$.
   c. $y' = -20y + 20\sin t + \cos t, \quad 0 \le t \le 2, \quad y(0) = 1$, with $h = 0.25$; actual solution $y(t) = \sin t + e^{-20t}$.
   d. $y' = \dfrac{50}{y} - 50y, \quad 0 \le t \le 1, \quad y(0) = \sqrt{2}$, with $h = 0.1$; actual solution $y(t) = (1 + e^{-100t})^{1/2}$.
2. Repeat Exercise 1 using the Runge-Kutta fourth-order method.
3. Repeat Exercise 1 using the Adams fourth-order predictor-corrector method.
4. Repeat Exercise 1 using the Trapezoidal Algorithm. Use $TOL = 10^{-5}$.
5. Solve the following stiff initial-value problem using the fourth-order Runge-Kutta method with (a) $h = 0.1$ and (b) $h = 0.025$.

$$u_1' = 32u_1 + 66u_2 + \frac{2}{3}t + \frac{2}{3}, \quad 0 \le t \le 0.5, \quad u_1(0) = \frac{1}{3};$$

$$u_2' = -66u_1 - 133u_2 - \frac{1}{3}t - \frac{1}{3}, \quad 0 \le t \le 0.5, \quad u_2(0) = \frac{1}{3}.$$

   Compare the results to the actual solution,

$$u_1(t) = \frac{2}{3}t + \frac{2}{3}e^{-t} - \frac{1}{3}e^{-100t} \quad \text{and} \quad u_2(t) = -\frac{1}{3}t - \frac{1}{3}e^{-t} + \frac{2}{3}e^{-100t}.$$

6. Show that the fourth-order Runge-Kutta method,

$$\begin{aligned}
k_1 &= hf(t_i, w_i),\\
k_2 &= hf(t_i + h/2, w_i + k_1/2),\\
k_3 &= hf(t_i + h/2, w_i + k_2/2),\\
k_4 &= hf(t_i + h, w_i + k_3),\\
w_{i+1} &= w_i + \frac{1}{6}(k_1 + 2k_2 + 2k_3 + k_4),
\end{aligned}$$

   when applied to the differential equation $y' = \lambda y$, can be written in the form

$$w_{i+1} = \left(1 + h\lambda + \frac{1}{2}(h\lambda)^2 + \frac{1}{6}(h\lambda)^3 + \frac{1}{24}(h\lambda)^4\right) w_i.$$

7. Discuss consistency, stability, and convergence for the Implicit Trapezoidal method

$$w_{i+1} = w_i + \frac{h}{2}[f(t_{i+1}, w_{i+1}) + f(t_i, w_i)], \quad \text{for } i = 0, 1, \ldots, N-1,$$

   with $w_0 = \alpha$ applied to the differential equation

$$y' = f(t, y), \quad a \le t \le b, \quad y(a) = \alpha.$$

8. The Backward Euler one-step method is defined by

$$w_{i+1} = w_i + hf(t_{i+1}, w_{i+1}), \quad \text{for } i = 0, \ldots, N-1.$$

a. Show that $Q(h\lambda) = 1/(1 - h\lambda)$ for the Backward Euler method.

b. Apply the Backward Euler method to the differential equations given in Exercise 1. Use Newton's method to solve for w_{i+1}.

9. **a.** Show that the Implicit Trapezoidal method (5.67) is A-stable.

b. Show that the Backward Euler method described in Exercise 8 is A-stable.

5.12 Survey of Methods and Software

In this chapter we have considered methods to approximate the solutions to initial-value problems for ordinary differential equations. We began with a discussion of the most elementary numerical technique, Euler's method. This procedure was not sufficiently accurate to be of use in applications, but it illustrated the general behavior of the more powerful techniques, without the accompanying algebraic difficulties. The Taylor methods were then considered as generalizations of Euler's method. They were found to be accurate but cumbersome because of the need to determine extensive partial derivatives of the defining function of the differential equation. The Runge-Kutta formulas simplified the Taylor methods, while not significantly increasing the error. To this point we had considered only one-step methods, techniques that use only data at the most recently computed point.

Multistep methods were discussed in Section 5.6, where explicit methods of Adams-Bashforth type and implicit methods of Adams-Moulton type were considered. These lead to predictor-corrector methods, which use an explicit method, such as an Adams-Bashforth, to predict the solution and then apply a corresponding implicit method, like an Adams-Moulton, to correct the approximation.

Section 5.9 illustrated how these techniques can be used to solve higher order initial-value problems and systems of initial-value problems.

These one- and multistep methods serve as an introduction to numerical methods for ordinary differential equations since the more accurate adaptive methods are based on these relatively uncomplicated techniques. In particular, we saw in Section 5.5 that the Runge-Kutta-Fehlberg method is a one-step procedure that seeks to select mesh spacing to keep the local error of the approximation under control. The Variable Step-Size Predictor-Corrector method presented in Section 5.7 is based on the four-step Adams-Bashforth method and three-step Adams-Moulton method. It also changes the step size to keep the local error within a given tolerance. The Extrapolation method discussed in Section 5.8 is based on a modification of the Midpoint method and incorporates extrapolation to maintain a desired accuracy of approximation.

The final topic in the chapter concerned the difficulty that is inherent in the approximation of the solution to a stiff equation, a differential equation whose exact solution contains a portion of the form $e^{-\lambda t}$, where λ is a positive constant. Special caution must be taken with problems of this type, or the results can be overwhelmed by roundoff error.

Methods of the Runge-Kutta-Fehlberg type are generally sufficient for nonstiff problems where moderate accuracy is required. The extrapolation procedures are recommended for nonstiff problems where high accuracy is required. Finally, extensions of the Implicit Trapezoidal method to variable-order and variable step-size implicit Adams-type methods are used for stiff initial-value problems.

The IMSL Library includes three subroutines for approximating the solutions of initial-value problems. Each of the methods solves a system of m first-order equations in m variables. The equations are of the form

$$\frac{du_i}{dt} = f_i(t, u_1, u_2, \ldots, u_m), \quad \text{for } i = 1, 2, \ldots, m,$$

where $u_i(t_0)$ is given for each i. The variable step-size subroutine IVPRK is based on the Runge-Kutta-Verner fifth- and sixth-order methods described in Exercise 4 of Section 5.5. The subroutine IVPBS is an extrapolation procedure based on the Bulirsh-Stoer extrapolation method. This technique uses extrapolation by rational functions instead of the polynomial extrapolation we considered in Section 5.8, when we discussed Gragg's extrapolation. A subroutine of Adams type to be used for stiff equations is due to C. William Gear and is given by IVPAG. This method uses implicit multistep methods of order up to 12 and backward differentiation formulas of order up to 5.

The Runge-Kutta-type procedures contained in the NAG Library are called D02BAF and D02BBF and are based on the Merson form of the Runge-Kutta method. A variable-order and variable step-size Adams method is contained in the procedures D02CAF and D02CBF. Variable-order, variable-step backward-difference formula methods for stiff systems are contained in the procedures D02EAF and D02EBF. Other routines incorporate the same methods but iterate until a component of the solution attains a given value or until a function of the solution is zero. The NAG Library contains numerous subroutines for the numerical solution of initial-value problems.

There are many books specializing in the numerical solution of initial-value problems. Two classics are by Henrici [He1] and Gear [Ge1]. Other books that survey the field are by Botha and Pinder [BP], Ortega and Poole [OP], Golub and Ortega [GO], Shampine [Sh], and Dormand [Do]. Two books by Hairer, Nörsett, and Warner provide comprehensive discussions on nonstiff [HNW1] and stiff [HNW2] problems. The book by Burrage [Bur] describes parallel and sequential methods.

CHAPTER

6

Direct Methods for Solving Linear Systems

■ ■ ■

Kirchhoff's laws of electrical circuits state that the net flow of current through each junction of a circuit is zero and that the net voltage drop around each closed loop of the circuit is zero. Suppose that a potential of V volts is applied between the points A and G in the circuit below and that i_1, i_2, i_3, i_4, and i_5 represent current flow as shown in the diagram. Using G as a reference point, Kirchhoff's laws imply that the currents satisfy the following system of linear equations:

$$
\begin{aligned}
5i_1 + 5i_2 &= V, \\
i_3 - i_4 - i_5 &= 0, \\
2i_4 - 3i_5 &= 0, \\
i_1 - i_2 - i_3 &= 0, \\
5i_2 - 7i_3 - 2i_4 &= 0.
\end{aligned}
$$

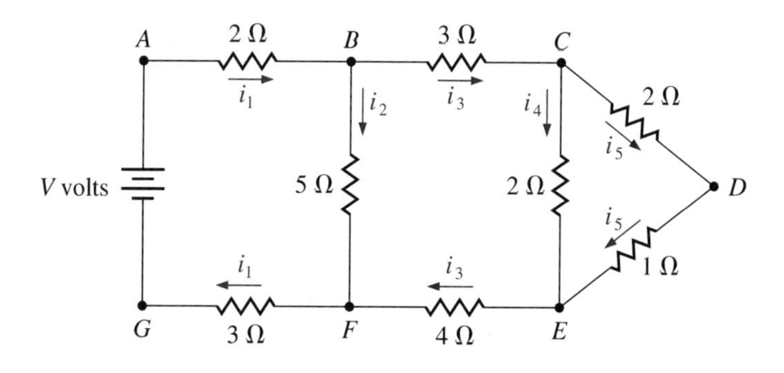

The solution of systems of this type will be considered in this chapter. This application is discussed in Exercise 23 of Section 6.6.

Linear systems of equations are associated with many problems in engineering and science, as well as with applications of mathematics to the social sciences and the quantitative study of business and economic problems.

In this chapter, direct techniques are considered to solve the linear system

$$
\begin{aligned}
E_1 &: \quad a_{11}x_1 + a_{12}x_2 + \cdots + a_{1n}x_n = b_1, \\
E_2 &: \quad a_{21}x_1 + a_{22}x_2 + \cdots + a_{2n}x_n = b_2, \\
&\quad\vdots \\
E_n &: \quad a_{n1}x_1 + a_{n2}x_2 + \cdots + a_{nn}x_n = b_n,
\end{aligned}
\tag{6.1}
$$

for $x_1, \ldots, x_n$, given the a_{ij} for each $i, j = 1, 2, \ldots, n$ and b_i, for each $i = 1, 2, \ldots, n$. Direct techniques are methods that give an answer in a fixed number of steps, subject only to roundoff errors. In the presentation we shall also introduce some elementary notions from the subject of linear algebra.

Methods of approximating the solution to linear systems by iterative methods will be discussed in Chapter 7.

6.1 Linear Systems of Equations

We use three operations to simplify the linear system given in (6.1):

1. Equation E_i can be multiplied by any nonzero constant λ and the resulting equation used in place of E_i. This operation is denoted $(\lambda E_i) \rightarrow (E_i)$.
2. Equation E_j can be multiplied by any constant λ, added to equation E_i, and the resulting equation used in place of E_i. This operation is denoted $(E_i + \lambda E_j) \rightarrow (E_i)$.
3. Equations E_i and E_j can be transposed in order. This operation is denoted $(E_i) \leftrightarrow (E_j)$.

By a sequence of the operations just given, a linear system can be transformed to a more easily solved linear system with the same solutions.

EXAMPLE 1 The four equations

$$
\begin{aligned}
E_1 &: \quad x_1 + x_2 \phantom{{}- x_3} + 3x_4 = 4, \\
E_2 &: \quad 2x_1 + x_2 - x_3 + x_4 = 1, \\
E_3 &: \quad 3x_1 - x_2 - x_3 + 2x_4 = -3, \\
E_4 &: \quad -x_1 + 2x_2 + 3x_3 - x_4 = 4,
\end{aligned}
\tag{6.2}
$$

will be solved for x_1, x_2, x_3, and x_4. We first use equation E_1 to eliminate the unknown x_1 from E_2, E_3, and E_4 by performing $(E_2 - 2E_1) \to (E_2)$, $(E_3 - 3E_1) \to (E_3)$, and $(E_4 + E_1) \to (E_4)$. The resulting system is

$$\begin{aligned}
E_1 &: x_1 + x_2 \qquad\quad + 3x_4 = 4,\\
E_2 &: \quad\; - x_2 - x_3 - 5x_4 = -7,\\
E_3 &: \quad\; - 4x_2 - x_3 - 7x_4 = -15,\\
E_4 &: \quad\;\; 3x_2 + 3x_3 + 2x_4 = 8,
\end{aligned}$$

where, for simplicity, the new equations are again labeled E_1, E_2, E_3, and E_4.

In the new system, E_2 is used to eliminate x_2 from E_3 and E_4 by performing $(E_3 - 4E_2) \to (E_3)$ and $(E_4 + 3E_2) \to (E_4)$, resulting in the system

$$\begin{aligned}
E_1 &: x_1 + x_2 \qquad\quad + 3x_4 = 4,\\
E_2 &: \quad\; - x_2 - x_3 - 5x_4 = -7,\\
E_3 &: \qquad\qquad 3x_3 + 13x_4 = 13,\\
E_4 &: \qquad\qquad\quad - 13x_4 = -13.
\end{aligned} \tag{6.3}$$

The system of equations (6.3) is now in **triangular** (or **reduced) form** and can be solved for the unknowns by a **backward-substitution process**. Noting that E_4 implies $x_4 = 1$, E_3 can be solved for x_3 to give

$$x_3 = \frac{1}{3}(13 - 13x_4) = \frac{1}{3}(13 - 13) = 0.$$

Continuing, E_2 gives

$$x_2 = -(-7 + 5x_4 + x_3) = -(-7 + 5 + 0) = 2$$

and E_1 gives

$$x_1 = 4 - 3x_4 - x_2 = 4 - 3 - 2 = -1.$$

The solution to (6.3), and to (6.2), is therefore, $x_1 = -1$, $x_2 = 2$, $x_3 = 0$, and $x_4 = 1$. ■

When performing the calculations of Example 1 we did not need to write out the full equations at each step or to carry the variables x_1, x_2, x_3, and x_4 through the calculations, since they always remained in the same column. The only variation from system to system occurred in the coefficients of the unknowns and in the values on the right side of the equations. For this reason, a linear system is often replaced by a *matrix*, which contains all the information about the system that is necessary to determine its solution, but in a compact form.

Definition 6.1 An n **by** m **matrix** is a rectangular array of elements with n rows and m columns in which not only is the value of an element important, but also its position in the array. ■

The notation for an $n \times m$ (n by m) matrix will be a capital letter such as A for the matrix and lowercase letters with double subscripts, such as a_{ij}, to refer to the entry at the intersection of the ith row and jth column; that is,

$$A = (a_{ij}) = \begin{bmatrix} a_{11} & a_{12} & \cdots & a_{1m} \\ a_{21} & a_{22} & \cdots & a_{2m} \\ \vdots & \vdots & & \vdots \\ a_{n1} & a_{n2} & \cdots & a_{nm} \end{bmatrix}.$$

The $1 \times n$ matrix

$$A = [a_{11} \quad a_{12} \quad \cdots \quad a_{1n}]$$

is called an ***n*-dimensional row vector**, and an $n \times 1$ matrix

$$A = \begin{bmatrix} a_{11} \\ a_{21} \\ \vdots \\ a_{n1} \end{bmatrix}$$

is called an ***n*-dimensional column vector**. Usually the unnecessary subscripts are omitted for vectors, and a boldface lowercase letter is used for notation. Thus,

$$\mathbf{x} = \begin{bmatrix} x_1 \\ x_2 \\ \vdots \\ x_n \end{bmatrix}$$

denotes a column vector, and

$$\mathbf{y} = [y_1 \quad y_2 \quad \cdots \quad y_n]$$

denotes a row vector.

An n by $(n+1)$ matrix can be used to represent the linear system

$$\begin{aligned} a_{11}x_1 + a_{12}x_2 + \cdots + a_{1n}x_n &= b_1, \\ a_{21}x_1 + a_{22}x_2 + \cdots + a_{2n}x_n &= b_2, \\ \vdots \qquad\quad \vdots \qquad\qquad\quad \vdots \quad &\quad \vdots \\ a_{n1}x_1 + a_{n2}x_2 + \cdots + a_{nn}x_n &= b_n, \end{aligned}$$

by first constructing

$$A = \begin{bmatrix} a_{11} & a_{12} & \cdots & a_{1n} \\ a_{21} & a_{22} & \cdots & a_{2n} \\ \vdots & \vdots & & \vdots \\ a_{n1} & a_{n2} & \cdots & a_{nn} \end{bmatrix} \quad \text{and} \quad \mathbf{b} = \begin{bmatrix} b_1 \\ b_2 \\ \vdots \\ b_n \end{bmatrix}$$

and then combining these matrices to form the **augmented matrix**

$$[A, \mathbf{b}] = \left[\begin{array}{cccc:c} a_{11} & a_{12} & \cdots & a_{1n} & b_1 \\ a_{21} & a_{22} & \cdots & a_{2n} & b_2 \\ \vdots & \vdots & & \vdots & \vdots \\ a_{n1} & a_{n2} & \cdots & a_{nn} & b_n \end{array}\right],$$

where the vertical dotted line is used to separate the coefficients of the unknowns from the values on the right-hand side of the equations.

Repeating the operations involved in Example 1 with the matrix notation results in first considering the augmented matrix

$$\left[\begin{array}{rrrr:r} 1 & 1 & 0 & 3 & 4 \\ 2 & 1 & -1 & 1 & 1 \\ 3 & -1 & -1 & 2 & -3 \\ -1 & 2 & 3 & -1 & 4 \end{array}\right].$$

Performing the operations as described in that example produces the matrices

$$\left[\begin{array}{rrrr:r} 1 & 1 & 0 & 3 & 4 \\ 0 & -1 & -1 & -5 & -7 \\ 0 & -4 & -1 & -7 & -15 \\ 0 & 3 & 3 & 2 & 8 \end{array}\right] \quad \text{and} \quad \left[\begin{array}{rrrr:r} 1 & 1 & 0 & 3 & 4 \\ 0 & -1 & -1 & -5 & -7 \\ 0 & 0 & 3 & 13 & 13 \\ 0 & 0 & 0 & -13 & -13 \end{array}\right].$$

The final matrix can now be transformed into its corresponding linear system, and solutions for x_1, x_2, x_3, and x_4, can be obtained. The procedure involved in this process is called **Gaussian elimination with backward substitution**.

The general Gaussian elimination procedure applied to the linear system

(6.4)

$$\begin{aligned} E_1 &: \quad a_{11}x_1 + a_{12}x_2 + \cdots + a_{1n}x_n = b_1, \\ E_2 &: \quad a_{21}x_1 + a_{22}x_2 + \cdots + a_{2n}x_n = b_2, \\ &\ \ \vdots \qquad \vdots \qquad\quad \vdots \qquad\qquad\quad \vdots \qquad \vdots \\ E_n &: \quad a_{n1}x_1 + a_{n2}x_2 + \cdots + a_{nn}x_n = b_n, \end{aligned}$$

is handled in a similar manner. First form the augmented matrix $\tilde{A}$:

(6.5)
$$\tilde{A} = [A, \mathbf{b}] = \begin{bmatrix} a_{11} & a_{12} & \cdots & a_{1n} & \vdots & a_{1,n+1} \\ a_{21} & a_{22} & \cdots & a_{2n} & \vdots & a_{2,n+1} \\ \vdots & \vdots & & \vdots & \vdots & \vdots \\ a_{n1} & a_{n2} & \cdots & a_{nn} & \vdots & a_{n,n+1} \end{bmatrix},$$

where A denotes the matrix formed by the coefficients. The entries in the $(n+1)$st column are the values of $\mathbf{b}$; that is, $a_{i,n+1} = b_i$ for each $i = 1, 2, \ldots, n$.

Provided $a_{11} \neq 0$, the operations corresponding to $(E_j - (a_{j1}/a_{11})E_1) \rightarrow (E_j)$ are performed for each $j = 2, 3, \ldots, n$ to eliminate the coefficient of x_1 in each of these rows. Although the entries in rows $2, 3, \ldots, n$ are expected to change, for ease of notation we again denote the entry in the ith row and the jth column by a_{ij}. With this in mind, we follow a sequential procedure for $i = 2, 3, \ldots, n-1$ and perform the operation $(E_j - (a_{ji}/a_{ii})E_i) \rightarrow (E_j)$ for each $j = i+1, i+2, \ldots, n$, provided $a_{ii} \neq 0$. This eliminates (that is, changes the coefficient to zero) x_i in each row below the ith for all values of $i = 1, 2, \ldots, n-1$. The resulting matrix has the form:

$$\tilde{\tilde{A}} = \begin{bmatrix} a_{11} & a_{12} & \cdots & a_{1n} & \vdots & a_{1,n+1} \\ 0 & a_{22} & \cdots & a_{2n} & \vdots & a_{2,n+1} \\ \vdots & \ddots & \ddots & \vdots & \vdots & \vdots \\ 0 & \cdots & 0 & a_{nn} & \vdots & a_{n,n+1} \end{bmatrix},$$

where, except in the first row, the values of a_{ij} are not expected to agree with those in the original matrix $\tilde{A}$. This matrix represents a linear system with the same solution set as the original system (6.4). Since the new linear system is triangular,

$$\begin{aligned} a_{11}x_1 + a_{12}x_2 + \cdots + a_{1n}x_n &= a_{1,n+1}, \\ a_{22}x_2 + \cdots + a_{2n}x_n &= a_{2,n+1}, \\ \ddots \qquad \vdots \quad &\quad \vdots \\ a_{nn}x_n &= a_{n,n+1}, \end{aligned}$$

backward substitution can be performed. Solving the nth equation for x_n gives

$$x_n = \frac{a_{n,n+1}}{a_{nn}}.$$

Solving the $(n-1)$st equation for x_{n-1} and using the known x_n yields

$$x_{n-1} = \frac{a_{n-1,n+1} - a_{n-1,n}x_n}{a_{n-1,n-1}},$$

and continuing this process, we obtain for each $i = n-1, n-2, \ldots, 2, 1$,

$$x_i = \frac{a_{i,n+1} - a_{i,n}x_n - a_{i,n-1}x_{n-1} - \cdots - a_{i,i+1}x_{i+1}}{a_{ii}} = \frac{a_{i,n+1} - \sum_{j=i+1}^{n} a_{ij}x_j}{a_{ii}}.$$

The Gaussian elimination procedure can be presented more precisely, although more intricately, by forming a sequence of augmented matrices $\tilde{A}^{(1)}, \tilde{A}^{(2)}, \ldots, \tilde{A}^{(n)}$, where $\tilde{A}^{(1)}$ is the matrix $\tilde{A}$ given in (6.5) and $\tilde{A}^{(k)}$ for each $k = 2, 3, \ldots, n$ has entries $a_{ij}^{(k)}$, where:

$$a_{ij}^{(k)} = \begin{cases} a_{ij}^{(k-1)}, & \text{when } i = 1, 2, \ldots, k-1 \text{ and } j = 1, 2, \ldots, n+1, \\ 0, & \text{when } i = k, k+1, \ldots, n \text{ and } j = 1, 2, \ldots, k-1, \\ a_{ij}^{(k-1)} - \dfrac{a_{i,k-1}^{(k-1)}}{a_{k-1,k-1}^{(k-1)}} a_{k-1,j}^{(k-1)}, & \text{when } i = k, k+1, \ldots, n \text{ and } j = k, k+1, \ldots, n+1. \end{cases}$$

Thus,

$$\textbf{(6.6)} \quad \tilde{A}^{(k)} = \left[\begin{array}{cccccccc:c} a_{11}^{(1)} & a_{12}^{(1)} & a_{13}^{(1)} & \cdots & a_{1,k-1}^{(1)} & a_{1k}^{(1)} & \cdots & a_{1n}^{(1)} & a_{1,n+1}^{(1)} \\ 0 & a_{22}^{(2)} & a_{23}^{(2)} & \cdots & a_{2,k-1}^{(2)} & a_{2k}^{(2)} & \cdots & a_{2n}^{(2)} & a_{2,n+1}^{(2)} \\ \vdots & & \ddots & & \vdots & \vdots & & \vdots & \vdots \\ \vdots & & & \ddots & a_{k-1,k-1}^{(k-1)} & a_{k-1,k}^{(k-1)} & \cdots & a_{k-1,n}^{(k-1)} & a_{k-1,n+1}^{(k-1)} \\ \vdots & & & & 0 & a_{kk}^{(k)} & \cdots & a_{kn}^{(k)} & a_{k,n+1}^{(k)} \\ \vdots & & & & \vdots & \vdots & & \vdots & \vdots \\ 0 & \cdots & \cdots & \cdots & 0 & a_{nk}^{(k)} & \cdots & a_{nn}^{(k)} & a_{n,n+1}^{(k)} \end{array}\right]$$

represents the equivalent linear system for which the variable x_{k-1} has just been eliminated from equations $E_k, E_{k+1}, \ldots, E_n$.

The procedure will fail if one of the elements $a_{11}^{(1)}, a_{22}^{(2)}, a_{33}^{(3)}, \ldots, a_{n-1,n-1}^{(n-1)}, a_{nn}^{(n)}$ is zero because, in this case, the step

$$\left(E_i - \frac{a_{i,k}^{(k)}}{a_{kk}^{(k)}} E_k\right) \to E_i$$

either cannot be performed (this occurs if one of $a_{11}^{(1)}, \ldots, a_{n-1,n-1}^{(n-1)}$ is zero), or the backward substitution cannot be accomplished (in the case $a_{nn}^{(n)} = 0$). This does not necessarily mean that the system has no solution, but rather that the technique for finding the solution must be altered. An illustration is given in the following example.

EXAMPLE 2 Consider the linear system

$$\begin{aligned} E_1 &: x_1 - x_2 + 2x_3 - x_4 = -8, \\ E_2 &: 2x_1 - 2x_2 + 3x_3 - 3x_4 = -20, \\ E_3 &: x_1 + x_2 + x_3 = -2, \\ E_4 &: x_1 - x_2 + 4x_3 + 3x_4 = 4. \end{aligned}$$

The augmented matrix is

$$\tilde{A} = \tilde{A}^{(1)} = \begin{bmatrix} 1 & -1 & 2 & -1 & \vdots & -8 \\ 2 & -2 & 3 & -3 & \vdots & -20 \\ 1 & 1 & 1 & 0 & \vdots & -2 \\ 1 & -1 & 4 & 3 & \vdots & 4 \end{bmatrix},$$

and performing the operations

$$(E_2 - 2E_1) \to (E_2),\ (E_3 - E_1) \to (E_3), \quad \text{and} \quad (E_4 - E_1) \to (E_4),$$

gives

$$\tilde{A}^{(2)} = \begin{bmatrix} 1 & -1 & 2 & -1 & \vdots & -8 \\ 0 & 0 & -1 & -1 & \vdots & -4 \\ 0 & 2 & -1 & 1 & \vdots & 6 \\ 0 & 0 & 2 & 4 & \vdots & 12 \end{bmatrix}.$$

Since $a_{22}^{(2)}$, called the **pivot element**, is zero, the procedure cannot continue in its present form. But the operation $(E_i) \leftrightarrow (E_j)$ is permitted, so a search is made of the elements $a_{32}^{(2)}$ and $a_{42}^{(2)}$ for the first nonzero element. Since $a_{32}^{(2)} \neq 0$, the operation $(E_2) \leftrightarrow (E_3)$ is performed to obtain a new matrix,

$$\tilde{A}^{(2)'} = \begin{bmatrix} 1 & -1 & 2 & -1 & \vdots & -8 \\ 0 & 2 & -1 & 1 & \vdots & 6 \\ 0 & 0 & -1 & -1 & \vdots & -4 \\ 0 & 0 & 2 & 4 & \vdots & 12 \end{bmatrix}.$$

Since x_2 is already eliminated from E_3 and E_4, $\tilde{A}^{(3)}$ will be $\tilde{A}^{(2)'}$, and the computations continue with the operation $(E_4 + 2E_3) \to (E_4)$, giving

$$\tilde{A}^{(4)} = \begin{bmatrix} 1 & -1 & 2 & -1 & \vdots & -8 \\ 0 & 2 & -1 & 1 & \vdots & 6 \\ 0 & 0 & -1 & -1 & \vdots & -4 \\ 0 & 0 & 0 & 2 & \vdots & 4 \end{bmatrix}.$$

Finally, the backward substitution is applied:

$$\begin{aligned} x_4 &= \frac{4}{2} = 2, \\ x_3 &= \frac{[-4 - (-1)x_4]}{-1} = 2, \\ x_2 &= \frac{[6 - x_4 - (-1)x_3]}{2} = 3, \\ x_1 &= \frac{[-8 - (-1)x_4 - 2x_3 - (-1)x_2]}{1} = -7. \end{aligned}$$

■

Example 2 illustrates what is done if $a_{kk}^{(k)} = 0$ for some $k = 1, 2, \ldots, n-1$. The kth column of $\tilde{A}^{(k-1)}$ from the kth row to the nth row is searched for the first nonzero entry. If $a_{pk}^{(k)} \neq 0$ for some p,with $k+1 \leq p \leq n$, then the operation $(E_k) \leftrightarrow (E_p)$ is performed to obtain $\tilde{A}^{(k-1)'}$. The procedure can then be continued to form $\tilde{A}^{(k)}$, and so on. If $a_{pk}^{(k)} = 0$ for each p, it can be shown (see Theorem 6.17 in Section 6.4) that the linear system does not have a unique solution and the procedure stops. Finally, if $a_{nn}^{(n)} = 0$, the linear system does not have a unique solution, and again the procedure stops. Algorithm 6.1 summarizes Gaussian elimination with backward substitution. The algorithm incorporates pivoting when one of the pivots $a_{kk}^{(k)}$ is zero by interchanging the kth row with the pth row, where p is the smallest integer greater than k for which a $a_{pk}^{(k)}$ is nonzero.

ALGORITHM **6.1**

Gaussian Elimination with Backward Substitution

To solve the $n \times n$ linear system

$$
\begin{array}{llllll}
E_1: & a_{11}x_1 & + a_{12}x_2 & + \cdots & + a_{1n}x_n & = a_{1,n+1} \\
E_2: & a_{21}x_1 & + a_{22}x_2 & + \cdots & + a_{2n}x_n & = a_{2,n+1} \\
\vdots & \vdots & \vdots & & \vdots & \vdots \\
E_n: & a_{n1}x_1 & + a_{n2}x_2 & + \cdots & + a_{nn}x_n & = a_{n,n+1}
\end{array}
$$

INPUT number of unknowns and equations n; augmented matrix $A = (a_{ij})$, where $1 \leq i \leq n$ and $1 \leq j \leq n+1$.

OUTPUT solution $x_1, x_2, \ldots, x_n$ or message that the linear system has no unique solution.

Step 1 For $i = 1, \ldots, n-1$ do Steps 2–4. *(Elimination process.)*

Step 2 Let p be the smallest integer with $i \leq p \leq n$ and $a_{pi} \neq 0$.
If no integer p can be found
then OUTPUT ('no unique solution exists');
STOP.

Step 3 If $p \neq i$ then perform $(E_p) \leftrightarrow (E_i)$.

Step 4 For $j = i+1, \ldots, n$ do Steps 5 and 6.

Step 5 Set $m_{ji} = a_{ji}/a_{ii}$.

Step 6 Perform $(E_j - m_{ji}E_i) \rightarrow (E_j)$;

Step 7 If $a_{nn} = 0$ then OUTPUT ('no unique solution exists');
STOP.

Step 8 Set $x_n = a_{n,n+1}/a_{nn}$. *(Start backward substitution.)*

Step 9 For $i = n-1, \ldots, 1$ set $x_i = \left[a_{i,n+1} - \sum_{j=i+1}^{n} a_{ij}x_j\right] \Big/ a_{ii}$.

Step 10 OUTPUT $(x_1, \ldots, x_n)$; *(Procedure completed successfully.)*
STOP.

To define matrices and perform Gaussian elimination using Maple, you must access the linear algebra library using the command

```
>with(linalg);
```

To define the matrix $\tilde{A}^{(1)}$ of Example 2, which we will call AA, use the command

```
>AA:=matrix(4,5,[1,-1,2,-1,-8,2,-2,3,-3,-20,1,1,1,0,-2,
                1,-1,4,3,4]);
```

The first two parameters, 4 and 5, give the number of rows and columns, respectively, and the last parameter is a list of the entries of $\tilde{A}^{(1)} \equiv AA$. The function `addrow(AA,i,j,m)` performs the operation $(E_j + mE_i) \rightarrow (E_j)$ and the function `swaprow(AA,i,j)` performs the operation $(E_i) \leftrightarrow (E_j)$. So, the sequence of operations

```
>AA:=addrow(AA,1,2,-2);
>AA:=addrow(AA,1,3,-1);
>AA:=addrow(AA,1,4,-1);
>AA:=swaprow(AA,2,3);
>AA:=addrow(AA,3,4,2);
```

gives the reduction to $\tilde{A}^{(4)}$, which is again called AA. Alternatively, the single command `AA:=gausselim(AA);` returns the reduced matrix. The final operation

```
>x:=backsub(AA);
```

produces the solution $x := [-7, 3, 2, 2]$.

EXAMPLE 3 The purpose of this example is to show what can happen if Algorithm 6.1 fails. The computations will be done simultaneously on two linear systems:

$$\begin{aligned} x_1 + x_2 + x_3 &= 4, \\ 2x_1 + 2x_2 + x_3 &= 6, \\ x_1 + x_2 + 2x_3 &= 6, \end{aligned} \quad \text{and} \quad \begin{aligned} x_1 + x_2 + x_3 &= 4, \\ 2x_1 + 2x_2 + x_3 &= 4, \\ x_1 + x_2 + 2x_3 &= 6. \end{aligned}$$

These systems produce matrices

$$\tilde{A} = \begin{bmatrix} 1 & 1 & 1 & \vdots & 4 \\ 2 & 2 & 1 & \vdots & 6 \\ 1 & 1 & 2 & \vdots & 6 \end{bmatrix} \quad \text{and} \quad \tilde{A} = \begin{bmatrix} 1 & 1 & 1 & \vdots & 4 \\ 2 & 2 & 1 & \vdots & 4 \\ 1 & 1 & 2 & \vdots & 6 \end{bmatrix}.$$

Since $a_{11} = 1$, we perform $(E_2 - 2E_1) \rightarrow (E_2)$ and $(E_3 - E_1) \rightarrow (E_3)$ to produce

$$\tilde{A} = \begin{bmatrix} 1 & 1 & 1 & \vdots & 4 \\ 0 & 0 & -1 & \vdots & -2 \\ 0 & 0 & 1 & \vdots & 2 \end{bmatrix} \quad \text{and} \quad \tilde{A} = \begin{bmatrix} 1 & 1 & 1 & \vdots & 4 \\ 0 & 0 & -1 & \vdots & -4 \\ 0 & 0 & 1 & \vdots & 2 \end{bmatrix}.$$

At this point, $a_{22} = a_{32} = 0$. The algorithm requires that the procedure be halted, and no solution to either system is obtained. Writing the equations for each system gives

$$\begin{aligned} x_1 + x_2 + \quad x_3 &= 4, \\ -x_3 &= -2, \\ x_3 &= 2, \end{aligned} \quad \text{and} \quad \begin{aligned} x_1 + x_2 + \quad x_3 &= 4, \\ -x_3 &= -4, \\ x_3 &= 2. \end{aligned}$$

The first linear system has an infinite number of solutions; $x_3 = 2$, $x_2 = 2 - x_1$, and x_1 arbitrary. The second system leads to the contradiction $x_3 = 2$ and $x_3 = 4$, so no solution exists. In each case, however, there is no *unique* solution, as we conclude from Algorithm 6.1. ■

Although Algorithm 6.1 can be viewed as the construction of the augmented matrices $\tilde{A}^{(1)}, \ldots, \tilde{A}^{(n)}$, the computations can be performed in a computer using only one n by $(n+1)$ array for storage. At each step we simply replace the previous value of a_{ij} by the new one. In addition, we can store the multipliers m_{ji} in the locations of a_{ji} since a_{ji} has the value zero for each $i = 1, 2, \ldots, n-1$, and $j = i+1, i+2, \ldots, n$. Thus, A can be overwritten by the multipliers below the main diagonal and by the nonzero entries of $\tilde{A}^{(n)}$ on and above the main diagonal. These values can be used to solve other linear systems involving the original matrix A, as we will see in Section 6.5.

Both the amount of time required to complete the calculations and the subsequent roundoff error depend on the number of floating-point arithmetic operations that need to be performed to solve a routine problem. In general, the amount of time required to perform a multiplication or division on a computer is about the same and is considerably greater than that required to perform an addition or subtraction. The actual differences in execution time, however, depend on the particular computing system being used. To demonstrate the counting operations for a given method, we will count the operations required to solve a typical linear system of n equations in n unknowns using Algorithm 6.1. We will keep the count of the additions/subtractions separate from the count of the multiplications/divisions because of the time differential.

No arithmetic operations are performed until Steps 5 and 6 in the algorithm. Step 5 requires that $(n-i)$ divisions be performed. The replacement of the equation E_j by $(E_j - m_{ji}E_i)$ in Step 6 requires that m_{ji} be multiplied by each term in E_i, resulting in a total of $(n-i)(n-i+1)$ multiplications. After this is completed, each term of the resulting equation is subtracted from the corresponding term in E_j. This requires $(n-i)(n-i+1)$ subtractions. For each $i = 1, 2, \ldots, n-1$, the operations required in Steps 5 and 6 are as follows.

Multiplications/divisions

$$(n-i) + (n-i)(n-i+1) = (n-i)(n-i+2).$$

Additions/subtractions

$$(n-i)(n-i+1).$$

The total number of operations required by these steps is obtained by summing the operation counts for each i. Recalling that

$$\sum_{j=1}^{m} 1 = m, \quad \sum_{j=1}^{m} j = \frac{m(m+1)}{2}, \quad \text{and} \quad \sum_{j=1}^{m} j^2 = \frac{m(m+1)(2m+1)}{6},$$

we have the following operation counts.

Multiplications/divisions

$$\begin{aligned}\sum_{i=1}^{n-1}(n-i)(n-i+2) &= \sum_{i=1}^{n-1}(n^2 - 2ni + i^2 + 2n - 2i)\\ &= (n^2+2n)\sum_{i=1}^{n-1} 1 - 2(n+1)\sum_{i=1}^{n-1} i + \sum_{i=1}^{n-1} i^2\\ &= \frac{2n^3 + 3n^2 - 5n}{6}.\end{aligned}$$

Additions/subtractions

$$\begin{aligned}\sum_{i=1}^{n-1}(n-i)(n-i+1) &= \sum_{i=1}^{n-1}(n^2 - 2ni + i^2 + n - i)\\ &= (n^2+n)\sum_{i=1}^{n-1} 1 - (2n+1)\sum_{i=1}^{n-1} i + \sum_{i=1}^{n-1} i^2\\ &= \frac{n^3 - n}{3}.\end{aligned}$$

The only other steps in Algorithm 6.1 that involve arithmetic operations are those required for backward substitution, Steps 8 and 9. Step 8 requires one division. Step 9 requires $(n-i)$ multiplications and $(n-i-1)$ additions for each summation term and then one subtraction and one division. The total number of operations in Steps 8 and 9 is as follows.

Multiplications/divisions

$$1 + \sum_{i=1}^{n-1}((n-i)+1) = \frac{n^2+n}{2}.$$

Additions/subtractions

$$\sum_{i=1}^{n-1}((n-i-1)+1) = \frac{n^2-n}{2}.$$

The total number of arithmetic operations in Algorithm 6.1 is, therefore:

Multiplications/divisions

$$\frac{2n^3+3n^2-5}{6}+\frac{n^2+n}{2}=\frac{n^3}{3}+n^2-\frac{n}{3}.$$

Additions/subtractions

$$\frac{n^3-n}{3}+\frac{n^2-n}{2}=\frac{n^3}{3}+\frac{n^2}{2}-\frac{5n}{6}.$$

For large n the total number of multiplications and divisions is approximately $n^3/3$, as is the total number of additions and subtractions. Thus the amount of computation and the time required increases with n in proportion to n^3, as shown in Table 6.1.

Table 6.1

n	Multiplications/Divisions	Additions/Subtractions
3	17	11
10	430	375
50	44,150	42,875
100	343,300	338,250

EXERCISE SET 6.1

1. For each of the following linear systems, obtain a solution by graphical methods, if possible. Explain the results from a geometrical standpoint.

   a. $x_1 + 2x_2 = 3,$ $x_1 - x_2 = 0.$
   b. $x_1 + 2x_2 = 0,$ $x_1 - x_2 = 0.$
   c. $x_1 + 2x_2 = 3,$ $2x_1 + 4x_2 = 6.$
   d. $x_1 + 2x_2 = 3,$ $-2x_1 - 4x_2 = 6.$
   e. $x_1 + 2x_2 = 0,$ $2x_1 + 4x_2 = 0.$
   f. $2x_1 + x_2 = -1,$ $x_1 + x_2 = 2,$ $x_1 - 3x_2 = 5,$
   g. $2x_1 + x_2 = -1,$ $4x_1 + 2x_2 = -2,$ $x_1 - 3x_2 = 5.$
   h. $2x_1 + x_2 + x_3 = 1,$ $2x_1 + 4x_2 - x_3 = -1.$

2. Use Gaussian elimination with backward substitution and two-digit rounding arithmetic to solve the following linear systems. Do not reorder the equations. (The exact solution to each system is $x_1 = 1, x_2 = -1, x_3 = 3$.)

   a. $4x_1 - x_2 + x_3 = 8,$ $2x_1 + 5x_2 + 2x_3 = 3,$ $x_1 + 2x_2 + 4x_3 = 11.$
   b. $4x_1 + x_2 + 2x_3 = 9,$ $2x_1 + 4x_2 - x_3 = -5,$ $x_1 + x_2 - 3x_3 = -9,$

3. Use the Gaussian Elimination Algorithm to solve the following linear systems, if possible, and determine whether row interchanges are necessary:

a.
$$\begin{aligned} x_1 - x_2 + 3x_3 &= 2, \\ 3x_1 - 3x_2 + x_3 &= -1, \\ x_1 + x_2 &= 3. \end{aligned}$$

b.
$$\begin{aligned} 2x_1 - 1.5x_2 + 3x_3 &= 1, \\ -x_1 + 2x_3 &= 3, \\ 4x_1 - 4.5x_2 + 5x_3 &= 1. \end{aligned}$$

c.
$$\begin{aligned} 2x_1 &= 3, \\ x_1 + 1.5x_2 &= 4.5, \\ -3x_2 + 0.5x_3 &= -6.6, \\ 2x_1 - 2x_2 + x_3 + x_4 &= 0.8. \end{aligned}$$

d.
$$\begin{aligned} x_1 - \tfrac{1}{2}x_2 + x_3 &= 4, \\ 2x_1 - x_2 - x_3 + x_4 &= 5, \\ x_1 + x_2 &= 2, \\ x_1 - \tfrac{1}{2}x_2 + x_3 + x_4 &= 5. \end{aligned}$$

e.
$$\begin{aligned} x_1 + x_2 + x_4 &= 2, \\ 2x_1 + x_2 - x_3 + x_4 &= 1, \\ 4x_1 - x_2 - 2x_3 + 2x_4 &= 0, \\ 3x_1 - x_2 - x_3 + 2x_4 &= -3. \end{aligned}$$

f.
$$\begin{aligned} x_1 + x_2 + x_4 &= 2, \\ 2x_1 + x_2 - x_3 + x_4 &= 1, \\ -x_1 + 2x_2 + 3x_3 - x_4 &= 4, \\ 3x_1 - x_2 - x_3 + 2x_4 &= -3. \end{aligned}$$

4. Use the Gaussian Elimination Algorithm and single-precision arithmetic on a computer to solve the following linear systems.

a.
$$\begin{aligned} \frac{1}{4}x_1 + \frac{1}{5}x_2 + \frac{1}{6}x_3 &= 9, \\ \frac{1}{3}x_1 + \frac{1}{4}x_2 + \frac{1}{5}x_3 &= 8, \\ \frac{1}{2}x_1 + x_2 + 2x_3 &= 8. \end{aligned}$$

b.
$$\begin{aligned} 3.333x_1 + 15920x_2 - 10.333x_3 &= 15913, \\ 2.222x_1 + 16.71x_2 + 9.612x_3 &= 28.544, \\ 1.5611x_1 + 5.1791x_2 + 1.6852x_3 &= 8.4254. \end{aligned}$$

c.
$$\begin{aligned} x_1 + \frac{1}{2}x_2 + \frac{1}{3}x_3 + \frac{1}{4}x_4 &= \frac{1}{6}, \\ \frac{1}{2}x_1 + \frac{1}{3}x_2 + \frac{1}{4}x_3 + \frac{1}{5}x_4 &= \frac{1}{7}, \\ \frac{1}{3}x_1 + \frac{1}{4}x_2 + \frac{1}{5}x_3 + \frac{1}{6}x_4 &= \frac{1}{8}, \\ \frac{1}{4}x_1 + \frac{1}{5}x_2 + \frac{1}{6}x_3 + \frac{1}{7}x_4 &= \frac{1}{9}. \end{aligned}$$

d.
$$\begin{aligned} 2x_1 + x_2 - x_3 + x_4 - 3x_5 &= 7, \\ x_1 + 2x_3 - x_4 + x_5 &= 2, \\ -2x_2 - x_3 + x_4 - x_5 &= -5, \\ 3x_1 + x_2 - 4x_3 + 5x_5 &= 6, \\ x_1 - x_2 - x_3 - x_4 + x_5 &= 3. \end{aligned}$$

5. Given the linear system

$$\begin{aligned} 2x_1 - 6\alpha x_2 &= 3, \\ 3\alpha x_1 - x_2 &= \frac{3}{2}. \end{aligned}$$

a. Find value(s) of α for which the system has no solutions.

b. Find value(s) of α for which the system has an infinite number of solutions.

c. Assuming a unique solution exists for a given α, find the solution.

6. Given the linear system

$$\begin{aligned} x_1 - x_2 + \alpha x_3 &= -2, \\ -x_1 + 2x_2 - \alpha x_3 &= 3, \\ \alpha x_1 + x_2 + x_3 &= 2. \end{aligned}$$

a. Find value(s) of α for which the system has no solutions.

b. Find value(s) of α for which the system has an infinite number of solutions.

c. Assuming a unique solution exists for a given α, find the solution.

7. Show that the operations

a. $(\lambda E_i) \longrightarrow (E_i)$ **b.** $(E_i + \lambda E_j) \longrightarrow (E_i)$ **c.** $(E_i) \leftrightarrow (E_j)$

do not change the solution set of a linear system.

8. **Gauss-Jordan Method** This method, used to solve the linear system (6.4), can be described as follows. Use the ith equation to eliminate not only x_i from the equations $E_{i+1}, E_{i+2}, \ldots, E_n$, as was done in the Gaussian elimination method, but also from $E_1, E_2, \ldots, E_{i-1}$. Upon reducing $[A, \mathbf{b}]$ to:

$$\left[\begin{array}{cccc:c} a_{11}^{(1)} & 0 & \cdots & 0 & a_{1,n+1}^{(1)} \\ 0 & a_{22}^{(2)} & \ddots & \vdots & a_{2,n+1}^{(2)} \\ \vdots & \ddots & \ddots & 0 & \vdots \\ 0 & \cdots & 0 & a_{nn}^{(n)} & a_{n,n+1}^{(n)} \end{array}\right],$$

the solution is obtained by setting

$$x_i = \frac{a_{i,n+1}^{(i)}}{a_{ii}^{(i)}},$$

for each $i = 1, 2, \ldots, n$. This procedure circumvents the backward substitution in the Gaussian elimination. Construct an algorithm for the Gauss-Jordan procedure patterned after that of Algorithm 6.1.

9. Use the Gauss-Jordan method and two-digit rounding arithmetic to solve the systems in Exercise 2.

10. Repeat Exercise 4 using the Gauss-Jordan method.

11. **a.** Show that the Gauss-Jordan method requires

$$\frac{n^3}{2} + n^2 - \frac{n}{2} \quad \text{multiplications/divisions}$$

and

$$\frac{n^3}{2} - \frac{n}{2} \quad \text{additions/subtractions.}$$

b. Make a table comparing the required operations for the Gauss-Jordan and Gaussian elimination methods for $n = 3, 10, 50, 100$. Which method requires less computation?

12. Consider the following Gaussian-elimination-Gauss-Jordan hybrid method for solving the system (6.4). First, apply the Gaussian-elimination technique to reduce the system to triangular form. Then use the nth equation to eliminate the coefficients of x_n in each of the first $n-1$ rows. After this is completed use the $(n-1)$st equation to eliminate the coefficients of x_{n-1} in the first $n-2$ rows, etc. The system will eventually appear as the reduced system in Exercise 8.

a. Show that this method requires

$$\frac{n^3}{3} + \frac{3}{2}n^2 - \frac{5}{6}n \quad \text{multiplications/divisions}$$

and

$$\frac{n^3}{3} + \frac{n^2}{2} - \frac{5}{6}n \quad \text{additions/subtractions.}$$

b. Make a table comparing the required operations for the Gaussian elimination, Gauss-Jordan, and hybrid methods for $n = 3, 10, 50, 100$.

13. Use the hybrid method described in Exercise 12 and two-digit rounding arithmetic to solve the systems in Exercise 2.

14. Repeat Exercise 4 using the method described in Exercise 12.

15. Suppose that in a biological system there are n species of animals and m sources of food. Let x_j represent the population of the jth species for each $j = 1, \ldots, n$; b_i represent the available daily supply of the ith food; and a_{ij} represent the amount of the ith food consumed on the average by a member of the jth species. The linear system

$$\begin{aligned} a_{11}x_1 + a_{12}x_2 + \cdots + a_{1n}x_n &= b_1, \\ a_{21}x_1 + a_{22}x_2 + \cdots + a_{2n}x_n &= b_2, \\ \vdots \qquad \vdots \qquad\qquad \vdots \quad &\quad \vdots \\ a_{m1}x_1 + a_{m2}x_2 + \cdots + a_{mn}x_n &= b_m \end{aligned}$$

represents an equilibrium where there is a daily supply of food to precisely meet the average daily consumption of each species.

a. Let

$$A = (a_{ij}) = \begin{bmatrix} 1 & 2 & 0 & 3 \\ 1 & 0 & 2 & 2 \\ 0 & 0 & 1 & 1 \end{bmatrix},$$

$\mathbf{x} = (x_j) = [1000, 500, 350, 400]$, and $\mathbf{b} = (b_i) = [3500, 2700, 900]$. Is there sufficient food to satisfy the average daily consumption?

b. What is the maximum number of animals of each species that could be individually added to the system with the supply of food still meeting the consumption?

c. If species 1 became extinct, how much of an individual increase of each of the remaining species could be supported?

d. If species 2 became extinct, how much of an individual increase of each of the remaining species could be supported?

16. A Fredholm integral equation of the second kind is an equation of the form

$$u(x) = f(x) + \int_a^b K(x,t)u(t)\,dt,$$

where a and b and the functions f and K are given. To approximate the function u on the interval $[a, b]$, a partition $x_0 = a < x_1 < \cdots < x_{m-1} < x_m = b$ is selected and the equations

$$u(x_i) = f(x_i) + \int_a^b K(x_i, t)u(t)\,dt, \qquad \text{for each } i = 0, \ldots, m,$$

are solved for $u(x_0), u(x_1), \ldots, u(x_m)$. The integrals are approximated using quadrature formulas based on the nodes $x_0, \ldots, x_m$. In our problem, $a = 0$, $b = 1$, $f(x) = x^2$, and $K(x,t) = e^{|x-t|}$.

a. Show that the linear system

$$u(0) = f(0) + \frac{1}{2}[K(0,0)u(0) + K(0,1)u(1)],$$

$$u(1) = f(1) + \frac{1}{2}[K(1,0)u(0) + K(1,1)u(1)]$$

must be solved when the Trapezoidal rule is used.

b. Set up and solve the linear system that results when the Composite Trapezoidal rule is used with $n = 4$.

c. Repeat part (b) using the Composite Simpson's rule.

6.2 Pivoting Strategies

In deriving Algorithm 6.1 we found that a row interchange is needed when one of the pivot elements $a_{kk}^{(k)}$ is zero. This row interchange has the form $(E_k) \leftrightarrow (E_p)$, where p is the smallest integer greater than k with $a_{pk}^{(k)} \neq 0$. To reduce roundoff error it is often necessary to perform row interchanges even when the pivot elements are not zero.

If $a_{kk}^{(k)}$ is small in magnitude compared to $a_{jk}^{(k)}$, the multiplier

$$m_{jk} = \frac{a_{jk}^{(k)}}{a_{kk}^{(k)}}$$

will have magnitude much larger than 1. Roundoff error introduced in the computation of one of the terms $a_{kl}^{(k)}$ will be multiplied by m_{jk} when computing $a_{jl}^{(k+1)}$, which may compound the original error. Also, when performing the backward substitution for

$$x_k = \frac{a_{k,n+1}^{(k)} - \sum_{j=k+1}^{n} a_{kj}^{(k)}}{a_{kk}^{(k)}}$$

with a small value of $a_{kk}^{(k)}$, any error in the numerator can be dramatically increased because of the division by $a_{kk}^{(k)}$. An illustration of this difficulty is given in the following example.

EXAMPLE 1 The linear system

$$\begin{aligned} E_1 &: 0.003000x_1 + 59.14x_2 = 59.17 \\ E_2 &: \quad 5.291x_1 - 6.130x_2 = 46.78, \end{aligned}$$

has the exact solution $x_1 = 10.00$ and $x_2 = 1.000$. To illustrate the difficulties of roundoff error, Gaussian elimination will be performed on this system using four-digit rounding arithmetic.

The first pivot element is a small number, $a_{11}^{(1)} = 0.003000$, and its associated multiplier,

$$m_{21} = \frac{5.291}{0.003000} = 1763.6\overline{6},$$

rounds to the large number 1764. Performing $(E_2 - m_{21}E_1) \rightarrow (E_2)$ and the appropriate rounding gives

$$\begin{aligned} 0.003000x_1 + 59.14x_2 &\approx 59.17 \\ -104300x_2 &\approx -104400, \end{aligned}$$

instead of the precise values,

$$\begin{aligned} 0.003000x_1 + 59.14x_2 &= 59.17 \\ -104309.37\overline{6}x_2 &= -104309.37\overline{6}. \end{aligned}$$

The disparity in the magnitudes of $m_{21}a_{13}$ and a_{23} has introduced roundoff error, but the roundoff error has not yet been propagated. Backward substitution yields

$$x_2 \approx 1.001,$$

which is a close approximation to the actual value, $x_2 = 1.000$. However, because of the small pivot $a_{11} = 0.003000$,

$$x_1 \approx \frac{59.17 - (59.14)(1.001)}{0.003000} = -10.00$$

contains the small error of 0.001 multiplied by

$$\frac{59.14}{0.003000}.$$

This ruins the approximation to the actual value $x_1 = 10.00$ (See Figure 6.1.) ■

Figure 6.1

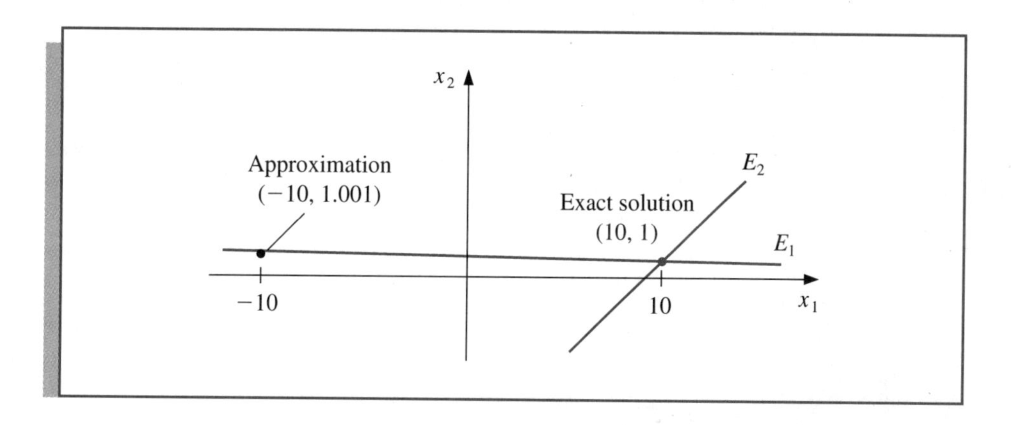

Example 1 illustrates that difficulties can arise when the pivot element $a_{kk}^{(k)}$ is small relative to the entries $a_{ij}^{(k)}$ for $k \le i \le n$ and $k \le j \le n$. To avoid this problem we use pivoting by selecting a larger element $a_{pq}^{(k)}$ for the pivot and interchanging the kth and pth rows, followed by the interchange of the kth and qth columns, if necessary. The simplest strategy is to select the element in the same column that is below the diagonal and has the largest absolute value; that is, we determine the smallest $p \ge k$ such that

$$|a_{pk}^{(k)}| = \max_{k \le i \le n} |a_{ik}^{(k)}|$$

and perform $(E_k) \leftrightarrow (E_p)$.

EXAMPLE 2 Reconsider the system

$$E_1 : 0.003000x_1 + 59.14x_2 = 59.17,$$
$$E_2 : \quad 5.291x_1 - 6.130x_2 = 46.78.$$

The pivoting procedure just described results in first finding

$$\max\left\{|a_{11}^{(1)}|, |a_{21}^{(1)}|\right\} = \max\{|0.003000|, |5.291|\} = |5.291| = |a_{21}^{(1)}|.$$

The operation $(E_2) \leftrightarrow (E_1)$ is performed to give the system

$$\begin{aligned} E_1 : \quad 5.291x_1 - 6.130x_2 &= 46.78, \\ E_2 : 0.003000x_1 + 59.14x_2 &= 59.17. \end{aligned}$$

The multiplier for this system is

$$m_{21} = \frac{a_{21}^{(1)}}{a_{11}^{(1)}} = 0.0005670,$$

and the operation $(E_2 - m_{21}E_1) \rightarrow (E_2)$ reduces the system to

$$\begin{aligned} 5.291x_1 - 6.130x_2 &= 46.78, \\ 59.14x_2 &\approx 59.14. \end{aligned}$$

The four-digit answers resulting from the backward substitution are the correct values $x_1 = 10.00$ and $x_2 = 1.000$. ■

This technique is called **partial pivoting**, or maximal column pivoting, and is detailed in Algorithm 6.2. The actual row interchanging is simulated in the algorithm by interchanging the values of *NROW* in Step 5.

ALGORITHM **6.2**

Gaussian Elimination with Partial Pivoting

To solve the $n \times n$ linear system

$$\begin{aligned} E_1 : \quad & a_{11}x_1 + a_{12}x_2 + \cdots + a_{1n}x_n = a_{1,n+1} \\ E_2 : \quad & a_{21}x_1 + a_{22}x_2 + \cdots + a_{2n}x_n = a_{2,n+1} \\ \vdots \quad & \quad \vdots \qquad\quad \vdots \qquad\qquad\quad \vdots \qquad\quad \vdots \\ E_n : \quad & a_{n1}x_1 + a_{n2}x_2 + \cdots + a_{nn}x_n = a_{n,n+1} \end{aligned}$$

INPUT number of unknowns and equations n; augmented matrix $A = (a_{ij})$ where $1 \leq i \leq n$ and $1 \leq j \leq n + 1$.

OUTPUT solution $x_1, \ldots, x_n$ or message that the linear system has no unique solution.

Step 1 For $i = 1, \ldots, n$ set $NROW(i) = i$. (*Initialize row pointer.*)

Step 2 For $i = 1, \ldots, n - 1$ do Steps 3–6. (*Elimination process.*)

Step 3 Let p be the smallest integer with $i \leq p \leq n$ and
$|a(NROW(p), i)| = \max_{i \leq j \leq n} |a(NROW(j), i)|$.
(Notation: $a(NROW(i), j) \equiv a_{NROW_i, j}$.*)*

Step 4 If $a(NROW(p), i) = 0$ then OUTPUT ('no unique solution exists');
STOP.

Step 5 If $NROW(i) \neq NROW(p)$ then set $NCOPY = NROW(i)$;
$NROW(i) = NROW(p)$;
$NROW(p) = NCOPY$.
(Simulated row interchange.)

Step 6 For $j = i + 1, \ldots, n$ do Steps 7 and 8.

Step 7 Set $m(NROW(j), i) = a(NROW(j), i)/a(NROW(i), i)$.

Step 8 Perform $(E_{NROW(j)} - m(NROW(j), i) \cdot E_{NROW(i)}) \rightarrow (E_{NROW(j)})$.

Step 9 If $a(NROW(n), n) = 0$ then OUTPUT ('no unique solution exists');
STOP.

Step 10 Set $x_n = a(NROW(n), n + 1)/a(NROW(n), n)$.
(Start backward substitution.)

Step 11 For $i = n - 1, \ldots, 1$

$$\text{set } x_i = \frac{a(NROW(i), n+1) - \sum_{j=i+1}^{n} a(NROW(i), j) \cdot x_j}{a(NROW(i), i)}.$$

Step 12 OUTPUT $(x_1, \ldots, x_n)$; *(Procedure completed successfully.)*
STOP.

Each multiplier m_{ji} in the partial pivoting algorithm has magnitude less than or equal to 1. Although this strategy is sufficient for most linear systems, situations do arise when it is inadequate.

EXAMPLE 3 The linear system

$$E_1 : 30.00x_1 + 591400x_2 = 591700,$$
$$E_2 : 5.291x_1 - 6.130x_2 = 46.78,$$

is the same as that in Examples 1 and 2 except that all the entries in the first equation have been multiplied by 10^4. The procedure described in Algorithm 6.2 with four-digit arithmetic would lead to the same results as obtained in Example 1. The maximal value in the first column is 30.00, and the multiplier

$$m_{21} = \frac{5.291}{30.00} = 0.1764$$

leads to the system

$$30.00x_1 + 591400x_2 \approx 591700,$$
$$-104300x_2 \approx -104400,$$

which has the same inaccurate solutions as in Example 1: $x_2 \approx 1.001$ and $x_1 \approx -10.00$. ■

Scaled partial pivoting, also called scaled-column pivoting, is appropriate for the system in Example 3. It places the element in the pivot position that is largest relative to the entries in its row. The first step in this procedure is to define for each row a scale factor s_i by

$$s_i = \max_{j=1,2,\ldots,n} |a_{ij}|.$$

If for some i we have $s_i = 0$, then the system has no unique solution since all entries in the ith row are zero. Assuming that this is not the case, the appropriate row interchange to place zeros in the first column is determined by choosing the least integer k with

$$\frac{|a_{k1}|}{s_k} = \max_{j=1,2,\ldots,n} \frac{|a_{j1}|}{s_j}$$

and performing $(E_1) \leftrightarrow (E_k)$. This ensures that the largest element in each row has a relative magnitude 1 before the comparison for row interchange is performed. The scaling is done only for comparison purposes, so the division to determine the scaled pivots produces no roundoff error in solving the system.

Applying scaled partial pivoting to Example 3 gives

$$s_1 = \max\{|30.00|, |591400|\} = 591400,$$

and

$$s_2 = \max\{|5.291|, |-6.130|\} = 6.130.$$

Consequently,

$$\frac{|a_{11}|}{s_1} = \frac{30.00}{591400} = 0.5073 \times 10^{-4}, \qquad \frac{|a_{21}|}{s_2} = \frac{5.291}{6.130} = 0.8631,$$

and the interchange $(E_1) \leftrightarrow (E_2)$ is made.

Applying Gaussian elimination to the new system

$$5.291x_1 - 6.130x_2 = 46.78$$
$$30.00x_1 + 591400x_2 = 591700$$

produces the correct results: $x_1 = 10.00$ and $x_2 = 1.000$.

Algorithm 6.3 implements scaled partial pivoting.

ALGORITHM 6.3

Gaussian Elimination with Scaled Partial Pivoting

The only steps in this algorithm that differ from those of Algorithm 6.2 are:

Step 1 For $i = 1, \ldots, n$ set $s_i = \max_{1 \le j \le n} |a_{ij}|$;
if $s_i = 0$ then OUTPUT ('no unique solution exists');
STOP.
set $NROW(i) = i$.

Step 2 For $i = 1, \ldots, n-1$ do Steps 3–6. (*Elimination process.*)

Step 3 Let p be the smallest integer with $i \le p \le n$ and

$$\frac{|a(NROW(p), i)|}{s(NROW(p))} = \max_{i \le j \le n} \frac{|a(NROW(j), i)|}{s(NROW(j))}.$$

The next example illustrates scaled partial pivoting using Maple with finite-digit rounding arithmetic.

EXAMPLE 4 Solve the linear system using three-digit rounding arithmetic.

$$\begin{aligned} 2.11x_1 - 4.21x_2 + 0.921x_3 &= 2.01, \\ 4.01x_1 + 10.2x_2 - 1.12x_3 &= -3.09, \\ 1.09x_1 + 0.987x_2 + 0.832x_3 &= 4.21. \end{aligned}$$

To obtain three-digit rounding arithmetic, enter

```
>Digits:=3;
```

We have $s_1 = 4.21$, $s_2 = 10.2$, and $s_3 = 1.09$. So

$$\frac{|a_{11}|}{s_1} = \frac{2.11}{4.21} = 0.501, \quad \frac{|a_{21}|}{s_2} = \frac{4.01}{10.2} = 0.393, \quad \text{and} \quad \frac{|a_{31}|}{s_3} = \frac{1.09}{1.09} = 1.$$

The augmented matrix AA is defined by

```
>AA:=matrix(3,4,[2.11,-4.21,0.921,2.01,4.01,10.2,-1.12,-3.09,1.09,
                 0.987,0.832,4.21]);
```

which gives

$$AA := \begin{bmatrix} 2.11 & -4.21 & .921 & 2.01 \\ 4.01 & 10.2 & -1.12 & -3.09 \\ 1.09 & .987 & .832 & 4.21 \end{bmatrix}.$$

Since $|a_{31}|/s_3$ is largest, we perform $(E_1) \leftrightarrow (E_3)$ using

```
>AA:=swaprow(AA,1,3);
```

to obtain

$$AA := \begin{bmatrix} 1.09 & .987 & .832 & 4.21 \\ 4.01 & 10.2 & -1.12 & -3.09 \\ 2.11 & -4.21 & .921 & 2.01 \end{bmatrix}.$$

Computing multipliers gives

```
>m21:=4.01/1.09;
```

$$m21 := 3.68$$

```
>m31:=2.11/1.09;
```

$$m31 := 1.94$$

We perform the first two eliminations using

```
>AA:=addrow(AA,1,2,-m21);
```

and

```
>AA:=addrow(AA,1,3,-m31);
```

to obtain

$$AA := \begin{bmatrix} 1.09 & .987 & .832 & 4.21 \\ 0 & 6.57 & -4.18 & -18.6 \\ 0 & -6.12 & -.689 & -6.16 \end{bmatrix}.$$

Since

$$\frac{|a_{22}|}{s_2} = \frac{6.57}{10.2} = 0.644 < \frac{|a_{32}|}{s_3} = \frac{6.12}{4.21} = 1.45,$$

we perform

```
>AA:=swaprow(AA,2,3);
```

giving

$$AA := \begin{bmatrix} 1.09 & .987 & .832 & 4.21 \\ 0 & -6.12 & -.689 & -6.16 \\ 0 & 6.57 & -4.18 & -18.6 \end{bmatrix}.$$

The multiplier m_{32} is computed by

```
>m32:=6.57/(-6.12);
```

$$m_{32} := -1.07.$$

The elimination step

```
>AA:=addrow(AA,2,3,-m32);
```

gives

$$AA := \begin{bmatrix} 1.09 & .987 & .832 & 4.21 \\ 0 & -6.12 & -.689 & -6.16 \\ 0 & .02 & -4.92 & -25.2 \end{bmatrix}.$$

We cannot use `backsub` because of the entry .02 in the (3, 2) position. This entry is nonzero due to rounding, but we can remedy this minor problem using the command

```
>AA[3,2]:=0;
```

which replaces the entry .02 with a 0. To see this enter

```
>evalm(AA);
```

which displays the matrix AA. Finally,

```
>x:=backsub(AA);
```

gives the solution

$$x := \begin{bmatrix} -.431 & .430 & 5.12 \end{bmatrix}.$$

■

The first additional computations required for scaled partial pivoting result from the determination of the scale factors; there are $(n-1)$ comparisons for each of the n rows, for a total of

$$n(n-1) \text{ comparisons.}$$

To determine the correct first interchange, n divisions are performed, followed by $n-1$ comparisons. So, the first interchange determination adds

$$n \text{ divisions and } (n-1) \text{ comparisons.}$$

Since the scaling factors are computed only once, the second step requires

$$(n-1) \text{ divisions and } (n-2) \text{ comparisons.}$$

We proceed in a similar manner until there are zeros below the main diagonal in all but nth row. The final step requires that we perform

$$2 \text{ divisions and } 1 \text{ comparison.}$$

As a consequence, scaled partial pivoting adds a total of

(6.7) $$n(n-1) + \sum_{k=1}^{n-1} k = n(n-1) + \frac{(n-1)n}{2} = \frac{3}{2}n(n-1) \quad \text{comparisons}$$

and

$$\sum_{k=2}^{n} k = \sum_{k=1}^{n} k - 1 = \frac{n(n+1)}{2} - 1 \quad \text{divisions}$$

to the Gaussian elimination procedure. The time required to perform a comparison is about the same as an addition/subtraction. Since the total time to perform the basic Gaussian elimination procedure is $O(n^3/3)$ multiplications/divisions and $O(n^3/3)$ additions/subtractions, scaled partial pivoting does not add significantly to the computational time required to solve a system for large values of n.

To emphasize the importance of choosing the scale factors only once, consider the amount of additional computation that would be required if the procedure were modified so that new scale factors were determined each time a row interchange decision was to be made. In this case, the term $n(n-1)$ in Eq. (6.7) would be replaced by

$$\sum_{k=2}^{n} k(k-1) = \frac{1}{3}n(n^2-1).$$

As a consequence, this pivoting technique would add $O(n^3/3)$ comparisons, in addition to the $[n(n+1)/2] - 1$ divisions. If a system warrants this type of pivoting, **complete** (or **maximal**) **pivoting** should instead be used. Complete pivoting at the kth step searches all the entries a_{ij}, for $i = k, k+1, \ldots, n$ and $j = k, k+1, \ldots, n$, to find the entry with the largest magnitude. Both row and column interchanges are performed to bring this entry to the pivot position. The first step of total pivoting requires that $n^2 - 1$ comparisons be performed, the second step requires $(n-1)^2 - 1$ comparisons, and so on. Hence the total additional time required to incorporate complete pivoting into Gaussian elimination is

$$\sum_{k=2}^{n} (k^2 - 1) = \frac{n(n-1)(2n+5)}{6}$$

comparisons. This figure is comparable to the number required for the modified scaled-column pivoting technique, but no divisions are required. Complete pivoting is, consequently, the strategy recommended for systems where accuracy is essential and the amount of execution time needed for this method can be justified.

EXERCISE SET 6.2

1. Find the row interchanges that are required to solve the following linear systems using Algorithm 6.1.

   a. $$\begin{aligned} x_1 - 5x_2 + x_3 &= 7 \\ 10x_1 \quad + 20x_3 &= 6 \\ 5x_1 \quad - x_3 &= 4 \end{aligned}$$

   b. $$\begin{aligned} x_1 + x_2 - x_3 &= 1 \\ x_1 + x_2 + 4x_3 &= 2 \\ 2x_1 - x_2 + 2x_3 &= 3 \end{aligned}$$

   c. $$\begin{aligned} 2x_1 - 3x_2 + 2x_3 &= 5 \\ -4x_1 + 2x_2 - 6x_3 &= 14 \\ 2x_1 + 2x_2 + 4x_3 &= 8 \end{aligned}$$

   d. $$\begin{aligned} x_2 + x_3 &= 6 \\ x_1 - 2x_2 - x_3 &= 4 \\ x_1 - x_2 + x_3 &= 5 \end{aligned}$$

2. Repeat Exercise 1 using Algorithm 6.2.
3. Repeat Exercise 1 using Algorithm 6.3.
4. Repeat Exercise 1 using complete pivoting.

5. Use Gaussian elimination and three-digit chopping arithmetic to solve the following linear systems, and compare the approximations to the actual solution.

a. $0.03x_1 + 58.9x_2 = 59.2$
$5.31x_1 - 6.10x_2 = 47.0$
Actual solution $x_1 = 10$, $x_2 = 1$

b. $58.9x_1 + 0.03x_2 = 59.2$
$-6.10x_1 + 5.31x_2 = 47.0$
Actual solution $x_1 = 1$, $x_2 = 10$

c. $3.03x_1 - 12.1x_2 + 14x_3 = -119$
$-3.03x_1 + 12.1x_2 - 7x_3 = 120$
$6.11x_1 - 14.2x_2 + 21x_3 = -139$
Actual solution $x_1 = 0$, $x_2 = 10$, $x_3 = \frac{1}{7}$

d. $3.3330x_1 + 15920x_2 + 10.333x_3 = 7953$
$2.2220x_1 + 16.710x_2 + 9.6120x_3 = 0.965$
$-1.5611x_1 + 5.1792x_2 - 1.6855x_3 = 2.714$
Actual solution $x_1 = 1$, $x_2 = 0.5$, $x_3 = -1$

e. $1.19x_1 + 2.11x_2 - 100x_3 + x_4 = 1.12$
$14.2x_1 - 0.122x_2 + 12.2x_3 - x_4 = 3.44$
$100x_2 - 99.9x_3 + x_4 = 2.15$
$15.3x_1 + 0.110x_2 - 13.1x_3 - x_4 = 4.16$
Actual solution $x_1 = 0.17682530$, $x_2 = 0.01269269$, $x_3 = -0.02065405$, $x_4 = -1.18260870$

f.
$$\pi x_1 - ex_2 + \sqrt{2}x_3 - \sqrt{3}x_4 = \sqrt{11}$$
$$\pi^2 x_1 + ex_2 - e^2 x_3 + \frac{3}{7}x_4 = 0$$
$$\sqrt{5}x_1 - \sqrt{6}x_2 + x_3 - \sqrt{2}x_4 = \pi$$
$$\pi^3 x_1 + e^2 x_2 - \sqrt{7}x_3 + \frac{1}{9}x_4 = \sqrt{2}$$
Actual solution $x_1 = 0.78839378$, $x_2 = -3.12541367$, $x_3 = 0.16759660$, $x_4 = 4.55700252$

6. Repeat Exercise 5 using three-digit rounding arithmetic.

7. Repeat Exercise 5 using Gaussian elimination with partial pivoting.

8. Repeat Exercise 6 using Gaussian elimination with partial pivoting.

9. Repeat Exercise 5 using Gaussian elimination with scaled partial pivoting.

10. Repeat Exercise 6 using Gaussian elimination with scaled partial pivoting.

11. Repeat Exercise 5 using Algorithm 6.1 with single-precision computer arithmetic.

12. Repeat Exercise 5 using Algorithm 6.2 with single-precision computer arithmetic.

13. Repeat Exercise 5 using Algorithm 6.3 with single-precision computer arithmetic.

14. Construct an algorithm for the complete pivoting procedure discussed in the text.

15. Use the complete pivoting algorithm developed in Exercise 14 to obtain solutions to

a. Exercise 5 **b.** Exercise 6 **c.** Exercise 11

16. Suppose that

$$2x_1 + x_2 + 3x_3 = 1$$
$$4x_1 + 6x_2 + 8x_3 = 5$$
$$6x_1 + \alpha x_2 + 10x_3 = 5$$

with $|\alpha| < 10$. For which of the following values of α will there be no row interchange required when solving this system using scaled partial pivoting?

a. $\alpha = 6$ **b.** $\alpha = 9$ **c.** $\alpha = -3$

6.3 Linear Algebra and Matrix Inversion

Matrices were introduced in Section 6.1 as a convenient method for expressing and manipulating a linear system. In this section we consider some algebra associated with matrices and show how it can be used to solve problems involving linear systems.

Definition 6.2 Two matrices A and B are **equal** if they have the same size, say $n \times m$, and if $a_{ij} = b_{ij}$ for each $i = 1, 2, \ldots, n$ and $j = 1, 2, \ldots, m$. ■

This definition means, for example, that

$$\begin{bmatrix} 2 & -1 & 7 \\ 3 & 1 & 0 \end{bmatrix} \neq \begin{bmatrix} 2 & 3 \\ -1 & 1 \\ 7 & 0 \end{bmatrix},$$

since they differ in dimension.

Two important operations performed on matrices are the sum of two matrices and the multiplication of a matrix by a real number.

Definition 6.3 If A and B are both $n \times m$ matrices, then the **sum** of A and B, denoted $A + B$, is the $n \times m$ matrix whose entries are $a_{ij} + b_{ij}$, for each $i = 1, 2, \ldots, n$ and $j = 1, 2, \ldots, m$. ■

Definition 6.4 If A is an $n \times m$ matrix and λ is a real number, then the **scalar multiplication** of λ and A, denoted λA, is the $n \times m$ matrix whose entries are λa_{ij}, for each $i = 1, 2, \ldots, n$ and $j = 1, 2, \ldots, m$. ■

EXAMPLE 1 Let

$$A = \begin{bmatrix} 2 & -1 & 7 \\ 3 & 1 & 0 \end{bmatrix}, \quad B = \begin{bmatrix} 4 & 2 & -8 \\ 0 & 1 & 6 \end{bmatrix}$$

and $\lambda = -2$. Then

$$A + B = \begin{bmatrix} 2+4 & -1+2 & 7-8 \\ 3+0 & 1+1 & 0+6 \end{bmatrix} = \begin{bmatrix} 6 & 1 & -1 \\ 3 & 2 & 6 \end{bmatrix},$$

and

$$\lambda A = \begin{bmatrix} -2(2) & -2(-1) & -2(7) \\ -2(3) & -2(\ 1) & -2(0) \end{bmatrix} = \begin{bmatrix} -4 & 2 & -14 \\ -6 & -2 & 0 \end{bmatrix}$$ ■

Let O denote a matrix all of whose entries are zero and $-A$ be the matrix whose entries are $-a_{ij}$. We have the following general properties for matrix addition and scalar multiplication. These properties are sufficient to classify the set of all $n \times m$ matrices with real entries as a **vector space** over the field of real numbers. (See [ND, pp. 107–109].)

Theorem 6.5 Let A, B, and C be $n \times m$ matrices and λ and μ be real numbers. The following properties of addition and scalar multiplication hold:

a. $A + B = B + A$

b. $(A + B) + C = A + (B + C)$,

c. $A + O = O + A = A$,

d. $A + (-A) = -A + A = O$,

e. $\lambda(A + B) = \lambda A + \lambda B$,

f. $(\lambda + \mu)A = \lambda A + \mu A$,

g. $\lambda(\mu A) = (\lambda\mu)A$,

h. $1A = A$.

All these properties follow from similar results about the real numbers. ■

Definition 6.6 Let A be an $n \times m$ matrix and B an $m \times p$ matrix. The **matrix product** of A and B, denoted AB, is an $n \times p$ matrix C whose entries c_{ij} are given by

$$c_{ij} = \sum_{k=1}^{m} a_{ik}b_{kj} = a_{i1}b_{1j} + a_{i2}b_{2j} + \cdots + a_{im}b_{mj},$$

for each $i = 1, 2, \ldots n$, and $j = 1, 2, \ldots, p$. ■

The computation of c_{ij} can be viewed as the multiplication of the entries of the ith row of A with corresponding entries in the jth column of B, followed by a summation. That is,

$$[a_{i1}, a_{i2}, \ldots, a_{im}] \begin{bmatrix} b_{1j} \\ b_{2j} \\ \vdots \\ b_{mj} \end{bmatrix} = c_{ij},$$

so

$$c_{ij} = a_{i1}b_{1j} + a_{i2}b_{2j} + \cdots + a_{im}b_{mj} = \sum_{k=1}^{m} a_{ik}b_{kj}.$$

This explains why the number of columns of A must equal the number of rows of B for the product AB to be defined.

The following example should serve to further clarify the matrix multiplication process.

EXAMPLE 2 Let

$$A = \begin{bmatrix} 2 & 1 & -1 \\ 3 & 1 & 2 \\ 0 & -2 & -3 \end{bmatrix}, \quad B = \begin{bmatrix} 3 & 2 \\ -1 & 1 \\ 6 & 4 \end{bmatrix},$$

$$C = \begin{bmatrix} 2 & 1 & 0 \\ -1 & 3 & 2 \end{bmatrix}, \text{ and } D = \begin{bmatrix} 1 & -1 & 1 \\ 2 & -1 & 2 \\ 3 & 0 & 3 \end{bmatrix}.$$

Then,

$$AD = \begin{bmatrix} 1 & -3 & 1 \\ 11 & -4 & 11 \\ -13 & 2 & -13 \end{bmatrix} \neq \begin{bmatrix} -1 & -2 & -6 \\ 1 & -3 & -10 \\ 6 & -3 & -12 \end{bmatrix} = DA.$$

Further,

$$BC = \begin{bmatrix} 4 & 9 & 4 \\ -3 & 2 & 2 \\ 8 & 18 & 8 \end{bmatrix} \quad \text{and} \quad CB = \begin{bmatrix} 5 & 5 \\ 6 & 9 \end{bmatrix}$$

are not even the same size.

Finally,

$$AB = \begin{bmatrix} -1 & 1 \\ 20 & 15 \\ -16 & -14 \end{bmatrix},$$

but BA cannot be computed.

Definition 6.7 A **square** matrix has the same number of rows as columns. A **diagonal** matrix is a square matrix $D = (d_{ij})$ with $d_{ij} = 0$ whenever $i \neq j$. The **identity matrix of order** n, $I_n = (\delta_{ij})$, is a diagonal matrix with entries

$$\delta_{ij} = \begin{cases} 1, & \text{if } i = j, \\ 0, & \text{if } i \neq j. \end{cases}$$

When the size of I_n is clear, this matrix is generally written simply as I. ■

Definition 6.8 An **upper-triangular** $n \times n$ matrix $U = (u_{ij})$ has, for each $j = 1, 2, \ldots, n$, the entries

$$u_{ij} = 0, \quad \text{for each } i = j+1, j+2, \ldots, n;$$

and a **lower-triangular** matrix $L = (l_{ij})$ has, for each $j = 1, 2, \ldots, n$, the entries

$$l_{ij} = 0, \quad \text{for each } i = 1, 2, \ldots, j-1.$$ ■

EXAMPLE 3 The identity matrix of order three is

$$I_3 = \begin{bmatrix} 1 & 0 & 0 \\ 0 & 1 & 0 \\ 0 & 0 & 1 \end{bmatrix}.$$

If A is any 3×3 matrix, then

$$AI_3 = \begin{bmatrix} a_{11} & a_{12} & a_{13} \\ a_{21} & a_{22} & a_{23} \\ a_{31} & a_{32} & a_{33} \end{bmatrix} \begin{bmatrix} 1 & 0 & 0 \\ 0 & 1 & 0 \\ 0 & 0 & 1 \end{bmatrix} = \begin{bmatrix} a_{11} & a_{12} & a_{13} \\ a_{21} & a_{22} & a_{23} \\ a_{31} & a_{32} & a_{33} \end{bmatrix} = A.$$ ■

The identity matrix I_n commutes with any $n \times n$ matrix A; that is, the order of multiplication does not matter, $I_nA = A = AI_n$. In Example 2 it was seen that the property $AB = BA$ is not generally true for matrix multiplication. Some of the properties involving matrix multiplication that do hold are presented in the next theorem.

Theorem 6.9 Let A be an $n \times m$ matrix, B an $m \times k$ matrix, C a $k \times p$ matrix, D an $m \times k$ matrix, and λ a real number. The following properties hold:

a. $A(BC) = (AB)C$;

b. $A(B + D) = AB + AD$;

c. $I_m B = B$ and $BI_k = B$;

d. $\lambda(AB) = (\lambda A)B = A(\lambda B)$.

Proof The verification of the property in part (a) is presented to show the method involved. The other parts can be shown in a similar manner.

To show that $A(BC) = (AB)C$, compute the i,j-entry of each side of the equation. BC is an $m \times p$ matrix with i,j-entry

$$(BC)_{ij} = \sum_{l=1}^{k} b_{il} c_{lj}$$

Thus, $A(BC)$ is an $n \times p$ matrix with entries

$$[A(BC)]_{ij} = \sum_{s=1}^{m} a_{is}(BC)_{sj} = \sum_{s=1}^{m} a_{is} \left(\sum_{l=1}^{k} b_{sl} c_{lj} \right) = \sum_{s=1}^{m} \sum_{l=1}^{k} a_{is} b_{sl} c_{lj}.$$

Similarly, AB is an $n \times k$ matrix with entries

$$(AB)_{ij} = \sum_{s=1}^{m} a_{is} b_{sj},$$

so $(AB)C$ is an $n \times p$ matrix with entries

$$[(AB)C]_{ij} = \sum_{l=1}^{k} (AB)_{il} c_{lj} = \sum_{l=1}^{k} \left(\sum_{s=1}^{m} a_{is} b_{sl} \right) c_{lj} = \sum_{l=1}^{k} \sum_{s=1}^{m} a_{is} b_{sl} c_{lj}.$$

Interchanging the order of summation on the right side gives

$$[(AB)C]_{ij} = \sum_{s=1}^{m} \sum_{l=1}^{k} a_{is} b_{sl} c_{lj} = [A(BC)]_{ij},$$

for each $i = 1, 2, \ldots, n$ and $j = 1, 2, \ldots, p$. So $A(BC) = (AB)C$. ■ ■ ■

The linear system

$$\begin{aligned}
a_{11}x_1 + a_{12}x_2 + \cdots + a_{1n}x_n &= b_1, \\
a_{21}x_1 + a_{22}x_2 + \cdots + a_{2n}x_n &= b_2, \\
\vdots \qquad \vdots \qquad\qquad \vdots \quad &\quad \vdots \\
a_{n1}x_1 + a_{n2}x_2 + \cdots + a_{nn}x_n &= b_n,
\end{aligned}$$

can be viewed as the matrix equation

$$A\mathbf{x} = \mathbf{b},$$

where

$$A = \begin{bmatrix} a_{11} & a_{12} & \cdots & a_{1n} \\ a_{21} & a_{22} & \cdots & a_{2n} \\ \vdots & \vdots & & \vdots \\ a_{n1} & a_{n2} & \cdots & a_{nn} \end{bmatrix}, \quad \mathbf{x} = \begin{bmatrix} x_1 \\ x_2 \\ \vdots \\ x_n \end{bmatrix}, \quad \text{and} \quad \mathbf{b} = \begin{bmatrix} b_1 \\ b_2 \\ \vdots \\ b_n \end{bmatrix}.$$

Related to the linear systems is the **inverse of a matrix**.

Definition 6.10 An $n \times n$ matrix A is said to be **nonsingular** if an $n \times n$ matrix A^{-1} exists with $AA^{-1} = A^{-1}A = I$. The matrix A^{-1} is called the **inverse** of A. A matrix without an inverse is called **singular**. ■

The following properties regarding matrix inverses follow from Definition 6.10. The proofs of these results are considered in Exercise 5.

Theorem 6.11 For any nonsingular $n \times n$ matrix A:

a. A^{-1} is unique.

b. A^{-1} is nonsingular and $(A^{-1})^{-1} = A$.

c. If B is also a nonsingular $n \times n$ matrix, then

$$(AB)^{-1} = B^{-1}A^{-1}.$$ ■

EXAMPLE 4 Let

$$A = \begin{bmatrix} 1 & 2 & -1 \\ 2 & 1 & 0 \\ -1 & 1 & 2 \end{bmatrix} \quad \text{and} \quad B = \begin{bmatrix} -\frac{2}{9} & \frac{5}{9} & -\frac{1}{9} \\ \frac{4}{9} & -\frac{1}{9} & \frac{2}{9} \\ -\frac{1}{3} & \frac{1}{3} & \frac{1}{3} \end{bmatrix}.$$

Then

$$AB = \begin{bmatrix} 1 & 2 & -1 \\ 2 & 1 & 0 \\ -1 & 1 & 2 \end{bmatrix} \cdot \begin{bmatrix} -\frac{2}{9} & \frac{5}{9} & -\frac{1}{9} \\ \frac{4}{9} & -\frac{1}{9} & \frac{2}{9} \\ -\frac{1}{3} & \frac{1}{3} & \frac{1}{3} \end{bmatrix} = \begin{bmatrix} 1 & 0 & 0 \\ 0 & 1 & 0 \\ 0 & 0 & 1 \end{bmatrix} = I_3.$$

In a similar manner, $BA = I_3$, so A and B are nonsingular with $B = A^{-1}$ and $A = B^{-1}$.

If we have the inverse of A, we can easily solve a linear system of the form $A\mathbf{x} = \mathbf{b}$. Suppose, for example, we want to solve

$$\begin{aligned} x_1 + 2x_2 - x_3 &= 2, \\ 2x_1 + x_2 \phantom{{}+2x_3} &= 3, \\ -x_1 + x_2 + 2x_3 &= 4. \end{aligned}$$

First, convert the system to the matrix equation

$$\begin{bmatrix} 1 & 2 & -1 \\ 2 & 1 & 0 \\ -1 & 1 & 2 \end{bmatrix} \begin{bmatrix} x_1 \\ x_2 \\ x_3 \end{bmatrix} = \begin{bmatrix} 2 \\ 3 \\ 4 \end{bmatrix},$$

and then multiply both sides by the inverse

$$\begin{bmatrix} -\frac{2}{9} & \frac{5}{9} & -\frac{1}{9} \\ \frac{4}{9} & -\frac{1}{9} & \frac{2}{9} \\ -\frac{1}{3} & \frac{1}{3} & \frac{1}{3} \end{bmatrix} \left(\begin{bmatrix} 1 & 2 & -1 \\ 2 & 1 & 0 \\ -1 & 1 & 2 \end{bmatrix} \begin{bmatrix} x_1 \\ x_2 \\ x_3 \end{bmatrix} \right) = \begin{bmatrix} -\frac{2}{9} & \frac{5}{9} & -\frac{1}{9} \\ \frac{4}{9} & -\frac{1}{9} & \frac{2}{9} \\ -\frac{1}{3} & \frac{1}{3} & \frac{1}{3} \end{bmatrix} \begin{bmatrix} 2 \\ 3 \\ 4 \end{bmatrix} = \begin{bmatrix} \frac{7}{9} \\ \frac{13}{9} \\ \frac{5}{3} \end{bmatrix},$$

so

$$\begin{bmatrix} \frac{7}{9} \\ \frac{13}{9} \\ \frac{5}{3} \end{bmatrix} = \left(\begin{bmatrix} -\frac{2}{9} & \frac{5}{9} & -\frac{1}{9} \\ \frac{4}{9} & -\frac{1}{9} & \frac{2}{9} \\ -\frac{3}{9} & \frac{3}{9} & \frac{3}{9} \end{bmatrix} \begin{bmatrix} 1 & 2 & -1 \\ 2 & 1 & 0 \\ -1 & 1 & 2 \end{bmatrix} \right) \begin{bmatrix} x_1 \\ x_2 \\ x_3 \end{bmatrix} = I_3 \begin{bmatrix} x_1 \\ x_2 \\ x_3 \end{bmatrix} = \begin{bmatrix} x_1 \\ x_2 \\ x_3 \end{bmatrix}.$$

This gives the solution $x_1 = 7/9$, $x_2 = 13/9$, and $x_3 = 5/3$. ■

Although it is easy to solve a linear system of the form $A\mathbf{x} = \mathbf{b}$ if A^{-1} is known, it is not computationally efficient to determine A^{-1} in order to solve the system. (See Exercise 8.) Even so, it is useful from a conceptual standpoint to describe a method for determining the inverse of a matrix.

To find a method of computing A^{-1} assuming its existence, let us look again at matrix multiplication. Let B_j be the jth column of the $n \times n$ matrix B,

$$B_j = \begin{bmatrix} b_{1j} \\ b_{2j} \\ \vdots \\ b_{nj} \end{bmatrix}.$$

If $AB = C$, then the jth column of C is given by the product

$$\begin{bmatrix} c_{1j} \\ c_{2j} \\ \vdots \\ c_{nj} \end{bmatrix} = C_j = AB_j = \begin{bmatrix} a_{11} & a_{12} & \cdots & a_{1n} \\ a_{21} & a_{22} & \cdots & a_{2n} \\ \vdots & \vdots & & \vdots \\ a_{n1} & a_{n2} & \cdots & a_{nn} \end{bmatrix} \begin{bmatrix} b_{1j} \\ b_{2j} \\ \vdots \\ b_{nj} \end{bmatrix} = \begin{bmatrix} \sum_{k=1}^n a_{1k}b_{kj} \\ \sum_{k=1}^n a_{2k}b_{kj} \\ \vdots \\ \sum_{k=1}^n a_{nk}b_{kj} \end{bmatrix}.$$

Suppose that A^{-1} exists and that $A^{-1} = B = (b_{ij})$. Then $AB = I$ and

$$AB_j = \begin{bmatrix} 0 \\ \vdots \\ 0 \\ 1 \\ 0 \\ \vdots \\ 0 \end{bmatrix}, \quad \text{where the value 1 appears in the } j\text{th row.}$$

To find B we must solve n linear systems in which the jth column of the inverse is the solution of the linear system with right-hand side the jth column of I. The next example demonstrates the method involved.

EXAMPLE 5 To determine the inverse of the matrix

$$A = \begin{bmatrix} 1 & 2 & -1 \\ 2 & 1 & 0 \\ -1 & 1 & 2 \end{bmatrix},$$

let us first consider the product AB, where B is an arbitrary 3×3 matrix.

$$AB = \begin{bmatrix} 1 & 2 & -1 \\ 2 & 1 & 0 \\ -1 & 1 & 2 \end{bmatrix} \begin{bmatrix} b_{11} & b_{12} & b_{13} \\ b_{21} & b_{22} & b_{23} \\ b_{31} & b_{32} & b_{33} \end{bmatrix}$$

$$= \begin{bmatrix} b_{11} + 2b_{21} - b_{31} & b_{12} + 2b_{22} - b_{32} & b_{13} + 2b_{23} - b_{33} \\ 2b_{11} + b_{21} & 2b_{12} + b_{22} & 2b_{13} + b_{23} \\ -b_{11} + b_{21} + 2b_{31} & -b_{12} + b_{22} + 2b_{32} & -b_{13} + b_{23} + 2b_{33} \end{bmatrix}.$$

If $B = A^{-1}$, then $AB = I$, so we must have

$$\begin{aligned} b_{11} + 2b_{21} - b_{31} &= 1, & b_{12} + 2b_{22} - b_{32} &= 0, & b_{13} + 2b_{23} - b_{33} &= 0, \\ 2b_{11} + b_{21} \phantom{+ 2b_{31}} &= 0, & 2b_{12} + b_{22} \phantom{+ 2b_{32}} &= 1, & 2b_{13} + b_{23} \phantom{+ 2b_{33}} &= 0, \\ -b_{11} + b_{21} + 2b_{31} &= 0, & -b_{12} + b_{22} + 2b_{32} &= 0, & -b_{13} + b_{23} + 2b_{33} &= 1. \end{aligned}$$

Notice that the coefficients in each of the systems of equations are the same, the only change in the systems occurs on the right side of the equations. As a consequence, Gaussian elimination can be performed on the larger augmented matrix, formed by combining the matrices for each of the systems:

$$\left[\begin{array}{rrr:rrr} 1 & 2 & -1 & 1 & 0 & 0 \\ 2 & 1 & 0 & 0 & 1 & 0 \\ -1 & 1 & 2 & 0 & 0 & 1 \end{array}\right].$$

First, $(E_2 - 2E_1) \rightarrow (E_2)$ and $(E_3 + E_1) \rightarrow (E_3)$ followed by $(E_3 + E_2) \rightarrow (E_3)$ produces

$$\left[\begin{array}{rrr:rrr} 1 & 2 & -1 & 1 & 0 & 0 \\ 0 & -3 & 2 & -2 & 1 & 0 \\ 0 & 3 & 1 & 1 & 0 & 1 \end{array}\right] \quad \text{and} \quad \left[\begin{array}{rrr:rrr} 1 & 2 & -1 & 1 & 0 & 0 \\ 0 & -3 & 2 & -2 & 1 & 0 \\ 0 & 0 & 3 & -1 & 1 & 1 \end{array}\right].$$

Backward substitution is performed on each of the three augmented matrices,

$$\left[\begin{array}{rrr:r} 1 & 2 & -1 & 1 \\ 0 & -3 & 2 & -2 \\ 0 & 0 & 3 & -1 \end{array}\right], \quad \left[\begin{array}{rrr:r} 1 & 2 & -1 & 0 \\ 0 & -3 & 2 & 1 \\ 0 & 0 & 3 & 1 \end{array}\right], \quad \left[\begin{array}{rrr:r} 1 & 2 & -1 & 0 \\ 0 & -3 & 2 & 0 \\ 0 & 0 & 3 & 1 \end{array}\right],$$

to give

$$\begin{aligned} b_{11} &= -\frac{2}{9}, & b_{12} &= \frac{5}{9}, & & & b_{13} &= -\frac{1}{9}, \\ b_{21} &= \frac{4}{9}, & b_{22} &= -\frac{1}{9}, & &\text{and} & b_{23} &= \frac{2}{9}, \\ b_{31} &= -\frac{1}{3}, & b_{32} &= \frac{1}{3}, & & & b_{33} &= \frac{1}{3}. \end{aligned}$$

As shown in Example 4, these are the entries of A^{-1}:

$$A^{-1} = \begin{bmatrix} -\frac{2}{9} & \frac{5}{9} & -\frac{1}{9} \\ \frac{4}{9} & -\frac{1}{9} & \frac{2}{9} \\ -\frac{1}{3} & \frac{1}{3} & \frac{1}{3} \end{bmatrix}$$

■

In the last example we illustrated the computation of A^{-1}. As we saw in that example, it is convenient to set up the larger augmented matrix,

$$\left[A \;\vdots\; I\right].$$

Upon performing the elimination in accordance with Algorithm 6.1, we obtain an augmented matrix of the form

$$\left[U \;\vdots\; Y\right],$$

where U is an upper-triangular matrix and Y is the matrix obtained by performing the same operations on the identity I that were performed to take A into U.

Gaussian elimination with backward substitution requires $4n^3/3 - n/3$ multiplications/divisions and $4n^3/3 - 3n^2/2 + n/6$ additions/subtractions to solve the n linear systems (see Exercise 8(c)). Special care can be taken in the implementation to note the operations that need *not* be performed, as, for example, a multiplication when one of the multipliers is known to be unity or a subtraction when the subtrahend is known to be zero. The number of multiplications/divisions required can then be reduced to n^3 and the number of additions/subtractions reduced to $n^3 - 2n^2 + n$ (see Exercise 8(d)).

Another important matrix associated with a given matrix A is its *transpose*, denoted A^t.

Definition 6.12 The **transpose** of an $n \times m$ matrix $A = (a_{ij})$ is the $m \times n$ matrix A^t, where for each i, the ith column of A^t is the same as the ith row of A, that is, $A^t = (a_{ji})$. A square matrix A is **symmetric** if $A = A^t$. ■

For example, the matrices

$$A = \begin{bmatrix} 7 & 2 & 0 \\ 3 & 5 & -1 \\ 0 & 5 & -6 \end{bmatrix}, \quad B = \begin{bmatrix} 2 & 4 & 7 \\ 3 & -5 & -1 \end{bmatrix}, \quad C = \begin{bmatrix} 6 & 4 & -3 \\ 4 & -2 & 0 \\ -3 & 0 & 1 \end{bmatrix}$$

have transposes

$$A^t = \begin{bmatrix} 7 & 3 & 0 \\ 2 & 5 & 5 \\ 0 & -1 & -6 \end{bmatrix}, \quad B^t = \begin{bmatrix} 2 & 3 \\ 4 & -5 \\ 7 & -1 \end{bmatrix}, \quad C^t = \begin{bmatrix} 6 & 4 & -3 \\ 4 & -2 & 0 \\ -3 & 0 & 1 \end{bmatrix}.$$

The matrix C is symmetric since $C^t = C$, but the matrices A and B are not.

The proof of the next result follows directly from the definition of the transpose.

Theorem 6.13 The following operations involving the transpose of a matrix hold whenever the operation is possible:

a. $(A^t)^t = A$,

b. $(A + B)^t = A^t + B^t$.

c. $(AB)^t = B^t A^t$,

d. If A^{-1} exists, $(A^{-1})^t = (A^t)^{-1}$. ■

Maple can be used to perform these arithmetic operations. Matrix addition is done with `matadd(A,B)` or `evalm(A+B)`. Scalar multiplication is defined by `scalarmul(A,c)` or `evalm(c*A)`.

Matrix multiplication is done using `multiply(A,B)` or `evalm(A&*B)`. Matrix transposition is achieved with `transpose(A)` and matrix inversion with `inverse(A)`.

EXERCISE SET 6.3

1. Determine which of the following matrices are nonsingular and compute the inverse of these matrices:

a. $\begin{bmatrix} 4 & 2 & 6 \\ 3 & 0 & 7 \\ -2 & -1 & -3 \end{bmatrix}$ **b.** $\begin{bmatrix} 1 & 2 & 0 \\ 2 & 1 & -1 \\ 3 & 1 & 1 \end{bmatrix}$

c. $\begin{bmatrix} 4 & 0 & 0 \\ 0 & 0 & 0 \\ 0 & 0 & 3 \end{bmatrix}$ **d.** $\begin{bmatrix} 1 & 1 & -1 & 1 \\ 1 & 2 & -4 & -2 \\ 2 & 1 & 1 & 5 \\ -1 & 0 & -2 & -4 \end{bmatrix}$

e. $\begin{bmatrix} 4 & 0 & 0 & 0 \\ 6 & 7 & 0 & 0 \\ 9 & 11 & 1 & 0 \\ 5 & 4 & 1 & 1 \end{bmatrix}$ **f.** $\begin{bmatrix} 2 & 0 & 1 & 2 \\ 1 & 1 & 0 & 2 \\ 2 & -1 & 3 & 1 \\ 3 & -1 & 4 & 3 \end{bmatrix}$

2. Consider the four 3×3 linear systems having the same coefficient matrix:

$$\begin{aligned} 2x_1 - 3x_2 + x_3 &= 2, \\ x_1 + x_2 - x_3 &= -1, \\ -x_1 + x_2 - 3x_3 &= 0, \end{aligned} \qquad \begin{aligned} 2x_1 - 3x_2 + x_3 &= 6, \\ x_1 + x_2 - x_3 &= 4, \\ -x_1 + x_2 - 3x_3 &= 5, \end{aligned}$$

$$\begin{aligned} 2x_1 - 3x_2 + x_3 &= 0, \\ x_1 + x_2 - x_3 &= 1, \\ -x_1 + x_2 - 3x_3 &= -3, \end{aligned} \qquad \begin{aligned} 2x_1 - 3x_2 + x_3 &= -1, \\ x_1 + x_2 - x_3 &= 0, \\ -x_1 + x_2 - 3x_3 &= 0. \end{aligned}$$

a. Solve the linear systems by applying Gaussian elimination to the augmented matrix

$$\left[\begin{array}{rrr:rrrr} 2 & -3 & 1 & 2 & 6 & 0 & -1 \\ 1 & 1 & -1 & -1 & 4 & 1 & 0 \\ -1 & 1 & -3 & 0 & 5 & -3 & 0 \end{array}\right].$$

b. Solve the linear systems by finding and multiplying by the inverse of

$$A = \begin{bmatrix} 2 & -3 & 1 \\ 1 & 1 & -1 \\ -1 & 1 & -3 \end{bmatrix}.$$

c. Which method requires more operations?

3. Repeat Exercise 2 using the linear systems

$$\begin{aligned} x_1 - x_2 + 2x_3 - x_4 &= 6, \\ x_1 \qquad - x_3 + x_4 &= 4, \\ 2x_1 + x_2 + 3x_3 - 4x_4 &= -2, \\ -x_2 + x_3 - x_4 &= 5, \end{aligned} \qquad \begin{aligned} x_1 - x_2 + 2x_3 - x_4 &= 1, \\ x_1 \qquad - x_3 + x_4 &= 1, \\ 2x_1 + x_2 + 3x_3 - 4x_4 &= 2, \\ -x_2 + x_3 - x_4 &= -1. \end{aligned}$$

4. Prove the following statements or provide counterexamples to show they are not true.

a. The product of two symmetric matrices is symmetric.

b. The inverse of a nonsingular symmetric matrix is a nonsingular symmetric matrix.

c. If A and B are $n \times n$ matrices, then $(AB)^t = A^t B^t$.

5. The following statements are needed to prove Theorem 6.11.

a. Show that if A^{-1} exists, it is unique.

b. Show that if A is nonsingular, then $(A^{-1})^{-1} = A$.

c. Show that if A and B are nonsingular $n \times n$ matricies, then $(AB)^{-1} = B^{-1}A^{-1}$.

6. Prove Theorem 6.5.

7. **a.** Show that the product of two $n \times n$ lower triangular matrices is lower triangular.

b. Show that the product of two $n \times n$ upper triangular matrices is upper triangular.

c. Show that the inverse of a nonsingular $n \times n$ lower triangular matrix is lower triangular.

8. Suppose m linear systems

$$A\mathbf{x}^{(p)} = \mathbf{b}^{(p)}, \quad p = 1, 2, \ldots, m,$$

are to be solved, each with the $n \times n$ coefficient matrix A.

a. Show that Gaussian elimination with backward substitution applied to the augmented matrix

$$[A : \; \mathbf{b}^{(1)}\mathbf{b}^{(2)} \cdots \mathbf{b}^{(m)}]$$

requires

$$\frac{1}{3}n^3 + mn^2 - \frac{1}{3}n \quad \text{multiplications/divisions}$$

and

$$\frac{1}{3}n^3 + mn^2 - \frac{1}{2}n^2 - mn + \frac{1}{6}n \quad \text{additions/subtractions.}$$

b. Show that the Gauss-Jordan method (see Exercise 8, Section 6.1) applied to the augmented matrix

$$\left[A : \ \mathbf{b}^{(1)}\mathbf{b}^{(2)}\cdots\mathbf{b}^{(m)}\right]$$

requires

$$\frac{1}{2}n^3 + mn^2 - \frac{1}{2}n \quad \text{multiplications/divisions}$$

and

$$\frac{1}{2}n^3 + (m-1)n^2 + \left(\frac{1}{2} - m\right) n \quad \text{additions/subtractions.}$$

c. For the special case

$$\mathbf{b}^{(p)} = \begin{bmatrix} 0 \\ \vdots \\ 0 \\ 1 \\ \vdots \\ 0 \end{bmatrix} \begin{matrix} \\ \\ \\ \leftarrow p\text{th row} \\ \\ \\ \end{matrix}$$

for each $p = 1, \ldots, m$, with $m = n$, the solution $\mathbf{x}^{(p)}$ is the pth column of A^{-1}. Show that Gaussian elimination with backward substitution requires

$$\frac{4}{3}n^3 - \frac{1}{3}n \quad \text{multiplications/divisions}$$

and

$$\frac{4}{3}n^3 - \frac{3}{2}n^2 + \frac{1}{6}n \quad \text{additions/subtractions}$$

for this application, and that the Gauss-Jordan method requires

$$\frac{3}{2}n^3 - \frac{1}{2}n \quad \text{multiplications/divisions}$$

and

$$\frac{3}{2}n^3 - 2n^2 + \frac{1}{2}n \quad \text{additions/subtractions}$$

d. Construct an algorithm using Gaussian elimination to find A^{-1}, but do not perform multiplications when one of the multipliers is known to be unity, and do not perform additions/subtractions when one of the elements involved is known to be zero. Show that the required computations are reduced to n^3 multiplications/divisions and $n^3 - 2n^2 + n$ additions/subtractions.

e. Show that solving the linear system $A\mathbf{x} = \mathbf{b}$, when A^{-1} is known, requires n^2 multiplications/divisions and $n^2 - n$ additions/subtractions.

f. Show that solving m linear systems $A\mathbf{x}^{(p)} = \mathbf{b}^{(p)}$, for $p = 1, 2, \ldots, m$, by the method $\mathbf{x}^{(p)} = A^{-1}\mathbf{b}^{(p)}$ requires mn^2 multiplications and $m(n^2 - n)$ additions, if A^{-1} is known.

g. Let A be an $n \times n$ matrix. Compare the number of operations required to solve n linear systems involving A by Gaussian elimination with backward substitution and by first inverting A and then multiplying $A\mathbf{x} = \mathbf{b}$ by A^{-1}, for $n = 3, 10, 50, 100$. Is it ever advantageous to compute A^{-1} for the purpose of solving linear systems?

9. Use the algorithm developed in Exercise 8(d) to find the inverses of the nonsingular matrices in Exercise 1.

10. It is often useful to partition matrices into a collection of submatrices. For example, the matrices

$$A = \begin{bmatrix} 1 & 2 & -1 \\ 3 & -4 & -3 \\ 6 & 5 & 0 \end{bmatrix} \quad \text{and} \quad B = \begin{bmatrix} 2 & -1 & 7 & 0 \\ 3 & 0 & 4 & 5 \\ -2 & 1 & -3 & 1 \end{bmatrix}$$

can be partitioned into

$$\left[\begin{array}{cc:c} 1 & 2 & -1 \\ 3 & -4 & -3 \\ \hdashline 6 & 5 & 0 \end{array}\right] = \left[\begin{array}{c:c} A_{11} & A_{12} \\ \hdashline A_{21} & A_{22} \end{array}\right]$$

and

$$\left[\begin{array}{ccc:c} 2 & -1 & 7 & 0 \\ 3 & 0 & 4 & 5 \\ \hdashline -2 & 1 & -3 & 1 \end{array}\right] = \left[\begin{array}{c:c} B_{11} & B_{12} \\ \hdashline B_{21} & B_{22} \end{array}\right]$$

a. Show that the product of A and B in this case is

$$AB = \left[\begin{array}{c:c} A_{11}B_{11} + A_{12}B_{21} & A_{11}B_{12} + A_{12}B_{22} \\ \hdashline A_{21}B_{11} + A_{22}B_{21} & A_{21}B_{12} + A_{22}B_{22} \end{array}\right]$$

b. If B were instead partitioned into

$$B = \left[\begin{array}{ccc:c} 2 & -1 & 7 & 0 \\ \hdashline 3 & 0 & 4 & 5 \\ -2 & 1 & -3 & 1 \end{array}\right] = \left[\begin{array}{c:c} B_{11} & B_{12} \\ \hdashline B_{21} & B_{22} \end{array}\right],$$

would the result in part (a) hold?

c. Make a conjecture concerning the conditions necessary for the result in part (a) to hold in the general case.

11. In a paper entitled "Population Waves," Bernadelli [Ber] (see also [Se]) hypothesizes a type of simplified beetle that has a natural life span of 3 years. The female of this species has a survival rate of $\frac{1}{2}$ in the first year of life, has a survival rate of $\frac{1}{3}$ from the second to third years, and gives birth to an average of six new females before expiring at the end of the third year. A matrix can be used to show the contribution an individual female beetle makes, in a probabilistic sense, to the female population of the species by letting a_{ij} in the matrix $A = (a_{ij})$ denote the contribution that a single female beetle of age j will make to the next year's female population of age i; that is,

$$A = \begin{bmatrix} 0 & 0 & 6 \\ \frac{1}{2} & 0 & 0 \\ 0 & \frac{1}{3} & 0 \end{bmatrix}.$$

a. The contribution that a female beetle makes to the population 2 years hence is determined from the entries of A^2, of 3 years hence from A^3, and so on. Construct A^2 and A^3, and try to make a general statement about the contribution of a female beetle to the population in n years' time for any positive integral value of n.

b. Use your conclusions from part (a) to describe what will occur in future years to a population of these beetles that initially consists of 6000 female beetles in each of the three age groups.

c. Construct A^{-1} and describe its significance regarding the population of this species.

12. The study of food chains is an important topic in the determination of the spread and accumulation of environmental pollutants in living matter. Suppose that a food chain has three links. The first link consists of vegetation of types $v_1, v_2, \ldots, v_n$, which provide all the food requirements for herbivores of species $h_1, h_2, \ldots, h_m$ in the second link. The third link consists of carnivorous animals $c_1, c_2, \ldots, c_k$, which depend entirely on the herbivores in the second link for their food supply. The coordinate a_{ij} of the matrix

$$A = \begin{bmatrix} a_{11} & a_{12} & \cdots & a_{1m} \\ a_{21} & a_{22} & \cdots & a_{2m} \\ \vdots & \vdots & & \vdots \\ a_{n1} & a_{n2} & \cdots & a_{nm} \end{bmatrix}$$

represents the total number of plants of type v_i eaten by the herbivores in the species h_j, whereas b_{ij} in

$$B = \begin{bmatrix} b_{11} & b_{12} & \cdots & b_{1k} \\ b_{21} & b_{22} & \cdots & b_{2k} \\ \vdots & \vdots & & \vdots \\ b_{m1} & b_{m2} & \cdots & b_{mk} \end{bmatrix}$$

describes the number of herbivores in species h_i that are devoured by the animals of type c_j.

a. Show that the number of plants of type v_i that eventually end up in the animals of species c_j is given by the entry in the ith row and jth column of the matrix AB.

b. What physical significance is associated with the matrices A^{-1}, B^{-1}, and $(AB)^{-1} = B^{-1}A^{-1}$?

13. In Section 3.5 we found that the parametric form $(x(t), y(t))$ of the cubic Hermite polynomials through $(x(0), y(0)) = (x_0, y_0)$ and $(x(1), y(1)) = (x_1, y_1)$ with guide points $(x_0 + \alpha_0, y_0 + \beta_0)$ and $(x_1 - \alpha_1, y_1 - \beta_1)$, respectively, are given by

$$x(t) = [2(x_0 - x_1) + (\alpha_0 + \alpha_1)]t^3 + [3(x_1 - x_0) - \alpha_1 - 2\alpha_0]t^2 + \alpha_0 t + x_0$$

and

$$y(t) = [2(y_0 - y_1) + (\beta_0 + \beta_1)]t^3 + [3(y_1 - y_0) - \beta_1 - 2\beta_0]t^2 + \beta_0 t + y_0.$$

The Bézier cubic polynomials have the form

$$\hat{x}(t) = [2(x_0 - x_1) + 3(\alpha_0 + \alpha_1)]t^3 + [3(x_1 - x_0) - 3(\alpha_1 + 2\alpha_0)]t^2 + 3\alpha_0 t + x_0$$

and

$$\hat{y}(t) = [2(y_0 - y_1) + 3(\beta_0 + \beta_1)]t^3 + [3(y_1 - y_0) - 3(\beta_1 + 2\beta_0)]t^2 + 3\beta_0 t + y_0.$$

a. Show that the matrix

$$A = \begin{bmatrix} 7 & 4 & 4 & 0 \\ -6 & -3 & -6 & 0 \\ 0 & 0 & 3 & 0 \\ 0 & 0 & 0 & 1 \end{bmatrix}$$

maps the Hermite polynomial coefficients onto the Bézier polynomial coefficients.

b. Determine a matrix B that maps the Bézier polynomial coefficients onto the Hermite polynomial coefficients.

14. Consider the 2×2 linear system $(A + iB)(\mathbf{x} + i\mathbf{y}) = \mathbf{c} + i\mathbf{d}$ with complex entries in component form:

$$(a_{11} + ib_{11})(x_1 + iy_1) + (a_{12} + ib_{12})(x_2 + iy_2) = c_1 + id_1,$$
$$(a_{21} + ib_{21})(x_1 + iy_1) + (a_{22} + ib_{22})(x_2 + iy_2) = c_2 + id_2.$$

a. Use the properties of complex numbers to convert this system to the equivalent 4×4 real linear system

$$A\mathbf{x} - B\mathbf{y} = \mathbf{c},$$
$$B\mathbf{x} + A\mathbf{y} = \mathbf{d}.$$

b. Solve the linear system

$$(1 - 2i)(x_1 + iy_1) + (3 + 2i)(x_2 + iy_2) = 5 + 2i,$$
$$(2 + i)(x_1 + iy_1) + (4 + 3i)(x_2 + iy_2) = 4 - i.$$

6.4 The Determinant of a Matrix

The *determinant* of a matrix is a fundamental concept of linear algebra that provides existence and uniqueness results for linear systems of equations. We will denote the determinant of a matrix A by $\det A$, but it is also common to use the notation $|A|$.

Definition 6.14

a. If $A = [a]$ is a 1×1 matrix, then $\det A = a$.

b. If A is an $n \times n$ matrix, the **minor** M_{ij} is the determinant of the $(n-1) \times (n-1)$ submatrix of A obtained by deleting the ith row and jth column of the matrix A.

c. The **cofactor** A_{ij} associated with M_{ij} is defined by $A_{ij} = (-1)^{i+j} M_{ij}$.

d. The **determinant** of the $n \times n$ matrix A, when $n > 1$, is given either by

$$\det A = \sum_{j=1}^{n} a_{ij} A_{ij} = \sum_{j=1}^{n} (-1)^{i+j} a_{ij} M_{ij} \quad \text{for any } i = 1, 2, \ldots, n,$$

or by

$$\det A = \sum_{i=1}^{n} a_{ij} A_{ij} = \sum_{i=1}^{n} (-1)^{i+j} a_{ij} M_{ij} \quad \text{for any } j = 1, 2, \ldots, n. \qquad \blacksquare$$

It can be shown that to calculate the determinant of a general $n \times n$ matrix by this definition requires $O(n!)$ multiplications/divisions and additions/subtractions. Even for relatively small values of n, the number of calculations becomes unwieldy.

It appears that there are $2n$ different definitions of $\det A$, depending on which row or column is chosen. However, all definitions give the same numerical result. The flexibility in the definition is used in the following example. It is most convenient to compute $\det A$ across the row or down the column with the most zeros.

EXAMPLE 1 Let

$$A = \begin{bmatrix} 2 & -1 & 3 & 0 \\ 4 & -2 & 7 & 0 \\ -3 & -4 & 1 & 5 \\ 6 & -6 & 8 & 0 \end{bmatrix}.$$

To compute $\det A$, it is easiest to expand about the fourth column:

$$\det A = a_{14}A_{14} + a_{24}A_{24} + a_{34}A_{34} + a_{44}A_{44} = 5A_{34} = -5M_{34}.$$

Eliminating the third row and the fourth column gives

$$\det A = -5 \det \begin{bmatrix} 2 & -1 & 3 \\ 4 & -2 & 7 \\ 6 & -6 & 8 \end{bmatrix}.$$

Now expanding along the top row gives

$$\det A = -5\left\{2 \det \begin{bmatrix} -2 & 7 \\ -6 & 8 \end{bmatrix} - (-1) \det \begin{bmatrix} 4 & 7 \\ 6 & 8 \end{bmatrix} + 3 \det \begin{bmatrix} 4 & -2 \\ 6 & -6 \end{bmatrix}\right\} = -30. \quad \blacksquare$$

The determinant of a matrix is computed in Maple by command `det(A);`.

The following properties are useful in relating linear systems and Gaussian elimination to determinants. These are proved in any standard linear algebra text. (See, for example, [ND, pp. 200–201].)

Theorem 6.15 Suppose A is an $n \times n$ matrix:

a. If any row or column of A has only zero entries, then $\det A = 0$.

b. If $\tilde{A}$ is obtained from A by the operation $(E_i) \leftrightarrow (E_j)$, with $i \neq j$, then $\det \tilde{A} = -\det A$.

c. If A has two rows the same or two columns the same, then $\det A = 0$.

d. If $\tilde{A}$ is obtained from A by the operation $(\lambda E_i) \longrightarrow (E_i)$, then $\det \tilde{A} = \lambda \det A$.

e. If $\tilde{A}$ is obtained from A by the operation $(E_i + \lambda E_j) \longrightarrow (E_i)$ with $i \neq j$, then $\det \tilde{A} = \det A$.

f. If B is also an $n \times n$ matrix, then $\det AB = \det A \det B$.

g. $\det A^t = \det A$.

h. When A^{-1} exists, $\det A^{-1} = (\det A)^{-1}$. ■

To evaluate the determinant of an arbitrary matrix can require considerable manipulation. A matrix in triangular form, however, has an easily calculated determinant.

Theorem 6.16 If $A = (a_{ij})$ is an $n \times n$ matrix that is either upper triangular or lower triangular (or diagonal), then $\det A = \prod_{i=1}^{n} a_{ii}$. ■

Theorem 6.16 follows from expanding the matrix and each submatrix about either the first row or first column.

The problem of computing the determinant of a matrix can be simplified by first reducing the matrix to triangular form, and then using Theorem 6.16 to find the determinant of the triangular matrix.

EXAMPLE 2 Let

$$A = \begin{bmatrix} 1 & 1 & 0 & 3 \\ 2 & 1 & -1 & 1 \\ -1 & 2 & 3 & -1 \\ 3 & -1 & -1 & 2 \end{bmatrix}.$$

The matrix A will first be reduced to an upper-triangular matrix. Perform $(E_2 - 2E_1) \to (E_2)$, $(E_3 + E_1) \to (E_3)$, and $(E_4 - 3E_1) \to (E_4)$ to obtain

$$\tilde{A}_1 = \begin{bmatrix} 1 & 1 & 0 & 3 \\ 0 & -1 & -1 & -5 \\ 0 & 3 & 3 & 2 \\ 0 & -4 & -1 & -7 \end{bmatrix}.$$

By Theorem 6.15(e), $\det A = \det \tilde{A}_1$. Form $\tilde{A}_2$ from $\tilde{A}_1$ by the operations $(E_3 + 3E_2) \to (E_3)$ and $(E_4 - 4E_2) \to (E_4)$:

$$\tilde{A}_2 = \begin{bmatrix} 1 & 1 & 0 & 3 \\ 0 & -1 & -1 & -5 \\ 0 & 0 & 0 & -13 \\ 0 & 0 & 3 & 13 \end{bmatrix}.$$

Then $\det \tilde{A}_2 = \det \tilde{A}_1 = \det A$. Let $\tilde{A}_3$ be formed from $\tilde{A}_2$ by $(E_3) \leftrightarrow (E_4)$:

$$\tilde{A}_3 = \begin{bmatrix} 1 & 1 & 0 & 3 \\ 0 & -1 & -1 & -5 \\ 0 & 0 & 3 & 13 \\ 0 & 0 & 0 & -13 \end{bmatrix}.$$

Since A_3 is an upper-triangular matrix, Theorem 6.16 implies that $\det \tilde{A}_3 = (1)(-1)(3)(-13) = 39$, and since $\tilde{A}_3$ was formed from $\tilde{A}_2$ by a row interchange,

$$\det A = \det \tilde{A}_2 = -\det \tilde{A}_3 = -39.$$

■

We now present the key result relating nonsingularity, Gaussian elimination, linear systems, and determinants. The proof of this theorem is not difficult, but it is laborious and is not presented here.

Theorem 6.17 The following statements are equivalent for any $n \times n$ matrix A:

a. The equation $A\mathbf{x} = \mathbf{0}$ has the unique solution $\mathbf{x} = \mathbf{0}$.

b. The system $A\mathbf{x} = \mathbf{b}$ has a unique solution for any n-dimensional column vector $\mathbf{b}$.

c. The matrix A is nonsingular; that is, A^{-1} exists.

d. $\det A \neq 0$.

e. Gaussian elimination with row interchanges can be performed on the system $A\mathbf{x} = \mathbf{b}$ for any n-dimensional column vector $\mathbf{b}$. ■

EXERCISE SET 6.4

1. Use Definition 6.14 to compute the determinants of the following matrices:

   a. $\begin{bmatrix} 1 & 2 & 0 \\ 2 & 1 & -1 \\ 3 & 1 & 1 \end{bmatrix}$ **b.** $\begin{bmatrix} 4 & 0 & 1 \\ 2 & 1 & 0 \\ 2 & 2 & 3 \end{bmatrix}$

   c. $\begin{bmatrix} 1 & 1 & -1 & 1 \\ 1 & 2 & -4 & -2 \\ 2 & 1 & 1 & 5 \\ -1 & 0 & -2 & -4 \end{bmatrix}$ **d.** $\begin{bmatrix} 2 & 0 & 1 & 2 \\ 1 & 1 & 0 & 2 \\ 2 & -1 & 3 & 1 \\ 3 & -1 & 4 & 3 \end{bmatrix}$

2. Repeat Exercise 1 using the method of Example 2.

3. Compute $\det A$, $\det B$, $\det AB$, and $\det BA$ for

$$A = \begin{bmatrix} 4 & 6 & 1 & -1 \\ 2 & 1 & 0 & \frac{1}{2} \\ 3 & 0 & 0 & 1 \\ 1 & -1 & 1 & 1 \end{bmatrix} \quad \text{and} \quad B = \begin{bmatrix} 1 & 2 & 3 & 4 \\ 0 & 2 & -1 & 1 \\ 0 & 0 & 3 & 2 \\ 0 & 0 & 0 & -1 \end{bmatrix}.$$

4. Let A be a 3×3 matrix. Show that if $\tilde{A}$ is the matrix obtained from A using any of the operations

$$(E_1) \leftrightarrow (E_2), \quad (E_1) \leftrightarrow (E_3), \quad \text{or} \quad (E_2) \leftrightarrow (E_3),$$

   then $\det \tilde{A} = -\det A$.

5. Find all values of α that make the following matrix singular.

$$A = \begin{bmatrix} 1 & -1 & \alpha \\ 2 & 2 & 1 \\ 0 & \alpha & -\frac{3}{2} \end{bmatrix}.$$

6. Find all values of α that make the following matrix singular.

$$A = \begin{bmatrix} 1 & 2 & -1 \\ 1 & \alpha & 1 \\ 2 & \alpha & -1 \end{bmatrix}.$$

7. Find all values of α so that the following linear system has no solutions.

$$\begin{aligned} 2x_1 - x_2 + 3x_3 &= 5, \\ 4x_1 + 2x_2 + 2x_3 &= 6, \\ -2x_1 + \alpha x_2 + 3x_3 &= 4. \end{aligned}$$

8. Find all values of α so that the following linear system has an infinite number of solutions.

$$\begin{aligned} 2x_1 - x_2 + 3x_3 &= 5, \\ 4x_1 + 2x_2 + 2x_3 &= 6, \\ -2x_1 + \alpha x_2 + 3x_3 &= 1. \end{aligned}$$

9. Use mathematical induction to show that when $n > 1$, the evaluation of the determinant of an $n \times n$ matrix using the definition requires $n! \sum_{k=1}^{n-1} \frac{1}{k!}$ multiplications/divisions and $n! - 1$ additions/subtractions.

10. Prove that AB is nonsingular if and only if both A and B are nonsingular.

11. The solution by **Cramer's rule** to the linear system

$$\begin{aligned} a_{11}x_1 + a_{12}x_2 + a_{13}x_3 &= b_1, \\ a_{21}x_1 + a_{22}x_2 + a_{23}x_3 &= b_2, \\ a_{31}x_1 + a_{32}x_2 + a_{33}x_3 &= b_3, \end{aligned}$$

has

$$x_1 = \frac{1}{D} \det \begin{bmatrix} b_1 & a_{12} & a_{13} \\ b_2 & a_{22} & a_{23} \\ b_3 & a_{32} & a_{33} \end{bmatrix} \equiv \frac{D_1}{D},$$

$$x_2 = \frac{1}{D} \det \begin{bmatrix} a_{11} & b_1 & a_{13} \\ a_{21} & b_2 & a_{23} \\ a_{31} & b_3 & a_{33} \end{bmatrix} \equiv \frac{D_2}{D},$$

and

$$x_3 = \frac{1}{D} \det \begin{bmatrix} a_{11} & a_{12} & b_1 \\ a_{21} & a_{22} & b_2 \\ a_{31} & a_{32} & b_3 \end{bmatrix} \equiv \frac{D_3}{D},$$

where

$$D = \det \begin{bmatrix} a_{11} & a_{12} & a_{13} \\ a_{21} & a_{22} & a_{23} \\ a_{31} & a_{32} & a_{33} \end{bmatrix}.$$

a. Find the solution to the linear system

$$\begin{aligned} 2x_1 + 3x_2 - x_3 &= 4, \\ x_1 - 2x_2 + x_3 &= 6, \\ x_1 - 12x_2 + 5x_3 &= 10 \end{aligned}$$

by Cramer's rule.

b. Show that the linear system

$$\begin{aligned} 2x_1 + 3x_2 - x_3 &= 4, \\ x_1 - 2x_2 + x_3 &= 6, \\ -x_1 - 12x_2 + 5x_3 &= 9 \end{aligned}$$

does not have a solution. Compute D_1, D_2, and D_3.

c. Show that the linear system

$$\begin{aligned} 2x_1 + 3x_2 - x_3 &= 4, \\ x_1 - 2x_2 + x_3 &= 6, \\ -x_1 - 12x_2 + 5x_3 &= 10 \end{aligned}$$

has an infinite number of solutions. Compute D_1, D_2, and D_3.

d. Prove that if a 3×3 linear system with $D = 0$ has solutions, then $D_1 = D_2 = D_3 = 0$.

e. Determine the number of multiplications/divisions and additions/subtractions required for Cramer's rule on a 3×3 system.

12. **a.** Generalize Cramer's rule to an $n \times n$ linear system.

b. Use the result in Exercise 9 to determine the number of multiplications/divisions and additions/subtractions required for Cramer's rule on an $n \times n$ system.

6.5 Matrix Factorization

Gaussian elimination is the principal tool in the direct solution of linear systems of equations, so it should be no surprise that it appears in other guises. In this section we will see that the steps used to solve a system of the form $A\mathbf{x} = \mathbf{b}$ can also be used to factor a matrix into a product of matrices. The factorization is particularly useful when it has the form $A = LU$, where L is lower triangular and U is upper triangular. Not all matrices can be factored in this way, but many can that frequently occur in the application of numerical techniques.

In Section 6.1 we found that Gaussian elimination applied to an arbitrary linear system $A\mathbf{x} = \mathbf{b}$ requires $O(n^3/3)$ arithmetic operations to determine $\mathbf{x}$. If A has been factored into the triangular form $A = LU$, then we can solve for $\mathbf{x}$ more easily by using a two-step process. First we let $\mathbf{y} = U\mathbf{x}$ and solve the system $L\mathbf{y} = \mathbf{b}$ for $\mathbf{y}$. Since L is triangular, determining $\mathbf{y}$ from this equation requires only $O(n^2)$ operations, the same as the backward substitution portion of Gaussian elimination. Once $\mathbf{y}$ is known, the triangular system $U\mathbf{x} = \mathbf{y}$ requires only an additional $O(n^2)$ operations to determine the solution $\mathbf{x}$. This fact means that the number of operations needed to solve the system $A\mathbf{x} = \mathbf{b}$ is reduced from $O(n^3/3)$ to $O(n^2)$. In systems greater than 100 by 100, this result can reduce the amount of calculation by more than 97%. Not surprisingly, the reductions resulting from matrix factorization do not come free; determining the specific matrices L and U requires $O(n^3/3)$ operations. But once the factorization is determined, any system involving the matrix A can be solved in the simplified manner.

To examine which matrices have an LU factorization and find how it is determined, first suppose that Gaussian elimination can be performed on the system $A\mathbf{x} = \mathbf{b}$ without row interchanges. Using the notation in Section 6.1, this is equivalent to having nonzero pivot elements $a_{ii}^{(i)}$ for each $i = 1, 2, \ldots, n$.

The first step in the Gaussian elimination process consists of performing, for each $j = 2, 3, \ldots, n$, the operations

(6.8)
$$(E_j - m_{j,1}E_1) \to (E_j), \quad \text{where} \quad m_{j,1} = \frac{a_{j1}^{(1)}}{a_{11}^{(1)}}.$$

These operations transform the system into one in which all the entries in the first column below the diagonal are zero.

The system of operations in (6.8) can be viewed in another way. It is simultaneously accomplished by multiplying the original matrix A on the left by the matrix

$$
M^{(1)} = \begin{bmatrix} 1 & 0 & \cdots & \cdots & 0 \\ -m_{21} & 1 & \ddots & & \vdots \\ \vdots & 0 & \ddots & \ddots & \vdots \\ \vdots & \vdots & \ddots & \ddots & 0 \\ -m_{n1} & 0 & \cdots & 0 & 1 \end{bmatrix}
$$

This is called the **first Gaussian transformation matrix**. We denote the product of this matrix with $A^{(1)} \equiv A$ by $A^{(2)}$ and with $\mathbf{b}$ by $\mathbf{b}^{(2)}$, so

$$A^{(2)}\mathbf{x} = M^{(1)}A\mathbf{x} = M^{(1)}\mathbf{b} = \mathbf{b}^{(2)}.$$

In a similar manner we construct $M^{(2)}$, the identity matrix with the entries below the diagonal in the second column replaced by the negatives of the multipliers

$$m_{j,2} = \frac{a_{j2}^{(2)}}{a_{22}^{(2)}}.$$

The product of this matrix with $A^{(2)}$ has zeros below the diagonal in the first two columns, and we let

$$A^{(3)}\mathbf{x} = M^{(2)}A^{(2)}\mathbf{x} = M^{(2)}M^{(1)}A\mathbf{x} = M^{(2)}M^{(1)}\mathbf{b} = \mathbf{b}^{(3)}.$$

In general, with $A^{(k)}\mathbf{x} = \mathbf{b}^{(k)}$ already formed, multiply by the ***k*th Gaussian transformation matrix**

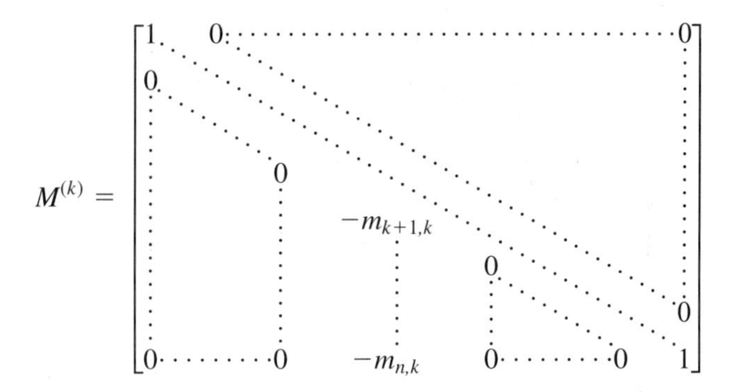

to obtain

$$\textbf{(6.9)}\quad A^{(k+1)}\mathbf{x} = M^{(k)}A^{(k)}\mathbf{x} = M^{(k)}\cdots M^{(1)}A\mathbf{x} = M^{(k)}\mathbf{b}^{(k)} = \mathbf{b}^{(k+1)} = M^{(k)}\cdots M^{(1)}\mathbf{b}.$$

The process ends with the formation of $A^{(n)}\mathbf{x} = \mathbf{b}^{(n)}$, where $A^{(n)}$ is the upper triangular matrix

$$
A^{(n)} = \begin{bmatrix} a_{11}^{(1)} & a_{12}^{(1)} & \cdots & \cdots & a_{1n}^{(1)} \\ 0 & a_{22}^{(2)} & \ddots & & \vdots \\ \vdots & \ddots & \ddots & \ddots & a_{n-1,n}^{(n-1)} \\ 0 & \cdots & \cdots & 0 & a_{nn}^{(n)} \end{bmatrix},
$$

given by

$$A^{(n)} = M^{(n-1)}M^{(n-2)}\cdots M^{(1)}A.$$

The process we have outlined forms one-half of the matrix factorization $A = LU$. We let U be the upper triangular matrix $A^{(n)}$. To determine the complementary lower triangular matrix L, first recall the multiplication of $A^{(k)}\mathbf{x} = \mathbf{b}^{(k)}$ by the Gaussian transformation $M^{(k)}$ used to obtain (6.9):

$$A^{(k+1)}\mathbf{x} = M^{(k)}A^{(k)}\mathbf{x} = M^{(k)}\mathbf{b}^{(k)} = \mathbf{b}^{(k+1)},$$

where $M^{(k)}$ generates the row operations

$$(E_j - m_{j,k}E_k) \rightarrow (E_j), \quad \text{for } j = k+1, \ldots, n.$$

To reverse the effects of this transformation and return to $A^{(k)}$ requires that the operations $(E_j + m_{j,k}E_k) \rightarrow (E_j)$ be performed for each $j = k+1, \ldots, n$. This is equivalent to multiplying by the inverse of the matrix $M^{(k)}$, the matrix

$$L^{(k)} = [M^{(k)}]^{-1} = \begin{bmatrix} 1 & 0 & \cdots & \cdots & \cdots & \cdots & 0 \\ 0 & \ddots & \ddots & & & & \vdots \\ \vdots & \ddots & 0 & \ddots & & & \vdots \\ \vdots & & \vdots & m_{k+1,k} & \ddots & & \vdots \\ \vdots & & \vdots & \vdots & 0 & \ddots & 0 \\ \vdots & & \vdots & \vdots & \vdots & \ddots & 0 \\ 0 & \cdots & 0 & m_{n,k} & 0 & \cdots\ 0 & 1 \end{bmatrix}.$$

The lower-triangular matrix L in the factorization of A is the product of the matrices $L^{(k)}$:

$$L = L^{(1)}L^{(2)}\cdots L^{(n-1)} = \begin{bmatrix} 1 & 0 & \cdots & 0 \\ m_{21} & 1 & \ddots & \vdots \\ \vdots & \ddots & \ddots & 0 \\ m_{n1} & \cdots & m_{n,n-1} & 1 \end{bmatrix},$$

since the product of L with the upper-triangular matrix $U = M^{(n-1)}\cdots M^{(2)}M^{(1)}A$ gives

$$\begin{aligned} LU &= L^{(1)}L^{(2)}\cdots L^{(n-3)}L^{(n-2)}L^{(n-1)} \cdot M^{(n-1)}M^{(n-2)}M^{(n-3)}\cdots M^{(2)}M^{(1)}A \\ &= [M^{(1)}]^{-1}[M^{(2)}]^{-1}\cdots[M^{(n-2)}]^{-1}[M^{(n-1)}]^{-1} \cdot M^{(n-1)}M^{(n-2)}\cdots M^{(2)}M^{(1)}A = A. \end{aligned}$$

Theorem 6.18 follows from these observations.

Theorem 6.18 If Gaussian elimination can be performed on the linear system $A\mathbf{x} = \mathbf{b}$ without row interchanges, then the matrix A can be factored into the product of a lower-triangular matrix L and an upper-triangular matrix U,

$$A = LU,$$

where

$$U = \begin{bmatrix} a_{11}^{(1)} & a_{12}^{(1)} & \cdots & a_{1n}^{(1)} \\ 0 & a_{22}^{(2)} & \ddots & \vdots \\ \vdots & \ddots & \ddots & a_{n-1,n}^{(n-1)} \\ 0 & \cdots & 0 & a_{nn}^{(n)} \end{bmatrix} \quad \text{and} \quad L = \begin{bmatrix} 1 & 0 & \cdots & 0 \\ m_{21} & 1 & \ddots & \vdots \\ \vdots & \ddots & \ddots & 0 \\ m_{n1} & \cdots & m_{n,n-1} & 1 \end{bmatrix}. \quad \blacksquare$$

EXAMPLE 1 The linear system

$$\begin{aligned} x_1 + x_2 + 3x_4 &= 4 \\ 2x_1 + x_2 - x_3 + x_4 &= 1 \\ 3x_1 - x_2 - x_3 + 2x_4 &= -3 \\ -x_1 + 2x_2 + 3x_3 - x_4 &= 4 \end{aligned}$$

was considered in Section 6.1. The sequence of operations $(E_2 - 2E_1) \to (E_2)$, $(E_3 - 3E_1) \to (E_3)$, $(E_4 - (-1)E_1) \to (E_4)$, $(E_3 - 4E_2) \to (E_3)$, $(E_4 - (-3)E_2) \to (E_4)$ converts the system to the triangular system

$$\begin{aligned} x_1 + x_2 + 3x_4 &= 4, \\ -x_2 - x_3 - 5x_4 &= -7, \\ 3x_3 + 13x_4 &= 13, \\ -13x_4 &= -13. \end{aligned}$$

The multipliers m_{ij} and the upper triangular matrix produce the factorization

$$A = \begin{bmatrix} 1 & 1 & 0 & 3 \\ 2 & 1 & -1 & 1 \\ 3 & -1 & -1 & 2 \\ -1 & 2 & 3 & -1 \end{bmatrix} = \begin{bmatrix} 1 & 0 & 0 & 0 \\ 2 & 1 & 0 & 0 \\ 3 & 4 & 1 & 0 \\ -1 & -3 & 0 & 1 \end{bmatrix} \begin{bmatrix} 1 & 1 & 0 & 3 \\ 0 & -1 & -1 & -5 \\ 0 & 0 & 3 & 13 \\ 0 & 0 & 0 & -13 \end{bmatrix} = LU.$$

This factorization permits us to easily solve any system involving the matrix A. For example, to solve

$$A\mathbf{x} = LU\mathbf{x} = \begin{bmatrix} 1 & 0 & 0 & 0 \\ 2 & 1 & 0 & 0 \\ 3 & 4 & 1 & 0 \\ -1 & -3 & 0 & 1 \end{bmatrix} \begin{bmatrix} 1 & 1 & 0 & 3 \\ 0 & -1 & -1 & -5 \\ 0 & 0 & 3 & 13 \\ 0 & 0 & 0 & -13 \end{bmatrix} \begin{bmatrix} x_1 \\ x_2 \\ x_3 \\ x_4 \end{bmatrix} = \begin{bmatrix} 8 \\ 7 \\ 14 \\ -7 \end{bmatrix},$$

we first introduce the substitution $\mathbf{y} = U\mathbf{x}$. Then $L\mathbf{y} = \mathbf{b}$, that is,

$$LU\mathbf{x} = L\mathbf{y} = \begin{bmatrix} 1 & 0 & 0 & 0 \\ 2 & 1 & 0 & 0 \\ 3 & 4 & 1 & 0 \\ -1 & -3 & 0 & 1 \end{bmatrix} \begin{bmatrix} y_1 \\ y_2 \\ y_3 \\ y_4 \end{bmatrix} = \begin{bmatrix} 8 \\ 7 \\ 14 \\ -7 \end{bmatrix}.$$

This system is solved for $\mathbf{y}$ by a simple forward-substitution process:

$$\begin{aligned} y_1 &= 8, & & \\ 2y_1 + y_2 &= 7, & \text{so } y_2 &= -9 \\ 3y_1 + 4y_2 + y_3 &= 14, & \text{so } y_3 &= 26 \\ -y_1 - 3y_3 + y_4 &= -7, & \text{so } y_4 &= -26. \end{aligned}$$

We then solve $U\mathbf{x} = \mathbf{y}$ for $\mathbf{x}$, the solution of the original system; that is,

$$\begin{bmatrix} 1 & 1 & 0 & 3 \\ 0 & -1 & -1 & -5 \\ 0 & 0 & 3 & 13 \\ 0 & 0 & 0 & -13 \end{bmatrix} \begin{bmatrix} x_1 \\ x_2 \\ x_3 \\ x_4 \end{bmatrix} = \begin{bmatrix} 8 \\ -9 \\ 26 \\ -26 \end{bmatrix}.$$

Using backward substitution we obtain $x_4 = 2$, $x_3 = 0$, $x_2 = -1$, $x_1 = 3$. ■

The factorization used in Example 1 is called *Doolittle's method* and requires that $l_{11} = l_{22} = l_{33} = l_{44} = 1$, which results in the factorization described in Theorem 6.18. In Section 6.6, we consider *Crout's method*, a factorization which requires the diagonal elements of U to be one, and *Choleski's method*, which requires that $l_{ii} = u_{ii}$ for each i.

A general procedure for factoring matrices into a product of triangular matrices is contained in Algorithm 6.4. Although new matrices L and U are constructed, the generated values can replace the corresponding entries of A that are no longer needed.

Algorithm 6.4 permits either the diagonal of L or the diagonal of U to be specified.

ALGORITHM 6.4

LU Factorization

To factor the $n \times n$ matrix $A = (a_{ij})$ into the product of the lower-triangular matrix $L = (l_{ij})$ and the upper-triangular matrix $U = (u_{ij})$; that is, $A = LU$, where the main diagonal of either L or U consists of all ones:

INPUT dimension n; the entries a_{ij}, $1 \le i, j \le n$ of A; the diagonal $l_{11} = \cdots = l_{nn} = 1$ of L or the diagonal $u_{11} = \cdots = u_{nn} = 1$ of U.

OUTPUT the entries l_{ij}, $1 \le j \le i$, $1 \le i \le n$ of L and the entries, u_{ij}, $i \le j \le n$, $1 \le i \le n$ of U.

Step 1 Select l_{11} and u_{11} satisfying $l_{11}u_{11} = a_{11}$.
If $l_{11}u_{11} = 0$ then OUTPUT ('Factorization impossible');
STOP.

Step 2 For $j = 2, \ldots, n$ set $u_{1j} = a_{1j}/l_{11}$; (*First row of U.*)
$l_{j1} = a_{j1}/u_{11}$. (*First column of L.*)

Step 3 For $i = 2, \ldots, n-1$ do Steps 4 and 5.

Step 4 Select l_{ii} and u_{ii} satisfying $l_{ii}u_{ii} = a_{ii} - \sum_{k=1}^{i-1} l_{ik}u_{ki}$.

If $l_{ii}u_{ii} = 0$ then OUTPUT ('Factorization impossible');
STOP.

Step 5 For $j = i+1, \ldots, n$

set $u_{ij} = \frac{1}{l_{ii}}\left[a_{ij} - \sum_{k=1}^{i-1} l_{ik}u_{kj}\right]$; (*ith row of U.*)

$l_{ji} = \frac{1}{u_{ii}}\left[a_{ji} - \sum_{k=1}^{i-1} l_{jk}u_{ki}\right]$. (*ith column of L.*)

Step 6 Select l_{nn} and u_{nn} satisfying $l_{nn}u_{nn} = a_{nn} - \sum_{k=1}^{n-1} l_{nk}u_{kn}$.

(*Note: If $l_{nn}u_{nn} = 0$, then $A = LU$ but A is singular.*)

Step 7 OUTPUT (l_{ij} for $j = 1, \ldots, i$ and $i = 1, \ldots, n$);
OUTPUT (u_{ij} for $j = i, \ldots, n$ and $i = 1, \ldots, n$);
STOP.

Once the matrix factorization is complete, the solution to a linear system of the form $A\mathbf{x} = LU\mathbf{x} = \mathbf{b}$ is found by first letting $\mathbf{y} = U\mathbf{x}$ and determining $\mathbf{y}$ from the equations

$$y_1 = \frac{b_1}{l_{11}}$$

and, for each $i = 2, 3, \ldots, n$,

$$y_i = \frac{1}{l_{ii}}\left[b_i - \sum_{j=1}^{i-1} l_{ij}y_j\right].$$

Once $\mathbf{y}$ is found, the upper-triangular system $U\mathbf{x} = \mathbf{y}$ is solved for $\mathbf{x}$ by backward substitution.

In the previous discussion we assumed that $A\mathbf{x} = \mathbf{b}$ can be solved using Gaussian elimination without row interchanges. From a practical standpoint, this factorization is useful only when row interchanges are not required to control the roundoff error resulting from the use of finite-digit arithmetic. Fortunately, many systems we encounter when using approximation methods are of this type, but it is interesting to see the modifications that must be made when row interchanges are required. We begin the discussion with the introduction of a class of matrices that are used to rearrange, or permute, rows of a given matrix.

An $n \times n$ **permutation matrix** P is obtained by rearranging the rows of the identity matrix. This gives a matrix with precisely one nonzero entry in each row and in each column. The nonzero entries are all 1s.

EXAMPLE 2 The matrix

$$P = \begin{bmatrix} 1 & 0 & 0 \\ 0 & 0 & 1 \\ 0 & 1 & 0 \end{bmatrix}$$

is a 3×3 permutation matrix. For any 3×3 matrix A, multiplying on the left by P has the effect of interchanging the second and third rows of A:

$$PA = \begin{bmatrix} 1 & 0 & 0 \\ 0 & 0 & 1 \\ 0 & 1 & 0 \end{bmatrix} \begin{bmatrix} a_{11} & a_{12} & a_{13} \\ a_{21} & a_{22} & a_{23} \\ a_{31} & a_{32} & a_{33} \end{bmatrix} = \begin{bmatrix} a_{11} & a_{12} & a_{13} \\ a_{31} & a_{32} & a_{33} \\ a_{21} & a_{22} & a_{23} \end{bmatrix}.$$

Similarly, multiplying A on the right by P interchanges the second and third columns of A. ■

There are two useful properties of permutation matrices that relate to Gaussian elimination. The first of these is illustrated in the previous example and states that if $k_1, \ldots, k_n$ is a permutation of the integers $1, \ldots, n$ and the permutation matrix $P = (p_{ij})$ is defined by

$$p_{ij} = \begin{cases} 1, & \text{if } j = k_i, \\ 0, & \text{otherwise,} \end{cases}$$

then PA permutes the rows of A, that is,

$$PA = \begin{bmatrix} a_{k_1,1} & a_{k_1,2} & \cdots & a_{k_1,n} \\ a_{k_2,1} & a_{k_2,2} & \cdots & a_{k_2,n} \\ \vdots & \vdots & & \vdots \\ a_{k_n,1} & a_{k_n,2} & \cdots & a_{k_n,n} \end{bmatrix}.$$

The second is that if P is a permutation matrix, then P^{-1} exists and $P^{-1} = P^t$.

At the end of Section 6.4 we saw that for any nonsingular matrix A, the linear system $A\mathbf{x} = \mathbf{b}$ can be solved by Gaussian elimination, with the possibility of row interchanges. If we knew the row interchanges that were required to solve the system by Gaussian elimination, we could arrange the original equations in an order that would ensure that no row interchanges are needed. Hence there is a rearrangement of the equations in the system that permits Gaussian elimination to proceed *without* row interchanges. This implies that for any nonsingular matrix A, a permutation matrix P exists for which the system

$$PA\mathbf{x} = P\mathbf{b}$$

can be solved without row interchanges. But this matrix PA can be factored into

$$PA = LU,$$

where L is lower triangular and U is upper triangular. Since $P^{-1} = P^t$, we have the factorization

$$A = P^{-1}LU = (P^tL)U.$$

However, the matrix P^tL is not lower triangular unless $P = I$.

EXAMPLE 3 Since $a_{11} = 0$, the matrix

$$A = \begin{bmatrix} 0 & 0 & -1 & 1 \\ 1 & 1 & -1 & 2 \\ 1 & 1 & 0 & 3 \\ 1 & 2 & -1 & 3 \end{bmatrix}$$

does not have an LU factorization. However, using the row interchange $(E_1) \leftrightarrow (E_2)$, followed by $(E_3 - E_1) \rightarrow E_3$ and $(E_4 - E_1) \rightarrow E_4$, produces

$$\begin{bmatrix} 1 & 1 & -1 & 2 \\ 0 & 0 & -1 & 1 \\ 0 & 0 & 1 & 1 \\ 0 & 1 & 0 & 1 \end{bmatrix}.$$

Then the row interchange $(E_2) \leftrightarrow (E_4)$, followed by $(E_4 + E_3) \rightarrow E_4$, gives the matrix

$$U = \begin{bmatrix} 1 & 1 & -1 & 2 \\ 0 & 1 & 0 & 1 \\ 0 & 0 & 1 & 1 \\ 0 & 0 & 0 & 2 \end{bmatrix}.$$

The permutation matrix associated with the row interchange $(E_1) \leftrightarrow (E_2)$ followed by the row interchange $(E_2) \leftrightarrow (E_4)$ is

$$P = \begin{bmatrix} 0 & 1 & 0 & 0 \\ 0 & 0 & 0 & 1 \\ 0 & 0 & 1 & 0 \\ 1 & 0 & 0 & 0 \end{bmatrix}.$$

Gaussian elimination can be performed on PA without row interchanges using the operations $(E_2 - E_1) \rightarrow (E_2)$, $(E_3 - E_1) \rightarrow (E_3)$, and $(E_4 + E_3) \rightarrow (E_4)$. This produces the LU factorization of PA,

$$PA = \begin{bmatrix} 1 & 1 & -1 & 2 \\ 1 & 2 & -1 & 3 \\ 1 & 1 & 0 & 3 \\ 0 & 0 & -1 & 2 \end{bmatrix} = \begin{bmatrix} 1 & 0 & 0 & 0 \\ 1 & 1 & 0 & 0 \\ 1 & 0 & 1 & 0 \\ 0 & 0 & -1 & 1 \end{bmatrix} \begin{bmatrix} 1 & 1 & -1 & 2 \\ 0 & 1 & 0 & 1 \\ 0 & 0 & 1 & 1 \\ 0 & 0 & 0 & 2 \end{bmatrix} = LU.$$

Multiplying by $P^{-1} = P^t$ produces the factorization

$$A = P^{-1}LU = (P^tL)U = \begin{bmatrix} 0 & 0 & -1 & 1 \\ 1 & 0 & 0 & 0 \\ 1 & 0 & 1 & 0 \\ 1 & 1 & 0 & 0 \end{bmatrix} \begin{bmatrix} 1 & 1 & -1 & 2 \\ 0 & 1 & 0 & 1 \\ 0 & 0 & 1 & 1 \\ 0 & 0 & 0 & 2 \end{bmatrix}.$$ ■

EXERCISE SET 6.5

1. Solve the following linear systems:

   a. $\begin{bmatrix} 1 & 0 & 0 \\ 2 & 1 & 0 \\ -1 & 0 & 1 \end{bmatrix} \begin{bmatrix} 2 & 3 & -1 \\ 0 & -2 & 1 \\ 0 & 0 & 3 \end{bmatrix} \begin{bmatrix} x_1 \\ x_2 \\ x_3 \end{bmatrix} = \begin{bmatrix} 2 \\ -1 \\ 1 \end{bmatrix}$

   b. $\begin{bmatrix} 2 & 0 & 0 \\ -1 & 1 & 0 \\ 3 & 2 & -1 \end{bmatrix} \begin{bmatrix} 1 & 1 & 1 \\ 0 & 1 & 2 \\ 0 & 0 & 1 \end{bmatrix} \begin{bmatrix} x_1 \\ x_2 \\ x_3 \end{bmatrix} = \begin{bmatrix} -1 \\ 3 \\ 0 \end{bmatrix}$

2. Factor the following matrices into the LU decomposition using the LU Factorization Algorithm with $l_{ii} = 1$ for all i.

   a. $\begin{bmatrix} 2 & -1 & 1 \\ 3 & 3 & 9 \\ 3 & 3 & 5 \end{bmatrix}$

   b. $\begin{bmatrix} 1.012 & -2.132 & 3.104 \\ -2.132 & 4.096 & -7.013 \\ 3.104 & -7.013 & 0.014 \end{bmatrix}$

   c. $\begin{bmatrix} 2 & 0 & 0 & 0 \\ 1 & 1.5 & 0 & 0 \\ 0 & -3 & 0.5 & 0 \\ 2 & -2 & 1 & 1 \end{bmatrix}$

   d. $\begin{bmatrix} 2.1756 & 4.0231 & -2.1732 & 5.1967 \\ -4.0231 & 6.0000 & 0 & 1.1973 \\ -1.0000 & -5.2107 & 1.1111 & 0 \\ 6.0235 & 7.0000 & 0 & -4.1561 \end{bmatrix}$

3. Modify the LU Factorization Algorithm so that it can be used to solve a linear system, and then solve the following linear systems.

   a. $$\begin{aligned} 2x_1 - x_2 + x_3 &= -1, \\ 3x_1 + 3x_2 + 9x_3 &= 0, \\ 3x_1 + 3x_2 + 5x_3 &= 4. \end{aligned}$$

   b. $$\begin{aligned} 1.012x_1 - 2.132x_2 + 3.104x_3 &= 1.984, \\ -2.132x_1 + 4.096x_2 - 7.013x_3 &= -5.049, \\ 3.104x_1 - 7.013x_2 + 0.014x_3 &= -3.895. \end{aligned}$$

   c. $$\begin{aligned} 2x_1 \qquad\qquad\qquad &= 3, \\ x_1 + 1.5x_2 \qquad\qquad &= 4.5, \\ - 3x_2 + 0.5x_3 \qquad &= -6.6, \\ 2x_1 - 2x_2 + x_3 + x_4 &= 0.8. \end{aligned}$$

   d. $$\begin{aligned} 2.1756x_1 + 4.0231x_2 - 2.1732x_3 + 5.1967x_4 &= 17.102, \\ -4.0231x_1 + 6.0000x_2 \qquad\qquad + 1.1973x_4 &= -6.1593, \\ -1.0000x_1 - 5.2107x_2 + 1.1111x_3 \qquad\qquad &= 3.0004, \\ 6.0235x_1 + 7.0000x_2 \qquad\qquad - 4.1561x_4 &= 0.0000. \end{aligned}$$

4. Consider the following matrices. Find the permutation matrix P so that PA can be factored into the product LU, where L is lower triangular with 1s on its diagonal and U is upper triangular for the following matrices.

a. $A = \begin{bmatrix} 1 & 2 & -1 \\ 2 & 4 & 0 \\ 0 & 1 & -1 \end{bmatrix}$ **b.** $A = \begin{bmatrix} 0 & 1 & 1 \\ 1 & -2 & -1 \\ 1 & -1 & 1 \end{bmatrix}$

c. $A = \begin{bmatrix} 1 & 1 & -1 & 0 \\ 1 & 1 & 4 & 3 \\ 2 & -1 & 2 & 4 \\ 2 & -1 & 2 & 3 \end{bmatrix}$ **d.** $A = \begin{bmatrix} 0 & 1 & 1 & 2 \\ 0 & 1 & 1 & -1 \\ 1 & 2 & -1 & 3 \\ 1 & 1 & 2 & 0 \end{bmatrix}$

5. Obtain factorizations of the form $A = P^tLU$ for the following matrices.

a. $A = \begin{bmatrix} 0 & 2 & 3 \\ 1 & 1 & -1 \\ 0 & -1 & 1 \end{bmatrix}$ **b.** $A = \begin{bmatrix} 1 & 2 & -1 \\ 1 & 2 & 3 \\ 2 & -1 & 4 \end{bmatrix}$

c. $A = \begin{bmatrix} 1 & -2 & 3 & 0 \\ 3 & -6 & 9 & 3 \\ 2 & 1 & 4 & 1 \\ 1 & -2 & 2 & -2 \end{bmatrix}$ **d.** $A = \begin{bmatrix} 1 & -2 & 3 & 0 \\ 1 & -2 & 3 & 1 \\ 1 & -2 & 2 & -2 \\ 2 & 1 & 3 & -1 \end{bmatrix}$

6. Suppose $A = P^tLU$, where P is a permutation matrix, L is a lower-triangular matrix with ones on the diagonal, and U is an upper-triangular matrix.

a. Count the number of operations needed to compute P^tLU for a given matrix A.

b. Show that if P contains k row interchanges, then

$$\det P = \det P^t = (-1)^k.$$

c. Use $\det A = \det P^t \det L \det U = (-1)^k \det U$ to count the number of operations for determining $\det A$ by factoring.

d. Compute $\det A$ and count the number of operations when

$$A = \begin{bmatrix} 0 & 2 & 1 & 4 & -1 & 3 \\ 1 & 2 & -1 & 3 & 4 & 0 \\ 0 & 1 & 1 & -1 & 2 & -1 \\ 2 & 3 & -4 & 2 & 0 & 5 \\ 1 & 1 & 1 & 3 & 0 & 2 \\ -1 & -1 & 2 & -1 & 2 & 0 \end{bmatrix}.$$

7. **a.** Show that the LU Factorization Algorithm requires $\frac{1}{3}n^3 - \frac{1}{3}n$ multiplications/divisions and $\frac{1}{3}n^3 - \frac{1}{2}n^2 + \frac{1}{6}n$ additions/subtractions.

b. Show that solving $L\mathbf{y} = \mathbf{b}$, where L is a lower-triangular matrix with $l_{ii} = 1$ for all i, requires $\frac{1}{2}n^2 - \frac{1}{2}n$ multiplications/divisions and $\frac{1}{2}n^2 - \frac{1}{2}n$ additions/subtractions.

c. Show that solving $A\mathbf{x} = \mathbf{b}$ by first factoring A into $A = LU$ and then solving $L\mathbf{y} = \mathbf{b}$ and $U\mathbf{x} = \mathbf{y}$ requires the same number of operations as the Gaussian Elimination Algorithm 6.1.

d. Count the number of operations required to solve m linear systems $A\mathbf{x}^{(k)} = \mathbf{b}^{(k)}$ for $k = 1, \ldots, m$ by first factoring A and then using the method of part (c) m times.

6.6 Special Types of Matrices

We will now turn attention to two classes of matrices for which Gaussian elimination can be performed without row interchanges. The first class is described in the following definition.

Definition 6.19 The $n \times n$ matrix A is said to be **strictly diagonally dominant** when

$$|a_{ii}| > \sum_{\substack{j=1,\\ j\neq i}}^{n} |a_{ij}|$$

holds for each $i = 1, 2, \ldots, n$. ■

EXAMPLE 1 Consider the matrices

$$A = \begin{bmatrix} 7 & 2 & 0 \\ 3 & 5 & -1 \\ 0 & 5 & -6 \end{bmatrix} \quad \text{and} \quad B = \begin{bmatrix} 6 & 4 & -3 \\ 4 & -2 & 0 \\ -3 & 0 & 1 \end{bmatrix}.$$

The nonsymmetric matrix A is strictly diagonally dominant, since $|7| > |2| + |0|$, $|5| > |3| + |-1|$, and $|-6| > |0| + |5|$. The symmetric matrix B is not strictly diagonally dominant since, for example, in the third row the absolute value of the diagonal element is $|6| < |4| + |-3| = 7$. It is interesting to note that A^t is not strictly diagonally dominant, nor, of course, is $B^t = B$. ■

The following theorem was used in Section 3.4 to ensure that there are unique solutions to the linear systems needed to determine cubic spline interpolants.

Theorem 6.20 A strictly diagonally dominant matrix A is nonsingular. Moreover, in this case, Gaussian elimination can be performed on any linear system of the form $A\mathbf{x} = \mathbf{b}$ to obtain its unique solution without row or column interchanges, and the computations are stable with respect to the growth of roundoff errors.

Proof We first use proof by contradiction to show that A is nonsingular. Consider the linear system described by $A\mathbf{x} = \mathbf{0}$, and suppose that a nonzero solution $\mathbf{x} = (x_i)$ to this system exists. Let k be an index for which

$$0 < |x_k| = \max_{1 \le j \le n} |x_j|.$$

Since $\sum_{j=1}^{n} a_{ij}x_j = 0$ for each $i = 1, 2, \ldots, n$, we have, when $i = k$,

$$a_{kk}x_k = -\sum_{\substack{j=1,\\ j\neq k}}^{n} a_{kj}x_j.$$

This implies that

$$|a_{kk}||x_k| \le \sum_{\substack{j=1,\\ j\neq k}}^{n} |a_{kj}||x_j|,$$

or

$$|a_{kk}| \leq \sum_{\substack{j=1,\\ j\neq k}}^{n} |a_{kj}| \frac{|x_j|}{|x_k|} \leq \sum_{\substack{j=1,\\ j\neq k}}^{n} |a_{kj}|$$

This inequality contradicts the strict diagonal dominance of A. Consequently, the only solution to $A\mathbf{x} = \mathbf{0}$ is $\mathbf{x} = \mathbf{0}$, a condition shown in Theorem 6.17 to be equivalent to the nonsingularity of A.

To prove that Gaussian elimination can be performed without row interchanges, we will show that each of the matrices $A^{(2)}, A^{(3)}, \ldots, A^{(n)}$ generated by the Gaussian elimination process (and described in Section 6.5) is strictly diagonally dominant.

Since A is strictly diagonally dominant, $a_{11} \neq 0$ and $A^{(2)}$ can be formed. For each $i = 2, 3, \ldots, n$,

$$a_{ij}^{(2)} = a_{ij}^{(1)} - \frac{a_{1j}^{(1)} a_{i1}^{(1)}}{a_{11}^{(1)}}, \quad \text{for} \quad 2 \leq j \leq n.$$

Since $a_{i1}^{(2)} = 0$,

$$\begin{aligned}
\sum_{\substack{j=2\\ j\neq i}}^{n} |a_{ij}^{(2)}| &= \sum_{\substack{j=2\\ j\neq i}}^{n} \left| a_{ij}^{(1)} - \frac{a_{1j}^{(1)} a_{i1}^{(1)}}{a_{11}^{(1)}} \right| \\
&\leq \sum_{\substack{j=2\\ j\neq i}}^{n} |a_{ij}^{(1)}| + \sum_{\substack{j=2\\ j\neq i}}^{n} \left| \frac{a_{1j}^{(1)} a_{i1}^{(1)}}{a_{11}^{(1)}} \right| \\
&< |a_{ii}^{(1)}| - |a_{i1}^{(1)}| + \frac{|a_{i1}^{(1)}|}{|a_{11}^{(1)}|} \sum_{\substack{j=2\\ j\neq i}}^{n} |a_{1j}^{(1)}| \\
&< |a_{ii}^{(1)}| - |a_{i1}^{(1)}| + \frac{|a_{i1}^{(1)}|}{|a_{11}^{(1)}|} \left(|a_{11}^{(1)}| - |a_{1i}^{(1)}| \right) \\
&= |a_{ii}^{(1)}| - \frac{|a_{i1}^{(1)}| |a_{1i}^{(1)}|}{|a_{11}^{(1)}|} \\
&\leq \left| a_{ii}^{(1)} - \frac{a_{i1}^{(1)} a_{1i}^{(1)}}{a_{11}^{(1)}} \right| = |a_{ii}^{(2)}|.
\end{aligned}$$

Thus, the strict diagonal dominance is established for rows $2, \ldots, n$. Since the first row of $A^{(2)}$ and A are the same, $A^{(2)}$ is strictly diagonally dominant.

This process is continued until the upper-triangular and strictly diagonally dominant $A^{(n)}$ is obtained. This implies that all the diagonal elements are nonzero, so Gaussian elimination can be performed without row interchanges.

The demonstration of stability for this procedure can be found in [We]. ■ ■ ■

The next special class of matrices is called *positive definite*.

Definition 6.21 A matrix A is **positive definite** if it is symmetric and if $\mathbf{x}^t A\mathbf{x} > 0$ for every n-dimensional column vector $\mathbf{x} \neq \mathbf{0}$. ■

Not all authors require symmetry of a positive definite matrix. For example, Golub and Van Loan [GV], a standard reference in matrix methods, requires only that $\mathbf{x}^t A\mathbf{x} > 0$ for each nonzero vector $\mathbf{x}$. Matrices we call positive definite are denoted symmetric positive definite in [GV]. Keep this discrepancy in mind if you are using material from other sources.

To be precise, Definition 6.21 should specify that the 1×1 matrix generated by the operation $\mathbf{x}^t A\mathbf{x}$ has a positive value for its only entry, since the operation is performed as follows:

$$\mathbf{x}^t A\mathbf{x} = [x_1, x_2, \ldots, x_n] \begin{bmatrix} a_{11} & a_{12} & \cdots & a_{1n} \\ a_{21} & a_{22} & \cdots & a_{2n} \\ \vdots & \vdots & & \vdots \\ a_{n1} & a_{n2} & \cdots & a_{nn} \end{bmatrix} \begin{bmatrix} x_1 \\ x_2 \\ \vdots \\ x_n \end{bmatrix}$$

$$= [x_1, x_2, \ldots, x_n] \begin{bmatrix} \sum_{j=1}^{n} a_{1j}x_j \\ \sum_{j=1}^{n} a_{2j}x_j \\ \vdots \\ \sum_{j=1}^{n} a_{nj}x_j \end{bmatrix} = \left[\sum_{i=1}^{n} \sum_{j=1}^{n} a_{ij}x_i x_j \right].$$

EXAMPLE 2 The matrix

$$A = \begin{bmatrix} 2 & -1 & 0 \\ -1 & 2 & -1 \\ 0 & -1 & 2 \end{bmatrix}$$

is positive definite, for suppose $\mathbf{x}$ is any three-dimensional column vector. Then

$$\mathbf{x}^t A\mathbf{x} = [x_1, x_2, x_3] \begin{bmatrix} 2 & -1 & 0 \\ -1 & 2 & -1 \\ 0 & -1 & 2 \end{bmatrix} \begin{bmatrix} x_1 \\ x_2 \\ x_3 \end{bmatrix}$$

$$= [x_1, x_2, x_3] \begin{bmatrix} 2x_1 - x_2 \\ -x_1 + 2x_2 - x_3 \\ -x_2 + 2x_3 \end{bmatrix}$$

$$= 2x_1^2 - 2x_1x_2 + 2x_2^2 - 2x_2x_3 + 2x_3^2.$$

Rearranging the terms gives

$$\begin{aligned}\mathbf{x}^t A\mathbf{x} &= x_1^2 + (x_1^2 - 2x_1x_2 + x_2^2) + (x_2^2 - 2x_2x_3 + x_3^2) + x_3^2 \\ &= x_1^2 + (x_1 - x_2)^2 + (x_2 - x_3)^2 + x_3^2\end{aligned}$$

and

$$x_1^2 + (x_1 - x_2)^2 + (x_2 - x_3)^2 + x_3^2 > 0,$$

unless $x_1 = x_2 = x_3 = 0$. ■

It should be clear from Example 2 that using the definition to determine whether a matrix is positive definite can be difficult. Fortunately, there are more easily verified criteria, which are presented in Chapter 9, for identifying members of this important class. The next result provides some conditions that can be used to eliminate certain matrices from consideration.

Theorem 6.22 If A is an $n \times n$ positive definite matrix, then

a. A is nonsingular

b. $a_{ii} > 0$ for each $i = 1, 2, \ldots, n$;

c. $\max_{1\le k,j\le n} |a_{kj}| \le \max_{1\le i\le n} |a_{ii}|$;

d. $(a_{ij})^2 < a_{ii}a_{jj}$ for each $i \neq j$.

Proof

a. If $\mathbf{x} \neq \mathbf{0}$ is a vector that satisfies $A\mathbf{x} = \mathbf{0}$, then $\mathbf{x}^t A\mathbf{x} = 0$. This contradicts the assumption that A is positive definite. Consequently, $A\mathbf{x} = \mathbf{0}$ has only the zero solution, and A is nonsingular.

b. For a given i, let $\mathbf{x} = (x_j)$ be defined by $x_i = 1$ and $x_j = 0$, if $j \neq i$. Since $\mathbf{x} \neq \mathbf{0}$,

$$0 < \mathbf{x}^t A\mathbf{x} = a_{ii}.$$

c. For $k \neq j$, define $\mathbf{x} = (x_i)$ by

$$x_i = \begin{cases} 0, & \text{if } i \neq j \text{ and } i \neq k, \\ 1, & \text{if } i = j, \\ -1, & \text{if } i = k. \end{cases}$$

Since $\mathbf{x} \neq \mathbf{0}$,

$$0 < \mathbf{x}^t A\mathbf{x} = a_{jj} + a_{kk} - a_{jk} - a_{kj}.$$

But $A^t = A$, so $a_{jk} = a_{kj}$ and

$$2a_{kj} < a_{jj} + a_{kk}. \tag{6.10}$$

Now define $\mathbf{z} = (z_i)$ by

$$z_i = \begin{cases} 0, & \text{if } i \neq j \text{ and } i \neq k, \\ 1, & \text{if } i = j \text{ or } i = k. \end{cases}$$

Then $\mathbf{z}^t A\mathbf{z} > 0$, so

(6.11)
$$-2a_{kj} < a_{kk} + a_{jj}.$$

Equations (6.10) and (6.11) imply that for each $k \neq j$,

$$|a_{kj}| < \frac{a_{kk} + a_{jj}}{2} \leq \max_{1\leq i\leq n} |a_{ii}|, \quad \text{so} \quad \max_{1\leq k,j\leq n} |a_{kj}| \leq \max_{1\leq i\leq n} |a_{ii}|.$$

d. For $i \neq j$, define $\mathbf{x} = (x_k)$ by

$$x_k = \begin{cases} 0, & \text{if } k \neq j \text{ and } k \neq i, \\ \alpha, & \text{if } k = i, \\ 1, & \text{if } k = j, \end{cases}$$

where α represents an arbitrary real number. Since $\mathbf{x} \neq \mathbf{0}$,

$$0 < \mathbf{x}^t A\mathbf{x} = a_{ii}\alpha^2 + 2a_{ij}\alpha + a_{jj}.$$

As a quadratic polynomial in α with no real roots, the discriminant of $P(\alpha) = a_{ii}\alpha^2 + 2a_{ij}\alpha + a_{jj}$ must be negative. Thus,

$$4a_{ij}^2 - 4a_{ii}a_{jj} < 0 \quad \text{and} \quad a_{ij}^2 < a_{ii}a_{jj}.$$ ■ ■ ■

Although Theorem 6.22 provides some important conditions that must be true of positive definite matrices, it does not ensure that a matrix satisfying these conditions is positive definite.

The following notion will be used to provide a necessary and sufficient condition.

Definition 6.23 A **leading principal submatrix** of a matrix A is a matrix of the form

$$A_k = \begin{bmatrix} a_{11} & a_{12} & \cdots & a_{1k} \\ a_{21} & a_{22} & \cdots & a_{2k} \\ \vdots & \vdots & & \vdots \\ a_{k1} & a_{k2} & \cdots & a_{kk} \end{bmatrix}$$

for some $1 \leq k \leq n$. ■

A proof of the following result can be found in [Stew2, p. 250].

Theorem 6.24 A symmetric matrix A is positive definite if and only if each of its leading principal submatrices has a positive determinant. ■

EXAMPLE 3 In Example 2 we used the definition to show that the symmetric matrix

$$A = \begin{bmatrix} 2 & -1 & 0 \\ -1 & 2 & -1 \\ 0 & -1 & 2 \end{bmatrix}$$

is positive definite. To confirm this using Theorem 6.24 note that

$$\det A_1 = \det[2] = 2 > 0,$$

$$\det A_2 = \det \begin{bmatrix} 2 & -1 \\ -1 & 2 \end{bmatrix} = 4 - 1 = 3 > 0,$$

and

$$\det A_3 = \det \begin{bmatrix} 2 & -1 & 0 \\ -1 & 2 & -1 \\ 0 & -1 & 2 \end{bmatrix} = 2 \det \begin{bmatrix} 2 & -1 \\ -1 & 2 \end{bmatrix} - (-1) \det \begin{bmatrix} -1 & -1 \\ 0 & 2 \end{bmatrix}$$
$$= 2(4 - 1) + (-2 + 0) = 4 > 0.$$ ■

Maple also has a useful command to determine the positive definiteness of a matrix. The command

```
>definite(A,positive_def);
```

returns *true* or *false* as an indication. Symmetry is required for a *true* result to be produced.

The next result extends part (a) of Theorem 6.22 and parallels the strictly diagonally dominant results presented in Theorem 6.20. We will not give a proof of this theorem, since it requires introducing terminology and results that are not needed for any other purpose. The development and proof can be found in [We, pp. 120 ff].

Theorem 6.25 The symmetric matrix A is positive definite if and only if Gaussian elimination without row interchanges can be performed on the linear system $A\mathbf{x} = \mathbf{b}$ with all pivot elements positive. Moreover, in this case, the computations are stable with respect to the growth of roundoff errors. ■

Some interesting facts that are uncovered in constructing the proof of Theorem 6.25 are presented in the following corollaries.

Corollary 6.26 The matrix A is positive definite if and only if A can be factored in the form LDL^t, where L is lower triangular with 1s on its diagonal and D is a diagonal matrix with positive diagonal entries. ■

Corollary 6.27 The matrix A is positive definite if and only if A can be factored in the form LL^t, where L is lower triangular with nonzero diagonal entries. ■

The matrix L in Corollary 6.27 is not the same as the matrix in Corollary 6.26. A relationship between them is presented in Exercise 26.

Algorithm 6.5 is based on the *LU* Factorization Algorithm 6.4 and obtains the LDL^t factorization described in Corollary 6.26.

ALGORITHM 6.5

LDL^t Factorization

To factor the positive definite $n \times n$ matrix A into the form LDL^t, where L is a lower triangular matrix with 1s along the diagonal and D is a diagonal matrix with positive entries on the diagonal:

INPUT the dimension n; entries a_{ij}, for $1 \le i, j \le n$ of A.

OUTPUT the entries l_{ij}, for $1 \le j < i$ and $1 \le i \le n$ of L, and d_i, for $1 \le i \le n$ of D.

Step 1 For $i = 1, \dots, n$ do Steps 2–4.

Step 2 For $j = 1, \dots, i-1$, set $v_j = l_{ij}d_j$.

Step 3 Set $d_i = a_{ii} - \sum_{j=1}^{i-1} l_{ij}v_j$.

Step 4 For $j = i+1, \dots, n$ set $l_{ji} = (a_{ji} - \sum_{k=1}^{i-1} l_{jk}v_k)/d_i$.

Step 5 OUTPUT (l_{ij} for $j = 1, \dots, i-1$ and $i = 1, \dots, n$);
OUTPUT (d_i for $i = 1, \dots, n$);
STOP.

Corollary 6.26 has a counterpart when A is symmetric but not necessarily positive definite. This result is widely applied since symmetric matrices are common and easily recognized.

Corollary 6.28 Let A be a symmetric $n \times n$ matrix for which Gaussian elimination can be applied without row interchanges. Then A can be factored into LDL^t, where L is lower triangular with 1s on its diagonal and D is the diagonal matrix with $a_{11}^{(1)}, \dots, a_{nn}^{(n)}$ on its diagonal. ■

Algorithm 6.5 is easily modified to factor the symmetric matrices described in Corollary 6.28. It simply requires adding a check to ensure that the diagonal elements are nonzero.

Choleski's Algorithm 6.6 produces the LL^t factorization described in Corollary 6.27.

ALGORITHM 6.6

Choleski's

To factor the positive definite $n \times n$ matrix A into LL^t, where L is lower triangular:

INPUT the dimension n; entries a_{ij}, for $1 \le i, j \le n$ of A.

OUTPUT the entries l_{ij}, for $1 \le j \le i$ and $1 \le i \le n$ of L. *(The entries of $U = L^t$ are $u_{ij} = l_{ji}$, for $i \le j \le n$ and $1 \le i \le n$.)*

Step 1 Set $l_{11} = \sqrt{a_{11}}$.

Step 2 For $j = 2, \dots, n$, set $l_{j1} = a_{j1}/l_{11}$.

Step 3 For $i = 2, \dots, n-1$ do Steps 4 and 5.

Step 4 Set $l_{ii} = \left[a_{ii} - \sum_{k=1}^{i-1} l_{ik}^2\right]^{1/2}$.

Step 5 For $j = i+1, \ldots, n$

set $l_{ji} = \frac{1}{l_{ii}} \left[a_{ji} - \sum_{k=1}^{i-1} l_{jk} l_{ik}\right]$.

Step 6 Set $l_{nn} = \left[a_{nn} - \sum_{k=1}^{n-1} l_{nk}^2\right]^{1/2}$.

Step 7 OUTPUT (l_{ij} for $j = 1, \ldots, i$ and $i = 1, \ldots, n$);
STOP.

EXAMPLE 4 The matrix

$$A = \begin{bmatrix} 4 & -1 & 1 \\ -1 & 4.25 & 2.75 \\ 1 & 2.75 & 3.5 \end{bmatrix}$$

is positive definite. The factorization LDL^t of A given in Algorithm 6.5 is

$$A = LDL^t = \begin{bmatrix} 1 & 0 & 0 \\ -0.25 & 1 & 0 \\ 0.25 & 0.75 & 1 \end{bmatrix} \begin{bmatrix} 4 & 0 & 0 \\ 0 & 4 & 0 \\ 0 & 0 & 1 \end{bmatrix} \begin{bmatrix} 1 & -0.25 & 0.25 \\ 0 & 1 & 0.75 \\ 0 & 0 & 1 \end{bmatrix},$$

and Choleski's Algorithm 6.6 produces the factorization

$$A = LL^t = \begin{bmatrix} 2 & 0 & 0 \\ -0.5 & 2 & 0 \\ 0.5 & 1.5 & 1 \end{bmatrix} \begin{bmatrix} 2 & -0.5 & 0.5 \\ 0 & 2 & 1.5 \\ 0 & 0 & 1 \end{bmatrix}.$$ ■

The LDL^t factorization described in Algorithm 6.5 requires $n^3/6 + n^2 - 7n/6$ multiplications/divisions and $n^3/6 - n/6$ additions/subtractions. The LL^t Choleski factorization of a positive definite matrix requires only $n^3/6 + n^2/2 - 2n/3$ multiplications/divisions and $n^3/6 - n/6$ additions/subtractions. The computational advantage of Choleski's factorization is misleading, however, since it requires extracting n square roots. However, the number of operations required for computing the n square roots is a linear factor of n and will decrease in significance as n increases.

Algorithm 6.5 provides a stable method for factoring a positive definite matrix into the form $A = LDL^t$, but it must be modified to solve the linear system $A\mathbf{x} = \mathbf{b}$. To do this, we delete the STOP statement from Step 5 in the algorithm and add the following steps to solve the lower triangular system $L\mathbf{y} = \mathbf{b}$:

Step 6 Set $y_1 = b_1$.

Step 7 For $i = 2, \ldots, n$ set $y_i = b_i - \sum_{j=1}^{i-1} l_{ij} y_j$.

The linear system $D\mathbf{z} = \mathbf{y}$ can then be solved by

Step 8 For $i = 1, \ldots, n$ set $z_i = y_i/d_i$.

Finally, the upper-triangular system $L^t\mathbf{x} = \mathbf{z}$ is solved with the steps given by

Step 9 Set $x_n = z_n$.

Step 10 For $i = n-1, \ldots, 1$ set $x_i = z_i - \sum_{j=i+1}^{n} l_{ji}x_j$.

Step 11 OUTPUT (x_i for $i = 1, \ldots, n$);
STOP.

The additional operations required to solve the linear system are shown in Table 6.2.

Table 6.2

Step	Multiplications/Divisions	Additions/Subtractions
6	0	0
7	$n(n-1)/2$	$n(n-1)/2$
8	n	0
9	0	0
10	$n(n-1)/2$	$n(n-1)/2$
Total	n^2	$n^2 - n$

If the Choleski factorization given in Algorithm 6.6 is preferred, the additional steps for solving the system $A\mathbf{x} = \mathbf{b}$ are as follows. First delete the STOP statement from Step 7. Then, add

Step 8 Set $y_1 = b_1/l_{11}$.

Step 9 For $i = 2, \ldots, n$ set $y_i = \left(b_i - \sum_{j=1}^{i-1} l_{ij}y_j\right) \Big/ l_{ii}$.

Step 10 Set $x_n = y_n/l_{nn}$.

Step 11 For $i = n-1, \ldots, 1$ set $x_i = \left(y_i - \sum_{j=i+1}^{n} l_{ji}x_j\right) \Big/ l_{ii}$.

Step 12 OUTPUT (x_i for $i = 1, \ldots, n$);
STOP.

Steps 8–12 require $n^2 + n$ multiplications/divisions and $n^2 - n$ additions/ subtractions.

The last class of matrices considered are called *band matrices*. In most applications, the band matrices are also strictly diagonally dominant or positive definite. This combination of properties is very useful.

Definition 6.29 An $n \times n$ matrix is called a **band matrix** if integers p and q, with $1 < p, q < n$, exist having the property that $a_{ij} = 0$ whenever $i + p \le j$ or $j + q \le i$. The **bandwidth** of a band matrix is defined as $w = p + q - 1$. ■

For example, the matrix

$$A = \begin{bmatrix} 7 & 2 & 0 \\ 3 & 5 & -1 \\ 0 & -5 & -6 \end{bmatrix}$$

is a band matrix with $p = q = 2$ and band width 3.

The definition of band matrix forces those matrices to concentrate all their nonzero entries about the diagonal. Two special cases of band matrices that occur often in practice have $p = q = 2$ and $p = q = 4$. The matrix of bandwidth 3, occurring when $p = q = 2$, has already been encountered in connection with the study of cubic spline approximations in Section 3.4. These matrices are called **tridiagonal**, since they have the form

$$A = \begin{bmatrix} a_{11} & a_{12} & 0 & \cdots & \cdots & 0 \\ a_{21} & a_{22} & a_{23} & \ddots & & \vdots \\ 0 & a_{32} & a_{33} & a_{34} & \ddots & \vdots \\ \vdots & \ddots & \ddots & \ddots & \ddots & 0 \\ \vdots & & \ddots & \ddots & \ddots & a_{n-1,n} \\ 0 & \cdots & \cdots & 0 & a_{n,n-1} & a_{nn} \end{bmatrix}.$$

Tridiagonal matrices are also considered in Chapter 11 in connection with the study of piecewise linear approximations to boundary-value problems. The case of $p = q = 4$ will be used for the solution of boundary-value problems when the approximating functions assume the form of cubic splines.

The factorization algorithms can be simplified considerably in the case of band matrices because a large number of zeros appear in these matrices in regular patterns. It is particularly interesting to observe the form the Crout or Doolittle method assumes in this case.

To illustrate the situation, suppose a tridiagonal matrix A can be factored into the triangular matrices L and U. Since A has only $(3n - 2)$ nonzero entries, there are only $(3n - 2)$ conditions to be applied to determine the entries of L and U, provided, of course, that the zero entries of A are also obtained. Suppose that the matrices can be found in the form

$$L = \begin{bmatrix} l_{11} & 0 & \cdots & \cdots & 0 \\ l_{21} & l_{22} & \ddots & & \vdots \\ 0 & \ddots & \ddots & \ddots & \vdots \\ \vdots & \ddots & \ddots & \ddots & 0 \\ 0 & \cdots & 0 & l_{n,n-1} & l_{nn} \end{bmatrix} \quad \text{and} \quad U = \begin{bmatrix} 1 & u_{12} & 0 & \cdots & 0 \\ 0 & 1 & \ddots & \ddots & \vdots \\ \vdots & \ddots & \ddots & \ddots & 0 \\ \vdots & & \ddots & \ddots & u_{n-1,n} \\ 0 & \cdots & \cdots & 0 & 1 \end{bmatrix}.$$

There are $(2n - 1)$ undetermined entries of L and $(n - 1)$ undetermined entries of U, which totals the number of conditions, $(3n - 2)$. The zero entries of A are obtained automatically.

The multiplication involved with $A = LU$ gives, in addition to the zero entries,

$$a_{11} = l_{11};$$

(6.12) $$a_{i,i-1} = l_{i,i-1}, \quad \text{for each } i = 2, 3, \ldots, n;$$

(6.13) $$a_{ii} = l_{i,i-1}u_{i-1,i} + l_{ii}, \quad \text{for each } i = 2, 3, \ldots, n;$$

and

(6.14) $$a_{i,i+1} = l_{ii}u_{i,i+1}, \quad \text{for each } i = 1, 2, \ldots, n-1.$$

A solution to this system is found by first using Eq. (6.12) to obtain all the nonzero off-diagonal terms in L and then using Eqs. (6.13) and (6.14) to alternately obtain the remainder of the entries in U and L. These can be stored in the corresponding entries of A.

Algorithm 6.7 solves an $n \times n$ system of linear equations whose coefficient matrix is tridiagonal. This algorithm requires only $(5n - 4)$ multiplications/divisions and $(3n - 3)$ additions/subtractions. Consequently, it has considerable computational advantage over the methods that do not consider the tridiagonality of the matrix.

ALGORITHM 6.7

Crout Factorization for Tridiagonal Linear Systems

To solve the $n \times n$ linear system

$$\begin{aligned}
E_1 &: \quad a_{11}x_1 + a_{12}x_2 &&= a_{1,n+1},\\
E_2 &: \quad a_{21}x_1 + a_{22}x_2 + a_{23}x_3 &&= a_{2,n+1},\\
&\vdots &&\vdots\\
E_{n-1} &: \quad a_{n-1,n-2}x_{n-2} + a_{n-1,n-1}x_{n-1} + a_{n-1,n}x_n &&= a_{n-1,n+1},\\
E_n &: \quad a_{n,n-1}x_{n-1} + a_{nn}x_n &&= a_{n,n+1},
\end{aligned}$$

which is assumed to have a unique solution:

INPUT the dimension n; the entries of A.

OUTPUT the solution $x_1, \ldots, x_n$.

(Steps 1–3 set up and solve $L\mathbf{z} = \mathbf{b}$.)

Step 1 Set $l_{11} = a_{11}$;
$u_{12} = a_{12}/l_{11}$;
$z_1 = a_{1,n+1}/l_{11}$.

Step 2 For $i = 2, \ldots, n-1$ set $l_{i,i-1} = a_{i,i-1}$; *(ith row of L.)*
$l_{ii} = a_{ii} - l_{i,i-1}u_{i-1,i}$;
$u_{i,i+1} = a_{i,i+1}/l_{ii}$; *((i + 1)th column of U.)*
$$z_i = \frac{1}{l_{ii}}[a_{i,n+1} - l_{i,i-1}z_{i-1}].$$

Step 3 Set $l_{n,n-1} = a_{n,n-1}$; *(nth row of L.)*
$l_{nn} = a_{nn} - l_{n,n-1}u_{n-1,n}$.
$$z_n = \frac{1}{l_{nn}}[a_{n,n+1} - l_{n,n-1}z_{n-1}].$$

(Steps 4 and 5 solve $U\mathbf{x} = \mathbf{z}$.)

Step 4 Set $x_n = z_n$.

Step 5 For $i = n-1, \ldots, 1$ set $x_i = z_i - u_{i,i+1}x_{i+1}$.

Step 6 OUTPUT $(x_1, \ldots, x_n)$;
STOP.

EXAMPLE 5 To illustrate the procedure for tridiagonal matrices, consider the tridiagonal system of equations

$$\begin{aligned} 2x_1 - x_2 \qquad\qquad &= 1, \\ -x_1 + 2x_2 - x_3 \qquad &= 0, \\ - x_2 + 2x_3 - x_4 &= 0, \\ - x_3 + 2x_4 &= 1, \end{aligned}$$

whose augmented matrix is

$$\begin{bmatrix} 2 & -1 & 0 & 0 & \vdots & 1 \\ -1 & 2 & -1 & 0 & \vdots & 0 \\ 0 & -1 & 2 & -1 & \vdots & 0 \\ 0 & 0 & -1 & 2 & \vdots & 1 \end{bmatrix}.$$

The Crout Factorization Algorithm produces the factorization

$$\begin{bmatrix} 2 & -1 & 0 & 0 \\ -1 & 2 & -1 & 0 \\ 0 & -1 & 2 & -1 \\ 0 & 0 & -1 & 2 \end{bmatrix} = \begin{bmatrix} 2 & 0 & 0 & 0 \\ -1 & \frac{3}{2} & 0 & 0 \\ 0 & -1 & \frac{4}{3} & 0 \\ 0 & 0 & -1 & \frac{5}{4} \end{bmatrix} \begin{bmatrix} 1 & -\frac{1}{2} & 0 & 0 \\ 0 & 1 & -\frac{2}{3} & 0 \\ 0 & 0 & 1 & -\frac{3}{4} \\ 0 & 0 & 0 & 1 \end{bmatrix} = LU.$$

Solving the system $L\mathbf{z} = \mathbf{b}$ gives $\mathbf{z} = (\frac{1}{2}, \frac{1}{3}, \frac{1}{4}, 1)^t$, and the solution of $U\mathbf{x} = \mathbf{z}$ is $\mathbf{x} = (1, 1, 1, 1)^t$. ■

The Crout Factorization Algorithm can be applied whenever $l_{ii} \neq 0$ for each $i = 1, 2, \ldots, n$. Two conditions, either of which ensure that this is true, are that the coefficient matrix of the system is positive definite or that it is strictly diagonally dominant. An additional condition that ensures this algorithm can be applied is given in the next theorem, whose proof is discussed in Exercise 22.

Theorem 6.30 Suppose that $A = (a_{ij})$ is tridiagonal with $a_{i,i-1}a_{i,i+1} \neq 0$, for each $i = 2, 3, \ldots, n-1$. If $|a_{11}| > |a_{12}|$, $|a_{ii}| \geq |a_{i,i-1}| + |a_{i,i+1}|$, for each $i = 2, 3, \ldots, n-1$, and $|a_{nn}| > |a_{n,n-1}|$, then A is nonsingular and the values of l_{ii} described in the Crout Factorization Algorithm are nonzero for each $i = 1, 2, \ldots, n$. ■

EXERCISE SET 6.6

1. Determine which of the following matrices are (i) symmetric, (ii) singular, (iii) strictly diagonally dominant, (iv) positive definite.

   a. $\begin{bmatrix} 2 & 1 \\ 1 & 3 \end{bmatrix}$ **b.** $\begin{bmatrix} -2 & 1 \\ 1 & -3 \end{bmatrix}$

   c. $\begin{bmatrix} 2 & 1 & 0 \\ 0 & 3 & 0 \\ 1 & 0 & 4 \end{bmatrix}$ **d.** $\begin{bmatrix} 2 & 1 & 0 \\ 0 & 3 & 2 \\ 1 & 2 & 4 \end{bmatrix}$

e. $\begin{bmatrix} 4 & 2 & 6 \\ 3 & 0 & 7 \\ -2 & -1 & -3 \end{bmatrix}$ **f.** $\begin{bmatrix} 2 & -1 & 0 \\ -1 & 4 & 2 \\ 0 & 2 & 2 \end{bmatrix}$

g. $\begin{bmatrix} 4 & 0 & 0 & 0 \\ 6 & 7 & 0 & 0 \\ 9 & 11 & 1 & 0 \\ 5 & 4 & 1 & 1 \end{bmatrix}$ **h.** $\begin{bmatrix} 2 & 3 & 1 & 2 \\ -2 & 4 & -1 & 5 \\ 3 & 7 & 1.5 & 1 \\ 6 & -9 & 3 & 7 \end{bmatrix}$

2. Use the LDL^t Factorization Algorithm to find a factorizaton of the form $A = LDL^t$ for the following matrices:

a. $A = \begin{bmatrix} 2 & -1 & 0 \\ -1 & 2 & -1 \\ 0 & -1 & 2 \end{bmatrix}$ **b.** $A = \begin{bmatrix} 4 & 1 & 1 & 1 \\ 1 & 3 & -1 & 1 \\ 1 & -1 & 2 & 0 \\ 1 & 1 & 0 & 2 \end{bmatrix}$

c. $A = \begin{bmatrix} 4 & 1 & -1 & 0 \\ 1 & 3 & -1 & 0 \\ -1 & -1 & 5 & 2 \\ 0 & 0 & 2 & 4 \end{bmatrix}$ **d.** $A = \begin{bmatrix} 6 & 2 & 1 & -1 \\ 2 & 4 & 1 & 0 \\ 1 & 1 & 4 & -1 \\ -1 & 0 & -1 & 3 \end{bmatrix}$

3. Use Choleski's Algorithm to find a factorization of the form $A = LL^t$ for the matrices in Exercise 2.

4. Modify the LDL^t Factorization Algorithm as suggested in the text so that it can be used to solve linear systems. Use the modified algorithm to solve the following linear systems.

a. $$\begin{aligned} 2x_1 - x_2 &= 3, \\ -x_1 + 2x_2 - x_3 &= -3, \\ - x_2 + 2x_3 &= 1. \end{aligned}$$

b. $$\begin{aligned} 4x_1 + x_2 + x_3 + x_4 &= 0.65, \\ x_1 + 3x_2 - x_3 + x_4 &= 0.05, \\ x_1 - x_2 + 2x_3 &= 0, \\ x_1 + x_2 + 2x_4 &= 0.5, \end{aligned}$$

c. $$\begin{aligned} 4x_1 + x_2 - x_3 &= 7, \\ x_1 + 3x_2 - x_3 &= 8, \\ -x_1 - x_2 + 5x_3 + 2x_4 &= -4, \\ 2x_3 + 4x_4 &= 6, \end{aligned}$$

d. $$\begin{aligned} 6x_1 + 2x_2 + x_3 - x_4 &= 0, \\ 2x_1 + 4x_2 + x_3 &= 7, \\ x_1 + x_2 + 4x_3 - x_4 &= -1, \\ -x_1 - x_3 + 3x_4 &= -2. \end{aligned}$$

5. Modify Choleski's Algorithm as suggested in the text so that it can be used to solve linear systems, and use the modified algorithm to solve the linear systems in Exercise 4.

6. Use Crout factorization for tridiagonal systems to solve the following linear systems.

a. $$\begin{aligned} x_1 - x_2 &= 0, \\ -2x_1 + 4x_2 - 2x_3 &= -1, \\ - x_2 + 2x_3 &= 1.5. \end{aligned}$$

b. $$\begin{aligned} 3x_1 + x_2 &= -1, \\ 2x_1 + 4x_2 + x_3 &= 7, \\ 2x_2 + 5x_3 &= 9, \end{aligned}$$

c. $$\begin{aligned} 2x_1 - x_2 &= 3, \\ -x_1 + 2x_2 - x_3 &= -3, \\ - x_2 + 2x_3 &= 1. \end{aligned}$$

d. $$\begin{aligned} 0.5x_1 + 0.25x_2 &= 0.35, \\ 0.35x_1 + 0.8x_2 + 0.4x_3 &= 0.77, \\ 0.25x_2 + x_3 + 0.5x_4 &= -0.5, \\ x_3 - 2x_4 &= -2.25. \end{aligned}$$

7. Let A be the 10×10 tridiagonal matrix given by $a_{ii} = 2, a_{i,i+1} = a_{i,i-1} = -1$, for each $i = 2, \ldots, 9$, and $a_{11} = a_{10,10} = 2, a_{12} = a_{10,9} = -1$. Let $\mathbf{b}$ be the ten-dimensional column vector given by $b_1 = b_{10} = 1$ and $b_i = 0$ for each $i = 2, 3, \ldots, 9$. Solve $A\mathbf{x} = \mathbf{b}$ using the Crout factorization for tridiagonal systems.

8. Modify the LDL^t factorization to factor a symmetric matrix A. [*Note:* The factorization may not always be possible.] Apply the new algorithm to the following matrices:

a. $A = \begin{bmatrix} 3 & -3 & 6 \\ -3 & 2 & -7 \\ 6 & -7 & 13 \end{bmatrix}$ **b.** $A = \begin{bmatrix} 3 & -6 & 9 \\ -6 & 14 & -20 \\ 9 & -20 & 29 \end{bmatrix}$

c. $A = \begin{bmatrix} -1 & 2 & 0 & 1 \\ 2 & -3 & 2 & -1 \\ 0 & 2 & 5 & 6 \\ 1 & -1 & 6 & 12 \end{bmatrix}$ **d.** $A = \begin{bmatrix} 2 & -2 & 4 & -4 \\ -2 & 3 & -4 & 5 \\ 4 & -4 & 10 & -10 \\ -4 & 5 & -10 & 14 \end{bmatrix}$

9. Which of the symmetric matrices in Exercise 8 are positive definite?

10. Find α so that $A = \begin{bmatrix} \alpha & 1 & -1 \\ 1 & 2 & 1 \\ -1 & 1 & 4 \end{bmatrix}$ is positive definite.

11. Find α so that $A = \begin{bmatrix} 2 & \alpha & -1 \\ \alpha & 2 & 1 \\ -1 & 1 & 4 \end{bmatrix}$ is positive definite.

12. Find α and $\beta > 0$ so that the matrix

$$A = \begin{bmatrix} 4 & \alpha & 1 \\ 2\beta & 5 & 4 \\ \beta & 2 & \alpha \end{bmatrix}$$

is strictly diagonally dominant.

13. Find $\alpha > 0$ and $\beta > 0$ so that the matrix

$$A = \begin{bmatrix} 3 & 2 & \beta \\ \alpha & 5 & \beta \\ 2 & 1 & \alpha \end{bmatrix}$$

is strictly diagonally dominant.

14. Suppose that A and B are strictly diagonally dominant $n \times n$ matrices.

a. Is $-A$ strictly diagonally dominant?

b. Is A^t strictly diagonally dominant?

c. Is $A + B$ strictly diagonally dominant?

d. Is A^2 strictly diagonally dominant?

e. Is $A - B$ strictly diagonally dominant?

15. Suppose that A and B are positive definite $n \times n$ matrices.

a. Is $-A$ positive definite?

b. Is A^t positive definite?

c. Is $A + B$ positive definite?

d. Is A^2 positive definite?

e. Is $A - B$ positive definite?

16. Let

$$A = \begin{bmatrix} 1 & 0 & -1 \\ 0 & 1 & 1 \\ -1 & 1 & \alpha \end{bmatrix}.$$

Find all values of α for which

a. A is singular.

b. A is strictly diagonally dominant.

c. A is symmetric.

d. A is positive definite.

17. Let

$$A = \begin{bmatrix} \alpha & 1 & 0 \\ \beta & 2 & 1 \\ 0 & 1 & 2 \end{bmatrix}$$

Find all values of α and β for which

a. A is singular.

b. A is strictly diagonally dominant.

c. A is symmetric.

d. A is positive definite.

18. Suppose A and B commute, that is, $AB = BA$. Must A^t and B^t also commute?

19. Construct a matrix A that is nonsymmetric but for which $\mathbf{x}^t A\mathbf{x} > 0$ for all $\mathbf{x} \neq \mathbf{0}$.

20. Show that Gaussian elimination can be performed on A without row interchanges if and only if all leading principal submatrices of A are nonsingular. [*Hint*: Partition each matrix in the equation

$$A^{(k)} = M^{(k-1)}M^{(k-2)}\cdots M^{(1)}A$$

vertically between the kth and $(k+1)$st columns and horizontally between the kth and $(k+1)$st rows (see Exercise 10 of Section 6.3). Show that the nonsingularity of the leading principal submatrix of A is equivalent to $a_{k,k}^{(k)} \neq 0$.]

21. Tridiagonal matrices are usually labeled by using the notation

$$A = \begin{bmatrix} a_1 & c_1 & 0 & \cdots & \cdots & 0 \\ b_2 & a_2 & c_2 & \ddots & & \vdots \\ 0 & b_3 & \ddots & \ddots & \ddots & 0 \\ \vdots & \ddots & \ddots & \ddots & \ddots & c_{n-1} \\ 0 & \cdots & \cdots & 0 & b_n & a_n \end{bmatrix}$$

to emphasize that it is not necessary to consider all the matrix entries. Rewrite the Crout Factorization Algorithm using this notation, and change the notation of the l_{ij} and u_{ij} in a similar manner.

22. Prove Theorem 6.30. [*Hint*: Show that $|u_{i,i+1}| < 1$ for each $i = 1, 2, \ldots, n-1$, and that $|l_{ii}| > 0$ for each $i = 1, 2, \ldots, n$. Deduce that $\det A = \det L \cdot \det U \neq 0$.]

23. Suppose $V = 5.5$ volts in the lead example of this chapter. By reordering the equations, a tridiagonal linear system can be formed. Use the Crout Factorization Algorithm to find the solution of the modified system.

24. Construct the operation count for solving an $n \times n$ linear system using the Crout Factorization Algorithm.

25. In a paper by Dorn and Burdick [DoB], it is reported that the average wing length that resulted from mating three mutant varieties of fruit flies (*Drosophila melanogaster*) can be expressed in the symmetric matrix form

$$A = \begin{bmatrix} 1.59 & 1.69 & 2.13 \\ 1.69 & 1.31 & 1.72 \\ 2.13 & 1.72 & 1.85 \end{bmatrix},$$

where a_{ij} denotes the average wing length of an offspring resulting from the mating of a male of type i with a female of type j.

a. What physical significance is associated with the symmetry of this matrix?

b. Is this matrix positive definite? If so, prove it; if not, find a nonzero vector $\mathbf{x}$ for which $\mathbf{x}^t A\mathbf{x} \leq 0$.

26. Suppose that the positive definite matrix A has the Cholesky factorization $A = LL^t$ and also the factorization $A = \hat{L}D\hat{L}^t$, where D is the diagonal matrix with positive diagonal entries d_{11}, $d_{22}, \ldots, d_{nn}$. Let $D^{1/2}$ be the diagonal matrix with diagonal entries $\sqrt{d_{11}}, \sqrt{d_{22}}, \ldots, \sqrt{d_{nn}}$.

a. Show that $D = D^{1/2}D^{1/2}$.

b. Show that $A = (\hat{L}D^{1/2})(\hat{L}D^{1/2})^t$.

6.7 Survey of Methods and Software

In this chapter we have looked at direct methods for solving linear systems. A linear system consists of n equations in n unknowns expressed in matrix notation as $A\mathbf{x} = \mathbf{b}$. These techniques use a finite sequence of arithmetic operations to determine the exact solution of the system subject only to roundoff error. We found that the linear system $A\mathbf{x} = \mathbf{b}$ has a unique solution if and only if A^{-1} exists, which is equivalent to $\det A \neq 0$. The solution of the linear system is the vector $\mathbf{x} = A^{-1}\mathbf{b}$.

Pivoting techniques were introduced to minimize the effects of roundoff error, which can dominate the solution when using direct methods. We described partial pivoting, scaled partial pivoting, and complete pivoting. We recommend the partial or scaled partial pivoting methods for most problems since these decrease the effects of roundoff error without adding much extra computation. Total pivoting should be used if roundoff error is suspected to be large. In Section 7.4 we will see some procedures for estimating this roundoff error.

Gaussian elimination with minor modifications was shown to yield a factorization of the matrix A into LU, where L is lower triangular with 1s on the diagonal and U is upper triangular. This process is called Crout factorization. Not all nonsingular matrices can be factored this way, but a permutation of the rows will always give a factorization of the form $PA = LU$, where P is the permutation matrix used to rearrange the rows of A. The advantage of the factorization is that the work is reduced when solving linear systems $A\mathbf{x} = \mathbf{b}$ with the same coefficient matrix A and different vectors $\mathbf{b}$.

When the matrix A is positive definite, factorizations take a simpler form. For example, the Choleski factorization has the form $A = LL^t$, where L is lower triangular. Positive definite matrices can also be factored in the form $A = LDL^t$, where L is lower triangular with 1s on the diagonal and D is diagonal. With these factorizations, manipulations involving A can be simplified. If A is tridiagonal, the LU factorization takes a particularly simple form, with L having 1s on the main diagonal and 0s elsewhere, except on the diagonal immediately below the main diagonal. In addition, U has its only nonzero entries on the main diagonal and one diagonal above.

The direct methods are the methods of choice for most linear systems. For tridiagonal, banded, and positive definite matrices, the special methods are recommended. For the general case, Gaussian elimination or LU factorization methods, which allow pivoting, are recommended. In these cases, the effects of roundoff error should be monitored. In Section 7.4 we discuss estimating errors in direct methods.

Large linear systems with primarily zero entries occurring in regular patterns can be solved efficiently using an iterative procedure such as those discussed in Chapter 7. Systems of this type arise naturally, for example, when finite-difference techniques are used to solve boundary-value problems, a common application in the numerical solution of partial-differential equations.

It can be very difficult to solve a large linear system that has primarily nonzero entries or one where the zero entries are not in a predictable pattern. The matrix associated with the system can be placed in secondary storage in partitioned form and portions read into main memory only as needed for calculation. Methods that require secondary storage can be either iterative or direct, but they generally require techniques from the fields of data structures and graph theory. A study of the problems involved in theory and implementation is beyond the scope of this text. The reader is referred to [BuR] and [RW] for a discussion of the current techniques.

The software for matrix operations and the direct solution of linear systems implemented in IMSL and NAG is based on LAPACK, a subroutine package in the public domain. There is excellent documentation available with it and from the books written about it. We will focus on several of the subroutines that are available in all three sources.

Accompanying LAPACK is a set of lower-level operations called Basic Linear Algebra Subprograms (BLAS). Level 1 of BLAS generally consists of vector operations with input data and operation counts of $O(n)$. Level 2 consists of the matrix-vector operations with input data and operation counts of $O(n^2)$. For example, in Level 1, the subroutine SCOPY overwrites a vector $\mathbf{y}$ with a vector $\mathbf{x}$; SSCAL computes a scalar a times a vector $\mathbf{x}$; SAXPY adds a scalar times a vector to a vector; SDOT computes the inner, or scalar, product of two vectors; SNRM2 computes the Euclidean norm of a vector by a method similar to that discussed in Section 1.4; and ISAMAX computes the index of the vector component that gives the maximum absolute value of all the components. In Level 2, MMULT computes the product of a matrix and a vector.

The subroutines in LAPACK for solving linear systems first factor the matrix A. The factorization depends on the type of matrix in the following way:

1. General matrix $PA = LU$;
2. Positive definite matrix $A = LL^t$;
3. Symmetric matrix $A = LDL^t$;
4. Tridiagonal matrix $A = LU$ (in banded form).

The subroutine STRTRS solves a triangular linear system when the matrix is either upper or lower triangular.

The subroutine SGETRF factors PA into LU as a preliminary operation to the subroutine SGETRS, which then computes the solution to $A\mathbf{x} = \mathbf{b}$. The subroutine SGETRI is used to construct the inverse of a matrix A and to calculate the determinant of A once A has been factored via SGETRF.

The Choleski factorization of a positive definite matrix A is obtained with the subroutine SPOTRF. The linear system $A\mathbf{x} = \mathbf{b}$ can then be solved using the subroutine SPOTRS. Inverses and determinants of positive definite matrices, given the Choleski factorization, can be computed using SPOTRI. If A is symmetric, the LDL^t factorization is found using SSYTRI. Linear systems can then be solved using SSISL. If inverses or determinants are desired, SSIDI can be used.

Many of the subroutines in LINPACK, and its successor LAPACK, can be implemented using MATLAB. A nonsingular matrix A is factored using the command

$$[LUP] = lu(A)$$

into the form $PA = LU$, where P is the permutation matrix defined by performing partial pivoting to solve a linear system involving A. If the nonsingular matrix A and the vector $\mathbf{b}$ have been defined in MATLAB, the command

$$x = A\backslash b$$

solves the linear system by first using the $PA = LU$ factoring command. Then it solves the lower-triangular system $L\mathbf{z} = \mathbf{b}$ for $\mathbf{z}$ using its command,

$$z = L\backslash b$$

This is followed by a solution to the upper-triangular system $U\mathbf{x} = \mathbf{z}$ using the command

$$x = U\backslash z$$

Other MATLAB commands include computing the inverse, transpose, and determinant of matrix A by issuing the commands $inv(A)$, A', and $\det(A)$, respectively.

In addition to the command `gausselim`, Maple uses the commands `gaussjord` and `solve` to solve linear systems. The command `gaussjord` uses the Gauss-Jordan method (see Exercise 8 in Section 6.1). The command `solve` uses a variety of techniques, depending on the equation or system of equations to be solved.

The IMSL Library includes counterparts to almost all the LAPACK subroutines and some extensions as well. They are named with regard to the tasks they perform as follows:

1. The first three letters of the name are used.
   - **a.** LSL: solves a linear system
   - **b.** LFT: factors a coefficient matrix
   - **c.** LFS: solves a linear system given factors from LFT
   - **d.** LFD: calculates the determinants of given factors
   - **e.** LIN: computes the inverse of given factors
2. The last two letters determine the type of matrix involved.
   - **a.** RG: real, general
   - **b.** RT: real triangular
   - **c.** DS: real positive definite
   - **d.** SF: real symmetric
   - **e.** RB: real banded

For example, the routine LFTDS factors a real positive definite matrix.

The NAG Library has many subroutines for direct methods of solving linear systems similar to those in LAPACK and IMSL. For example, the subroutine F04AEF solves linear systems using Crout factorization. The subroutine F04ATF solves a single linear system using Crout factorization, as in F04AEF. The subroutine F04EAF solves a single linear system where the matrix is real and tridiagonal, and F04ASF solves a system when the matrix is real and positive definite. Inverse matrices can be computed by F01AAF for an arbitrary real matrix and by F01ACF if the matrix is positive definite. A determinant can he computed using F03AAF. Factorizations can be obtained using F01BTF for the LU factorization of a real matrix and using F01LEF for a tridiagonal matrix. Linear systems can then be solved using F04AYF. Choleski's factorization of a positive definite matrix can be obtained using F01BXF, and a linear system can then be solved using F04AZF. The NAG library also includes the lower-level matrix-vector manipulations.

Further information on the numerical solution of linear systems and matrices can be found in Golub and Van Loan [GV], Forsythe and Moler [FM], and Stewart [Stew1]. The use of direct techniques for solving large sparse systems is discussed in detail in George and Liu [GL] and in Pissanetzky [Pi]. Coleman and Van Loan [CV] consider the use of BLAS, LINPACK, and MATLAB.

CHAPTER

7 Iterative Techniques in Matrix Algebra

■ ■ ■

Trusses are lightweight structures capable of carrying heavy loads. In bridge design, the individual members of the truss are connected with rotatable pin joints that permit forces to be transferred from one member of the truss to another. The accompanying figure shows a truss that is held stationary at the lower left endpoint ①, is permitted to move horizontally at the lower right endpoint ④, and has pin joints at ①, ②, ③, and ④. A load of 10,000 newtons (N) is placed at the joint ③, and the forces on the members of the truss have magnitudes given by f_1, f_2, f_3, f_4, and f_5, as shown. The stationary support member has both a horizontal force F_1 and a vertical force F_2, but the movable support member has only the vertical force F_3.

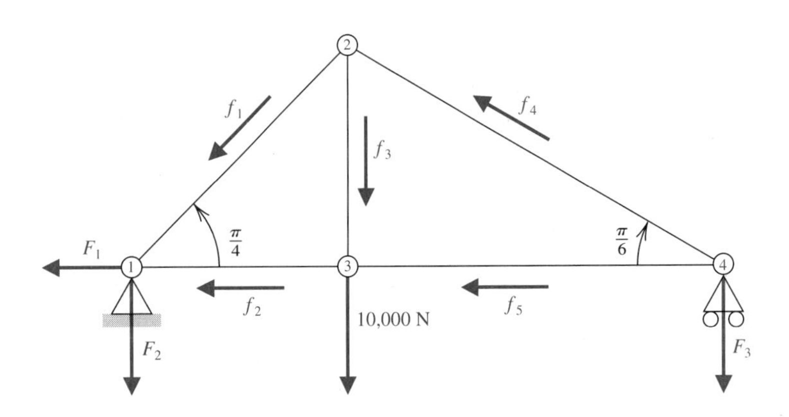

If the truss is in static equilibrium, the forces at each joint must add to the zero vector, so the sum of the horizontal and vertical components of

the forces at each joint must be zero. This produces the system of linear equations shown in the accompanying table. An 8 × 8 matrix describing this system has 46 zero entries and only 18 nonzero entries. Matrices with a high percentage of zero entries are called *sparse* and are often solved using iterative, rather than direct, techniques. The iterative solution to this system is considered in Exercise 16 of Section 7.3.

Joint	Horizontal Component	Vertical Component
①	$-F_1 + \frac{\sqrt{2}}{2} f_1 + f_2 = 0$	$\frac{\sqrt{2}}{2} f_1 - F_2 = 0$
②	$-\frac{\sqrt{2}}{2} f_1 + \frac{\sqrt{3}}{2} f_4 = 0$	$-\frac{\sqrt{2}}{2} f_1 - f_3 + \frac{1}{2} f_4 = 0$
③	$-f_2 + f_5 = 0$	$f_3 - 10,000 = 0$
④	$-\frac{\sqrt{3}}{2} f_4 - f_5 = 0$	$\frac{1}{2} f_4 - F_3 = 0$

The methods presented in Chapter 6 used direct techniques to solve a system of $n \times n$ linear equations of the form $Ax = b$. In this chapter, we present iterative methods to solve a system of this type.

7.1 Norms of Vectors and Matrices

In Chapter 2 we described iterative techniques for finding roots of equations of the form $f(x) = 0$. An initial approximation (or approximations) was found, and new approximations were then determined based on how well the previous approximations satisfied the equation. To discuss iterative methods for solving linear systems, we first need a means for measuring the distance between n-dimensional column vectors to determine whether a sequence of such vectors converges to a solution of the system. In actuality, this measure is also needed when the solution is obtained by the direct methods presented in Chapter 6. Those methods required a large number of arithmetic operations, and using finite-digit arithmetic leads only to an approximation to an actual solution of the system.

Let $\mathbb{R}^n$ denote the set of all n-dimensional column vectors with real-number components. To define a distance in $\mathbb{R}^n$, we use the notion of a norm.

Definition 7.1 A **vector norm** on $\mathbb{R}^n$ is a function, $\|\cdot\|$, from $\mathbb{R}^n$ into $\mathbb{R}$ with the following properties:

(i) $\|\mathbf{x}\| \geq 0$ for all $\mathbf{x} \in \mathbb{R}^n$,

(ii) $\|\mathbf{x}\| = 0$ if and only if $\mathbf{x} = (0, 0, \ldots, 0)^t \equiv \mathbf{0}$,

(iii) $\|\alpha\mathbf{x}\| = |\alpha|\|\mathbf{x}\|$ for all $\alpha \in \mathbb{R}$ and $\mathbf{x} \in \mathbb{R}^n$,

(iv) $\|\mathbf{x} + \mathbf{y}\| \leq \|\mathbf{x}\| + \|\mathbf{y}\|$ for all $\mathbf{x}, \mathbf{y} \in \mathbb{R}^n$. ■

We will need only two specific norms on $\mathbb{R}^n$, although a third norm on $\mathbb{R}^n$ is presented in Exercise 2.

Since vectors in $\mathbb{R}^n$ are column vectors, it is convenient to use the transpose notation presented in Section 6.3 when a vector is represented in terms of its components. For example, the vector

$$\mathbf{x} = \begin{bmatrix} x_1 \\ x_2 \\ \vdots \\ x_n \end{bmatrix}$$

will be written $\mathbf{x} = (x_1, x_2, \ldots, x_n)^t$.

Definition 7.2 The l_2 and l_∞ norms for the vector $\mathbf{x} = (x_1, x_2, \ldots, x_n)^t$ are defined by

$$\|\mathbf{x}\|_2 = \left\{\sum_{i=1}^{n} x_i^2\right\}^{1/2} \quad \text{and} \quad \|\mathbf{x}\|_\infty = \max_{1\leq i\leq n} |x_i|.$$ ■

The l_2 norm is called the **Euclidean norm** of the vector $\mathbf{x}$ since it represents the usual notion of distance from the origin in case $\mathbf{x}$ is in $\mathbb{R}^1 \equiv \mathbb{R}$, $\mathbb{R}^2$, or $\mathbb{R}^3$. For example, the l_2 norm of the vector $\mathbf{x} = (x_1, x_2, x_3)^t$ denotes the length of the straight line joining the points (0, 0, 0) and (x_1, x_2, x_3). Figure 7.1 shows the boundary of those vectors in $\mathbb{R}^2$ and $\mathbb{R}^3$ that have l_2 norm less than 1. Figure 7.2 is a similar illustration for the l_∞ norm.

Figure 7.1

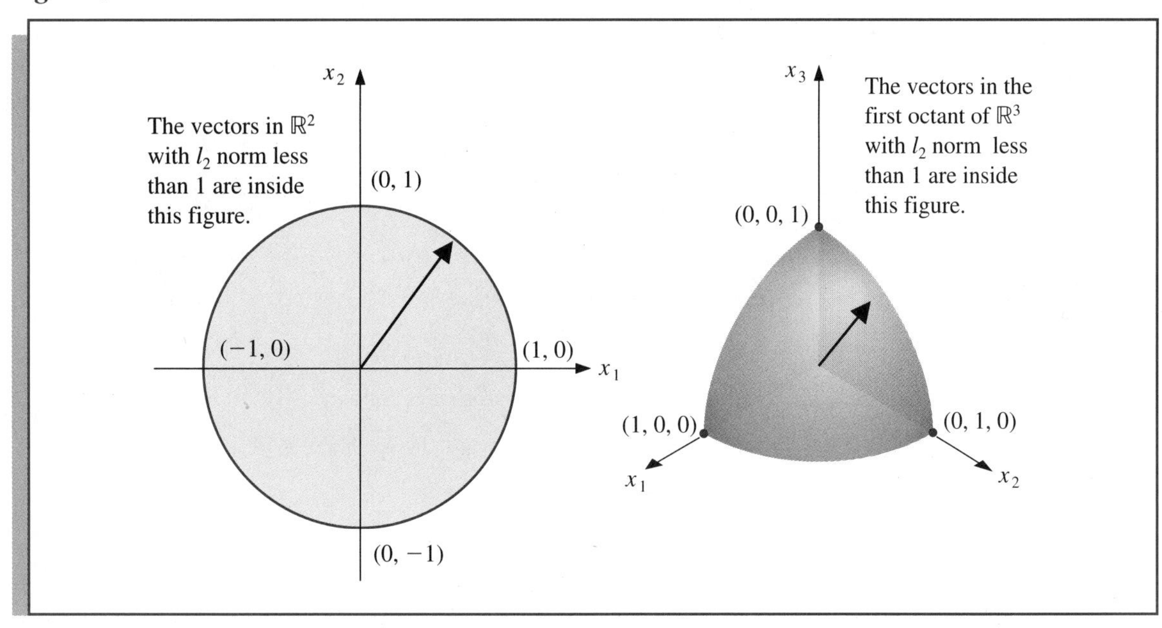

Figure 7.2

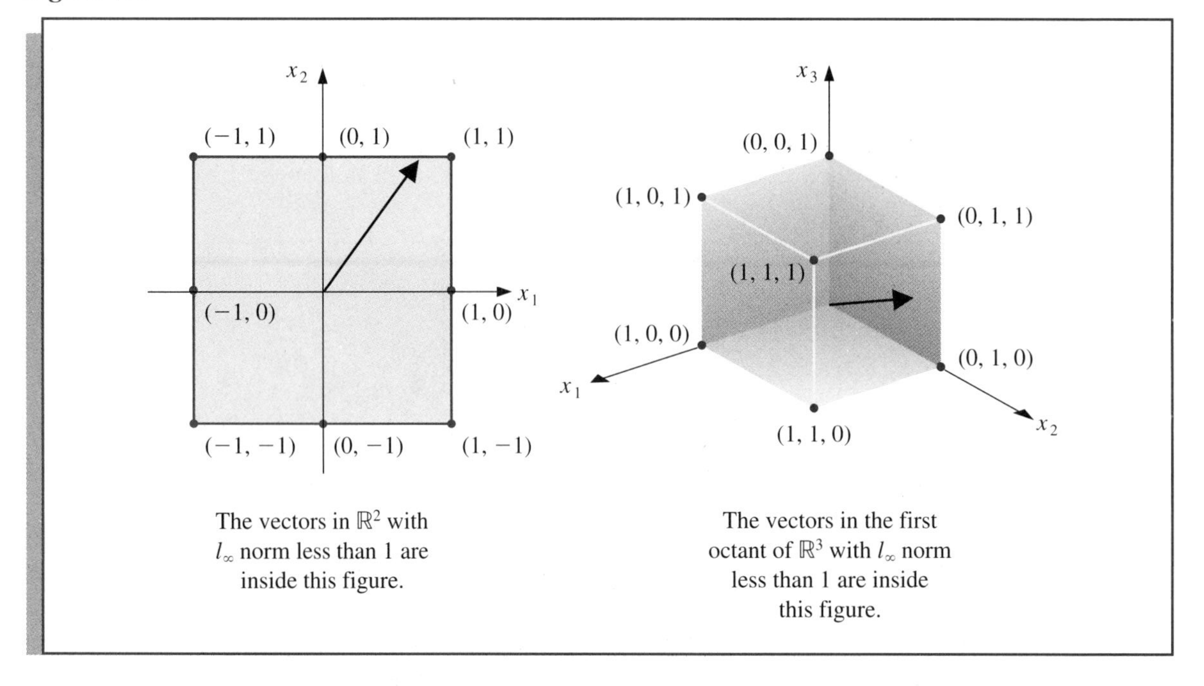

The vectors in $\mathbb{R}^2$ with l_∞ norm less than 1 are inside this figure.

The vectors in the first octant of $\mathbb{R}^3$ with l_∞ norm less than 1 are inside this figure.

EXAMPLE 1 The vector $\mathbf{x} = (-1, 1, -2)^t$ in $\mathbb{R}^3$ has norms

$$\|\mathbf{x}\|_2 = \sqrt{(-1)^2 + (1)^2 + (-2)^2} = \sqrt{6}$$

and

$$\|\mathbf{x}\|_\infty = \max\{|-1|, |1|, |-2|\} = 2. \qquad \blacksquare$$

It is easy to show that the properties in Definition 7.1 hold for the l_∞ norm since they follow from similar results for absolute values. For example, if $\mathbf{x} = (x_1, x_2, \ldots, x_n)^t$ and $\mathbf{y} = (y_1, y_2, \ldots, y_n)^t$, then

$$\|\mathbf{x} + \mathbf{y}\|_\infty = \max_{1\le i\le n} |x_i + y_i| \le \max_{1\le i\le n} \{|x_i| + |y_i|\} \le \max_{1\le i\le n} |x_i| + \max_{1\le i\le n} |y_i| = \|\mathbf{x}\|_\infty + \|\mathbf{y}\|_\infty.$$

To show that

$$\|\mathbf{x}\|_2 = \left\{\sum_{i=1}^n x_i^2\right\}^{1/2}$$

is a norm is more difficult. The first three properties follow easily. To show that

$$\|\mathbf{x} + \mathbf{y}\|_2 \le \|\mathbf{x}\|_2 + \|\mathbf{y}\|_2 \quad \text{for each } \mathbf{x}, \mathbf{y} \in \mathbb{R}_n,$$

we need a famous inequality.

Theorem 7.3 **(Cauchy-Buniakowsky-Schwarz Inequality)**

For each $\mathbf{x} = (x_1, x_2, \ldots, x_n)^t$ and $\mathbf{y} = (y_1, y_2, \ldots, y_n)^t$ in $\mathbb{R}^n$,

(7.1)
$$\sum_{i=1}^{n} |x_i y_i| \leq \left\{\sum_{i=1}^{n} x_i^2\right\}^{1/2} \left\{\sum_{i=1}^{n} y_i^2\right\}^{1/2}.$$

Proof If $\mathbf{y} = \mathbf{0}$ or $\mathbf{x} = \mathbf{0}$, the result is immediate since both sides of the inequality are zero.

Suppose $\mathbf{y} \neq \mathbf{0}$ and $\mathbf{x} \neq \mathbf{0}$. For each $\lambda \in \mathbb{R}$,

$$0 \leq \|\mathbf{x} + \lambda \mathbf{y}\|_2^2 = \sum_{i=1}^{n} (x_i + \lambda y_i)^2 = \sum_{i=1}^{n} x_i^2 - 2\lambda \sum_{i=1}^{n} x_i y_i + \lambda^2 \sum_{i=1}^{n} y_i^2,$$

so

$$2\lambda \sum_{i=1}^{n} x_i y_i \leq \sum_{i=1}^{n} x_i^2 + \lambda^2 \sum_{i=1}^{n} y_i^2 = \|\mathbf{x}\|_2^2 + \lambda^2 \|\mathbf{y}\|_2^2.$$

Since $\|\mathbf{x}\|_2 > 0$ and $\|\mathbf{y}\|_2 > 0$, we can let $\lambda = \|\mathbf{x}\|_2 / \|\mathbf{y}\|_2$ to give

$$\left(2\frac{\|\mathbf{x}\|_2}{\|\mathbf{y}\|_2}\right)\left(\sum_{i=1}^{n} x_i y_i\right) \leq \|\mathbf{x}\|_2^2 + \frac{\|\mathbf{x}\|_2^2}{\|\mathbf{y}\|_2^2}\|\mathbf{y}\|_2^2 = 2\|\mathbf{x}\|_2^2,$$

and divide by λ to produce

$$2\sum_{i=1}^{n} x_i y_i \leq 2\|\mathbf{x}\|_2^2 \frac{\|\mathbf{y}\|_2}{\|\mathbf{x}\|_2} = 2\|\mathbf{x}\|_2\|\mathbf{y}\|_2.$$

Thus,

$$\sum_{i=1}^{n} x_i y_i \leq \|\mathbf{x}\|_2\|\mathbf{y}\|_2.$$

Replace x_i by $-x_i$ whenever $x_i y_i < 0$ and call the new vector $\tilde{\mathbf{x}} = (\tilde{x}_i)$. Then $\|\tilde{\mathbf{x}}\|_2 = \|\mathbf{x}\|_2$ and

$$\sum_{i=1}^{n} |x_i y_i| = \sum_{i=1}^{n} \tilde{x}_i y_i \leq \|\tilde{\mathbf{x}}\|_2\|\mathbf{y}\|_2 = \|\mathbf{x}\|_2\|\mathbf{y}\|_2 = \left\{\sum_{i=1}^{n} x_i^2\right\}^{1/2} \left\{\sum_{i=1}^{n} y_i^2\right\}^{1/2}. \quad \blacksquare \blacksquare \blacksquare$$

With this result we see that for each $\mathbf{x}, \mathbf{y} \in \mathbb{R}^n$,

$$\|\mathbf{x} + \mathbf{y}\|_2^2 = \sum_{i=1}^{n} (x_i + y_i)^2 = \sum_{i=1}^{n} x_i^2 + 2\sum_{i=1}^{n} x_i y_i + \sum_{i=1}^{n} y_i^2,$$

which gives the final norm property

$$\|\mathbf{x} + \mathbf{y}\|_2 \leq \left[\|\mathbf{x}\|_2^2 + 2\|\mathbf{x}\|_2\|\mathbf{y}\|_2 + \|\mathbf{y}\|_2^2\right]^{1/2} = \|\mathbf{x}\|_2 + \|\mathbf{y}\|_2.$$

Since the norm of a vector gives a measure for the distance between an arbitrary vector and the zero vector, the **distance between two vectors** can be defined as the norm of the difference of the vectors.

Definition 7.4 If $\mathbf{x} = (x_1, x_2, \ldots, x_n)^t$ and $\mathbf{y} = (y_1, y_2, \ldots, y_n)^t$ are vectors in $\mathbb{R}^n$, the l_2 and l_∞ distances between $\mathbf{x}$ and $\mathbf{y}$ are defined by

$$\|\mathbf{x} - \mathbf{y}\|_2 = \left\{\sum_{i=1}^{n}(x_i - y_i)^2\right\}^{1/2} \quad \text{and} \quad \|\mathbf{x} - \mathbf{y}\|_\infty = \max_{1\le i\le n} |x_i - y_i|. \quad \blacksquare$$

EXAMPLE 2 The linear system

$$\begin{aligned} 3.3330x_1 + 15920x_2 - 10.333x_3 &= 15913, \\ 2.2220x_1 + 16.710x_2 + 9.6120x_3 &= 28.544, \\ 1.5611x_1 + 5.1791x_2 + 1.6852x_3 &= 8.4254 \end{aligned}$$

has solution $(x_1, x_2, x_3)^t = (1.0000, 1.0000, 1.0000)^t$. If Gaussian elimination is performed in five-digit rounding arithmetic using partial pivoting (Algorithm 6.2), the solution obtained is

$$\tilde{\mathbf{x}} = (\tilde{x}_1, \tilde{x}_2, \tilde{x}_3)^t = (1.2001, 0.99991, 0.92538)^t.$$

Measurements of $\mathbf{x} - \tilde{\mathbf{x}}$ are given by

$$\begin{aligned} \|\mathbf{x} - \tilde{\mathbf{x}}\|_\infty &= \max\{|1.0000 - 1.2001|, |1.0000 - 0.99991|, |1.0000 - 0.92538|\} \\ &= \max\{0.2001, 0.00009, 0.07462\} = 0.2001 \end{aligned}$$

and

$$\begin{aligned} \|\mathbf{x} - \tilde{\mathbf{x}}\|_2 &= [(1.0000 - 1.2001)^2 + (1.0000 - 0.99991)^2 + (1.0000 - 0.92538)^2]^{1/2} \\ &= [(0.2001)^2 + (0.00009)^2 + (0.07462)^2]^{1/2} = 0.21356. \end{aligned}$$

Although the components $\tilde{x}_2$ and $\tilde{x}_3$ are good approximations to x_2 and x_3, the component $\tilde{x}_1$ is a poor approximation to x_1, and $|x_1 - \tilde{x}_1|$ dominates the norms. ■

The concept of distance in $\mathbb{R}^n$ is also used to define a limit of a sequence of vectors in this space.

Definition 7.5 A sequence $\{\mathbf{x}^{(k)}\}_{k=1}^{\infty}$ of vectors in $\mathbb{R}^n$ is said to **converge** to $\mathbf{x}$ with respect to the norm $\|\cdot\|$ if, given any $\varepsilon > 0$, there exists an integer $N(\varepsilon)$ such that

$$\|\mathbf{x}^{(k)} - \mathbf{x}\| < \varepsilon \quad \text{for all } k \ge N(\varepsilon). \quad \blacksquare$$

Theorem 7.6 The sequence of vectors $\{\mathbf{x}^{(k)}\}$ converges to $\mathbf{x}$ in $\mathbb{R}^n$ with respect to $\|\cdot\|_\infty$ if and only if $\lim_{k\to\infty} x_i^{(k)} = x_i$ for each $i = 1, 2, \ldots, n$.

Proof Suppose $\{\mathbf{x}^{(k)}\}$ converges to $\mathbf{x}$ with respect to $\|\cdot\|_\infty$. Given any $\varepsilon > 0$, there exists an integer $N(\varepsilon)$ such that for all $k \ge N(\varepsilon)$,

$$\max_{1\le i\le n} |x_i^{(k)} - x_i| = \|\mathbf{x}^{(k)} - \mathbf{x}\|_\infty < \varepsilon.$$

This result implies that $|x_i^{(k)} - x_i| < \varepsilon$ for each $i = 1, 2, \ldots, n$, so $\lim_{k\to\infty} x_i^{(k)} = x_i$ for each i.

Conversely, suppose that $\lim_{k\to\infty} x_i^{(k)} = x_i$ for every $i = 1, 2, \ldots, n$. For a given $\varepsilon > 0$, let $N_i(\varepsilon)$ for each i represent an integer with the property that

$$|x_i^{(k)} - x_i| < \varepsilon$$

whenever $k \geq N_i(\varepsilon)$. Define $N(\varepsilon) = \max_{1\leq i\leq n} N_i(\varepsilon)$. If $k \geq N(\varepsilon)$, then $|x_i^{(k)} - x_i| < \varepsilon$ for each i and

$$\max_{1\leq i\leq n} |x_i^{(k)} - x_i| = \|\mathbf{x}^{(k)} - \mathbf{x}\|_\infty < \varepsilon.$$

This implies that $\{\mathbf{x}^{(k)}\}$ converges to $\mathbf{x}$. ■ ■ ■

EXAMPLE 3 Let $\mathbf{x}^{(k)} \in \mathbb{R}^4$ be defined by

$$\mathbf{x}^{(k)} = (x_1^{(k)}, x_2^{(k)}, x_3^{(k)}, x_4^{(k)})^t = \left(1, 2 + \frac{1}{k}, \frac{3}{k^2}, e^{-k} \sin k\right)^t.$$

Since $\lim_{k\to\infty} 1 = 1$, $\lim_{k\to\infty}(2 + 1/k) = 2$, $\lim_{k\to\infty} 3/k^2 = 0$, and $\lim_{k\to\infty} e^{-k} \sin k = 0$, Theorem 7.6 implies that the sequence $\{\mathbf{x}^{(k)}\}$ converges to $(1, 2, 0, 0)^t$ with respect to $\|\cdot\|_\infty$. ■

To show directly that the sequence in Example 3 converges to $(1, 2, 0, 0)^t$ with respect to the l_2 norm is quite complicated. It is easier to prove the next result and apply it to this special case.

Theorem 7.7 For each $\mathbf{x} \in \mathbb{R}^n$,

$$\|\mathbf{x}\|_\infty \leq \|\mathbf{x}\|_2 \leq \sqrt{n}\|\mathbf{x}\|_\infty.$$

Proof Let x_j be a coordinate of $\mathbf{x}$ such that $\|\mathbf{x}\|_\infty = \max_{1\leq i\leq n} |x_i| = |x_j|$. Then

$$\|\mathbf{x}\|_\infty^2 = |x_j|^2 = x_j^2 \leq \sum_{i=1}^{n} x_i^2 \leq \sum_{i=1}^{n} x_j^2 = nx_j^2 = n\|\mathbf{x}\|_\infty^2.$$

Thus,

$$\|\mathbf{x}\|_\infty \leq \left\{\sum_{i=1}^{n} x_i^2\right\}^{1/2} = \|\mathbf{x}\|_2 \leq \sqrt{n}\|\mathbf{x}\|_\infty.$$ ■ ■ ■

Figure 7.3 illustrates this result when $n = 2$.

Figure 7.3

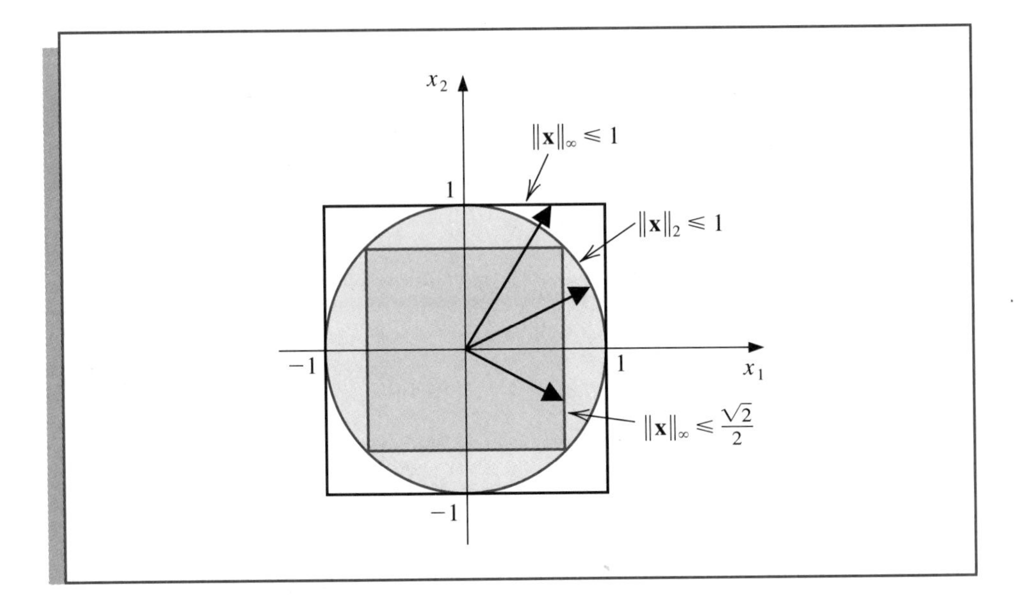

EXAMPLE 4 In Example 3, we found that the sequence $\{\mathbf{x}^{(k)}\}$, defined by

$$\mathbf{x}^{(k)} = \left(1, 2 + \frac{1}{k}, \frac{3}{k^2}, e^{-k} \sin k\right)^t,$$

converges to $\mathbf{x} = (1, 2, 0, 0)^t$ with respect to $\|\cdot\|_\infty$. Given any $\varepsilon > 0$, there exists an integer $N(\varepsilon/2)$ with the property that

$$\|\mathbf{x}^{(k)} - \mathbf{x}\|_\infty < \frac{\varepsilon}{2}$$

whenever $k \geq N(\varepsilon/2)$. By Theorem 7.7, this implies that

$$\|\mathbf{x}^{(k)} - \mathbf{x}\|_2 < \sqrt{4}\|\mathbf{x}^{(k)} - \mathbf{x}\|_\infty < 2(\varepsilon/2) = \varepsilon$$

when $k \geq N(\varepsilon/2)$. So $\{\mathbf{x}^{(k)}\}$ converges to $\mathbf{x}$ with respect to $\|\cdot\|_2$. ■

It can be shown that all norms on $\mathbb{R}^n$ are equivalent with respect to convergence; that is, if $\|\cdot\|$ and $\|\cdot\|'$ are any two norms on $\mathbb{R}^n$ and $\{\mathbf{x}^{(k)}\}_{k=1}^\infty$ has the limit $\mathbf{x}$ with respect to $\|\cdot\|$, then $\{\mathbf{x}^{(k)}\}_{k=1}^\infty$ also has the limit $\mathbf{x}$ with respect to $\|\cdot\|'$. The proof of this fact for the general case can be found in [Or2, p. 8]. The case for the norms $\|\cdot\|_2$ and $\|\cdot\|_\infty$ follows from Theorem 7.7.

In the subsequent sections of this and later chapters, we will need methods for determining the distance between $n \times n$ matrices. This again requires the use of the norm concept.

Definition 7.8 A **matrix norm** on the set of all $n \times n$ matrices is a real-valued function, $\|\cdot\|$, defined on this set, satisfying for all $n \times n$ matrices A and B and all real numbers α:

(i) $\|A\| \geq 0$.
(ii) $\|A\| = 0$, if and only if A is O, the matrix with all zero entries.
(iii) $\|\alpha A\| = |\alpha|\|A\|$.
(iv) $\|A + B\| \leq \|A\| + \|B\|$.
(v) $\|AB\| \leq \|A\|\|B\|$. ∎

A **distance between $n \times n$ matrices** A and B with respect to this matrix norm is $\|A - B\|$.

Although matrix norms can be obtained in various ways, the only norms we consider are those that are natural consequences of the vector norms l_2 and l_∞.

The following theorem is not difficult to show, and its proof is left as Exercise 13.

Theorem 7.9 If $\|\cdot\|$ is a vector norm on $\mathbb{R}^n$, then

$$\|A\| = \max_{\|\mathbf{x}\|=1} \|A\mathbf{x}\|$$

is a matrix norm. ∎

This is called the **natural**, or **induced**, **matrix norm** associated with the vector norm. In this text, all matrix norms will be assumed to be natural matrix norms unless specified otherwise.

The following corollary is often used to bound a value of $\|A\mathbf{x}\|$.

Corollary 7.10 For any vector $\mathbf{x} \neq \mathbf{0}$, matrix A, and any natural norm $\|\cdot\|$, we have

$$\|A\mathbf{x}\| \leq \|A\| \cdot \|\mathbf{x}\|.$$

Proof First note that for $\mathbf{x} \neq \mathbf{0}$, the vector $\mathbf{x}/\|\mathbf{x}\|$ has length 1. So by Theorem 7.9 we have

$$\left\| A\left(\frac{\mathbf{x}}{\|\mathbf{x}\|}\right) \right\| \leq \|A\|.$$

But $\|\mathbf{x}\|$ is a nonzero real number, which implies that

$$A\left(\frac{\mathbf{x}}{\|\mathbf{x}\|}\right) = \frac{1}{\|\mathbf{x}\|} A\mathbf{x}.$$

Hence

$$\frac{1}{\|\mathbf{x}\|}\|A\mathbf{x}\| = \left\|\frac{1}{\|\mathbf{x}\|}A\mathbf{x}\right\| = \left\| A\left(\frac{\mathbf{x}}{\|\mathbf{x}\|}\right) \right\| \leq \|A\|,$$

which implies that

$$\|A\mathbf{x}\| \leq \|A\| \cdot \|\mathbf{x}\|.$$ ∎ ∎ ∎

The measure given to a matrix under a natural norm describes how the matrix stretches unit vectors relative to that norm. The largest amount of stretch is the norm of the matrix.

The matrix norms we will consider have the forms

$$\|A\|_\infty = \max_{\|\mathbf{x}\|_\infty=1} \|A\mathbf{x}\|_\infty, \quad \text{the } l_\infty \text{ norm,}$$

and

$$\|A\|_2 = \max_{\|\mathbf{x}\|_2=1} \|A\mathbf{x}\|_2, \quad \text{the } l_2 \text{ norm.}$$

An illustration of these norms when $n = 2$ is shown in Figures 7.4 and 7.5.

Figure 7.4

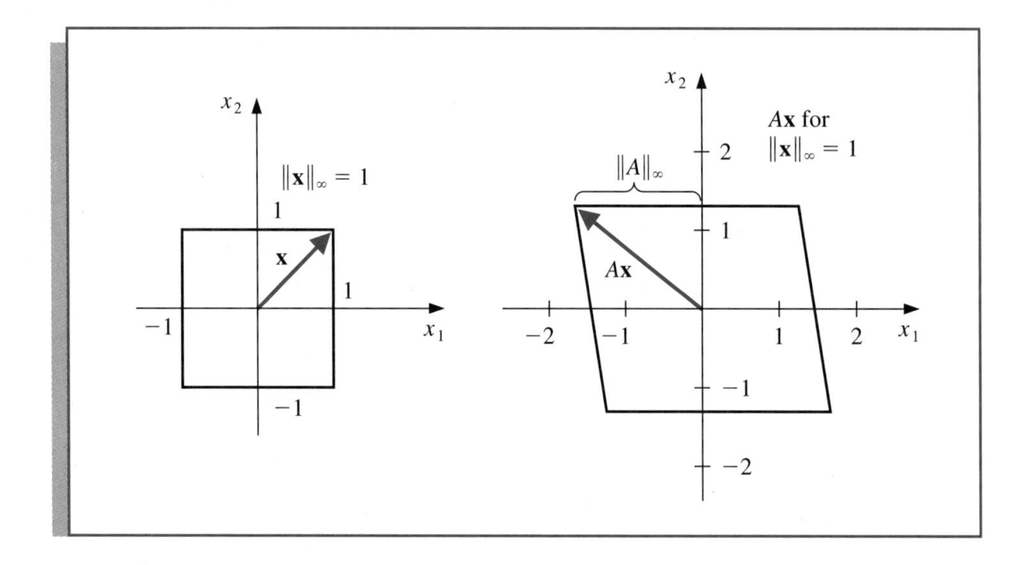

Figure 7.5

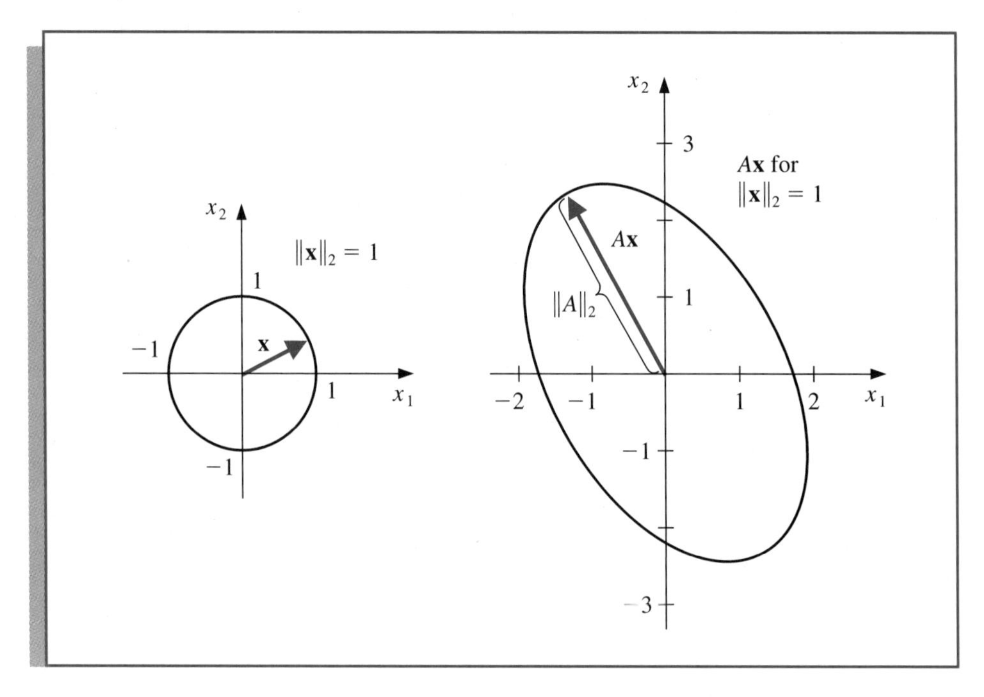

The l_∞ norm of a matrix has an interesting representation with respect to the entries of the matrix.

Theorem 7.11 If $A = (a_{ij})$ is an $n \times n$ matrix, then

$$\|A\|_\infty = \max_{1 \le i \le n} \sum_{j=1}^{n} |a_{ij}|.$$

Proof First we show that $\|A\|_\infty \le \max_{1\le i\le n} \sum_{j=1}^n |a_{ij}|$. Let $\mathbf{x}$ be an n-dimensional column vector with $1 = \|\mathbf{x}\|_\infty = \max_{1\le i\le n} |x_i|$. Since $A\mathbf{x}$ is also an n-dimensional column vector,

$$\|A\mathbf{x}\|_\infty = \max_{1\le i\le n} |(A\mathbf{x})_i| = \max_{1\le i\le n} \left| \sum_{j=1}^n a_{ij}x_j \right| \le \max_{1\le i\le n} \sum_{j=1}^n |a_{ij}| \max_{1\le j\le n} |x_j|.$$

Since $\|\mathbf{x}\|_\infty = 1$, we have

$$\|A\mathbf{x}\|_\infty \le \max_{1\le i\le n} \sum_{j=1}^n |a_{ij}| \|\mathbf{x}\|_\infty = \max_{1\le i\le n} \sum_{j=1}^n |a_{ij}|.$$

Consequently,

$$\|A\|_\infty = \max_{\|\mathbf{x}\|_\infty = 1} \|A\mathbf{x}\|_\infty \le \max_{1\le i\le n} \sum_{j=1}^n |a_{ij}|. \tag{7.2}$$

Now we need to show the opposite inequality, that $\|A\|_\infty \ge \max_{1\le i\le n} \sum_{j=1}^n |a_{ij}|$. Let p be an integer with

$$\sum_{j=1}^n |a_{pj}| = \max_{1\le i\le n} \sum_{j=1}^n |a_{ij}|,$$

and $\mathbf{x}$ be the vector with components

$$x_j = \begin{cases} 1, & \text{if } a_{pj} \ge 0, \\ -1, & \text{if } a_{pj} < 0. \end{cases}$$

Then $\|\mathbf{x}\|_\infty = 1$ and $a_{pj}x_j = |a_{pj}|$, for all $j = 1, 2, \ldots, n$, so

$$\|A\mathbf{x}\|_\infty = \max_{1\le i\le n} \left| \sum_{j=1}^n a_{ij}x_j \right| \ge \left| \sum_{j=1}^n a_{pj}x_j \right| = \left| \sum_{j=1}^n |a_{pj}| \right| = \max_{1\le i\le n} \sum_{j=1}^n |a_{ij}|.$$

This result implies that

$$\|A\|_\infty = \max_{\|\mathbf{x}\|_\infty = 1} \|A\mathbf{x}\|_\infty \ge \max_{1\le i\le n} \sum_{j=1}^n |a_{ij}|,$$

which, together with Inequality (7.2), gives

$$\|A\|_\infty = \max_{1\le i\le n} \sum_{j=1}^{n} |a_{ij}|. \qquad \blacksquare\;\blacksquare\;\blacksquare$$

EXAMPLE 5 If

$$A = \begin{bmatrix} 1 & 2 & -1 \\ 0 & 3 & -1 \\ 5 & -1 & 1 \end{bmatrix},$$

then

$$\sum_{j=1}^{3} |a_{1j}| = |1| + |2| + |-1| = 4,$$

$$\sum_{j=1}^{3} |a_{2j}| = |0| + |3| + |-1| = 4,$$

and

$$\sum_{j=1}^{3} |a_{3j}| = |5| + |-1| + |1| = 7;$$

so

$$\|A\|_\infty = \max\{4, 4, 7\} = 7. \qquad \blacksquare$$

In the next section, we will discover an alternative method for finding the l_2 norm of a matrix.

EXERCISE SET 7.1

1. Find $\|\mathbf{x}\|_\infty$ and $\|\mathbf{x}\|_2$ for the following vectors.
   a. $\mathbf{x} = (3, -4, 0, \frac{3}{2})^t$
   b. $\mathbf{x} = (2, 1, -3, 4)^t$
   c. $\mathbf{x} = (\sin k, \cos k, 2^k)^t$ for a fixed positive integer k
   d. $\mathbf{x} = (4/(k+1), 2/k^2, k^2 e^{-k})^t$ for a fixed positive integer k
2. a. Verify that the function $\|\cdot\|_1$, defined on $\mathbb{R}^n$ by

   $$\|\mathbf{x}\|_1 = \sum_{i=1}^{n} |x_i|,$$

   is a norm on $\mathbb{R}^n$.
   b. Find $\|\mathbf{x}\|_1$ for the vectors given in Exercise 1.
3. Prove that the following sequences are convergent and find their limits.
   a. $\mathbf{x}^{(k)} = (1/k, e^{1-k}, -2/k^2)^t$
   b. $\mathbf{x}^{(k)} = \left(e^{-k}\cos k, k\sin\frac{1}{k}, 3 + k^{-2}\right)^t$

c. $\mathbf{x}^{(k)} = (ke^{-k^2}, (\cos k)/k, \sqrt{k^2+k}-k)^t$

d. $\mathbf{x}^{(k)} = (e^{1/k}, (k^2+1)/(1-k^2), (1/k^2)(1+3+5+\cdots+(2k-1)))^t$

4. Find $\|\cdot\|_\infty$ for the following matrices.

a. $\begin{bmatrix} 10 & 15 \\ 0 & 1 \end{bmatrix}$ **b.** $\begin{bmatrix} 10 & 0 \\ 15 & 1 \end{bmatrix}$

c. $\begin{bmatrix} 2 & -1 & 0 \\ -1 & 2 & -1 \\ 0 & -1 & 2 \end{bmatrix}$ **d.** $\begin{bmatrix} 4 & -1 & 7 \\ -1 & 4 & 0 \\ -7 & 0 & 4 \end{bmatrix}$

5. The following linear systems $A\mathbf{x} = \mathbf{b}$ have $\mathbf{x}$ as the actual solution and $\tilde{\mathbf{x}}$ as an approximate solution. Compute $\|\mathbf{x}-\tilde{\mathbf{x}}\|_\infty$ and $\|A\tilde{\mathbf{x}}-\mathbf{b}\|_\infty$.

a. $\frac{1}{2}x_1 + \frac{1}{3}x_2 = \frac{1}{63}$,
$\frac{1}{3}x_1 + \frac{1}{4}x_2 = \frac{1}{168}$,
$\mathbf{x} = \left(\frac{1}{7}, -\frac{1}{6}\right)^t$,
$\tilde{\mathbf{x}} = (0.142, -0.166)^t$.

b. $x_1 + 2x_2 + 3x_3 = 1$,
$2x_1 + 3x_2 + 4x_3 = -1$,
$3x_1 + 4x_2 + 6x_3 = 2$,
$\mathbf{x} = (0, -7, 5)^t$,
$\tilde{\mathbf{x}} = (-0.33, -7.9, 5.8)^t$.

c. $x_1 + 2x_2 + 3x_3 = 1$,
$2x_1 + 3x_2 + 4x_3 = -1$,
$3x_1 + 4x_2 + 6x_3 = 2$,
$\mathbf{x} = (0, -7, 5)^t$,
$\tilde{\mathbf{x}} = (-0.2, -7.5, 5.4)^t$.

d. $0.04x_1 + 0.01x_2 - 0.01x_3 = 0.06$,
$0.2x_1 + 0.5x_2 - 0.2x_3 = 0.3$,
$x_1 + 2x_2 + 4x_3 = 11$,
$\mathbf{x} = (1.827586, 0.6551724, 1.965517)^t$,
$\tilde{\mathbf{x}} = (1.8, 0.64, 1.9)^t$.

6. The matrix norm $\|\cdot\|_1$, defined by $\|A\|_1 = \max_{\|\mathbf{x}\|_1=1} \|A\mathbf{x}\|_1$, can be computed using the formula

$$\|A\|_1 = \max_{1\le j\le n} \sum_{i=1}^{n} |a_{ij}|,$$

where the vector norm $\|\cdot\|_1$ is defined in Exercise 2. Find $\|\cdot\|_1$ for the matrices in Exercise 4.

7. Show by example that $\|\cdot\|_{\textcircled{\infty}}$, defined by $\|A\|_{\textcircled{\infty}} = \max_{1\le i,j\le n} |a_{ij}|$, does not define a matrix norm.

8. Show that $\|\cdot\|_{\textcircled{1}}$ defined by

$$\|A\|_{\textcircled{1}} = \sum_{i=1}^{n}\sum_{j=1}^{n} |a_{ij}|$$

is a matrix norm. Find $\|\cdot\|_{\textcircled{1}}$ for the matrices in Exercise 4.

9. **a.** The Frobenius norm (which is not a natural norm) is defined for an $n \times n$ matrix A by

$$\|A\|_F = \left(\sum_{i=1}^{n}\sum_{j=1}^{n} |a_{ij}|^2\right)^{1/2}.$$

Show that $\|\cdot\|_F$ is a matrix norm.

b. Find $\|\cdot\|_F$ for the matrices in Exercise 4.

c. For any matrix A, show that $\|A\|_2 \le \|A\|_F \le n^{1/2}\|A\|_2$.

10. In Exercise 9 the Frobenius norm of a matrix was defined. Show that for any $n \times n$ matrix A and vector $\mathbf{x}$ in $\mathbb{R}^n$, $\|A\mathbf{x}\|_2 \le \|A\|_F\|\mathbf{x}\|_2$.

11. Let S be a positive definite $n \times n$ matrix. For any $\mathbf{x}$ in $\mathbb{R}^n$ define $\|\mathbf{x}\| = (\mathbf{x}^t S\mathbf{x})^{1/2}$. Show that this defines a norm on $\mathbb{R}^n$. [*Hint*: First show that $\mathbf{x}^t S\mathbf{y} = \mathbf{y}^t S\mathbf{x} \leq (\mathbf{x}^t S\mathbf{x})^{1/2}(\mathbf{y}^t S\mathbf{y})^{1/2}$.]
12. Let S be a real and nonsingular matrix and let $\|\cdot\|$ be any norm on $\mathbb{R}^n$. Define $\|\cdot\|'$ by $\|\mathbf{x}\|' = \|S\mathbf{x}\|$. Show that $\|\cdot\|'$ is also a norm on $\mathbb{R}^n$.
13. Prove that if $\|\cdot\|$ is a vector norm on $\mathbb{R}^n$, then $\|A\| = \max_{\|\mathbf{x}\|=1} \|A\mathbf{x}\|$ is a matrix norm.
14. The following excerpt from the Mathematics Magazine [Sz] gives an alternative way to prove the Cauchy-Buniakowsky-Schwarz Inequality.
    a. Show that when $\mathbf{x} \neq \mathbf{0}$ and $\mathbf{y} \neq \mathbf{0}$ we have

$$\frac{\sum_{i=1}^n x_i y_i}{\left(\sum_{i=1}^n x_i^2\right)^{1/2}\left(\sum_{i=1}^n y_i^2\right)^{1/2}} = 1 - \frac{1}{2}\sum_{i=1}^n \left(\frac{x_i}{\left(\sum_{j=1}^n x_j^2\right)^{1/2}} - \frac{y_i}{\left(\sum_{j=1}^n y_j^2\right)^{1/2}} \right)^2.$$

    b. Use the result in part (a) to show that

$$\sum_{i=1}^n x_i y_i \leq \left(\sum_{i=1}^n x_i^2\right)^{1/2}\left(\sum_{i=1}^n y_i^2\right)^{1/2}.$$

7.2 Eigenvalues and Eigenvectors

The last result in Section 7.1 gave us a method for determining the l_∞ norm of a matrix that does not require applying the definition. In this section, we will see that the l_2 norm of a matrix can also be determined without referring directly to the definition. To develop this technique, we introduce the notions of eigenvalues and eigenvectors.

Definition 7.12 If A is a square matrix, the polynomial defined by

$$p(\lambda) = \det(A - \lambda I)$$

is called the **characteristic polynomial** of A. ∎

It is not difficult to show (see Exercise 7) that p is an nth-degree polynomial and, consequently, has at most n distinct zeros, some of which may be complex. If λ is a zero of p, then, since $\det(A - \lambda I) = 0$, Theorem 6.17 in Section 6.4 implies that the linear system defined by $(A - \lambda I)\mathbf{x} = \mathbf{0}$ has a solution other than $\mathbf{x} = \mathbf{0}$. We wish to study the zeros of p and the nonzero solutions corresponding to these systems.

Definition 7.13 If p is the characteristic polynomial of the matrix A, the zeros of p are called **eigenvalues**, or characteristic values, of the matrix A. If λ is a eigenvalue of A and $\mathbf{x} \neq \mathbf{0}$ has the property that $(A - \lambda I)\mathbf{x} = \mathbf{0}$, then $\mathbf{x}$ is called an **eigenvector**, or characteristic vector, of A corresponding to the eigenvalue λ. ∎

If $\mathbf{x}$ is an eigenvector associated with the eigenvalue λ, then $A\mathbf{x} = \lambda\mathbf{x}$, so the matrix A takes the vector $\mathbf{x}$ into a scalar multiple of itself. If λ is real and $\lambda > 1$, then A has the effect of stretching $\mathbf{x}$ by a factor of λ, as illustrated in Figure 7.6(a). If $0 < \lambda < 1$, then

A shrinks $\mathbf{x}$ by a factor of λ (see Figure 7.6(b)). When $\lambda < 0$, the effects are similar (see Figure 7.6(c) and (d)), although the direction of $A\mathbf{x}$ is reversed.

Figure 7.6

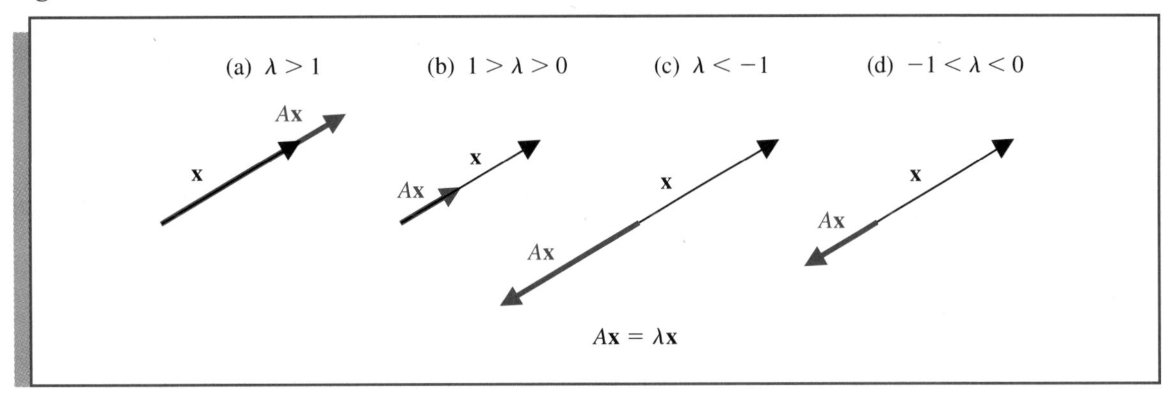

EXAMPLE 1 Let

$$A = \begin{bmatrix} 1 & 0 & 2 \\ 0 & 1 & -1 \\ -1 & 1 & 1 \end{bmatrix}.$$

To compute the eigenvalues of A, consider

$$p(\lambda) = \det(A - \lambda I) = \det \begin{bmatrix} 1-\lambda & 0 & 2 \\ 0 & 1-\lambda & -1 \\ -1 & 1 & 1-\lambda \end{bmatrix} = (1-\lambda)(\lambda^2 - 2\lambda + 4).$$

The eigenvalues of A are the solutions of $p(\lambda) = 0$, which are $\lambda_1 = 1$, $\lambda_2 = 1 + \sqrt{3}i$, and $\lambda_3 = 1 - \sqrt{3}i$.

An eigenvector $\mathbf{x}$ of A associated with λ_1 is a solution of the system $(A - \lambda_1 I)\mathbf{x} = \mathbf{0}$:

$$\begin{bmatrix} 0 & 0 & 2 \\ 0 & 0 & -1 \\ -1 & 1 & 0 \end{bmatrix} \begin{bmatrix} x_1 \\ x_2 \\ x_3 \end{bmatrix} = \begin{bmatrix} 0 \\ 0 \\ 0 \end{bmatrix}.$$

Thus,

$$2x_3 = 0, \quad -x_3 = 0, \quad \text{and} \quad -x_1 + x_2 = 0,$$

which implies that

$$x_3 = 0, \quad x_2 = x_1, \quad \text{and} \quad x_1 \text{ is arbitrary.}$$

The choice $x_1 = 1$ produces the eigenvector $(1, 1, 0)^t$ corresponding to the eigenvalue $\lambda_1 = 1$.

Since λ_2 and λ_3 are complex numbers, their corresponding eigenvectors are also complex. To find an eigenvector for λ_2, we solve the system

$$\begin{bmatrix} 1-(1+\sqrt{3}i) & 0 & 2 \\ 0 & 1-(1+\sqrt{3}i) & -1 \\ -1 & 1 & 1-(1+\sqrt{3}i) \end{bmatrix} \begin{bmatrix} x_1 \\ x_2 \\ x_3 \end{bmatrix} = \begin{bmatrix} 0 \\ 0 \\ 0 \end{bmatrix}.$$

Using complex arithmetic, it can be shown that one solution to this system is the vector

$$\left(-\frac{2\sqrt{3}}{3}i, \frac{\sqrt{3}}{3}i, 1\right)^t.$$

In a similar manner, the vector

$$\left(\frac{2\sqrt{3}}{3}i, -\frac{\sqrt{3}}{3}i, 1\right)^t$$

is an eigenvector corresponding to the eigenvalue $\lambda_3 = 1 - \sqrt{3}i$. ■

Maple provides the function `Eigenvals` to compute eigenvalues and, optionally, eigenvectors of a matrix. For the example we enter the following:

```
>with(linalg);
>A:=matrix(3,3,[1,0,2,0,1,-1,-1,1,1]);
>evalf(Eigenvals(A));
```

$$[1.000000000 + 1.732050807I, \quad 1.000000000 - 1.732050807I, \quad 1.000000000]$$

This computes the eigenvalues

$$\lambda_2 = 1 + \sqrt{3}i, \; \lambda_3 = 1 - \sqrt{3}i, \; \lambda_1 = 1.$$

To compute both the eigenvalues and eigenvectors, use

```
>evalf(Eigenvals(A,B));
```

The eigenvalues are computed and displayed as before, and the eigenvectors are indicated in the columns of B. If the eigenvalues are all real, each column of B gives an eigenvector. However, for our example we display B using

```
>evalm(B);
```

$$B = \begin{bmatrix} 1.154700538 & .6324555321\ 10^{-10} & .7453559925 \\ -.5773502680 & .1264911064\ 10^{-9} & .7453559926 \\ -.2581988896\ 10^{-19} & 1.000000000 & -.72572776\ 10^{-11} \end{bmatrix}$$

The first two columns correspond to the real and imaginary parts of the eigenvectors corresponding to eigenvalues λ_2 and λ_3. Thus, an eigenvector for λ_2 is

$$\begin{bmatrix} 1.154700538 \\ -.5773502680 \\ -.2581988896\ 10^{-19} \end{bmatrix} + \begin{bmatrix} .6324555321\ 10^{-10} \\ .1264911064\ 10^{-9} \\ 1.000000000 \end{bmatrix} i \approx \begin{bmatrix} 1.154700538 \\ -.5773502680 \\ 0 \end{bmatrix} + \begin{bmatrix} 0 \\ 0 \\ 1 \end{bmatrix} i,$$

that is,

$$(1.154700538, -0.5773502680, i)^t = \left(\frac{2\sqrt{3}}{3}, -\frac{\sqrt{3}}{3}, i \right)^t.$$

Since any multiple of an eigenvector is also an eigenvector, multiplying each coordinate by $-i$ gives the eigenvector

$$\left(-\frac{2\sqrt{3}}{3} i, \frac{\sqrt{3}}{3} i, 1 \right)^t.$$

Similarly, we can obtain the eigenvector

$$\left(\frac{2\sqrt{3}}{3} i, -\frac{\sqrt{3}}{3} i, 1 \right)^t$$

for the eigenvalue λ_3. Since λ_1 is real, the third column of B is an eigenvector corresponding to λ_1. We can obtain the eigenvector $(1, 1, 0)^t$ from the third column by setting the third coordinate to zero and dividing by the first coordinate.

The notions of eigenvalues and eigenvectors are introduced here for a specific computational convenience, but these concepts arise frequently in the study of physical systems. In fact, they are of sufficient interest that Chapter 9 is devoted to their numerical approximation.

Definition 7.14 The **spectral radius** $\rho(A)$ of a matrix A is defined by

$$\rho(A) = \max |\lambda|, \quad \text{where } \lambda \text{ is an eigenvalue of } A.$$

(Recall that for complex $\lambda = \alpha + \beta i$, we have $|\lambda| = (\alpha^2 + \beta^2)^{1/2}$.) ■

For the matrix considered in Example 1,

$$\rho(A) = \max\{1, |1 + \sqrt{3}i|, |1 - \sqrt{3}i|\} = \max\{1, 2, 2\} = 2.$$

The spectral radius is closely related to the norm of a matrix, as shown in the following theorem.

Theorem 7.15 If A is an $n \times n$ matrix, then

(i) $\|A\|_2 = [\rho(A^t A)]^{1/2}$,

(ii) $\rho(A) \leq \|A\|$, for any natural norm $\|\cdot\|$.

Proof The proof of part (i) requires more information concerning eigenvalues than we presently have available. For the details involved in the proof, see [Or2, p. 21].

To prove part (ii), suppose λ is an eigenvalue of A with eigenvector $\mathbf{x}$ where $\|\mathbf{x}\| = 1$. (Exercise 6 ensures that such an eigenvector exists.) Since $A\mathbf{x} = \lambda\mathbf{x}$, for any natural norm

$$|\lambda| = |\lambda| \cdot \|\mathbf{x}\| = \|\lambda\mathbf{x}\| = \|A\mathbf{x}\| \leq \|A\|\|\mathbf{x}\| = \|A\|.$$

Thus,

$$\rho(A) = \max|\lambda| \leq \|A\|.$$

▪ ▪ ▪

An interesting and useful result, which is similar to part (ii) of Theorem 7.15, is that for any matrix A and any $\varepsilon > 0$, there exists a natural norm $\|\cdot\|$ with the property that $\rho(A) < \|A\| < \rho(A) + \varepsilon$. Consequently, $\rho(A)$ is the greatest lower bound for the natural norms on A. The proof of this result can be found in [Or2, p. 23].

EXAMPLE 2 If

$$A = \begin{bmatrix} 1 & 1 & 0 \\ 1 & 2 & 1 \\ -1 & 1 & 2 \end{bmatrix},$$

then

$$A^tA = \begin{bmatrix} 1 & 1 & -1 \\ 1 & 2 & 1 \\ 0 & 1 & 2 \end{bmatrix} \begin{bmatrix} 1 & 1 & 0 \\ 1 & 2 & 1 \\ -1 & 1 & 2 \end{bmatrix} = \begin{bmatrix} 3 & 2 & -1 \\ 2 & 6 & 4 \\ -1 & 4 & 5 \end{bmatrix}.$$

To calculate $\rho(A^tA)$, we need the eigenvalues of A^tA. If

$$0 = \det(A^tA - \lambda I) = \det \begin{bmatrix} 3-\lambda & 2 & -1 \\ 2 & 6-\lambda & 4 \\ -1 & 4 & 5-\lambda \end{bmatrix}$$

$$= -\lambda^3 + 14\lambda^2 - 42\lambda = -\lambda(\lambda^2 - 14\lambda + 42),$$

then

$$\lambda = 0 \quad \text{or} \quad \lambda = 7 \pm \sqrt{7},$$

so

$$\|A\|_2 = \sqrt{\rho(A^tA)} = \sqrt{\max\{0, 7-\sqrt{7}, 7+\sqrt{7}\}} = \sqrt{7+\sqrt{7}} \approx 3.106.$$

▪

Matrix norms can be computed in Maple using the command `norm`. The following commands illustrate the procedure for the l_2 norm:

```
>with(linalg);
>A:=matrix(3,3,[1,1,0,1,2,1,-1,1,2]);
>norm(A,2);
```

$$\sqrt{7+\sqrt{7}}$$

To determine the l_∞ norm of A, replace the last command with `norm(A,infinity);`.

In studying iterative matrix techniques, it is of particular importance to know when powers of a matrix become small (that is, when all the entries approach zero). Matrices of this type are called **convergent**.

Definition 7.16 We call an $n \times n$ matrix A **convergent** if

$$\lim_{k\to\infty}(A^k)_{ij} = 0, \quad \text{for each } i = 1, 2, \ldots, n \text{ and } j = 1, 2, \ldots, n. \qquad \blacksquare$$

EXAMPLE 3 Let

$$A = \begin{bmatrix} \frac{1}{2} & 0 \\ \frac{1}{4} & \frac{1}{2} \end{bmatrix}.$$

Computing powers of A, we obtain:

$$A^2 = \begin{bmatrix} \frac{1}{4} & 0 \\ \frac{1}{4} & \frac{1}{4} \end{bmatrix}, \quad A^3 = \begin{bmatrix} \frac{1}{8} & 0 \\ \frac{3}{16} & \frac{1}{8} \end{bmatrix}, \quad A^4 = \begin{bmatrix} \frac{1}{16} & 0 \\ \frac{1}{8} & \frac{1}{16} \end{bmatrix},$$

and, in general,

$$A^k = \begin{bmatrix} (\frac{1}{2})^k & 0 \\ \frac{k}{2^{k+1}} & (\frac{1}{2})^k \end{bmatrix}.$$

Since

$$\lim_{k\to\infty}\left(\frac{1}{2}\right)^k = 0 \quad \text{and} \quad \lim_{k\to\infty}\frac{k}{2^{k+1}} = 0,$$

A is a convergent matrix. $\blacksquare$

The convergent matrix A in Example 3 has $\rho(A) = \frac{1}{2}$, since $\frac{1}{2}$ is the only eigenvalue of A. This illustrates the important connection that exists between the spectral radius of a matrix and the convergence of the matrix, as detailed in the following result.

Theorem 7.17 The following statements are equivalent.

(i) A is a convergent matrix.

(ii) $\lim_{n\to\infty}\|A^n\| = 0$, for some natural norm.

(iii) $\lim_{n\to\infty}\|A^n\| = 0$, for all natural norms.

(iv) $\rho(A) < 1$.

(v) $\lim_{n\to\infty} A^n\mathbf{x} = \mathbf{0}$, for every $\mathbf{x}$. ■

The proof of this theorem can be found in [IK, p. 14].

EXERCISE SET 7.2

1. Compute the eigenvalues and associated eigenvectors of the following matrices.

   a. $\begin{bmatrix} 2 & -1 \\ -1 & 2 \end{bmatrix}$ **b.** $\begin{bmatrix} 0 & 1 \\ 1 & 1 \end{bmatrix}$

   c. $\begin{bmatrix} 0 & \frac{1}{2} \\ \frac{1}{2} & 0 \end{bmatrix}$ **d.** $\begin{bmatrix} 1 & 1 \\ -2 & -2 \end{bmatrix}$

   e. $\begin{bmatrix} 2 & 1 & 0 \\ 1 & 2 & 0 \\ 0 & 0 & 3 \end{bmatrix}$ **f.** $\begin{bmatrix} -1 & 2 & 0 \\ 0 & 3 & 4 \\ 0 & 0 & 7 \end{bmatrix}$

   g. $\begin{bmatrix} 2 & 1 & 1 \\ 2 & 3 & 2 \\ 1 & 1 & 2 \end{bmatrix}$ **h.** $\begin{bmatrix} 3 & 2 & -1 \\ 1 & -2 & 3 \\ 2 & 0 & 4 \end{bmatrix}$

2. Find the spectral radius for each matrix in Exercise 1.

3. Which of the matrices in Exercise 1 are convergent?

4. Show that

   $$A_1 = \begin{bmatrix} 1 & 0 \\ \frac{1}{4} & \frac{1}{2} \end{bmatrix}$$

   is not convergent, but

   $$A_2 = \begin{bmatrix} \frac{1}{2} & 0 \\ 16 & \frac{1}{2} \end{bmatrix}$$

   is convergent.

5. Find $\|\cdot\|_2$ for the matrices in Exercise 1.

6. Show that if λ is a eigenvalue of a matrix A and $\|\cdot\|$ is a vector norm, then an eigenvector $\mathbf{x}$ associated with λ exists with $\|\mathbf{x}\| = 1$.

7. Show that the characteristic polynomial $p(\lambda) = \det(A - \lambda I)$ for the $n \times n$ matrix A is an nth-degree polynomial. [*Hint*: Expand $\det(A - \lambda I)$ along the first row and use mathematical induction on n.]

8. **a.** Show that if A is an $n \times n$ matrix, then

   $$\det A = \prod_{i=1}^{n} \lambda_i,$$

   where $\lambda_1, \ldots, \lambda_n$ are the eigenvalues of A. [*Hint*: Consider $p(0)$.]

   b. Show that A is singular if and only if $\lambda = 0$ is an eigenvalue of A.

9. Show that if A is symmetric, then $\|A\|_2 = \rho(A)$.

10. Let λ be an eigenvalue of the $n \times n$ matrix A and $\mathbf{x} \neq \mathbf{0}$ be an associated eigenvector.

a. Show that λ is also an eigenvalue of A^t.

b. Show for any integer $k \geq 1$ that λ^k is an eigenvalue of A^k with eigenvector $\mathbf{x}$.

c. Show that if A^{-1} exists, then $1/\lambda$ is an eigenvalue of A^{-1} with eigenvector $\mathbf{x}$.

d. Generalize parts (b) and (c) to $(A^{-1})^k$ for integers $k \geq 2$.

e. Given the polynomial $q(x) = q_0 + q_1 x + \cdots + q_k x^k$, define $q(A)$ to be the matrix $q(A) = q_0 I + q_1 A + \cdots + q_k A^k$. Show that $q(\lambda)$ is an eigenvalue of $q(A)$ with eigenvector $\mathbf{x}$.

f. Let $\alpha \neq \lambda$ be given. Show that if $A - \alpha I$ is a nonsingular, then $1/(\lambda - \alpha)$ is an eigenvalue of $(A - \alpha I)^{-1}$ with eigenvector $\mathbf{x}$.

11. In Exercise 11 of Section 6.3, we assumed that the contribution a female beetle of a certain type made to the future years' beetle population could be expressed in terms of the matrix

$$A = \begin{bmatrix} 0 & 0 & 6 \\ \frac{1}{2} & 0 & 0 \\ 0 & \frac{1}{3} & 0 \end{bmatrix},$$

where the entry in the ith row and jth column represents the probabilistic contribution of a beetle of age j onto the next year's female population of age i.

a. Does the matrix A have any real eigenvalues? If so, determine them and any associated eigenvectors.

b. If a sample of this species was needed for laboratory test purposes that would have a constant proportion in each age group from year to year, what criteria could be imposed on the initial population to ensure that this requirement would be satisfied?

12. Find matrices A and B for which $\rho(A + B) > \rho(A) + \rho(B)$. (This shows that $\rho(A)$ cannot be a matrix norm.)

13. Show that if $||\cdot||$ is any natural norm, then $(1/||A^{-1}||) \leq |\lambda| \leq ||A||$ for any eigenvalue λ of the nonsingular matrix A.

7.3 Iterative Techniques for Solving Linear Systems

An iterative technique to solve the linear system $A\mathbf{x} = \mathbf{b}$ starts with an initial approximation $\mathbf{x}^{(0)}$ to the solution $\mathbf{x}$ and generates a sequence of vectors $\{\mathbf{x}^{(k)}\}_{k=0}^{\infty}$ that converges to $\mathbf{x}$. Iterative techniques involve a process that converts the system $A\mathbf{x} = \mathbf{b}$ into an equivalent system of the form $\mathbf{x} = T\mathbf{x} + \mathbf{c}$ for some fixed matrix T and vector $\mathbf{c}$. After the initial vector $\mathbf{x}^{(0)}$ is selected, the sequence of approximate solution vectors is generated by computing

$$\mathbf{x}^{(k)} = T\mathbf{x}^{(k-1)} + \mathbf{c}$$

for each $k = 1, 2, 3, \ldots$. This result should be reminiscent of the fixed-point iteration studied in Chapter 2.

Iterative techniques are seldom used for solving linear systems of small dimension since the time required for sufficient accuracy exceeds that required for direct techniques such as the Gaussian elimination method. For large systems with a high percentage of zero entries, however, these techniques are efficient in terms of both computer storage and computational time. Systems of this type arise frequently in circuit analysis and in the numerical solution of boundary-value problems and partial-differential equations.

EXAMPLE 1 The linear system $A\mathbf{x} = \mathbf{b}$ given by

$$\begin{aligned}
E_1: &\quad 10x_1 - x_2 + 2x_3 &= 6,\\
E_2: &\quad -x_1 + 11x_2 - x_3 + 3x_4 &= 25,\\
E_3: &\quad 2x_1 - x_2 + 10x_3 - x_4 &= -11,\\
E_4: &\quad 3x_2 - x_3 + 8x_4 &= 15
\end{aligned}$$

has the unique solution $\mathbf{x} = (1, 2, -1, 1)^t$. To convert $A\mathbf{x} = \mathbf{b}$ to the form $\mathbf{x} = T\mathbf{x} + \mathbf{c}$, solve equation E_i for x_i for each $i = 1, 2, 3, 4$, to obtain

$$\begin{aligned}
x_1 &= \frac{1}{10}x_2 - \frac{1}{5}x_3 + \frac{3}{5},\\
x_2 &= \frac{1}{11}x_1 + \frac{1}{11}x_3 - \frac{3}{11}x_4 + \frac{25}{11},\\
x_3 &= -\frac{1}{5}x_1 + \frac{1}{10}x_2 + \frac{1}{10}x_4 - \frac{11}{10},\\
x_4 &= -\frac{3}{8}x_2 + \frac{1}{8}x_3 + \frac{15}{8}.
\end{aligned}$$

To write this in the form $\mathbf{x} = T\mathbf{x} + \mathbf{c}$, we use

$$T = \begin{bmatrix} 0 & \frac{1}{10} & -\frac{1}{5} & 0 \\ \frac{1}{11} & 0 & \frac{1}{11} & -\frac{3}{11} \\ -\frac{1}{5} & \frac{1}{10} & 0 & \frac{1}{10} \\ 0 & -\frac{3}{8} & \frac{1}{8} & 0 \end{bmatrix} \quad \text{and} \quad \mathbf{c} = \begin{bmatrix} \frac{3}{5} \\ \frac{25}{11} \\ -\frac{11}{10} \\ \frac{15}{8} \end{bmatrix}.$$

For an initial approximation, we let $\mathbf{x}^{(0)} = (0, 0, 0, 0)^t$. Then $\mathbf{x}^{(1)}$ is given by

$$\begin{aligned}
x_1^{(1)} &= \frac{1}{10}x_2^{(0)} - \frac{1}{5}x_3^{(0)} + \frac{3}{5} = 0.6000,\\
x_2^{(1)} &= \frac{1}{11}x_1^{(0)} + \frac{1}{11}x_3^{(0)} - \frac{3}{11}x_4^{(0)} + \frac{25}{11} = 2.2727,\\
x_3^{(1)} &= -\frac{1}{5}x_1^{(0)} + \frac{1}{10}x_2^{(0)} + \frac{1}{10}x_4^{(0)} - \frac{11}{10} = -1.1000,\\
x_4^{(1)} &= -\frac{3}{8}x_2^{(0)} + \frac{1}{8}x_3^{(0)} + \frac{15}{8} = 1.8750.
\end{aligned}$$

Additional iterates, $\mathbf{x}^{(k)} = (x_1^{(k)}, x_2^{(k)}, x_3^{(k)}, x_4^{(k)})^t$, are generated in a similar manner and are presented in Table 7.1.

Table 7.1

k	0	1	2	3	4	5	6	7	8	9	10
$x_1^{(k)}$	0.0000	0.6000	1.0473	0.9326	1.0152	0.9890	1.0032	0.9981	1.0006	0.9997	1.0001
$x_2^{(k)}$	0.0000	2.2727	1.7159	2.0533	1.9537	2.0114	1.9922	2.0023	1.9987	2.0004	1.9998
$x_3^{(k)}$	0.0000	−1.1000	−0.8052	−1.0493	−0.9681	−1.0103	−0.9945	−1.0020	−0.9990	−1.0004	−0.9998
$x_4^{(k)}$	0.0000	1.8750	0.8852	1.1309	0.9739	1.0214	0.9944	1.0036	0.9989	1.0006	0.9998

The decision to stop after ten iterations was based on the criterion

$$\frac{\|\mathbf{x}^{(10)} - \mathbf{x}^{(9)}\|_\infty}{\|\mathbf{x}^{(10)}\|_\infty} = \frac{8.0 \times 10^{-4}}{1.9998} < 10^{-3}.$$

In fact, $\|\mathbf{x}^{(10)} - \mathbf{x}\|_\infty = 0.0002$. ■

The method of Example 1 is called the **Jacobi iterative method**. It consists of solving the ith equation in $A\mathbf{x} = \mathbf{b}$ for x_i to obtain (provided $a_{ii} \neq 0$)

$$x_i = \sum_{\substack{j=1 \\ j\neq i}}^{n} \left(-\frac{a_{ij}x_j}{a_{ii}} \right) + \frac{b_i}{a_{ii}}, \qquad \text{for } i = 1, 2, \ldots, n$$

and generating each $x_i^{(k)}$ from components of $\mathbf{x}^{(k-1)}$ for $k \geq 1$ by

$$x_i^{(k)} = \frac{\sum_{\substack{j=1 \\ j\neq i}}^{n} \left(-a_{ij}x_j^{(k-1)} \right) + b_i}{a_{ii}}, \qquad \text{for } i = 1, 2, \ldots, n. \tag{7.3}$$

The method is written in the form $\mathbf{x}^{(k)} = T\mathbf{x}^{(k-1)} + \mathbf{c}$ by splitting A into its diagonal and off-diagonal parts. To see this, let D be the diagonal matrix whose diagonal is the same as A, $-L$ be the strictly lower-triangular part of A, and $-U$ be the strictly upper-triangular part of A. With this notation, A is split into

$$A = \begin{bmatrix} a_{11} & a_{12} & \cdots & a_{1n} \\ a_{21} & a_{22} & \cdots & a_{2n} \\ \vdots & \vdots & & \vdots \\ a_{n1} & a_{n2} & \cdots & a_{nn} \end{bmatrix}$$

$$= \begin{bmatrix} a_{11} & 0 & \cdots & 0 \\ 0 & a_{22} & \ddots & \vdots \\ \vdots & \ddots & \ddots & 0 \\ 0 & \cdots & 0 & a_{nn} \end{bmatrix} - \begin{bmatrix} 0 & \cdots & \cdots & 0 \\ -a_{21} & \ddots & & \vdots \\ \vdots & \ddots & \ddots & \vdots \\ -a_{n1} & \cdots & -a_{n,n-1} & 0 \end{bmatrix} - \begin{bmatrix} 0 & -a_{12} & \cdots & -a_{1n} \\ \vdots & \ddots & \ddots & \vdots \\ \vdots & & \ddots & -a_{n-1,n} \\ 0 & \cdots & \cdots & 0 \end{bmatrix}$$

$$= D \qquad\qquad - L \qquad\qquad - U.$$

The equation $A\mathbf{x} = \mathbf{b}$, or $(D - L - U)\mathbf{x} = \mathbf{b}$, is then transformed into

$$D\mathbf{x} = (L + U)\mathbf{x} + \mathbf{b}$$

and, finally,

$$\mathbf{x} = D^{-1}(L + U)\mathbf{x} + D^{-1}\mathbf{b}.$$

This results in the matrix form of the Jacobi iterative technique:

(7.4) $$\mathbf{x}^{(k)} = D^{-1}(L + U)\mathbf{x}^{(k-1)} + D^{-1}\mathbf{b}, \quad k = 1, 2, \ldots.$$

Introducing the notation $T_j = D^{-1}(L + U)$ and $\mathbf{c}_j = D^{-1}\mathbf{b}$, the Jacobi technique has the form

(7.5) $$\mathbf{x}^{(k)} = T_j\mathbf{x}^{(k-1)} + \mathbf{c}_j.$$

In practice, Eq. (7.3) is used in computation and Eq. (7.5) for theoretical purposes.

Algorithm 7.1 implements the Jacobi iterative technique.

ALGORITHM 7.1

Jacobi Iterative

To solve $A\mathbf{x} = \mathbf{b}$ given an initial approximation $\mathbf{x}^{(0)}$:

INPUT the number of equations and unknowns n; the entries a_{ij}, $1 \le i, j \le n$ of the matrix A; the entries b_i, $1 \le i \le n$ of $\mathbf{b}$; the entries XO_i, $1 \le i \le n$ of $\mathbf{XO} = \mathbf{x}^{(0)}$; tolerance TOL; maximum number of iterations N.

OUTPUT the approximate solution $x_1, \ldots, x_n$ or a message that the number of iterations was exceeded.

Step 1 Set $k = 1$.

Step 2 While $(k \le N)$ do Steps 3–6.

Step 3 For $i = 1, \ldots, n$

$$\text{set } x_i = \frac{-\sum_{\substack{j=1 \\ j \ne i}}^{n} (a_{ij}XO_j) + b_i}{a_{ii}}.$$

Step 4 If $||\mathbf{x} - \mathbf{XO}|| < TOL$ then OUTPUT $(x_1, \ldots, x_n)$;
(*Procedure completed successfully.*)
STOP.

Step 5 Set $k = k + 1$.

Step 6 For $i = 1, \ldots, n$ set $XO_i = x_i$.

Step 7 OUTPUT ('Maximum number of iterations exceeded');
(*Procedure completed unsuccessfully.*)
STOP.

Step 3 of the algorithm requires that $a_{ii} \neq 0$ for each $i = 1, 2, \ldots, n$. If one of the a_{ii} entries is zero and the system is nonsingular, a reordering of the equations can be performed so that no $a_{ii} = 0$. To speed convergence, the equations should be arranged so that a_{ii} is as large as possible. This subject is discussed in more detail later in this chapter.

Another possible stopping criterion in Step 4 is to iterate until

$$\frac{\|\mathbf{x}^{(k)} - \mathbf{x}^{(k-1)}\|}{\|\mathbf{x}^{(k)}\|}$$

is smaller than some prescribed tolerance $\varepsilon > 0$. For this purpose, any convenient norm can be used, the usual being the l_∞ norm.

An improvement in Algorithm 7.1 is suggested by an analysis of Eq. (7.3). To compute $x_i^{(k)}$, the components of $\mathbf{x}^{(k-1)}$ are used. Since, for $i > 1$, $x_1^{(k)}, \ldots, x_{i-1}^{(k)}$ have already been computed and are likely to be better approximations to the actual solutions $x_1, \ldots, x_{i-1}$ than $x_1^{(k-1)}, \ldots, x_{i-1}^{(k-1)}$, it seems more reasonable to compute $x_i^{(k)}$ using these most recently calculated values; that is,

$$x_i^{(k)} = \frac{-\sum_{j=1}^{i-1}(a_{ij}x_j^{(k)}) - \sum_{j=i+1}^{n}(a_{ij}x_j^{(k-1)}) + b_i}{a_{ii}}, \tag{7.6}$$

for each $i = 1, 2, \ldots, n$, instead of Eq. (7.3). This modification is called the **Gauss-Seidel iterative technique** and is illustrated in the following example.

EXAMPLE 2 The linear system given by

$$\begin{aligned}
10x_1 - x_2 + 2x_3 &= 6, \\
-x_1 + 11x_2 - x_3 + 3x_4 &= 25, \\
2x_1 - x_2 + 10x_3 - x_4 &= -11, \\
3x_2 - x_3 + 8x_4 &= 15
\end{aligned}$$

was solved in Example 1 by the Jacobi iterative method. Incorporating Eq. (7.6) into Algorithm 7.1 gives the equations to be used for each $k = 1, 2, \ldots,$

$$\begin{aligned}
x_1^{(k)} &= \frac{1}{10}x_2^{(k-1)} - \frac{1}{5}x_3^{(k-1)} + \frac{3}{5}, \\
x_2^{(k)} &= \frac{1}{11}x_1^{(k)} + \frac{1}{11}x_3^{(k-1)} - \frac{3}{11}x_4^{(k-1)} + \frac{25}{11}, \\
x_3^{(k)} &= -\frac{1}{5}x_1^{(k)} + \frac{1}{10}x_2^{(k)} + \frac{1}{10}x_4^{(k-1)} - \frac{11}{10}, \\
x_4^{(k)} &= -\frac{3}{8}x_2^{(k)} + \frac{1}{8}x_3^{(k)} + \frac{15}{8}.
\end{aligned}$$

Letting $\mathbf{x}^{(0)} = (0, 0, 0, 0)^t$, we generate the iterates in Table 7.2.

Table 7.2

k	0	1	2	3	4	5
$x_1^{(k)}$	0.0000	0.6000	1.030	1.0065	1.0009	1.0001
$x_2^{(k)}$	0.0000	2.3272	2.037	2.0036	2.0003	2.0000
$x_3^{(k)}$	0.0000	−0.9873	−1.014	−1.0025	−1.0003	−1.0000
$x_4^{(k)}$	0.0000	0.8789	0.9844	0.9983	0.9999	1.0000

Since

$$\frac{\|\mathbf{x}^{(5)} - \mathbf{x}^{(4)}\|_\infty}{\|\mathbf{x}^{(5)}\|_\infty} = \frac{0.0008}{2.000} = 4 \times 10^{-4},$$

$\mathbf{x}^{(5)}$ is accepted as a reasonable approximation to the solution. Note that Jacobi's method in Example 1 required twice the iterations for the same degree of accuracy. ■

To write the Gauss-Seidel method in matrix form, multiply both sides of Eq. (7.6) by a_{ii} and collect all kth iterate terms to give

$$a_{i1}x_1^{(k)} + a_{i2}x_2^{(k)} + \cdots + a_{ii}x_i^{(k)} = -a_{i,i+1}x_{i+1}^{(k-1)} - \cdots - a_{in}x_n^{(k-1)} + b_i,$$

for each $i = 1, 2, \ldots, n$. Writing all n equations gives

$$\begin{aligned}
a_{11}x_1^{(k)} \qquad\qquad\qquad\qquad &= -a_{12}x_2^{(k-1)} - a_{13}x_3^{(k-1)} - \cdots - a_{1n}x_n^{(k-1)} + b_1,\\
a_{21}x_1^{(k)} + a_{22}x_2^{(k)} \qquad\qquad &= \qquad\qquad -a_{23}x_3^{(k-1)} - \cdots - a_{2n}x_n^{(k-1)} + b_2,\\
&\vdots\\
a_{n1}x_1^{(k)} + a_{n2}x_2^{(k)} + \cdots + a_{nn}x_n^{(k)} &= \qquad\qquad\qquad\qquad\qquad\qquad b_n;
\end{aligned}$$

and it follows that the matrix form of the Gauss-Seidel method is

$$(D - L)\mathbf{x}^{(k)} = U\mathbf{x}^{(k-1)} + \mathbf{b}$$

or

(7.7) $$\mathbf{x}^{(k)} = (D - L)^{-1}U\mathbf{x}^{(k-1)} + (D - L)^{-1}\mathbf{b}, \quad \text{for each } k = 1, 2, \ldots.$$

Letting $T_g = (D - L)^{-1}U$ and $\mathbf{c}_j = (D - L)^{-1}\mathbf{b}$, the Gauss-Seidel technique has the form

(7.8) $$\mathbf{x}^{(k)} = T_g\mathbf{x}^{(k-1)} + \mathbf{c}_g.$$

For the lower-triangular matrix $D - L$ to be nonsingular, it is necessary and sufficient that $a_{ii} \neq 0$ for each $i = 1, 2, \ldots, n$.

Algorithm 7.2 implements the Gauss-Seidel method.

ALGORITHM **7.2**

Gauss-Seidel Iterative

To solve $A\mathbf{x} = \mathbf{b}$ given an initial approximation $\mathbf{x}^{(0)}$:

INPUT the number of equations and unknowns n; the entries a_{ij}, $1 \le i, j \le n$ of the matrix A; the entries b_i, $1 \le i \le n$ of $\mathbf{b}$; the entries XO_i, $1 \le i \le n$ of $\mathbf{XO} = \mathbf{x}^{(0)}$; tolerance TOL; maximum number of iterations N.

OUTPUT the approximate solution $x_1, \ldots, x_n$ or a message that the number of iterations was exceeded.

Step 1 Set $k = 1$.

Step 2 While ($k \le N$) do Steps 3–6.

Step 3 For $i = 1, \ldots, n$

$$\text{set } x_i = \frac{-\sum_{j=1}^{i-1} a_{ij}x_j - \sum_{j=i+1}^{n} a_{ij}XO_j + b_i}{a_{ii}}.$$

Step 4 If $||\mathbf{x} - \mathbf{XO}|| < TOL$ then OUTPUT $(x_1, \ldots, x_n)$;
(*Procedure completed successfully.*)
STOP.

Step 5 Set $k = k + 1$.

Step 6 For $i = 1, \ldots, n$ set $XO_i = x_i$.

Step 7 OUTPUT ('Maximum number of iterations exceeded');
(*Procedure completed unsuccessfully.*)
STOP.

The comments following Algorithm 7.1 regarding reordering and stopping criteria also apply to the Gauss-Seidel Algorithm 7.2.

The results of Examples 1 and 2 appear to imply that the Gauss-Seidel method is superior to the Jacobi method. This is generally, but not always, true. There are linear systems for which the Jacobi method converges and the Gauss-Seidel method does not and others for which the Gauss-Seidel method converges and the Jacobi method does not. (See [Var, p. 74].)

To study the convergence of general iteration techniques, we consider the formula

$$\mathbf{x}^{(k)} = T\mathbf{x}^{(k-1)} + \mathbf{c}, \quad \text{for each } k = 1, 2, \ldots,$$

where $\mathbf{x}^{(0)}$ is arbitrary.

Lemma 7.18 If the spectral radius $\rho(T)$ satisfies $\rho(T) < 1$, then $(I - T)^{-1}$ exists, and

$$(I - T)^{-1} = I + T + T^2 + \cdots.$$

Proof Since $T\mathbf{x} = \lambda\mathbf{x}$ is true precisely when $(I - T)\mathbf{x} = (1 - \lambda)\mathbf{x}$ we have λ as an eigenvalue of T precisely when $1 - \lambda$ is an eigenvalue of $I - T$. But $|\lambda| \le \rho(T) < 1$, so $\lambda = 1$ is not an eigenvalue of T, and 0 cannot be an eigenvalue of $I - T$. Hence $I - T$ is nonsingular.

Let $S_m = I + T + T^2 + \cdots + T^m$. Then

$$(I - T)S_m = (1 + T + T^2 + \cdots + T^m) - (T + T^2 + \cdots + T^{m+1}) = I - T^{m+1}.$$

Since $\rho(T) < 1$, the result at the end of Section 7.2 implies that T is convergent and

$$\lim_{m\to\infty} (I - T)S_m = \lim_{m\to\infty} (I - T^{m+1}) = I.$$

Thus, $(I - T)^{-1} = \lim_{m\to\infty} S_m = I + T + T^2 + \cdots$. ■ ■ ■

Theorem 7.19 For any $\mathbf{x}^{(0)} \in \mathbb{R}^n$, the sequence $\{\mathbf{x}^{(k)}\}_{k=0}^{\infty}$ defined by

(7.9) $$\mathbf{x}^{(k)} = T\mathbf{x}^{(k-1)} + \mathbf{c}, \quad \text{for each } k \ge 1,$$

converges to the unique solution of $\mathbf{x} = T\mathbf{x} + \mathbf{c}$ if and only if $\rho(T) < 1$.

Proof First assume that $\rho(T) < 1$. From Eq. (7.9),

$$\begin{aligned}
\mathbf{x}^{(k)} &= T\mathbf{x}^{(k-1)} + \mathbf{c} \\
&= T(T\mathbf{x}^{(k-2)} + \mathbf{c}) + \mathbf{c} \\
&= T^2\mathbf{x}^{(k-2)} + (T + I)\mathbf{c} \\
&\ \ \vdots \\
&= T^k\mathbf{x}^{(0)} + (T^{k-1} + \cdots + T + I)\mathbf{c}.
\end{aligned}$$

Since $\rho(T) < 1$, the matrix T is convergent and

$$\lim_{k\to\infty} T^k\mathbf{x}^{(0)} = \mathbf{0}.$$

Lemma 7.18 implies that

$$\lim_{k\to\infty} \mathbf{x}^{(k)} = \lim_{k\to\infty} T^k\mathbf{x}^{(0)} + \lim_{k\to\infty} \left(\sum_{j=0}^{k-1} T^j\right)\mathbf{c} = \mathbf{0} + (I - T)^{-1}\mathbf{c} = (I - T)^{-1}\mathbf{c}.$$

Since $\mathbf{x} = T\mathbf{x} + \mathbf{c}$ implies that $(I - T)\mathbf{x} = \mathbf{c}$, the sequence $\{\mathbf{x}^{(k)}\}$ converges to the unique solution to the equation, the vector $\mathbf{x} = (I - T)^{-1}\mathbf{c}$.

To prove the converse, we show that for any $\mathbf{z} \in \mathbb{R}^n$ we have $\lim_{k\to\infty} T^k\mathbf{z} = \mathbf{0}$.

Let $\mathbf{x}$ be the unique solution to the equation $\mathbf{x} = T\mathbf{x}$, that is, Eq. (7.9) with $\mathbf{c} = \mathbf{0}$. For $\mathbf{x}^{(0)} = \mathbf{x} - \mathbf{z}$, we have

$$\lim_{k\to\infty} T^k\mathbf{z} = \lim_{k\to\infty} T^k(\mathbf{x} - \mathbf{x}^{(0)}) = \lim_{k\to\infty} T^{k-1}(T\mathbf{x} - T\mathbf{x}^{(0)}) = \lim_{k\to\infty} T^{k-1}(\mathbf{x} - \mathbf{x}^{(1)}).$$

Continuing in this manner, we have

$$\lim_{k\to\infty} T^k \mathbf{z} = \lim_{k\to\infty} T^{k-1}(\mathbf{x} - \mathbf{x}^{(1)}) = \lim_{k\to\infty} T^{k-2}(\mathbf{x} - \mathbf{x}^{(2)}) = \cdots = \lim_{k\to\infty}(\mathbf{x} - \mathbf{x}^{(k)}) = \mathbf{0}.$$

Since $\mathbf{z} \in \mathbb{R}^n$ was arbitrary, Theorem 7.17 implies that T is a convergent matrix and that $\rho(T) < 1$. ■ ■ ■

The proof of the following corollary is similar to the proofs in Corollary 2.4. It is considered in Exercise 9.

Corollary 7.20 If $\|T\| < 1$ for any natural matrix norm and $\mathbf{c}$ is a given vector, then the sequence $\{\mathbf{x}^{(k)}\}_{k=0}^{\infty}$ defined by $\mathbf{x}^{(k)} = T\mathbf{x}^{(k-1)} + \mathbf{c}$ converges, for any $\mathbf{x}^{(0)} \in \mathbb{R}^n$, to a vector $\mathbf{x} \in \mathbb{R}^n$, and the following error bounds hold:

(i) $\|\mathbf{x} - \mathbf{x}^{(k)}\| \le \|T\|^k \|\mathbf{x}^{(0)} - \mathbf{x}\|$.

(ii) $\|\mathbf{x} - \mathbf{x}^{(k)}\| \le \dfrac{\|T\|^k}{1 - \|T\|} \|\mathbf{x}^{(1)} - \mathbf{x}^{(0)}\|$. ■

We have seen that the Jacobi and Gauss-Seidel iterative techniques can be written

$$\mathbf{x}^{(k)} = T_j \mathbf{x}^{(k-1)} + \mathbf{c}_j \quad \text{and} \quad \mathbf{x}^{(k)} = T_g \mathbf{x}^{(k-1)} + \mathbf{c}_g,$$

using the matrices

$$T_j = D^{-1}(L + U) \quad \text{and} \quad T_g = (D - L)^{-1} U.$$

If $\rho(T_j)$ or $\rho(T_g)$ is less than 1, then the corresponding sequence $\{\mathbf{x}^{(k)}\}_{k=0}^{\infty}$ will converge to the solution $\mathbf{x}$ of $A\mathbf{x} = \mathbf{b}$. For example, the Jacobi scheme has

$$\mathbf{x}^{(k)} = D^{-1}(L + U)\mathbf{x}^{(k-1)} + D^{-1}\mathbf{b},$$

and, if $\{\mathbf{x}^{(k)}\}_{k=0}^{\infty}$ converges to $\mathbf{x}$, then

$$\mathbf{x} = D^{-1}(L + U)\mathbf{x} + D^{-1}\mathbf{b}$$

This implies that

$$D\mathbf{x} = (L + U)\mathbf{x} + \mathbf{b} \quad \text{and} \quad (D - L - U)\mathbf{x} = \mathbf{b}.$$

Since $D - L - U = A$, the solution $\mathbf{x}$ satisfies $A\mathbf{x} = \mathbf{b}$.

We can now give easily verified sufficiency conditions for convergence of the Jacobi and Gauss-Seidel methods. (To prove convergence for the Jacobi scheme, see Exercise 10, and for the Gauss-Seidel scheme, see [Or2, p. 120].)

Theorem 7.21 If A is strictly diagonally dominant, then for any choice of $\mathbf{x}^{(0)}$, both the Jacobi and Gauss-Seidel methods give sequences $\{\mathbf{x}^{(k)}\}_{k=0}^{\infty}$ that converge to the unique solution of $A\mathbf{x} = \mathbf{b}$. ■

The relationship of the rapidity of convergence to the spectral radius of the iteration matrix T can be seen from Corollary 7.20. Since the inequalities hold for any natural matrix norm, it follows from the statement after Theorem 7.15 that

(7.10) $$\|\mathbf{x}^{(k)} - \mathbf{x}\| \approx \rho(T)^k \|\mathbf{x}^{(0)} - \mathbf{x}\|.$$

Thus, it is desirable to select the iterative technique with minimal $\rho(T) < 1$ for a particular system $A\mathbf{x} = \mathbf{b}$.

No general results exist to tell which of the two techniques, Jacobi or Gauss-Seidel, will be most successful for an arbitrary linear system. In special cases, however, the answer is known, as is demonstrated in the following theorem. The proof of this result can be found in [Y, pp. 120–127].

Theorem 7.22 **(Stein-Rosenberg)**

If $a_{ij} \le 0$ for each $i \ne j$ and $a_{ii} > 0$ for each $i = 1, 2, \ldots, n$, then one and only one of the following statements holds:

a. $0 \le \rho(T_g) < \rho(T_j) < 1$.
b. $1 < \rho(T_j) < \rho(T_g)$.
c. $\rho(T_j) = \rho(T_g) = 0$.
d. $\rho(T_j) = \rho(T_g) = 1$. ■

For the special case described in Theorem 7.22, we see from part (a) that when one method gives convergence, then both give convergence, and the Gauss-Seidel method converges faster than the Jacobi method. Part (b) indicates that when one method diverges then both diverge, and the divergence is more pronounced for the Gauss-Seidel method.

Since the rate of convergence of a procedure depends on the spectral radius of the matrix associated with the method, one way to select a procedure to accelerate convergence is to choose a method whose associated matrix has minimal spectral radius. Before describing a procedure for selecting such a method, we need to introduce a new means of measuring the amount by which an approximation to the solution to a linear system differs from the true solution to the system. The method makes use of the vector described in the following definition.

Definition 7.23 Suppose $\tilde{\mathbf{x}} \in \mathbb{R}^n$ is an approximation to the solution of the linear system defined by $A\mathbf{x} = \mathbf{b}$. The **residual vector** for $\tilde{\mathbf{x}}$ with respect to this system is $\mathbf{r} = \mathbf{b} - A\tilde{\mathbf{x}}$. ■

In procedures such as the Jacobi or Gauss-Seidel methods, a residual vector is associated with each calculation of an approximation component to the solution vector. The object of the method is to generate a sequence of approximations that will cause the associated residual vectors to converge rapidly to zero. Suppose we let

$$\mathbf{r}_i^{(k)} = (r_{1i}^{(k)}, r_{2i}^{(k)}, \ldots, r_{ni}^{(k)})^t$$

denote the residual vector for the Gauss-Seidel method corresponding to the approximate solution vector $\mathbf{x}_i^{(k)}$ defined by

$$\mathbf{x}_i^{(k)} = (x_1^{(k)}, x_2^{(k)}, \ldots, x_{i-1}^{(k)}, x_i^{(k-1)}, \ldots, x_n^{(k-1)})^t.$$

The mth component of $\mathbf{r}_i^{(k)}$ is

$$r_{mi}^{(k)} = b_m - \sum_{j=1}^{i-1} a_{mj}x_j^{(k)} - \sum_{j=i}^{n} a_{mj}x_j^{(k-1)}, \tag{7.11}$$

or, equivalently,

$$r_{mi}^{(k)} = b_m - \sum_{j=1}^{i-1} a_{mj}x_j^{(k)} - \sum_{j=i+1}^{n} a_{mj}x_j^{(k-1)} - a_{mi}x_i^{(k-1)},$$

for each $m = 1, 2, \ldots, n$.

In particular, the ith component of $\mathbf{r}_i^{(k)}$ is

$$r_{ii}^{(k)} = b_i - \sum_{j=1}^{i-1} a_{ij}x_j^{(k)} - \sum_{j=i+1}^{n} a_{ij}x_j^{(k-1)} - a_{ii}x_i^{(k-1)};$$

so

$$a_{ii}x_i^{(k-1)} + r_{ii}^{(k)} = b_i - \sum_{j=1}^{i-1} a_{ij}x_j^{(k)} - \sum_{j=i+1}^{n} a_{ij}x_j^{(k-1)}. \tag{7.12}$$

Recall, however, that in the Gauss-Seidel method, $x_i^{(k)}$ is chosen to be

$$x_i^{(k)} = \frac{1}{a_{ii}}\left[b_i - \sum_{j=1}^{i-1} a_{ij}x_j^{(k)} - \sum_{j=i+1}^{n} a_{ij}x_j^{(k-1)}\right], \tag{7.13}$$

so Eq. (7.12) can be rewritten as

$$a_{ii}x_i^{(k-1)} + r_{ii}^{(k)} = a_{ii}x_i^{(k)}.$$

Consequently, the Gauss-Seidel method can be characterized as choosing $x_i^{(k)}$ to satisfy

$$x_i^{(k)} = x_i^{(k-1)} + \frac{r_{ii}^{(k)}}{a_{ii}}. \tag{7.14}$$

We can derive another connection between the residual vectors and the Gauss-Seidel technique. Consider the residual vector $\mathbf{r}_{i+1}^{(k)}$, associated with the vector $\mathbf{x}_{i+1}^{(k)} = (x_1^{(k)}, \ldots, x_i^{(k)}, x_{i+1}^{(k-1)}, \ldots, x_n^{(k-1)})^t$. By (7.11), the ith component of $\mathbf{r}_{i+1}^{(k)}$ is

$$\begin{aligned} r_{i,i+1}^{(k)} &= b_i - \sum_{j=1}^{i} a_{ij}x_j^{(k)} - \sum_{j=i+1}^{n} a_{ij}x_j^{(k-1)} \\ &= b_i - \sum_{j=1}^{i-1} a_{ij}x_j^{(k)} - \sum_{j=i+1}^{n} a_{ij}x_j^{(k-1)} - a_{ii}x_i^{(k)}. \end{aligned}$$

Equation (7.13) implies that $r_{i,i+1}^{(k)} = 0$. In a sense, then, the Gauss-Seidel technique is also characterized by choosing $x_i^{(k)}$ in such a way that the ith component of $\mathbf{r}_{i+1}^{(k)}$ is zero.

Reducing one coordinate of the residual vector to zero, however, is not generally the most efficient way to reduce the overall size of the vector $\mathbf{r}_{i+1}^{(k)}$. Instead, we need to choose $x_i^{(k)}$ so that $\|\mathbf{r}_{i+1}^{(k)}\|$ is small. Modifying the Gauss-Seidel procedure as given by Eq. (7.14) to

(7.15)
$$x_i^{(k)} = x_i^{(k-1)} + \omega \frac{r_{ii}^{(k)}}{a_{ii}}$$

for certain choices of positive ω reduces the norm of the residual vector and leads to significantly faster convergence.

Methods involving Eq. (7.15) are called **relaxation methods**. For choices of ω with $0 < \omega < 1$, the procedures are called **under-relaxation methods** and can be used to obtain convergence of some systems that are not convergent by the Gauss-Seidel method. For choices of ω with $1 < \omega$, the procedures are called **over-relaxation methods**, which are used to accelerate the convergence for systems that are convergent by the Gauss-Seidel technique. These methods are abbreviated **SOR**, for **Successive Over-Relaxation**, and are particularly useful for solving the linear systems that occur in the numerical solution of certain partial-differential equations.

Before illustrating the advantages of the SOR method, we note that by using Eq. (7.11) with $m = i$, Eq. (7.15) can be reformulated for calculation purposes to

$$x_i^{(k)} = (1-\omega)x_i^{(k-1)} + \frac{\omega}{a_{ii}}\left[b_i - \sum_{j=1}^{i-1} a_{ij}x_j^{(k)} - \sum_{j=i+1}^{n} a_{ij}x_j^{(k-1)}\right].$$

To determine the matrix of the SOR method, we rewrite this as

$$a_{ii}x_i^{(k)} + \omega\sum_{j=1}^{i-1} a_{ij}x_j^{(k)} = (1-\omega)a_{ii}x_i^{(k-1)} - \omega\sum_{j=i+1}^{n} a_{ij}x_j^{(k-1)} + \omega b_i$$

so

$$(D - \omega L)\mathbf{x}^{(k)} = [(1-\omega)D + \omega U]\mathbf{x}^{(k-1)} + \omega\mathbf{b}$$

or

(7.16)
$$\mathbf{x}^{(k)} = (D-\omega L)^{-1}[(1-\omega)D + \omega U]\mathbf{x}^{(k-1)} + \omega(D-\omega L)^{-1}\mathbf{b}.$$

If we let $T_\omega = (D-\omega L)^{-1}[(1-\omega)D + \omega U]$ and $\mathbf{c}_\omega = \omega(D-\omega L)^{-1}\mathbf{b}$ we can express the SOR technique in the form

(7.17)
$$\mathbf{x}^{(k)} = T_\omega \mathbf{x}^{(k-1)} + \mathbf{c}_\omega.$$

EXAMPLE 3 The linear system $A\mathbf{x} = \mathbf{b}$ given by

$$\begin{aligned} 4x_1 + 3x_2 \quad\quad &= 24, \\ 3x_1 + 4x_2 - \ x_3 &= 30, \\ - \ x_2 + 4x_3 &= -24, \end{aligned}$$

has the solution $(3, 4, -5)^t$. The Gauss-Seidel method and the SOR method with $\omega = 1.25$ will be used to solve this system, using $\mathbf{x}^{(0)} = (1, 1, 1)^t$ for both methods. The equations for the Gauss-Seidel method are

$$\begin{aligned} x_1^{(k)} &= -0.75x_2^{(k-1)} + 6, \\ x_2^{(k)} &= -0.75x_1^{(k)} + 0.25x_3^{(k-1)} + 7.5, \\ x_3^{(k)} &= 0.25x_2^{(k)} - 6, \end{aligned}$$

and the equations for the SOR method with $\omega = 1.25$ are

$$\begin{aligned} x_1^{(k)} &= -0.25x_1^{(k-1)} - 0.9375x_2^{(k-1)} + 7.5, \\ x_2^{(k)} &= -0.9375x_1^{(k)} - 0.25x_2^{(k-1)} + 0.3125x_3^{(k-1)} + 9.375, \\ x_3^{(k)} &= 0.3125x_2^{(k)} - 0.25x_3^{(k-1)} - 7.5. \end{aligned}$$

The first seven iterates for each method are listed in Tables 7.3 and 7.4. For the iterates to be accurate to seven decimal places, the Gauss-Seidel method requires 34 iterations, as opposed to 14 iterations for the over-relaxation method with $\omega = 1.25$. ■

Table 7.3 **Gauss-Seidel**

k	0	1	2	3	4	5	6	7
$x_1^{(k)}$	1	5.250000	3.1406250	3.0878906	3.0549316	3.0343323	3.0214577	3.0134110
$x_2^{(k)}$	1	3.812500	3.8828125	3.9267578	3.9542236	3.9713898	3.9821186	3.9888241
$x_3^{(k)}$	1	−5.046875	−5.0292969	−5.0183105	−5.0114441	−5.0071526	−5.0044703	−5.0027940

Table 7.4 **SOR with $\omega = 1.25$**

k	0	1	2	3	4	5	6	7
$x_1^{(k)}$	1	6.312500	2.6223145	3.1333027	2.9570512	3.0037211	2.9963276	3.0000498
$x_2^{(k)}$	1	3.5195313	3.9585266	4.0102646	4.0074838	4.0029250	4.0009262	4.0002586
$x_3^{(k)}$	1	−6.6501465	−4.6004238	−5.0966863	−4.9734897	−5.0057135	−4.9982822	−5.0003486

The obvious question to ask is how the appropriate value of ω is chosen. Although no complete answer to this question is known for the general $n \times n$ linear system, the following results can be used in certain situations.

Theorem 7.24 **(Kahan)**

If $a_{ii} \neq 0$ for each $i = 1, 2, \ldots, n$, then $\rho(T_\omega) \geq |\omega - 1|$. This implies that the SOR method can converge only if $0 < \omega < 2$. ■

The proof of this theorem is considered in Exercise 11. The proof of the next two results can be found in [Or2, pp. 123–133]. These results will be used in Chapter 12.

Theorem 7.25 **(Ostrowski-Reich)**

If A is a positive definite matrix and $0 < \omega < 2$, then the SOR method converges for any choice of initial approximate vector $\mathbf{x}^{(0)}$. ■

Theorem 7.26 If A is positive definite and tridiagonal, then $\rho(T_g) = [\rho(T_j)]^2 < 1$, and the optimal choice of ω for the SOR method is

$$\omega = \frac{2}{1 + \sqrt{1 - [\rho(T_j)]^2}}.$$

With this choice of ω, we have $\rho(T_\omega) = \omega - 1$. ■

EXAMPLE 4 In Example 3 the matrix was given by

$$A = \begin{bmatrix} 4 & 3 & 0 \\ 3 & 4 & -1 \\ 0 & -1 & 4 \end{bmatrix}.$$

This matrix is positive definite and tridiagonal, so Theorem 7.26 applies. Since

$$T_j = D^{-1}(L + U) = \begin{bmatrix} \frac{1}{4} & 0 & 0 \\ 0 & \frac{1}{4} & 0 \\ 0 & 0 & \frac{1}{4} \end{bmatrix} \begin{bmatrix} 0 & -3 & 0 \\ -3 & 0 & 1 \\ 0 & 1 & 0 \end{bmatrix} = \begin{bmatrix} 0 & -0.75 & 0 \\ -0.75 & 0 & 0.25 \\ 0 & 0.25 & 0 \end{bmatrix},$$

we have

$$T_j - \lambda I = \begin{bmatrix} -\lambda & -0.75 & 0 \\ -0.75 & -\lambda & 0.25 \\ 0 & 0.25 & -\lambda \end{bmatrix},$$

so

$$\det(T_j - \lambda I) = -\lambda(\lambda^2 - 0.625).$$

Thus,

$$\rho(T_j) = \sqrt{0.625}$$

and

$$\omega = \frac{2}{1+\sqrt{1-[\rho(T_j)]^2}} = \frac{2}{1+\sqrt{1-0.625}} \approx 1.24.$$

This explains the rapid convergence obtained in Example 1 by using $\omega = 1.25$. ■

We close this section with Algorithm 7.3 for the SOR method.

ALGORITHM 7.3

SOR

To solve $A\mathbf{x} = \mathbf{b}$ given the parameter ω and an initial approximation $\mathbf{x}^{(0)}$:

INPUT the number of equations and unknowns n; the entries a_{ij}, $1 \le i, j \le n$, of the matrix A; the entries b_i, $1 \le i \le n$, of $\mathbf{b}$; the entries XO_i, $1 \le i \le n$, of $\mathbf{XO} = \mathbf{x}^{(0)}$; the parameter ω; tolerance TOL; maximum number of iterations N.

OUTPUT the approximate solution $x_1, \ldots, x_n$ or a message that the number of iterations was exceeded.

Step 1 Set $k = 1$.

Step 2 While ($k \le N$) do Steps 3–6.

Step 3 For $i = 1, \ldots, n$

$$\text{set } x_i = (1-\omega)XO_i + \frac{\omega\left(-\sum_{j=1}^{i-1} a_{ij}x_j - \sum_{j=i+1}^{n} a_{ij}XO_j + b_i\right)}{a_{ii}}.$$

Step 4 If $||\mathbf{x} - \mathbf{XO}|| < TOL$ then OUTPUT $(x_1, \ldots, x_n)$;
(*Procedure completed successfully.*)
STOP.

Step 5 Set $k = k + 1$.

Step 6 For $i = 1, \ldots, n$ set $XO_i = x_i$.

Step 7 OUTPUT ('Maximum number of iterations exceeded');
(*Procedure completed unsuccessfully.*)
STOP.

EXERCISE SET 7.3

1. Find the first two iterations of the Jacobi method for the following linear systems, using $\mathbf{x}^{(0)} = \mathbf{0}$:

a.
$$\begin{aligned} 3x_1 - x_2 + x_3 &= 1, \\ 3x_1 + 6x_2 + 2x_3 &= 0, \\ 3x_1 + 3x_2 + 7x_3 &= 4. \end{aligned}$$

b.
$$\begin{aligned} 10x_1 - x_2 &= 9, \\ -x_1 + 10x_2 - 2x_3 &= 7, \\ -2x_2 + 10x_3 &= 6. \end{aligned}$$

c.
$$\begin{aligned} 10x_1 + 5x_2 &= 6, \\ 5x_1 + 10x_2 - 4x_3 &= 25, \\ -4x_2 + 8x_3 - x_4 &= -11, \\ -x_3 + 5x_4 &= -11. \end{aligned}$$

d.
$$\begin{aligned} 4x_1 + x_2 - x_3 + x_4 &= -2, \\ x_1 + 4x_2 - x_3 - x_4 &= -1, \\ -x_1 - x_2 + 5x_3 + x_4 &= 0, \\ x_1 - x_2 + x_3 + 3x_4 &= 1. \end{aligned}$$

e.
$$\begin{aligned} 4x_1 + x_2 + x_3 + x_5 &= 6,\\ -x_1 - 3x_2 + x_3 + x_4 &= 6,\\ 2x_1 + x_2 + 5x_3 - x_4 - x_5 &= 6,\\ -x_1 - x_2 - x_3 + 4x_4 &= 6,\\ 2x_2 - x_3 + x_4 + 4x_5 &= 6. \end{aligned}$$

f.
$$\begin{aligned} 4x_1 - x_2 - x_4 &= 0,\\ -x_1 + 4x_2 - x_3 - x_5 &= 5,\\ -x_2 + 4x_3 - x_6 &= 0,\\ -x_1 + 4x_4 - x_5 &= 6,\\ -x_2 - x_4 + 4x_5 - x_6 &= -2,\\ -x_3 - x_5 + 4x_6 &= 6. \end{aligned}$$

2. Repeat Exercise 1 using the Gauss-Seidel method.

3. Use the Jacobi method to solve the linear systems in Exercise 1, with $TOL = 10^{-3}$ in the l_∞ norm.

4. Repeat Exercise 3 using the Gauss-Seidel Algorithm.

5. Find the first two iterations of the SOR method with $\omega = 1.1$ for the following linear systems, using $\mathbf{x}^{(0)} = \mathbf{0}$:

a.
$$\begin{aligned} 3x_1 - x_2 + x_3 &= 1,\\ 3x_1 + 6x_2 + 2x_3 &= 0,\\ 3x_1 + 3x_2 + 7x_3 &= 4. \end{aligned}$$

b.
$$\begin{aligned} 10x_1 - x_2 &= 9,\\ -x_1 + 10x_2 - 2x_3 &= 7,\\ -2x_2 + 10x_3 &= 6. \end{aligned}$$

c.
$$\begin{aligned} 10x_1 + 5x_2 &= 6,\\ 5x_1 + 10x_2 - 4x_3 &= 25,\\ -4x_2 + 8x_3 - x_4 &= -11,\\ -x_3 + 5x_4 &= -11. \end{aligned}$$

d.
$$\begin{aligned} 4x_1 + x_2 - x_3 + x_4 &= -2,\\ x_1 + 4x_2 - x_3 - x_4 &= -1,\\ -x_1 - x_2 + 5x_3 + x_4 &= 0,\\ x_1 - x_2 + x_3 + 3x_4 &= 1. \end{aligned}$$

e.
$$\begin{aligned} 4x_1 + x_2 + x_3 + x_5 &= 6,\\ -x_1 - 3x_2 + x_3 + x_4 &= 6,\\ 2x_1 + x_2 + 5x_3 - x_4 - x_5 &= 6,\\ -x_1 - x_2 - x_3 + 4x_4 &= 6,\\ 2x_2 - x_3 + x_4 + 4x_5 &= 6. \end{aligned}$$

f.
$$\begin{aligned} 4x_1 - x_2 - x_4 &= 0,\\ -x_1 + 4x_2 - x_3 - x_5 &= 5,\\ -x_2 + 4x_3 - x_6 &= 0,\\ -x_1 + 4x_4 - x_5 &= 6,\\ -x_2 - x_4 + 4x_5 - x_6 &= -2,\\ -x_3 - x_5 + 4x_6 &= 6. \end{aligned}$$

6. Repeat Exercise 5 using $\omega = 1.3$.

7. Use the SOR method with $\omega = 1.2$ to solve the linear systems in Exercise 5 with a tolerance $TOL = 10^{-3}$ in the l_∞ norm.

8. Determine which matrices in Exercise 5 are tridiagonal and positive definite. Repeat Exercise 7 for these matrices using the optimal choice of ω.

9. **a.** Prove that

$$\|\mathbf{x}^{(k)} - \mathbf{x}\| \le \|T\|^k \, \|\mathbf{x}^{(0)} - \mathbf{x}\| \quad \text{and} \quad \|\mathbf{x}^{(k)} - \mathbf{x}\| \le \frac{\|T\|^k}{1 - \|T\|}\|\mathbf{x}^{(1)} - \mathbf{x}^{(0)}\|,$$

where T is an $n \times n$ matrix with $\|T\| < 1$ and

$$\mathbf{x}^{(k)} = T\mathbf{x}^{(k-1)} + \mathbf{c}, \quad k = 1, 2, \ldots,$$

with $\mathbf{x}^{(0)}$ arbitrary, $\mathbf{c} \in \mathbb{R}^n$, and $\mathbf{x} = T\mathbf{x} + \mathbf{c}$.

b. Apply the bounds to Exercise 1, when possible, using the l_∞ norm.

10. Show that is A is strictly diagonally dominant, then $\|T_j\|_\infty < 1$.

11. Prove Theorem 7.24. [*Hint*: If $\lambda_1, \ldots, \lambda_n$ are eigenvalues of T_ω, then $\det T_\omega = \prod_{i=1}^n \lambda_i$. Since $\det D^{-1} = \det(D - \omega L)^{-1}$ and the determinant of a product of matrices is the product of the determinants of the factors, the result follows from the definition of T_ω.]

12. Suppose that an object can be at any one of $n+1$ equally-spaced points $x_0, x_1, \ldots, x_n$ on a line. When an object is at location x_i, it is equally likely to move to either x_{i-1} or x_{i+1} and cannot directly move to any other location. Consider the probabilities $\{P_i\}_{i=0}^n$ that an object starting at location x_i will reach the left endpoint x_0 before reaching the right endpoint x_n. Clearly, $P_0 = 1$ and $P_n = 0$. Since the object can move to x_i only from x_{i-1} or x_{i+1} and does so with probability $\frac{1}{2}$ for each of these locations,

$$P_i = \frac{1}{2}P_{i-1} + \frac{1}{2}P_{i+1}, \quad \text{for each } i = 1, 2, \ldots, n-1.$$

a. Show that

$$\begin{bmatrix} 1 & -\frac{1}{2} & 0 & \cdots & \cdots & 0 \\ -\frac{1}{2} & 1 & -\frac{1}{2} & \ddots & & \vdots \\ 0 & -\frac{1}{2} & 1 & \ddots & \ddots & \vdots \\ \vdots & \ddots & \ddots & \ddots & \ddots & 0 \\ \vdots & & \ddots & -\frac{1}{2} & 1 & -\frac{1}{2} \\ 0 & \cdots & \cdots & 0 & -\frac{1}{2} & 1 \end{bmatrix} \begin{bmatrix} P_1 \\ P_2 \\ \vdots \\ P_{n-1} \end{bmatrix} = \begin{bmatrix} \frac{1}{2} \\ 0 \\ \vdots \\ 0 \end{bmatrix}.$$

b. Solve this system using $n = 10$, 50, and 100.

c. Change the probabilities to α and $1 - \alpha$ for movement to the left and right, respectively, and derive the linear system similar to the one in part (a).

d. Repeat part (b) with $\alpha = \frac{1}{3}$.

13. Use all the applicable methods in this section to find solutions to the linear system of $A\mathbf{x} = \mathbf{b}$ to within 10^{-5} in the l_∞ norm.

a.

$$a_{i,j} = \begin{cases} 4, & \text{when } j = i \text{ and } i = 1, 2, \ldots, 16, \\ -1, & \text{when } \begin{cases} j = i+1 \text{ and } i = 1, 2, 3, 5, 6, 7, 9, 10, 11, 13, 14, 15, \\ j = i-1 \text{ and } i = 2, 3, 4, 6, 7, 8, 10, 11, 12, 14, 15, 16, \\ j = i+4 \text{ and } i = 1, 2, \ldots, 12, \\ j = i-4 \text{ and } i = 5, 6, \ldots, 16, \end{cases} \\ 0, & \text{otherwise} \end{cases}$$

and

$$\mathbf{b} = (1.902207, 1.051143, 1.175689, 3.480083, 0.819600, -0.264419, -0.412789, 1.175689, 0.913337, -0.150209, -0.264419, 1.051143, 1.966694, 0.913337, 0.819600, 1.902207)^t$$

b.

$$a_{i,j} = \begin{cases} 4, & \text{when } j = i \text{ and } i = 1, 2, \ldots, 25, \\ -1, & \text{when } \begin{cases} j = i+1 \text{ and } i = \begin{cases} 1, 2, 3, 4, 6, 7, 8, 9, 11, 12, 13, 14, \\ 16, 17, 18, 19, 21, 22, 23, 24, \end{cases} \\ j = i-1 \text{ and } i = \begin{cases} 2, 3, 4, 5, 7, 8, 9, 10, 12, 13, 14, 15, \\ 17, 18, 19, 20, 22, 23, 24, 25, \end{cases} \\ j = i+5 \text{ and } i = 1, 2, \ldots, 20, \\ j = i-5 \text{ and } i = 6, 7, \ldots, 25, \end{cases} \\ 0, & \text{otherwise} \end{cases}$$

and

$$\mathbf{b} = (1, 0, -1, 0, 2, 1, 0, -1, 0, 2, 1, 0, -1, 0, 2, 1, 0, -1, 0, 2, 1, 0, -1, 0, 2)^t$$

c.

$$a_{i,j} = \begin{cases} 2i, & \text{when } j = i \text{ and } i = 1, 2, \ldots, 40, \\ -1, & \text{when } \begin{cases} j = i+1 \text{ and } i = 1, 2, \ldots, 39, \\ j = i-1 \text{ and } i = 2, 3, \ldots, 40, \end{cases} \\ 0, & \text{otherwise} \end{cases}$$

and $b_i = 1.5i - 6$ for each $i = 1, 2, \ldots, 40$

d.

$$a_{i,j} = \begin{cases} 2i, & \text{when } j = i \text{ and } i = 1, 2, \ldots, 80, \\ 0.5i, & \text{when } \begin{cases} j = i+2 \text{ and } i = 1, 2, \ldots, 78, \\ j = i-2 \text{ and } i = 3, 4, \ldots, 80, \end{cases} \\ 0.25i, & \text{when } \begin{cases} j = i+4 \text{ and } i = 1, 2, \ldots, 76, \\ j = i-4 \text{ and } i = 5, 6, \ldots, 80, \end{cases} \\ 0, & \text{otherwise} \end{cases}$$

and $b_i = \pi$ for each $i = 1, 2, \ldots, 80$

14. Suppose that A is positive definite.

a. Show that we can write $A = D - L - L^t$, where D is diagonal with $d_{ii} > 0$ for each $1 \le i \le n$ and L is lower triangular. Further, show that $D - L$ is nonsingular.

b. Let $T_g = (D - L)^{-1}L^t$ and $P = A - T_g^t A T_g$. Show that P is symmetric.

c. Show that T_g can also be written as $T_g = I - (D - L)^{-1}A$.

d. Let $Q = (D - L)^{-1}A$. Show that $T_g = I - Q$ and $P = Q^t[AQ^{-1} - A + (Q^t)^{-1}A]Q$.

e. Show that $P = Q^t DQ$ and P is positive definite.

f. Let λ be an eigenvalue of T_g with eigenvector $\mathbf{x} \neq \mathbf{0}$. Use part (b) to show that $\mathbf{x}^t P\mathbf{x} > 0$ implies that $|\lambda| < 1$.

g. Show that T_g is convergent and prove that the Gauss-Seidel method converges.

15. Extend the method of proof in Exercise 14 to the SOR method with $0 < \omega < 2$.

16. The forces on the bridge truss described in the opening to this chapter satisfy the equations in the following table :

Joint	Horizontal Component	Vertical Component
①	$-F_1 + \frac{\sqrt{2}}{2} f_1 + f_2 = 0$	$\frac{\sqrt{2}}{2} f_1 - F_2 = 0$
②	$-\frac{\sqrt{2}}{2} f_1 + \frac{\sqrt{3}}{2} f_4 = 0$	$-\frac{\sqrt{2}}{2} f_1 - f_3 + \frac{1}{2} f_4 = 0$
③	$-f_2 + f_5 = 0$	$f_3 - 10{,}000 = 0$
④	$-\frac{\sqrt{3}}{2} f_4 - f_5 = 0$	$\frac{1}{2} f_4 - F_3 = 0$

This linear system can be placed in the matrix form

$$\begin{bmatrix} -1 & 0 & 0 & \frac{\sqrt{2}}{2} & 1 & 0 & 0 & 0 \\ 0 & -1 & 0 & \frac{\sqrt{2}}{2} & 0 & 0 & 0 & 0 \\ 0 & 0 & -1 & 0 & 0 & 0 & \frac{1}{2} & 0 \\ 0 & 0 & 0 & -\frac{\sqrt{2}}{2} & 0 & -1 & \frac{1}{2} & 0 \\ 0 & 0 & 0 & 0 & -1 & 0 & 0 & 1 \\ 0 & 0 & 0 & 0 & 0 & 1 & 0 & 0 \\ 0 & 0 & 0 & -\frac{\sqrt{2}}{2} & 0 & 0 & \frac{\sqrt{3}}{2} & 0 \\ 0 & 0 & 0 & 0 & 0 & 0 & -\frac{\sqrt{3}}{2} & -1 \end{bmatrix} \begin{bmatrix} F_1 \\ F_2 \\ F_3 \\ f_1 \\ f_2 \\ f_3 \\ f_4 \\ f_5 \end{bmatrix} = \begin{bmatrix} 0 \\ 0 \\ 0 \\ 0 \\ 0 \\ 10{,}000 \\ 0 \\ 0 \end{bmatrix}.$$

a. Explain why the system of equations was reordered.

b. Approximate the solution of the resulting linear system to within 10^{-2} in the l_∞ norm using as initial approximation the vector all of whose entries are 1s and (i) the Gauss-Seidel method, (ii) the Jacobi method, and (iii) the SOR method with $\omega = 1.25$.

7.4 Error Estimates and Iterative Refinement

It seems intuitively reasonable that if $\tilde{\mathbf{x}}$ is an approximation to the solution $\mathbf{x}$ of $A\mathbf{x} = \mathbf{b}$ and the residual vector $\mathbf{r} = \mathbf{b} - A\tilde{\mathbf{x}}$ has the property that $\|\mathbf{r}\|$ is small, then $\|\mathbf{x} - \tilde{\mathbf{x}}\|$ would be small as well. This is often the case, but certain systems, which occur frequently in practice, fail to have this property.

EXAMPLE 1 The linear system $A\mathbf{x} = \mathbf{b}$ given by

$$\begin{bmatrix} 1 & 2 \\ 1.0001 & 2 \end{bmatrix} \begin{bmatrix} x_1 \\ x_2 \end{bmatrix} = \begin{bmatrix} 3 \\ 3.0001 \end{bmatrix}$$

has the unique solution $\mathbf{x} = (1, 1)^t$. The poor approximation $\tilde{\mathbf{x}} = (3, 0)^t$ has the residual vector

$$\mathbf{r} = \mathbf{b} - A\tilde{\mathbf{x}} = \begin{bmatrix} 3 \\ 3.0001 \end{bmatrix} - \begin{bmatrix} 1 & 2 \\ 1.0001 & 2 \end{bmatrix} \begin{bmatrix} 3 \\ 0 \end{bmatrix} = \begin{bmatrix} 0 \\ -0.0002 \end{bmatrix},$$

so $\|\mathbf{r}\|_\infty = 0.0002$. Although the norm of the residual vector is small, the approximation $\tilde{\mathbf{x}} = (3, 0)^t$ is obviously quite poor; in fact, $\|\mathbf{x} - \tilde{\mathbf{x}}\|_\infty = 2$. ■

The difficulty in Example 1 is explained quite simply by noting that the solution to the system represents the intersection of the lines

$$l_1: \quad x_1 + 2x_2 = 3 \quad \text{and} \quad l_2: \quad 1.0001x_1 + 2x_2 = 3.0001.$$

The point (3, 0) lies on l_1, and the lines are nearly parallel. This implies that (3, 0) also lies close to l_2, even though it differs significantly from the intersection point (1, 1). (See Figure 7.7.)

Figure 7.7

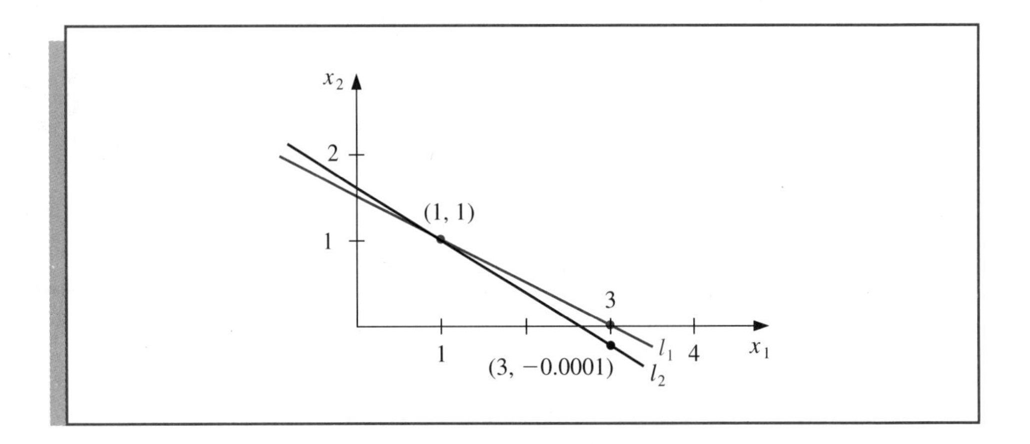

Example 1 was clearly constructed to show the difficulties that can—and, in fact do—arise. Had the lines not been nearly coincident, we would expect a small residual vector to imply an accurate approximation.

In the general situation, we cannot rely on the geometry of the system to give an indication of when problems might occur. We can, however, obtain this information by considering the norms of the matrix A and its inverse.

Theorem 7.27 Suppose that $\tilde{\mathbf{x}}$ is an approximation to the solution of $A\mathbf{x} = \mathbf{b}$, A is a nonsingular matrix, and $\mathbf{r}$ is the residual vector for $\tilde{\mathbf{x}}$. Then for any natural norm,

$$\|\mathbf{x} - \tilde{\mathbf{x}}\| \leq \|\mathbf{r}\| \cdot \|A^{-1}\|$$

and

$$\frac{\|\mathbf{x} - \tilde{\mathbf{x}}\|}{\|\mathbf{x}\|} \leq \|A\| \cdot \|A^{-1}\| \frac{\|\mathbf{r}\|}{\|\mathbf{b}\|}, \quad \text{provided } \mathbf{x} \neq \mathbf{0} \text{ and } \mathbf{b} \neq \mathbf{0}. \tag{7.18}$$

Proof Since $\mathbf{r} = \mathbf{b} - A\tilde{\mathbf{x}} = A\mathbf{x} - A\tilde{\mathbf{x}}$ and A is nonsingular, $\mathbf{x} - \tilde{\mathbf{x}} = A^{-1}\mathbf{r}$, Corollary 7.10 in Section 7.1 implies that

$$\|\mathbf{x} - \tilde{\mathbf{x}}\| = \|A^{-1}\mathbf{r}\| \leq \|A^{-1}\|\|\mathbf{r}\|.$$

Moreover, since $\mathbf{b} = A\mathbf{x}$, we have $\|\mathbf{b}\| \leq \|A\|\|\mathbf{x}\|$, so

$$\frac{1}{\|\mathbf{x}\|} \leq \frac{\|A\|}{\|\mathbf{b}\|}, \quad \text{and} \quad \frac{\|\mathbf{x} - \tilde{\mathbf{x}}\|}{\|\mathbf{x}\|} \leq \frac{\|A\|\|A^{-1}\|}{\|\mathbf{b}\|}\|\mathbf{r}\|.$$

▪ ▪ ▪

The inequalities in Theorem 7.27 imply that the quantities $\|A^{-1}\|$ and $\|A\|\|A^{-1}\|$ provide an indication of the connection between the residual vector and the accuracy of the approximation. In general, the relative error $\|\mathbf{x} - \tilde{\mathbf{x}}\|/\|\mathbf{x}\|$ is of most interest, and, by Inequality (7.18), this error is bounded by the product of $\|A\|\|A^{-1}\|$ with the relative residual for this approximation, $\|\mathbf{r}\|/\|\mathbf{b}\|$. Any convenient norm can be used for this approximation; the only requirement is that it be used consistently throughout.

Definition 7.28 The **condition number** of the nonsingular matrix A relative to a norm $\|\cdot\|$ is

$$K(A) = \|A\| \cdot \|A^{-1}\|.$$

■

With this notation, the inequalities in Theorem 7.27 become

$$\|\mathbf{x} - \tilde{\mathbf{x}}\| \leq K(A)\frac{\|\mathbf{r}\|}{\|A\|}$$

and

$$\frac{\|\mathbf{x} - \tilde{\mathbf{x}}\|}{\|\mathbf{x}\|} \leq K(A)\frac{\|\mathbf{r}\|}{\|\mathbf{b}\|}.$$

For any nonsingular matrix A and natural norm $\|\cdot\|$,

$$1 = \|I\| = \|A \cdot A^{-1}\| \leq \|A\| \cdot \|A^{-1}\| = K(A).$$

A matrix A is **well-conditioned** if $K(A)$ is close to 1, and is **ill-conditioned** when $K(A)$ is significantly greater than 1. Condition in this context refers to the relative security that a small residual vector implies a correspondingly accurate approximate solution.

EXAMPLE 2 The matrix for the system considered in Example 1 was

$$A = \begin{bmatrix} 1 & 2 \\ 1.0001 & 2 \end{bmatrix},$$

which has $\|A\|_\infty = 3.0001$. This norm would not be considered large. However,

$$A^{-1} = \begin{bmatrix} -10000 & 10000 \\ 5000.5 & -5000 \end{bmatrix}, \quad \text{so} \quad \|A^{-1}\|_\infty = 20000,$$

and, for the infinity norm, $K(A) = (20000)(3.0001) = 60002$. The size of the condition number for this example should certainly keep us from making hasty accuracy decisions based on the residual of an approximation. ■

In Maple the condition number K_∞ for the matrix in Example 2 can be computed as follows:

```
>with(linalg);
>A:=matrix(2,2,[1,2,1.0001,2]);
>cond(A);
```

$$60002.00000$$

Although the condition number of a matrix depends totally on the norms of the matrix and its inverse, in practice the calculation of the inverse is subject to roundoff error and is dependent on the accuracy with which the calculations are performed. If the operations involve arithmetic with t significant digits of accuracy, the approximate condition number for the matrix A is the norm of the matrix times the norm of the approximation to the inverse of A, which is obtained using t-digit arithmetic. In fact, this condition number also depends on the method used to calculate the inverse of A.

If we assume that the approximate solution to the linear system $A\mathbf{x} = \mathbf{b}$ is being determined using t-digit arithmetic and Gaussian elimination, it can be shown (see [FM, pp. 49–51]) that the residual vector $\mathbf{r}$ for the approximation $\tilde{\mathbf{x}}$ has

$$\|\mathbf{r}\| \approx 10^{-t}\|A\|\|\tilde{\mathbf{x}}\|. \tag{7.19}$$

From this approximation, an estimate for the effective condition number in t-digit arithmetic can be obtained without inverting the matrix A. In actuality, this approximation assumes that all the arithmetic operations in the Gaussian elimination technique are performed using t-digit arithmetic but that the operations needed to determine the residual are done in double-precision (that is, $2t$-digit) arithmetic. This technique does not add significantly to the computational effort and eliminates much of the loss of accuracy (see Section 1.2) involved with the subtraction of the nearly equal numbers that occur in the calculation of the residual.

The approximation for the t-digit condition number $K(A)$ comes from consideration of the linear system

$$A\mathbf{y} = \mathbf{r}.$$

The solution to this system can be readily approximated since the multipliers for the Gaussian elimination method have already been calculated. In fact $\tilde{\mathbf{y}}$, the approximate solution of $A\mathbf{y} = \mathbf{r}$, satisfies

$$\tilde{\mathbf{y}} \approx A^{-1}\mathbf{r} = A^{-1}(\mathbf{b} - A\tilde{\mathbf{x}}) = A^{-1}\mathbf{b} - A^{-1}A\tilde{\mathbf{x}} = \mathbf{x} - \tilde{\mathbf{x}}; \tag{7.20}$$

and

$$\mathbf{x} \approx \tilde{\mathbf{x}} + \tilde{\mathbf{y}}.$$

So $\tilde{\mathbf{y}}$ is an estimate of the error produced when $\tilde{\mathbf{x}}$ approximates the solution $\mathbf{x}$ to the original system. Equations (7.19) and (7.20) imply that

$$\|\tilde{\mathbf{y}}\| \approx \|\mathbf{x} - \tilde{\mathbf{x}}\| = \|A^{-1}\mathbf{r}\| \le \|A^{-1}\|\|\mathbf{r}\| \approx \|A^{-1}\| \left(10^{-t}\|A\|\|\tilde{\mathbf{x}}\|\right) = 10^{-t}\|\tilde{\mathbf{x}}\|K(A).$$

This gives an approximation for the condition number involved with solving the system $A\mathbf{x} = \mathbf{b}$ using Gaussian elimination and the t-digit type of arithmetic just described:

$$K(A) \approx \frac{\|\tilde{\mathbf{y}}\|}{\|\tilde{\mathbf{x}}\|}10^{t}. \tag{7.21}$$

EXAMPLE 3 The linear system given by

$$\begin{bmatrix} 3.3330 & 15920 & -10.333 \\ 2.2220 & 16.710 & 9.6120 \\ 1.5611 & 5.1791 & 1.6852 \end{bmatrix} \begin{bmatrix} x_1 \\ x_2 \\ x_3 \end{bmatrix} = \begin{bmatrix} 15913 \\ 28.544 \\ 8.4254 \end{bmatrix}$$

has the exact solution $\mathbf{x} = (1, 1, 1)^t$.

Using Gaussian elimination and five-digit rounding arithmetic leads successively to the augmented matrices

$$\begin{bmatrix} 3.3330 & 15920 & -10.333 & \vdots & 15913 \\ 0 & -10596 & 16.501 & \vdots & 10580 \\ 0 & -7451.4 & 6.5250 & \vdots & -7444.9 \end{bmatrix}$$

and

$$\begin{bmatrix} 3.3330 & 15920 & -10.333 & \vdots & 15913 \\ 0 & -10596 & 16.501 & \vdots & -10580 \\ 0 & 0 & -5.0790 & \vdots & -4.7000 \end{bmatrix}.$$

The approximate solution to this system is

$$\tilde{\mathbf{x}} = (1.2001, 0.99991, 0.92538)^t.$$

The residual vector corresponding to $\tilde{\mathbf{x}}$ is computed in double precision to be

$$\begin{aligned} \mathbf{r} &= \mathbf{b} - A\tilde{\mathbf{x}} \\ &= \begin{bmatrix} 15913 \\ 28.544 \\ 8.4254 \end{bmatrix} - \begin{bmatrix} 3.3330 & 15920 & -10.333 \\ 2.2220 & 16.710 & 9.6120 \\ 1.5611 & 5.1791 & 1.6852 \end{bmatrix} \begin{bmatrix} 1.20001 \\ 0.99991 \\ 0.92538 \end{bmatrix} \\ &= \begin{bmatrix} -0.00518 \\ 0.27413 \\ -0.18616 \end{bmatrix}; \end{aligned}$$

so

$$\|\mathbf{r}\|_\infty = 0.27413.$$

The estimate for the condition number given in the preceding discussion is obtained by first solving the system $A\mathbf{y} = \mathbf{r}$ for $\tilde{\mathbf{y}}$:

$$\begin{bmatrix} 3.3330 & 15920 & -10.333 \\ 2.2220 & 16.710 & 9.6120 \\ 1.5611 & 5.1791 & 1.6852 \end{bmatrix} \begin{bmatrix} y_1 \\ y_2 \\ y_3 \end{bmatrix} = \begin{bmatrix} -0.00518 \\ 0.27413 \\ -0.18616 \end{bmatrix}.$$

This implies that $\tilde{\mathbf{y}} = (-0.20008, 8.9987 \times 10^{-5}, 0.074607)^t$. Using the estimate in Eq. (7.21) gives

$$K(A) \approx \frac{\|\tilde{\mathbf{y}}\|_\infty}{\|\tilde{\mathbf{x}}\|_\infty} 10^5 = \frac{(0.20008)10^5}{1.2001} = 16672. \tag{7.22}$$

To determine the *exact* condition number of A, we first must construct A^{-1}. Using five-digit rounding arithmetic for the calculations gives the approximation:

$$A^{-1} = \begin{bmatrix} -1.1701 \times 10^{-4} & -1.4983 \times 10^{-1} & 8.5416 \times 10^{-1} \\ 6.2782 \times 10^{-5} & 1.2124 \times 10^{-4} & -3.0662 \times 10^{-4} \\ -8.6631 \times 10^{-5} & 1.3846 \times 10^{-1} & -1.9689 \times 10^{-1} \end{bmatrix}.$$

Theorem 7.11 implies that $\|A^{-1}\|_\infty = 1.0041$ and $\|A\|_\infty = 15934$.

As a consequence, the ill-conditioned matrix A has

$$K(A) = (1.0041)(15934) = 15999.$$

The estimate in (7.22) is quite close to $K(A)$ and requires considerably less computational effort.

Since the actual solution $\mathbf{x} = (1, 1, 1)^t$ is known for this system, we can calculate both.

$$\|\mathbf{x} - \tilde{\mathbf{x}}\|_\infty = 0.2001 \quad \text{and} \quad \frac{\|\mathbf{x} - \tilde{\mathbf{x}}\|_\infty}{\|\mathbf{x}\|_\infty} = \frac{0.2001}{1} = 0.2001.$$

The error bounds given in Theorem 7.27 for these values are

$$\|\mathbf{x} - \tilde{\mathbf{x}}\|_\infty \le K(A)\frac{\|\mathbf{r}\|_\infty}{\|A\|_\infty} = \frac{(15999)(0.27413)}{15934} = 0.27525$$

and

$$\frac{\|\mathbf{x} - \tilde{\mathbf{x}}\|_\infty}{\|\mathbf{x}\|_\infty} \le K(A)\frac{\|\mathbf{r}\|_\infty}{\|\mathbf{b}\|_\infty} = \frac{(15999)(0.27413)}{15913} = 0.27561. \quad \blacksquare$$

In Eq. (7.20), we used the estimate $\tilde{\mathbf{y}} \approx \mathbf{x} - \tilde{\mathbf{x}}$, where $\tilde{\mathbf{y}}$ is the approximate solution to the system $A\mathbf{y} = \mathbf{r}$. In general, $\tilde{\mathbf{x}} + \tilde{\mathbf{y}}$ is a more accurate approximation to the solution of the linear system $A\mathbf{x} = \mathbf{b}$ than the original approximation $\tilde{\mathbf{x}}$. The method using this assumption is called **iterative refinement**, or iterative improvement, and consists of performing iterations on the system whose right-hand side is the residual vector for successive approximations until satisfactory accuracy results.

If the process is applied using t-digit arithmetic and if $K_\infty(A) \approx 10^q$, then after k iterations of iterative refinement the solution has approximately the smaller of t and $k(t-q)$ correct digits. If the system is well-conditioned, one or two iterations will indicate that the solution is accurate. There may also be significant improvement on ill-conditioned systems unless the matrix A is so ill-conditioned that $K_\infty(A) > 10^t$. In that situation increased precision should be used for the calculations.

Algorithm 7.4 implements the Iterative Refinement technique.

ALGORITHM 7.4

Iterative Refinement

To approximate the solution to the linear system $A\mathbf{x} = \mathbf{b}$:

INPUT the number of equations and unknowns n; the entries a_{ij}, $1 \le i, j \le n$ of the matrix A; the entries b_i, $1 \le i \le n$ of $\mathbf{b}$; the maximum number of iterations N; tolerance TOL; number of digits of precision t.

OUTPUT the approximation $\mathbf{xx} = (xx_i, \ldots, xx_n)^t$ or a message that the number of iterations was exceeded, and an approximation *COND* to $K_\infty(A)$.

Step 0 Solve the system $A\mathbf{x} = \mathbf{b}$ for $x_1, \ldots, x_n$ by Gaussian elimination saving the multipliers m_{ji}, $j = i+1, i+2, \ldots, n$, $i = 1, 2, \ldots, n-1$ and noting row interchanges.

Step 1 Set $k = 1$.

Step 2 While $(k \leq N)$ do Steps 3–9.

Step 3 For $i = 1, 2, \ldots, n$ (*Calculate* $\mathbf{r}$.)

$$\text{set } r_i = b_i - \sum_{j=1}^{n} a_{ij} x_j.$$

(*Perform the computations in double-precision arithmetic.*)

Step 4 Solve the linear system $A\mathbf{y} = \mathbf{r}$ by using Gaussian elimination in the same order as in Step 0.

Step 5 For $i = 1, \ldots, n$ set $xx_i = x_i + y_i$.

Step 6 If $k = 1$ then set $COND = \dfrac{\|\mathbf{y}\|_\infty}{\|\mathbf{xx}\|_\infty} 10^t$.

Step 7 If $\|\mathbf{x} - \mathbf{xx}\|_\infty < TOL$ then OUTPUT ($\mathbf{xx}$);
OUTPUT (*COND*);
(*Procedure completed successfully.*)
STOP.

Step 8 Set $k = k + 1$.

Step 9 For $i = 1, \ldots, n$ set $x_i = xx_i$.

Step 10 OUTPUT ('Maximum number of iterations exceeded');
OUTPUT (*COND*);
(*Procedure completed unsuccessfully.*)
STOP.

If t-digit arithmetic is used, a recommended stopping procedure in Step 7 is to iterate until $|y_i^{(k)}| \leq 10^{-t}$, for each $i = 1, 2, \ldots, n$.

EXAMPLE 4 In Example 3 we found the approximation to the problem we have been considering, using five-digit arithmetic and Gaussian elimination, to be

$$\tilde{\mathbf{x}}^{(1)} = (1.2001, 0.99991, 0.92538)^t$$

and the solution to $A\mathbf{y} = \mathbf{r}^{(1)}$ to be

$$\tilde{\mathbf{y}}^{(1)} = (-0.20008, 8.9987 \times 10^{-5}, 0.074607)^t$$

By Step 5 in Algorithm 7.4,

$$\tilde{\mathbf{x}}^{(2)} = \tilde{\mathbf{x}}^{(1)} + \tilde{\mathbf{y}}^{(1)} = (1.0000, 1.0000, 0.99999)^t.$$

and the actual error in this approximation is

$$\|\mathbf{x} - \tilde{\mathbf{x}}^{(2)}\|_\infty = 1 \times 10^{-5}.$$

Using the suggested stopping technique for the algorithm, we compute $\mathbf{r}^{(2)} = \mathbf{b} - A\tilde{\mathbf{x}}^{(2)}$ and solve the system $A\mathbf{y}^{(2)} = \mathbf{r}^{(2)}$, which gives

$$\tilde{\mathbf{y}}^{(2)} = (1.5002 \times 10^{-9}, 2.0951 \times 10^{-10}, 1.0000 \times 10^{-5})^t.$$

Since $\|\tilde{\mathbf{y}}^{(2)}\|_\infty \leq 10^{-5}$, we conclude that

$$\tilde{\mathbf{x}}^{(3)} = \tilde{\mathbf{x}}^{(2)} + \tilde{\mathbf{y}}^{(2)} = (1.0000, 1.0000, 1.0000)^t$$

is sufficiently accurate, which is certainly correct. ■

Throughout this section it has been assumed that in the linear system $A\mathbf{x} = \mathbf{b}$, the matrix A and the vector $\mathbf{b}$ can be represented exactly. Realistically, the entries a_{ij} and b_j will be altered or perturbed by an amount δa_{ij} and δb_j, causing the linear system

$$(A + \delta A)\mathbf{x} = \mathbf{b} + \delta\mathbf{b}$$

to be solved in place of $A\mathbf{x} = \mathbf{b}$. Normally, if $\|\delta A\|$ and $\|\delta\mathbf{b}\|$ are small (on the order of 10^{-t}), the t-digit arithmetic should yield a solution $\tilde{\mathbf{x}}$ for which $\|\mathbf{x} - \tilde{\mathbf{x}}\|$ is correspondingly small. However, in the case of ill-conditioned systems, we have seen that even if A and $\mathbf{b}$ are represented exactly, roundoff errors can cause $\|\mathbf{x} - \tilde{\mathbf{x}}\|$ to be large. The following theorem relates the perturbations of linear systems to the condition number of a matrix. The proof of this result can be found in [Or2, p. 33].

Theorem 7.29 Suppose A is nonsingular and

$$\|\delta A\| < \frac{1}{\|A^{-1}\|}.$$

The solution $\tilde{\mathbf{x}}$ to $(A + \delta A)\tilde{\mathbf{x}} = \mathbf{b} + \delta\mathbf{b}$ approximates the solution $\mathbf{x}$ of $A\mathbf{x} = \mathbf{b}$ with error estimate

$$\frac{\|\mathbf{x} - \tilde{\mathbf{x}}\|}{\|\mathbf{x}\|} \leq \frac{K(A)}{1 - K(A)(\|\delta A\|/\|A\|)}\left(\frac{\|\delta\mathbf{b}\|}{\|\mathbf{b}\|} + \frac{\|\delta A\|}{\|A\|}\right). \tag{7.23}$$

■

The estimate in inequality (7.23) states that if the matrix A is well-conditioned (that is, $K(A)$ is not too large), then small changes in A and $\mathbf{b}$ produce correspondingly small changes in the solution $\mathbf{x}$. If, on the other hand, A is ill-conditioned, then small changes in A and $\mathbf{b}$ may produce large changes in $\mathbf{x}$.

The theorem is independent of the particular numerical procedure used to solve $A\mathbf{x} = \mathbf{b}$. It can be shown, by means of Wilkinson's backward error analysis (see [Wil1] or [Wil2]), that if Gaussian elimination with pivoting is used to solve $A\mathbf{x} = \mathbf{b}$ in t-digit arithmetic, the numerical solution $\tilde{\mathbf{x}}$ is the actual solution of a linear system:

$$(A + \delta A)\tilde{\mathbf{x}} = \mathbf{b}, \quad \text{where } \|\delta A\|_\infty \le f(n)10^{1-t} \max_{i,j,k} |a_{ij}^{(k)}|.$$

Wilkinson found in practice that $f(n) \approx n$ and, at worst, $f(n) \le 1.01(n^3 + 3n^2)$.

EXERCISE SET 7.4

1. Compute the condition numbers of the following matrices relative to $\|\cdot\|_\infty$.

a. $\begin{bmatrix} \frac{1}{2} & \frac{1}{3} \\ \frac{1}{3} & \frac{1}{4} \end{bmatrix}$

b. $\begin{bmatrix} 3.9 & 1.6 \\ 6.8 & 2.9 \end{bmatrix}$

c. $\begin{bmatrix} 1 & 2 \\ 1.0001 & 2 \end{bmatrix}$

d. $\begin{bmatrix} 1.003 & 58.09 \\ 5.550 & 321.8 \end{bmatrix}$

e. $\begin{bmatrix} 1 & -1 & -1 \\ 0 & 1 & -1 \\ 0 & 0 & -1 \end{bmatrix}$

f. $\begin{bmatrix} 0.04 & 0.01 & -0.01 \\ 0.2 & 0.5 & -0.2 \\ 1 & 2 & 4 \end{bmatrix}$

2. The following linear systems $A\mathbf{x} = \mathbf{b}$ have $\mathbf{x}$ as the actual solution and $\tilde{\mathbf{x}}$ as an approximate solution. Using the results of Exercise 1, compute $\|\mathbf{x} - \tilde{\mathbf{x}}\|_\infty$ and $K_\infty(A)\dfrac{\|\mathbf{b} - A\tilde{\mathbf{x}}\|_\infty}{\|A\|_\infty}$.

a.
$$\tfrac{1}{2}x_1 + \tfrac{1}{3}x_2 = \tfrac{1}{63},$$
$$\tfrac{1}{3}x_1 + \tfrac{1}{4}x_2 = \tfrac{1}{168},$$
$$\mathbf{x} = \left(\tfrac{1}{7}, -\tfrac{1}{6}\right)^t,$$
$$\tilde{\mathbf{x}} = (0.142, -0.166)^t.$$

b.
$$3.9x_1 + 1.6x_2 = 5.5,$$
$$6.8x_1 + 2.9x_2 = 9.7,$$
$$\mathbf{x} = (1, 1)^t,$$
$$\tilde{\mathbf{x}} = (0.98, 1.1)^t.$$

c.
$$x_1 + 2x_2 = 3,$$
$$1.0001x_1 + 2x_2 = 3.0001,$$
$$\mathbf{x} = (1, 1)^t,$$
$$\tilde{\mathbf{x}} = (0.96, 1.02)^t$$

d.
$$1.003x_1 + 58.09x_2 = 68.12,$$
$$5.550x_1 + 321.8\ x_2 = 377.3,$$
$$\mathbf{x} = (10, 1)^t,$$
$$\tilde{\mathbf{x}} = (-10, 1)^t$$

e.
$$x_1 - x_2 - x_3 = 2\pi,$$
$$x_2 - x_3 = 0,$$
$$-x_3 = \pi.$$
$$\mathbf{x} = (0, -\pi, -\pi)^t,$$
$$\tilde{\mathbf{x}} = (-0.1, -3.15, -3.14)^t$$

f.
$$0.04x_1 + 0.01x_2 - 0.01x_3 = 0.06,$$
$$0.2x_1 + 0.5\ x_2 - 0.2\ x_3 = 0.3,$$
$$x_1 + 2\ x_2 + 4\ x_3 = 11,$$
$$\mathbf{x} = (1.827586, 0.6551724, 1.965517)^t,$$
$$\tilde{\mathbf{x}} = (1.8, 0.64, 1.9)^t$$

3. The linear system

$$A\mathbf{x} = \begin{bmatrix} 1 & 2 \\ 1.0001 & 2 \end{bmatrix} \begin{bmatrix} x_1 \\ x_2 \end{bmatrix} = \begin{bmatrix} 3 \\ 3.0001 \end{bmatrix}$$

has solution $(1, 1)^t$. Change A slightly to

$$\begin{bmatrix} 1 & 2 \\ 0.9999 & 2 \end{bmatrix}$$

and consider the linear system

$$\begin{bmatrix} 1 & 2 \\ 0.9999 & 2 \end{bmatrix} \begin{bmatrix} x_1 \\ x_2 \end{bmatrix} = \begin{bmatrix} 3 \\ 3.0001 \end{bmatrix}.$$

Compute the new solution using five-digit rounding arithmetic and compare the actual error to the estimate (7.23). Is A ill-conditioned?

4. The linear system $A\mathbf{x} = \mathbf{b}$ given by

$$\begin{bmatrix} 1 & 2 \\ 1.00001 & 2 \end{bmatrix} \begin{bmatrix} x_1 \\ x_2 \end{bmatrix} = \begin{bmatrix} 3 \\ 3.00001 \end{bmatrix}$$

has solution $(1, 1)^t$. Use seven-digit rounding arithmetic to find the solution of the perturbed system

$$\begin{bmatrix} 1 & 2 \\ 1.000011 & 2 \end{bmatrix} \begin{bmatrix} x_1 \\ x_2 \end{bmatrix} = \begin{bmatrix} 3.00001 \\ 3.00003 \end{bmatrix},$$

and compare the actual error to the estimate (7.23). Is A ill-conditioned?

5. **a.** Use single precision arithmetic on a computer to solve the following linear system using the Gaussian Elimination with Backward Substitution Algorithm 6.1.

$$\begin{aligned}
\frac{1}{3}x_1 - \frac{1}{3}x_2 - \frac{1}{3}x_3 - \frac{1}{3}x_4 - \frac{1}{3}x_5 &= 1 \\
\frac{1}{3}x_2 - \frac{1}{3}x_3 - \frac{1}{3}x_4 - \frac{1}{3}x_5 &= 0 \\
\frac{1}{3}x_3 - \frac{1}{3}x_4 - \frac{1}{3}x_5 &= -1 \\
\frac{1}{3}x_4 - \frac{1}{3}x_5 &= 2 \\
\frac{1}{3}x_5 &= 7
\end{aligned}$$

b. Compute the condition number of the matrix for the system relative to $\|\cdot\|_\infty$.

c. Find the exact solution to the linear system.

6. The $n \times n$ *Hilbert* matrix $H^{(n)}$ defined by

$$H_{ij}^{(n)} = \frac{1}{i + j - 1}, \quad 1 \le i, j \le n$$

is an ill-conditioned matrix that arises in solving the normal equations for the coefficients of the least-squares polynomial (see Example 1 of Section 8.2).

a. Show that

$$[H^{(4)}]^{-1} = \begin{bmatrix} 16 & -120 & 240 & -140 \\ -120 & 1200 & -2700 & 1680 \\ 240 & -2700 & 6480 & -4200 \\ -140 & 1680 & -4200 & 2800 \end{bmatrix},$$

and compute $K_\infty(H^{(4)})$.

b. Show that

$$[H^{(5)}]^{-1} = \begin{bmatrix} 25 & -300 & 1050 & -1400 & 630 \\ -300 & 4800 & -18900 & 26880 & -12600 \\ 1050 & -18900 & 79380 & -117600 & 56700 \\ -1400 & 26880 & -117600 & 179200 & -88200 \\ 630 & -12600 & 56700 & -88200 & 44100 \end{bmatrix},$$

and compute $K_\infty(H^{(5)})$.

c. Solve the linear system

$$H^{(4)}\begin{bmatrix}x_1\\x_2\\x_3\\x_4\end{bmatrix} = \begin{bmatrix}1\\0\\0\\1\end{bmatrix}$$

using five-digit rounding arithmetic, and compare the actual error to that estimated in (7.23).

7. Show that if B is singular, then

$$\frac{1}{K(A)} \le \frac{||A - B||}{||A||}.$$

[*Hint*: There exists a vector with $||\mathbf{x}|| = 1$, such that $B\mathbf{x} = \mathbf{0}$. Derive the estimate using $||A\mathbf{x}|| \ge ||\mathbf{x}|| \,/\, ||A^{-1}||$.]

8. Using Exercise 7, estimate the condition numbers for the following matrices:

a. $\begin{bmatrix}1 & 2\\1.0001 & 2\end{bmatrix}$ **b.** $\begin{bmatrix}3.9 & 1.6\\6.8 & 2.9\end{bmatrix}$

9. Use four-digit rounding arithmetic to compute the inverse H^{-1} of the 3×3 Hilbert matrix H and then compute $\hat{H} = (H^{-1})^{-1}$. Determine $||H - \hat{H}||_\infty$.

7.5 Survey of Methods and Software

In this chapter we have studied iterative techniques to approximate the solution of linear systems. We began with the Jacobi method and the Gauss-Seidel method to introduce the iterative methods. Both methods require an arbitrary initial approximation $\mathbf{x}^{(0)}$ and generate a sequence of vectors $\mathbf{x}^{(i+1)}$ using an equation of the form

$$\mathbf{x}^{(i+1)} = T\mathbf{x}^{(i)} + \mathbf{c}.$$

It was noted that the method will converge if and only if the spectral radius of the iteration matrix $\rho(T) < 1$, and the smaller the spectral radius, the faster the convergence. Analysis of the residual vectors of the Gauss-Seidel technique led to the SOR iterative method, which involves a parameter ω to speed convergence.

These iterative methods and modifications are used extensively in the solution of linear systems which arise in the numerical solution of boundary value problems and partial differential equations (see Chapters 11 and 12). These systems are often very large, on the order of more than 10000 equations in 10000 unknowns, and are sparse with their nonzero entries in predictable positions. The iterative methods are also useful for other large sparse systems and are easily adapted for efficient use on parallel computers.

The concepts of condition number and poorly conditioned matrices were introduced in the last section of the chapter. Many of the subroutines for solving a linear system or for factoring a matrix into an *LU* factorization include checks for ill-conditioned matrices and also give an estimate of the condition number.

The packages LINPACK and LAPACK contain only direct methods for the solution of linear systems. Neither the IMSL Library nor the NAG Library contains subroutines for the iterative solution of linear systems since neither library focuses on methods for boundary value problems or partial differential equations that require the solution of large sparse systems. However, the public domain packages ITPACK, SLAP, and SPARSPAK contain iterative methods.

The subroutine SGETRF in LAPACK factors the real matrix A into an LU factorization and gives the row ordering for the permutation matrix P, where $PA = LU$. The subroutine SGECON also gives the condition number of A. LAPACK has other subroutines for special matrices, for example, SPOTRF performs the Choleski factorization of a positive definite matrix A and estimates its condition number.

The IMSL Library has subroutines that estimate the condition number. For example, the subroutine LFCRG computes an LU factorization $PA = LU$ of the matrix A and also gives an estimate of the condition number. The NAG Library has similar subroutines.

LAPACK, LINPACK, the IMSL Library, and the NAG Library have subroutines that improve on a solution to a linear system that is poorly conditioned. The subroutines test the condition number and then use iterative refinement to obtain the most accurate solution possible given the precision of the computer.

More information on the use of iterative methods for solving linear systems can be found in Varga [Var], Young [Y], Hageman and Young [HY], and in the recent book by Axelsson [Ax]. Iterative methods for large sparse systems are discussed in Barrett et al [Barr], Hackbusch [Hac], and Saad [Sa2].

CHAPTER

8 Approximation Theory

■ ■ ■

Hooke's law states that when a force is applied to a spring constructed of uniform material, the length of the spring is a linear function of the force applied. We can write the linear function as $F(l) = k(l - E)$, where $F(l)$ represents the force required to stretch the spring l units, the constant E represents the length of the spring with no force applied, and the constant k is the spring constant.

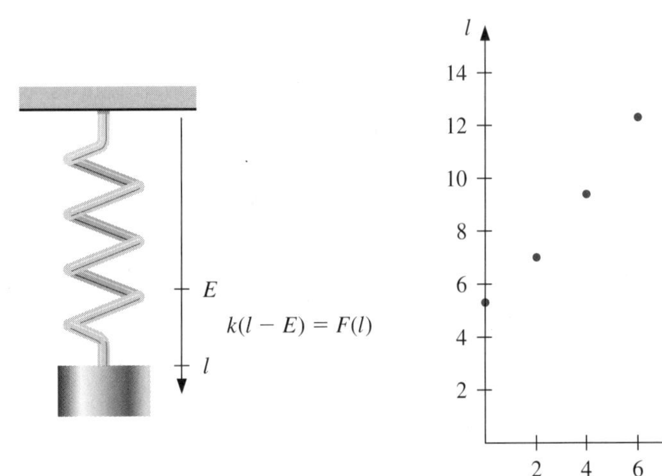

Suppose we want to determine the spring constant for a spring that has initial length 5.3 in. We consecutively apply forces of 2, 4, and 6 lb to the spring and find that its length increases to 7.0, 9.4, and 12.3 in., respectively. A quick examination shows that the points (0, 5.3), (2, 7.0), (4, 9.4), and (6, 12.3) do not quite lie in a straight line. Although we could simply use one random pair of these data points to approximate the spring constant, it would seem more reasonable to find the line that *best*

approximates all the data points to determine the constant. This type of approximation will be considered in this chapter, and this spring application can be found in Exercise 7 of Section 8.1.

The study of approximation theory involves two general types of problems. One problem arises when a function is given explicitly, but we wish to find a "simpler" type of function, such as a polynomial, that can be used to determine approximate values of the given function. The other problem in approximation theory is concerned with fitting functions to given data and finding the "best" function in a certain class that can be used to represent the data.

Both problems have been touched upon in Chapter 3. The Taylor polynomial of degree n about the number x_0 is an excellent approximation to an $(n+1)$-times differentiable function f in a small neighborhood of x_0. The Lagrange interpolating polynomials, or, more generally, osculatory polynomials, were discussed both as approximating polynomials and as polynomials to fit certain data. Cubic splines were also discussed in that chapter. In this chapter, limitations to these techniques are considered and other avenues of approach discussed.

8.1 Discrete Least Squares Approximation

Consider the problem of estimating the values of a function at nontabulated points, given the experimental data in Table 8.1.

Table 8.1

x_i	y_i
1	1.3
2	3.5
3	4.2
4	5.0
5	7.0
6	8.8
7	10.1
8	12.5
9	13.0
10	15.6

Interpolation requires a function that assumes the value of y_i at x_i for each $i = 1, 2, \ldots, 10$. Figure 8.1 shows a graph of the values in Table 8.1. From this graph, it appears that the actual relationship between x and y is linear and that no line precisely fits the data because of errors in the data collection procedure.

In this case, it is unreasonable to require that the approximating function agree exactly with the given data. In fact, such a function would introduce oscillations that were not

Figure 8.1

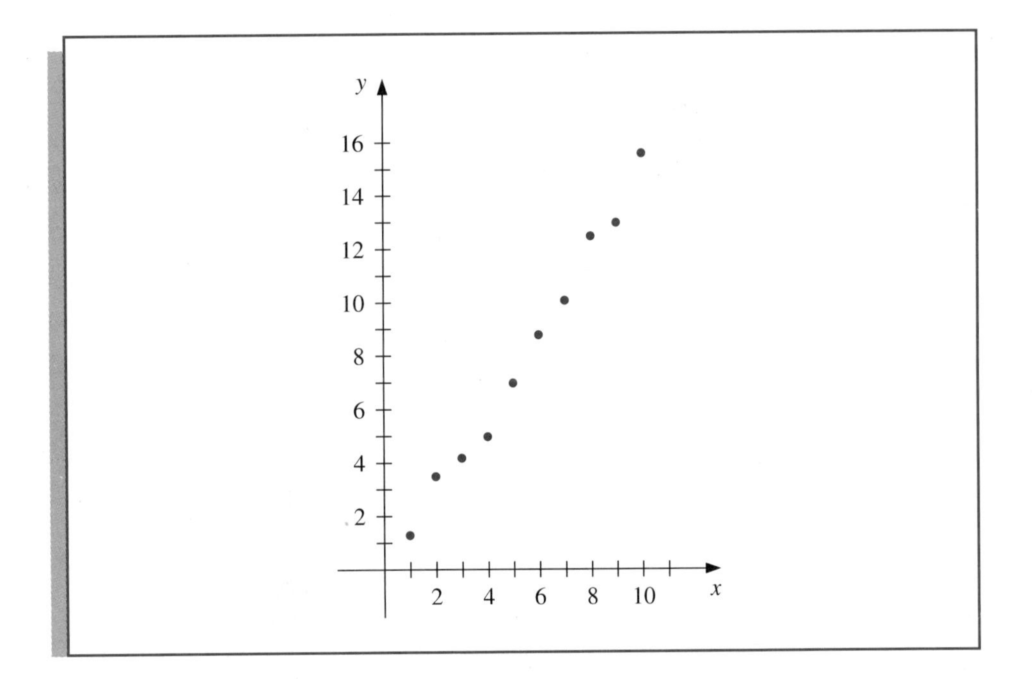

present originally. A better approach for a problem of this type would be to find the "best" (in some sense) line that can be used as an approximating function, even though it might not agree precisely with the data at any point.

Let $ax_i + b$ denote the ith value on the approximating line and y_i be the ith given y-value. The problem of finding the equation of the best linear approximation in the absolute sense requires that values of a and b be found to minimize

$$\max_{i=1,2,\ldots,10} \{|y_i - (ax_i + b)|\}.$$

This is commonly called a **minimax** problem and cannot be handled by elementary techniques.

Another approach to determining the best linear approximation involves finding values of a and b to minimize

$$\sum_{i=1}^{10} |y_i - (ax_i + b)|.$$

This quantity is called the **absolute deviation**. To minimize the absolute deviation, it is necessary that:

$$0 = \frac{\partial}{\partial a} \sum_{i=1}^{10} |y_i - (ax_i + b)| \quad \text{and} \quad 0 = \frac{\partial}{\partial b} \sum_{i=1}^{10} |y_i - (ax_i + b)|.$$

The difficulty with this procedure is that the absolute value function is not differentiable at zero, and solutions to this pair of equations may not necessarily be obtainable.

The **least squares** approach to this problem involves determining the best approximating line when the error involved is the sum of the squares of the differences between the y-values on the approximating line and the given y-values. Hence, constants a and b must be found that minimize the least squares error:

$$\sum_{i=1}^{10} [y_i - (ax_i + b)]^2.$$

The least squares method is the most convenient procedure for determining best linear approximations, but there are also important theoretical considerations that favor this method. The minimax approach generally assigns too much weight to a bit of data that is badly in error. The method using absolute deviation simply averages the error at the various points and does not give sufficient weight to a point that is considerably out of line with the approximation. The least squares approach puts substantially more weight on a point that is out of line with the rest of the data but will not allow that point to completely dominate the approximation. An additional reason for considering the least squares approach involves the study of the statistical distribution of error. (See [Lar, pp. 463–481].)

The general problem of fitting the best least squares line to a collection of data $\{(x_i, y_i)\}_{i=1}^{m}$ involves minimizing $\sum_{i=1}^{m} [y_i - (ax_i + b)]^2$ with respect to the parameters a and b. For a minimum to occur, it is necessary that

$$0 = \frac{\partial}{\partial a} \sum_{i=1}^{m} [y_i - (ax_i + b)]^2 = 2 \sum_{i=1}^{m} (y_i - ax_i - b)(-x_i)$$

and

$$0 = \frac{\partial}{\partial b} \sum_{i=1}^{m} [y_i - (ax_i + b)]^2 = 2 \sum_{i=1}^{m} (y_i - ax_i - b)(-1).$$

These equations simplify to the **normal equations**:

$$a \sum_{i=1}^{m} x_i^2 + b \sum_{i=1}^{m} x_i = \sum_{i=1}^{m} x_i y_i \quad \text{and} \quad a \sum_{i=1}^{m} x_i + b \cdot m = \sum_{i=1}^{m} y_i.$$

The solution to this system of equations is

(8.1)
$$a = \frac{m \left(\sum_{i=1}^{m} x_i y_i \right) - \left(\sum_{i=1}^{m} x_i \right) \left(\sum_{i=1}^{m} y_i \right)}{m \left(\sum_{i=1}^{m} x_i^2 \right) - \left(\sum_{i=1}^{m} x_i \right)^2}$$

and

$$
\textbf{(8.2)} \qquad b = \frac{\left(\sum_{i=1}^{m} x_i^2\right)\left(\sum_{i=1}^{m} y_i\right) - \left(\sum_{i=1}^{m} x_i y_i\right)\left(\sum_{i=1}^{m} x_i\right)}{m\left(\sum_{i=1}^{m} x_i^2\right) - \left(\sum_{i=1}^{m} x_i\right)^2}.
$$

EXAMPLE 1 Consider the data presented in Table 8.1. To find the least squares line approximating this data, extend the table and sum the columns, as shown in the third and fourth columns of Table 8.2.

Table 8.2

x_i	y_i	x_i^2	$x_i y_i$	$P(x_i) = 1.538x_i - 0.360$
1	1.3	1	1.3	1.18
2	3.5	4	7.0	2.72
3	4.2	9	12.6	4.25
4	5.0	16	20.0	5.79
5	7.0	25	35.0	7.33
6	8.8	36	52.8	8.87
7	10.1	49	70.7	10.41
8	12.5	64	100.0	11.94
9	13.0	81	117.0	13.48
10	15.6	100	156.0	15.02
55	81.0	385	572.4	$E = \Sigma_{i=1}^{10}(y_i - P(x_i))^2 \approx 2.34$

The normal equations (8.1) and (8.2) imply that

$$
a = \frac{10(572.4) - 55(81)}{10(385) - (55)^2} = 1.538
$$

and

$$
b = \frac{385(81) - 55(572.4)}{10(385) - (55)^2} = -0.360.
$$

The graph of this line and the data points are shown in Figure 8.2. The approximate values given by the least squares technique at the data points are in Table 8.2. ■

The general problem of approximating a set of data, $\{(x_i, y_i) | i = 1, 2, \ldots, m\}$, with an algebraic polynomial $P_n(x) = \Sigma_{k=0}^{n} a_k x^k$ of degree $n < m - 1$ using the least squares procedure is handled in a similar manner. It requires choosing the constants $a_0, a_1, \ldots, a_n$ to minimize the least squares error

Figure 8.2

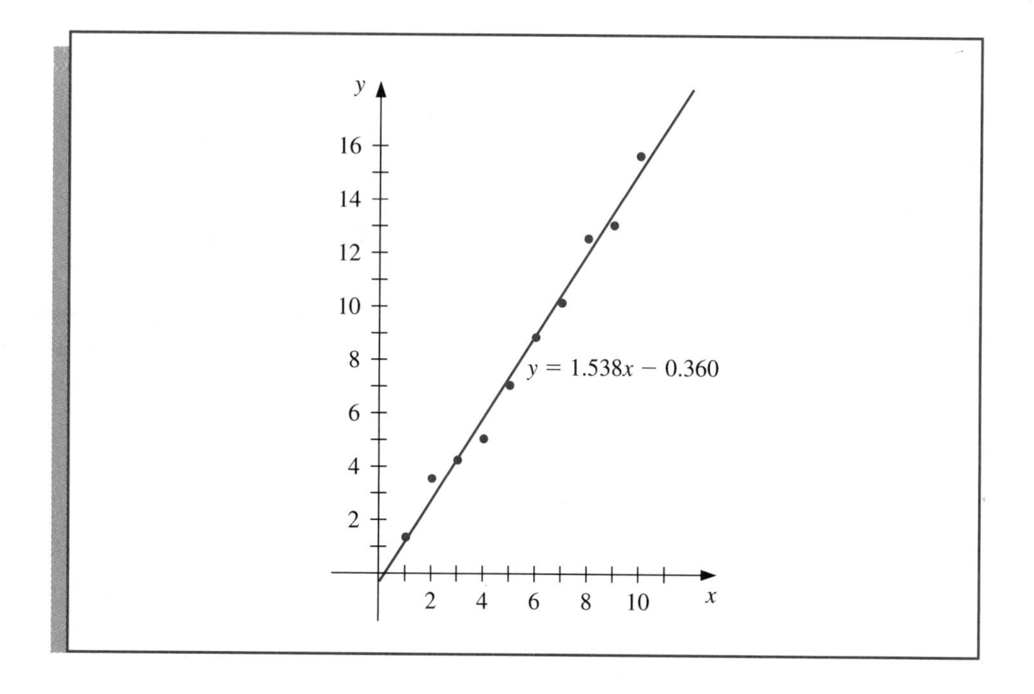

$$
\begin{aligned}
E &= \sum_{i=1}^{m} (y_i - P_n(x_i))^2 \\
&= \sum_{i=1}^{m} y_i^2 - 2\sum_{i=1}^{m} P_n(x_i) y_i + \sum_{i=1}^{m} (P_n(x_i))^2 \\
&= \sum_{i=1}^{m} y_i^2 - 2\sum_{i=1}^{m} \left(\sum_{j=0}^{n} a_j x_i^j\right) y_i + \sum_{i=1}^{m} \left(\sum_{j=0}^{n} a_j x_i^j\right)^2 \\
&= \sum_{i=1}^{m} y_i^2 - 2\sum_{j=0}^{n} a_j \left(\sum_{i=1}^{m} y_i x_i^j\right) + \sum_{j=0}^{n}\sum_{k=0}^{n} a_j a_k \left(\sum_{i=1}^{m} x_i^{j+k}\right).
\end{aligned}
$$

As in the linear case, for E to be minimized, it is necessary that $\partial E/\partial a_j = 0$ for each $j = 0, 1, \ldots, n$. Thus, for each j,

$$
0 = \frac{\partial E}{\partial a_j} = -2\sum_{i=1}^{m} y_i x_i^j + 2\sum_{k=0}^{n} a_k \sum_{i=1}^{m} x_i^{j+k}.
$$

This gives $n + 1$ **normal equations** in the $n + 1$ unknowns a_j,

(8.3)
$$
\sum_{k=0}^{n} a_k \sum_{i=1}^{m} x_i^{j+k} = \sum_{i=1}^{m} y_i x_i^j, \quad j = 0, 1, \ldots, n.
$$

It is helpful to write the equations as follows:

$$
\begin{aligned}
a_0 \sum_{i=1}^{m} x_i^0 + a_1 \sum_{i=1}^{m} x_i^1 \quad + a_2 \sum_{i=1}^{m} x_i^2 \quad + \cdots + a_n \sum_{i=1}^{m} x_i^n \quad &= \sum_{i=1}^{m} y_i x_i^0, \\
a_0 \sum_{i=1}^{m} x_i^1 + a_1 \sum_{i=1}^{m} x_i^2 \quad + a_2 \sum_{i=1}^{m} x_i^3 \quad + \cdots + a_n \sum_{i=1}^{m} x_i^{n+1} &= \sum_{i=1}^{m} y_i x_i^1, \\
\vdots \qquad\qquad\qquad\qquad\qquad\qquad\qquad\qquad & \qquad \vdots \\
a_0 \sum_{i=1}^{m} x_i^n + a_1 \sum_{i=1}^{m} x_i^{n+1} + a_2 \sum_{i=1}^{m} x_i^{n+2} + \cdots + a_n \sum_{i=1}^{m} x_i^{2n} &= \sum_{i=1}^{m} y_i x_i^n.
\end{aligned}
$$

It can be shown (see Exercise 14) that the normal equations have a unique solution provided that the x_i are distinct.

EXAMPLE 2 Fit the data in Table 8.3 with the discrete least squares polynomial of degree two. For this problem, $n = 2$, $m = 5$, and the three normal equations are

$$
\begin{aligned}
5a_0 + \quad 2.5a_1 + \quad 1.875a_2 &= 8.7680, \\
2.5a_0 + \quad 1.875a_1 + 1.5625a_2 &= 5.4514, \\
1.875a_0 + 1.5625a_1 + 1.3828a_2 &= 4.4015.
\end{aligned}
$$

Table 8.3

i	1	2	3	4	5
x_i	0.00	0.25	0.50	0.75	1.00
y_i	1.0000	1.2840	1.6487	2.1170	2.7183

The solution to this system is

$$a_0 = 1.0052, \quad a_1 = 0.8641, \quad a_2 = 0.8437.$$

Thus, the least squares polynomial of degree two fitting the preceding data is $P_2(x) = 1.0052 + 0.8641x + 0.8437x^2$, whose graph is shown in Figure 8.3. At the given values of x_i we have the approximations shown in Table 8.4.

The total error,

$$\sum_{i=1}^{5} (y_i - P(x_i))^2 = 2.76 \times 10^{-4},$$

is the least that can be obtained by using a polynomial of degree at most two. ■

Figure 8.3

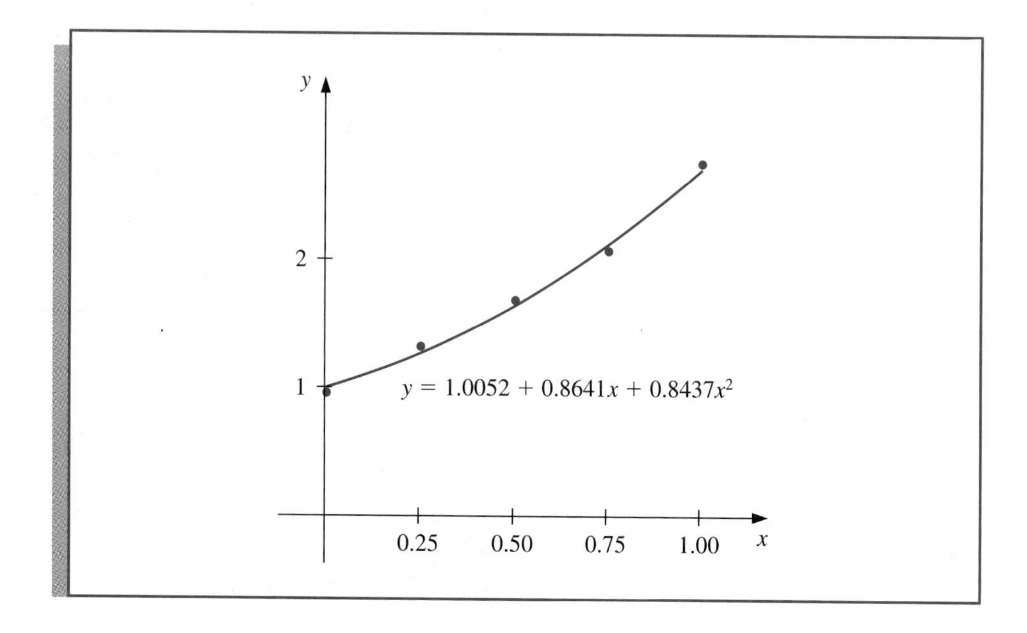

Table 8.4

i	1	2	3	4	5
x_i	0.00	0.25	0.50	0.75	1.00
y_i	1.0000	1.2840	1.6487	2.1170	2.7183
$P(x_i)$	1.0052	1.2740	1.6482	2.1279	2.7130
$y_i - P(x_i)$	−0.0052	0.0100	0.0005	−0.0109	0.0053

Occasionally it is appropriate to assume that the data are exponentially related. This requires the approximating function to be of the form

(8.4) $$y = be^{ax}$$

or

(8.5) $$y = bx^a,$$

for some constants a and b. The difficulty with applying the least squares procedure in a situation of this type comes from attempting to minimize

$$E = \sum_{i=1}^{m} (y_i - be^{ax_i})^2, \quad \text{in the case of Eq. (8.4),}$$

or

$$E = \sum_{i=1}^{m} (y_i - bx_i^a)^2, \quad \text{in the case of Eq. (8.5).}$$

The normal equations associated with these procedures are obtained from either

$$0 = \frac{\partial E}{\partial b} = 2\sum_{i=1}^{m}(y_i - be^{ax_i})(-e^{ax_i})$$

and

$$0 = \frac{\partial E}{\partial a} = 2\sum_{i=1}^{m}(y_i - be^{ax_i})(-bx_i e^{ax_i}), \quad \text{the case of Eq. (8.4),}$$

or

$$0 = \frac{\partial E}{\partial b} = 2\sum_{i=1}^{m}(y_i - bx_i^a)(-x_i^a)$$

and

$$0 = \frac{\partial E}{\partial a} = 2\sum_{i=1}^{m}(y_i - bx_i^a)(-b(\ln x_i)x_i^a), \quad \text{the case of Eq. (8.5).}$$

No exact solution to either of these systems can generally be found.

The method that is commonly used when the data are suspected to be exponentially related is to consider the logarithm of the approximating equation:

$$\ln y = \ln b + ax, \quad \text{in the case of Eq. (8.4),}$$

and

$$\ln y = \ln b + a \ln x, \quad \text{in the case of Eq. (8.5).}$$

In either case, a linear problem now appears, and solutions for $\ln b$ and a can be obtained by appropriately modifying the normal equations (8.1) and (8.2).

However, the approximation obtained in this manner is *not* the least squares approximation for the original problem, and this approximation can in some cases differ significantly from the least squares approximation to the original problem. The application in Exercise 13 describes such a problem. This application will be reconsidered as Exercise 7 in Section 10.3, where the exact solution to the exponential least squares problem is approximated by using methods suitable for solving nonlinear systems of equations.

EXAMPLE 3 Consider the collection of data in the first three columns of Table 8.5. If x_i is graphed with $\ln y_i$, the data appear to have a linear relation, so it is reasonable to assume an approximation of the form

$$y = be^{ax} \quad \text{or} \quad \ln y = \ln b + ax.$$

Extending the table and summing the appropriate columns gives the remaining data in Table 8.5.

Table 8.5

i	x_i	y_i	$\ln y_i$	x_i^2	$x_i \ln y_i$
1	1.00	5.10	1.629	1.0000	1.629
2	1.25	5.79	1.756	1.5625	2.195
3	1.50	6.53	1.876	2.2500	2.814
4	1.75	7.45	2.008	3.0625	3.514
5	2.00	8.46	2.135	4.0000	4.270
	7.50		9.404	11.875	14.422

Using the normal equations (8.1) and (8.2),

$$a = \frac{(5)(14.422) - (7.5)(9.404)}{(5)(11.875) - (7.5)^2} = 0.5056$$

and

$$\ln b = \frac{(11.875)(9.404) - (14.422)(7.5)}{(5)(11.875) - (7.5)^2} = 1.122.$$

Since $b = e^{1.122} = 3.071$, the approximation assumes the form

$$y = 3.071e^{0.5056x},$$

which, at the data points, gives the values in Table 8.6. (See Figure 8.4.) ■

Figure 8.4

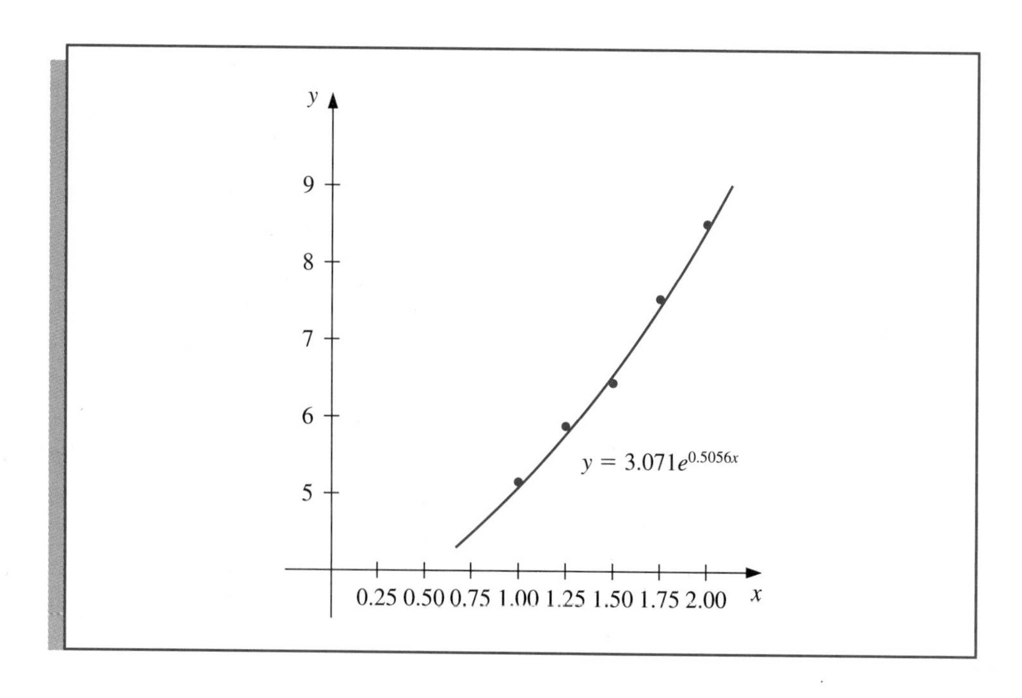

Table 8.6

i	x_i	y_i	$3.071e^{0.5056x_i}$
1	1.00	5.10	5.09
2	1.25	5.79	5.78
3	1.50	6.53	6.56
4	1.75	7.45	7.44
5	2.00	8.46	8.44

EXERCISE SET 8.1

1. Compute the linear least squares polynomial and total error E for the data of Example 2.
2. Compute the least squares polynomial of degree two for the data of Example 1 and compare the total error E for the two methods.
3. Find the least squares polynomials of degrees one, two, and three for the data in the following table. Compute the error E in each case. Graph the data and the polynomials.

x_i	1.0	1.1	1.3	1.5	1.9	2.1
y_i	1.84	1.96	2.21	2.45	2.94	3.18

4. Find the least squares polynomials of degrees one, two, and three for the data in the following table. Compute the error E in each case. Graph the data and the polynomials.

x_i	0	0.15	0.31	0.5	0.6	0.75
y_i	1.0	1.004	1.031	1.117	1.223	1.422

5. Given the data

x_i	4.0	4.2	4.5	4.7	5.1	5.5	5.9	6.3	6.8	7.1
y_i	102.56	113.18	130.11	142.05	167.53	195.14	224.87	256.73	299.50	326.72

   a. Construct the least squares polynomial of degree one and compute the error.
   b. Construct the least squares polynomial of degree two and compute the error.
   c. Construct the least squares polynomial of degree three and compute the error.
   d. Construct the least squares approximation of the form be^{ax} and compute the error.
   e. Construct the least squares approximation of the form bx^a and compute the error.
   f. What form of relationship between the data do you think holds?

6. Repeat Exercise 5 for the following data.

x_i	0.2	0.3	0.6	0.9	1.1	1.3	1.4	1.6
y_i	0.050446	0.098426	0.33277	0.72660	1.0972	1.5697	1.8487	2.5015

7. In the lead example of this chapter, an experiment was described to determine the spring constant k in Hooke's law:

$$F(l) = k(l - E).$$

The function F is the force required to stretch the spring l units, where the constant $E = 5.3$ in. is the length of the unstretched spring.

a. Suppose measurements are made of the length l in inches for applied weights $F(l)$ in pounds, as given in the following table.

$F(l)$	l
2	7.0
4	9.4
6	12.3

Find the least squares approximation for k.

b. More measurements are made, giving additional data:

$F(l)$	l
3	8.3
5	11.3
8	14.4
10	15.9

Compute the new least squares approximation for k. Then compare the total error in parts (a) and (b).

8. The following list contains homework grades and the final-examination grades for 30 numerical analysis students. Find the equation of the least squares line for this data, and use this line to determine the homework grade required to predict minimal A (90%) and D (60%) grades on the final.

Homework	Final	Homework	Final
302	45	323	83
325	72	337	99
285	54	337	70
339	54	304	62
334	79	319	66
322	65	234	51
331	99	337	53
279	63	351	100
316	65	339	67
347	99	343	83
343	83	314	42
290	74	344	79
326	76	185	59
233	57	340	75
254	45	316	45

9. The following table lists the college grade-point averages of 20 mathematics and computer science majors, together with the scores that these students received on the mathematics portion of the ACT (American College Testing Program) test while in high school. Plot these data, and find the equation of the least squares line for this data.

ACT score	Grade-point average	ACT score	Grade-point average
28	3.84	29	3.75
25	3.21	28	3.65
28	3.23	27	3.87
27	3.63	29	3.75
28	3.75	21	1.66
33	3.20	28	3.12
28	3.41	28	2.96
29	3.38	26	2.92
23	3.53	30	3.10
27	2.03	24	2.81

10. The following set of data, presented to the Senate Antitrust Subcommittee, shows the comparative crash-survivability characteristics of cars in various classes. Find the least squares line that approximates these data. (The table shows the percent of accident-involved vehicles in which the most severe injury was fatal or serious.)

Type	Average Weight	Percent Occurrence
1. Domestic luxury regular	4800 lb	3.1
2. Domestic intermediate regular	3700 lb	4.0
3. Domestic economy regular	3400 lb	5.2
4. Domestic compact	2800 lb	6.4
5. Foreign compact	1900 lb	9.6

11. To determine a relationship between the number of fish and the number of species of fish in samples taken for a portion of the Great Barrier Reef, P. Sale and R. Dybdahl [SD] fit a linear least squares polynomial to the following collection of data, which were collected in samples over a 2-year period. Let x be the number of fish in the sample and y be the number of species in the sample.

x	y	x	y	x	y
13	11	29	12	60	14
15	10	30	14	62	21
16	11	31	16	64	21
21	12	36	17	70	24
22	12	40	13	72	17
23	13	42	14	100	23
25	13	55	22	130	34

Determine the linear least squares polynomial for these data.

12. To determine a functional relationship between the attenuation coefficient and the thickness of a sample of taconite, V. P. Singh [Si] fit a collection of data by using a linear least squares polynomial. The following collection of data is taken from a graph in that paper. Find the linear least squares polynomial fitting these data.

Thickness (cm)	Attenuation coefficient (dB/cm)
0.040	26.5
0.041	28.1
0.055	25.2
0.056	26.0
0.062	24.0
0.071	25.0
0.071	26.4
0.078	27.2
0.082	25.6
0.090	25.0
0.092	26.8
0.100	24.8
0.105	27.0
0.120	25.0
0.123	27.3
0.130	26.9
0.140	26.2

13. In a paper dealing with the efficiency of energy utilization of the larvae of the modest sphinx moth (*Pachysphinx modesta*), L. Schroeder [Schr1] used the following data to determine a relation between W, the live weight of the larvae in grams, and R, the oxygen consumption of the larvae in milliliters/hour. For biological reasons, it is assumed that a relationship in the form of $R = bW^a$ exists between W and R.

a. Find the logarithmic linear least squares polynomial by using

$$\ln R = \ln b + a \ln W.$$

b. Compute the error associated with the approximation in part (a):

$$E = \sum_{i=1}^{37} (R_i - bW_i^a)^2.$$

c. Modify the logarithmic least squares equation in part (a) by adding the quadratic term $c(\ln W_i)^2$, and determine the logarithmic quadratic least squares polynomial.

d. Determine the formula for and compute the error associated with the approximation in part (c).

W	R	W	R	W	R	W	R	W	R
0.017	0.154	0.025	0.23	0.020	0.181	0.020	0.180	0.025	0.234
0.087	0.296	0.111	0.357	0.085	0.260	0.119	0.299	0.233	0.537
0.174	0.363	0.211	0.366	0.171	0.334	0.210	0.428	0.783	1.47
1.11	0.531	0.999	0.771	1.29	0.87	1.32	1.15	1.35	2.48
1.74	2.23	3.02	2.01	3.04	3.59	3.34	2.83	1.69	1.44
4.09	3.58	4.28	3.28	4.29	3.40	5.48	4.15	2.75	1.84
5.45	3.52	4.58	2.96	5.30	3.88			4.83	4.66
5.96	2.40	4.68	5.10					5.53	6.94

14. Show that the normal equations (8.3) resulting from discrete least squares approximation yield a symmetric and nonsingular matrix and hence have a unique solution. [*Hint*: Let $A = (a_{ij})$,

where

$$a_{ij} = \sum_{k=1}^{m} x_k^{i+j-2}$$

and $x_1, x_2, \ldots, x_m$ are distinct with $n < m - 1$. Suppose A is singular and that $\mathbf{c} \neq \mathbf{0}$ is such that $\mathbf{c}^t A\mathbf{c} = 0$. Show that the nth-degree polynomial whose coefficients are the coordinates of $\mathbf{c}$ has more than n roots, and use this to establish a contradiction.]

8.2 Orthogonal Polynomials and Least Squares Approximation

The previous section considered the problem of least squares approximation to fit a collection of data. The other approximation problem mentioned in the introduction concerns the approximation of functions.

Suppose $f \in C[a, b]$ and that a polynomial $P_n(x)$ of degree at most n is required that will minimize the error

$$\int_a^b [f(x) - P_n(x)]^2\, dx.$$

To determine a least squares approximating polynomial, that is, a polynomial to minimize this expression, let

$$P_n(x) = a_n x^n + a_{n-1}x^{n-1} + \cdots + a_1 x + a_0 = \sum_{k=0}^{n} a_k x^k,$$

and define, as shown in Figure 8.5,

$$E(a_0, a_1, \ldots, a_n) = \int_a^b \left(f(x) - \sum_{k=0}^{n} a_k x^k \right)^2 dx.$$

Figure 8.5

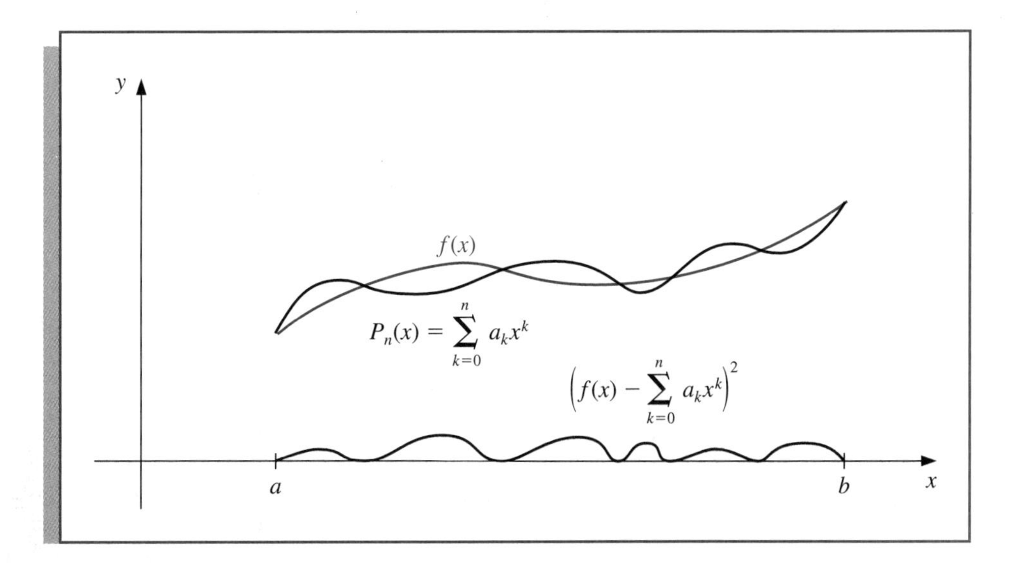

The problem is to find real coefficients $a_0, a_1, \ldots, a_n$ that will minimize E. A necessary condition for the numbers $a_0, a_1, \ldots, a_n$ to minimize E is that

$$\frac{\partial E}{\partial a_j} = 0 \quad \text{for each } j = 0, 1, \ldots, n.$$

Since

$$E = \int_a^b [f(x)]^2\,dx - 2\sum_{k=0}^{n} a_k \int_a^b x^k f(x)\,dx + \int_a^b \left(\sum_{k=0}^{n} a_k x^k\right)^2 dx,$$

we have

$$\frac{\partial E}{\partial a_j} = -2\int_a^b x^j f(x)\,dx + 2\sum_{k=0}^{n} a_k \int_a^b x^{j+k}\,dx.$$

Hence to find $P_n(x)$ the $(n+1)$ linear **normal equations**

(8.6)
$$\sum_{k=0}^{n} a_k \int_a^b x^{j+k}\,dx = \int_a^b x^j f(x)\,dx, \quad \text{for each } j = 0, 1, \ldots, n,$$

must be solved for the $(n+1)$ unknowns a_j. It can be shown that the normal equations always have a unique solution provided $f \in C[a, b]$. (See Exercise 15.)

EXAMPLE 1 Find the least squares approximating polynomial of degree two for the function $f(x) = \sin \pi x$ on the interval $[0, 1]$. The normal equations for $P_2(x) = a_2 x^2 + a_1 x + a_0$ are

$$a_0 \int_0^1 1\,dx + a_1 \int_0^1 x\,dx + a_2 \int_0^1 x^2\,dx = \int_0^1 \sin \pi x\,dx,$$

$$a_0 \int_0^1 x\,dx + a_1 \int_0^1 x^2\,dx + a_2 \int_0^1 x^3\,dx = \int_0^1 x \sin \pi x\,dx,$$

$$a_0 \int_0^1 x^2\,dx + a_1 \int_0^1 x^3\,dx + a_2 \int_0^1 x^4\,dx = \int_0^1 x^2 \sin \pi x\,dx.$$

Performing the integration yields

$$a_0 + \frac{1}{2}a_1 + \frac{1}{3}a_2 = \frac{2}{\pi}, \quad \frac{1}{2}a_0 + \frac{1}{3}a_1 + \frac{1}{4}a_2 = \frac{1}{\pi}, \quad \frac{1}{3}a_0 + \frac{1}{4}a_1 + \frac{1}{5}a_2 = \frac{\pi^2 - 4}{\pi^3}.$$

These three equations in three unknowns can be solved to obtain

$$a_0 = \frac{12\pi^2 - 120}{\pi^3} \approx -0.050465 \quad \text{and} \quad a_1 = -a_2 = \frac{720 \quad 60\pi^2}{\pi^3} \approx 4.12251.$$

Consequently, the least squares polynomial approximation of degree two for $f(x) = \sin \pi x$ on $[0, 1]$ is $P_2(x) = -4.12251x^2 + 4.12251x - 0.050465$. (See Figure 8.6.) ■

Figure 8.6

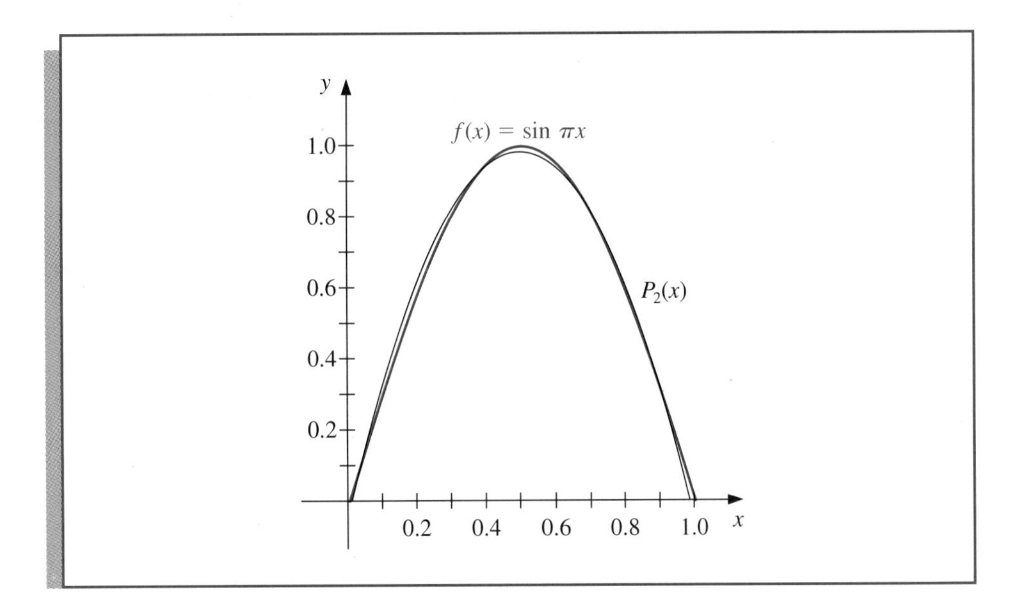

Example 1 illustrates the difficulty in obtaining a least squares polynomial approximation. An $(n + 1) \times (n + 1)$ linear system must be solved for the coefficients $a_0, \ldots, a_n$ of $P_n(x)$. The coefficients in the linear system are of the form

$$\int_a^b x^{j+k}\,dx = \frac{b^{j+k+1} - a^{j+k+1}}{j + k + 1},$$

a linear system that does not have an easily computed numerical solution. The matrix in the linear system is known as a **Hilbert matrix**. This ill-conditioned matrix is a classic example for demonstrating roundoff error difficulties. (See Exercise 6 of Section 7.4.) Another disadvantage is similar to the situation that occurred when the Lagrange polynomials were first introduced in Section 3.1. The calculations that were performed in obtaining the best nth-degree polynomial, $P_n(x)$, do not lessen the amount of work required to obtain $P_{n+1}(x)$, the polynomial of next higher degree.

A different technique to obtain least squares approximations will now be considered. This turns out to be computationally efficient, and once $P_n(x)$ is known, it is easy to determine $P_{n+1}(x)$. To facilitate the discussion, we need some new concepts.

Definition 8.1 The set of functions $\{\phi_0, \ldots, \phi_n\}$ is said to be **linearly independent** on $[a, b]$ if, whenever

$$c_0\phi_0(x) + c_1\phi_1(x) + \cdots + c_n\phi_n(x) = 0, \quad \text{for all } x \in [a, b],$$

then $c_0 = c_1 = \cdots = c_n = 0$.

Otherwise the set of functions is said to be **linearly dependent**. ■

Theorem 8.2 If $\phi_j(x)$ is a polynomial of degree j for each $j = 0, 1, \ldots, n$, then $\{\phi_0, \ldots, \phi_n\}$ is linearly independent on any interval $[a, b]$.

Proof Suppose $c_0, \ldots, c_n$ are real numbers for which

$$P(x) = c_0\phi_0(x) + c_1\phi_1(x) + \cdots + c_n\phi_n(x) = 0, \quad \text{for all } x \in [a, b].$$

Since the polynomial $P(x)$ vanishes on $[a, b]$, the coefficients of all the powers of x are zero. In particular, the coefficient of x^n is zero. Since $c_n\phi_n(x)$ is the only term in $P(x)$ that contains x^n, we must have $c_n = 0$ and

$$P(x) = \sum_{j=0}^{n-1} c_j\phi_j(x).$$

In this representation of $P(x)$, the only term that contains a power of x^{n-1} is $c_{n-1}\phi_{n-1}(x)$, so this term must also be zero and

$$P(x) = \sum_{j=0}^{n-2} c_j\phi_j(x).$$

In like manner, the remaining constants $c_{n-2}, c_{n-3}, \ldots, c_1, c_0$ are all zero, which implies that $\{\phi_0, \phi_1, \ldots, \phi_n\}$ is linearly independent. ■ ■ ■

EXAMPLE 2 Let $\phi_0(x) = 2$, $\phi_1(x) = x - 3$, and $\phi_2(x) = x^2 + 2x + 7$. By Theorem 8.2, $\{\phi_0, \phi_1, \phi_2\}$ is linearly independent on any interval $[a, b]$. Suppose $Q(x) = a_0 + a_1x + a_2x^2$. We will show that there exist constants c_0, c_1, and c_2 such that $Q(x) = c_0\phi_0(x) + c_1\phi_1(x) + c_2\phi_2(x)$. Note that

$$1 = \frac{1}{2}\phi_0(x), \quad x = \phi_1(x) + 3 = \phi_1(x) + \frac{3}{2}\phi_0(x),$$

and

$$\begin{aligned} x^2 &= \phi_2(x) - 2x - 7 = \phi_2(x) - 2\left[\phi_1(x) + \frac{3}{2}\phi_0(x)\right] - 7\left[\frac{1}{2}\phi_0(x)\right] \\ &= \phi_2(x) - 2\phi_1(x) - \frac{13}{2}\phi_0(x). \end{aligned}$$

Hence,

$$\begin{aligned} Q(x) &= a_0\left[\frac{1}{2}\phi_0(x)\right] + a_1\left[\phi_1(x) + \frac{3}{2}\phi_0(x)\right] + a_2\left[\phi_2(x) - 2\phi_1(x) - \frac{13}{2}\phi_0(x)\right] \\ &= \left(\frac{1}{2}a_0 + \frac{3}{2}a_1 - \frac{13}{2}a_2\right)\phi_0(x) + [a_1 - 2a_2]\phi_1(x) + a_2\phi_2(x). \end{aligned}$$ ■

The situation illustrated in Example 2 holds in a much more general setting. Let $\prod_n$ be **the set of all polynomials of degree at most n**. The following result is used extensively in many applications of linear algebra. Its proof is considered in Exercise 13.

Theorem 8.3 If $\{\phi_0(x), \phi_1(x), \ldots, \phi_n(x)\}$ is a collection of linearly independent polynomials in $\prod_n$, then any polynomial in $\prod_n$ can be written uniquely as a linear combination of $\phi_0(x), \phi_1(x), \ldots, \phi_n(x)$. ■

To discuss general function approximation requires the introduction of the notions of weight functions and orthogonality.

Definition 8.4 An integrable function w is called a **weight function** on the interval I if $w(x) \geq 0$ for all x in I, but $w(x) \not\equiv 0$ on any subinterval of I. ■

The purpose of a weight function is to assign varying degrees of importance to approximations on certain portions of the interval. For example, the weight function

$$w(x) = \frac{1}{\sqrt{1 - x^2}}$$

places less emphasis near the center of the interval $(-1, 1)$ and more emphasis when $|x|$ is near one (see Figure 8.7). This weight function is used in the next section.

Figure 8.7

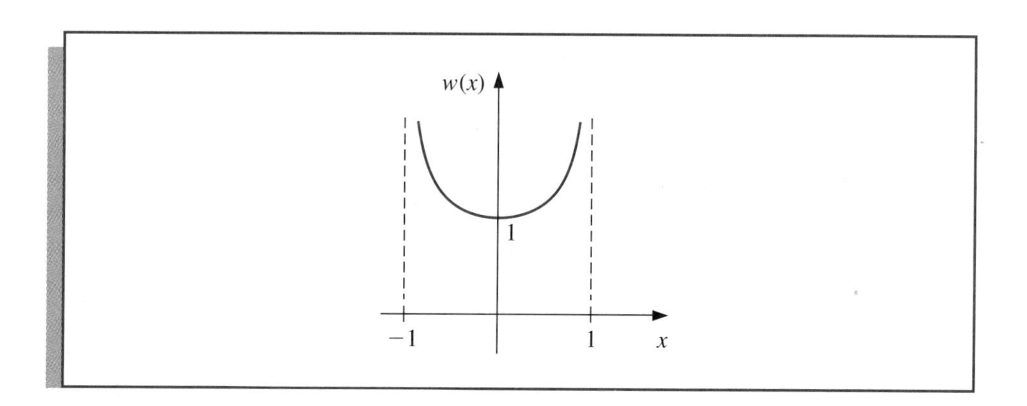

Suppose $\{\phi_0, \phi_1, \ldots, \phi_n\}$ is a set of linearly independent functions on $[a, b]$, w is a weight function for $[a, b]$, and, for $f \in C[a, b]$, a linear combination

$$\sum_{k=0}^{n} a_k \phi_k(x)$$

is sought to minimize the error

$$E(a_0, \ldots, a_n) = \int_a^b w(x) \left[f(x) - \sum_{k=0}^{n} a_k \phi_k(x) \right]^2 dx.$$

This problem reduces to the situation considered at the beginning of this section in the special case when $w(x) \equiv 1$ and $\phi_k(x) = x^k$ for each $k = 0, 1, \ldots, n$.

The normal equations associated with this problem are derived from the fact that for each $j = 0, 1, \ldots, n$,

$$0 = \frac{\partial E}{\partial a_j} = 2\int_a^b w(x)\left[f(x) - \sum_{k=0}^{n} a_k\phi_k(x)\right]\phi_j(x)\,dx.$$

The system of normal equations can be written

$$\int_a^b w(x)f(x)\phi_j(x)\,dx = \sum_{k=0}^{n} a_k \int_a^b w(x)\phi_k(x)\phi_j(x)\,dx, \quad \text{for } j = 0, 1, \ldots, n.$$

If the functions $\phi_0, \phi_1, \ldots, \phi_n$ can be chosen so that

$$\int_a^b w(x)\phi_k(x)\phi_j(x)\,dx = \begin{cases} 0, & \text{when } j \neq k, \\ \alpha_j > 0, & \text{when } j = k \end{cases} \tag{8.7}$$

then the normal equations reduce to

$$\int_a^b w(x)f(x)\phi_j(x)\,dx = a_j\int_a^b w(x)[\phi_j(x)]^2\,dx = a_j\alpha_j$$

for each $j = 0, 1, \ldots, n$, and

$$a_j = \frac{1}{\alpha_j}\int_a^b w(x)f(x)\phi_j(x)\,dx.$$

Hence the least squares approximation problem is greatly simplified when the functions $\phi_0, \phi_1, \ldots, \phi_n$ are chosen to satisfy Eq. (8.7). The remainder of this section is devoted to studying collections of this type.

Definition 8.5 $\{\phi_0, \phi_1, \ldots, \phi_n\}$ is said to be an **orthogonal set of functions** for the interval $[a, b]$ with respect to the weight function w if

$$\int_a^b w(x)\phi_j(x)\phi_k(x)\,dx = \begin{cases} 0, & \text{when } j \neq k, \\ \alpha_k > 0, & \text{when } j = k. \end{cases}$$

If, in addition, $\alpha_k = 1$ for each $k = 0, 1, \ldots, n$, the set is said to be **orthonormal**. ■

This definition, together with the remarks preceding it, produces the following theorem.

Theorem 8.6 If $\{\phi_0, \ldots, \phi_n\}$ is an orthogonal set of functions on an interval $[a, b]$ with respect to the weight function w, then the least squares approximation to f on $[a, b]$ with respect to w is

$$\sum_{k=0}^{n} a_k\phi_k(x),$$

where, for each $k = 0, 1, \ldots, n$,

$$a_k = \frac{\int_a^b w(x)\phi_k(x)f(x)\,dx}{\int_a^b w(x)[\phi_k(x)]^2\,dx} = \frac{1}{\alpha_k}\int_a^b w(x)\phi_k(x)f(x)\,dx. \qquad \blacksquare$$

For the remainder of this section, only orthogonal sets of polynomials will be considered. The next theorem, which is based on the **Gram-Schmidt process**, describes how to construct orthogonal polynomials on $[a, b]$ with respect to a weight function w.

Theorem 8.7 The set of polynomial functions $\{\phi_0, \phi_1, \ldots, \phi_n\}$ defined in the following way is orthogonal on $[a, b]$ with respect to the weight function w.

$$\phi_0(x) \equiv 1, \quad \phi_1(x) = x - B_1, \quad \text{for each } x \text{ in } [a, b],$$

where

$$B_1 = \frac{\int_a^b xw(x)[\phi_0(x)]^2\,dx}{\int_a^b w(x)[\phi_0(x)]^2\,dx},$$

and when $k \geq 2$,

$$\phi_k(x) = (x - B_k)\phi_{k-1}(x) - C_k\phi_{k-2}(x), \quad \text{for each } x \text{ in } [a, b],$$

where

$$B_k = \frac{\int_a^b xw(x)[\phi_{k-1}(x)]^2\,dx}{\int_a^b w(x)[\phi_{k-1}(x)]^2\,dx}$$

and

$$C_k = \frac{\int_a^b xw(x)\phi_{k-1}(x)\phi_{k-2}(x)\,dx}{\int_a^b w(x)[\phi_{k-2}(x)]^2\,dx}. \qquad \blacksquare$$

Theorem 8.7 provides a recursive procedure for constructing a set of orthogonal polynomials. The proof of this theorem follows by applying mathematical induction to the degree of the polynomial $\phi_n(x)$.

Corollary 8.8 For any $n > 0$, the set of polynomial functions $\{\phi_0, \ldots, \phi_n\}$ given in Theorem 8.7 is linearly independent on $[a, b]$ and

$$\int_a^b w(x)\phi_n(x)Q_k(x)\,dx = 0,$$

for any polynomial $Q_k(x)$ of degree $k < n$.

Proof Since $\phi_n(x)$ is a polynomial of degree n, Theorem 8.2 implies that $\{\phi_0, \ldots, \phi_n\}$ is a linearly independent set.

Let $Q_k(x)$ be a polynomial of degree k. By Theorem 8.3 there exist numbers $c_0, \ldots, c_k$ such that

$$Q_k(x) = \sum_{j=0}^{k} c_j \phi_j(x).$$

Thus,

$$\int_a^b w(x) Q_k(x) \phi_n(x)\, dx = \sum_{j=0}^{k} c_j \int_a^b w(x) \phi_j(x) \phi_n(x)\, dx = \sum_{j=0}^{k} c_j \cdot 0 = 0,$$

since ϕ_n is orthogonal to ϕ_j for each $j = 0, 1, \ldots, k$. ▪ ▪ ▪

EXAMPLE 3 The set of **Legendre polynomials**, $\{P_n(x)\}$, is orthogonal on $[-1, 1]$ with respect to the weight function $w(x) \equiv 1$. The classical definition of the Legendre polynomials requires that $P_n(1) = 1$ for each n, and a recursive relation is used to generate the polynomials when $n \geq 2$. This normalization will not be needed in our discussion, and the least squares approximating polynomials generated in either case are essentially the same. Using the recursive procedure of Theorem 8.7 with $P_0(x) \equiv 1$ gives

$$B_1 = \frac{\int_{-1}^{1} x\, dx}{\int_{-1}^{1} dx} = 0 \quad \text{and} \quad P_1(x) = (x - B_1)P_0(x) = x.$$

Figure 8.8

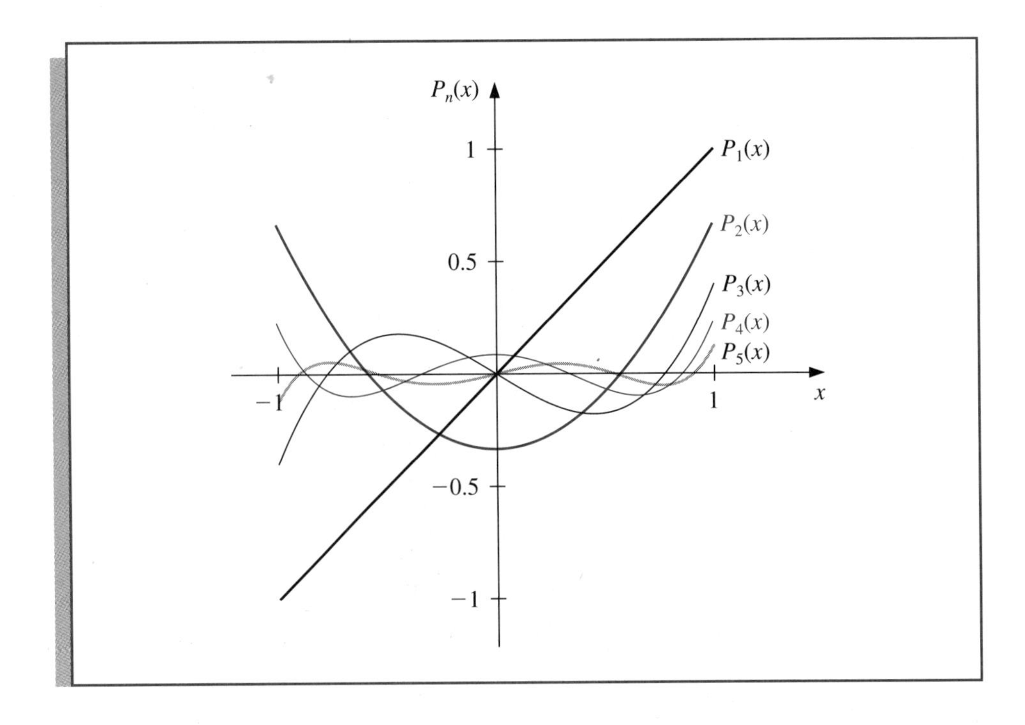

Also,

$$B_2 = \frac{\int_{-1}^{1} x^3\,dx}{\int_{-1}^{1} x^2\,dx} = 0 \quad \text{and} \quad C_2 = \frac{\int_{-1}^{1} x^2\,dx}{\int_{-1}^{1} 1\,dx} = \frac{1}{3},$$

so

$$P_2(x) = (x - B_2)P_1(x) - C_2P_0(x) = (x - 0)x - \frac{1}{3} \cdot 1 = x^2 - \frac{1}{3}.$$

Higher-degree Legendre polynomials are derived in the same manner. The next three are $P_3(x) = x^3 - \frac{3}{5}x$, $P_4(x) = x^4 - \frac{6}{7}x^2 + \frac{3}{35}$, and $P_5(x) = x^5 - \frac{10}{9}x^3 + \frac{5}{21}x$. Figure 8.8 shows the graphs of these polynomials. ■

The Legendre polynomials were mentioned in Section 4.7, where their roots were used as the nodes in Gaussian quadrature.

EXERCISE SET 8.2

1. Find the linear least squares polynomial approximation to $f(x)$ on the indicated interval if
   a. $f(x) = x^2 + 3x + 2$, $[0, 1]$;
   b. $f(x) = x^3$, $[0, 2]$;
   c. $f(x) = \frac{1}{x}$, $[1, 3]$;
   d. $f(x) = e^x$, $[0, 2]$;
   e. $f(x) = \frac{1}{2}\cos x + \frac{1}{3}\sin 2x$, $[0, 1]$;
   f. $f(x) = x \ln x$, $[1, 3]$.
2. Find the least squares polynomial approximation of degree two to the functions and intervals in Exercise 1.
3. Find the linear least squares polynomial approximation on the interval $[-1, 1]$ for the following functions.
   a. $f(x) = x^2 - 2x + 3$
   b. $f(x) = x^3$
   c. $f(x) = \dfrac{1}{x + 2}$
   d. $f(x) = e^x$
   e. $f(x) = \frac{1}{2}\cos x + \frac{1}{3}\sin 2x$
   f. $f(x) = \ln(x + 2)$
4. Find the least squares polynomial approximation of degree two on the interval $[-1, 1]$ for the functions in Exercise 3.
5. Compute the error E for the approximations in Exercise 3.
6. Compute the error E for the approximations in Exercise 4.
7. Use the Gram-Schmidt process to construct $\phi_0(x)$, $\phi_1(x)$, $\phi_2(x)$, and $\phi_3(x)$ for the following intervals.
   a. $[0, 1]$
   b. $[0, 2]$
   c. $[1, 3]$
8. Repeat Exercise 1 using the results of Exercise 7.
9. Repeat Exercise 2 using the results of Exercise 7.
10. Obtain the least squares approximation polynomial of degree three for the functions in Exercise 1 using the results of Exercise 7.

11. Use the Gram-Schmidt procedure to calculate L_1, L_2, and L_3, where $\{L_0(x), L_1(x), L_2(x), L_3(x)\}$ is an orthogonal set of polynomials on $(0, \infty)$ with respect to the weight functions $w(x) = e^{-x}$ and $L_0(x) \equiv 1$. The polynomials obtained from this procedure are called the **Laguerre polynomials**.
12. Use the Laguerre polynomials calculated in Exercise 11 to compute the least squares polynomials of degree one, two, and three on the interval $(0, \infty)$ with respect to the weight function $w(x) = e^{-x}$ for the following functions:

    a. $f(x) = x^2$ **b.** $f(x) = e^{-x}$ **c.** $f(x) = x^3$ **d.** $f(x) = e^{-2x}$
13. Suppose $\{\phi_0(x), \phi_1(x), \ldots, \phi_n(x)\}$ is any linearly independent set in $\prod_n$. Show that for any element $Q(x) \in \prod_n$ there exist unique constants $c_0, c_1, \ldots, c_n$ such that

$$Q(x) = \sum_{k=0}^{n} c_k \phi_k(x).$$

14. Show that if $\{\phi_0, \phi_1, \ldots, \phi_n\}$ is an orthogonal set of functions on $[a, b]$ with respect to the weight function w, then $\{\phi_0, \phi_1, \ldots, \phi_n\}$ is a linearly independent set.
15. Show that the normal equations (8.6) have a unique solution. [*Hint*: Show that the only solution for the function $f(x) \equiv 0$ is $a_j = 0, j = 0, 1, \ldots, n$. Multiply Eq. (8.6) by a_j and sum over all j. Interchange the integral sign and the summation sign to obtain $\int_a^b [P(x)]^2 dx = 0$. Thus, $P(x) \equiv 0$, so $a_j = 0$ for $j = 0, \ldots, n$. Hence, the coefficient matrix is nonsingular, and there is a unique solution to Eq. (8.6).]

8.3 Chebyshev Polynomials and Economization of Power Series

The Chebyshev polynomials $\{T_n(x)\}$ are orthogonal on $(-1, 1)$ with respect to the weight function $w(x) = (1 - x^2)^{-1/2}$. Although they can be derived by the method in the previous section, it is easier to give their definition and then show that they satisfy the required orthogonality properties.

For $x \in [-1, 1]$, define

(8.8) $$T_n(x) = \cos[n \arccos x], \quad \text{for each } n \geq 0.$$

Introducing the substitution $\theta = \arccos x$ changes this equation to

$$T_n(\theta(x)) \equiv T_n(\theta) = \cos(n\theta), \quad \text{where } \theta \in [0, \pi].$$

A recurrence relation is derived by noting that

$$T_{n+1}(\theta) = \cos(n\theta) \cos \theta - \sin(n\theta) \sin \theta$$

and

$$T_{n-1}(\theta) = \cos(n\theta) \cos \theta + \sin(n\theta) \sin \theta.$$

so

$$T_{n+1}(\theta) = 2 \cos(n\theta) \cos \theta - T_{n-1}(\theta).$$

Returning to the variable x gives

(8.9) $$T_{n+1}(x) = 2xT_n(x) - T_{n-1}(x), \quad \text{for each } n \geq 1.$$

Since

$$T_0(x) = \cos(0 \cdot \arccos x) = 1 \quad \text{and} \quad T_1(x) = \cos(1 \arccos x) = x,$$

the recurrence relation implies that $T_n(x)$ is a polynomial of degree n with leading coefficient 2^{n-1}. The next three Chebyshev polynomials are

$$T_2(x) = 2xT_1(x) - T_0(x) = 2x^2 - 1,$$
$$T_3(x) = 2xT_2(x) - T_1(x) = 4x^3 - 3x,$$

and

$$T_4(x) = 2xT_3(x) - T_2(x) = 8x^4 - 8x^2 + 1.$$

The graphs of T_1, T_2, T_3, and T_4 are shown in Figure 8.9. To show the orthogonality of the Chebyshev polynomials, consider

$$\int_{-1}^{1} \frac{T_n(x)T_m(x)}{\sqrt{1-x^2}}\,dx = \int_{-1}^{1} \frac{\cos(n \arccos x)\cos(m \arccos x)}{\sqrt{1-x^2}}\,dx.$$

Figure 8.9

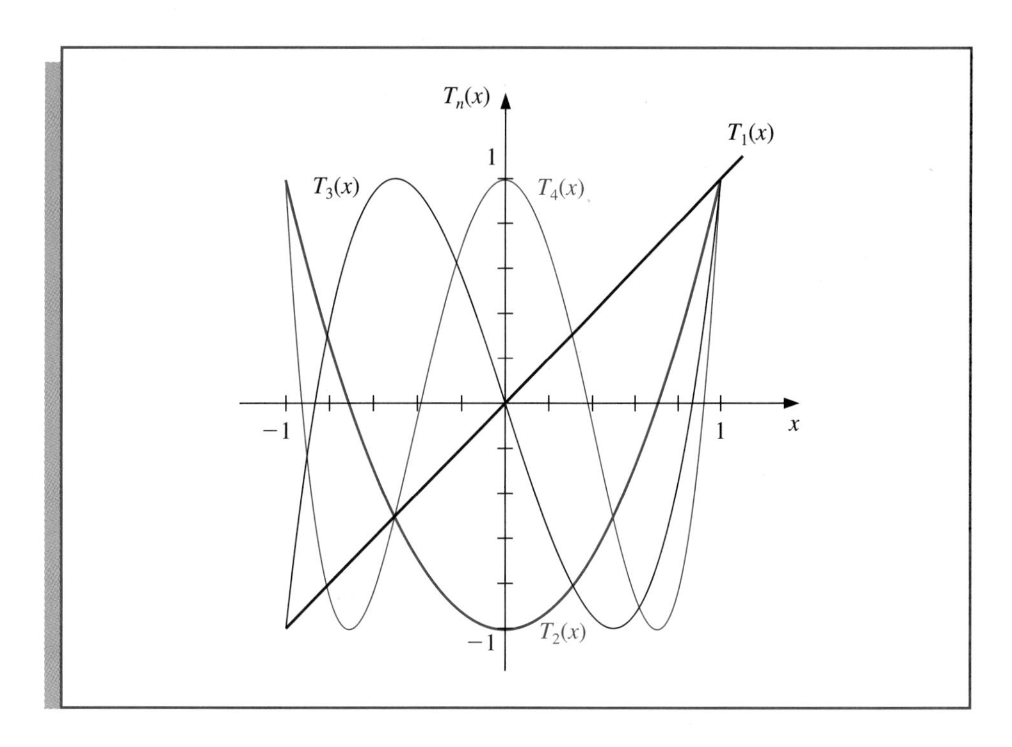

Reintroducing the substitution $\theta = \arccos x$ gives

$$d\theta = -\frac{1}{\sqrt{1-x^2}}\,dx$$

and

$$\int_{-1}^{1} \frac{T_n(x)T_m(x)}{\sqrt{1-x^2}}\,dx = -\int_{\pi}^{0} \cos(n\theta)\cos(m\theta)\,d\theta = \int_{0}^{\pi} \cos(n\theta)\cos(m\theta)\,d\theta.$$

Suppose $n \neq m$. Since $\cos(n\theta)\cos(m\theta) = \frac{1}{2}[\cos(n+m)\theta + \cos(n-m)\theta]$, we have

$$\begin{aligned}\int_{-1}^{1} \frac{T_n(x)T_m(x)}{\sqrt{1-x^2}}\,dx &= \frac{1}{2}\int_{0}^{\pi} \cos((n+m)\theta)\,d\theta + \frac{1}{2}\int_{0}^{\pi} \cos((n-m)\theta)\,d\theta \\ &= \left[\frac{1}{2(n+m)}\sin((n+m)\theta) + \frac{1}{2(n-m)}\sin((n-m)\theta)\right]_{0}^{\pi} \\ &= 0.\end{aligned}$$

By a similar technique, it can also be shown that when $n = m$,

(8.10)
$$\int_{-1}^{1} \frac{[T_n(x)]^2}{\sqrt{1-x^2}}\,dx = \frac{\pi}{2} \quad \text{for each } n \geq 1.$$

The Chebyshev polynomials are used to minimize approximation error. We will see how they are used to solve two problems of this type:

1. An optimal placing of interpolating points to minimize the error in Lagrange interpolation;
2. A means of reducing the degree of an approximating polynomial with minimal loss of accuracy.

The next result concerns the zeros and extreme points of T_n.

Theorem 8.9 The Chebyshev polynomial $T_n(x)$ of degree $n \geq 1$ has n simple zeros in $[-1, 1]$ at

$$\bar{x}_k = \cos\left(\frac{2k-1}{2n}\pi\right), \quad \text{for each } k = 1, 2, \ldots, n.$$

Moreover, T_n assumes its absolute extrema for each $k = 0, 1, \ldots, n$ at

$$\bar{x}'_k = \cos\left(\frac{k\pi}{n}\right), \quad \text{with} \quad T_n(\bar{x}'_k) = (-1)^k.$$

Proof If we let

$$\bar{x}_k = \cos\left(\frac{2k-1}{2n}\pi\right), \quad \text{for } k = 1, 2, \ldots, n,$$

then

$$T_n(\bar{x}_k) = \cos(n \arccos \bar{x}_k) = \cos\left(n \arccos\left(\cos\left(\frac{2k-1}{2n}\pi\right)\right)\right)$$
$$= \cos\left(\frac{2k-1}{2}\pi\right) = 0,$$

and each $\bar{x}_k$ is a distinct zero of T_n. Since $T_n(x)$ is a polynomial of degree n, all zeros of T_n must be of this form.

To show the second part, first note that

$$T_n'(x) = \frac{d}{dx}[\cos(n \arccos x)] = \frac{n \sin(n \arccos x)}{\sqrt{1-x^2}},$$

and that, when $k = 1, 2, \ldots, n-1$,

$$T_n'(\bar{x}_k') = \frac{n \sin\left(n \arccos\left(\cos\left(\frac{k\pi}{n}\right)\right)\right)}{\sqrt{1-\left[\cos\left(\frac{k\pi}{n}\right)\right]^2}} = \frac{n \sin(k\pi)}{\sin\left(\frac{k\pi}{n}\right)} = 0.$$

Since $T_n(x)$ is a polynomial of degree n, its derivative $T_n'(x)$ is a polynomial of degree $(n-1)$, and all the zeros of T_n' occur at these $n-1$ points. The only other possibilities for extrema of T_n occur at the endpoints of the interval $[-1, 1]$; that is, $\bar{x}_0' = 1$ and $\bar{x}_n' = -1$. Since for any $k = 0, 1, \ldots, n$, we have

$$T_n(\bar{x}_k') = \cos\left(n \arccos\left(\cos\left(\frac{k\pi}{n}\right)\right)\right) = \cos(k\pi) = (-1)^k,$$

a maximum occurs at each even value of k and a minimum at each odd value. ■ ■ ■

The monic Chebyshev polynomials (polynomials with leading coefficient 1) $\tilde{T}_n(x)$ are derived from the Chebyshev polynomial $T_n(x)$ by dividing by the leading coefficient 2^{n-1}. Hence,

(8.11) $$\tilde{T}_0(x) = 1 \quad \text{and} \quad \tilde{T}_n(x) = \frac{1}{2^{n-1}}T_n(x), \quad \text{for each } n \geq 1.$$

The recurrence relationship satisfied by the Chebyshev polynomials implies that

(8.12) $$\tilde{T}_2(x) = x\tilde{T}_1(x) - \frac{1}{2}\tilde{T}_0(x) \quad \text{and}$$
$$\tilde{T}_{n+1}(x) = x\tilde{T}_n(x) - \frac{1}{4}\tilde{T}_{n-1}(x), \quad \text{for each } n \geq 2.$$

The graphs of $\tilde{T}_1$, $\tilde{T}_2$, $\tilde{T}_3$, $\tilde{T}_4$, and $\tilde{T}_5$ are shown in Figure 8.10.

Figure 8.10

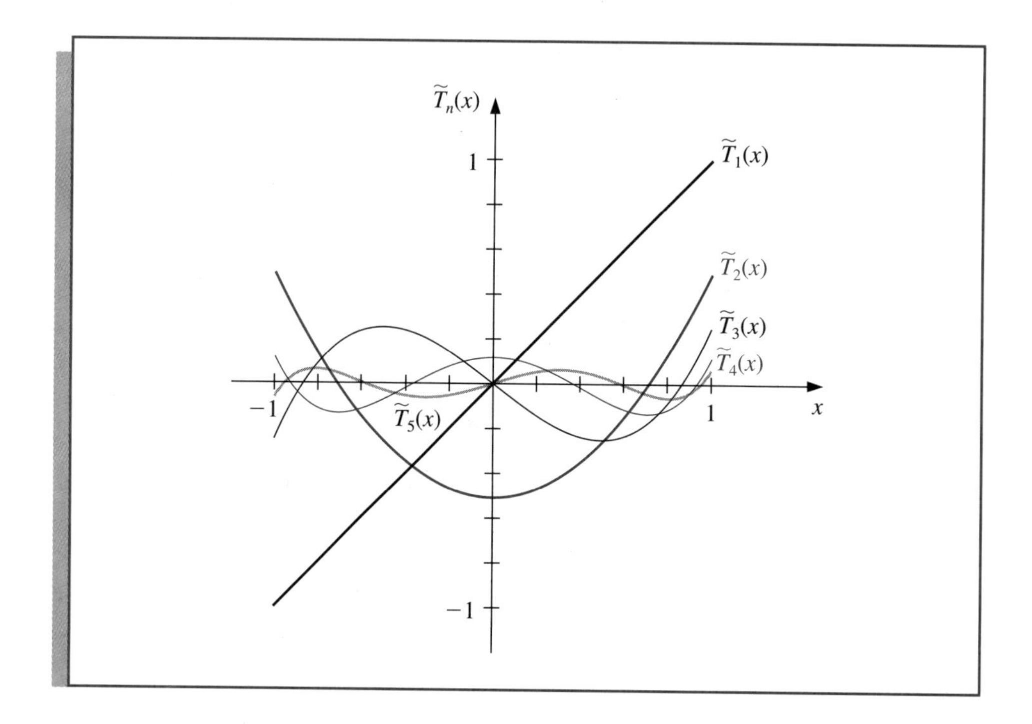

Because $\tilde{T}_n$ is just a multiple of T_n, Theorem 8.9 implies that the zeros of $\tilde{T}_n$ also occur at

$$\bar{x}_k = \cos\left(\frac{2k-1}{2n}\pi\right), \quad \text{for each } k = 1, 2, \ldots, n,$$

and the extreme values of $\tilde{T}_n(x)$, for $n \geq 1$, occur at

$$\textbf{(8.13)} \quad \bar{x}'_k = \cos\left(\frac{k\pi}{n}\right), \quad \text{with} \quad \tilde{T}_n(\bar{x}'_k) = \frac{(-1)^k}{2^{n-1}}, \quad \text{for each } k = 0, 1, 2, \ldots, n.$$

Let $\tilde{\Pi}_n$ denote **the set of all monic polynomials of degree n**. The relation expressed in Eq. (8.13) leads to an important minimization property that distinguishes $\tilde{T}_n(x)$ from the other members of $\tilde{\Pi}_n$.

Theorem 8.10 The polynomials of the form $\tilde{T}_n(x)$, when $n \geq 1$, have the property that

$$\frac{1}{2^{n-1}} = \max_{x\in[-1,1]} |\tilde{T}_n(x)| \leq \max_{x\in[-1,1]} |P_n(x)|, \quad \text{for all } P_n(x) \in \tilde{\Pi}_n.$$

Moreover, equality can occur only if $P_n \equiv \tilde{T}_n$.

Proof Suppose $P_n(x) \in \tilde{\Pi}_n$ and

$$\max_{x\in[-1,1]} |P_n(x)| \leq \frac{1}{2^{n-1}} = \max_{x\in[-1,1]} |\tilde{T}_n(x)|.$$

Let $Q = \tilde{T}_n - P_n$. Since $\tilde{T}_n(x)$ and $P_n(x)$ are both monic polynomials of degree n, $Q(x)$ is a polynomial of degree at most $(n-1)$. Moreover, at the extreme points $\bar{x}'_k$ of $\tilde{T}_n$,

$$Q(\bar{x}'_k) = \tilde{T}_n(\bar{x}'_k) - P_n\left(\bar{x}'_k\right) = \frac{(-1)^k}{2^{n-1}} - P_n\left(\bar{x}'_k\right).$$

Since

$$|P_n\left(\bar{x}'_k\right)| \le \frac{1}{2^{n-1}}, \quad \text{for each } k = 0, 1, \ldots, n,$$

we have

$$Q\left(\bar{x}'_k\right) \le 0 \quad \text{when } k \text{ is odd} \quad \text{and} \quad Q\left(\bar{x}'_k\right) \ge 0 \quad \text{when } k \text{ is even.}$$

Since Q is continuous, the Intermediate Value Theorem implies that the polynomial $Q(x)$ has at least one zero between $\bar{x}'_j$ and $\bar{x}'_{j+1}$ for each $j = 0, 1, \ldots, n-1$. Thus Q has at least n zeros in the interval $[-1, 1]$. But the degree of $Q(x)$ is less than n, so $Q \equiv 0$. This implies that $P_n \equiv \tilde{T}_n$. ■ ■ ■

This theorem can be used to answer the question of where to place interpolating nodes to minimize the error in Lagrange interpolation. Theorem 3.3 applied to the interval $[-1, 1]$ states that, if $x_0, \ldots, x_n$ are distinct numbers in the interval $[-1, 1]$ and $f \in C^{n+1}[-1, 1]$, then, for each $x \in [-1, 1]$, a number $\xi(x)$ exists in $(-1, 1)$ with

$$f(x) - P(x) = \frac{f^{(n+1)}(\xi(x))}{(n+1)!}(x - x_0)(x - x_1) \cdots (x - x_n),$$

where $P(x)$ denotes the Lagrange interpolating polynomial. Generally, there is no control over $\xi(x)$, so to minimize the error by shrewd placement of the nodes $x_0, \ldots, x_n$, we find $x_0, \ldots, x_n$ to minimize the quantity

$$|(x - x_0)(x - x_1) \cdots (x - x_n)|$$

throughout the interval $[-1, 1]$. Since $(x - x_0)(x - x_1) \cdots (x - x_n)$ is a monic polynomial of degree $(n + 1)$, we have just seen that the minimum is obtained if and only if

$$(x - x_0)(x - x_1) \cdots (x - x_n) = \tilde{T}_{n+1}(x).$$

The maximum value of $|(x - x_0)(x - x_1) \cdots (x - x_n)|$ is minimized when x_k is chosen to be the $(k + 1)$st zero of $\tilde{T}_{n+1}$, for each $k = 0, 1, \ldots, n$; that is, when x_k is

$$\bar{x}_{k+1} = \cos \frac{2k + 1}{2(n + 1)}\pi.$$

Since $\max_{x\in[-1,1]} |\tilde{T}_{n+1}(x)| = 2^{-n}$, this also implies that

$$\frac{1}{2^n} = \max_{x\in[-1,1]} |(x-\bar{x}_1)\cdots(x-\bar{x}_{n+1})| \le \max_{x\in[-1,1]} |(x-x_0)\cdots(x-x_n)|$$

for any choice of $x_0, x_1, \ldots, x_n$ in the interval $[-1, 1]$.

The next corollary follows from this discussion.

Corollary 8.11 If $P(x)$ is the interpolating polynomial of degree at most n with nodes at the roots of $T_{n+1}(x)$, then

$$\max_{x\in[-1,1]} |f(x) - P(x)| \le \frac{1}{2^n(n+1)!} \max_{x\in[-1,1]} |f^{(n+1)}(x)|, \quad \text{for each } f \in C^{n+1}[-1,1]. \quad \blacksquare$$

This technique for choosing points to minimize the interpolating error is extended to a general closed interval $[a, b]$ by using the change of variables

$$\tilde{x} = \frac{1}{2}[(b-a)x + a + b]$$

to shift the numbers $\bar{x}_k$ in the interval $[-1, 1]$ into the corresponding number $\tilde{x}_k$ in the interval $[a, b]$.

EXAMPLE 1 Let $f(x) = xe^x$ on $[0, 1.5]$. Two interpolation polynomials of degree at most three will be constructed. First, the equally spaced nodes $x_0 = 0$, $x_1 = 0.5$, $x_2 = 1$, and $x_3 = 1.5$ are used to give

$$L_0(x) = -1.3333x^3 + 4.0000x^2 - 3.6667x + 1,$$

$$L_1(x) = 4.0000x^3 - 10.000x^2 + 6.0000x,$$

$$L_2(x) = -4.0000x^3 + 8.0000x^2 - 3.0000x,$$

$$L_3(x) = 1.3333x^3 - 2.000x^2 + 0.66667x.$$

For the values listed in the first two columns of Table 8.7, the first polynomial is given by

$$P_3(x) = 1.3875x^3 + 0.057570x^2 + 1.2730x.$$

Table 8.7

x	$f(x) = xe^x$	$\tilde{x}$	$f(\tilde{x}) = \tilde{x}e^{\tilde{x}}$
$x_0 = 0.0$	0.00000	$\tilde{x}_0 = 1.44291$	6.10783
$x_1 = 0.5$	0.824361	$\tilde{x}_1 = 1.03701$	2.92517
$x_2 = 1.0$	2.71828	$\tilde{x}_2 = 0.46299$	0.73560
$x_3 = 1.5$	6.72253	$\tilde{x}_3 = 0.05709$	0.060444

For the second interpolating polynomial, shift the zeros $\bar{x}_k = \cos((2k+1)/8)\pi$, for $k = 0, 1, 2, 3$, of $\tilde{T}_4$ from $[-1, 1]$ to $[0, 1.5]$, using the linear transformation

$$\tilde{x}_k = 0.75 + 0.75\bar{x}_k$$

to obtain

$$\tilde{x}_0 = 1.44291, \quad \tilde{x}_1 = 1.03701, \quad \tilde{x}_2 = 0.46299, \quad \text{and} \quad \tilde{x}_3 = 0.05709.$$

The Lagrange coefficient polynomials for this set of nodes are then computed as:

$$\tilde{L}_0(x) = 1.8142x^3 - 2.8249x^2 + 1.0264x - 0.049728,$$
$$\tilde{L}_1(x) = -4.3799x^3 + 8.5977x^2 - 3.4026x + 0.16705,$$
$$\tilde{L}_2(x) = 4.3799x^3 - 11.112x^2 + 7.1738x - 0.37415,$$
$$\tilde{L}_3(x) = -1.8142x^3 + 5.3390x^2 - 4.7976x + 1.2568.$$

The functional values required for these polynomials are given in the last two columns of Table 8.7. The interpolation polynomial of degree at most three is given by

$$\tilde{P}_3(x) = 1.3811x^3 + 0.044652x^2 + 1.3031x - 0.014352.$$

For comparison, Table 8.8 lists various values of x, together with the values of $f(x)$, $P_3(x)$, and $\tilde{P}_3(x)$. It can be seen from this table that, although the error using $P_3(x)$ is less than using $\tilde{P}_3(x)$ near the middle of the table, the maximum error involved with using $\tilde{P}_3(x)$ is 0.0190 compared to a maximum error of 0.0290 when using $P_3(x)$. (See Figure 8.11 on page 504.) ■

Table 8.8

x	$f(x) = xe^x$	$P_3(x)$	$\lvert xe^x - P_3(x)\rvert$	$\tilde{P}_3(x)$	$\lvert xe^x - \tilde{P}_3(x)\rvert$
0.15	0.1743	0.1969	0.0226	0.1868	0.0125
0.25	0.3210	0.3435	0.0225	0.3358	0.0148
0.35	0.4967	0.5121	0.0154	0.5064	0.0097
0.65	1.245	1.233	0.012	1.231	0.014
0.75	1.588	1.572	0.016	1.571	0.017
0.85	1.989	1.976	0.013	1.974	0.015
1.15	3.632	3.650	0.018	3.644	0.012
1.25	4.363	4.391	0.028	4.382	0.019
1.35	5.208	5.237	0.029	5.224	0.016

Chebyshev polynomials can also be used to reduce the degree of an approximating polynomial with a minimal loss of accuracy. Because the Chebyshev polynomials have a minimum maximum-absolute value that is spread uniformly on an interval, they can be used to reduce the degree of an approximation polynomial without exceeding the error tolerance.

Consider approximating an arbitrary nth-degree polynomial

$$P(x) = a_n x^n + a_{n-1}x^{n-1} + \cdots + a_1 x + a_0$$

Figure 8.11

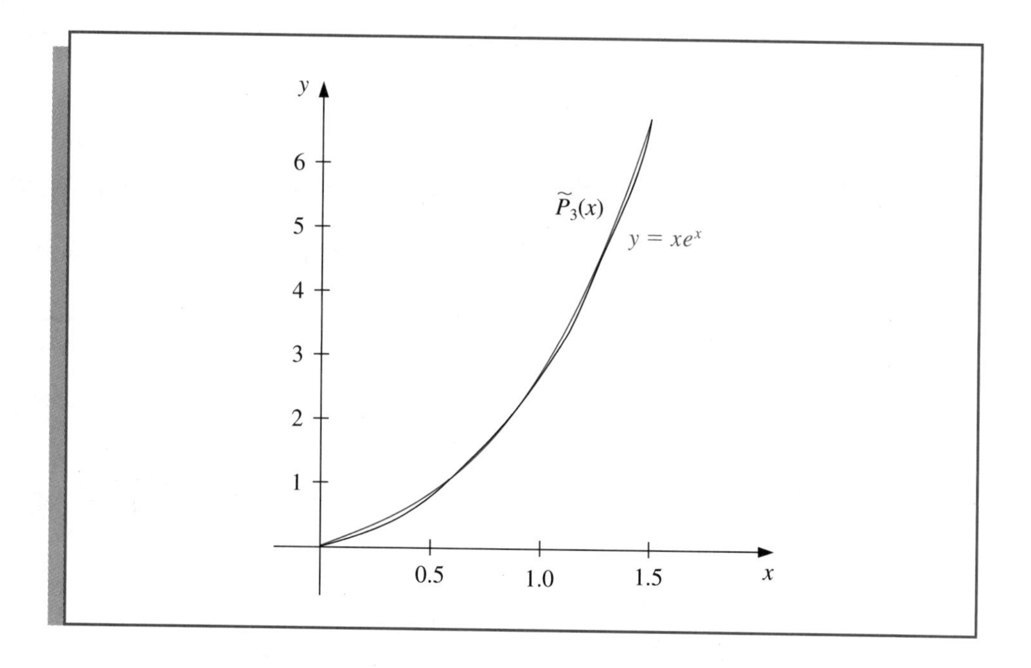

on $[-1, 1]$ with a polynomial of degree at most $n-1$. The object is to choose $P_{n-1}(x)$ in $\prod_{n-1}$ so that

$$\max_{x\in[-1,1]} |P(x) - P_{n-1}(x)|$$

is as small as possible.

We first note that $(P(x) - P_{n-1}(x))/a_n$ is a monic polynomial of degree n. Applying Theorem 8.10 gives

$$\max_{x\in[-1,1]} \left| \frac{1}{a_n}(P_n(x) - P_{n-1}(x)) \right| \geq \frac{1}{2^{n-1}}.$$

Equality occurs precisely when

$$\frac{1}{a_n}(P_n(x) - P_{n-1}(x)) = \tilde{T}_n(x).$$

This means that we should choose

$$P_{n-1}(x) = P(x) - a_n\tilde{T}_n(x),$$

and with this choice we have the minimum value of

$$\max_{x\in[-1,1]} |P(x) - P_{n-1}(x)| = |a_n| \max_{x\in[-1,1]} \left| \frac{1}{a_n}(P(x) - P_{n-1}(x)) \right| = \frac{|a_n|}{2^{n-1}}.$$

EXAMPLE 2 The function $f(x) = e^x$ will be approximated on the interval $[-1, 1]$ by the fourth Maclaurin polynomial

$$P_4(x) = 1 + x + \frac{x^2}{2} + \frac{x^3}{6} + \frac{x^4}{24},$$

which has truncation error

$$|R_4(x)| = \frac{|f^{(5)}(\xi(x))||x^5|}{120} \leq \frac{e}{120} \approx 0.023, \quad \text{for } -1 \leq x \leq 1.$$

Suppose that an error of 0.05 is tolerable and that we would like to reduce the degree of the approximating polynomial while staying within this bound.

The polynomial of degree three or less that best uniformly approximates $P_4(x)$ on $[-1, 1]$ is

$$\begin{aligned} P_3(x) &= P_4(x) - a_4\tilde{T}_4(x) \\ &= 1 + x + \frac{x^2}{2} + \frac{x^3}{6} + \frac{x^4}{24} - \frac{1}{24}\left(x^4 - x^2 + \frac{1}{8}\right) \\ &= \frac{191}{192} + x + \frac{13}{24}x^2 + \frac{1}{6}x^3. \end{aligned}$$

With this choice we have

$$|P_4(x) - P_3(x)| = |a_4\tilde{T}_4(x)| \leq \frac{1}{24} \cdot \frac{1}{2^3} = \frac{1}{192} \leq 0.0053.$$

Adding this error bound to the bound for the Maclaurin truncation error gives

$$0.023 + 0.0053 = 0.0283,$$

which is still within the permissible error of 0.05.

The polynomial of degree two or less that best uniformly approximates $P_3(x)$ on $[-1, 1]$ is

$$\begin{aligned} P_2(x) &= P_3(x) - \frac{1}{6}\tilde{T}_3(x) \\ &= \frac{191}{192} + x + \frac{13}{24}x^2 + \frac{1}{6}x^3 - \frac{1}{6}\left(x^3 - \frac{3}{4}x\right) = \frac{191}{192} + \frac{9}{8}x + \frac{13}{24}x^2. \end{aligned}$$

However,

$$|P_3(x) - P_2(x)| = \left|\frac{1}{6}\tilde{T}_3(x)\right| = \frac{1}{6}\left(\frac{1}{2}\right)^2 = \frac{1}{24} \approx 0.042,$$

which—when added to the already accumulated error bound of 0.0283—exceeds the tolerance of 0.05. Consequently, the polynomial of least degree that best approximates e^x on

$[-1, 1]$ with an error bound of less than 0.05 is

$$P_3(x) = \frac{191}{192} + x + \frac{13}{24}x^2 + \frac{1}{6}x^3.$$

Table 8.9 lists the function and the approximating polynomials at various points in $[-1, 1]$. Note that the tabulated entries for P_2 are well within the tolerance of 0.05, even though the error bound for $P_2(x)$ exceeded the tolerance. ■

Table 8.9

x	e^x	$P_4(x)$	$P_3(x)$	$P_2(x)$	$\lvert e^x - P_2(x)\rvert$
−0.75	0.47237	0.47412	0.47917	0.45573	0.01664
−0.25	0.77880	0.77881	0.77604	0.74740	0.03140
0.00	1.00000	1.00000	0.99479	0.99479	0.00521
0.25	1.28403	1.28402	1.28125	1.30990	0.02587
0.75	2.11700	2.11475	2.11979	2.14323	0.02623

EXERCISE SET 8.3

1. Use the zeros of $\tilde{T}_3$ to construct an interpolating polynomial of degree two for the following functions on the interval $[-1, 1]$.
   a. $f(x) = e^x$
   b. $f(x) = \sin x$
   c. $f(x) = \ln(x + 2)$
   d. $f(x) = x^4$
2. Find bounds for the maximum error of the approximations in Exercise 1 on the interval $[-1, 1]$.
3. Use the zeros of $\tilde{T}_4$ to construct an interpolating polynomial of degree three for the functions in Exercise 1.
4. Repeat Exercise 2 for the approximations computed in Exercise 3.
5. Use the zeros of $\tilde{T}_3$ and transformations of the given interval to construct an interpolating polynomial of degree two for the following functions.
   a. $f(x) = \frac{1}{x}$, $[1, 3]$
   b. $f(x) = e^{-x}$, $[0, 2]$
   c. $f(x) = \frac{1}{2}\cos x + \frac{1}{3}\sin 2x$, $[0, 1]$
   d. $f(x) = x \ln x$, $[1, 3]$
6. Find the sixth Maclaurin polynomial for xe^x, and use Chebyshev economization to obtain a lesser-degree polynomial approximation while keeping the error less than 0.01 on $[-1, 1]$.
7. Find the sixth Maclaurin polynomial for $\sin x$ and use Chebyshev economization to obtain a lesser-degree polynomial approximation while keeping the error less than 0.01 on $[-1, 1]$.
8. Show that for any positive integers i and j with $i > j$, we have $T_i(x)T_j(x) = \frac{1}{2}[T_{i+j}(x) + T_{i-j}(x)]$.
9. Show that for each Chebyshev polynomial $T_n(x)$, we have

$$\int_{-1}^{1} \frac{[T_n(x)]^2}{\sqrt{1-x^2}}\,dx = \frac{\pi}{2}.$$

8.4 Rational Function Approximation

The class of algebraic polynomials has some distinct advantages for use in approximation. There is a sufficient number of polynomials to approximate any continuous function on a closed interval to within an arbitrary tolerance, polynomials are easily evaluated at arbitrary values, and the derivatives and integrals of polynomials exist and are easily determined. The disadvantages of using polynomials for approximation is their tendency for oscillation. This often causes error bounds in polynomial approximation to significantly exceed the average approximation error since error bounds are determined by the maximum approximation error. We now consider methods that spread the approximation error more evenly over the approximation interval. These techniques involve rational functions.

A **rational function** r of degree N has the form

$$r(x) = \frac{p(x)}{q(x)},$$

where $p(x)$ and $q(x)$ are polynomials whose degrees sum to N.

Since every polynomial is a rational function (simply let $q(x) \equiv 1$), approximation by rational functions gives results that are no worse than approximation by polynomials. However, rational functions whose numerator and denominator have the same or nearly the same degree generally produce approximation results superior to polynomial methods for the same amount of computation effort. (This statement is based on the assumption that the amount of computation effort required for division is approximately the same as for multiplication.) Rational functions have the added advantage of permitting efficient approximation of functions that have infinite discontinuities near, but outside, the interval of approximation. Polynomial approximation is generally unacceptable in this situation.

Suppose r is a rational function of degree $N = n + m$ of the form

$$r(x) = \frac{p(x)}{q(x)} = \frac{p_0 + p_1 x + \cdots + p_n x^n}{q_0 + q_1 x + \cdots + q_m x^m}$$

that is used to approximate a function f on a closed interval I containing zero. For r to be defined at zero requires that $q_0 \neq 0$. We can assume that $q_0 = 1$, for if this is not the case we simply replace $p(x)$ by $p(x)/q_0$ and $q(x)$ by $q(x)/q_0$. Consequently, there are $N + 1$ parameters $q_1, q_2, \ldots, q_m, p_0, p_1, \ldots, p_n$ available for the approximation of f by r.

The **Padé approximation technique** chooses the $N + 1$ parameters so that $f^{(k)}(0) = r^{(k)}(0)$ for each $k = 0, 1, \ldots, N$. Padé approximation is the extension of Taylor polynomial approximation to rational functions. In fact, when $n = N$ and and $m = 0$, the Padé approximation is just the Nth Maclaurin polynomial.

Consider the difference

$$f(x) - r(x) = f(x) - \frac{p(x)}{q(x)} = \frac{f(x)q(x) - p(x)}{q(x)} = \frac{f(x)\sum_{i=0}^{m} q_i x^i - \sum_{i=0}^{n} p_i x^i}{q(x)}$$

and suppose f has the Maclaurin series expansion $f(x) = \sum_{i=0}^{\infty} a_i x^i$. Then

$$f(x) - r(x) = \frac{\sum_{i=0}^{\infty} a_i x^i \sum_{i=0}^{m} q_i x^i - \sum_{i=0}^{n} p_i x^i}{q(x)}. \tag{8.14}$$

The object is to choose the constants $q_1, q_2, \ldots, q_m$ and $p_0, p_1, \ldots, p_n$ so that

$$f^{(k)}(0) - r^{(k)}(0) = 0, \quad \text{for each } k = 0, 1, \ldots, N.$$

In Section 2.4 (see, in particular, Exercise 10) we found that this is equivalent to $f - r$ having a zero of multiplicity $N + 1$ at $x = 0$. As a consequence, we choose $q_1, q_2, \ldots, q_m$ and $p_0, p_1, \ldots, p_n$ so that the numerator on the right side of Eq. (8.14),

(8.15) $$(a_0 + a_1 x + \cdots)(1 + q_1 x + \cdots + q_m x^m) - (p_0 + p_1 x + \cdots + p_n x^n),$$

has no terms of degree less than or equal to N. To simplify notation, we define $p_{n+1} = p_{n+2} = \cdots = p_N = 0$ and $q_{m+1} = q_{m+2} = \cdots = q_N = 0$. We can then express the coefficient of x^k in expression (8.15) as

$$\left(\sum_{i=0}^{k} a_i q_{k-i} \right) - p_k.$$

So, the rational function for Padé approximation results from the solution of the $N + 1$ linear equations

$$\sum_{i=0}^{k} a_i q_{k-i} = p_k, \quad k = 0, 1, \ldots, N$$

in the $N + 1$ unknowns $q_1, q_2, \ldots, q_m, p_0, p_1, \ldots, p_n$.

EXAMPLE 1 The Maclaurin series expansion for e^{-x} is

$$\sum_{i=0}^{\infty} \frac{(-1)^i}{i!} x^i.$$

To find the Padé approximation to e^{-x} of degree five with $n = 3$ and $m = 2$ requires choosing p_0, p_1, p_2, p_3, q_1, and q_2 so that the coefficients of x^k for $k = 0, 1, \ldots, 5$ are zero in the expression

$$\left(1 - x + \frac{x^2}{2} - \frac{x^3}{6} + \cdots \right)(1 + q_1 x + q_2 x^2) - (p_0 + p_1 x + p_2 x^2 + p_3 x^3).$$

Expanding and collecting terms produces

$$\begin{aligned}
x^5&: \quad -\frac{1}{120} + \frac{1}{24} q_1 - \frac{1}{6} q_2 = 0; & x^2&: \quad \frac{1}{2} - q_1 + q_2 = p_2;\\
x^4&: \quad \frac{1}{24} - \frac{1}{6} q_1 + \frac{1}{2} q_2 = 0; & x^1&: \quad -1 + q_1 = p_1;\\
x^3&: \quad -\frac{1}{6} + \frac{1}{2} q_1 - q_2 = p_3; & x^0&: \quad 1 = p_0.
\end{aligned}$$

The solution to this system is

$$p_0 = 1,\ p_1 = -\frac{3}{5},\ p_2 = \frac{3}{20},\ p_3 = -\frac{1}{60},\ q_1 = \frac{2}{5},\ \text{and } q_2 = \frac{1}{20},$$

so the Padé approximation is

$$r(x) = \frac{1 - \frac{3}{5}x + \frac{3}{20}x^2 - \frac{1}{60}x^3}{1 + \frac{2}{5}x + \frac{1}{20}x^2}.$$

Table 8.10 lists values of $r(x)$ and $P_5(x)$, the fifth Maclaurin polynomial. The Padé approximation is clearly superior in this example. ■

Table 8.10

x	e^{-x}	$P_5(x)$	$\lvert e^{-x} - P_5(x)\rvert$	$r(x)$	$\lvert e^{-x} - r(x)\rvert$
0.2	0.81873075	0.81873067	8.64×10^{-8}	0.81873075	7.55×10^{-9}
0.4	0.67032005	0.67031467	5.38×10^{-6}	0.67031963	4.11×10^{-7}
0.6	0.54881164	0.54875200	5.96×10^{-5}	0.54880763	4.00×10^{-6}
0.8	0.44932896	0.44900267	3.26×10^{-4}	0.44930966	1.93×10^{-5}
1.0	0.36787944	0.36666667	1.21×10^{-3}	0.36781609	6.33×10^{-5}

Maple can be used to compute a Padé approximation. We first compute the Maclaurin series with the call

```
>series(exp(-x),x);
```

to obtain

$$1 - x + \frac{1}{2}x^2 - \frac{1}{6}x^3 + \frac{1}{24}x^4 - \frac{1}{120}x^5 + O(x^6)$$

The Padé approximation with $n = 3$ and $m = 2$ is computed using the command

```
>g:=convert(",ratpoly,3,2);
```

where the double quote refers to the result of the preceding calculation, namely, the series. The result is

$$g := \frac{1 - \frac{3}{5}x + \frac{3}{20}x^2 - \frac{1}{60}x^3}{1 + \frac{2}{5}x + \frac{1}{20}x^2}$$

We can then compute $g(0.8)$ by entering

```
>evalf(subs(x=0.8,g));
```

to get .4493096647.

Algorithm 8.1 implements the Padé approximation technique.

ALGORITHM 8.1

Padé Rational Approximation

To obtain the rational approximation

$$r(x) = \frac{p(x)}{q(x)} = \frac{\sum_{i=0}^{n} p_i x^i}{\sum_{j=0}^{m} q_j x^j}$$

for a given function $f(x)$:

INPUT nonnegative integers m and n.

OUTPUT coefficients $q_0, q_1, \ldots, q_m$ and $p_0, p_1, \ldots, p_n$.

Step 1 Set $N = m + n$.

Step 2 For $i = 0, 1, \ldots, N$ set $a_i = \dfrac{f^{(i)}(0)}{i!}$.

(The coefficients of the Maclaurin polynomial are $a_0, \ldots, a_N$, which can be input instead of calculated.)

Step 3 Set $q_0 = 1$;
$p_0 = a_0$.

Step 4 For $i = 1, 2, \ldots, N$ do Steps 5–10. (*Set up a linear system with matrix B.*)

Step 5 For $j = 1, 2, \ldots, i - 1$
if $j \le n$ then set $b_{i,j} = 0$.

Step 6 If $i \le n$ then set $b_{i,i} = 1$.

Step 7 For $j = i + 1, i + 2, \ldots, N$ set $b_{i,j} = 0$.

Step 8 For $j = 1, 2, \ldots, i$
if $j \le m$ then set $b_{i,n+j} = -a_{i-j}$.

Step 9 For $j = n + i + 1, n + i + 2, \ldots, N$ set $b_{i,j} = 0$.

Step 10 Set $b_{i,N+1} = a_i$.

(*Steps 11–22 solve the linear system using partial pivoting.*)

Step 11 For $i = n + 1, n + 2, \ldots, N - 1$ do Steps 12–18.

Step 12 Let k be the smallest integer with $i \le k \le N$ and $|b_{k,i}| = \max_{i \le j \le N} |b_{j,i}|$.
(*Find pivot element.*)

Step 13 If $b_{k,i} = 0$ then OUTPUT ("The system is singular ");
STOP.

Step 14 If $k \ne i$ then (*Interchange row i and row k.*)
for $j = i, i + 1, \ldots, N + 1$ set

$$b_{COPY} = b_{i,j};$$
$$b_{i,j} = b_{k,j};$$
$$b_{k,j} = b_{COPY}.$$

Step 15 For $j = i+1, i+2, \ldots, N$ do Steps 16–18. (*Perform elimination.*)

Step 16 Set $xm = \dfrac{b_{j,i}}{b_{i,i}}$.

Step 17 For $k = i+1, i+2, \ldots, N+1$
set $b_{j,k} = b_{j,k} - xm \cdot b_{i,k}$.

Step 18 Set $b_{j,i} = 0$.

Step 19 If $b_{N,N} = 0$ then OUTPUT ("The system is singular");
STOP.

Step 20 If $m > 0$ then set $q_m = \dfrac{b_{N,N+1}}{b_{N,N}}$. (*Start backward substitution.*)

Step 21 For $i = N-1, N-2, \ldots, n+1$ set $q_{i-n} = \dfrac{b_{i,N+1} - \sum_{j=i+1}^{N} b_{i,j}q_{j-n}}{b_{i,i}}$.

Step 22 For $i = n, n-1, \ldots, 1$ set $p_i = b_{i,N+1} - \sum_{j=n+1}^{N} b_{i,j}q_{j-n}$.

Step 23 OUTPUT $(q_0, q_1, \ldots, q_m, p_0, p_1, \ldots, p_n)$;
STOP. (*Procedure completed successfully.*)

It is interesting to compare the number of arithmetic operations required for calculations of $P_5(x)$ and $r(x)$ in Example 1. Using nested multiplication, $P_5(x)$ can be expressed as

$$P_5(x) = 1 - x\left(1 - x\left(\frac{1}{2} - x\left(\frac{1}{6} - x\left(\frac{1}{24} - \frac{1}{120}x\right)\right)\right)\right).$$

Assuming that the coefficients of $1, x, x^2, x^3, x^4$, and x^5 are represented as decimals, a single calculation of $P_5(x)$ in nested form requires five multiplications and five additions/subtractions.

Using nested multiplication, $r(x)$ is expressed as

$$r(x) = \frac{1 - x\left(\frac{3}{5} - x\left(\frac{3}{20} - \frac{1}{60}x\right)\right)}{1 + x\left(\frac{2}{5} + \frac{1}{20}x\right)},$$

so a single calculation of $r(x)$ requires five multiplications, five additions/subtractions, and one division. Hence, computational effort appears to favor the polynomial approximation. However, by reexpressing $r(x)$ by continued division, we can write

$$\begin{aligned} r(x) &= \frac{1 - \frac{3}{5}x + \frac{3}{20}x^2 - \frac{1}{60}x^3}{1 + \frac{2}{5}x + \frac{1}{20}x^2} \\ &= \frac{-\frac{1}{3}x^3 + 3x^2 - 12x + 20}{x^2 + 8x + 20} \\ &= -\frac{1}{3}x + \frac{17}{3} + \frac{\left(-\frac{152}{3}x - \frac{280}{3}\right)}{x^2 + 8x + 20} \end{aligned}$$

$$= -\frac{1}{3}x + \frac{17}{3} + \frac{-\frac{152}{3}}{\left(\frac{x^2+8x+20}{x+(35/19)}\right)},$$

or

(8.16)
$$r(x) = -\frac{1}{3}x + \frac{17}{3} + \frac{-\frac{152}{3}}{\left(x + \frac{117}{19} + \frac{3125/361}{x+(35/19)}\right)}.$$

Written in this form, a single calculation of $r(x)$ requires one multiplication, five additions/subtractions, and two divisions. If the amount of computation required for division is approximately the same as for multiplication, the computational effort required for an evaluation of $P_5(x)$ significantly exceeds that required for an evaluation of $r(x)$.

Expressing a rational function approximation in a form such as Eq. (8.16) is called **continued-fraction** approximation. This is a classical approximation technique of current interest because of the computational efficiency of this representation. It is, however, a specialized technique, and one we will not discuss further. A rather extensive treatment of this subject and of rational approximation in general can be found in [RR, pp. 285–322].

Although the rational-function approximation in Example 1 gave results superior to the polynomial approximation of the same degree, the approximation has a wide variation in accuracy. The approximation at 0.2 is accurate to within 8×10^{-9}, while at 1.0 the approximation and the function agree only to within 7×10^{-5}. This accuracy variation is expected because the Padé approximation is based on a Taylor polynomial representation of e^{-x}, and the Taylor representation has a wide variation of accuracy in $[0.2, 1.0]$.

To obtain more uniformly accurate rational-function approximations we use Chebyshev polynomials, a class that exhibits more uniform behavior. The general Chebyshev rational-function approximation method proceeds in the same manner as Padé approximation, except that each x^k term in the Padé approximation is replaced by the kth-degree Chebyshev polynomial $T_k(x)$.

Suppose we want to approximate the function f by an Nth-degree rational function r written in the form

$$r(x) = \frac{\sum_{k=0}^{n} p_k T_k(x)}{\sum_{k=0}^{m} q_k T_k(x)}, \quad \text{where } N = n + m \text{ and } q_0 = 1.$$

Writing $f(x)$ in a series involving Chebyshev polynomials gives

$$f(x) - r(x) = \sum_{k=0}^{\infty} a_k T_k(x) - \frac{\sum_{k=0}^{n} p_k T_k(x)}{\sum_{k=0}^{m} q_k T_k(x)},$$

or

(8.17)
$$f(x) - r(x) = \frac{\sum_{k=0}^{\infty} a_k T_k(x) \sum_{k=0}^{m} q_k T_k(x) - \sum_{k=0}^{n} p_k T_k(x)}{\sum_{k=0}^{m} q_k T_k(x)}.$$

The coefficients $q_1, q_2, \ldots, q_m$ and $p_0, p_1, \ldots, p_n$ are chosen so that the numerator on the right-hand side of this equation has zero coefficients for $T_k(x)$ when $k = 0, 1, \ldots, N$. This implies that

$$(a_0T_0(x) + a_1T_1(x) + \cdots)(T_0(x) + q_1T_1(x) + \cdots + q_mT_m(x))$$
$$-(p_0T_0(x) + p_1T_1(x) + \cdots + p_nT_n(x))$$

has no terms of degree less than or equal to N.

Two problems arise with the Chebyshev procedure that make it more difficult to implement than the Padé method. One occurs because the product of the polynomial $q(x)$ and the series for $f(x)$ involves products of Chebyshev polynomials. This problem is resolved by making use of the relationship

(8.18) $$T_i(x)T_j(x) = \frac{1}{2}\left[T_{i+j}(x) + T_{|i-j|}(x)\right].$$

(See Exercise 8 of Section 8.3.) The other problem is more difficult to resolve and involves the computation of the Chebyshev series for $f(x)$. In theory, this is not difficult, for if

$$f(x) = \sum_{k=0}^{\infty} a_k T_k(x),$$

then the orthogonality of the Chebyshev polynomials implies that

$$a_0 = \frac{1}{\pi}\int_{-1}^{1} \frac{f(x)}{\sqrt{1-x^2}}\,dx \quad \text{and} \quad a_k = \frac{2}{\pi}\int_{-1}^{1} \frac{f(x)T_k(x)}{\sqrt{1-x^2}}\,dx, \quad \text{where } k \geq 1.$$

Practically, however, these integrals can seldom be evaluated in closed form, and a numerical integration technique is required for each evaluation.

EXAMPLE 2 The first five terms of the Chebyshev expansion for e^{-x} are

$$\tilde{P}_5(x) = 1.266066T_0(x) - 1.130318T_1(x) + 0.271495T_2(x) - 0.044337T_3(x)$$
$$+ 0.005474T_4(x) - 0.000543T_5(x).$$

To determine the Chebyshev rational approximation of degree five with $n = 3$ and $m = 2$ requires choosing p_0, p_1, p_2, p_3, q_1, and q_2 so that for $k = 0, 1, 2, 3, 4$, and 5 the coefficients of $T_k(x)$ are zero in the expansion

$$\tilde{P}_5(x)[T_0(x) + q_1T_1(x) + q_2T_2(x)] - [p_0T_0(x) + p_1T_1(x) + p_2T_2(x) + p_3T_3(x)].$$

Using the relation (8.18) and collecting terms gives the equations

$$
\begin{aligned}
T_0 &: \quad 1.266066 - 0.565159q_1 + 0.1357485q_2 = p_0, \\
T_1 &: \quad -1.130318 + 1.401814q_1 - 0.587328q_2 = p_1, \\
T_2 &: \quad 0.271495 - 0.587328q_1 + 1.268803q_2 = p_2, \\
T_3 &: \quad -0.044337 + 0.138485q_1 - 0.565431q_2 = p_3, \\
T_4 &: \quad 0.005474 - 0.022440q_1 + 0.135748q_2 = 0, \\
T_5 &: \quad -0.000543 + 0.002737q_1 - 0.022169q_2 = 0.
\end{aligned}
$$

The solution to this system produces the rational function

$$r_T(x) = \frac{1.055265T_0(x) - 0.613016T_1(x) + 0.077478T_2(x) - 0.004506T_3(x)}{T_0(x) + 0.378331T_1(x) + 0.022216T_2(x)}.$$

We found at the beginning of Section 8.3 that $T_0(x) = 1$, $T_1(x) = x$, $T_2(x) = 2x^2 - 1$, $T_3(x) = 4x^3 - 3x$. Using these to convert to an expression involving powers of x gives

$$r_T(x) = \frac{0.977787 - 0.599499x + 0.154956x^2 - 0.018022x^3}{0.977784 + 0.378331x + 0.044432x^2}.$$

Table 8.11 lists values of $r_T(x)$ and, for comparison purposes, the values of $r(x)$ obtained in Example 1. Note that the approximation given by $r(x)$ is superior to that of $r_T(x)$ for $x = 0.2$ and 0.4, but that the maximum error for $r(x)$ is 6.33×10^{-5} compared to 9.13×10^{-6} for $r_T(x)$. ■

Table 8.11

x	e^{-x}	$r(x)$	$\lvert e^{-x} - r(x)\rvert$	$r_T(x)$	$\lvert e^{-x} - r_T(x)\rvert$
0.2	0.81873075	0.81873075	7.55×10^{-9}	0.81872510	5.66×10^{-6}
0.4	0.67032005	0.67031963	4.11×10^{-7}	0.67031310	6.95×10^{-6}
0.6	0.54881164	0.54880763	4.00×10^{-6}	0.54881292	1.28×10^{-6}
0.8	0.44932896	0.44930966	1.93×10^{-5}	0.44933809	9.13×10^{-6}
1.0	0.36787944	0.36781609	6.33×10^{-5}	0.36787155	7.89×10^{-6}

The Chebyshev approximation can be generated using Algorithm 8.2.

ALGORITHM 8.2

Chebyshev Rational Approximation

To obtain the rational approximation

$$r_T(x) = \frac{\sum_{k=0}^{n} p_k T_k(x)}{\sum_{k=0}^{m} q_k T_k(x)}$$

for a given function $f(x)$:

INPUT nonnegative integers m and n.

OUTPUT coefficients $q_0, q_1, \ldots, q_m$ and $p_0, p_1, \ldots, p_n$.

Step 1 Set $N = m + n$.

Step 2 Set $a_0 = \frac{2}{\pi}\int_0^{\pi} f(\cos\theta)d\theta$; *(The coefficient a_0 is doubled for computational efficiency.)*

For $k = 1, 2, \ldots, N + m$ set

$$a_k = \frac{2}{\pi}\int_0^{\pi} f(\cos\theta)\cos k\theta d\theta.$$

(The integrals can be evaluated using a numerical integration procedure or the coefficients can be input directly.)

Step 3 Set $q_0 = 1$.

Step 4 For $i = 0, 1, \ldots, N$ do Steps 5–9. *(Set up a linear system with matrix B.)*

Step 5 For $j = 0, 1, \ldots, i$
if $j \le n$ then set $b_{i,j} = 0$.

Step 6 If $i \le n$ then set $b_{i,i} = 1$.

Step 7 For $j = i + 1, i + 2, \ldots, n$ set $b_{i,j} = 0$.

Step 8 For $j = n + 1, n + 2, \ldots, N$

if $i \ne 0$ then set $b_{i,j} = -\frac{1}{2}(a_{i+j-n} + a_{|i-j+n|})$
else set $b_{i,j} = -\frac{1}{2}a_{j-n}$.

Step 9 If $i \ne 0$ then set $b_{i,N+1} = a_i$
else set $b_{i,N+1} = \frac{1}{2}a_i$.

(Steps 10–21 solve the linear system using partial pivoting.)

Step 10 For $i = n + 1, n + 2, \ldots, N - 1$ do Steps 11–17.

Step 11 Let k be the smallest integer with $i \le k \le N$ and $|b_{k,i}| = \max_{i \le j \le N} |b_{j,i}|$. *(Find pivot element.)*

Step 12 If $b_{k,i} = 0$ then OUTPUT ("The system is singular");
STOP.

Step 13 If $k \ne i$ then *(Interchange row i and row k.)*
for $j = i, i + 1, \ldots, N + 1$ set

$$b_{COPY} = b_{i,j};$$
$$b_{i,j} = b_{k,j};$$
$$b_{k,j} = b_{COPY}.$$

Step 14 For $j = i + 1, i + 2, \ldots, N$ do Steps 15–17. *(Perform elimination.)*

Step 15 Set $xm = \frac{b_{j,i}}{b_{i,i}}$.

Step 16 For $k = i + 1, i + 2, \ldots, N + 1$

set $b_{j,k} = b_{j,k} - xm \cdot b_{i,k}$.

Step 17 Set $b_{j,i} = 0$.

Step 18 If $b_{N,N} = 0$ then OUTPUT ("The system is singular");
STOP.

Step 19 If $m > 0$ then set $q_m = \dfrac{b_{N,N+1}}{b_{N,N}}$. (*Start backward substitution.*)

Step 20 For $i = N-1, N-2, \ldots, n+1$ set $q_{i-n} = \dfrac{b_{i,N+1} - \sum_{j=i+1}^{N} b_{i,j} q_{j-n}}{b_{i,i}}$.

Step 21 For $i = n, n-1, \ldots, 0$ set $p_i = b_{i,N+1} - \sum_{j=n+1}^{N} b_{i,j} q_{j-n}$.

Step 22 OUTPUT $(q_0, q_1, \ldots, q_m, p_0, p_1, \ldots, p_n)$;
STOP. (*Procedure completed successfully.*)

We can obtain both the Chebyshev series expansion and the Chebyshev rational approximation using Maple. To make the Chebyshev polynomials accessible to Maple, enter the command

```
>with(orthopoly,T);
```

The procedure to compute the Chebyshev series as an approximation is

```
>g:=chebyshev(exp(-x),x,0.000001);
```

where the third parameter specifies the required accuracy. The result is

$$g := 1.266065878\,T(0,x) - 1.130318208\,T(1,x) + .2714953396\,T(2,x)$$
$$- .04433684985\,T(3,x) + .005474240443\,T(4,x) - .0005429263119\,T(5,x)$$
$$+ .00004497732296\,T(6,x) - .3198436462\;10^{-5}\,T(7,x)$$

and we can evaluate $g(0.8)$ using

```
>evalf(subs(x=0.8,g));
```

to obtain .4493288893.

To get the Chebyshev rational approximation we first need to clear Maple's memory with the command

```
>restart;
```

Then we start again with the Chebyshev series

```
>chebyshev(exp(-x),x,0.000001);
```

and enter

```
>g:=convert(",ratpoly,3,2);
```

resulting in

$$g := (1.050531166\,T(0,x) - .6016362117\,T(1,x) + .07417897134\,T(2,x) - .004109558332\,T(3,x))/(T(0,x) + .3870509569\,T(1,x) + .02365167318\,T(2,x))$$

Since we have cleared Maple's memory, we need to reenter the command

```
>with(orthopoly,T);
```

so that we can evaluate $g(0.8)$ by

```
>evalf(subs(x=0.8,g));
```

to get .4493317581.

The Chebyshev method does not produce the best rational function approximation in the sense of the approximation whose maximum approximation error is minimal. The method can, however, be used as a starting point for an iterative method known as the second Remes' algorithm that converges to the best approximation. A discussion of the techniques involved with this procedure and an improvement on this algorithm can be found in [RR, pp. 292–305], or in [Po, pp. 90–92].

EXERCISE SET 8.4

1. Determine all Padé approximations for $f(x) = e^{2x}$ of degree two. Compare the results at $x_i = 0.2i$, for $i = 1, 2, 3, 4, 5$, with the actual values $f(x_i)$.
2. Determine all Padé approximations for $f(x) = x \ln(x + 1)$ of degree three. Compare the results at $x_i = 0.2i$, for $i = 1, 2, 3, 4, 5$, with the actual values $f(x_i)$.
3. Determine the Padé approximation of degree five with $n = 2$ and $m = 3$ for $f(x) = e^x$. Compare the results at $x_i = 0.2i$, for $i = 1, 2, 3, 4, 5$, with those from the fifth Maclaurin polynomial.
4. Repeat Exercise 3 using instead the Padé approximation of degree five with $n = 3$ and $m = 2$. Compare the results at each x_i with those computed in Exercise 3.
5. Determine the Padé approximation of degree six with $n = m = 3$ for $f(x) = \sin x$. Compare the results at $x_i = 0.1i$, for $i = 0, 1, \ldots, 5$, with the exact results and with the results of the sixth Maclaurin polynomial.
6. Determine the Padé approximations of degree six with (a) $n = 2, m = 4$ and (b) $n = 4, m = 2$ for $f(x) = \sin x$. Compare the results at each x_i to those obtained in Exercise 5.
7. Table 8.10 lists results of the Padé approximation of degree five with $n = 3$ and $m = 2$, the fifth Maclaurin polynomial, and the exact values of $f(x) = e^{-x}$ when $x_i = 0.2i$, for $i = 1, 2, 3, 4$, and 5. Compare these results with those produced from the other Padé approximations of degree five.

   a. $n = 0, m = 5$ **b.** $n = 1, m = 4$

   c. $n = 3, m = 2$ **d.** $n = 4, m = 1$
8. Express the following rational functions in continued-fraction form:

   a. $\dfrac{x^2 + 3x + 2}{x^2 - x + 1}$ **b.** $\dfrac{4x^2 + 3x - 7}{2x^3 + x^2 - x + 5}$

   c. $\dfrac{2x^3 - 3x^2 + 4x - 5}{x^2 + 2x + 4}$ **d.** $\dfrac{2x^3 + x^2 - x + 3}{3x^3 + 2x^2 - x + 1}$

9. Find all the Chebyshev rational approximations of degree two for $f(x) = e^{-x}$. Which give the best approximations to $f(x) = e^{-x}$ at $x = 0.25, 0.5$, and 1?

10. Find all the Chebyshev rational approximations of degree three for $f(x) = \cos x$. Which give the best approximations to $f(x) = \cos x$ at $x = \pi/4$ and $\pi/3$?

11. Find the Chebyshev rational approximation of degree four with $n = m = 2$ for $f(x) = \sin x$. Compare the results at $x_i = 0.1i$, for $i = 0, 1, 2, 3, 4, 5$ from this approximation with those obtained in Exercise 5 using a sixth-degree Padé approximation.

12. Find all Chebyshev rational approximations of degree five for $f(x) = e^x$. Compare the results at $x_i = 0.2i$, for $i = 1, 2, 3, 4, 5$ from this approximation with those obtained in Exercises 3 and 4.

13. To accurately approximate $f(x) = e^x$ for the inclusion in a mathematical library, we first restrict the domain of f. Given a real number x, divide by $\ln \sqrt{10}$ to obtain the relation

$$x = M \cdot \ln \sqrt{10} + s,$$

where M is an integer and s is a real number satisfying $|s| \le \frac{1}{2} \ln \sqrt{10}$.

a. Show that $e^x = e^s \cdot 10^{M/2}$.

b. Construct a rational function approximation for e^s using $n = m = 3$. Estimate the error when $0 \le |s| \le \frac{1}{2} \ln \sqrt{10}$.

c. Design an implementation of e^x using the results of part (a) and (b) and the approximations

$$\frac{1}{\ln \sqrt{10}} = 0.8685889638 \quad \text{and} \quad \sqrt{10} = 3.162277660.$$

14. To accurately approximate $\sin x$ and $\cos x$ for inclusion in a mathematical library, we first restrict their domains. Given a real number x, divide by π to obtain the relation

$$|x| = M\pi + s, \quad \text{where } M \text{ is an integer and } |s| \le \frac{\pi}{2}.$$

a. Show that $\sin x = \text{sign}(x) \cdot (-1)^M \cdot \sin s$, where $\text{sign}(x) = |x|/x$, when $x \ne 0$.

b. Construct a rational approximation to $\sin s$ using $n = m = 4$. Estimate the error when $0 \le |s| \le \frac{\pi}{2}$.

c. Design an implementation of $\sin x$ using parts (a) and (b).

d. Repeat part (c) for $\cos x$ using the fact that $\cos x = \sin(x + \frac{\pi}{2})$.

8.5 Trigonometric Polynomial Approximation

The use of series of sine and cosine functions to represent arbitrary functions had its beginnings in the 1750s with the study of the motion of a vibrating string. This problem was considered by Jean d'Alembert and then taken up by the foremost mathematician of the time, Leonhard Euler. But it was Daniel Bernoulli who first advocated the use of the infinite sums of sine and cosines as a solution to the problem, sums that we now know as Fourier series. In the early part of the 19th century, Jean Baptiste Joseph Fourier used these series to study the flow of heat and developed quite a complete theory of the subject.

The first observation in the development of Fourier series is that, for each positive integer n, the set of functions $\{\phi_0, \phi_1, \ldots, \phi_{2n-1}\}$, where

$$\phi_0(x) = \frac{1}{2},$$

$$\phi_k(x) = \cos kx, \quad \text{for each } k = 1, 2, \ldots, n,$$

and

$$\phi_{n+k}(x) = \sin kx, \quad \text{for each } k = 1, 2, \ldots, n-1,$$

is an orthogonal set on $[-\pi, \pi]$ with respect to $w(x) \equiv 1$. This orthogonality follows from the fact that, for every integer j, the integrals of $\sin jx$ and $\cos jx$ over $[-\pi, \pi]$ are 0, and we can rewrite products of sine and cosine functions as sums by using the three trigonometric identities

(8.19)

$$\sin t_1 \sin t_2 = \frac{1}{2}[\cos(t_1 - t_2) - \cos(t_1 + t_2)],$$

$$\cos t_1 \cos t_2 = \frac{1}{2}[\cos(t_1 - t_2) + \cos(t_1 + t_2)],$$

$$\sin t_1 \cos t_2 = \frac{1}{2}[\sin(t_1 - t_2) + \sin(t_1 + t_2)].$$

Let $\mathcal{T}_n$ denote the set of all linear combinations of the functions $\phi_0, \phi_1, \ldots, \phi_{2n-1}$. This set is called the set of **trigonometric polynomials** of degree less than or equal to n. (Some sources include an additional function in the set, $\phi_{2n}(x) = \sin nx$. We will not follow this convention.)

For a function $f \in C[-\pi, \pi]$, we want to find the *continuous least squares* approximation by functions in $\mathcal{T}_n$ in the form

$$S_n(x) = \frac{a_0}{2} + a_n \cos nx + \sum_{k=1}^{n-1}(a_k \cos kx + b_k \sin kx).$$

Since the set of functions $\{\phi_0, \phi_1, \ldots, \phi_{2n-1}\}$ is orthogonal on $[-\pi, \pi]$ with respect to $w(x) \equiv 1$, it follows from Theorem 8.6 that the appropriate selection of coefficients is

$$a_k = \frac{1}{\pi}\int_{-\pi}^{\pi} f(x)\cos kx\,dx, \quad \text{for each } k = 0, 1, 2, \ldots, n$$

and

$$b_k = \frac{1}{\pi}\int_{-\pi}^{\pi} f(x)\sin kx\,dx, \quad \text{for each } k = 1, 2, \ldots, n-1.$$

The limit of $S_n(x)$ when $n \to \infty$ is called the **Fourier series** of f.

EXAMPLE 1 To determine the trigonometric polynomial from $\mathcal{T}_n$ that approximates

$$f(x) = |x|, \quad \text{for } -\pi < x < \pi,$$

requires finding

$$a_0 = \frac{1}{\pi}\int_{-\pi}^{\pi} |x|\,dx = -\frac{1}{\pi}\int_{-\pi}^{0} x\,dx + \frac{1}{\pi}\int_{0}^{\pi} x\,dx = \frac{2}{\pi}\int_{0}^{\pi} x\,dx = \pi,$$

$$a_k = \frac{1}{\pi}\int_{-\pi}^{\pi} |x|\cos kx\,dx = \frac{2}{\pi}\int_{0}^{\pi} x\cos kx\,dx = \frac{2}{\pi k^2}\left[(-1)^k - 1\right],$$

for each $k = 1, 2, \ldots, n$, and

$$b_k = \frac{1}{\pi}\int_{-\pi}^{\pi} |x| \sin kx \, dx = 0, \quad \text{for each } k = 1, 2, \ldots, n-1.$$

That the b_k's are all 0 follows from the fact that $g(x) = |x| \sin kx$ is an odd function for each k, and the integral of any odd function over any interval of the form $[-a, a]$ is 0. (See Exercises 13 and 14.) The trigonometric polynomial from $\mathcal{T}_n$ approximating f is, therefore,

$$S_n(x) = \frac{\pi}{2} + \frac{2}{\pi}\sum_{k=1}^{n} \frac{(-1)^k - 1}{k^2} \cos kx.$$

The first few trigonometric polynomials for $f(x) = |x|$ are shown in Figure 8.12.

Figure 8.12

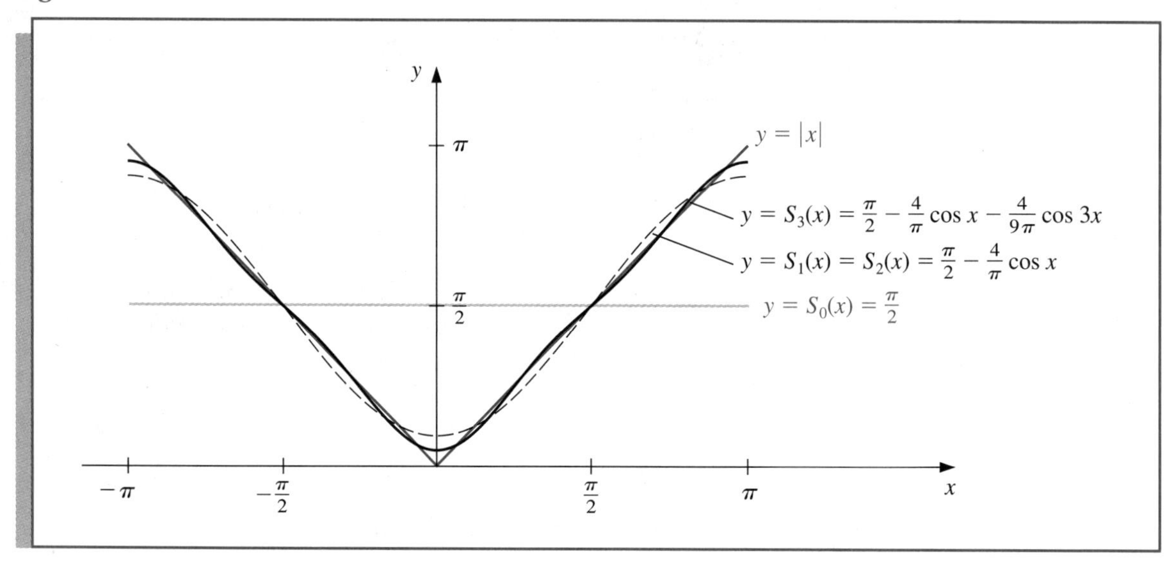

The Fourier series for f is

$$S(x) = \lim_{n\to\infty} S_n(x) = \frac{\pi}{2} + \frac{2}{\pi}\sum_{k=1}^{\infty} \frac{(-1)^k - 1}{k^2} \cos kx.$$

Since $|\cos kx| \le 1$ for every k and x, the series converges and $S(x)$ exists for all real numbers x. ■

There is a discrete analog that is useful for the *discrete least squares* approximation and the interpolation of large amounts of data.

Suppose that a collection of $2m$ paired data points $\{(x_j, y_j)\}_{j=0}^{2m-1}$ is given, with the first elements in the pairs equally partitioning a closed interval. For convenience, we assume that the interval is $[-\pi, \pi]$, so, as shown in Figure 8.13,

(8.20) $$x_j = -\pi + \left(\frac{j}{m}\right)\pi, \quad \text{for each } j = 0, 1, \ldots, 2m-1.$$

Figure 8.13

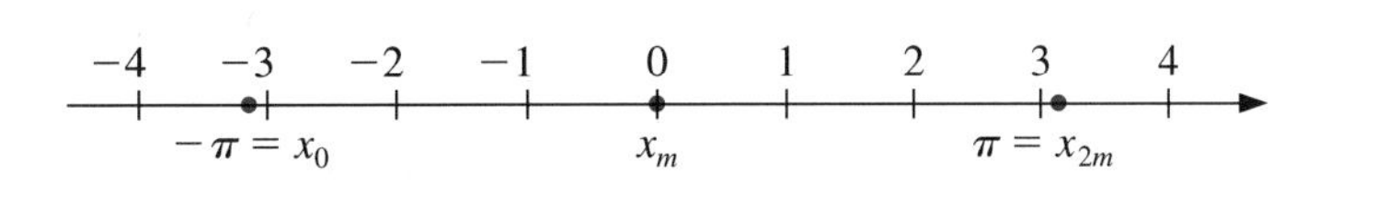

If this is not the case, a simple linear transformation could be used to translate the data into this form.

The goal in the discrete case is to determine the trigonometric polynomial $S_n(x)$ in $\mathcal{T}_n$ that will minimize

$$E(S_n) = \sum_{j=0}^{2m-1} [y_j - S_n(x_j)]^2.$$

To do this we need to choose the constants $a_0, a_1, \ldots, a_n, b_1, b_2, \ldots, b_{n-1}$ so that

(8.21) $$E(S_n) = \sum_{j=0}^{2m-1} \left\{ y_j - \left[\frac{a_0}{2} + a_n \cos nx_j + \sum_{k=1}^{n-1} (a_k \cos kx_j + a_{n+k} \sin kx_j) \right] \right\}^2$$

is a minimum.

The determination of the constants is simplified by the fact that the set $\{\phi_0, \phi_1, \ldots, \phi_{2n-1}\}$ is also orthogonal with respect to summation over the equally spaced points $\{x_j\}_{j=0}^{2m-1}$ in $[-\pi, \pi]$. By this we mean that for each $k \neq l$,

(8.22) $$\sum_{j=0}^{2m-1} \phi_k(x_j)\phi_l(x_j) = 0.$$

To show this orthogonality, we use the following lemma.

Lemma 8.12 If the integer r is not a multiple of $2m$, then

$$\sum_{j=0}^{2m-1} \cos rx_j = 0 \quad \text{and} \quad \sum_{j=0}^{2m-1} \sin rx_j = 0.$$

Moreover, if r is not a multiple of m, then

$$\sum_{j=0}^{2m-1} (\cos rx_j)^2 = m \quad \text{and} \quad \sum_{j=0}^{2m-1} (\sin rx_j)^2 = m.$$

Proof If i denotes the complex number with $i^2 = -1$, then Euler's formula,

(8.23) $$e^{iz} = \cos z + i \sin z,$$

gives

$$\sum_{j=0}^{2m-1} \cos rx_j + i \sum_{j=0}^{2m-1} \sin rx_j = \sum_{j=0}^{2m-1} [\cos rx_j + i \sin rx_j] = \sum_{j=0}^{2m-1} e^{irx_j}.$$

But

$$e^{irx_j} = e^{ir(-\pi + j\pi/m)} = e^{-ir\pi} \cdot e^{irj\pi/m},$$

so

$$\sum_{j=0}^{2m-1} \cos rx_j + i \sum_{j=0}^{2m-1} \sin rx_j = e^{-ir\pi} \sum_{j=0}^{2m-1} e^{irj\pi/m}.$$

Since $\sum_{j=0}^{2m-1} e^{irj\pi/m}$ is a geometric series with first term 1 and ratio $e^{ir\pi/m} \neq 1$, we have

$$\sum_{j=0}^{2m-1} e^{irj\pi/m} = \frac{1 - (e^{ir\pi/m})^{2m}}{1 - e^{ir\pi/m}} = \frac{1 - e^{2ir\pi}}{1 - e^{ir\pi/m}}.$$

But $e^{2ir\pi} = \cos 2r\pi + i \sin 2r\pi = 1$, so

$$\sum_{j=0}^{2m-1} \cos rx_j + i \sum_{j=0}^{2m-1} \sin rx_j = e^{-ir\pi} \sum_{j=0}^{2m-1} e^{irj\pi/m} = 0.$$

This implies that

$$\sum_{j=0}^{2m-1} \cos rx_j = 0 \quad \text{and} \quad \sum_{j=0}^{2m-1} \sin rx_j = 0.$$

If r is not a multiple of m, these sums imply that both

$$\begin{aligned}\sum_{j=0}^{2m-1} (\cos rx_j)^2 &= \sum_{j=0}^{2m-1} \frac{1}{2}[1 + \cos 2rx_j] \\ &= \frac{1}{2}\left[\sum_{j=0}^{2m-1} 1 + \sum_{j=0}^{2m-1} \cos 2rx_j\right] = \frac{1}{2}(2m + 0) = m\end{aligned}$$

and, similarly, that

$$\sum_{j=0}^{2m-1} (\sin rx_j)^2 = m.$$

▪ ▪ ▪

We can now show the orthogonality stated in (8.22). Consider, for example, the case

$$\sum_{j=0}^{2m-1} \phi_k(x_j)\phi_{n+l}(x_j) = \sum_{j=0}^{2m-1} (\cos kx_j)(\sin lx_j).$$

Since

$$\cos kx_j \sin lx_j = \frac{1}{2}[\sin(l+k)x_j + \sin(l-k)x_j],$$

and $(l+k)$ and $(l-k)$ are both integers that are not multiples of $2m$, Lemma 8.12 implies that

$$\sum_{j=0}^{2m-1} (\cos kx_j)(\sin lx_j) = \frac{1}{2}\left[\sum_{j=0}^{2m-1} \sin(l+k)x_j + \sum_{j=0}^{2m-1} \sin(l-k)x_j\right] = \frac{1}{2}[0+0] = 0.$$

This technique is used to show that the orthogonality condition is satisfied for any pair of the functions and is used to produce the following result.

Theorem 8.13 The constants in the summation

$$S_n(x) = \frac{a_0}{2} + a_n \cos nx + \sum_{k=1}^{n-1}(a_k \cos kx + b_k \sin kx)$$

that minimize the least squares sum

$$E(a_0, \ldots, a_n, b_1, \ldots, b_{n-1}) = \sum_{j=0}^{2m-1} (y_j - S_n(x_j))^2$$

are

$$a_k = \frac{1}{m}\sum_{j=0}^{2m-1} y_j \cos kx_j, \quad \text{for each } k = 0, 1, \ldots, n,$$

and

$$b_k = \frac{1}{m}\sum_{j=0}^{2m-1} y_j \sin kx_j, \quad \text{for each } k = 1, 2, \ldots, n-1.$$

■

The theorem is proved by setting the partial derivatives of E with respect to the a_k's and the b_k's to zero, as was done in Sections 8.1 and 8.2, and applying the orthogonality to simplify the equations. For example,

$$0 = \frac{\partial E}{\partial b_k} = 2\sum_{j=0}^{2m-1} [y_j - S_n(x_j)](-\sin kx_j),$$

so

$$0 = \sum_{j=0}^{2m-1} y_j \sin kx_j - \frac{a_0}{2} \sum_{j=0}^{2m-1} \sin kx_j - a_n \sum_{j=0}^{2m-1} \sin kx_j \cos nx_j$$

$$- \sum_{l=1}^{n-1} a_l \sum_{j=0}^{2m-1} \sin kx_j \cos lx_j - \sum_{\substack{l=1,\\ l\neq k}}^{n-1} b_l \sum_{j=0}^{2m-1} \sin kx_j \sin lx_j - b_k \sum_{j=0}^{2m-1} (\sin kx_j)^2.$$

The orthogonality implies that all but the first and last sums on the right side are zero, and Lemma 8.12 states the final sum is m. Hence

$$b_k = \frac{1}{m} \sum_{j=0}^{2m-1} y_j \sin kx_j.$$

EXAMPLE 2 Let $f(x) = x^4 - 3x^3 + 2x^2 - \tan x(x-2)$. To find the discrete least squares approximation S_3 for the data $\{(x_j, y_j)\}_{j=0}^{9}$, where $x_j = j/5$ and $y_j = f(x_j)$, requires a transformation from $[0, 2]$ to $[-\pi, \pi]$. It is easily verified that the required linear transformation is

$$z_j = \pi(x_j - 1),$$

and that the translated data is of the form

$$\left\{ \left(z_j, f\left(1 + \frac{z_j}{\pi}\right) \right) \right\}_{j=0}^{9}.$$

The least squares trigonometric polynomial is, consequently,

$$S_3(z) = \frac{a_0}{2} + a_3 \cos 3z + \sum_{k=1}^{2} (a_k \cos kz + b_k \sin kz),$$

where

$$a_k = \frac{1}{5} \sum_{j=0}^{9} f\left(1 + \frac{z_j}{\pi}\right) \cos kz_j, \quad \text{for } k = 0, 1, 2, 3,$$

and

$$b_k = \frac{1}{5} \sum_{j=0}^{9} f\left(1 + \frac{z_j}{\pi}\right) \sin kz_j, \quad \text{for } k = 1, 2.$$

Evaluating these sums produces the approximation

$$S_3(z) = 0.76201 + 0.77177 \cos z + 0.017423 \cos 2z + 0.0065673 \cos 3z$$
$$- 0.38676 \sin z + 0.047806 \sin 2z.$$

Converting back to the variable x gives

$$S_3(x) = 0.76201 + 0.77177\cos\pi(x-1) + 0.017423\cos 2\pi(x-1) + 0.0065673\cos 3\pi(x-1) - 0.38676\sin\pi(x-1) + 0.047806\sin 2\pi(x-1).$$

Table 8.12 lists values of $f(x)$ and $S_3(x)$. ■

Table 8.12

x	$f(x)$	$S_3(x)$	$\lvert f(x) - S_3(x)\rvert$
0.125	0.26440	0.24060	2.38×10^{-2}
0.375	0.84081	0.85154	1.07×10^{-2}
0.625	1.36150	1.36248	9.74×10^{-4}
0.875	1.61282	1.60406	8.75×10^{-3}
1.125	1.36672	1.37566	8.94×10^{-3}
1.375	0.71697	0.71545	1.52×10^{-3}
1.625	0.07909	0.06929	9.80×10^{-3}
1.875	−0.14576	−0.12302	2.27×10^{-2}

EXERCISE SET 8.5

1. Find the continuous least squares trigonometric polynomial $S_2(x)$ for $f(x) = x^2$ on $[-\pi, \pi]$.
2. Find the continuous least squares trigonometric polynomial $S_3(x)$ for $f(x) = x$ on $[-\pi, \pi]$.
3. Find the continuous least squares trigonometric polynomial $S_3(x)$ for $f(x) = e^x$ on $[-\pi, \pi]$.
4. Find the general continuous least squares trigonometric polynomial $S_n(x)$ for $f(x) = e^x$ on $[-\pi, \pi]$.
5. Find the general continuous least squares trigonometric polynomial $S_n(x)$ for

$$f(x) = \begin{cases} 0, & \text{if } -\pi < x \le 0, \\ 1, & \text{if } 0 < x < \pi. \end{cases}$$

6. Find the general continuous least squares trigonometric polynomial $S_n(x)$ in for

$$f(x) = \begin{cases} -1, & \text{if } -\pi < x < 0, \\ 1, & \text{if } 0 \le x \le \pi. \end{cases}$$

7. Determine the discrete least squares trigonometric polynomial $S_n(x)$ on the interval $[-\pi, \pi]$ for the following functions, using the given values of m and n:
   - **a.** $f(x) = \cos 2x, \quad m = 4, n = 2$
   - **b.** $f(x) = \cos 3x, \quad m = 4, n = 2$
   - **c.** $f(x) = \sin\frac{1}{2}x + 2\cos\frac{1}{3}x, \quad m = 6, n = 3$
   - **d.** $f(x) = x^2\cos x, \quad m = 6, n = 3$
8. Compute the error $E(S_n)$ for each of the functions in Exercise 7.
9. Determine the discrete least squares trigonometric polynomial $S_3(x)$, using $m = 4$ for $f(x) = e^x\cos 2x$ on the interval $[-\pi, \pi]$. Compute the error $E(S_3)$.

10. Repeat Exercise 9 using $m = 8$. Compare the values of the approximating polynomials with the values of f at the points $\xi_j = -\pi + 0.2j\pi$, for $0 \leq j \leq 10$. Which approximation is better?

11. Let $f(x) = 2\tan x - \sec 2x$, for $2 \leq x \leq 4$. Determine the discrete least squares trigonometric polynomials $S_n(x)$, using the values of n and m as follows, and compute the error in each case.

a. $n = 3, \quad m = 6$ **b.** $n = 4, \quad m = 6$

12. **a.** Determine the discrete least squares trigonometric polynomial $S_4(x)$, using $m = 16$, for $f(x) = x^2 \sin x$ on the interval $[0, 1]$.

b. Compute $\int_0^1 S_4(x)\,dx$.

c. Compare the integral in part (b) to $\int_0^1 x^2 \sin x\,dx$.

13. Show that for any continuous odd function f defined on the interval $[-a, a]$, we have $\int_{-a}^{a} f(x)\,dx = 0$.

14. Show that for any continuous even function f defined on the interval $[-a, a]$, we have $\int_{-a}^{a} f(x)\,dx = 2\int_0^a f(x)\,dx$.

15. Show that the functions $\phi_0(x) = 1/2, \phi_1(x) = \cos x, \ldots, \phi_n(x) = \cos nx, \phi_{n+1}(x) = \sin x, \ldots, \phi_{2n-1}(x) = \sin(n-1)x$, are orthogonal on $[-\pi, \pi]$ with respect to $w(x) \equiv 1$.

16. In Example 1 the Fourier series was determined for $f(x) = |x|$. Use this series and the assumption that it represents f at zero to find the value of the convergent infinite series $\sum_{k=0}^{\infty} \frac{1}{(2k+1)^2}$.

8.6 Fast Fourier Transforms

In the second half of Section 8.5 we determined the form of the discrete least squares polynomial of degree n on the $2m - 1$ data points $\{(x_j, y_j)\}_{j=0}^{2m-1}$, where $x_j = -\pi + (j/m)\pi$, for each $j = 0, 1, \ldots, 2m - 1$.

The *interpolatory* trigonometric polynomial in $\mathcal{T}_m$ on these $2m$ data points is almost the same as the least squares polynomial. This is because the least squares trigonometric polynomial minimizes the error term

$$E(S_m) = \sum_{j=0}^{2m-1} \left(y_j - S_m(x_j)\right)^2$$

and this error is zero, hence minimized, when the $S_m(x_j) = y_j$ for each $j = 0, 1, \ldots, 2m-1$. There is a modification that is needed to the form of the polynomial, however, if we want the coefficients to assume the same form as in the least squares case. In Lemma 8.12 we found that if r is not a multiple of m, then

$$\sum_{j=0}^{2m-1} (\cos rx_j)^2 = m.$$

Interpolation requires computing

$$\sum_{j=0}^{2m-1} (\cos mx_j)^2,$$

which (see Exercise 8) has the value $2m$. This requires the interpolatory polynomial to be written as

(8.24)
$$S_m(x) = \frac{a_0 + a_m \cos mx}{2} + \sum_{k=1}^{m-1} (a_k \cos kx + b_k \sin kx)$$

if we want the form of the constants a_k and b_k to agree with those of the discrete least squares polynomial. So we want the constants to be

(8.25)
$$a_k = \frac{1}{m} \sum_{j=0}^{2m-1} y_j \cos kx_j, \quad \text{for each } k = 0, 1, \ldots, m$$

and

(8.26)
$$b_k = \frac{1}{m} \sum_{j=0}^{2m-1} y_j \sin kx_j \quad \text{for each } k = 1, 2, \ldots, m - 1.$$

The interpolation of large amounts of equally spaced data by trigonometric polynomials can produce very accurate results. It is the appropriate approximation technique used in areas such as those involving digital filters, antenna field patterns, quantum mechanics, optics, and many areas involving simulation problems. Until the middle of the 1960s, however, the method had not been extensively applied due to the number of arithmetic calculations required for the determination of the constants in the approximation. The interpolation of $2m$ data points by the direct-calculation technique requires approximately $(2m)^2$ multiplications and $(2m)^2$ additions. The approximation of many thousands of data points is not unusual in areas requiring trigonometric interpolation, so the direct methods for evaluating the constants require multiplication and addition operations numbering in the millions. The roundoff error associated with this number of calculations generally dominates the approximation.

In 1965 a paper by J. W Cooley and J. W Tukey in the journal *Mathematics of Computation* [CT] described a different method of calculating the constants in the interpolating trigonometric polynomial. This method requires only $O(m \log_2 m)$ multiplications and $O(m \log_2 m)$ additions, provided m is chosen in an appropriate manner. For a problem with thousands of data points, this reduces the number of calculations to thousands, compared to millions for the direct technique. The method had actually been discovered a number of years before the Cooley-Tukey paper appeared but had gone largely unnoticed. ([Brigh, pp. 8–9], contains a short, but interesting, historical summary of the method.)

The method described by Cooley and Tukey is known either as the **Cooley-Tukey algorithm** or the **fast Fourier transform (FFT) algorithm** and has led to a revolution in the use of interpolatory trigonometric polynomials. The method consists of organizing the problem so that the number of data points being used can be easily factored, particularly into powers of two.

Instead of directly evaluating the constants a_k and b_k, the fast Fourier transform procedure computes the complex coefficients c_k in

(8.27)
$$\frac{1}{m}\sum_{k=0}^{2m-1} c_k e^{ikx},$$

where

(8.28)
$$c_k = \sum_{j=0}^{2m-1} y_j e^{ik\pi j/m}, \quad \text{for each } k = 0, 1, \ldots, 2m-1.$$

Once the constants c_k have been determined, a_k and b_k can be recovered using the fact that, for each $k = 0, 1, \ldots, m$,

$$\begin{aligned}
\frac{1}{m}c_k(-1)^k = \frac{1}{m}c_k e^{-i\pi k} &= \frac{1}{m}\sum_{j=0}^{2m-1} y_j e^{ik\pi j/m} e^{-i\pi k} = \frac{1}{m}\sum_{j=0}^{2m-1} y_j e^{ik(-\pi+(\pi j/m))} \\
&= \frac{1}{m}\sum_{j=0}^{2m-1} y_j \left(\cos k\left(-\pi + \frac{\pi j}{m}\right) + i \sin k\left(-\pi + \frac{\pi j}{m}\right)\right) \\
&= \frac{1}{m}\sum_{j=0}^{2m-1} y_j(\cos kx_j + i \sin kx_j).
\end{aligned}$$

So

(8.29)
$$a_k + ib_k = \frac{(-1)^k}{m} c_k.$$

For notational convenience, b_0 and b_m are added to the collection, but both are zero and do not contribute to the resulting sum.

The operation-reduction feature of the fast Fourier transform results from calculating the coefficients c_k in clusters and uses as a basic relation the fact that for any integer n,

$$e^{n\pi i} = \cos n\pi + i \sin n\pi = (-1)^n.$$

Suppose $m = 2^p$ for some positive integer p. For each $k = 0, 1, \ldots, m-1$,

$$c_k + c_{m+k} = \sum_{j=0}^{2m-1} y_j e^{ik\pi j/m} + \sum_{j=0}^{2m-1} y_j e^{i(m+k)\pi j/m} = \sum_{j=0}^{2m-1} y_j e^{ik\pi j/m}(1 + e^{\pi i j}).$$

But

$$1 + e^{i\pi j} = \begin{cases} 2, & \text{if } j \text{ is even,} \\ 0, & \text{if } j \text{ is odd.} \end{cases}$$

As a consequence, there are only m nonzero terms to be summed. If j is replaced by $2j$ in the index of the sum, we can write the sum as

$$c_k + c_{m+k} = 2\sum_{j=0}^{m-1} y_{2j} e^{ik\pi(2j)/m};$$

that is,

$$c_k + c_{m+k} = 2\sum_{j=0}^{m-1} y_{2j} e^{ik\pi j/(m/2)}. \tag{8.30}$$

In a similar manner,

$$c_k - c_{m+k} = 2e^{ik\pi/m}\sum_{j=0}^{m-1} y_{2j+1} e^{ik\pi j/(m/2)}. \tag{8.31}$$

Since c_k and c_{m+k} can both be recovered from Eqs. (8.30) and (8.31), these relations determine all the coefficients c_k. Note that the sums in Eqs. (8.30) and (8.31) are of the same form as the sum in Eq. (8.28), except that the index m has been replaced by $m/2$.

There are $2m$ coefficients $c_0, c_1, \ldots, c_{2m-1}$ to be calculated. Using the basic formula (8.28) requires $2m$ complex multiplications per coefficient, for a total of $(2m)^2 = 4m^2$ operations. Equation (8.30) requires m complex multiplications for each $k = 0, 1, \ldots, m-1$, and (8.31) requires $m+1$ complex multiplications for each $k = 0, 1, \ldots, m-1$. Using these equations to compute $c_0, c_1, \ldots, c_{2m-1}$ reduces the number of complex multiplications to

$$m \cdot m + m(m+1) = 2m^2 + m.$$

Since the sums in (8.30) and (8.31) have the same form as the original and m is a power of 2, the reduction technique can be reapplied to the sums in (8.30) and (8.31). Each of these is replaced by two sums from $j = 0$ to $j = (m/2) - 1$. This reduces the $2m^2$ portion of the sum to

$$2\left[\frac{m}{2} \cdot \frac{m}{2} + \frac{m}{2} \cdot \left(\frac{m}{2} + 1\right)\right] = m^2 + m.$$

So a total of $(m^2 + m) + m = m^2 + 2m$ complex multiplications are now needed.

Applying the technique one more time gives us 4 sums each with $m/4$ terms and reduces the m^2 portion of this total to

$$4\left[\left(\frac{m}{4}\right)^2 + \frac{m}{4}\left(\frac{m}{4} + 1\right)\right] = \frac{m^2}{2} + m$$

for a new total of $(m^2/2) + 3m$ complex multiplications. Repeating the process r times reduces the total number of required complex multiplications to

$$\frac{m^2}{2^{r-2}} + mr.$$

The process is complete when $r = p + 1$, since $m = 2^p$ and $2m = 2^{p+1}$. As a consequence, after $r = p+1$ reductions of this type, the number of complex multiplications is reduced to

$$\frac{(2^p)^2}{2^{p-1}} + m(p+1) = 2m + pm + m = 3m + m\log_2 m = O(m\log_2 m).$$

Because of the way the calculations are arranged, the number of required complex additions is comparable. To illustrate the significance of this reduction, suppose we have $m = 2^{10} = 1024$. The direct calculation would require

$$(2m)^2 = (2048)^2 \approx 4{,}200{,}000.$$

The fast Fourier transform procedure reduces the number of calculations to

$$3(1024) + 1024\log_2 1024 \approx 13{,}300.$$

EXAMPLE 1 Consider the fast Fourier transform technique applied to $8 = 2^3$ data points $\{(x_j, y_j)\}_{j=0}^{7}$, where $x_j = -\pi + j\pi/4$, for each $j = 0, 1, \ldots, 7$. In this case $2m = 8$, so $m = 4 = 2^2$ and $p = 2$.

From (8.24) we have

$$S_4(x) = \frac{a_0 + a_4\cos 4x}{2} + \sum_{k=1}^{3}(a_k\cos kx + b_k\sin kx),$$

where

$$a_k = \frac{1}{4}\sum_{j=0}^{7} y_j\cos kx_j \quad \text{and} \quad b_k = \frac{1}{4}\sum_{j=0}^{7} y_j\sin kx_j, \quad k = 0, 1, 2, 3, 4.$$

Define

$$F(x) = \frac{1}{4}\sum_{j=0}^{7} c_k e^{ikx},$$

where

$$c_k = \sum_{j=0}^{7} y_j e^{ik\pi j/4}, \quad \text{for } k = 0, 1, \ldots, 7.$$

Then by (8.29), for $k = 0, 1, 2, 3, 4$,

$$\frac{1}{4}c_k e^{-ik\pi} = a_k + ib_k.$$

By direct calculation, the complex constants c_k are given by

$$c_0 = y_0 + y_1 + y_2 + y_3 + y_4 + y_5 + y_6 + y_7;$$

$$c_1 = y_0 + \left((i+1)/\sqrt{2}\right) y_1 + iy_2 + \left((i-1)/\sqrt{2}\right) y_3 - y_4$$
$$- \left((i+1)/\sqrt{2}\right) y_5 - iy_6 - \left((i-1)/\sqrt{2}\right) y_7;$$
$$c_2 = y_0 + iy_1 - y_2 - iy_3 + y_4 + iy_5 - y_6 - iy_7;$$
$$c_3 = y_0 + \left((i-1)/\sqrt{2}\right) y_1 - iy_2 + \left((i+1)/\sqrt{2}\right) y_3 - y_4$$
$$- \left((i-1)/\sqrt{2}\right) y_5 + iy_6 - \left((i+1)/\sqrt{2}\right) y_7;$$
$$c_4 = y_0 - y_1 + y_2 - y_3 + y_4 - y_5 + y_6 - y_7;$$
$$c_5 = y_0 - \left((i+1)/\sqrt{2}\right) y_1 + iy_2 - \left((i-1)/\sqrt{2}\right) y_3 - y_4$$
$$+ \left((i+1)/\sqrt{2}\right) y_5 - iy_6 + \left((i-1)/\sqrt{2}\right) y_7;$$
$$c_6 = y_0 - iy_1 - y_2 + iy_3 + y_4 - iy_5 - y_6 + iy_7;$$
$$c_7 = y_0 - \left((i-1)/\sqrt{2}\right) y_1 - iy_2 - \left((i+1)/\sqrt{2}\right) y_3 - y_4$$
$$+ \left((i-1)/\sqrt{2}\right) y_5 + iy_6 + \left((i+1)/\sqrt{2}\right) y_7.$$

Because of the small size of the collection of data points, many of the coefficients of the y_j in these equations are 1 or -1. This frequency will decrease in a larger application, so to count the computational operations accurately, multiplication by 1 or -1 will be included, even though it would not be necessary in this example. With this understanding, 64 multiplications/divisions and 56 additions/subtractions are required for the direct computation of $c_0, c_1, \ldots, c_7$.

To apply the fast Fourier transform procedure with $r = 1$, we first define

$$d_0 = \frac{1}{2}(c_0 + c_4) = y_0 + y_2 + y_4 + y_6;$$
$$d_1 = \frac{1}{2}(c_0 - c_4) = y_1 + y_3 + y_5 + y_7;$$
$$d_2 = \frac{1}{2}(c_1 + c_5) = y_0 + iy_2 - y_4 - iy_6;$$
$$d_3 = \frac{1}{2}(c_1 - c_5) = \left((i+1)/\sqrt{2}\right)(y_1 + iy_3 - y_5 - iy_7);$$
$$d_4 = \frac{1}{2}(c_2 + c_6) = y_0 - y_2 + y_4 - y_6;$$
$$d_5 = \frac{1}{2}(c_2 - c_6) = i(y_1 - y_3 + y_5 - y_7);$$
$$d_6 = \frac{1}{2}(c_3 + c_7) = y_0 - iy_2 - y_4 + iy_6;$$
$$d_7 = \frac{1}{2}(c_3 - c_7) = \left((i-1)/\sqrt{2}\right)(y_1 - iy_3 - y_5 + iy_7).$$

We then define, for $r = 2$,

$$
\begin{aligned}
e_0 &= \frac{1}{2}(d_0 + d_4) = y_0 + y_4;\\
e_1 &= \frac{1}{2}(d_0 - d_4) = y_2 + y_6;\\
e_2 &= \frac{1}{2}(id_1 + d_5) = i(y_1 + y_5);\\
e_3 &= \frac{1}{2}(id_1 - d_5) = i(y_3 + y_7);\\
e_4 &= \frac{1}{2}(d_2 + d_6) = y_0 - y_4;\\
e_5 &= \frac{1}{2}(d_2 - d_6) = i(y_2 - y_6);\\
e_6 &= \frac{1}{2}(id_3 + d_7) = \left((i-1)/\sqrt{2}\right)(y_1 - y_5);\\
e_7 &= \frac{1}{2}(id_3 - d_7) = i\left((i-1)/\sqrt{2}\right)(y_3 - y_7).
\end{aligned}
$$

Finally, for $r = p + 1 = 3$, we define

$$
\begin{aligned}
f_0 &= \frac{1}{2}(e_0 + e_4) = y_0;\\
f_1 &= \frac{1}{2}(e_0 - e_4) = y_4;\\
f_2 &= \frac{1}{2}(ie_1 + e_5) = iy_2;\\
f_3 &= \frac{1}{2}(ie_1 - e_5) = iy_6;\\
f_4 &= \frac{1}{2}\left(\left((i+1)/\sqrt{2}\right)e_2 + e_6\right) = \left((i-1)/\sqrt{2}\right)y_1;\\
f_5 &= \frac{1}{2}\left(\left((i+1)/\sqrt{2}\right)e_2 - e_6\right) = \left((i-1)/\sqrt{2}\right)y_5;\\
f_6 &= \frac{1}{2}\left(\left((i-1)/\sqrt{2}\right)e_3 + e_7\right) = \left(-(i+1)/\sqrt{2}\right)y_3;\\
f_7 &= \frac{1}{2}\left(\left((i-1)/\sqrt{2}\right)e_3 - e_7\right) = \left(-(i+1)/\sqrt{2}\right)y_7.
\end{aligned}
$$

The $c_0, \ldots, c_7$, $d_0, \ldots, d_7$, $e_0, \ldots, e_7$, and $f_0, \ldots, f_7$ are independent of the particular data points; they depend only on the fact that $m = 4$. For each m there is a unique set of constants $\{c_k\}_{k=0}^{2m-1}$, $\{d_k\}_{k=0}^{2m-1}$, $\{e_k\}_{k=0}^{2m-1}$, and $\{f_k\}_{k=0}^{2m-1}$. This portion of the work is not needed for a particular application. Only the calculations that follow are required:

1. $f_0 = y_0$; $f_1 = y_4$; $f_2 = iy_2$; $f_3 = iy_6$;

$f_4 = \left((i-1)/\sqrt{2}\right) y_1$; $f_5 = \left((i-1)/\sqrt{2}\right) y_5$; $f_6 = \left(-(i+1)/\sqrt{2}\right) y_3$;

$f_7 = \left(-(i+1)/\sqrt{2}\right) y_7$.

2. $e_0 = f_0 + f_1$; $e_1 = -i(f_2 + f_3)$; $e_2 = \left((-i+1)/\sqrt{2}\right)(f_4 + f_5)$;

$e_3 = \left((-i+1)/\sqrt{2}\right)(f_6 + f_7)$; $e_4 = f_0 - f_1$; $e_5 = f_2 - f_3$;

$e_6 = f_4 - f_5$; $e_7 = f_6 - f_7$.

3. $d_0 = e_0 + e_1$; $d_1 = -i(e_2 + e_3)$; $d_2 = e_4 + e_5$; $d_3 = -i(e_6 + e_7)$;

$d_4 = e_0 - e_1$; $d_5 = e_2 - e_3$; $d_6 = e_4 - e_5$; $d_7 = e_6 - e_7$.

4. $c_0 = d_0 + d_1$; $c_1 = d_2 + d_3$; $c_2 = d_4 + d_5$; $c_3 = d_6 + d_7$;

$c_4 = d_0 - d_1$; $c_5 = d_2 - d_3$; $c_6 = d_4 - d_5$; $c_7 = d_6 - d_7$.

Computing the constants $c_0, c_1, \ldots, c_7$ in this manner requires the number of operations shown in Table 8.13. Note again that multiplication by 1 or -1 has been included in the count, even though this does not require computational effort.

Table 8.13

Step	Multiplications/divisions	Additions/subtractions
(1)	8	0
(2)	8	8
(3)	8	8
(4)	0	8
Total	24	24

The lack of multiplications/divisions in Step 4 reflects the fact that for any m, the coefficients $\{c_k\}_{k=0}^{2m-1}$ are computed from $\{d_k\}_{k=0}^{2m-1}$ in the same manner:

$$c_k = d_{2k} + d_{2k+1}$$

and

$$c_{k+m} = d_{2k} - d_{2k+1},$$

for $k = 0, 1, \ldots, m-1$, so no complex multiplication is involved.

In summary, the direct computation of the coefficients $c_0, c_1, \ldots, c_7$ requires 64 multiplications/divisions and 56 additions/subtractions. The fast Fourier transform technique reduces the computations to 24 multiplications/divisions and 24 additions/subtractions. ■

Algorithm 8.3 performs the fast Fourier transform when $m = 2^p$ for some positive integer p. Modifications of the technique can be made when m takes other forms.

ALGORITHM 8.3

Fast Fourier Transform

To compute the coefficients in the summation

$$\frac{1}{m}\sum_{k=0}^{2m-1} c_k e^{ikx} = \frac{1}{m}\sum_{k=0}^{2m-1} c_k(\cos kx + i\sin kx), \quad \text{where } i = \sqrt{-1},$$

for the data $\{(x_j, y_j)\}_{j=0}^{2m-1}$ where $m = 2^p$ and $x_j = -\pi + j\pi/m$ for $j = 0, 1, \ldots, 2m-1$:

INPUT $m, p; y_0, y_1, \ldots, y_{2m-1}$.

OUTPUT complex numbers $c_0, \ldots, c_{2m-1}$; real numbers $a_0, \ldots, a_m; b_1, \ldots, b_{m-1}$.

Step 1 Set $M = m$;
$q = p$;
$\zeta = e^{\pi i/m}$.

Step 2 For $j = 0, 1, \ldots, 2m-1$ set $c_j = y_j$.

Step 3 For $j = 1, 2, \ldots, M$ set $\xi_j = \zeta^j$;
$\xi_{j+M} = -\xi_j$.

Step 4 Set $K = 0$;
$\xi_0 = 1$.

Step 5 For $L = 1, 2, \ldots, p+1$ do Steps 6–12.

Step 6 While $K < 2m-1$ do Steps 7–11.

Step 7 For $j = 1, 2, \ldots, M$ do Steps 8–10.

Step 8 Let $K = k_p \cdot 2^p + k_{p-1} \cdot 2^{p-1} + \cdots + k_1 \cdot 2 + k_0$; (*Decompose k.*)
set $K_1 = K/2^q = k_p \cdot 2^{p-q} + \cdots + k_{q+1} \cdot 2 + k_q$;
$K_2 = k_q \cdot 2^p + k_{q+1} \cdot 2^{p-1} + \cdots + k_p \cdot 2^q$.

Step 9 Set $\eta = c_{K+M}\xi_{K_2}$;
$c_{K+M} = c_K - \eta$;
$c_K = c_K + \eta$.

Step 10 Set $K = K + 1$.

Step 11 Set $K = K + M$.

Step 12 Set $K = 0$;
$M = M/2$;
$q = q - 1$.

Step 13 While $K < 2m-1$ do Steps 14–16.

Step 14 Let $K = k_p \cdot 2^p + k_{p-1} \cdot 2^{p-1} + \cdots + k_1 \cdot 2 + k_0$; (*Decompose k.*)
set $j = k_0 \cdot 2^p + k_1 \cdot 2^{p-1} + \cdots + k_{p-1} \cdot 2 + k_p$.

Step 15 If $j > K$ then interchange c_j and c_k.

Step 16 Set $K = K + 1$.

Step 17 Set $a_0 = c_0/m$;
$a_m = \text{Re}(e^{-i\pi m}c_m/m)$.

Step 18 For $j = 1, \ldots, m-1$ set $a_j = \text{Re}(e^{-i\pi j}c_j/m)$;
$b_j = \text{Im}(e^{-i\pi j}c_j/m)$.

Step 19 OUTPUT $(c_0, \ldots, c_{2m-1}; a_0, \ldots, a_m; b_1, \ldots, b_{m-1})$;
STOP.

EXAMPLE 2 Let $f(x) = x^4 - 3x^3 + 2x^2 - \tan x(x-2)$. To determine the trigonometric interpolating polynomial of degree four for the data $\{(x_j, y_j)\}_{j=0}^{7}$ where $x_j = j/4$ and $y_j = f(x_j)$ requires a transformation of the interval $[0, 2]$ to $[-\pi, \pi]$. The linear translation is given by

$$z_j = \pi(x_j - 1),$$

so that the input data to Algorithm 8.3 are

$$\left\{z_j, f\left(1 + \frac{z_j}{\pi}\right)\right\}_{j=0}^{7}.$$

The interpolating polynomial in z is

$$S_4(z) = 0.761979 + 0.771841 \cos z + 0.0173037 \cos 2z + 0.00686304 \cos 3z$$
$$- 0.000578545 \cos 4z - 0.386374 \sin z + 0.0468750 \sin 2z - 0.0113738 \sin 3z.$$

The trigonometric polynomial $S_4(x)$ on $[0, 2]$ is obtained by substituting $z = \pi(x-1)$ into $S_4(z)$. The graphs of $y = f(x)$ and $y = S_4(x)$ are shown in Figure 8.14. Values of $f(x)$ and $S_4(x)$ are given in Table 8.14. ■

Table 8.14

x	$f(x)$	$S_4(x)$	$\lvert f(x) - S_4(x)\rvert$
0.125	0.26440	0.25001	1.44×10^{-2}
0.375	0.84081	0.84647	5.66×10^{-3}
0.625	1.36150	1.35824	3.27×10^{-3}
0.875	1.61282	1.61515	2.33×10^{-3}
1.125	1.36672	1.36471	2.02×10^{-3}
1.375	0.71697	0.71931	2.33×10^{-3}
1.625	0.07909	0.07496	4.14×10^{-3}
1.875	−0.14576	−0.13301	1.27×10^{-2}

More details on the verification of the validity of the fast Fourier transform procedure can be found in [Ham], which presents the method from a mathematical approach, or in [Brac], where the presentation is based on methods more likely to be familiar to engineers. [AHU, pp. 252–269], is a good reference for a discussion of the computational aspects of the method. Modification of the procedure for the case when m is not a power of 2 can be found in [Win]. A presentation of the techniques and related material from the point of view of applied abstract algebra is given in [Lau, pp. 438–465].

Figure 8.14

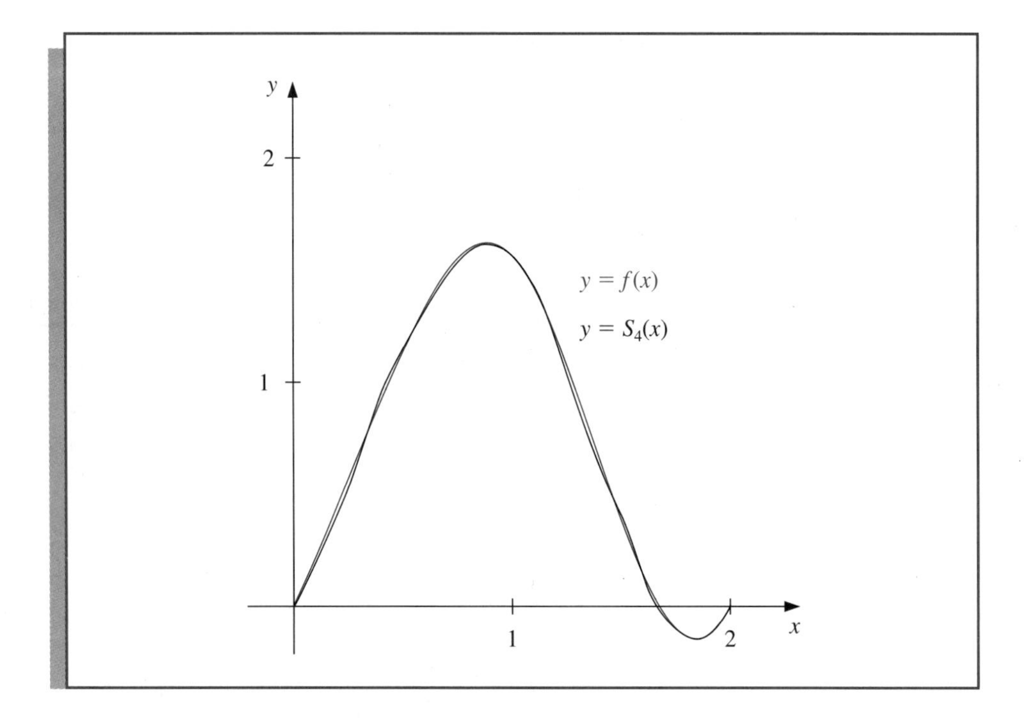

EXERCISE SET 8.6

1. Determine the trigonometric interpolating polynomial $S_2(x)$ of degree two on $[-\pi, \pi]$ for the following functions, and graph $f(x) - S_2(x)$:

   a. $f(x) = \pi(x - \pi)$ **b.** $f(x) = x(\pi - x)$

   c. $f(x) = |x|$ **d.** $f(x) = \begin{cases} -1, & -\pi \le x \le 0 \\ 1, & 0 < x \le \pi \end{cases}$

2. Determine the trigonometric interpolating polynomial of degree four for $f(x) = x(\pi - x)$ on the interval $[-\pi, \pi]$ using

   a. Direct calculation;

   b. The Fast Fourier Transform Algorithm.

3. Use the Fast Fourier Transform Algorithm to compute the trigonometric interpolating polynomial of degree four on $[-\pi, \pi]$ for the following functions.

   a. $f(x) = \pi(x - \pi)$ **b.** $f(x) = |x|$

   c. $f(x) = \cos \pi x - 2 \sin \pi x$ **d.** $f(x) = x \cos x^2 + e^x \cos e^x$

4. **a.** Determine the trigonometric interpolating polynomial $S_4(x)$ of degree four for $f(x) = x^2 \sin x$ on the interval $[0, 1]$.

   b. Compute $\int_0^1 S_4(x)\,dx$.

   c. Compare the integral in part (b) to $\int_0^1 x^2 \sin x\,dx$.

5. Use the approximations obtained in Exercise 3 to approximate the following integrals, and compare your results to the actual values.

a. $\int_{-\pi}^{\pi} \pi(x - \pi)\,dx$ **b.** $\int_{-\pi}^{\pi} |x|\,dx$

c. $\int_{-\pi}^{\pi} (\cos \pi x - 2 \sin \pi x)\,dx$ **d.** $\int_{-\pi}^{\pi} (x \cos x^2 + e^x \cos e^x)\,dx$

6. Use the Fast Fourier Transform Algorithm to determine the trigonometric interpolating polynomial of degree 16 for $f(x) = x^2 \cos x$ on $[-\pi, \pi]$.

7. Use the Fast Fourier Transform Algorithm to determine the trigonometric interpolating polynomial of degree 64 for $f(x) = x^2 \cos x$ on $[-\pi, \pi]$.

8. Use a trigonometric identity to show that $\sum_{j=0}^{2m-1} (\cos mx_j)^2 = 2m$.

9. Show that $c_0, \ldots, c_{2m-1}$ in Algorithm 8.3 are given by

$$\begin{bmatrix} c_0 \\ c_1 \\ c_2 \\ \vdots \\ c_{2m-1} \end{bmatrix} = \begin{bmatrix} 1 & 1 & 1 & \cdots & 1 \\ 1 & \zeta & \zeta^2 & \cdots & \zeta^{2m-1} \\ 1 & \zeta^2 & \zeta^4 & \cdots & \zeta^{4m-2} \\ \vdots & \vdots & \vdots & & \vdots \\ 1 & \zeta^{2m-1} & \zeta^{4m-2} & \cdots & \zeta^{(2m-1)^2} \end{bmatrix} \begin{bmatrix} y_0 \\ y_1 \\ y_2 \\ \vdots \\ y_{2m-1} \end{bmatrix},$$

where $\zeta = e^{\pi i/m}$.

10. In the discussion preceding Algorithm 8.3, an example for $m = 4$ was explained. Define vectors $\mathbf{c}$, $\mathbf{d}$, $\mathbf{e}$, $\mathbf{f}$, and $\mathbf{y}$ as

$$\mathbf{c} = (c_0, c_1, \ldots, c_7)^t,$$
$$\mathbf{d} = (d_0, d_1, \ldots, d_7)^t,$$
$$\mathbf{e} = (e_0, e_1, \ldots, e_7)^t,$$
$$\mathbf{f} = (f_0, f_1, \ldots, f_7)^t,$$
$$\mathbf{y} = (y_0, y_1, \ldots, y_7)^t.$$

Find matrices A, B, C, and D so that $\mathbf{c} = A\mathbf{d}$, $\mathbf{d} = B\mathbf{e}$, $\mathbf{e} = C\mathbf{f}$, and $\mathbf{f} = D\mathbf{y}$.

8.7 Survey of Methods and Software

In this chapter we have considered approximating data and functions with elementary functions. The elementary functions used were polynomials, rational functions, and trigonometric polynomials. We considered two types of approximations, discrete and continuous. Discrete approximations arise when approximating a finite set of data with an elementary function. Continuous approximations are used when the function to be approximated is known.

Discrete least squares techniques are recommended when the function is specified by giving a set of data that may not exactly represent the function. Least squares fit of data can take the form of a linear or other polynomial approximation or even an exponential form. These approximations are computed by solving sets of normal equations, as given in Section 8.1.

If the data are periodic, a trigonometric least squares fit may be appropriate. Because of the orthogonality of the trigonometric basis functions, the least squares trigonometric approximation does not require the solution of a linear system. For large amounts of periodic data, interpolation by trigonometric polynomials is also recommended. An efficient method of computing the trigonometric interpolating polynomial is given by the fast Fourier transform.

When the function to be approximated can be evaluated at any required argument, the approximations seek to minimize an integral instead of a sum. The continuous least squares polynomial approximations were considered in Section 8.2. Efficient computation of least squares polynomials lead to orthogonal sets of polynomials, such as the Legendre and Chebyshev polynomials. Approximation by rational functions was studied in Section 8.4, where Padé approximation as a generalization of the Maclaurin polynomial was presented and the extension to Chebyshev rational approximation was introduced. Both methods allow a more uniform method of approximation than polynomials. Continuous least squares approximation by trigonometric functions was discussed in Section 8.5, especially as it relates to Fourier series.

The IMSL Library provides a number of routines for approximation. The subroutine RLINE gives the least squares line for a set of data points. The subroutine returns statistics such as means and variances. Polynomial least squares approximations can be obtained with the subroutine RCURV. The subroutine FNLSQ computes the discrete least squares approximation for a user's choice of basis functions. The subroutine BSLSQ computes a least squares cubic spline approximation. RATCH computes the rational weighted Chebyshev approximation to a continuous functions on an interval $[a, b]$. The subroutine FFTCB computes the fast Fourier transform for a given set of data and is similar to Algorithm 8.3.

The NAG Library has many subroutines for function approximation. Least squares polynomial approximation is given in the subroutine E02ADF. This subroutine is quite versatile in that it computes least squares polynomials for varying degrees and gives their least squares errors. It uses Chebyshev polynomials to minimize roundoff error and enhance accuracy.

The routine E02AEF can be used to evaluate the approximation obtained by E02ADF. NAG also supplies the routine E02BAF to compute least squares cubic spline fits, E02GAF to compute the best L_1 linear fit and E02GCF to compute the best L_∞ fit. The routine E02RAF computes the Padé approximation. The NAG Library also includes many routines for fast Fourier transforms, one of which is C06ECF.

For further information on the general theory of approximation see Powell [Po], Davis [Da], or Cheney [Ch]. A good reference for methods of least squares is Lawson and Hanson [LH], and information about Fourier transforms can be found in Van Loan [Van] and in Briggs and Hanson [BH].

CHAPTER

9 Approximating Eigenvalues

■ ■ ■

The longitudinal vibrations of an elastic bar of local stiffness $p(x)$ and density $\rho(x)$ are described by the partial differential equation

$$\rho(x)\frac{\partial^2 v}{\partial t^2}(x,t) = \frac{\partial}{\partial x}\left[p(x)\frac{\partial v}{\partial x}(x,t)\right],$$

where $v(x,t)$ is the mean longitudinal displacement of a section of the bar from its equilibrium position x at time t. The vibrations can be written as a sum of simple harmonic vibrations:

$$v(x,t) = \sum_{k=0}^{\infty} c_k u_k(x) \cos\sqrt{\lambda_k}(t - t_0),$$

where

$$\frac{d}{dx}\left[p(x)\frac{du_k}{dx}\right] + \lambda_k \rho(x) u_k = 0.$$

If the bar has length l and is fixed at its ends, then this differential equation holds for $0 < x < l$ and $v(0) = v(l) = 0$. A system of these differential equations is called a Sturm-Liouville system, and the numbers λ_k are eigenvalues with corresponding eigenfunctions $u_k(x)$.

Suppose the bar is 1 m long with uniform stiffness $p(x) = p$ and uniform density $\rho(x) = \rho$. To approximate u and λ let $h = 0.2$. Then $x_j = 0.2j$, for $0 \le j \le 5$, and we can use the centered-difference formula (4.5) in Section 4.1 to approximate the first derivatives. This gives the linear system $A\mathbf{w} = -0.04\frac{\rho}{p}\lambda\mathbf{w}$, where

$$\begin{bmatrix} 2 & -1 & 0 & 0 \\ -1 & 2 & -1 & 0 \\ 0 & -1 & 2 & -1 \\ 0 & 0 & -1 & 2 \end{bmatrix} \begin{bmatrix} w_1 \\ w_2 \\ w_3 \\ w_4 \end{bmatrix} = -0.04\frac{\rho}{p}\lambda \begin{bmatrix} w_1 \\ w_2 \\ w_3 \\ w_4 \end{bmatrix}.$$

In this system, $w_j \approx u(x_j)$, for $1 \leq j \leq 4$, and $w_0 = w_5 = 0$. The four eigenvalues of A approximate the eigenvalues of the Sturm-Liouville system. It is the approximation of eigenvalues that we will consider in this chapter. A Sturm-Liouville application is discussed in Exercise 11 of Section 9.4.

9.1 Linear Algebra and Eigenvalues

The solution to many physical problems requires the calculation, or at least estimation, of the eigenvalues and corresponding eigenvectors of a matrix. We have seen that an $n \times n$ matrix A has precisely n (not necessarily distinct) eigenvalues that are the roots of the polynomial $p(\lambda) = \det(A - \lambda I)$. Theoretically, the eigenvalues of A are obtained by finding the n roots of $p(\lambda)$; then the associated linear systems are solved to determine the corresponding eigenvectors. In practice, $p(\lambda)$ is difficult to obtain, and except for small values of n, it is difficult to determine the roots of an nth-degree polynomial. Approximation techniques are needed for finding eigenvalues and eigenvectors.

Before considering further results concerning eigenvalues and eigenvectors, we need some definitions and results from linear algebra. All the general results that will be needed in the remainder of this chapter are listed here for ease of reference. The proofs of the results that are not given can be found in most standard texts on linear algebra (see, for example, [ND]). The first definition parallels the definition for the linear independence of functions described in Section 8.2.

Definition 9.1 Let $\{\mathbf{v}^{(1)}, \mathbf{v}^{(2)}, \mathbf{v}^{(3)}, \ldots, \mathbf{v}^{(k)}\}$ be a set of vectors. The set is **linearly independent** if whenever

$$\mathbf{0} = \alpha_1 \mathbf{v}^{(1)} + \alpha_2 \mathbf{v}^{(2)} + \alpha_3 \mathbf{v}^{(3)} + \cdots + \alpha_k \mathbf{v}^{(k)},$$

then $\alpha_1 = \alpha_2 = \alpha_3 = \cdots = \alpha_k = 0$.

Otherwise the set of vectors is **linearly dependent**. ■

Note that any set of vectors containing the zero vector is linearly dependent.

Theorem 9.2 If $\{\mathbf{v}^{(1)}, \mathbf{v}^{(2)}, \mathbf{v}^{(3)}, \ldots, \mathbf{v}^{(n)}\}$ is a set of n linearly independent vectors in $\mathbb{R}^n$, then any vector $\mathbf{x} \in \mathbb{R}^n$ can be written uniquely as

$$\mathbf{x} = \beta_1 \mathbf{v}^{(1)} + \beta_2 \mathbf{v}^{(2)} + \beta_3 \mathbf{v}^{(3)} + \cdots + \beta_n \mathbf{v}^{(n)}$$

for some collection of constants $\beta_1, \beta_2, \ldots, \beta_n$.

Proof Suppose that A is the matrix whose columns are the vectors $\mathbf{v}^{(1)}, \mathbf{v}^{(2)}, \cdots, \mathbf{v}^{(n)}$. Then the set $\{\mathbf{v}^{(1)}, \mathbf{v}^{(2)}, \cdots, \mathbf{v}^{(n)}\}$ is linearly independent if and only if the matrix equation $A\boldsymbol{\alpha} = \mathbf{0}$ has the unique solution $\boldsymbol{\alpha} = \mathbf{0}$. But by Theorem 9.2, this is equivalent to the statement that for any vector $\mathbf{x} \in \mathbb{R}^n$, $A\boldsymbol{\beta} = \mathbf{x}$ has a unique solution. This is, in turn, equivalent to the statement that for any $\mathbf{x} \in \mathbb{R}^n$,

$$\mathbf{x} = \beta_1 \mathbf{v}^{(1)} + \beta_2 \mathbf{v}^{(2)} + \cdots + \beta_n \mathbf{v}^{(n)}$$

for some unique set of constants $\beta_1, \beta_2, \ldots, \beta_n$. ■ ■ ■

Any collection of n linearly independent vectors in $\mathbb{R}^n$ is called a **basis** for $\mathbb{R}^n$.

EXAMPLE 1 Let $\mathbf{v}^{(1)} = (1, 0, 0)^t$, $\mathbf{v}^{(2)} = (-1, 1, 1)^t$, and $\mathbf{v}^{(3)} = (0, 4, 2)^t$. If α_1, α_2, and α_3 are numbers with

$$\mathbf{0} = \alpha_1 \mathbf{v}^{(1)} + \alpha_2 \mathbf{v}^{(2)} + \alpha_3 \mathbf{v}^{(3)},$$

then

$$\begin{aligned}(0, 0, 0)^t &= \alpha_1(1, 0, 0)^t + \alpha_2(-1, 1, 1)^t + \alpha_3(0, 4, 2)^t \\ &= (\alpha_1 - \alpha_2, \alpha_2 + 4\alpha_3, \alpha_2 + 2\alpha_3)^t,\end{aligned}$$

so

$$\alpha_1 - \alpha_2 = 0, \qquad \alpha_2 + 4\alpha_3 = 0, \qquad \text{and} \qquad \alpha_2 + 2\alpha_3 = 0.$$

The only solution to this system is $\alpha_1 = \alpha_2 = \alpha_3 = 0$, so the set $\{\mathbf{v}^{(1)}, \mathbf{v}^{(2)}, \mathbf{v}^{(3)}\}$ is linearly independent in $\mathbb{R}^3$ and is a basis for $\mathbb{R}^3$.

Any vector $\mathbf{x} = (x_1, x_2, x_3)^t$ in $\mathbb{R}^3$ can be written as

$$\mathbf{x} = \beta_1 \mathbf{v}^{(1)} + \beta_2 \mathbf{v}^{(2)} + \beta_3 \mathbf{v}^{(3)}$$

by choosing

$$\beta_1 = x_1 - x_2 + 2x_3, \qquad \beta_2 = 2x_3 - x_2, \qquad \text{and} \qquad \beta_3 = \frac{1}{2}(x_2 - x_3).$$ ■

The next result will be used in the following section to develop the Power method for approximating eigenvalues. A proof of this result is considered in Exercise 8.

Theorem 9.3 If A is a matrix and $\lambda_1, \ldots, \lambda_k$ are distinct eigenvalues of A with associated eigenvectors $\mathbf{x}^{(1)}, \mathbf{x}^{(2)}, \ldots, \mathbf{x}^{(k)}$, then $\{\mathbf{x}^{(1)}, \mathbf{x}^{(2)}, \ldots, \mathbf{x}^{(k)}\}$ is linearly independent. ■

In Section 8.2 we considered orthogonal and orthonormal sets of functions. Vectors with these properties are defined in a similar manner.

Definition 9.4 A set of vectors $\{\mathbf{v}^{(1)}, \mathbf{v}^{(2)}, \ldots, \mathbf{v}^{(n)}\}$ is called **orthogonal** if $(\mathbf{v}^{(i)})^t \mathbf{v}^{(j)} = 0$ for all $i \neq j$. If, in addition, $(\mathbf{v}^{(i)})^t \mathbf{v}^{(i)} = 1$ for all $i = 1, 2, \ldots, n$, then the set is **orthonormal**. ■

Since $\mathbf{x}^t \mathbf{x} = \|\mathbf{x}\|_2^2$, a set of orthogonal vectors $\{\mathbf{v}^{(1)}, \mathbf{v}^{(2)}, \ldots, \mathbf{v}^{(n)}\}$ is orthonormal if and only if

$$\|\mathbf{v}^{(i)}\|_2 = 1 \qquad \text{for each } i = 1, 2, \ldots, n.$$

EXAMPLE 2 The vectors $\mathbf{v}^{(1)} = (0, 4, 2)^t$, $\mathbf{v}^{(2)} = (-1, -\frac{1}{5}, \frac{2}{5})^t$, and $\mathbf{v}^{(3)} = (\frac{1}{6}, -\frac{1}{6}, \frac{1}{3})^t$ form an orthogonal set. The l_2 norms of these vectors are $\|\mathbf{v}^{(1)}\|_2 = 2\sqrt{5}$, $\|\mathbf{v}^{(2)}\|_2 = \frac{\sqrt{30}}{5}$, and $\|\mathbf{v}^{(3)}\|_2 = \frac{\sqrt{6}}{6}$. As a consequence, the vectors

$$\mathbf{u}^{(1)} = \frac{\mathbf{v}^{(1)}}{\|\mathbf{v}^{(1)}\|_2} = \left(0, \frac{2\sqrt{5}}{5}, \frac{\sqrt{5}}{5}\right)^t,$$

$$\mathbf{u}^{(2)} = \frac{\mathbf{v}^{(2)}}{\|\mathbf{v}^{(2)}\|_2} = \left(-\frac{\sqrt{30}}{6}, -\frac{\sqrt{30}}{30}, \frac{\sqrt{30}}{15}\right)^t,$$

and

$$\mathbf{u}^{(3)} = \frac{\mathbf{v}^{(3)}}{\|\mathbf{v}^{(3)}\|_2} = \left(\frac{\sqrt{6}}{6}, -\frac{\sqrt{6}}{6}, \frac{\sqrt{6}}{3}\right)^t$$

form an orthonormal set, since they inherit orthogonality from $\mathbf{v}^{(1)}$, $\mathbf{v}^{(2)}$, and $\mathbf{v}^{(3)}$, and, in addition,

$$\|\mathbf{u}^{(1)}\|_2 = \|\mathbf{u}^{(2)}\|_2 = \|\mathbf{u}^{(3)}\|_2 = 1.$$

∎

The proof of the next result is considered in Exercise 5.

Theorem 9.5 An orthogonal set of vectors that does not contain the zero vector is linearly independent. ∎

The terminology in the next definition follows from the fact that the columns of an orthogonal matrix form an orthogonal, in fact orthonormal, set of vectors. (See Exercise 6.)

Definition 9.6 A matrix P is said to be an **orthogonal matrix** if $P^{-1} = P^t$. ∎

Recall that the permutation matrices discussed in Section 6.5 have this property, so they are orthogonal.

EXAMPLE 3 The orthogonal matrix P formed from the orthonormal set of vectors found in Example 2 is

$$P = [\mathbf{u}^{(1)}, \mathbf{u}^{(2)}, \mathbf{u}^{(3)}] = \begin{bmatrix} 0 & -\frac{\sqrt{30}}{6} & \frac{\sqrt{6}}{6} \\ \frac{2\sqrt{5}}{5} & -\frac{\sqrt{30}}{30} & -\frac{\sqrt{6}}{6} \\ \frac{\sqrt{5}}{5} & \frac{\sqrt{30}}{15} & \frac{\sqrt{6}}{3} \end{bmatrix}.$$

Note that

$$PP^t = \begin{bmatrix} 0 & \frac{-\sqrt{30}}{6} & \frac{\sqrt{6}}{6} \\ \frac{2\sqrt{5}}{5} & -\frac{\sqrt{30}}{30} & -\frac{\sqrt{6}}{6} \\ \frac{\sqrt{5}}{5} & \frac{\sqrt{30}}{15} & \frac{\sqrt{6}}{3} \end{bmatrix} \cdot \begin{bmatrix} 0 & \frac{2\sqrt{5}}{5} & \frac{\sqrt{5}}{5} \\ -\frac{\sqrt{30}}{6} & -\frac{\sqrt{30}}{30} & \frac{\sqrt{30}}{15} \\ \frac{\sqrt{6}}{6} & -\frac{\sqrt{6}}{6} & \frac{\sqrt{6}}{3} \end{bmatrix} = \begin{bmatrix} 1 & 0 & 0 \\ 0 & 1 & 0 \\ 0 & 0 & 1 \end{bmatrix}.$$

It is also true that $P^tP = I$, so $P^t = P^{-1}$. ∎

Definition 9.7 Two matrices A and B are said to be **similar** if a nonsingular matrix S exists with $A = S^{-1}BS$. ■

Theorem 9.8 Suppose A and B are similar matrices and λ is an eigenvalue of A with associated eigenvector $\mathbf{x}$. Then λ is also an eigenvalue of B, and if $A = S^{-1}BS$, then $S\mathbf{x}$ is an eigenvector associated with λ for the matrix B.

Proof Suppose that $\mathbf{x} \neq \mathbf{0}$ is such that

$$S^{-1}BS\mathbf{x} = A\mathbf{x} = \lambda\mathbf{x}.$$

Multiplying on the left by the matrix S gives

$$BS\mathbf{x} = \lambda S\mathbf{x}.$$

Since $\mathbf{x} \neq \mathbf{0}$ and S is nonsingular, $S\mathbf{x} \neq \mathbf{0}$. Hence, $S\mathbf{x}$ is an eigenvector of B corresponding to its eigenvalue λ. ■ ■ ■

The determination of eigenvalues is easy for a triangular matrix A, for in this case λ is a solution to the equation

$$0 = \det(A - \lambda I) = \prod_{i=1}^{n}(a_{ii} - \lambda)$$

if and only if $\lambda = a_{ii}$ for some i. The next result describes a relationship, called a **similarity transformation**, between arbitrary matrices and triangular matrices.

Theorem 9.9 **(Schur)**
Let A be an arbitrary matrix. A nonsingular matrix U exists with the property that

$$T = U^{-1}AU,$$

where T is an upper-triangular matrix whose diagonal entries consist of the eigenvalues of A. ■

The matrix U whose existence is ensured in Theorem 9.9 satisfies the condition $\|U\mathbf{x}\|_2 = \|\mathbf{x}\|_2$ for any vector $\mathbf{x}$. Matrices with this property are called **unitary matrices**. Although we will not make use of this norm-preserving property, it does significantly increase the application of the theorem.

Theorem 9.9 is an existence theorem that ensures that the triangular matrix T exists but without providing a constructive means for finding T. The proof of the theorem requires a knowledge of the eigenvalues of A. In most instances, then, the similarity transformation is too difficult to determine. The following restriction of Theorem 9.9 to symmetric matrices reduces the complication, since in this case the transformation matrix is orthogonal.

Theorem 9.10 If A is a symmetric matrix and D is a diagonal matrix whose diagonal entries are the eigenvalues of A, then there exists an orthogonal matrix P such that $D = P^{-1}AP = P^tAP$. ■

The following corollaries to Theorem 9.10 demonstrate some of the interesting properties of symmetric matrices.

Corollary 9.11 If A is a symmetric $n \times n$ matrix, then the eigenvalues of A are real numbers, and there exist n eigenvectors of A that form an orthonormal set.

Proof If $P = (p_{ij})$ and $D = (d_{ij})$ are the matrices specified in Theorem 9.10, then

$$D = P^{-1}AP \quad \text{implies that} \quad AP = PD.$$

Let $\mathbf{v} = (p_{1i}, p_{2i}, \ldots, p_{ni})^t$ be the ith column of P. Then

$$A\mathbf{v} = d_{ii}\mathbf{v},$$

and the n columns of P are eigenvectors of A that form an orthonormal set.

Multiplying this equation on the left by $\mathbf{v}^t$ gives

$$\mathbf{v}^t A\mathbf{v} = d_{ii}\mathbf{v}^t\mathbf{v}.$$

Since $\mathbf{v}^t A\mathbf{v}$ and $\mathbf{v}^t\mathbf{v}$ are real numbers and $\mathbf{v}^t\mathbf{v} \neq 0$, the eigenvalue d_{ii} is also a real number.

▪ ▪ ▪

Recall from Section 6.6 that a symmetric matrix A is called positive definite if for all nonzero vectors $\mathbf{x}$ we have $\mathbf{x}^t A\mathbf{x} > 0$. The following theorem characterizes positive definite matrices in terms of eigenvalues. This eigenvalue property makes positive definite matrices important in applications.

Theorem 9.12 A symmetric matrix A is positive definite if and only if all the eigenvalues of A are positive.

Proof First suppose that A is positive definite and that λ is an eigenvalue of A with associated eigenvector $\mathbf{x}$. Then

$$0 < \mathbf{x}^t A\mathbf{x} = \lambda\mathbf{x}^t\mathbf{x} = \lambda\|\mathbf{x}\|_2^2,$$

so $\lambda > 0$. Hence, every eigenvalue of a positive definite matrix is positive.

To show the converse, suppose that A is symmetric with positive eigenvalues. By Corollary 9.11, there exist n eigenvectors of A, $\mathbf{v}^{(1)}, \mathbf{v}^{(2)}, \ldots, \mathbf{v}^{(n)}$, that form an orthonormal and, by Theorem 9.5, linearly independent set. Hence, for any nonzero vector $\mathbf{x}$ there exists a unique set of constants $\beta_1, \beta_2, \ldots, \beta_n$ that are not all zero and for which

$$\mathbf{x} = \sum_{i=1}^{n} \beta_i \mathbf{v}^{(i)}.$$

Multiplying by $\mathbf{x}^t A$ gives

$$\mathbf{x}^t A\mathbf{x} = \mathbf{x}^t\left(\sum_{i=1}^{n} \beta_i A\mathbf{v}^{(i)}\right) = \mathbf{x}^t\left(\sum_{i=1}^{n} \beta_i \lambda_i \mathbf{v}^{(i)}\right) = \sum_{j=1}^{n}\sum_{i=1}^{n} \beta_j \beta_i \lambda_i (\mathbf{v}^{(j)})^t \mathbf{v}^{(i)}$$

But the vectors $\mathbf{v}^{(1)}, \mathbf{v}^{(2)}, \ldots, \mathbf{v}^{(n)}$ form an orthonormal set, so

$$(\mathbf{v}^{(j)})^t\mathbf{v}^{(i)} = \begin{cases} 0, & \text{if } i \neq j, \\ 1, & \text{if } i = j. \end{cases}$$

This, together with the fact that the λ_i are all positive, implies that

$$\mathbf{x}^t A\mathbf{x} = \sum_{j=1}^{n}\sum_{i=1}^{n} \beta_j\beta_i\lambda_i(\mathbf{v}^{(j)})^t\mathbf{v}^{(i)} = \sum_{i=1}^{n} \lambda_i\beta_i^2 > 0.$$

Hence, A is positive definite. ■ ■ ■

The final result of the section concerns bounds for the approximation of eigenvalues.

Theorem 9.13 **(Gerschgorin Circle Theorem)**

Let A be an $n \times n$ matrix and R_i denote the circle in the complex plane with center a_{ii} and radius $\sum_{j=1, j\neq i}^{n} |a_{ij}|$; that is,

$$R_i = \left\{ z \in C \;\middle|\; |z - a_{ii}| \leq \sum_{\substack{j=1,\\ j\neq i}}^{n} |a_{ij}| \right\},$$

where C denotes the complex plane. The eigenvalues of A are contained within $R = \cup_{i=1}^{n} R_i$. Moreover, the union of any k of these circles that do not intersect the remaining $(n-k)$ contains precisely k (counting multiplicities) of the eigenvalues.

Proof Suppose that λ is an eigenvalue of A with associated eigenvector $\mathbf{x}$, where $\|\mathbf{x}\|_\infty = 1$. Since $A\mathbf{x} - \lambda\mathbf{x} = \mathbf{0}$, the equivalent component representation is

$$\sum_{j=1}^{n} a_{ij}x_j = \lambda x_i, \quad \text{for each } i = 1, 2, \ldots, n.$$

If k is an integer with $|x_k| = \|\mathbf{x}\|_\infty = 1$, this equation, with $i = k$, implies that

$$\sum_{j=1}^{n} a_{kj}x_j = \lambda x_k.$$

Thus

$$\sum_{\substack{j=1,\\ j\neq k}}^{n} a_{kj}x_j = \lambda x_k - a_{kk}x_k = (\lambda - a_{kk})x_k,$$

and

$$|\lambda - a_{kk}|\,|x_k| = |(\lambda - a_{kk})x_k| = \left|\sum_{\substack{j=1,\\ j\neq k}}^{n} a_{kj}x_j\right| \leq \sum_{\substack{j=1,\\ j\neq k}}^{n} |a_{kj}||x_j|.$$

Since $|x_k| \geq |x_j|$ for all $j = 1, 2, \ldots, n$,

$$|\lambda - a_{kk}| \leq \sum_{\substack{j=1,\\ j\neq k}}^{n} |a_{kj}| \left|\frac{x_j}{x_k}\right| \leq \sum_{\substack{j=1,\\ j\neq k}}^{n} |a_{kj}|.$$

Thus $\lambda \in R_k$, which proves the first assertion in the theorem. The second part of this theorem requires a clever continuity argument. A quite readable proof is contained in [Or2, p. 48]. ▪ ▪ ▪

EXAMPLE 4 For the matrix

$$A = \begin{bmatrix} 4 & 1 & 1 \\ 0 & 2 & 1 \\ -2 & 0 & 9 \end{bmatrix},$$

the circles in the Gerschgorin Theorem are (see Figure 9.1)

$$R_1 = \{z \in C \mid |z - 4| \leq 2\},$$
$$R_2 = \{z \in C \mid |z - 2| \leq 1\},$$

and

$$R_3 = \{z \in C \mid \; |z - 9| \leq 2\}.$$

Since R_1 and R_2 are disjoint from R_3, there must be precisely two eigenvalues within $R_1 \cup R_2$ and one within R_3. Moreover, since $\rho(A) = \max_{1\leq i\leq 3} |\lambda_i|$, we have $7 \leq \rho(A) \leq 11$. ■

Figure 9.1

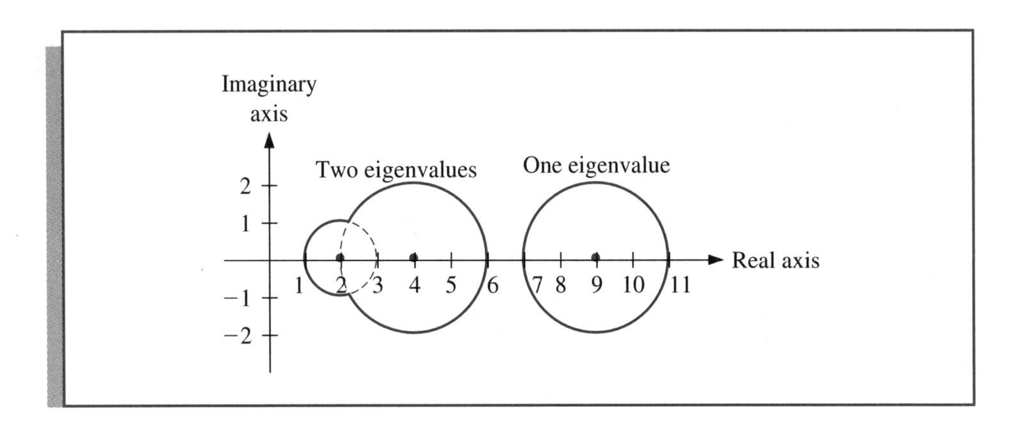

EXERCISE SET 9.1

1. Find the eigenvalues and associated eigenvectors for the following 3×3 matrices. Is there a set of three linearly independent eigenvectors?

a. $A = \begin{bmatrix} 2 & -3 & 6 \\ 0 & 3 & -4 \\ 0 & 2 & -3 \end{bmatrix}$ b. $A = \begin{bmatrix} 1 & 0 & 0 \\ -1 & 0 & 1 \\ -1 & -1 & 2 \end{bmatrix}$

c. $A = \begin{bmatrix} 2 & 0 & 1 \\ 0 & 2 & 0 \\ 1 & 0 & 2 \end{bmatrix}$ d. $A = \begin{bmatrix} 2 & -1 & -1 \\ -1 & 2 & -1 \\ -1 & -1 & 2 \end{bmatrix}$

e. $A = \begin{bmatrix} 1 & 1 & 1 \\ 1 & 1 & 0 \\ 1 & 0 & 1 \end{bmatrix}$ f. $A = \begin{bmatrix} 2 & 1 & 1 \\ 1 & 2 & 1 \\ 1 & 1 & 2 \end{bmatrix}$

2. The matrices in Exercise 1(c), (d), (e), and (f) are symmetric.

a. Are any positive definite?

b. Consider the positive definite matrices in part (a). Construct an orthogonal matrix P for which $P^tAP = D$, a diagonal matrix, using the eigenvectors found in Exercise 1.

3. Use the Gerschgorin Circle Theorem to determine bounds for the eigenvalues of the following matrices.

a. $\begin{bmatrix} 1 & 0 & 0 \\ -1 & 0 & 1 \\ -1 & -1 & 2 \end{bmatrix}$ b. $\begin{bmatrix} 4 & -1 & 0 \\ -1 & 4 & -1 \\ -1 & -1 & 4 \end{bmatrix}$

c. $\begin{bmatrix} 3 & 2 & 1 \\ 2 & 3 & 0 \\ 1 & 0 & 3 \end{bmatrix}$ d. $\begin{bmatrix} 4.75 & 2.25 & -0.25 \\ 2.25 & 4.75 & 1.25 \\ -0.25 & 1.25 & 4.75 \end{bmatrix}$

e. $\begin{bmatrix} -4 & 0 & 1 & 3 \\ 0 & -4 & 2 & 1 \\ 1 & 2 & -2 & 0 \\ 3 & 1 & 0 & -4 \end{bmatrix}$ f. $\begin{bmatrix} 1 & 0 & -1 & 1 \\ 2 & 2 & -1 & 1 \\ 0 & 1 & 3 & -2 \\ 1 & 0 & 1 & 4 \end{bmatrix}$

4. Show that any four vectors in $\mathbb{R}^3$ are linearly dependent.

5. Let $\{\mathbf{v}_1, \ldots, \mathbf{v}_k\}$ be a set of k orthogonal nonzero vectors. Show that $\{\mathbf{v}_1, \ldots, \mathbf{v}_k\}$ is a linearly independent set.

6. Let P be an orthogonal matrix. Show that the columns of P form an orthonormal set of vectors. Also, show that $\|P\|_2 = 1$ and $\|P^t\|_2 = 1$.

7. Let $\{\mathbf{v}_1, \ldots, \mathbf{v}_n\}$ be a set of orthonormal nonzero vectors in $\mathbb{R}^n$ and $\mathbf{x} \in \mathbb{R}^n$. Show that

$$\mathbf{x} = \sum_{k=1}^{n} c_k \mathbf{v}_k, \quad \text{where} \quad c_k = \mathbf{v}_k^t \mathbf{x}.$$

8. Show that if A is an $n \times n$ matrix with n distinct eigenvalues, then A has n linearly independent eigenvectors.

9. In Exercise 25 of Section 6.6, a symmetric matrix

$$A = \begin{bmatrix} 1.59 & 1.69 & 2.13 \\ 1.69 & 1.31 & 1.72 \\ 2.13 & 1.72 & 1.85 \end{bmatrix}$$

was used to describe the average wing lengths of fruit flies that were offspring resulting from the mating of three mutants of the flies. The entry a_{ij} represents the average wing length of a fly that is the offspring of a male fly of type i and a female fly of type j.

a. Find the eigenvalues and associated eigenvectors of this matrix.

b. Use Theorem 9.12 to answer the question posed in part (b) of Exercise 25 Section 6.6, that is, is this matrix positive definite?

10. A **persymmetric matrix** is a matrix that is symmetric about both diagonals; that is, an $N \times N$ matrix $A = (a_{ij})$ is persymmetric if $a_{ij} = a_{ji} = a_{N+1-i,N+1-j}$ for all $i = 1, 2, \ldots, N$ and

$j = 1, 2, \ldots, N$. A number of problems in communication theory have solutions that involve the eigenvalues and eigenvectors of matrices that are in persymmetric form. For example, the eigenvector corresponding to the minimal eigenvalue of the 4×4 persymmetric matrix

$$A = \begin{bmatrix} 2 & -1 & 0 & 0 \\ -1 & 2 & -1 & 0 \\ 0 & -1 & 2 & -1 \\ 0 & 0 & -1 & 2 \end{bmatrix}$$

gives the unit energy-channel impulse response for a given error sequence of length 2, and subsequently the minimum weight of any possible error sequence.

a. Use the Gerschgorin Circle Theorem to show that if A is the matrix given above and λ is its minimal eigenvalue, then $|\lambda - 4| = \rho(A - 4I)$, where ρ denotes the spectral radius.

b. Find the minimal eigenvalue of the matrix A by finding all the eigenvalues of $A - 4I$ and computing its spectral radius. Then find the corresponding eigenvector.

c. Use the Gerschgorin Circle Theorem to show that if λ is the minimal eigenvalue of the matrix

$$B = \begin{bmatrix} 3 & -1 & -1 & 1 \\ -1 & 3 & -1 & -1 \\ -1 & -1 & 3 & -1 \\ 1 & -1 & -1 & 3 \end{bmatrix},$$

then $|\lambda - 6| = \rho(B - 6I)$.

d. Repeat part (b) using the matrix B and the result in part (c).

9.2 The Power Method

In this section we derive the iterative **Power method** for approximating eigenvalues. To apply the Power method, we assume that the $n \times n$ matrix A has n eigenvalues $\lambda_1, \lambda_2, \ldots, \lambda_n$ with an associated collection of linearly independent eigenvectors $\{\mathbf{v}^{(1)}, \mathbf{v}^{(2)}, \mathbf{v}^{(3)}, \ldots, \mathbf{v}^{(n)}\}$. Moreover, we assume A has precisely one eigenvalue that is largest in magnitude.

Let $\lambda_1, \lambda_2, \ldots, \lambda_n$ denote the eigenvalues of A with $|\lambda_1| > |\lambda_2| \geq |\lambda_3| \geq \cdots \geq |\lambda_n|$. If $\mathbf{x}$ is any vector in $\mathbb{R}^n$, the fact that $\{\mathbf{v}^{(1)}, \mathbf{v}^{(2)}, \mathbf{v}^{(3)}, \ldots, \mathbf{v}^{(n)}\}$ is linearly independent implies that constants $\beta_1, \beta_2, \ldots, \beta_n$ exist with

$$\mathbf{x} = \sum_{j=1}^{n} \beta_j \mathbf{v}^{(j)}.$$

Multiplying both sides of this equation by $A, A^2, \ldots, A^k$, we obtain:

$$A\mathbf{x} = \sum_{j=1}^{n} \beta_j A\mathbf{v}^{(j)} = \sum_{j=1}^{n} \beta_j \lambda_j \mathbf{v}^{(j)},$$

$$A^2\mathbf{x} = \sum_{j=1}^{n} \beta_j \lambda_j A\mathbf{v}^{(j)} = \sum_{j=1}^{n} \beta_j \lambda_j^2 \mathbf{v}^{(j)},$$

$$\vdots$$

$$A^k\mathbf{x} = \sum_{j=1}^{n} \beta_j \lambda_j^k \mathbf{v}^{(j)}.$$

If λ_1^k is factored from each term on the right side of the last equation, then

$$A^k\mathbf{x} = \lambda_1^k \sum_{j=1}^{n} \beta_j \left(\frac{\lambda_j}{\lambda_1}\right)^k \mathbf{v}^{(j)}.$$

Since $|\lambda_1| > |\lambda_j|$ for all $j = 2, 3, \ldots, n$, we have $\lim_{k\to\infty}(\lambda_j/\lambda_1)^k = 0$, and

(9.1) $$\lim_{k\to\infty} A^k\mathbf{x} = \lim_{k\to\infty} \lambda_1^k \beta_1 \mathbf{v}^{(1)}.$$

This gives us the way to proceed to find λ_1 and an associated eigenvector, but we cannot use the sequence in (9.1) directly since it converges to zero if $|\lambda_1| < 1$ and diverges if $|\lambda_1| > 1$, provided, of course, that $\beta_1 \neq 0$.

Advantage can be made of the relationship expressed in Eq. (9.1) by scaling the powers of $A^k\mathbf{x}$ in an appropriate manner to ensure that the limit in Eq. (9.1) is finite and nonzero. The scaling begins by choosing $\mathbf{x}$ to be a unit vector $\mathbf{x}^{(0)}$ relative to $\|\cdot\|_\infty$ and a component $x_{p_0}^{(0)}$ of $\mathbf{x}^{(0)}$ with

$$x_{p_0}^{(0)} = 1 = \|\mathbf{x}^{(0)}\|_\infty.$$

Let $\mathbf{y}^{(1)} = A\mathbf{x}^{(0)}$, and define $\mu^{(1)} = y_{p_0}^{(1)}$.

With this notation,

$$\mu^{(1)} = y_{p_0}^{(1)} = \frac{y_{p_0}^{(1)}}{x_{p_0}^{(0)}} = \frac{\beta_1\lambda_1 v_{p_0}^{(1)} + \sum_{j=2}^{n}\beta_j\lambda_j v_{p_0}^{(j)}}{\beta_1 v_{p_0}^{(1)} + \sum_{j=2}^{n}\beta_j v_{p_0}^{(j)}} = \lambda_1\left[\frac{\beta_1 v_{p_0}^{(1)} + \sum_{j=2}^{n}\beta_j(\lambda_j/\lambda_1)v_{p_0}^{(j)}}{\beta_1 v_{p_0}^{(1)} + \sum_{j=2}^{n}\beta_j v_{p_0}^{(j)}}\right].$$

Then let p_1 be the least integer such that

$$|y_{p_1}^{(1)}| = \|\mathbf{y}^{(1)}\|_\infty$$

and define $\mathbf{x}^{(1)}$ by

$$\mathbf{x}^{(1)} = \frac{1}{y_{p_1}^{(1)}}\mathbf{y}^{(1)} = \frac{1}{y_{p_1}^{(1)}}A\mathbf{x}^{(0)}.$$

Then

$$x_{p_1}^{(1)} = 1 = \|\mathbf{x}^{(1)}\|_\infty.$$

Now define

$$\mathbf{y}^{(2)} = A\mathbf{x}^{(1)} = \frac{1}{y_{p_1}^{(1)}}A^2\mathbf{x}^{(0)}$$

and

$$\mu^{(2)} = y_{p_1}^{(2)} = \frac{y_{p_1}^{(2)}}{x_{p_1}^{(1)}} = \frac{\left[\beta_1 \lambda_1^2 v_{p_1}^{(1)} + \sum_{j=2}^n \beta_j \lambda_j^2 v_{p_1}^{(j)}\right] \Big/ y_{p_1}^{(1)}}{\left[\beta_1 \lambda_1 v_{p_1}^{(1)} + \sum_{j=2}^n \beta_j \lambda_j v_{p_1}^{(j)}\right] \Big/ y_{p_1}^{(1)}}$$

$$= \lambda_1 \left[\frac{\beta_1 v_{p_1}^{(1)} + \sum_{j=2}^n \beta_j (\lambda_j/\lambda_1)^2 v_{p_1}^{(j)}}{\beta_1 v_{p_1}^{(1)} + \sum_{j=2}^n \beta_j (\lambda_j/\lambda_1) v_{p_1}^{(j)}}\right].$$

Let p_2 be the smallest integer with

$$|y_{p_2}^{(2)}| = \|\mathbf{y}^{(2)}\|_\infty$$

and define

$$\mathbf{x}^{(2)} = \frac{1}{y_{p_2}^{(2)}} \mathbf{y}^{(2)} = \frac{1}{y_{p_2}^{(2)}} A\mathbf{x}^{(1)} = \frac{1}{y_{p_2}^{(2)} y_{p_1}^{(1)}} A^2 \mathbf{x}^{(0)}.$$

In a similar manner, define sequences of vectors $\{\mathbf{x}^{(m)}\}_{m=0}^\infty$ and $\{\mathbf{y}^{(m)}\}_{m=1}^\infty$, and a sequence of scalars $\{\mu^{(m)}\}_{m=1}^\infty$ inductively by

$$\mathbf{y}^{(m)} = A\mathbf{x}^{(m-1)},$$

(9.2)
$$\mu^{(m)} = y_{p_{m-1}}^{(m)} = \lambda_1 \left[\frac{\beta_1 v_{p_{m-1}}^{(1)} + \sum_{j=2}^n (\lambda_j/\lambda_1)^m \beta_j v_{p_{m-1}}^{(j)}}{\beta_1 v_{p_{m-1}}^{(1)} + \sum_{j=2}^n (\lambda_j/\lambda_1)^{m-1} \beta_j v_{p_{m-1}}^{(j)}}\right],$$

and

$$\mathbf{x}^{(m)} = \frac{\mathbf{y}^{(m)}}{y_{p_m}^{(m)}} = \frac{A^m \mathbf{x}^{(0)}}{\prod_{k=1}^m y_{p_k}^{(k)}},$$

where at each step p_m is used to represent the smallest integer for which

$$|y_{p_m}^{(m)}| = \|\mathbf{y}^{(m)}\|_\infty.$$

By examining Eq. (9.2), we see that since $|\lambda_j/\lambda_1| < 1$ for each $j = 2, 3, \ldots, n$, $\lim_{m\to\infty} \mu^{(m)} = \lambda_1$, provided that $\mathbf{x}^{(0)}$ is chosen so that $\beta_1 \neq 0$. Moreover, the sequence of vectors $\{\mathbf{x}^{(m)}\}_{m=0}^\infty$ converges to an eigenvector associated with λ_1 that has l_∞-norm one.

The Power method has the disadvantage that it is unknown at the outset whether or not the matrix has a single dominant eigenvalue. Nor is it known how $\mathbf{x}^{(0)}$ should be chosen so as to ensure that its representation in terms of the eigenvectors of the matrix will contain a nonzero contribution from the eigenvector associated with the dominant eigenvalue, should it exist.

Algorithm 9.1 implements the Power method.

ALGORITHM **9.1**

Power Method

To approximate the dominant eigenvalue and an associated eigenvector of the $n \times n$ matrix A given a nonzero vector $\mathbf{x}$:

INPUT dimension n; matrix A; vector $\mathbf{x}$; tolerance TOL; maximum number of iterations N.

OUTPUT approximate eigenvalue μ; approximate eigenvector $\mathbf{x}$ (with $\|\mathbf{x}\|_\infty = 1$) or a message that the maximum number of iterations was exceeded.

Step 1 Set $k = 1$.

Step 2 Find the smallest integer p with $1 \leq p \leq n$ and $|x_p| = \|\mathbf{x}\|_\infty$.

Step 3 Set $\mathbf{x} = \dfrac{1}{x_p}\mathbf{x}$.

Step 4 While ($k \leq N$) do Steps 5–11.

Step 5 Set $\mathbf{y} = A\mathbf{x}$.

Step 6 Set $\mu = y_p$.

Step 7 Find the smallest integer p with $1 \leq p \leq n$ and $|y_p| = \|\mathbf{y}\|_\infty$.

Step 8 If $y_p = 0$ then OUTPUT ('Eigenvector', $\mathbf{x}$);
OUTPUT ('A has the eigenvalue 0, select a new vector $\mathbf{x}$ and restart');
STOP.

Step 9 Set $ERR = \left\|\mathbf{x} - \dfrac{1}{y_p}\mathbf{y}\right\|_\infty$;

$$\mathbf{x} = \frac{1}{y_p}\mathbf{y}.$$

Step 10 If $ERR < TOL$ then OUTPUT (μ, $\mathbf{x}$);
(*Procedure completed successfully.*)
STOP.

Step 11 Set $k = k + 1$.

Step 12 OUTPUT ('Maximum number of iterations exceeded');
(*Procedure completed unsuccessfully.*)
STOP.

Choosing, in Step 7, the smallest integer p_m for which $|y_{p_m}^{(m)}| = \|\mathbf{y}^{(m)}\|_\infty$ will generally ensure that this index eventually becomes invariant. The rate at which $\{\mu^{(m)}\}_{m=1}^{\infty}$ converges to λ_1 is determined by the ratios $|\lambda_j/\lambda_1|^m$, for $j = 2, 3, \ldots, n$, and in particular by $|\lambda_2/\lambda_1|^m$. The rate of convergence is $O(|\lambda_2/\lambda_1|^m)$ (see [IK, p. 148]), so there is a constant k such that for large m,

$$|\mu^{(m)} - \lambda_1| \approx k\left|\frac{\lambda_2}{\lambda_1}\right|^m.$$

This implies that

$$\lim_{m\to\infty} \frac{|\mu^{(m+1)} - \lambda_1|}{|\mu^{(m)} - \lambda_1|} \approx \left|\frac{\lambda_2}{\lambda_1}\right| < 1.$$

The sequence $\{\mu^{(m)}\}$ converges linearly to λ_1, so the Aitken's Δ^2 procedure discussed in Section 2.5 can be used to speed the convergence. Implementing the Δ^2 procedure in Algorithm 9.1 is accomplished by modifying the algorithm as follows:

Step 1 Set $k = 1$;
$\mu_0 = 0$;
$\mu_1 = 0$.

Step 6 Set $\mu = y_p$;

$$\hat{\mu} = \mu_0 - \frac{(\mu_1 - \mu_0)^2}{\mu - 2\mu_1 + \mu_0}.$$

Step 10 If $ERR < TOL$ and $k \geq 4$ then OUTPUT $(\hat{\mu}, \mathbf{x})$;
STOP.

Step 11 Set $k = k + 1$;
$\mu_0 = \mu_1$;
$\mu_1 = \mu$.

It is not necessary for the matrix to have distinct eigenvalues for the Power method to converge. If the unique dominant eigenvalue, λ_1, has multiplicity r greater than 1 and $\mathbf{v}^{(1)}, \mathbf{v}^{(2)}, \ldots, \mathbf{v}^{(r)}$ are linearly independent eigenvectors associated with λ_1, the procedure will still converge to λ_1. The sequence of vectors $\{\mathbf{x}^{(m)}\}_{m=0}^{\infty}$ will, in this case, converge to an eigenvector of λ_1 of norm one that is a linear combination of $\mathbf{v}^{(1)}, \mathbf{v}^{(2)}, \ldots, \mathbf{v}^{(r)}$, and depends on the choice of the initial vector $\mathbf{x}^{(0)}$.

EXAMPLE 1 The matrix

$$A = \begin{bmatrix} -4 & 14 & 0 \\ -5 & 13 & 0 \\ -1 & 0 & 2 \end{bmatrix}$$

has eigenvalues $\lambda_1 = 6$, $\lambda_2 = 3$, and $\lambda_3 = 2$. The Power method described in Algorithm 9.1 will, consequently, converge. Let $\mathbf{x}^{(0)} = (1, 1, 1)^t$, then

$$\mathbf{y}^{(1)} = A\mathbf{x}^{(0)} = (10, 8, 1)^t,$$

so

$$\|\mathbf{y}^{(1)}\|_\infty = 10, \quad \mu^{(1)} = y_1^{(1)} = 10, \quad \text{and} \quad \mathbf{x}^{(1)} = \frac{\mathbf{y}^{(1)}}{10} = (1, 0.8, 0.1)^t.$$

Continuing in this manner leads to the values in Table 9.1, where $\hat{\mu}^{(m)}$ represents the sequence generated by the Aitken's Δ^2 procedure. An approximation to the dominant eigenvalue, 6, at this stage is $\hat{\mu}^{(10)} = 6.000000$ with approximate unit eigenvector $(1, 0.714316, -0.249895)^t$. ■

Table 9.1

m	$(\mathbf{x}^{(m)})^t$	$\mu^{(m)}$	$\hat{\mu}^{(m)}$
0	$(1, 1, 1)$		
1	$(1, 0.8, 0.1)$	10	6.266667
2	$(1, 0.75, -0.111)$	7.2	6.062473
3	$(1, 0.730769, -0.188803)$	6.5	6.015054
4	$(1, 0.722200, -0.220850)$	6.230769	6.004202
5	$(1, 0.718182, -0.235915)$	6.111000	6.000855
6	$(1, 0.716216, -0.243095)$	6.054546	6.000240
7	$(1, 0.715247, -0.246588)$	6.027027	6.000058
8	$(1, 0.714765, -0.248306)$	6.013453	6.000017
9	$(1, 0.714525, -0.249157)$	6.006711	6.000003
10	$(1, 0.714405, -0.249579)$	6.003352	6.000000
11	$(1, 0.714346, -0.249790)$	6.001675	
12	$(1, 0.714316, -0.249895)$	6.000837	

When A is symmetric, a variation in the choice of the vectors $\mathbf{x}^{(m)}$, $\mathbf{y}^{(m)}$, and scalars $\mu^{(m)}$ can be made to significantly improve the rate of convergence of the sequence $\{\mu^{(m)}\}_{m=1}^{\infty}$ to the dominant eigenvalue λ_1. In fact, although the rate of convergence of the general Power method is $O(|\lambda_2/\lambda_1|^m)$, the rate of convergence of the modified procedure given in Algorithm 9.2 for symmetric matrices is $O(|\lambda_2/\lambda_1|^{2m})$. (See [IK, pp. 149 ff].) The sequence $\{\mu^{(m)}\}$ is still linearly convergent, so Aitken's Δ^2 procedure can again be applied.

ALGORITHM 9.2

Symmetric Power Method

To approximate the dominant eigenvalue and an associated eigenvector of the $n \times n$ symmetric matrix A, given a nonzero vector $\mathbf{x}$:

INPUT dimension n; matrix A; vector $\mathbf{x}$; tolerance TOL; maximum number of iterations N.

OUTPUT approximate eigenvalue μ; approximate eigenvector $\mathbf{x}$ (with $\|\mathbf{x}\|_2 = 1$) or a message that the maximum number of iterations was exceeded.

Step 1 Set $k = 1$;
$\mathbf{x} = \mathbf{x}/\|\mathbf{x}\|_2$.

Step 2 While ($k \leq N$) do Steps 3–8.

Step 3 Set $\mathbf{y} = A\mathbf{x}$.

Step 4 Set $\mu = \mathbf{x}^t\mathbf{y}$.

Step 5 If $\|\mathbf{y}\|_2 = 0$, then OUTPUT ('Eigenvector', $\mathbf{x}$);
OUTPUT ('A has eigenvalue 0, select new vector $\mathbf{x}$ and restart');
STOP.

Step 6 Set $ERR = \left\|\mathbf{x} - \dfrac{\mathbf{y}}{\|\mathbf{y}\|_2}\right\|_2$;
$\mathbf{x} = \mathbf{y}/\|\mathbf{y}\|_2$.

Step 7 If $ERR < TOL$ then OUTPUT $(\mu, \mathbf{x})$;
(*Procedure completed successfully.*)
STOP.

Step 8 Set $k = k + 1$.

Step 9 OUTPUT ('Maximum number of iterations exceeded');
(*Procedure completed unsuccessfully.*)
STOP.

EXAMPLE 2 The matrix

$$A = \begin{bmatrix} 4 & -1 & 1 \\ -1 & 3 & -2 \\ 1 & -2 & 3 \end{bmatrix}$$

Table 9.2

m	$(\mathbf{y}^{(m)})^t$	$\mu^{(m)}$	$\hat{\mu}^{(m)}$	$(\mathbf{x}^{(m)})^t$ with $\Vert\mathbf{x}^{(m)}\Vert_\infty = 1$
0				$(1, 0, 0)$
1	$(4, -1, 1)$	4		$(1, -0.25, 0.25)$
2	$(4.5, -2.25, 2.25)$	4.5	7	$(1, -0.5, 0.5)$
3	$(5, -3.5, 3.5)$	5	$6.\dot{2}$	$(1, -0.7, 0.7)$
4	$(5.4, -4.5, 4.5)$	5.4	6.047617	$(1, -8.33\bar{3}, 0.833\bar{3})$
5	$(5.66\bar{6}, -5.166\bar{6}, 5.166\bar{6})$	$5.66\bar{6}$	6.011767	$(1, -0.911765, 0.911765)$
6	$(5.823529, -5.558824, 5.558824)$	5.823529	6.002931	$(1, -0.954545, 0.954545)$
7	$(5.909091, -5.772727, 5.772727)$	5.909091	6.000733	$(1, -0.976923, 0.976923)$
8	$(5.953846, -5.884615, 5.884615)$	5.953846	6.000184	$(1, -0.988372, 0.988372)$
9	$(5.976744, -5.941861, 5.941861)$	5.976744		$(1, -0.994163, 0.994163)$
10	$(5.988327, -5.970817, 5.970817)$	5.988327		$(1, -0.997076, 0.997076)$

Table 9.3

m	$(\mathbf{y}^{(m)})^t$	$\mu^{(m)}$	$\hat{\mu}^{(m)}$	$(\mathbf{x}^{(m)})^t$ with $\Vert\mathbf{x}^{(m)}\Vert_2 = 1$
0	$(1, 0, 0)$			$(1, 0, 0)$
1	$(4, -1, 1)$	4	7	$(0.942809, -0.235702, 0.235702)$
2	$(4.242641, -2.121320, 2.121320$	5	6.047619	$(0.816497, -0.408248, 0.408248)$
3	$(4.082483, -2.857738, 2.857738)$	5.666667	6.002932	$(0.710669, -0.497468, 0.497468)$
4	$(3.837613, -3.198011, 3.198011)$	5.909091	6.000183	$(0.646997, -0.539164, 0.539164)$
5	$(3.666314, -3.342816, 3.342816)$	5.976744	6.000012	$(0.612836, -0.558763, 0.558763)$
6	$(3.568871, -3.406650, 3.406650)$	5.994152	6.000000	$(0.595247, -0.568190, 0.568190)$
7	$(3.517370, -3.436200, 3.436200)$	5.998536	6.000000	$(0.586336, -0.572805, 0.572805)$
8	$(3.490952, -3.450359, 3.450359)$	5.999634		$(0.581852, -0.575086, 0.575086)$
9	$(3.477580, -3.457283, 3.457283)$	5.999908		$(0.579603, -0.576220, 0.576220)$
10	$(3.470854, -3.460706, 3.460706)$	5.999977		$(0.578477, -0.576786, 0.576786)$

is symmetric with eigenvalues $\lambda_1 = 6$, $\lambda_2 = 3$, and $\lambda_3 = 1$. Listing the results of the first ten iterations of the Power methods presented in Algorithms 9.1 and 9.2 with $\mathbf{y}^{(0)} = \mathbf{x}^{(0)} = (1, 0, 0)^t$ demonstrates the significant improvement obtained by using the latter algorithm. Using Algorithm 9.1 we have the results listed in Table 9.2. The results listed in Table 9.3 are obtained using Algorithm 9.2. Aitken's Δ^2 procedure is used in both cases to improve convergence to the dominant eigenvalue. No value of $\hat{\mu}^{(1)}$ is given in Table 9.2 because the calculation leads to a fraction with a zero denominator. ■

The following gives an error bound for approximating the eigenvalues of a symmetric matrix.

Theorem 9.14 If A is an $n \times n$ symmetric matrix with eigenvalues $\lambda_1, \lambda_2, \ldots, \lambda_n$ and $\|A\mathbf{x} - \lambda\mathbf{x}\|_2 < \varepsilon$ for some vector $\mathbf{x}$ with $\|\mathbf{x}\|_2 = 1$ and real number λ, then

$$\min_{1 \le j \le n} |\lambda_j - \lambda| < \varepsilon.$$

Proof Suppose that $\mathbf{v}^{(1)}, \mathbf{v}^{(2)}, \ldots, \mathbf{v}^{(n)}$ form an orthonormal set of eigenvectors of A associated, respectively, with the eigenvalues $\lambda_1, \lambda_2, \ldots, \lambda_n$. By Theorems 9.5 and 9.2, $\mathbf{x}$ can be expressed, for some unique set of constants $\beta_1, \beta_2, \ldots, \beta_n$, as

$$\mathbf{x} = \sum_{j=1}^{n} \beta_j \mathbf{v}^{(j)}.$$

Thus,

$$\|A\mathbf{x} - \lambda\mathbf{x}\|_2^2 = \left\| \sum_{j=1}^{n} \beta_j(\lambda_j - \lambda)\mathbf{v}^{(j)} \right\|_2^2 = \sum_{j=1}^{n} |\beta_j|^2 |\lambda_j - \lambda|^2 \ge \min_{1 \le j \le n} |\lambda_j - \lambda|^2 \sum_{j=1}^{n} |\beta_j|^2.$$

But

$$\sum_{j=1}^{n} |\beta_j|^2 = \|\mathbf{x}\|_2^2 = 1,$$

so

$$\varepsilon \ge \|A\mathbf{x} - \lambda\mathbf{x}\|_2 > \min_{1 \le j \le n} |\lambda_j - \lambda|.$$ ■ ■ ■

The **Inverse Power method** is a modification of the Power method that gives faster convergence. It is used to determine the eigenvalue of A that is closest to a specified number q.

Assume that the matrix A has eigenvalues $\lambda_1, \ldots, \lambda_n$ with linearly independent eigenvectors $\mathbf{v}^{(1)}, \ldots, \mathbf{v}^{(n)}$. Consider the matrix $(A - qI)^{-1}$, where $q \ne \lambda_i$ for $i = 1, 2, \ldots, n$. The eigenvalues of $(A - qI)^{-1}$ are

$$\frac{1}{\lambda_1 - q}, \quad \frac{1}{\lambda_2 - q}, \quad \ldots, \quad \frac{1}{\lambda_n - q}$$

with eigenvectors $\mathbf{v}^{(1)}, \mathbf{v}^{(2)}, \ldots, \mathbf{v}^{(n)}$ (See Exercise 10 of Section 7.2.). Applying the Power method to $(A - qI)^{-1}$ gives

$$\mathbf{y}^{(m)} = (A - qI)^{-1}\mathbf{x}^{(m-1)},$$

(9.3)
$$\mu^{(m)} = y^{(m)}_{p_{m-1}} = \frac{y^{(m)}_{p_{m-1}}}{x^{(m-1)}_{p_{m-1}}} = \frac{\displaystyle\sum_{j=1}^{n} \beta_j \frac{1}{(\lambda_j - q)^m} v^{(j)}_{p_{m-1}}}{\displaystyle\sum_{j=1}^{n} \beta_j \frac{1}{(\lambda_j - q)^{m-1}} v^{(j)}_{p_{m-1}}},$$

and

$$\mathbf{x}^{(m)} = \frac{\mathbf{y}^{(m)}}{y^{(m)}_{p_m}},$$

where, at each step, p_m represents the smallest integer for which $|y^{(m)}_{p_m}| = \|\mathbf{y}^{(m)}\|_\infty$. The sequence $\{\mu^{(m)}\}$ in Eq. (9.3) converges to

$$\frac{1}{|\lambda_k - q|} = \max_{1 \le i \le n} \frac{1}{|\lambda_i - q|},$$

and λ_k is the eigenvalue of A closest to q.

With k known, Eq. (9.3) can be written as

(9.4)
$$\mu^{(m)} = \frac{1}{\lambda_k - q} \left[\frac{\beta_k v^{(k)}_{p_{m-1}} + \displaystyle\sum_{\substack{j=1 \\ j \ne k}}^{n} \beta_j \left[\frac{\lambda_k - q}{\lambda_j - q}\right]^m v^{(j)}_{p_{m-1}}}{\beta_k v^{(k)}_{p_{m-1}} + \displaystyle\sum_{\substack{j=1 \\ j \ne k}}^{n} \beta_j \left[\frac{\lambda_k - q}{\lambda_j - q}\right]^{m-1} v^{(j)}_{p_{m-1}}} \right].$$

Thus the choice of q determines the convergence, provided that $1/(\lambda_k - q)$ is a unique dominant eigenvalue of $(A - qI)^{-1}$ (although it may be a multiple eigenvalue). The closer q is to an eigenvalue λ_k of A, the faster the convergence since the convergence is of order

$$O\left(\left|\frac{(\lambda - q)^{-1}}{(\lambda_k - q)^{-1}}\right|^m\right) = O\left(\left|\frac{(\lambda_k - q)}{(\lambda - q)}\right|^m\right),$$

where λ represents the eigenvalue of A that is second closest to q.

The determination of $\mathbf{y}^{(m)}$ is obtained from the equation

$$(A - qI)\mathbf{y}^{(m)} = \mathbf{x}^{(m-1)}.$$

In general, Gaussian elimination with pivoting is used to solve this system.

Although the Inverse Power method requires the solution of an $n \times n$ system at each step, the multipliers can be saved to reduce the computation. The selection of q can be based on the Gerschgorin Circle Theorem or on any other means of localizing an eigenvalue.

Algorithm 9.3 computes q from an initial approximation $\mathbf{x}^{(0)}$ to an eigenvector by

$$q = \frac{\mathbf{x}^{(0)t} A\mathbf{x}^{(0)}}{\mathbf{x}^{(0)t}\mathbf{x}^{(0)}}.$$

This choice of q results from the observation that if $\mathbf{x}$ is an eigenvector of A with respect to the eigenvalue λ, then $A\mathbf{x} = \lambda\mathbf{x}$. So, $\mathbf{x}^t A\mathbf{x} = \lambda\mathbf{x}^t\mathbf{x}$ and

$$\lambda = \frac{\mathbf{x}^t A\mathbf{x}}{\mathbf{x}^t\mathbf{x}} = \frac{\mathbf{x}^t A\mathbf{x}}{\|\mathbf{x}\|_2^2}$$

If q is close to an eigenvalue, the convergence will be quite rapid, but a pivoting technique should be used in Step 6 to avoid contamination by roundoff error.

Algorithm 9.3 is often used to approximate an eigenvector when an approximate eigenvalue q is known.

ALGORITHM 9.3

Inverse Power Method

To approximate an eigenvalue and an associated eigenvector of the $n \times n$ matrix A given a nonzero vector $\mathbf{x}$:

INPUT dimension n; matrix A; vector $\mathbf{x}$; tolerance TOL; maximum number of iterations N.

OUTPUT approximate eigenvalue μ; approximate eigenvector $\mathbf{x}$ (with $\|\mathbf{x}\|_\infty = 1$) or a message that the maximum number of iterations was exceeded.

Step 1 Set $q = \dfrac{\mathbf{x}^t A\mathbf{x}}{\mathbf{x}^t\mathbf{x}}$.

Step 2 Set $k = 1$.

Step 3 Find the smallest integer p with $1 \le p \le n$ and $|x_p| = \|\mathbf{x}\|_\infty$.

Step 4 Set $\mathbf{x} = \dfrac{1}{x_p}\mathbf{x}$.

Step 5 While ($k \le N$) do Steps 6–12.

Step 6 Solve the linear system $(A - qI)\mathbf{y} = \mathbf{x}$.

Step 7 If the system does not have a unique solution, then
OUTPUT ('q is an eigenvalue', q);
STOP.

Step 8 Set $\mu = y_p$.

Step 9 Find the smallest integer p with $1 \le p \le n$ and $|y_p| = \|\mathbf{y}\|_\infty$.

Step 10 Set $ERR = \left\|\mathbf{x} - \dfrac{1}{y_p}\mathbf{y}\right\|_\infty$;

$$\mathbf{x} = \frac{1}{y_p}\mathbf{y}.$$

Step 11 If $ERR < TOL$ then set $\mu = (1/\mu) + q$;
OUTPUT (μ, $\mathbf{x}$);
(*Procedure completed successfully.*)
STOP.

Step 12 Set $k = k + 1$.

Step 13 OUTPUT ('Maximum number of iterations exceeded');
(*Procedure completed unsuccessfully.*)
STOP.

Since the convergence of the Inverse Power Method is linear, Aitken Δ^2 procedure can again be used to speed convergence. The following example illustrates the fast convergence of the Inverse Power method if q is close to an eigenvalue.

EXAMPLE 3 The matrix

$$A = \begin{bmatrix} -4 & 14 & 0 \\ -5 & 13 & 0 \\ -1 & 0 & 2 \end{bmatrix}$$

was considered in Example 1. Algorithm 9.1 gave the approximation $\mu^{(12)} = 6.000837$ using $\mathbf{x}^{(0)} = (1, 1, 1)^t$. With $\mathbf{x}^{(0)} = (1, 1, 1)^t$, we have

$$q = \frac{\mathbf{x}^{(0)t} A \mathbf{x}^{(0)}}{\mathbf{x}^{(0)t} \mathbf{x}^{(0)}} = \frac{19}{3} = 6.333333.$$

The results of applying Algorithm 9.3 and Aitken's Δ^2 method are listed in Table 9.4. ■

Table 9.4

m	$\mathbf{x}^{(m)t}$	$\mu^{(m)}$	$\hat{\mu}^{(m)}$
0	(1, 1, 1)		
1	(1, 0.720727, −0.194042)	6.183183	6.000116
2	(1, 0.715518, −0.245052)	6.017244	6.000004
3	(1, 0.714409, −0.249522)	6.001719	6.000004
4	(1, 0.714298, −0.249953)	6.000175	6.000003
5	(1, 0.714287, −0.250000)	6.000021	
6	(1, 0.714286, −0.249999)	6.000005	

If A is symmetric, then for any real number q, $(A - qI)^{-1}$ is also symmetric, so the Symmetric Power method, Algorithm 9.2, can be applied to $(A - qI)^{-1}$ to speed the convergence to

$$O\left(\left|\frac{\lambda_k - q}{\lambda - q}\right|^{2m}\right).$$

Numerous techniques are available for obtaining approximations to the other eigenvalues of a matrix once an approximation to the dominant eigenvalue has been computed. We will restrict our presentation to **deflation techniques**.

Deflation techniques involve forming a new matrix B whose eigenvalues are the same as those of A, except that the dominant eigenvalue of A is replaced by the eigenvalue 0 in B. The following result justifies the procedure. The proof of this theorem can be found in [Wil2, p. 596].

Theorem 9.15 Suppose $\lambda_1, \lambda_2, \ldots, \lambda_n$ are eigenvalues of A with associated eigenvectors $\mathbf{v}^{(1)}, \mathbf{v}^{(2)}, \ldots, \mathbf{v}^{(n)}$ and that λ_1 has multiplicity one. Let $\mathbf{x}$ be a vector with $\mathbf{x}^t\mathbf{v}^{(1)} = 1$. Then the matrix

$$B = A - \lambda_1 \mathbf{v}^{(1)}\mathbf{x}^t$$

has eigenvalues $0, \lambda_2, \lambda_3, \ldots, \lambda_n$ with associated eigenvectors $\mathbf{v}^{(1)}, \mathbf{w}^{(2)}, \mathbf{w}^{(3)}, \ldots, \mathbf{w}^{(n)}$, where $\mathbf{v}^{(i)}$ and $\mathbf{w}^{(i)}$ are related by the equation

(9.5)
$$\mathbf{v}^{(i)} = (\lambda_i - \lambda_1)\mathbf{w}^{(i)} + \lambda_1 \left(\mathbf{x}^t\mathbf{w}^{(i)}\right) \mathbf{v}^{(1)},$$

for each $i = 2, 3, \ldots, n$. ■

There are many choices of the vector $\mathbf{x}$ that could be used in Theorem 9.15. **Wielandt deflation** proceeds from defining

(9.6)
$$\mathbf{x} = \frac{1}{\lambda_1 v_i^{(1)}} \begin{bmatrix} a_{i1} \\ a_{i2} \\ \vdots \\ a_{in} \end{bmatrix},$$

where $v_i^{(1)}$ is a coordinate of $\mathbf{v}^{(1)}$ that is nonzero, and the values $a_{i1}, a_{i2}, \ldots, a_{in}$ are the entries in the ith row of A.

With this definition,

$$\mathbf{x}^t\mathbf{v}^{(1)} = \frac{1}{\lambda_1 v_i^{(1)}}[a_{i1}, a_{i2}, \ldots, a_{in}]\left(v_1^{(1)}, v_2^{(1)}, \ldots, v_n^{(1)}\right)^t = \frac{1}{\lambda_1 v_i^{(1)}} \sum_{j=1}^{n} a_{ij}v_j^{(1)},$$

where the sum is the ith coordinate of the product $A\mathbf{v}^{(1)}$. Since $A\mathbf{v}^{(1)} = \lambda_1\mathbf{v}^{(1)}$, this implies that

$$\mathbf{x}^t\mathbf{v}^{(1)} = \frac{1}{\lambda_1 v_i^{(1)}}\left(\lambda_1 v_i^{(1)}\right) = 1,$$

so $\mathbf{x}$ satisfies the hypotheses of Theorem 9.15. Moreover (see Exercise 12), the ith row of $B = A - \lambda_1\mathbf{v}^{(1)}\mathbf{x}^t$ consists entirely of zero entries.

If $\lambda \neq 0$ is an eigenvalue with associated eigenvector $\mathbf{w}$, the relation $B\mathbf{w} = \lambda\mathbf{w}$ implies that the ith coordinate of $\mathbf{w}$ must also be zero. Consequently the ith column of the matrix B makes no contribution to the product $B\mathbf{w} = \lambda\mathbf{w}$. Thus, the matrix B can be replaced by an

$(n-1)\times(n-1)$ matrix B' obtained by deleting the ith row and column from B. The matrix B' has eigenvalues $\lambda_2, \lambda_3, \ldots, \lambda_n$. If $|\lambda_2| > |\lambda_3|$, the Power method is reapplied to the matrix B' to determine this new dominant eigenvalue and an eigenvector, $\mathbf{w}^{(2)'}$, associated with λ_2, with respect to the matrix B'. To find the associated eigenvector $\mathbf{w}^{(2)}$ for the matrix B, insert a zero coordinate between the coordinates $w_{i-1}^{(2)'}$ and $w_i^{(2)'}$ of the $(n-1)$-dimensional vector $\mathbf{w}^{(2)'}$ and then calculate $\mathbf{v}^{(2)}$ by the use of Eq. (9.5).

EXAMPLE 4 From Example 2, we know that the matrix

$$A = \begin{bmatrix} 4 & -1 & 1 \\ -1 & 3 & -2 \\ 1 & -2 & 3 \end{bmatrix}$$

has eigenvalues $\lambda_1 = 6$, $\lambda_2 = 3$, and $\lambda_3 = 1$. Assuming that the dominant eigenvalue $\lambda_1 = 6$ and associated unit eigenvector $\mathbf{v}^{(1)} = (1, -1, 1)^t$ have been calculated, the procedure just outlined for obtaining λ_2 proceeds as follows:

$$\mathbf{x} = \frac{1}{6}\begin{bmatrix} 4 \\ -1 \\ 1 \end{bmatrix} = \left(\frac{2}{3}, -\frac{1}{6}, \frac{1}{6}\right)^t,$$

$$\mathbf{v}^{(1)}\mathbf{x}^t = \begin{bmatrix} 1 \\ -1 \\ 1 \end{bmatrix}\left[\tfrac{2}{3}, \ -\tfrac{1}{6}, \ \tfrac{1}{6}\right] = \begin{bmatrix} \frac{2}{3} & -\frac{1}{6} & \frac{1}{6} \\ -\frac{2}{3} & \frac{1}{6} & -\frac{1}{6} \\ \frac{2}{3} & -\frac{1}{6} & \frac{1}{6} \end{bmatrix},$$

and

$$B = A - \lambda_1\mathbf{v}^{(1)}\mathbf{x}^t = \begin{bmatrix} 4 & -1 & 1 \\ -1 & 3 & -2 \\ 1 & -2 & 3 \end{bmatrix} - 6\begin{bmatrix} \frac{2}{3} & -\frac{1}{6} & \frac{1}{6} \\ -\frac{2}{3} & \frac{1}{6} & -\frac{1}{6} \\ \frac{2}{3} & -\frac{1}{6} & \frac{1}{6} \end{bmatrix}$$

$$= \begin{bmatrix} 0 & 0 & 0 \\ 3 & 2 & -1 \\ -3 & -1 & 2 \end{bmatrix}.$$

Deleting the first row and column gives

$$B' = \begin{bmatrix} 2 & -1 \\ -1 & 2 \end{bmatrix},$$

which has eigenvalues $\lambda_2 = 3$ and $\lambda_3 = 1$. For $\lambda_2 = 3$ the eigenvector $\mathbf{w}^{(2)'}$ can be obtained by solving the second-order linear system

$$(B' - 3I)\mathbf{w}^{(2)'} = \mathbf{0},$$

resulting in

$$\mathbf{w}^{(2)'} = (1, -1)^t.$$

Thus, $\mathbf{w}^{(2)} = (0, 1, -1)^t$ and, from Eq. (9.5), we have

$$\mathbf{v}^{(2)} = (3-6)(0, 1, -1)^t + 6\left[\left(\frac{2}{3}, -\frac{1}{6}, \frac{1}{6}\right)(0, 1, -1)^t\right](1, -1, 1)^t = (-2, -1, 1)^t. \quad \blacksquare$$

Although this deflation process can be used to find approximations to all of the eigenvalues and eigenvectors of a matrix, the process is susceptible to roundoff error. If this technique is used to find all the eigenvalues of a matrix, the approximations obtained should be used as starting values for the Inverse Power method applied to the original matrix. This will ensure that the approximations converge to eigenvalues of the original matrix, not to those of the reduced matrix, which likely contains errors. Techniques based on similarity transformations are presented in the next two sections. These methods are generally preferable when approximations to all the eigenvalues are needed.

We close this section with Algorithm 9.4, which calculates the second most dominant eigenvalue and associated eigenvector for a matrix, once the dominant eigenvalue and associated eigenvector have been determined.

ALGORITHM 9.4

Wielandt Deflation

To approximate the second most dominant eigenvalue and an associated eigenvector of the $n \times n$ matrix A given an approximation λ to the dominant eigenvalue, an approximation $\mathbf{v}$ to a corresponding eigenvector, and a vector $\mathbf{x} \in \mathbb{R}^{n-1}$:

INPUT dimension n; matrix A; approximate eigenvalue λ with eigenvector $\mathbf{v} \in \mathbb{R}^n$; vector $\mathbf{x} \in \mathbb{R}^{n-1}$, tolerance TOL, maximum number of iterations N.

OUTPUT approximate eigenvalue μ; approximate eigenvector $\mathbf{u}$ or a message that the method fails.

Step 1 Let i be the smallest integer with $1 \le i \le n$ and $|v_i| = \max_{1\le j\le n} |v_j|$.

Step 2 If $i \ne 1$ then
  for $k = 1, \ldots, i-1$
    for $j = 1, \ldots, i-1$

$$\text{set } b_{kj} = a_{kj} - \frac{v_k}{v_i} a_{ij}.$$

Step 3 If $i \ne 1$ and $i \ne n$ then
  for $k = i, \ldots, n-1$
    for $j = 1, \ldots, i-1$

$$\text{set } b_{kj} = a_{k+1,j} - \frac{v_{k+1}}{v_i} a_{ij};$$

$$b_{jk} = a_{j,k+1} - \frac{v_j}{v_i} a_{i,k+1}.$$

Step 4 If $i \ne n$ then
  for $k = i, \ldots, n-1$
    for $j = i, \ldots, n-1$

$$\text{set } b_{kj} = a_{k+1,j+1} - \frac{v_{k+1}}{v_i} a_{i,j+1}.$$

Step 5 Perform the power method on the $(n-1) \times (n-1)$ matrix $B' = (b_{kj})$ with $\mathbf{x}$ as initial approximation.

Step 6 If the method fails, then OUTPUT ('Method fails');
STOP
else let μ be the approximate eigenvalue and
$\mathbf{w}' = (w'_1, \ldots, w'_{n-1})^t$ the approximate eigenvector.

Step 7 If $i \neq 1$ then for $k = 1, \ldots, i-1$ set $w_k = w'_k$.

Step 8 Set $w_i = 0$.

Step 9 If $i \neq n$ then for $k = i+1, \ldots, n$ set $w_k = w'_{k-1}$.

Step 10 For $k = 1, \ldots, n$

$$\text{set } u_k = (\mu - \lambda)w_k + \left(\sum_{j=1}^{n} a_{ij}w_j\right)\frac{v_k}{v_i}.$$

(*Compute the eigenvector using Eq.* (9.5).)

Step 11 OUTPUT $(\mu, \mathbf{u})$; (*Procedure completed successfully.*)
STOP.

EXERCISE SET 9.2

1. Find the first three iterations obtained by the Power method applied to the following matrices.

a. $\begin{bmatrix} 2 & 1 & 1 \\ 1 & 2 & 1 \\ 1 & 1 & 2 \end{bmatrix}$;
Use $\mathbf{x}^{(0)} = (1, -1, 2)^t$.

b. $\begin{bmatrix} 1 & 1 & 1 \\ 1 & 1 & 0 \\ 1 & 0 & 1 \end{bmatrix}$;
Use $\mathbf{x}^{(0)} = (-1, 0, 1)^t$.

c. $\begin{bmatrix} 1 & -1 & 0 \\ -2 & 4 & -2 \\ 0 & -1 & 2 \end{bmatrix}$;
Use $\mathbf{x}^{(0)} = (-1, 2, 1)^t$.

d. $\begin{bmatrix} 4 & 1 & 1 & 1 \\ 1 & 3 & -1 & 1 \\ 1 & -1 & 2 & 0 \\ 1 & 1 & 0 & 2 \end{bmatrix}$;
Use $\mathbf{x}^{(0)} = (1, -2, 0, 3)^t$.

e. $\begin{bmatrix} 5 & -2 & -\frac{1}{2} & \frac{3}{2} \\ -2 & 5 & \frac{3}{2} & -\frac{1}{2} \\ -\frac{1}{2} & \frac{3}{2} & 5 & -2 \\ \frac{3}{2} & -\frac{1}{2} & -2 & 5 \end{bmatrix}$;
Use $\mathbf{x}^{(0)} = (1, 1, 0, -3)^t$.

f. $\begin{bmatrix} -4 & 0 & \frac{1}{2} & \frac{1}{2} \\ \frac{1}{2} & -2 & 0 & \frac{1}{2} \\ \frac{1}{2} & \frac{1}{2} & 0 & 0 \\ 0 & 1 & 1 & 4 \end{bmatrix}$;
Use $\mathbf{x}^{(0)} = (0, 0, 0, 1)^t$.

2. Repeat Exercise 1 using the Inverse Power method.

3. Find the first three iterations obtained by the Symmetric Power method applied to the following matrices.

a. $\begin{bmatrix} 2 & 1 & 1 \\ 1 & 2 & 1 \\ 1 & 1 & 2 \end{bmatrix}$;
Use $\mathbf{x}^{(0)} = (1, -1, 2)^t$.

b. $\begin{bmatrix} 1 & 1 & 1 \\ 1 & 1 & 0 \\ 1 & 0 & 1 \end{bmatrix}$;
Use $\mathbf{x}^{(0)} = (-1, 0, 1)^t$.

c. $\begin{bmatrix} 4.75 & 2.25 & -0.25 \\ 2.25 & 4.75 & 1.25 \\ -0.25 & 1.25 & 4.75 \end{bmatrix}$; Use $\mathbf{x}^{(0)} = (0, 1, 0)^t$.

d. $\begin{bmatrix} 4 & 1 & -1 & 0 \\ 1 & 3 & -1 & 0 \\ -1 & -1 & 5 & 2 \\ 0 & 0 & 2 & 4 \end{bmatrix}$; Use $\mathbf{x}^{(0)} = (0, 1, 0, 0)^t$.

e. $\begin{bmatrix} 4 & 1 & 1 & 1 \\ 1 & 3 & -1 & 1 \\ 1 & -1 & 2 & 0 \\ 1 & 1 & 0 & 2 \end{bmatrix}$; Use $\mathbf{x}^{(0)} = (1, 0, 0, 0)^t$.

f. $\begin{bmatrix} 5 & -2 & -\frac{1}{2} & \frac{3}{2} \\ -2 & 5 & \frac{3}{2} & -\frac{1}{2} \\ -\frac{1}{2} & \frac{3}{2} & 5 & -2 \\ \frac{3}{2} & -\frac{1}{2} & -2 & 5 \end{bmatrix}$; Use $\mathbf{x}^{(0)} = (1, 1, 0, -3)^t$.

4. Develop an algorithm to incorporate the Inverse Power method into the Symmetric Power method. Repeat Exercise 3 using the new algorithm.

5. Use the Power method and Wielandt deflation to approximate the two most dominant eigenvalues for the matrices in Exercise 1. Iterate until a tolerance of 10^{-4} is achieved or until the number of iterations exceeds 25.

6. Repeat Exercise 5 using Aitken's Δ^2 technique and the Power method for the first eigenvalue.

7. Use the Symmetric Power method to compute the largest eigenvalue (in absolute value) of the matrices given in Exercise 3. Iterate until a tolerance of 10^{-4} is achieved or until the number of iterations exceeds 25.

8. Repeat Exercise 6 using the Inverse Power method for the first eigenvalue.

9. Repeat Exercise 7 using the Inverse Power method.

10. **Annihilation Technique** Suppose the $n \times n$ matrix A has eigenvalues $\lambda_1, \ldots, \lambda_n$ ordered by

$$|\lambda_1| > |\lambda_2| > |\lambda_3| \geq \cdots \geq |\lambda_n|$$

with linearly independent eigenvectors $\mathbf{v}^{(1)}, \mathbf{v}^{(2)}, \ldots, \mathbf{v}^{(n)}$.

a. Show that if the Power method is applied with an initial vector $\mathbf{x}^{(0)}$ given by

$$\mathbf{x}^{(0)} = \beta_2 \mathbf{v}^{(2)} + \beta_3 \mathbf{v}^{(3)} + \cdots + \beta_n \mathbf{v}^{(n)},$$

then the sequence $\{\mu^{(m)}\}$ described in Algorithm 9.1 will converge to λ_2.

b. Show that for any vector $\mathbf{x} = \sum_{i=1}^{n} \beta_i \mathbf{v}^{(i)}$, the vector $\mathbf{x}^{(0)} = (A - \lambda_1 I)\mathbf{x}$ satisfies the property given in part (a).

c. Obtain an approximation to λ_2 for the matrices in Exercise 1.

d. Show that this method can be continued to find λ_3 using $\mathbf{x}^{(0)} = (A - \lambda_2 I)(A - \lambda_1 I)\mathbf{x}$.

11. **Hotelling Deflation** Assume that the largest eigenvalue λ_1 in magnitude and an associated eigenvector $\mathbf{v}^{(1)}$ have been obtained for the $n \times n$ symmetric matrix A. Show that the matrix

$$B = A - \frac{\lambda_1}{(\mathbf{v}^{(1)})^t \mathbf{v}^{(1)}} \mathbf{v}^{(1)} \left(\mathbf{v}^{(1)}\right)^t$$

has the same eigenvalues $\lambda_2, \ldots, \lambda_n$ as A, except that B has eigenvalue 0 with eigenvector $\mathbf{v}^{(1)}$ instead of eigenvalue λ_1. Use this deflation method to find λ_2 for each matrix in Exercise 3. Theoretically, this method can be continued to find more eigenvalues, but roundoff error soon makes the effort worthless.

12. Show that the ith row of $B = A - \lambda_1 \mathbf{v}^{(1)} \mathbf{x}^t$ is zero, where λ_1 is the largest eigenvalue of A in absolute value, $\mathbf{v}^{(1)}$ is the associated eigenvector of A for λ_1, and $\mathbf{x}$ is the vector defined in Eq. (9.6).

13. Following along the line of Exercise 11 in Section 6.3 and Exercise 11 in Section 7.2, suppose that a species of beetle has a life span of 4 years, and that a female in the first year has a survival rate of $\frac{1}{2}$, in the second year a survival rate of $\frac{1}{4}$, and in the third year a survival rate of $\frac{1}{8}$. Suppose additionally that a female gives birth, on the average, to two new females in the third year and to four new females in the fourth year. The matrix describing a single female's contribution in one year to the female population in the succeeding year is

$$A = \begin{bmatrix} 0 & 0 & 2 & 4 \\ \frac{1}{2} & 0 & 0 & 0 \\ 0 & \frac{1}{4} & 0 & 0 \\ 0 & 0 & \frac{1}{8} & 0 \end{bmatrix},$$

where again the entry in the ith row and jth column denotes the probabilistic contribution that a female of age j makes on the next year's female population of age i.

a. Use the Gerschgorin Circle Theorem to determine a region in the complex plane containing all the eigenvalues of A.

b. Use the Power method to determine the dominant eigenvalue of the matrix and its associated eigenvector.

c. Use Algorithm 9.4 to determine any remaining eigenvalues and eigenvectors of A.

d. Find the eigenvalues of A by using the characteristic polynomial of A and the Newton-Raphson method.

e. What is your long-range prediction for the population of these beetles?

14. A linear dynamical system can be represented by the equations

$$\frac{d\mathbf{x}}{dt} = A(t)\mathbf{x}(t) + B(t)\mathbf{u}(t), \quad \mathbf{y}(t) = C(t)\mathbf{x}(t) + D(t)\mathbf{u}(t),$$

where A is an $n \times n$ variable matrix, B is an $n \times r$ variable matrix, C is an $m \times n$ variable matrix, D is an $m \times r$ variable matrix, $\mathbf{x}$ is an n-dimensional vector variable, $\mathbf{y}$ is an m-dimensional vector variable, and $\mathbf{u}$ is an r-dimensional vector variable. For the system to be stable, the matrix A must have all its eigenvalues with nonpositive real part for all t.

a. Is the system stable if

$$A(t) = \begin{bmatrix} -1 & 2 & 0 \\ -2.5 & -7 & 4 \\ 0 & 0 & -5 \end{bmatrix}?$$

b. Is the system stable if

$$A(t) = \begin{bmatrix} -1 & 1 & 0 & 0 \\ 0 & -2 & 1 & 0 \\ 0 & 0 & -5 & 1 \\ -1 & -1 & -2 & -3 \end{bmatrix}?$$

15. The $(m-1) \times (m-1)$ tridiagonal matrix

$$A = \begin{bmatrix} 1+2\alpha & -\alpha & 0 & \cdots & \cdots & 0 \\ -\alpha & 1+2\alpha & -\alpha & \ddots & & \vdots \\ 0 & \ddots & \ddots & \ddots & \ddots & 0 \\ \vdots & & \ddots & \ddots & \ddots & -\alpha \\ 0 & \cdots & \cdots & 0 & -\alpha & 1+2\alpha \end{bmatrix}$$

is involved in the Backward Difference method to solve the heat equation. (See Section 12.2.) For the stability of the method we need $\rho(A^{-1}) < 1$. Let $m = 11$. Approximate $\rho(A^{-1})$ for each of the following.

a. $\alpha = \frac{1}{4}$ **b.** $\alpha = \frac{1}{2}$ **c.** $\alpha = \frac{3}{4}$

When is the method stable?

16. The eigenvalues of the matrix A in Exercise 15 are

$$\lambda_i = 1 + 4\alpha\left(\sin\frac{\pi i}{2m}\right)^2, \quad \text{for } i = 1, \ldots, m-1.$$

Compare the approximation in Exercise 15 to the actual value of $\rho(A^{-1})$. Again, when is the method stable?

17. The $(m-1) \times (m-1)$ matrices A and B given by

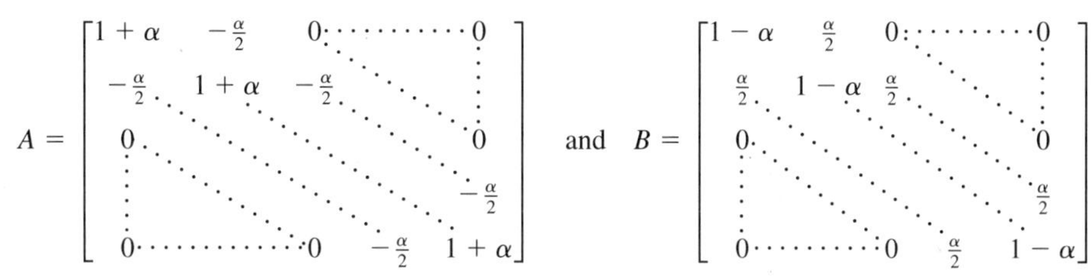

$$A = \begin{bmatrix} 1+\alpha & -\frac{\alpha}{2} & 0 & \cdots & 0 \\ -\frac{\alpha}{2} & 1+\alpha & -\frac{\alpha}{2} & \ddots & \vdots \\ 0 & \ddots & \ddots & \ddots & 0 \\ \vdots & & \ddots & \ddots & -\frac{\alpha}{2} \\ 0 & \cdots & 0 & -\frac{\alpha}{2} & 1+\alpha \end{bmatrix} \quad \text{and} \quad B = \begin{bmatrix} 1-\alpha & \frac{\alpha}{2} & 0 & \cdots & 0 \\ \frac{\alpha}{2} & 1-\alpha & \frac{\alpha}{2} & \ddots & \vdots \\ 0 & \ddots & \ddots & \ddots & 0 \\ \vdots & & \ddots & \ddots & \frac{\alpha}{2} \\ 0 & \cdots & 0 & \frac{\alpha}{2} & 1-\alpha \end{bmatrix}$$

are involved in the Crank-Nicolson method to solve the heat equation (See Section 12.2). With $m = 11$, approximate $\rho(A^{-1}B)$ for each of the following.

a. $\alpha = \frac{1}{4}$ **b.** $\alpha = \frac{1}{2}$ **c.** $\alpha = \frac{3}{4}$

9.3 Householder's Method

Householder's method is used to find a symmetric tridiagonal matrix B that is similar to a given symmetric matrix A. Theorem 9.10 implies that A is similar to a diagonal matrix D since an orthogonal matrix Q exists with the property that $D = Q^{-1}AQ = Q^tAQ$. Because the matrix Q (and consequently D) is generally difficult to compute, Householder's method offers a compromise. After Householder's method has been implemented, efficient methods such as the QR algorithm can be used for accurate approximation of the eigenvalues of the resulting symmetric tridiagonal matrix.

Definition 9.16 Let $\mathbf{w} \in \mathbb{R}^n$ with $\mathbf{w}^t\mathbf{w} = 1$. The $n \times n$ matrix

$$P = I - 2\mathbf{w}\mathbf{w}^t$$

is called a **Householder transformation**. ■

Householder transformations are used to selectively zero out blocks of entries in vectors or columns of matrices in a manner that is extremely stable with respect to roundoff error. (See [Wil2, pp. 152–162], for further discussion.) Properties of Householder transformations are given in the following theorem.

Theorem 9.17 If $P = I - 2\mathbf{w}\mathbf{w}^t$ is a Householder transformation, then P is symmetric and orthogonal, so $P^{-1} = P$.

Proof Since

$$(\mathbf{w}\mathbf{w}^t)^t = (\mathbf{w}^t)^t\mathbf{w}^t = \mathbf{w}\mathbf{w}^t,$$

it follows that

$$P^t = (I - 2\mathbf{w}\mathbf{w}^t)^t = I - 2\mathbf{w}\mathbf{w}^t = P.$$

Further, since $\mathbf{w}^t\mathbf{w} = 1$,

$$PP^t = (I - 2\mathbf{w}\mathbf{w}^t)(I - 2\mathbf{w}\mathbf{w}^t) = I - 2\mathbf{w}\mathbf{w}^t - 2\mathbf{w}\mathbf{w}^t + 4\mathbf{w}\mathbf{w}^t\mathbf{w}\mathbf{w}^t$$
$$= I - 4\mathbf{w}\mathbf{w}^t + 4\mathbf{w}\mathbf{w}^t = I,$$

so

$$P^{-1} = P^t = P. \qquad ▪ ▪ ▪$$

Householder's method begins by determining a transformation $P^{(1)}$ so that $A^{(2)} = P^{(1)}AP^{(1)}$ has

(9.7) $$a_{j1}^{(2)} = 0, \quad \text{for each } j = 3, 4, \ldots, n.$$

By symmetry, this also implies that $a_{1j}^{(2)} = 0$, for each $j = 3, 4, \ldots, n$.

The vector $\mathbf{w} = (w_1, w_2, \ldots, w_n)^t$ is chosen so that $\mathbf{w}^t\mathbf{w} = 1$, Eq. (9.7) holds, and $a_{11}^{(2)} = a_{11}$. This choice imposes n conditions on the n unknowns $w_1, w_2, \ldots, w_n$.

Setting $w_1 = 0$ ensures that $a_{11}^{(2)} = a_{11}$. We want

$$P^{(1)} = I - 2\mathbf{w}\mathbf{w}^t$$

to satisfy

(9.8) $$P^{(1)}(a_{11}, a_{21}, a_{31}, \ldots, a_{n1})^t = (a_{11}, \alpha, 0, \ldots, 0)^t,$$

where α will be chosen later. To simplify notation, let

$$\hat{\mathbf{w}} = (w_2, w_3, \ldots, w_n)^t \in \mathbb{R}^{n-1}, \ \hat{\mathbf{y}} = (a_{21}, a_{31}, \ldots, a_{n1})^t \in \mathbb{R}^{n-1},$$

and $\hat{P}$ be the $(n-1) \times (n-1)$ Householder transformation

$$\hat{P} = I_{n-1} - 2\hat{\mathbf{w}}\hat{\mathbf{w}}^t.$$

Eq. (9.8) then becomes

$$P^{(1)} \begin{bmatrix} a_{11} \\ a_{21} \\ a_{31} \\ \vdots \\ a_{n1} \end{bmatrix} = \left[\begin{array}{c|ccc} 1 & 0 & \cdots & 0 \\ \hline 0 & & & \\ \vdots & & \hat{P} & \\ 0 & & & \end{array}\right] \cdot \begin{bmatrix} a_{11} \\ \cdots \\ \hat{\mathbf{y}} \end{bmatrix} = \begin{bmatrix} a_{11} \\ \cdots \\ \hat{P}\hat{\mathbf{y}} \end{bmatrix} = \begin{bmatrix} a_{11} \\ \alpha \\ 0 \\ \vdots \\ 0 \end{bmatrix}$$

with

(9.9) $$\hat{P}\hat{\mathbf{y}} = (I_{n-1} - 2\hat{\mathbf{w}}\hat{\mathbf{w}}^t)\hat{\mathbf{y}} = \hat{\mathbf{y}} - 2(\hat{\mathbf{w}}^t\hat{\mathbf{y}})\hat{\mathbf{w}} = (\alpha, 0, \ldots, 0)^t.$$

Let $r = \hat{\mathbf{w}}^t\hat{\mathbf{y}}$. Then

$$(\alpha, 0, \ldots, 0)^t = (a_{21} - 2rw_2, a_{31} - 2rw_3, \ldots, a_{n1} - 2rw_n)^t,$$

and we can determine all of the w_i if we can determine α and r. Equating components gives

$$\alpha = a_{21} - 2rw_2$$

and

$$0 = a_{j1} - 2rw_j, \quad \text{for each } j = 3, \ldots, n.$$

Thus,

(9.10) $$2rw_2 = a_{21} - \alpha$$

and

(9.11) $$2rw_j = a_{j1}, \quad \text{for each } j = 3, \ldots, n.$$

Squaring both sides of each of the equations and adding gives

$$4r^2 \sum_{j=2}^{n} w_j^2 = (a_{21} - \alpha)^2 + \sum_{j=3}^{n} a_{j1}^2.$$

Since $\mathbf{w}^t\mathbf{w} = 1$ and $w_1 = 0$, we have $\Sigma_{j=2}^{n} w_j^2 = 1$ and

(9.12) $$4r^2 = \sum_{j=2}^{n} a_{j1}^2 - 2\alpha a_{21} + \alpha^2.$$

From Eq. (9.9) and the fact that P is orthogonal, we have

$$\alpha^2 = (\alpha, 0, \ldots, 0)(\alpha, 0, \ldots, 0)^t = (\hat{P}\hat{\mathbf{y}})^t\hat{P}\hat{\mathbf{y}} = \hat{\mathbf{y}}^t\hat{P}^t\hat{P}\hat{\mathbf{y}} = \hat{\mathbf{y}}^t\hat{\mathbf{y}} = \sum_{j=2}^{n} a_{j1}^2,$$

so

$$\alpha = \pm\left(\sum_{j=2}^{n} a_{j1}^2\right)^{1/2}$$

which, when substituted into (9.12), gives

$$2r^2 = \sum_{j=2}^{n} a_{j1}^2 \pm a_{21}\left(\sum_{j=2}^{n} a_{j1}^2\right)^{1/2}.$$

To ensure that $2r^2 = 0$ only if $a_{21} = a_{31} = \cdots = a_{n1} = 0$, we choose the sign so that

$$2r^2 = \sum_{j=2}^{n} a_{j1}^2 + |a_{21}|\left(\sum_{j=2}^{n} a_{j1}^2\right)^{1/2},$$

that is, we choose

$$\alpha = -(\text{sign } a_{21})\left(\sum_{j=2}^{n} a_{j1}^2\right)^{1/2}.$$

With these choices of α and $2r^2$, we solve Eqs. (9.10) and (9.11) to obtain

$$w_2 = \frac{a_{21} - \alpha}{2r}$$

and

$$w_j = \frac{a_{j1}}{2r}, \quad \text{for each } j = 3, \ldots, n.$$

To summarize the choice of $P^{(1)}$, we have

$$\alpha = -(\text{sign } a_{21})\left(\sum_{j=2}^{n} a_{j1}^2\right)^{1/2},$$

$$r = \left(\frac{1}{2}\alpha^2 - \frac{1}{2}a_{21}\alpha\right)^{1/2},$$

$$w_1 = 0,$$

$$w_2 = \frac{a_{21} - \alpha}{2r},$$

and

$$w_j = \frac{a_{j1}}{2r}, \quad \text{for each } j = 3, \ldots, n.$$

With this choice,

$$A^{(2)} = P^{(1)}AP^{(1)} = \begin{bmatrix} a_{11}^{(2)} & a_{12}^{(2)} & 0 & \cdots & 0 \\ a_{21}^{(2)} & a_{22}^{(2)} & a_{23}^{(2)} & \cdots & a_{2n}^{(2)} \\ 0 & a_{32}^{(2)} & a_{33}^{(2)} & \cdots & a_{3n}^{(2)} \\ \vdots & \vdots & \vdots & & \vdots \\ 0 & a_{n2}^{(2)} & a_{n3}^{(2)} & \cdots & a_{nn}^{(2)} \end{bmatrix}.$$

Having found $P^{(1)}$ and $A^{(2)}$, the process is repeated to find $P^{(2)} = I - 2\mathbf{w}^{(2)}\,[\mathbf{w}^{(2)}]^t$, where $[\mathbf{w}^{(2)}]^t\mathbf{w}^{(2)} = 1$ and $w_1^{(2)} = w_2^{(2)} = 0$, with the property that

$$A^{(3)} = P^{(2)}A^{(2)}P^{(2)} = \begin{bmatrix} a_{11}^{(3)} & a_{12}^{(3)} & 0 & 0 & \cdots & 0 \\ a_{21}^{(3)} & a_{22}^{(3)} & a_{23}^{(3)} & 0 & \cdots & 0 \\ 0 & a_{32}^{(3)} & a_{33}^{(3)} & a_{34}^{(3)} & \cdots & a_{3n}^{(3)} \\ \vdots & 0 & a_{43}^{(3)} & a_{44}^{(3)} & \cdots & a_{4n}^{(3)} \\ \vdots & \vdots & \vdots & \vdots & & \vdots \\ 0 & 0 & a_{n3}^{(3)} & a_{n4}^{(3)} & \cdots & a_{nn}^{(3)} \end{bmatrix}.$$

Continuing in this manner, the tridiagonal and symmetric matrix $A^{(n-1)}$ is formed, where

$$A^{(n-1)} = P^{(n-2)}P^{(n-3)} \cdots P^{(1)}AP^{(1)} \cdots P^{(n-3)}P^{(n-2)}.$$

Algorithm 9.5 performs Householder's method as described here, although the actual matrix multiplications are circumvented.

ALGORITHM **9.5**

Householder's

To obtain a symmetric tridiagonal matrix $A^{(n-1)}$ similar to the symmetric matrix $A = A^{(1)}$, construct the following matrices $A^{(2)}, A^{(3)}, \ldots, A^{(n-1)}$, where $A^{(k)} = (a_{ij}^{(k)})$ for each $k = 1, 2, \ldots, n-1$:

INPUT dimension n; matrix A.

OUTPUT $A^{(n-1)}$. *(At each step, A can be overwritten.)*

Step 1 For $k = 1, 2, \ldots, n-2$ do Steps 2–14.

Step 2 Set

$$q = \sum_{j=k+1}^{n} \left(a_{jk}^{(k)}\right)^2.$$

Step 3 If $a_{k+1,k}^{(k)} = 0$ then set $\alpha = -q^{1/2}$

$$\text{else set } \alpha = -\frac{q^{1/2} a_{k+1,k}^{(k)}}{|a_{k+1,k}^{(k)}|}$$

Step 4 Set $RSQ = \alpha^2 - \alpha a_{k+1,k}^{(k)}$. (*Note:* $RSQ = 2r^2$)

Step 5 Set $v_k = 0$; (*Note:* $v_1 = \cdots = v_{k-1} = 0$, *but are not needed.*)
$v_{k+1} = a_{k+1,k}^{(k)} - \alpha$;
For $j = k+2, \ldots, n$ set $v_j = a_{jk}^{(k)}$.

$$\left(\textit{Note: } \mathbf{w} = \left(\frac{1}{\sqrt{2RSQ}}\right)\mathbf{v} = \frac{1}{2r}\mathbf{v}.\right)$$

Step 6 For $j = k, k+1, \ldots, n$ set $u_j = \left(\frac{1}{RSQ}\right) \sum_{i=k+1}^{n} a_{ji}^{(k)} v_i$.

$$\left(\textit{Note: } \mathbf{u} = \left(\frac{1}{RSQ}\right) A^{(k)}\mathbf{v} = \frac{1}{2r^2}A^{(k)}\mathbf{v} = \frac{1}{r}A^{(k)}\mathbf{w}.\right)$$

Step 7 Set $PROD = \sum_{i=k+1}^{n} v_i u_i$.

$$\left(\textit{Note: } PROD = \mathbf{v}^t\mathbf{u} = \frac{1}{2r^2}\mathbf{v}^t A^{(k)}\mathbf{v}.\right)$$

Step 8 For $j = k, k+1, \ldots, n$ set $z_j = u_j - \left(\frac{PROD}{2RSQ}\right) v_j$.

$$\left(\textit{Note: } \mathbf{z} = \mathbf{u} - \frac{1}{2RSQ}\mathbf{v}^t\mathbf{u}\mathbf{v} = \mathbf{u} - \frac{1}{4r^2}\mathbf{v}^t\mathbf{u}\mathbf{v}\right.$$
$$\left. = \mathbf{u} - \mathbf{w}\mathbf{w}^t\mathbf{u} = \frac{1}{r}A^{(k)}\mathbf{w} - \mathbf{w}\mathbf{w}^t\frac{1}{r}A^{(k)}\mathbf{w}.\right)$$

Step 9 For $l = k+1, k+2, \ldots, n-1$ do Steps 10 and 11.
(*Note: Compute* $A^{(k+1)} = A^{(k)} - \mathbf{v}\mathbf{z}^t - \mathbf{z}\mathbf{v}^t = (I - 2\mathbf{w}\mathbf{w}^t)A^{(k)}(I - 2\mathbf{w}\mathbf{w}^t)$.)

Step 10 For $j = l+1, \ldots, n$ set
$a_{jl}^{(k+1)} = a_{jl}^{(k)} - v_l z_j - v_j z_l$;
$a_{lj}^{(k+1)} = a_{jl}^{(k+1)}$.

Step 11 Set $a_{ll}^{(k+1)} = a_{ll}^{(k)} - 2v_l z_l$.

Step 12 Set $a_{nn}^{(k+1)} = a_{nn}^{(k)} - 2v_n z_n$.

Step 13 For $j = k+2, \ldots, n$ set $a_{kj}^{(k+1)} = a_{jk}^{(k+1)} = 0$.

Step 14 Set $a_{k+1,k}^{(k+1)} = a_{k+1,k}^{(k)} - v_{k+1} z_k$;
$a_{k,k+1}^{(k+1)} = a_{k+1,k}^{(k+1)}$.
(*Note: The other elements of* $A^{(k+1)}$ *are the same as* $A^{(k)}$.)

Step 15 OUTPUT ($A^{(n-1)}$);
(*The process is complete.* $A^{(n-1)}$ *is symmetric, tridiagonal, and similar to A.*)
STOP.

EXAMPLE 1 The 4×4 matrix

$$A = \begin{bmatrix} 4 & 1 & -2 & 2 \\ 1 & 2 & 0 & 1 \\ -2 & 0 & 3 & -2 \\ 2 & 1 & -2 & -1 \end{bmatrix}$$

is symmetric. For the first application of a Householder transformation,

$$q = \sum_{j=2}^{4}(a_{j1})^2 = (1)^2 + (-2)^2 + (2)^2 = 9,$$

and since $a_{21} \neq 0$,

$$\alpha = \frac{-3 \cdot 1}{|1|} = -3, \quad RSQ = (3)^2 - (-3)(1) = 12, \quad \mathbf{v} = (0, 4, -2, 2)^t,$$

$$\mathbf{u} = \frac{1}{12}\begin{bmatrix} 4 & 1 & -2 & 2 \\ 1 & 2 & 0 & 1 \\ -2 & 0 & 3 & -2 \\ 2 & 1 & -2 & -1 \end{bmatrix}\begin{bmatrix} 0 \\ 4 \\ -2 \\ 2 \end{bmatrix} = \begin{bmatrix} 1 \\ \frac{5}{6} \\ -\frac{5}{6} \\ \frac{1}{2} \end{bmatrix},$$

$$PROD = 6,$$

and

$$\mathbf{z} = \left(1, -\frac{1}{6}, -\frac{1}{3}, 0\right)^t.$$

So

$$A^{(2)} = \begin{bmatrix} 4 & -3 & 0 & 0 \\ -3 & \frac{10}{3} & 1 & \frac{4}{3} \\ 0 & 1 & \frac{5}{3} & -\frac{4}{3} \\ 0 & \frac{4}{3} & -\frac{4}{3} & -1 \end{bmatrix}.$$

Note that

$$P^{(1)} = I - \frac{1}{RSQ}\mathbf{v}\mathbf{v}^t = \begin{bmatrix} 1 & 0 & 0 & 0 \\ 0 & -\frac{1}{3} & \frac{2}{3} & -\frac{2}{3} \\ 0 & \frac{2}{3} & \frac{2}{3} & \frac{1}{3} \\ 0 & -\frac{2}{3} & \frac{1}{3} & \frac{2}{3} \end{bmatrix}.$$

Continuing to the second iteration,

$$q = \frac{25}{9}, \ \alpha = -\frac{5}{3}, \ RSQ = \frac{40}{9}, \ \mathbf{v} = \left(0, 0, \frac{8}{3}, \frac{4}{3}\right)^t, \ \mathbf{u} = \left(0, 1, \frac{3}{5}, -\frac{11}{10}\right)^t,$$

$$PROD = \frac{2}{15}, \ \mathbf{z} = \left(0, 1, \frac{14}{25}, -\frac{28}{25}\right)^t,$$

and the symmetric tridiagonal matrix is

$$A^{(3)} = \begin{bmatrix} 4 & -3 & 0 & 0 \\ -3 & \frac{10}{3} & -\frac{5}{3} & 0 \\ 0 & -\frac{5}{3} & -\frac{33}{25} & \frac{68}{75} \\ 0 & 0 & \frac{68}{75} & \frac{149}{75} \end{bmatrix}.$$

Note also that

$$P^{(2)} = \begin{bmatrix} 1 & 0 & 0 & 0 \\ 0 & 1 & 0 & 0 \\ 0 & 0 & -\frac{3}{5} & -\frac{4}{5} \\ 0 & 0 & -\frac{4}{5} & \frac{3}{5} \end{bmatrix}.$$ ∎

To apply Householder's Algorithm to an arbitrary $n \times n$ matrix, the following modifications must be made to account for a possible lack of symmetry.

Step 6 For $j = 1, 2, \ldots, n$ set $u_j = \frac{1}{RSQ} \sum_{i=k+1}^{n} a_{ji}^{(k)} v_i;$

$$y_j = \frac{1}{RSQ} \sum_{i=k+1}^{n} a_{ij}^{(k)} v_i.$$

Step 8 For $j = 1, 2, \ldots, n$ set $z_j = u_j - \frac{PROD}{RSQ} v_j.$

Step 9 For $l = k+1, k+2, \ldots, n$ do Steps 10 and 11.

Step 10 For $j = 1, 2, \ldots, k$ set $a_{jl}^{(k+1)} = a_{jl}^{(k)} - z_j v_l;$

$$a_{lj}^{(k+1)} = a_{lj}^{(k)} - y_j v_l.$$

Step 11 For $j = k+1, \ldots, n$ set $a_{jl}^{(k+1)} = a_{jl}^{(k)} - z_j v_l - y_l v_j.$

After these steps are modified, delete Steps 12 through 14 and output $A^{(n-1)}$.

The resulting matrix $A^{(n-1)}$ will not be tridiagonal unless the original matrix A is symmetric. It will, however, have only zero entries below the lower subdiagonal. A matrix of this type is called *upper Hessenberg*. That is, $H = (h_{ij})$ is **upper Hessenberg** if $h_{ij} = 0$ for all $i \geq j + 2$.

In the next section, we will examine how the QR algorithm can then be applied to $A^{(n-1)}$ to determine the eigenvalues of $A^{(n-1)}$, which are the same as those of the original matrix A.

EXERCISE SET 9.3

1. Use Householder's method to place the following matrices in tridiagonal form.

a. $\begin{bmatrix} 12 & 10 & 4 \\ 10 & 8 & -5 \\ 4 & -5 & 3 \end{bmatrix}$

b. $\begin{bmatrix} 2 & -1 & -1 \\ -1 & 2 & -1 \\ -1 & -1 & 2 \end{bmatrix}$

c. $\begin{bmatrix} 1 & 1 & 1 \\ 1 & 1 & 0 \\ 1 & 0 & 1 \end{bmatrix}$

d. $\begin{bmatrix} 4.75 & 2.25 & -0.25 \\ 2.25 & 4.75 & 1.25 \\ -0.25 & 1.25 & 4.75 \end{bmatrix}$

2. Use Householder's method to place the following matrices in tridiagonal form.

a. $\begin{bmatrix} 4 & -1 & -1 & 0 \\ -1 & 4 & 0 & -1 \\ -1 & 0 & 4 & -1 \\ 0 & -1 & -1 & 4 \end{bmatrix}$

b. $\begin{bmatrix} 5 & -2 & -0.5 & 1.5 \\ -2 & 5 & 1.5 & -0.5 \\ -0.5 & 1.5 & 5 & -2 \\ 1.5 & -0.5 & -2 & 5 \end{bmatrix}$

c. $\begin{bmatrix} 8 & 0.25 & 0.5 & 2 & -1 \\ 0.25 & -4 & 0 & 1 & 2 \\ 0.5 & 0 & 5 & 0.75 & -1 \\ 2 & 1 & 0.75 & 5 & -0.5 \\ -1 & 2 & -1 & -0.5 & 6 \end{bmatrix}$

d. $\begin{bmatrix} 2 & -1 & -1 & 0 & 0 \\ -1 & 3 & 0 & -2 & 0 \\ -1 & 0 & 4 & 2 & 1 \\ 0 & -2 & 2 & 8 & 3 \\ 0 & 0 & 1 & 3 & 9 \end{bmatrix}$

3. Modify Householder's Algorithm 9.5 to compute similar upper Hessenberg matrices for the following matrices.

a. $\begin{bmatrix} 2 & -1 & 3 \\ 2 & 0 & 1 \\ -2 & 1 & 4 \end{bmatrix}$

b. $\begin{bmatrix} -1 & 2 & 3 \\ 2 & 3 & -2 \\ 3 & 1 & -1 \end{bmatrix}$

c. $\begin{bmatrix} 5 & -2 & -3 & 4 \\ 0 & 4 & 2 & -1 \\ 1 & 3 & -5 & 2 \\ -1 & 4 & 0 & 3 \end{bmatrix}$

d. $\begin{bmatrix} 4 & -1 & -1 & -1 \\ -1 & 4 & 0 & -1 \\ -1 & -1 & 4 & -1 \\ -1 & -1 & -1 & 4 \end{bmatrix}$

9.4 The QR Algorithm

The deflation methods discussed in Section 9.2 are not generally suitable for calculating all the eigenvalues of a matrix because of the growth of roundoff error. In this section we consider the QR Algorithm, a matrix reduction technique used to simultaneously determine all the eigenvalues of a symmetric matrix.

For the QR Algorithm to be effective, the symmetric matrix must be in tridiagonal form. If the matrix is not in this form, the first step is to apply Householder's Algorithm since this produces a symmetric, tridiagonal matrix similar to the given symmetric matrix.

In the remainder of this section it will be assumed that the symmetric matrix for which the eigenvalues are to be calculated is tridiagonal. If we let A denote a matrix of this type, we can simplify the notation somewhat by labeling the entries of A as follows:

(9.13)
$$A = \begin{bmatrix} a_1 & b_2 & 0 & \cdots & 0 \\ b_2 & a_2 & b_3 & \ddots & \vdots \\ 0 & b_3 & a_3 & \ddots & 0 \\ \vdots & \ddots & \ddots & \ddots & b_n \\ 0 & \cdots & 0 & b_n & a_n \end{bmatrix}.$$

If $b_2 = 0$ or $b_n = 0$, then the 1×1 matrix $[a_1]$ or $[a_n]$ immediately produces an eigenvalue a_1 or a_n of A.

When $b_j = 0$ for some j, where $2 < j < n$, the problem can be reduced to considering, instead of A, the smaller matrices

(9.14)
$$\begin{bmatrix} a_1 & b_2 & 0 & \cdots & 0 \\ b_2 & a_2 & b_3 & \ddots & \vdots \\ 0 & b_3 & a_3 & \ddots & 0 \\ \vdots & \ddots & \ddots & \ddots & b_{j-1} \\ 0 & \cdots & 0 & b_{j-1} & a_{j-1} \end{bmatrix} \quad \text{and} \quad \begin{bmatrix} a_j & b_{j+1} & 0 & \cdots & 0 \\ b_{j+1} & a_{j+1} & b_{j+2} & \ddots & \vdots \\ 0 & b_{j+2} & a_{j+2} & \ddots & 0 \\ \vdots & \ddots & \ddots & \ddots & b_n \\ 0 & \cdots & 0 & b_n & a_n \end{bmatrix}.$$

If none of the b_j are zero, the QR method proceeds by forming a sequence of matrices $A = A^{(1)}, A^{(2)}, A^{(3)}, \ldots,$ as follows:

1. $A^{(1)} = A$ is factored as a product $A^{(1)} = Q^{(1)}R^{(1)}$, where $Q^{(1)}$ is orthogonal and $R^{(1)}$ is upper triangular.
2. $A^{(2)}$ is defined as $A^{(2)} = R^{(1)}Q^{(1)}$.

In general, for $i \geq 2$, $A^{(i)}$ is factored as a product $A^{(i)} = Q^{(i)}R^{(i)}$ of an orthogonal matrix $Q^{(i)}$ and an upper triangular matrix $R^{(i)}$. Then $A^{(i+1)}$ is defined by the product of $R^{(i)}$ and $Q^{(i)}$ in the reverse direction $A^{(i+1)} = R^{(i)}Q^{(i)}$. Since $Q^{(i)}$ is orthogonal,

(9.15)
$$A^{(i+1)} = R^{(i)}Q^{(i)} = (Q^{(i)t}A^{(i)})Q^{(i)} = Q^{(i)t}A^{(i)}Q^{(i)},$$

and $A^{(i+1)}$ is symmetric with the same eigenvalues as $A^{(i)}$. By the way we define $R^{(i)}$ and $Q^{(i)}$ we can also ensure that $A^{(i+1)}$ is tridiagonal.

Continuing by induction, $A^{(i+1)}$ has the same eigenvalues as the original matrix A. The success of the procedure is a result of the fact that $A^{(i+1)}$ tends to a diagonal matrix with the eigenvalues of A along the diagonal.

To describe the construction of the factoring matrices $Q^{(i)}$ and $R^{(i)}$, we need the notion of a *rotation matrix*.

Definition 9.18 A **rotation matrix** P differs from the identity matrix in at most four elements. These four elements are of the form

$$p_{ii} = p_{jj} = \cos\theta \quad \text{and} \quad p_{ij} = -p_{ji} = \sin\theta$$

for some θ and some $i \neq j$. ■

It is easy to show (see Exercise 6) that, for any rotation matrix P, the matrix AP differs from A only in the ith and jth columns and the matrix PA differs from A only in the ith and jth rows. For any $i \neq j$, the angle θ can be chosen so that the product PA has a zero entry for $(PA)_{ij}$. In addition, every rotation matrix P is orthogonal, since the definition implies that $PP^t = I$.

The factorization of $A^{(1)}$ into $A^{(1)} = Q^{(1)}R^{(1)}$ uses a product of $n-1$ rotation matrices of this type to construct

$$R^{(1)} = P_n P_{n-1} \cdots P_2 A^{(1)}.$$

We first choose the rotation matrix P_2 to have

$$p_{11} = p_{22} = \cos\theta_2 \quad \text{and} \quad p_{12} = -p_{21} = \sin\theta_2,$$

where

$$\sin\theta_2 = \frac{b_2}{\sqrt{b_2^2 + a_1^2}} \quad \text{and} \quad \cos\theta_2 = \frac{a_1}{\sqrt{b_2^2 + a_1^2}}.$$

Then the matrix

$$A_2^{(1)} = P_2 A^{(1)}$$

has a zero in the $(2, 1)$ position, that is, in the second row and first column, since the $(2, 1)$ entry in $A_2^{(1)}$ is

$$(-\sin\theta_2)a_1 + (\cos\theta_2)b_2 = \frac{-b_2 a_1}{\sqrt{b_2^2 + a_1^2}} + \frac{a_1 b_2}{\sqrt{b_2^2 + a_1^2}} = 0.$$

Since the multiplication $P_2 A^{(1)}$ affects both rows 1 and 2 of $A^{(1)}$, the new matrix does not necessarily retain zero entries in positions $(1, 3), (1, 4), \ldots,$ and $(1, n)$. However, $A^{(1)}$ is tridiagonal, so the $(1, 4), \ldots, (1, n)$ entries of $A_2^{(1)}$ are zero. Only the $(1, 3)$-entry, the element in the first row and third column, can become nonzero.

In general, the matrix P_k is chosen so that the $(k, k-1)$-entry in $A_k^{(1)} = P_k A_{k-1}^{(1)}$ is zero, which results in the $(k-1, k+1)$-entry becoming nonzero. The matrix $A_k^{(1)}$ has the form

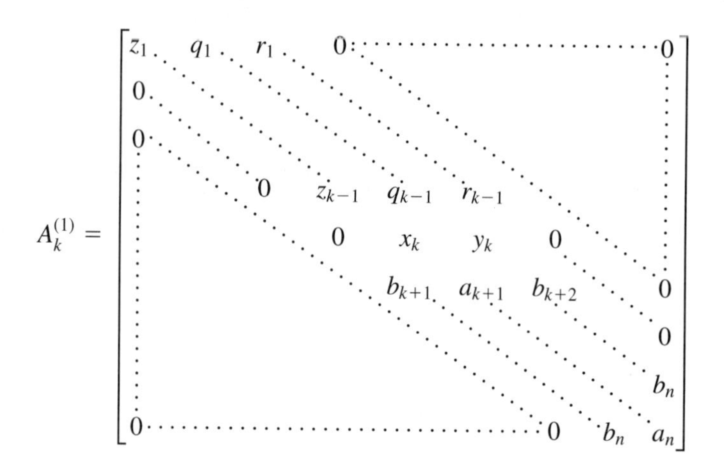

and P_{k+1} has the form

(9.16)

$$P_{k+1} = \left[\begin{array}{c|cc|c} I_{k-1} & & O & O \\ \hline & c_{k+1} & s_{k+1} & \\ O & & & O \\ & -s_{k+1} & c_{k+1} & \\ \hline O & & O & I_{n-k-1} \end{array}\right] \begin{array}{l} \\ \leftarrow \text{row } k \\ \\ \\ \\ \end{array}$$

$\uparrow$ column k

The constants $c_{k+1} = \cos\theta_{k+1}$ and $s_{k+1} = \sin\theta_{k+1}$ in P_{k+1} are chosen so that the $(k+1, k)$-entry in $A_{k+1}^{(1)}$ is zero; that is,

$$s_{k+1}x_k - c_{k+1}b_{k+1} = 0.$$

Since $c_{k+1}^2 + s_{k+1}^2 = 1$, the solution to this equation is

$$s_{k+1} = \frac{b_{k+1}}{\sqrt{b_{k+1}^2 + x_k^2}} \quad \text{and} \quad c_{k+1} = \frac{x_k}{\sqrt{b_{k+1}^2 + x_k^2}},$$

and $A_{k+1}^{(1)}$ has the form

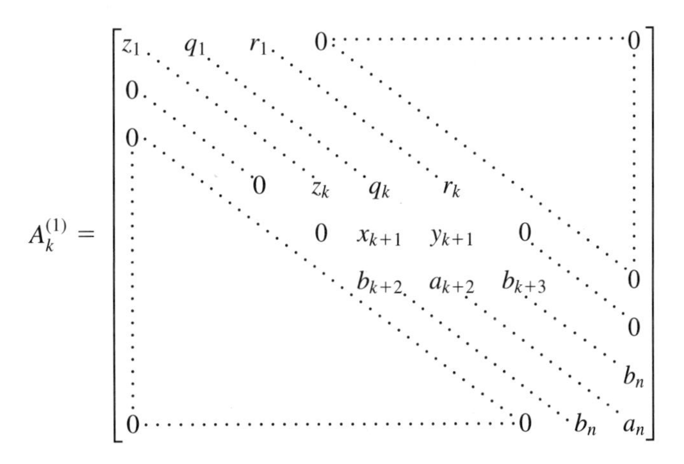

Proceeding with this construction in the sequence $P_2, \ldots, P_n$ produces the upper triangular matrix

$$R^{(1)} \equiv A_n^{(1)} = \begin{bmatrix} z_1 & q_1 & r_1 & 0 & \cdots & 0 \\ 0 & \ddots & \ddots & \ddots & \ddots & \vdots \\ \vdots & & \ddots & \ddots & \ddots & 0 \\ \vdots & & & \ddots & \ddots & r_{n-2} \\ \vdots & & & & z_{n-1} & q_{n-1} \\ 0 & \cdots & \cdots & \cdots & 0 & x_n \end{bmatrix}.$$

The other half of the QR factorization is the matrix

$$Q^{(1)} = P_2^t P_3^t \cdots P_n^t,$$

since the orthogonality of the rotation matrices implies that

$$Q^{(1)}R^{(1)} = (P_2^t P_3^t \cdots P_n^t) \cdot (P_n \cdots P_3 P_2)A^{(1)} = A^{(1)}.$$

The matrix $Q^{(1)}$ is orthogonal since

$$(Q^{(1)})^t Q^{(1)} = (P_2^t P_3^t \cdots P_n^t)^t (P_2^t P_3^t \cdots P_n^t) = P_n \cdots P_3 P_2 P_2^t P_3^t \cdots P_n^t = I.$$

In addition, $Q^{(1)}$ is an upper-Hessenberg matrix. To see why this is true, you can follow the steps in Exercises 7 and 8.

As a consequence, $A^{(2)} = R^{(1)}Q^{(1)}$ is also an upper-Hessenberg matrix, since multiplying $Q^{(1)}$ on the left by the upper triangular matrix $R^{(1)}$ does not affect the entries in the lower triangle. This implies that $A^{(2)}$ is in fact tridiagonal, since we already know that it is symmetric.

The entries off the diagonal of $A^{(2)}$ will generally be smaller in magnitude than the corresponding entries of $A^{(1)}$, so $A^{(2)}$ is closer to being a diagonal matrix then is $A^{(1)}$. The process is repeated to construct $A^{(3)}, A^{(4)}, \cdots$.

If the eigenvalues of A have distinct moduli and are ordered by $|\lambda_1| > |\lambda_2| > \cdots > |\lambda_n|$, the QR method converges (see [Fr]); that is, $A^{(i+1)}$ tends to a diagonal matrix. The rate of convergence of an off-diagonal entry $b_{j+1}^{(i+1)}$ to zero depends on the ratio $|\lambda_{j+1}/\lambda_j|$. Thus, the convergence of $a_j^{(i+1)}$ to the eigenvalue λ_j can be slow if $|\lambda_{j+1}/\lambda_j|$ is close to unity.

To accelerate this convergence, a shifting technique is employed similar to that used with the Inverse Power method in Section 9.2. A constant s is selected close to an eigenvalue of A. This modifies the factorization in Eq. (9.15) to choosing $Q^{(i)}$ and $R^{(i)}$ so that

(9.17) $$A^{(i)} - sI = Q^{(i)}R^{(i)},$$

and, correspondingly, the matrix $A^{(i+1)}$ is defined to be

(9.18) $$A^{(i+1)} = R^{(i)}Q^{(i)} + sI.$$

With this modification, the rate of convergence of $b_{j+1}^{(i+1)}$ to zero depends on the ratio $|(\lambda_{j+1} - s)/(\lambda_j - s)|$, which can result in a significant improvement over the original rate of convergence of $a_j^{(i+1)}$ to λ_j if s is close to λ_{j+1} but not close to λ_j.

In the listing of the QR Algorithm, we change s at each step so that when A has eigenvalues of distinct modulus, $b_n^{(i+1)}$ converges to zero faster than $b_j^{(i+1)}$ for any integer j less than n. When $b_n^{(i+1)}$ is sufficiently small, we assume that $\lambda_n \approx a_n^{(i+1)}$, delete the nth row and column of the matrix, and proceed in the same manner to find an approximation to λ_{n-1}. The process is continued until an approximation has been determined for each eigenvalue.

The algorithm incorporates the shifting technique by choosing at the ith step the shifting constant s_i, where s_i is the eigenvalue closest to $a_n^{(i)}$ of the matrix

$$E^{(i)} = \begin{bmatrix} a_{n-1}^{(i)} & b_n^{(i)} \\ b_n^{(i)} & a_n^{(i)} \end{bmatrix}.$$

This shift translates the eigenvalues of A by a factor s_i. With this shifting technique, the convergence is usually cubic. (See [WR, p. 270].) The algorithm accumulates these shifts until $b_n^{(i+1)} \approx 0$ and then adds the shifts to $a_n^{(i+1)}$ to approximate the eigenvalue λ_n.

If A has eigenvalues of the same modulus, $b_j^{(i+1)}$ may tend to zero for some $j \neq n$ at a faster rate than $b_n^{(i+1)}$. In this case, the matrix-splitting technique described in (9.14) can be employed to reduce the problem to one involving a pair of matrices of reduced order. Algorithm 9.6 implements the QR method in this manner.

ALGORITHM 9.6

QR

To obtain the eigenvalues of the symmetric, tridiagonal $n \times n$ matrix

$$A \equiv A_1 = \begin{bmatrix} a_1^{(1)} & b_2^{(1)} & 0 & \cdots & \cdots & 0 \\ b_2^{(1)} & a_2^{(1)} & \ddots & \ddots & & \vdots \\ 0 & \ddots & \ddots & \ddots & \ddots & \vdots \\ \vdots & \ddots & \ddots & \ddots & \ddots & 0 \\ \vdots & & \ddots & \ddots & \ddots & b_n^{(1)} \\ 0 & \cdots & \cdots & 0 & b_n^{(1)} & a_n^{(1)} \end{bmatrix}$$

INPUT $n; a_1^{(1)}, \ldots, a_n^{(1)}, b_2^{(1)}, \ldots, b_n^{(1)}$; tolerance TOL; maximum number of iterations M.

OUTPUT eigenvalues of A, or recommended splitting of A, or a message that the maximum number of iterations was exceeded.

Step 1 Set $k = 1$;
$SHIFT = 0$. (*Accumulated shift.*)

Step 2 While $k \leq M$ do Steps 3–19.
(Steps 3–7 test for success.)

Step 3 If $|b_n^{(k)}| \leq TOL$ then set $\lambda = a_n^{(k)} + SHIFT$;
OUTPUT (λ);
set $n = n - 1$.

Step 4 If $|b_2^{(k)}| \leq TOL$ then set $\lambda = a_1^{(k)} + SHIFT$;
OUTPUT (λ);
set $n = n - 1$;
$a_1^{(k)} = a_2^{(k)}$;
for $j = 2, \ldots, n$
set $a_j^{(k)} = a_{j+1}^{(k)}$;
$b_j^{(k)} = b_{j+1}^{(k)}$.

Step 5 If $n = 0$ then
STOP.

Step 6 If $n = 1$ then
set $\lambda = a_1^{(k)} + SHIFT$;
OUTPUT (λ);
STOP.

Step 7 For $j = 3, \ldots, n - 1$
if $|b_j^{(k)}| \leq TOL$ then
OUTPUT ('split into', $a_1^{(k)}, \ldots, a_{j-1}^{(k)}, b_2^{(k)}, \ldots, b_{j-1}^{(k)}$,
'and',
$a_j^{(k)}, \ldots, a_n^{(k)}, b_{j+1}^{(k)}, \ldots, b_n^{(k)}, SHIFT$);
STOP.

Step 8 (*Compute shift.*)

Set $b = -(a_{n-1}^{(k)} + a_n^{(k)})$;

$c = a_n^{(k)} a_{n-1}^{(k)} - \left[b_n^{(k)}\right]^2$;

$d = (b^2 - 4c)^{1/2}$.

Step 9 If $b > 0$ then set $\mu_1 = -2c/(b + d)$;
$\mu_2 = -(b + d)/2$
else set $\mu_1 = (d - b)/2$;
$\mu_2 = 2c/(d - b)$.

Step 10 If $n = 2$ then set $\lambda_1 = \mu_1 + SHIFT$;
$\lambda_2 = \mu_2 + SHIFT$;
OUTPUT (λ_1, λ_2);
STOP.

Step 11 Choose s so that $|s - a_n^{(k)}| = \min\{|\mu_1 - a_n^{(k)}|, |\mu_2 - a_n^{(k)}|\}$.

Step 12 (*Accumulate shift.*)
Set $SHIFT = SHIFT + s$.

Step 13 (*Perform shift.*)
For $j = 1, \ldots, n$, set $d_j = a_j^{(k)} - s$.

Step 14 (*Steps* 14 *and* 15 *compute* $R^{(k)}$.)
Set $x_1 = d_1$;
$y_1 = b_2$.

Step 15 For $j = 2, \ldots, n$

set $z_{j-1} = \left\{x_{j-1}^2 + \left[b_j^{(k)}\right]^2\right\}^{1/2}$;

$c_j = \dfrac{x_{j-1}}{z_{j-1}}$;

$s_j = \dfrac{b_j^{(k)}}{z_{j-1}}$;

$q_{j-1} = c_j y_{j-1} + s_j d_j$;
$x_j = -s_j y_{j-1} + c_j d_j$;
If $j \neq n$ then set $r_{j-1} = s_j b_{j+1}^{(k)}$;
$y_j = c_j b_{j+1}^{(k)}$.

$\left(A_j^{(k)} = P_j A_{j-1}^{(k)}\right.$ *has just been computed and* $\left.R^{(k)} = A_n^{(k)}.\right)$

Step 16 $\left(\text{\textit{Steps} 16–18 \textit{compute} } A^{(k+1)}\right)$.)
Set $z_n = x_n$;

$a_1^{(k+1)} = s_2 q_1 + c_2 z_1$;

$b_2^{(k+1)} = s_2 z_2$.

Step 17 For $j = 2, 3, \ldots, n - 1$
set $a_j^{(k+1)} = s_{j+1} q_j + c_j c_{j+1} z_j$;

$b_{j+1}^{(k+1)} = s_{j+1} z_{j+1}$.

Step 18 Set $a_n^{(k+1)} = c_n z_n$.

Step 19 Set $k = k + 1$.

Step 20 OUTPUT ('Maximum number of iterations exceeded');
(*Procedure completed unsuccessfully.*)
STOP.

EXAMPLE 1 Let

$$A = \begin{bmatrix} 3 & 1 & 0 \\ 1 & 3 & 1 \\ 0 & 1 & 3 \end{bmatrix} = \begin{bmatrix} a_1^{(1)} & b_2^{(1)} & 0 \\ b_2^{(1)} & a_2^{(1)} & b_3^{(1)} \\ 0 & b_3^{(1)} & a_3^{(1)} \end{bmatrix}.$$

To find the acceleration parameter for shifting we need the eigenvalues of

$$\begin{bmatrix} a_2^{(1)} & b_3^{(1)} \\ b_3^{(1)} & a_3^{(1)} \end{bmatrix} = \begin{bmatrix} 3 & 1 \\ 1 & 3 \end{bmatrix},$$

which are $\mu_1 = 4$ and $\mu_2 = 2$. The choice of eigenvalue closest to $a_3^{(1)} = 3$ is arbitrary, and we choose $\mu_2 = 2$ so that *SHIFT* $= 2$. With the notation of Algorithm 9.6,

$$\begin{bmatrix} d_1 & b_2^{(1)} & 0 \\ b_2^{(1)} & d_2 & b_3^{(1)} \\ 0 & b_3^{(1)} & d_3 \end{bmatrix} = \begin{bmatrix} 1 & 1 & 0 \\ 1 & 1 & 1 \\ 0 & 1 & 1 \end{bmatrix}.$$

Continuing the computation gives

$$x_1 = 1, \quad y_1 = 1, \quad z_1 = \sqrt{2}, \quad c_2 = \frac{\sqrt{2}}{2}, \quad s_2 = \frac{\sqrt{2}}{2},$$

$$q_1 = \sqrt{2}, \quad x_2 = 0, \quad r_1 = \frac{\sqrt{2}}{2}, \quad \text{and} \quad y_2 = \frac{\sqrt{2}}{2}.$$

So

$$A_2^{(1)} = \begin{bmatrix} \sqrt{2} & \sqrt{2} & \frac{\sqrt{2}}{2} \\ 0 & 0 & \frac{\sqrt{2}}{2} \\ 0 & 1 & 1 \end{bmatrix}.$$

Further,

$$z_2 = 1, \quad c_3 = 0, \quad s_3 = 1, \quad q_2 = 1, \quad \text{and} \quad x_3 = -\frac{\sqrt{2}}{2}.$$

So

$$R^{(1)} = A_3^{(1)} = \begin{bmatrix} \sqrt{2} & \sqrt{2} & \frac{\sqrt{2}}{2} \\ 0 & 1 & 1 \\ 0 & 0 & -\frac{\sqrt{2}}{2} \end{bmatrix}.$$

To compute $A^{(2)}$, we have

$$z_3 = -\frac{\sqrt{2}}{2}, \quad a_1^{(2)} = 2, \quad b_2^{(2)} = \frac{\sqrt{2}}{2}, \quad a_2^{(2)} = 1, \quad b_3^{(2)} = -\frac{\sqrt{2}}{2}, \quad \text{and} \quad a_3^{(2)} = 0.$$

So

$$A^{(2)} = R^{(1)}Q^{(1)} = \begin{bmatrix} 2 & \frac{\sqrt{2}}{2} & 0 \\ \frac{\sqrt{2}}{2} & 1 & -\frac{\sqrt{2}}{2} \\ 0 & -\frac{\sqrt{2}}{2} & 0 \end{bmatrix}.$$

One iteration of the QR method is complete. Since neither $b_2^{(2)} = \sqrt{2}/2$ nor $b_3^{(2)} = -\sqrt{2}/2$ is small, another iteration of the QR Algorithm will be performed. This iteration gives $SHIFT = 2 + (1 - \sqrt{3})/2 \approx 1.6339746$ and

$$A^{(3)} = \begin{bmatrix} 2.6720277 & 0.37597448 & 0 \\ 0.37597448 & 1.4736080 & 0.030396964 \\ 0 & 0.030396964 & -0.047559530 \end{bmatrix}.$$

If $b_3^{(3)} = 0.030396964$ is sufficiently small, then the approximation to the eigenvalue λ_3 is $a_3{}^{(3)} + SHIFT = 1.5864151$. Deleting the third row and column gives

$$A^{(3)} = \begin{bmatrix} 2.6720277 & 0.37597448 \\ 0.37597448 & 1.4736080 \end{bmatrix},$$

which has eigenvalues $\mu_1 = 2.7802140$ and $\mu_2 = 1.3654218$. Adding on the shifting factor gives the approximations

$$\lambda_1 \approx 4.4141886 \quad \text{and} \quad \lambda_2 \approx 2.9993964.$$

The actual eigenvalues of the matrix A are 4.41420, 3.00000, and 1.58579, so the QR Algorithm gave four digits of accuracy in only two iterations. ■

A similar procedure can be used to find approximations to the eigenvalues of a nonsymmetric $n \times n$ matrix. The matrix is first reduced to a similar upper-Hessenberg matrix H using the Householder Algorithm for nonsymmetric matrices.

The factoring process assumes the following form. First

(9.19) $$H \equiv H^{(1)} = Q^{(1)}R^{(1)}.$$

Then $H^{(2)}$ is defined by

$$H^{(2)} = R^{(1)}Q^{(1)} \tag{9.20}$$

and factored into

$$H^{(2)} = Q^{(2)}R^{(2)}. \tag{9.21}$$

The method of factoring proceeds with the same aim as the QR Algorithm. The matrices are chosen to introduce zeros at appropriate entries of the matrix, and a shifting procedure is used similar to that in the QR method. However, the shifting is somewhat more complicated for nonsymmetric matrices since complex eigenvalues with the same modulus can occur. The shifting process modifies the calculations in Eqs. (9.19), (9.20), and (9.21) to obtain the double QR method $H^{(1)} - s_1 I = Q^{(1)}R^{(1)}$, $H^{(2)} = R^{(1)}Q^{(1)} + s_1 I$, $H^{(2)} - s_2 I = Q^{(2)}R^{(2)}$, and $H^{(3)} = R^{(2)}Q^{(2)} + s_2 I$, where s_1 and s_2 are complex conjugates and $H^{(1)}, H^{(2)}, \ldots$ are real upper-Hessenberg matrices.

A complete description of the QR method can be found in [Wil2]. Detailed algorithms and programs for this method and most other commonly employed methods are given in [WR]. We refer the reader to these works if the method we have discussed does not give satisfactory results.

The QR method can be performed in a manner that will produce the eigenvectors of a matrix as well as its eigenvalues, but Algorithm 9.6 has not been designed to accomplish this. If the eigenvectors of a symmetric matrix are needed as well as the eigenvalues, we suggest either using the Inverse Power method after Algorithm 9.6 has been employed or using a more powerful technique such as those listed in [WR], methods that are designed expressly for this purpose.

EXERCISE SET 9.4

1. Apply two iterations of the QR Algorithm to the following matrices.

   a. $\begin{bmatrix} 2 & -1 & 0 \\ -1 & 2 & -1 \\ 0 & -1 & 2 \end{bmatrix}$ b. $\begin{bmatrix} 3 & 1 & 0 \\ 1 & 4 & 2 \\ 0 & 2 & 1 \end{bmatrix}$

   c. $\begin{bmatrix} 4 & -1 & 0 \\ -1 & 3 & -1 \\ 0 & -1 & 2 \end{bmatrix}$ d. $\begin{bmatrix} 1 & 1 & 0 & 0 \\ 1 & 2 & -1 & 0 \\ 0 & -1 & 3 & 1 \\ 0 & 0 & 1 & 4 \end{bmatrix}$

   e. $\begin{bmatrix} -2 & 1 & 0 & 0 \\ 1 & -3 & -1 & 0 \\ 0 & -1 & 1 & 1 \\ 0 & 0 & 1 & 3 \end{bmatrix}$ f. $\begin{bmatrix} 0.5 & 0.25 & 0 & 0 \\ 0.25 & 0.8 & 0.4 & 0 \\ 0 & 0.4 & 0.6 & 0.1 \\ 0 & 0 & 0.1 & 1 \end{bmatrix}$

2. Use the QR Algorithm to determine, to within 10^{-5}, all the eigenvalues of the following matrices.

   a. $\begin{bmatrix} 2 & -1 & 0 \\ -1 & -1 & -2 \\ 0 & -2 & 3 \end{bmatrix}$ b. $\begin{bmatrix} 3 & 1 & 0 \\ 1 & 4 & 2 \\ 0 & 2 & 3 \end{bmatrix}$

c. $\begin{bmatrix} 4 & 2 & 0 & 0 & 0 \\ 2 & 4 & 2 & 0 & 0 \\ 0 & 2 & 4 & 2 & 0 \\ 0 & 0 & 2 & 4 & 2 \\ 0 & 0 & 0 & 2 & 4 \end{bmatrix}$ **d.** $\begin{bmatrix} 5 & -1 & 0 & 0 & 0 \\ -1 & 4.5 & 0.2 & 0 & 0 \\ 0 & 0.2 & 1 & -0.4 & 0 \\ 0 & 0 & -0.4 & 3 & 1 \\ 0 & 0 & 0 & 1 & 3 \end{bmatrix}$

3. Use the QR Algorithm to determine, to within 10^{-5}, all the eigenvalues for the matrices given in Exercise 1.

4. Use the Inverse Power method to determine, to within 10^{-5}, the eigenvectors of the matrices in Exercise 1.

5. **a.** Show that the rotation matrix $\begin{bmatrix} \cos\theta & -\sin\theta \\ \sin\theta & \cos\theta \end{bmatrix}$ applied to the vector $\mathbf{x} = (x_1, x_2)^t$ has the geometric effect of rotating $\mathbf{x}$ through the angle θ without changing its magnitude with respect to $\|\cdot\|_2$.

b. Show that the magnitude of $\mathbf{x}$ with respect to $\|\cdot\|_\infty$ can be changed by a rotation matrix.

6. Let P be the rotation matrix with $p_{ii} = p_{jj} = \cos\theta$, $p_{ij} = -p_{ji} = \sin\theta$ for $j < i$. Show that for any $n \times n$ matrix A:

$$(AP)_{pq} = \begin{cases} a_{pq}, & \text{if } q \neq i, j, \\ (\cos\theta)a_{pj} + (\sin\theta)a_{pi}, & \text{if } q = j, \\ (\cos\theta)a_{pi} - (\sin\theta)a_{pj}, & \text{if } q = i. \end{cases}$$

$$(PA)_{pq} = \begin{cases} a_{pq}, & \text{if } p \neq i, j, \\ (\cos\theta)a_{jq} - (\sin\theta)a_{iq}, & \text{if } p = j, \\ (\sin\theta)a_{jq} + (\cos\theta)a_{iq}, & \text{if } p = i. \end{cases}$$

7. Show that the product of an upper triangular matrix (on the left) and an upper Hessenberg matrix produces an upper Hessenberg matrix.

8. Let P_k denote a rotation matrix of the form given in (9.16).

a. Show that $P_2^t P_3^t$ differs from an upper triangular matrix only in at most the (2, 1) and (3, 2) positions.

b. Assume that $P_2^t P_3^t \cdots P_k^t$ differs from an upper triangular matrix only in at most the $(2, 1), (3, 2), \ldots, (k, k-1)$ positions. Show that $P_2^t P_3^t \cdots P_k^t P_{k+1}^t$ differs from an upper triangular matrix only in at most the $(2, 1), (3, 2), \ldots, (k, k-1), (k+1, k)$ positions.

c. Show that the matrix $P_2^t P_3^t \cdots P_n^t$ is upper Hessenberg.

9. **Jacobi's method** for a symmetric matrix A is described by

$$A_1 = A,$$

$$A_2 = P_1 A_1 P_1^t.$$

and, in general,

$$A_{i+1} = P_i A_i P_i^t.$$

The matrix A_{i+1} tends to a diagonal matrix, where P_i is a rotation matrix chosen to eliminate a large off-diagonal element in A_i. If a_{jk} and a_{kj} are to be set to zero where $j \neq k$, then if $a_{jj} \neq a_{kk}$,

$$(P_i)_{jj} = (P_i)_{kk} = \sqrt{\frac{1}{2}\left(1 + \frac{b}{\sqrt{c^2 + b^2}}\right)},$$

$$(P_i)_{kj} = \frac{c}{2(P_i)_{jj}\sqrt{c^2 + b^2}} = -(P_i)_{jk},$$

where

$$c = 2a_{jk}\text{sign}(a_{jj} - a_{kk}) \quad \text{and} \quad b = |a_{jj} - a_{kk}|,$$

or if $a_{jj} = a_{kk}$,

$$(P_i)_{jj} = (P_i)_{kk} = \frac{\sqrt{2}}{2},$$

and

$$(P_i)_{kj} = -(P_i)_{jk} = \frac{\sqrt{2}}{2}.$$

Develop an algorithm to implement Jacobi's method by setting $a_{21} = 0$. Then set $a_{31}, a_{32}, a_{41}, a_{42}, a_{43}, \ldots, a_{n,1}, \ldots, a_{n,n-1}$ in turn to zero. This is repeated until a matrix A_k is computed with

$$\sum_{i=1}^{n} \sum_{\substack{j=1 \\ j \neq i}}^{n} |a_{ij}^{(k)}|$$

sufficiently small. The eigenvalues of A can then be approximated by the diagonal entries of A_k.

10. Repeat Exercise 3 using the Jacobi method.

11. In the lead example of this chapter, the linear system $A\mathbf{w} = -0.04\frac{\rho}{p}\lambda\mathbf{w}$ must be solved for $\mathbf{w}$ and λ in order to approximate the eigenvalues λ_k of the Strum-Liouville system.

a. Find all four eigenvalues $\mu_1, \ldots, \mu_4$ of the matrix

$$A = \begin{bmatrix} 2 & -1 & 0 & 0 \\ -1 & 2 & -1 & 0 \\ 0 & -1 & 2 & -1 \\ 0 & 0 & -1 & 2 \end{bmatrix}$$

to within 10^{-5}.

b. Approximate the eigenvalues $\lambda_1, \ldots, \lambda_4$ of the system in terms of ρ and p.

12. The $(m-1) \times (m-1)$ tridiagonal matrix

$$A = \begin{bmatrix} 1-2\alpha & \alpha & 0 & \cdots & 0 \\ \alpha & 1-2\alpha & \alpha & \ddots & \vdots \\ 0 & \ddots & \ddots & \ddots & 0 \\ \vdots & \ddots & \ddots & \ddots & \alpha \\ 0 & \cdots & 0 & \alpha & 1-2\alpha \end{bmatrix}$$

is involved in the Forward Difference method to solve the heat equation (see Section 12.2). For the stability of the method we need $\rho(A) < 1$. Let $m = 11$. Approximate the eigenvalues of A for each of the following.

a. $\alpha = \frac{1}{4}$ **b.** $\alpha = \frac{1}{2}$ **c.** $\alpha = \frac{3}{4}$

When is the method stable?

13. The eigenvalues of the matrix A in Exercise 12 are

$$\lambda_i = 1 - 4\alpha\left(\sin\frac{\pi i}{2m}\right)^2, \quad \text{for } i = 1, \ldots, m-1.$$

Compare the approximation in Exercise 12 to the actual eigenvalue. Again, when is the method stable?

9.5 Survey of Methods and Software

This chapter discussed the approximation of eigenvalues and eigenvectors. The Gerschgorin circles give a crude approximation to the location of the eigenvalues of a matrix. The Power method can be used to find the dominant eigenvalue and an associated eigenvector for an arbitrary matrix A. If A is symmetric, the Symmetric Power method gives faster convergence to the dominant eigenvalue and an associated eigenvector. The Inverse Power method will find the eigenvalue closest to a given value and an associated eigenvector. This method is often used to refine an approximate eigenvalue and to compute an eigenvector once an eigenvalue has been found by some other technique. Deflation methods, such as Wielandt deflation, obtain other eigenvalues once the dominant eigenvalue is known. These methods are used if only a few eigenvalues are required since they are susceptible to roundoff error. The Inverse Power method should be used to improve the accuracy of approximate eigenvalues obtained from a deflation technique.

Methods based on similarity transformations, such as Householder's method, are used to convert a symmetric matrix into a similar matrix that is tridiagonal (or upper Hessenberg if the matrix is not symmetric). Techniques such as the QR method can then be applied to the tridiagonal (or upper-Hessenberg) matrix to obtain approximations to all the eigenvalues. The associated eigenvectors can be found by using an iterative method, such as the Inverse Power method, applied to the eigenvalues obtained from the QR method. We restricted our study to symmetric matrices and presented the QR method only to compute eigenvalues for the symmetric case.

The subroutines in the IMSL and NAG libraries are based on those contained in EISPACK and LAPACK, packages that were discussed in Section 1.4. In general, the subroutines transform a matrix into the appropriate form for the QR method or one of its modifications, such as the QL method. The subroutines approximate all the eigenvalues and can approximate an associated eigenvector for each eigenvalue. There are special routines that find all the eigenvalues within an interval or region or find only the largest or smallest eigenvalue. Subroutines are also available to approximate the accuracy of the eigenvalue and the sensitivity of the process to roundoff error. The LAPACK routine SGEBAL prepares a real nonsymmetric matrix A for further processing. It tries to use permutation matrices to transform A to a similar block upper triangular form. Similarity transformations are used to balance the rows and columns in norm. The routine SGEHRD can then be used to convert A to a similar upper Hessenberg matrix H. The matrix H is then reduced via SHSEQR to Schur form STS^t, where S is orthogonal and the diagonal of T holds the eigenvalues of A. STREVC can then be used to obtain the corresponding eigenvectors.

The LAPACK routine SSYTRD is used to reduce a real symmetric matrix A to a similar tridiagonal matrix via Householder's method. The routine SSTEQR uses an implicity shifted QR algorithm to obtain all the eigenvalues and eigenvectors of A. The IMSL subroutine EVLRG produces all eigenvalues of A in increasing order of magnitude. This subroutine first balances the matrix A using a version of the EISPACK routine BALANC, so that the sums of the magnitudes of the entries in each row and in each column are approximately the same. This leads to greater stability in the ensuing computations. EVLRG next performs orthogonal similarity transformations, such as in Householder's method, to reduce A to a similar upper Hessenberg matrix. This portion is similar to the EISPACK subroutine ORTHES. Finally, the shifted QR algorithm is performed to obtain all the eigenvalues. This part is similar to the subroutine HQR in EISPACK. The IMSL subroutine EVCRG is the same as EVRLG, except that corresponding eigenvectors are computed. The subroutine EVLSF computes the eigenvalues of the real symmetric matrix A. The matrix A is first reduced to tridiagonal form using a modification of the EISPACK routine TRED2. Then the eigenvalues are computed using a modification of the EISPACK routine IMTQL2, which is a variation of the QR method called the implicit QL method. The subroutine EVCSF is the same as EVLSF except that the eigenvectors are also calculated. Finally, EVLRH and EVCRH compute all eigenvalues of the upper Hessenberg matrix A and, additionally, EVCRH computes the eigenvectors. These subroutines are based on the subroutines HQR and HQR2, respectively, in EISPACK.

The NAG library has similar subroutines based on the EISPACK routines. The subroutine F02AFF computes the eigenvalues of a real matrix. The subroutine F02AGF is the same as F02AFF except the eigenvectors are also calculated. The matrix is first balanced and then is reduced to upper-Hessenberg form for the QR method. The subroutine F02AAF is used on a real symmetric matrix. The eigenvalues, which are real, are computed in increasing order of magnitude. If the eigenvectors are also needed, the subroutine F02ABF can be applied. In either case, the matrix is reduced to tridiagonal form using Householder's method and the eigenvalues are then computed using the QL algorithm. The subroutine F01AGF implements Householder's algorithm directly for symmetric matrices to produce a similar tridiagonal symmetric matrix. Routines are also available in the NAG library for directly balancing real matrices, recovering eigenvectors if a matrix was first balanced, and performing other operations on special types of matrices.

The Maple procedure `Eigenvals(A)`; computes the eigenvalues of A by first balancing and then transforming A to tridiagonal or upper Hessenberg form. The QR method is then applied to obtain all eigenvalues and eigenvectors. The tridiagonal form as in Algorithm 9.6 is used for a symmetric matrix.

The books by Wilkinson [Wil2] and Wilkinson and Reinsch [WR] are classics in the study of eigenvalue problems. Stewart [Stew1] is also a good source of information on the general problem and Parlett [Par] considers the symmetric problem. A study of the nonsymmetric problem can be found in Saad [Sa1].

CHAPTER

10 Numerical Solutions of Nonlinear Systems of Equations

■ ■ ■

The amount of pressure required to sink a large heavy object into soft, homogeneous soil lying above a hard base soil can be predicted by the amount of pressure required to sink smaller objects in the same soil. Specifically, the amount of pressure p to sink a circular plate of radius r a distance d in the soft soil, where the hard base soil lies a distance $D > d$ below the surface, can be approximated by an equation of the form

$$p = k_1 e^{k_2 r} + k_3 r,$$

where k_1, k_2, and k_3 are constants depending on d and the consistency of the soil but not on the radius of the plate.

To determine the minimal size of plate required to sustain a large load, three small plates with differing radii are sunk to the same distance, and the loads required for this sinkage are shown in the accompanying figure.

This produces the three nonlinear equations

$$m_1 = k_1 e^{k_2 r_1} + k_3 r_1$$
$$m_2 = k_1 e^{k_2 r_2} + k_3 r_2$$
$$m_3 = k_1 e^{k_2 r_3} + k_3 r_3$$

in the three unknowns k_1, k_2, and k_3. Numerical approximation methods are usually needed for solving systems of equations when the equations are not linear. Exercise 8 of Section 10.2 concerns an application of the type described here.

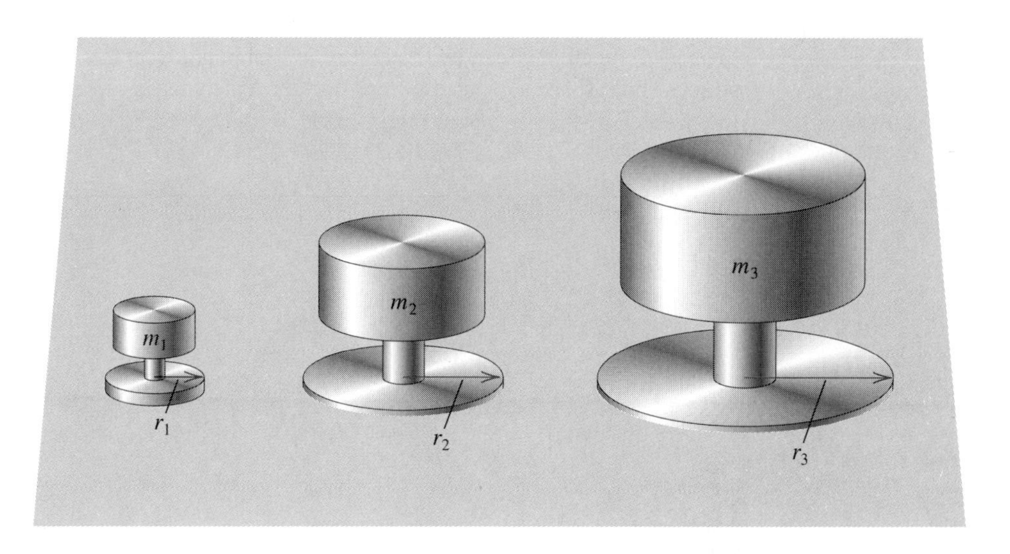

In Chapter 2 we considered the problem of approximating solutions to a single nonlinear equation of the form $f(x) = 0$. In this chapter we consider generalizations of those techniques that enable us to approximate the solutions of systems of nonlinear equations.

Fixed-point methods dominated the study in Chapter 2, with the most frequently applied techniques being based on Newton's method. This will also be the case in this chapter, although Section 10.4 considers an alternative method that provides more reliable but slower convergence.

Most of the proofs of the theoretical results in this chapter are omitted since they involve methods that are usually studied in advanced calculus. A good general reference for this material is Ortega's book entitled *Numerical Analysis—A Second Course* [Or2]. A more complete reference is [OR].

10.1 Fixed Points for Functions of Several Variables

A system of nonlinear equations has the form

$$\begin{aligned} f_1(x_1, x_2, \ldots, x_n) &= 0, \\ f_2(x_1, x_2, \ldots, x_n) &= 0, \\ \vdots \qquad & \quad \vdots \\ f_n(x_1, x_2, \ldots, x_n) &= 0, \end{aligned} \tag{10.1}$$

where each function f_i can be thought of as mapping a vector $\mathbf{x} = (x_1, x_2, \ldots, x_n)^t$ of the n-dimensional space $\mathbb{R}^n$ into the real line $\mathbb{R}$. A geometric representation of a nonlinear system when $n = 2$ is given in Figure 10.1.

Figure 10.1

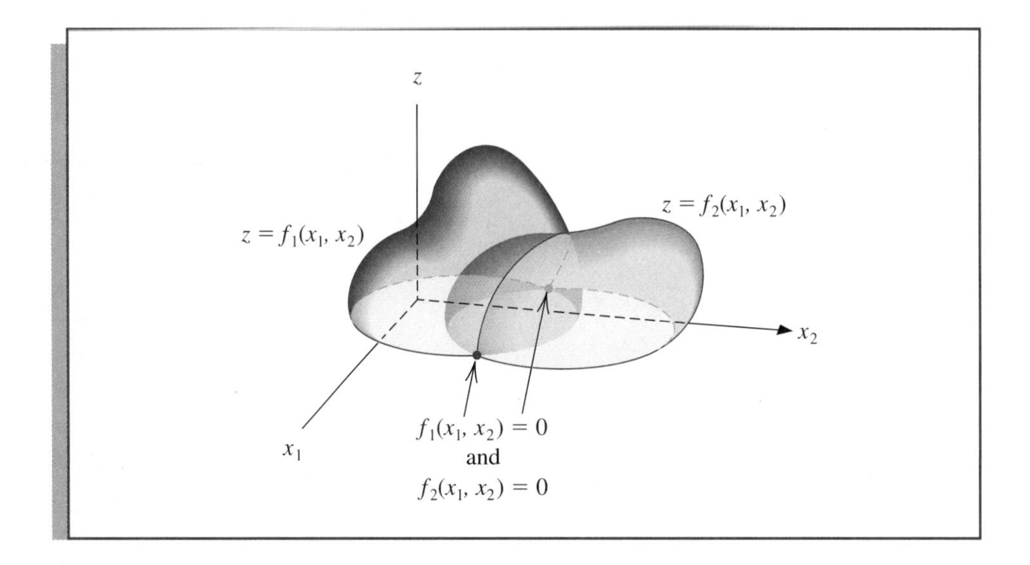

This system of n nonlinear equations in n unknowns can alternatively be represented by defining a function $\mathbf{F}$, mapping $\mathbb{R}^n$ into $\mathbb{R}^n$ by

$$\mathbf{F}(x_1, x_2, \ldots, x_n) = (f_1(x_1, x_2, \ldots, x_n), f_2(x_1, x_2, \ldots, x_n), \ldots, f_n(x_1, x_2, \ldots, x_n))^t.$$

If vector notation is used to represent the variables $x_1, x_2, \ldots, x_n$, system (10.1) assumes the form

$$\mathbf{F}(\mathbf{x}) = \mathbf{0}. \qquad \textbf{(10.2)}$$

The functions $f_1, f_2, \ldots, f_n$ are the **coordinate functions** of $\mathbf{F}$.

EXAMPLE 1 The three by three nonlinear system

$$3x_1 - \cos(x_2 x_3) - \frac{1}{2} = 0,$$

$$x_1^2 - 81(x_2 + 0.1)^2 + \sin x_3 + 1.06 = 0,$$

$$e^{-x_1 x_2} + 20x_3 + \frac{10\pi - 3}{3} = 0$$

can be placed in the form (10.2) by defining the three functions f_1, f_2, and f_3 from $\mathbb{R}^3$ to $\mathbb{R}$ as

$$f_1(x_1, x_2, x_3) = 3x_1 - \cos(x_2 x_3) - \frac{1}{2},$$

$$f_2(x_1, x_2, x_3) = x_1^2 - 81(x_2 + 0.1)^2 + \sin x_3 + 1.06,$$

$$f_3(x_1, x_2, x_3) = e^{-x_1 x_2} + 20x_3 + \frac{10\pi - 3}{3}$$

and $\mathbf{F}$ from $\mathbb{R}^3 \to \mathbb{R}^3$ by

$$\begin{aligned}\mathbf{F}(\mathbf{x}) &= \mathbf{F}(x_1, x_2, x_3)\\ &= (f_1(x_1, x_2, x_3), f_2(x_1, x_2, x_3), f_3(x_1, x_2, x_3))^t\\ &= \left(3x_1 - \cos(x_2 x_3) - \frac{1}{2}, x_1^2 - 81(x_2 + 0.1)^2 + \sin x_3 + 1.06,\right.\\ &\qquad \left. e^{-x_1 x_2} + 20x_3 + \frac{10\pi - 3}{3}\right)^t.\end{aligned}$$

■

Before discussing the solution of a system given in the form (10.1) or (10.2), we need some results concerning continuity and differentiability of functions from $\mathbb{R}^n$ into $\mathbb{R}^n$. Although this study could be presented directly (see Exercise 10), we use an alternative method that allows us to present the more theoretically difficult concepts of *limits* and *continuity* in terms of functions from $\mathbb{R}^n$ into $\mathbb{R}$.

Definition 10.1 Let f be a function defined on a set $D \subset \mathbb{R}^n$ and mapping into $\mathbb{R}$. The function f is said to have the **limit** L at $\mathbf{x}_0$, written

$$\lim_{\mathbf{x} \to \mathbf{x}_0} f(\mathbf{x}) = L,$$

if, given any number $\varepsilon > 0$, a number $\delta > 0$ exists with the property that

$$|f(\mathbf{x}) - L| < \varepsilon$$

whenever $\mathbf{x} \in D$ and

$$0 < ||\mathbf{x} - \mathbf{x}_0|| < \delta.$$

■

The existence of a limit is independent of the particular vector norm being used (see Section 7.1).

Definition 10.2 Let f be a function from a set $D \subset \mathbb{R}^n$ into $\mathbb{R}$. The function f is **continuous** at $\mathbf{x}_0 \in D$ provided $\lim_{\mathbf{x} \to \mathbf{x}_0} f(\mathbf{x})$ exists and

$$\lim_{\mathbf{x} \to \mathbf{x}_0} f(\mathbf{x}) = f(\mathbf{x}_0).$$

Moreover, f is **continuous on a set** D if f is continuous at every point of D. This concept is expressed by writing $f \in C(D)$. ■

We can now define the limit and continuity concepts for functions from $\mathbb{R}^n$ into $\mathbb{R}^n$ by considering the coordinate functions from $\mathbb{R}^n$ into $\mathbb{R}$.

Definition 10.3 Let $\mathbf{F}$ be a function from $D \subset \mathbb{R}^n$ into $\mathbb{R}^n$ of the form

$$\mathbf{F}(\mathbf{x}) = (f_1(\mathbf{x}), f_2(\mathbf{x}), \ldots, f_n(\mathbf{x}))^t,$$

where f_i is a mapping from $\mathbb{R}^n$ into $\mathbb{R}$ for each i. We define

$$\lim_{\mathbf{x}\to\mathbf{x}_0} \mathbf{F}(\mathbf{x}) = \mathbf{L} = (L_1, L_2, \ldots, L_n)^t$$

if and only if $\lim_{\mathbf{x}\to\mathbf{x}_0} f_i(\mathbf{x}) = L_i$ for each i. ■

The function $\mathbf{F}$ is *continuous* at $\mathbf{x}_0 \in D$ provided $\lim_{\mathbf{x}\to\mathbf{x}_0} \mathbf{F}(\mathbf{x})$ exists and $\lim_{\mathbf{x}\to\mathbf{x}_0} \mathbf{F}(\mathbf{x}) = \mathbf{F}(\mathbf{x}_0)$. In addition, $\mathbf{F}$ is continuous on the set D if $\mathbf{F}$ is continuous at each $\mathbf{x}$ in D. This concept is expressed by writing $\mathbf{F} \in C(D)$.

For functions from $\mathbb{R}$ into $\mathbb{R}$, continuity can often be shown by demonstrating that the function is differentiable (see Theorem 1.6). Although this theorem generalizes to functions of several variables, the derivative (or total derivative) of a function of several variables is quite involved and will not be presented here. Instead we state the following theorem, which relates the continuity of a function of n variables at a point to the partial derivatives of the function at the point.

Theorem 10.4 Let f be a function from $D \subset \mathbb{R}^n$ into $\mathbb{R}$ and $\mathbf{x}_0 \in D$. If constants $\delta > 0$ and $K > 0$ exist with

$$\left|\frac{\partial f(\mathbf{x})}{\partial x_j}\right| \le K, \quad \text{for each } j = 1, 2, \ldots, n,$$

whenever $\|\mathbf{x} - \mathbf{x}_0\| < \delta$ and $\mathbf{x} \in D$, then f is continuous at $\mathbf{x}_0$. ■

In Chapter 2, an iterative process for solving an equation $f(x) = 0$ was developed by first transforming the equation into one of the form $x = g(x)$. The function g is defined to have fixed points precisely at solutions to the original equation. A similar procedure will be investigated for functions from $\mathbb{R}^n$ into $\mathbb{R}^n$.

Definition 10.5 A function $\mathbf{G}$ from $D \subset \mathbb{R}^n$ into $\mathbb{R}^n$ has a **fixed point** at $\mathbf{p} \in D$ if $\mathbf{G}(\mathbf{p}) = \mathbf{p}$. ■

The following theorem extends the Fixed-Point Theorem 2.3 to the n-dimensional case. This theorem is a special case of the Contraction Mapping Theorem, and its proof can be found in [Or2, p. 153].

Theorem 10.6 Let $D = \{(x_1, x_2, \ldots, x_n)^t \mid a_i \le x_i \le b_i$ for each $i = 1, 2, \ldots, n\}$ for some collection of constants $a_1, a_2, \ldots, a_n$ and $b_1, b_2, \ldots, b_n$. Suppose $\mathbf{G}$ is a continuous function from $D \subset \mathbb{R}^n$ into $\mathbb{R}^n$ with the property that $\mathbf{G}(\mathbf{x}) \in D$ whenever $\mathbf{x} \in D$. Then $\mathbf{G}$ has a fixed point in D.

Suppose, in addition, that G has continuous partial derivatives and a constant $K < 1$ exists with

$$\left|\frac{\partial g_i(\mathbf{x})}{\partial x_j}\right| \le \frac{K}{n}, \quad \text{whenever } \mathbf{x} \in D,$$

for each $j = 1, 2, \ldots, n$ and each component function g_i. Then the sequence $\{\mathbf{x}^{(k)}\}_{k=0}^{\infty}$ defined by an arbitrarily selected $\mathbf{x}^{(0)}$ in D and generated by

$$\mathbf{x}^{(k)} = G(\mathbf{x}^{(k-1)}), \quad \text{for each } k \ge 1,$$

converges to the unique fixed point $\mathbf{p} \in D$ and

$$\text{(10.3)} \qquad \|\mathbf{x}^{(k)} - \mathbf{p}\|_\infty \le \frac{K^k}{1-K}\|\mathbf{x}^{(1)} - \mathbf{x}^{(0)}\|_\infty. \qquad \blacksquare$$

EXAMPLE 2 Consider the nonlinear system from Example 1 given by

$$3x_1 - \cos(x_2 x_3) - \frac{1}{2} = 0,$$
$$x_1^2 - 81(x_2 + 0.1)^2 + \sin x_3 + 1.06 = 0,$$
$$e^{-x_1 x_2} + 20x_3 + \frac{10\pi - 3}{3} = 0.$$

If the ith equation is solved for x_i, the system is changed into the fixed-point problem

$$x_1 = \frac{1}{3}\cos(x_2 x_3) + \frac{1}{6},$$
$$\text{(10.4)} \qquad x_2 = \frac{1}{9}\sqrt{x_1^2 + \sin x_3 + 1.06} - 0.1,$$
$$x_3 = -\frac{1}{20}e^{-x_1 x_2} - \frac{10\pi - 3}{60}.$$

Let $\mathbf{G}: \mathbb{R}^3 \to \mathbb{R}^3$ be defined by $\mathbf{G}(\mathbf{x}) = (g_1(\mathbf{x}), g_2(\mathbf{x}), g_3(\mathbf{x}))^t$, where

$$g_1(x_1, x_2, x_3) = \frac{1}{3}\cos(x_2 x_3) + \frac{1}{6},$$
$$g_2(x_1, x_2, x_3) = \frac{1}{9}\sqrt{x_1^2 + \sin x_3 + 1.06} - 0.1,$$
$$g_3(x_1, x_2, x_3) = -\frac{1}{20}e^{-x_1 x_2} - \frac{10\pi - 3}{60}.$$

Theorems 10.4 and 10.6 will be used to show that $\mathbf{G}$ has a unique fixed point in

$$D = \{(x_1, x_2, x_3)^t \mid -1 \le x_i \le 1, \text{ for each } i = 1, 2, 3\}.$$

For $\mathbf{x} = (x_1, x_2, x_3)^t$ in D,

$$|g_1(x_1, x_2, x_3)| \le \frac{1}{3}|\cos(x_2x_3)| + \frac{1}{6} \le 0.50,$$

$$|g_2(x_1, x_2, x_3)| = \left|\frac{1}{9}\sqrt{x_1^2 + \sin x_3 + 1.06} - 0.1\right|$$

$$= \frac{1}{9}\sqrt{1 + \sin 1 + 1.06} - 0.1 < 0.09,$$

and

$$|g_3(x_1, x_2, x_3)| = \frac{1}{20}e^{-x_1x_2} + \frac{10\pi - 3}{60} \le \frac{1}{20}e + \frac{10\pi - 3}{60} < 0.61;$$

so $-1 \le g_i(x_1, x_2, x_3) \le 1$, for each $i = 1, 2, 3$. Thus, $\mathbf{G}(\mathbf{x}) \in D$ whenever $\mathbf{x} \in D$.

Finding bounds for the partial derivatives on D gives

$$\left|\frac{\partial g_1}{\partial x_1}\right| = 0, \quad \left|\frac{\partial g_2}{\partial x_2}\right| = 0, \quad \text{and} \quad \left|\frac{\partial g_3}{\partial x_3}\right| = 0,$$

as well as

$$\left|\frac{\partial g_1}{\partial x_2}\right| \le \frac{1}{3}|x_3||\sin x_2x_3| \le \frac{1}{3}\sin 1 < 0.281,$$

$$\left|\frac{\partial g_1}{\partial x_3}\right| \le \frac{1}{3}|x_2||\sin x_2x_3| \le \frac{1}{3}\sin 1 < 0.281,$$

$$\left|\frac{\partial g_2}{\partial x_1}\right| = \frac{|x_1|}{9\sqrt{x_1^2 + \sin x_3 + 1.06}} < \frac{1}{9\sqrt{0.218}} < 0.238,$$

$$\left|\frac{\partial g_2}{\partial x_3}\right| = \frac{|\cos x_3|}{18\sqrt{x_1^2 + \sin x_3 + 1.06}} < \frac{1}{18\sqrt{0.218}} < 0.119,$$

$$\left|\frac{\partial g_3}{\partial x_1}\right| = \frac{|x_2|}{20}e^{-x_1x_2} \le \frac{1}{20}e < 0.14,$$

and

$$\left|\frac{\partial g_3}{\partial x_2}\right| = \frac{|x_1|}{20}e^{-x_1x_2} \le \frac{1}{20}e < 0.14.$$

Since the partial derivatives of g_1, g_2, and g_3 are bounded on D, Theorem 10.4 implies that these functions are continuous on D. Consequently, G is continuous on D. Moreover, for every $\mathbf{x} \in D$

$$\left|\frac{\partial g_i(\mathbf{x})}{\partial x_j}\right| \le 0.281, \quad \text{for each } i = 1, 2, 3 \text{ and } j = 1, 2, 3,$$

and the condition in the second part of Theorem 10.6 holds with $K = 3(0.281) = 0.843$.

In the same manner it can also be shown that $\partial g_i/\partial x_j$ is continuous on D for each $i = 1, 2, 3$ and $j = 1, 2, 3$. (This is considered in Exercise 3.) Consequently, $\mathbf{G}$ has a unique fixed point in D and the nonlinear system has a solution in D.

Note that G having a unique solution in D does not imply that the solution to the original system is unique on this domain, since the solution for x_2 in (10.4) involved the choice of the principal square root. Exercise 7(d) examines the situation that occurs if, instead, the negative square root is chosen in this step.

To approximate the fixed point $\mathbf{p}$, we choose $\mathbf{x}^{(0)} = (0.1, 0.1, -0.1)^t$. The sequence of vectors generated by

$$x_1^{(k)} = \frac{1}{3}\cos x_2^{(k-1)}x_3^{(k-1)} + \frac{1}{6},$$

$$x_2^{(k)} = \frac{1}{9}\sqrt{\left(x_1^{(k-1)}\right)^2 + \sin x_3^{(k-1)} + 1.06} - 0.1,$$

$$x_3^{(k)} = -\frac{1}{20}e^{-x_1^{(k-1)}x_2^{(k-1)}} - \frac{10\pi - 3}{60}$$

converges to the unique solution of (10.4). In this example the sequence was generated until

$$\|\mathbf{x}^{(k)} - \mathbf{x}^{(k-1)}\|_\infty < 10^{-5}.$$

The results are given in Table 10.1.

Table 10.1

k	$x_1^{(k)}$	$x_2^{(k)}$	$x_3^{(k)}$	$\|\mathbf{x}^{(k)} - \mathbf{x}^{(k-1)}\|_\infty$
0	0.10000000	0.10000000	−0.10000000	
1	0.49998333	0.00944115	−0.52310127	0.423
2	0.49999593	0.00002557	−0.52336331	9.4×10^{-3}
3	0.50000000	0.00001234	−0.52359814	2.3×10^{-4}
4	0.50000000	0.00000003	−0.52359847	1.2×10^{-5}
5	0.50000000	0.00000002	−0.52359877	3.1×10^{-7}

Using the error bound (10.3) with $K = 0.843$ gives

$$\|\mathbf{x}^{(5)} - \mathbf{p}\|_\infty \le \frac{(0.843)^5}{1 - 0.843}(0.423) < 1.15,$$

which does not indicate the true accuracy of $\mathbf{x}^{(5)}$ because of the inaccurate initial approximation. The actual solution is

$$\mathbf{p} = \left(0.5, 0, -\frac{\pi}{6}\right)^t \approx (0.5, 0, -0.5235987757)^t,$$

so the true error is

$$\|\mathbf{x}^{(5)} - \mathbf{p}\|_\infty \leq 2 \times 10^{-8}.$$

One way to accelerate convergence of the fixed-point iteration is to use the latest estimates $x_1^{(k)}, \ldots, x_{i-1}^{(k)}$ instead of $x_1^{(k-1)}, \ldots, x_{i-1}^{(k-1)}$ to compute $x_i^{(k)}$, as in the Gauss-Seidel method for linear systems. The component equations then become

$$x_1^{(k)} = \frac{1}{3}\cos\left(x_2^{(k-1)}x_3^{(k-1)}\right) + \frac{1}{6},$$

$$x_2^{(k)} = \frac{1}{9}\sqrt{\left(x_1^{(k)}\right)^2 + \sin x_3^{(k-1)} + 1.06} - 0.1,$$

$$x_3^{(k)} = -\frac{1}{20}e^{-x_1^{(k)}x_2^{(k)}} - \frac{10\pi - 3}{60}.$$

With $\mathbf{x}^{(0)} = (0.1, 0.1, -0.1)^t$, the results of these calculations are listed in Table 10.2.

The iterate $\mathbf{x}^{(4)}$ is accurate to within 10^{-7} in the l_∞ norm; so the convergence was indeed accelerated for this problem by using the Seidel method. It should be remarked, however, that the Seidel method does not *always* accelerate the convergence. ■

Table 10.2

k	$x_1^{(k)}$	$x_2^{(k)}$	$x_3^{(k)}$	$\|\mathbf{x}^{(k)} - \mathbf{x}^{(k-1)}\|_\infty$
0	0.10000000	0.10000000	−0.10000000	
1	0.49998333	0.02222979	−0.52304613	0.423
2	0.49997747	0.00002815	−0.52359807	2.2×10^{-2}
3	0.50000000	0.00000004	−0.52359877	2.8×10^{-5}
4	0.50000000	0.00000000	−0.52359877	3.8×10^{-8}

Maple provides the function `fsolve` to solve systems of equations. The fixed-point problem of Example 2 can be solved with the following commands:

```
>g1:=x1=(2*cos(x2*x3)+1)/6:
>g2:=x2=sqrt(x1^2+sin(x3)+1.06)/9-0.1:
>g3:=x3=-(3*exp(-x1*x2)+10*Pi-3)/60:
>fsolve({g1,g2,g3},{x1,x2,x3},{x1=-1..1,x2=-1..1,x3=-1..1});
```

The first three commands define the system, and the last command involves the procedure `fsolve`. The answer displayed is

$$\{x3 = -.5235987758,\ x1 = .5000000000,\ x2 = -.2102454409\ 10^{-10}\}$$

In general, `fsolve(eqns,vars,options)` solves the system of equations represented by the parameter `eqns` for the variables represented by the parameter `vars` under optional parameters represented by `options`. Under `options` we specify a region in which the routine is required to search for a solution. This specification is not mandatory, and Maple determines its own search space if the options are omitted.

EXERCISE SET 10.1

1. Show that the function $\mathbf{F} : \mathbb{R}^3 \to \mathbb{R}^3$ defined by

$$\mathbf{F}(x_1, x_2, x_3) = \left(x_1 + 2x_3, x_1 \cos x_2, x_2^2 + x_3\right)^t$$

is a continuous at each point of $\mathbb{R}^3$.

2. Give an example of a function $\mathbf{F} : \mathbb{R}^2 \to \mathbb{R}^2$ that is continuous at each point of $\mathbb{R}^2$ except at $(1, 0)$.

3. Show that the first partial derivatives in Example 2 are continuous on D.

4. The nonlinear system

$$-x_1(x_1 + 1) + 2x_2 = 18$$
$$(x_1 - 1)^2 + (x_2 - 6)^2 = 25$$

has two solutions.

   a. Approximate the solutions graphically.
   b. Use the approximations from part (a) as initial approximations for an appropriate functional iteration.
   c. Determine the solutions to within 10^{-5} in the l_∞ norm.

5. The nonlinear system

$$x_1^2 - 10x_1 + x_2^2 + 8 = 0,$$
$$x_1x_2^2 + x_1 - 10x_2 + 8 = 0,$$

can be transformed into the fixed-point problem

$$x_1 = g_1(x_1, x_2) = \frac{x_1^2 + x_2^2 + 8}{10},$$
$$x_2 = g_2(x_1, x_2) = \frac{x_1x_2^2 + x_1 + 8}{10}.$$

   a. Use Theorem 10.6 to show that $\mathbf{G} = (g_1, g_2)^t : D \subset \mathbb{R}^2 \to \mathbb{R}^2$ has a unique fixed point in

$$D = \{(x_1, x_2)^t \mid 0 \le x_1, x_2 \le 1.5\}.$$

   b. Apply functional iteration to approximate the solution.
   c. Does the Seidel method accelerate convergence?

6. The nonlinear system

$$5x_1^2 - x_2^2 = 0$$
$$x_2 - 0.25(\sin x_1 + \cos x_2) = 0$$

has a solution near $\left(\frac{1}{4}, \frac{1}{4}\right)^t$.

   a. Find a function $\mathbf{G}$ and a set D in $\mathbb{R}^2$ such that $\mathbf{G} : D \to \mathbb{R}^2$ and $\mathbf{G}$ has a unique fixed point in D.
   b. Apply functional iteration to approximate the solution to within 10^{-5} in the l_∞ norm.
   c. Does the Seidel method accelerate convergence?

7. Use Theorem 10.6 to show that $\mathbf{G} : D \subset \mathbb{R}^3 \to \mathbb{R}^3$ has a fixed point in D. Apply functional iteration to approximate the solution to within 10^{-5} using $\|\cdot\|_\infty$.

a.

$$\mathbf{G}(x_1, x_2, x_3) = \left(\frac{\cos(x_2 x_3) + 0.5}{3}, \frac{1}{25}\sqrt{x_1^2 + 0.3125} - 0.03, -\frac{1}{20}e^{-x_1 x_2} - \frac{10\pi - 3}{60} \right)^t;$$

$$D = \{(x_1, x_2, x_3)^t \mid -1 \le x_i \le 1, i = 1, 2, 3\}$$

b.

$$\mathbf{G}(x_1, x_2, x_3) = \left(\frac{13 - x_2^2 + 4x_3}{15}, \frac{11 + x_3 - x_1^2}{10}, \frac{22 + x_2^3}{25} \right);$$

$$D = \{(x_1, x_2, x_3)^t \mid 0 \le x_i \le 1.5, i = 1, 2, 3\}$$

c.

$$\mathbf{G}(x_1, x_2, x_3) = (1 - \cos(x_1 x_2 x_3), 1 - (1 - x_1)^{1/4} - 0.05x_3^2 + 0.15x_3, x_1^2 + 0.1x_2^2 - 0.01x_2 + 1)^t;$$

$$D = \{(x_1, x_2, x_3)^t \mid -0.1 \le x_1 \le 0.1, -0.1 \le x_2 \le 0.3, 0.5 \le x_3 \le 1.1\}$$

d.

$$\mathbf{G}(x_1, x_2, x_3) = \left(\frac{1}{3}\cos(x_2 x_3) + \frac{1}{6}, -\frac{1}{9}\sqrt{x_1^2 + \sin x_3 + 1.06} - 0.1, -\frac{1}{20}e^{-x_1 x_2} - \frac{10\pi - 3}{60} \right)^t;$$

$$D = \{(x_1, x_2, x_3)^t \mid -1 \le x_i \le 1, i = 1, 2, 3\}$$

8. Use the Seidel method to approximate the fixed points in Exercise 7 to within 10^{-5} using $\|\cdot\|_\infty$.

9. Use functional iteration to find solutions to the following nonlinear systems, accurate to within 10^{-5}, using $\|\cdot\|_\infty$.

a.
$$x_1^2 + x_2^2 - x_1 = 0,$$
$$x_1^2 - x_2^2 - x_2 = 0.$$

b.
$$3x_1^2 - x_2^2 = 0,$$
$$3x_1 x_2^2 - x_1^3 - 1 = 0.$$

c.
$$x_1^2 + x_2 - 37 = 0,$$
$$x_1 - x_2^2 - 5 = 0,$$
$$x_1 + x_2 + x_3 - 3 = 0.$$

d.
$$x_1^2 + 2x_2^2 - x_2 - 2x_3 = 0,$$
$$x_1^2 - 8x_2^2 + 10x_3 = 0,$$
$$\frac{x_1^2}{7x_2 x_3} - 1 = 0.$$

10. Show that a function $\mathbf{F} : D \subset \mathbb{R}^n \to \mathbb{R}^n$ is continuous at $\mathbf{x}_0 \in D$ precisely when, given any number $\epsilon > 0$, a number $\delta > 0$ can be found with the property that

$$\|\mathbf{F}(\mathbf{x}) - \mathbf{F}(\mathbf{x}_0)\| < \epsilon$$

whenever $\mathbf{x} \in D$ and $\|\mathbf{x} - \mathbf{x}_0\| < \delta$.

11. In Exercise 8 of Section 5.9, we considered the problem of predicting the population of two species that compete for the same food supply. In the problem, we made the assumption that the populations could be predicted by solving the system of equations

$$\frac{dx_1(t)}{dt} = x_1(t)(4 - 0.0003x_1(t) - 0.0004x_2(t))$$

and

$$\frac{dx_2(t)}{dt} = x_2(t)(2 - 0.0002x_1(t) - 0.0001x_2(t)).$$

In this exercise, we would like to consider the problem of determining equilibrium populations of the two species. The mathematical criteria that must be satisfied in order for the populations to be at equilibrium is that, simultaneously,

$$\frac{dx_1(t)}{dt} = 0 \quad \text{and} \quad \frac{dx_2(t)}{dt} = 0.$$

This occurs when the first species is extinct and the second species has a population of 20,000 or when the second species is extinct and the first species has a population of 13,333. Can an equilibrium occur in any other situation?

10.2 Newton's Method

The problem in Example 2 of the previous section is transformed into a convergent fixed-point problem by algebraically solving the three equations for the three variables x_1, x_2, and x_3. It is, however, rather unusual for this technique to be successful. In this section, we consider an algorithmic procedure to perform the transformation in a more general situation.

To construct the algorithm that led to an appropriate fixed-point method in the one-dimensional case, we found a function ϕ with the property that

$$g(x) = x - \phi(x)f(x)$$

gives quadratic convergence to the fixed point p of the function g (see Section 2.4). From this condition Newton's method evolved by choosing $\phi(x) = 1/f'(x)$, assuming that $f'(x) \neq 0$.

Using a similar approach in the n-dimensional case involves a matrix

$$\textbf{(10.5)} \qquad A(\mathbf{x}) = \begin{bmatrix} a_{11}(\mathbf{x}) & a_{12}(\mathbf{x}) & \cdots & a_{1n}(\mathbf{x}) \\ a_{21}(\mathbf{x}) & a_{22}(\mathbf{x}) & \cdots & a_{2n}(\mathbf{x}) \\ \vdots & \vdots & & \vdots \\ a_{n1}(\mathbf{x}) & a_{n2}(\mathbf{x}) & \cdots & a_{nn}(\mathbf{x}) \end{bmatrix},$$

where each of the entries $a_{ij}(\mathbf{x})$ is a function from $\mathbb{R}^n$ into $\mathbb{R}$. This requires that $A(\mathbf{x})$ be found so that

$$\mathbf{G}(\mathbf{x}) = \mathbf{x} - A(\mathbf{x})^{-1}\mathbf{F}(\mathbf{x})$$

gives quadratic convergence to the solution of $\mathbf{F}(\mathbf{x}) = \mathbf{0}$, assuming that $A(\mathbf{x})$ is nonsingular at the fixed point $\mathbf{p}$ of $\mathbf{G}$.

The following theorem parallels Theorem 2.8 in Section 2.4. Its proof requires being able to express $\mathbf{G}$ in terms of its Taylor series in n variables about $\mathbf{p}$.

Theorem 10.7 Suppose $\mathbf{p}$ is a solution of $\mathbf{G}(\mathbf{x}) = \mathbf{x}$ for some function $\mathbf{G} = (g_1, g_2, \ldots, g_n)^t$ mapping $\mathbb{R}^n$ into $\mathbb{R}^n$. If a number $\delta > 0$ exists with the property that

(i) $\partial g_i/\partial x_j$ is continuous on $N_\delta = \{\mathbf{x} \mid \|\mathbf{x} - \mathbf{p}\| < \delta\}$ for each $i = 1, 2, \ldots, n$ and $j = 1, 2, \ldots, n$;

(ii) $\partial^2 g_i(\mathbf{x})/(\partial x_j \partial x_k)$ is continuous, and $|\partial^2 g_i(\mathbf{x})/(\partial x_j \partial x_k)| \leq M$ for some constant M whenever $\mathbf{x} \in N_\delta$, for each $i = 1, 2, \ldots, n$, $j = 1, 2, \ldots, n$, and $k = 1, 2, \ldots, n$;

(iii) $\partial g_i(\mathbf{p})/\partial x_k = 0$ for each $i = 1, 2, \ldots, n$ and $k = 1, 2, \ldots, n$,

then a number $\hat{\delta} \leq \delta$ exists such that the sequence generated by $\mathbf{x}^{(k)} = \mathbf{G}(\mathbf{x}^{(k-1)})$ converges quadratically to $\mathbf{p}$ for any choice of $\mathbf{x}^{(0)}$ provided that $\|\mathbf{x}^{(0)} - \mathbf{p}\| < \hat{\delta}$. Moreover,

$$\|\mathbf{x}^{(k)} - \mathbf{p}\|_\infty \leq \frac{n^2 M}{2}\|\mathbf{x}^{(k-1)} - \mathbf{p}\|_\infty^2, \quad \text{for each } k \geq 1. \qquad \blacksquare$$

To use Theorem 10.7, suppose that $A(\mathbf{x})$ is an $n \times n$ matrix of functions from $\mathbb{R}^n$ into $\mathbb{R}$ in the form of Eq. (10.5), where the specific entries will be chosen later. Assume, moreover, that $A(\mathbf{x})$ is nonsingular near a solution $\mathbf{p}$ of $\mathbf{F}(\mathbf{x}) = \mathbf{0}$, and let $b_{ij}(\mathbf{x})$ denote the entry of $A(\mathbf{x})^{-1}$ in the ith row and jth column.

Since $\mathbf{G}(\mathbf{x}) = \mathbf{x} - A(\mathbf{x})^{-1}\mathbf{F}(\mathbf{x})$, we have $g_i(\mathbf{x}) = x_i - \sum_{j=1}^n b_{ij}(\mathbf{x}) f_j(\mathbf{x})$ and

$$\frac{\partial g_i(\mathbf{x})}{\partial x_k} = \begin{cases} 1 - \displaystyle\sum_{j=1}^n \left(b_{ij}(\mathbf{x}) \frac{\partial f_j}{\partial x_k}(\mathbf{x}) + \frac{\partial b_{ij}}{\partial x_k}(\mathbf{x}) f_j(\mathbf{x}) \right), & \text{if } i = k, \\ -\displaystyle\sum_{j=1}^n \left(b_{ij}(\mathbf{x}) \frac{\partial f_j}{\partial x_k}(\mathbf{x}) + \frac{\partial b_{ij}}{\partial x_k}(\mathbf{x}) f_j(\mathbf{x}) \right), & \text{if } i \neq k. \end{cases}$$

Theorem 10.7 implies that we need $\partial g_i(\mathbf{p})/\partial x_k = 0$ for each $i = 1, 2, \ldots, n$ and $k = 1, 2, \ldots, n$. This means that for $i = k$,

$$0 = 1 - \sum_{j=1}^n b_{ij}(\mathbf{p}) \frac{\partial f_j}{\partial x_i}(\mathbf{p}),$$

so

$$\sum_{j=1}^n b_{ij}(\mathbf{p}) \frac{\partial f_j}{\partial x_i}(\mathbf{p}) = 1. \tag{10.6}$$

When $k \neq i$,

$$0 = -\sum_{j=1}^n b_{ij}(\mathbf{p}) \frac{\partial f_j}{\partial x_k}(\mathbf{p}),$$

so

$$\sum_{j=1}^n b_{ij}(\mathbf{p}) \frac{\partial f_j}{\partial x_k}(\mathbf{p}) = 0. \tag{10.7}$$

Defining the matrix $J(\mathbf{x})$ by

(10.8)
$$J(\mathbf{x}) = \begin{bmatrix} \dfrac{\partial f_1(\mathbf{x})}{\partial x_1} & \dfrac{\partial f_1(\mathbf{x})}{\partial x_2} & \cdots & \dfrac{\partial f_1(\mathbf{x})}{\partial x_n} \\ \dfrac{\partial f_2(\mathbf{x})}{\partial x_1} & \dfrac{\partial f_2(\mathbf{x})}{\partial x_2} & \cdots & \dfrac{\partial f_2(\mathbf{x})}{\partial x_n} \\ \vdots & \vdots & & \vdots \\ \dfrac{\partial f_n(\mathbf{x})}{\partial x_1} & \dfrac{\partial f_n(\mathbf{x})}{\partial x_2} & \cdots & \dfrac{\partial f_n(\mathbf{x})}{\partial x_n} \end{bmatrix},$$

we see that conditions (10.6) and (10.7) require

$$A(\mathbf{p})^{-1}J(\mathbf{p}) = I, \quad \text{the identity matrix,}$$

so

$$A(\mathbf{p}) = J(\mathbf{p}).$$

An appropriate choice for $A(\mathbf{x})$ is, consequently, $A(\mathbf{x}) = J(\mathbf{x})$ since condition (iii) in Theorem 10.7 is then satisfied.

The function $\mathbf{G}$ is defined by

$$\mathbf{G}(\mathbf{x}) = \mathbf{x} - J(\mathbf{x})^{-1}\mathbf{F}(\mathbf{x}),$$

and the functional iteration procedure evolves from selecting $\mathbf{x}^{(0)}$ and generating, for $k \geq 1$,

(10.9)
$$\mathbf{x}^{(k)} = \mathbf{G}(\mathbf{x}^{(k-1)}) = \mathbf{x}^{(k-1)} - J(\mathbf{x}^{(k-1)})^{-1}\mathbf{F}(\mathbf{x}^{(k-1)}),$$

This is called **Newton's method for nonlinear systems**, and is generally expected to give quadratic convergence, provided that a sufficiently accurate starting value is known and $J(\mathbf{p})^{-1}$ exists.

The matrix $J(\mathbf{x})$ is called the **Jacobian** matrix and has a number of applications in analysis. It might, in particular, be familiar to the reader due to its application in the multiple integration of a function of several variables over a region that requires a change of variables to be performed.

The weakness in Newton's method arises from the need to compute and invert the matrix $J(\mathbf{x})$ at each step. In practice, explicit computation of $J(\mathbf{x})^{-1}$ is avoided by performing the operation in a two-step manner. First, a vector $\mathbf{y}$ is found that will satisfy $J(\mathbf{x}^{(k-1)})\mathbf{y} = -\mathbf{F}(\mathbf{x}^{(k-1)})$. After this has been accomplished, the new approximation, $\mathbf{x}^{(k)}$, is obtained by adding $\mathbf{y}$ to $\mathbf{x}^{(k-1)}$. Algorithm 10.1 uses this two-step procedure.

ALGORITHM **10.1**

Newton's Method for Systems

To approximate the solution of the nonlinear system $\mathbf{F}(\mathbf{x}) = \mathbf{0}$ given an initial approximation $\mathbf{x}$:

INPUT number n of equations and unknowns; initial approximation $\mathbf{x} = (x_1, \ldots, x_n)^t$, tolerance TOL; maximum number of iterations N.

OUTPUT approximate solution $\mathbf{x} = (x_1, \ldots, x_n)^t$ or a message that the number of iterations was exceeded.

Step 1 Set $k = 1$.

Step 2 While ($k \leq N$) do Steps 3–7.

Step 3 Calculate $\mathbf{F}(\mathbf{x})$ and $J(\mathbf{x})$, where $J(\mathbf{x})_{i,j} = (\partial f_i(\mathbf{x})/\partial x_j)$ for $1 \leq i, j \leq n$.

Step 4 Solve the $n \times n$ linear system $J(\mathbf{x})\mathbf{y} = -\mathbf{F}(\mathbf{x})$.

Step 5 Set $\mathbf{x} = \mathbf{x} + \mathbf{y}$.

Step 6 If $||\mathbf{y}|| < TOL$ then OUTPUT ($\mathbf{x}$);
(*Procedure completed successfully.*)
STOP.

Step 7 Set $k = k + 1$.

Step 8 OUTPUT ('Maximum number of iterations exceeded');
(*Procedure completed unsuccessfully.*)
STOP.

EXAMPLE 1 The nonlinear system

$$3x_1 - \cos(x_2x_3) - \frac{1}{2} = 0,$$
$$x_1^2 - 81(x_2 + 0.1)^2 + \sin x_3 + 1.06 = 0,$$
$$e^{-x_1x_2} + 20x_3 + \frac{10\pi - 3}{3} = 0$$

was shown, in Example 2 of Section 10.1, to have an approximate solution at $(0.5, 0, -0.52359877)^t$. Newton's method will be used to obtain this approximation when the initial approximation is $\mathbf{x}^{(0)} = (0.1, 0.1, -0.1)^t$.

The Jacobian matrix $J(\mathbf{x})$ for this system is

$$J(x_1, x_2, x_3) = \begin{bmatrix} 3 & x_3 \sin x_2x_3 & x_2 \sin x_2x_3 \\ 2x_1 & -162(x_2 + 0.1) & \cos x_3 \\ -x_2e^{-x_1x_2} & -x_1e^{-x_1x_2} & 20 \end{bmatrix}$$

and

$$\begin{bmatrix} x_1^{(k)} \\ x_2^{(k)} \\ x_3^{(k)} \end{bmatrix} = \begin{bmatrix} x_1^{(k-1)} \\ x_2^{(k-1)} \\ x_3^{(k-1)} \end{bmatrix} + \begin{bmatrix} y_1^{(k-1)} \\ y_2^{(k-1)} \\ y_3^{(k-1)} \end{bmatrix},$$

where

$$\begin{bmatrix} y_1^{(k-1)} \\ y_2^{(k-1)} \\ y_3^{(k-1)} \end{bmatrix} = -\left(J\left(x_1^{(k-1)}, x_2^{(k-1)}, x_3^{(k-1)}\right)\right)^{-1} \mathbf{F}\left(x_1^{(k-1)}, x_2^{(k-1)}, x_3^{(k-1)}\right).$$

Thus, at the kth step, the linear system $J(\mathbf{x}^{(k-1)})\mathbf{y}^{(k-1)} = -\mathbf{F}(\mathbf{x}^{(k-1)})$ must be solved, where

$$J(\mathbf{x}^{(k-1)}) = \begin{bmatrix} 3 & x_3^{(k-1)} \sin x_2^{(k-1)} x_3^{(k-1)} & x_2^{(k-1)} \sin x_2^{(k-1)} x_3^{(k-1)} \\ 2x_1^{(k-1)} & -162(x_2^{(k-1)} + 0.1) & \cos x_3^{(k-1)} \\ -x_2^{(k-1)} e^{-x_1^{(k-1)} x_2^{(k-1)}} & -x_1^{(k-1)} e^{-x_1^{(k-1)} x_2^{(k-1)}} & 20 \end{bmatrix},$$

$$\mathbf{y}^{(k-1)} = \begin{bmatrix} y_1^{(k-1)} \\ y_2^{(k-1)} \\ y_3^{(k-1)} \end{bmatrix},$$

and

$$\mathbf{F}(\mathbf{x}^{(k-1)}) = \begin{bmatrix} 3x_1^{(k-1)} - \cos x_2^{(k-1)} x_3^{(k-1)} - \frac{1}{2} \\ (x_1^{(k-1)})^2 - 81(x_2^{(k-1)} + 0.1)^2 + \sin x_3^{(k-1)} + 1.06 \\ e^{-x_1^{(k-1)} x_2^{(k-1)}} + 20x_3^{(k-1)} + \frac{10\pi - 3}{3} \end{bmatrix}.$$

The results using this iterative procedure are shown in Table 10.3.

Table 10.3

k	$x_1^{(k)}$	$x_2^{(k)}$	$x_3^{(k)}$	$\lVert \mathbf{x}^{(k)} - \mathbf{x}^{(k-1)} \rVert_\infty$
0	0.10000000	0.10000000	−0.10000000	
1	0.50003702	0.01946686	−0.52152047	0.422
2	0.50004593	0.00158859	−0.52355711	1.79×10^{-2}
3	0.50000034	0.00001244	−0.52359845	1.58×10^{-3}
4	0.50000000	0.00000000	−0.52359877	1.24×10^{-5}
5	0.50000000	0.00000000	−0.52359877	0

The previous example illustrates that Newton's method can converge very rapidly once an approximation is obtained that is near the true solution. However, it is not always easy to determine starting values that will lead to a solution, and the method is comparatively expensive to employ. In the next section, we consider a method for overcoming the latter weakness. Good starting values can usually be found by the method that will be discussed in Section 10.4.

Initial approximation to the solutions of 2×2 and often 3×3 nonlinear systems can also be obtained using the graphing facilities of Maple. The nonlinear system

$$x_1^2 - x_2^2 + 2x_2 = 0,$$
$$2x_1 + x_2^2 - 6 = 0.$$

has two solutions, $(0.625204094, 2.179355825)^t$ and $(2.109511920, -1.334532188)^t$. To use Maple we first define the 2 equations

```
>eq1:=x1^2-x2^2+2*x2=0;
>eq2:=2*x1+x2^2-6=0;
```

To obtain a graph of the two equations for $-3 \leq x_1, x_2 \leq 3$, enter the commands

```
>with(plots);
>implicitplot({eq1,eq2},x1=-3..3,x2=-3..3);
```

From the graph we are able to estimate that there are solutions near $(0.64, 2.2)^t$ and $(2.1, -1.3)^t$. This gives us good starting values for Newton's method.

The problem is more difficult in three dimensions. Consider the nonlinear system

$$2x_1 - 3x_2 + x_3 - 4 = 0,$$
$$2x_1 + x_2 - x_3 + 4 = 0,$$
$$x_1^2 + x_2^2 + x_3^2 - 4 = 0.$$

Define three equations using the Maple commands

```
>eq1:=2*x1-3*x2+x3-4=0;
>eq2:=2*x1+x2-x3+4=0;
>eq3:=x1^2+x2^2+x3^2-4=0;
```

The third equation describes a sphere of radius 2 and center (0, 0, 0), so `x1`, `x2`, and `x3` are in $[-2, 2]$. The Maple commands to obtain the graph in this case is

```
>with(plots);
>implicitplot3d({eq1,eq2,eq3},x1=-2..2,x2=-2..2,x3=-2..2);
```

Various three-dimensional plotting options are available in Maple for isolating a solution to the nonlinear system. For example, we can rotate the graph to better view the sections of the surfaces. Then we can zoom into regions where the intersections lie and alter the display form of the axes for a more accurate view of the intersection's coordinates. For this problem a reasonable initial approximation is $(x_1, x_2, x_3) = (-0.5, -1.5, 1.5)$.

EXERCISE SET 10.2

1. Use Newton's method with $\mathbf{x}^{(0)} = \mathbf{0}$ to compute $\mathbf{x}^{(2)}$ for each of the following nonlinear systems.

   a. $$4x_1^2 - 20x_1 + \frac{1}{4}x_2^2 + 8 = 0,$$
   $$\frac{1}{2}x_1x_2^2 + 2x_1 - 5x_2 + 8 = 0.$$

   b. $$\sin(4\pi x_1 x_2) - 2x_2 - x_1 = 0,$$
   $$\left(\frac{4\pi - 1}{4\pi}\right)(e^{2x_1} - e) + 4ex_2^2 - 2ex_1 = 0.$$

   c. $$3x_1 - \cos(x_2x_3) - \frac{1}{2} = 0,$$
   $$4x_1^2 - 625x_2^2 + 2x_2 - 1 = 0,$$
   $$e^{-x_1x_2} + 20x_3 + \frac{10\pi - 3}{3} = 0.$$

   d. $$x_1^2 + x_2 - 37 = 0,$$
   $$x_1 - x_2^2 - 5 = 0,$$
   $$x_1 + x_2 + x_3 - 3 = 0.$$

2. Use the graphing facilities in Maple to approximate solutions to the following nonlinear systems.

   a. $$x_1(1 - x_1) + 4x_2 = 12,$$
   $$(x_1 - 2)^2 + (2x_2 - 3)^2 = 25.$$

   b. $$5x_1^2 - x_2^2 = 0,$$
   $$x_2 - 0.25(\sin x_1 + \cos x_2) = 0.$$

c. $15x_1 + x_2^2 - 4x_3 = 13,$
$x_1^2 + 10x_2 - x_3 = 11,$
$x_2^3 - 25x_3 = -22.$

d. $10x_1 - 2x_2^2 + x_2 - 2x_3 - 5 = 0,$
$8x_2^2 + 4x_3^2 - 9 = 0,$
$8x_2x_3 + 4 = 0.$

3. Use Newton's method to find a solution to the following nonlinear systems with the given initial approximation. Iterate until $\|\mathbf{x}^{(k)} - \mathbf{x}^{(k-1)}\|_\infty < 10^{-6}$.

a. $3x_1^2 - x_2^2 = 0,$
$3x_1x_2^2 - x_1^3 - 1 = 0.$
Use $\mathbf{x}^{(0)} = (1, 1)^t$.

b. $\ln\left(x_1^2 + x_2^2\right) - \sin(x_1x_2) = \ln 2 + \ln \pi,$
$e^{x_1 - x_2} + \cos(x_1x_2) = 0.$
Use $\mathbf{x}^{(0)} = (2, 2)^t$.

c. $x_1^3 + x_1^2x_2 - x_1x_3 + 6 = 0,$
$e^{x_1} + e^{x_2} - x_3 = 0,$
$x_2^2 - 2x_1x_3 = 4.$
Use $\mathbf{x}^{(0)} = (-1, -2, 1)^t$.

d. $6x_1 - 2\cos(x_2x_3) - 1 = 0,$
$9x_2 + \sqrt{x_1^2 + \sin x_3 + 1.06} + 0.9 = 0,$
$60x_3 + 3e^{-x_1x_2} + 10\pi - 3 = 0.$
Use $\mathbf{x}^{(0)} = (0, 0, 0)^t$.

4. Use the answers obtained in Exercise 2 as initial approximations to Newton's method. Iterate until $\|\mathbf{x}^{(k)} - \mathbf{x}^{(k-1)}\|_\infty < 10^{-6}$.

5. The nonlinear system

$$3x_1 - \cos(x_2x_3) - \frac{1}{2} = 0,$$
$$x_1^2 - 625x_2^2 - \frac{1}{4} = 0,$$
$$e^{-x_1x_2} + 20x_3 + \frac{10\pi - 3}{3} = 0$$

has a singular Jacobian matrix at the solution. Apply Newton's method with $\mathbf{x}^{(0)} = (1, 1 - 1)^t$. Note that convergence may be slow or may not occur within a reasonable number of iterations.

6. The nonlinear system

$$4x_1 - x_2 + x_3 = x_1x_4,$$
$$-x_1 + 3x_2 - 2x_3 = x_2x_4,$$
$$x_1 - 2x_2 + 3x_3 = x_3x_4,$$
$$x_1^2 + x_2^2 + x_3^2 = 1$$

has three solutions. Use Newton's method to approximate all three solutions. Iterate until $\|\mathbf{x}^{(k)} - \mathbf{x}^{(k-1)}\|_\infty < 10^{-5}$.

7. C. Chiarella, W. Charlton, and A. W. Roberts [CCR], in calculating the shape of a gravity-flow discharge chute that will minimize transit time of discharged granular particles, solve the following equations by Newton's method:

(i) $f_n(\theta_1, \ldots, \theta_N) = \dfrac{\sin \theta_{n+1}}{v_{n+1}}(1-\mu w_{n+1}) - \dfrac{\sin \theta_n}{v_n}(1-\mu w_n) = 0$, for each $n = 1, 2, \ldots, N-1$.

(ii) $f_N(\theta_1, \ldots, \theta_N) = \Delta y \sum_{i=1}^N \tan \theta_i - X = 0$, where

a. $v_n^2 = v_0^2 - 2gn\Delta y - 2\mu\Delta y \sum_{j=1}^n \dfrac{1}{\cos \theta_j},$ for each $n = 1, 2, \ldots, N$, and

b. $w_n = -\Delta y v_n \sum_{i=1}^N \dfrac{1}{v_i^3 \cos \theta_i},$ for each $n = 1, 2, \ldots, N$.

The constant v_0 is the initial velocity of the granular material, X is the x-coordinate of the end of the chute, μ is the friction force, N is the number of chute segments, and g is the gravitational constant. The variable θ_i is the angle of the ith chute segment from the vertical as shown in the following figure, and v_i is the particle velocity in the ith chute segment. Solve (i) and (ii) for $\boldsymbol{\theta} = (\theta_1, \ldots, \theta_N)^t$ with $\mu = 0, X = 2, \Delta y = 0.2, N = 20, v_0 = 0$, and $g = -32.17$ ft/s^2, where the values for v_n and w_n can be obtained directly from (a) and (b). Iterate until $||\boldsymbol{\theta}^{(k)} - \boldsymbol{\theta}^{(k-1)}||_\infty < 10^{-2}$.

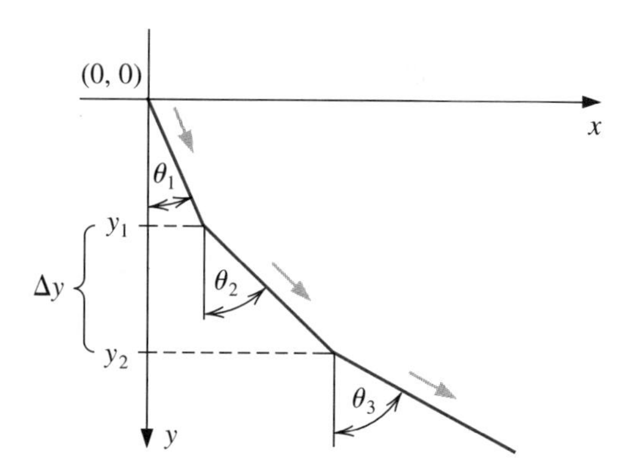

8. The amount of pressure required to sink a large, heavy object in a soft homogeneous soil that lies above a hard base soil can be predicted by the amount of pressure required to sink smaller objects in the same soil. Specifically, the amount of pressure p required to sink a circular plate of radius r a distance d in the soft soil, where the hard base soil lies a distance $D > d$ below the surface, can be approximated by an equation of the form

$$p = k_1 e^{k_2 r} + k_3 r,$$

where k_1, k_2, and k_3 are constants, with $k_2 > 0$, depending on d and the consistency of the soil but not on the radius of the plate. (See [Bek, pp. 89–94].)

a. Find the values of k_1, k_2, and k_3 if we assume that a plate of radius 1 in. requires a pressure of 10 lb/in.^{2} to sink 1 ft in a muddy field, a plate of radius 2 in. requires a pressure of 12 lb/in.^{2} to sink 1 ft, and a plate of radius 3 in. requires a pressure of 15 lb/in.^{2} to sink this distance (assuming that the mud is more than 1 ft deep).

b. Use your calculations from part (a) to predict the minimal size of circular plate that would be required to sustain a load of 500 lb on this field with sinkage of less than 1 ft.

9. An interesting biological experiment (see [Schr2]) concerns the determination of the maximum water temperature, X_M, at which various species of hydra can survive without shortened life expectancy. One approach to the solution of this problem uses a weighted least squares fit of the form $f(x) = y = a/(x - b)^c$ to a collection of experimental data. The x-values of the data refer to water temperature. The constant b is the asymptote of the graph of f and as such is an approximation to X_M.

a. Show that choosing a, b, and c to minimize

$$\sum_{i=1}^{n} \left[w_i y_i - \frac{a}{(x_i - b)^c} \right]^2$$

reduces to solving the nonlinear system

$$a = \left(\sum_{i=1}^{n} \frac{w_i y_i}{(x_i - b)^c}\right) \Big/ \left(\sum_{i=1}^{n} \frac{1}{(x_i - b)^{2c}}\right),$$

$$0 = \sum_{i=1}^{n} \frac{w_i y_i}{(x_i - b)^c} \cdot \sum_{i=1}^{n} \frac{1}{(x_i - b)^{2c+1}} - \sum_{i=1}^{n} \frac{w_i y_i}{(x_i - b)^{c+1}} \cdot \sum_{i=1}^{n} \frac{1}{(x_i - b)^{2c}},$$

$$0 = \sum_{i=1}^{n} \frac{w_i y_i}{(x_i - b)^c} \cdot \sum_{i=1}^{n} \frac{\ln(x_i - b)}{(x_i - b)^{2c}} - \sum_{i=1}^{n} \frac{w_i y_i \ln(x_i - b)}{(x_i - b)^c} \cdot \sum_{i=1}^{n} \frac{1}{(x_i - b)^{2c}}.$$

b. Solve the nonlinear system for the species with the following data. Use the weights $w_i = \ln y_i$.

i	1	2	3	4
y_i	2.40	3.80	4.75	21.60
x_i	31.8	31.5	31.2	30.2

10.3 Quasi-Newton Methods

A significant weakness of Newton's method for solving systems of nonlinear equations lies in the requirement that, at each iteration, a Jacobian matrix be computed and an $n \times n$ linear system solved that involves this matrix. To illustrate the magnitude of this weakness, let us consider the amount of computation associated with one iteration of Newton's method. The Jacobian matrix associated with a system of n nonlinear equations written in the form $\mathbf{F}(\mathbf{x}) = \mathbf{0}$ requires that the n^2 partial derivatives of the n component functions of $\mathbf{F}$ be determined and evaluated. In most situations, the exact evaluation of the partial derivatives is inconvenient, although the problem has been made more tractable with the widespread use of symbolic computation systems, such as Maple.

When the exact evaluation is not practical, we can use finite difference approximations to the partial derivatives. For example,

(10.10)
$$\frac{\partial f_j}{\partial x_k}(\mathbf{x}^{(i)}) \approx \frac{f_j(\mathbf{x}^{(i)} + \mathbf{e}_k h) - f_j(\mathbf{x}^{(i)})}{h},$$

where h is small in absolute value and $\mathbf{e}_k$ is the vector whose only nonzero entry is a 1 in the kth coordinate. This approximation, however, still requires that at least n^2 scalar functional evaluations be performed to approximate the Jacobian and does not decrease the amount of calculation, in general $O(n^3)$, required for solving the linear system involving this approximate Jacobian. The total computational effort for just one iteration of Newton's method is consequently at least $n^2 + n$ scalar functional evaluations (n^2 for the evaluation of the Jacobian matrix and n for the evaluation of $\mathbf{F}$) together with $O(n^3)$ arithmetic operations to solve the linear system. This amount of computational effort is extensive, except for relatively small values of n and easily evaluated scalar functions.

In this section, we consider a generalization of the Secant method to systems of nonlinear equations, a technique known as **Broyden's method** (see [Broy]). The method requires only n scalar functional evaluations per iteration and also reduces the number of arithmetic calculations to $O(n^2)$. It belongs to a class of methods known as *least-change secant updates* that produce algorithms called **quasi-Newton**. These methods replace the Jacobian matrix in Newton's method with an approximation matrix that is updated at each iteration. The disadvantage of the methods is that the quadratic convergence of Newton's method is lost, being replaced in general by a convergence called **superlinear**, which implies that

$$\lim_{i\to\infty} \frac{\|\mathbf{x}^{(i+1)} - \mathbf{p}\|}{\|\mathbf{x}^{(i)} - \mathbf{p}\|} = 0,$$

where $\mathbf{p}$ denotes the solution to $\mathbf{F}(\mathbf{x}) = \mathbf{0}$ and $\mathbf{x}^{(i)}$ and $\mathbf{x}^{(i+1)}$ are consecutive approximations to $\mathbf{p}$.

In most applications the reduction to superlinear convergence is a more than acceptable trade-off for the decrease in the amount of computation. An additional disadvantage of quasi-Newton methods is that, unlike Newton's method, they are not self-correcting. Newton's method will generally correct for roundoff error with successive iterations, but unless special safeguards are incorporated, Broyden's method will not.

Suppose that an initial approximation $\mathbf{x}^{(0)}$ is given to the solution $\mathbf{p}$ of $\mathbf{F}(\mathbf{x}) = \mathbf{0}$. We calculate the next approximation $\mathbf{x}^{(1)}$ in the same manner as Newton's method, or, if it is inconvenient to determine $J(\mathbf{x}^{(0)})$ exactly, we use the difference equations given by (10.10) to approximate the partial derivatives. To compute $\mathbf{x}^{(2)}$, however, we depart from Newton's method and examine the Secant method for a single nonlinear equation. The Secant method uses the approximation

$$f'(x_1) \approx \frac{f(x_1) - f(x_0)}{x_1 - x_0}$$

as a replacement for $f'(x_1)$ in Newton's method. For nonlinear systems, $\mathbf{x}^{(1)} - \mathbf{x}^{(0)}$ is a vector, and the corresponding quotient is undefined. However, the method proceeds similarly in that we replace the matrix $J(\mathbf{x}^{(1)})$ in Newton's method by a matrix A_1 with the property that

$$A_1(\mathbf{x}^{(1)} - \mathbf{x}^{(0)}) = \mathbf{F}(\mathbf{x}^{(1)}) - \mathbf{F}(\mathbf{x}^{(0)}). \tag{10.11}$$

Any nonzero vector in $\mathbb{R}^n$ can be written as the sum of a multiple of $\mathbf{x}^{(1)} - \mathbf{x}^{(0)}$ and a multiple of a vector in the orthogonal complement of $\mathbf{x}^{(1)} - \mathbf{x}^{(0)}$. So to uniquely define the matrix A_1, we need to specify how it acts on the orthogonal complement of $\mathbf{x}^{(1)} - \mathbf{x}^{(0)}$. Since no information is available about the change in $\mathbf{F}$ in a direction orthogonal to $\mathbf{x}^{(1)} - \mathbf{x}^{(0)}$, we require that

$$A_1\mathbf{z} = J\left(\mathbf{x}^{(0)}\right)\mathbf{z}, \quad \text{whenever } \left(\mathbf{x}^{(1)} - \mathbf{x}^{(0)}\right)^t \mathbf{z} = 0. \tag{10.12}$$

This condition specifies that any vector orthogonal to $\mathbf{x}^{(1)} - \mathbf{x}^{(0)}$ is unaffected by the update from $J(\mathbf{x}^{(0)})$, which was used to compute $\mathbf{x}^{(1)}$, to A_1, which is used in the determination of $\mathbf{x}^{(2)}$.

Conditions (10.11) and (10.12) uniquely define A_1 (see [DM]) as

$$A_1 = J\left(\mathbf{x}^{(0)}\right) + \frac{\left[\mathbf{F}\left(\mathbf{x}^{(1)}\right) - \mathbf{F}\left(\mathbf{x}^{(0)}\right) - J\left(\mathbf{x}^{(0)}\right)\left(\mathbf{x}^{(1)} - \mathbf{x}^{(0)}\right)\right]\left(\mathbf{x}^{(1)} - \mathbf{x}^{(0)}\right)^t}{||\mathbf{x}^{(1)} - \mathbf{x}^{(0)}||_2^2}.$$

It is this matrix that is used in place of $J(\mathbf{x}^{(1)})$ to determine $\mathbf{x}^{(2)}$ as

$$\mathbf{x}^{(2)} = \mathbf{x}^{(1)} - A_1^{-1}\mathbf{F}\left(\mathbf{x}^{(1)}\right).$$

The method is then repeated to determine $\mathbf{x}^{(3)}$, using A_1 in place of $A_0 \equiv J(\mathbf{x}^{(0)})$ and with $\mathbf{x}^{(2)}$ and $\mathbf{x}^{(1)}$ in place of $\mathbf{x}^{(1)}$ and $\mathbf{x}^{(0)}$. In general, once $\mathbf{x}^{(i)}$ has been determined, $\mathbf{x}^{(i+1)}$ is computed by

(10.13) $$A_i = A_{i-1} + \frac{\mathbf{y}_i - A_{i-1}\mathbf{s}_i}{||\mathbf{s}_i||_2^2}\mathbf{s}_i^t,$$

and

(10.14) $$\mathbf{x}^{(i+1)} = \mathbf{x}^{(i)} - A_i^{-1}\mathbf{F}\left(\mathbf{x}^{(i)}\right),$$

where the notation $\mathbf{y}_i = \mathbf{F}\left(\mathbf{x}^{(i)}\right) - \mathbf{F}\left(\mathbf{x}^{(i-1)}\right)$ and $\mathbf{s}_i = \mathbf{x}^{(i)} - \mathbf{x}^{(i-1)}$ is introduced into (10.13) to simplify the equations.

If the method is performed as outlined in Eqs. (10.13) and (10.14), the number of scalar functional evaluations is reduced from $n^2 + n$ to n (those required for evaluating $\mathbf{F}\left(\mathbf{x}^{(i)}\right)$), but $O(n^3)$ calculations are still required to solve the associated $n \times n$ linear system (see Step 4 in Algorithm 10.1)

(10.15) $$A_i\mathbf{y}_i = -\mathbf{F}\left(\mathbf{x}^{(i)}\right).$$

Employing the method in this form would not be justified because of the reduction to superlinear convergence from the quadratic convergence of Newton's method.

A considerable improvement can be incorporated, however, by employing a matrix inversion formula of Sherman and Morrison (see, for example, [DM, p. 55]). This result states that if A is a nonsingular matrix and $\mathbf{x}$ and $\mathbf{y}$ are vectors, then $A + \mathbf{x}\mathbf{y}^t$ is nonsingular provided that $\mathbf{y}^t A^{-1}\mathbf{x} \neq -1$, and

(10.16) $$(A + \mathbf{x}\mathbf{y}^t)^{-1} = A^{-1} - \frac{A^{-1}\mathbf{x}\mathbf{y}^t A^{-1}}{1 + \mathbf{y}^t A^{-1}\mathbf{x}}.$$

This formula permits A_i^{-1} to be computed directly from A_{i-1}^{-1}, eliminating the need for a matrix inversion with each iteration. By letting $A = A_{i-1}$, $\mathbf{x} = (\mathbf{y}_i - A_{i-1}\mathbf{s}_i)/||\mathbf{s}_i||_2^2$, and

$\mathbf{y} = \mathbf{s}_i$, Eq. (10.13) together with Eq. (10.16) implies that

$$A_i^{-1} = \left(A_{i-1} + \frac{\mathbf{y}_i - A_{i-1}\mathbf{s}_i}{||\mathbf{s}_i||_2^2}\mathbf{s}_i^t\right)^{-1}$$

$$= A_{i-1}^{-1} - \frac{A_{i-1}^{-1}\left(\frac{\mathbf{y}_i - A_{i-1}\mathbf{s}_i}{||\mathbf{s}_i||_2^2}\mathbf{s}_i^t\right)A_{i-1}^{-1}}{1 + \mathbf{s}_i^t A_{i-1}^{-1}\left(\frac{\mathbf{y}_i - A_{i-1}\mathbf{s}_i}{||\mathbf{s}_i||_2^2}\right)}$$

$$= A_{i-1}^{-1} - \frac{(A_{i-1}^{-1}\mathbf{y}_i - \mathbf{s}_i)\mathbf{s}_i^t A_{i-1}^{-1}}{||\mathbf{s}_i||_2^2 + \mathbf{s}_i^t A_{i-1}^{-1}\mathbf{y}_i - ||\mathbf{s}_i||_2^2},$$

so

(10.17)
$$A_i^{-1} = A_{i-1}^{-1} + \frac{(\mathbf{s}_i - A_{i-1}^{-1}\mathbf{y}_i)\mathbf{s}_i^t A_{i-1}^{-1}}{\mathbf{s}_i^t A_{i-1}^{-1}\mathbf{y}_i}.$$

This computation involves only matrix-vector multiplication at each step and therefore requires only $O(n^2)$ arithmetic calculations. The calculation of A_i is bypassed, as is the necessity of solving the linear system (10.15). Algorithm 10.2 follows directly from this construction, incorporating (10.17) into the iterative technique (10.14).

ALGORITHM 10.2

Broyden

To approximate the solution of the nonlinear system $\mathbf{F}(\mathbf{x}) = \mathbf{0}$ given an initial approximation $\mathbf{x}$:

INPUT number n of equations and unknowns; initial approximation $\mathbf{x} = (x_1, \ldots, x_n)^t$; tolerance TOL; maximum number of iterations N.

OUTPUT approximate solution $\mathbf{x} = (x_1, \ldots, x_n)^t$ or a message that the number of iterations was exceeded.

Step 1 Set $A_0 = J(\mathbf{x})$ where $J(\mathbf{x})_{i,j} = \partial f_i(\mathbf{x})/\partial x_j$ for $1 \le i, j \le n$;
$\mathbf{v} = \mathbf{F}(\mathbf{x})$. (*Note:* $\mathbf{v} = \mathbf{F}(\mathbf{x}^{(0)})$.)

Step 2 Set $A = A_0^{-1}$. (*Use Gaussian elimination.*)

Step 3 Set $\mathbf{s} = -A\mathbf{v}$; (*Note:* $\mathbf{s} = \mathbf{s}_1$.)
$\mathbf{x} = \mathbf{x} + \mathbf{s}$; (*Note:* $\mathbf{x} = \mathbf{x}^{(1)}$.)
$k = 2$.

Step 4 While ($k \le N$) do Steps 5–13.

Step 5 Set $\mathbf{w} = \mathbf{v}$; (*Save* $\mathbf{v}$.)
$\mathbf{v} = \mathbf{F}(\mathbf{x})$; (*Note:* $\mathbf{v} = \mathbf{F}(\mathbf{x}^{(k)})$.)
$\mathbf{y} = \mathbf{v} - \mathbf{w}$. (*Note:* $\mathbf{y} = \mathbf{y}_k$.)

Step 6 Set $\mathbf{z} = -A\mathbf{y}$. (*Note:* $\mathbf{z} = -A_{k-1}^{-1}\mathbf{y}_k$.)

Step 7 Set $p = -\mathbf{s}^t\mathbf{z}$. (*Note:* $p = \mathbf{s}_k^t A_{k-1}^{-1}\mathbf{y}_k$.)

Step 8 Set $\mathbf{u}^t = \mathbf{s}^t A$.

Step 9 Set $A = A + \frac{1}{p}(\mathbf{s} + \mathbf{z})\mathbf{u}^t$. (*Note:* $A = A_k^{-1}$.)

Step 10 Set $\mathbf{s} = -A\mathbf{v}$. (*Note:* $\mathbf{s} = -A_k^{-1}\mathbf{F}(\mathbf{x}^{(k)})$.)

Step 11 Set $\mathbf{x} = \mathbf{x} + \mathbf{s}$. (*Note:* $\mathbf{x} = \mathbf{x}^{(k+1)}$.)

Step 12 If $||\mathbf{s}|| < TOL$ then OUTPUT ($\mathbf{x}$);
(*Procedure completed successfully.*)
STOP.

Step 13 Set $k = k + 1$.

Step 14 OUTPUT ('Maximum number of iterations exceeded');
(*Procedure completed unsuccessfully.*)
STOP.

EXAMPLE 1 The nonlinear system

$$3x_1 - \cos(x_2x_3) - \frac{1}{2} = 0,$$

$$x_1^2 - 81(x_2 + 0.1)^2 + \sin x_3 + 1.06 = 0,$$

$$e^{-x_1x_2} + 20x_3 + \frac{10\pi - 3}{3} = 0$$

was solved by Newton's method in Example 1 of Section 10.2. The Jacobian matrix for this system is

$$J(x_1, x_2, x_3) = \begin{bmatrix} 3 & x_3 \sin x_2x_3 & x_2 \sin x_2x_3 \\ 2x_1 & -162(x_2 + 0.1) & \cos x_3 \\ -x_2e^{-x_1x_2} & -x_1e^{-x_1x_2} & 20 \end{bmatrix}.$$

With $\mathbf{x}^{(0)} = (0.1, 0.1, -0.1)^t$ we have

$$\mathbf{F}\left(\mathbf{x}^{(0)}\right) = \begin{bmatrix} -1.199950 \\ -2.269833 \\ 8.462025 \end{bmatrix}.$$

Since

$$A_0 = J\left(x_1^{(0)}, x_2^{(0)}, x_3^{(0)}\right)$$
$$= \begin{bmatrix} 3 & 9.999833 \times 10^{-4} & -9.999833 \times 10^{-4} \\ 0.2 & -32.4 & 0.9950042 \\ -9.900498 \times 10^{-2} & -9.900498 \times 10^{-2} & 20 \end{bmatrix},$$

we have

$$A_0^{-1} = J\left(x_1^{(0)}, x_2^{(0)}, x_3^{(0)}\right)^{-1}$$

$$= \begin{bmatrix} 0.3333332 & 1.023852 \times 10^{-5} & 1.615701 \times 10^{-5} \\ 2.108607 \times 10^{-3} & -3.086883 \times 10^{-2} & 1.535836 \times 10^{-3} \\ 1.660520 \times 10^{-3} & -1.527577 \times 10^{-4} & 5.000768 \times 10^{-2} \end{bmatrix}.$$

So

$$\mathbf{x}^{(1)} = \mathbf{x}^{(0)} - A_0^{-1}\mathbf{F}\left(\mathbf{x}^{(0)}\right) = \begin{bmatrix} 0.4998697 \\ 1.946685 \times 10^{-2} \\ -0.5215205 \end{bmatrix},$$

$$\mathbf{F}\left(\mathbf{x}^{(1)}\right) = \begin{bmatrix} -3.394465 \times 10^{-4} \\ -0.3443879 \\ 3.188238 \times 10^{-2} \end{bmatrix},$$

$$\mathbf{y}_1 = \mathbf{F}\left(\mathbf{x}^{(1)}\right) - \mathbf{F}\left(\mathbf{x}^{(0)}\right) = \begin{bmatrix} 1.199611 \\ 1.925445 \\ -8.430143 \end{bmatrix},$$

$$\mathbf{s}_1 = \begin{bmatrix} 0.3998697 \\ -8.053315 \times 10^{-2} \\ -0.4215204 \end{bmatrix},$$

$$\mathbf{s}_1^t A_0^{-1}\mathbf{y}_1 = 0.3424604,$$

$$A_1^{-1} = A_0^{-1} + (1/0.3424604)[(\mathbf{s}_1 - A_0^{-1}\mathbf{y}_1)\mathbf{s}_1^t A_0^{-1}]$$

$$= \begin{bmatrix} 0.3333781 & 1.11050 \times 10^{-5} & 8.967344 \times 10^{-6} \\ -2.021270 \times 10^{-3} & -3.094849 \times 10^{-2} & 2.196906 \times 10^{-3} \\ 1.022214 \times 10^{-3} & -1.650709 \times 10^{-4} & 5.010986 \times 10^{-2} \end{bmatrix},$$

and

$$\mathbf{x}^{(2)} = \mathbf{x}^{(1)} - A_1^{-1}\mathbf{F}\left(\mathbf{x}^{(1)}\right) = \begin{bmatrix} 0.4999863 \\ 8.737833 \times 10^{-3} \\ -0.5231746 \end{bmatrix}.$$

Additional iterations are listed in Table 10.4. ∎

Table 10.4

k	$x_1^{(k)}$	$x_2^{(k)}$	$x_3^{(k)}$	$\|\mathbf{x}^{(k)} - \mathbf{x}^{(k-1)}\|_2$
3	0.5000066	8.672157×10^{-4}	−0.5236918	7.88×10^{-3}
4	0.5000003	6.083352×10^{-5}	−0.5235954	8.12×10^{-4}
5	0.5000000	-1.448889×10^{-6}	−0.5235989	6.24×10^{-5}
6	0.5000000	6.059030×10^{-9}	−0.5235988	1.50×10^{-6}

Procedures are also available that maintain quadratic convergence but significantly reduce the number of required functional evaluations. Methods of this type were originally proposed by Brown [Brow,K]. A survey and comparison of some commonly used methods of this type can be found in [MC]. In general, however, these methods are much more difficult to implement efficiently than Broyden's method.

EXERCISE SET 10.3

1. Use Broyden's method with $\mathbf{x}^{(0)} = \mathbf{0}$ to compute $\mathbf{x}^{(2)}$ for each of the following nonlinear systems.

a.
$$4x_1^2 - 20x_1 + \frac{1}{4}x_2^2 + 8 = 0,$$
$$\frac{1}{2}x_1x_2^2 + 2x_1 - 5x_2 + 8 = 0.$$

b.
$$\sin(4\pi x_1x_2) - 2x_2 - x_1 = 0,$$
$$\left(\frac{4\pi - 1}{4\pi}\right)\left(e^{2x_1} - e\right) + 4ex_2^2 - 2ex_1 = 0.$$

c.
$$3x_1 - \cos(x_2x_3) - \frac{1}{2} = 0,$$
$$4x_1^2 - 625x_2^2 + 2x_2 - 1 = 0,$$
$$e^{-x_1x_2} + 20x_3 + \frac{10\pi - 3}{3} = 0.$$

d.
$$x_1^2 + x_2 - 37 = 0,$$
$$x_1 - x_2^2 - 5 = 0,$$
$$x_1 + x_2 + x_3 - 3 = 0.$$

2. Use Broyden's method to approximate solutions to the nonlinear systems in Exercise 1. Iterate until $\|\mathbf{x}^{(k)} - \mathbf{x}^{(k-1)}\|_\infty < 10^{-6}$. The initial approximations $\mathbf{x}^{(0)}$ in Exercise 1 may not lead to convergence. If not, use a different value of $\mathbf{x}^{(0)}$.

3. Use Broyden's method to find a solution to the following nonlinear systems. Iterate until $\|\mathbf{x}^{(k)} - \mathbf{x}^{(k-1)}\|_\infty < 10^{-6}$.

a.
$$3x_1^2 - x_2^2 = 0,$$
$$3x_1x_2^2 - x_1^3 - 1 = 0.$$
Use $\mathbf{x}^{(0)} = (1, 1)^t$.

b.
$$\ln\left(x_1^2 + x_2^2\right) - \sin(x_1x_2) = \ln 2 + \ln \pi,$$
$$e^{x_1 - x_2} + \cos(x_1x_2) = 0.$$
Use $\mathbf{x}^{(0)} = (2, 2)^t$.

c.
$$x_1^3 + x_1^2x_2 - x_1x_3 + 6 = 0,$$
$$e^{x_1} + e^{x_2} - x_3 = 0,$$
$$x_2^2 - 2x_1x_3 = 4.$$
Use $\mathbf{x}^{(0)} = (-1, -2, 1)^t$.

d.
$$6x_1 - 2\cos(x_2x_3) - 1 = 0,$$
$$9x_2 + \sqrt{x_1^2 + \sin x_3 + 1.06} + 0.9 = 0,$$
$$60x_3 + 3e^{-x_1x_2} + 10\pi - 3 = 0.$$
Use $\mathbf{x}^{(0)} = (0, 0, 0)^t$.

4. Use Broyden's method to approximate solutions to the following nonlinear systems. Iterate until $\|\mathbf{x}^{(k)} - \mathbf{x}^{(k-1)}\|_\infty < 10^{-6}$.

a.
$$x_1(1 - x_1) + 4x_2 = 12,$$
$$(x_1 - 2)^2 + (2x_2 - 3)^2 = 25.$$

b.
$$5x_1^2 - x_2^2 = 0,$$
$$x_2 - 0.25(\sin x_1 + \cos x_2) = 0.$$

c.
$$15x_1 + x_2^2 - 4x_3 = 13,$$
$$x_1^2 + 10x_2 - x_3 = 11,$$
$$x_2^3 - 25x_3 = -22.$$

d.
$$10x_1 - 2x_2^2 + x_2 - 2x_3 - 5 = 0,$$
$$8x_2^2 + 4x_3^2 - 9 = 0,$$
$$8x_2x_3 + 4 = 0.$$

5. The nonlinear system

$$3x_1 - \cos(x_2 x_3) - \frac{1}{2} = 0,$$

$$x_1^2 - 625x_2^2 - \frac{1}{4} = 0,$$

$$e^{-x_1 x_2} + 20x_3 + \frac{10\pi - 3}{3} = 0$$

has a singular Jacobian matrix at the solution. Apply Broyden's method with $\mathbf{x}^{(0)} = (1, 1 - 1)^t$. Note that convergence may be slow or may not occur within a reasonable number of iterations.

6. The nonlinear system

$$4x_1 - x_2 + x_3 = x_1 x_4,$$

$$-x_1 + 3x_2 - 2x_3 = x_2 x_4,$$

$$x_1 - 2x_2 + 3x_3 = x_3 x_4,$$

$$x_1^2 + x_2^2 + x_3^2 = 1$$

has three solutions. Use Broyden's method to approximate all three solutions. Iterate until $\|\mathbf{x}^{(k)} - \mathbf{x}^{(k-1)}\|_\infty < 10^{-5}$.

7. Exercise 13 of Section 8.1 dealt with determining an exponential least squares relationship of the form $R = bw^a$ to approximate a collection of data relating the weight and respiration rule of *Modest sphinx* moths. In that exercise, the problem was converted to a log-log relationship, and in part (c), a quadratic term was introduced in an attempt to improve the approximation. Instead of converting the problem, determine the constants a and b that minimize $\sum_{i=1}^{n}(R_i - bw_i^a)^2$ for the data listed in Exercise 13 of 8.1. Compute the error associated with this approximation, and compare this to the error of the previous approximations for this problem.

8. Show that if $\mathbf{0} \neq \mathbf{y} \in \mathbb{R}^n$ and $\mathbf{z} \in \mathbb{R}^n$, then $\mathbf{z} = \mathbf{z}_1 + \mathbf{z}_2$, where $\mathbf{z}_1 = \dfrac{\mathbf{y}^t\mathbf{z}}{\|\mathbf{y}\|_2^2}\mathbf{y}$ is parallel to $\mathbf{y}$ and $\mathbf{z}_2$ is orthogonal to $\mathbf{y}$.

9. Show that if $\mathbf{u}, \mathbf{v} \in \mathbb{R}^n$, then $\det(I + \mathbf{u}\mathbf{v}^t) = 1 + \mathbf{v}^t\mathbf{u}$.

10. **a.** Use the result in Exercise 9 to show that if A^{-1} exists and $\mathbf{x}, \mathbf{y} \in \mathbb{R}^n$, then $(A + \mathbf{x}\mathbf{y}^t)^{-1}$ exists if and only if $\mathbf{y}^t A^{-1}\mathbf{x} \neq -1$.

    b. By multiplying on the right by $A + \mathbf{x}\mathbf{y}^t$, show that when $\mathbf{y}^t A^{-1}\mathbf{x} \neq -1$ we have

$$(A + \mathbf{x}\mathbf{y}^t)^{-1} = A^{-1} - \frac{A^{-1}\mathbf{x}\mathbf{y}^t A^{-1}}{1 + \mathbf{y}^t A^{-1}\mathbf{x}}.$$

10.4 Steepest Descent Techniques

The advantage of the Newton and quasi-Newton methods for solving systems of nonlinear equations is their speed of convergence once a sufficiently accurate approximation is known. A weakness of these methods is that an accurate initial approximation to the solution is needed to ensure convergence. The **Steepest Descent** method considered in this section converges only linearly to the solution, but it will usually converge even for poor initial approximations. As a consequence, this method is used to find sufficiently accurate starting approximations for the Newton-based techniques, in the same way the Bisection method is used for a single equation.

The method of Steepest Descent determines a local minimum for a multivariable function of the form $g : \mathbb{R}^n \to \mathbb{R}$. Although the method is valuable quite apart from the application as a starting method for solving nonlinear systems, we restrict our discussion to that situation. (Some other applications are considered in the exercises.)

The connection between the minimization of a function from $\mathbb{R}^n$ to $\mathbb{R}$ and the solution of a system of nonlinear equations is due to the fact that a system of the form

$$\begin{aligned} f_1(x_1, x_2, \ldots, x_n) &= 0, \\ f_2(x_1, x_2, \ldots, x_n) &= 0, \\ \vdots \qquad & \quad \vdots \\ f_n(x_1, x_2, \ldots, x_n) &= 0, \end{aligned}$$

has a solution at $\mathbf{x} = (x_1, x_2, \ldots, x_n)^t$ precisely when the function g defined by

$$g(x_1, x_2, \ldots, x_n) = \sum_{i=1}^{n} [f_i(x_1, x_2, \ldots, x_n)]^2$$

has the minimal value zero.

The method of Steepest Descent for finding a local minimum for an arbitrary function g from $\mathbb{R}^n$ into $\mathbb{R}$ can be intuitively described as follows:

1. Evaluate g at an initial approximation $\mathbf{x}^{(0)} = \left(x_1^{(0)}, x_2^{(0)}, \ldots, x_n^{(0)}\right)^t$.
2. Determine a direction from $\mathbf{x}^{(0)}$ that results in a decrease in the value of g.
3. Move an appropriate amount in this direction and call the new vector $\mathbf{x}^{(1)}$.
4. Repeat steps 1 through 3 with $\mathbf{x}^{(0)}$ replaced by $\mathbf{x}^{(1)}$.

Before describing how to choose the correct direction and the appropriate distance to move in this direction, we need to review some results from calculus. The Extreme Value Theorem implies that a differentiable single-variable function can have a relative minimum only when the derivative is zero. To extend this result to multivariable functions, we need the following definition.

Definition 10.8 If $g : \mathbb{R}^n \to \mathbb{R}$, the **gradient** of g at $\mathbf{x} = (x_1, x_2, \ldots, x_n)^t$ is denoted $\nabla g(\mathbf{x})$ and defined by

$$\nabla g(\mathbf{x}) = \left(\frac{\partial g}{\partial x_1}(\mathbf{x}), \frac{\partial g}{\partial x_2}(\mathbf{x}), \ldots, \frac{\partial g}{\partial x_n}(\mathbf{x})\right)^t. \qquad \blacksquare$$

The gradient for a multivariable function is analogous to the derivative of a single-variable function in the sense that a differentiable multivariable function can have a relative minimum at $\mathbf{x}$ only when the gradient is zero.

The gradient has another important property connected with the minimization of multivariable functions. Suppose that $\mathbf{v} = (v_1, v_2, \ldots, v_n)^t$ is a unit vector in $\mathbb{R}^n$; that is,

$$||\mathbf{v}||_2^2 = \sum_{i=1}^{n} v_i^2 = 1.$$

The directional derivative of g at $\mathbf{x}$ in the direction of $\mathbf{v}$ is defined by

$$D_{\mathbf{v}} g(\mathbf{x}) = \lim_{h \to 0} \frac{1}{h} [g(\mathbf{x} + h\mathbf{v}) - g(\mathbf{x})] = \mathbf{v} \cdot \nabla g(\mathbf{x}).$$

The directional derivative of g at $\mathbf{x}$ in the direction of $\mathbf{v}$ measures the change in the value of the function g relative to the change in the variable in the direction of $\mathbf{v}$. (Figure 10.2 gives an illustration when g is a function of two variables.)

Figure 10.2

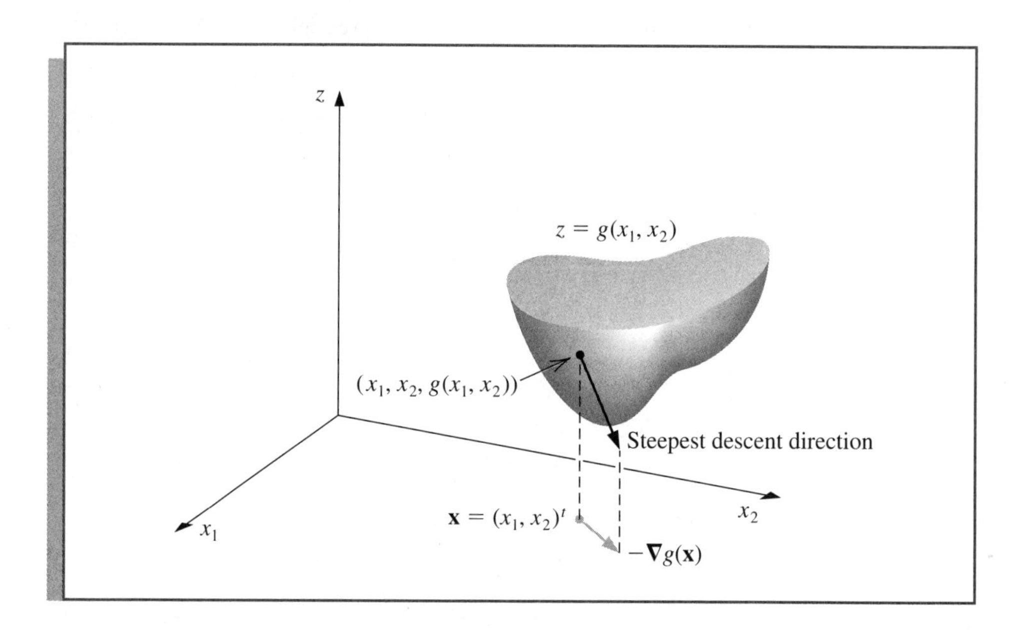

A standard result from the calculus of multivariable functions states that, when g is differentiable, the direction that produces the maximum value for the directional derivative occurs when $\mathbf{v}$ is chosen to be parallel to $\nabla g(\mathbf{x})$, provided that $\nabla g(\mathbf{x}) \neq \mathbf{0}$. As a consequence, the direction of greatest decrease in the value of g at $\mathbf{x}$ is the direction given by $-\nabla g(\mathbf{x})$. Since the object is to reduce $g(\mathbf{x})$ to its minimal value zero, an appropriate choice for $\mathbf{x}^{(1)}$ is

(10.18) $$\mathbf{x}^{(1)} = \mathbf{x}^{(0)} - \alpha \nabla g\left(\mathbf{x}^{(0)}\right), \quad \text{for some constant } \alpha > 0.$$

The problem now reduces to choosing α so that $g\left(\mathbf{x}^{(1)}\right)$ will be significantly less than

$g(\mathbf{x}^{(0)})$. To determine an appropriate choice for the value α, we consider the single-variable function

$$h(\alpha) = g\left(\mathbf{x}^{(0)} - \alpha \nabla g\left(\mathbf{x}^{(0)}\right)\right). \tag{10.19}$$

The value of α that minimizes h is the value needed for Eq. (10.18).

Finding a minimal value for h directly would require differentiating h and then solving a root-finding problem to determine the critical points of h. This procedure is generally too costly. Instead, we choose three numbers $\alpha_1 < \alpha_2 < \alpha_3$ that, we hope, are close to where the minimum value of $h(\alpha)$ occurs. Then we construct the quadratic polynomial $P(x)$ that interpolates h at α_1, α_2, and α_3. We define $\hat{\alpha}$ in $[\alpha_1, \alpha_3]$ so that $P(\hat{\alpha})$ is a minimum in $[\alpha_1, \alpha_3]$ and use $P(\hat{\alpha})$ to approximate the minimal value of $h(\alpha)$. Then $\hat{\alpha}$ is used to determine the new iterate for approximating the minimal value of g:

$$\mathbf{x}^{(1)} = \mathbf{x}^{(0)} - \hat{\alpha} \nabla g\left(\mathbf{x}^{(0)}\right).$$

Since $g\left(\mathbf{x}^{(0)}\right)$ is available, we first choose $\alpha_1 = 0$ to minimize the computation. Next a number α_3 is found with $h(\alpha_3) < h(\alpha_1)$. (Since α_1 does not minimize h, such a number α_3 does exist.) Finally, α_2 is chosen to be $\alpha_3/2$.

The minimum value $\hat{\alpha}$ of P on $[\alpha_1, \alpha_3]$ occurs either at the only critical point of P or at the right endpoint α_3, for, by assumption, $P(\alpha_3) = h(\alpha_3) < h(\alpha_1) = P(\alpha_1)$. The critical point is easily determined since P is a quadratic polynomial.

Algorithm 10.3 applies the method of Steepest Descent to approximate the minimal value of $g(\mathbf{x})$. To begin an iteration, the value 0 is assigned to α_1 and the value 1 is assigned to α_3. If $h(\alpha_3) \geq h(\alpha_1)$, then successive divisions of α_3 by 2 are performed and the value of α_3 is reassigned until $h(\alpha_3) < h(\alpha_1)$ and $\alpha_3 = 2^{-k}$ for some value of k.

To employ the method to approximate the solution to the system

$$\begin{aligned} f_1(x_1, x_2, \ldots, x_n) &= 0, \\ f_2(x_1, x_2, \ldots, x_n) &= 0, \\ \vdots \qquad & \quad \vdots \\ f_n(x_1, x_2, \ldots, x_n) &= 0, \end{aligned}$$

we replace the function g with $\sum_{i=1}^{n} f_i^2$.

ALGORITHM 10.3

Steepest Descent

To approximate a solution $\mathbf{p}$ to the minimization problem

$$g(\mathbf{p}) = \min_{\mathbf{x} \in \mathbb{R}^n} g(\mathbf{x})$$

given an initial approximation $\mathbf{x}$:

INPUT number n of variables; initial approximation $\mathbf{x} = (x_1, \ldots, x_n)^t$; tolerance TOL; maximum number of iterations N.

OUTPUT approximate solution $\mathbf{x} = (x_1, \ldots, x_n)^t$ or a message of failure.

Step 1 Set $k = 1$.

Step 2 While ($k \leq N$) do Steps 3–15.

Step 3 Set $g_1 = g(x_1, \ldots, x_n)$; (*Note:* $g_1 = g\left(\mathbf{x}^{(k)}\right)$.)
$\mathbf{z} = \nabla g(x_1, \ldots, x_n)$; (*Note:* $\mathbf{z} = \nabla g\left(\mathbf{x}^{(k)}\right)$.)
$z_0 = ||\mathbf{z}||_2$.

Step 4 If $z_0 = 0$ then OUTPUT ('Zero gradient');
OUTPUT $(x_1, \ldots, x_n, g_1)$;
(*Procedure completed, may have a minimum.*)
STOP.

Step 5 Set $\mathbf{z} = \mathbf{z}/z_0$; (*Make* $\mathbf{z}$ *a unit vector.*)
$\alpha_1 = 0$;
$\alpha_3 = 1$;
$g_3 = g(\mathbf{x} - \alpha_3\mathbf{z})$.

Step 6 While ($g_3 \geq g_1$) do Steps 7 and 8.

Step 7 Set $\alpha_3 = \alpha_3/2$;
$g_3 = g(\mathbf{x} - \alpha_3\mathbf{z})$.

Step 8 If $\alpha_3 < TOL/2$ then
OUTPUT ('No likely improvement');
OUTPUT $(x_1, \ldots, x_n, g_1)$;
(*Procedure completed, may have a minimum.*)
STOP.

Step 9 Set $\alpha_2 = \alpha_3/2$;
$g_2 = g(\mathbf{x} - \alpha_2\mathbf{z})$.

Step 10 Set $h_1 = (g_2 - g_1)/\alpha_2$;
$h_2 = (g_3 - g_2)/(\alpha_3 - \alpha_2)$;
$h_3 = (h_2 - h_1)/\alpha_3$.
(*Note: Newton's forward divided-difference formula is used to find the quadratic* $P(\alpha) = g_1 + h_1\alpha + h_3\alpha(\alpha - \alpha_2)$ *that interpolates* $h(\alpha)$ *at* $\alpha = 0$, $\alpha = \alpha_2$, $\alpha = \alpha_3$.)

Step 11 Set $\alpha_0 = 0.5(\alpha_2 - h_1/h_3)$; (*Critical point of P occurs at* α_0.)
$g_0 = g(\mathbf{x} - \alpha_0\mathbf{z})$.

Step 12 Find α from $\{\alpha_0, \alpha_3\}$ so that $g = g(\mathbf{x} - \alpha\mathbf{z}) = \min\{g_0, g_3\}$.

Step 13 Set $\mathbf{x} = \mathbf{x} - \alpha\mathbf{z}$.

Step 14 If $|g - g_1| < TOL$ then
OUTPUT $(x_1, \ldots, x_n, g)$;
(*Procedure completed successfully.*)
STOP.

Step 15 Set $k = k + 1$.

Step 16 OUTPUT ('Maximum iterations exceeded');
(*Procedure completed unsuccessfully.*)
STOP.

EXAMPLE 1 To find a reasonable initial approximation to the solution of the nonlinear system

$$f_1(x_1, x_2, x_3) = 3x_1 - \cos(x_2 x_3) - \frac{1}{2} = 0,$$

$$f_2(x_1, x_2, x_3) = x_1^2 - 81(x_2 + 0.1)^2 + \sin x_3 + 1.06 = 0,$$

$$f_3(x_1, x_2, x_3) = e^{-x_1 x_2} + 20x_3 + \frac{10\pi - 3}{3} = 0,$$

we use Algorithm 10.3 with $TOL = 0.05$, $N = 10$, and $\mathbf{x}^{(0)} = (0, 0, 0)^t$.

Let $g(x_1, x_2, x_3) = [f_1(x_1, x_2, x_3)]^2 + [f_2(x_1, x_2, x_3)]^2 + [f_3(x_1, x_2, x_3)]^2$; then

$$\begin{aligned}\nabla g(x_1, x_2, x_3) \equiv \nabla g(\mathbf{x}) = \Bigg(& 2f_1(\mathbf{x})\frac{\partial f_1}{\partial x_1}(\mathbf{x}) + 2f_2(\mathbf{x})\frac{\partial f_2}{\partial x_1}(\mathbf{x}) + 2f_3(\mathbf{x})\frac{\partial f_3}{\partial x_1}(\mathbf{x}),\\ & 2f_1(\mathbf{x})\frac{\partial f_1}{\partial x_2}(\mathbf{x}) + 2f_2(\mathbf{x})\frac{\partial f_2}{\partial x_2}(\mathbf{x}) + 2f_3(\mathbf{x})\frac{\partial f_3}{\partial x_2}(\mathbf{x}),\\ & 2f_1(\mathbf{x})\frac{\partial f_1}{\partial x_3}(\mathbf{x}) + 2f_2(\mathbf{x})\frac{\partial f_2}{\partial x_3}(\mathbf{x}) + 2f_3(\mathbf{x})\frac{\partial f_3}{\partial x_3}(\mathbf{x})\Bigg)^t\\ = {} & 2\mathbf{J}(\mathbf{x})^t\mathbf{F}(\mathbf{x}).\end{aligned}$$

For $\mathbf{x}^{(0)} = (0, 0, 0)^t$, we have

$$g(\mathbf{x}^{(0)}) = 111.975, \quad z_0 = ||\mathbf{z}||_2 = 419.554,$$
$$\mathbf{z} = (-0.0214514, -0.0193062, 0.999583)^t,$$

and

$$\begin{aligned}
&\alpha_1 = 0, && g_1 = 111.975, && && \\
&\alpha_2 = 0.5, && g_2 = 2.53557, && h_1 = -218.878, && \\
&\alpha_3 = 1, && g_3 = 93.5649, && h_2 = 182.059, && h_3 = 400.937,
\end{aligned}$$

so $P(\alpha) = 111.975 - 218.878\alpha + 400.937\alpha(\alpha - 0.5)$. Hence,

$$\alpha_0 = 0.5\left(\alpha_2 - \frac{h_1}{h_3}\right) = 0.522959, \quad \text{and} \quad g_0 = g(\mathbf{x}^{(0)} - \alpha_0\mathbf{z}) = 2.32762.$$

Since $g_0 = 2.32762 < 93.5649 = g_3$, we have $\alpha = \alpha_0 = 0.522959$,

$$\mathbf{x}^{(1)} = \mathbf{x}^{(0)} - \alpha\mathbf{z} = (0.0112182, 0.0100964, -0.522741)^t,$$

and

$$g(\mathbf{x}^{(1)}) = 2.32762.$$

Table 10.5 contains the remainder of the results.

Table 10.5

k	$x_1^{(k)}$	$x_2^{(k)}$	$x_3^{(k)}$	$g(x_1^{(k)}, x_2^{(k)}, x_3^{(k)})$
2	0.137860	−0.205453	−0.522059	1.27406
3	0.266959	0.00551102	−0.558494	1.06813
4	0.272734	−0.00811751	−0.522006	0.468309
5	0.308689	−0.0204026	−0.533112	0.381087
6	0.314308	−0.0147046	−0.520923	0.318837
7	0.324267	−0.00852549	−0.528431	0.287024

A solution to the nonlinear system is given in Example 2 of Section 10.1 as $\mathbf{p} = (0.5, 0, -0.5235988)^t$. The results here would be adequate as initial approximations for Newton's and Broyden's methods. One of these quicker converging techniques would be appropriate at this stage, since 70 iterations of the Steepest Descent method are required to find $\|\mathbf{x}^{(k)} - \mathbf{p}\|_\infty < 0.01$. ■

There are many variations of the method of Steepest Descent, some of which involve more intricate methods for determining the value of α that will produce a minimum for the single-variable function h defined in Eq. (10.19). Other techniques use a multidimensional Taylor polynomial to replace the original multivariable function g and minimize the polynomial instead of g. Although there are advantages to some of these methods over the procedure discussed here, all the Steepest Descent methods are, in general, linearly convergent and converge independent of the starting approximation. In some instances, however, the methods may converge to something other than the absolute minimum of the function g.

A more complete discussion of Steepest Descent methods can be found in [OR] or [RR].

EXERCISE SET 10.4

1. Use the method of Steepest Descent with $TOL = 0.05$ to approximate the solutions of the following nonlinear systems.

   a. $4x_1^2 - 20x_1 + \frac{1}{4}x_2^2 + 8 = 0,$

   $\frac{1}{2}x_1x_2^2 + 2x_1 - 5x_2 + 8 = 0.$

   b. $3x_1^2 - x_2^2 = 0,$

   $3x_1x_2^2 - x_1^3 - 1 = 0.$

   c. $\ln(x_1^2 + x_2^2) - \sin(x_1x_2) = \ln 2 + \ln \pi,$

   $e^{x_1 - x_2} + \cos(x_1x_2) = 0.$

   d. $\sin(4\pi x_1x_2) - 2x_2 - x_1 = 0,$

   $\left(\frac{4\pi - 1}{4\pi}\right)(e^{2x_1} - e) + 4ex_2^2 - 2ex_1 = 0.$

2. Use the results in Exercise 1 and Newton's method to approximate the solutions of the nonlinear systems in Exercise 1 to within 10^{-6}.

3. Use the method of Steepest Descent with $TOL = 0.05$ to approximate the solutions of the following nonlinear systems.

a. $$15x_1 + x_2^2 - 4x_3 = 13,$$ $$x_1^2 + 10x_2 - x_3 = 11,$$ $$x_2^3 - 25x_3 = -22.$$

b. $$10x_1 - 2x_2^2 + x_2 - 2x_3 - 5 = 0,$$ $$8x_2^2 + 4x_3^2 - 9 = 0,$$ $$8x_2x_3 + 4 = 0.$$

c. $$x_1^3 + x_1^2x_2 - x_1x_3 + 6 = 0,$$ $$e^{x_1} + e^{x_2} - x_3 = 0,$$ $$x_2^2 - 2x_1x_3 = 4.$$

d. $$x_1 + \cos(x_1x_2x_3) - 1 = 0,$$ $$(1 - x_1)^{1/4} + x_2 + 0.05x_3^2 - 0.15x_3 - 1 = 0,$$ $$-x_1^2 - 0.1x_2^2 + 0.01x_2 + x_3 - 1 = 0.$$

4. Use the results of Exercise 3 and Newton's method to approximate the solutions to within 10^{-6} for the nonlinear systems in Exercise 3.

5. Use the method of Steepest Descent to approximate minima to within 0.005 for the following functions.

a. $g(x_1, x_2) = \cos(x_1 + x_2) + \sin x_1 + \cos x_2$

b. $g(x_1, x_2) = 100(x_1^2 - x_2)^2 + (1 - x_1)^2$

c. $g(x_1, x_2, x_3) = x_1^2 + 2x_2^2 + x_3^2 - 2x_1x_2 + 2x_1 - 2.5x_2 - x_3 + 2$

d. $g(x_1, x_2, x_3) = x_1^4 + 2x_2^4 + 3x_3^4 + 1.01$

6. **a.** Show that the quadratic polynomial

$$P(\alpha) = g_1 + h_1\alpha + h_3\alpha(\alpha - \alpha_2)$$

interpolates the function h defined in (10.19):

$$h(\alpha) = g\left(\mathbf{x}^{(0)} - \alpha\nabla g\left(\mathbf{x}^{(0)}\right)\right)$$

at $\alpha = 0$, α_2, and α_3.

b. Show that a critical point of P occurs at

$$\alpha_0 = 0.5\left(\alpha_2 - \frac{h_1}{h_3}\right).$$

10.5 Survey of Methods and Software

In this chapter we considered methods to approximate solutions to nonlinear systems

$$f_1(x_1, x_2, \ldots, x_n) = 0,$$
$$f_2(x_1, x_2, \ldots, x_n) = 0,$$
$$\vdots \qquad \vdots$$
$$f_n(x_1, x_2, \ldots, x_n) = 0,$$

Newton's method for systems requires a good initial approximation $\left(x_1^{(0)}, x_2^{(0)}, \ldots, x_n^{(0)}\right)^t$ and generates a sequence

$$\mathbf{x}^{(k)} = \mathbf{x}^{(k-1)} - J\left(\mathbf{x}^{(k-1)}\right)^{-1} \mathbf{F}\left(\mathbf{x}^{(k-1)}\right),$$

which converges rapidly to a solution $\mathbf{p}$ if $\mathbf{x}^{(0)}$ is sufficiently close to $\mathbf{p}$. However, Newton's method requires evaluating, or approximating, n^2 partial derivatives and solving an n by n linear system at each step.

Broyden's method reduces the amount of computation at each step without significantly degrading the speed of convergence. This technique replaces the Jacobian matrix J with a matrix A_{k-1} whose inverse is directly determined at each step. This reduces the arithmetic computations from $O(n^3)$ to $O(n^2)$. Moreover, the only scalar function evaluations required are in evaluating the f_i, saving n^2 scalar function evaluations per step. Broyden's method also requires a good initial approximation.

The Steepest Descent method was presented as a way to obtain good initial approximations for Newton and Broyden's methods. Although Steepest Descent does not give a rapidly convergent sequence, it does not require a good initial approximation. The Steepest Descent method approximates a minimum of a multivariable function g. For our application we chose

$$g(x_1, x_2, \ldots, x_n) = \sum_{i=1}^{n} [f_i(x_1, x_2, \ldots, x_n)]^2.$$

The minimum of g is zero, which occurs when the functions f_i are simultaneously zero.

Homotopy and continuation methods are also used for nonlinear systems and are the subject of current research. (See [AG].) In these methods a given problem

$$\mathbf{F}(\mathbf{x}) = \mathbf{0}$$

is embedded in a one-parameter family of problems using a parameter λ assuming values in [0, 1]. The original problem corresponds to $\lambda = 1$ and a problem with a known solution corresponds to $\lambda = 0$. For example, the set of problems

$$G(\mathbf{x}, \lambda) = \lambda\mathbf{F}(\mathbf{x}) + (1 - \lambda)\mathbf{F}(\mathbf{x}_0) = \mathbf{0}, \quad 0 \le \lambda \le 1,$$

for fixed $\mathbf{x}_0 \in \mathbb{R}^n$ forms a homotopy. When $\lambda = 0$, the solution is $\mathbf{x}(\lambda = 0) = \mathbf{x}_0$. The solution to the original problem corresponds to $\mathbf{x}(\lambda = 1)$. A continuation method attempts to determine $\mathbf{x}(\lambda = 1)$ by solving the sequence of problems corresponding to $\lambda_0 = 0 < \lambda_1 < \lambda_2 < \cdots < \lambda_m = 1$. The initial approximation to the solution of

$$\lambda_i\mathbf{F}(\mathbf{x}) + (1 - \lambda_i)\mathbf{F}(\mathbf{x}_0) = \mathbf{0}$$

would be the solution $\mathbf{x}(\lambda = \lambda_{i-1})$ to the problem

$$\lambda_{i-1}\mathbf{F}(\mathbf{x}) + (1 - \lambda_{i-1})\mathbf{F}(\mathbf{x}_0) = \mathbf{0}.$$

The methods in the IMSL and NAG libraries are based on two subroutines HYBRDI and HYBRDJ contained in MINPACK, a public-domain package. Both methods use the

Levenberg-Marquardt method, which is a weighted average of Newton's method and the Steepest Descent method. The weight is biased toward the Steepest Descent method until convergence is detected, at which time the weight is shifted toward the more rapidly convergent Newton's method. The subroutine HYBRDI uses a finite-difference approximation to the Jacobian, and HYBRDJ requires a user-supplied subroutine to compute the Jacobian.

The IMSL subroutine NEQNF solves a nonlinear system without a user-supplied Jacobian. The subroutine NEQNJ is similar to NEQNF, except that the user must supply a subroutine to calculate the Jacobian.

In the NAG Library C05NBF is similar to HYBRDI. The subroutine C05PBF is similar to C05NBF except that the user must supply a subroutine to compute the Jacobian. Subroutine C05PBF is based on HYBRDJ in the MINPACK package. NAG also contains other modifications of the Levenberg-Marquardt method.

A comprehensive treatment of methods for solving nonlinear systems of equations can be found in Ortega and Rheinbolt [OR] and in Dennis and Schnabel [DenS]. Recent developments on iterative methods can be found in Argyros and Szidarovszky [AS], and information on the use of continuation methods is available in Allgower and Georg [AG].

CHAPTER

11

Boundary-Value Problems for Ordinary Differential Equations

■ ■ ■

A common problem in civil engineering concerns the deflection of a beam of rectangular cross section subject to uniform loading while the ends of the beam are supported so that they undergo no deflection.

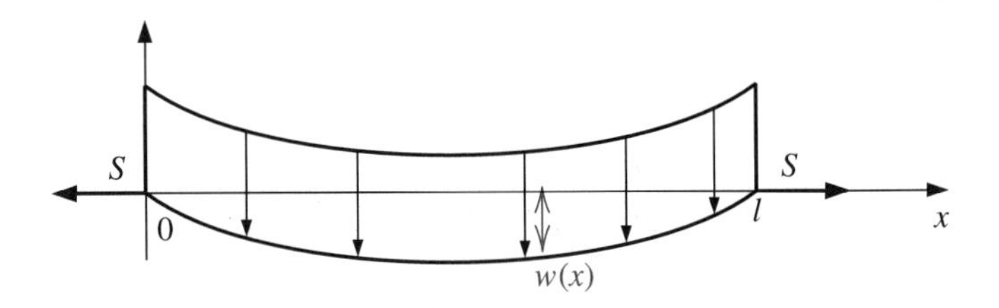

The differential equation approximating the physical situation is of the form

$$\frac{d^2w}{dx^2} = \frac{S}{EI}w + \frac{qx}{2EI}(x - l),$$

where $w = w(x)$ is the deflection a distance x from the left end of the beam, and l, q, E, S, and I represent, respectively, the length of the beam, the intensity of the uniform load, the modulus of elasticity, the stress at the endpoints, and the central moment of inertia. Associated with this differential equation are two boundary conditions given by the assumption that no deflection occurs at the ends of the beam,

$$w(0) = w(l) = 0.$$

When the beam is of uniform thickness, the product EI is constant, and the exact solution is easily obtained. In many applications, however, the thickness is not uniform, so the moment of inertia I is a function of x, and approximation techniques are required. Problems of this type are considered in Exercises 7 of Section 11.3 and 6 of Section 11.4.

Methods for finding approximate solutions to differential equations, studied in Chapter 5, require that all conditions imposed on the differential equation occur at an initial point. For a second-order equation, we need to know both $w(0)$ and $w'(0)$, which is not the case in this problem. New techniques are required for handling problems when the conditions imposed are of a boundary-value rather than an initial-value type.

Physical problems that are position-dependent rather than time-dependent are often described in terms of differential equations with conditions imposed at more than one point. The two-point boundary-value problems in this chapter involve a second-order differential equation of the form

(11.1) $$y'' = f(x, y, y'), \quad a \le x \le b,$$

together with the boundary conditions

(11.2) $$y(a) = \alpha \quad \text{and} \quad y(b) = \beta.$$

11.1 The Linear Shooting Method

The following theorem gives general conditions that ensure that the solution to a second-order boundary value problem exists and is unique. The proof of this theorem can be found in [K,H].

Theorem 11.1 Suppose the function f in the boundary-value problem

$$y'' = f(x, y, y'), \quad a \le x \le b, \quad y(a) = \alpha, \quad y(b) = \beta,$$

is continuous on the set

$$D = \{(x, y, y') \mid a \le x \le b, -\infty < y < \infty, \ -\infty < y' < \infty\},$$

and that $\partial f/\partial y$ and $\partial f/\partial y'$ are also continuous on D. If

(i) $\dfrac{\partial f}{\partial y}(x, y, y') > 0$ for all $(x, y, y') \in D$, and

(ii) A constant M exists, with

$$\left|\frac{\partial f}{\partial y'}(x, y, y')\right| \le M, \quad \text{for all } (x, y, y') \in D,$$

then the boundary-value problem has a unique solution. ■

EXAMPLE 1 The boundary-value problem

$$y'' + e^{-xy} + \sin y' = 0, \quad 1 \le x \le 2, \quad y(1) = y(2) = 0,$$

has

$$f(x, y, y') = -e^{-xy} - \sin y'.$$

Since

$$\frac{\partial f}{\partial y}(x, y, y') = xe^{-xy} > 0 \quad \text{and} \quad \left|\frac{\partial f}{\partial y'}(x, y, y')\right| = |-\cos y'| \le 1,$$

this problem has a unique solution. ■

When $f(x, y, y')$ has the form

$$f(x, y, y') = p(x)y' + q(x)y + r(x),$$

the differential equation

$$y'' = f(x, y, y')$$

is **linear**. Problems of this type frequently occur, and in this situation Theorem 11.1 can be simplified.

Corollary 11.2 If the linear boundary-value problem

$$y'' = p(x)y' + q(x)y + r(x), \quad a \le x \le b, \quad y(a) = \alpha, \quad y(b) = \beta$$

satisfies

(i) $p(x)$, $q(x)$, and $r(x)$ are continuous on $[a, b]$,
(ii) $q(x) > 0$ on $[a, b]$,

then the problem has a unique solution. ■

To approximate the unique solution guaranteed by the satisfaction of the hypotheses of Corollary 11.2, let us first consider the initial-value problems

(11.3) $$y'' = p(x)y' + q(x)y + r(x), \quad a \le x \le b, \quad y(a) = \alpha, \quad y'(a) = 0,$$

and

(11.4) $$y'' = p(x)y' + q(x)y, \quad a \le x \le b, \quad y(a) = 0, \quad y'(a) = 1.$$

Theorem 5.16 in Section 5.9 ensures that under the hypotheses in Corollary 11.2, both problems have a unique solution. If $y_1(x)$ denotes the solution to (11.3) and $y_2(x)$ denotes the solution to (11.4), it is not difficult to verify that

$$y(x) = y_1(x) + \frac{\beta - y_1(b)}{y_2(b)} y_2(x) \tag{11.5}$$

is the unique solution to our boundary-value problem, provided, of course that $y_2(b) \neq 0$. (That $y_2(b) = 0$ is in conflict with the hypotheses of Corollary 11.2 is considered in Exercise 8.)

The Shooting method for linear equations is based on the replacement of the linear boundary-value problem by the two initial-value problems (11.3) and (11.4). Numerous methods are available from Chapter 5 for approximating the solutions $y_1(x)$ and $y_2(x)$, and once these approximations are available, the solution to the boundary-value problem is approximated using Eq. (11.5). Graphically, the method has the appearance shown in Figure 11.1.

Figure 11.1

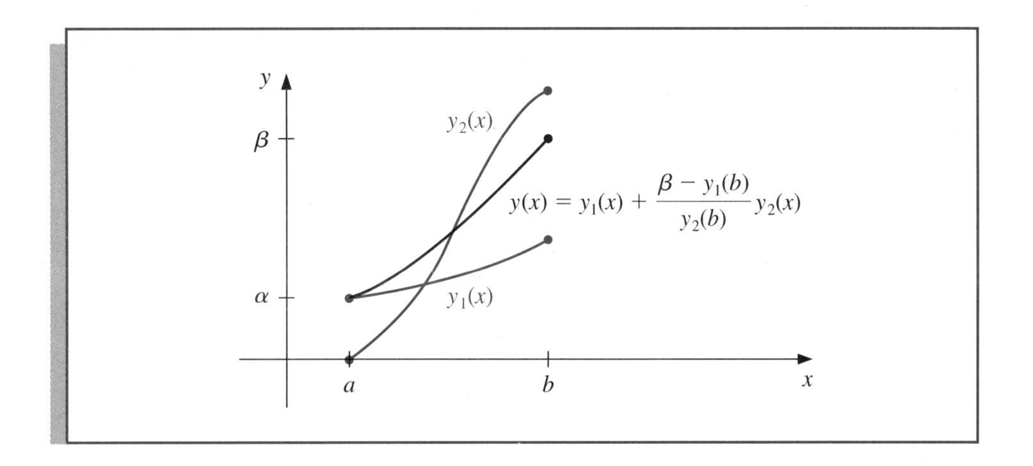

Algorithm 11.1 uses the fourth-order Runge-Kutta technique to find the approximations to $y_1(x)$ and $y_2(x)$, but any other technique for approximating the solutions to initial-value problems can be substituted into Step 4.

The algorithm has the additional feature of obtaining approximations for the derivative of the solution to the boundary-value problem as well as to the solution of the problem itself. The use of the algorithm is not restricted to those problems for which the hypotheses of Corollary 11.2 can be verified; it gives satisfactory results for many problems that do not satisfy these hypotheses.

ALGORITHM 11.1

Linear Shooting

To approximate the solution of the boundary-value problem

$$-y'' + p(x)y' + q(x)y + r(x) = 0, \quad a \leq x \leq b, \quad y(a) = \alpha, \quad y(b) = \beta:$$

(*Note: Equations* (11.3) *and* (11.4) *are written as first-order systems and solved.*)

INPUT endpoints a, b; boundary conditions α, β; number of subintervals N.

OUTPUT approximations $w_{1,i}$ to $y(x_i)$; $w_{2,i}$ to $y'(x_i)$ for each $i = 0, 1, \ldots, N$.

Step 1 Set $h = (b - a)/N$;
$u_{1,0} = \alpha$;
$u_{2,0} = 0$;
$v_{1,0} = 0$;
$v_{2,0} = 1$.

Step 2 For $i = 0, \ldots, N - 1$ do Steps 3 and 4.
(*The Runge-Kutta method for systems is used in Steps* 3 *and* 4.)

Step 3 Set $x = a + ih$.

Step 4 Set $k_{1,1} = hu_{2,i}$;

$$k_{1,2} = h\left[p(x)u_{2,i} + q(x)u_{1,i} + r(x)\right];$$
$$k_{2,1} = h\left[u_{2,i} + \tfrac{1}{2}k_{1,2}\right];$$
$$k_{2,2} = h\left[p(x + h/2)\left(u_{2,i} + \tfrac{1}{2}k_{1,2}\right) + q(x + h/2)\left(u_{1,i} + \tfrac{1}{2}k_{1,1}\right) + r(x + h/2)\right];$$
$$k_{3,1} = h\left[u_{2,i} + \tfrac{1}{2}k_{2,2}\right];$$
$$k_{3,2} = h\left[p(x + h/2)\left(u_{2,i} + \tfrac{1}{2}k_{2,2}\right) + q(x + h/2)(u_{1,i} + \tfrac{1}{2}k_{2,1}) + r(x + h/2)\right];$$
$$k_{4,1} = h\left[u_{2,i} + k_{3,2}\right];$$
$$k_{4,2} = h\left[p(x + h)(u_{2,i} + k_{3,2}) + q(x + h)(u_{1,i} + k_{3,1}) + r(x + h)\right];$$
$$u_{1,i+1} = u_{1,i} + \tfrac{1}{6}\left[k_{1,1} + 2k_{2,1} + 2k_{3,1} + k_{4,1}\right];$$
$$u_{2,i+1} = u_{2,i} + \tfrac{1}{6}\left[k_{1,2} + 2k_{2,2} + 2k_{3,2} + k_{4,2}\right];$$
$$k'_{1,1} = hv_{2,i};$$
$$k'_{1,2} = h\left[p(x)v_{2,i} + q(x)v_{1,i}\right];$$
$$k'_{2,1} = h\left[v_{2,i} + \tfrac{1}{2}k'_{1,2}\right];$$
$$k'_{2,2} = h\left[p(x + h/2)\left(v_{2,i} + \tfrac{1}{2}k'_{1,2}\right) + q(x + h/2)\left(v_{1,i} + \tfrac{1}{2}k'_{1,1}\right)\right];$$
$$k'_{3,1} = h\left[v_{2,i} + \tfrac{1}{2}k'_{2,2}\right];$$
$$k'_{3,2} = h\left[p(x + h/2)\left(v_{2,i} + \tfrac{1}{2}k'_{2,2}\right) + q(x + h/2)\left(v_{1,i} + \tfrac{1}{2}k'_{2,1}\right)\right];$$
$$k'_{4,1} = h\left[v_{2,i} + k'_{3,2}\right];$$
$$k'_{4,2} = h\left[p(x + h)(v_{2,i} + k'_{3,2}) + q(x + h)(v_{1,i} + k'_{3,1})\right];$$
$$v_{1,i+1} = v_{1,i} + \tfrac{1}{6}\left[k'_{1,1} + 2k'_{2,1} + 2k'_{3,1} + k'_{4,1}\right];$$
$$v_{2,i+1} = v_{2,i} + \tfrac{1}{6}\left[k'_{1,2} + 2k'_{2,2} + 2k'_{3,2} + k'_{4,2}\right].$$

Step 5 Set $w_{1,0} = \alpha$;

$$w_{2,0} = \frac{\beta - u_{1,N}}{v_{1,N}};$$

OUTPUT $(a, w_{1,0}, w_{2,0})$.

Step 6 For $i = 1, \ldots, N$
set $W1 = u_{1,i} + w_{2,0}v_{1,i}$;
$W2 = u_{2,i} + w_{2,0}v_{2,i}$;
$x = a + ih$;
OUTPUT $(x, W1, W2)$. (*Output is* $x_i, w_{1,i}, w_{2,i}$.)

Step 7 STOP. (*Process is complete.*)

EXAMPLE 2 The boundary-value problem

$$y'' = -\frac{2}{x}y' + \frac{2}{x^2}y + \frac{\sin(\ln x)}{x^2}, \quad 1 \le x \le 2, \quad y(1) = 1, \quad y(2) = 2$$

has the exact solution

$$y = c_1 x + \frac{c_2}{x^2} - \frac{3}{10}\sin(\ln x) - \frac{1}{10}\cos(\ln x),$$

where

$$c_2 = \frac{1}{70}[8 - 12\sin(\ln 2) - 4\cos(\ln 2)] \approx -0.03920701320$$

and

$$c_1 = \frac{11}{10} - c_2 \approx 1.1392070132.$$

Applying Algorithm 11.1 to this problem requires approximating the solutions to the initial-value problems

$$y_1'' = -\frac{2}{x}y_1' + \frac{1}{x^2}y_1 + \frac{\sin(\ln x)}{x^2}, \quad 1 \le x \le 2, \quad y_1(1) = 1, \quad y_1'(1) = 0,$$

and

$$y_2'' = -\frac{2}{x}y_2' + \frac{2}{x^2}y_2, \quad 1 \le x \le 2, \quad y_2(1) = 0, \quad y_2'(1) = 1.$$

The results of the calculations, using Algorithm 11.1 with $N = 10$ and $h = 0.1$, are given in Table 11.1. The value listed as $u_{1,i}$ approximates $y_1(x_i)$, $v_{1,i}$ approximates $y_2(x_i)$, and w_i approximates

$$y(x_i) = y_1(x_i) + \frac{2 - y_1(2)}{y_2(2)}y_2(x_i). \qquad \blacksquare$$

The accuracy found in Table 11.1 is expected because the fourth-order Runge-Kutta method gives $O(h^4)$ accuracy to the solutions of the initial-value problems. Unfortunately,

Table 11.1

x_i	$u_{1,i}$	$v_{1,i}$	w_i	$y(x_i)$	$\lvert y(x_i) - w_i \rvert$
1.0	1.00000000	0.00000000	1.00000000	1.00000000	
1.1	1.00896058	0.09117986	1.09262917	1.09262930	1.43×10^{-7}
1.2	1.03245472	0.16851175	1.18708471	1.18708484	1.34×10^{-7}
1.3	1.06674375	0.23608704	1.28338227	1.28338236	9.78×10^{-8}
1.4	1.10928795	0.29659067	1.38144589	1.38144595	6.02×10^{-8}
1.5	1.15830000	0.35184379	1.48115939	1.48115942	3.06×10^{-8}
1.6	1.21248372	0.40311695	1.58239245	1.58239246	1.08×10^{-8}
1.7	1.27087454	0.45131840	1.68501396	1.68501396	5.43×10^{-10}
1.8	1.33273851	0.49711137	1.78889854	1.78889853	5.05×10^{-9}
1.9	1.39750618	0.54098928	1.89392951	1.89392951	4.41×10^{-9}
2.0	1.46472815	0.58332538	2.00000000	2.00000000	

because of roundoff errors, there can be problems hidden in this technique. If $y_1(x)$ rapidly increases as x goes from a to b, then $u_{1,N} \approx y_1(b)$ will be large. Should β be small in magnitude compared to $u_{1,N}$, the term $w_{2,0} = (\beta - u_{1,N})/v_{1,N}$ will be approximately $-u_{1,N}/v_{1,N}$. The computations in Step 6 then become

$$W1 = u_{1,i} + w_{2,0}v_{1,i} \approx u_{1,i} - \left(\frac{u_{1,N}}{v_{1,N}}\right) v_{1,i},$$

$$W2 = u_{2,i} + w_{2,0}v_{2,i} \approx u_{2,i} - \left(\frac{u_{1,N}}{v_{1,N}}\right) v_{2,i},$$

which allows a possibility of a loss of significant digits due to cancellation. However, since $u_{1,i}$ is an approximation to $y_1(x_i)$, the behavior of y_1 can easily be monitored, and if $u_{1,i}$ increases rapidly from a to b, the shooting technique can be employed backward, that is, solving instead the initial-value problems

$$y'' = p(x)y' + q(x)y + r(x), \quad a \leq x \leq b, \quad y(b) = \beta, \quad y'(b) = 0,$$

and

$$y'' = p(x)y' + q(x)y, \quad a \leq x \leq b, \quad y(b) = 0, \quad y'(b) = 1.$$

If this reverse shooting technique still gives cancellation of significant digits and if increased precision does not yield greater accuracy, other techniques must be used, such as those presented later in this chapter. In general, however, if $u_{1,i}$ and $v_{1,i}$ are $O(h^n)$ approximations to $y_1(x_i)$ and $y_2(x_i)$, respectively, for each $i = 0, 1, \ldots, N$, then $w_{1,i}$ will be an $O(h^n)$ approximation to $y(x_i)$. In particular,

$$|w_{1,i} - y(x_i)| \leq Kh^n \left| 1 + \frac{v_{1,i}}{v_{1,N}} \right|,$$

for some constant K (see [IK, p. 426]).

EXERCISE SET 11.1

1. The boundary-value problem

$$y'' = 4(y - x), \quad 0 \le x \le 1, \quad y(0) = 0, \quad y(1) = 2,$$

has the solution $y(x) = e^2(e^4 - 1)^{-1}(e^{2x} - e^{-2x}) + x$. Use the Linear Shooting method to approximate the solution and compare the results to the actual solution.

   a. With $h = \frac{1}{3}$; b. With $h = \frac{1}{4}$.

2. The boundary-value problem

$$y'' = y' + 2y + \cos x, \quad 0 \le x \le \frac{\pi}{2}, \quad y(0) = -0.3, \quad y\left(\frac{\pi}{2}\right) = -0.1$$

has the solution $y(x) = -\frac{1}{10}[\sin x + 3\cos x]$. Use the Linear Shooting method to approximate the solution and compare the results to the actual solution.

   a. With $h = \frac{\pi}{4}$; b. With $h = \frac{\pi}{6}$.

3. Use the Linear Shooting method to approximate the solution to the following boundary-value problems.

   a. $y'' = -3y' + 2y + 2x + 3, \quad 0 \le x \le 1, \ y(0) = 2, \ y(1) = 1;$ use $h = 0.1$.

   b. $y'' = -\frac{4}{x}y' - \frac{2}{x^2}y - \frac{2\ln x}{x^2}, \quad 1 \le x \le 2, \ y(1) = -\frac{1}{2}, \ y(2) = \ln 2;$ use $h = 0.05$.

   c. $y'' = -(x+1)y' + 2y + (1 - x^2)e^{-x}, \quad 0 \le x \le 1, \ y(0) = -1, \ y(1) = 0;$ use $h = 0.1$.

   d. $y'' = \frac{y'}{x} + \frac{3y}{x^2} + \frac{\ln x}{x} - 1, \quad 1 \le x \le 2, \ y(1) = y(2) = 0;$ use $h = 0.1$.

4. Although $q(x) < 0$ in the following boundary-value problems, unique solutions exist and are given. Use the Linear Shooting Algorithm to approximate the solutions to the following problems and compare the results to the actual solutions.

   a. $y'' + y = 0, \quad 0 \le x \le \frac{\pi}{4}, \ y(0) = 1, \ y\left(\frac{\pi}{4}\right) = 1;$ use $h = \frac{\pi}{20}$; actual solution

$$y(x) = \cos x + \frac{2 - \sqrt{2}}{\sqrt{2}} \sin x.$$

   b. $y'' + 4y = \cos x, \quad 0 \le x \le \frac{\pi}{4}, \ y(0) = 0, \ y\left(\frac{\pi}{4}\right) = 0;$ use $h = \frac{\pi}{20}$; actual solution

$$y(x) = -\frac{1}{3}\cos 2x - \frac{\sqrt{2}}{6}\sin 2x + \frac{1}{3}\cos x.$$

   c. $y'' = -\frac{4}{x}y' - \frac{2}{x^2}y + \frac{2}{x^2}\ln x, \quad 1 \le x \le 2, \ y(1) = \frac{1}{2}, \ y(2) = \ln 2;$ use $h = 0.05$; actual solution $y(x) = \frac{4}{x} - \frac{2}{x^2} + \ln x - \frac{3}{2}$.

   d. $y'' = 2y' - y + xe^x - x, \quad 0 \le x \le 2, \ y(0) = 0, \ y(2) = -4;$ use $h = 0.2$; actual solution $y(x) = \frac{1}{6}x^3e^x - \frac{5}{3}xe^x + 2e^x - x - 2$.

5. Use the Linear Shooting Algorithm to approximate the solution $y = e^{-10x}$ to the boundary-value problem

$$y'' = 100y, \quad 0 \le x \le 1, \quad y(0) = 1, \quad y(1) = e^{-10}.$$

Use $h = 0.1$ and 0.05.

6. Write the second-order initial-value problems (11.3) and (11.4) as first-order systems, and derive the equations necessary to solve the systems using the fourth-order Runge-Kutta method for systems.

7. Let u represent the electrostatic potential between two concentric metal spheres of radii R_1 and R_2, with $R_1 < R_2$, such that the potential of the inner sphere is kept constant at V_1 volts and the potential of the outer sphere is 0 volts. The potential in the region between the two spheres is governed by Laplace's equation, which, in this particular application, reduces to

$$\frac{d^2u}{dr^2} + \frac{2}{r}\frac{du}{dr} = 0, \quad R_1 \le r \le R_2, \quad u(R_1) = V_1, \quad u(R_2) = 0.$$

Suppose $R_1 = 2$ in., $R_2 = 4$ in., and $V_1 = 110$ volts.

   a. Approximate $u(3)$ using the Linear Shooting Algorithm.

   b. Compare the results of part (a) with the actual potential $u(3)$, where

$$u(r) = \frac{V_1 R_1}{r}\left(\frac{R_2 - r}{R_2 - R_1}\right).$$

8. Show that if y_2 is the solution to $y'' = p(x)y' + q(x)y$ and $y_2(a) = y_2(b) = 0$, then $y_2 \equiv 0$.

9. Consider the boundary-value problem

$$y'' + y = 0, \quad 0 \le x \le b, \quad y(0) = 0, \quad y(b) = B.$$

Find choices for b and B so that the boundary-value problem has

   a. No solution;

   b. Exactly one solution;

   c. Infinitely many solutions.

10. Attempt to apply Exercise 9 to the boundary-value problem

$$y'' - y = 0, \quad 0 \le x \le b, \quad y(0) = 0, \quad y(b) = B.$$

What happens? How do both problems relate to Corollary 11.2?

11.2 The Shooting Method for Nonlinear Problems

The shooting technique for the nonlinear second-order boundary-value problem

$$y'' = f(x, y, y'), \quad a \le x \le b, \quad y(a) = \alpha, \quad y(b) = \beta, \tag{11.6}$$

is similar to the linear case, except that the solution to a nonlinear problem cannot be expressed as a linear combination of the solutions to two initial-value problems. Instead, we need to use the solutions to a *sequence* of initial-value problems of the form

$$y'' = f(x, y, y'), \quad a \le x \le b, \quad y(a) = \alpha, \quad y'(a) = t, \tag{11.7}$$

involving a parameter t, to approximate the solution to the boundary-value problem. We do this by choosing the parameters $t = t_k$ so that

$$\lim_{k\to\infty} y(b, t_k) = y(b) = \beta,$$

where $y(x, t_k)$ denotes the solution to the initial-value problem (11.7) with $t = t_k$ and $y(x)$ denotes the solution to the boundary-value problem (11.6).

Figure 11.2

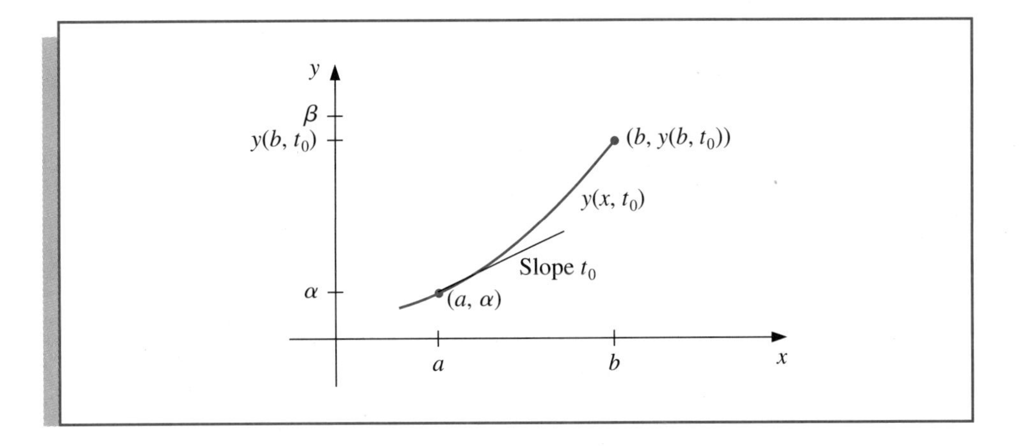

This technique is called a "shooting" method, by analogy to the procedure of firing objects at a stationary target. (See Figure 11.2.) We start with a parameter t_0 that determines the initial elevation at which the object is fired from the point (a, α) and along the curve described by the solution to the initial-value problem:

$$y'' = f(x, y, y'), \quad a \le x \le b, \quad y(a) = \alpha, \quad y'(a) = t_0.$$

If $y(b, t_0)$ is not sufficiently close to β, we correct our approximation by choosing elevations t_1, t_2, and so on, until $y(b, t_k)$ is sufficiently close to "hitting" β. (See Figure 11.3.)

Figure 11.3

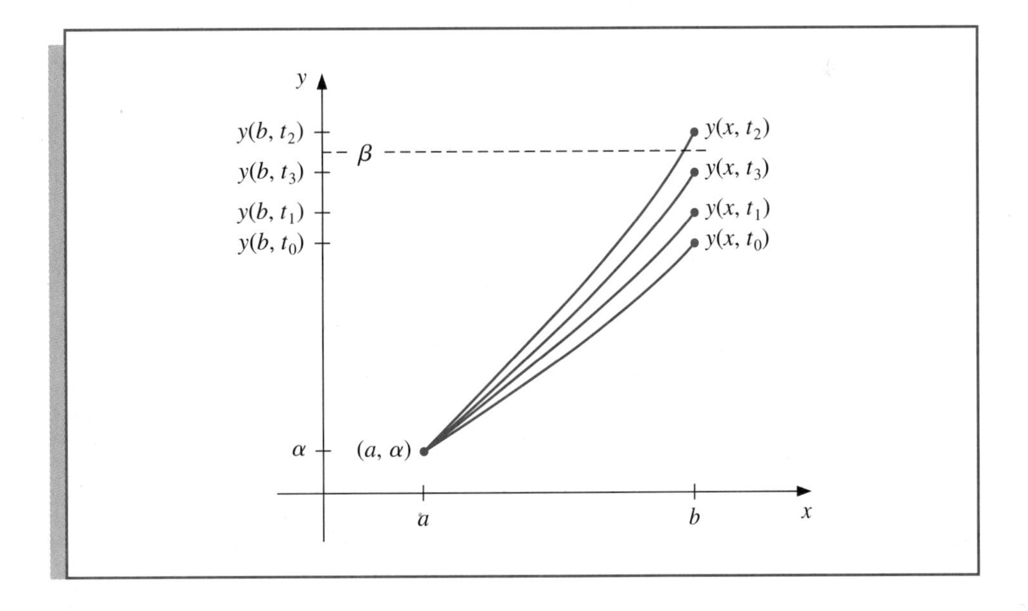

To determine the parameters t_k, suppose a boundary-value problem of the form (11.6) satisfies the hypotheses of Theorem 11.1. If $y(x, t)$ denotes the solution to the initial-value problem (11.7), the problem is to determine t so that

(11.8)
$$y(b, t) - \beta = 0.$$

This is a nonlinear equation of the type considered in Chapter 2, so a number of methods are available.

To use the Secant method to solve the problem, we need to choose initial approximations t_0 and t_1 and then generate the remaining terms of the sequence by

$$t_k = t_{k-1} - \frac{(y(b, t_{k-1}) - \beta)(t_{k-1} - t_{k-2})}{y(b, t_{k-1}) - y(b, t_{k-2})}, \qquad k = 2, 3, \ldots.$$

To use the more powerful Newton's method to generate the sequence $\{t_k\}$, only one initial approximation, t_0, is needed. However, the iteration has the form

(11.9)
$$t_k = t_{k-1} - \frac{y(b, t_{k-1}) - \beta}{\dfrac{dy}{dt}(b, t_{k-1})},$$

and requires the knowledge of $(dy/dt)(b, t_{k-1})$. This presents a difficulty since an explicit representation for $y(b, t)$ is not known; we know only the values $y(b, t_0), y(b, t_1), \ldots, y(b, t_{k-1})$.

Suppose we rewrite the initial-value problem (11.7), emphasizing that the solution depends on both x and t as

(11.10)
$$y''(x, t) = f(x, y(x, t), y'(x, t)), \quad a \le x \le b, \quad y(a, t) = \alpha, \quad y'(a, t) = t.$$

We have retained the prime notation to indicate differentiation with respect to x. Since we need to determine $(dy/dt)(b, t)$ when $t = t_{k-1}$, we first take the partial derivative of (11.10) with respect to t. This implies that

$$\begin{aligned}\frac{\partial y''}{\partial t}(x, t) &= \frac{\partial f}{\partial t}(x, y(x, t), y'(x, t))\\ &= \frac{\partial f}{\partial x}(x, y(x, t), y'(x, t))\frac{\partial x}{\partial t} + \frac{\partial f}{\partial y}(x, y(x, t), y'(x, t))\frac{\partial y}{\partial t}(x, t)\\ &\quad + \frac{\partial f}{\partial y'}(x, y(x, t), y'(x, t))\frac{\partial y'}{\partial t}(x, t).\end{aligned}$$

Since x and t are independent, $\partial x/\partial t = 0$ and

(11.11)
$$\frac{\partial y''}{\partial t}(x, t) = \frac{\partial f}{\partial y}(x, y(x, t), y'(x, t))\frac{\partial y}{\partial t}(x, t) + \frac{\partial f}{\partial y'}(x, y(x, t), y'(x, t))\frac{\partial y'}{\partial t}(x, t)$$

for $a \le x \le b$. The initial conditions give

$$\frac{\partial y}{\partial t}(a, t) = 0 \quad \text{and} \quad \frac{\partial y'}{\partial t}(a, t) = 1.$$

If we simplify the notation by using $z(x, t)$ to denote $(\partial y/\partial t)(x, t)$ and assume that the order of differentiation of x and t can be reversed, (11.11) with the initial conditions becomes the initial-value problem

$$z''(x, t) = \frac{\partial f}{\partial y}(x, y, y')z(x, t) + \frac{\partial f}{\partial y'}(x, y, y')z'(x, t), \quad a \le x \le b, \quad z(a, t) = 0, \quad z'(a, t) = 1.$$

(11.12)

Newton's method therefore requires that two initial-value problems be solved for each iteration, (11.10) and (11.12). Then from Eq. (11.9),

(11.13)
$$t_k = t_{k-1} - \frac{y(b, t_{k-1}) - \beta}{z(b, t_{k-1})}.$$

Of course, none of these initial-value problems is solved exactly; the solutions are approximated by one of the methods discussed in Chapter 5. Algorithm 11.2 uses the fourth-order Runge-Kutta method to approximate both solutions required by Newton's method. A similar procedure for the Secant method is considered in Exercise 4.

ALGORITHM 11.2

Nonlinear Shooting

To approximate the solution of the nonlinear boundary-value problem

$$y'' = f(x, y, y'), \quad a \le x \le b, \quad y(a) = \alpha, \quad y(b) = \beta:$$

(*Note: Equations* (11.10) *and* (11.12) *are written as first-order systems and solved.*)

INPUT endpoints a, b; boundary conditions α, β; number of subintervals $N \ge 2$; tolerance TOL; maximum number of iterations M.

OUTPUT approximations $w_{1,i}$ to $y(x_i)$; $w_{2,i}$ to $y'(x_i)$ for each $i = 0, 1, \ldots, N$ or a message that the maximum number of iterations was exceeded.

Step 1 Set $h = (b - a)/N$;
$k = 1$;
$TK = (\beta - \alpha)/(b - a)$. (*Note: TK could also be input.*)

Step 2 While ($k \le M$) do Steps 3–10.

Step 3 Set $w_{1,0} = \alpha$;
$w_{2,0} = TK$;
$u_1 = 0$;
$u_2 = 1$.

Step 4 For $i = 1, \ldots, N$ do Steps 5 and 6.
(*The Runge-Kutta method for systems is used in Steps* 5 *and* 6.)

Step 5 Set $x = a + (i - 1)h$.

Step 6 Set $k_{1,1} = hw_{2,i-1}$;
$k_{1,2} = hf(x, w_{1,i-1}w_{2,i-1})$;

$$k_{2,1} = h\left(w_{2,i-1} + \tfrac{1}{2}k_{1,2}\right);$$
$$k_{2,2} = hf\left(x + h/2, w_{1,i-1} + \tfrac{1}{2}k_{1,1}, w_{2,i-1} + \tfrac{1}{2}k_{1,2}\right);$$
$$k_{3,1} = h\left(w_{2,i-1} + \tfrac{1}{2}k_{2,2}\right);$$
$$k_{3,2} = hf\left(x + h/2, w_{1,i-1} + \tfrac{1}{2}k_{2,1}, w_{2,i-1} + \tfrac{1}{2}k_{2,2}\right);$$
$$k_{4,1} = h(w_{2,i-1} + k_{3,2});$$
$$k_{4,2} = hf(x + h, w_{1,i-1} + k_{3,1}, w_{2,i-1} + k_{3,2});$$
$$w_{1,i} = w_{1,i-1} + (k_{1,1} + 2k_{2,1} + 2k_{3,1} + k_{4,1})/6;$$
$$w_{2,i} = w_{2,i-1} + (k_{1,2} + 2k_{2,2} + 2k_{3,2} + k_{4,2})/6;$$
$$k'_{1,1} = hu_2;$$
$$k'_{1,2} = h[f_y(x, w_{1,i-1}, w_{2,i-1})u_1 + f_{y'}(x, w_{1,i-1}, w_{2,i-1})u_2];$$
$$k'_{2,1} = h\left[u_2 + \tfrac{1}{2}k'_{1,2}\right];$$
$$k'_{2,2} = h\left[f_y(x + h/2, w_{1,i-1}, w_{2,i-1})\left(u_1 + \tfrac{1}{2}k'_{1,1}\right) + f_{y'}(x + h/2, w_{1,i-1}, w_{2,i-1})\left(u_2 + \tfrac{1}{2}k'_{1,2}\right)\right];$$
$$k'_{3,1} = h\left(u_2 + \tfrac{1}{2}k'_{2,2}\right);$$
$$k'_{3,2} = h\left[f_y(x + h/2, w_{1,i-1}, w_{2,i-1})\left(u_1 + \tfrac{1}{2}k'_{2,1}\right) + f_{y'}(x + h/2, w_{1,i-1}, w_{2,i-1})\left(u_2 + \tfrac{1}{2}k'_{2,2}\right)\right];$$
$$k'_{4,1} = h(u_2 + k'_{3,2});$$
$$k'_{4,2} = h\left[f_y(x + h, w_{1,i-1}, w_{2,i-1})\left(u_1 + k'_{3,1}\right) + f_{y'}(x + h, w_{1,i-1}, w_{2,i-1})\left(u_2 + k'_{3,2}\right)\right];$$
$$u_1 = u_1 + \tfrac{1}{6}[k'_{1,1} + 2k'_{2,1} + 2k'_{3,1} + k'_{4,1}];$$
$$u_2 = u_2 + \tfrac{1}{6}[k'_{1,2} + 2k'_{2,2} + 2k'_{3,2} + k'_{4,2}].$$

Step 7 If $|w_{1,N} - \beta| \leq TOL$ then do Steps 8 and 9.

Step 8 For $i = 0, 1, \ldots, N$
set $x = a + ih$;
OUTPUT $(x, w_{1,i}, w_{2,i})$.

Step 9 (*Procedure is complete.*)
STOP.

Step 10 Set $TK = TK - \left(\dfrac{w_{1,N} - \beta}{u_1}\right)$; (*Newton's method is used to compute TK.*)
$k = k + 1$.

Step 11 OUTPUT ('Maximum number of iterations exceeded');
(*Procedure completed unsuccessfully.*)
STOP.

In Step 7, the best approximation to β we can expect for $w_{1,N}(t_k)$ is $O(h^n)$, if the approximation method selected for Step 6 gives $O(h^n)$ rate of convergence.

The value $t_0 = TK$ selected in Step 1 is the slope of the straight line through (a, α) and (b, β). If the problem satisfies the hypotheses of Theorem 11.1, any choice of t_0 will give convergence; but given a good choice of t_0, the convergence will improve and the procedure will work for many problems that do not satisfy these hypotheses.

EXAMPLE 1 Consider the boundary-value problem

$$y'' = \frac{1}{8}(32 + 2x^3 - yy'), \quad 1 \le x \le 3, \quad y(1) = 17, \quad y(3) = \frac{43}{3},$$

which has the exact solution $y(x) = x^2 + 16/x$.

Applying the Shooting method given in Algorithm 11.2 to this problem requires approximating the initial-value problems

$$y'' = \frac{1}{8}(32 + 2x^3 - yy'), \quad 1 \le x \le 3, \quad y(1) = 17, \quad y'(1) = t_k,$$

and

$$z'' = \frac{\partial f}{\partial y} z + \frac{\partial f}{\partial y'} z' = -\frac{1}{8}(y'z + yz'), \quad 1 \le x \le 3, \quad z(1) = 0, \quad z'(1) = 1,$$

at each step in the iteration.

Table 11.2

x_i	$w_{1,i}$	$y(x_i)$	$\lvert w_{1,i} - y(x_i)\rvert$
1.0	17.000000	17.000000	
1.1	15.755495	15.755455	4.06×10^{-5}
1.2	14.773389	14.773333	5.60×10^{-5}
1.3	13.997752	13.997692	5.94×10^{-5}
1.4	13.388629	13.388571	5.71×10^{-5}
1.5	12.916719	12.916667	5.23×10^{-5}
1.6	12.560046	12.560000	4.64×10^{-5}
1.7	12.301805	12.301765	4.02×10^{-5}
1.8	12.128923	12.128889	3.14×10^{-5}
1.9	12.031081	12.031053	2.84×10^{-5}
2.0	12.000023	12.000000	2.32×10^{-5}
2.1	12.029066	12.029048	1.84×10^{-5}
2.2	12.112741	12.112727	1.40×10^{-5}
2.3	12.246532	12.246522	1.01×10^{-5}
2.4	12.426673	12.426667	6.68×10^{-6}
2.5	12.650004	12.650000	3.61×10^{-6}
2.6	12.913847	12.913846	9.17×10^{-7}
2.7	13.215924	13.215926	1.43×10^{-6}
2.8	13.554282	13.554286	3.46×10^{-6}
2.9	13.927236	13.927241	5.21×10^{-6}
3.0	14.333327	14.333333	6.69×10^{-6}

If the stopping technique

$$|w_{1,N}(t_k) - y(3)| \leq 10^{-5}$$

is used, this problem requires four iterations and $t_4 = -14.000203$. The results obtained for this value of t are shown in Table 11.2. ■

Although Newton's method used with the shooting technique requires the solution of an additional initial-value problem, it will generally be faster than the Secant method. Both methods are only locally convergent, since they require good initial approximations. For a general discussion of the convergence of the shooting techniques for nonlinear problems, the reader is referred to the excellent book by Keller [K,H]. In that reference, more general boundary conditions are discussed. It is also noted that the shooting technique for nonlinear problems is sensitive to round-off errors, especially if the solution $y(x)$ and $z(x, t)$ are rapidly increasing functions on $[a, b]$.

EXERCISE SET 11.2

1. Use the Nonlinear Shooting Algorithm with $h = 0.5$ to approximate the solution to the boundary-value problem

$$y'' = -(y')^2 - y + \ln x, \quad \text{for} \quad 1 \leq x \leq 2, \quad \text{where} \quad y(1) = 0 \quad \text{and} \quad y(2) = \ln 2.$$

Compare your results to the actual solution $y(x) = \ln x$.

2. Use the Nonlinear Shooting Algorithm with $h = 0.25$ to approximate the solution to the boundary-value problem

$$y'' = 2y^3, \quad \text{for} \quad 1 \leq x \leq 2, \quad \text{where} \quad y(1) = \frac{1}{4} \quad \text{and} \quad y(2) = \frac{1}{5}.$$

Compare your results to the actual solution $y(x) = 1/(x + 3)$.

3. Use the Nonlinear Shooting method with $TOL = 10^{-4}$ to approximate the solution to the following boundary-value problems. The actual solution is given for comparison to your results.

   a. $y'' = y^3 - yy'$, $1 \leq x \leq 2$, $y(1) = \frac{1}{2}$, $y(2) = \frac{1}{3}$; use $h = 0.1$ and compare the results to $y(x) = (x + 1)^{-1}$.

   b. $y'' = 2y^3 - 6y - 2x^3$, $1 \leq x \leq 2$, $y(1) = 2$, $y(2) = \frac{5}{2}$; use $h = 0.1$ and compare the results to $y(x) = x + x^{-1}$.

   c. $y'' = y' + 2(y - \ln x)^3 - x^{-1}$, $1 \leq x \leq 2$, $y(1) = 1$, $y(2) = \frac{1}{2} + \ln 2$; use $h = 0.1$ and compare the results to $y(x) = x^{-1} + \ln x$.

   d. $y'' = [x^2(y')^2 - 9y^2 + 4x^6]/x^5$, $1 \leq x \leq 2$, $y(1) = 0$, $y(2) = \ln 256$; use $h = 0.05$ and compare the results to $y(x) = x^3 \ln x$.

4. Change Algorithm 11.2 to incorporate the Secant method instead of Newton's method. Use $t_0 = (\beta - \alpha)/(b - a)$ and $t_1 = t_0 + (\beta - y(b, t_0))/(b - a)$.

5. Repeat Exercise 3(a) and 3(c) using the Secant algorithm derived in Exercise 4, and compare the number of iterations required for the two methods.

6. The Van der Pol equation

$$y'' - \mu(y^2 - 1)y' + y = 0, \quad \mu > 0$$

governs the flow of current in a vacuum tube with three internal elements. Let $\mu = \frac{1}{2}$, $y(0) = 0$, and $y(2) = 1$. Approximate the solution $y(t)$ for $t = 0.2i$, where $1 \leq i \leq 9$.

11.3 Finite-Difference Methods for Linear Problems

Although the Shooting methods can be used for both linear and nonlinear boundary-value problems, they often present problems of instability. The methods in this section have better stability characteristics, but they generally require more work to obtain a specified accuracy.

Methods involving finite differences for solving boundary-value problems replace each of the derivatives in the differential equation by an appropriate difference-quotient approximation of the type considered in Section 4.1. The difference quotient is chosen to maintain a specified order of truncation error. However, the parameter h cannot be chosen too small because of the instability of the finite-difference approximations to derivatives.

The linear second-order boundary-value problem,

$$\textbf{(11.14)} \qquad y'' = p(x)y' + q(x)y + r(x), \quad a \leq x \leq b, \quad y(a) = \alpha, \quad y(b) = \beta.$$

requires that difference-quotient approximations be used to approximate both y' and y''. First, we select an integer $N > 0$ and divide the interval $[a, b]$ into $(N + 1)$ equal subintervals whose endpoints are the mesh points $x_i = a + ih$, for $i = 0, 1, \ldots, N + 1$, where $h = (b - a)/(N + 1)$. Choosing the constant h in this manner facilitates the application of a matrix algorithm from Chapter 6, which solves a linear system involving an $N \times N$ matrix.

At the interior mesh points, x_i, for $i = 1, 2, \ldots, N$, the differential equation to be approximated is

$$\textbf{(11.15)} \qquad y''(x_i) = p(x_i)y'(x_i) + q(x_i)y(x_i) + r(x_i).$$

Expanding y in a third Taylor polynomial about x_i evaluated at x_{i+1} and x_{i-1}, we have

$$y(x_{i+1}) = y(x_i + h) = y(x_i) + hy'(x_i) + \frac{h^2}{2}y''(x_i) + \frac{h^3}{6}y'''(x_i) + \frac{h^4}{24}y^{(4)}(\xi_i^+),$$

for some ξ_i^+ in (x_i, x_{i+1}), and

$$y(x_{i-1}) = y(x_i - h) = y(x_i) - hy'(x_i) + \frac{h^2}{2}y''(x_i) - \frac{h^3}{6}y'''(x_i) + \frac{h^4}{24}y^{(4)}(\xi_i^-),$$

for some ξ_i^- in (x_{i-1}, x_i), assuming $y \in C^4[x_{i-1}, x_{i+1}]$. If these equations are added, the terms involving $y'(x_i)$ and $y'''(x_i)$ are eliminated, and a simple algebraic manipulation gives

$$y''(x_i) = \frac{1}{h^2}[y(x_{i+1}) - 2y(x_i) + y(x_{i-1})] - \frac{h^2}{24}[y^{(4)}(\xi_i^+) + y^{(4)}(\xi_i^-)].$$

The Intermediate Value Theorem can be used to simplify this further to

(11.16)
$$y''(x_i) = \frac{1}{h^2}[y(x_{i+1}) - 2y(x_i) + y(x_{i-1})] - \frac{h^2}{12}y^{(4)}(\xi_i),$$

for some ξ_i in (x_{i-1}, x_{i+1}). This is called the **centered-difference formula** for $y''(x_i)$.

A centered-difference formula for $y'(x_i)$ is obtained in a similar manner (the details are considered in Section 4.1), resulting in

(11.17)
$$y'(x_i) = \frac{1}{2h}[y(x_{i+1}) - y(x_{i-1})] - \frac{h^2}{6}y'''(\eta_i),$$

for some η_i in (x_{i-1}, x_{i+1}).

The use of these centered-difference formulas in Eq. (11.15) results in the equation

$$\frac{y(x_{i+1}) - 2y(x_i) + y(x_{i-1})}{h^2} = p(x_i)\left[\frac{y(x_{i+1}) - y(x_{i-1})}{2h}\right] + q(x_i)y(x_i)$$
$$+ r(x_i) - \frac{h^2}{12}\left[2p(x_i)y'''(\eta_i) - y^{(4)}(\xi_i)\right].$$

A Finite-Difference method with truncation error of order $O(h^2)$ results by using this equation together with the boundary conditions $y(a) = \alpha$ and $y(b) = \beta$ to define

$$w_0 = \alpha, \qquad w_{N+1} = \beta$$

and

(11.18)
$$\left(\frac{2w_i - w_{i+1} - w_{i-1}}{h^2}\right) + p(x_i)\left(\frac{w_{i+1} - w_{i-1}}{2h}\right) + q(x_i)w_i = -r(x_i),$$

for each $i = 1, 2, \ldots, N$.

In the form we will consider, Eq. (11.18) is rewritten as

$$-\left(1 + \frac{h}{2}p(x_i)\right)w_{i-1} + (2 + h^2q(x_i))w_i - \left(1 - \frac{h}{2}p(x_i)\right)w_{i+1} = -h^2r(x_i),$$

and the resulting system of equations is expressed in the tridiagonal $N \times N$ matrix form

(11.19)
$$A\mathbf{w} = \mathbf{b}, \quad \text{where}$$

$$A = \begin{bmatrix} 2 + h^2 q(x_1) & -1 + \frac{h}{2}p(x_1) & 0 & \cdots & \cdots & 0 \\ -1 - \frac{h}{2}p(x_2) & 2 + h^2 q(x_2) & -1 + \frac{h}{2}p(x_2) & \ddots & & \vdots \\ 0 & \ddots & \ddots & \ddots & \ddots & 0 \\ \vdots & & \ddots & \ddots & \ddots & -1 + \frac{h}{2}p(x_{N-1}) \\ 0 & \cdots & \cdots & 0 & -1 - \frac{h}{2}p(x_N) & 2 + h^2 q(x_N) \end{bmatrix},$$

$$\mathbf{w} = \begin{bmatrix} w_1 \\ w_2 \\ \vdots \\ w_{N-1} \\ w_N \end{bmatrix}, \quad \text{and} \quad \mathbf{b} = \begin{bmatrix} -h^2 r(x_1) + \left(1 + \frac{h}{2}p(x_1)\right) w_0 \\ -h^2 r(x_2) \\ \vdots \\ -h^2 r(x_{N-1}) \\ -h^2 r(x_N) + \left(1 - \frac{h}{2}p(x_N)\right) w_{N+1} \end{bmatrix}.$$

The following theorem gives conditions under which the tridiagonal linear system (11.19) has a unique solution. Its proof is a consequence of Theorem 6.30 and is considered in Exercise 9.

Theorem 11.3 Suppose that p, q, and r are continuous on $[a, b]$. If $q(x) \geq 0$ on $[a, b]$, then the tridiagonal linear system (11.19) has a unique solution provided that $h < 2/L$, where $L = \max_{a \leq x \leq b} |p(x)|$. ■

It should be noted that the hypotheses of Theorem 11.3 guarantee a unique solution to the boundary-value problem (11.14), but they do not guarantee that $y \in C^4[a, b]$. We need to establish that $y^{(4)}$ is continuous on $[a, b]$ to ensure that the truncation error has order $O(h^2)$.

Algorithm 11.3 implements the Linear Finite-Difference method.

ALGORITHM 11.3

Linear Finite-Difference

To approximate the solution of the boundary-value problem

$$y'' = p(x)y' + q(x)y + r(x), \quad a \leq x \leq b, \quad y(a) = \alpha, \quad y(b) = \beta:$$

INPUT endpoints a, b; boundary conditions α, β; integer $N \geq 2$.

OUTPUT approximations w_i to $y(x_i)$ for each $i = 0, 1, \ldots, N + 1$.

Step 1 Set $h = (b - a)/(N + 1)$;
$x = a + h$;
$a_1 = 2 + h^2 q(x)$;
$b_1 = -1 + (h/2)p(x)$;
$d_1 = -h^2 r(x) + (1 + (h/2)p(x))\alpha$.

Step 2 For $i = 2, \ldots, N-1$
set $x = a + ih$;
$a_i = 2 + h^2 q(x)$;
$b_i = -1 + (h/2)p(x)$;
$c_i = -1 - (h/2)p(x)$;
$d_i = -h^2 r(x)$.

Step 3 Set $x = b - h$;
$a_N = 2 + h^2 q(x)$;
$c_N = -1 - (h/2)p(x)$;
$d_N = -h^2 r(x) + (1 - (h/2)p(x))\beta$.

Step 4 Set $l_1 = a_1$; (*Steps* 4–8 *solve a tridiagonal linear system using Algorithm* 6.7.)
$u_1 = b_1/a_1$;
$z_1 = d_1/l_1$.

Step 5 For $i = 2, \ldots, N-1$ set $l_i = a_i - c_i u_{i-1}$;
$u_i = b_i/l_i$;
$z_i = (d_i - c_i z_{i-1})/l_i$.

Step 6 Set $l_N = a_N - c_N u_{N-1}$;
$z_N = (d_N - c_N z_{N-1})/l_N$.

Step 7 Set $w_0 = \alpha$;
$w_{N+1} = \beta$.
$w_N = z_N$.

Step 8 For $i = N-1, \ldots, 1$ set $w_i = z_i - u_i w_{i+1}$.

Step 9 For $i = 0, \ldots, N+1$ set $x = a + ih$;
OUTPUT (x, w_i).

Step 10 STOP. (*Procedure is complete.*)

EXAMPLE 1 Algorithm 11.3 will be used to approximate the solution to the linear boundary-value problem

$$y'' = -\frac{2}{x}y' + \frac{2}{x^2}y + \frac{\sin(\ln x)}{x^2}, \quad 1 \le x \le 2, \quad y(1) = 1, \quad y(2) = 2,$$

which was also approximated by the Shooting method in Example 2 of Section 11.1. For this example, we will use $N = 9$, so $h = 0.1$ and we have the same spacing as in Example 2 of Section 11.1. The results are listed in Table 11.3.

Note that these results are considerably less accurate than those obtained in Example 2 of Section 11.1. This is because the method used in that example involved a Runge-Kutta technique with truncation error of order $O(h^4)$, whereas the difference method used here has truncation error of order $O(h^2)$. ■

To obtain a difference method with greater accuracy, we can proceed in a number of ways. Using fifth-order Taylor series for approximating $y''(x_i)$ and $y'(x_i)$ results in a truncation error term involving h^4. However, this process requires using multiples not only

Table 11.3

x_i	w_i	$y(x_i)$	$\lvert w_i - y(x_i)\rvert$
1.0	1.00000000	1.00000000	
1.1	1.09260052	1.09262930	2.88×10^{-5}
1.2	1.18704313	1.18708484	4.17×10^{-5}
1.3	1.28333687	1.28338236	4.55×10^{-5}
1.4	1.38140205	1.38144595	4.39×10^{-5}
1.5	1.48112026	1.48115942	3.92×10^{-5}
1.6	1.58235990	1.58239246	3.26×10^{-5}
1.7	1.68498902	1.68501396	2.49×10^{-5}
1.8	1.78888175	1.78889853	1.68×10^{-5}
1.9	1.89392110	1.89392951	8.41×10^{-6}
2.0	2.00000000	2.00000000	

of $y(x_{i+1})$ and $y(x_{i-1})$, but also $y(x_{i+2})$ and $y(x_{i-2})$ in the approximation formulas for $y''(x_i)$ and $y'(x_i)$. This leads to difficulty at $i = 0$ and $i = N$. Moreover, the resulting system of equations analogous to (11.19) is not in tridiagonal form, and the solution to the system requires many more calculations.

Instead of attempting to obtain a difference method with a higher-order truncation error in this manner, it is generally more satisfactory to consider a reduction in step size. In addition, it can be shown (see, for example, [K,H, p. 81]) that Richardson's extrapolation can be used effectively for this method since the error term is expressed in even powers of h with coefficients independent of h, provided y is sufficiently differentiable.

EXAMPLE 2 Applying Richardson's extrapolation to approximate the solution to the boundary-value problem

$$y'' = -\frac{2}{x}y' + \frac{2}{x^2}y + \frac{\sin(\ln x)}{x^2}, \quad 1 \le x \le 2, \quad y(1) = 1, \quad y(2) = 2,$$

with $h = 0.1, 0.05$, and 0.025 gives the results listed in Table 11.4. The first extrapolation is

$$\text{Ext}_{1i} = \frac{4w_i(h = 0.05) - w_i(h = 0.1)}{3};$$

the second extrapolation is

$$\text{Ext}_{2i} = \frac{4w_i(h = 0.025) - w_i(h = 0.05)}{3};$$

and the final extrapolation is

$$\text{Ext}_{3i} = \frac{16\text{Ext}_{2i} - \text{Ext}_{1i}}{15}.$$

Table 11.4

x_i	$w_i(h = 0.1)$	$w_i(h = 0.05)$	$w_i(h = 0.025)$	Ext_{1i}	Ext_{2i}	Ext_{3i}
1.0	1.00000000	1.00000000	1.00000000	1.00000000	1.00000000	1.00000000
1.1	1.09260052	1.09262207	1.09262749	1.09262925	1.09262930	1.09262930
1.2	1.18704313	1.18707436	1.18708222	1.18708477	1.18708484	1.18708484
1.3	1.28333687	1.28337094	1.28337950	1.28338230	1.28338236	1.28338236
1.4	1.38140205	1.38143493	1.38144319	1.38144589	1.38144595	1.38144595
1.5	1.48112026	1.48114959	1.48115696	1.48115937	1.48115941	1.48115942
1.6	1.58235990	1.58238429	1.58239042	1.58239242	1.58239246	1.58239246
1.7	1.68498902	1.68500770	1.68501240	1.68501393	1.68501396	1.68501396
1.8	1.78888175	1.78889432	1.78889748	1.78889852	1.78889853	1.78889853
1.9	1.89392110	1.89392740	1.89392898	1.89392950	1.89392951	1.89392951
2.0	2.00000000	2.00000000	2.00000000	2.00000000	2.00000000	2.00000000

All the results of EXT_{3i} are correct to the decimal places listed. In fact, if sufficient digits are maintained, this approximation gives results that agree with the exact solution with maximum error of 6.3×10^{-11} at the mesh points, an impressive improvement. ■

EXERCISE SET 11.3

1. The boundary-value problem

$$y'' = 4(y - x), \quad 0 \le x \le 1, \quad y(0) = 0, \quad y(1) = 2$$

has the solution $y(x) = e^2(e^4 - 1)^{-1}(e^{2x} - e^{-2x}) + x$. Use the Linear Finite-Difference Algorithm to approximate the solution and compare the results to the actual solution.

   a. With $h = \frac{1}{3}$; b. With $h = \frac{1}{4}$.

2. The boundary-value problem

$$y'' = y' + 2y + \cos x, \quad 0 \le x \le \frac{\pi}{2}, \quad y(0) = -0.3, \quad y\left(\frac{\pi}{2}\right) = -0.1$$

has the solution $y(x) = -\frac{1}{10}[\sin x + 3\cos x]$. Use the Linear Finite-Difference Algorithm to approximate the solution and compare the results to the actual solution.

   a. With $h = \frac{\pi}{4}$; b. With $h = \frac{\pi}{6}$.

3. Use the Linear Finite-Difference Algorithm to approximate the solution to the following boundary-value problems.

   a. $y'' = -3y' + 2y + 2x + 3, \quad 0 \le x \le 1, \ y(0) = 2, \ y(1) = 1$; use $h = 0.1$.

   b. $y'' = -\frac{4}{x}y' + \frac{2}{x^2}y - \frac{2}{x^2}\ln x, \quad 1 \le x \le 2, \ y(1) = -\frac{1}{2}, \ y(2) = \ln 2$; use $h = 0.05$.

   c. $y'' = -(x+1)y' + 2y + (1 - x^2)e^{-x}, \quad 0 \le x \le 1, \ y(0) = -1, \ y(1) = 0$; use $h = 0.1$.

   d. $y'' = \frac{y'}{x} + \frac{3y}{x^2} + \frac{\ln x}{x} - 1, \quad 1 \le x \le 2, \ y(1) = y(2) = 0$; use $h = 0.1$.

4. Although $q(x) < 0$ in the following boundary-value problems, unique solutions exist and are given. Use the Linear Finite-Difference Algorithm to approximate the solutions and compare the results to the actual solutions.

a. $y'' + y = 0, \quad 0 \le x \le \frac{\pi}{4}, \; y(0) = 1, \; y\left(\frac{\pi}{4}\right) = 1$; use $h = \frac{\pi}{20}$; actual solution $y(x) = \cos x + (\sqrt{2} - 1)\sin x$.

b. $y'' + 4y = \cos x, \quad 0 \le x \le \frac{\pi}{4}, \; y(0) = 0, \; y\left(\frac{\pi}{4}\right) = 0$; use $h = \frac{\pi}{20}$; actual solution $y(x) = -\frac{1}{3}\cos 2x - \frac{\sqrt{2}}{6}\sin 2x + \frac{1}{3}\cos x$.

c. $y'' = -\frac{4}{x}y' - \frac{2}{x^2}y + \frac{2\ln x}{x^2}, \quad 1 \le x \le 2, \; y(1) = \frac{1}{2}, \; y(2) = \ln 2$; use $h = 0.05$; actual solution $y(x) = \frac{4}{x} - \frac{2}{x^2} + \ln x - \frac{3}{2}$.

d. $y'' = 2y' - y + xe^x - x, \quad 0 \le x \le 2, \; y(0) = 0, \; y(2) = -4$; use $h = 0.2$; actual solution $y(x) = \frac{1}{6}x^3e^x - \frac{5}{3}xe^x + 2e^x - x - 2$.

5. Use the Linear Finite-Difference Algorithm to approximate the solution $y(x) = e^{-10x}$ to the boundary-value problem

$$y'' = 100y, \quad 0 \le x \le 1, \; y(0) = 1, \; y(1) = e^{-10}.$$

Use $h = 0.1$ and 0.05. Can you explain the consequences?

6. Repeat Exercise 3(a) and (b) using the extrapolation discussed in Example 2.

7. The lead example of this chapter concerned the deflection of a beam with supported ends subject to uniform loading. The boundary-value problem governing this physical situation is

$$\frac{d^2w}{dx^2} = \frac{S}{EI}w + \frac{qx}{2EI}(x - l), \quad 0 < x < l,$$

with boundary conditions $w(0) = 0$ and $w(l) = 0$.

Suppose the beam is a W10-type steel I-beam with the following characteristics: length $l = 120$ in., intensity of uniform load $q = 100$ lb/ft, modulus of elasticity $E = 3.0 \times 10^7$ lb/in.^{2}, stress at ends $S = 1000$ lb, and central moment of inertia $I = 625$ in.^{4}.

a. Approximate the deflection $w(x)$ of the beam every 6 in.

b. The actual relationship is given by

$$w(x) = c_1e^{ax} + c_2e^{-ax} + b(x - l)x + c,$$

where $c_1 = 7.7042537 \times 10^4$, $c_2 = 7.9207462 \times 10^4$, $a = 2.3094010 \times 10^{-4}$, $b = -4.1666666 \times 10^{-3}$, and $c = -1.5625 \times 10^5$. Is the maximum error on the interval within 0.2 in.?

c. State law requires that $\max_{0<x<l} w(x) < 1/300$. Does this beam meet state code?

8. The deflection of a uniformly loaded, long rectangular plate under an axial tension force is governed by a second-order differential equation. Let S represent the axial force and q the intensity of the uniform load. The deflection W along the elemental length is given by

$$W''(x) - \frac{S}{D}W(x) = \frac{-ql}{2D}x + \frac{q}{2D}x^2, \quad 0 \le x \le l, \; W(0) = W(l) = 0,$$

where l is the length of the plate and D is the flexual rigidity of the plate. Let $q = 200$ lb/in.^{2}, $S = 100$ lb/in., $D = 8.8 \times 10^7$ lb/in., and $l = 50$ in. Approximate the deflection at 1-in. intervals.

9. Prove Theorem 11.3. [*Hint*: To use Theorem 6.30, first show that $|hp(x_i)/2| < 1$ implies that $|1 + hp(x_i)/2| + |1 - hp(x_i)/2| = 2$.]

10. Show that if $y \in C^6[a, b]$ and if $w_0, w_1, \ldots, w_{N+1}$ satisfy Eq. (11.18), then

$$w_i - y(x_i) = Ah^2 + O(h^4),$$

where A is independent of h, provided $q(x) \geq w > 0$ on $[a, b]$ for some w.

11.4 Finite-Difference Methods for Nonlinear Problems

For the general nonlinear boundary-value problem

$$y'' = f(x, y, y'), \quad a \leq x \leq b, \quad y(a) = \alpha, \quad y(b) = \beta,$$

the difference method is similar to the method applied to linear problems in Section 11.3. Here, however, the system of equations will not be linear, so an iterative process is required to solve it.

For the development of the procedure, we assume throughout that f satisfies the following conditions:

1. f and the partial derivatives $f_y \equiv \partial f/\partial y$ and $f_{y'} \equiv \partial f/\partial y'$ are all continuous on

$$D = \{(x, y, y') \mid a \leq x \leq b, \quad -\infty < y < \infty, -\infty < y' < \infty\};$$

2. $f_y(x, y, y') \geq \delta$ on D for some $\delta > 0$;
3. Constants k and L exist, with

$$k = \max_{(x,y,y')\in D} |f_y(x, y, y')|, \quad L = \max_{(x,y,y')\in D} |f_{y'}(x, y, y')|.$$

This ensures, by Theorem 11.1, that a unique solution exists.

As in the linear case, we divide $[a, b]$ into $(N + 1)$ equal subintervals of width $h = (b - a)/(N + 1)$ whose endpoints are at $x_i = a + ih$, for $i = 0, 1, \ldots, N + 1$. Assuming that the exact solution has a bounded fourth derivative allows us to replace $y''(x_i)$ and $y'(x_i)$ in each of the equations

$$y''(x_i) = f(x_i, y(x_i), y'(x_i))$$

by the appropriate centered-difference formula given in Eqs. (11.16) and (11.17). This gives, for each $i = 1, 2, \ldots, N$,

$$\frac{y(x_{i+1}) - 2y(x_i) + y(x_{i-1})}{h^2} = f\left(x_i, y(x_i), \frac{y(x_{i+1}) - y(x_{i-1})}{2h} - \frac{h^2}{6}y'''(\eta_i)\right) + \frac{h^2}{12}y^{(4)}(\xi_i),$$

for some ξ_i and η_i in the interval (x_{i-1}, x_{i+1}).

As in the linear case, the difference method results when the error terms are deleted and the boundary conditions are employed:

$$w_0 = \alpha, \quad w_{N+1} = \beta,$$

and

$$-\frac{w_{i+1} - 2w_i + w_{i-1}}{h^2} + f\left(x_i, w_i, \frac{w_{i+1} - w_{i-1}}{2h}\right) = 0,$$

for each $i = 1, 2, \ldots, N$.

The $N \times N$ nonlinear system obtained from this method,

(11.20)

$$\begin{aligned}
2w_1 - w_2 + h^2 f\left(x_1, w_1, \frac{w_2 - \alpha}{2h}\right) - \alpha &= 0,\\
-w_1 + 2w_2 - w_3 + h^2 f\left(x_2, w_2, \frac{w_3 - w_1}{2h}\right) &= 0,\\
&\vdots\\
-w_{N-2} + 2w_{N-1} - w_N + h^2 f\left(x_{N-1}, w_{N-1}, \frac{w_N - w_{N-2}}{2h}\right) &= 0,\\
-w_{N-1} + 2w_N + h^2 f\left(x_N, w_N, \frac{\beta - w_{N-1}}{2h}\right) - \beta &= 0.
\end{aligned}$$

has a unique solution provided that $h < 2/L$, as shown in [K,H, p. 86].

We use Newton's method for nonlinear systems, discussed in Section 10.2, to approximate the solution to this system. A sequence of iterates $\{(w_1^{(k)}, w_2^{(k)}, \ldots, w_N^{(k)})^t\}$ is generated that converges to the solution of system (11.20), provided that the initial approximation $(w_1^{(0)}, w_2^{(0)}, \ldots, w_N^{(0)})^t$ is sufficiently close to the solution $(w_1, w_2, \ldots, w_N)^t$, and that the Jacobian matrix for the system is nonsingular. For system (11.20), the Jacobian matrix $J(w_1, \ldots, w_N)$ is tridiagonal with ij-th entry

(11.21)

$$J(w_1, \ldots, w_N)_{ij} = \begin{cases} -1 + \dfrac{h}{2} f_{y'}\left(x_i, w_i, \dfrac{w_{i+1} - w_{i-1}}{2h}\right), & \text{for } i = j - 1 \text{ and } j = 2, \ldots, N,\\[2ex] 2 + h^2 f_y\left(x_i, w_i, \dfrac{w_{i+1} - w_{i-1}}{2h}\right), & \text{for } i = j \text{ and } j = 1, \ldots, N,\\[2ex] -1 - \dfrac{h}{2} f_{y'}\left(x_i, w_i, \dfrac{w_{i+1} - w_{i-1}}{2h}\right), & \text{for } i = j + 1 \text{ and } j = 1, \ldots, N - 1, \end{cases}$$

where $w_0 = \alpha$ and $w_{N+1} = \beta$.

Newton's method for nonlinear systems requires that at each iteration the $N \times N$ linear system

$$\begin{aligned}
J(w_1, \ldots, w_N)(v_1, \ldots, v_n)^t = -\Bigg(&2w_1 - w_2 - \alpha + h^2 f\left(x_1, w_1, \frac{w_2 - \alpha}{2h}\right),\\
&-w_1 + 2w_2 - w_3 + h^2 f\left(x_2, w_2, \frac{w_3 - w_1}{2h}\right), \ldots,
\end{aligned}$$

$$-w_{N-2} + 2w_{N-1} - w_N + h^2 f\left(x_{N-1}, w_{N-1}, \frac{w_N - w_{N-2}}{2h}\right),$$

$$\left.-w_{N-1} + 2w_N + h^2 f\left(x_N, w_N, \frac{\beta - w_{N-1}}{2h}\right) - \beta\right)^t$$

be solved for $v_1, v_2, \ldots, v_N$, since

$$w_i^{(k)} = w_i^{(k-1)} + v_i, \quad \text{for each } i = 1, 2, \ldots, N.$$

Since J is tridiagonal, this is not as formidable a problem as it might at first appear. The Crout Factorization Algorithm for Tridiagonal Systems 6.7 can be applied. The process is detailed in Algorithm 11.4.

ALGORITHM 11.4

Nonlinear Finite-Difference

To approximate the solution to the nonlinear boundary-value problem

$$y'' = f(x, y, y'), \quad a \le x \le b, \quad y(a) = \alpha, \quad y(b) = \beta:$$

INPUT endpoints a, b; boundary conditions α, β; integer $N \ge 2$; tolerance TOL; maximum number of iterations M.

OUTPUT approximations w_i to $y(x_i)$ for each $i = 0, 1, \ldots, N+1$ or a message that the maximum number of iterations was exceeded.

Step 1 Set $h = (b - a)/(N + 1)$;
$w_0 = \alpha$;
$w_{N+1} = \beta$.

Step 2 For $i = 1, \ldots, N$ set $w_i = \alpha + i\left(\dfrac{\beta - \alpha}{b - a}\right)h$.

Step 3 Set $k = 1$.

Step 4 While $k \le M$ do Steps 5–16.

Step 5 Set $x = a + h$;
$t = (w_2 - \alpha)/(2h)$;
$a_1 = 2 + h^2 f_y(x, w_1, t)$;
$b_1 = -1 + (h/2) f_{y'}(x, w_1, t)$;
$d_1 = -(2w_1 - w_2 - \alpha + h^2 f(x, w_1, t))$.

Step 6 For $i = 2, \ldots, N - 1$
set $x = a + ih$;
$t = (w_{i+1} - w_{i-1})/(2h)$;
$a_i = 2 + h^2 f_y(x, w_i, t)$;
$b_i = -1 + (h/2) f_{y'}(x, w_i, t)$;
$c_i = -1 - (h/2) f_{y'}(x, w_i, t)$;
$d_i = -(2w_i - w_{i+1} - w_{i-1} + h^2 f(x, w_i, t))$.

Step 7 Set $x = b - h$;
$t = (\beta - w_{N-1})/(2h)$;
$a_N = 2 + h^2 f_y(x, w_N, t)$;
$c_N = -1 - (h/2) f_{y'}(x, w_N, t)$;
$d_N = -(2w_N - w_{N-1} - \beta + h^2 f(x, w_N, t))$.

Step 8 Set $l_1 = a_1$; (*Steps* 8–12 *solve a tridiagonal linear system using Algorithm* 6.7.)
$u_1 = b_1/a_1$;
$z_1 = d_1/l_1$.

Step 9 For $i = 2, \ldots, N-1$ set $l_i = a_i - c_i u_{i-1}$;
$u_i = b_i/l_i$;
$z_i = (d_i - c_i z_{i-1})/l_i$.

Step 10 Set $l_N = a_N - c_N u_{N-1}$;
$z_N = (d_N - c_N z_{N-1})/l_N$.

Step 11 Set $v_N = z_N$;
$w_N = w_N + v_N$.

Step 12 For $i = N-1, \ldots, 1$ set $v_i = z_i - u_i v_{i+1}$;
$w_i = w_i + v_i$.

Step 13 If $\|\mathbf{v}\| \le TOL$ then do Steps 14 and 15.

Step 14 For $i = 0, \ldots, N+1$ set $x = a + ih$;
OUTPUT (x, w_i).

Step 15 STOP. (*Procedure completed successfully.*)

Step 16 Set $k = k + 1$.

Step 17 OUTPUT ('Maximum number of iterations exceeded');
(*Procedure completed unsuccessfully.*)
STOP.

It can be shown (see [IK, p. 433]) that this Nonlinear Finite-Difference method is of order $O(h^2)$.

Since a good initial approximation is required when the satisfaction of conditions (1), (2), and (3) given at the beginning of this presentation cannot be verified, an upper bound for k should be specified and, if exceeded, a new initial approximation or a reduction in step size considered. The initial approximations $w_i^{(0)}$ to w_i, for each $i = 1, 2, \ldots, N$, are obtained in Step 2 by passing a straight line through (a, α) and (b, β) and evaluating at x_i.

EXAMPLE 1 Applying Algorithm 11.4, with $h = 0.1$, to the nonlinear boundary-value problem

$$y'' = \frac{1}{8}(32 + 2x^3 - yy'), \quad 1 \le x \le 3, \quad y(1) = 17, \quad y(3) = \frac{43}{3}$$

gives the results in Table 11.5. The stopping procedure used in this example was to iterate until values of successive iterates differed by less than 10^{-8}. This was accomplished with four iterations. Note that the problem in this example is the same as that considered for the Nonlinear Shooting method, Example 1 of Section 11.2. ■

Table 11.5

x_i	w_i	$y(x_i)$	$\lvert w_i - y(x_i)\rvert$
1.0	17.000000	17.000000	
1.1	15.754503	15.755455	9.520×10^{-4}
1.2	14.771740	14.773333	1.594×10^{-3}
1.3	13.995677	13.997692	2.015×10^{-3}
1.4	13.386297	13.388571	2.275×10^{-3}
1.5	12.914252	12.916667	2.414×10^{-3}
1.6	12.557538	12.560000	2.462×10^{-3}
1.7	12.299326	12.301765	2.438×10^{-3}
1.8	12.126529	12.128889	2.360×10^{-3}
1.9	12.028814	12.031053	2.239×10^{-3}
2.0	11.997915	12.000000	2.085×10^{-3}
2.1	12.027142	12.029048	1.905×10^{-3}
2.2	12.111020	12.112727	1.707×10^{-3}
2.3	12.245025	12.246522	1.497×10^{-3}
2.4	12.425388	12.426667	1.278×10^{-3}
2.5	12.648944	12.650000	1.056×10^{-3}
2.6	12.913013	12.913846	8.335×10^{-4}
2.7	13.215312	13.215926	6.142×10^{-4}
2.8	13.553885	13.554286	4.006×10^{-4}
2.9	13.927046	13.927241	1.953×10^{-4}
3.0	14.333333	14.333333	

Table 11.6

x_i	$w_i(h = 0.05)$	$w_i(h = 0.025)$	Ext_{1i}	Ext_{2i}	Ext_{3i}
1.0	17.00000000	17.00000000	17.00000000	17.00000000	17.00000000
1.1	15.75521721	15.75539525	15.75545543	15.75545460	15.75545455
1.2	14.77293601	14.77323407	14.77333479	14.77333342	14.77333333
1.3	13.99718996	13.99756680	13.99769413	13.99769242	13.99769231
1.4	13.38800424	13.38842973	13.38857346	13.38857156	13.38857143
1.5	12.91606471	12.91651628	12.91666881	12.91666680	12.91666667
1.6	12.55938618	12.55984665	12.56000217	12.56000014	12.56000000
1.7	12.30115670	12.30161280	12.30176684	12.30176484	12.30176471
1.8	12.12830042	12.12874287	12.12899094	12.12888902	12.12888889
1.9	12.03049438	12.03091316	12.03105457	12.03105275	12.03105263
2.0	11.99948020	11.99987013	12.00000179	12.00000011	12.00000000
2.1	12.02857252	12.02892892	12.02902924	12.02904772	12.02904762
2.2	12.11230149	12.11262089	12.11272872	12.11272736	12.11272727
2.3	12.24614846	12.24642848	12.24652299	12.24652182	12.24652174
2.4	12.42634789	12.42658702	12.42666773	12.42666673	12.42666667
2.5	12.64973666	12.64993420	12.65000086	12.65000005	12.65000000
2.6	12.91363828	12.91379422	12.91384683	12.91384620	12.91384615
2.7	13.21577275	13.21588765	13.21592641	13.21592596	13.21592593
2.8	13.55418579	13.55426075	13.55428603	13.55428573	13.55428571
2.9	13.92719268	13.92722921	13.92724153	13.92724139	13.92724138
3.0	14.33333333	14.33333333	14.33333333	14.33333333	14.33333333

Richardson's extrapolation procedure can also be used for the Nonlinear Finite-Difference method. Table 11.6 lists the results when this method is applied to our example using $h = 0.1, 0.05$, and 0.025, with four iterations in each case. The notation is the same as in Example 2 of Section 11.3, and the values of EXT_{3i} are all accurate to the places listed, with an actual maximum error of 3.68×10^{-10}. The values of $w_i(h = 0.1)$ are omitted from the table, since they were listed previously.

EXERCISE SET 11.4

1. Use the Nonlinear Finite-Difference Algorithm with $h = 0.5$ to approximate the solution to the boundary-value problem

$$y'' = -(y')^2 - y + \ln x, \quad 1 \le x \le 2, \quad y(1) = 0, \quad y(2) = \ln 2.$$

Compare your results to the actual solution $y(x) = \ln x$.

2. Use the Nonlinear Finite-Difference Algorithm with $h = 0.25$ to approximate the solution to the boundary-value problem

$$y'' = 2y^3, \quad 1 \le x \le 2, \quad y(1) = \frac{1}{4}, \quad y(2) = \frac{1}{5}.$$

Compare your results to the actual solution $y(x) = 1/(x + 3)$.

3. Use the Nonlinear Finite-Difference Algorithm with $TOL = 10^{-4}$ to approximate the solution to the following boundary-value problems. The actual solution is given for comparison to your results.

a. $y'' = y^3 - yy', \quad 1 \le x \le 2, \ y(1) = \frac{1}{2}, \ y(2) = \frac{1}{3}$; use $h = 0.1$ and compare the results to $y(x) = (x + 1)^{-1}$.

b. $y'' = 2y^3 - 6y - 2x^3, \quad 1 \le x \le 2, \ y(1) = 2, \ y(2) = \frac{5}{2}$; use $h = 0.1$ and compare the results to $y(x) = x + x^{-1}$.

c. $y'' = y' + 2(y - \ln x)^3 - x^{-1}, \quad 1 \le x \le 2, \ y(1) = 1, \ y(2) = \frac{1}{2} + \ln 2$; use $h = 0.1$ and compare the results to $y(x) = x^{-1} + \ln x$.

d. $y'' = \left(x^2(y')^2 - 9y^2 + 4x^6\right)/x^5, \quad 1 \le x \le 2, \ y(1) = 0, \ y(2) = \ln 256$; use $h = 0.05$ and compare the results to $y(x) = x^3 \ln x$.

4. Repeat Exercise 3(a) and (b) using extrapolation.

5. Show that the hypotheses listed at the beginning of the section ensure the nonsingularity of the Jacobian matrix J for $h < 2/L$.

6. In Exercise 7 of Section 11.3 the deflection of beam with supported ends subject to uniform loading was approximated. Using a more appropriate representation of curvature gives the differential equation

$$\left[1 + (w'(x))^2\right]^{-3/2} w''(x) = \frac{S}{El}w(x) + \frac{qx}{2El}(x - l), \quad \text{for } 0 < x < l.$$

Approximate the deflection $w(x)$ of the beam every 6 in. and compare the results to those of Exercise 7 of Section 11.3.

11.5 The Rayleigh-Ritz Method

In the finite-difference approach to approximating the solution to boundary-value problems, we replaced the continuous operation of differentiation with the discrete operation of finite differences. The Rayleigh-Ritz method attacks the problem from a different approach. The boundary-value problem is first reformulated as a problem of choosing, from the set of all sufficiently differentiable functions satisfying the boundary conditions, the function that minimizes a certain integral. Then the set of feasible functions is reduced in size, to result in an approximation to the solution to the minimization problem and (as a consequence) an approximation to the solution to the boundary-value problem.

To describe the Rayleigh-Ritz method we consider approximating the solution to a linear two-point boundary-value problem from beam stress analysis. This boundary-value problem is described by the differential equation

(11.22)
$$-\frac{d}{dx}\left(p(x)\frac{dy}{dx}\right) + q(x)y = f(x), \quad \text{for } 0 \le x \le 1,$$

with the boundary conditions

(11.23)
$$y(0) = y(1) = 0.$$

This differential equation describes the deflection $y(x)$ of a beam of length 1 with variable cross section represented by $q(x)$. The deflection is due to the added stresses $p(x)$ and $f(x)$.

In the discussion that follows we assume that $p \in C^1[0, 1]$ and $q, f \in C[0, 1]$. Further, we assume that there exists a constant $\delta > 0$ such that

$$p(x) \ge \delta, \quad \text{and that} \quad q(x) \ge 0, \quad \text{for each } x \text{ in } [0, 1].$$

These assumptions are sufficient to guarantee that the boundary-value problem given in (11.22) and (11.23) has a unique solution (see [BSW]).

As is the case in many boundary-value problems that describe physical phenomena, the solution to the beam equation satisfies a **variational** property. The variational principle for the beam equation is fundamental to the development of the Rayleigh-Ritz method and characterizes the solution to the beam equation as the function that minimizes a certain integral over all functions in $C_0^2[0, 1]$, the set of those functions u in $C^2[0, 1]$ with the property that $u(0) = u(1) = 0$. The following theorem gives the characterization. The proof of this theorem, while not difficult, is lengthy. It can be found in [Schul, pp. 88–89].

Theorem 11.4 Let $p \in C^1[0, 1]$, $q, f \in C[0, 1]$, and

$$p(x) \geq \delta > 0, \quad q(x) \geq 0, \quad \text{for } 0 \leq x \leq 1.$$

The function $y \in C_0^2[0, 1]$ is the unique solution to the differential equation

$$\text{(11.24)} \qquad -\frac{d}{dx}\left(p(x)\frac{dy}{dx}\right) + q(x)y = f(x), \quad 0 \leq x \leq 1,$$

if and only if y is the unique function in $C_0^2[0, 1]$ that minimizes the integral

$$\text{(11.25)} \qquad I[u] = \int_0^1 \left\{p(x)[u'(x)]^2 + q(x)[u(x)]^2 - 2f(x)u(x)\right\}\, dx. \qquad \blacksquare$$

The Rayleigh-Ritz method approximates the solution y by minimizing the integral, not over all the functions in $C_0^2[0, 1]$, but over a smaller set of functions consisting of linear combinations of certain basis functions $\phi_1, \phi_2, \ldots, \phi_n$.The basis functions are linearly independent and satisfy

$$\phi_i(0) = \phi_i(1) = 0, \qquad \text{for each } i = 1, 2, \ldots, n.$$

An approximation $\phi(x) = \sum_{i=1}^n c_i\phi_i(x)$ to the solution $y(x)$ of Eq. (11.24) is then obtained by finding constants $c_1, c_2, \ldots, c_n$ to minimize $I\left[\sum_{i=1}^n c_i\phi_i\right]$.

From Eq. (11.25),

$$\text{(11.26)} \quad I[\phi] = I\left[\sum_{i=1}^n c_i\phi_i\right]$$

$$= \int_0^1 \left\{p(x)\left[\sum_{i=1}^n c_i\phi_i'(x)\right]^2 + q(x)\left[\sum_{i=1}^n c_i\phi_i(x)\right]^2 - 2f(x)\sum_{i=1}^n c_i\phi_i(x)\right\} dx,$$

and, for a minimum to occur, it is necessary when considering I as a function of $c_1, c_2, \ldots, c_n$, to have

$$\text{(11.27)} \qquad \frac{\partial I}{\partial c_j} = 0, \qquad \text{for each } j = 1, 2, \ldots, n.$$

Differentiating (11.26) gives

$$\frac{\partial I}{\partial c_j} = \int_0^1 \left\{2p(x)\sum_{i=1}^n c_i\phi_i'(x)\phi_j'(x) + 2q(x)\sum_{i=1}^n c_i\phi_i(x)\phi_j(x) - 2f(x)\phi_j(x)\right\} dx,$$

and substituting into Eq. (11.27) yields

(11.28)
$$0 = \sum_{i=1}^{n} \left[\int_0^1 \{p(x)\phi_i'(x)\phi_j'(x) + q(x)\phi_i(x)\phi_j(x)\}\, dx \right] c_i - \int_0^1 f(x)\phi_j(x)\, dx,$$

for each $j = 1, 2, \ldots, n$.

The equations described in Eq. (11.28) produce an $n \times n$ linear system $A\mathbf{c} = \mathbf{b}$ in the variables $c_1, c_2, \ldots, c_n$, where the symmetric matrix A is given by

$$a_{ij} = \int_0^1 \left[p(x)\phi_i'(x)\phi_j'(x) + q(x)\phi_i(x)\phi_j(x)\right] dx$$

and $\mathbf{b}$ is defined by

$$b_i = \int_0^1 f(x)\phi_i(x)\, dx.$$

The most elementary choice of basis functions involves piecewise linear polynomials. The first step is to form a partition of $[0, 1]$ by choosing points $x_0, x_1, \ldots, x_{n+1}$ with

$$0 = x_0 < x_1 < \cdots < x_n < x_{n+1} = 1.$$

Letting $h_i = x_{i+1} - x_i$ for each $i = 0, 1, \ldots, n$, we define the basis functions $\phi_1(x)$, $\phi_2(x)$, $\ldots$, $\phi_n(x)$ by

(11.29)
$$\phi_i(x) = \begin{cases} 0, & 0 \le x \le x_{i-1}, \\ \dfrac{x - x_{i-1}}{h_{i-1}}, & x_{i-1} < x \le x_i, \\ \dfrac{x_{i+1} - x}{h_i}, & x_i < x \le x_{i+1}, \\ 0, & x_{i+1} < x \le 1, \end{cases}$$

for each $i = 1, 2, \ldots, n$. (See Figure 11.4.)

Figure 11.4

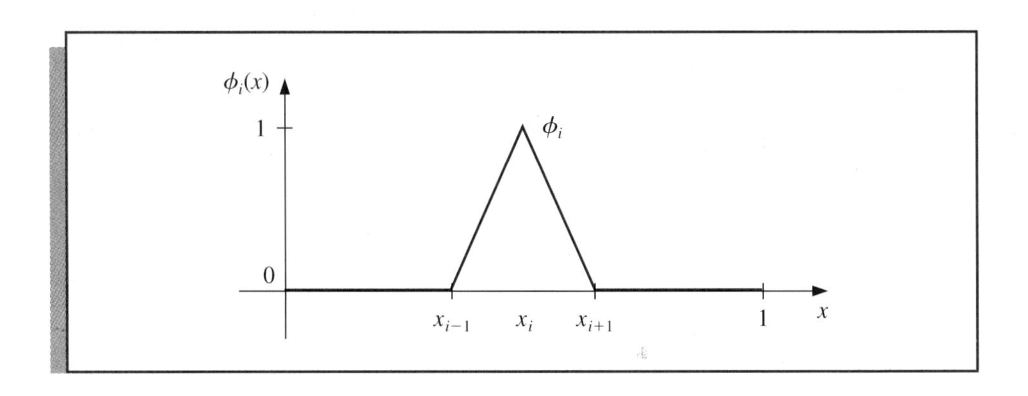

Since the functions ϕ_i are piecewise linear, the derivatives ϕ_i', while not continuous, are constant on the open subinterval (x_j, x_{j+1}) for each $j = 0, 1, \ldots, n$. Thus, we have

(11.30)
$$\phi_i'(x) = \begin{cases} 0, & 0 < x < x_{i-1}, \\ \dfrac{1}{h_{i-1}}, & x_{i-1} < x < x_i, \\ -\dfrac{1}{h_i}, & x_i < x < x_{i+1}, \\ 0, & x_{i+1} < x < 1, \end{cases}$$

for each $i = 1, 2, \ldots, n$.

Because ϕ_i and ϕ_i' are nonzero only on (x_{i-1}, x_{i+1}),

$$\phi_i(x)\phi_j(x) \equiv 0 \qquad \text{and} \qquad \phi_i'(x)\phi_j'(x) \equiv 0,$$

except when j is $i-1$, i, or $i+1$. As a consequence, the linear system given by (11.28) reduces to an $n \times n$ tridiagonal linear system. The nonzero entries in A are

$$\begin{aligned} a_{ii} &= \int_0^1 \left\{p(x)[\phi_i'(x)]^2 + q(x)[\phi_i(x)]^2\right\} dx \\ &= \left(\frac{1}{h_{i-1}}\right)^2 \int_{x_{i-1}}^{x_i} p(x)\,dx + \left(\frac{-1}{h_i}\right)^2 \int_{x_i}^{x_{i+1}} p(x)\,dx \\ &\quad + \left(\frac{1}{h_{i-1}}\right)^2 \int_{x_{i-1}}^{x_i} (x - x_{i-1})^2 q(x)\,dx + \left(\frac{1}{h_i}\right)^2 \int_{x_i}^{x_{i+1}} (x_{i+1} - x)^2 q(x)\,dx, \end{aligned}$$

for each $i = 1, 2, \ldots, n$;

$$\begin{aligned} a_{i,i+1} &= \int_0^1 \left\{p(x)\phi_i'(x)\phi_{i+1}'(x) + q(x)\phi_i(x)\phi_{i+1}(x)\right\} dx \\ &= -\left(\frac{1}{h_i}\right)^2 \int_{x_i}^{x_{i+1}} p(x)\,dx + \left(\frac{1}{h_i}\right)^2 \int_{x_i}^{x_{i+1}} (x_{i+1} - x)(x - x_i) q(x)\,dx, \end{aligned}$$

for each $i = 1, 2, \ldots, n-1$; and

$$\begin{aligned} a_{i,i-1} &= \int_0^1 \left\{p(x)\phi_i'(x)\phi_{i-1}'(x) + q(x)\phi_i(x)\phi_{i-1}(x)\right\} dx \\ &= -\left(\frac{1}{h_{i-1}}\right)^2 \int_{x_{i-1}}^{x_i} p(x)\,dx + \left(\frac{1}{h_{i-1}}\right)^2 \int_{x_{i-1}}^{x_i} (x_i - x)(x - x_{i-1}) q(x)\,dx, \end{aligned}$$

for each $i = 2, \ldots, n$. The entries in $\mathbf{b}$ are

$$b_i = \int_0^1 f(x)\phi_i(x)\,dx = \frac{1}{h_{i-1}} \int_{x_{i-1}}^{x_i} (x - x_{i-1}) f(x)\,dx + \frac{1}{h_i} \int_{x_i}^{x_{i+1}} (x_{i+1} - x) f(x)\,dx,$$

for each $i = 1, 2, \ldots, n$.

There are six types of integrals to be evaluated:

$$Q_{1,i} = \left(\frac{1}{h_i}\right)^2 \int_{x_i}^{x_{i+1}} (x_{i+1} - x)(x - x_i)q(x)\,dx, \quad \text{for each } i = 1, 2, \ldots, n-1,$$

$$Q_{2,i} = \left(\frac{1}{h_{i-1}}\right)^2 \int_{x_{i-1}}^{x_i} (x - x_{i-1})^2 q(x)\,dx, \quad \text{for each } i = 1, 2, \ldots, n,$$

$$Q_{3,i} = \left(\frac{1}{h_i}\right)^2 \int_{x_i}^{x_{i+1}} (x_{i+1} - x)^2 q(x)\,dx, \quad \text{for each } i = 1, 2, \ldots, n,$$

$$Q_{4,i} = \left(\frac{1}{h_{i-1}}\right)^2 \int_{x_{i-1}}^{x_i} p(x)\,dx, \quad \text{for each } i = 1, 2, \ldots, n+1,$$

$$Q_{5,i} = \frac{1}{h_{i-1}} \int_{x_{i-1}}^{x_i} (x - x_{i-1})f(x)\,dx, \quad \text{for each } i = 1, 2, \ldots, n,$$

and

$$Q_{6,i} = \frac{1}{h_i} \int_{x_i}^{x_{i+1}} (x_{i+1} - x)f(x)\,dx, \quad \text{for each } i = 1, 2, \ldots, n.$$

The matrix A and the $\mathbf{b}$ in the linear system $A\mathbf{c} = \mathbf{b}$ have the entries

$$\begin{aligned} a_{i,i} &= Q_{4,i} + Q_{4,i+1} + Q_{2,i} + Q_{3,i}, && \text{for each } i = 1, 2, \ldots, n, \\ a_{i,i+1} &= -Q_{4,i+1} + Q_{1,i}, && \text{for each } i = 1, 2, \ldots, n-1, \\ a_{i,i-1} &= -Q_{4,i} + Q_{1,i-1}, && \text{for each } i = 2, 3, \ldots, n, \end{aligned}$$

and

$$b_i = Q_{5,i} + Q_{6,i}, \quad \text{for each } i = 1, 2, \ldots, n.$$

The entries in $\mathbf{c}$ are the unknown coefficients $c_1, c_2, \ldots, c_n$, from which the Rayleigh-Ritz approximation ϕ, given by $\phi(x) = \sum_{i=1}^n c_i\phi_i(x)$, is constructed.

A practical difficulty with this method is the necessity of evaluating $6n$ integrals. The integrals can be evaluated either directly or by a quadrature formula such as Simpson's method. An alternative approach for the integral evaluation is to approximate each of the functions p, q, and f with its piecewise linear interpolating polynomial and then integrate the approximation. Consider, for example, the integral $Q_{1,i}$. The piecewise linear interpolation of q is

$$P_q(x) = \sum_{i=0}^{n+1} q(x_i)\phi_i(x),$$

where $\phi_1, \ldots, \phi_n$ are defined in (11.29) and

$$\phi_0(x) = \begin{cases} \dfrac{x_1 - x}{x_1}, & 0 \le x \le x_1 \\ 0, & \text{elsewhere} \end{cases} \quad \text{and} \quad \phi_{n+1}(x) = \begin{cases} \dfrac{x - x_n}{1 - x_n}, & x_n \le x \le 1 \\ 0, & \text{elsewhere.} \end{cases}$$

Since the interval of integration is $[x_i, x_{i+1}]$, P_q reduces to

$$P_q(x) = q(x_i)\phi_i(x) + q(x_{i+1})\phi_{i+1}(x).$$

This is the first-degree interpolating polynomial studied in Section 3.1. By Theorem 3.3,

$$|q(x) - P_q(x)| = O\left(h_i^2\right), \quad x_i \le x \le x_{i+1},$$

if $q \in C^2[x_i, x_{i+1}]$. For $i = 1, 2, \ldots, n-1$, the approximation to $Q_{1,i}$ is obtained by integrating the approximation to the integrand

$$\begin{aligned} Q_{1,i} &= \left(\frac{1}{h_i}\right)^2 \int_{x_i}^{x_{i+1}} (x_{i+1} - x)(x - x_i)q(x)\,dx \\ &\approx \left(\frac{1}{h_i}\right)^2 \int_{x_i}^{x_{i+1}} (x_{i+1} - x)(x - x_i)\left[\frac{q(x_i)(x_{i+1} - x)}{h_i} + \frac{q(x_{i+1})(x - x_i)}{h_i}\right] dx \\ &= \frac{h_i}{12}[q(x_i) + q(x_{i+1})]. \end{aligned}$$

Further, if $q \in C^2[x_i, x_{i+1}]$, then

$$\left| Q_{1,i} - \frac{h_i}{12}[q(x_i) + q(x_{i+1})] \right| = O\left(h_i^3\right).$$

Approximations to the other integrals are derived in a similar manner and are given by

$$Q_{2,i} \approx \frac{h_{i-1}}{12}[3q(x_i) + q(x_{i-1})], \quad Q_{3,i} \approx \frac{h_i}{12}[3q(x_i) + q(x_{i+1})],$$
$$Q_{4,i} \approx \frac{h_{i-1}}{2}[p(x_i) + p(x_{i-1})], \quad Q_{5,i} \approx \frac{h_{i-1}}{6}[2f(x_i) + f(x_{i-1})],$$

and

$$Q_{6,i} \approx \frac{h_i}{6}[2f(x_i) + f(x_{i+1})].$$

Algorithm 11.5 sets up the tridiagonal linear system and incorporates the Crout Factorization Algorithm 6.7 to solve the system. The integrals $Q_{1,i}, \ldots, Q_{6,i}$ can be computed by one of the methods mentioned previously.

ALGORITHM 11.5

Piecewise Linear Rayleigh-Ritz

To approximate the solution to the boundary-value problem

$$-\frac{d}{dx}\left(p(x)\frac{dy}{dx}\right) + q(x)y = f(x), \quad 0 \le x \le 1, \quad y(0) = y(1) = 0,$$

with the piecewise linear function

$$\phi(x) = \sum_{i=1}^{n} c_i \phi_i(x):$$

INPUT integer $n \ge 1$; points $x_0 = 0 < x_1 < \cdots < x_n < x_{n+1} = 1$.

OUTPUT coefficients $c_1, \ldots, c_n$.

Step 1 For $i = 0, \ldots, n$ set $h_i = x_{i+1} - x_i$.

Step 2 For $i = 1, \ldots, n$ define the piecewise linear basis ϕ_i by

$$\phi_i(x) = \begin{cases} 0, & 0 \le x \le x_{i-1}, \\ \dfrac{x - x_{i-1}}{h_{i-1}}, & x_{i-1} < x \le x_i, \\ \dfrac{x_{i+1} - x}{h_i}, & x_i < x \le x_{i+1}, \\ 0, & x_{i+1} < x \le 1. \end{cases}$$

Step 3 For each $i = 1, 2, \ldots, n-1$ compute $Q_{1,i}, Q_{2,i}, Q_{3,i}, Q_{4,i}, Q_{5,i}, Q_{6,i}$;
Compute $Q_{2,n}, Q_{3,n}, Q_{4,n}, Q_{4,n+1}, Q_{5,n}, Q_{6,n}$.

Step 4 For each $i = 1, 2, \ldots, n-1$, set $\alpha_i = Q_{4,i} + Q_{4,i+1} + Q_{2,i} + Q_{3,i}$;
$\beta_i = Q_{1,i} - Q_{4,i+1}$;
$b_i = Q_{5,i} + Q_{6,i}$.

Step 5 Set $\alpha_n = Q_{4,n} + Q_{4,n+1} + Q_{2,n} + Q_{3,n}$;
$b_n = Q_{5,n} + Q_{6,n}$.

Step 6 Set $a_1 = \alpha_1$; (*Steps* 6–10 *solve a symmetric tridiagonal linear system using Algorithm* 6.7.)
$\zeta_1 = \beta_1/\alpha_1$;
$z_1 = b_1/a_1$.

Step 7 For $i = 2, \ldots, n-1$ set $a_i = \alpha_i - \beta_{i-1}\zeta_{i-1}$;
$\zeta_i = \beta_i/a_i$;
$z_i = (b_i - \beta_{i-1}z_{i-1})/a_i$.

Step 8 Set $a_n = \alpha_n - \beta_{n-1}\zeta_{n-1}$;
$z_n = (b_n - \beta_{n-1}z_{n-1})/a_n$.

Step 9 Set $c_n = z_n$;
OUTPUT (c_n).

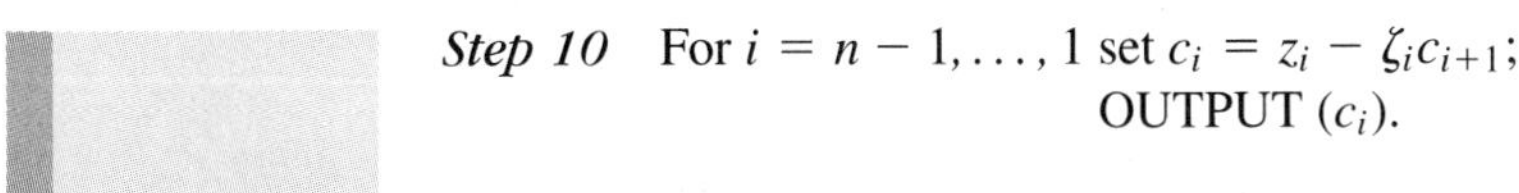

Step 10 For $i = n-1, \ldots, 1$ set $c_i = z_i - \zeta_i c_{i+1}$;
OUTPUT (c_i).

Step 11 STOP. (*Procedure is complete.*)

The following example uses Algorithm 11.5. Because of the elementary nature of this example, the integrals in Steps 3, 4, and 5 were found directly.

EXAMPLE 1 Consider the boundary-value problem

$$-y'' + \pi^2 y = 2\pi^2 \sin(\pi x), \quad 0 \le x \le 1, \quad y(0) = y(1) = 0.$$

Let $h_i = h = 0.1$, so that $x_i = 0.1i$ for each $i = 0, 1, \ldots, 9$. The integrals are

$$Q_{1,i} = 100 \int_{0.1i}^{0.1i+0.1} (0.1i + 0.1 - x)(x - 0.1i)\pi^2 \, dx = \frac{\pi^2}{60},$$

$$Q_{2,i} = 100 \int_{0.1i-0.1}^{0.1i} (x - 0.1i + 0.1)^2 \pi^2 \, dx = \frac{\pi^2}{30},$$

$$Q_{3,i} = 100 \int_{0.1i}^{0.1i+0.1} (0.1i + 0.1 - x)^2 \pi^2 \, dx = \frac{\pi^2}{30},$$

$$Q_{4,i} = 100 \int_{0.1i-0.1}^{0.1i} dx = 10,$$

$$\begin{aligned} Q_{5,i} &= 10 \int_{0.1i-0.1}^{0.1i} (x - 0.1i + 0.1) 2\pi^2 \sin \pi x \, dx \\ &= -2\pi \cos 0.1\pi i + 20[\sin(0.1\pi i) - \sin((0.1i - 0.1)\pi)], \end{aligned}$$

and

$$\begin{aligned} Q_{6,i} &= 10 \int_{0.1i}^{0.1i+0.1} (0.1i + 0.1 - x) 2\pi^2 \sin \pi x \, dx \\ &= 2\pi \cos 0.1\pi i - 20[\sin((0.1i + 0.1)\pi) - \sin(0.1\pi i)]. \end{aligned}$$

The linear system $A\mathbf{c} = \mathbf{b}$ has

$$a_{i,i} = 20 + \frac{\pi^2}{15}, \qquad \text{for each } i = 1, 2, \ldots, 9,$$

$$a_{i,i+1} = -10 + \frac{\pi^2}{60}, \qquad \text{for each } i = 1, 2, \ldots, 8,$$

$$a_{i,i-1} = -10 + \frac{\pi^2}{60}, \qquad \text{for each } i = 2, 3, \ldots, 9,$$

and

$$b_i = 40\sin(0.1\pi i)[1 - \cos 0.1\pi], \qquad \text{for each } i = 1, 2, \ldots, 9.$$

The solution to the tridiagonal linear system is

$$c_9 = 0.31029,\ c_8 = 0.59020,\ c_7 = 0.81234,$$
$$c_6 = 0.95496,\ c_5 = 1.00411,\ c_4 = 0.95496,$$
$$c_3 = 0.81234,\ c_2 = 0.59020,\ c_1 = 0.31029.$$

The piecewise linear approximation is

$$\phi(x) = \sum_{i=1}^{9} c_i \phi_i(x).$$

The actual solution to the boundary-value problem is

$$y(x) = \sin \pi x.$$

Table 11.7 lists the error in the approximation at x_i for each $i = 1, \ldots, 9$. ∎

Table 11.7

i	x_i	$\phi(x_i)$	$y(x_i)$	$\|\phi(x_i) - y(x_i)\|$
1	0.1	0.31029	0.30902	0.00127
2	0.2	0.59020	0.58779	0.00241
3	0.3	0.81234	0.80902	0.00332
4	0.4	0.95496	0.95106	0.00390
5	0.5	1.00411	1.00000	0.00411
6	0.6	0.95496	0.95106	0.00390
7	0.7	0.81234	0.80902	0.00332
8	0.8	0.59020	0.58779	0.00241
9	0.9	0.31029	0.30902	0.00127

It can be shown that the tridiagonal matrix A given by the piecewise linear basis functions is positive definite (see Exercise 12), so, by Theorem 6.25, the linear system is stable with respect to roundoff error. Under the hypotheses presented at the beginning of this section, we have

$$|\phi(x) - y(x)| = O(h^2), \quad \text{for each } x \text{ in } [0, 1].$$

A proof of this result can be found in [Schul, pp. 103–104].

The use of piecewise linear basis functions results in an approximate solution to Eqs. (11.22) and (11.23) that is continuous but not differentiable on [0, 1]. A more complicated set of basis functions is required to construct an approximation that belongs to $C_0^2[0, 1]$. These basis functions are similar to the cubic interpolatory splines discussed in Section 3.4.

Recall that the cubic *interpolatory* spline S on the five nodes x_0, x_1, x_2, x_3, and x_4 for a function f is defined by:

a. S is a cubic polynomial, denoted by S_j, on $[x_j, x_{j+1}]$, for each $j = 0, 1, 2, 3$. (*This gives* 16 *selectable constants for S*, 4 *for each cubic*.)

b. $S(x_j) = f(x_j)$, for $j = 0, 1, 2, 3, 4$ (5 *specified conditions*).

c. $S_{j+1}(x_{j+1}) = S_j(x_{j+1})$, for $j = 0, 1, 2$ (3 *specified conditions*).

d. $S'_{j+1}(x_{j+1}) = S'_j(x_{j+1})$, for $j = 0, 1, 2$ (3 *specified conditions*).

e. $S''_{j+1}(x_{j+1}) = S''_j(x_{j+1})$, for $j = 0, 1, 2$ (3 *specified conditions*).

f. One of the following boundary conditions is satisfied:

(i) Free: $S''(x_0) = S''(x_4) = 0$ (2 *specified conditions*).

(ii) Clamped: $S'(x_0) = f'(x_0)$ and $S'(x_4) = f'(x_4)$ (2 *specified conditions*).

Since uniqueness of solution requires the number of constants in (a), 16, to equal the number of conditions in (b) through (f), only one of the boundary conditions in (f) can be specified for the interpolatory cubic splines.

The cubic spline functions we will use for our basis functions are called **B-splines**, or *bell-shaped splines*. These differ from interpolatory splines in that both sets of boundary conditions in (f) are satisfied. This requires the relaxation of two of the conditions in (b) through (e). Since the spline must have two continuous derivatives on $[x_0, x_4]$, we delete from the description of the interpolatory splines two of the interpolation conditions. In particular, we modify condition (b) to

b. $S(x_j) = f(x_j)$ for $j = 0, 2, 4$.

The basic B-spline S defined next and shown in Figure 11.5 uses the equally spaced nodes $x_0 = -2$, $x_1 = -1$, $x_2 = 0$, $x_3 = 1$, and $x_4 = 2$. It satisfies the interpolatory conditions

b. $S(x_0) = 0, \quad S(x_2) = 1, \quad S(x_4) = 0;$

as well as both sets of conditions

$$\text{(i) } S''(x_0) = S''(x_4) = 0 \quad \text{and} \quad \text{(ii) } S'(x_0) = S'(x_4) = 0.$$

Figure 11.5

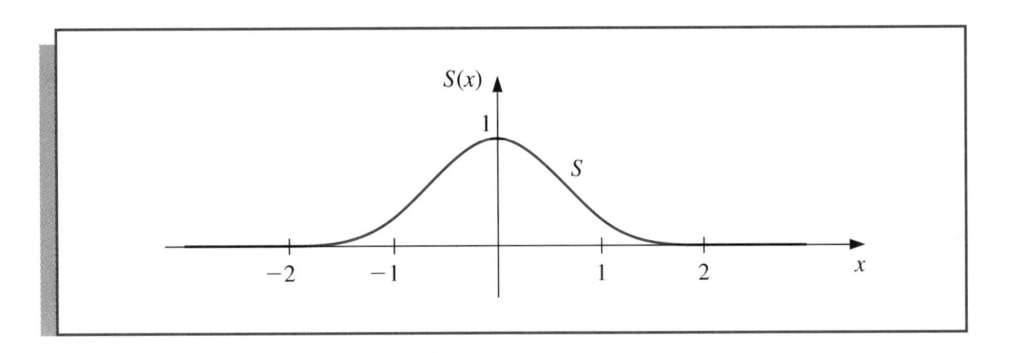

As a consequence, $S \in C^2(-\infty, \infty)$.

(11.31)
$$S(x) = \begin{cases} 0, & x \le -2, \\ \frac{1}{4}(2+x)^3, & -2 \le x \le -1, \\ \frac{1}{4}\left[(2+x)^3 - 4(1+x)^3\right], & -1 < x \le 0, \\ \frac{1}{4}\left[(2-x)^3 - 4(1-x)^3\right], & 0 < x \le 1, \\ \frac{1}{4}(2-x)^3, & 1 < x \le 0, \\ 0, & 2 < x. \end{cases}$$

To construct the basis functions ϕ_i in $C_0^2[0, 1]$ we first partition $[0, 1]$ by choosing a positive integer n and defining $h = 1/(n+1)$. This produces the equally spaced nodes $x_i = ih$, for each $i = 0, 1, \ldots, n+1$. We then define the basis functions $\{\phi_i\}_{i=0}^{n+1}$ as

$$\phi_i(x) = \begin{cases} S\left(\frac{x}{h}\right) - 4S\left(\frac{x+h}{h}\right), & i = 0, \\ S\left(\frac{x-h}{h}\right) - S\left(\frac{x+h}{h}\right), & i = 1, \\ S\left(\frac{x-ih}{h}\right), & 2 \le i \le n-1, \\ S\left(\frac{x-nh}{h}\right) - S\left(\frac{x-(n+2)h}{h}\right), & i = n, \\ S\left(\frac{x-(n+1)h}{h}\right) - 4S\left(\frac{x-(n+2)h}{h}\right), & i = n+1. \end{cases}$$

It is not difficult to show that $\{\phi_i\}_{i=0}^{n+1}$ is a linearly independent set of cubic splines satisfying $\phi_i(0) = \phi_i(1) = 0$, for each $i = 0, 1, \ldots, n, n+1$ (see Exercise 11). The graphs of ϕ_i, for $2 \le i \le n-1$, are shown in Figure 11.6 and the graphs of ϕ_0, ϕ_1, ϕ_n, and ϕ_{n+1} are in Figure 11.7.

Since $\phi_i(x)$ and $\phi_i'(x)$ are nonzero only for $x_{i-2} \le x \le x_{i+2}$, the matrix in the Rayleigh-Ritz approximation is a band matrix with bandwidth at most seven:

(11.32)
$$A = \begin{bmatrix} a_{00} & a_{01} & a_{02} & a_{03} & 0 & \cdots & \cdots & \cdots & 0 \\ a_{10} & a_{11} & a_{12} & a_{13} & a_{14} & \ddots & & & \vdots \\ a_{20} & a_{21} & a_{22} & a_{23} & a_{24} & a_{25} & \ddots & & \vdots \\ a_{30} & a_{31} & a_{32} & a_{33} & a_{34} & a_{35} & a_{36} & \ddots & \vdots \\ 0 & \ddots & \ddots & \ddots & \ddots & \ddots & \ddots & \ddots & 0 \\ \vdots & \ddots & \ddots & \ddots & \ddots & \ddots & \ddots & \ddots & a_{n-2,n+1} \\ \vdots & & \ddots & \ddots & \ddots & \ddots & \ddots & \ddots & a_{n-1,n+1} \\ \vdots & & & \ddots & \ddots & \ddots & \ddots & \ddots & a_{n,n+1} \\ 0 & \cdots & \cdots & \cdots & 0 & a_{n+1,n-2} & a_{n+1,n-1} & a_{n+1,n} & a_{n+1,n+1} \end{bmatrix},$$

where

$$a_{ij} = \int_0^1 \left\{p(x)\phi_i'(x)\phi_j'(x) + q(x)\phi_i(x)\phi_j(x)\right\}\,dx,$$

Figure 11.6

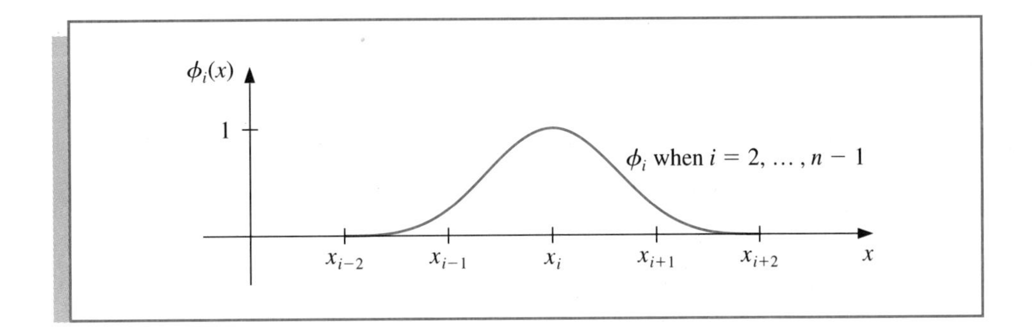

Figure 11.7

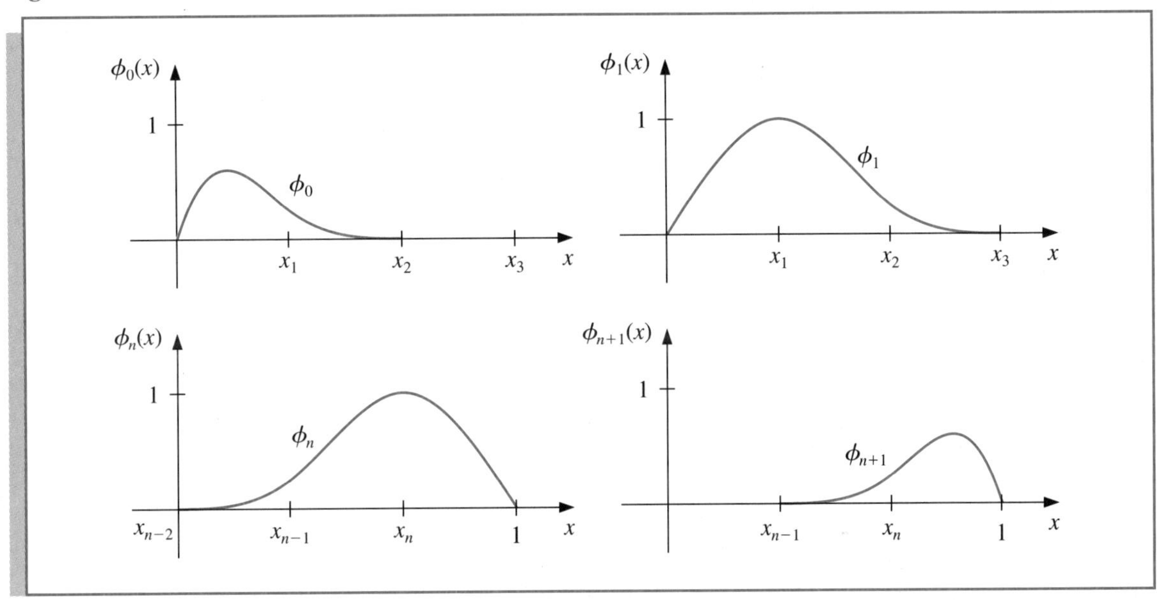

for each $i, j = 0, 1, \ldots, n+1$. The matrix A is positive definite (see Exercise 13), so the linear system (11.32) can be solved by Choleski's Algorithm 6.6 or by Gaussian elimination. Algorithm 11.6 details the construction of the cubic spline approximation $\phi(x)$ by the Rayleigh-Ritz method for the boundary-value problem (11.22) and (11.23) given at the beginning of this section.

ALGORITHM **11.6**

Cubic Spline Rayleigh-Ritz

To approximate the solution to the boundary-value problem

$$-\frac{d}{dx}\left(p(x)\frac{dy}{dx}\right) + q(x)y = f(x), \quad 0 \le x \le 1, \quad y(0) = y(1) = 0$$

with the sum of cubic splines

$$\phi(x) = \sum_{i=0}^{n+1} c_i \phi_i(x):$$

INPUT integer $n \geq 1$.

OUTPUT coefficients $c_0, \ldots, c_{n+1}$.

Step 1 Set $h = 1/(n+1)$.

Step 2 For $i = 0, \ldots, n+1$ set $x_i = ih$.
Set $x_{-2} = x_{-1} = 0$; $x_{n+2} = x_{n+3} = 1$.

Step 3 Define the function S by

$$S(x) = \begin{cases} 0, & x \leq -2, \\ \frac{1}{4}(2+x)^3, & -2 < x \leq -1, \\ \frac{1}{4}\left[(2+x)^3 - 4(1+x)^3\right], & -1 < x \leq 0, \\ \frac{1}{4}\left[(2-x)^3 - 4(1-x)^3\right], & 0 < x \leq 1, \\ \frac{1}{4}(2-x)^3, & 1 < x \leq 2, \\ 0, & 2 < x \end{cases}$$

Step 4 Define the cubic spline basis $\{\phi_i\}_{i=0}^{n+1}$ by

$$\phi_0(x) = S\left(\frac{x}{h}\right) - 4S\left(\frac{x+h}{h}\right),$$

$$\phi_1(x) = S\left(\frac{x-x_1}{h}\right) - S\left(\frac{x+h}{h}\right),$$

$$\phi_i(x) = S\left(\frac{x-x_i}{h}\right), \text{ for } i = 2, \ldots, n-1,$$

$$\phi_n(x) = S\left(\frac{x-x_n}{h}\right) - S\left(\frac{x-(n+2)h}{h}\right),$$

$$\phi_{n+1}(x) = S\left(\frac{x-x_{n+1}}{h}\right) - 4S\left(\frac{x-(n+2)h}{h}\right).$$

Step 5 For $i = 0, \ldots, n+1$ do Steps 6–9.
(*Note: The integrals in Steps* 6 *and* 9 *can be evaluated using a numerical integration procedure.*)

Step 6 For $j = i, i+1, \ldots, \min\{i+3, n+1\}$
set $L = \max\{x_{j-2}, 0\}$;
$U = \min\{x_{i+2}, 1\}$;

$$a_{ij} = \int_L^U \left[p(x)\phi_i'(x)\phi_j'(x) + q(x)\phi_i(x)\phi_j(x)\right] dx;$$

if $i \neq j$, then set $a_{ji} = a_{ij}$. (*Since A is symmetric.*)

Step 7 If $i \geq 4$ then for $j = 0, \ldots, i-4$ set $a_{ij} = 0$.

Step 8 If $i \le n-3$ then for $j = i+4, \ldots, n+1$ set $a_{ij} = 0$.

Step 9 Set $L = \max\{x_{i-2}, 0\}$;
$U = \min\{x_{i+2}, 1\}$;

$$b_i = \int_L^U f(x)\phi_i(x)\,dx.$$

Step 10 Solve the linear system $A\mathbf{c} = \mathbf{b}$, where $A = (a_{ij})$, $\mathbf{b} = (b_0, \ldots, b_{n+1})^t$ and $\mathbf{c} = (c_0, \ldots, c_{n+1})^t$.

Step 11 For $i = 0, \ldots, n+1$
OUTPUT (c_i).

Step 12 STOP. (*Procedure is complete.*)

EXAMPLE 2 Consider the boundary-value problem

$$-y'' + \pi^2 y = 2\pi^2 \sin(\pi x), \quad 0 \le x \le 1, \quad y(0) = y(1) = 0.$$

In Example 1 we let $h = 0.1$ and generated approximations using piecewise linear basis functions. Table 11.8 lists the results obtained by applying the B-splines as detailed in Algorithm 11.6 with this same choice of nodes. ■

Table 11.8

i	c_i	x_i	$\phi(x_i)$	$y(x_i)$	$\lvert y(x_i) - \phi(x_i)\rvert$
0	$0.50964361 \times 10^{-5}$	0	0.00000000	0.00000000	0.00000000
1	0.20942608	0.1	0.30901644	0.30901699	0.00000055
2	0.39835678	0.2	0.58778549	0.58778525	0.00000024
3	0.54828946	0.3	0.80901687	0.80901699	0.00000012
4	0.64455358	0.4	0.95105667	0.95105652	0.00000015
5	0.67772340	0.5	1.00000002	1.00000000	0.00000020
6	0.64455370	0.6	0.95105713	0.95105652	0.00000061
7	0.54828951	0.7	0.80901773	0.80901699	0.00000074
8	0.39835730	0.8	0.58778690	0.58778525	0.00000165
9	0.20942593	0.9	0.30901810	0.30901699	0.00000111
10	$0.74931285 \times 10^{-5}$	1.0	0.00000000	0.00000000	0.00000000

It is recommended that the integrations in Steps 6 and 9 be performed in two steps. First, construct cubic spline interpolatory polynomials for p, q, and f using the methods presented in Section 3.4. Then approximate the integrands by products of cubic splines or derivatives of cubic splines. The integrands are now piecewise polynomials and can be integrated exactly on each subinterval, and then summed. This leads to accurate approximations of the integrals.

The hypotheses assumed at the beginning of this section are sufficient to guarantee that

$$\left\{\int_0^1 |y(x) - \phi(x)|^2\,dx\right\}^{1/2} = O(h^4), \quad 0 \le x \le 1.$$

For a proof of this result, see [Schul, pp. 107–108].

B-splines can also be defined for unequally spaced nodes, but the details are more complicated. A presentation of the technique can be found in [Schul, p. 73]. Another commonly used basis is the piecewise cubic Hermite polynomials. For an excellent presentation of this method, again see [Schul, pp. 24ff].

Other methods that receive considerable attention are Galerkin, or "weak form," methods. For the boundary-value problem we have been considering,

$$-\frac{d}{dx}\left(p(x)\frac{dy}{dx}\right) + q(x)y = f(x), \quad y(0) = y(1) = 0, \quad 0 \le x \le 1,$$

under the assumptions listed at the beginning of this section, the Galerkin and Rayleigh-Ritz methods are both determined by Eq. (11.28). This is not the case for an arbitrary boundary-value problem, however. A treatment of the similarities and differences in the two methods and a discussion of the wide application of the Galerkin method can be found in [Schul] and in [SF].

Another popular technique for solving boundary-value problems is the **method of collocation**. This procedure begins by selecting a set of basis functions $\{\phi_1, \ldots, \phi_N\}$, a set of numbers $\{x_i, \ldots, x_n\}$ in $[0, 1]$, and requiring that an approximation

$$\sum_{i=1}^{N} c_i \phi_i(x)$$

satisfy the differential equation at each of the numbers x_j for $1 \le j \le n$. If, in addition, it is required that $\phi_i(0) = \phi_i(1) = 0$ for $1 \le i \le N$, the boundary conditions are automatically satisfied. Much attention in the literature has been given to the choice of the numbers $\{x_j\}$ and the basis functions $\{\phi_i\}$. One popular choice is to let the ϕ_i be the basis functions for spline functions relative to a partition of $[0, 1]$, and to let the nodes $\{x_j\}$ be the Gaussian points or roots of certain orthogonal polynomials, transformed to the proper subinterval. A comparison of various collocation methods and finite difference methods is contained in [Ru]. The conclusion is that the collocation methods using higher-degree splines are competitive with finite-difference techniques using extrapolation. Other references for collocation methods are [DebS] and [LR].

EXERCISE SET 11.5

1. Use the Piecewise Linear Algorithm to approximate the solution to the boundary-value problem

$$y'' + \frac{\pi^2}{4}y = \frac{\pi^2}{16}\cos\frac{\pi}{4}x, \quad 0 \le x \le 1, \quad y(0) = y(1) = 0$$

using $x_0 = 0, x_1 = 0.3, x_2 = 0.7, x_3 = 1$ and compare the results to the actual solution $y(x) = -\frac{1}{3}\cos\frac{\pi}{2}x - \frac{\sqrt{2}}{6}\sin\frac{\pi}{2}x + \frac{1}{3}\cos\frac{\pi}{4}x$.

2. Use the Piecewise Linear Algorithm to approximate the solution to the boundary-value problem

$$-\frac{d}{dx}(xy') + 4y = 4x^2 - 8x + 1, \quad 0 \le x \le 1, \quad y(0) = y(1) = 0$$

using $x_0 = 0, x_1 = 0.4, x_2 = 0.8, x_3 = 1$ and compare the results to the actual solution $y(x) = x^2 - x$.

3. Use the Piecewise Linear Algorithm to approximate the solutions to the following boundary-value problems and compare the results to the actual solution:

a. $-x^2y'' - 2xy' + 2y = -4x^2$, $0 \le x \le 1$, $y(0) = y(1) = 0$; use $h = 0.1$; actual solution $y(x) = x^2 - x$.

b. $-\frac{d}{dx}(e^x y') + e^x y = x + (2 - x)e^x$, $0 \le x \le 1$, $y(0) = y(1) = 0$; use $h = 0.1$; actual solution $y(x) = (x - 1)(e^{-x} - 1)$.

c. $-\frac{d}{dx}(e^{-x} y') + e^{-x} y = (x - 1) - (x + 1)e^{-(x-1)}$, $0 \le x \le 1$, $y(0) = y(1) = 0$; use $h = 0.05$; actual solution $y(x) = x(e^x - e)$.

d. $-(x+1)y'' - y' + (x+2)y = [2 - (x+1)^2]e \ln 2 - 2e^x$, $0 \le x \le 1$, $y(0) = y(1) = 0$; use $h = 0.05$; actual solution $y(x) = e^x \ln(x + 1) - (e \ln 2)x$.

4. Use the Cubic Spline Algorithm with $n = 3$ to approximate the solution to the following boundary-value problems and compare the results to the actual solutions given in Exercise 1:

a. $y'' + \frac{\pi^2}{4} y = \frac{\pi^2}{16} \cos \frac{\pi}{4} x$, $0 \le x \le 1$, $y(0) = 0$, $y(1) = 0$

b. $-\frac{d}{dx}(xy') + 4y = 4x^2 - 8x + 1$, $0 \le x \le 1$, $y(0) = 0$, $y(1) = 0$

5. Repeat Exercise 3 using the Cubic Spline Algorithm.

6. Show that the boundary-value problem

$$-\frac{d}{dx}(p(x)y') + q(x)y = f(x), \quad 0 \le x \le 1, \quad y(0) = \alpha, \quad y(1) = \beta,$$

can be transformed by the change of variable

$$z = y - \beta x - (1 - x)\alpha$$

into the form

$$-\frac{d}{dx}(p(x)z') + q(x)z = F(x), \quad 0 \le x \le 1, \quad z(0) = 0, \quad z(1) = 0.$$

7. Use Exercise 6 and the Piecewise Linear Algorithm with $n = 9$ to approximate the solution to the boundary-value problem

$$-y'' + y = x, \quad 0 \le x \le 1, \quad y(0) = 1, \quad y(1) = 1 + e^{-1}.$$

8. Repeat Exercise 7 using the Cubic Spline Algorithm.

9. Show that the boundary-value problem

$$-\frac{d}{dx}(p(x)y') + q(x)y = f(x), \quad a \le x \le b, \quad y(a) = \alpha, \quad y(b) = \beta,$$

can be transformed into the form

$$-\frac{d}{dw}(p(w)z') + q(w)z = F(w), \quad 0 \le w \le 1, \quad z(0) = 0, \quad z(1) = 0,$$

by a method similar to that given in Exercise 6.

10. Show that $\{\phi_i\}_{i=1}^n$ is a linearly independent set of functions on $[0, 1]$ for the functions ϕ_i defined in Eq. (11.29).

11. Repeat Exercise 10 using the cubic spline basis $\{\phi_i\}_{i=0}^{n+1}$.

12. Show that the matrix given by the piecewise linear basis functions is positive definite. [*Hint*: Use the definition.]

13. Show that the matrix A in (11.32) is positive definite.

11.6 Survey of Methods and Software

In this chapter we discussed methods for approximating solutions to boundary-value problems. For the linear boundary-value problem

$$y'' = p(x)y' + q(x)y + r(x), \quad \text{for} \quad a \le x \le b, \quad \text{where} \quad y(a) = \alpha \quad \text{and} \quad y(b) = \beta,$$

we considered both a linear shooting method and a finite-difference method to approximate the solution. The shooting method uses an initial-value technique to solve the problems

$$y'' = p(x)y' + q(x)y + r(x), \quad \text{where} \quad y(a) = \alpha \quad \text{and} \quad y'(a) = 0,$$

and

$$y'' = p(x)y' + q(x)y, \quad \text{where} \quad y(a) = 0 \quad \text{and} \quad y'(a) = 1.$$

A weighted average of these solutions produces a solution to the linear boundary-value problem.

In the finite-difference method, we replaced y'' and y' with difference approximations and solved a linear system. Although the approximations may not be as accurate as the shooting method, there is less sensitivity to roundoff error. Higher-order difference methods are available, or extrapolation can be used to improve accuracy.

For the nonlinear boundary problem

$$y'' = f(x, y, y'), \quad \text{for} \quad a \le x \le b, \quad \text{where} \quad y(a) = \alpha \quad \text{and} \quad y(b) = \beta,$$

we also presented two methods. The nonlinear shooting method requires the solution of the initial-value problem

$$y'' = f(x, y, y'), \quad \text{where} \quad y(a) = \alpha \quad \text{and} \quad y'(a) = t,$$

for an initial choice of t. We improved the choice by using Newton's method to approximate the solution, t, to $y(b, t) = \beta$. This method required solving two initial-value problems at each iteration. The accuracy is dependent on the choice of method for solving the initial-value problems.

The finite-difference method for the nonlinear equation requires the replacement of y'' and y' by difference quotients, which results in a nonlinear system. This system is solved using Newton's method. Higher-order differences or extrapolation can be used to improve accuracy. Finite-difference methods tend to be less sensitive to roundoff error than shooting methods.

The Rayleigh-Ritz-Galerkin method was illustrated by approximating the solution to the boundary-value problem

$$-\frac{d\left(p(x)\frac{dy}{dx}\right)}{dx} + q(x)y = f(x), \quad \text{for} \quad 0 \le x \le 1, \quad \text{where} \quad y(0) = y(1) = 0.$$

A piecewise linear approximation or a cubic spline approximation can be obtained.

Most of the material concerning second-order boundary-value problems can be extended to problems with boundary conditions of the form

$$\alpha_1 y(a) + \beta_1 y'(a) = \alpha \quad \text{and} \quad \alpha_2 y(b) + \beta_2 y'(b) = \beta,$$

where $|\alpha_1| + |\beta_1| \neq 0$ and $|\alpha_2| + |\beta_2| \neq 0$, but some of the techniques become quite complicated. The reader who is interested in problems of this type is advised to consider a book specializing in boundary-value problems, such as [K,H].

We mention only two of the many methods in the IMSL Library for solving boundary-value problems. The subroutine BVPFD is based on finite differences and BVPMS is based on multiple shooting using IVPRK, a Runge-Kutta-Verner method for initial-value problems. Both methods can be used for systems of parameterized boundary-value problems.

The NAG Library also has a multitude of subroutines for solving boundary-value problems. The subroutine D02HAF is a shooting method using the Runge-Kutta-Merson initial-value method in conjunction with Newton's method. The subroutine D02GAF uses the finite-difference method with Newton's method to solve the nonlinear system. The subroutine D02GBF is a linear finite-difference method, and D02JAF is a method based on collocation.

Further information on the general problems involved with the numerical solution to two-point boundary-value problems can be found in Keller [K,H] and Bailey, Shampine and Waltman [BSW]. Roberts and Shipman [RS] focuses on the shooting methods for the two-point boundary-value problem, and Pryce [Pr] restricts attention to Sturm-Liouville problems. The book by Ascher, Mattheij, and Russell [AMR] has a comprehensive presentation of multiple shooting and parallel shooting methods.

CHAPTER

12

Numerical Solutions to Partial-Differential Equations

■ ■ ■

A body is called *isotropic* if the thermal conductivity at each point in the body is independent of the direction of heat flow through the point. The temperature, $u \equiv u(x, y, z, t)$ in an isotropic body can be found by solving the partial-differential equation

$$\frac{\partial}{\partial x}\left(k\frac{\partial u}{\partial x}\right) + \frac{\partial}{\partial y}\left(k\frac{\partial u}{\partial y}\right) + \frac{\partial}{\partial z}\left(k\frac{\partial u}{\partial z}\right) = c\rho\frac{\partial u}{\partial t},$$

where k, c, and ρ are functions of (x, y, z) and represent, respectively, the thermal conductivity, specific heat, and density of the body at the point (x, y, z).

When k, c, and ρ are constants, this equation is known as the simple three-dimensional heat equation and is expressed as

$$\frac{\partial^2 u}{\partial x^2} + \frac{\partial^2 u}{\partial y^2} + \frac{\partial^2 u}{\partial z^2} = \frac{c\rho}{k}\frac{\partial u}{\partial t}.$$

If the boundary of the body is relatively simple, the solution to this equation can be found using Fourier series.

In most situations where k, c, and ρ are not constant or when the boundary is irregular, the solution to the partial-differential equation must be obtained by approximation techniques. An introduction to techniques of this type is presented in this chapter.

Physical situations involving more than one independent variable are expressed using equations involving partial derivatives. In this chapter, we present a brief introduction to some of the techniques available for approximating the solution to partial-differential equations involving two variables by showing how these techniques can be applied to certain standard physical problems. We limit our treatment to problems of this type because most of the more advanced techniques require a stronger background in analysis than this book assumes.

In Section 12.1 we consider an **elliptic** partial-differential equation known as the **Poisson equation**:

$$\frac{\partial^2 u}{\partial x^2}(x, y) + \frac{\partial^2 u}{\partial y^2}(x, y) = f(x, y).$$

We assume in this equation that the function f describes the input to the problem on a plane region R whose boundary we denote by S. Equations of this type arise naturally in the study of various time-independent physical problems such as the steady-state distribution of heat in a plane region, the potential energy of a point in a plane acted on by gravitational forces in the plane, and two-dimensional steady-state problems involving incompressible fluids.

To obtain a unique solution to the Poisson equation, additional constraints are placed on the solution. For example, the study of the steady-state distribution of heat in a plane region requires that $f(x, y) \equiv 0$, resulting in a simplification of the Poisson equation to

$$\frac{\partial^2 u}{\partial x^2}(x, y) + \frac{\partial^2 u}{\partial y^2}(x, y) = 0,$$

which is called **Laplace's equation**. If the temperature within the region is determined by the temperature distribution on the boundary of the region, the constraints are called the **Dirichlet boundary conditions**. These are given by

$$u(x, y) = g(x, y),$$

for all (x, y) on S, the boundary of the region R. (See Figure 12.1.)

Figure 12.1

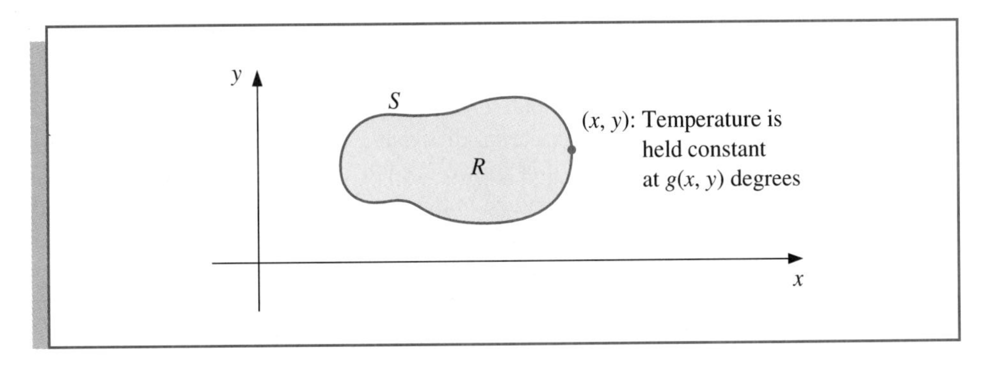

In Section 12.2, we consider the numerical solution to a problem involving a parabolic partial-differential equation of the form

$$\frac{\partial u}{\partial t}(x, t) = \alpha^2 \frac{\partial^2 u}{\partial x^2}(x, t).$$

The physical problem considered here concerns the flow of heat along a rod of length l (see Figure 12.2), which is assumed to have a uniform temperature within each cross-sectional element. This condition requires the rod to be perfectly insulated on its lateral surface. The constant α is determined by the heat-conductive properties of the material of which the rod is composed and is assumed to be independent of the position in the rod.

Figure 12.2

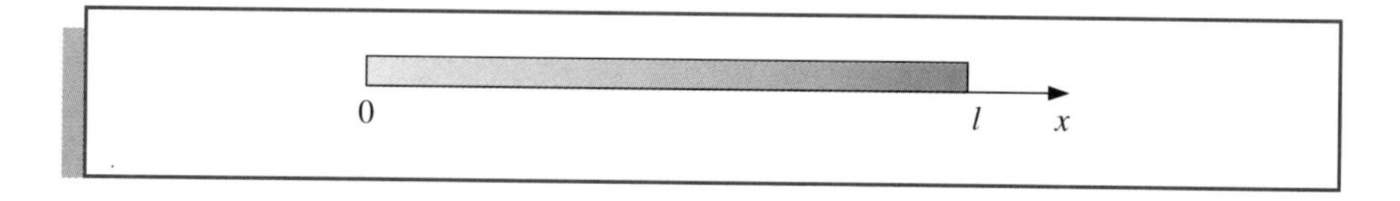

One of the typical sets of constraints for a heat-flow problem of this type is to specify the initial heat distribution in the rod,

$$u(x, 0) = f(x),$$

and describe the behavior at the ends of the rod. For example, if the ends of the rod are held at constant temperatures U_1 and U_2, the boundary conditions have the form

$$u(0, t) = U_1 \quad \text{and} \quad u(l, t) = U_2,$$

and the heat distribution in the rod approaches the limiting temperature distribution

$$\lim_{t\to\infty} u(x, t) = U_1 + \frac{U_2 - U_1}{l}x.$$

If, instead, the rod is insulated so that no heat flows through the ends, the boundary conditions are

$$\frac{\partial u}{\partial x}(0, t) = 0 \quad \text{and} \quad \frac{\partial u}{\partial x}(l, t) = 0$$

and result in a constant temperature in the rod as the limiting case.

The parabolic partial-differential equation is also of importance in the study of gas diffusion; in fact, it is known in some circles as the **diffusion equation**.

The problem studied in Section 12.3 is the one-dimensional **wave equation** and is an example of a **hyperbolic** partial-differential equation. Suppose an elastic string of length l is stretched between two supports at the same horizontal level (see Figure 12.3). If the string is set in motion so that it vibrates in a vertical plane, the vertical displacement $u(x, t)$ of a point x at time t satisfies the partial-differential equation

$$\alpha^2 \frac{\partial^2 u}{\partial x^2}(x, t) = \frac{\partial^2 u}{\partial t^2}(x, t), \quad 0 < x < l, \quad 0 < t,$$

provided that damping effects are neglected and the amplitude is not too large. To impose constraints on this problem, assume that the initial position and velocity of the string are

Figure 12.3

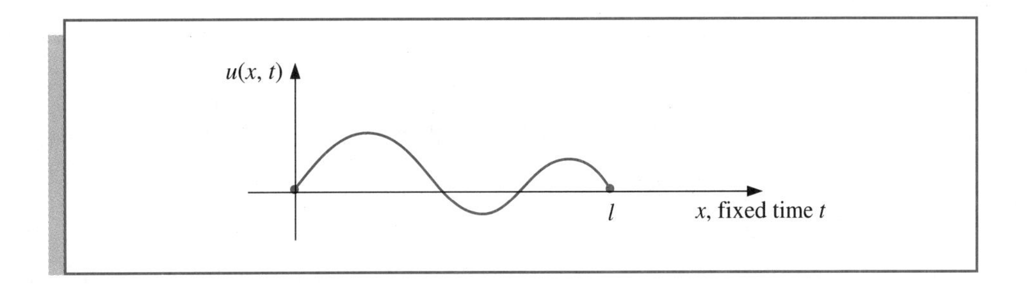

given by

$$u(x, 0) = f(x) \quad \text{and} \quad \frac{\partial u}{\partial t}(x, 0) = g(x), \quad 0 \le x \le l,$$

and use the fact that the endpoints are fixed. This implies that $u(0, t) = 0$ and $u(l, t) = 0$.

Other physical problems involving the hyperbolic partial-differential equation occur in the study of vibrating beams with one or both ends clamped and in the transmission of electricity in a long transmission line where there is some leakage of current to the ground.

12.1 Elliptic Partial-Differential Equations

The *elliptic* partial-differential equation we consider is the Poisson equation,

$$\nabla^2 u(x, y) \equiv \frac{\partial^2 u}{\partial x^2}(x, y) + \frac{\partial^2 u}{\partial y^2}(x, y) = f(x, y) \tag{12.1}$$

on $(x, y) \in R$ and

$$u(x, y) = g(x, y) \quad \text{for } (x, y) \in S,$$

where

$$R = \{(x, y) | a < x < b, c < y < d\}$$

and S denotes the boundary of R. For this discussion, we assume that both f and g are continuous on their domains, and a unique solution is ensured.

The method used is an adaptation of the finite-difference method for boundary-value problems, which was discussed in Section 11.3. The first step is to choose integers n and m and define step sizes h and k by $h = (b - a)/n$ and $k = (d - c)/m$. Partitioning the interval $[a, b]$ into n equal parts of width h and the interval $[c, d]$ into m equal parts of width k (see Figure 12.4) provides a grid on the rectangle R by drawing vertical and horizontal lines through the points with coordinates (x_i, y_j), where

$$x_i = a + ih, \quad \text{for each } i = 0, 1, \ldots, n,$$

and

$$y_j = c + jk, \quad \text{for each } j = 0, 1, \ldots, m.$$

Figure 12.4

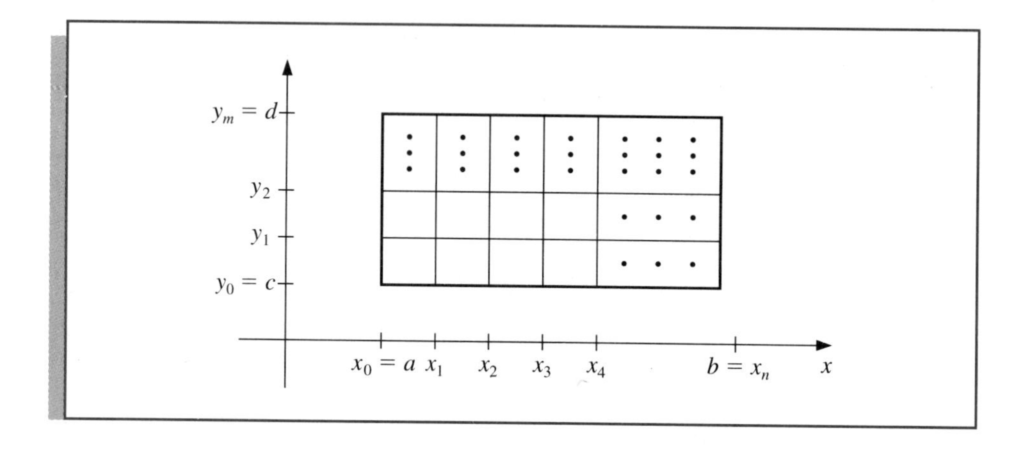

The lines $x = x_i$ and $y = y_j$ are **grid lines**, and their intersections are the **mesh points** of the grid. For each mesh point in the interior of the grid, (x_i, y_j), for $i = 1, 2, \ldots, n-1$ and $j = 1, 2, \ldots, m-1$, we use the Taylor series in the variable x about x_i to generate the central-difference formula

$$\textbf{(12.2)} \qquad \frac{\partial^2 u}{\partial x^2}(x_i, y_j) = \frac{u(x_{i+1}, y_j) - 2u(x_i, y_j) + u(x_{i-1}, y_j)}{h^2} - \frac{h^2}{12}\frac{\partial^4 u}{\partial x^4}(\xi_i, y_j),$$

where $\xi_i \in (x_{i-1}, x_{i+1})$. We also use the Taylor series in the variable y about y_j to generate the central-difference formula

$$\textbf{(12.3)} \qquad \frac{\partial^2 u}{\partial y^2}(x_i, y_j) = \frac{u(x_i, y_{j+1}) - 2u(x_i, y_j) + u(x_i, y_{j-1})}{k^2} - \frac{k^2}{12}\frac{\partial^4 u}{\partial y^4}(x_i, \eta_j),$$

where $\eta_j \in (y_{j-1}, y_{j+1})$.

Using these formulas in Eq. (12.1) allows us to express the Poisson equation at the points (x_i, y_j) as

$$\frac{u(x_{i+1}, y_j) - 2u(x_i, y_j) + u(x_{i-1}, y_j)}{h^2} + \frac{u(x_i, y_{j+1}) - 2u(x_i, y_j) + u(x_i, y_{j-1})}{k^2}$$
$$= f(x_i, y_j) + \frac{h^2}{12}\frac{\partial^4 u}{\partial x^4}(\xi_i, y_j) + \frac{k^2}{12}\frac{\partial^4 u}{\partial y^4}(x_i, \eta_j),$$

for each $i = 1, 2, \ldots, n-1$ and $j = 1, 2, \ldots, m-1$ and the boundary conditions as

$$u(x_0, y_j) = g(x_0, y_j), \quad \text{for each } j = 0, 1, \ldots, m,$$
$$u(x_n, y_j) = g(x_n, y_j), \quad \text{for each } j = 0, 1, \ldots, m,$$

$$u(x_i, y_0) = g(x_i, y_0), \quad \text{for each } i = 1, 2, \ldots, n-1,$$
$$u(x_i, y_m) = g(x_i, y_m), \quad \text{for each } i = 1, 2, \ldots, n-1.$$

In difference-equation form, this results in the **Central-Difference method**, with local truncation error of order $O(h^2 + k^2)$:

(12.4)
$$2\left[\left(\frac{h}{k}\right)^2 + 1\right] w_{ij} - (w_{i+1,j} + w_{i-1,j}) - \left(\frac{h}{k}\right)^2 (w_{i,j+1} + w_{i,j-1}) = -h^2 f(x_i, y_j),$$

for each $i = 1, 2, \ldots, n-1$ and $j = 1, 2, \ldots, m-1$, and

(12.5)
$$\begin{aligned} w_{0j} &= g(x_0, y_j), && \text{for each } j = 0, 1, \ldots, m, \\ w_{nj} &= g(x_n, y_j), && \text{for each } j = 0, 1, \ldots, m, \\ w_{i0} &= g(x_i, y_0), && \text{for each } i = 1, 2, \ldots, n-1, \\ w_{im} &= g(x_i, y_m), && \text{for each } i = 1, 2, \ldots, n-1, \end{aligned}$$

where w_{ij} approximates $u(x_i, y_j)$.

The typical equation in (12.4) involves approximations to $u(x, y)$ at the points

$$(x_{i-1}, y_j), \quad (x_i, y_j), \quad (x_{i+1}, y_j), \quad (x_i, y_{j-1}), \quad \text{and} \quad (x_i, y_{j+1}).$$

Reproducing the portion of the grid where these points are located (see Figure 12.5) shows that each equation involves approximations in a star-shaped region about (x_i, y_j).

Figure 12.5

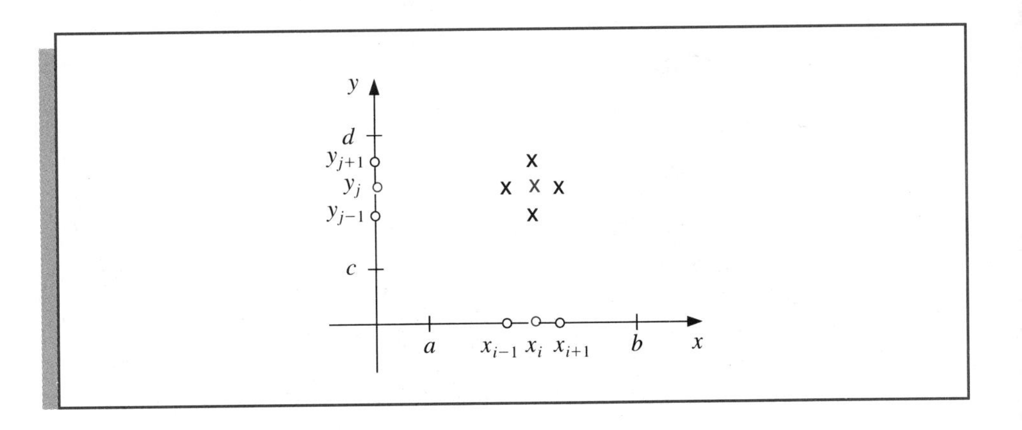

If we use the information from the boundary conditions (12.5) whenever appropriate in the system given by (12.4), that is, at all points (x_i, y_j) adjacent to a boundary mesh point, we have an $(n-1)(m-1)$ by $(n-1)(m-1)$ linear system with the unknowns being the approximations $w_{i,j}$ to $u(x_i, y_j)$ at the interior mesh points.

The linear system involving these unknowns is expressed for matrix calculations more efficiently if a relabeling of the interior mesh points is introduced. A recommended labeling of these points (see [Var, p. 187]) is to let

$$P_l = (x_i, y_j) \quad \text{and} \quad w_l = w_{i,j},$$

where $l = i + (m - 1 - j)(n - 1)$, for each $i = 1, 2, \ldots, n - 1$ and $j = 1, 2, \ldots, m - 1$. This labels the mesh points consecutively from left to right and top to bottom. For example, with $n = 4$ and $m = 5$ the relabeling results in a grid whose points are shown in Figure 12.6. Labeling the points in this manner ensures that the system needed to determine the $w_{i,j}$ is a banded matrix with band width at most $2n - 1$.

Figure 12.6

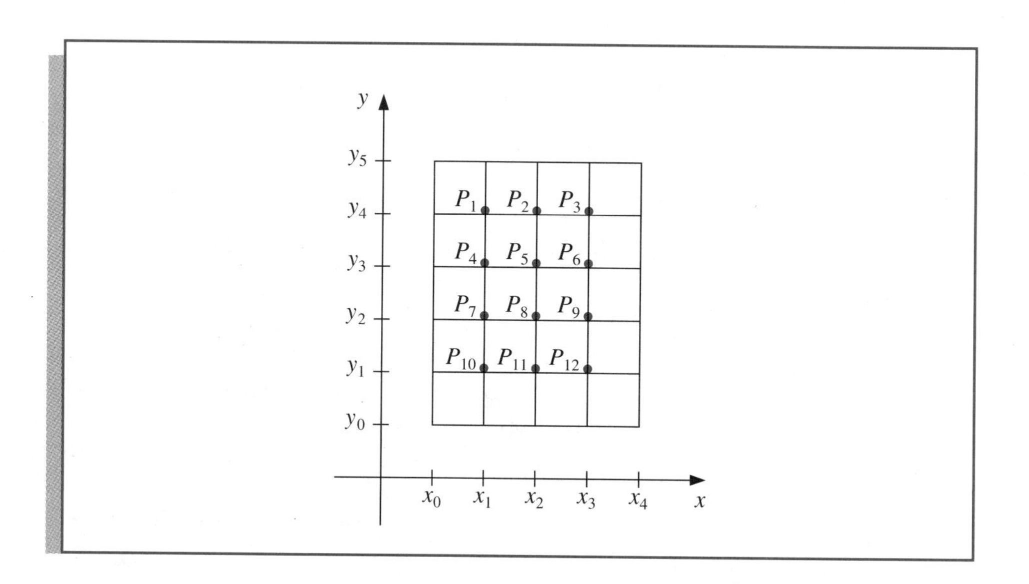

EXAMPLE 1 Consider the problem of determining the steady-state heat distribution in a thin square metal plate with dimensions 0.5 m by 0.5 m. Two adjacent boundaries are held at 0°C, while the heat on the other boundaries increases linearly from 0°C at one corner to 100°C where the sides meet. If we place the sides with the zero boundary conditions along the x- and y-axes, the problem is expressed as

$$\frac{\partial^2 u}{\partial x^2}(x, y) + \frac{\partial^2 u}{\partial y^2}(x, y) = 0$$

for (x, y) in the set $R = \{(x, y) \mid 0 < x < 0.5,\ 0 < y < 0.5\}$, with the boundary conditions

$$u(0, y) = 0, \quad u(x, 0) = 0, \quad u(x, 0.5) = 200x, \quad u(0.5, y) = 200y.$$

If $n = m = 4$, the problem has the grid given in Figure 12.7, and the difference equation (12.4) is

$$4w_{i,j} - w_{i+1,j} - w_{i-1,j} - w_{i,j-1} - w_{i,j+1} = 0$$

for each $i = 1, 2, 3$ and $j = 1, 2, 3$.

Expressing this in terms of the relabeled interior grid points $w_i = u(P_i)$ implies that the equations at the points P_i are:

$$\begin{aligned}
P_1:&\quad 4w_1 - w_2 - w_4 = w_{0,3} + w_{1,4},\\
P_2:&\quad 4w_2 - w_3 - w_1 - w_5 = w_{2,4},\\
P_3:&\quad 4w_3 - w_2 - w_6 = w_{4,3} + w_{3,4},\\
P_4:&\quad 4w_4 - w_5 - w_1 - w_7 = w_{0,2},\\
P_5:&\quad 4w_5 - w_6 - w_4 - w_2 - w_8 = 0,\\
P_6:&\quad 4w_6 - w_5 - w_3 - w_9 = w_{4,2},\\
P_7:&\quad 4w_7 - w_8 - w_4 = w_{0,1} + w_{1,0},\\
P_8:&\quad 4w_8 - w_9 - w_7 - w_5 = w_{2,0},\\
P_9:&\quad 4w_9 - w_8 - w_6 = w_{3,0} + w_{4,1},
\end{aligned}$$

where the right sides of the equations are obtained from the boundary conditions.

Figure 12.7

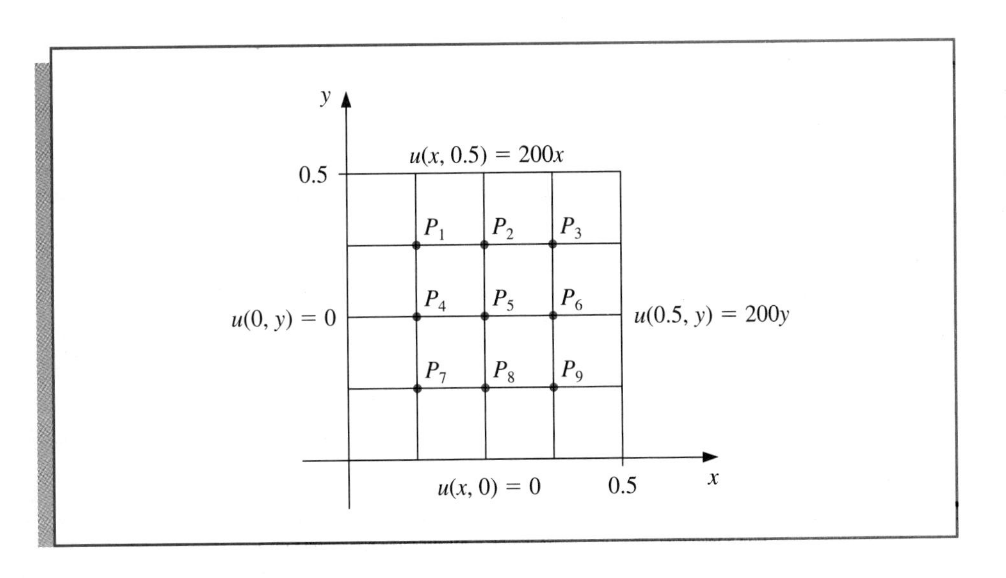

The boundary conditions imply that

$$w_{1,0} = w_{2,0} = w_{3,0} = w_{0,1} = w_{0,2} = w_{0,3} = 0,$$

$$w_{1,4} = w_{4,1} = 25, \quad w_{2,4} = w_{4,2} = 50, \quad \text{and} \quad w_{3,4} = w_{4,3} = 75.$$

The linear system associated with this problem has the form

$$\begin{bmatrix} 4 & -1 & 0 & -1 & 0 & 0 & 0 & 0 & 0 \\ -1 & 4 & -1 & 0 & -1 & 0 & 0 & 0 & 0 \\ 0 & -1 & 4 & 0 & 0 & -1 & 0 & 0 & 0 \\ -1 & 0 & 0 & 4 & -1 & 0 & -1 & 0 & 0 \\ 0 & -1 & 0 & -1 & 4 & -1 & 0 & -1 & 0 \\ 0 & 0 & -1 & 0 & -1 & 4 & 0 & 0 & -1 \\ 0 & 0 & 0 & -1 & 0 & 0 & 4 & -1 & 0 \\ 0 & 0 & 0 & 0 & -1 & 0 & -1 & 4 & -1 \\ 0 & 0 & 0 & 0 & 0 & -1 & 0 & -1 & 4 \end{bmatrix} \begin{bmatrix} w_1 \\ w_2 \\ w_3 \\ w_4 \\ w_5 \\ w_6 \\ w_7 \\ w_8 \\ w_9 \end{bmatrix} = \begin{bmatrix} 25 \\ 50 \\ 150 \\ 0 \\ 0 \\ 50 \\ 0 \\ 0 \\ 25 \end{bmatrix}.$$

The values of $w_1, w_2, \ldots, w_9$, found by applying the Gauss-Seidel method to this matrix, are given in Table 12.1.

Table 12.1

i	1	2	3	4	5	6	7	8	9
w_i	18.75	37.50	56.25	12.50	25.00	37.50	6.25	12.50	18.75

These answers are exact, since the true solution, $u(x, y) = 400xy$, has

$$\frac{\partial^4 u}{\partial x^4} = \frac{\partial^4 u}{\partial y^4} \equiv 0,$$

and so the truncation error is zero at each step. ■

The problem we considered in Example 1 has the same mesh size, 0.125, on each axis and requires solving only a 9×9 linear system. This simplifies the situation and does not introduce the computational problems that are present when the system is larger. Algorithm 12.1 uses the Gauss-Seidel iterative method for solving the linear system that is produced and permits unequal mesh sizes on the axes.

ALGORITHM **12.1**

Poisson Equation Finite-Difference

To approximate the solution to the Poisson equation

$$\frac{\partial^2 u}{\partial x^2}(x, y) + \frac{\partial^2 u}{\partial y^2}(x, y) = f(x, y), \quad a \le x \le b, \quad c \le y \le d,$$

subject to the boundary conditions

$$u(x, y) = g(x, y) \quad \text{if } x = a \text{ or } x = b \quad \text{and} \quad c \le y \le d$$

and

$$u(x, y) = g(x, y) \quad \text{if } y = c \text{ or } y = d \quad \text{and} \quad a \le x \le b:$$

INPUT endpoints a, b, c, d; integers $m \geq 3$, $n \geq 3$; tolerance TOL; maximum number of iterations N.

OUTPUT approximations $w_{i,j}$ to $u(x_i, y_j)$ for each $i = 1, \ldots, n-1$ and for each $j = 1, \ldots, m-1$ or a message that the maximum number of iterations was exceeded.

Step 1 Set $h = (b-a)/n$;
$k = (d-c)/m$.

Step 2 For $i = 1, \ldots, n-1$ set $x_i = a + ih$. (*Steps* 2 *and* 3 *construct mesh points.*)

Step 3 For $j = 1, \ldots, m-1$ set $y_j = c + jk$.

Step 4 For $i = 1, \ldots, n-1$
for $j = 1, \ldots, m-1$ set $w_{i,j} = 0$.

Step 5 Set $\lambda = h^2/k^2$;
$\mu = 2(1+\lambda)$;
$l = 1$.

Step 6 While $l \leq N$ do Steps 7–20. (*Steps* 7–20 *perform Gauss-Seidel iterations.*)

Step 7 Set $z = (-h^2 f(x_1, y_{m-1}) + g(a, y_{m-1}) + \lambda g(x_1, d) + \lambda w_{1,m-2} + w_{2,m-1})/\mu$;
$NORM = |z - w_{1,m-1}|$;
$w_{1,m-1} = z$.

Step 8 For $i = 2, \ldots, n-2$
set $z = (-h^2 f(x_i, y_{m-1}) + \lambda g(x_i, d) + w_{i-1,m-1}$
$+ w_{i+1,m-1} + \lambda w_{i,m-2})/\mu$;
if $|w_{i,m-1} - z| > NORM$ then set $NORM = |w_{i,m-1} - z|$;
set $w_{i,m-1} = z$.

Step 9 Set $z = (-h^2 f(x_{n-1}, y_{m-1}) + g(b, y_{m-1}) + \lambda g(x_{n-1}, d)$
$+ w_{n-2,m-1} + \lambda w_{n-1,m-2})/\mu$;
if $|w_{n-1,m-1} - z| > NORM$ then set $NORM = |w_{n-1,m-1} - z|$;
set $w_{n-1,m-1} = z$.

Step 10 For $j = m-2, \ldots, 2$ do Steps 11, 12, and 13.

Step 11 Set $z = (-h^2 f(x_1, y_j) + g(a, y_j) + \lambda w_{1,j+1} + \lambda w_{1,j-1} + w_{2,j})/\mu$;
if $|w_{1,j} - z| > NORM$ then set $NORM = |w_{1,j} - z|$;
set $w_{1,j} = z$.

Step 12 For $i = 2, \ldots, n-2$
set $z = (-h^2 f(x_i, y_j) + w_{i-1,j} + \lambda w_{i,j+1} + w_{i+1,j} + \lambda w_{i,j-1})/\mu$;
if $|w_{i,j} - z| > NORM$ then set $NORM = |w_{i,j} - z|$;
set $w_{i,j} = z$.

Step 13 Set $z = (-h^2 f(x_{n-1}, y_j) + g(b, y_j) + w_{n-2,j}$
$+ \lambda w_{n-1,j+1} + \lambda w_{n-1,j-1})/\mu$;
if $|w_{n-1,j} - z| > NORM$ then set $NORM = |w_{n-1,j} - z|$;
set $w_{n-1,j} = z$.

Step 14 Set $z = (-h^2 f(x_1, y_1) + g(a, y_1) + \lambda g(x_1, c) + \lambda w_{1,2} + w_{2,1})/\mu$;
if $|w_{1,1} - z| > NORM$ then set $NORM = |w_{1,1} - z|$;
set $w_{1,1} = z$.

Step 15 For $i = 2, \ldots, n-2$
set $z = (-h^2 f(x_i, y_1) + \lambda g(x_i, c) + w_{i-1,1} + \lambda w_{i,2} + w_{i+1,1})/\mu$;
if $|w_{i,1} - z| > NORM$ then set $NORM = |w_{i,1} - z|$;
set $w_{i,1} = z$.

Step 16 Set $z = (-h^2 f(x_{n-1}, y_1) + g(b, y_1) + \lambda g(x_{n-1}, c) + w_{n-2,1} + \lambda w_{n-1,2})/\mu$;
if $|w_{n-1,1} - z| > NORM$ then set $NORM = |w_{n-1,1} - z|$;
set $w_{n-1,1} = z$.

Step 17 If $NORM \le TOL$ then do Steps 18 and 19.

Step 18 For $i = 1, \ldots, n-1$
for $j = 1, \ldots, m-1$ OUTPUT $(x_i, y_j, w_{i,j})$.

Step 19 STOP. (*Procedure completed successfully.*)

Step 20 Set $l = l + 1$.

Step 21 OUTPUT ('Maximum number of iterations exceeded');
(*Procedure completed unsuccessfully.*)
STOP.

Although the Gauss-Seidel iterative procedure is incorporated into Algorithm 12.1 for simplicity, it is advisable to use a direct technique such as Gaussian elimination when the system is small, on the order of 100 or less, since the positive definiteness ensures stability with respect to roundoff errors. In particular, a generalization of the Crout Factorization Algorithm 6.7, see [Var, p. 196], is efficient for solving this system since the matrix is in symmetric-block tridiagonal form

$$\begin{bmatrix} A_1 & C_1 & 0 & \cdots & \cdots & 0 \\ C_1 & A_2 & C_2 & \ddots & & \vdots \\ 0 & C_2 & \ddots & \ddots & \ddots & \vdots \\ \vdots & \ddots & \ddots & \ddots & \ddots & 0 \\ \vdots & & \ddots & \ddots & \ddots & C_{m-1} \\ 0 & \cdots & \cdots & 0 & C_{m-1} & A_{m-1} \end{bmatrix},$$

with square blocks of size $(n-1)$ by $(n-1)$.

For large systems, an iterative method should be used—specifically, the SOR method discussed in Algorithm 7.3. The choice of ω that is optimal in this situation comes from the fact that when A is decomposed into its diagonal D and upper- and lower-triangular parts U and L,

$$A = D - L - U,$$

and B is the matrix for the Jacobi method,

$$B = D^{-1}(L + U),$$

then the spectral radius of B is (see [Var, p. 203])

$$\rho(B) = \frac{1}{2}\left[\cos\left(\frac{\pi}{m}\right) + \cos\left(\frac{\pi}{n}\right)\right].$$

The value of ω to be used is consequently,

$$\omega = \frac{2}{1 + \sqrt{1 - [\rho(B)]^2}} = \frac{4}{2 + \sqrt{4 - \left[\cos\left(\frac{\pi}{m}\right) + \cos\left(\frac{\pi}{n}\right)\right]^2}}.$$

A block technique can be incorporated into the algorithm for faster convergence of the SOR procedure. For a presentation of the technique involved see [Var, pp. 194–199].

EXAMPLE 2 Consider Poisson's equation

$$\frac{\partial^2 u}{\partial x^2}(x, y) + \frac{\partial^2 u}{\partial y^2}(x, y) = xe^y, \quad 0 < x < 2, \quad 0 < y < 1,$$

Table 12.2

i	j	x_i	y_j	$w_{i,j}^{(61)}$	$u(x_i, y_j)$	$\lvert u(x_i, y_j) - w_{i,j}^{(61)}\rvert$
1	1	0.3333	0.2000	0.40726	0.40713	1.30×10^{-4}
1	2	0.3333	0.4000	0.49748	0.49727	2.08×10^{-4}
1	3	0.3333	0.6000	0.60760	0.60737	2.23×10^{-4}
1	4	0.3333	0.8000	0.74201	0.74185	1.60×10^{-4}
2	1	0.6667	0.2000	0.81452	0.81427	2.55×10^{-4}
2	2	0.6667	0.4000	0.99496	0.99455	4.08×10^{-4}
2	3	0.6667	0.6000	1.2152	1.2147	4.37×10^{-4}
2	4	0.6667	0.8000	1.4840	1.4837	3.15×10^{-4}
3	1	1.0000	0.2000	1.2218	1.2214	3.64×10^{-4}
3	2	1.0000	0.4000	1.4924	1.4918	5.80×10^{-4}
3	3	1.0000	0.6000	1.8227	1.8221	6.24×10^{-4}
3	4	1.0000	0.8000	2.2260	2.2255	4.51×10^{-4}
4	1	1.3333	0.2000	1.6290	1.6285	4.27×10^{-4}
4	2	1.3333	0.4000	1.9898	1.9891	6.79×10^{-4}
4	3	1.3333	0.6000	2.4302	2.4295	7.35×10^{-4}
4	4	1.3333	0.8000	2.9679	2.9674	5.40×10^{-4}
5	1	1.6667	0.2000	2.0360	2.0357	3.71×10^{-4}
5	2	1.6667	0.4000	2.4870	2.4864	5.84×10^{-4}
5	3	1.6667	0.6000	3.0375	3.0369	6.41×10^{-4}
5	4	1.6667	0.8000	3.7097	3.7092	4.89×10^{-4}

with the boundary conditions

$$u(0, y) = 0, \quad u(2, y) = 2e^y, \quad 0 \le y \le 1,$$

$$u(x, 0) = x, \quad u(x, 1) = ex, \quad 0 \le x \le 2.$$

We will use Algorithm 12.1 to approximate the exact solution $u(x, y) = xe^y$ with $n = 6$ and $m = 5$. The stopping criterion for the Gauss-Seidel method in the Step 17 requires that

$$|w_{ij}^{(l)} - w_{ij}^{(l-1)}| \le 10^{-10},$$

for each $i = 1, \ldots, 5$ and $j = 1, \ldots, 4$. So the solution to the difference equation was accurately obtained, and the procedure stopped at $l = 61$. The results, along with the correct values, are presented in Table 12.2 on page 681. ■

EXERCISE SET 12.1

1. Use Algorithm 12.1 to approximate the solution to the elliptic partial-differential equation

$$\frac{\partial^2 u}{\partial x^2} + \frac{\partial^2 u}{\partial y^2} = 4, \quad 0 < x < 1, \quad 0 < y < 2;$$

$$u(x, 0) = x^2, \quad u(x, 2) = (x - 2)^2, \quad 0 \le x \le 1;$$
$$u(0, y) = y^2, \quad u(1, y) = (y - 1)^2, \quad 0 \le y \le 2.$$

Use $h = k = \frac{1}{2}$ and compare the results to the actual solution $u(x, y) = (x - y)^2$.

2. Use Algorithm 12.1 to approximate the solution to the elliptic partial-differential equation

$$\frac{\partial^2 u}{\partial x^2} + \frac{\partial^2 u}{\partial y^2} = 0, \quad 1 < x < 2, \quad 0 < y < 1;$$

$$u(x, 0) = 2 \ln x, \quad u(x, 1) = \ln(x^2 + 1), \quad 1 \le x \le 2;$$
$$u(1, y) = \ln(y^2 + 1), \quad u(2, y) = \ln(y^2 + 4), \quad 0 \le y \le 1.$$

Use $h = k = \frac{1}{3}$ and compare the results to the actual solution $u(x, y) = \ln(x^2 + y^2)$.

3. Approximate the solutions to the following elliptic partial-differential equations, using Algorithm 12.1:

   a. $$\frac{\partial^2 u}{\partial x^2} + \frac{\partial^2 u}{\partial y^2} = 0, \quad 0 < x < 1, \quad 0 < y < 1;$$

   $$u(x, 0) = 0, \quad u(x, 1) = x, \quad 0 \le x \le 1;$$
   $$u(0, y) = 0, \quad u(1, y) = y, \quad 0 \le y \le 1.$$

   Use $h = k = 0.2$ and compare the results with the solution $u(x, y) = xy$.

   b. $$\frac{\partial^2 u}{\partial x^2} + \frac{\partial^2 u}{\partial y^2} = -\cos(x + y) - \cos(x - y), \quad 0 < x < \pi, \quad 0 < y < \frac{\pi}{2};$$

   $$u(0, y) = \cos y, \quad u(\pi, y) = -\cos y, \quad 0 \le y \le \frac{\pi}{2};$$

   $$u(x, 0) = \cos x, \quad u\left(x, \frac{\pi}{2}\right) = 0, \quad 0 \le x \le \pi.$$

   Use $h = \pi/5$ and $k = \pi/10$ and compare the results with the solution $u(x, y) = \cos x \cos y$.

c. $\dfrac{\partial^2 u}{\partial x^2} + \dfrac{\partial^2 u}{\partial y^2} = (x^2 + y^2)e^{xy}, \quad 0 < x < 2, \quad 0 < y < 1;$

$$u(0, y) = 1, \quad u(2, y) = e^{2y}, \quad 0 \le y \le 1;$$
$$u(x, 0) = 1, \quad u(x, 1) = e^{x}, \quad 0 \le x \le 2.$$

Use $h = 0.2$ and $k = 0.1$, and compare the results with the solution $u(x, y) = e^{xy}$.

d. $\dfrac{\partial^2 u}{\partial x^2} + \dfrac{\partial^2 u}{\partial y^2} = \dfrac{x}{y} + \dfrac{y}{x}, \quad 1 < x < 2, \quad 1 < y < 2;$

$$u(x, 1) = x \ln x, \quad u(x, 2) = x \ln(4x^2), \quad 1 \le x \le 2;$$
$$u(1, y) = y \ln y, \quad u(2, y) = 2y \ln(2y), \quad 1 \le y \le 2.$$

Use $h = k = 0.1$ and compare the results with the solution $u(x, y) = xy \ln xy$.

4. Repeat Exercise 3(a) using extrapolation with $h_0 = 0.2$, $h_1 = h_0/2$, and $h_2 = h_0/4$.

5. Construct an algorithm similar to Algorithm 12.1, except use the SOR method with optimal ω instead of the Gauss-Seidel method for solving the linear system.

6. Repeat Exercise 3, using the algorithm constructed in Exercise 5.

7. A coaxial cable is made of a 0.1-in. square inner conductor and a 0.5-in. square outer conductor. The potential at a point in the cross section of the cable is described by Laplace's equation. Suppose the inner conductor is kept at 0 volts and the outer conductor is kept at 110 volts. Find the potential between the two conductors by placing a grid with horizontal mesh spacing $h = 0.1$ in. and vertical mesh spacing $k = 0.1$ in. on the region

$$D = \{(x, y) \mid 0 \le x \le 0.5,\ 0 \le y \le 0.5\}.$$

Approximate the solution to Laplace's equation at each grid point, and use the two sets of boundary conditions to derive a linear system to be solved by the Gauss-Seidel method.

8. A 6-cm $\times$ 5-cm rectangular silver plate has heat being uniformly generated at each point at the rate $q = 1.5$ cal/cm^3·s. Let x represent the distance along the edge of the plate of length 6 cm and y be the distance along the edge of the plate of length 5 cm. Suppose the temperature u along the edges is kept at the following temperatures:

$$u(x, 0) = x(6 - x), \quad u(x, 5) = 0, \quad 0 \le x \le 6,$$
$$u(0, y) = y(5 - y), \quad u(6, y) = 0, \quad 0 \le y \le 5,$$

where the origin lies at a corner of the plate with coordinates $(0, 0)$ and the edges lie along the positive x- and y-axes. The steady-state temperature $u = u(x, y)$ satisfies Poisson's equation:

$$\frac{\partial^2 u}{\partial x^2}(x, y) + \frac{\partial^2 u}{\partial y^2}(x, y) = -\frac{q}{K}, \quad 0 < x < 6, \quad 0 < y < 5,$$

where K, the thermal conductivity, is 1.04 cal/cm·deg·s. Approximate the temperature $u(x, y)$ using Algorithm 12.1 with $h = 0.4$ and $k = \frac{1}{3}$.

12.2 Parabolic Partial-Differential Equations

The parabolic partial-differential equation we study is the heat, or diffusion, equation

(12.6)
$$\frac{\partial u}{\partial t}(x, t) = \alpha^2 \frac{\partial^2 u}{\partial x^2}(x, t), \quad 0 < x < l, \quad t > 0,$$

subject to the conditions

$$u(0, t) = u(l, t) = 0, \quad t > 0,$$

and

$$u(x, 0) = f(x), \quad 0 \leq x \leq l.$$

The approach we use to approximate the solution to this problem involves finite differences and is similar to the method used in Section 12.1.

First select an integer $m > 0$ and a time step $k > 0$, and let $h = l/m$. The grid points for this situation are (x_i, t_j), where $x_i = ih$ for $i = 0, 1, \ldots, m$, and $t_j = jk$, for $j = 0, 1, \ldots$.

We obtain the difference method by using the Taylor series in t to form the difference quotient

(12.7)
$$\frac{\partial u}{\partial t}(x_i, t_j) = \frac{u(x_i, t_j + k) - u(x_i, t_j)}{k} - \frac{k}{2}\frac{\partial^2 u}{\partial t^2}(x_i, \mu_j),$$

for some $\mu_j \in (t_j, t_{j+1})$, and the Taylor series in x to form the difference quotient

(12.8)
$$\frac{\partial^2 u}{\partial x^2}(x_i, t_j) = \frac{u(x_i + h, t_j) - 2u(x_i, t_j) + u(x_i - h, t_j)}{h^2} - \frac{h^2}{12}\frac{\partial^4 u}{\partial x^4}(\xi_i, t_j),$$

where $\xi_i \in (x_{i-1}, x_{i+1})$.

The partial-differential equation (12.6) implies that at the interior gridpoint (x_i, t_j), for each $i = 1, 2, \ldots, m - 1$ and $j = 1, 2, \ldots$, we have

$$\frac{\partial u}{\partial t}(x_i, t_j) - \alpha^2 \frac{\partial^2 u}{\partial x^2}(x_i, t_j) = 0,$$

so the difference method using the difference quotients (12.7) and (12.8) is

(12.9)
$$\frac{w_{i,j+1} - w_{ij}}{k} - \alpha^2 \frac{w_{i+1,j} - 2w_{ij} + w_{i-1,j}}{h^2} = 0,$$

where w_{ij} approximates $u(x_i, t_j)$.

The local truncation error for this difference equation is

(12.10)
$$\tau_{ij} = \frac{k}{2}\frac{\partial^2 u}{\partial t^2}(x_i, \mu_j) - \alpha^2 \frac{h^2}{12}\frac{\partial^4 u}{\partial x^4}(\xi_i, t_j).$$

Solving Eq. (12.9) for $w_{i,j+1}$ gives

(12.11)
$$w_{i,j+1} = \left(1 - \frac{2\alpha^2 k}{h^2}\right) w_{ij} + \alpha^2 \frac{k}{h^2}(w_{i+1,j} + w_{i-1,j}),$$

for each $i = 1, 2, \ldots, (m - 1)$ and $j = 1, 2, \ldots$. Since the initial condition $u(x, 0) = f(x)$, for each $0 \leq x \leq 1$, implies that $w_{i,0} = f(x_i)$, for each $i = 0, 1, \ldots, m$, these values can be used in Eq. (12.11) to find the value of $w_{i,1}$ for each $i = 1, 2, \ldots, m - 1$. The additional conditions $u(0, t) = 0$ and $u(l, t) = 0$ imply that $w_{0,1} = w_{m,1} = 0$, so all the entries of the form $w_{i,1}$ can be determined. If the procedure is reapplied once all the approximations $w_{i,1}$ are known, the values of $w_{i,2}, w_{i,3}, \ldots, w_{i,m-1}$ can be obtained in a similar manner.

The explicit nature of the difference method implies that the $(m-1)$ by $(m-1)$ matrix associated with this system can be written in the tridiagonal form

$$A = \begin{bmatrix} (1-2\lambda) & \lambda & 0 & \cdots & 0 \\ \lambda & (1-2\lambda) & \lambda & \ddots & \vdots \\ 0 & \ddots & \ddots & \ddots & 0 \\ \vdots & & \ddots & \ddots & \lambda \\ 0 & \cdots & 0 & \lambda & (1-2\lambda) \end{bmatrix},$$

where $\lambda = \alpha^2(k/h^2)$. If we let

$$\mathbf{w}^{(0)} = (f(x_1), f(x_2), \ldots, f(x_{m-1}))^t$$

and

$$\mathbf{w}^{(j)} = (w_{1j}, w_{2j}, \ldots, w_{m-1,j})^t, \quad \text{for each } j = 1, 2, \ldots,$$

then the approximate solution is given by

$$\mathbf{w}^{(j)} = A\mathbf{w}^{(j-1)}, \quad \text{for each } j = 1, 2, \ldots.$$

This is known as the **Forward-Difference method**. If the solution to the partial-differential equation has four continuous partial derivatives in x and two in t, then Eq. (12.10) implies that the method is of order $O(k + h^2)$.

EXAMPLE 1 Consider the heat equation

$$\frac{\partial u}{\partial t}(x, t) - \frac{\partial^2 u}{\partial x^2}(x, t) = 0, \quad 0 < x < 1, \quad 0 \le t,$$

with boundary conditions

$$u(0, t) = u(1, t) = 0, \quad 0 < t,$$

and initial conditions

$$u(x, 0) = \sin \pi x, \quad 0 \le x \le 1.$$

The solution to this problem is

$$u(x, t) = e^{-\pi^2 t} \sin \pi x.$$

The solution at $t = 0.5$ will be approximated using the Forward-Difference method, first with $h = 0.1$, $k = 0.0005$, and $\lambda = 0.05$ and then with $h = 0.1$, $k = 0.01$, and $\lambda = 1$. The results are presented in Table 12.3. ■

Table 12.3

x_i	$u(x_i, 0.5)$	$w_{i,1000}$ $k = 0.0005$	$\lvert u(x_i, 0.5) - w_{i,1000}\rvert$	$w_{i,50}$ $k = 0.01$	$\lvert u(x_i, 0.5) - w_{i,50}\rvert$
0.0	0	0		0	
0.1	0.00222241	0.00228652	6.411×10^{-5}	8.19876×10^{7}	8.199×10^{7}
0.2	0.00422728	0.00434922	1.219×10^{-4}	-1.55719×10^{8}	1.557×10^{8}
0.3	0.00581836	0.00598619	1.678×10^{-4}	2.13833×10^{8}	2.138×10^{8}
0.4	0.00683989	0.00703719	1.973×10^{-4}	-2.50642×10^{8}	2.506×10^{8}
0.5	0.00719188	0.00739934	2.075×10^{-4}	2.62685×10^{8}	2.627×10^{8}
0.6	0.00683989	0.00703719	1.973×10^{-4}	-2.49015×10^{8}	2.490×10^{8}
0.7	0.00581836	0.00598619	1.678×10^{-4}	2.11200×10^{8}	2.112×10^{8}
0.8	0.00422728	0.00434922	1.219×10^{-4}	-1.53086×10^{8}	1.531×10^{8}
0.9	0.00222241	0.00228652	6.511×10^{-5}	8.03604×10^{7}	8.036×10^{7}
1.0	0	0		0	

A truncation error of order $O(k + h^2)$ is expected in Example 1. Although this is obtained with $h = 0.1$ and $k = 0.0005$, it certainly is not when $h = 0.1$ and $k = 0.01$. To explain the difficulty, we must look at the stability of the Forward-Difference method.

If an error $\mathbf{e}^{(0)} = (e_1^{(0)}, e_2^{(0)}, \ldots, e_{m-1}^{(0)})^t$ is made in representing the initial data $\mathbf{w}^{(0)} = (f(x_1), f(x_2), \ldots, f(x_{m-1}))^t$ (or in any particular step, the choice of the initial step is simply for convenience), an error of $A\mathbf{e}^{(0)}$ propagates in $\mathbf{w}^{(1)}$, since

$$\mathbf{w}^{(1)} = A(\mathbf{w}^{(0)} + \mathbf{e}^{(0)}) = A\mathbf{w}^{(0)} + A\mathbf{e}^{(0)}.$$

This process continues. At the nth time step, the error in $\mathbf{w}^{(n)}$ due to $\mathbf{e}^{(0)}$ is $A^n\mathbf{e}^{(0)}$. The method is consequently stable if and only if these errors do not grow as n increases, that is, if and only if for any initial error $\mathbf{e}^{(0)}$ we have $\|A^n\mathbf{e}^{(0)}\| \le \|\mathbf{e}^{(0)}\|$ for all n. This implies that $\|A^n\| \le 1$, a condition that, by Theorem 7.15, requires that the spectral radius $\rho(A^n) = (\rho(A))^n \le 1$. The Forward-Difference method is therefore stable only if $\rho(A) \le 1$.

The eigenvalues of A can be shown (see Exercise 7) to be

$$\mu_i = 1 - 4\lambda\left(\sin\frac{i\pi}{2m}\right)^2, \quad \text{for each } i = 1, 2, \ldots, m-1.$$

The condition for stability consequently reduces to determining whether

$$\rho(A) = \max_{1 \le i \le m-1}\left|1 - 4\lambda\left(\sin\frac{i\pi}{2m}\right)^2\right| \le 1,$$

which simplifies to

$$0 \le \lambda\left(\sin\frac{i\pi}{2m}\right)^2 \le \frac{1}{2}, \quad \text{for each } i = 1, 2, \ldots, m-1.$$

Since stability requires that this inequality condition hold as $h \to 0$, or, equivalently, as $m \to \infty$, the fact that

$$\lim_{m\to\infty}\left[\sin\frac{(m-1)\pi}{2m}\right]^2 = 1$$

means that stability will occur only if $0 \le \lambda \le \frac{1}{2}$. Since $\lambda = \alpha^2(k/h^2)$, this inequality requires that h and k be chosen so that

$$\alpha^2\frac{k}{h^2} \le \frac{1}{2}.$$

This condition was satisfied in our example when $h = 0.1$ and $k = 0.0005$; but when k was increased to 0.01 with no corresponding increase in h, the ratio was

$$\frac{0.01}{(0.1)^2} = 1 > \frac{1}{2},$$

and stability problems became apparent.

Consistent with the terminology of Chapter 5, we call the Forward-Difference method **conditionally stable** and remark that the method converges to the solution of Eq. (12.6) with rate of convergence $O(k + h^2)$, provided

$$\alpha^2\frac{k}{h^2} \le \frac{1}{2}$$

and the required continuity conditions on the solution are met. (For a detailed proof of this fact, see [IK, pp. 502–505].)

To obtain a method that is **unconditionally stable**, we consider an implicit-difference method that results from using the backward-difference quotient for $(\partial u/\partial t)(x_i, t_j)$ in the form

$$\frac{\partial u}{\partial t}(x_i, t_j) = \frac{u(x_i, t_j) - u(x_i, t_{j-1})}{k} + \frac{k}{2}\frac{\partial^2 u}{\partial t^2}(x_i, \mu_j),$$

where μ_j in (t_{j-1}, t_j). Substituting this equation, together with Eq. (12.8) for $\partial^2 u/\partial x^2$, into the partial-differential equation gives

$$\frac{u(x_i, t_j) - u(x_i, t_{j-1})}{k} - \alpha^2\frac{u(x_{i+1}, t_j) - 2u(x_i, t_j) + u(x_{i-1}, t_j)}{h^2}$$
$$= -\frac{k}{2}\frac{\partial^2 u}{\partial t^2}(x_i, \mu_j) - \alpha^2\frac{h^2}{12}\frac{\partial^4 u}{\partial x^4}(\xi_i, t_j),$$

for some $\xi_i \in (x_{i-1}, x_{i+1})$. The **Backward-Difference method** that results is

(12.12) $$\frac{w_{ij} - w_{i,j-1}}{k} - \alpha^2\frac{w_{i+1,j} - 2w_{ij} + w_{i-1,j}}{h^2} = 0,$$

for each $i = 1, 2, \ldots, m-1$, and $j = 1, 2, \ldots$.

The Backward-Difference method involves, at a typical step, the mesh points

$$(x_i, t_j), \quad (x_i, t_{j-1}), \quad (x_{i-1}, t_j), \quad \text{and} \quad (x_{i+1}, t_j),$$

and, in grid form, involves approximations at the points marked with ×'s in Figure 12.8.

Figure 12.8

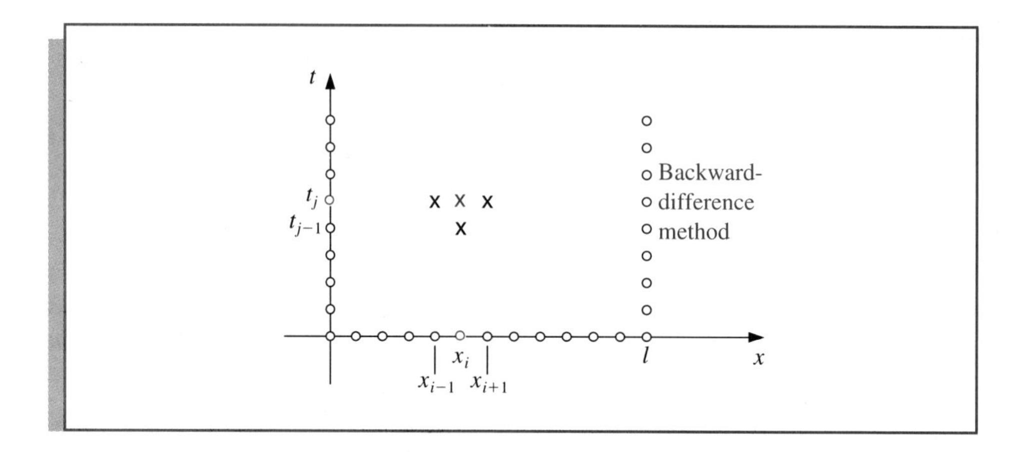

Since the boundary and initial conditions associated with the problem give information at the circled mesh points, the figure shows that no explicit procedures can be used to solve Eq. (12.12). In the Forward-Difference method (see Figure 12.9), approximations at

$$(x_{i-1}, t_j), \quad (x_i, t_j), \quad (x_i, t_{j+1}), \quad \text{and} \quad (x_{i+1}, t_j),$$

Figure 12.9

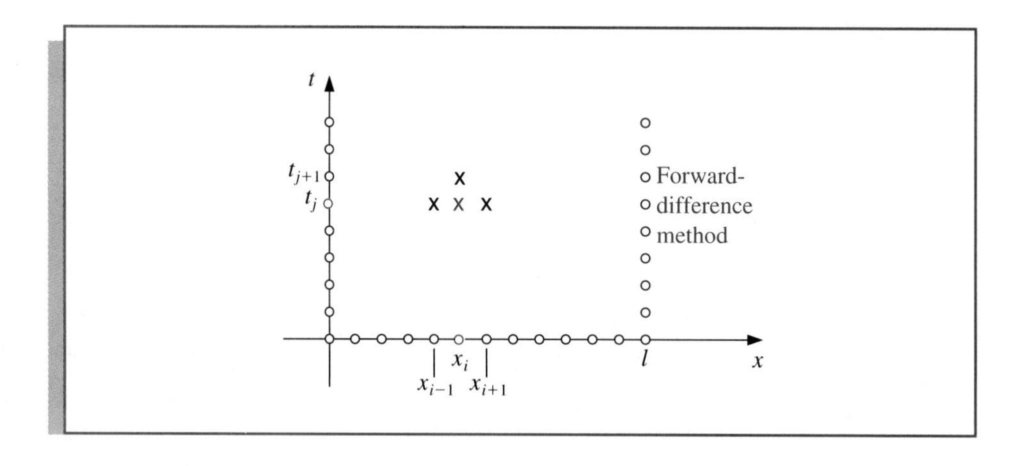

were used, so an explicit method for finding the approximations, based on the information from the initial and boundary conditions, was available.

If we again let λ denote the quantity $\alpha^2(k/h^2)$, the Backward-Difference method becomes

$$(1+2\lambda)w_{ij} - \lambda w_{i+1,j} - \lambda w_{i-1,j} = w_{i,j-1},$$

for each $i = 1, 2, \ldots, m-1$, and $j = 1, 2, \ldots$. Using the knowledge that $w_{i,0} = f(x_i)$ for each $i = 1, 2, \ldots, m-1$ and $w_{m,j} = w_{0,j} = 0$ for each $j = 1, 2, \ldots$, this difference method has the matrix representation:

$$\begin{bmatrix} (1+2\lambda) & -\lambda & 0 & \cdots & 0 \\ -\lambda & \ddots & \ddots & \ddots & \vdots \\ 0 & \ddots & \ddots & \ddots & 0 \\ \vdots & \ddots & \ddots & \ddots & -\lambda \\ 0 & \cdots & 0 & -\lambda & (1+2\lambda) \end{bmatrix} \begin{bmatrix} w_{1,j} \\ w_{2,j} \\ \vdots \\ w_{m-1,j} \end{bmatrix} = \begin{bmatrix} w_{1,j-1} \\ w_{2,j-1} \\ \vdots \\ w_{m-1,j-1} \end{bmatrix}, \tag{12.13}$$

or $A\mathbf{w}^{(j)} = \mathbf{w}^{(j-1)}$ for each $j = 1, 2, \ldots$. Since $\lambda > 0$, the matrix A is positive definite and strictly diagonally dominant, as well as being tridiagonal. We can use either the Crout Factorization for Tridiagonal Linear Systems Algorithm 6.7 or the SOR Algorithm 7.3, to solve this system. Algorithm 12.2 solves (12.13) using Crout factorization, which is an acceptable method unless m is large. In this algorithm we assume, for stopping purposes, that a bound is given for t.

ALGORITHM **12.2**

Heat Equation Backward-Difference

To approximate the solution to the parabolic partial-differential equation

$$\frac{\partial u}{\partial t}(x,t) - \alpha^2\frac{\partial^2 u}{\partial x^2}(x,t) = 0, \quad 0 < x < l, \quad 0 < t < T,$$

subject to the boundary conditions

$$u(0,t) = u(l,t) = 0, \quad 0 < t < T,$$

and the initial conditions

$$u(x,0) = f(x), \quad 0 \le x \le l:$$

INPUT endpoint l; maximum time T; constant α; integers $m \ge 3$, $N \ge 1$.

OUTPUT approximations $w_{i,j}$ to $u(x_i, t_j)$ for each $i = 1, \ldots, m-1$ and $j = 1, \ldots, N$.

Step 1 Set $h = l/m$;
$k = T/N$;
$\lambda = \alpha^2 k/h^2$.

Step 2 For $i = 1, \ldots, m-1$ set $w_i = f(ih)$. (*Initial values.*)
(*Steps 3–11 solve a tridiagonal linear system using Algorithm 6.7.*)

Step 3 Set $l_1 = 1 + 2\lambda$;
$u_1 = -\lambda/l_1$.

Step 4 For $i = 2, \ldots, m-2$ set $l_i = 1 + 2\lambda + \lambda u_{i-1}$;
$u_i = -\lambda/l_i$.

Step 5 Set $l_{m-1} = 1 + 2\lambda + \lambda u_{m-2}$.

Step 6 For $j = 1, \ldots, N$ do Steps 7–11.

Step 7 Set $t = jk$; (*Current t_j.*)
$z_1 = w_1/l_1$.

Step 8 For $i = 2, \ldots, m-1$ set $z_i = (w_i + \lambda z_{i-1})/l_i$.

Step 9 Set $w_{m-1} = z_{m-1}$.

Step 10 For $i = m-2, \ldots, 1$ set $w_i = z_i - u_i w_{i+1}$.

Step 11 OUTPUT (t); (*Note: $t = t_j$.*)
For $i = 1, \ldots, m-1$ set $x = ih$;
OUTPUT (x, w_i). (*Note*: $w_i = w_{i,j}$.)

Step 12 STOP. (*Procedure is complete.*)

EXAMPLE 2 The Backward-Difference method (Algorithm 12.2) with $h = 0.1$ and $k = 0.01$ will be used to approximate the solution to the heat equation

$$\frac{\partial u}{\partial t}(x,t) - \frac{\partial^2 u}{\partial x^2}(x,t) = 0, \quad 0 < x < 1, \quad 0 < t,$$

subject to the constraints

$$u(0,t) = u(1,t) = 0, \quad 0 < t, \quad u(x,0) = \sin \pi x, \quad 0 \leq x \leq 1,$$

which was considered in Example 1. To demonstrate the unconditional stability of the Backward-Difference method, we again compare $w_{i,50}$ to $u(x_i, 0.5)$, where $i = 0, 1, \ldots, 10$.

The results listed in Table 12.4 should be compared with the fifth and sixth columns of Table 12.3. ■

The Backward-Difference method does not have the stability problems of the Forward-Difference method. This can be seen by analyzing the eigenvalues of the matrix A. For the Backward-Difference method (see Exercise 8), the eigenvalues are

$$\mu_i = 1 + 4\lambda\left(\sin\frac{i\pi}{2m}\right)^2, \quad \text{for each } i = 1, 2, \ldots, m-1;$$

and since $\lambda > 0$, we have $\mu_i > 1$ for all $i = 1, 2, \ldots, m-1$. This implies that A^{-1} exists since zero is not an eigenvalue of A. An error $\mathbf{e}^{(0)}$ in the initial data produces an error $(A^{-1})^n \mathbf{e}^{(0)}$ at the nth step. Since the eigenvalues of A^{-1} are the reciprocals of the eigenvalues of A, the spectral radius of A^{-1} is bounded above by 1 and the method is stable,

Table 12.4

x_i	$w_{i,50}$	$u(x_i, 0.5)$	$\lvert w_{i,50} - u(x_i, 0.5)\rvert$
0.0	0	0	
0.1	0.00289802	0.00222241	6.756×10^{-4}
0.2	0.00551236	0.00422728	1.285×10^{-3}
0.3	0.00758711	0.00581836	1.769×10^{-3}
0.4	0.00891918	0.00683989	2.079×10^{-3}
0.5	0.00937818	0.00719188	2.186×10^{-3}
0.6	0.00891918	0.00683989	2.079×10^{-3}
0.7	0.00758711	0.00581836	1.769×10^{-3}
0.8	0.00551236	0.00422728	1.285×10^{-3}
0.9	0.00289802	0.00222241	6.756×10^{-4}
1.0	0	0	

independent of the choice of $\lambda = \alpha^2(k/h^2)$. In the terminology of Chapter 5, we call the Backward-Difference method an **unconditionally stable** method. The local truncation error for the method is of order $O(k + h^2)$, provided the solution of the differential equation satisfies the usual differentiability conditions. In this case, the method converges to the solution of the partial-differential equation with this same rate of convergence (see [IK, p. 508]).

The weakness in the Backward-Difference method results from the fact that the local truncation error has a portion with order $O(k)$, requiring that time intervals be made much smaller than spatial intervals. It would clearly be desirable to have a procedure with local truncation error of order $O(k^2 + h^2)$. The first step in this direction is to use a difference equation that has $O(k^2)$ error for $u_t(x, t)$ instead of those we have used previously, whose error was $O(k)$. This can be done by using the Taylor series in t for the function $u(x, t)$ at the point (x_i, t_j) and evaluating at (x_i, t_{j+1}) and (x_i, t_{j-1}) to obtain the Central-Difference formula

$$\frac{\partial u}{\partial t}(x_i, t_j) = \frac{u(x_i, t_{j+1}) - u(x_i, t_{j-1})}{2k} + \frac{k^2}{6}\frac{\partial^3 u}{\partial t^3}(x_i, \mu_j),$$

where $\mu_j \in (t_{j-1}, t_{j+1})$. The difference method that results from substituting this and the usual difference quotient for $\partial^2 u/\partial x^2$, Eq. (12.8), into the differential equation is called **Richardson's method** and is given by

(12.14)
$$\frac{w_{i,j+1} - w_{i,j-1}}{2k} - \alpha^2 \frac{w_{i+1,j} - 2w_{ij} + w_{i-1,j}}{h^2} = 0.$$

Richardson's method has local truncation error of order $O(k^2 + h^2)$, but unfortunately it also has serious stability problems (see Exercise 6).

A more rewarding method is derived by averaging the Forward-Difference method at the jth step in t,

$$\frac{w_{i,j+1} - w_{i,j}}{k} - \alpha^2 \frac{w_{i+1,j} - 2w_{i,j} + w_{i-1,j}}{h^2} = 0,$$

which has local truncation error

$$\tau_F = \frac{k}{2}\frac{\partial^2 u}{\partial t^2}(x_i, \mu_j) + O(h^2),$$

and the Backward-Difference method at the $(j+1)$st step in t,

$$\frac{w_{i,j+1} - w_{i,j}}{k} - \alpha^2 \frac{w_{i+1,j+1} - 2w_{i,j+1} + w_{i-1,j+1}}{h^2} = 0,$$

which has local truncation error

$$\tau_B = -\frac{k}{2}\frac{\partial^2 u}{\partial t^2}(x_i, \hat{u}_j) + O(h^2).$$

If we assume that

$$\frac{\partial^2 u}{\partial t^2}(x_i, \hat{\mu}_j) \approx \frac{\partial^2 u}{\partial t^2}(x_i, \mu_j),$$

then the averaged-difference method,

$$\frac{w_{i,j+1} - w_{ij}}{k} - \frac{\alpha^2}{2}\left[\frac{w_{i+1,j} - 2w_{i,j} + w_{i-1,j}}{h^2} + \frac{w_{i+1,j+1} - 2w_{i,j+1} + w_{i-1,j+1}}{h^2}\right] = 0,$$

has local truncation error of order $O(k^2 + h^2)$, provided, of course, that the usual differentiability conditions are satisfied. This is known as the **Crank-Nicolson method** and is represented in the matrix form

(12.15) $$A\mathbf{w}^{(j+1)} = B\mathbf{w}^{(j)}, \quad \text{for each } j = 0, 1, 2, \ldots,$$

where

$$\lambda = \alpha^2 \frac{k}{h^2}, \quad \mathbf{w}^{(j)} = (w_{1,j}, w_{2,j}, \ldots, w_{m-1,j})^t,$$

and the matrices A and B are given by:

$$A = \begin{bmatrix} (1+\lambda) & -\frac{\lambda}{2} & 0 & \cdots & 0 \\ -\frac{\lambda}{2} & \ddots & \ddots & \ddots & \vdots \\ 0 & \ddots & \ddots & \ddots & 0 \\ \vdots & \ddots & \ddots & \ddots & -\frac{\lambda}{2} \\ 0 & \cdots & 0 & -\frac{\lambda}{2} & (1+\lambda) \end{bmatrix}$$

and

$$B = \begin{bmatrix} (1-\lambda) & \frac{\lambda}{2} & 0 & \cdots & \cdots & 0 \\ \frac{\lambda}{2} & \ddots & \ddots & \ddots & & \vdots \\ 0 & \ddots & \ddots & \ddots & \ddots & 0 \\ \vdots & & \ddots & \ddots & \ddots & \frac{\lambda}{2} \\ 0 & \cdots & \cdots & 0 & \frac{\lambda}{2} & (1-\lambda) \end{bmatrix}$$

Since A is a positive definite, strictly diagonally dominant, and tridiagonal matrix, it is nonsingular. Either the Crout Factorization for Tridiagonal Linear System Algorithm 6.7 or the SOR Algorithm 7.3 can be used to obtain $\mathbf{w}^{(j+1)}$ from $\mathbf{w}^{(j)}$, for each $j = 0, 1, 2, \ldots$. Algorithm 12.3 incorporates Crout factorization into the Crank-Nicolson technique. As in Algorithm 12.2, a finite length for the time interval must be specified to determine a stopping procedure.

ALGORITHM 12.3

Crank-Nicolson

To approximate the solution to the parabolic partial-differential equation

$$\frac{\partial u}{\partial t}(x,t) - \alpha^2 \frac{\partial^2 u}{\partial x^2}(x,t) = 0, \quad 0 < x < l, \quad 0 < t < T,$$

subject to the boundary conditions

$$u(0,t) = u(l,t) = 0, \quad 0 < t < T,$$

and the initial conditions

$$u(x,0) = f(x), \quad 0 \le x \le l:$$

INPUT endpoint l; maximum time T; constant α; integers $m \ge 3$, $N \ge 1$.

OUTPUT approximations $w_{i,j}$ to $u(x_i, t_j)$ for each $i = 1, \ldots, m-1$ and $j = 1, \ldots, N$.

Step 1 Set $h = l/m$;
$k = T/N$;
$\lambda = \alpha^2 k/h^2$;
$w_m = 0$.

Step 2 For $i = 1, \ldots, m-1$ set $w_i = f(ih)$. (*Initial values.*)
(*Steps* 3–11 *solve a tridiagonal linear system using Algorithm* 6.7.)

Step 3 Set $l_1 = 1 + \lambda$;
$u_1 = -\lambda/(2l_1)$.

Step 4 For $i = 2, \ldots, m-2$ set $l_i = 1 + \lambda + \lambda u_{i-1}/2$;
$u_i = -\lambda/(2l_i)$.

Step 5 Set $l_{m-1} = 1 + \lambda + \lambda u_{m-2}/2$.

Step 6 For $j = 1, \ldots, N$ do Steps 7–11.

Step 7 Set $t = jk$; (*Current* t_j.)

$$z_1 = \left[(1-\lambda)w_1 + \frac{\lambda}{2}w_2\right] \Big/ l_1.$$

Step 8 For $i = 2, \ldots, m-1$ set

$$z_i = \left[(1-\lambda)w_i + \frac{\lambda}{2}(w_{i+1} + w_{i-1} + z_{i-1})\right] \Big/ l_i.$$

Step 9 Set $w_{m-1} = z_{m-1}$.

Step 10 For $i = m-2, \ldots, 1$ set $w_i = z_i - u_i w_{i+1}$.

Step 11 OUTPUT (t); (*Note:* $t = t_j$.)
For $i = 1, \ldots, m-1$ set $x = ih$;
OUTPUT (x, w_i). (*Note:* $w_i = w_{i,j}$.)

Step 12 STOP. (*Procedure is complete.*)

The verification that the Crank-Nicolson method is unconditionally stable and has order of convergence $O(k^2 + h^2)$ can be found in [IK, pp. 508–512].

EXAMPLE 3 The Crank-Nicolson method will be used to approximate the solution to the problem in Examples 1 and 2, consisting of the equation

$$\frac{\partial u}{\partial t}(x,t) - \frac{\partial^2 u}{\partial x^2}(x,t) = 0, \quad 0 < x < 1, \quad 0 < t,$$

subject to the conditions

$$u(0,t) = u(1,t) = 0, \quad 0 < t,$$

Table 12.5

x_i	$w_{i,50}$	$u(x_i, 0.5)$	$\lvert w_{i,50} - u(x_i, 0.5)\rvert$
0.0	0	0	
0.1	0.00230512	0.00222241	8.271×10^{-5}
0.2	0.00438461	0.00422728	1.573×10^{-4}
0.3	0.00603489	0.00581836	2.165×10^{-4}
0.4	0.00709444	0.00683989	2.546×10^{-4}
0.5	0.00745954	0.00719188	2.677×10^{-4}
0.6	0.00709444	0.00683989	2.546×10^{-4}
0.7	0.00603489	0.00581836	2.165×10^{-4}
0.8	0.00438461	0.00422728	1.573×10^{-4}
0.9	0.00230512	0.00222241	8.271×10^{-5}
1.0	0	0	

and

$$u(x, 0) = \sin \pi x, \quad 0 \le x \le 1.$$

The choices $m = 10$, $h = 0.1$, $N = 50$, $k = 0.01$, and $\lambda = 1$ are used in Algorithm 12.3, as they were in the previous examples. The results in Table 12.5 indicate the increase in accuracy of the Crank-Nicolson method over the Backward-Difference method, the best of the two previously discussed techniques. ■

EXERCISE SET 12.2

1. Approximate the solution to the following partial-differential equations using the Backward-Difference Algorithm.

   a. $$\frac{\partial u}{\partial t} - \frac{\partial^2 u}{\partial x^2} = 0, \quad 0 < x < 2,\ 0 < t;$$
   $$u(0, t) = u(2, t) = 0, \quad 0 < t,$$
   $$u(x, 0) = \sin \frac{\pi}{2} x, \quad 0 \le x \le 2.$$

   Use $m = 4$, $T = 0.1$, and $N = 2$ and compare your answers to the actual solution $u(x, t) = e^{-(\pi^2/4)t} \sin \frac{\pi}{2} x$.

   b. $$\frac{\partial u}{\partial t} - \frac{1}{16} \frac{\partial^2 u}{\partial x^2} = 0, \quad 0 < x < 1,\ 0 < t;$$
   $$u(0, t) = u(1, t) = 0, \quad 0 < t,$$
   $$u(x, 0) = 2 \sin 2\pi x, \quad 0 \le x \le 1.$$

   Use $m = 3$, $T = 0.1$, and $N = 2$ and compare your answers to the actual solution $u(x, t) = 2e^{-(\pi^2/4)t} \sin 2\pi x$.

2. Repeat Exercise 1 using the Crank-Nicolson Algorithm.

3. Use the Forward-Difference method to approximate the solution to the following parabolic partial-differential equations.

   a. $$\frac{\partial u}{\partial t} - \frac{\partial^2 u}{\partial x^2} = 0, \quad 0 < x < 2,\ 0 < t;$$
   $$u(0, t) = u(2, t) = 0, \quad 0 < t,$$
   $$u(x, 0) = \sin 2\pi x, \quad 0 \le x \le 2.$$

   Use $h = 0.4$ and $k = 0.1$, and compare your answers at $t = 0.5$ to the actual solution $u(x, t) = e^{-4\pi^2 t} \sin 2\pi x$. Then use $h = 0.4$ and $k = 0.05$, and compare the answers.

   b. $$\frac{\partial u}{\partial t} - \frac{\partial^2 u}{\partial x^2} = 0, \quad 0 < x < \pi,\ 0 < t;$$
   $$u(0, t) = u(\pi, t) = 0, \quad 0 < t,$$
   $$u(x, 0) = \sin x, \quad 0 \le x \le \pi.$$

   Use $h = \pi/10$ and $k = 0.05$ and compare your answers to the actual solution $u(x, t) = e^{-t} \sin x$ at $t = 0.5$.

c. $\dfrac{\partial u}{\partial t} - \dfrac{4}{\pi^2}\dfrac{\partial^2 u}{\partial x^2} = 0, \quad 0 < x < 4, \ 0 < t;$

$$u(0,t) = u(4,t) = 0, \quad 0 < t,$$

$$u(x,0) = \sin\frac{\pi}{4}x\left(1 + 2\cos\frac{\pi}{4}x\right), \quad 0 \le x \le 4.$$

Use $h = 0.2$ and $k = 0.04$. Compare your answers to the actual solution $u(x,t) = e^{-t}\sin\frac{\pi}{2}x + e^{-t/4}\sin\frac{\pi}{4}x$ at $t = 0.4$.

d. $\dfrac{\partial u}{\partial t} - \dfrac{1}{\pi^2}\dfrac{\partial^2 u}{\partial x^2} = 0, \quad 0 < x < 1, \ 0 < t;$

$$u(0,t) = u(1,t) = 0, \quad 0 < t,$$

$$u(x,0) = \cos\pi\left(x - \frac{1}{2}\right), \quad 0 \le x \le 1.$$

Use $h = 0.1$ and $k = 0.04$. Compare your answers to the actual solution $u(x,t) = e^{-t}\cos\pi(x - \frac{1}{2})$ at $t = 0.4$.

4. Repeat Exercise 3 using the Backward-Difference Algorithm.

5. Repeat Exercise 3 using the Crank-Nicolson Algorithm.

6. Repeat Exercise 3 using Richardson's method.

7. Show that the eigenvalues for the $(m-1)$ by $(m-1)$ tridiagonal matrix A given by

$$a_{ij} = \begin{cases} \lambda, & j = i-1 \text{ or } j = i+1, \\ 1 - 2\lambda, & j = i, \\ 0, & \text{otherwise} \end{cases}$$

are

$$\mu_i = 1 - 4\lambda\left(\sin\frac{i\pi}{2m}\right)^2, \quad \text{for each } i = 1, 2, \ldots, m-1,$$

with corresponding eigenvectors $\mathbf{v}^{(i)}$, where $v_j^{(i)} = \sin\dfrac{ij\pi}{m}$.

8. Show that the $(m-1)$ by $(m-1)$ tridiagonal matrix A given by

$$a_{ij} = \begin{cases} -\lambda, & j = i-1 \text{ or } j = i+1, \\ 1 + 2\lambda, & j = i, \\ 0, & \text{otherwise} \end{cases}$$

where $\lambda > 0$ is positive definite and diagonally dominant and has eigenvalues

$$\mu_i = 1 + 4\lambda\left(\sin\frac{i\pi}{2m}\right)^2, \quad \text{for each } i = 1, 2, \ldots, m-1,$$

with corresponding eigenvectors $\mathbf{v}^{(i)}$, where $v_j^{(i)} = \sin\dfrac{ij\pi}{m}$.

9. Modify Algorithms 12.2 and 12.3 to include the parabolic partial-differential equation

$$\frac{\partial u}{\partial t} - \frac{\partial^2 u}{\partial x^2} = F(x), \quad 0 < x < l, \ 0 < t;$$

$$u(0,t) = u(l,t) = 0, \quad 0 < t,$$

$$u(x,0) = f(x), \quad 0 \le x \le l.$$

10. Use the results of Exercise 9 to approximate the solution to

$$\frac{\partial u}{\partial t} - \frac{\partial^2 u}{\partial x^2} = 2, \quad 0 < x < 1, \ 0 < t;$$

$$u(0, t) = u(1, t) = 0, \quad 0 < t,$$

$$u(x, 0) = \sin \pi x + x(1 - x),$$

with $h = 0.1$ and $k = 0.01$. Compare your answer to the actual solution $u(x, t) = e^{-\pi^2 t} \sin \pi x + x(1 - x)$ at $t = 0.25$.

11. Change Algorithms 12.2 and 12.3 to accommodate the partial-differential equation

$$\frac{\partial u}{\partial t} - \alpha^2 \frac{\partial^2 u}{\partial x^2} = 0, \quad 0 < x < l, \ 0 < t;$$

$$u(0, t) = \phi(t), \ u(l, t) = \Psi(t), \quad 0 < t;$$

$$u(x, 0) = f(x), \quad 0 \le x \le l,$$

where $f(0) = \phi(0)$ and $f(l) = \Psi(0)$.

12. The temperature $u(x, t)$ of a long, thin rod of constant cross section and homogeneous conducting material is governed by the one-dimensional heat equation. If heat is generated in the material, for example by resistance to current or nuclear reaction, the heat equation becomes

$$\frac{\partial^2 u}{\partial x^2} + \frac{Kr}{\rho C} = K \frac{\partial u}{\partial t}, \quad 0 < x < l, \quad 0 < t,$$

where l is the length, ρ is the density, C is the specific heat, and K is the thermal diffusivity of the rod. The function $r = r(x, t, u)$ represents the heat generated per unit volume. Suppose that

$$l = 1.5 \text{ cm}, \qquad K = 1.04 \text{ cal/cm} \cdot \text{deg} \cdot \text{s},$$
$$\rho = 10.6 \text{ g/cm}^3, \quad C = 0.056 \text{ cal/g} \cdot \text{deg},$$

and

$$r(x, t, u) = 5.0 \text{ cal/cm}^3 \cdot \text{s}.$$

If the ends of the rod are kept at 0°C, then

$$u(0, t) = u(l, t) = 0, \quad t > 0.$$

Suppose the initial temperature distribution is given by

$$u(x, 0) = \sin \frac{\pi x}{l}, \quad 0 \le x \le l.$$

Use the results of Exercise 9 to approximate the temperature distribution with $h = 0.15$ and $k = 0.0225$.

13. Sagar and Payne [SP] analyze the stress-strain relationships and material properties of a cylinder alternately subjected to heating and cooling and consider the equation

$$\frac{\partial^2 T}{\partial r^2} + \frac{1}{r}\frac{\partial T}{\partial r} = \frac{1}{4K}\frac{\partial T}{\partial t}, \quad \frac{1}{2} < r < 1, \ 0 < T,$$

where $T = T(r, t)$ is the temperature, r is the radial distance from the center of the cylinder, t is time, and K is a diffusivity coefficient.

a. Find approximations to $T(r, 10)$ for a cylinder with outside radius 1, given the initial and boundary conditions:

$$T(1, t) = 100 + 40t, \quad 0 \le t \le 10,$$

$$T\left(\frac{1}{2}, t\right) = t, \quad 0 \le t \le 10,$$

$$T(r, 0) = 200(r - 0.5), \quad 0.5 \le r \le 1.$$

Use a modification of the Backward-Difference method with $K = 0.1$, $k = 0.5$, and $h = \Delta r = 0.1$.

b. Using the temperature distribution of part (a), to calculate the strain I by approximating the integral

$$I = \int_{0.5}^{1} \alpha T(r, t) r \, dr,$$

where $\alpha = 10.7$ and $t = 10$. Use the Composite Trapezoidal method with $n = 5$.

12.3 Hyperbolic Partial-Differential Equations

In this section, we consider the numerical solution to the wave equation, an example of a hyperbolic partial-differential equation. The wave equation is given by the differential equation

(12.16)
$$\frac{\partial^2 u}{\partial t^2}(x, t) - \alpha^2 \frac{\partial^2 u}{\partial x^2}(x, t) = 0, \quad 0 < x < l, \quad t > 0,$$

subject to the conditions

$$u(0, t) = u(l, t) = 0, \quad t > 0,$$
$$u(x, 0) = f(x), \quad 0 \leq x \leq l,$$

and

$$\frac{\partial u}{\partial t}(x, 0) = g(x), \quad 0 \leq x \leq l,$$

where α is a constant. To set up the finite-difference method, select an integer $m > 0$ and time-step size $k > 0$. With $h = l/m$, the mesh points (x_i, t_j) are

$$x_i = ih, \quad \text{for each } i = 0, 1, \ldots, m,$$

and

$$t_j = jk, \quad \text{for each } j = 0, 1, \ldots.$$

At any interior mesh point (x_i, t_j), the wave equation becomes

(12.17)
$$\frac{\partial^2 u}{\partial t^2}(x_i, t_j) - \alpha^2 \frac{\partial^2 u}{\partial x^2}(x_i, t_j) = 0.$$

The difference method is obtained using the centered-difference quotient for the second partial derivatives given by

$$\frac{\partial^2 u}{\partial t^2}(x_i, t_j) = \frac{u(x_i, t_{j+1}) - 2u(x_i, t_j) + u(x_i, t_{j-1})}{k^2} - \frac{k^2}{12} \frac{\partial^4 u}{\partial t^4}(x_i, \mu_j),$$

where $\mu_j \in (t_{j-1}, t_{j+1})$ and

$$\frac{\partial^2 u}{\partial x^2}(x_i, t_j) = \frac{u(x_{i+1}, t_j) - 2u(x_i, t_j) + u(x_{i-1}, t_j)}{h^2} - \frac{h^2}{12}\frac{\partial^4 u}{\partial x^4}(\xi_i, t_j),$$

where $\xi_i \in (x_{i-1}, x_{i+1})$. Substituting these into Eq. (12.17) gives

$$\frac{u(x_i, t_{j+1}) - 2u(x_i, t_j) + u(x_i, t_{j-1})}{k^2} - \alpha^2 \frac{u(x_{i+1}, t_j) - 2u(x_i, t_j) + u(x_{i-1}, t_j)}{h^2}$$
$$= \frac{1}{12}\left[k^2 \frac{\partial^4 u}{\partial t^4}(x_i, \mu_j) - \alpha^2 h^2 \frac{\partial^4 u}{\partial x^4}(\xi_i, t_j)\right].$$

Neglecting the error term

$$\tau_{i,j} = \frac{1}{12}\left[k^2 \frac{\partial^4 u}{\partial t^4}(x_i, \mu_j) - \alpha^2 h^2 \frac{\partial^4 u}{\partial x^4}(\xi_i, t_j)\right]$$

leads to the difference equation

$$\frac{w_{i,j+1} - 2w_{i,j} + w_{i,j-1}}{k^2} - \alpha^2 \frac{w_{i+1,j} - 2w_{i,j} + w_{i-1,j}}{h^2} = 0.$$

If $\lambda = \alpha k/h$, we can write the difference equation as

$$w_{i,j+1} - 2w_{i,j} + w_{i,j-1} - \lambda^2 w_{i+1,j} + 2\lambda^2 w_{i,j} - \lambda^2 w_{i-1,j} = 0$$

and solve for $w_{i,j+1}$, the most advanced time-step approximation, to obtain

(12.18)
$$w_{i,j+1} = 2(1 - \lambda^2)w_{i,j} + \lambda^2(w_{i+1,j} + w_{i-1,j}) - w_{i,j-1}.$$

This equation holds for each $i = 1, 2, \ldots, m-1$ and $j = 1, 2, \ldots$. The boundary conditions give

(12.19)
$$w_{0,j} = w_{m,j} = 0, \quad \text{for each } j = 1, 2, 3, \ldots,$$

and the initial condition implies that

(12.20)
$$w_{i,0} = f(x_i), \quad \text{for each } i = 1, 2, \ldots, m - 1.$$

Writing this set of equations in matrix form gives

$$\begin{bmatrix} w_{1,j+1} \\ w_{2,j+1} \\ \vdots \\ w_{m-1,j+1} \end{bmatrix} = \begin{bmatrix} 2(1-\lambda^2) & \lambda^2 & 0 & \cdots & 0 \\ \lambda^2 & 2(1-\lambda^2) & \lambda^2 & \ddots & \vdots \\ 0 & \ddots & \ddots & \ddots & 0 \\ \vdots & & \ddots & \ddots & \lambda^2 \\ 0 & \cdots & 0 & \lambda^2 & 2(1-\lambda^2) \end{bmatrix} \begin{bmatrix} w_{1,j} \\ w_{2,j} \\ \vdots \\ w_{m-1,j} \end{bmatrix} - \begin{bmatrix} w_{1,j-1} \\ w_{2,j-1} \\ \vdots \\ w_{m-1,j-1} \end{bmatrix}.$$

(12.21)

Figure 12.10

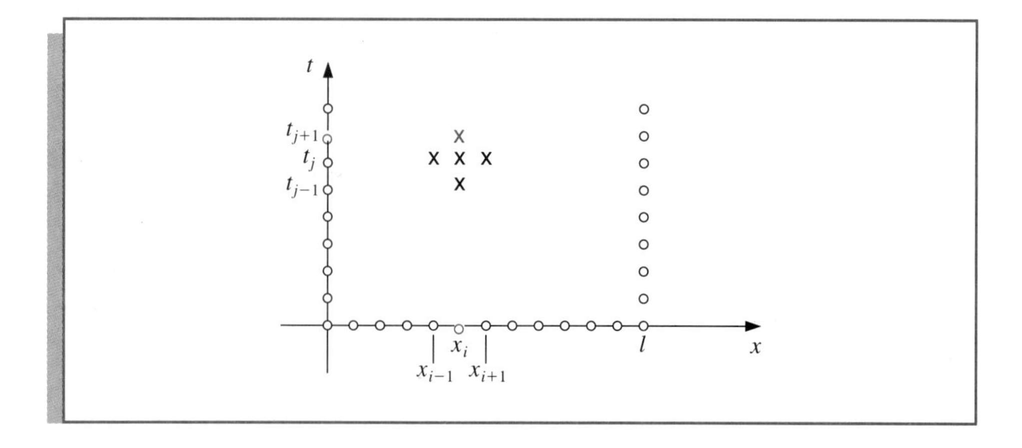

Equations (12.18) and (12.19) imply that the $(j+1)$st time step requires values from the jth and $(j-1)$st time steps. (See Figure 12.10.) This produces a minor starting problem since values for $j=0$ are given by Eq. (12.20), but values for $j=1$, which are needed in Eq. (12.18) to compute $w_{i,2}$, must be obtained from the initial-velocity condition

$$\frac{\partial u}{\partial t}(x,0)=g(x), \quad 0\le x\le l.$$

One approach is to replace $\partial u/\partial t$ by a forward-difference approximation,

(12.22) $$\frac{\partial u}{\partial t}(x_i,0)=\frac{u(x_i,t_1)-u(x_i,0)}{k}-\frac{k}{2}\frac{\partial^2 u}{\partial t^2}(x_i,\tilde{\mu}_i), \quad 0<\tilde{\mu}_i<t_1.$$

Solving for $u(x_i,t_1)$ gives

$$\begin{aligned}u(x_i,t_1)&=u(x_i,0)+k\frac{\partial u}{\partial t}(x_i,0)+\frac{k^2}{2}\frac{\partial^2 u}{\partial t^2}(x_i,\tilde{\mu}_i)\\&=u(x_i,0)+kg(x_i)+\frac{k^2}{2}\frac{\partial^2 u}{\partial t^2}(x_i,\tilde{\mu}_i).\end{aligned}$$

As a consequence,

(12.23) $$w_{i,1}=w_{i,0}+kg(x_i), \quad \text{for each } i=1,\ldots,m-1.$$

From Eq. (12.22), however, we can see that this result gives an approximation that has local truncation error of only $O(k)$. A better approximation to $u(x_i,0)$ can be obtained rather easily, particularly when the second derivative of f at x_i can be determined. Using a second Taylor polynomial in t for u at $(x_i,0)$ we can write

(12.24) $$\frac{u(x_i,t_1)-u(x_i,0)}{k}=\frac{\partial u}{\partial t}(x_i,0)+\frac{k}{2}\frac{\partial^2 u}{\partial t^2}(x_i,0)+\frac{k^2}{6}\frac{\partial^3 u}{\partial t^3}(x_i,\hat{\mu}_i)$$

for some $\hat{\mu}_i$ with $0 < \hat{\mu}_i < t_1$. Suppose the wave equation also holds on the initial line; that is,

$$\frac{\partial^2 u}{\partial t^2}(x_i, 0) - \alpha^2 \frac{\partial^2 u}{\partial x^2}(x_i, 0) = 0, \quad \text{for each } i = 0, 1, \ldots, m.$$

If f'' exists, then

$$\frac{\partial^2 u}{\partial t^2}(x_i, 0) = \alpha^2 \frac{\partial^2 u}{\partial x^2}(x_i, 0) = \alpha^2 \frac{d^2 f}{dx^2}(x_i) = \alpha^2 f''(x_i).$$

Substituting into Eq. (12.24) and solving for $u(x_i, t_1)$ gives

$$u(x_i, t_1) = u(x_i, 0) + kg(x_i) + \frac{\alpha^2 k^2}{2} f''(x_i) + \frac{k^3}{6} \frac{\partial^3 u}{\partial t^3}(x_i, \hat{\mu}_i)$$

and

(12.25)
$$w_{i,1} = w_{i,0} + kg(x_i) + \frac{\alpha^2 k^2}{2} f''(x_i).$$

This is an approximation with local truncation error $O(k^2)$ for each $i = 1, 2, \ldots, m-1$.

If $f''(x_i)$ is not readily available, we can use the difference equation in Eq. (4.8) to write

$$f''(x_i) = \frac{f(x_{i+1}) - 2f(x_i) + f(x_{i-1})}{h^2} - \frac{h^2}{12} f^{(4)}(\tilde{\xi}_i),$$

for some $\tilde{\xi}_i$ in (x_{i-1}, x_{i+1}), provided that $f \in C^4[0, 1]$. This implies that the approximation becomes

$$\frac{u(x_i, t_1) - u(x_i, 0)}{k} = g(x_i) + \frac{k\alpha^2}{2h^2}[f(x_{i+1}) - 2f(x_i) + f(x_{i-1})] + O(k^2 + h^2 k)$$

or, letting $\lambda = \dfrac{k\alpha}{h}$,

$$\begin{aligned} u(x_i, t_1) &= u(x_i, 0) + kg(x_i) + \frac{\lambda^2}{2}[f(x_{i+1}) - 2f(x_i) + f(x_{i-1})] + O(k^3 + h^2 k^2) \\ &= (1 - \lambda^2) f(x_i) + \frac{\lambda^2}{2} f(x_{i+1}) + \frac{\lambda^2}{2} f(x_{i-1}) + kg(x_i) + O(k^3 + h^2 k^2). \end{aligned}$$

Thus, the difference equation

(12.26)
$$w_{i,1} = (1 - \lambda^2) f(x_i) + \frac{\lambda^2}{2} f(x_{i+1}) + \frac{\lambda^2}{2} f(x_{i-1}) + kg(x_i)$$

can be used to find $w_{i,1}$ for each $i = 1, 2, \ldots, m-1$.

Algorithm 12.4 uses Eq. (12.26) to approximate $w_{i,1}$, although Eq. (12.23) could also be used. It is assumed that there is an upper bound for the value of t to be used in the stopping technique.

ALGORITHM 12.4

Wave Equation Finite-Difference

To approximate the solution to the wave equation

$$\frac{\partial^2 u}{\partial t^2}(x, t) - \alpha^2 \frac{\partial^2 u}{\partial x^2}(x, t) = 0, \quad 0 < x < l, \quad 0 < t < T,$$

subject to the boundary conditions

$$u(0, t) = u(l, t) = 0, \quad 0 < t < T,$$

and the initial conditions

$$u(x, 0) = f(x), \quad 0 \leq x \leq l,$$
$$\frac{\partial u}{\partial t}(x, 0) = g(x), \quad 0 \leq x \leq l:$$

INPUT endpoint l; maximum time T; constant α; integers $m \geq 2$, $N \geq 2$.

OUTPUT approximations $w_{i,j}$ to $u(x_i, t_j)$ for each $i = 0, \ldots, m$ and $j = 0, \ldots, N$.

Step 1 Set $h = l/m$;
$k = T/N$;
$\lambda = k\alpha/h$.

Step 2 For $j = 1, \ldots, N$ set $w_{0,j} = 0$;
$w_{m,j} = 0$;

Step 3 Set $w_{0,0} = f(0)$;
$w_{m,0} = f(l)$.

Step 4 For $i = 1, \ldots, m-1$ (*Initialize for $t = 0$ and $t = k$.*)
set $w_{i,0} = f(ih)$;
$$w_{i,1} = (1 - \lambda^2) f(ih) + \frac{\lambda^2}{2}[f((i+1)h) + f((i-1)h)] + kg(ih).$$

Step 5 For $j = 1, \ldots, N-1$ (*Perform matrix multiplication.*)
for $i = 1, \ldots, m-1$
set $w_{i,j+1} = 2(1 - \lambda^2) w_{i,j} + \lambda^2 (w_{i+1,j} + w_{i-1,j}) + w_{i,j-1}$.

Step 6 For $j = 0, \ldots, N$
set $t = jk$;
for $i = 0, \ldots, m$
set $x = ih$;
OUTPUT $(x, t, w_{i,j})$.

Step 7 STOP. (*Procedure is complete.*)

EXAMPLE 1 Consider the hyperbolic problem

$$\frac{\partial^2 u}{\partial t^2}(x,t) - 4\frac{\partial^2 u}{\partial x^2}(x,t) = 0, \quad 0 < x < 1, \quad 0 < t,$$

with boundary conditions $u(0,t) = u(1,t) = 0$, for $0 < t$, and initial conditions

$$u(x,0) = \sin \pi x, \; 0 \le x \le 1, \quad \text{and} \quad \frac{\partial u}{\partial t}(x,0) = 0, \quad 0 \le x \le 1.$$

It is easily verified that the solution to this problem is

$$u(x,t) = \sin \pi x \cos 2\pi t.$$

The Finite-Difference Algorithm 12.4 is used in this example with $m = 10$, $T = 1$, and $N = 20$, which implies that $h = 0.1$, $k = 0.05$, and $\lambda = 1$. Table 12.6 lists the results of the approximation $w_{i,N}$ for $i = 0, 1, \ldots, 10$. The values listed in the table are correct to the places given. ■

Table 12.6

x_i	$w_{i,20}$
0.0	0.0000000000
0.1	0.3090169944
0.2	0.5877852523
0.3	0.8090169944
0.4	0.9510565163
0.5	1.0000000000
0.6	0.9510565163
0.7	0.8090169944
0.8	0.5877852523
0.9	0.3090169944
1.0	0.0000000000

The results of the example were very accurate, more so than the truncation error $O(k^2 + h^2)$ would lead us to believe. The explanation for this lies in the fact that the true solution to the equation is infinitely differentiable. When this is the case, using Taylor series gives

$$\frac{u(x_{i+1},t_j) - 2u(x_i,t_j) + u(x_{i-1},t_j)}{h^2}$$

$$= \frac{\partial^2 u}{\partial x^2}(x_i,t_j) + 2\left[\frac{h^2}{4!}\frac{\partial^4 u}{\partial x^4}(x_i,t_j) + \frac{h^4}{6!}\frac{\partial^6 u}{\partial x^6}(x_i,t_j) + \cdots\right]$$

and

$$\frac{u(x_i, t_{j+1}) - 2u(x_i, t_j) + u(x_i, t_{j-1})}{k^2}$$
$$= \frac{\partial^2 u}{\partial t^2}(x_i, t_j) + 2\left[\frac{k^2}{4!}\frac{\partial^4 u}{\partial t^4}(x_i, t_j) + \frac{k^4}{6!}\frac{\partial^6 u}{\partial t^6}(x_i, t_j) + \cdots\right]$$

Since $u(x, t)$ satisfies the partial-differential equation,

(12.27)
$$\frac{u(x_i, t_{j+1}) - 2u(x_i, t_j) + u(x_i, t_{j-1})}{k^2} - \alpha^2 \frac{u(x_{i+1}, t_j) - 2u(x_i, t_j) + u(x_{i-1}, t_j)}{h^2}$$
$$= 2\left[\frac{1}{4!}\left(k^2 \frac{\partial^4 u}{\partial t^4}(x_i, t_j) - \alpha^2 h^2 \frac{\partial^4 u}{\partial x^4}(x_i, t_j)\right) + \frac{1}{6!}\left(k^4 \frac{\partial^6 u}{\partial t^6}(x_i, t_j) - \alpha^2 h^4 \frac{\partial^6 u}{\partial x^6}(x_i, t_j)\right) + \cdots\right].$$

However, by differentiating the wave equation,

$$k^2 \frac{\partial^4}{\partial t^4}(x_i, t_j) = k^2 \frac{\partial^2}{\partial t^2}\left[\alpha^2 \frac{\partial^2 u}{\partial x^2}(x_i, t_j)\right] = \alpha^2 k^2 \frac{\partial^2}{\partial x^2}\left[\frac{\partial^2 u}{\partial t^2}(x_i, t_j)\right]$$
$$= \alpha^2 k^2 \frac{\partial^2}{\partial x^2}\left[\alpha^2 \frac{\partial^2 u}{\partial x^2}(x_i, t_j)\right] = \alpha^4 k^2 \frac{\partial^4 u}{\partial x^4}(x_i, t_j),$$

we see that since $\lambda^2 = \dfrac{\alpha^2 k^2}{h^2} = 1$,

$$\frac{1}{4!}\left[k^2 \frac{\partial^4 u}{\partial t^4}(x_i, t_j) - \alpha^2 h^2 \frac{\partial^4 u}{\partial x^4}(x_i, t_j)\right] = \frac{\alpha^2}{4!}[\alpha^2 k^2 - h^2]\frac{\partial^4 u}{\partial x^4}(x_i, t_j) = 0.$$

Continuing in this manner, all the terms on the right-hand side of (12.27) are zero, implying a zero local truncation error. The only errors in Example 1 are those due to the approximation of $w_{i,1}$ and to roundoff.

As in the case of the Forward-Difference method for the heat equation, the Explicit Finite-Difference method for the wave equation has stability problems. In fact, it is necessary that $\lambda = \alpha k/h \leq 1$ for the method to be stable. (See [IK, p. 489].) The explicit method given in Algorithm 12.4, with $\lambda \leq 1$, is $O(h^2 + k^2)$ convergent if f and g are sufficiently differentiable. For verification of this, see [IK, p. 491].

Although we will not discuss them, there are implicit methods that are unconditionally stable. A discussion of these methods can be found in [Am, p. 199], [Mi], or [Sm,G].

EXERCISE SET 12.3

1. Approximate the solution to the wave equation

$$\frac{\partial^2 u}{\partial t^2} - \frac{\partial^2 u}{\partial x^2} = 0, \quad 0 < x < 1, \quad 0 < t;$$

$$u(0, t) = u(1, t) = 0, \quad 0 < t,$$

$$u(x, 0) = \sin \pi x, \quad 0 \le x \le 1,$$

$$\frac{\partial u}{\partial t}(x, 0) = 0, \quad 0 \le x \le 1,$$

using the Finite-Difference Algorithm with $m = 4$, $N = 4$, and $T = 1.0$ and compare your results to the actual solution $u(x, t) = \cos \pi t \sin \pi x$ at $t = 1.0$.

2. Approximate the solution to the wave equation

$$\frac{\partial^2 u}{\partial t^2} - \frac{1}{16\pi^2} \frac{\partial^2 u}{\partial x^2} = 0, \quad 0 < x < 0.5,\ 0 < t;$$

$$u(0, t) = u(0.5, t) = 0, \quad 0 < t,$$

$$u(x, 0) = 0, \quad 0 \le x \le 0.5,$$

$$\frac{\partial u}{\partial t}(x, 0) = \sin 4\pi x, \quad 0 \le x \le 0.5,$$

using the Finite-Difference Algorithm with $m = 4$, $N = 4$ and $T = 0.5$ and compare your results to the actual solution $u(x, t) = \sin t \sin 4\pi x$ at $t = 0.5$.

3. Approximate the solution to the wave equation

$$\frac{\partial^2 u}{\partial t^2} - \frac{\partial^2 u}{\partial x^2} = 0, \quad 0 < x < \pi,\ 0 < t;$$

$$u(0, t) = u(\pi, t) = 0, \quad 0 < t,$$

$$u(x, 0) = \sin x, \quad 0 \le x \le \pi,$$

$$\frac{\partial u}{\partial t}(x, 0) = 0, \quad 0 \le x \le \pi,$$

using the Finite-Difference Algorithm with $h = \pi/10$ and $k = 0.05$, with $h = \pi/20$ and $k = 0.1$, and then with $h = \pi/20$ and $k = 0.05$. Compare your results to the actual solution $u(x, t) = \cos t \sin x$ at $t = 0.5$.

4. Repeat Exercise 3, using in Step 4 of Algorithm 12.4 the approximation

$$w_{i,1} = w_{i,0} + kg(x_i), \quad \text{for each } i = 1, \ldots, m-1.$$

5. Approximate the solution to the wave equation

$$\frac{\partial^2 u}{\partial t^2} - \frac{\partial^2 u}{\partial x^2} = 0, \quad 0 < x < 1,\ 0 < t;$$

$$u(0, t) = u(1, t) = 0, \quad 0 < t,$$

$$u(x, 0) = \sin 2\pi x, \quad 0 \le x \le 1,$$

$$\frac{\partial u}{\partial t}(x, 0) = 2\pi \sin 2\pi x, \quad 0 \le x \le 1,$$

using Algorithm 12.4 with $h = 0.1$ and $k = 0.1$. Compare your results to the actual solution $u(x, t) = \sin 2\pi x(\cos 2\pi t + \sin 2\pi t)$, at $t = 0.3$.

6. Approximate the solution to the wave equation at $t = 0.5$

$$\frac{\partial^2 u}{\partial t^2} - \frac{\partial^2 u}{\partial x^2} = 0, \quad 0 < x < 1,\ 0 < t;$$

$$u(0,t) = u(1,t) = 0, \quad 0 < t,$$

$$u(x,0) = \begin{cases} 1, & 0 \le x \le \frac{1}{2}, \\ -1, & \frac{1}{2} < x \le 1, \end{cases}$$

$$\frac{\partial u}{\partial t}(x,0) = 0, \quad 0 \le x \le 1.$$

using Algorithm 12.4 with $h = 0.1$ and $k = 0.1$.

7. The air pressure $p(x,t)$ in an organ pipe is governed by the wave equation

$$\frac{\partial^2 p}{\partial x^2} = \frac{1}{c^2}\frac{\partial^2 p}{\partial t^2}, \quad 0 < x < l,\ 0 < t,$$

where l is the length of the pipe and c is a physical constant. If the pipe is open, the boundary conditions are given by

$$p(0,t) = p_0 \quad \text{and} \quad p(l,t) = p_0.$$

If the pipe is closed at the end where $x = l$, the boundary conditions are

$$p(0,t) = p_0 \quad \text{and} \quad \frac{\partial p}{\partial x}(l,t) = 0.$$

Assume that $c = 1$, $l = 1$, and the initial conditions are

$$p(x,0) = p_0 \cos 2\pi x, \quad \text{and} \quad \frac{\partial p}{\partial t}(x,0) = 0, \quad 0 \le x \le 1.$$

a. Approximate the pressure for an open pipe with $p_0 = 0.9$ at $x = \frac{1}{2}$ for $t = 0.5$ and $t = 1$, using Algorithm 12.4 with $h = k = 0.1$.

b. Modify Algorithm 12.4 for the closed-pipe problem with $p_0 = 0.9$, and approximate $p(0.5, 0.5)$ and $p(0.5, 1)$ using $h = k = 0.1$.

8. In an electric transmission line of length l that carries alternating current of high frequency (called a "lossless" line), the voltage V and current i are described by

$$\frac{\partial^2 V}{\partial x^2} = LC\frac{\partial^2 V}{\partial t^2}, \quad 0 < x < l,\ 0 < t,$$

$$\frac{\partial^2 i}{\partial x^2} = LC\frac{\partial^2 i}{\partial t^2}, \quad 0 < x < l,\ 0 < t,$$

where L is the inductance per unit length and C is the capacitance per unit length. Suppose the line is 200 ft long and the constants C and L are given by

$$C = 0.1 \text{ farads/ft} \quad \text{and} \quad L = 0.3 \text{ henries/ft}.$$

Suppose the voltage and current also satisfy

$$V(0,t) = V(200,t) = 0, \quad 0 < t,$$

$$V(x,0) = 110 \sin\frac{\pi x}{200}, \quad 0 \le x \le 200,$$

$$\frac{\partial V}{\partial t}(x,0) = 0, \quad 0 \le x \le 200,$$

$$i(0, t) = i(200, t) = 0, \quad 0 < t,$$

$$i(x, 0) = 5.5 \cos \frac{\pi x}{200}, \quad 0 \le x \le 200,$$

and

$$\frac{\partial i}{\partial t}(x, 0) = 0, \quad 0 \le x \le 200.$$

Approximate the voltage and current at $t = 0.2$ and $t = 0.5$ using Algorithm 12.4 with $h = 10$ and $k = 0.1$.

12.4 An Introduction to the Finite-Element Method

The **Finite-Element method** is similar to the Rayleigh-Ritz method introduced in Section 11.5 for approximating the solution to two-point boundary-value problems. It was originally developed for use in civil engineering but is now used for approximating the solutions to partial-differential equations that arise in all areas of applied mathematics.

One advantage of the Finite-Element method over finite-difference methods is the relative ease with which the boundary conditions of the problem are handled. Many physical problems have boundary conditions involving derivatives and irregularly shaped boundaries. Boundary conditions of this type are difficult to handle using finite-difference techniques since each boundary condition involving a derivative must be approximated by a difference quotient at the grid points, and irregular shaping of the boundary makes placing the grid points difficult. The Finite-Element method includes the boundary conditions as integrals in a functional that is being minimized, so the construction procedure is independent of the particular boundary conditions of the problem.

In our discussion, we consider the partial-differential equation

$$\text{(12.28)} \qquad \frac{\partial}{\partial x}\left(p(x, y)\frac{\partial u}{\partial x}\right) + \frac{\partial}{\partial y}\left(q(x, y)\frac{\partial u}{\partial y}\right) + r(x, y)u(x, y) = f(x, y),$$

with $(x, y) \in \mathcal{D}$, where $\mathcal{D}$ is a plane region with boundary S.

Boundary conditions of the form

$$\text{(12.29)} \qquad u(x, y) = g(x, y)$$

are imposed on a portion S_1 of the boundary. On the remainder of the boundary, S_2, the solution $u(x, y)$ is required to satisfy

$$\text{(12.30)} \quad p(x, y)\frac{\partial u}{\partial x}(x, y)\cos\theta_1 + q(x, y)\frac{\partial u}{\partial y}(x, y)\cos\theta_2 + g_1(x, y)u(x, y) = g_2(x, y),$$

where θ_1 and θ_2 are the direction angles of the outward normal to the boundary at the point (x, y). (See Figure 12.11.)

Figure 12.11

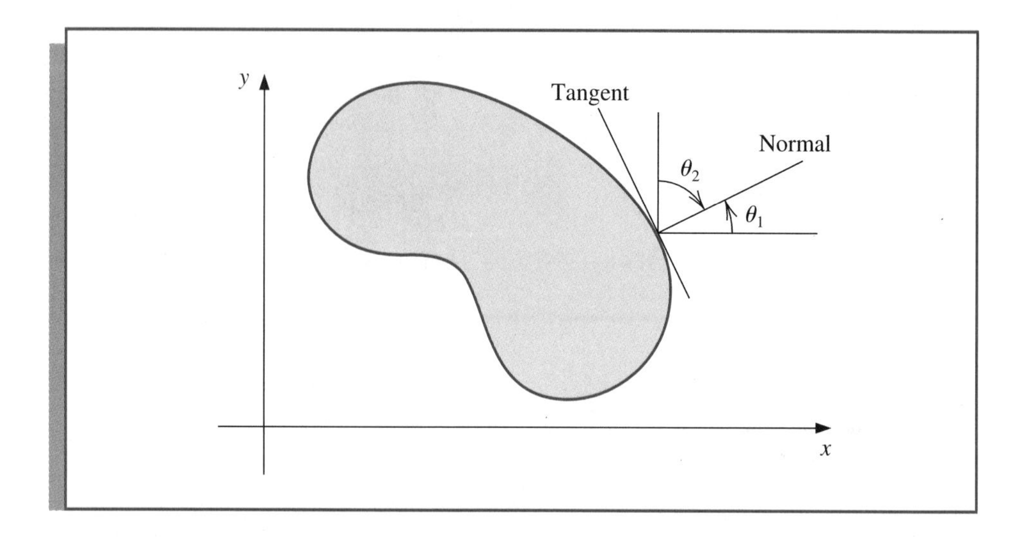

Physical problems in the areas of solid mechanics and elasticity have associated partial-differential equations similar to Eq. (12.28). The solution to a problem of this type typically minimizes a certain functional, involving integrals, over a class of functions determined by the problem.

Suppose p, q, r, and f are all continuous in $\mathcal{D} \cup S$, that p and q have continuous first partial derivatives, and that g_1 and g_2 are continuous on S_2. Suppose, in addition, that $p(x, y) > 0, q(x, y) > 0, r(x, y) \leq 0$, and $g_1(x, y) > 0$. Then a solution to Eq. (12.28) uniquely minimizes the functional

(12.31)

$$I[w] = \iint_{\mathcal{D}} \left\{ \frac{1}{2}\left[p(x, y)\left(\frac{\partial w}{\partial x}\right)^2 + q(x, y)\left(\frac{\partial w}{\partial y}\right)^2 - r(x, y)w^2 \right] + f(x, y)w \right\} dx\, dy$$

$$+ \int_{S_2} \{-g_2(x, y)w + \frac{1}{2} g_1(x, y)w^2\}\, dS$$

over all functions w satisfying Eq. (12.29) on S_1 that are twice continuously differentiable. The Finite-Element method approximates this solution by minimizing the functional I over a smaller class of functions, just as the Rayleigh-Ritz method did for the boundary-value problem considered in Section 11.5.

The first step is to divide the region into a finite number of sections, or elements, of a regular shape, either rectangles or triangles. (See Figure 12.12.)

The set of functions used for approximation is generally a set of piecewise polynomials of fixed degree in x and y, and the approximation requires that the polynomials be pieced together in such a manner that the resulting function is continuous with an integrable or continuous first or second derivative on the entire region. Polynomials of linear type in x and y,

$$\phi(x, y) = a + bx + cy,$$

Figure 12.12

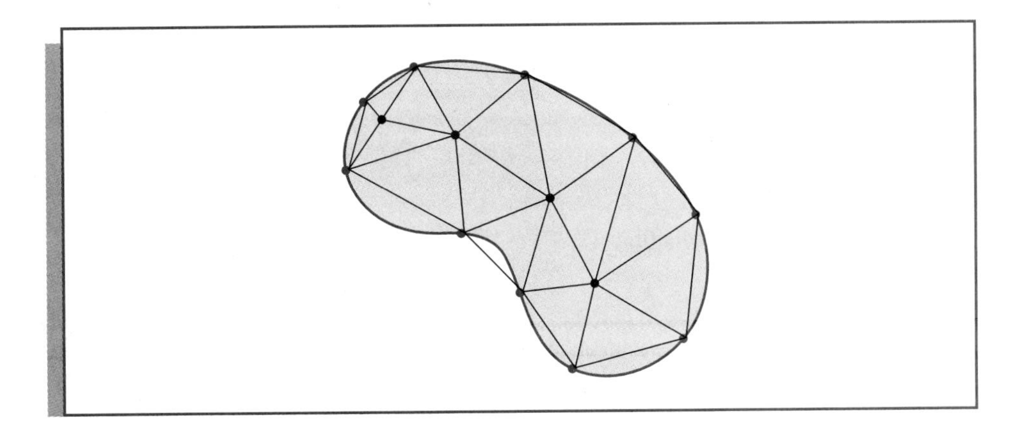

are commonly used with triangular elements, whereas polynomials of bilinear type in x and y,

$$\phi(x, y) = a + bx + cy + dxy,$$

are used with rectangular elements.

For our discussion, suppose that the region $\mathcal{D}$ has been subdivided into triangular elements. The collection of triangles is denoted D, and the vertices of these triangles are called *nodes*. The method seeks an approximation of the form

$$\phi(x, y) = \sum_{i=1}^{m} \gamma_i \phi_i(x, y),$$

where $\phi_1, \phi_2, \ldots, \phi_m$ are linearly independent piecewise-linear polynomials and $\gamma_1, \gamma_2, \ldots, \gamma_m$ are constants. Some of these constants, say, $\gamma_{n+1}, \gamma_{n+2}, \ldots, \gamma_m$, are used to ensure that the boundary condition

$$\phi(x, y) = g(x, y)$$

is satisfied on S_1, and the remaining constants, $\gamma_1, \gamma_2, \ldots, \gamma_n$, are used to minimize the functional $I\left[\sum_{i=1}^{m} \gamma_i \phi_i\right]$.

From Eq. (12.31), the functional is of the form

(12.32)

$$I[\phi] = I\left[\sum_{i=1}^{m} \gamma_i \phi_i\right] = \iint_{\mathcal{D}} \left(\frac{1}{2}\left\{p(x, y)\left[\sum_{i=1}^{m} \gamma_i \frac{\partial \phi_i}{\partial x}(x, y)\right]^2 + q(x, y)\left[\sum_{i=1}^{m} \gamma_i \frac{\partial \phi_i}{\partial y}(x, y)\right]^2 \right.\right.$$

$$\left.\left. - r(x, y)\left[\sum_{i=1}^{m} \gamma_i \phi_i(x, y)\right]^2\right\} + f(x, y) \sum_{i=1}^{m} \gamma_i \phi_i(x, y)\right) dy\, dx$$

$$+ \int_{S_2} \left\{ - g_2(x, y) \sum_{i=1}^{m} \gamma_i \phi_i(x, y) + \frac{1}{2} g_1(x, y) \left[\sum_{i=1}^{m} \gamma_i \phi_i(x, y)\right]^2 \right\} dS.$$

For a minimum to occur, considering I as a function of $\gamma_1, \gamma_2, \ldots, \gamma_n$, it is necessary to have

$$\frac{\partial I}{\partial \gamma_j} = 0, \quad \text{for each } j = 1, 2, \ldots, n.$$

Differentiating (12.32) gives

$$\begin{aligned}\frac{\partial I}{\partial \gamma_j} = \iint_{\mathcal{D}} \Bigg\{ & p(x,y) \sum_{i=1}^{m} \gamma_i \frac{\partial \phi_i}{\partial x}(x,y) \frac{\partial \phi_j}{\partial x}(x,y) + q(x,y) \sum_{i=1}^{m} \gamma_i \frac{\partial \phi_i}{\partial y}(x,y) \frac{\partial \phi_j}{\partial y}(x,y) \\ & - r(x,y) \sum_{i=1}^{m} \gamma_i \phi_i(x,y) \phi_j(x,y) + f(x,y) \phi_j(x,y) \Bigg\}\, dx\, dy \\ & + \int_{S_2} \Bigg\{ - g_2(x,y) \phi_j(x,y) + g_1(x,y) \sum_{i=1}^{m} \gamma_i \phi_i(x,y) \phi_j(x,y) \Bigg\}\, dS,\end{aligned}$$

so

$$\begin{aligned}0 = \sum_{i=1}^{m} \Bigg[\iint_{\mathcal{D}} \Bigg\{ & p(x,y) \frac{\partial \phi_i}{\partial x}(x,y) \frac{\partial \phi_j}{\partial x}(x,y) + q(x,y) \frac{\partial \phi_i}{\partial y}(x,y) \frac{\partial \phi_j}{\partial y}(x,y) \\ & - r(x,y) \phi_i(x,y) \phi_j(x,y) \Bigg\}\, dx\, dy + \int_{S_2} g_1(x,y) \phi_i(x,y) \phi_j(x,y)\, dS \Bigg] \gamma_i \\ & + \iint_{\mathcal{D}} f(x,y) \phi_j(x,y)\, dx\, dy - \int_{S_2} g_2(x,y) \phi_j(x,y)\, dS,\end{aligned}$$

for each $j = 1, 2, \ldots, n$. This set of equations can be written as a linear system:

$$A\mathbf{c} = \mathbf{b},$$

where $\mathbf{c} = (\gamma_1, \ldots, \gamma_n)^t$ and where $A = (\alpha_{ij})$ and $\mathbf{b} = (\beta_1, \ldots, \beta_n)^t$ are defined by

$$\textbf{(12.33)} \quad \begin{aligned}\alpha_{ij} = \iint_{\mathcal{D}} \Bigg[& p(x,y) \frac{\partial \phi_i}{\partial x}(x,y) \frac{\partial \phi_j}{\partial x}(x,y) + q(x,y) \frac{\partial \phi_i}{\partial y}(x,y) \frac{\partial \phi_j}{\partial y}(x,y) \\ & - r(x,y) \phi_i(x,y) \phi_j(x,y) \Bigg]\, dx\, dy + \int_{S_2} g_1(x,y) \phi_i(x,y) \phi_j(x,y)\, dS,\end{aligned}$$

for each $i = 1, 2, \ldots, n$ and $j = 1, 2, \ldots, m$, and

$$\textbf{(12.34)}\quad \beta_i = -\iint_{\mathcal{D}} f(x, y)\phi_i(x, y)\,dx\,dy + \int_{S_2} g_2(x, y)\phi_i(x, y)\,dS - \sum_{k=n+1}^{m} \alpha_{ik}\gamma_k,$$

for each $i = 1, \ldots, n$.

The particular choice of basis functions is important since the appropriate choice can often make the matrix A positive definite and banded. For the second-order problem (12.28), we assume that $\mathcal{D}$ is polygonal and that S, is a contiguous set of straight lines, so that $\mathcal{D} = D$. To begin the procedure, we divide the region D into a collection of triangles $T_1, T_2, \ldots, T_m$, with the ith triangle having three vertices, or nodes, denoted

$$V_j^{(i)} = \left(x_j^{(i)}, y_j^{(i)}\right), \quad \text{for } j = 1, 2, 3.$$

To simplify the notation, we write $V_j^{(i)}$ simply as $V_j = (x_j, y_j)$ when working with the fixed triangle T_i. With each vertex V_j we associate a linear polynomial

$$N_j^{(i)} \equiv N_j = a_j + b_j x + c_j y, \quad \text{where} \quad N_j^{(i)}(x_k, y_k) = \begin{cases} 1, & \text{if } j = k, \\ 0, & \text{if } j \neq k. \end{cases}$$

This produces linear systems of the form

$$\begin{bmatrix} 1 & x_1 & y_1 \\ 1 & x_2 & y_2 \\ 1 & x_3 & y_3 \end{bmatrix} \begin{bmatrix} a_j \\ b_j \\ c_j \end{bmatrix} = \begin{bmatrix} 0 \\ 1 \\ 0 \end{bmatrix},$$

with the element one occurring in the jth row in the vector on the right (here $j = 2$).

Let $E_1, \ldots, E_n$ be a labeling of the nodes lying in $D \cup S$ in a left-to-right, top-to-bottom fashion. With each node E_k, we associate a function ϕ_k that is linear on each triangle, has the value 1 at E_k, and is 0 at each of the other nodes. This choice makes ϕ_k identical to $N_j^{(i)}$ on triangle T_i when the node E_k is the vertex denoted $V_j^{(i)}$.

EXAMPLE 1 Suppose that a finite-element problem contains the triangles T_1 and T_2 shown in Figure 12.13. The linear function $N_1^{(1)}(x, y)$ that assumes the value 1 at $(1, 1)$ and 0 at both $(0, 0)$ and $(-1, 2)$ satisfies

$$a_1^{(1)} + b_1^{(1)}(1) \quad + c_1^{(1)}(1) = 1,$$

$$a_1^{(1)} + b_1^{(1)}(-1) + c_1^{(1)}(2) = 0,$$

and

$$a_1^{(1)} + b_1^{(1)}(0) \quad + c_1^{(1)}(0) = 0,$$

Figure 12.13

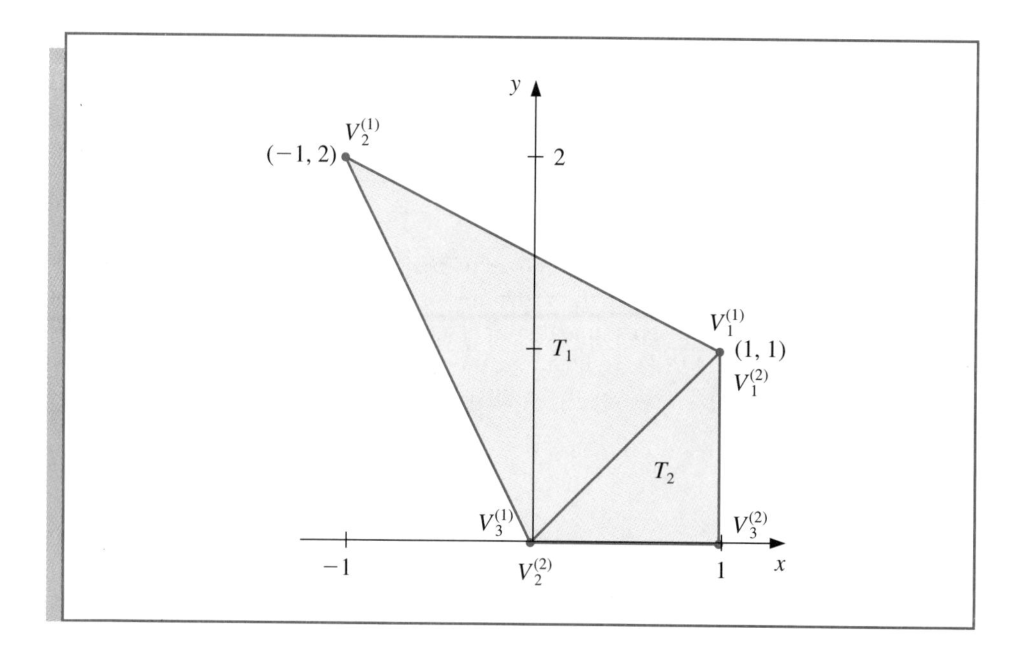

So $a_1^{(1)} = 0, b_1^{(1)} = \frac{2}{3}, c_1^{(1)} = \frac{1}{3}$, and

$$N_1^{(1)}(x, y) = \frac{2}{3}x + \frac{1}{3}y.$$

In a similar manner, the linear function $N_1^{(2)}(x, y)$ that assumes the value 1 at (1, 1) and 0 at both (0, 0) and (1, 0) satisfies

$$a_1^{(2)} + b_1^{(2)}(1) + c_1^{(2)}(1) = 1,$$
$$a_1^{(2)} + b_1^{(2)}(0) + c_1^{(2)}(0) = 0,$$

and

$$a_1^{(2)} + b_1^{(2)}(1) + c_1^{(2)}(0) = 0,$$

so $a_1^{(2)} = 0, b_1^{(2)} = 0$, and $c_1^{(2)} = 1$. As a consequence, $N_1^{(2)}(x, y) = y$. Note that on the common boundary of T_1 and T_2, $N_1^{(1)}(x, y) = N_1^{(2)}(x, y)$ since $y = x$. ■

Consider Figure 12.14, the upper left portion of the region shown in Figure 12.12. We will generate the entries in the matrix A that correspond to the nodes shown in this figure.

For simplicity, we assume that E_1 is not one of the nodes on S_2. The relationship between the nodes and the vertices of the triangles for this portion is

$$E_1 = V_3^{(1)} = V_1^{(2)}, \quad E_4 = V_2^{(2)}, \quad E_3 = V_2^{(1)} = V_3^{(2)}, \quad \text{and} \quad E_2 = V_1^{(1)}.$$

Figure 12.14

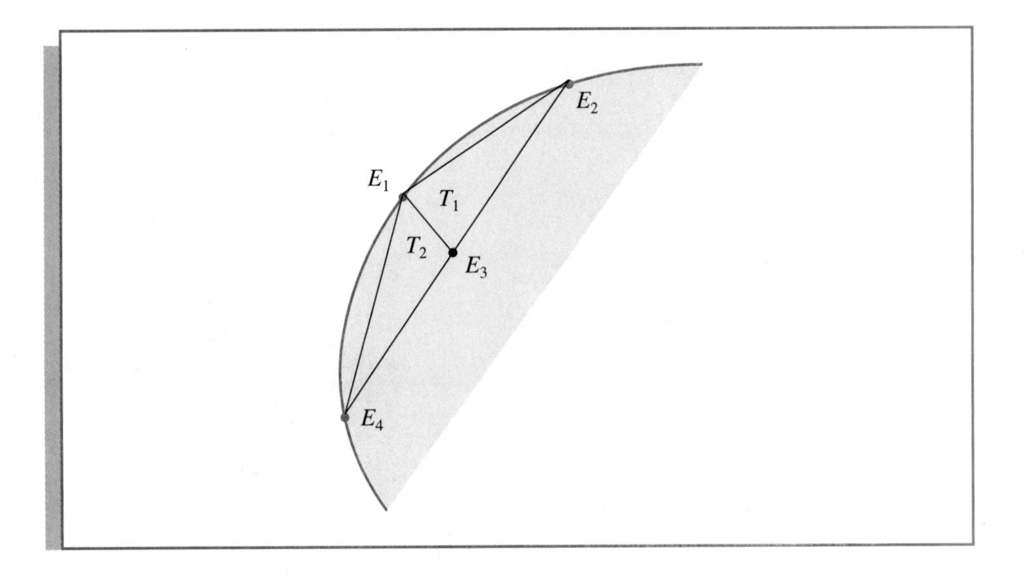

Since ϕ_1 and ϕ_3 are both nonzero on T_1 and T_2, the entries $\alpha_{1,3} = \alpha_{3,1}$ are computed by

$$\alpha_{1,3} = \iint_D \left[p \frac{\partial \phi_1}{\partial x} \frac{\partial \phi_3}{\partial x} + q \frac{\partial \phi_1}{\partial y} \frac{\partial \phi_3}{\partial y} - r\phi_1\phi_3 \right] dx\, dy$$

$$= \iint_{T_1} \left[p \frac{\partial \phi_1}{\partial x} \frac{\partial \phi_3}{\partial x} + q \frac{\partial \phi_1}{\partial y} \frac{\partial \phi_3}{\partial y} - r\phi_1\phi_3 \right] dx\, dy$$

$$+ \iint_{T_2} \left[p \frac{\partial \phi_1}{\partial x} \frac{\partial \phi_3}{\partial x} + q \frac{\partial \phi_1}{\partial y} \frac{\partial \phi_3}{\partial y} - r\phi_1\phi_3 \right] dx\, dy.$$

On triangle T_1,

$$\phi_1(x, y) = N_3^{(1)} = a_3^{(1)} + b_3^{(1)}x + c_3^{(1)}y$$

and

$$\phi_3(x, y) = N_2^{(1)} = a_2^{(1)} + b_2^{(1)}x + c_2^{(1)}y,$$

so

$$\frac{\partial \phi_1}{\partial x} = b_3^{(1)}, \quad \frac{\partial \phi_1}{\partial y} = c_3^{(1)}, \quad \frac{\partial \phi_3}{\partial x} = b_2^{(1)}, \quad \text{and} \quad \frac{\partial \phi_3}{\partial y} = c_2^{(1)}.$$

Similarly, on T_2,

$$\phi_1(x, y) = N_1^{(2)} = a_1^{(2)} + b_1^{(2)}x + c_1^{(2)}y$$

and

$$\phi_3(x, y) = N_3^{(2)} = a_3^{(2)} + b_3^{(2)}x + c_3^{(2)}y,$$

so

$$\frac{\partial \phi_1}{\partial x} = b_1^{(2)}, \quad \frac{\partial \phi_1}{\partial y} = c_1^{(2)}, \quad \frac{\partial \phi_3}{\partial x} = b_3^{(2)}, \quad \text{and} \quad \frac{\partial \phi_3}{\partial y} = c_3^{(2)}.$$

Thus,

$$\begin{aligned}
\alpha_{1,3} = {} & b_3^{(1)}b_2^{(1)} \iint_{T_1} p\,dx\,dy + c_3^{(1)}c_2^{(1)} \iint_{T_1} q\,dx\,dy \\
& - \iint_{T_1} r(a_3^{(1)} + b_3^{(1)}x + c_3^{(1)}y)(a_2^{(1)} + b_2^{(1)}x + c_2^{(1)}y)\,dx\,dy \\
& + b_1^{(2)}b_3^{(2)} \iint_{T_2} p\,dx\,dy + c_1^{(2)}c_3^{(2)} \iint_{T_2} q\,dx\,dy \\
& - \iint_{T_2} r(a_1^{(2)} + b_1^{(2)}x + c_1^{(2)}y)(a_3^{(2)} + b_3^{(2)}x + c_3^{(2)}y)\,dx\,dy.
\end{aligned}$$

All the double integrals over D reduce to double integrals over triangles. The usual procedure is to compute all possible integrals over the triangles and accumulate them into the correct entry α_{ij} in A.

Similarly, the double integrals of the form

$$\iint_D f(x, y)\phi_i(x, y)\,dx\,dy$$

are computed over triangles and then accumulated into the correct entry β_i of the vector **b**. For example,

$$\begin{aligned}
-\iint_D f(x, y)\phi_1(x, y)\,dx\,dy = {} & -\iint_{T_1} f(x, y)\left[a_3^{(1)} + b_3^{(1)}x + c_3^{(1)}y\right]\,dx\,dy \\
& -\iint_{T_2} f(x, y)\left[a_1^{(2)} + b_1^{(2)}x + c_1^{(2)}y\right]\,dx\,dy.
\end{aligned}$$

Part of β_i is contributed by ϕ_1 restricted to T_1 and the remainder by ϕ_1 restricted to T_2, since E_1 is a vertex of both T_1 and T_2. In addition, nodes that lie on S_2 have line integrals added to their entries in A and **b**.

Algorithm 12.5 performs the Finite-Element method on a second-order elliptic differential equation. The algorithm sets all values of the matrix A and vector **b** initially to 0 and, after all the integrations have been performed on all the triangles, adds these values to the appropriate entries in A and **b**.

ALGORITHM **12.5**

Finite-Element

To approximate the solution to the partial-differential equation

$$\frac{\partial}{\partial x}\left(p(x, y)\frac{\partial u}{\partial x}\right) + \frac{\partial}{\partial y}\left(q(x, y)\frac{\partial u}{\partial y}\right) + r(x, y)u = f(x, y), \quad (x, y) \in D$$

subject to the boundary conditions

$$u(x, y) = g(x, y), \quad (x, y) \in S_1$$

and

$$p(x, y)\frac{\partial u}{\partial x}(x, y)\cos\theta_1 + q(x, y)\frac{\partial u}{\partial y}(x, y)\cos\theta_2 + g_1(x, y)u(x, y) = g_2(x, y), \quad (x, y) \in S_2,$$

where $S_1 \cup S_2$ is the boundary of D, and θ_1 and θ_2 are the direction angles of the normal to the boundary:

Step 0 Divide the region D into triangles $T_1, \ldots, T_M$ such that: $T_1, \ldots, T_K$ are the triangles with no edges on S_1 or S_2;
 (*Note:* $K = 0$ *implies that no triangle is interior to* D.)
 $T_{K+1}, \ldots, T_N$ are the triangles with at least one edge on S_2; $T_{N+1}, \ldots, T_M$ are the remaining triangles.
 (*Note:* $M = N$ *implies that all triangles have edges on* S_2.)
 Label the three vertices of the triangle T_i by
 $\left(x_1^{(i)}, y_1^{(i)}\right)$, $\left(x_2^{(i)}, y_2^{(i)}\right)$, and $\left(x_3^{(i)}, y_3^{(i)}\right)$.
 Label the nodes (vertices) $E_1, \ldots, E_m$ where
 $E_1, \ldots, E_n$ are in $D \cup S_2$ and $E_{n+1}, \ldots, E_m$ are on S_1.
 (*Note:* $n = m$ *implies that* S_1 *contains no nodes.*)

INPUT integers K, N, M, n, m; vertices $\left(x_1^{(i)}, y_1^{(i)}\right)$, $\left(x_2^{(i)}, y_2^{(i)}\right)$, $\left(x_3^{(i)}, y_3^{(i)}\right)$ for each $i = 1, \ldots, M$; nodes E_j for each $j = 1, \ldots, m$.

(*Note: All that is needed is a means of corresponding a vertex* $\left(x_k^{(i)}, y_k^{(i)}\right)$ *to a node* $E_j = (x_j, y_j)$.)

OUTPUT constants $\gamma_1, \ldots, \gamma_m$; $a_j^{(i)}, b_j^{(i)}, c_j^{(i)}$ for each $j = 1, 2, 3$ and $i = 1, \ldots, M$.

Step 1 For $l = n + 1, \ldots, m$ set $\gamma_l = g(x_l, y_l)$. (*Note:* $E_l = (x_l, y_l)$.)

Step 2 For $i = 1, \ldots, n$
 set $\beta_i = 0$;
 for $j = 1, \ldots, n$ set $\alpha_{i,j} = 0$.

Step 3 For $i = 1, \ldots, M$

$$\text{set } \Delta_i = \det\begin{vmatrix} 1 & x_1^{(i)} & y_1^{(i)} \\ 1 & x_2^{(i)} & y_2^{(i)} \\ 1 & x_3^{(i)} & y_3^{(i)} \end{vmatrix};$$

$$a_1^{(i)} = \frac{x_2^{(i)} y_3^{(i)} - y_2^{(i)} x_3^{(i)}}{\Delta_i}; \qquad b_1^{(i)} = \frac{y_2^{(i)} - y_3^{(i)}}{\Delta_i}; \qquad c_1^{(i)} = \frac{x_3^{(i)} - x_2^{(i)}}{\Delta_i};$$

$$a_2^{(i)} = \frac{x_3^{(i)} y_1^{(i)} - y_3^{(i)} x_1^{(i)}}{\Delta_i}; \qquad b_2^{(i)} = \frac{y_3^{(i)} - y_1^{(i)}}{\Delta_i}; \qquad c_2^{(i)} = \frac{x_1^{(i)} - x_3^{(i)}}{\Delta_i};$$

$$a_3^{(i)} = \frac{x_1^{(i)} y_2^{(i)} - y_1^{(i)} x_2^{(i)}}{\Delta_i}; \qquad b_3^{(i)} = \frac{y_1^{(i)} - y_2^{(i)}}{\Delta_i}; \qquad c_3^{(i)} = \frac{x_2^{(i)} - x_1^{(i)}}{\Delta_i};$$

for $j = 1, 2, 3$
  define $N_j^{(i)}(x, y) = a_j^{(i)} + b_j^{(i)} x + c_j^{(i)} y$.

Step 4 For $i = 1, \ldots, M$ *(The integrals in Steps 4 and 5 can be evaluated using numerical integration.)*
  for $j = 1, 2, 3$
    for $k = 1, \ldots, j$ *(Compute all double integrals over the triangles.)*

$$\text{set } z_{j,k}^{(i)} = b_j^{(i)} b_k^{(i)} \iint_{T_i} p(x, y)\, dx\, dy + c_j^{(i)} c_k^{(i)} \iint_{T_i} q(x, y)\, dx\, dy - \iint_{T_i} r(x, y) N_j^{(i)}(x, y) N_k^{(i)}(x, y)\, dx\, dy;$$

$$\text{set } H_j^{(i)} = -\iint_{T_i} f(x, y) N_j^{(i)}(x, y)\, dx\, dy.$$

Step 5 For $i = K + 1, \ldots, N$ *(Compute all line integrals.)*
  for $j = 1, 2, 3$
    for $k = 1, \ldots, j$

$$\text{set } J_{j,k}^{(i)} = \int_{S_2} g_1(x, y) N_j^{(i)}(x, y) N_k^{(i)}(x, y)\, dS;$$

$$\text{set } I_j^{(i)} = \int_{S_2} g_2(x, y) N_j^{(i)}(x, y)\, dS.$$

Step 6 For $i = 1, \ldots, M$ do Steps 7–12. *(Assembling the integrals over each triangle into the linear system.)*

Step 7 For $k = 1, 2, 3$ do Steps 8–12.

Step 8 Find l so that $E_l = \left(x_k^{(i)}, y_k^{(i)}\right)$.

Step 9 If $k > 1$ then for $j = 1, \ldots, k - 1$ do Steps 10, 11.

Step 10 Find t so that $E_t = \left(x_j^{(i)}, y_j^{(i)}\right)$.

Step 11 If $l \leq n$ then
  if $t \leq n$ then set $\alpha_{lt} = \alpha_{lt} + z_{k,j}^{(i)}$;
    $\alpha_{tl} = \alpha_{tl} + z_{k,j}^{(i)}$
  else set $\beta_l = \beta_l - \gamma_t z_{k,j}^{(i)}$
else
  if $t \leq n$ then set $\beta_t = \beta_t - \gamma_l z_{k,j}^{(i)}$.

Step 12 If $l \leq n$ then set $a_{ll} = \alpha_{ll} + z_{k,k}^{(i)}$;
  $\beta_l = \beta_l + H_k^{(i)}$.

Step 13 For $i = K+1, \ldots, N$ do Steps 14–19. *(Assembling the line integrals into the linear system.)*

Step 14 For $k = 1, 2, 3$ do Steps 15–19.

Step 15 Find l so that $E_l = \left(x_k^{(i)}, y_k^{(i)}\right)$.

Step 16 If $k > 1$ then for $j = 1, \ldots, k-1$ do Steps 17, 18.

Step 17 Find t so that $E_t = \left(x_j^{(i)}, y_j^{(i)}\right)$.

Step 18 If $l \le n$ then

if $t \le n$ then set $\alpha_{lt} = \alpha_{lt} + J_{k,j}^{(i)}$;

$\alpha_{tl} = \alpha_{tl} + J_{k,j}^{(i)}$

else set $\beta_l = \beta_l - \gamma_t J_{k,j}^{(i)}$

else

if $t \le n$ then set $\beta_t = \beta_t - \gamma_l J_{k,j}^{(i)}$.

Step 19 If $l \le n$ then set $\alpha_{ll} = \alpha_{ll} + J_{k,k}^{(i)}$;

$\beta_l = \beta_l + I_k^{(i)}$.

Step 20 Solve the linear system $A\mathbf{c} = \mathbf{b}$ where $A = (\alpha_{l,t})$, $\mathbf{b} = (\beta_l)$ and $\mathbf{c} = (\gamma_t)$ for $1 \le l \le n$ and $1 \le t \le n$.

Step 21 OUTPUT $(\gamma_1, \ldots, \gamma_m)$.

(For each $k = 1, \ldots, m$ let $\phi_k = N_j^{(i)}$ on T_i if $E_k = \left(x_j^{(i)}, y_j^{(i)}\right)$.

Then $\phi(x, y) = \sum_{k=1}^{m} \gamma_k \phi_k(x, y)$ approximates $u(x, y)$ on $D \cup S_1 \cup S_2$.)

Step 22 For $i = 1, \ldots, M$

for $j = 1, 2, 3$ OUTPUT $\left(a_j^{(i)}, b_j^{(i)}, c_j^{(i)}\right)$.

Step 23 STOP. *(Procedure is complete.)*

EXAMPLE 2 The temperature, $u(x, y)$, in a two-dimensional region D satisfies Laplace's equation

$$\frac{\partial^2 u}{\partial x^2}(x, y) + \frac{\partial^2 u}{\partial y^2}(x, y) = 0 \quad \text{on } D.$$

Consider the region D shown in Figure 12.15 and suppose that the following boundary conditions are given:

$$u(x, y) = 4, \qquad \text{for } (x, y) \in L_6 \text{ and } (x, y) \in L_7,$$

$$\frac{\partial u}{\partial n}(x, y) = x, \qquad \text{for } (x, y) \in L_2 \text{ and } (x, y) \in L_4,$$

$$\frac{\partial u}{\partial n}(x, y) = y, \qquad \text{for } (x, y) \in L_5,$$

$$\frac{\partial u}{\partial n}(x, y) = \frac{x + y}{\sqrt{2}}, \qquad \text{for } (x, y) \in L_1 \text{ and } (x, y) \in L_3,$$

Figure 12.15

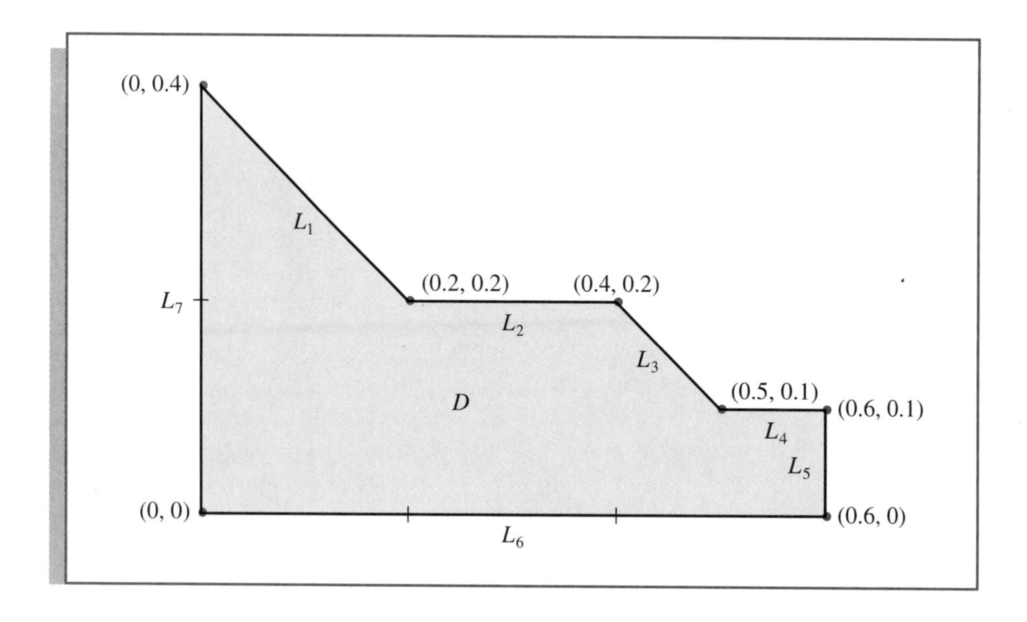

where $\partial u/\partial n$ denotes the directional derivative in the direction of the normal to the boundary of the region D at the point (x, y).

We first subdivide D into triangles with the labeling suggested in Step 0 of the algorithm. For this example, $S_1 = L_6 \cup L_7$ and $S_2 = L_1 \cup L_2 \cup L_3 \cup L_4 \cup L_5$. The labeling of triangles is shown in Figure 12.16.

Figure 12.16

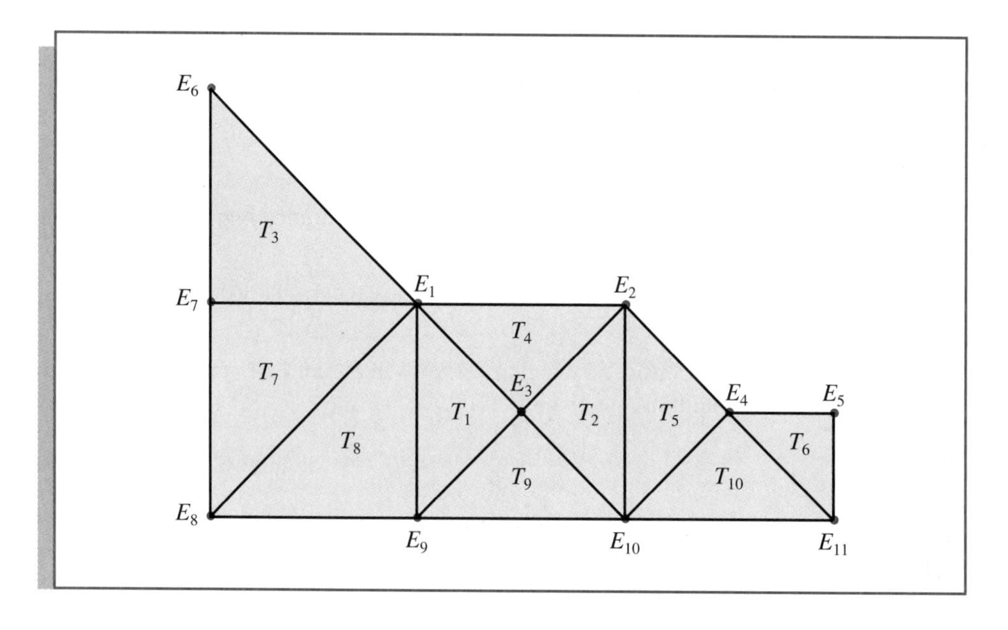

The boundary condition $u(x, y) = 4$ on L_6 and L_7 implies that $\gamma_t = 4$ when $t = 6, 7, \ldots, 11$. To determine the values of γ_l for $l = 1, 2, \ldots, 5$, apply the remaining steps of the algorithm and generate the matrix

$$A = \begin{bmatrix} 2.5 & 0 & -1 & 0 & 0 \\ 0 & 1.5 & -1 & -0.5 & 0 \\ -1 & -1 & 4 & 0 & 0 \\ 0 & -0.5 & 0 & 2.5 & -0.5 \\ 0 & 0 & 0 & -0.5 & 1 \end{bmatrix}$$

and the vector

$$\mathbf{b} = \begin{bmatrix} 6.066\bar{6} \\ 0.063\bar{3} \\ 8.0000 \\ 6.056\bar{6} \\ 2.031\bar{6} \end{bmatrix}.$$

The solution to the equation $A\mathbf{c} = \mathbf{b}$ is

$$\mathbf{c} = \begin{bmatrix} \gamma_1 \\ \gamma_2 \\ \gamma_3 \\ \gamma_4 \\ \gamma_5 \end{bmatrix} = \begin{bmatrix} 4.0383 \\ 4.0782 \\ 4.0291 \\ 4.0496 \\ 4.0565 \end{bmatrix},$$

which gives the following approximation to the solution of Laplace's equation and the boundary conditions on the respective triangles:

$$\begin{aligned}
T_1: &\quad \phi(x, y) = 4.0383(1 - 5x + 5y) + 4.0291(-2 + 10x) + 4(2 - 5x - 5y),\\
T_2: &\quad \phi(x, y) = 4.0782(-2 + 5x + 5y) + 4.0291(4 - 10x) + 4(-1 + 5x - 5y),\\
T_3: &\quad \phi(x, y) = 4(-1 + 5y) + 4(2 - 5x - 5y) + 4.0383(5x),\\
T_4: &\quad \phi(x, y) = 4.0383(1 - 5x + 5y) + 4.0782(-2 + 5x + 5y) + 4.0291(2 - 10y),\\
T_5: &\quad \phi(x, y) = 4.0782(2 - 5x + 5y) + 4.0496(-4 + 10x) + 4(3 - 5x - 5y),\\
T_6: &\quad \phi(x, y) = 4.0496(6 - 10x) + 4.0565(-6 + 10x + 10y) + 4(1 - 10y),\\
T_7: &\quad \phi(x, y) = 4(-5x + 5y) + 4.0383(5x) + 4(1 - 5y),\\
T_8: &\quad \phi(x, y) = 4.0383(5y) + 4(1 - 5x) + 4(5x - 5y),\\
T_9: &\quad \phi(x, y) = 4.0291(10y) + 4(2 - 5x - 5y) + 4(-1 + 5x - 5y),\\
T_{10}: &\quad \phi(x, y) = 4.0496(10y) + 4(3 - 5x - 5y) + 4(-2 + 5x - 5y).
\end{aligned}$$

The actual solution to the boundary-value problem is $u(x, y) = xy + 4$. Table 12.7 on the following page compares the value of u to the value of ϕ at E_i, for each $i = 1, \ldots, 5$.

■

Typically, the error for elliptic second-order problems of the type (12.28) with smooth coefficient functions is $O(h^2)$, where h is the maximum diameter of the triangular elements.

Table 12.7

x	y	$\phi(x, y)$	$u(x, y)$	$\lvert\phi(x, y) - u(x, y)\rvert$
0.2	0.2	4.0383	4.04	0.0017
0.4	0.2	4.0782	4.08	0.0018
0.3	0.1	4.0291	4.03	0.0009
0.5	0.1	4.0496	4.05	0.0004
0.6	0.1	4.0565	4.06	0.0035

Piecewise bilinear basis functions on rectangular elements are also expected to give $O(h^2)$ results, where h is the maximum diagonal length of the rectangular elements. Other classes of basis functions can be used to give $O(h^4)$ results, but the construction is more complex. Efficient error theorems for finite-element methods are difficult to state and apply because the accuracy of the approximation depends on the continuity properties of the solution and the regularity of the boundary.

The Finite-Element method can also be applied to parabolic and hyperbolic partial-differential equations, but the minimization procedure is more difficult. A good survey on the advantages and techniques of the Finite-Element method applied to various physical problems can be found in a paper by [Fi]. For a more extensive discussion, refer to [SF], [ZM], or [AB].

EXERCISE SET 12.4

1. Use Algorithm 12.5 to approximate the solution to the following partial-differential equation (see the figure):

$$\frac{\partial}{\partial x}\left(y^2\frac{\partial u}{\partial x}(x, y)\right) + \frac{\partial}{\partial y}\left(y^2\frac{\partial u}{\partial y}(x, y)\right) - yu(x, y) = -x, \quad (x, y) \in D,$$

$$u(x, 0.5) = 2x, \quad 0 \le x \le 0.5, \quad u(0, y) = 0, \quad 0.5 \le y \le 1,$$

$$y^2\frac{\partial u}{\partial x}(x, y)\cos\theta_1 + y^2\frac{\partial u}{\partial y}(x, y)\cos\theta_2 = \frac{\sqrt{2}}{2}(y - x) \quad \text{for } (x, y) \in S_2.$$

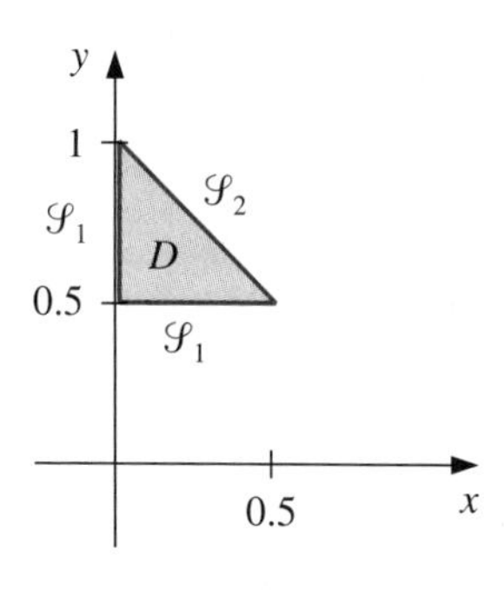

Let $M = 2$; T_1 have vertices (0, 0.5), (0.25, 0.75), (0, 1); and T_2 have vertices (0, 0.5), (0.5, 0.5), and (0.25, 0.75).

2. Repeat Exercise 1, using instead the triangles

$$T_1: \quad (0, 0.75), (0, 1), (0.25, 0.75);$$
$$T_2: \quad (0.25, 0.5), (0.25, 0.75), (0.5, 0.5);$$
$$T_3: \quad (0, 0.5), (0, 0.75), (0.25, 0.75);$$
$$T_4: \quad (0, 0.5), (0.25, 0.5), (0.25, 0.75).$$

3. Approximate the solution to the partial-differential equation

$$\frac{\partial^2 u}{\partial x^2}(x, y) + \frac{\partial^2 u}{\partial y^2}(x, y) - 12.5\pi^2 u(x, y) = -25\pi^2 \sin\frac{5\pi}{2}x \sin\frac{5\pi}{2}y, \quad 0 < x,\ y < 0.4,$$

subject to the Dirichlet boundary condition

$$u(x, y) = 0,$$

using the Finite-Element Algorithm with the elements given in the accompanying figure. Compare the approximate solution to the actual solution

$$u(x, y) = \sin\frac{5\pi}{2}x \sin\frac{5\pi}{2}y$$

at the interior vertices and at the points $(0.125, 0.125)$, $(0.125, 0.25)$, $(0.25, 0.125)$, and $(0.25, 0.25)$.

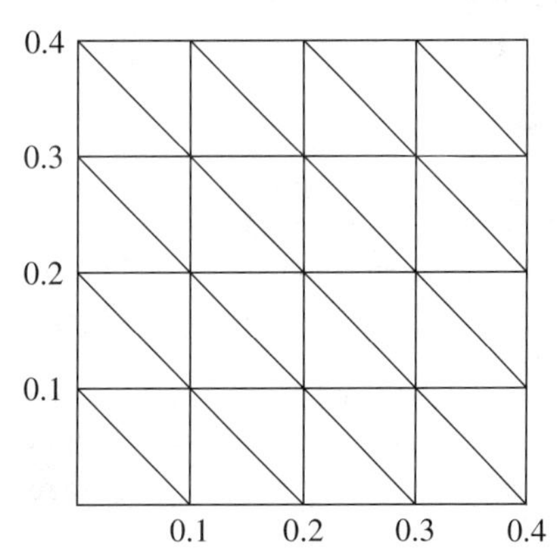

4. Repeat Exercise 3 with

$$f(x, y) = -25\pi^2 \cos\frac{5\pi}{2}x \cos\frac{5\pi}{2}y,$$

using the Neumann boundary condition

$$\frac{\partial u}{\partial n}(x, y) = 0.$$

The actual solution for this problem is

$$u(x, y) = \cos\frac{5\pi}{2}x \cos\frac{5\pi}{2}y.$$

5. A silver plate in the shape of a trapezoid (see the accompanying figure) has heat being uniformly generated at each point at the rate $q = 1.5\ \text{cal/cm}^3\cdot\text{s}$. The steady-state temperature $u(x, y)$ of the plate satisfies the Poisson equation

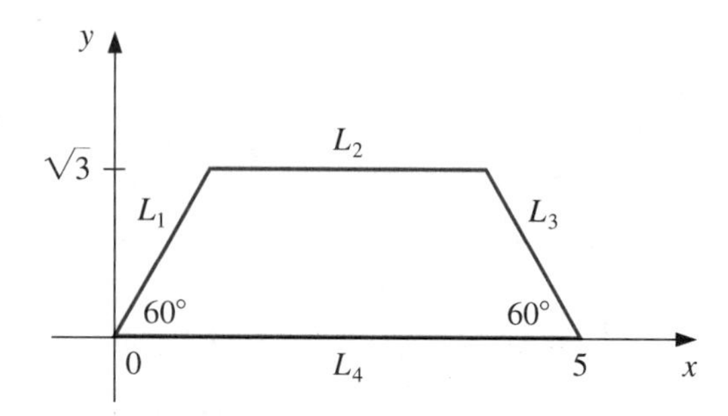

$$\frac{\partial^2 u}{\partial x^2}(x, y) + \frac{\partial^2 u}{\partial y^2}(x, y) = \frac{-q}{k},$$

where k, the thermal conductivity, is 1.04 cal/cm·deg·s. Assume that the temperature is held at 15°C on L_2, that heat is lost on the slanted edges L_1 and L_3 according to the boundary condition $\partial u/\partial n = 4$, and that no heat is lost on L_4, that is, $\partial u/\partial n = 0$. Approximate the temperature of the plate at (1, 0), (4, 0), and $(5/2, \sqrt{3}/2)$ by using Algorithm 12.5.

12.5 Survey of Methods and Software

In this chapter, methods to approximate solutions to partial-differential equations were considered. We restricted our attention to Poisson's equation as an example of an elliptic partial-differential equation, the heat or diffusion equation as an example of a parabolic partial-differential equation, and the wave equation as an example of a hyperbolic partial-differential equation. Finite-difference approximations were discussed for these three examples.

Poisson's equation on a rectangle required the solution of a large sparse linear system, for which iterative techniques, such as the SOR method, are recommended. Four finite-difference methods were presented for the heat equation. The Forward-Difference and Richardson's methods had stability problems, so the Backward-Difference method and the Crank-Nicolson methods were introduced. Although a tridiagonal linear system must be solved at each time step with these implicit methods, they are more stable than the explicit Forward-Difference and Richardson's methods. The Finite-Difference method for the wave equation is explicit and can also have stability problems for certain choice of time and space discretizations.

In the last section of the chapter we presented an introduction to the Finite-Element method for a self-adjoint elliptic partial-differential equation on a polygonal domain. Although our methods will work adequately for the problems and examples in the textbook, more powerful generalizations and modifications of these techniques are required for commercial applications.

We consider two subroutines from the IMSL Library. The subroutine MOLCH is used to solve the partial-differential equation

$$\frac{\partial u}{\partial t} = F\left(x, t, u, \frac{\partial u}{\partial x}, \frac{\partial^2 u}{\partial x^2}\right)$$

with boundary conditions

$$\alpha(x, t)u(x, t) + \beta(x, t)\frac{\partial u}{\partial x}(x, t) = \gamma(x, t).$$

The method is based on collocation at Gaussian points on the x-axis for each value of t and uses cubic Hermite splines as basis functions.

The subroutine FPS2H is used to solve Poisson's equation on a rectangle. The method of solution is based on a choice of second- or fourth-order finite differences on a uniform mesh.

The NAG Library has a number of subroutines for partial-differential equations. The subroutine D03EAF is used for Laplace's equation on an arbitrary domain in the xy-plane. The subroutine D03PAF is used to solve a single parabolic partial-differential equation by the method of lines.

There are specialized packages, such as NASTRAN, consisting of codes for the Finite-Element method. These packages are popular in engineering applications. General codes for partial-differential equations are difficult to write because of the problem of specifying domains other than common geometrical figures. Research in the area of solution of partial-differential equations is currently very active.

We have only presented a small sample of the many techniques used for approximating the solutions to the problems involving partial-differential equations. Further information on the general topic can be found in Lapidus and Pinder [LP], Twizell [Tw], and the recent book by Morton and Mayers [MM]. Software information can be found in Rice and Boisvert [RB] and in Bank [Ban].

Books that focus on finite-difference methods include Strikwerda [Stri], Thomas [Th], and Shashkov and Steinberg [ShS]. Strang and Fix [SF] and Zienkiewicz and Morgan [ZM] are good sources for information on the finite-element method. Time-dependent equations are treated in Schiesser [Schi] and in Gustafsson, Kreiss, and Oliger [GKO], and Birkhoff and Lynch [BL] and Roache [Ro] discuss the solution to elliptic problems.

Multigrid methods use coarse grid approximations and iterative techniques to provide approximations on finer grids. References on these techniques include Briggs [Brigg], McCormick [Mc], and Bramble [Bram].

Bibliography

[AHU] Aho, A. V., J. E. Hopcroft, and J. D. Ullman, *The design and analysis of computer algorithms*, Addison-Wesley, Reading, MA, 1974, 470 pp. QA76.6.A36 *535*

[AG] Allgower, E. and K. Georg, *Numerical continuation methods: an introduction*, Springer-Verlag, New York, 1990, 388 pp. QA377.A56 *622, 623*

[Am] Ames, W. F., *Numerical methods for partial differential equations*, (Third edition), Academic Press, New York, 1992, 451 pp. QA374.A46 *704*

[An] Anderson, E., et al., *LAPACK user's guide*, (Second edition), SIAM Publications, Philadelphia, PA, 1995, 325 pp. QA76.73.F25 L36 *44*

[AS] Argyros, I. K. and F. Szidarovszky, *The theory and applications of iteration methods*, CRC Press, Boca Raton, FL, 1993, 355 pp. QA297.8.A74 *623*

[AMR] Ascher, U. M., R. M. M. Mattheij, and R. B. Russell, *Numerical solution of boundary value problems for ordinary differential equations*, Prentice-Hall, Englewood Cliffs, NJ, 1988, 595 pp. QA379.A83 *669*

[Ax] Axelsson, O., *Iterative solution methods*, Cambridge University Press, New York, 1994, 654 pp. QA297.8.A94 *472*

[AB] Axelsson, O. and V. A. Barker, *Finite element solution of boundary value problems: theory and computation*, Academic Press, Orlando, FL, 1984, 432 pp. QA379.A9 *720*

[Ba1] Bailey, N. T. J., *The mathematical approach to biology and medicine*, John Wiley & Sons, New York, 1967, 269 pp. QH324.B28 *293*

[Ba2] Bailey, N. T. J., *The mathematical theory of epidemics*, Hafner, New York, 1957, 194 pp. RA625.B3 *294*

[BSW] Bailey, P. B., L. F. Shampine, and P. E. Waltman, *Nonlinear two-point boundary-value problems*, Academic Press, New York, 1968, 171 pp. QA372.B27 *652, 669*

[Ban] Bank, R. E., *PLTMG, A software package for solving elliptic partial differential equations: Users' Guide 7.0*, SIAM Publications, Philadelphia, PA, 1994, 128 pp. QA377.B26 *723*

[Barr] Barrett, R., et al., *Templates for the solution of linear systems: building blocks for iterative methods*, SIAM Publications, Philadelphia, PA, 1994, 112 pp. QA297.8.T45 *472*

[Bart] Bartle, R. G., *The elements of real analysis*, (Second edition), John Wiley & Sons, New York, 1976, 480 pp. QA300.B29 *105, 123*

[Bek] Bekker, M. G., *Introduction to terrain vehicle systems*, University of Michigan Press, Ann Arbor, MI, 1969, 846 pp. TL243.B39 *78, 605*

[Ber] Bernadelli, H., *Population waves*, Journal of the Burma Research Society **31** (1941), 1–18, DS527.B85 *387*

[BD] Birkhoff, G. and C. De Boor, *Error bounds for spline interpolation*, Journal of Mathematics and Mechanics **13** (1964), 827–836, QA1.J975 *154*

[BL] Birkhoff, G. and R. E. Lynch, *Numerical solution of elliptic problems*, SIAM Publications, Philadelphia, PA, 1984, 319 pp. QA377.B672 *723*

[BiR] Birkhoff, G. and G. Rota, *Ordinary differential equations*, John Wiley & Sons, New York, 1978, 342 pp. QA372.B58 *255, 257, 319*

[BP] Botha, J. F. and G. F. Pinder, *Fundamental concepts in the numerical solution of differential equations*, Wiley-Interscience, New York, 1983, 202 pp. QA374.B74 *349*

[Brac] Bracewell, R., *The Fourier transform and its application*, (Second edition), McGraw-Hill, New York, 1978, 444 pp. QA403.5.B7 *535*

[Bram] Bramble, J.H., *Multigrid methods*, John Wiley & Sons, New York, 1993, 161 pp. QA377.B73 *723*

[Bre] Brent, R., *Algorithms for minimization without derivatives*, Prentice-Hall, Englewood Cliffs, NJ, 1973, 195 pp. QA402.5.B74 *102, 103*

[Brigg] Briggs, W. L., *A multigrid tutorial*, SIAM Publications, Philadelphia, PA, 1987, 88 pp. QA377.B75 *723*

[BH] Briggs, W. L. and V. E. Henson, *The DFT: an owner's manual for the discrete Fourier transform*, SIAM Publications, Philadelphia, PA, 1995, 434 pp. QA403.5.B75 *538*

[Brigh] Brigham, E. O., *The fast Fourier transform*, Prentice-Hall, Englewood Cliffs, NJ, 1974, 252 pp. QA403.B74 *527*

[Brow,K] Brown, K. M., *A quadratically convergent Newton-like method based upon Gaussian elimination*, SIAM Journal on Numerical Analysis **6**, No. 4 (1969), 560–569, QA297.A1S2 *613*

[Brow,W] Brown, W. S., *A simple but realistic model of floating point computation*, ACM transactions of Mathematical Software **7** (1981), 445–480, QA76.A8 *41, 43*

[Broy] Broyden, C. G., *A class of methods for solving nonlinear simultaneous equations*, Mathematics of Computation **19** (1965), 577–593, QA1.M4144 *608*

[BS1] Bulirsch R. and J. Stoer, *Numerical treatment of ordinary differential equations by extrapolation methods*, Numerische Mathematik **8** (1966), 1–13, QA241.N9 *316*

[BS2] Bulirsch, R. and J. Stoer, *Fehlerabschätzungen und extrapolation mit rationalen Funktionen bei Verfahren von Richardson-typus*, Numerische Mathematik **6** (1964), 413–427, QA241.N9 *316*

[BS3] Bulirsch, R. and J. Stoer, *Asymptotic upper and lower bounds for results of extrapolation methods*, Numerische Mathematik **8** (1966), 93–104, QA241.N9 *316*

[BuR] Bunch, J. R. and D. J. Rose (eds.), *Sparse matrix computations* (Proceedings of a conference held at Argonne National Laboratories, September 9–11, 1975), Academic Press, New York, 1976, 453 pp. QA188.S9 *420*

[BFR] Burden, R. L., J. D. Faires, and A. C. Reynolds, *Numerical Analysis*, (Second edition), Prindle, Weber & Schmidt, Boston, MA, 1981, 598 pp. QA297.B84 *96*

[Bur] Burrage, K., 1995, *Parallel and sequential methods for ordinary differential equations*, Oxford University Press, New York, 446 pp. QA372.B883 *349*

[But] Butcher, J. C., *The non-existence of ten-stage eighth-order explicit Runge-Kutta methods*, BIT **25** (1985), 521–542, QA76.N62 *282*

[CF] Chaitin-Chatelin, F. and Fraysse, V., *Lectures on finite precision computations*, SIAM Publications, Philadelphia, PA, 1996, 235 pp. QA297.C417 *46*

[CCR] Chiarella, C., W. Charlton, and A. W. Roberts, *Optimum chute profiles in gravity flow of granular materials: a discrete segment solution method*, Transactions of the ASME, Journal of Engineering for Industry Series B **97** (1975), 10–13, TJ1.A712 *605*

[Ch] Cheney, E. W., *Introduction to approximation theory*, McGraw-Hill, New York, 1966, 259 pp. QA221.C47 *538*

[CC] Clenshaw, C. W. and C. W. Curtis, *A method for numerical integration on an automatic computer*, Numerische Mathematik **2** (1960), 197–205, QA241.N9 *251*

[CW] Cody, W. J. and W. Waite, *Software manual for the elementary functions*, Prentice-Hall, Englewood Cliffs, NJ, 1980, 269 pp. QA331.C635 *46*

[CV] Coleman, T. F. and C. Van Loan, *Handbook for matrix computations*, SIAM Publications, Philadelphia, PA, 1988, 264 pp. QA188.C65 *44, 422*

[CT] Cooley, J. W. and J. W. Tukey, *An algorithm for the machine calculation of complex Fourier series*, Mathematics of Computation **19**, No. 90 (1965), 297–301, QA1.M4144 *527*

[Co] Cowell, W. (ed.), *Sources and development of mathematical software*, Prentice-Hall, Englewood Cliffs, NJ, 1984, 404 pp. QA76.95.S68 *43*

[DaB] Dahlquist, G. and Å. Björck (Translated by N. Anderson), *Numerical methods*, Prentice-Hall, Englewood Cliffs, NJ, 1974, 573 pp. QA297.D3313 *86*

[Da] Davis, P. J., *Interpolation and Approximation*, Dover, New York, 1975, 393 pp. QA221.D33 *538*

[DR] Davis, P. J. and P. Rabinowitz, *Methods of numerical integration*, Academic Press, New York, 1975, 459 pp. QA299.3.D28 *251*

[Deb] De Boor, C., *A practical guide to splines*, Springer-Verlag, New York, 1978, 392 pp. QA1.A647 vol. 27 *154, 165, 166*

[DebS] De Boor, C. and B. Swartz, *Collocation at Gaussian points*, SIAM Journal on Numerical Analysis **10**, No. 4 (1973), 582–606, QA297.A1S2 *666*

[DM] Dennis, J. E., Jr. and J. J. Moré, *Quasi-Newton methods, motivation and theory*, SIAM Review **19**, No. 1 (1977), 46–89, QA1.S2 *609*

[DenS] Dennis, J. E., Jr. and R. B. Schnabel, *Numerical methods for unconstrained optimization and nonlinear equations*, Prentice-Hall, Englewood Cliffs, NJ, 1983, 378 pp. QA402.5.D44 *623*

[Di] Dierckx, P., *Curve and surface fitting with splines*, Oxford University Press, New York, 1993, 285 pp. QA297.6.D54 *166*

[DBMS] Dongarra, J. J., J. R. Bunch, C. B. Moler, and G. W. Stewart, *LINPACK users guide*, SIAM Publications, Philadephia, PA, 1979, 367 pp. QA214.L56 *44*

[DRW] Dongarra, J. J., T. Rowan, and R. Wade, *Software distributions using Xnetlib*, ACM Transactions on Mathematical Software **21**, No. 1 (1995), 79–88 QA76.6.A8 *44*

[DW] Dongarra, J. and D. W. Walker, *Software libraries for linear algebra computation on high performance computers*, SIAM Review **37**, No. 2 (1995), 151–180 QA1.S2 *46*

[Do] Dormand, J. R., *Numerical methods for differential equations: a computational approach*, CRC Press, Boca Raton, FL, 1996, 368 pp. QA372.D67 *349*

[DoB] Dorn, G. L. and A. B. Burdick, *On the recombinational structure of complementation relationships in the m-dy complex of the Drosophila melanogaster*, Genetics **47** (1962), 503–518, QH431.G43 *418*

[E] Engels, H., *Numerical Quadrature and Cubature*, Academic Press, New York, 1980, 441 pp. QA299.3.E5 *251*

[Fe] Fehlberg, E., *Klassische Runge-Kutta Formeln vierter und niedrigerer Ordnung mit Schrittweiten-Kontrolle und ihre Anwendung auf Wärmeleitungsprobleme*, Computing **6** (1970), 61–71, QA76.C777 *289*

[Fi] Fix, G., *A survey of numerical methods for selected problems in continuum mechanics*, Proceedings of a Conference on Numerical Methods of Ocean Circulation, National Academy of Sciences (1975), 268–283, Q11.N26 *720*

[FM] Forsythe, G. E. and C. B. Moler, *Computer solution of linear algebraic systems*, Prentice-Hall, Englewood Cliffs, NJ, 1967, 148 pp. QA297.F57 *422, 464*

[Fr] Francis, J. G. F., *The QR transformation*, Computer Journal **4** (1961–2), Part I, 265–271; Part II, 332–345, QA76.C57 *578*

[Fu] Fulks, W., *Advanced Calculus*, (Third edition), John Wiley & Sons, New York, 1978, 731 pp. QA303.F954 *8, 9, 276*

[Gar] Garbow, B. S., et al., *Matrix eigensystem routines: EISPACK guide extension*, Springer-Verlag, New York, 1977, 343 pp. QA193.M38 *44*

[Ge1] Gear, C. W., *Numerical initial-value problems in ordinary differential equations*, Prentice-Hall, Englewood Cliffs, NJ, 1971, 253 pp. QA372.G4 *328, 331, 342, 349*

[Ge2] Gear, C. W., *Numerical solution of ordinary differential equations: Is there anything left to do?*, SIAM Review **23**, No. 1 (1981), 10–24, QA1.S2 *346*

[GL] George, A. and J. W. Liu, *Computer solution of large sparse positive definite systems*, Prentice-Hall, Englewood Cliffs, NJ, 1981, 324pp. QA188.G46 *422*

[Go] Goldberg, D., *What every scientist should know about floating-point arithmetic*, ACM Computing Surveys **23**, No. 1 (1991), 5–48, QA76.5.A1 *46*

[GO] Golub, G. H. and Ortega, J. M., *Scientific computing: an introduction with parallel computing*, Academic Press, Boston, MA, 1993, 442 pp. QA76.58.G64 *46, 349*

[GV] Golub, G. H. and C. F. Van Loan, *Matrix computations*, (Second edition), John Hopkins University Press, Baltimore, MD, 1989, 642 pp. QA188.G65 *406, 422*

[Gr] Gragg, W. B., *On extrapolation algorithms for ordinary initial-value problems*, SIAM Journal on Numerical Analysis **2** (1965), 384–403, QA297.A1S2 *312, 316*

[GKO] Gustafsson, B., H. Kreiss, and J. Oliger, *Time dependent problems and difference methods*, John Wiley & Sons, New York, 1995, 642 pp. QA374.G974 *723*

[Hac] Hackbusch, W., *Iterative solution of large sparse systems of equations*, Springer-Verlag, New York, 1994, 429 pp. QA1.A647 vol. 95 *472*

[HY] Hageman, L. A. and D. M. Young, *Applied iterative methods*, Academic Press, New York, 1981, 386 pp. QA297.8.H34 *472*

[HNW1] Hairer, E., S. P. Nörsett, and G. Wanner, *Solving ordinary differential equations. Vol. 1: Nonstiff equations*, Springer-Verlag, New York, 1987, QA372.H16 *328, 349*

[HNW2] Hairer, E., S. P. Nörsett, and G. Wanner, *Solving ordinary differential equations. Vol. 2: Stiff and differential-algebraic problems*, Springer-Verlag, New York, 1991, QA372.H16 *349*

[Ham] Hamming, R. W., *Numerical methods for scientists and engineers*, (Second edition), McGraw-Hill, New York, 1973, 721 pp. QA297.H28 *535*

[He1] Henrici, P., *Discrete variable methods in ordinary differential equations*, John Wiley & Sons, New York, 1962, 407 pp. QA372.H48 *349*

[He2] Henrici, P., *Elements of numerical analysis*, John Wiley & Sons, New York, 1964, 328 pp. QA297.H54 *90, 337*

[Hild] Hildebrand, F. B., *Introduction to numerical analysis*, (Second edition), McGraw-Hill, New York, 1974, 669 pp. QA297.H54 *123, 125, 130*

[Hill] Hill, F. S., Jr., *Computer graphics*, Macmillan, New York, 1990, 754 pp. T385.H549 *164*

[Ho] Householder, A. S., *The numerical treatment of a single nonlinear equation*, McGraw-Hill, New York, 1970, 216 pp. QA218.H68 *102, 103*

[IK] Issacson, E. and H. B. Keller, *Analysis of numerical methods*, John Wiley & Sons, New York, 1966, 541 pp. QA297.I8 *90, 194, 196, 334, 337, 442, 551, 553, 630, 649, 687, 691, 694, 704*

[JT] Jenkins, M. A. and J. F. Traub, *A three-stage algorithm for real polynomials using quadratic iteration*, SIAM Journal on Numerical Analysis **7**, No. 4 (1970), 545–566, QA297.A1S2 *102*

[Joh] Johnston, R. L., *Numerical methods: a software approach*, John Wiley & Sons, New York, 1982, 276 pp. QA297.J64 *213*

[Joy] Joyce, D. C., *Survey of extrapolation processes in numerical analysis*, SIAM Review **13**, No. 4 (1971), 435–490, QA1.S2 *180*

[K,H] Keller, H. B., *Numerical methods for two-point boundary-value problems*, Blaisdell, Waltham, MA, 1968, 184 pp. QA372.K42 *625, 638, 643, 647, 669*

[K,J] Keller, J. B., *Probability of a shutout in racquetball*, SIAM Review **26**, No. 2 (1984), 267–268, QA1.S2 *78*

[Ko] Köckler, N., *Numerical methods and scientific computing: using software libraries for problem solving*, Oxford University Press, New York, 1994, 328 pp. TA345.K653 *46*

[Lam] Lambert, J. D., *The initial value problem for ordinary differential equations. The state of art in numerical analysis* (D. Jacobs, ed.), Academic Press, New York, 1977, 451–501 pp. QA297.C646 *346*

[LP] Lapidus, L. and G. F. Pinder, *Numerical solution of partial differential equations in science and engineering*, John Wiley & Sons, New York, 1982, 677 pp. Q172.L36 *723*

[Lar] Larson, H. J., *Introduction to probability theory and statistical inference*, (Third edition), John Wiley & Sons, New York, 1982, 637 pp. QA273.L352 *476*

[Lau] Laufer, H. B., *Discrete mathematics and applied modern algebra*, PWS-Kent Publishing, Boston, MA, 1984, 538 pp. QA161.L38 *536*

[LH] Lawson, C. L. and Hanson, R. J., *Solving least squares problems*, SIAM Publications, Philadelphia, PA, 1995, 337 pp. QA275.L38 *538*

[LR] Lucas, T. R. and G. W. Reddien, Jr., *Some collocation methods for nonlinear boundary value problems*, SIAM Journal on Numerical Analysis **9**, No. 2 (1972), 341–356, QA297.A1S2 *666*

[Ma] Mano, M. M., *Computer system architecture*, Prentice-Hall, Englewood Cliffs, NJ, 1982, 531 pp. QA76.9A73 M36 *21*

[Mc] McCormick, S. F., *Multigrid methods*, SIAM Publications, Philadelphia, PA, 1987, 282 pp. QA374.M84 *723*

[Mi] Mitchell, A. R., *Computation methods in partial differential equations*, John Wiley & Sons, New York, 1969, 255 pp. QA374.M68 *704*

[Mo] Moler, C. B., *Demonstration of a matrix laboratory. Lecture notes in Mathematics* (J. P. Hennart, ed.), Springer-Verlag, Berlin, 1982, 84–98 *45*

[MC] Moré J. J. and M. Y. Cosnard, *Numerical solution of nonlinear equations*, ACM Transactions on Mathematical Software **5**, No. 1 (1979), 64–85, QA76.6.A8 *613*

[MM] Morton, K. W. and D. F. Mayers, *Numerical solution of partial differential equations: an introduction*, Cambridge University Press, New York, 1994, 227 pp. QA377.M69 *723*

[Mu] Müller, D. E., *A method for solving algebraic equations using an automatic computer*, Mathematical Tables and Other Aids to Computation **10** (1956), 208–215, QA47.M29 *96*

[ND] Noble, B. and J. W. Daniel, *Applied linear algebra*, (Second edition), Prentice-Hall, Englewood Cliffs, NJ, 1977, 477 pp. QA184.N6 *376, 390, 540*

[Or1] Ortega, J. M., *Introduction to parallel and vector solution of linear systems*, Plenum Press, New York, 1988, 305 pp. QA218.O78 *46*

[Or2] Ortega, J. M., *Numerical analysis; a second course*, Academic Press, New York, 1972, 201 pp. QA297.O78 *430, 440, 451, 456, 468, 546, 589, 592*

[OP] Ortega, J. M. and W. G. Poole, Jr., *An introduction to numerical methods for differential equations*, Pitman Publishing, Marshfield, MA, 1981, 329 pp. QA371.O65 *349*

[OR] Ortega, J. M. and W. C. Rheinboldt, *Iterative solution of nonlinear equations in several variables*, Academic Press, New York, 1970, 572 pp. QA297.8.O77 *589, 620, 623*

[Os] Ostrowski, A. M., *Solution of equations and systems of equations*, (Second edition), Academic Press, New York, 1966, 338 pp. QA3.P8 vol. 9 *103*

[Par] Parlett, B., *The symmetric eigenvalue problem*, Prentice-Hall, Englewood Cliffs, NJ, 1980, 348 pp. QA188.P37 *587*

[Pat] Patterson, T. N. L., *The optimum addition of points to quadrature formulae*, Mathematics of Computation **22**, No. 104 (1968), 847–856, QA1.M4144 *250*

[PF] Phillips, C. and T. L. Freeman, *Parallel numerical algorithms*, Prentice-Hall, New York, 1992, 315 pp. QA76.9.A43 F74 *46*

[Ph] Phillips, J., *The NAG Library: a beginner's guide*, Clarendon Press, Oxford, 1986, 245 pp. QA297.P35 *45*

[PDUK] Piessens, R., E. de Doncker-Kapenga, C. W. Überhuber, and D. K. Kahaner, *QUADPACK: a subroutine package for automatic integration*, Springer-Verlag, New York, 1983, 301 pp. QA299.3.Q36 *250*

[Pi] Pissanetzky, S., *Sparse matrix technology*, Academic Press, New York, 1984, 321 pp. QA188.P57 *422*

[Po] Powell, M. J. D., *Approximation theory and methods*, Cambridge University Press, Cambridge, 1981, 339 pp. QA221.P65 *139, 140, 166, 517, 538*

[Pr] Pryce, J. D., *Numerical solution of Sturm-Liouville problems*, Oxford University Press, New York, 1993, 322 pp. QA379.P79 *669*

[RR] Ralston, A. and P. Rabinowitz, *A first course in numerical analysis*, (Second edition), McGraw-Hill, New York, 1978, 556 pp. QA297.R3 *211, 512, 517, 620*

[Ra] Rashevsky, N., *Looking at history through mathematics*, Massachusetts Institute of Technology Press, Cambridge, MA, 1968, 199 pp. D16.25.R3 *268*

[RB] Rice, J. R. and R. F. Boisvert, *Solving elliptic problems using ELLPACK*, Springer-Verlag, New York, 1985, 497 pp. QA377.R53 *723*

[RG] Richardson, L. F. and J. A. Gaunt, *The deferred approach to the limit*, Philosophical Transactions of the Royal Society of London **226A** (1927), 299–361, Q41.L82 *180*

[Ro] Roache, P. J., *Elliptic marching methods and domain decomposition*, CRC Press, Boca Raton, FL, 1995, 190 pp. QA377.R63 *723*

[RS] Roberts, S. and J. Shipman, *Two-point boundary value problems: shooting methods*, Elsevier, New York, 1972, 269 pp. QA372.R76 *669*

[RW] Rose, D. J. and R. A. Willoughby (eds.), *Sparse matrices and their applications* (Proceedings of a conference held at IBM Research, New York, September 9–10, 1971. 215 pp.), Plenum Press, New York, 1972, QA263.S94 *420*

[Ru] Russell, R. D., *A comparison of collocation and finite differences for two-point boundary value problems*, SIAM Journal on Numerical Analysis **14**, No. 1 (1977), 19–39, QA297.A1S2 *666*

[Sa1] Saad, Y., *Numerical methods for large eigenvalue problems*, Halsted Press, New York, 1992, 346 pp. QA188.S18 *587*

[Sa2] Saad, Y., *Iterative methods for sparse linear systems*, PWS-Kent Publishing, Boston, MA, 1996, 447 pp. QA188.S17 *472*

[SaS] Saff, E. B. and A. D. Snider, *Fundamentals of complex analysis for mathematics, science, and engineering*, Prentice-Hall, Englewood Cliffs, NJ, 1976, 444 pp. QA300.S18 *92*

[SP] Sagar, V. and D. J. Payne, *Incremental collapse of thick-walled circular cylinders under steady axial tension and torsion loads and cyclic transient heating*, Journal of the Mechanics and Physics of Solids **21**, No. 1 (1975), 39–54, TA350.J68 *697*

[SD] Sale, P. F. and R. Dybdahl, *Determinants of community structure for coral-reef fishes in experimental habitat*, Ecology **56** (1975), 1343–1355, QH540.E3 *485*

[Sche] Schendel, U., *Introduction to numerical methods for parallel computers*, (Translated by B.W. Conolly), Halsted Press, New York, 1984, 151 pp. QA297.S3813 *46*

[Schi] Schiesser, W. E., *Computational mathematics in engineering and applied science: ODE's, DAE's, and PDE's*, CRC Press, Boca Raton, FL, 1994, 587 pp. TA347.D45 S34 *723*

[Scho] Schoenberg, I. J., *Contributions to the problem of approximation of equidistant data by analytic functions*, Quarterly of Applied Mathematics **4**, (1946), Part A, 45–99; Part B, 112–141, QA1.A26 *166*

[Schr1] Schroeder, L. A., *Energy budget of the larvae of the moth Pachysphinx modesta*, Oikos **24** (1973), 278–281, QH540.O35 *486*

[Schr2] Schroeder, L. A., *Thermal tolerances and acclimation of two species of hydras*, Limnology and Oceanography **26**, No. 4 (1981), 690–696, GC1.L5 *605*

[Schul] Schultz, M. H., *Spline analysis*, Prentice-Hall, Englewood Cliffs, NJ, 1973, 156 pp. QA211.S33 *153, 166, 653, 660, 665, 666*

[Schum] Schumaker, L. L., *Spline functions: basic theory*, Wiley-Interscience, New York, 1981, 553 pp. QA224.S33 *166*

[Se] Searle, S. R., *Matrix algebra for the biological sciences*, John Wiley & Sons, New York, 1966 296 pp. QH324.S439 *387*

[SH] Secrist, D. A. and R. W. Hornbeck, *An analysis of heat transfer and fade in disk brakes*, Transactions of the ASME, Journal of Engineering for Industry Series B **98** No. 2 (1976), 385–390, TJ1.A712 *207*

[Sh] Shampine, L. F., *Numerical solution of ordinary differential equations*, Chapman & Hall, New York, 1994, 484 pp. QA372.S417 *349*

[SGe] Shampine, L. F. and C. W. Gear, *A user's view of solving stiff ordinary differential equations*, SIAM Review **21**, No. 1 (1979), 1–17, QA1.S2 *376*

[ShS] Shashkov, M. and S. Steinberg, *Conservative finite-difference methods on general grids*, CRC Press, Boca Raton, FL, 1996, 359 pp. QA431.S484 *723*

[SW] Simon, B. and R. M. Wilson, *Supercalculators on the PC*, Notices of the American Mathematical Society **35**, No. 7 (1988), 978–1001, QA1.N6 *45*

[Si] Singh, V. P., *Investigations of attentuation and internal friction of rocks by ultrasonics*, International Journal of Rock Mechanics and Mining Sciences (1976), 69–72, TA706.I45 *485*

[SJ] Sloan, I. H. and S. Joe, *Lattice methods for multiple integration*, Oxford University Press, New York, 1994, 239 pp. QA311.S56 *251*

[Sm,B] Smith, B. T., et al., *Matrix eigensystem routines: EISPACK guide*, (Second edition), Springer-Verlag, New York, 1976, 551 pp. QA193.M37 *43*

[Sm,G] Smith, G. D., *Numerical solution of partial differential equations*, Oxford University Press, New York, 1965, 179 pp. QA377.S59 *704*

[Stet] Stetter, H. J., *Analysis of discretization methods for ordinary differential equations. From tracts in natural philosophy*, Springer-Verlag, New York, 1973, 388 pp. QA372.S84 *316*

[Stew1] Stewart, G. W., *Afternotes on numerical analysis*, SIAM Publications, Philadelphia, PA, 1996, 200 pp. QA297.S785 *422, 587*

[Stew2] Stewart, G. W., *Introduction to matrix computations*, Academic Press, New York, 1973, 441 pp. QA188.S7 *408*

[SF] Strang, W. G. and G. J. Fix, *An analysis of the finite element method*, Prentice-Hall, Englewood Cliffs, NJ, 1973, 306 pp. TA335.S77 *666, 720, 723*

[Stri] Strikwerda, J. C., *Finite difference schemes and partial differential equations*, Brooks/Cole Publishing, Pacific Grove, CA, 1989, 386 pp. QA374.S88 *723*

[Stro] Stroud, A. H., *Approximate calculation of multiple integrals*, Prentice-Hall, Englewood Cliffs, NJ, 1971, 431 pp. QA311.S85 *251*

[StS] Stroud, A. H. and D. Secrest, *Gaussian quadrature formulas*, Prentice-Hall, Englewood Cliffs, NJ, 1966, 374 pp. QA299.4.G4 S7 *226, 251*

[Sz] Szüsz, P., *Math bite*, Mathematics Magazine **68**, No. 2, 1995, 97, QA1.N28 *436*

[Th] Thomas, J. W., *Numerical partial differential equations*, Springer-Verlag, New York, 1995, 445 pp. QA377.T495 *723*

[Tr] Traub, J. F., *Iterative methods for the solution of equations*, Prentice-Hall, Englewood Cliffs, NJ, 1964, 310 pp. QA297.T7 *103*

[Tw] Twizell, E. H., *Computational methods for partial differential equations*, Ellis Horwood Ltd., Chichester, West Sussex, England, 1984, 276 pp. QA377.T95 *723*

[Van] Van Loan, C. F., *Computational frameworks for the fast Fourier transform*, SIAM Publications, Philadelphia, PA, 1992, 273 pp. QA403.5.V35 *538*

[Var] Varga, R. S., *Matrix iterative analysis*, Prentice-Hall, Englewood Cliffs, NJ, 1962, 322 pp. QA263.V3 *449, 472, 676, 680, 681*

[We] Wendroff, B., *Theoretical numerical analysis*, Academic Press, New York, 1966, 239 pp. QA297.W43 *405, 409*

[Wil1] Wilkinson, J. H., *Rounding errors in algebraic processes*, Prentice-Hall, Englewood Cliffs, NJ, 1963, 161 pp. QA76.5.W53 *468*

[Wil2] Wilkinson, J. H., *The algebraic eigenvalue problem*, Clarendon Press, Oxford, 1965, 662 pp. QA218.W5 *468, 559, 565, 583, 587*

[WR] Wilkinson, J. H. and C. Reinsch (eds.), *Handbook for automatic computation. Vol. 2: Linear algebra*, Springer-Verlag, New York, 1971, 439 pp. QA251.W67 *43, 578, 583, 587*

[Win] Winograd, S., *On computing the discrete Fourier transform*, Mathematics of Computation **32** (1978), 175–199, QA1.M4144 *536*

[Y] Young, D. M., *Iterative solution of large linear systems*, Academic Press, New York, 1971, 570 pp. QA195.Y68 *452, 472*

[YG] Young, D. M. and R. T. Gregory, *A survey of numerical mathematics. Vol. 1*, Addison-Wesley, Reading, MA, 1972, 533 pp. QA297.Y63 *102*

[ZM] Zienkiewicz, O. C. and K. Morgan, *Finite elements and approximation*, John Wiley & Sons, New York, 1983, 328 pp. QA297.5.Z53 *720, 723*

Answers to Selected Exercises

Exercise Set 1.1

1. For each part, $f \in C[a, b]$ on the given interval. Since $f(a)$ and $f(b)$ are of opposite sign, the Intermediate Value Theorem implies a number c exists with $f(c) = 0$.

3. For each part, $f \in C[a, b]$, f' exists on (a, b), and $f(a) = f(b) = 0$. Rolle's Theorem implies that a number c exists in (a, b) with $f'(c) = 0$. For part (d), we can use $[a, b] = [-1, 0]$ or $[a, b] = [0, 2]$.

5. For $x < 0$, $f(x) < 2x + k < 0$, provided that $x < -\frac{1}{2}k$. Similarly, for $x > 0$, $f(x) > 2x + k > 0$, provided that $x > -\frac{1}{2}k$. By the Intermediate Value Theorem, there exists a number c with $f(c) = 0$. If $f(c) = 0$ and $f(c') = 0$ for some $c' \neq c$, then by Theorem 1.7, there exists a number p between c and c' with $f'(p) = 0$. But $f'(x) = 3x^2 + 2 > 0$ for all x.

7. a. $P_2(x) = 0$

b. $R_2(0.5) = 0.125$; actual error $= 0.125$

c. $P_2(x) = 1 + 3(x - 1) + 3(x - 1)^2$

d. $R_2(0.5) = -0.125$; actual error $= -0.125$

9. Since

$$P_2(x) = 1 + x \quad \text{and} \quad R_2(x) = \frac{-2e^{\xi}(\sin \xi + \cos \xi)}{6}x^3$$

for some ξ between x and 0, we have the following:

a. $P_2(0.5) = 1.5$ and $|f(0.5) - P_2(0.5)| \leq 0.0532$

b. $|f(x) - P_2(x)| \leq 1.252$

c. $\int_0^1 f(x)\,dx \approx 1.5$

d. $|\int_0^1 f(x)\,dx - \int_0^1 P_2(x)\,dx| \leq \int_0^1 |R_2(x)|\,dx \leq 0.313$, and the actual error is 0.122.

11. $P_3(x) = (x - 1)^2 - \frac{1}{2}(x - 1)^3$

a. $P_3(0.5) = 0.312500$, $f(0.5) = 0.346574$. An error bound is $0.291\overline{6}$, and the actual error is 0.034074.

b. $|f(x) - P_3(x)| \leq 0.291\overline{6}$ on $[0.5, 1.5]$

c. $\int_{0.5}^{1.5} P_3(x)\,dx = 0.08\overline{3}$, $\int_{0.5}^{1.5}(x - 1)\ln x\,dx = 0.088020$

d. An error bound is $0.058\overline{3}$, and the actual error is 4.687×10^{-3}.

13. $P_4(x) = x + x^3$

a. $|f(x) - P_4(x)| \leq 0.012405$

b. $\int_0^{0.4} P_4(x)\,dx = 0.0864$, $\int_0^{0.4} xe^{x^2}\,dx = 0.086755$

c. 8.27×10^{-4}

d. $P_4'(0.2) = 1.12$, $f'(0.2) = 1.124076$. The actual error is 4.076×10^{-3}.

15. Since $42^\circ = 7\pi/30$ radians, use $x_0 = \pi/4$. Then

$$\left|R_n\left(\frac{7\pi}{30}\right)\right| \leq \frac{(\frac{\pi}{4} - \frac{7\pi}{30})^{n+1}}{(n + 1)!} < \frac{(0.053)^{n+1}}{(n + 1)!}.$$

For $|R_n(\frac{7\pi}{30})| < 10^{-6}$ it suffices to take $n = 3$. To 7 digits, $\cos 42^\circ = 0.7431448$ and $P_3(42^\circ) = P_3(\frac{7\pi}{30}) = 0.7431446$, so the actual error is 2×10^{-7}.

17. a. $P_3(x) = \ln(3) + \frac{2}{3}(x - 1) + \frac{1}{9}(x - 1)^2 - \frac{10}{81}(x - 1)^3$

b. $\max_{0 \leq x \leq 1} |f(x) - P_3(x)| = |f(0) - P_3(0)| = 0.02663366$

c. $\tilde{P}_3(x) = \ln(2) + \frac{1}{2}x^2$

d. $\max_{0\le x\le 1} |f(x) - \tilde{P}_3(x)| = |f(1) - \tilde{P}_3(1)| = 0.09453489$

e. $\tilde{P}_3(0)$ approximates $f(0)$ better than $\tilde{P}_3(1)$ approximates $f(1)$.

19. $P_n(x) = \sum_{k=0}^{n} \frac{1}{k!}x^k, \; n \ge 7$

21. A bound for the maximum error is 0.0026.

23. Since $R_2(1) = \frac{1}{6}e^{\xi}$ for some ξ in (0, 1), we have $|E - R_2(1)| = \frac{1}{6}|1 - e^{\xi}| \le \frac{1}{6}(e - 1)$.

25. a. Let x_0 be any number in $[a, b]$. Given $\epsilon > 0$, let $\delta = \epsilon/L$. If $|x - x_0| < \delta$ and $a \le x \le b$, then $|f(x) - f(x_0)| \le L|x - x_0| < \epsilon$.

b. Using the Mean Value Theorem we have

$$|f(x_2) - f(x_1)| = |f'(\xi)||x_2 - x_1|$$

for some ξ between x_1 and x_2, so

$$|f(x_2) - f(x_1)| \le L|x_2 - x_1|.$$

c. One example is $f(x) = x^{1/3}$ on [0, 1].

Exercise Set 1.2

1.	Absolute Error	Relative Error
a.	0.001264	4.025×10^{-4}
b.	7.346×10^{-6}	2.338×10^{-6}
c.	2.818×10^{-4}	1.037×10^{-4}
d.	2.136×10^{-4}	1.510×10^{-4}
e.	2.647×10^{1}	1.202×10^{-3}
f.	1.454×10^{1}	1.050×10^{-2}
g.	420	1.042×10^{-2}
h.	3.343×10^{3}	9.213×10^{-3}

3. The largest intervals are **a.** (149.85, 150.15) **b.** (899.1, 900.9) **c.** (1498.5, 1501.5) **d.** (89.91, 90.09)

5.	Approximation	Absolute Error	Relative Error
a.	134	0.079	5.90×10^{-4}
b.	133	0.499	3.77×10^{-3}
c.	2.00	0.327	0.195
d.	1.67	0.003	1.79×10^{-3}
e.	1.80	0.154	0.0786
f.	−15.1	0.0546	3.60×10^{-3}
g.	0.286	2.86×10^{-4}	10^{-3}
h.	0.00	0.0215	1.00

7.	Approximation	Absolute Error	Relative Error
a.	133	0.921	6.88×10^{-3}
b.	132	0.501	3.78×10^{-3}
c.	1.00	0.673	0.402
d.	1.67	0.003	1.79×10^{-3}
e.	3.55	1.60	0.817
f.	−15.2	0.0454	0.00299
g.	0.284	0.00171	0.00600
h.	0	0.02150	1

9.	Approximation	Absolute Error	Relative Error
a.	3.14557613	3.983×10^{-3}	1.268×10^{-3}
b.	3.14162103	2.838×10^{-5}	9.032×10^{-6}

11. a. $\lim_{x\to 0} \frac{x\cos x - \sin x}{x - \sin x} = \lim_{x\to 0} \frac{-x\sin x}{1-\cos x} = \lim_{x\to 0} \frac{-\sin x - x\cos x}{\sin x} = \lim_{x\to 0} \frac{-2\cos x + x\sin x}{\cos x} = -2$

b. -1.941

c. $\frac{x(1-\frac{1}{2}x^2) - (x - \frac{1}{6}x^3)}{x - (x - \frac{1}{6}x^3)} = -2$

d. The relative error in part (b) is 0.029. The relative error in part (c) is 0.00050.

13.	x_1	Absolute Error	Relative Error	x_2	Absolute Error	Relative Error
a.	92.26	0.01542	1.672×10^{-4}	0.005419	6.273×10^{-7}	1.157×10^{-4}
b.	0.005421	1.264×10^{-6}	2.333×10^{-4}	-92.26	4.580×10^{-3}	4.965×10^{-5}
c.	10.98	6.875×10^{-3}	6.257×10^{-4}	0.001149	7.566×10^{-8}	6.584×10^{-5}
d.	-0.001149	7.566×10^{-8}	6.584×10^{-5}	-10.98	6.875×10^{-3}	6.257×10^{-4}

15. The machine numbers are equivalent to **a.** 2707 **b.** -2707 **c.** 0.0174560546875 **d.** 0.0174560584130

17. b. The first formula gives -0.00658 and the second formula gives -0.0100. The true three-digit value is -0.0116.

19. The approximate solutions to the systems are **a.** $x = 2.451$, $y = -1.635$ **b.** $x = 507.7$, $y = 82.00$

21. a. In nested form we have

$$f(x) = (((1.01e^x - 4.62)e^x - 3.11)e^x + 12.2)e^x - 1.99.$$

b. -6.79 **c.** -7.07

23. a. $n = 77$ **b.** $n = 35$

25. a. $m = 17$

b. $\binom{m}{k} = \frac{m!}{k!(m-k)!} = \frac{m(m-1)\cdots(m-k-1)(m-k)!}{k!(m-k)!} = \left(\frac{m}{k}\right)\left(\frac{m-1}{k-1}\right)\cdots\left(\frac{m-k-1}{1}\right)$

c. $m = 181707$

d. 2, 597, 000; actual error 1960; relative error 7.541×10^{-4}

27. a. 124.03 **b.** 124.03 **c.** -124.03 **d.** -124.03 **e.** 0.0065 **f.** 0.0065 **g.** -0.0065 **h.** -0.0065

Exercise Set 1.3

1. a. The approximate sums are 1.53 and 1.54, respectively. The actual value is 1.549. Significant roundoff error occurs earlier with the first method.

3. a. 2000 terms **b.** 20,000,000,000 terms

5. 3 terms

7. The rates of convergence are **a.** $O(h^2)$ **b.** $O(h)$ **c.** $O(h^2)$ **d.** $O(h)$

13. a. If $\frac{|\alpha_n - \alpha|}{(1/n^p)} \le K$, then $|\alpha_n - \alpha| \le K(1/n^p) \le K(1/n^q)$ since $0 < q < p$. Thus, $\frac{|\alpha_n - \alpha|}{(1/n^p)} \le K$ and $\{\alpha_n\}_{n=1}^{\infty} \to \alpha$ with rate of convergence $O(1/n^p)$.

b.

n	$1/n$	$1/n^2$	$1/n^3$	$1/n^4$
5	0.2	0.04	0.008	0.0016
10	0.1	0.01	0.001	0.0001
50	0.02	0.0004	8×10^{-6}	1.6×10^{-7}
100	0.01	10^{-4}	10^{-6}	10^{-8}

The most rapid convergence rate is $O(1/n^4)$.

17. a. 354224848179261915075

b. .3542248538 10^{21}

c. The result in part (a) is computed using exact integer arithmetic, and the result in part (b) is computed using 10-digit rounding arithmetic.

d. The result in part (a) required traversing a loop 98 times.

e. The result is the same as the result in part (a).

Exercise Set 2.1

1. $p_3 = 0.625$

3. The Bisection method gives **a.** $p_7 = 0.5859$ **b.** $p_8 = 3.002$ **c.** $p_7 = 3.419$

5. The Bisection method gives $p_9 = 4.4932$.

7. The Bisection method gives

a. $p_{17} = 0.641182$

b. $p_{17} = 0.257530$

c. For the interval $[-3, -2]$ we have $p_{17} = -2.191307$, and for the interval $[-1, 0]$ we have $p_{17} = -0.798164$.

d. For the interval $[0.2, 0.3]$ we have $p_{14} = 0.297528$, and for the interval $[1.2, 1.3]$ we have $p_{14} = 1.256622$.

9. a. 2 **b.** -2 **c.** -1 **d.** 1

11. The third root of 25 is approximately $p_{14} = 2.92401$, using $[2, 3]$.

13. A bound is $n \geq 14$, and $p_{14} = 1.32477$.

15. Since $\lim_{n\to\infty}(p_n - p_{n-1}) = \lim_{n\to\infty} 1/n = 0$, the difference in the terms goes to zero. However, p_n is the nth term of the divergent harmonic series, so $\lim_{n\to\infty} p_n = \infty$.

17. The depth of the water is 0.838 ft.

Exercise Set 2.2

1. For the value of x under consideration we have

a. $x = (3 + x - 2x^2)^{(1/4)} \Leftrightarrow x^4 = 3 + x - 2x^2 \Leftrightarrow f(x) = 0$

b. $x = \left(\dfrac{x + 3 - x^4}{2}\right)^{1/2} \Leftrightarrow 2x^2 = x + 3 - x^4 \Leftrightarrow f(x) = 0$

c. $x = \left(\dfrac{x + 3}{x^2 + 2}\right)^{1/2} \Leftrightarrow x^2(x^2 + 2) = x + 3 \Leftrightarrow f(x) = 0$

d. $x = \dfrac{3x^4 + 2x^2 + 3}{4x^3 + 4x - 1} \Leftrightarrow 4x^4 + 4x^2 - x = 3x^4 + 2x^2 + 3 \Leftrightarrow f(x) = 0$

3. The order in descending speed of convergence is (b), (d), (a). The sequence in (c) does not converge.

5. With $g(x) = (3x^2 + 3)^{1/4}$ and $p_0 = 1$, $p_6 = 1.94332$ is accurate to within 0.01.

7. Since $g'(x) = \frac{1}{4}\cos\frac{x}{2}$, g is continuous and g' exists on $[0, 2\pi]$. Further, $g'(x) = 0$ only when $x = \pi$, so that $g(0) = g(2\pi) = \pi \leq g(x) \leq g(\pi) = \pi + \frac{1}{2}$ and $|g'(x)| \leq \frac{1}{4}$ for $0 \leq x \leq 2\pi$. Theorem 2.2 implies that a unique fixed point p exists in $[0, 2\pi]$. With $k = \frac{1}{4}$ and $p_0 = \pi$, we have $p_1 = \pi + \frac{1}{2}$. Corollary 2.4 implies that

$$|p_n - p| \leq \frac{k^n}{1-k}|p_1 - p_0| = \frac{2}{3}\left(\frac{1}{4}\right)^n.$$

For the bound to be less than 0.1, we need $n \geq 4$. However, $p_3 = 3.626996$ is accurate to within 0.01.

9. For $p_0 = 1.0$ and $g(x) = 0.5(x + \frac{3}{x})$, we have $\sqrt{3} \approx p_4 = 1.73205$.

11. a. With $[0, 1]$ and $p_0 = 0$, we have $p_9 = 0.257531$.

b. With $[2.5, 3.0]$ and $p_0 = 2.5$, we have $p_{17} = 2.690650$.
c. With $[0.25, 1]$ and $p_0 = 0.25$, we have $p_{14} = 0.909999$.
d. With $[0.3, 0.7]$ and $p_0 = 0.3$, we have $p_{39} = 0.469625$.
e. With $[0.3, 0.6]$ and $p_0 = 0.3$, we have $p_{48} = 0.448059$.
f. With $[0, 1]$ and $p_0 = 0$, we have $p_6 = 0.704812$.

13. For $g(x) = (2x^2 - 10\cos x)/3x$, we have the following:

$$p_0 = 3 \Rightarrow p_8 = 3.16193; \quad p_0 = -3 \Rightarrow p_8 = -3.16193.$$

For $g(x) = \arccos(-0.1x^2)$, we have the following:

$$p_0 = 1 \Rightarrow p_{11} = 1.96882; \quad p_0 = -1 \Rightarrow p_{11} = -1.96882.$$

15. With $g(x) = \frac{1}{\pi}\arcsin(-\frac{x}{2}) + 2$, we have $p_5 = 1.683855$.

17. One of many examples is $g(x) = \sqrt{2x - 1}$ on $\left[\frac{1}{2}, 1\right]$.

21. Replace the second sentence in the proof with: "Since g satisfies a Lipschitz condition on $[a, b]$ with a Lipschitz constant $L < 1$, we have, for each n,

$$|p_n - p| = |g(p_{n-1}) - g(p)| \leq L|p_{n-1} - p|."$$

The rest of the proof is the same, with k replaced by L.

23. With $g(t) = \dfrac{501.0625 - 201.0625e^{-0.4t}}{80.425}$ and $p_0 = 5.0$, $p_3 = 6.0028$ is within 0.01 s of the actual time.

Exercise Set 2.3

1. $p_2 = 2.60714$

3. **a.** 2.45454 **b.** 2.44444 **c.** Part (b) is better

5. **a.** For $p_0 = 2$, we have $p_5 = 2.69065$
b. For $p_0 = -3$, we have $p_3 = -2.87939$
c. For $p_0 = 0$, we have $p_4 = 0.73909$
d. For $p_0 = 0$, we have $p_3 = 0.96434$

7. Using the endpoints of the intervals as p_0 and p_1, we have
i. **a.** $p_{11} = 2.69065$ **b.** $p_7 = -2.87939$ **c.** $p_6 = 0.73909$ **d.** $p_5 = 0.96433$
ii. **a.** $p_{16} = 2.69060$ **b.** $p_6 = -2.87938$ **c.** $p_7 = 0.73908$ **d.** $p_6 = 0.96433$

9. For $p_0 = 1$, we have $p_5 = 0.589755$. The point has the coordinates $(0.589755, 0.347811)$.

11. The equation of the tangent line is

$$y - f(p_{n-1}) = f'(p_{n-1})(x - p_{n-1}).$$

To complete this problem, set $y = 0$ and solve for $x = p_n$.

13. **a.** For $p_0 = -1$ and $p_1 = 0$, we have $p_{17} = -0.04065850$, and for $p_0 = 0$ and $p_1 = 1$ we have $p_9 = 0.9623984$.
b. For $p_0 = -1$ and $p_1 = 0$, we have $p_5 = -0.04065929$, and for $p_0 = 0$ and $p_1 = 1$ we have $p_{12} = -0.04065929$.
c. For $p_0 = -0.5$, we have $p_5 = -0.04065929$, and for $p_0 = 0.5$ we have $p_{21} = 0.9623989$.

15. This formula involves the subtraction of nearly equal numbers in both the numerator and denominator if p_{n-1} and p_{n-2} are nearly equal.

17. 7

19. For $f(x) = \ln(x^2 + 1) - e^{0.4x} \cos \pi x$, we have the following roots

a. For $p_0 = -0.5$, we have $p_3 = -0.4341431$

b. For $p_0 = 0.5$, we have $p_3 = 0.4506567$
For $p_0 = 1.5$, we have $p_3 = 1.7447381$.
For $p_0 = 2.5$, we have $p_5 = 2.2383198$.
For $p_0 = 3.5$, we have $p_4 = 3.7090412$.

c. The initial approximation $n - 0.5$ is quite reasonable.

d. For $p_0 = 24.5$ we have $p_2 = 24.4998870$

21. The two numbers are approximately 6.512849 and 13.487151.

23. The borrower can afford to pay at most 8.10%.

25. a. `solve(3^(3*x+1)-7*5^(2*x),x)` and `fsolve(3^(3*x+1)-7*5^(2*x),x)` both fail.

b. `plot(3^(3*x+1)-7*5^(2*x),x=a..b)` generally yields no useful information. However, $a = 10.5$ and $b = 11.5$ in the plot command show that $f(x)$ has a root near $x = 11$.

c. With $p_0 = 11$, $p_5 = 11.0094386442681716$ is accurate to 10^{-16}.

d. $p = \ln \frac{3}{7} / \ln \frac{25}{27}$

27. We have $P_L = 265816$, $c = -0.75658125$, and $k = 0.045017502$. The 1980 population is $P(30) = 222{,}248{,}320$, and the 2000 population is $P(50) = 246{,}200{,}780$.

29. Using $p_0 = 0.5$ and $p_1 = 0.9$, the Secant method gives $p_5 = 0.842$.

Exercise Set 2.4

1. a. For $p_0 = 0.5$, we have $p_{13} = 0.567135$.

b. For $p_0 = -1.5$, we have $p_{23} = -1.414325$.

c. For $p_0 = 0.5$, we have $p_{22} = 0.641166$.

d. For $p_0 = -0.5$, we have $p_{23} = -0.183274$.

3. Newton's method with $p_0 = -0.5$ gives $p_{13} = -0.169607$. Modified Newton's method in Eq. (2.10) with $p_0 = -0.5$ gives $p_{11} = -0.169607$.

5. For $k > 0$,

$$\lim_{n\to\infty} \frac{|p_{n+1} - 0|}{|p_n - 0|} = \lim_{n\to\infty} \frac{1/(n+1)^k}{1/n^k} = \lim_{n\to\infty} \left(\frac{n}{n+1}\right)^k = 1,$$

so the convergence is linear. We need to have $N > 10^{m/k}$.

7. Typical examples are **a.** $p_n = 10^{-3^n}$ and **b.** $p_n = 10^{-\alpha^n}$.

9. This follows from the fact that

$$\lim_{n\to\infty} \frac{|(b-a)/2^{n+1}|}{|(b-a)/2^n|} = \frac{1}{2}.$$

11. If $\dfrac{|p_{n+1} - p|}{|p_n - p|^3} = 0.75$ and $|p_0 - p| = 0.5$, then

$$|p_n - p| = (0.75)^{(3^n-1)/2} |p_0 - p|^{3^n}.$$

To have $|p_n - p| \le 10^{-8}$ requires that $n \ge 3$.

Exercise Set 2.5

1. The results are listed in the following table.

	a	**b**	**c**	**d**
$\hat{p}_0$	0.258684	0.907859	0.548101	0.731385
$\hat{p}_1$	0.257613	0.909568	0.547915	0.736087
$\hat{p}_2$	0.257536	0.909917	0.547847	0.737653
$\hat{p}_3$	0.257531	0.909989	0.547823	0.738469
$\hat{p}_4$	0.257530	0.910004	0.547814	0.738798
$\hat{p}_5$	0.257530	0.910007	0.547810	0.738958

3. $p_0^{(1)} = 0.826427$

5. $p_1^{(0)} = 1.5$ or $p_1^{(0)} = 0$

7. For $g(x) = \sqrt{1 + \frac{1}{x}}$ and $p_0 = 1$, we have $p_3 = 1.32472$.

9. For $g(x) = 0.5(x + \frac{3}{x})$ and $p_0 = 0.5$, we have $p_4 = 1.73205$.

11. Aitken's Δ^2 method gives **a.** $\hat{p}_{10} = 0.0\overline{45}$ **b.** $\hat{p}_2 = 0.0363$.

13.

$$\frac{|p_{n+1} - p_n|}{|p_n - p|} = \frac{|p_{n+1} - p + p - p_n|}{|p_n - p|} = \left|\frac{p_{n+1} - p}{p_n - p} - 1\right|.$$

So

$$\lim_{n\to\infty} \frac{|p_{n+1} - p_n|}{|p_n - p|} = \lim_{n\to\infty} \left|\frac{p_{n+1} - p}{p_n - p} - 1\right| = 1.$$

15. a. *Hint:* First show that $p_n - p = [-e^{\xi}/(n+1)!]x^{n+1}$, where ξ is between 0 and 1.

b.

n	p_n	$\hat{p}_n$
0	1	3
1	2	2.75
2	2.5	$2.7\overline{2}$
3	$2.\overline{6}$	2.71875
4	$2.708\overline{3}$	$2.718\overline{3}$
5	$2.71\overline{6}$	2.7182870
6	$2.7180\overline{5}$	2.7182823
7	2.7182539	2.7182818
8	2.7182787	2.7182818
9	2.7182815	
10	2.7182818	

Exercise Set 2.6

1. a. For $p_0 = 1$, we have $p_{22} = 2.69065$.

b. For $p_0 = 1$, we have $p_5 = 0.53209$; for $p_0 = -1$, we have $p_3 = -0.65270$, and for $p_0 = -3$, we have $p_3 = -2.87939$.

c. For $p_0 = 1$, we have $p_5 = 1.32472$.

d. For $p_0 = 1$, we have $p_4 = 1.12412$; and for $p_0 = 0$, we have $p_8 = -0.87605$.

e. For $p_0 = 0$, we have $p_6 = -0.47006$; for $p_0 = -1$, we have $p_4 = -0.88533$; and for $p_0 = -3$, we have $p_4 = -2.64561$.

f. For $p_0 = 0$, we have $p_{10} = 1.49819$.

3. The following table lists the initial approximation and the roots.

	p_0	p_1	p_2	Approximate Roots	Complex Conjugate Roots
a	-1	0	1	$p_7 = -0.34532 - 1.31873i$	$-0.34532 + 1.31873i$
	0	1	2	$p_6 = 2.69065$	
b	0	1	2	$p_6 = 0.53209$	
	1	2	3	$p_9 = -0.65270$	
	-2	-3	-2.5	$p_4 = -2.87939$	
c	0	1	2	$p_5 = 1.32472$	
	-2	-1	0	$p_7 = -0.66236 - 0.56228i$	$-0.66236 + 0.56228i$
d	0	1	2	$p_5 = 1.12412$	
	2	3	4	$p_{12} = -0.12403 + 1.74096i$	$-0.12403 - 1.74096i$
	-2	0	-1	$p_5 = -0.87605$	
e	0	1	2	$p_{10} = -0.88533$	
	1	0	-0.5	$p_5 = -0.47006$	
	-1	-2	-3	$p_5 = -2.64561$	
f	0	1	2	$p_6 = 1.49819$	
	-1	-2	-3	$p_{10} = -0.51363 - 1.09156i$	$-0.51363 + 1.09156i$
	1	0	-1	$p_8 = 0.26454 - 1.32837i$	$0.26454 + 1.32837i$

5. a. The roots are 1.244, 8.847, and -1.091, and the critical points are 0 and 6.

b. The roots are 0.5798, 1.521, 2.332, and -2.432, and the critical points are 1, 2.001, and -1.5.

7. Let $c_1 = (2 + \frac{2}{9}\sqrt{129})^{-1/3}$ and $c_2 = (2 + \frac{2}{9}\sqrt{129})^{1/3}$. The roots are $c_2 - \frac{4}{3}c_1$, $-\frac{1}{2}c_2 + \frac{2}{3}c_1 + \frac{1}{2}\sqrt{3}(c_2 + \frac{4}{3}c_1)i$, and $-\frac{1}{2}c_2 + \frac{2}{3}c_1 - \frac{1}{2}\sqrt{3}(c_2 + \frac{4}{3}c_1)i$.

11. The minimal material is approximately 573.64895 cm^2.

Exercise Set 3.1

1. a. $P_1(x) = -0.148878x + 1$; $P_2(x) = -0.452592x^2 - 0.0131009x + 1$; $P_1(0.45) = 0.933005$; $|f(0.45) - P_1(0.45)| = 0.032558$; $P_2(0.45) = 0.902455$; $|f(0.45) - P_2(0.45)| = 0.002008$

b. $P_1(x) = 0.467251x + 1$; $P_2(x) = -0.0780026x^2 + 0.490652x + 1$; $P_1(0.45) = 1.210263$; $|f(0.45) - P_1(0.45)| = 0.006104$ $P_2(0.45) = 1.204998$; $|f(0.45) - P_2(0.45)| = 0.000839$

c. $P_1(x) = 0.874548x$; $P_2(x) = -0.268961x^2 + 0.955236x$ $P_1(0.45) = 0.393546$; $|f(0.45) - P_1(0.45)| = 0.0212983$ $P_2(0.45) = 0.375392$; $|f(0.45) - P_2(0.45)| = 0.003828$

d. $P_1(x) = 1.031121x$; $P_2(x) = 0.615092x^2 + 0.846593x$ $P_1(0.45) = 0.464004$; $|f(0.45) - P_1(0.45)| = 0.019051$; $P_2(0.45) = 0.505523$; $|f(0.45) - P_2(0.45)| = 0.022468$

3. a.

n	$x_0, x_1, \ldots, x_n$	$P_n(8.4)$
1	8.3, 8.6	17.87833
2	8.3, 8.6, 8.7	17.87716
3	8.3, 8.6, 8.7, 8.1	17.87714

b.

n	$x_0, x_1, \ldots, x_n$	$P_n(-1/3)$
1	-0.5, -0.25	0.21504167
2	-0.5, -0.25, 0.0	0.16988889
3	-0.5, -0.25, 0.0, -0.75	0.17451852

c.

n	$x_0, x_1, \ldots, x_n$	$P_n(0.25)$
1	0.2, 0.3	-0.13869287
2	0.2, 0.3, 0.4	-0.13259734
3	0.2, 0.3, 0.4, 0.1	-0.13277477

d.

n	$x_0, x_1, \ldots, x_n$	$P_n(0.9)$
1	0.8, 1.0	0.44086280
2	0.8, 1.0, 0.7	0.43841352
3	0.8, 1.0, 0.7, 0.6	0.44198500

5. $\sqrt{3} \approx P_4\left(\frac{1}{2}\right) = 1.708\overline{3}$

7. a.

n	Actual Error	Error Bound
1	0.00118	0.00120
2	1.367×10^{-5}	1.452×10^{-5}

b.

n	Actual Error	Error Bound
1	4.0523×10^{-2}	4.5153×10^{-2}
2	4.6296×10^{-3}	4.6296×10^{-3}

c.

n	Actual Error	Error Bound
1	5.9210×10^{-3}	6.0971×10^{-3}
2	1.7455×10^{-4}	1.8128×10^{-4}

d.

n	Actual Error	Error Bound
1	2.7296×10^{-3}	1.4080×10^{-2}
2	5.1789×10^{-3}	9.2215×10^{-3}

9. $y = 4.25$

11. $f(1.09) \approx 0.2826$. The actual error is 4.3×10^{-5}, and an error bound is 7.4×10^{-6}. The discrepancy is due to the fact that the data are given to only four decimal places and only four-digit arithmetic is used.

13. $P_2 = f(0.7) = 6.4$

15. a. $P_2(x) = -11.22388889x^2 + 3.810500000x + 1$, and an error bound is 0.11371294.

b. $P_2(x) = -0.1306344167x^2 + 0.8969979335x - 0.63249693$, and an error bound is 9.45762×10^{-4}.

c. $P_3(x) = 0.1970056667x^3 - 1.06259055x^2 + 2.532453189x - 1.666868305$, and an error bound is 10^{-4}.

d. $P_3(x) = -0.07932x^3 - 0.545506x^2 + 1.0065992x + 1$, and an error bound is 1.591376×10^{-3}.

17. The largest possible step size is 0.004291932, so 0.004 would be a reasonable choice.

19. $P_{0,1,2,3}(2.5) = 2.875$

21. The first ten terms of the sequence are 0.038462, 0.333671, 0.116605, −0.371760, −0.0548919, 0.605935, 0.190249, −0.513353, −0.0668173, and 0.448335. Since $f(1 + \sqrt{10}) = 0.0545716$, the sequence does not appear to converge.

25. a. Sample 1: $P_6(x) = 6.67 - 42.6434x + 16.1427x^2 - 2.09464x^3 + 0.126902x^4 - 0.00367168x^5 + 0.0000409458x^6$;
Sample 2:
$P_6(x) = 6.67 - 5.67821x + 2.91281x^2 - 0.413799x^3 + 0.0258413x^4 - 0.000752546x^5 + 0.00000836160x^6$

b. Sample 1: 42.71 mg; Sample 2: 19.42 mg

27. Since $g(x) = g(x_0) = 0$, there exists a number ξ_1 between x and x_0 for which $g'(\xi_1) = 0$. Also, $g'(x_0) = 0$, so there exists a number ξ_2 between x_0 and ξ_1 for which $g''(\xi_2) = 0$. The process is continued by induction to show that a number ξ_{n+1} between x_0 and ξ_n exists with $g^{(n+1)}(\xi_{n+1}) = 0$. The error formula for Taylor polynomials follows.

29. a. (i) $B_3(x) = x$ **(ii)** $B_3(x) = 1$

Exercise Set 3.2

1. a. $P_1(x) = 16.9441 + 3.1041(x - 8.1)$; $P_1(8.4) = 17.87533$; $P_2(x) = P_1(x) + 0.06(x - 8.1)(x - 8.3)$; $P_2(8.4) = 17.87713$; $P_3(x) = P_2(x) - 0.00208333(x - 8.1)(x - 8.3)(x - 8.6)$; $P_3(8.4) = 17.87714$

b. $P_1(x) = -0.1769446 + 1.9069687(x - 0.6)$; $P_1(0.9) = 0.395146$; $P_2(x) = P_1(x) + 0.959224(x - 0.6)(x - 0.7)$; $P_2(0.9) = 0.4526995$; $P_3(x) = P_2(x) - 1.785741(x - 0.6)(x - 0.7)(x - 0.8)$; $P_3(0.9) = 0.4419850$

3. In the following equations we have $s = \frac{1}{h}(x - x_n)$.

a. $P_1(s) = 1.101 + 0.7660625s$; $f(-\frac{1}{3}) \approx P_1(-\frac{4}{3}) = 0.07958333$;
$P_2(s) = P_1(s) + 0.406375s(s + 1)/2$; $f(-\frac{1}{3}) \approx P_2(-\frac{4}{3}) = 0.1698889$;
$P_3(s) = P_2(s) + 0.09375s(s + 1)(s + 2)/6$; $f(-\frac{1}{3}) \approx P_3(-\frac{4}{3}) = 0.1745185$

b. $P_1(s) = 0.2484244 + 0.2418235s$; $f(0.25) \approx P_1(-1.5) = -0.1143108$; $P_2(s) = P_1(s) - 0.04876419s(s + 1)/2$; $f(0.25) \approx P_2(-1.5) = -0.1325973$; $P_3(s) = P_2(s) - 0.00283891s(s + 1)(s + 2)/6$; $f(0.25) \approx P_3(-1.5) = -0.1327748$

5. a. $f(0.05) \approx 1.05126$ **b.** $f(0.65) \approx 1.91555$ **c.** $f(0.43) \approx 1.53725$

7. a. $P(-2) = Q(-2) = -1$, $P(-1) = Q(-1) = 3$, $P(0) = Q(0) = 1$, $P(1) = Q(1) = -1$, $P(2) = Q(2) = 3$

b. The format of the polynomial is not unique. If $P(x)$ and $Q(x)$ are expanded, they are identical. There is only one interpolating polynomial of degree less than or equal to four for the given data. However, it can be expressed in an infinite number of ways.

9. The coefficient of x^2 is 3.5.

11. The approximation to $f(0.3)$ should be increased by 5.9375.

13. $f[x_0] = f(x_0) = 1$, $f[x_1] = f(x_1) = 3$, $f[x_0, x_1] = 5$

15. Since $f[x_2] = f[x_0] + f[x_0, x_1](x_2 - x_0) + a_2(x_2 - x_0)(x_2 - x_1)$,

$$a_2 = \frac{f[x_2] - f[x_0]}{(x_2 - x_0)(x_2 - x_1)} - \frac{f[x_0, x_1]}{(x_2 - x_1)}.$$

This simplifies to $f[x_0, x_1, x_2]$.

17. Let $\tilde{P}(x) = f[x_{i_0}] + \sum_{k=1}^{n} f[x_{i_0}, \ldots, x_{i_k}](x - x_{i_0}) \cdots (x - x_{i_k})$ and $\hat{P}(x) = f[x_0] + \sum_{k=1}^{n} f[x_0, \ldots, x_k](x - x_0) \cdots (x - x_k)$. The polynomial $\tilde{P}(x)$ interpolates $f(x)$ at the nodes $x_{i_0}, \ldots, x_{i_n}$ and the polynomial $\hat{P}(x)$ interpolates $f(x)$ at the nodes $x_0, \ldots, x_n$. Since both sets of nodes are the same and the interpolating polynomial is unique, we have $\tilde{P}(x) = \hat{P}(x)$. The coefficient of x^n in $\tilde{P}(x)$ is $f[x_{i_0}, \ldots, x_{i_n}]$ and the coefficient of x^n in $\hat{P}(x)$ is $f[x_0, \ldots, x_n]$. Thus, $f[x_{i_0}, \ldots, x_{i_n}] = f[x_0, \ldots, x_n]$.

Exercise Set 3.3

1. The coefficients for the polynomials in divided-difference form are given in the following tables. For example, the polynomial in part (a) is

$$H_3(x) = 17.56492 + 3.116256(x - 8.3) + 0.05948(x - 8.3)^2 - 0.00202222(x - 8.3)^2(x - 8.6).$$

a	b	c	d
17.56492	0.022363362	−0.02475	−0.62049958
3.116256	2.1691753	0.751	3.5850208
0.05948	0.01558225	2.751	−2.1989182
−0.00202222	−3.2177925	1	−0.490447
		0	0.037205
		0	0.040475
			−0.0025277777
			0.0029629628

3. a. We have $\sin 0.34 \approx H_5(0.34) = 0.33349$.

b. The formula gives an error bound of 3.05×10^{-14}, but the actual error is 2.91×10^{-6}. The discrepancy is due to the fact that the data are given to only five decimal places.

c. We have $\sin 0.34 \approx H_7(0.34) = 0.33350$. Although the error bound is now 5.4×10^{-20}, the accuracy of the given data dominates the calculations. This result is actually less accurate than the approximation in part (b), since $\sin 0.34 = 0.333487$.

5. For 2(a) we have an error bound of 5.9×10^{-8}. The error bound for 2(c) is 0 since $f^{(n)}(x) \equiv 0$ for $n > 3$.

7. The Hermite polynomial generated from these data is

$$\begin{aligned} H_9(x) = {} & 75x + 0.222222x^2(x-3) - 0.0311111x^2(x-3)^2 - 0.00644444x^2(x-3)^2(x-5) \\ & + 0.00226389x^2(x-3)^2(x-5)^2 - 0.000913194x^2(x-3)^2(x-5)^2(x-8) \\ & + 0.000130527x^2(x-3)^2(x-5)^2(x-8)^2 - 0.0000202236x^2(x-3)^2(x-5)^2(x-8)^2(x-13). \end{aligned}$$

a. The Hermite polynomial predicts a position of $H_9(10) = 743$ ft and a speed of $H_9'(10) = 48$ ft/s. Although the position approximation is reasonable, the low speed prediction is suspect.

b. To find the first time the speed exceeds 55 mi/h $= 80.\overline{6}$ ft/s, we solve for the smallest value of t in the equation $80.\overline{6} = H_9'(x)$. This gives $x \approx 5.6488092$.

c. The estimated maximum speed is $H_9'(12.37187) = 119.423$ ft/s ≈ 81.425 mi/h.

9. Let

$$H(x) = f[z_0] + f[z_0, z_1](x - x_0) + f[z_0, z_1, z_2](x - x_0)^2 + f[z_0, z_1, z_2, z_3](x - x_0)^2(x - x_1).$$

Substituting $f[z_0] = f(x_0)$, $f[z_0, z_1] = f'(x_0)$,

$$f[z_0, z_1, z_2] = \frac{f(x_1) - f(x_0) - f'(x_0)(x_1 - x_0)}{x_1 - x_0}$$

$$f[z_0, z_1, z_2, z_3] = \frac{f'(x_1)(x_1 - x_0) - 2f(x_1) + 2f(x_0) + f'(x_0)(x_1 - x_0)}{(x_1 - x_0)^3}$$

into $H(x)$ and simplifying gives

$$\begin{aligned} H(x) = f(x_0) + f'(x_0)(x - x_0) &+ \frac{f(x_1) - f(x_0) - f'(x_0)(x_1 - x_0)}{(x_1 - x_0)^2}(x - x_0)^2 \\ &+ \frac{f'(x_1)(x_1 - x_0) - 2f(x_1) + 2f(x_0) + f'(x_0)(x_1 - x_0)}{(x_1 - x_0)^3}(x - x_0)^2(x - x_1). \end{aligned}$$

Thus, $H(x_0) = f(x_0)$, and

$$\begin{aligned} H(x_1) &= f(x_0) + f'(x_0)(x_1 - x_0) + [f(x_1) - f(x_0) - f'(x_0)(x_1 - x_0)] \\ &= f(x_1). \end{aligned}$$

Further,

$$\begin{aligned} H'(x) = f'(x_0) &+ 2\frac{f(x_1) - f(x_0) - f'(x_0)(x_1 - x_0)}{(x_1 - x_0)^2}(x - x_0) \\ &+ \frac{f'(x_1)(x_1 - x_0) - 2f(x_1) + 2f(x_0) + f'(x_0)(x_1 - x_0)}{(x_1 - x_0)^3}[2(x - x_0)(x - x_1) + (x - x_0)^2] \end{aligned}$$

so

$$\begin{aligned} H'(x_0) &= f'(x_0) \quad \text{and} \\ H'(x_1) &= f'(x_0) + \frac{2f(x_1)}{x_1 - x_0} - \frac{2f(x_0)}{x_1 - x_0} - 2f'(x_0) + f'(x_1) - \frac{2f(x_1)}{x_1 - x_0} + \frac{2f(x_0)}{x_1 - x_0} + f'(x_0) \\ &= f'(x_1). \end{aligned}$$

Thus, H satisfies the requirements of the cubic Hermite polynomial H_3 and the uniqueness of H_3 implies $H_3 = H$.

Exercise Set 3.4

1. $S(x) = x$ on $[0, 2]$

3. The equations of the respective free cubic splines are given by

$$S(x) = S_i(x) = a_i + b_i(x - x_i) + c_i(x - x_i)^2 + d_i(x - x_i)^3,$$

for x in $[x_i, x_{i+1}]$, where the coefficients are given in the following tables.

a.

i	a_i	b_i	c_i	d_i
0	17.564920	3.13410000	0.00000000	0.00000000

b.

i	a_i	b_i	c_i	d_i
0	0.22363362	2.17229175	0.00000000	0.00000000

c.

i	a_i	b_i	c_i	d_i
0	−0.02475000	1.03237500	0.00000000	6.50200000
1	0.33493750	2.25150000	4.87650000	−6.50200000

d.

i	a_i	b_i	c_i	d_i
0	−0.62049958	3.45508693	0.00000000	−8.9957933
1	−0.28398668	3.18521313	−2.69873800	−0.94630333
2	0.00660095	2.61707643	−2.98262900	9.9420966

5. The equations of the respective clamped cubic splines are given by

$$s(x) = s_i(x) = a_i + b_i(x - x_i) + c_i(x - x_i)^2 + d_i(x - x_i)^3,$$

for x in $[x_i, x_{i+1}]$, where the coefficients are given in the following tables.

a.

i	a_i	b_i	c_i	d_i
0	17.564920	3.1162560	0.0600867	−0.00202222

b.

i	a_i	b_i	c_i	d_i
0	0.22363362	2.1691753	0.65914075	−3.2177925

c.

i	a_i	b_i	c_i	d_i
0	−0.02475000	0.75100000	2.5010000	1.0000000
1	0.33493750	2.18900000	3.2510000	1.0000000

d.

i	a_i	b_i	c_i	d_i
0	−0.62049958	3.5850208	−2.1498407	−0.49077413
1	−0.28398668	3.1403294	−2.2970730	−0.47458360
2	0.006600950	2.6666773	−2.4394481	−0.44980146

7. $a = 2, b = -1, c = -3, d = 1$

9. $B = \frac{1}{4}, D = \frac{1}{4}, b = -\frac{1}{2}, d = \frac{1}{4}$

11. The equation of the spline is

$$S(x) = S_i(x) = a_i + b_i(x - x_i) + c_i(x - x_i)^2 + d_i(x - x_i)^3$$

on the interval $[x_i, x_{i+1}]$, where the coefficients are given in the following table.

x_i	a_i	b_i	c_i	d_i
0	1.0	−0.7573593	0.0	−6.627417
0.25	0.7071068	−2.0	−4.970563	6.627417
0.5	0.0	−3.242641	0.0	6.627417
0.75	−0.7071068	−2.0	4.970563	−6.627417

$\int_0^1 S(x)\,dx = 0.000000$, $S'(0.5) = -3.24264$, and $S''(0.5) = 0.0$

13. The equation of the spline is

$$s(x) = s_i(x) = a_i + b_i(x - x_i) + c_i(x - x_i)^2 + d_i(x - x_i)^3$$

on the interval $[x_i, x_{i+1}]$, where the coefficients are given in the following table.

x_i	a_i	b_i	c_i	d_i
0	1.0	0.0	−5.193321	2.028118
0.25	0.7071068	−2.216388	−3.672233	4.896310
0.5	0.0	−3.134447	0.0	4.896310
0.75	−0.7071068	−2.216388	3.672233	2.028118

$\int_0^1 s(x)\,dx = 0.000000$, $s'(0.5) = -3.13445$, and $s''(0.5) = 0.0$

15. Let $f(x) = a + bx + cx^2 + dx^3$. Clearly, f satisfies properties (a), (c), (d), (e) of Definition 3.10 and f interpolates itself for any choice of $x_0, \ldots, x_n$. Since (ii) of (f) in Definition 3.10 holds, f must be its own clamped cubic spline. However, $f''(x) = 2c + 6dx$ can be zero only at $x = -c/3d$. Thus, part (i) of (f) in Definition 3.10 cannot hold at two values x_0 and x_n. Thus, f cannot be a natural cubic spline.

17. The piecewise linear approximation to f is given by

$$F(x) = \begin{cases} 20(e^{0.1} - 1)x + 1, & \text{for } x \text{ in } [0, 0.05] \\ 20(e^{0.2} - e^{0.1})x + 2e^{0.1} - e^{0.2}, & \text{for } x \text{ in } (0.05, 1]. \end{cases}$$

We have

$$\int_0^{0.1} F(x)\,dx = 0.1107936 \quad \text{and} \quad \int_0^{0.1} f(x)\,dx = 0.1107014.$$

21. **a.** On $[0, 0.05]$ we have $s(x) = 1.000000 + 1.999999x + 1.998302x^2 + 1.401310x^3$, and on $(0.05, 0.1]$ we have $s(x) = 1.105170 + 2.210340(x - 0.05) + 2.208498(x - 0.05)^2 + 1.548758(x - 0.05)^3$.

b. $\int_0^{0.1} s(x)\,dx = 0.110701$

c. 1.6×10^{-7}

d. On $[0, 0.05]$ we have $S(x) = 1 + 2.04811x + 22.12184x^3$, and on $(0.05, 0.1]$ we have $S(x) = 1.105171 + 2.214028(x - 0.05) + 3.318277(x - 0.05)^2 - 22.12184(x - 0.05)^3$. $S(0.02) = 1.041139$ and $S(0.02) = 1.040811$.

23.

$$S(x) = \begin{cases} 2x - x^2, & 0 \le x \le 1 \\ 1 + (x - 1)^2, & 1 \le x \le 2 \end{cases}$$

25. The spline has the equation $s(x) = s_i(x) = a_i + b_i(x - x_i) + c_i(x - x_i)^2 + d_i(x - x_i)^3$ on $[x_i, x_{i+1}]$, where the coefficients are given in the following table.

x_i	a_i	b_i	c_i	d_i
0	0	75	−0.659292	0.219764
3	225	76.9779	1.31858	−0.153761
5	383	80.4071	0.396018	−0.177237
8	623	77.9978	−1.19912	0.0799115

The spline predicts a position of $s(10) = 774.84$ ft and a speed of $s'(10) = 74.16$ ft/s. To maximize the speed, we find the single critical points of $s'(x)$ and compare the values of $s(x)$ at these points and the endpoints. We find that max $s'(x) = s'(5.7448) = 80.7$ ft/s $= 55.02$ mi/h. The speed 55 mi/h was first exceeded at approximately 5.5 s.

27. The equation of the spline is

$$S(x) = S_i(x) = a_i + b_i(x - x_i) + c_i(x - x_i)^2 + d_i(x - x_i)^3$$

on the interval $[x_i, x_{i+1}]$, where the coefficients are given in the following table.

	Sample 1				Sample 2			
x_i	a_i	b_i	c_i	d_i	a_i	b_i	c_i	d_i
0	6.67	−0.44687	0	0.06176	6.67	1.6629	0	−0.00249
6	17.33	6.2237	1.1118	−0.27099	16.11	1.3943	−0.04477	−0.03251
10	42.67	2.1104	−2.1401	0.28109	18.89	−0.52442	−0.43490	0.05916
13	37.33	−3.1406	0.38974	−0.01411	15.00	−1.5365	0.09756	0.00226
17	30.10	−0.70021	0.22036	−0.02491	10.56	−0.64732	0.12473	−0.01113
20	29.31	−0.05069	−0.00386	0.00016	9.44	−0.19955	0.02453	−0.00102

29. The three natural splines have equations of the form

$$S_i(x) = a_i + b_i(x - x_i) + c_i(x - x_i)^2 + d_i(x - x_i)^3$$

on $[x_i, x_{i+1}]$ where the values of the coefficients are given in the following tables.

Spline 1

i	x_i	$a_i = f(x_i)$	b_i	c_i	d_i
0	1	3.0	0.786	0.0	−0.086
1	2	3.7	0.529	−0.257	0.034
2	5	3.9	−0.086	0.052	0.334
3	6	4.2	1.019	1.053	−0.572
4	7	5.7	1.408	−0.664	0.156
5	8	6.6	0.547	−0.197	0.024
6	10	7.1	0.049	−0.052	−0.003
7	13	6.7	−0.342	−0.078	0.007
8	17	4.5			

Spline 2

i	x_i	$a_i = f(x_i)$	b_i	c_i	d_i
0	17	4.5	1.106	0.0	−0.030
1	20	7.0	0.289	−0.272	0.025
2	23	6.1	−0.660	−0.044	0.204
3	24	5.6	−0.137	0.567	−0.230
4	25	5.8	0.306	−0.124	−0.089
5	27	5.2	−1.263	−0.660	0.314
6	27.7	4.1			

Spline 3

i	x_i	$a_i = f(x_i)$	b_i	c_i	d_i
0	27.7	4.1	0.749	0.0	−0.910
1	28	4.3	0.503	−0.819	0.116
2	29	4.1	−0.787	−0.470	0.157
3	30	3.0			

Exercise Set 3.5

1. a. $x(t) = -10t^3 + 14t^2 + t, \quad y(t) = -2t^3 + 3t^2 + t$

b. $x(t) = -10t^3 + 14.5t^2 + 0.5t, \quad y(t) = -3t^3 + 4.5t^2 + 0.5t$

c. $x(t) = -10t^3 + 14t^2 + t, \quad y(t) = -4t^3 + 5t^2 + t$

d. $x(t) = -10t^3 + 13t^2 + 2t, \quad y(t) = 2t$

3. a. $x(t) = -11.5t^3 + 15t^2 + 1.5t + 1, \quad y(t) = -4.25t^3 + 4.5t^2 + 0.75t + 1$

b. $x(t) = -6.25t^3 + 10.5t^2 + 0.75t + 1, \quad y(t) = -3.5t^3 + 3t^2 + 1.5t + 1$

c. For t between $(0, 0)$ and $(4, 6)$ we have

$$x(t) = -5t^3 + 7.5t^2 + 1.5t, \quad y(t) = -13.5t^3 + 18t^2 + 1.5t,$$

and for t between $(4, 6)$ and $(6, 1)$ we have

$$x(t) = -5.5t^3 + 6t^2 + 1.5t + 4, \quad y(t) = 4t^3 - 6t^2 - 3t + 6.$$

d. For t between $(0, 0)$ and $(2, 1)$ we have

$$x(t) = -5.5t^3 + 6t^2 + 1.5t, \quad y(t) = -0.5t^3 + 1.5t,$$

for t between $(2, 1)$ and $(4, 0)$ we have

$$x(t) = -4t^3 + 3t^2 + 3t + 2, \quad y(t) = -t^3 + 1,$$

and for t between $(4, 0)$ and $(6, -1)$ we have

$$x(t) = -8.5t^3 + 13.5t^2 - 3t + 4, \quad y(t) = -3.25t^3 + 5.25t^2 - 3t.$$

Exercise Set 4.1

1. From the forward-backward difference Formula (4.1) we have the following approximations:

a. $f'(0.5) \approx 0.8520$, $f'(0.6) \approx 0.8520$, $f'(0.7) \approx 0.7960$

b. $f'(0.0) \approx 3.7070$, $f'(0.2) \approx 3.1520$, $f'(0.4) \approx 3.1520$

3. For the endpoints of the tables we use Formula (4.4). The other approximations come from Formula (4.5)

a. $f'(1.1) \approx 17.769705$, $f'(1.2) \approx 22.193635$, $f'(1.3) \approx 27.107350$, $f'(1.4) \approx 32.150850$

b. $f'(8.1) \approx 3.092050$, $f'(8.3) \approx 3.116150$, $f'(8.5) \approx 3.139975$, $f'(8.7) \approx 3.163525$

c. $f'(2.9) \approx 5.101375$, $f'(3.0) \approx 6.654785$, $f'(3.1) \approx 8.216330$, $f'(3.2) \approx 9.786010$

d. $f'(2.0) \approx 0.13533150$, $f'(2.1) \approx -0.09989550$, $f'(2.2) \approx -0.3298960$, $f'(2.3) \approx -0.5546700$

5. The approximations and the formulas used are:

a. $f'(2.1) \approx 3.899344$ from (4.7) $\quad f'(2.2) \approx 2.876876$ from (4.7) $\quad f'(2.3) \approx 2.249704$ from (4.6)
$f'(2.4) \approx 1.837756$ from (4.6) $\quad f'(2.5) \approx 1.544210$ from (4.7) $\quad f'(2.6) \approx 1.355496$ from (4.7)

b. $f'(-3.0) \approx -5.877358$ from (4.7) $\quad f'(-2.8) \approx -5.468933$ from (4.7) $\quad f'(-2.6) \approx -5.059884$ from (4.6)
$f'(-2.4) \approx -4.650223$ from (4.6) $\quad f'(-2.2) \approx -4.239911$ from (4.7) $\quad f'(-2.0) \approx -3.828853$ from (4.7)

7. $f'(3) \approx \frac{1}{12}[f(1) - 8f(2) + 8f(4) - f(5)] = 0.21062$ with an error bound given by

$$\max_{1 \le x \le 5} \frac{|f^{(5)}(x)|h^4}{30} \le \frac{23}{30} = 0.7\overline{6}.$$

9. From the forward-backward formula (4.1) we have the following approximations:

a. $f'(0.5) \approx 0.852$, $f'(0.6) \approx 0.852$, $f'(0.7) \approx 0.7960$

b. $f'(0.0) \approx 3.707$, $f'(0.2) \approx 3.153$, $f'(0.4) \approx 3.153$

11. For the endpoints of the tables we use Formula (4.7). The other approximations come from Formula (4.6).

a. $f'(2.1) \approx 3.884 \quad f'(2.2) \approx 2.896 \quad f'(2.3) \approx 2.249 \quad f'(2.4) \approx 1.836 \quad f'(2.5) \approx 1.550 \quad f'(2.6) \approx 1.348$

b. $f'(-3.0) \approx -5.883$ $f'(-2.8) \approx -5.467$ $f'(-2.6) \approx -5.059$ $f'(-2.4) \approx -4.650$ $f'(-2.2) \approx -4.208$ $f'(-2.0) \approx -3.875$

13. The approximation is -4.8×10^{-9}. $f''(0.5) = 0$. The error bound is 0.35874. The method is very accurate since the function is symmetric about $x = 0.5$.

15. **a.** $f'(0.2) \approx -0.1951027$ **b.** $f'(1.0) \approx -1.541415$ **c.** $f'(0.6) \approx -0.6824175$

17. $f'(0.4) \approx -0.4249840$ and $f'(0.8) \approx -1.032772$.

19. Using three point formulas gives the following table:

Time	0	3	5	8	10	13
Speed	79	82.4	74.2	76.8	69.4	71.2

21. The approximations eventually become zero since the numerator becomes zero.

23. Since $e'(h) = (-\varepsilon/h^2) + (hM/3)$, we have $e'(h) = 0$ if and only if $h = \sqrt[3]{3\varepsilon/M}$. Also, $e'(h) < 0$ if $h < \sqrt[3]{3\varepsilon/M}$ and $e'(h) > 0$ if $h > \sqrt[3]{3\varepsilon/M}$, so an absolute minimum for $e(h)$ occurs at $h = \sqrt[3]{3\varepsilon/M}$.

Exercise Set 4.2

1. **a.** $f'(1) \approx 1.0000109$ **b.** $f'(0) \approx 2.0000000$ **c.** $f'(1.05) \approx 2.2751459$ **d.** $f'(2.3) \approx -19.646799$

3. **a.** $f'(1) \approx 1.001$ **b.** $f'(0) \approx 1.999$ **c.** $f'(1.05) \approx 2.283$ **d.** $f'(2.3) \approx -19.61$

5. $\int_0^{\pi} \sin x\,dx \approx 1.999999$

9. Let $N_2(h) = N\left(\frac{h}{3}\right) + \left(\frac{N(h/3)-N(h)}{2}\right)$ and $N_3(h) = N_2\left(\frac{h}{3}\right) + \left(\frac{N_2(h/3)-N_2(h)}{8}\right)$. Then $N_3(h)$ is an $O(h^3)$ approximation to M.

11. Let $N(h) = (1+h)^{1/h}$, $N_2(h) = 2N\left(\frac{h}{2}\right) - N(h)$, $N_3(h) = N_2\left(\frac{h}{2}\right) + \frac{1}{3}(N_2\left(\frac{h}{2}\right) - N_2(h))$.

a. $N(0.04) = 2.665836331$, $N(0.02) = 2.691588029$, $N(0.01) = 2.704813829$

b. $N_2(0.04) = 2.717339727$, $N_2(0.02) = 2.718039629$. The $O(h^3)$ approximation is $N_3(0.04) = 2.718272931$.

c. Yes, since the errors seem proportional to h for $N(h)$, to h^2 for $N_2(h)$, and to h^3 for $N_3(h)$.

15. **c.**

k	4	8	16	32	64	128	256	512
p_k	$2\sqrt{2}$	3.0614675	3.1214452	3.1365485	3.1403312	3.1412723	3.1415138	3.1415729
P_k	4	3.3137085	3.1825979	3.1517249	3.144184	3.1422236	3.1417504	3.1416321

Values of p_k and P_k are given in the following tables, together with the extrapolation results:

e.

4				
3.3137085	3.0849447			
3.1825979	3.1388943	3.1424910		
3.1517249	3.1414339	3.1416032	3.1415891	
3.1441184	3.1415829	3.1415928	3.1415926	3.1415927

2.8284271				
3.0614675	3.1391476			
3.1214452	3.1414377	3.1415904		
3.1365485	3.1415829	3.1415926	3.1415927	
3.1403312	3.1415921	3.1415927	3.1415927	3.1415927

Exercise Set 4.3

1. The Trapezoidal rule gives the following approximations. **a.** 0.265625 **b.** -0.2678571 **c.** 0.2280741 **d.** 0.1839397 **e.** -0.8666667 **f.** -0.6166667 **g.** 0.2180895 **h.** 4.1432597

3. Simpson's rule gives the following approximations. **a.** 0.1940104 **b.** -0.2670635 **c.** 0.1922453 **d.** 0.16240168 **e.** -0.7391053 **f.** -0.5518759 **g.** 0.1513826 **h.** 2.5836964

5. The Midpoint rule gives the following approximations. **a.** 0.1582031 **b.** −0.2666667 **c.** 0.1743309 **d.** 0.1516327 **e.** −0.6753247 **f.** −0.5194805 **g.** 0.1180292 **h.** 1.8039148

7. $f(1) = \frac{1}{2}$

9. The degree of precision is 3.

11. $c_0 = \frac{1}{3}$, $c_1 = \frac{4}{3}$, $c_2 = \frac{1}{3}$

13. $c_0 = \frac{1}{4}$, $c_1 = \frac{3}{4}$, $x_1 = \frac{2}{3}$ gives the highest degree of precision, 2.

15. The following approximations are obtained from Formula (4.23) through Formula (4.30), respectively.

a. 0.1024404, 0.1024598, 0.1024598, 0.1024598, 0.1024695, 0.1024663, 0.1024598, and 0.1024598

b. 0.7853982, 0.7853982, 0.7853982, 0.7853982, 0.7853982, 0.7853982, 0.7853982, and 0.7853982

c. 1.497171, 1.477536, 1.477529, 1.477523, 1.467719, 1.470981, 1.477512, and 1.477515

d. 4.950000, 2.740909, 2.563393, 2.385700, 1.636364, 1.767857, 2.074893, and 2.116379

e. 3.293182, 2.407901, 2.359772, 2.314751, 1.965260, 2.048634, 2.233251, and 2.249001

f. 0.5000000, 0.6958004, 0.7126032, 0.7306341, 0.7937005, 0.7834709, 0.7611137, and 0.7593572

17. The errors in Exercise 16 are 1.6×10^{-6}, 5.3×10^{-8}, -6.7×10^{-7}, -7.2×10^{-7}, and -1.3×10^{-6}, respectively.

19. If $E(x^k) = 0$ for all $k = 0, 1, \ldots, n$ and $E(x^{n+1}) \neq 0$, then with $p_{n+1}(x) = x^{n+1}$ we have a polynomial of degree $n + 1$ for which $E(p_{n+1}(x)) \neq 0$. Let $p(x) = a_n x^n + \cdots + a_1 x + a_0$ be any polynomial of degree less than or equal to n. Then $E(p(x)) = a_n E(x^n) + \cdots + a_1 E(x) + a_0 E(1) = 0$. Conversely if $E(p(x)) = 0$ for all polynomials of degree less than or equal to n, it follows that $E(x^k) = 0$ for all $k = 0, 1, \ldots, n$. Let $p_{n+1}(x) = a_{n+1}x^{n+1} + \cdots + a_0$ be a polynomial of degree $n + 1$ for which $E(p_{n+1}(x)) \neq 0$. Since $a_{n+1} \neq 0$ we have

$$x^{n+1} = \frac{1}{a_{n+1}} p_{n+1}(x) - \frac{a_n}{a_{n+1}} x^n - \cdots - \frac{a_0}{a_{n+1}}.$$

Then

$$\begin{aligned} E(x^{n+1}) &= \frac{1}{a_{n+1}} E(p_{n+1}(x)) - \frac{a_n}{a_{n+1}} E(x^n) - \cdots - \frac{a_0}{a_{n+1}} E(1) \\ &= \frac{1}{a_{n+1}} E(p_{n+1}(x)) \neq 0. \end{aligned}$$

Thus, the quadrature formula has degree of precision n.

Exercise Set 4.4

1. The Composite Trapezoidal rule approximations are **a.** 0.639900 **b.** 31.3653 **c.** 0.784241 **d.** −6.42872 **e.** −13.5760 **f.** 0.476977 **g.** 0.605498 **h.** 0.970926

3. The Composite Midpoint rule approximations are **a.** 0.633096 **b.** 11.1568 **c.** 0.786700 **d.** −6.11274 **e.** −14.9985 **f.** 0.478751 **g.** 0.602961 **h.** 0.947868

5. $\alpha = 1.5$

7. a. The Composite Trapezoidal rule requires $h < 0.000922295$ and $n \geq 2168$.

b. The Composite Simpson's rule requires $h < 0.037658$ and $n \geq 54$.

c. The Composite Midpoint rule requires $h < 0.00065216$ and $n \geq 3066$.

9. a. The Composite Trapezoidal rule requires $h < 0.04382$ and $n \geq 46$. The approximation is 0.405471.

b. The Composite Simpson's rule requires $h < 0.44267$ and $n \geq 6$. The approximation is 0.405466.

c. The Composite Midpoint rule requires $h < 0.03098$ and $n \geq 64$. The approximation is 0.405460.

11. a. Because the right and left limits at 0.1 and 0.2 for f, f', and f'' are the same, the functions are continuous on $[0, 0.3]$. However,

$$f'''(x) = \begin{cases} 6, & 0 \leq x \leq 0.1 \\ 12, & 0.1 < x \leq 0.2 \\ 12, & 0.2 < x \leq 0.3 \end{cases}$$

is discontinuous at $x = 0.1$.

b. We have 0.302506 with an error bound of 1.9×10^{-4}.

c. We have 0.302425, and the value of the actual integral is the same.

13. a. For the Composite Trapezoidal rule we have

$$E(f) = -\frac{h^3}{12}\sum_{j=1}^{n} f''(\xi_j) = -\frac{h^2}{12}\sum_{j=1}^{n} f''(\xi_j)h$$

$$= -\frac{h^2}{12}\sum_{j=1}^{n} f''(\xi_j)\Delta x_j,$$

where $\Delta x_j = x_{j+1} - x_j = h$ for each j. Since $\sum_{j=1}^{n} f''(\xi_j)\Delta x_j$ is a Riemann sum for $\int_a^b f''(x)\,dx = f'(b) - f'(a)$, we have

$$E(f) \approx -\frac{h^2}{12}[f'(b) - f'(a)].$$

b. For the Composite Midpoint rule we have

$$E(f) = \frac{h^3}{3}\sum_{j=1}^{n/2} f''(\xi_j) = \frac{h^2}{6}\sum_{j=1}^{n/2} f''(\xi_j)(2h).$$

But $\sum_{j=1}^{n/2} f''(\xi_j)(2h)$ is a Riemann sum for $\int_a^b f''(x)\,dx = f'(b) - f'(a)$, so

$$E(f) \approx \frac{h^2}{6}[f'(b) - f'(a)].$$

15. a. The estimate using the Composite Trapezoidal rule is $-\frac{1}{2}h^2 \ln 2 = -6.296 \times 10^{-6}$.

b. The estimate using the Composite Simpson's rule is $-\frac{1}{240}h^2 = -3.75 \times 10^{-6}$.

c. The estimate using the Composite Midpoint rule is $\frac{1}{6}h^2 \ln 2 = 6.932 \times 10^{-6}$.

17. The length is approximately 15.8655.

19. Composite Simpson's rule with $h = 0.25$ gives 2.61972 s.

21. The length is approximately 58.47082 using $n = 100$ in the Composite Simpson's rule.

Exercise Set 4.5

1. Romberg integration gives $R_{3,3}$ as follows: **a.** 0.1922593 **b.** 0.1606105 **c.** −0.1768200 **d.** 0.08875677 **e.** 2.5879685 **f.** −0.7341567 **g.** 0.6362135 **h.** 0.6426970

3. Romberg integration gives **a.** 0.19225936 with $n = 4$ **b.** 0.16060279 with $n = 5$ **c.** −0.17682002 with $n = 4$ **d.** 0.088755284 with $n = 5$ **e.** 2.5886286 with $n = 6$ **f.** −0.73396918 with $n = 6$ **g.** 0.63621335 with $n = 4$ **h.** 0.64269908 with $n = 5$

5. $R_{33} = 11.5246$

7. $f(2.5) \approx 0.43457$

9. $R_{31} = 5$

11. We have

$$R_{k,2} = \frac{4R_{k,1} - R_{k-1,1}}{3}$$

$$= \frac{1}{3}\left[R_{k-1,1} + 2h_{k-1}\sum_{i=1}^{2^{k-2}} f(a + (i - 1/2))h_{k-1})\right] \quad \text{from (4.32)}$$

$$= \frac{1}{3}\left[\frac{h_{k-1}}{2}(f(a) + f(b)) + h_{k-1}\sum_{i=1}^{2^{k-2}-1} f(a + ih_{k-1})\right.$$

$$+ 2h_{k-1}\sum_{i=1}^{2^{k-2}} f(a+(i-1/2)h_{k-1})\Bigg] \quad \text{from (4.31) with } k-1 \text{ instead of } k$$

$$= \frac{1}{3}\left[h_k(f(a)+f(b)) + 2h_k\sum_{i=1}^{2^{k-2}-1} f(a+2ih_k) + 4h_k\sum_{i=1}^{2^{k-2}} f(a+(2i-1)h)\right]$$

$$= \frac{h}{3}\left[f(a)+f(b)+2\sum_{i=1}^{M-1} f(a+2ih) + 4\sum_{i=1}^{M} f(a+(2i-1)h)\right],$$

where $h = h_k$ and $M = 2^{k-2}$.

13. Equation (4.32) follows from

$$R_{k,1} = \frac{h_k}{2}\left[f(a)+f(b)+2\sum_{i=1}^{2^{k-1}-1} f(a+ih_k)\right]$$

$$= \frac{h_k}{2}\left[f(a)+f(b)+2\sum_{i=1}^{2^{k-1}-1} f\left(a+\frac{i}{2}h_{k-1}\right)\right]$$

$$= \frac{h_k}{2}\left[f(a)+f(b)+2\sum_{i=1}^{2^{k-2}-1} f(a+ih_{k-1}) + 2\sum_{i=1}^{2^{k-2}} f(a+(i-1/2)h_{k-1})\right]$$

$$= \frac{1}{2}\left\{\frac{h_{k-1}}{2}\left[f(a)+f(b)+2\sum_{i=1}^{2^{k-2}-1} f(a+ih_{k-1})\right] + h_{k-1}\sum_{i=1}^{2^{k-2}} f(a+(i-1/2)h_{k-1})\right\}$$

$$= \frac{1}{2}\left[R_{k-1,1} + h_{k-1}\sum_{i=1}^{2^{k-2}} f(a+(i-1/2)h_{k-1})\right].$$

Exercise Set 4.6

1. Simpson's rule gives

a. $S(1, 1.5) = 0.19224530$, $S(1, 1.25) = 0.039372434$, $S(1.25, 1.5) = 0.15288602$, and the actual value is 0.19225935.

b. $S(0, 1) = 0.16240168$, $S(0, 0.5) = 0.028861071$, $S(0.5, 1) = 0.13186140$, and the actual value is 0.16060279.

c. $S(0, 0.35) = -0.17682156$, $S(0, 0.175) = -0.087724382$, $S(0.175, 0.35) = -0.089095736$, and the actual value is -0.17682002.

d. $S(0, \frac{\pi}{4}) = 0.087995669$, $S(0, \frac{\pi}{8}) = 0.0058315797$, $S(\frac{\pi}{8}, \frac{\pi}{4}) = 0.082877624$, and the actual value is 0.088755285.

e. $S(0, \frac{\pi}{4}) = 2.5836964$, $S(0, \frac{\pi}{8}) = 0.33088926$, $S(\frac{\pi}{8}, \frac{\pi}{4}) = 2.2568121$, and the actual value is 2.5886286.

f. $S(1, 1.6) = -0.73910533$, $S(1, 1.3) = -0.26141244$, $S(1.3, 1.6) = -0.47305351$, and the actual value is -0.73396917.

g. $S(3, 3.5) = 0.63623873$, $S(3, 3.25) = 0.32567095$, $S(3.25, 3.5) = 0.31054412$, and the actual value is 0.63621334.

h. $S(0, \frac{\pi}{4}) = 0.64326905$, $S(0, \frac{\pi}{8}) = 0.37315002$, $S(\frac{\pi}{8}, \frac{\pi}{4}) = 0.26958270$, and the actual value is 0.64269908.

3. Adaptive quadrature gives **a.** 108.555281 **b.** -1724.966983 **c.** -15.306308 **d.** -18.945949

5. Adaptive quadrature gives

$$\int_{0.1}^{2} \sin\frac{1}{x}\,dx = 1.1454 \quad \text{and} \quad \int_{0.1}^{2} \cos\frac{1}{x}\,dx = 0.67378.$$

7. $\int_0^{2\pi} u(t)\,dt \approx 0.00001$

9.

t	$c(t)$	$s(t)$
0.1	0.0999975	0.000523589
0.2	0.199921	0.00418759
0.3	0.299399	0.0141166
0.4	0.397475	0.0333568
0.5	0.492327	0.0647203
0.6	0.581061	0.110498
0.7	0.659650	0.172129
0.8	0.722844	0.249325
0.9	0.764972	0.339747
1.0	0.779880	0.438245

Exercise Set 4.7

1. Gaussian quadrature gives **a.** 0.1922687 **b.** 0.1594104 **c.** −0.1768190 **d.** 0.08926302 **e.** 2.5913247 **f.** −0.7307230 **g.** 0.6361966 **h.** 0.6423172

3. Gaussian quadrature gives **a.** 0.1922594 **b.** 0.1606028 **c.** −0.1768200 **d.** 0.08875529 **e.** 2.5886327 **f.** −0.7339604 **g.** 0.6362133 **h.** 0.6426991

5. $a = 1$, $b = 1$, $c = \frac{1}{3}$, $d = -\frac{1}{3}$

Exercise Set 4.8

1. Algorithm 4.4 with $n = m = 2$ gives **a.** 0.3115733 **b.** 0.2552526 **c.** 16.50864 **d.** 1.476684

3. Algorithm 4.4 with $n = 2$ and $m = 4$, $n = 4$ and $m = 2$, and $n = m = 3$ gives

a. 0.5119875, 0.5118533, 0.5118722

b. 1.718857, 1.718220, 1.718385

c. 1.001953, 1.000122, 1.000386

d. 0.7838542, 0.7833659, 0.7834362

e. −1.985611, −1.999182, −1.997353

f. 2.004596, 2.000879, 2.000980

g. 0.3084277, 0.3084562, 0.3084323

h. −22.61612, −19.85408, −20.14117

5. Algorithm 4.5 with $n = m = 2$ gives **a.** 0.3115733 **b.** 0.2552446 **c.** 16.50863 **d.** 1.488875

7. Algorithm 4.4 with $n = m = 3$, $n = 3$ and $m = 4$, $n = 4$ and $m = 3$, and $n = m = 4$ gives

a. 0.5118655, 0.5118445, 0.5118655, 0.5118445, 2.1×10^{-5}, 1.3×10^{-7}, 2.1×10^{-5}, 1.3×10^{-7}

b. 1.718163, 1.718302, 1.718139, 1.718277, 1.2×10^{-4}, 2.0×10^{-5}, 1.4×10^{-4}, 4.8×10^{-6}

c. 1.000000, 1.000000, 1.0000000, 1.000000, 0, 0, 0, 0

d. 0.7833333, 0.7833333, 0.7833333, 0.7833333, 0, 0, 0, 0

e. −1.991878, −2.000124, −1.991878, −2.000124, 8.1×10^{-3}, 1.2×10^{-4}, 8.1×10^{-3}, 1.2×10^{-4}

f. 2.001494, 2.000080, 2.001388, 1.999984, 1.5×10^{-3}, 8×10^{-5}, 1.4×10^{-3}, 1.6×10^{-5}

g. 0.3084151, 0.3084145, 0.3084246, 0.3084245, 10^{-5}, 5.5×10^{-7}, 1.1×10^{-5}, 6.4×10^{-7}

h. −12.74790, −21.21539, −11.83624, −20.30373, 7.0, 1.5, 7.9, 0.564

9. Algorithm 4.4 with $n = m = 7$ gives 0.1479103 and Algorithm 4.5 with $n = m = 4$ gives 0.1506823.

11. The approximation to the center of mass is $(\overline{x}, \overline{y})$, where $\overline{x} = 0.3806333$ and $\overline{y} = 0.3822558$.

13. The area is approximately 1.0402528.

15. Algorithm 4.6 with $n = m = p = 2$ gives the first listed value. The second is the exact result.

a. 5.204036, $e(e^{0.5} - 1)(e - 1)^2$

b. $0.08429784, \frac{1}{12}$
c. $0.08641975, \frac{1}{14}$
d. $0.09722222, \frac{1}{12}$
e. $7.103932, 2 + \frac{1}{2}\pi^2$
f. $1.428074, \frac{1}{2}(e^2 + 1) - e$

17. Algorithm 4.6 with $n = m = p = 4$ gives the first listed value. The second is from Algorithm 4.6 with $n = m = p = 5$.
a. 5.206447, 5.206447
b. 0.08333333,0.08333333
c. 0.07142857,0.07142857
d. 0.08333333,0.08333333
e. 6.934912,6.934801
f. 1.476207, 1.476246

19. The approximation 20.41887 requires 125 functional evaluations.

Exercise Set 4.9

1. The Composite Simpson's rule gives **a.** 0.5284163 **b.** 4.266654 **c.** 0.4329748 **d.** 0.8802210
3. The Composite Simpson's rule gives **a.** 0.4112649 **b.** 0.2440679 **c.** 0.05501681 **d.** 0.2903746
7. When $n = 2$, we have 0.9238795; and when $n = 3$, we have 0.9064405.
9. The escape velocity is approximately 6.9450 mi/s.

Exercise Set 5.1

1. a. Since $f(t, y) = y \cos t$, we have $(\partial f/\partial y)(t, y) = \cos t$ and f satisfies a Lipschitz condition in y on

$$D = \{(t, y) \mid 0 \le t \le 1, -\infty < y < \infty\} \quad \text{with } L = 1.$$

f is continuous on D, so there exists a unique solution. The solution is $y(t) = e^{\sin t}$.

b. $f(t, y) = \frac{2}{t}y + t^2e^t$, $\partial f/\partial y = 2/t$; f satisfies a Lipschitz condition in y on

$$D = \{(t, y) \mid 1 \le t \le 2, -\infty < y < \infty\} \quad \text{with } L = 2.$$

f is continuous on D, so there exists a unique solution, which is $y(t) = t^2(e^t - e)$.

c. $f(t, y) = -\frac{2}{t}y + t^2e^t$, $\frac{\partial f}{\partial y} = -\frac{2}{t}$; f satisfies a Lipschitz condition in y on

$$D = \{(t, y) \mid 1 \le t \le 2, -\infty < y < \infty\} \quad \text{with } L = 2.$$

f is continuous on D, so there exists a unique solution, which is

$$y(t) = (t^4e^t - 4t^3e^t + 12t^2e^t - 24te^t + 24e^t + (\sqrt{2} - 9)e)/t^2.$$

d. $f(t, y) = 4t^3y/(1 + t^4)$, $\partial f/\partial y = 4t^3/(1 + t^4)$ satisfies a Lipschitz condition on

$$D = \{(t, y) \mid 0 \le t \le 1, -\infty < y < \infty\} \quad \text{with } L = 2.$$

f is continuous on D, so there exists a unique solution, which is $y(t) = 1 + t^4$.

3. a. Differentiating $y^3t + yt = 2$ gives $3y^2y't + y^3 + y't + y = 0$. Solving for y' gives the original differential equation, and setting $t = 1$ and $y = 1$ verifies the initial condition. To approximate $y(2)$, use Newton's method to solve the equation $y^3 + y - 1 = 0$. This gives $y(2) \approx 0.6823278$.

b. Differentiating $y \sin t + t^2e^y + 2y - 1 = 0$ gives $y' \sin t + y \cos t + 2te^y + t^2e^yy' + 2y' = 0$. Solving for y' gives the original differential equation and setting $t = 1$ and $y = 0$ verifies the initial condition. To approximate $y(2)$, use Newton's method to solve the equation $(2 + \sin 2)y + 4e^y - 1 = 0$. This gives $y(2) \approx -0.4946599$.

5. Let (t_1, y_1) and (t_2, y_2) be in D, with $a \le t_1 \le b, a \le t_2 \le b, -\infty < y_1 < \infty$, and $-\infty < y_2 < \infty$. For $0 \le \lambda \le 1$, we have $(1-\lambda)a \le (1-\lambda)t_1 \le (1-\lambda)b$ and $\lambda a \le \lambda t_2 \le \lambda b$. Hence, $a = (1-\lambda)a + \lambda a \le (1-\lambda)t_1 + \lambda t_2 \le (1-\lambda)b + \lambda b = b$. Also, $-\infty < (1-\lambda)y_1 + \lambda y_2 < \infty$, so D is convex.

7. a. Since $y' = f(t, y(t))$,

$$\int_a^t y'(z)\,dz = \int_a^t f(z, y(z))\,dz.$$

So

$$y(t) - y(a) = \int_a^t f(z, y(z))\,dz$$

and

$$y(t) = \alpha + \int_a^t f(z, y(z))\,dz.$$

The iterative method follows from this equation.

b. We have $y_0(t) = 1$, $y_1(t) = 1 + \frac{1}{2}t^2$, $y_2(t) = 1 + \frac{1}{2}t^2 - \frac{1}{6}t^3$, and $y_3(t) = 1 + \frac{1}{2}t^2 - \frac{1}{6}t^3 + \frac{1}{24}t^4$.

c. We have $y(t) = 1 + \frac{1}{2}t^2 - \frac{1}{6}t^3 + \frac{1}{24}t^4 - \frac{1}{120}t^5 + \cdots$.

Exercise Set 5.2

1. Euler's method gives the approximations in the following tables.

a.

i	t_i	w_i	$y(t_i)$
1	0.500	0.0000000	0.2836165
2	1.000	1.1204223	3.2190993

b.

i	t_i	w_i	$y(t_i)$
1	2.500	2.0000000	1.8333333
2	3.000	2.6250000	2.5000000

c.

i	t_i	w_i	$y(t_i)$
1	1.250	2.7500000	2.7789294
2	1.500	3.5500000	3.6081977
3	1.750	4.3916667	4.4793276
4	2.000	5.2690476	5.3862944

d.

i	t_i	w_i	$y(t_i)$
1	0.250	1.2500000	1.3291498
2	0.500	1.6398053	1.7304898
3	0.750	2.0242547	2.0414720
4	1.000	2.2364573	2.1179795

3. Euler's method gives the approximations in the following tables.

a.

i	t_i	w_i	$y(t_i)$
2	1.200	1.0082645	1.0149523
4	1.400	1.0385147	1.0475339
6	1.600	1.0784611	1.0884327
8	1.800	1.1232621	1.1336536
10	2.000	1.1706516	1.1812322

b.

i	t_i	w_i	$y(t_i)$
2	1.400	0.4388889	0.4896817
4	1.800	1.0520380	1.1994386
6	2.200	1.8842608	2.2135018
8	2.600	3.0028372	3.6784753
10	3.000	4.5142774	5.8741000

c.

i	t_i	w_i	$y(t_i)$
2	0.400	−1.6080000	−1.6200510
4	0.800	−1.3017370	−1.3359632
6	1.200	−1.1274909	−1.1663454
8	1.600	−1.0491191	−1.0783314
10	2.000	−1.0181518	−1.0359724

d.

i	t_i	w_i	$y(t_i)$
2	0.2	0.1083333	0.1626265
4	0.4	0.1620833	0.2051118
6	0.6	0.3455208	0.3765957
8	0.8	0.6213802	0.6461052
10	1.0	0.9803451	1.0022460

5. Euler's method gives the approximations in the following table.

a.

i	t_i	w_i	$y(t_i)$
1	1.1	0.271828	0.345920
5	1.5	3.18744	3.96767
6	1.6	4.62080	5.70296
9	1.9	11.7480	14.3231
10	2.0	15.3982	18.6831

b. Linear interpolation gives the approximations in the following table.

t	Approximation	$y(t)$	Error
1.04	0.108731	0.119986	0.01126
1.55	3.90412	4.78864	0.8845
1.97	14.3031	17.2793	2.976

c. $h < 0.00064$

7. a. Euler's method produces the following approximation to $y(5) = 5.00674$.

	$h = 0.2$	$h = 0.1$	$h = 0.05$
w_N	5.00377	5.00515	5.00592

b. $h = \sqrt{2 \times 10^{-6}} \approx 0.0014142$.

9. a. $h = 10^{-n/2}$ **b.** The minimal error is $10^{-n/2}(e - 1) + 5e10^{-n-1}$.

c.

t	$w(h = 0.1)$	$w(h = 0.01)$	$y(t)$	Error $(n = 8)$
0.5	0.40951	0.39499	0.39347	1.5×10^{-4}
1.0	0.65132	0.63397	0.63212	3.1×10^{-4}

11. b. $w_{50} = 0.10430 \approx p(50)$ **c.** Since $p(t) = 1 - 0.99e^{-0.002t}$, $p(50) = 0.10421$.

Exercise Set 5.3

1. a.

t_i	w_i	$y(t_i)$
0.50	0.12500000	0.28361652
1.00	2.02323897	3.21909932

b.

t_i	w_i	$y(t_i)$
2.50	1.75000000	1.83333333
3.00	2.42578125	2.50000000

c.

t_i	w_i	$y(t_i)$
1.25	2.78125000	2.77892944
1.50	3.61250000	3.60819766
1.75	4.48541667	4.47932763
2.00	5.39404762	5.38629436

d.

t_i	w_i	$y(t_i)$
0.25	1.34375000	1.32914981
0.50	1.77218707	1.73048976
0.75	2.11067606	2.04147203
1.00	2.20164395	2.11797955

3. a.

i	t_i	Order 2 w_i	Order 4 w_i	$y(t_i)$
1	1.1	1.214999	1.215883	1.215886
2	1.2	1.465250	1.467561	1.467570

b.

i	t_i	Order 2 w_i	Order 4 w_i	$y(t_i)$
1	0.5	0.5000000	0.5156250	0.5158868
2	1.0	1.076858	1.091267	1.091818

c.

i	t_i	Order 2 w_i	Order 4 w_i	$y(t_i)$
1	1.5	−2.000000	−2.000000	−1.500000
2	2.0	−1.777776	−1.679012	−1.333333
3	2.5	−1.585732	−1.484493	−1.250000
4	3.0	−1.458882	−1.374440	−1.200000

d.

i	t_i	Order 2 w_i	Order 4 w_i	$y(t_i)$
1	0.25	1.093750	1.086426	1.087088
2	0.50	1.312319	1.288245	1.289805
3	0.75	1.538468	1.512576	1.513490
4	1.0	1.720480	1.701494	1.701870

5. a. Taylor's method of order two gives the results in the following table.

i	t_i	w_i	$y(t_i)$
1	1.1	0.3397852	0.3459199
5	1.5	3.910985	3.967666
6	1.6	5.643081	5.720962
9	1.9	14.15268	14.32308
10	2.0	18.46999	18.68310

b. Linear interpolation gives $y(1.04) \approx 0.1359139$, $y(1.55) \approx 4.777033$, and $y(1.97) \approx 17.17480$. Actual values are $y(1.04) = 0.1199875$, $y(1.55) = 4.788635$, and $y(1.97) = 17.27930$.

c. Taylor's method of order four gives the results in the following table.

i	t_i	w_i
1	1.1	0.3459127
5	1.5	3.967603
6	1.6	5.720875
9	1.9	14.32290
10	2.0	18.68287

d. Cubic Hermite interpolation gives $y(1.04) \approx 0.1199704$, $y(1.55) \approx 4.788527$, and $y(1.97) \approx 17.27904$.

7. a.

i	t_i	Order 2	Order 4
2	0.2	5.86595	5.86433
5	0.5	2.82145	2.81789
7	0.7	0.84926	0.84455
10	1.0	−2.08606	−2.09015

b. 0.8 s

Exercise Set 5.4

1. a.

t	Modified Euler	$y(t)$
0.5	0.5602111	0.2836165
1.0	5.3014898	3.2190993

b.

t	Modified Euler	$y(t)$
2.5	1.8125000	1.8333333
3.0	2.4815531	2.5000000

c.

t	Modified Euler	$y(t)$
1.25	2.7750000	2.7789294
1.50	3.6008333	3.6081977
1.75	4.4688294	4.4793276
2.00	5.3728586	5.3862944

d.

t	Modified Euler	$y(t)$
0.25	1.3199027	1.3291498
0.50	1.7070300	1.7304898
0.75	2.0053560	2.0414720
1.00	2.0770789	2.1179795

3. a.

t	Midpoint	$y(t)$
0.5	0.2646250	0.2836165
1.0	3.1300023	3.2190993

b.

t	Midpoint	$y(t)$
2.5	1.7812500	1.8333333
3.0	2.4550638	2.5000000

c.

t	Midpoint	$y(t)$
1.25	2.7777778	2.7789294
1.50	3.6060606	3.6081977
1.75	4.4763015	4.4793276
2.00	5.3824398	5.3862944

d.

t	Midpoint	$y(t)$
0.25	1.3337962	1.3291498
0.50	1.7422854	1.7304898
0.75	2.0596374	2.0414720
1.00	2.1385560	2.1179795

5. a. $1.0221167 \approx y(1.25) = 1.0219569$, $1.1640347 \approx y(1.93) = 1.1643901$

b. $1.9086500 \approx y(2.1) = 1.9249616$, $4.3105913 \approx y(2.75) = 4.3941697$

c. $-1.1461434 \approx y(1.3) = -1.1382768$, $-1.0454854 \approx y(1.93) = -1.0412665$

d. $0.3271470 \approx y(0.54) = 0.3140018$, $0.8967073 \approx y(0.94) = 0.8866318$

7. a. $1.0225530 \approx y(1.25) = 1.0219569$, $1.1646155 \approx y(1.93) = 1.1643901$

b. $1.9132167 \approx y(2.1) = 1.9249616$, $4.3246152 \approx y(2.75) = 4.3941697$

c. $-1.1441775 \approx y(1.3) = -1.1382768$, $-1.0447403 \approx y(1.93) = -1.0412665$

d. $0.3251049 \approx y(0.54) = 0.3140018$, $0.8945125 \approx y(0.94) = 0.8866318$

9. a. $1.0227863 \approx y(1.25) = 1.0219569$, $1.1649247 \approx y(1.93) = 1.1643901$

b. $1.9153749 \approx y(2.1) = 1.9249616$, $4.3312939 \approx y(2.75) = 4.3941697$

c. $-1.1432070 \approx y(1.3) = -1.1382768$, $-1.0443743 \approx y(1.93) = -1.0412665$

d. $0.3240839 \approx y(0.54) = 0.3140018$, $0.8934152 \approx y(0.94) = 0.8866318$

11. a. The Runge-Kutta method of order four gives the results in the following tables.

t	Runge-Kutta	$y(t)$
1.2	1.0149520	1.0149523
1.4	1.0475336	1.0475339
1.6	1.0884323	1.0884327
1.8	1.1336532	1.1336536
2.0	1.1812319	1.1812322

b.

t	Runge-Kutta	$y(t)$
1.4	0.4896842	0.4896817
1.8	1.1994320	1.1994386
2.2	2.2134693	2.2135018
2.6	3.6783790	3.6784753
3.0	5.8738386	5.8741000

c.

t	Runge-Kutta	$y(t)$
0.4	−1.6200576	−1.6200510
0.8	−1.3359824	−1.3359632
1.2	−1.1663735	−1.1663454
1.6	−1.0783582	−1.0783314
2.0	−1.0359922	−1.0359724

d.

t	Runge-Kutta	$y(t)$
0.2	0.1627655	0.1626265
0.4	0.2052405	0.2051118
0.6	0.3766981	0.3765957
0.8	0.6461896	0.6461052
1.0	1.0023207	1.0022460

15. In 0.2 s we have approximately 2099 units of KOH.

17. The appropriate constants are $\alpha_1 = \delta_1 = \alpha_2 = \delta_2 = \gamma_2 = \gamma_3 = \gamma_4 = \gamma_5 = \gamma_6 = \gamma_7 = \frac{1}{2}$ and $\alpha_3 = \delta_3 = 1$.

Exercise Set 5.5

1. The Runge-Kutta-Fehlberg Algorithm gives the results in the following tables.

a.

i	t_i	w_i	h_i	y_i
1	0.2093900	0.0298184	0.2093900	0.0298337
3	0.5610469	0.4016438	0.1777496	0.4016860
5	0.8387744	1.5894061	0.1280905	1.5894600
7	1.0000000	3.2190497	0.0486737	3.2190993

b.

i	t_i	w_i	h_i	y_i
1	2.2500000	1.4499988	0.2500000	1.4500000
2	2.5000000	1.8333332	0.2500000	1.8333333
3	2.7500000	2.1785718	0.2500000	2.1785714
4	3.0000000	2.5000005	0.2500000	2.5000000

c.

i	t_i	w_i	h_i	y_i
1	1.2500000	2.7789299	0.2500000	2.7789294
2	1.5000000	3.6081985	0.2500000	3.6081977
3	1.7500000	4.4793288	0.2500000	4.4793276
4	2.0000000	5.3862958	0.2500000	5.3862944

d.

i	t_i	w_i	h_i	y_i
1	0.2500000	1.3291478	0.2500000	1.3291498
2	0.5000000	1.7304857	0.2500000	1.7304898
3	0.7500000	2.0414669	0.2500000	2.0414720
4	1.0000000	2.1179750	0.2500000	2.1179795

3. The Runge-Kutta-Fehlberg Algorithm gives the results in the following tables.

a.

i	t_i	w_i	h_i	y_i
1	1.1101946	1.0051237	0.1101946	1.0051237
5	1.7470584	1.1213948	0.2180472	1.1213947
7	2.3994350	1.2795396	0.3707934	1.2795395
11	4.0000000	1.6762393	0.1014853	1.6762391

b.

i	t_i	w_i	h_i	y_i
4	1.5482238	0.7234123	0.1256486	0.7234119
7	1.8847226	1.3851234	0.1073571	1.3851226
10	2.1846024	2.1673514	0.0965027	2.1673499
16	2.6972462	4.1297939	0.0778628	4.1297904
21	3.0000000	5.8741059	0.0195070	5.8741000

c.

i	t_i	w_i	h_i	y_i
1	0.1633541	−1.8380836	0.1633541	−1.8380836
5	0.7585763	−1.3597623	0.1266248	−1.3597624
9	1.1930325	−1.1684827	0.1048224	−1.1684830
13	1.6229351	−1.0749509	0.1107510	−1.0749511
17	2.1074733	−1.0291158	0.1288897	−1.0291161
23	3.0000000	−1.0049450	0.1264618	−1.0049452

d.

i	t_i	w_i	h_i	y_i
1	0.3986051	0.3108201	0.3986051	0.3108199
3	0.9703970	0.2221189	0.2866710	0.2221186
5	1.5672905	0.1133085	0.3042087	0.1133082
8	2.0000000	0.0543454	0.0902302	0.0543455

5. a. The number of infectives is $y(30) \approx 80295.7$.

b. The limiting value for the number of infectives for this model is $\lim_{t\to\infty} y(t) = 100{,}000$.

Exercise Set 5.6

1. The Adams-Bashforth methods give the results in the following tables.

a.

t	2-step	3-step	4-step	5-step	$y(t)$
0.2	0.0268128	0.0268128	0.0268128	0.0268128	0.0268128
0.4	0.1200522	0.1507778	0.1507778	0.1507778	0.1507778
0.6	0.4153551	0.4613866	0.4960196	0.4960196	0.4960196
0.8	1.1462844	1.2512447	1.2961260	1.3308570	1.3308570
1.0	2.8241683	3.0360680	3.1461400	3.1854002	3.2190993

b.

t	2-step	3-step	4-step	5-step	$y(t)$
2.2	1.3666667	1.3666667	1.3666667	1.3666667	1.3666667
2.4	1.6750000	1.6857143	1.6857143	1.6857143	1.6857143
2.6	1.9632431	1.9794407	1.9750000	1.9750000	1.9750000
2.8	2.2323184	2.2488759	2.2423065	2.2444444	2.2444444
3.0	2.4884512	2.5051340	2.4980306	2.5011406	2.5000000

c.

t	2-step	3-step	4-step	5-step	$y(t)$
1.2	2.6187859	2.6187859	2.6187859	2.6187859	2.6187859
1.4	3.2734823	3.2710611	3.2710611	3.2710611	3.2710611
1.6	3.9567107	3.9514231	3.9520058	3.9520058	3.9520058
1.8	4.6647738	4.6569191	4.6582078	4.6580160	4.6580160
2.0	5.3949416	5.3848058	5.3866452	5.3862177	5.3862944

d.

t	2-step	3-step	4-step	5-step	$y(t)$
0.2	1.2529306	1.2529306	1.2529306	1.2529306	1.2529306
0.4	1.5986417	1.5712255	1.5712255	1.5712255	1.5712255
0.6	1.9386951	1.8827238	1.8750869	1.8750869	1.8750869
0.8	2.1766821	2.0844122	2.0698063	2.0789180	2.0789180
1.0	2.2369407	2.1115540	2.0998117	2.1180642	2.1179795

3. The Adams-Bashforth methods give the results in the following tables.

a.

t	2-step	3-step	4-step	5-step	$y(t)$
1.2	1.0161982	1.0149520	1.0149520	1.0149520	1.0149523
1.4	1.0497665	1.0468730	1.0477278	1.0475336	1.0475339
1.6	1.0910204	1.0875837	1.0887567	1.0883045	1.0884327
1.8	1.1363845	1.1327465	1.1340093	1.1334967	1.1336536
2.0	1.1840272	1.1803057	1.1815967	1.1810689	1.1812322

b.

t	2-step	3-step	4-step	5-step	$y(t)$
1.4	0.4867550	0.4896842	0.4896842	0.4896842	0.4896817
1.8	1.1856931	1.1982110	1.1990422	1.1994320	1.1994386
2.2	2.1753785	2.2079987	2.2117448	2.2134792	2.2135018
2.6	3.5849181	3.6617484	3.6733266	3.6777236	3.6784753
3.0	5.6491203	5.8268008	5.8589944	5.8706101	5.8741000

c.

t	2-step	3-step	4-step	5-step	$y(t)$
0.5	−1.5357010	−1.5381988	−1.5379372	−1.5378676	−1.5378828
1.0	−1.2374093	−1.2389605	−1.2383734	−1.2383693	−1.2384058
1.5	−1.0952910	−1.0950952	−1.0947925	−1.0948481	−1.0948517
2.0	−1.0366643	−1.0359996	−1.0359497	−1.0359760	−1.0359724

d.

t	2-step	3-step	4-step	5-step	$y(t)$
0.2	0.1739041	0.1627655	0.1627655	0.1627655	0.1626265
0.4	0.2144877	0.2026399	0.2066057	0.2052405	0.2051118
0.6	0.3822803	0.3747011	0.3787680	0.3765206	0.3765957
0.8	0.6491272	0.6452640	0.6487176	0.6471458	0.6461052
1.0	1.0037415	1.0020894	1.0064121	1.0073348	1.0022460

5. The Adams Fourth-order Predictor-Corrector Algorithm gives the results in the following tables.

a.

t	w	$y(t)$
1.2	1.0149520	1.0149523
1.4	1.0475227	1.0475339
1.6	1.0884141	1.0884327
1.8	1.1336331	1.1336536
2.0	1.1812112	1.1812322

b.

t	w	$y(t)$
1.4	0.4896842	0.4896817
1.8	1.1994245	1.1994386
2.2	2.2134701	2.2135018
2.6	3.6784144	3.6784753
3.0	5.8739518	5.8741000

c.

t	w	$y(t)$
0.5	−1.5378788	−1.5378828
1.0	−1.2384134	−1.2384058
1.5	−1.0948609	−1.0948517
2.0	−1.0359757	−1.0359724

d.

t	w	$y(t)$
0.2	0.1627655	0.1626265
0.4	0.2048557	0.2051118
0.6	0.3762804	0.3765957
0.8	0.6458949	0.6461052
1.0	1.0021372	1.0022460

7. a. With $h = 0.01$, the three-step Adams-Moulton method gives the values in the following table.

i	t_i	w_i
10	0.1	1.317218
20	0.2	1.784511

b. Newton's method will reduce the number of iterations per step from three to two, using the stopping criterion

$$|w_i^{(k)} - w_i^{(k-1)}| \leq 10^{-6}.$$

13. To derive Milne's method, integrate $y'(t) = f(t, y(t))$ on the interval $[t_{i-3}, t_{i+1}]$ to obtain

$$y(t_{i+1}) - y(t_{i-3}) = \int_{t_{i-3}}^{t_{i+1}} f(t, y(t))\, dt.$$

Using the open Newton-Cotes formula (4.29) we have

$$y(t_{i+1}) - y(t_{i-3}) = \frac{4h[2f(t_i, y(t_i)) - f(t_{i-1}, y(t_{i-1})) + 2f(t_{i-2}, y(t_{i-2}))]}{3} + \frac{14h^5 f^{(4)}(\xi, y(\xi))}{45}.$$

The difference equation becomes

$$w_{i+1} = w_{i-3} + \frac{h[8f(t_i, w_i) - 4f(t_{i-1}, w_{i-1}) + 8f(t_{i-2}, w_{i-2})]}{3}$$

with local truncation error

$$\tau_{i+1}(h) = \frac{14h^4 y^{(5)}(\xi)}{45}.$$

Exercise Set 5.7

1. The Adams Variable Step-Size Predictor-Corrector Algorithm gives the results in the following tables.

a.

i	t_i	w_i	h_i	y_i
1	0.04275596	0.00096891	0.04275596	0.00096887
5	0.22491460	0.03529441	0.05389076	0.03529359
12	0.60214994	0.50174348	0.05389076	0.50171761
17	0.81943926	1.45544317	0.04345786	1.45541453
22	0.99830392	3.19605697	0.03577293	3.19602842
26	1.00000000	3.21912776	0.00042395	3.21909932

b.

i	t_i	w_i	h_i	y_i
1	2.06250000	1.12132350	0.06250000	1.12132353
5	2.31250000	1.55059834	0.06250000	1.55059524
9	2.62471924	2.00923157	0.09360962	2.00922829
13	2.99915773	2.49895243	0.09360962	2.49894707
17	3.00000000	2.50000535	0.00021057	2.50000000

c.

i	t_i	w_i	h_i	y_i
1	1.06250000	2.18941363	0.06250000	2.18941366
4	1.25000000	2.77892931	0.06250000	2.77892944
8	1.85102559	4.84179835	0.15025640	4.84180141
12	2.00000000	5.38629105	0.03724360	5.38629436

d.

i	t_i	w_i	h_i	y_i
1	0.06250000	1.06817960	0.06250000	1.06817960
5	0.31250000	1.42861668	0.06250000	1.42861361
10	0.62500000	1.90768386	0.06250000	1.90767015
13	0.81250000	2.08668486	0.06250000	2.08666541
16	1.00000000	2.11800208	0.06250000	2.11797955

3. The following tables list representative results from the Adams Variable Step-Size Predictor-Corrector Algorithm.

a.

i	t_i	w_i	h_i	y_i
5	1.10431651	1.00463041	0.02086330	1.00463045
15	1.31294952	1.03196889	0.02086330	1.03196898
25	1.59408142	1.08714711	0.03122028	1.08714722
35	2.00846205	1.18327922	0.04824992	1.18327937
45	2.66272188	1.34525123	0.07278716	1.34525143
52	3.40193112	1.52940900	0.11107035	1.52940924
57	4.00000000	1.67623887	0.12174963	1.67623914

b.

i	t_i	w_i	h_i	y_i
5	1.18519603	0.20333499	0.03703921	0.20333497
15	1.55558810	0.73586642	0.03703921	0.73586631
25	1.92598016	1.48072467	0.03703921	1.48072442
35	2.29637222	2.51764797	0.03703921	2.51764743
45	2.65452689	3.92602442	0.03092051	3.92602332
55	2.94341188	5.50206466	0.02584049	5.50206279
61	3.00000000	5.87410206	0.00122679	5.87409998

c.

i	t_i	w_i	h_i	y_i
5	0.16854008	−1.83303780	0.03370802	−1.83303783
17	0.64833341	−1.42945306	0.05253230	−1.42945304
27	1.06742915	−1.21150951	0.04190957	−1.21150932
41	1.75380240	−1.05819340	0.06681937	−1.05819325
51	2.50124702	−1.01335240	0.07474446	−1.01335258
61	3.00000000	−1.00494507	0.01257155	−1.00494525

d.

i	t_i	w_i	h_i	y_i
5	0.28548652	0.32153668	0.05709730	0.32153674
15	0.85645955	0.24281066	0.05709730	0.24281095
20	1.35101725	0.15096743	0.09891154	0.15096772
25	1.66282314	0.09815109	0.06236118	0.09815137
29	1.91226786	0.06418555	0.06236118	0.06418579
33	2.00000000	0.05434530	0.02193303	0.05434551

5. The current after 2 s is approximately $i(2) = 8.693$ amperes.

Exercise Set 5.8

1. The Extrapolation Algorithm gives the results in the following tables.

a.

i	t_i	w_i	h	k	y_i
1	0.25	0.04543132	0.25	3	0.04543123
2	0.50	0.28361684	0.25	3	0.28361652
3	0.75	1.05257634	0.25	4	1.05257615
4	1.00	3.21909944	0.25	4	3.21909932

b.

i	t_i	w_i	h	k	y_i
1	2.25	1.44999987	0.25	3	1.45000000
2	2.50	1.83333321	0.25	3	1.83333333
3	2.75	2.17857133	0.25	3	2.17857143
4	3.00	2.49999993	0.25	3	2.50000000

c.

i	t_i	w_i	h	k	y_i
1	1.25	2.77892942	0.25	3	2.77892944
2	1.50	3.60819763	0.25	3	3.60819766
3	1.75	4.47932759	0.25	3	4.47932763
4	2.00	5.38629431	0.25	3	5.38629436

d.

i	t_i	w_i	h	k	y_i
1	0.25	1.32914981	0.25	3	1.32914981
2	0.50	1.73048976	0.25	3	1.73048976
3	0.75	2.04147203	0.25	3	2.04147203
4	1.00	2.11797954	0.25	3	2.11797955

3. The Extrapolation Algorithm gives the results in the following tables.

a.

i	t_i	w_i	h	k	y_i
1	1.50	1.06726237	0.50	4	1.06726235
2	2.00	1.18123223	0.50	3	1.18123222
3	2.50	1.30460372	0.50	3	1.30460371
4	3.00	1.42951608	0.50	3	1.42951607
5	3.50	1.55364771	0.50	3	1.55364770
6	4.00	1.67623915	0.50	3	1.67623914

b.

i	t_i	w_i	h	k	y_i
1	1.50	0.64387537	0.50	4	0.64387533
2	2.00	1.66128182	0.50	5	1.66128176
3	2.50	3.25801550	0.50	5	3.25801536
4	3.00	5.87410027	0.50	5	5.87409998

c.

i	t_i	w_i	h	k	y_i
1	0.50	−1.53788284	0.50	4	−1.53788284
2	1.00	−1.23840584	0.50	5	−1.23840584
3	1.50	−1.09485175	0.50	5	−1.09485175
4	2.00	−1.03597242	0.50	5	−1.03597242
5	2.50	−1.01338570	0.50	5	−1.01338570
6	3.00	−1.00494526	0.50	4	−1.00494525

d.

i	t_i	w_i	h	k	y_i
1	0.50	0.29875177	0.50	4	0.29875178
2	1.00	0.21662642	0.50	4	0.21662642
3	1.50	0.12458565	0.50	4	0.12458565
4	2.00	0.05434552	0.50	4	0.05434551

Exercise Set 5.9

1. The Runge-Kutta for Systems Algorithm gives the results in the following tables.

a.

t_i	w_{1i}	u_{1i}	w_{2i}	u_{2i}
0.200	2.12036583	2.12500839	1.50699185	1.51158743
0.400	4.44122776	4.46511961	3.24224021	3.26598528
0.600	9.73913329	9.83235869	8.16341700	8.25629549
0.800	22.67655977	23.00263945	21.34352778	21.66887674
1.000	55.66118088	56.73748265	56.03050296	57.10536209

b.

t_i	w_{1i}	u_{1i}	w_{2i}	u_{2i}
0.500	0.95671390	0.95672798	−1.08381950	−1.08383310
1.000	1.30654440	1.30655930	−0.83295364	−0.83296776
1.500	1.34416716	1.34418117	−0.56980329	−0.56981634
2.000	1.14332436	1.14333672	−0.36936318	−0.36937457

c.

t_i	w_{1i}	u_{1i}	w_{2i}	u_{2i}	w_{3i}	u_{3i}
0.5	0.70787076	0.70828683	−1.24988663	−1.25056425	0.39884862	0.39815702
1.0	−0.33691753	−0.33650854	−3.01764179	−3.01945051	−0.29932294	−0.30116868
1.5	−2.41332734	−2.41345688	−5.40523279	−5.40844686	−0.92346873	−0.92675778
2.0	−5.89479008	−5.89590551	−8.70970537	−8.71450036	−1.32051165	−1.32544426

d.

t_i	w_{1i}	u_{1i}	w_{2i}	u_{2i}	w_{3i}	u_{3i}
0.2	1.38165297	1.38165325	1.00800000	1.00800000	−0.61833075	−0.61833075
0.5	1.90753116	1.90753184	1.12500000	1.12500000	−0.09090565	−0.09090566
0.7	2.25503524	2.25503620	1.34300000	1.34000000	0.26343971	0.26343970
1.0	2.83211921	2.83212056	2.00000000	2.00000000	0.88212058	0.88212056

5. The Adams fourth-order predictor-corrector method for systems gives the results in the following tables.

a.

t_i	w_{1i}	$y(t_i)$
0.200	0.00015352	0.00015350
0.500	0.00743133	0.00743027
0.700	0.03300266	0.03299805
1.000	0.17134711	0.17132880

b.

t_i	w_{1i}	$y(t_i)$
1.200	0.96152437	0.96152583
1.500	0.77796798	0.77797237
1.700	0.59373213	0.59373830
2.000	0.27258055	0.27258872

c.

t_i	w_{1i}	$y(t_i)$
1.000	3.73186337	3.73170445
2.000	11.31462595	11.31452924
3.000	34.04548233	34.04517155

d.

t_i	w_{1i}	$y(t_i)$
1.200	0.27273759	0.27273791
1.500	1.08847933	1.08849259
1.700	2.04352376	2.04353642
2.000	4.36157310	4.36157780

7. The predicted number of prey, x_{1i}, and predators, x_{2i}, are given in the following table.

i	t_i	x_{1i}	x_{2i}
10	1.0	4393	1512
20	2.0	288	3175
30	3.0	32	2042
40	4.0	25	1258

Exercise Set 5.10

1. Let L be the Lipschitz constant for ϕ. Then

$$u_{i+1} - v_{i+1} = u_i - v_i + h[\phi(t_i, u_i, h) - \phi(t_i, v_i, h)],$$

so

$$|u_{i+1} - v_{i+1}| \leq (1 + hL)|u_i - v_i| \leq (1 + hL)^{i+1}|u_0 - v_0|.$$

3. By Exercise 17 in Section 5.4 we have

$$\phi(t, w, h) = \frac{1}{6}f(t, w) + \frac{1}{3}f\left(t + \frac{1}{2}h, w + \frac{1}{2}hf(t, w)\right) + \frac{1}{3}f\left(t + \frac{1}{2}h, w + \frac{1}{2}hf\left(t + \frac{1}{2}h, w + \frac{1}{2}hf(t, w)\right)\right)$$
$$+ \frac{1}{6}f\left(t + h, w + hf\left(t + \frac{1}{2}h, w + \frac{1}{2}hf\left(t + \frac{1}{2}h, w + \frac{1}{2}hf(t, w)\right)\right)\right),$$

so

$$\phi(t, w, 0) = \frac{1}{6}f(t, w) + \frac{1}{3}f(t, w) + \frac{1}{3}f(t, w) + \frac{1}{6}f(t, w) = f(t, w).$$

5. a. The local truncation error is $\tau_{i+1} = \frac{1}{4}h^3 y^{(4)}(\xi_i)$ for some ξ_i, where $t_{i-2} < \xi_i < t_{i+1}$.

b. The method is consistent but unstable and not convergent.

7. The method is unstable.

Exercise Set 5.11

1. Euler's method gives the results in the following tables.

a.

t_i	w_i	y_i
0.200	0.027182818	0.449328964
0.500	0.000027183	0.030197383
0.700	0.000000272	0.004991594
1.000	0.000000000	0.000335463

b.

t_i	w_i	y_i
0.200	0.373333333	0.046105213
0.500	−0.093333333	0.250015133
0.700	0.146666667	0.490000277
1.000	1.333333333	1.000000001

c.

t_i	w_i	y_i
0.500	16.47925	0.479470939
1.000	256.7930	0.841470987
1.500	4096.142	0.997494987
2.000	65523.12	0.909297427

d.

t_i	w_i	y_i
0.200	6.128259	1.000000001
0.500	−378.2574	1.000000000
0.700	−6052.063	1.000000000
1.000	387332.0	1.000000000

3. The Adams Fourth-Order Predictor-Corrector Algorithm gives the results in the following tables.

a.

t_i	w_i	y_i
0.200	0.4588119	0.4493290
0.500	−0.0112813	0.0301974
0.700	0.0013734	0.0049916
1.000	0.0023604	0.0003355

b.

t_i	w_i	y_i
0.200	0.0792593	0.0461052
0.500	0.1554027	0.2500151
0.700	0.5507445	0.4900003
1.000	0.7278557	1.0000000

c.

t_i	w_i	y_i
.500	188.3082	0.4794709
1.000	38932.03	0.8414710
1.500	9073607	0.9974950
2.000	2115741299	0.9092974

d.

t_i	w_i	y_i
0.200	−215.7459	1.000000001
0.500	−682637.0	1.000000000
0.700	−159172736	1.000000000
1.000	−566751172258	1.000000000

5. a.

t_i	w_{1i}	u_{1i}	w_{2i}	u_{2i}
0.100	−96.33011	0.66987648	193.6651	−0.33491554
0.200	−28226.32	0.67915383	56453.66	−0.33957692
0.300	−8214056	0.69387881	16428113	−0.34693941
0.400	−2390290586	0.71354670	4780581173	−0.35677335
0.500	−695574560790	0.73768711	1391149121600	−0.36884355

b.

t_i	w_{1i}	u_{1i}	w_{2i}	u_{2i}
0.100	0.61095960	0.66987648	−0.21708179	−0.33491554
0.200	0.66873489	0.67915383	−0.31873903	−0.33957692
0.300	0.69203679	0.69387881	−0.34325535	−0.34693941
0.400	0.71322103	0.71354670	−0.35612202	−0.35677335
0.500	0.73762953	0.73768711	−0.36872840	−0.36884355

9. a. The Trapezoidal method applied to the test equation gives

$$w_{j+1} = \frac{1 + \frac{h\lambda}{2}}{1 - \frac{h\lambda}{2}} w_j$$

so

$$Q(h\lambda) = \frac{2 + h\lambda}{2 - h\lambda}.$$

Thus, $|Q(h\lambda)| < 1$ whenever $\text{Re}(h\lambda) < 0$.

b. The Backward Euler method applied to the test equation gives

$$w_{j+1} = \frac{w_j}{1 - h\lambda}$$

so

$$Q(h\lambda) = \frac{1}{1 - h\lambda}.$$

Thus, $|Q(h\lambda)| < 1$ whenever $\mathrm{Re}(h\lambda) < 0$.

Exercise Set 6.1

1. **a.** Intersecting lines with solution $x_1 = x_2 = 1$.
b. Intersecting lines with solution $x_1 = x_2 = 0$.
c. One line, so there is an infinite number of solutions with $x_2 = \frac{3}{2} - \frac{1}{2}x_1$.
d. Parallel lines, so there is no solution.
e. One line, so there is an infinite number of solutions with $x_2 = -\frac{1}{2}x_1$.
f. Three lines in the plane that do not intersect at a common point.
g. Intersecting lines with solution $x_1 = \frac{2}{7}$ and $x_2 = -\frac{11}{7}$.
h. Two planes in space which intersect in a line with $x_1 = -\frac{5}{4}x_2$ and $x_3 = \frac{3}{2}x_2 + 1$.

3. Gaussian elimination gives the following solutions.
a. $x_1 = 1.1875$, $x_2 = 1.8125$, $x_3 = 0.875$ with one row interchange required.
b. $x_1 = -1$, $x_2 = 0$, $x_3 = 1$ with no interchange required.
c. $x_1 = 1.5$, $x_2 = 2$, $x_3 = -1.2$, $x_4 = 3$ with no interchange required.
d. $x_1 = \frac{22}{9}$, $x_2 = -\frac{4}{9}$, $x_3 = \frac{4}{3}$, $x_4 = 1$ with one row interchange required.
e. no unique solution.
f. $x_1 = -1$, $x_2 = 2$, $x_3 = 0$, $x_4 = 1$ with one row interchange required.

5. **a.** When $\alpha = -1/3$ there is no solution.
b. When $\alpha = 1/3$ there is an infinite number of solutions with $x_1 = x_2 + 1.5$ and x_2 is arbitrary.
c. If $\alpha \neq \pm 1/3$, then the unique solution is

$$x_1 = \frac{3}{2(1+3\alpha)} \quad \text{and} \quad x_2 = \frac{-3}{2(1+3\alpha)}.$$

9. The Gauss-Jordan method gives the following results.
a. $x_1 = 0.98$, $x_2 = -0.98$, $x_3 = 2.9$
b. $x_1 = 1.1$, $x_2 = -1.0$, $x_3 = 2.9$

11. **b.** The results for this exercise are listed in the following table. (The abbreviations M/D and A/S are used for multiplications/divisions and additions/subtractions, respectively.)

	Gaussian Elimination		Gauss-Jordan	
n	M/D	A/S	M/D	A/S
3	17	11	21	12
10	430	375	595	495
50	44150	42875	64975	62475
100	343300	338250	509950	499950

13. The Gaussian-Elimination–Gauss-Jordan hybrid method gives the following results.
a. $x_1 = 1.0$, $x_2 = -0.98$, $x_3 = 2.9$
b. $x_1 = 1.0$, $x_2 = -1.0$, $x_3 = 2.9$

15. **a.** There is sufficient food to satisfy the average daily consumption.
b. We could add 200 of species 1, or 150 of species 2, or 100 of species 3, or 100 of species 4.

c. Assuming none of the increases indicated in part (b) was selected, species 2 could be increased by 650, or species 3 could be increased by 150, or species 4 could be increased by 150.

d. Assuming none of the increases indicated in parts (b) or (c) was selected, species 3 could be increased by 150, or species 4 could be increased by 150.

Exercise Set 6.2

1. a. none **b.** Interchange rows 2 and 3 **c.** none **d.** Interchange rows 1 and 2

3. a. Interchange rows 1 and 3, then interchange rows 2 and 3.

b. Interchange rows 2 and 3.

c. Interchange rows 2 and 3.

d. Interchange rows 1 and 3, then interchange rows 2 and 3.

5. Gaussian elimination with three-digit chopping arithmetic gives the following results.

a. $x_1 = 30.0$, $x_2 = 0.990$

b. $x_1 = 1.00$, $x_2 = 9.98$

c. $x_1 = 0.00$, $x_2 = 10.0$, $x_3 = 0.142$

d. $x_1 = 12.0$, $x_2 = 0.492$, $x_3 = -9.78$

e. $x_1 = 0.206$, $x_2 = 0.0154$, $x_3 = -0.0156$, $x_4 = -0.716$

f. $x_1 = 0.828$, $x_2 = -3.32$, $x_3 = 0.153$, $x_4 = 4.91$

7. Gaussian elimination with partial pivoting and three-digit chopping arithmetic gives the following results.

a. $x_1 = 10.0$, $x_2 = 1.00$

b. $x_1 = 1.00$, $x_2 = 9.98$

c. $x_1 = -0.163$, $x_2 = 9.98$, $x_3 = 0.142$

d. $x_1 = 12.0$, $x_2 = 0.504$, $x_3 = -9.78$

e. $x_1 = 0.177$, $x_2 = -0.0072$, $x_3 = -0.0208$, $x_4 = -1.18$

f. $x_1 = 0.777$, $x_2 = -3.10$, $x_3 = 0.161$, $x_4 = 4.50$

9. Gaussian elimination with scaled partial pivoting and three-digit chopping arithmetic gives the following results.

a. $x_1 = 10.0$, $x_2 = 1.00$

b. $x_1 = 1.00$, $x_2 = 9.98$

c. $x_1 = -0.163$, $x_2 = 9.98$, $x_3 = 0.142$

d. $x_1 = 0.993$, $x_2 = 0.500$, $x_3 = -1.00$

e. $x_1 = 0.171$, $x_2 = 0.0102$, $x_3 = -0.0217$, $x_4 = -1.27$

f. $x_1 = 0.687$, $x_2 = -2.66$, $x_3 = 0.117$, $x_4 = 3.59$

11. The Gaussian Elimination with Backward Substitution Algorithm and single-precision arithmetic gives the following results.
For (1a) we have $x_1 = 10.000000$, $x_2 = 1.0000000$.
For (1b) we have $x_1 = 1.0000000$, $x_2 = 10.000000$.
For (1c) we have $x_1 = 0.0000000$, $x_2 = 10.000000$, $x_3 = 0.14285714$.
For (1d) we have $x_1 = 0.99999999$, $x_2 = 0.50000000$, $x_3 = -1.00000000$.
For (1e) we have $x_1 = 0.17682530$, $x_2 = 0.012692691$, $x_3 = -0.020654050$, $x_4 = -1.1826087$.
For (1f) we have $x_1 = 0.78838790$, $x_2 = -3.1253894$, $x_3 = 0.1675964$, $x_4 = 4.5569519$.

13. The Gaussian Elimination with Scaled Partial Pivoting Algorithm and single-precision arithmetic gives the following results.
For (1a) we have $x_1 = 10.000000$, $x_2 = 1.0000000$.
For (1b) we have $x_1 = 1.0000000$, $x_2 = 10.000000$.
For (1c) we have $x_1 = 0.0000000$, $x_2 = 10.000000$, $x_3 = 0.14285714$.
For (1d) we have $x_1 = 1.00000000$, $x_2 = 0.50000000$, $x_3 = -1.00000000$.

For (1e) we have $x_1 = 0.17682530$, $x_2 = 0.012692691$, $x_3 = -0.020654050$, $x_4 = -1.1826087$.
For (1f) we have $x_1 = 0.78838790$, $x_2 = -3.1253894$, $x_3 = 0.1675946$, $x_4 = 4.5569519$.

15. The total pivoting algorithm with single-precision arithmetic gives the following results.

a. For (1a) we have $x_1 = 9.98$, $x_2 = 1.00$.
For (1b) we have $x_1 = 1.00$, $x_2 = 9.98$.
For (1c) we have $x_1 = 0.0724$, $x_2 = 10.0$, $x_3 = 0.0952$.
For (1d) we have $x_1 = 0.982$, $x_2 = 0.500$, $x_3 = -0.994$.
For (1e) we have $x_1 = 0.161$, $x_2 = 0.0125$, $x_3 = -0.0232$, $x_4 = -1.42$.
For (1f) we have $x_1 = 0.719$, $x_2 = -2.86$, $x_3 = 0.146$, $x_4 = 4.00$.

b. For (2a) we have $x_1 = 10.0$, $x_2 = 1.00$.
For (2b) we have $x_1 = 1.00$, $x_2 = 10.0$.
For (2c) we have $x_1 = 0.00$, $x_2 = 10.0$, $x_3 = 0.143$.
For (2d) we have $x_1 = 1.01$, $x_2 = 0.501$, $x_3 = -1.00$.
For (2e) we have $x_1 = 0.179$, $x_2 = 0.0127$, $x_3 = -0.0203$, $x_4 = -1.15$.
For (2f) we have $x_1 = 0.874$, $x_2 = -3.49$, $x_3 = 0.192$, $x_4 = 5.33$.

c. For (7a) we have $x_1 = 10.000000$, $x_2 = 1.0000000$.
For (7b) we have $x_1 = 1.0000000$, $x_2 = 10.000000$.
For (7c) we have $x_1 = 0.0000000$, $x_2 = 10.000000$, $x_3 = 0.14285714$.
For (7d) we have $x_1 = 1.00000000$, $x_2 = 0.50000000$, $x_3 = -1.00000000$.
For (7e) we have $x_1 = 0.17682530$, $x_2 = 0.012692691$, $x_3 = -0.020654050$, $x_4 = -1.1826087$.
For (7f) we have $x_1 = 0.78838790$, $x_2 = -3.1253894$, $x_3 = 0.16759460$, $x_4 = 4.5569519$.

Exercise Set 6.3

1. a. The matrix is singular. **b.** $\begin{bmatrix} -\frac{1}{4} & \frac{1}{4} & \frac{1}{4} \\ \frac{5}{8} & -\frac{1}{8} & -\frac{1}{8} \\ \frac{1}{8} & -\frac{5}{8} & \frac{3}{8} \end{bmatrix}$ **c.** The matrix is singular. **d.** The matrix is singular.

e. $\begin{bmatrix} \frac{1}{4} & 0 & 0 & 0 \\ -\frac{3}{14} & \frac{1}{7} & 0 & 0 \\ \frac{3}{28} & -\frac{11}{7} & 1 & 0 \\ -\frac{1}{2} & 1 & -1 & 1 \end{bmatrix}$ **f.** $\begin{bmatrix} 1 & 0 & 1 & -1 \\ -1 & \frac{5}{3} & \frac{5}{3} & -1 \\ -1 & \frac{2}{3} & \frac{2}{3} & 0 \\ 0 & -\frac{1}{3} & -\frac{4}{3} & 1 \end{bmatrix}$

3. The solutions to the linear systems obtained in parts (a) and (b) are, from left to right,

$$3,\ -6,\ -2,\ -1 \quad \text{and} \quad 1,\ 1,\ 1,\ 1.$$

5. a. Suppose $\tilde{A}$ and $\hat{A}$ are both inverses of A. Then $A\tilde{A} = \tilde{A}A = I$ and $A\hat{A} = \hat{A}A = I$. Thus,

$$\tilde{A} = \tilde{A}I = \tilde{A}(A\hat{A}) = (\tilde{A}A)\hat{A} = I\hat{A} = \hat{A}.$$

b. $(AB)(B^{-1}A^{-1}) = A(BB^{-1})A^{-1} = AIA^{-1} = AA^{-1} = I$ and $(B^{-1}A^{-1})(AB) = B^{-1}(A^{-1}A)B = B^{-1}IB = B^{-1}B = I$, so $(AB)^{-1} = B^{-1}A^{-1}$ since there is only one inverse.

c. Since $A^{-1}A = AA^{-1} = I$, it follows that A^{-1} is nonsingular. Since the inverse is unique, we have $(A^{-1})^{-1} = A$.

7. a. If $C = AB$, where A and B are lower triangular, then $a_{ik} = 0$ if $k > i$ and $b_{kj} = 0$ if $k < j$. Thus,

$$c_{ij} = \sum_{k=1}^{n} a_{ik}b_{kj} = \sum_{k=j}^{i} a_{ik}b_{kj},$$

which will have the sum zero unless $j \leq i$. Hence C is lower triangular.

b. We have $a_{ik} = 0$ if $k < i$ and $b_{kj} = 0$ if $k > j$. The steps are similar to those in part (a).

c. Let L be a nonsingular lower triangular matrix. To obtain the ith column of L^{-1}, solve n linear systems of the form

$$\begin{bmatrix} l_{11} & 0 & \cdots & \cdots & 0 \\ l_{21} & l_{22} & \ddots & & \vdots \\ \vdots & \vdots & \ddots & \ddots & \vdots \\ l_{i1} & l_{i2} & \cdots & l_{ii} & \vdots \\ \vdots & \vdots & & \ddots & 0 \\ l_{n1} & l_{n2} & \cdots & \cdots & l_{nn} \end{bmatrix} \begin{bmatrix} x_1 \\ x_2 \\ \vdots \\ x_i \\ \vdots \\ x_n \end{bmatrix} = \begin{bmatrix} 0 \\ 0 \\ \vdots \\ 0 \\ 1 \\ 0 \\ \vdots \\ 0 \end{bmatrix},$$

where the 1 appears in the ith position, to obtain the ith column of L^{-1}.

9. The answers are the same as those in Exercise 1.

11. a.

$$A^2 = \begin{bmatrix} 0 & 2 & 0 \\ 0 & 0 & 3 \\ \frac{1}{6} & 0 & 0 \end{bmatrix}, \quad A^3 = \begin{bmatrix} 1 & 0 & 0 \\ 0 & 1 & 0 \\ 0 & 0 & 1 \end{bmatrix},$$

$A^4 = A, \quad A^5 = A^2, \quad A^6 = I, \ldots$

b.

	Year 1	Year 2	Year 3	Year 4
Age 1	6000	36000	12000	6000
Age 2	6000	3000	18000	6000
Age 3	6000	2000	1000	6000

c.

$$A^{-1} = \begin{bmatrix} 0 & 2 & 0 \\ 0 & 0 & 3 \\ \frac{1}{6} & 0 & 0 \end{bmatrix}.$$

The i, j-entry is the number of beetles of age i necessary to produce one beetle of age j.

13. a. We have

$$\begin{bmatrix} 7 & 4 & 4 & 0 \\ -6 & -3 & -6 & 0 \\ 0 & 0 & 3 & 0 \\ 0 & 0 & 0 & 1 \end{bmatrix} \begin{bmatrix} 2(x_0 - x_1) + \alpha_0 + \alpha_1 \\ 3(x_1 - x_0) - \alpha_1 - 2\alpha_0 \\ \alpha_0 \\ x_0 \end{bmatrix} = \begin{bmatrix} 2(x_0 - x_1) + 3\alpha_0 + 3\alpha_1 \\ 3(x_1 - x_0) - 3\alpha_1 - 6\alpha_0 \\ 3\alpha_0 \\ x_0 \end{bmatrix}$$

b. B

Exercise

1. The de ... 3

3. We ha

5. $\alpha = -$

7. $\alpha = -$

11. a. The

b. We h ... no solutions.

c. We ha ... solutions.

e. Crame ... ons/subtractions.

Exercise Se

1. a. $x_1 = -$

b. $x_1 = \frac{1}{2},$

3. a. $P = \begin{bmatrix} 1 \\ 0 \\ 0 \end{bmatrix}$ ...

b. $P = \begin{bmatrix} 0 \\ 1 \\ 0 \end{bmatrix}$ ...

c. $P = \begin{bmatrix} 1 & 0 & 0 & 0 \\ 0 & 0 & 1 & 0 \\ 0 & 1 & 0 & 0 \\ 0 & 0 & 0 & 1 \end{bmatrix}$

d. $P = \begin{bmatrix} 0 & 0 & 1 & 0 \\ 0 & 1 & 0 & 0 \\ 0 & 0 & 0 & 1 \\ 1 & 0 & 0 & 0 \end{bmatrix}$

5. a. $P^tLU = \begin{bmatrix} 0 & 1 & 0 \\ 1 & 0 & 0 \\ 0 & 0 & 1 \end{bmatrix} \begin{bmatrix} 1 & 0 & 0 \\ 0 & 1 & 0 \\ 0 & -\frac{1}{2} & 1 \end{bmatrix} \begin{bmatrix} 1 & 1 & -1 \\ 0 & 2 & 3 \\ 0 & 0 & \frac{5}{2} \end{bmatrix}$

b.

$$P^tLU = \begin{bmatrix} 1 & 0 & 0 \\ 0 & 0 & 1 \\ 0 & 1 & 0 \end{bmatrix} \begin{bmatrix} 1 & 0 & 0 \\ 2 & 1 & 0 \\ 1 & 0 & 1 \end{bmatrix} \begin{bmatrix} 1 & 2 & -1 \\ 0 & -5 & 6 \\ 0 & 0 & 4 \end{bmatrix}$$

c.

$$P^tLU = \begin{bmatrix} 1 & 0 & 0 & 0 \\ 0 & 0 & 0 & 1 \\ 0 & 1 & 0 & 0 \\ 0 & 0 & 1 & 0 \end{bmatrix} \begin{bmatrix} 1 & 0 & 0 & 0 \\ 2 & 1 & 0 & 0 \\ 1 & 0 & 1 & 0 \\ 3 & 0 & 0 & 1 \end{bmatrix} \begin{bmatrix} 1 & -2 & 3 & 0 \\ 0 & 5 & -2 & 1 \\ 0 & 0 & -1 & -2 \\ 0 & 0 & 0 & 3 \end{bmatrix}$$

d.

$$P^tLU = \begin{bmatrix} 1 & 0 & 0 & 0 \\ 0 & 0 & 0 & 1 \\ 0 & 0 & 1 & 0 \\ 0 & 1 & 0 & 0 \end{bmatrix} \begin{bmatrix} 1 & 0 & 0 & 0 \\ 2 & 1 & 0 & 0 \\ 1 & 0 & 1 & 0 \\ 1 & 0 & 0 & 1 \end{bmatrix} \begin{bmatrix} 1 & -2 & 3 & 0 \\ 0 & 5 & -3 & -1 \\ 0 & 0 & -1 & -2 \\ 0 & 0 & 0 & 1 \end{bmatrix}$$

7. c.

	Multiplications/Divisions	Additions/Subtractions
Factoring into LU	$\frac{1}{3}n^3 - \frac{1}{3}n$	$\frac{1}{3}n^3 - \frac{1}{2}n^2 + \frac{1}{6}n$
Solving $Ly = b$	$\frac{1}{2}n^2 - \frac{1}{2}n$	$\frac{1}{2}n^2 - \frac{1}{2}n$
Solving $Ux = y$	$\frac{1}{2}n^2 + \frac{1}{2}n$	$\frac{1}{2}n^2 - \frac{1}{2}n$
Total	$\frac{1}{3}n^3 + n^2 - \frac{1}{3}n$	$\frac{1}{3}n^3 + \frac{1}{2}n^2 - \frac{5}{6}n$

d.

	Multiplications/Divisions	Additions/Subtractions
Factoring into LU	$\frac{1}{3}n^3 - \frac{1}{3}n$	$\frac{1}{3}n^3 - \frac{1}{2}n^2 + \frac{1}{6}n$
Solving $Ly^{(k)} = b^{(k)}$	$(\frac{1}{2}n^2 - \frac{1}{2}n)m$	$(\frac{1}{2}n^2 - \frac{1}{2}n)m$
Solving $Ux^{(k)} = y^{(k)}$	$(\frac{1}{2}n^2 + \frac{1}{2}n)m$	$(\frac{1}{2}n^2 - \frac{1}{2}n)m$
Total	$\frac{1}{3}n^3 + mn^2 - \frac{1}{3}n$	$\frac{1}{3}n^3 + (m - \frac{1}{2})n^2 - (m - \frac{1}{6})n$

Exercise Set 6.6

1. **(i)** The symmetric matrices are in (a), (b), and (f).
(ii) The singular matrices are in (e) and (h).
(iii) The strictly diagonally dominant matrices are in (a), (b), (c), and (d).
(iv) The positive definite matrices are in (a) and (f).

3. Choleski's Algorithm gives the following results.

a.

$$L = \begin{bmatrix} 1.414213 & 0 & 0 \\ -0.7071069 & 1.224743 & 0 \\ 0 & -0.8164972 & 1.154699 \end{bmatrix}$$

b.

$$L = \begin{bmatrix} 2 & 0 & 0 & 0 \\ 0.5 & 1.658311 & 0 & 0 \\ 0.5 & -0.7537785 & 1.087113 & 0 \\ 0.5 & 0.4522671 & 0.08362442 & 1.240346 \end{bmatrix}$$

c.

$$L = \begin{bmatrix} 2 & 0 & 0 & 0 \\ 0.5 & 1.658311 & 0 & 0 \\ -0.5 & -0.4522671 & 2.132006 & 0 \\ 0 & 0 & 0.9380833 & 1.766351 \end{bmatrix}$$

d.

$$L = \begin{bmatrix} 2.449489 & 0 & 0 & 0 \\ 0.8164966 & 1.825741 & 0 & 0 \\ 0.4082483 & 0.3651483 & 1.923538 & 0 \\ -0.4082483 & 0.1825741 & -0.4678876 & 1.606574 \end{bmatrix}$$

5. The modified Choleski's algorithm gives the following results.

a. $x_1 = 1, x_2 = -1, x_3 = 0$

b. $x_1 = 0.2, x_2 = -0.2, x_3 = -0.2, x_4 = 0.25$

c. $x_1 = 1, x_2 = 2, x_3 = -1, x_4 = 2$

d. $x_1 = -0.85863874, x_2 = 2.4188482, x_3 = -0.95811518, x_4 = -1.2722513$

7. We have $x_i = 1$ for each $i = 1, \ldots, 10$.

9. Only the matrix in (d) is positive definite.

11. $-2 < \alpha < \frac{3}{2}$

13. $0 < \beta < 1$ and $3 < \alpha < 5 - \beta$

15. a. No, consider $\begin{bmatrix} 1 & 0 \\ 0 & 1 \end{bmatrix}$.

b. Yes, since $A = A^t$.

c. Yes, since $\mathbf{x}^t(A + B)\mathbf{x} = \mathbf{x}^t A\mathbf{x} + \mathbf{x}^t B\mathbf{x}$.

d. Yes, since $\mathbf{x}^t A^2\mathbf{x} = \mathbf{x}^t A^t A\mathbf{x} = (A\mathbf{x})^t(A\mathbf{x}) \geq 0$ and because A is nonsingular, equality holds only if $\mathbf{x} = \mathbf{0}$.

e. No, consider $A = \begin{bmatrix} 1 & 0 \\ 0 & 1 \end{bmatrix}$ and $B = \begin{bmatrix} 10 & 0 \\ 0 & 10 \end{bmatrix}$.

17. a. Since $\det A = 3\alpha - 2\beta$, A is singular if and only if $\alpha = 2\beta/3$. **b.** $|\alpha| > 1, |\beta| < 1$ **c.** $\beta = 1$
d. $\alpha > \frac{2}{3}, \beta = 1$

19. $A = \begin{bmatrix} 1.0 & 0.2 \\ 0.1 & 1.0 \end{bmatrix}$

23. $i_1 = 0.6785047$, $i_2 = 0.4214953$,
$i_3 = 0.2570093$, $i_4 = 0.1542056$,
$i_5 = 0.1028037$

25. a. Mating male i with female j produces offspring with the same wing characteristics as mating male j with female i.

b. No. Consider, for example, $\mathbf{x} = (1, 0, -1)^t$.

Exercise Set 7.1

1. a. We have $\|\mathbf{x}\|_\infty = 4$ and $\|\mathbf{x}\|_2 = 5.220153$.

b. We have $\|\mathbf{x}\|_\infty = 4$ and $\|\mathbf{x}\|_2 = 5.477226$.

c. We have $\|\mathbf{x}\|_\infty = 2^k$ and $\|\mathbf{x}\|_2 = (1 + 4^k)^{1/2}$.

d. We have $\|\mathbf{x}\|_\infty = 4/(k + 1)$ and $\|\mathbf{x}\|_2 = (16/(k + 1)^2 + 4/k^4 + k^4 e^{-2k})^{1/2}$.

3. a. We have $\lim_{k\to\infty} \mathbf{x}^{(k)} = (0, 0, 0)^t$.

b. We have $\lim_{k\to\infty} \mathbf{x}^{(k)} = (0, 1, 3)^t$.

c. We have $\lim_{k\to\infty} \mathbf{x}^{(k)} = (0, 0, \frac{1}{2})^t$.

d. We have $\lim_{k\to\infty} \mathbf{x}^{(k)} = (1, -1, 1)^t$.

5. a. We have $||\mathbf{x} - \hat{\mathbf{x}}||_\infty = 8.57 \times 10^{-4}$ and $||A\hat{\mathbf{x}} - \mathbf{b}||_\infty = 2.06 \times 10^{-4}$.

b. We have $||\mathbf{x} - \hat{\mathbf{x}}||_\infty = 0.90$ and $||A\hat{\mathbf{x}} - \mathbf{b}||_\infty = 0.27$.

c. We have $||\mathbf{x} - \hat{\mathbf{x}}||_\infty = 0.5$ and $||A\hat{\mathbf{x}} - \mathbf{b}||_\infty = 0.3$.

d. We have $||\mathbf{x} - \hat{\mathbf{x}}||_\infty = 6.55 \times 10^{-2}$, and $||A\hat{\mathbf{x}} - \mathbf{b}||_\infty = 0.32$.

7. Let $A = \begin{bmatrix} 1 & 1 \\ 0 & 1 \end{bmatrix}$ and $B = \begin{bmatrix} 1 & 0 \\ 1 & 1 \end{bmatrix}$. Then $||AB||_{\otimes} = 2$, but $||A||_{\otimes} \cdot ||B||_{\otimes} = 1$.

9. b. We have

(4a) $||A||_F = \sqrt{326}$

(4b) $||A||_F = \sqrt{326}$

(4c) $||A||_F = 4$

(4d) $||A||_F = \sqrt{148}$

Exercise Set 7.2

1. a. The eigenvalue $\lambda_1 = 3$ has the eigenvector $\mathbf{x}_1 = (1, -1)^t$, and the eigenvalue $\lambda_2 = 1$ has the eigenvector $\mathbf{x}_2 = (1, 1)^t$.

b. The eigenvalue $\lambda_1 = \frac{1+\sqrt{5}}{2}$ has the eigenvector $\mathbf{x} = \left(1, \frac{1+\sqrt{5}}{2}\right)^t$, and the eigenvalue $\lambda_2 = \frac{1-\sqrt{5}}{2}$ has the eigenvector $\mathbf{x} = \left(1, \frac{1-\sqrt{5}}{2}\right)^t$.

c. The eigenvalue $\lambda_1 = \frac{1}{2}$ has the eigenvector $\mathbf{x}_1 = (1, 1)^t$ and the eigenvalue $\lambda_2 = -\frac{1}{2}$ has the eigenvector $\mathbf{x}_2 = (1, -1)^t$.

d. The eigenvalue $\lambda_1 = 0$ has the eigenvector $\mathbf{x}_1 = (1, -1)^t$ and the eigenvalue $\lambda_2 = -1$ has the eigenvector $\mathbf{x}_2 = (1, -2)^t$.

e. The eigenvalue $\lambda_1 = \lambda_2 = 3$ has the eigenvectors $\mathbf{x}_1 = (0, 0, 1)^t$ and $\mathbf{x}_2 = (1, 1, 0)^t$, and the eigenvalue $\lambda_3 = 1$ has the eigenvector $\mathbf{x}_3 = (1, -1, 0)^t$.

f. The eigenvalue $\lambda_1 = 7$ has the eigenvector $\mathbf{x}_1 = (1, 4, 4)^t$, the eigenvalue $\lambda_2 = 3$ has the eigenvector $\mathbf{x}_2 = (1, 2, 0)^t$, and the eigenvalue $\lambda_3 = -1$ has the eigenvector $\mathbf{x}_3 = (1, 0, 0)^t$.

g. The eigenvalue $\lambda_1 = \lambda_2 = 1$ has the eigenvectors $\mathbf{x}_1 = (-1, 1, 0)^t$ and $\mathbf{x}_2 = (-1, 0, 1)^t$, and the eigenvalue $\lambda_3 = 5$ has the eigenvector $\mathbf{x}_3 = (1, 2, 1)^t$.

h. The eigenvalue $\lambda_1 = 3$ has the eigenvector $\mathbf{x}_1 = (-1, 1, 2)^t$, the eigenvalue $\lambda_2 = 4$ has the eigenvector $\mathbf{x}_2 = (0, 1, 2)^t$, and the eigenvalue $\lambda_3 = -2$ has the eigenvector $\mathbf{x} = (-3, 8, 1)^t$.

3. Only the matrix in (c) is convergent.

5. a. 3 **b.** 1.618034 **c.** 0.5 **d.** 3.162278 **e.** 3 **f.** 8.224257 **g.** 5.203527 **h.** 5.601152

9. Since $A^tA = A^2$ and $A\mathbf{x} = \lambda\mathbf{x}$, we have $A^2\mathbf{x} = \lambda^2\mathbf{x}$. Thus, $\rho(A^tA) = \rho(A^2) = [\rho(A)]^2$ and $||A||_2 = [\rho(A^tA)]^{1/2} = \rho(A)$.

11. a. We have the real eigenvalue $\lambda = 1$ with the eigenvector $\mathbf{x} = (6, 3, 1)^t$.

b. Choose any multiple of the vector $(6, 3, 1)^t$.

13. Let $A\mathbf{x} = \lambda\mathbf{x}$. Then $|\lambda|\ ||\mathbf{x}|| = ||A\mathbf{x}|| \leq ||A||\ ||\mathbf{x}||$, which implies $|\lambda| \leq ||A||$. Also, $(1/\lambda)\mathbf{x} = A^{-1}\mathbf{x}$ so $1/|\lambda| \leq ||A^{-1}||$ and $||A^{-1}||^{-1} \leq |\lambda|$.

Exercise Set 7.3

1. Two iterations of Jacobi's method gives the following results.

a. $(0.1428571, -0.3571429, 0.4285714)^t$

b. $(0.97, 0.91, 0.74)^t$

c. $(-0.65, 1.65, -0.4, -2.475)^t$

d. $(-0.5208333, -0.04166667, -0.2166667, 0.4166667)^t$

e. $(1.325, -1.6, 1.6, 1.675, 2.425)^t$

f. $(0.6875, 1.125, 0.6875, 1.375, 0.5625, 1.375)^t$

3. Jacobi's Algorithm gives the following results.

a. $\mathbf{x}^{(10)} = (0.03507839, -0.2369262, 0.6578015)^t$

b. $\mathbf{x}^{(6)} = (0.9957250, 0.9577750, 0.7914500)^t$

c. $\mathbf{x}^{(22)} = (-0.7975853, 2.794795, -0.2588888, -2.251879)^t$

d. $\mathbf{x}^{(14)} = (-0.7529267, 0.04078538, -0.2806091, 0.6911662)^t$

e. $\mathbf{x}^{(12)} = (0.7870883, -1.003036, 1.866048, 1.912449, 1.985707)^t$

f. $\mathbf{x}^{(17)} = (0.9996805, 1.999774, 0.9996805, 1.999840, 0.9995482, 1.999840)^t$

5. Two iterations of the SOR method give the following results.

a. $(0.05410079, -0.2115435, 0.6477159)^t$

b. $(0.9876790, 0.9784935, 0.7899328)^t$

c. $(-0.71885, 2.818822, -0.2809726, -2.235422)^t$

d. $(-0.6604902, 0.03700749, -0.2493513, 0.6561139)^t$

e. $(1.079675, -1.260654, 2.042489, 1.995373, 2.049536)^t$

f. $(0.8318750, 1.647766, 0.9189856, 1.791281, 0.8712129, 1.959155)^t$

7. The SOR Algorithm gives the following results.

a. $\mathbf{x}^{(12)} = (0.03488469, -0.2366474, 0.6579013)^t$

b. $\mathbf{x}^{(7)} = (0.9958341, 0.9579041, 0.7915756)^t$

c. $\mathbf{x}^{(8)} = (-0.7976009, 2.795288, -0.2588293, -2.251768)^t$

d. $\mathbf{x}^{(7)} = (-0.7534489, 0.04106617, -0.2808146, 0.6918049)^t$

e. $\mathbf{x}^{(10)} = (0.7866310, -1.002807, 1.866530, 1.912645, 1.989792)^t$

f. $\mathbf{x}^{(7)} = (0.9999442, 1.999934, 1.000033, 1.999958, 0.9999815, 2.000007)^t$

9. a. Subtract $\mathbf{x} = T\mathbf{x} + \mathbf{c}$ from $\mathbf{x}^{(k)} = T\mathbf{x}^{(k-1)} + \mathbf{c}$ to obtain $\mathbf{x}^{(k)} - \mathbf{x} = T(\mathbf{x}^{(k-1)} - \mathbf{x})$. Thus,

$$\|\mathbf{x}^{(k)} - \mathbf{x}\| \le \|T\|\,\|\mathbf{x}^{(k-1)} - \mathbf{x}\|.$$

Inductively, we have

$$\|\mathbf{x}^{(k)} - \mathbf{x}\| \le \|T\|^k\|\mathbf{x}^{(0)} - \mathbf{x}\|.$$

The remainder of the proof is similar to the proof of Corollary 2.4.

b. The last column has no entry when $\|T\|_\infty = 1$.

	$\|\mathbf{x}^{(2)} - \mathbf{x}\|_\infty$	$\|T\|_\infty$	$\|T\|_\infty^2\|\mathbf{x}^{(0)} - \mathbf{x}\|_\infty$	$\frac{\|T\|_\infty^2}{1-\|T\|_\infty}\|\mathbf{x}^{(1)} - \mathbf{x}^{(0)}\|_\infty$
1 (a)	0.22932	0.857143	0.48335	2.9388
1 (b)	0.051579	0.3	0.089621	0.11571
1 (c)	1.1453	0.9	2.2642	20.25
1 (d)	0.27511	1	0.75342	
1 (e)	0.59743	1	1.9897	
1 (f)	0.875	0.75	1.125	3.375

13.

a.

	Jacobi 49 iterations	Gauss-Seidel 28 iterations	SOR ($\omega = 1.3$) 13 iterations
x_1	0.93406183	0.93406917	0.93407584
x_2	0.97473885	0.97475285	0.97476180
x_3	1.10688692	1.10690302	1.10691093
x_4	1.42346150	1.42347226	1.42347591
x_5	0.85931331	0.85932730	0.85933633
x_6	0.80688119	0.80690725	0.80691961
x_7	0.85367746	0.85370564	0.85371536
x_8	1.10688692	1.10690579	1.10691075
x_9	0.87672774	0.87674384	0.87675177
x_{10}	0.80424512	0.80427330	0.80428301
x_{11}	0.80688119	0.80691173	0.80691989
x_{12}	0.97473885	0.97475850	0.97476265
x_{13}	0.93003466	0.93004542	0.93004899
x_{14}	0.87672774	0.87674661	0.87675155
x_{15}	0.85931331	0.85933296	0.85933709
x_{16}	0.93406183	0.93407462	0.93407672

b.

	Jacobi 60 iterations	Gauss-Seidel 35 iterations	SOR ($\omega = 1.2$) 23 iterations
x_1	0.39668038	0.39668651	0.39668915
x_2	0.07175540	0.07176830	0.07177348
x_3	−0.23080396	−0.23078609	−0.23077981
x_4	0.24549277	0.24550989	0.24551535
x_5	0.83405412	0.83406516	0.83406823
x_6	0.51497606	0.51498897	0.51499414
x_7	0.12116003	0.12118683	0.12119625
x_8	−0.24044414	−0.24040991	−0.24039898
x_9	0.37873579	0.37876891	0.37877812
x_{10}	1.09073364	1.09075392	1.09075899
x_{11}	0.54207872	0.54209658	0.54210286
x_{12}	0.13838259	0.13841682	0.13842774
x_{13}	−0.23083868	−0.23079452	−0.23078224
x_{14}	0.41919067	0.41923122	0.41924136
x_{15}	1.15015953	1.15018477	1.15019025
x_{16}	0.51497606	0.51499318	0.51499864
x_{17}	0.12116003	0.12119315	0.12120236
x_{18}	−0.24044414	−0.24040359	−0.24039345
x_{19}	0.37873579	0.37877365	0.37878188
x_{20}	1.09073364	1.09075629	1.09076069
x_{21}	0.39668038	0.39669142	0.39669449
x_{22}	0.07175540	0.07177567	0.07178074
x_{23}	−0.23080396	−0.23077872	−0.23077323
x_{24}	0.24549277	0.24551542	0.24551982
x_{25}	0.83405412	0.83406793	0.83407025

c.	Jacobi 15 iterations	Gauss-Seidel 9 iterations	SOR ($\omega = 1.1$) 8 iterations
x_1	−3.07611424	−3.07611739	−3.07611796
x_2	−1.65223176	−1.65223563	−1.65223579
x_3	−0.53282391	−0.53282528	−0.53282531
x_4	−0.04471548	−0.04471608	−0.04471609
x_5	0.17509673	0.17509661	0.17509661
x_6	0.29568226	0.29568223	0.29568223
x_7	0.37309012	0.37309011	0.37309011
x_8	0.42757934	0.42757934	0.42757934
x_9	0.46817927	0.46817927	0.46817927
x_{10}	0.49964748	0.49964748	0.49964748
x_{11}	0.52477026	0.52477026	0.52477026
x_{12}	0.54529835	0.54529835	0.54529835
x_{13}	0.56239007	0.56239007	0.56239007
x_{14}	0.57684345	0.57684345	0.57684345
x_{15}	0.58922662	0.58922662	0.58922662
x_{16}	0.59995522	0.59995522	0.59995522
x_{17}	0.60934045	0.60934045	0.60934045
x_{18}	0.61761997	0.61761997	0.61761997
x_{19}	0.62497846	0.62497846	0.62497846
x_{20}	0.63156161	0.63156161	0.63156161
x_{21}	0.63748588	0.63748588	0.63748588
x_{22}	0.64284553	0.64284553	0.64284553
x_{23}	0.64771764	0.64771764	0.64771764
x_{24}	0.65216585	0.65216585	0.65216585
x_{25}	0.65624320	0.65624320	0.65624320
x_{26}	0.65999423	0.65999423	0.65999423
x_{27}	0.66345660	0.66345660	0.66345660
x_{28}	0.66666242	0.66666242	0.66666242
x_{29}	0.66963919	0.66963919	0.66963919
x_{30}	0.67241061	0.67241061	0.67241061
x_{31}	0.67499722	0.67499722	0.67499722
x_{32}	0.67741692	0.67741692	0.67741691
x_{33}	0.67968535	0.67968535	0.67968535
x_{34}	0.68181628	0.68181628	0.68181628
x_{35}	0.68382184	0.68382184	0.68382184
x_{36}	0.68571278	0.68571278	0.68571278
x_{37}	0.68749864	0.68749864	0.68749864
x_{38}	0.68918652	0.68918652	0.68918652
x_{39}	0.69067718	0.69067718	0.69067718
x_{40}	0.68363346	0.68363346	0.68363346

d.	Jacobi 33 iterations	Gauss-Seidel 8 iterations	SOR ($\omega = 1.2$) 13 iterations
x_1	1.53873501	1.53873270	1.53873549
x_2	0.73142167	0.73141966	0.73142226
x_3	0.10797136	0.10796931	0.10797063
x_4	0.17328530	0.17328340	0.17328480
x_5	0.04055865	0.04055595	0.04055737
x_6	0.08525019	0.08524787	0.08524925
x_7	0.16645040	0.16644711	0.16644868
x_8	0.12198156	0.12197878	0.12198026
x_9	0.10125265	0.10124911	0.10125043
x_{10}	0.09045966	0.09045662	0.09045793
x_{11}	0.07203172	0.07202785	0.07202912
x_{12}	0.07026597	0.07026266	0.07026392
x_{13}	0.06875835	0.06875421	0.06875546
x_{14}	0.06324659	0.06324307	0.06324429
x_{15}	0.05971510	0.05971083	0.05971200
x_{16}	0.05571199	0.05570834	0.05570949
x_{17}	0.05187851	0.05187416	0.05187529
x_{18}	0.04924911	0.04924537	0.04924648
x_{19}	0.04678213	0.04677776	0.04677885
x_{20}	0.04448679	0.04448303	0.04448409
x_{21}	0.04246924	0.04246493	0.04246597
x_{22}	0.04053818	0.04053444	0.04053546
x_{23}	0.03877273	0.03876852	0.03876952
x_{24}	0.03718190	0.03717822	0.03717920
x_{25}	0.03570858	0.03570451	0.03570548
x_{26}	0.03435107	0.03434748	0.03434844
x_{27}	0.03309542	0.03309152	0.03309246
x_{28}	0.03192212	0.03191866	0.03191958
x_{29}	0.03083007	0.03082637	0.03082727
x_{30}	0.02980997	0.02980666	0.02980755
x_{31}	0.02885510	0.02885160	0.02885248
x_{32}	0.02795937	0.02795621	0.02795707
x_{33}	0.02711787	0.02711458	0.02711543
x_{34}	0.02632478	0.02632179	0.02632262
x_{35}	0.02557705	0.02557397	0.02557479
x_{36}	0.02487017	0.02486733	0.02486814
x_{37}	0.02420147	0.02419858	0.02419938
x_{38}	0.02356750	0.02356482	0.02356560
x_{39}	0.02296603	0.02296333	0.02296410
x_{40}	0.02239424	0.02239171	0.02239247
x_{41}	0.02185033	0.02184781	0.02184855
x_{42}	0.02133203	0.02132965	0.02133038
x_{43}	0.02083782	0.02083545	0.02083615
x_{44}	0.02036585	0.02036360	0.02036429
x_{45}	0.01991483	0.01991261	0.01991324

(continued)

	Jacobi 33 iterations	Gauss-Seidel 8 iterations	SOR ($\omega = 1.2$) 13 iterations
x_{46}	0.01948325	0.01948113	0.01948175
x_{47}	0.01907002	0.01906793	0.01906846
x_{48}	0.01867387	0.01867187	0.01867239
x_{49}	0.01829386	0.01829190	0.01829233
x_{50}	0.01792896	0.01792707	0.01792749
x_{51}	0.01757833	0.01757648	0.01757683
x_{52}	0.01724113	0.01723933	0.01723968
x_{53}	0.01691660	0.01691487	0.01691517
x_{54}	0.01660406	0.01660237	0.01660267
x_{55}	0.01630279	0.01630127	0.01630146
x_{56}	0.01601230	0.01601082	0.01601101
x_{57}	0.01573198	0.01573087	0.01573077
x_{58}	0.01546129	0.01546020	0.01546010
x_{59}	0.01519990	0.01519909	0.01519878
x_{60}	0.01494704	0.01494626	0.01494595
x_{61}	0.01470181	0.01470085	0.01470077
x_{62}	0.01446510	0.01446417	0.01446409
x_{63}	0.01423556	0.01423437	0.01423461
x_{64}	0.01401350	0.01401233	0.01401256
x_{65}	0.01380328	0.01380234	0.01380242
x_{66}	0.01359448	0.01359356	0.01359363
x_{67}	0.01338495	0.01338434	0.01338418
x_{68}	0.01318840	0.01318780	0.01318765
x_{69}	0.01297174	0.01297109	0.01297107
x_{70}	0.01278663	0.01278598	0.01278597
x_{71}	0.01270328	0.01270263	0.01270271
x_{72}	0.01252719	0.01252656	0.01252663
x_{73}	0.01237700	0.01237656	0.01237654
x_{74}	0.01221009	0.01220965	0.01220963
x_{75}	0.01129043	0.01129009	0.01129008
x_{76}	0.01114138	0.01114104	0.01114102
x_{77}	0.01217337	0.01217312	0.01217313
x_{78}	0.01201771	0.01201746	0.01201746
x_{79}	0.01542910	0.01542896	0.01542896
x_{80}	0.01523810	0.01523796	0.01523796

Exercise Set 7.4

1. The $\|\cdot\|_\infty$ condition number is **a.** 50 **b.** 241.37 **c.** 60,002 **d.** 339,866 **e.** 12 **h.** 198.17

3. The matrix is ill-conditioned since $K_\infty = 60002$. We have $\tilde{\mathbf{x}} = (-1.0000, 2.0000)^t$.

5. a. We have $\tilde{\mathbf{x}} = (188.9998, 92.99998, 45.00001, 27.00001, 21.00002)^t$.

b. The condition number is $K_\infty = 80$.

c. The exact solution is $\mathbf{x} = (189, 93, 45, 27, 21)^t$.

9. For the 3×3 Hilbert matrix H we have

$$\hat{H}^{-1} = \begin{bmatrix} 8.968 & -35.77 & 29.77 \\ -35.77 & 190.6 & -178.6 \\ 29.77 & -178.6 & 178.6 \end{bmatrix}, \qquad \hat{H} = \begin{bmatrix} 0.9799 & 0.4870 & 0.3238 \\ 0.4860 & 0.3246 & 0.2434 \\ 0.3232 & 0.2433 & 0.1949 \end{bmatrix}$$

and $\|H - \hat{H}\|_\infty = 0.04260$.

Exercise Set 8.1

1. The linear least-squares polynomial is $1.70784x + 0.89968$ with $E = 0.039198$.

3. The least-squares polynomials with their errors are, respectively, $0.6208950 + 1.219621x$, with $E = 2.719 \times 10^{-5}$; $0.5965807 + 1.253293x - 0.01085343x^2$, with $E = 1.801 \times 10^{-5}$; and $0.6290193 + 1.185010x + 0.03533252x^2 - 0.01004723x^3$, with $E = 1.741 \times 10^{-5}$.

5. **a.** The linear least-squares polynomial is $72.0845x - 194.138$, with error of 329.

b. The least-squares polynomial of degree two is $6.61821x^2 - 1.14352x + 1.23556$, with error of 1.44×10^{-3}.

c. The least-squares polynomial of degree three is $-0.0136742x^3 + 6.84557x^2 - 2.37919x + 3.42904$, with error of 5.27×10^{-4}.

d. The least-squares approximation of the form be^{ax} is $24.2588e^{0.372382x}$, with error of 418.

e. The least-squares approximation of the form bx^a is $6.23903x^{2.01954}$, with error of 0.00703.

7. **a.** $k = 0.8996$, $E(k) = 0.407$ **b.** $k = 0.9069$, $E(k) = 0.486$ Part (b) fits the total experimental data best.

9. Point average $= 0.101$(ACT score) $+0.487$

11. The linear least-squares polynomial gives $y \approx 0.17952x + 8.2084$.

13. **a.** $\ln R = \ln 1.304 + 0.5756 \ln W$

b. $E = 25.25$

c. $\ln R = \ln 1.051 + 0.7006 \ln W + 0.06695(\ln W)^2$

d. $E = \sum_{i=1}^{37} \left(R_i - bW_i^a e^{c(\ln W_i)^2}\right)^2 = 20.30$

Exercise Set 8.2

1. The linear least-squares approximations are

a. $P_1(x) = 1.833333 + 4x$

b. $P_1(x) = -1.600003 + 3.600003x$

c. $P_1(x) = 1.140981 - 0.2958375x$

d. $P_1(x) = 0.1945267 + 3.000001x$

e. $P_1(x) = 0.6109245 + 0.09167105x$

f. $P_1(x) = -1.861455 + 1.666667x$

3. The linear least-squares approximations on $[-1, 1]$ are

a. $P_1(x) = 3.333333 - 2x$

b. $P_1(x) = 0.6000025x$

c. $P_1(x) = 0.5493063 - 0.2958375x$

d. $P_1(x) = 1.175201 + 1.103639x$

e. $P_1(x) = 0.4207355 + 0.4353975x$

f. $P_1(x) = 0.6479184 + 0.5281226x$

5. The errors for the approximations in Exercise 3 are **a.** 0.177779 **b.** 0.0457206 **c.** 0.00484624 **d.** 0.0526541 **e.** 0.0153784 **f.** 0.00363453

7. The Gram-Schmidt process produces the following collections of polynomials;

a. $\phi_0(x) = 1$, $\phi_1(x) = x - 0.5$, $\phi_2(x) = x^2 - x + \frac{1}{6}$, and $\phi_3(x) = x^3 - 1.5x^2 + 0.6x - 0.05$

b. $\phi_0(x) = 1$, $\phi_1(x) = x - 1$, $\phi_2(x) = x^2 - 2x + \frac{2}{3}$, and $\phi_3(x) = x^3 - 3x^2 + \frac{12}{5}x - \frac{2}{5}$

c. $\phi_0(x) = 1$, $\phi_1(x) = x - 2$, $\phi_2(x) = x^2 - 4x + \frac{11}{3}$, and $\phi_3(x) = x^3 - 6x^2 + 11.4x - 6.8$

9. The least-squares polynomials of degree two are

a. $P_2(x) = 3.833333\phi_0(x) + 4\phi_1(x) + 0.9999998\phi_2(x)$

b. $P_2(x) = 2\phi_0(x) + 3.6\phi_1(x) + 3\phi_2(x)$

c. $P_2(x) = 0.5493061\phi_0(x) - 0.2958369\phi_1(x) + 0.1588785\phi_2(x)$

d. $P_2(x) = 3.194528\phi_0(x) + 3\phi_1(x) + 1.458960\phi_2(x)$

e. $P_2(x) = 0.6567600\phi_0(x) + 0.09167105\phi_1(x) - 0.7375118\phi_2(x)$

f. $P_2(x) = 1.471878\phi_0(x) + 1.666667\phi_1(x) + 0.2597705\phi_2(x)$

11. The Laguerre polynomials are $L_1(x) = x - 1$, $L_2(x) = x^2 - 4x + 2$ and $L_3(x) = x^3 - 9x^2 + 18x - 6$.

Exercise Set 8.3

1. The interpolating polynomials of degree two are

a. $P_2(x) = 2.377443 + 1.590534(x - 0.8660254) + 0.5320418(x - 0.8660254)x$

b. $P_2(x) = 0.7617600 + 0.8796047(x - 0.8660254)$

c. $P_2(x) = 1.052926 + 0.4154370(x - 0.8660254) - 0.1384262x(x - 0.8660254)$

d. $P_2(x) = 0.5625 + 0.649519(x - 0.8660254) + 0.75x(x - 0.8660254)$

3. The interpolating polynomials of degree three are

a.
$$\begin{aligned} P_3(x) = {} & 2.519044 + 1.945377(x - 0.9238795) \\ & + 0.7047420(x - 0.9238795)(x - 0.3826834) \\ & + 0.1751757(x - 0.9238795)(x - 0.3826834)(x + 0.3826834) \end{aligned}$$

b.
$$\begin{aligned} P_3(x) = {} & 0.7979459 + 0.7844380(x - 0.9238795) \\ & - 0.1464394(x - 0.9238795)(x - 0.3826834) \\ & - 0.1585049(x - 0.9238795)(x - 0.3826834)(x + 0.3826834) \end{aligned}$$

c.
$$\begin{aligned} P_3(x) = {} & 1.072911 + 0.3782067(x - 0.9238795) \\ & - 0.09799213(x - 0.9238795)(x - 0.3826834) \\ & + 0.04909073(x - 0.9238795)(x - 0.3826834)(x + 0.3826834) \end{aligned}$$

d.
$$\begin{aligned} P_3(x) = {} & 0.7285533 + 1.306563(x - 0.9238795) \\ & + 0.9999999(x - 0.9238795)(x - 0.3826834) \end{aligned}$$

5. The zeros of $\tilde{T}_3$ produce the following interpolating polynomials of degree two.

a. $P_2(x) = 0.3489153 - 0.1744576(x - 2.866025) + 0.1538462(x - 2.866025)(x - 2)$

b. $P_2(x) = 0.1547375 - 0.2461152(x - 1.866025) + 0.1957273(x - 1.866025)(x - 1)$

c. $P_2(x) = 0.6166200 - 0.2370869(x - 0.9330127) - 0.7427732(x - 0.9330127)(x - 0.5)$

d. $P_2(x) = 3.0177125 + 1.883800(x - 2.866025) + 0.2584625(x - 2.866025)(x - 2)$

7. The cubic polynomial $(383/384)x - (5/32)x^3$ approximates $\sin x$ with error at most 7.19×10^{-4}.

9. The change of variable $x = \cos\theta$ produces

$$\int_{-1}^{1} \frac{T_n^2(x)}{\sqrt{1-x^2}}\,dx = \int_{-1}^{1} \frac{[\cos(n \arccos x)]^2}{\sqrt{1-x^2}}\,dx = \int_0^{\pi} \cos^2(n\theta)\,dx = \frac{\pi}{2}.$$

Exercise Set 8.4

1. The Padé approximations of degree two for $f(x) = e^{2x}$ are

$$n = 2,\ m = 0 : r_{2,0}(x) = 1 + 2x + 2x^2$$

$$n = 1,\ m = 1 : r_{1,1}(x) = (1 + x)/(1 - x)$$

$$n = 0,\ m = 2 : r_{0,2}(x) = (1 - 2x + 2x^2)^{-1}$$

i	x_i	$f(x_i)$	$r_{2,0}(x_i)$	$r_{1,1}(x_i)$	$r_{0,2}(x_i)$
1	0.2	1.4918	1.4800	1.5000	1.4706
2	0.4	2.2255	2.1200	2.3333	1.9231
3	0.6	3.3201	2.9200	4.0000	1.9231
4	0.8	4.9530	3.8800	9.0000	1.4706
5	1.0	7.3891	5.0000	undefined	1.0000

3. $r_{2,3}(x) = (1 + \frac{2}{5}x + \frac{1}{20}x^2)/(1 - \frac{3}{5}x + \frac{3}{20}x^2 - \frac{1}{60}x^3)$

i	x_i	$f(x_i)$	$r_{2,3}(x_i)$
1	0.2	1.22140276	1.22140277
2	0.4	1.49182470	1.49182561
3	0.6	1.82211880	1.82213210
4	0.8	2.22554093	2.22563652
5	1.0	2.71828183	2.71875000

5. $r_{3,3}(x) = (x - \frac{7}{60}x^3)/(1 + \frac{1}{20}x^2)$

i	x_i	$f(x_i)$	MacLaurin polynomial of degree 6	$r_{3,3}(x_i)$
0	0.0	0.00000000	0.00000000	0.00000000
1	0.1	0.09983342	0.09966675	0.09938640
2	0.2	0.19866933	0.19733600	0.19709571
3	0.3	0.29552021	0.29102025	0.29246305
4	0.4	0.38941834	0.37875200	0.38483660
5	0.5	0.47942554	0.45859375	0.47357724

7. The Padé approximations of degree five are

a. $r_{0,5}(x) = (1 + x + \frac{1}{2}x^2 + \frac{1}{6}x^3 + \frac{1}{24}x^4 + \frac{1}{120}x^5)^{-1}$

b. $r_{1,4}(x) = (1 - \frac{1}{5}x)/(1 + \frac{4}{5}x + \frac{3}{10}x^2 + \frac{1}{15}x^3 + \frac{1}{120}x^4)$

c. $r_{3,2}(x) = (1 - \frac{3}{5}x + \frac{3}{20}x^2 - \frac{1}{60}x^3)/(1 + \frac{2}{5}x + \frac{1}{20}x^2)$

d. $r_{4,1}(x) = (1 - \frac{4}{5}x + \frac{3}{10}x^2 - \frac{1}{15}x^3 + \frac{1}{120}x^4)/(1 + \frac{1}{5}x)$

i	x_i	$f(x_i)$	$r_{0,5}(x_i)$	$r_{1,4}(x_i)$	$r_{2,3}(x_i)$	$r_{4,1}(x_i)$
1	0.2	0.81873075	0.81873081	0.81873074	0.81873075	0.81873077
2	0.4	0.67032005	0.67032276	0.67031942	0.67031963	0.67032099
3	0.6	0.54881164	0.54883296	0.54880635	0.54880763	0.54882143
4	0.8	0.44932896	0.44941181	0.44930678	0.44930966	0.44937931
5	1.0	0.36787944	0.36809816	0.36781609	0.36781609	0.36805556

9. $r_{T_{2,0}}(x) = (1.266066T_0(x) - 1.130318T_1(x) + 0.2714953T_2(x))/T_0(x)$

$r_{T_{1,1}}(x) = (0.9945705T_0(x) - 0.4569046T_1(x))/(T_0(x) + 0.48038745T_1(x))$

$r_{T_{0,2}}(x) = 0.7940220T_0(x)/(T_0(x) + 0.8778575T_1(x) + 0.1774266T_2(x))$

i	x_i	$f(x_i)$	$r_{T_{2,0}}(x_i)$	$r_{T_{1,1}}(x_i)$	$r_{T_{0,2}}(x_i)$
1	0.25	0.77880078	0.74592811	0.78595377	0.74610974
2	0.50	0.60653066	0.56515935	0.61774075	0.58807059
3	1.00	0.36787944	0.40724330	0.36319269	0.38633199

11. $r_{T_{2,2}}(x) = \dfrac{0.91747T_1(x)}{T_0(x) + 0.088914T_2(x)}$

i	x_i	$f(x_i)$	$r_{T_{2,2}}(x_i)$
0	0.00	0.00000000	0.00000000
1	0.10	0.09983342	0.09093843
2	0.20	0.19866933	0.18028797
3	0.30	0.29552021	0.26808992
4	0.40	0.38941834	0.35438412

13. a. $e^x = e^{M \ln \sqrt{10} + s} = e^{M \ln \sqrt{10}} e^s = e^{\ln 10(M/2)} e^s = 10^{M/2} e^s$

b. $e^s \approx \left(1 + \frac{1}{2}s + \frac{1}{10}s^2 + \frac{1}{120}s^3\right) \Big/ \left(1 - \frac{1}{2}s + \frac{1}{10}s^2 - \frac{1}{120}s^3\right)$ with $|\text{error}| \le 3.75 \times 10^{-7}$

c. Set $M = \text{round}(0.8685889638x)$, $s = x - M/(0.8685889638)$, and

$$\hat{f} = \left(1 + \frac{1}{2}s + \frac{1}{10}s^2 + \frac{1}{120}s^3\right) \Big/ \left(1 - \frac{1}{2}s + \frac{1}{10}s^2 - \frac{1}{120}s^3\right).$$

Then $f = (3.16227766)^M \hat{f}$ is the approximation.

Exercise Set 8.5

1. $S_2(x) = \frac{\pi^2}{3} - 4\cos x + \cos 2x$

3. $S_3(x) = 3.676078 - 3.676078\cos x + 1.470431\cos 2x - 0.7352156\cos 3x + 3.676078\sin x - 2.940862\sin 2x$

5. $S_n(x) = \frac{1}{2} + \frac{1}{\pi}\sum_{k=1}^{n-1} \frac{1-(-1)^k}{k}\sin kx$

7. The trigonometric least-squares polynomials are

a. $S_2(x) = \cos 2x$

b. $S_2(x) = 0$

c. $S_3(x) = 1.566453 + 0.5886815\cos x - 0.2700642\cos 2x + 0.2175679\cos 3x + 0.8341640\sin x - 0.3097866\sin 2x$

d. $S_3(x) = -2.046326 + 3.883872\cos x - 2.320482\cos 2x + 0.7310818\cos 3x$

9. The trigonometric least-squares polynomial is
$S_3(x) = -0.4968929 + 0.2391965\cos x + 1.515393\cos 2x + 0.2391965\cos 3x - 1.150649\sin x$ with error $E(S_3) = 7.271197$.

11. The trigonometric least-squares polynomials and their errors are

a. $S_3(x) = -0.08676065 - 1.446416\cos\pi(x-3) - 1.617554\cos 2\pi(x-3) + 3.980729\cos 3\pi(x-3) - 2.154320\sin\pi(x-3) + 3.907451\sin 2\pi(x-3)$;
$E(S_3) = 210.90453$

b. $S_3(x) = -0.0867607 - 1.446416\cos\pi(x-3) - 1.617554\cos 2\pi(x-3) + 3.980729\cos 3\pi(x-3) - 2.354088\cos 4\pi(x-3) - 2.154320\sin\pi(x-3) + 3.907451\sin 2\pi(x-3) - 1.166181\sin 3\pi(x-3)$;
$E(S_4) = 169.4943$

13. Let $f(-x) = -f(x)$. The integral $\int_{-a}^{0} f(x)\,dx$ under the change of variable $t = -x$ transforms to

$$-\int_a^0 f(-t)\,dt = \int_0^a f(-t)\,dt = -\int_0^a f(t)\,dt = -\int_0^a f(x)\,dx.$$

Thus,

$$\int_{-a}^{a} f(x)\,dx = \int_{-a}^{0} f(x)\,dx + \int_0^a f(x)\,dx = -\int_0^a f(x)\,dx + \int_0^a f(x)\,dx = 0.$$

Exercise Set 8.6

1. The trigonometric interpolating polynomials are

a. $S_2(x) = -12.33701 + 4.934802\cos x - 2.467401\cos 2x + 4.934802\sin x$

b. $S_2(x) = -6.168503 + 9.869604\cos x - 3.701102\cos 2x + 4.934802\sin x$

c. $S_2(x) = 1.570796 - 1.570796\cos x$

d. $S_2(x) = -0.5 - 0.5\cos 2x + \sin x$

3. The Fast Fourier Transform Algorithm gives the following trigonometric interpolating polynomials.

a. $S_4(x) = -11.10331 + 2.467401\cos x - 2.467401\cos 2x + 2.467401\cos 3x - 1.233701\cos 4x + 5.956833\sin x - 2.467401\sin 2x + 1.022030\sin 3x$

b. $S_4(x) = 1.570796 - 1.340759\cos x - 0.2300378\cos 3x$

c. $S_4(x) = -0.1264264 + 0.2602724\cos x - 0.3011140\cos 2x + 1.121372\cos 3x + 0.04589648\cos 4x - 0.1022190\sin x + 0.2754062\sin 2x - 2.052955\sin 3x$

d. $S_4(x) = -0.1526819 + 0.04754278\cos x + 0.6862114\cos 2x - 1.216913\cos 3x + 1.176143\cos 4x - 0.8179387\sin x + 0.1802450\sin 2x + 0.2753402\sin 3x$

5.

	Approximation	Actual
a.	−69.76415	−62.01255
b.	9.869602	9.869604
c.	−0.7943605	−0.2739383
d.	−0.9593287	−0.9557781

7. The b_j terms are all zero. The a_j terms are as follows:

$a_0 = -4.0008033$	$a_1 = 3.7906715$	$a_2 = -2.2230259$	$a_3 = 0.6258042$
$a_4 = -0.3030271$	$a_5 = 0.1813613$	$a_6 = -0.1216231$	$a_7 = 0.0876136$
$a_8 = -0.0663172$	$a_9 = 0.0520612$	$a_{10} = -0.0420333$	$a_{11} = 0.0347040$
$a_{12} = -0.0291807$	$a_{13} = 0.0249129$	$a_{14} = -0.0215458$	$a_{15} = 0.0188421$
$a_{16} = -0.0166380$	$a_{17} = 0.0148174$	$a_{18} = -0.0132962$	$a_{19} = 0.0120123$
$a_{20} = -0.0109189$	$a_{21} = 0.0099801$	$a_{22} = -0.0091683$	$a_{23} = 0.0084617$
$a_{24} = -0.0078430$	$a_{25} = 0.0072984$	$a_{26} = -0.0068167$	$a_{27} = 0.0063887$
$a_{28} = -0.0060069$	$a_{29} = 0.0056650$	$a_{30} = -0.0053578$	$a_{31} = 0.0050810$

$a_{32}=-0.0048308$	$a_{33}=0.0046040$	$a_{34}=-0.0043981$	$a_{35}=0.0042107$
$a_{36}=-0.0040398$	$a_{37}=0.0038837$	$a_{38}=-0.0037409$	$a_{39}=0.0036102$
$a_{40}=-0.0034903$	$a_{41}=0.0033803$	$a_{42}=-0.0032793$	$a_{43}=0.0031866$
$a_{44}=-0.0031015$	$a_{45}=0.0030233$	$a_{46}=-0.0029516$	$a_{47}=0.0028858$
$a_{48}=-0.0028256$	$a_{49}=0.0027705$	$a_{50}=-0.0027203$	$a_{51}=0.0026747$
$a_{52}=-0.0026333$	$a_{53}=0.0025960$	$a_{54}=-0.0025626$	$a_{55}=0.0025328$
$a_{56}=-0.0025066$	$a_{57}=0.0024837$	$a_{58}=-0.0024642$	$a_{59}=0.0024478$
$a_{60}=-0.0024345$	$a_{61}=0.0024242$	$a_{62}=-0.0024169$	$a_{63}=0.0024125$

Exercise Set 9.1

1. a. The eigenvalues and associated eigenvectors are $\lambda_1 = 2$, $\mathbf{v}^{(1)} = (1, 0, 0)^t$; $\lambda_2 = 1$, $\mathbf{v}^{(2)} = (0, 2, 1)^t$; and $\lambda_3 = -1$, $\mathbf{v}^{(3)} = (-1, 1, 1)^t$. Yes, the set is linearly independent.

b. The eigenvalues and associated eigenvectors are $\lambda_1 = \lambda_2 = \lambda_3 = 1$, $\mathbf{v}^{(1)} = \mathbf{v}^{(2)} = (1, 0, 1)^t$ and $\mathbf{v}^{(3)} = (0, 1, 1)$. No, the set is linearly independent.

c. The eigenvalues and associated eigenvectors are $\lambda_1 = 2$, $\mathbf{v}^{(1)} = (0, 1, 0)^t$; $\lambda_2 = 3$, $\mathbf{v}^{(2)} = (1, 0, 1)^t$; and $\lambda_3 = 1$, $\mathbf{v}^{(3)} = (1, 0, -1)^t$. Yes, the set is linearly independent.

d. The eigenvalues and associated eigenvectors are $\lambda_1 = \lambda_2 = 3$, $\mathbf{v}^{(1)} = (1, 0, -1)^t$, $\mathbf{v}^{(2)} = (0, 1, -1)^t$; and $\lambda_3 = 0$, $\mathbf{v}^{(3)} = (1, 1, 1)^t$. Yes, the set is linearly independent.

e. The eigenvalues and associated eigenvectors are $\lambda_1 = 1, \mathbf{v}^{(1)} = (0, -1, 1)^t$; $\lambda_2 = 1 + \sqrt{2}$, $\mathbf{v}^{(2)} = (\sqrt{2}, 1, 1)^t$; and $\lambda_3 = 1 - \sqrt{2}$, $\mathbf{v}^{(3)} = (-\sqrt{2}, 1, 1)^t$; Yes, the set is linearly independent.

f. The eigenvalues and associated eigenvectors are $\lambda_1 = 1$, $\mathbf{v}^{(1)} = (1, 0, -1)^t$; $\lambda_2 = 1$, $\mathbf{v}^{(2)} = (1, -1, 0)^t$; and $\lambda_3 = 4$, $\mathbf{v}^{(3)} = (1, 1, 1)^t$. Yes, the set is linearly independent.

3. a. The three eigenvalues are within $\{\lambda \mid |\lambda| \le 2\} \cup \{\lambda \mid |\lambda - 2| \le 2\}$.

b. The three eigenvalues are within $\{\lambda \mid |\lambda - 4| \le 2\}$.

c. The three real eigenvalues satisfy $0 \le \lambda \le 6$.

d. The three real eigenvalues satisfy $1.25 \le \lambda \le 8.25$.

e. The four real eigenvalues satisfy $-8 \le \lambda \le 1$.

f. The four eigenvalues are within $\{\lambda \mid |\lambda - 2| \le 4\}$.

5. If $c_1\mathbf{v}_1 + \cdots + c_k\mathbf{v}_k = \mathbf{0}$, then for any j, with $1 \le j \le k$, we have $c_1\mathbf{v}_j^t\mathbf{v}_1 + \cdots + c_k\mathbf{v}_j^t\mathbf{v}_k = \mathbf{0}$. But orthogonality gives $c_i\mathbf{v}_j^t\mathbf{v}_i = 0$ for $i \ne j$, so $c_j\mathbf{v}_j^t\mathbf{v}_j = 0$ and $c_j = 0$.

7. Since $\{\mathbf{v}_i\}_{i=1}^n$ is linearly independent in $\mathbb{R}^n$, there exist numbers $c_1, \ldots, c_n$ with

$$\mathbf{x} = c_1\mathbf{v}_1 + \cdots + c_n\mathbf{v}_n.$$

Hence, for any j,with $1 \le j \le n$,

$$\mathbf{v}_j^t\mathbf{x} = c_1\mathbf{v}_j^t\mathbf{v}_1 + \cdots + c_n\mathbf{v}_j^t\mathbf{v}_n = c_j\mathbf{v}_j^t\mathbf{v}_j = c_j.$$

9. a. The eigenvalues are $\lambda_1 = 5.307857563$, $\lambda_2 = -0.4213112993$, $\lambda_3 = -0.1365462647$ with associated eigenvectors $(0.59020967, 0.51643129, 0.62044441)^t$, $(0.77264234, -0.13876278, -0.61949069)^t$, and $(0.23382978, -0.84501102, 0.48091581)^t$, respectively.

b. A is not positive definite, since $\lambda_2 < 0$ and $\lambda_3 < 0$.

Exercise Set 9.2

1. The approximate eigenvalues and approximate eigenvectors are

a. $\mu^{(3)} = 3.666667$, $\mathbf{x}^{(3)} = (0.9772727, 0.9318182, 1)^t$

b. $\mu^{(3)} = 2.000000$, $\mathbf{x}^{(3)} = (1, 1, 0.5)^t$

c. $\mu^{(3)} = 5.000000$, $\mathbf{x}^{(3)} = (-0.2578947, 1, -0.2842105)^t$

d. $\mu^{(3)} = 5.038462$, $\mathbf{x}^{(3)} = (1, 0.2213741, 0.3893130, 0.4045802)^t$

e. $\mu^{(3)} = 7.531073$, $\mathbf{x}^{(3)} = (0.6886722, -0.6706677, -0.9219805, 1)^t$

f. $\mu^{(3)} = 4.106061$, $\mathbf{x}^{(3)} = (0.1254613, 0.08487085, 0.00922509, 1)^t$

3. The approximate eigenvalues and approximate eigenvectors are

a. $\mu^{(3)} = 3.959538$, $\mathbf{x}^{(3)} = (0.5816124, 0.5545606, 0.5951383)^t$

b. $\mu^{(3)} = 2.0000000$, $\mathbf{x}^{(3)} = (-0.6666667, -0.6666667, -0.3333333)^t$

c. $\mu^{(3)} = 7.189567$, $\mathbf{x}^{(3)} = (0.5995308, 0.7367472, 0.3126762)^t$

d. $\mu^{(3)} = 6.037037$, $\mathbf{x}^{(3)} = (0.5073714, 0.4878571, -0.6634857, -0.2536857)^t$

e. $\mu^{(3)} = 5.142562$, $\mathbf{x}^{(3)} = (0.8373051, 0.3701770, 0.1939022, 0.3525495)^t$

f. $\mu^{(3)} = 8.593142$, $\mathbf{x}^{(3)} = (-0.4134762, 0.4026664, 0.5535536, -0.6003962)^t$

5. The approximate eigenvalues and approximate eigenvectors are

a. $\lambda_1 \approx \mu^{(9)} = 3.999908$, $\mathbf{x}^{(9)} = (0.9999943, 0.9999828, 1)^t$

$\lambda_2 \approx \mu^{(1)} = 1.000000$, $\mathbf{x}^{(1)} = (-2.999908, 2.999908, 0)^t$

b. $\lambda_1 \approx \mu^{(13)} = 2.414214$, $\mathbf{x}^{(13)} = (1, 0.7071429, 0.7070707)^t$

$\lambda_2 \approx \mu^{(1)} = 1.000000$, $\mathbf{x}^{(1)} = (0, -1.414214, 1.414214)^t$

c. $\lambda_1 \approx \mu^{(9)} = 5.124749$, $\mathbf{x}^{(9)} = (-0.2424476, 1, -0.3199733)^t$

$\lambda_2 \approx \mu^{(6)} = 1.636734$, $\mathbf{x}^{(6)} = (1.783218, -1.135350, -3.124733)^t$

d. $\lambda_1 \approx \mu^{(24)} = 5.235861$, $\mathbf{x}^{(24)} = (1, 0.6178361, 0.1181667, 0.4999220)^t$

$\lambda_2 \approx \mu^{(10)} = 3.618177$, $\mathbf{x}^{(10)} = (0.7236390, -1.170573, 1.170675, -0.2763374)^t$

e. $\lambda_1 \approx \mu^{(17)} = 8.999667$, $\mathbf{x}^{(17)} = (0.9999085, -0.9999078, -0.9999993, 1)^t$

$\lambda_2 \approx \mu^{(21)} = 5.000051$, $\mathbf{x}^{(21)} = (1.999338, -1.999603, 1.999603, -2.000198)^t$

f. The method did not converge in 25 iterations, but $\lambda_1 \approx \mu^{(363)} = 4.105309$, $\mathbf{x}^{(363)} = (0.06286299, 0.08702754, 0.01824680, 1)^t$, $\lambda_2 \approx \mu^{(15)} = -4.024308$, $\mathbf{x}^{(15)} = (-8.151965, 2.100699, 0.7519080, -0.3554941)^t$

7. The approximate eigenvalues and approximate eigenvectors are

a. $\mu^{(8)} = 4.0000000$, $\mathbf{x}^{(8)} = (0.5773547, 0.5773282, 0.5773679)^t$

b. $\mu^{(13)} = 2.414214$, $\mathbf{x}^{(13)} = (-0.7071068, -0.5000255, -0.4999745)^t$

c. $\mu^{(16)} = 7.223663$, $\mathbf{x}^{(16)} = (0.6247845, 0.7204271, 0.3010466)^t$

d. $\mu^{(20)} = 7.086130$, $\mathbf{x}^{(20)} = (0.3325999, 0.2671862, 0.7590108, -0.4918246)^t$

e. $\mu^{(21)} = 5.236068$, $\mathbf{x}^{(21)} = (0.7795539, 0.4815996, 0.09214214, 0.3897016)^t$

f. $\mu^{(16)} = 9.0000000$, $\mathbf{x}^{(16)} = (-0.4999592, 0.4999584, 0.5000408, -0.5000416)^t$

9. The approximate eigenvalues and approximate eigenvectors are

a. $\mu^{(9)} = 1.000000$, $\mathbf{x}^{(9)} = (-0.1542994, 0.7715207, -0.6172095)^t$

b. $\mu^{(12)} = -0.4142136$, $\mathbf{x}^{(12)} = (-0.7071068, 0.4999894, 0.5000106)^t$

c. $\mu^{(6)} = 4.961699$, $\mathbf{x}^{(6)} = (-0.4812465, 0.05195336, 0.8750444)^t$

d. $\mu^{(14)} = 2.485863$, $\mathbf{x}^{(14)} = (-0.6096695, 0.6451951, -0.2779286, 0.3671268)^t$

e. $\mu^{(10)} = 3.618034$, $\mathbf{x}^{(10)} = (0.3958550, -0.6404796, 0.6404886, -0.1511924)^t$

f. $\mu^{(6)} = 4.0000000$, $\mathbf{x}^{(6)} = (-0.4999985, -0.5000015, -0.4999985, -0.5000015)^t$

11. The approximate eigenvalues and approximate eigenvectors are

a. $\mu^{(2)} = 1.000000$, $\mathbf{x}^{(2)} = (0.1542373, -0.7715828, 0.6171474)^t$

b. $\mu^{(13)} = 1.000000$, $\mathbf{x}^{(13)} = (0.00007432, -0.7070723, 0.7071413)^t$

c. $\mu^{(14)} = 4.961699$, $\mathbf{x}^{(14)} = (-0.4814472, 0.05180473, 0.8749428)^t$

d. $\mu^{(17)} = 4.428007$, $\mathbf{x}^{(17)} = (0.7194230, 0.4231908, 0.1153589, 0.5385466)^t$

e. $\mu^{(10)} = 3.618034$, $\mathbf{x}^{(10)} = (0.3956185, -0.6406258, 0.6404462, -0.1513711)^t$

f. The method did not converge in 25 iterations, but $\mu^{(31)} = 5.0000000$, $\mathbf{x}^{(31)} = (0.4999091, -0.5002392, 0.4997607, -0.50009009)^t$.

13. a. We have $|\lambda| \leq 6$ for all eigenvalues λ.

b. The approximate eigenvalue is $\mu^{(133)} = 0.69766854$, with the approximate eigenvector $\mathbf{x}^{(133)} = (1, 0.7166727, 0.2568099, 0.04601217)^t$.

d. The characteristic polynomial is $P(\lambda) = \lambda^4 - \frac{1}{4}\lambda - \frac{1}{16}$ and the eigenvalues are $\lambda_1 = 0.6976684972$, $\lambda_2 = -0.2301775942 + 0.56965884i$, $\lambda_3 = -0.2301775942 - 0.56965884i$, and $\lambda_4 = -0.237313308$.

e. The beetle population should approach zero since A is convergent.

15. Using the Inverse Power method with $\mathbf{x}^{(0)} = (1, 0, 0, 1, 0, 0, 1, 0, 0, 1)^t$ and $q = 0$ gives the following results:

a. $\mu^{(49)} = 1.0201926$, so $\rho(A^{-1}) \approx 1/\mu^{(49)} = 0.9802071$

b. $\mu^{(30)} = 1.0404568$, so $\rho(A^{-1}) \approx 1/\mu^{(30)} = 0.9611163$

c. $\mu^{(22)} = 1.0606974$, so $\rho(A^{-1}) \approx 1/\mu^{(22)} = 0.9427760$. The method appears to be stable for all α in $[\frac{1}{4}, \frac{3}{4}]$.

17. Forming $A^{-1}B$ and using the Power method with $\mathbf{x}^{(0)} = (1, 0, 0, 1, 0, 0, 1, 0, 0, 1)^t$ gives the following results:

a. The spectral radius is approximately $\mu^{(46)} = 0.9800021$.

b. The spectral radius is approximately $\mu^{(25)} = 0.9603543$.

c. The spectral radius is approximately $\mu^{(18)} = 0.9410754$.

Exercise Set 9.3

1. Householder's method produces the following tridiagonal matrices.

a. $\begin{bmatrix} 12.00000 & -10.77033 & 0.0 \\ -10.77033 & 3.862069 & 5.344828 \\ 0.0 & 5.344828 & 7.137931 \end{bmatrix}$ **b.** $\begin{bmatrix} 2.0000000 & 1.414214 & 0.0 \\ 1.414214 & 1.000000 & 0.0 \\ 0.0 & 0.0 & 3.0 \end{bmatrix}$

c. $\begin{bmatrix} 1.0000000 & -1.414214 & 0.0 \\ -1.414214 & 1.000000 & 0.0 \\ 0.0 & 0.0 & 1.000000 \end{bmatrix}$ **d.** $\begin{bmatrix} 4.750000 & -2.263846 & 0.0 \\ -2.263846 & 4.475610 & -1.219512 \\ 0.0 & -1.219512 & 5.024390 \end{bmatrix}$

3. Householder's method produces the following upper Hessenberg matrices.

a. $\begin{bmatrix} 2.0000000 & 2.8284271 & 1.4142136 \\ -2.8284271 & 1.0000000 & 2.0000000 \\ 0.0000000 & 2.0000000 & 3.0000000 \end{bmatrix}$ **b.** $\begin{bmatrix} -1.0000000 & -3.0655513 & 0.0000000 \\ -3.6055513 & -0.23076923 & 3.1538462 \\ 0.0000000 & 0.15384615 & 2.2307692 \end{bmatrix}$

c. $\begin{bmatrix} 5.0000000 & 4.9497475 & -1.4320780 & -1.5649769 \\ -1.4142136 & -2.0000000 & -2.4855515 & 1.8226448 \\ 0.0000000 & -5.4313902 & -1.4237288 & -2.6486542 \\ 0.0000000 & 0.0000000 & 1.5939865 & 5.4237288 \end{bmatrix}$

d. $\begin{bmatrix} 4.0000000 & 1.7320508 & 0.0000000 & 0.0000000 \\ 1.7320508 & 2.3333333 & 0.23570226 & 0.40824829 \\ 0.0000000 & -0.47140452 & 4.6666667 & -0.57735027 \\ 0.0000000 & 0.0000000 & 0.0000000 & 5.0000000 \end{bmatrix}$

Exercise Set 9.4

1. Two iterations of the QR Algorithm produces the following matrices.

a. $A^{(3)} = \begin{bmatrix} 0.6939977 & -0.3759745 & 0.0 \\ -0.3759745 & 1.892417 & -0.03039696 \\ 0.0 & -0.03039696 & 3.413585 \end{bmatrix}$

b. $A^{(3)} = \begin{bmatrix} 4.535466 & 1.212648 & 0.0 \\ 1.212648 & 3.533242 & 3.83 \times 10^{-7} \\ 0.0 & 3.83 \times 10^{-7} & -0.06870782 \end{bmatrix}$

c. $A^{(3)} = \begin{bmatrix} 4.679567 & -0.2969009 & 0.0 \\ -2.969009 & 3.052484 & -1.207346 \times 10^{-5} \\ 0.0 & -1.207346 \times 10^{-5} & 1.267949 \end{bmatrix}$

d. $A^{(3)} = \begin{bmatrix} 0.3862092 & 0.4423226 & 0.0 & 0.0 \\ 0.4423226 & 1.787694 & -0.3567744 & 0.0 \\ 0.0 & -0.3567744 & 3.080815 & 3.116382 \times 10^{-5} \\ 0.0 & 0.0 & 3.116382 \times 10^{-5} & 4.745281 \end{bmatrix}$

e. $A^{(3)} = \begin{bmatrix} -2.826365 & 1.130297 & 0.0 & 0.0 \\ 1.130297 & -2.429647 & -0.1734156 & 0.0 \\ 0.0 & -0.1734156 & 0.8172086 & 1.863997 \times 10^{-9} \\ 0.0 & 0.0 & 1.863997 \times 10^{-9} & 3.438803 \end{bmatrix}$

f. $A^{(3)} = \begin{bmatrix} 0.2763388 & 0.1454371 & 0.0 & 0.0 \\ 0.1454371 & 0.4543713 & 0.1020836 & 0.0 \\ 0.0 & 0.1020836 & 1.174446 & -4.36 \times 10^{-5} \\ 0.0 & 0.0 & -4.36 \times 10^{-5} & 0.9948441 \end{bmatrix}$

3. The matrices in Exercise 1 have the following eigenvalues, accurate to within 10^{-5}.

a. 3.414214, 2.000000, 0.58578644

b. −0.06870782, 5.346462, 2.722246

c. 1.267949, 4.732051, 3.000000

d. 4.745281, 3.177283, 1.822717, 0.2547188

e. 3.438803, 0.8275517, −1.488068, −3.778287

f. 0.9948440, 1.189091, 0.5238224, 0.1922421

5. a. Let

$$P = \begin{bmatrix} \cos\theta & -\sin\theta \\ \sin\theta & \cos\theta \end{bmatrix}$$

and $\mathbf{y} = P\mathbf{x}$. Show that $\|\mathbf{x}\|_2 = \|\mathbf{y}\|_2$. Use the relationship $x_1 + ix_2 = re^{i\alpha}$, where $r = \|\mathbf{x}\|_2$ and $\alpha = \tan^{-1}(x_2/x_1)$, and $y_1 + iy_2 = re^{i(\alpha+\theta)}$.

b. Let $\mathbf{x} = (1, 0)^t$ and $\theta = \pi/4$.

11. a. To within 10^{-5}, the eigenvalues are 2.618034, 3.618034, 1.381966, and 0.3819660.

b. In terms of p and ρ the eigenvalues are $-65.45085p/\rho$, $-90.45085p/\rho$, $-34.54915p/\rho$, and $-9.549150p/\rho$.

13. The actual eigenvalues are

a. When $\alpha = 1/4$ we have 0.97974649, 0.92062677, 0.82743037, 0.70770751, 0.57115742, 0.42884258, 0.29229249, 0.17256963, 0.07937323, and 0.02025351

b. When $\alpha = 1/2$ we have 0.95949297, 0.84125353, 0.65486073, 0.41541501, 0.14231484, −0.14231484, −0.41541501, −0.65486073, −0.84125353, and −0.95949297.

c. When $\alpha = 3/4$ we have 0.93923946, 0.76188030, 0.48229110, 0.12312252, −0.28652774, −0.71347226, −1.12312252, −1.48229110, −1.76188030, and −1.93923946. The method appears to be stable for $\alpha \le \frac{1}{2}$.

Exercise Set 10.1

1. Use Theorem 10.4.

3. Use Theorem 10.4 for each of the partial derivatives.

5. b. With $\mathbf{x}^{(0)} = (0, 0)^t$ and tolerance 10^{-5}, we have $\mathbf{x}^{(13)} = (0.9999973, 0.9999973)^t$.

c. With $\mathbf{x}^{(0)} = (0, 0)^t$ and tolerance 10^{-5}, we have $\mathbf{x}^{(11)} = (0.9999984, 0.9999991)^t$.

7. a. With $\mathbf{x}^{(0)} = (1, 1, 1)^t$ we have $\mathbf{x}^{(5)} = (5.0000000, 0.0000000, -0.5235988)^t$.

b. With $\mathbf{x}^{(0)} = (1, 1, 1)^t$ we have $\mathbf{x}^{(9)} = (1.0364011, 1.0857072, 0.93119113)^t$.

c. With $\mathbf{x}^{(0)} = (0, 0, 0.5)^t$ we have $\mathbf{x}^{(5)} = (0.00000000, 0.09999999, 1.0000000)^t$.

d. With $\mathbf{x}^{(0)} = (0, 0, 0)^t$ we have $\mathbf{x}^{(5)} = (0.49814471, -0.19960600, -0.52882595)^t$.

9. a. With $\mathbf{G}(\mathbf{x}) = \left(\sqrt{x_1 - x_2^2}, \sqrt{x_1^2 - x_2}\right)^t$ and $\mathbf{x}^{(0)} = (0.7, 0.4)^t$, we have $\mathbf{x}^{(14)} = (0.77184647, 0.41965131)^t$.

b. With $\mathbf{G}(\mathbf{x}) = \left(x/\sqrt{3}, \sqrt{(1 + x_1^3)/(3x_1)}\right)^t$ and $\mathbf{x}^{(0)} = (0.4, 0.7)^t$, we have $\mathbf{x}^{(20)} = (0.4999980, 0.8660221)^t$.

c. With $\mathbf{G}(\mathbf{x}) = \left(\sqrt{37 - x_2}, \sqrt{x_1 - 5}, 3 - x_1 - x_2\right)^t$ and $\mathbf{x}^{(0)} = (5, 1, -1)^t$ we have $\mathbf{x}^{(10)} = (6.0000002, 1.0000000, -3.9999971)^t$.

d. With $\mathbf{G}(\mathbf{x}) = \left(\sqrt{2x_3 + x_2 - 2x_2^2}, \sqrt{(10x_3 + x_1^2)/8}, x_1^2/(7x_2)\right)^t$ and $\mathbf{x}^{(0)} = (0.5, 0.5, 0)^t$, we have $\mathbf{x}^{(60)} = (0.5291548, 0.4000018, 0.09999853)^t$.

11. Yes, a stable solution occurs when $x_1 = 8000$ and $x_2 = 4000$.

Exercise Set 10.2

1. a. $\mathbf{x}^{(2)} = (0.4958936, 1.983423)^t$

b. $\mathbf{x}^{(2)} = (-0.5131616, -0.01837622)^t$

c. $\mathbf{x}^{(2)} = (0.5001667, 0.2508036, -0.5173874)^t$

d. $\mathbf{x}^{(2)} = (4.350877, 18.49123, -19.84211)^t$

3. a. $\mathbf{x}^{(5)} = (0.5000000, 0.8660254)^t$

b. $\mathbf{x}^{(6)} = (1.772454, 1.772454)^t$

c. $\mathbf{x}^{(5)} = (-1.456043, -1.664230, 0.4224934)^t$

d. $\mathbf{x}^{(4)} = (0.4981447, -0.1996059, -0.5288260)^t$

5. With $\mathbf{x}^{(0)} = (1, 1 - 1)^t$ and $TOL = 10^{-6}$, we have $\mathbf{x}^{(20)} = (0.5, 9.5 \times 10^{-7}, -0.5235988)^t$

7. With $\theta_i^{(0)} = 1$ for each $i = 1, 2, \ldots, 20$ the following results are obtained:

i	1	2	3	4	5	6	7	8	9	10
$\theta_i^{(5)}$	0.14062	0.19954	0.24522	0.28413	0.31878	0.35045	0.37990	0.40763	0.43398	0.45920
i	11	12	13	14	15	16	17	18	19	20
$\theta_i^{(5)}$	0.48348	0.50697	0.52980	0.55205	0.57382	0.59516	0.61615	0.63683	0.65726	0.67746

9. a. We have

$$\frac{\partial E}{\partial a} = 2\sum_{i=1}^{n}\left(w_i y_i - \frac{a}{(x_i - b)^c}\right)\left(\frac{1}{(x_i - b)^c}\right) = 0,$$

$$\frac{\partial E}{\partial b} = 2\sum_{i=1}^{n}\left(w_i y_i - \frac{a}{(x_i - b)^c}\right)\left(\frac{-ac}{(x_i - b)^{c+1}}\right) = 0,$$

and

$$\frac{\partial E}{\partial c} = 2\sum_{i=1}^{n}\left(w_i y_i - \frac{a}{(x_i - b)^c}\right)\ln(x_i - b)\left(\frac{-a}{(x_i - b)^c}\right) = 0.$$

Solving for a in the first equation and substituting into the second and third equations gives the linear system.

b. With $\mathbf{x}^{(0)} = (26.8, 8.3)^t = (b_0, c_0)^t$, we have $\mathbf{x}^{(7)} = (26.77021, 8.451831)^t$. Thus, $a = 2.217952 \times 10^6$, $b = 26.77021$, $c = 8.451831$, and

$$\sum_{i=1}^{n}\left(w_i y_i - \frac{a}{(x_i - b)^c}\right)^2 = 0.7821139.$$

Exercise Set 10.3

1. a. $\mathbf{x}^{(2)} = (0.4777920, 1.927557)^t$

b. $\mathbf{x}^{(2)} = (-0.3250070, -0.1386967)^t$

c. $\mathbf{x}^{(2)} = (0.5115893, 38.31494, 31.69089)^t$

d. $\mathbf{x}^{(2)} = (-67.00583, 35.06480, -123.3408)^t$

3. a. $\mathbf{x}^{(9)} = (0.5, 0.8660254)^t$

b. $\mathbf{x}^{(8)} = (1.772454, 1.772454)^t$

c. $\mathbf{x}^{(9)} = (-1.456043, -1.664231, 0.4224934)^t$

d. $\mathbf{x}^{(5)} = (0.4981447, -0.1996059, -0.5288260)^t$

5. With $\mathbf{x}^{(0)} = (1, 1 - 1)^t$, we have $\mathbf{x}^{(56)} = (0.5000591, 0.01057235, -0.5224818)^t$.

7. With $\mathbf{x}^{(0)} = (0.75, 1.25)^t$, we have $\mathbf{x}^{(4)} = (0.7501948, 1.184712)^t$. Thus, $a = 0.7501948$, $b = 1.184712$, and the error is 19.796.

Exercise Set 10.4

1. a. With $\mathbf{x}^{(0)} = (0, 0)^t$, we have $\mathbf{x}^{(11)} = (0.4943541, 1.948040)^t$.

b. With $\mathbf{x}^{(0)} = (1, 1)^t$, we have $\mathbf{x}^{(2)} = (0.4970073, 0.8644143)^t$.

c. With $\mathbf{x}^{(0)} = (2, 2)^t$, we have $\mathbf{x}^{(1)} = (1.736083, 1.804428)^t$.

d. With $\mathbf{x}^{(0)} = (0, 0)^t$, we have $\mathbf{x}^{(2)} = (-0.3610092, 0.05788368)^t$.

3. a. With $\mathbf{x}^{(0)} = (0, 0, 0)^t$, we have $\mathbf{x}^{(14)} = (1.043605, 1.064058, 0.9246118)^t$.

b. With $\mathbf{x}^{(0)} = (0, 0, 0)^t$, we have $\mathbf{x}^{(9)} = (0.4932739, 0.9863888, -0.5175964)^t$.

c. With $\mathbf{x}^{(0)} = (0, 0, 0)^t$, we have $\mathbf{x}^{(11)} = (-1.608296, -1.192750, 0.7205642)^t$.

d. With $\mathbf{x}^{(0)} = (0, 0, 0)^t$, we have $\mathbf{x}^{(1)} = (0, 0.00989056, 0.9890556)^t$.

5. a. With $\mathbf{x}^{(0)} = (0, 0)^t$, we have $\mathbf{x}^{(8)} = (3.136548, 0)^t$ and $\mathbf{g}(\mathbf{x}^{(8)}) = 0.005057848$.

b. With $\mathbf{x}^{(0)} = (0, 0)^t$, we have $\mathbf{x}^{(13)} = (0.6157412, 0.3768953)^t$ and $\mathbf{g}(\mathbf{x}^{(13)}) = 0.1481574$.

c. With $\mathbf{x}^{(0)} = (0, 0, 0)^t$, we have $\mathbf{x}^{(5)} = (-0.6633785, 0.3145720, 0.5000740)^t$ and $\mathbf{g}(\mathbf{x}^{(5)}) = 0.6921548$.

d. With $\mathbf{x}^{(0)} = (1, 1, 1)^t$, we have $\mathbf{x}^{(4)} = (0.04022273, 0.01592477, 0.01594401)^t$ and $\mathbf{g}(\mathbf{x}^{(4)}) = 1.010003$.

Exercise Set 11.1

1. The Linear Shooting Algorithm gives the results in the following tables.

a.

i	x_i	w_{1i}	$y(x_i)$
1	0.333333	0.5311664	0.5310687
2	0.666667	1.153515	1.153323

b.

i	x_i	w_{1i}	$y(x_i)$
1	0.25	0.3937095	0.3936767
2	0.50	0.8240948	0.8240271
3	0.75	1.337160	1.337086

3. The Linear Shooting Algorithm gives the results in the following tables.

a.

i	x_i	w_{1i}	$y(x_i)$
3	0.3	0.7833204	0.7831923
6	0.6	0.6023521	0.6022801
9	0.9	0.8568906	0.8568760

b.

i	x_i	w_{1i}	$y(x_i)$
5	1.25	0.1676179	0.1676243
10	1.50	0.4581901	0.4581935
15	1.75	0.6077718	0.6077740

c.

i	x_i	w_{1i}	$y(x_i)$
3	0.3	−0.5185754	−0.5185728
6	0.6	−0.2195271	−0.2195247
9	0.9	−0.0406577	−0.0406570

d.

i	x_i	w_{1i}	$y(x_i)$
3	1.3	0.0655336	0.06553420
6	1.6	0.0774590	0.07745947
9	1.9	0.0305619	0.03056208

5. The Linear Shooting Algorithm with $h = 0.05$ gives the following results.

i	x_i	w_{1i}
6	0.3	0.04990547
10	0.5	0.00676467
16	0.8	0.00033755

The Linear Shooting Algorithm with $h = 0.1$ gives the following results.

i	x_i	w_{1i}
3	0.3	0.05273437
5	0.5	0.00741571
8	0.8	0.00038976

7. **a.** The approximate potential is $u(3) \approx 36.66702$ using $h = 0.1$.
   b. The actual potential is $u(3) = 36.66667$.

9. **a.** There are no solutions if b is an integer multiple of π and $B \neq 0$.
   b. A unique solution exists whenever b is not an integer multiple of π.
   c. There is an infinite number of solutions if b is an multiple integer of π and $B = 0$.

Exercise Set 11.2

1. The Nonlinear Shooting Algorithm gives $w_1 = 0.405505 \approx \ln 1.5 = 0.405465$.

3. The Nonlinear Shooting Algorithm gives the results in the following tables.

   a. 4 iterations required:

i	x_i	w_{1i}	$y(x_i)$
3	1.3	0.4347934	0.4347826
6	1.6	0.3846363	0.3846154
9	1.9	0.3448586	0.3448276

   b. 6 iterations required:

i	x_i	w_{1i}	$y(x_i)$
3	1.3	2.069249	2.069231
6	1.6	2.225013	2.225000
9	1.9	2.426317	2.426316

   c. 4 iterations required:

i	x_i	w_{1i}	$y(x_i)$
3	1.3	1.031597	1.031595
6	1.6	1.095007	1.095004
9	1.9	1.168174	1.168170

d. To apply the algorithm we need to redefine the initial value of TK to be 2. 7 iterations required:

i	x_i	w_{1i}	$y(x_i)$
5	1.25	0.4358290	0.4358272
10	1.50	1.3684496	1.3684447
15	1.75	2.9992010	2.9991909

5. The algorithm gives the results in the following tables.

a. 3 iterations required:

i	x_i	w_{1i}	$y(x_i)$
3	1.3	0.4347720	0.4347826
6	1.6	0.3845947	0.3846154
9	1.9	0.3447969	0.3448276

b. To apply the algorithm we need to define the initial approximations for t_0 and t_1 to be -0.5 and 0.5. 15 iterations required:

i	x_i	w_{1i}	$y(x_i)$
3	1.3	2.0692491	2.0692308
6	1.6	2.2250137	2.2250000
9	1.9	2.4263174	2.4263158

c. 6 iterations required:

i	x_i	w_{1i}	$y(x_i)$
3	1.3	1.0315965	1.0315950
6	1.6	1.0950047	1.0950036
9	1.9	1.1681698	1.1681697

d. To apply the algorithm we need to define the initial approximations for t_0 and t_1 to be 0.5 and 1.81832. 7 iterations required:

i	x_i	w_{1i}	$y(x_i)$
5	1.25	0.4358261	0.4358273
10	1.50	1.3684417	1.3684447
15	1.75	2.9991849	2.9991909

Exercise Set 11.3

1. The Linear Finite-Difference Algorithm gives the results in the following tables.

a.

i	x_i	w_{1i}	$y(x_i)$
1	0.333333	0.5343259	0.5310687
2	0.666667	1.1579818	1.1533232

b.

i	x_i	w_{1i}	$y(x_i)$
1	0.25	0.3951247	0.3936767
2	0.50	0.8265306	0.8240271
3	0.75	1.3395692	1.3370861

3. The Linear Finite-Difference Algorithm gives the results in the following tables.

a.

i	x_i	w_{1i}	$y(x_i)$
2	0.2	1.018096	1.0221404
5	0.5	0.5942743	0.59713617
7	0.7	0.6514520	0.65290384

b.

i	x_i	w_{1i}	$y(x_i)$
5	1.25	0.16797186	0.16762427
10	1.50	0.45842388	0.45819349
15	1.75	0.60787334	0.60777401

c.

i	x_i	w_{1i}	$y(x_i)$
3	0.3	−0.5183084	−0.5185728
6	0.6	−0.2192657	−0.2195247
9	0.9	−0.04057484	−0.04065697

d.

i	x_i	w_{1i}	$y(x_i)$
3	1.3	0.0654387	0.0655342
6	1.6	0.0773936	0.0774595
9	1.9	0.0305465	0.0305621

5. The Linear Finite-Difference Algorithm gives the results in the following tables.

i	x_i	$w_i(h = 0.1)$
3	0.3	0.05572807
6	0.6	0.00310518
9	0.9	0.00016516

i	x_i	$w_i(h = 0.05)$
6	0.3	0.05132396
12	0.6	0.00263406
18	0.9	0.00013340

7. a. The approximate deflections are shown in the following table.

i	x_i	w_{1i}
5	30	0.0102808
10	60	0.0144277
15	90	0.0102808

b. Yes

c. Yes; maximum deflection occurs at $x = 60$. The exact solution is within tolerance, but the approximation is not.

Exercise Set 11.4

1. The Nonlinear Finite-Difference Algorithm gives the following results.

i	x_i	w_i	$y(x_i)$
1	1.5	0.4067967	0.4054651

3. The Nonlinear Finite-Difference Algorithm gives the results in the following tables.

a.

i	x_i	w_i	$y(x_i)$
3	1.3	0.4347972	0.4347826
6	1.6	0.3846286	0.3846154
9	1.9	0.3448316	0.3448276

b.

i	x_i	w_i	$y(x_i)$
3	1.3	2.0694081	2.0692308
6	1.6	2.2250937	2.2250000
9	1.9	2.4263387	2.4263158

c.

i	x_i	w_i	$y(x_i)$
3	1.3	1.031970	1.031595
6	1.6	1.095321	1.095004
9	1.9	1.168271	1.168170

d.

i	x_i	w_i	$y(x_i)$
5	1.25	0.4345979	0.4358273
10	1.50	1.3662119	1.3684447
15	1.75	2.9969339	2.9991909

Exercise Set 11.5

1. The Piecewise Linear Algorithm gives $\phi(x) = -0.07713271\phi_1(x) - 0.07442678\phi_2(x)$. Actual values are $y(x_1) = -0.07988545$ and $y(x_2) = -0.07712903$.

3. The Piecewise Linear Algorithm gives the results in the following tables.

a.

i	x_i	$\phi(x_i)$	$y(x_i)$
3	0.3	−0.212333	−0.21
6	0.6	−0.241333	−0.24
9	0.9	−0.090333	−0.09

b.

i	x_i	$\phi(x_i)$	$y(x_i)$
3	0.3	0.1815138	0.1814273
6	0.6	0.1805502	0.1804753
9	0.9	0.05936468	0.05934303

c.

i	x_i	$\phi(x_i)$	$y(x_i)$
5	0.25	−0.3585989	−0.3585641
10	0.50	−0.5348383	−0.5347803
15	0.75	−0.4510165	−0.4509614

d.

i	x_i	$\phi(x_i)$	$y(x_i)$a
5	0.25	−0.1846134	−0.1845204
10	0.50	−0.2737099	−0.2735857
15	0.75	−0.2285169	−0.2284204

5. The Cubic Spline Algorithm gives the results in the following tables.

a.

i	x_i	$\phi(x_i)$	$y(x_i)$
3	0.3	−0.2100000	−0.21
6	0.6	−0.2400000	−0.24
9	0.9	−0.0900000	−0.09

b.

i	x_i	$\phi(x_i)$	$y(x_i)$
3	0.3	0.1814269	0.1814273
6	0.6	0.1804753	0.1804754
9	0.9	0.05934321	0.05934303

c.

i	x_i	$\phi(x_i)$	$y(x_i)$
5	0.25	−0.3585639	−0.3585641
10	0.50	−0.5347779	−0.5347803
15	0.75	−0.4509109	−0.4509614

d.

i	x_i	$\phi(x_i)$	$y(x_i)$
5	0.25	−0.1845191	−0.1845204
10	0.50	−0.2735833	−0.2735857
15	0.75	−0.2284186	−0.2284204

7.

i	x_i	$\phi(x_i)$	$y(x_i)$
3	0.3	1.0408182	1.0408182
6	0.6	1.1065307	1.1065306
9	0.9	1.3065697	1.3065697

9. A change in variable $w = (x - a)/(b - a)$ gives the boundary value problem

$$-\frac{d}{dw}(p((b-a)w + a)y') + (b-a)^2 q((b-a)w + a)y = (b-a)^2 f((b-a)w + a),$$

where $0 < w < 1$, $y(0) = \alpha$, and $y(1) = \beta$. Then Exercise 6 can be used.

13. For $\mathbf{c} = (c_0, c_1, \ldots, c_{n+1})^t$ and $\phi(x) = \sum_{i=0}^{n+1} c_i\phi_i(x)$, we have

$$\mathbf{c}^t A\mathbf{c} = \int_0^1 p(x)[\phi'(x)]^2 + q(x)[\phi(x)]^2\, dx.$$

But $p(x) > 0$ and $q(x)[\phi(x)]^2 \geq 0$, so $\mathbf{c}^t A\mathbf{c} \geq 0$. It can be zero only if $\phi'(x) = 0$ on $[0, 1]$. However, $\{\phi_0', \phi_1', \ldots, \phi_{n+1}'\}$ is linearly independent, so $\mathbf{c}^t A\mathbf{c} = 0$ if and only if $\mathbf{c} = \mathbf{0}$.

Exercise Set 12.1

1. The Poisson Equation Finite-Difference Algorithm gives the following results.

i	j	x_i	y_j	$w_{i,j}$	$u(x_i, y_j)$
1	1	0.5	0.5	0.0	0
1	2	0.5	1.0	0.25	0.25
1	3	0.5	1.5	1.0	1

3. The Poisson Equation Finite-Difference Algorithm gives the following results.

a. 30 iterations required:

i	j	x_i	y_j	$w_{i,j}$	$u(x_i, y_j)$
2	2	0.4	0.4	0.1599988	0.16
2	4	0.4	0.8	0.3199988	0.32
4	2	0.8	0.4	0.3199995	0.32
4	4	0.8	0.8	0.6399996	0.64

b. 29 iterations required:

i	j	x_i	y_j	$w_{i,j}$	$u(x_i, y_j)$
2	1	1.256637	0.3141593	0.2951855	0.2938926
2	3	1.256637	0.9424778	0.1830822	0.1816356
4	1	2.513274	0.3141593	−0.7721948	−0.7694209
4	3	2.513274	0.9424778	−0.4785169	−0.4755283

c. 126 iterations required:

i	j	x_i	y_j	$w_{i,j}$	$u(x_i, y_j)$
4	3	0.8	0.3	1.2714468	1.2712492
4	7	0.8	0.7	1.7509419	1.7506725
8	3	1.6	0.3	1.6167917	1.6160744
8	7	1.6	0.7	3.0659184	3.0648542

d. 127 iterations required:

i	j	x_i	y_j	$w_{i,j}$	$u(x_i, y_j)$
2	2	1.2	1.2	0.5251533	0.5250861
4	4	1.4	1.4	1.3190830	1.3189712
6	6	1.6	1.6	2.4065150	2.4064186
8	8	1.8	1.8	3.8088995	3.8088576

7. The approximate potential at some typical points gives the following results.

i	j	x_i	y_j	$w_{i,j}$
1	4	0.1	0.4	88
2	1	0.2	0.1	66
4	2	0.4	0.2	66

Exercise Set 12.2

1. The Heat Equation Backward-Difference Algorithm gives the following results.

a.

i	j	x_i	t_j	w_{ij}	$u(x_i, t_j)$
1	1	0.5	0.05	0.632952	0.652037
2	1	1.0	0.05	0.895129	0.883937
3	1	1.5	0.05	0.632952	0.625037
1	2	0.5	0.1	0.566574	0.552493
2	2	1.0	0.1	0.801256	0.781344
3	2	1.5	0.1	0.566574	0.552493

b.

i	j	x_i	t_j	w_{ij}	$u(x_i, t_j)$
1	1	1/3	0.05	1.59728	1.53102
2	1	2/3	0.05	−1.59728	−1.53102
1	2	1/3	0.1	1.47300	1.35333
2	2	2/3	0.1	−1.47300	−1.35333

3. The Forward-Difference Algorithm gives the following results.

a. For $h = 0.4$ and $k = 0.1$:

i	j	x_i	t_j	w_{ij}	$u(x_i, t_j)$
2	5	0.8	0.5	3.035630	0
3	5	1.2	0.5	−3.035630	0
4	5	1.6	0.5	1.876122	0

For $h = 0.4$ and $k = 0.05$:

i	j	x_i	t_j	w_{ij}	$u(x_i, t_j)$
2	10	0.8	0.5	0	0
3	10	1.2	0.5	0	0
4	10	1.6	0.5	0	0

b. For $h = \frac{\pi}{10}$ and $k = 0.05$:

i	j	x_i	t_j	w_{ij}	$u(x_i, t_j)$
3	10	0.94247780	0.5	0.4864823	0.4906936
6	10	1.88495559	0.5	0.5718943	0.5768449
9	10	2.82743339	0.5	0.1858197	0.1874283

c. For $h = 0.2$ and $k = 0.04$:

i	j	x_i	t_j	w_{ij}	$u(x_i, t_j)$
4	10	0.8	0.4	1.166149	1.169362
8	10	1.6	0.4	1.252413	1.254556
12	10	2.4	0.4	0.4681813	0.4665473
16	10	3.2	0.4	−0.1027637	−0.1056622

d. For $h = 0.1$ and $k = 0.04$:

i	j	x_i	t_j	w_{ij}	$u(x_i, t_j)$
3	10	0.3	0.4	0.5397009	0.5423003
6	10	0.6	0.4	0.6344565	0.6375122
9	10	0.9	0.4	0.2061474	0.2071403

5. The Crank-Nicolson Algorithm gives the following results.

a. For $h = 0.4$ and $k = 0.1$:

i	j	x_i	t_j	w_{ij}	$u(x_i, t_j)$
2	5	0.8	0.5	8.2×10^{-7}	0
3	5	1.2	0.5	-8.2×10^{-7}	0
4	5	1.6	0.5	5.1×10^{-7}	0

For $h = 0.4$ and $k = 0.05$:

i	j	x_i	t_j	w_{ij}	$u(x_i, t_j)$
2	10	0.8	0.5	-2.6×10^{-6}	0
3	10	1.2	0.5	2.6×10^{-6}	0
4	10	1.6	0.5	-1.6×10^{-6}	0

b. For $h = \frac{\pi}{10}$ and $k = 0.05$:

i	j	x_i	t_j	w_{ij}	$u(x_i, t_j)$
3	10	0.94247780	0.5	0.4926589	0.4906936
6	10	1.88495559	0.5	0.5791553	0.5768449
9	10	2.82743339	0.5	0.1881790	0.1874283

c. For $h = 0.2$ and $k = 0.04$:

i	j	x_i	t_j	w_{ij}	$u(x_i, t_j)$
4	10	0.8	0.4	1.171532	1.169362
8	10	1.6	0.4	1.256005	1.254556
12	10	2.4	0.4	0.4654499	0.4665473
16	10	3.2	0.4	−0.1076139	−0.1056622

d. For $h = 0.1$ and $k = 0.04$:

i	j	x_i	t_j	w_{ij}	$u(x_i, t_j)$
3	10	0.3	0.4	0.5440532	0.5423003
6	10	0.6	0.4	0.6395728	0.6375122
9	10	0.9	0.4	0.2078098	0.2071403

9. To modify Algorithm 12.2, change the following:

Step 7 Set

$$t = jk;$$

$$z_1 = (w_1 + kF(h))/l_1.$$

Step 8 For $i = 2, \ldots, m-1$ set

$$z_i = (w_i + kF(ih) + \lambda z_{i-1})/l_i.$$

To modify Algorithm 12.3, change the following:
Step 7 Set

$$t = jk;$$

$$z_1 = \left[(1-\lambda)w_1 + \frac{\lambda}{2}w_2 + kF(h)\right]/l_1.$$

Step 8 For $i = 2, \ldots, m-1$ set

$$z_i = \left[(1-\lambda)w_i + \frac{\lambda}{2}(w_{i+1} + w_{i-1} + z_{i-1}) + kF(ih)\right]/l_i.$$

13. a. The approximate temperature at some typical points is given in the table.

i	j	r_i	t_j	$w_{i,j}$
1	20	0.6	10	137.6753
2	20	0.7	10	245.9678
3	20	0.8	10	340.2862
4	20	0.9	10	424.1537

b. The strain is approximately $I = 1242.537$.

Exercise Set 12.3

1. The Wave Equation Finite-Difference Algorithm gives the following results.

i	j	x_i	t_j	w_{ij}	$u(x_i, t_j)$
2	4	0.25	1.0	−0.7071068	−0.7071068
3	4	0.50	1.0	−1.0000000	−1.0000000
4	4	0.75	1.0	−0.7071068	−0.7071068

3. The Wave Equation Finite-Difference Algorithm with $h = \frac{\pi}{10}$ and $k = 0.05$ gives the following results.

i	j	x_i	t_j	w_{ij}	$u(x_i, t_j)$
2	10	$\frac{\pi}{5}$	0.5	0.5163933	0.5158301
5	10	$\frac{\pi}{2}$	0.5	0.8785407	0.8775826
8	10	$\frac{4\pi}{5}$	0.5	0.5163933	0.5158301

The Wave Equation Finite-Difference Algorithm with $h = \frac{\pi}{20}$ and $k = 0.1$ gives the following results.

i	j	x_i	t_j	w_{ij}
4	5	$\frac{\pi}{5}$	0.5	0.5159163
10	5	$\frac{\pi}{2}$	0.5	0.8777292
16	5	$\frac{4\pi}{5}$	0.5	0.5159163

The Wave Equation Finite-Difference Algorithm with $h = \frac{\pi}{20}$ and $k = 0.05$ gives the following results.

i	j	x_i	t_j	w_{ij}
4	10	$\frac{\pi}{5}$	0.5	0.5159602
10	10	$\frac{\pi}{2}$	0.5	0.8778039
16	10	$\frac{4\pi}{5}$	0.5	0.5159602

5. The Wave Equation Finite-Difference Algorithm gives the following results.

i	j	x_i	t_j	w_{ij}	$u(x_i, t_j)$
2	3	0.2	0.3	0.6729902	0.61061587
5	3	0.5	0.3	0	0
8	3	0.8	0.3	−0.6729902	−0.61061587

7. a. The air pressure for the open pipe is $p(0.5, 0.5) \approx 0.9$ and $p(0.5, 1.0) \approx 2.7$.
b. The air pressure for the closed pipe is $p(0.5, 0.5) \approx 0.9$ and $p(0.5, 1.0) \approx 0.9187927$.

Exercise Set 12.4

1. With $E_1 = (0.25, 0.75)$, $E_2 = (0, 1)$, $E_3 = (0.5, 0.5)$, and $E_4 = (0, 0.5)$, the basis functions are

$$\phi_1(x, y) = \begin{cases} 4x & \text{on } T_1 \\ -2 + 4y & \text{on } T_2 \end{cases}$$

$$\phi_2(x, y) = \begin{cases} -1 - 2x + 2y & \text{on } T_1 \\ 0 & \text{on } T_2 \end{cases}$$

$$\phi_3(x, y) = \begin{cases} 0 & \text{on } T_1 \\ 1 + 2x - 2y & \text{on } T_2 \end{cases}$$

$$\phi_4(x, y) = \begin{cases} 2 - 2x - 2y & \text{on } T_1 \\ 2 - 2x - 2y & \text{on } T_2 \end{cases}$$

and $\gamma_1 = 0.323825$, $\gamma_2 = 0$, $\gamma_3 = 1.0000$, and $\gamma_4 = 0$.

3. The Finite-Element Algorithm with $K = 8, N = 8, M = 32, n = 9, m = 25$, and $NL = 0$ gives the following results.

$$\gamma_1 = 0.511023$$
$$\gamma_2 = 0.720476$$
$$\gamma_3 = 0.507899$$
$$\gamma_4 = 0.720476$$
$$\gamma_5 = 1.01885$$
$$\gamma_6 = 0.720476$$
$$\gamma_7 = 0.507898$$
$$\gamma_8 = 0.720476$$
$$\gamma_9 = 0.511023$$
$$\gamma_i = 0, \quad \text{for } 10 \le i \le 25$$

$$u(0.125, 0.125) \approx 0.614187$$

$$u(0.125, 0.25) \approx 0.690343$$

$$u(0.25, 0.125) \approx 0.690343$$

$$u(0.25, 0.25) \approx 0.720476$$

(See the diagram.)

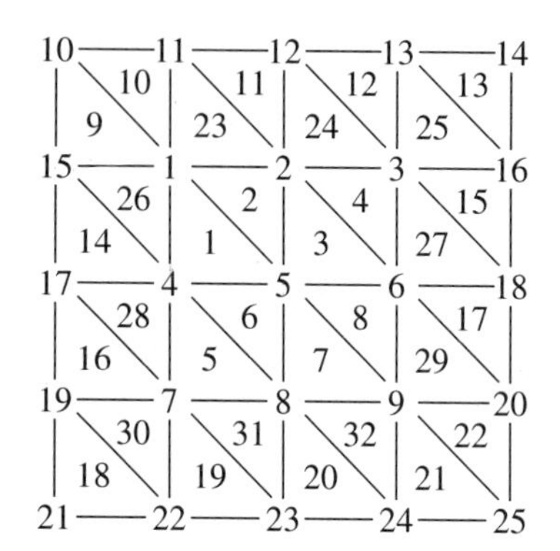

5. The Finite-Element Algorithm with $K = 0$, $N = 12$, $M = 32$, $n = 20$, $m = 27$, and $NL = 14$ gives the following results.

$\gamma_1 = 21.40335$	$\gamma_8 = 24.19855$	$\gamma_{15} = 20.23334$	$\gamma_{22} = 15$
$\gamma_2 = 19.87372$	$\gamma_9 = 24.16799$	$\gamma_{16} = 20.50056$	$\gamma_{23} = 15$
$\gamma_3 = 19.10019$	$\gamma_{10} = 27.55237$	$\gamma_{17} = 21.35070$	$\gamma_{24} = 15$
$\gamma_4 = 18.85895$	$\gamma_{11} = 25.11508$	$\gamma_{18} = 22.84663$	$\gamma_{25} = 15$
$\gamma_5 = 19.08533$	$\gamma_{12} = 22.92824$	$\gamma_{19} = 24.98178$	$\gamma_{26} = 15$
$\gamma_6 = 19.84115$	$\gamma_{13} = 21.39741$	$\gamma_{20} = 27.41907$	$\gamma_{27} = 15$
$\gamma_7 = 21.34694$	$\gamma_{14} = 20.52179$	$\gamma_{21} = 15$	

$$u(1, 0) \approx 22.92824$$

$$u(4, 0) \approx 22.84663$$

$$u\left(\frac{5}{2}, \frac{\sqrt{3}}{2}\right) \approx 18.85895$$

(See the diagram.)

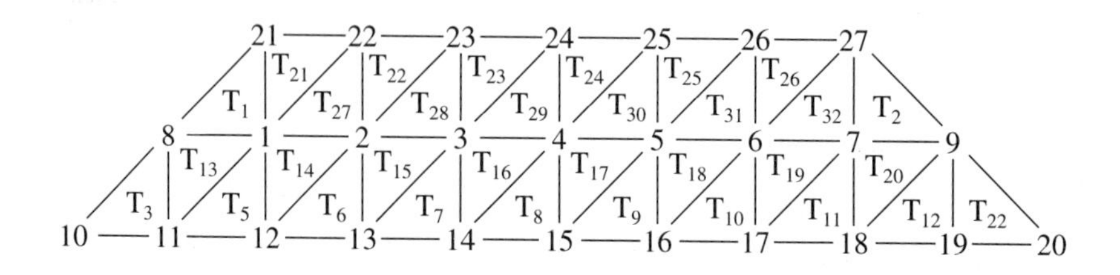

Index